CHILTON'S GUIDE TO BRAKES, STEERING and SUSPENSION 1980-87

President	Gary R. Ingersoll
Senior Vice President, Book Publishing & Research	Ronald A. Hoxter
Publisher	Kerry A. Freeman, S.A.E.
Editor-In-Chief	Dean F. Morgantini, S.A.E.
Managing Editor	David H. Lee, A.S.E., S.A.E.
Manager of Manufacturing	John J. Cantwell
Production Manager	W. Calvin Settle, Jr., S.A.E.
Senior Editor	Richard J. Rivele, S.A.E.
Senior Editor	Nick D'Andrea
Senior Editor	Ron Webb

CHILTON BOOK COMPANY

ONE OF THE *ABC* PUBLISHING COMPANIES,
A PART OF *CAPITAL CITIES/ABC, INC.*

Manufactured in USA
© 1988 Chilton Book Company
Chilton Way, Radnor, PA 19089
ISBN 0-8019-7819-X
Library of Congress Card Catalog No. 87-47939
4567890123   0987654321

SAFETY NOTICE

Proper service and repair procedures are vital to the safe, reliable operation of all motor vehicles, as well as the personal safety of those performing repairs. This manual outlines procedures for servicing and repairing vehicles using safe, effective methods. The procedures contain many NOTES, CAUTIONS and WARNINGS which should be followed along with standard safety procedures to eliminate the possibilty of personal injury or improper service which could damage the vehicle or compromise its safety.

It is important to note that the repair procedures and techniques, tools and parts for servicng motor vehicles, as well as the skill and experience of the individual performing the work vary widely. It is not possible to anticipate all of the conceivable ways or conditions under which vehicles may be serviced, or to provide cautions as to all of the possible hazards that may result. Standard and accepted safety precautions and equipment should be used when handling toxic or flammable fluids, and safety goggles or other protection should be used during cutting, grinding, chiseling, prying, or any other process that can cause material removal or projectiles.

Some procedures require the use of tools specially designed for a specific purpose. Before substituting another tool or procedure, you must be completely satisfied that neither your personal safety, nor the performance of the vehicle will be endangered

PART NUMBERS

Part numbers listed in this reference are not recomendations by Chilton for any product by brand name. They are references that can be used with interchange manuals and aftermarket supplier catalogs to locate each brand supplier's discrete part number.

Although information in this manual is based on industry sources and is complete as possible at the time of publication, the possibilty exists that some car manufacturers made later changes which could not be included here. While striving for total accuracy, Chilton Book Company cannot assume responsibility for any errors, changes or omissions that may occur in the compilation of this data.

No part of this publication may be reproduced, transmitted or stored in any form or by any means, electronic or mechanical, including photocopy, recording, or by information storage or retrieval system withiout prior written permission from the publisher.

Contents

Section 1
BRAKES

General Information	1-3
Domestic Vehicles	1-13
Import Vehicles	1-67
Light Trucks & Vans	1-105
Specifications	1-147
Domestic Cars	1-148
Domestic Light Trucks	1-156
Import Cars & Light Trucks	1-161
Caliper Design Illustrations	1-173

Section 2
STEERING AND SUSPENSION

General Information	2-3
Domestic Vehicles	2-37
Import Vehicles	2-185
Light Trucks & Vans	2-283
Specifications	2-365
Domestic Car	2-366
Domestic Light Trucks	2-378
Import Cars & Light Trucks	2-389
Adjustment Illustrations	2-406

Brakes
INDEX

GENERAL INFORMATION
Drum Brake	1-3
Brake System Bleeding	1-7
Flushing System	1-7
Disc Brake Service	1-8
Performance Diagnosis	1-11

DOMESTIC CARS

HYDRAULIC BRAKE SERVICE
Brake Actuating System	1-14
Brake Bleeding Sequence	1-23

FRONT DISC BRAKES
Applications & Specifications	1-23
Trouble Diagnosis	1-31
Rotors	1-29
Caliper & Pad Replacement	1-32 — 1-36

REAR DISC BRAKES
Caliper & Pad Replacement	1-42 — 1-46

DRUM BRAKES
Specifications	1-46
Trouble Diagnosis	1-49
Applications	1-50
Duo Servo Type	1-50
Ford Non-Servo Type	1-52
Chrysler Non-Servo Type	1-54

POWER BRAKES
Vacuum Operated Booster	1-55
Hydro Boost	1-58
Powermaster Unit	1-25

ANTI-LOCK BRAKES
Ford Motor Company	1-36
General Motors Corporation	1-63

IMPORT CARS

GENERAL INFORMATION
Hydraulic System	1-67
Cylinders & Valves	1-68
Power Brakes	1-74
Anti-Lock Brakes	1-74
Servicing Disc Brakes	1-75

DISC BRAKES
Applications	1-77
Pad Replacement	1-80 — 1-89

DRUM BRAKES
General Information	1-89
Applications	1-92
Servicing Drum Brakes	1-93 — 1-104

LIGHT TRUCKS & VANS

GENERAL INFORMATION
System Service	1-105
Trouble Diagnosis	1-105
Control Valves	1-107
Master Cylinder Service	1-108 — 1-114
Bleeding Brakes	1-114
Wheel Cylinders Replacement	1-116
Disc Brake Rotor	1-118

DRUM BRAKES
Brake Shoes Replacement	1-120 — 1-133
Wheel Cylinders Replacement	1-133 — 1-136

DISC BRAKES
Trouble Diagnosis	1-136
Pads & Caliper Replacement	1-136 — 1-146

BRAKE SPECIFICATIONS
Domestic Passenger Cars	1-147 — 1-161
Domestic Light Trucks	1-156 — 1-161
Imported Cars & Light Trucks	1-161 — 1-172
Caliper Design Illustrations	1-172 — 1-179

BRAKES
General Information

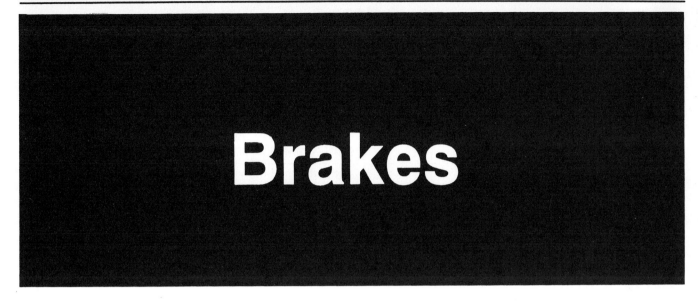

Brakes

DRUM BRAKE SERVICE

Most drum brakes are self energizing, with automatic adjusters. Utilization of the frictional force to increase the pressure of shoes against the drum is called *self-energizing* action. Utilization of force in one shoe to apply the opposite shoe is called *servo* action.

Brake Lining

Brake lining is made of asbestos impregnated with special compounds to bind the asbestos fibers together. Some linings are woven of asbestos threads and fine copper wire. With a few exceptions, most brake lining material is made from asbestos fibers ground up, pressed into shape and either riveted or bonded onto the brake shoe.

The primary shoe, sometimes called leading or forward brake shoe, is the shoe that faces toward the front of the car.

The secondary shoe, sometimes called the trailing or reverse brake shoe, faces the rear of the car.

Backing Plate

Thorough brake work starts at the brake backing plate. Check the brake area for any indication of lubricant leakage. If the leakage is due to brake fluid, replace or rebuild the wheel cylinder. If the leakage consists of wheel bearing grease, replace the inner bearing seal. It may be necessary to replace the axle bearing or seal. To check the backing plate mounting, tap the plate clockwise and counterclockwise. If movement occurs in either direction, remove the backing plate and check for worn bolts or elongated bolt holes. Replace worn parts. A loose backing plate can usually be detected by listening for a "clicking" sound when applying the brakes while the car is moved forward and backward.

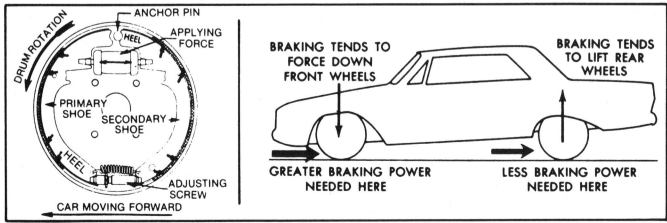

Action of the car and the car's brakes as the brakes are applied

1–3

BRAKES
General Information

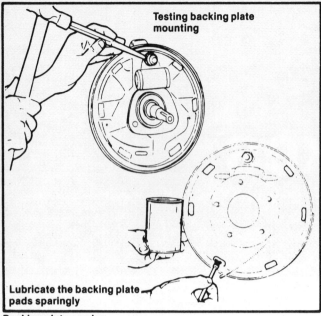

Backing plate service

Wheel Cylinder

Wheel cylinders should be inspected for leakage. Carefully inspect the boots. If they are torn, cut, heat cracked or show evidence of leakage, the wheel cylinder should be replaced or overhauled. Don't gamble. If the cylinder doesn't look healthy, replace or rebuild.

INSPECTION

1. Wash all parts in clean denatured alcohol. If alcohol is not available, use specified brake fluid. Dry with compressed air.
2. Replace scored pistons. Always replace the rubber cups and dust boots.
3. Inspect the cylinder bore for score marks or rust. If either condition is present, the cylinder bore must be honed. However, the cylinder should not be honed more than 0.003 inch beyond its original.
4. Check the bleeder hole to be sure that it is open.

ASSEMBLY

1. Apply a coating of heavy-duty brake fluid to all internal parts.
2. Thread the bleeder screw into the cylinder and tighten securely.
3. Insert the return spring, cups, and pistons into their respective positions in the cylinder bore. Place a boot over each end of the cylinder.

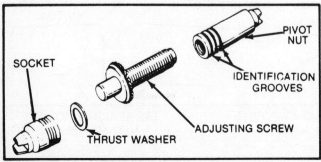

Typical adjusting screw assembly

Adjusting Screw Assembly

Disassemble the adjusting screw assembly. Using an electric wire brush, clean up the threads. Lubricate with brake fluid and reassemble the unit. Turn the threads all the way in by hand. If the threads bind at any point, replace the unit.

Installing Brake Shoes

1. Preassemble the brake shoes adjusting screw assembly and spring, plus the parking lever assembly (rear brakes only).
2. Spread the assembly; place it on the backing plate. Make sure that the wheel cylinder sockets are in the proper position.
3. Install the retainer pin and spring on both shoes.
4. Install the shoe guide.
5. Install the adjusting cable.
6. Install the parking link and spring (rear only).
7. Install the primary retracting spring.
8. Install the secondary retracting spring.

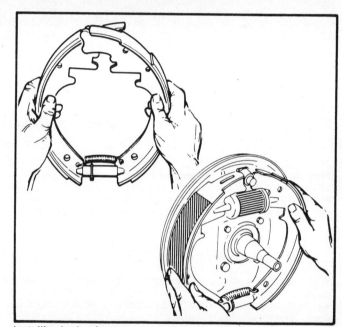

Installing brake shoes

Brake Drums

BRAKE DRUM TYPES

The *full-cast* drum has a cast iron web (back) of 3/16 to 1/4 inch thickness (passenger car sizes) whereas the *composite* drum has a steel web approximately 1/8 inch thick. These two types of drums, with few exceptions, are not interchangeable.

BRAKE DRUM DEPTH

Place a straightedge across the drum diameter on the open side. The actual drum depth is the measurement at a right angle from the straightedge to that part of the web which mates against the hub mounting flange.

ALUMINUM DRUMS

When replaced by other types, aluminum drums must be replaced in pairs.

BRAKES
General Information

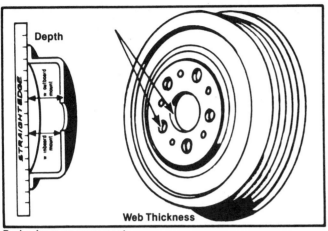
Brake drum measurement

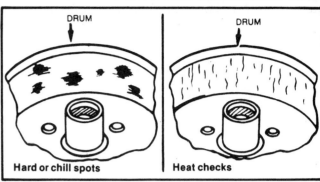
Brake drum inspection

METALLIC BRAKES

Drums designed for use with standard brake linings should not be used with metallic brakes.

BOLT CIRCLE

The circumference on which the centers of the wheel bolt holes are located around the drum-hub center is the bolt circle. It is shown as a double number (example: 6-5½). The first digit indicates the number of holes. The second number indicates the bolt circle diameter.

REMOVING TIGHT DRUMS

Difficulty removing a brake drum can be caused by shoes which are expanded beyond the drum's inner ridge, or shoes which have cut into and ridged the drum. In either case, back off the adjuster to obtain sufficient clearance for removal.

BRAKE DRUM INSPECTION

The condition of the brake drum surface is just as important as the surface to the brake lining. All drum surfaces should be clean, smooth, free from hard spots, heat checks, score marks and foreign matter imbedded in the drum surface. They should not be out of round, bellmouthed or barrel shaped. It is recommended that all drums be first checked with a drum micrometer to see if they are within oversize limits. If drum is within safe limits, even though the surface appears smooth, it should be turned not only to assure a true drum surface but also to remove any possible contamination in the surface from previous brake linings, road dusts, etc. Too much metal removed from a drum is unsafe and may result in:

1. Brake fade due to the thin drum being unable to absorb the heat generated.
2. Poor and erratic brake action due to distortion of drums.
3. Noise due to vibration caused by thin drums.

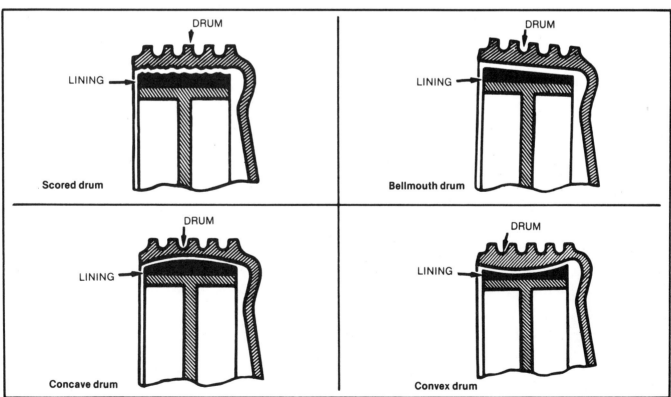
Checking brake drum wear

1-5

SECTION 1

BRAKES
General Information

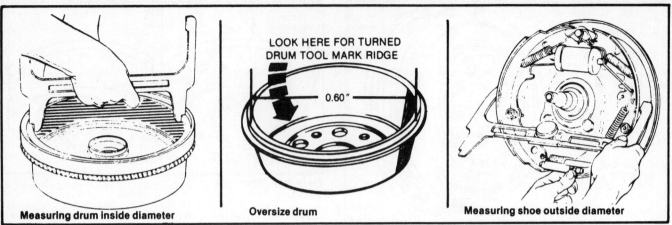

Checking brake drum-to-shoe clearance

4. A cracked or broken drum on a severe or very hard brake application.

Brake drum run-out should not exceed .005". Drums turned to more than .060" oversize are unsafe and should be replaced with new drums, except for some heavy ribbed drums which have an .080" limit. It is recommended that the diameters of the left and right drums on any one axle be within .010" of each other. In order to avoid erratic brake action when replacing drums, it is always good to replace the drums on both wheels at the same time.

If the drums are true, smooth up any slight scores by polishing with fine emery cloth. If deep scores or grooves are present which cannot be removed by this method, then the drum must be turned.

Adjusting Drum Brakes

PRELIMINARY ADJUSTMENT

1. Set a brake shoe adjustment gauge at .030 inch less than the brake drum diameter.
2. Center the gauge over the shoes at the greatest lining thickness and run out the adjuster until the new lining touches the gauge.
3. Install the drum.

NOTE: Tight clearance can aggravate normal seating problems.

Most service technicians prefer to set the initial brake adjustment with a gauge and then brake the vehicle backward and forward allowing the shoes and drums to seek the correct running clearance. Complete seating normally occurs within 1000 miles.

ROUTINE ADJUSTMENT

1. Use a brake adjusting tool to expand the brake shoes against the drum. Raising the tool handle turns the star wheel adjuster in the proper direction to expand the shoes. Turn the adjuster until a heavy drag is felt while turning the wheel.
2. Depress the brake pedal hard several times and recheck wheel drag. Continue to depress brake pedal and recheck drag until a true heavy drag is obtained.
3. Turn the star wheel adjuster in the opposite direction until the wheel turns freely.
4. Drive the car, braking forward and backward, to allow the self-adjusters to obtain the best running clearance.

NOTE: Exceptions to the preceding are General Motors' "H" body cars (Astre, Monza, Skyhawk, Starfire, Vega). These brakes are automatically adjusted when the parking brake is applied. After brake service, apply and release the parking brake until the brakes are correctly adjusted.

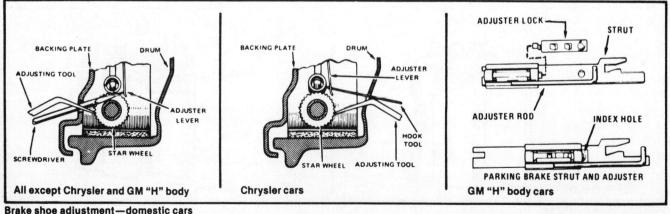

Brake shoe adjustment—domestic cars

BRAKES
General Information

BRAKE SYSTEM BLEEDING

If the master cylinder has been replaced it is more practical and safe to bleed most of the air out at the master cylinder. This can be done either on or off the car and prevents great masses of air from being passed through the system.

Manual Bleeding

1. Fill the master cylinder with new fluid of the correct type.
2. On cars with power brakes pump the brake pedal several times to remove all vacuum from the power unit.
3. Pump the brake pedal to pressurize the system and, while holding the pedal down, release the hydraulic pressure at the wheel cylinder bleeder valve. The pedal must be held depressed until the bleeder valve is closed to prevent air from entering the system.
4. Repeat until a steady, clear (no air bubbles) flow of fluid is seen at the wheel cylinder.

— CAUTION —
The bleeder valve at the wheel cylinder must be closed at the end of each stroke, and before the brake pedal is released, to insure that no air can enter the system. It is also important that the brake pedal be returned to the full up position so the piston in the master cylinder moves back enough to clear the bypass outlets.

Pressure Bleeding

Pressure bleeding equipment should be of the diaphragm type, placing a diaphragm between the pressurized air supply and the brake fluid. This prevents moisture and other contaminants from entering the hydraulic system.

NOTE: Front disc/rear drum equipped vehicles use a metering valve which closes off pressure to the front brakes under certain conditions. These systems contain manual release actuators which must be engaged to pressure bleed the front brakes.

1. Connect the tank hydraulic hose and adapter to the master cylinder.
2. Close hydraulic valve on the bleeder equipment.
3. Apply air pressure to the bleeder equipment.

— CAUTION —
Follow equipment manufacturer's recommendations for correct air pressure.

4. Open the valve to bleed air out of the pressure hose to the master cylinder.

NOTE: Never bleed this system using the secondary piston stopscrew on the bottom of many master cylinders.

5. Open the hydraulic valve and bleed each wheel cylinder. Bleed rear brake system first when bleeding both front and rear systems.

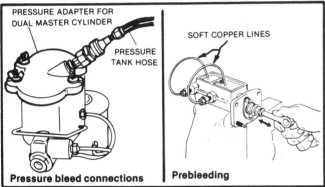

Bleeding the master cylinder

FLUSHING HYDRAULIC BRAKE SYSTEMS

Hydraulic brake systems must be totally flushed if the fluid becomes contaminated with water, dirt or other corrosive chemicals. To flush, simply bleed the entire system until *all* fluid has been replaced with the correct type of new fluid.

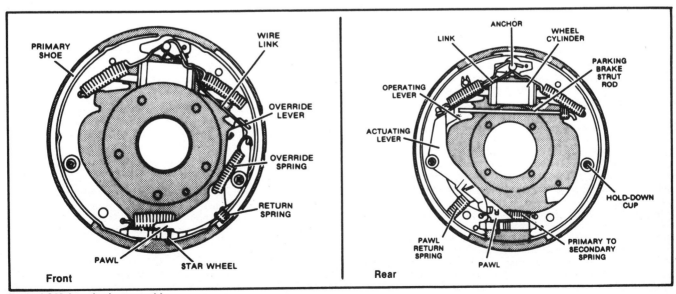

Typical GM drum brake assembly

1-7

SECTION 1

BRAKES
General Information

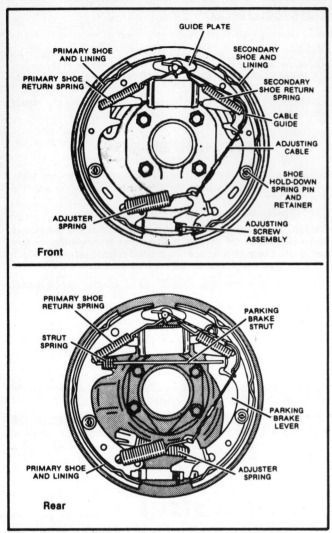

Typical AMC and Ford drum brake assembly

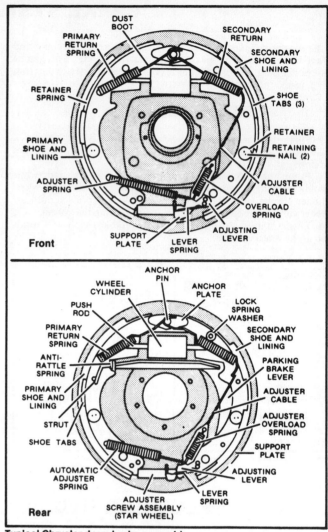

Typical Chrysler drum brake assembly

DISC BRAKE SERVICE

Caliper disc brakes can be divided into three types: the four-piston fixed-caliper type, the single-piston floating-caliper type, and the single piston sliding-caliper type.

In the four piston type (two in each side of the caliper) braking effect is achieved by hydraulically pushing both shoes against the disc sides.

With the single piston floating-caliper type, the inboard shoe is pushed hydraulically into contact with the disc; the reaction force thus generated is used to pull the outboard shoe into frictional contact (made possible by letting the caliper move slightly along the axle centerline).

In the sliding caliper (single piston) type, the caliper assembly slides along the machined surfaces of the anchor plate. A steel key located between the machined surfaces of the caliper and the machined surfaces of the anchor plate is held in place with either a retaining screw or two cotter pins. The caliper is held in place against the anchor plate with one or two support springs.

All disc brake systems are inherently self-adjusting and have no provision for manual adjustment.

Inspection

Disc pads (lining and shoe assemblies) should be replaced in axle sets (both wheels) when the lining on any pad is worn to $1/16$ in. at any point. *If lining is allowed to wear past $1/16$ in. minimum thickness severe damage to disc may result.*

NOTE: State inspection specifications take precedence over these general recommendations.

Note that disc pads in floating caliper type brakes may wear at any angle, and measurement should be made at the narrow end of the taper. Tapered linings should be replaced if the taper exceeds $1/8$ in. from end to end (the difference between the thickest and thinnest points).

---- **CAUTION** ----
To prevent costly paint damage, remove some brake fluid (don't re-use) from the reservoir and install the reservoir cover before replacing the disc pads. When replacing the pads, the piston is

BRAKES
General Information

depressed and fluid is forced back through the lines to squirt out of the fluid reservoir.

If the caliper is unbolted from the hub, do not let it dangle by the brake hose; it can be rested on a suspension member or wired onto the frame.

Servicing the Caliper Assembly

1. Raise the vehicle on a hoist and remove the front wheels.
2. Working on one side at a time only, disconnect the hydraulic inlet line from the caliper and plug the end. Remove the caliper mounting bolts or pins, and shims if used, and slide the caliper off the disc.
3. Remove the disc pads from the caliper. If the old ones are to be reused, mark them so that they can be reinstalled in their original positions.
4. Open the caliper bleed screw and drain the fluid. Clean the outside of the caliper and mount it in a vise with padded jaws.

CAUTION

When cleaning any brake components, use only brake fluid or denatured (Isopropyl) alcohol. Never use a mineral-based solvent, such as gasoline or paint thinner, since it will cause rubber parts to swell and quickly deteriorate.

---

5. Remove the bridge bolts, separate the caliper halves, and remove the two O-ring seals from the transfer holes.
6. Pry the lip on each piston dust boot from its groove and remove the piston assemblies and springs from the bores. If necessary, air pressure may be used to force the pistons out of the bores, using care to prevent them from popping out of control.
7. Remove the boots and seals from the pistons and clean the pistons in brake fluid. Blow out the caliper passages with an air hose.
8. Inspect the cylinder bores for scoring, pitting, or corrosion. Corrosion is a pitted or rough condition not to be confused with staining. Light rough spots may be removed by rotating crocus cloth, using finger pressure, in the bores. Do not polish with an in-and-out motion or use any other abrasive.
9. If the pistons are pitted, scored, or worn, they must be replaced. A corroded or deeply scored caliper should also be replaced.
10. Check the clearance of the pistons in the bores using a feeler gauge. Clearance should be 0.002–0.006 in. If there is excessive clearance, the caliper must be replaced.
11. Replace all rubber parts and lubricate with brake fluid. Install the seals and boots in the grooves in each piston. The seal should be installed in the groove closest to the closed end of the piston with the seal lips facing the closed end. The lip on the boot should be facing the seal.
12. Lubricate the piston and bore with brake fluid. Position the piston return spring, large coil first, in the piston bore.
13. Install the piston in the bore, taking great care to avoid damaging the seal lip as it passes the edge of the cylinder bore.
14. Compress the lip on the dust boot into the groove in the caliper. Be sure the boot is fully seated in the groove, as poor sealing will allow contaminants to ruin the bore.
15. Position the O-rings in the cavities around the caliper transfer holes, and fit the caliper halves together. Install the bridge bolts (lubricated with brake fluid) and be sure to torque to specification.
16. Install the disc pads in the caliper and remount the caliper on the hub. Connect the brake line to the caliper and bleed the brakes. Replace the wheels. Recheck the brake fluid level, check the brake pedal travel, and road test the vehicle.

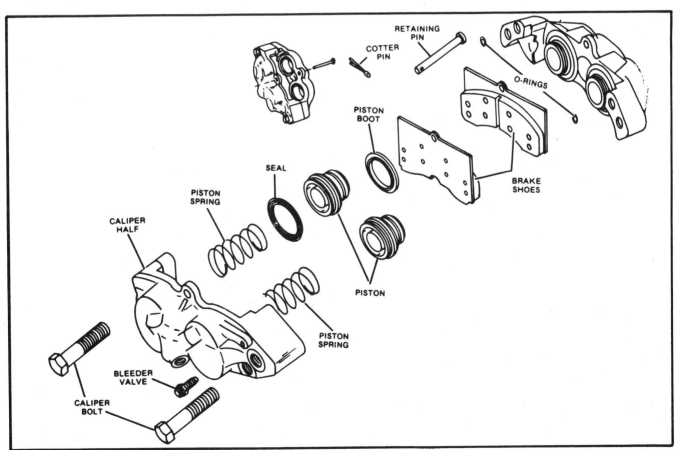

Typical four piston caliper

SECTION 1

BRAKES
General Information

BRAKE SYSTEM TUNE-UP PROCEDURE

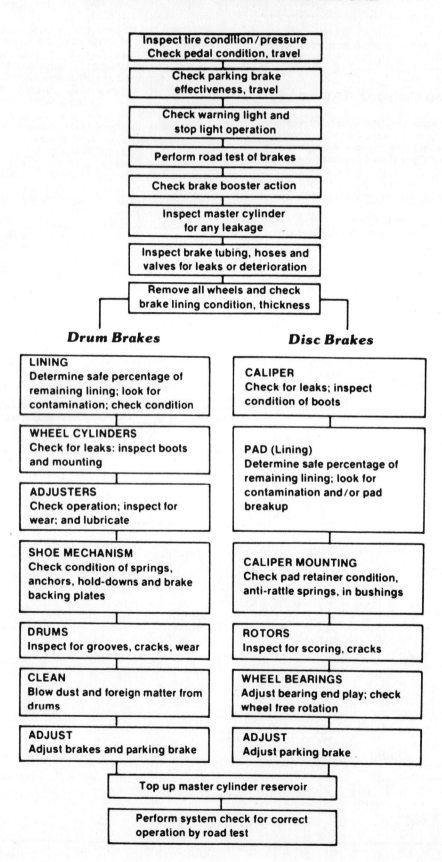

1-10

BRAKES
General Information

BRAKE PERFORMANCE DIAGNOSIS

The Condition	The Possible Cause	The Corrective Action
PEDAL GOES TO FLOOR	(a) Fluid low in reservoir. (b) Air in hydraulic brake system. (c) Improperly adjusted brake. (d) Leaking wheel cylinders. (e) Loose or broken brake lines. (f) Leaking or worn master cylinder. (g) Excessively worn brake lining.	(a) Fill and bleed master cylinder. (b) Fill and bleed hydraulic brake system. (c) Repair or replace self-adjuster as required. (d) Recondition or replace wheel cylinder and replace both brake shoes. (e) Tighten all brake fittings or replace brake line. (f) Recondition or replace master cylinder and bleed hydraulic system. (g) Reline and adjust brakes.
SPONGY BRAKE PEDAL	(a) Air in hydraulic system. (b) Improper brake fluid (low boiling point). (c) Excessively worn or cracked brake drums. (d) Broken pedal pivot bushing.	(a) Fill master cylinder and bleed hydraulic system. (b) Drain, flush and refill with brake fluid. (c) Replace all faulty brake drums. (d) Replace nylon pivot bushing.
BRAKES PULLING	(a) Contaminated lining. (b) Front end out of alignment. (c) Incorrect brake adjustment. (d) Unmatched brake lining. (e) Brake drums out of round. (f) Brake shoes distorted. (g) Restricted brake hose or line. (h) Broken rear spring.	(a) Replace contaminated brake lining. (b) Align front end. (c) Adjust brakes and check fluid. (d) Match primary, secondary with same type of lining on all wheels. (e) Grind or replace brake drums. (f) Replace faulty brake shoes. (g) Replace plugged hose or brake line. (h) Replace broken spring.
SQUEALING BRAKES	(a) Glazed brake lining. (b) Saturated brake lining. (c) Weak or broken brake shoe retaining spring. (d) Broken or weak brake shoe return spring. (e) Incorrect brake lining. (f) Distorted brake shoes. (g) Bent support plate. (h) Dust in brakes or scored brake drums.	(a) Cam grind or replace brake lining. (b) Replace saturated lining. (c) Replace retaining spring. (d) Replace return spring. (e) Install matched brake lining. (f) Replace brake shoes. (g) Replace support plate. (h) Blow out brake assembly with compressed air and grind brake drums.
CHIRPING BRAKES	(a) Out of round drum or eccentric axle flange pilot.	(a) Repair as necessary, and lubricate support plate contact areas (6 places).
DRAGGING BRAKES	(a) Incorrect wheel or parking brake adjustment. (b) Parking brakes engaged. (c) Weak or broken brake shoe return spring. (d) Brake pedal binding. (e) Master cylinder cup sticking. (f) Obstructed master cylinder relief port. (g) Saturated brake lining. (h) Bent or out of round brake drum.	(a) Adjust brake and check fluid. (b) Release parking brakes. (c) Replace brake shoe return spring. (d) Free up and lubricate brake pedal and linkage. (e) Recondition master cylinder. (f) Use compressed air and blow out relief port. (g) Replace brake lining. (h) Grind or replace faulty brake drum.
HARD PEDAL	(a) Brake booster inoperative. (b) Incorrect brake lining. (c) Restricted brake line or hose. (d) Frozen brake pedal linkage.	(a) Replace brake booster. (b) Install matched brake lining. (c) Clean out or replace brake line or hose. (d) Free up and lubricate brake linkage.
WHEEL LOCKS	(a) Contaminated brake lining. (b) Loose or torn brake lining. (c) Wheel cylinder cups sticking. (d) Incorrect wheel bearing adjustment.	(a) Reline both front or rear of all four brakes. (b) Replace brake lining. (c) Recondition or replace wheel cylinder. (d) Clean, pack and adjust wheel bearings.
BRAKES FADE (HIGH SPEED)	(a) Incorrect lining. (b) Overheated brake drums. (c) Incorrect brake fluid (low boiling temperature). (d) Saturated brake lining.	(e) Replace lining. (b) Inspect for dragging brakes. (c) Drain, flush, refill and bleed hydraulic brake system. (d) Reline both front or rear or all four brakes.
PEDAL PULSATES	(a) Bent or out of round brake drum.	(a) Grind or replace brake drums.
BRAKE CHATTER AND SHOE KNOCK	(a) Out of round brake drum. (b) Loose support plate. (c) Bent support plate. (d) Distorted brake shoes. (e) Machine grooves in contact face of brake drum. (Shoe Knock) (f) Contaminated brake lining.	(a) Grind or replace brake drums. (b) Tighten support plate bolts to proper specifications. (c) Replace support plate. (d) Replace brake shoes. (e) Grind or replace brake drum. (f) Replace either front or rear or all four linings
BRAKES DO NOT SELF ADJUST	(a) Adjuster screw frozen in thread. (b) Adjuster screw corroded at thrust washer. (c) Adjuster lever does not engage star wheel. (d) Adjuster installed on wrong wheel.	(a) Clean and free-up all thread areas. (b) Clean threads and replace thrust washer if necessary. (c) Repair, free up or replace adjusters as required. (d) Install correct adjuster parts.

SECTION 1

BRAKES
General Information

BRAKE PERFORMANCE DIAGNOSIS

The Condition	The Possible Cause	The Corrective Action
NOISE—Groan—Brake noise emanating when slowly releasing brakes (creep-groan).	(a) Not detrimental to function of disc brakes—no corrective action required. (Indicate to operator this noise may be eliminated by slightly increasing or decreasing brake pedal efforts.)	
RATTLE—Brake noise or rattle emanating at low speeds on rough roads, (front wheels only).	(a) Shoe anti-rattle spring missing or not properly positioned. (b) Excessive clearance between shoe and caliper.	(a) Install new anti-rattle spring or position properly. (b) Install new shoe and lining assemblies.
SCRAPING	(a) Mounting bolts too long. (b) Loose wheel bearings.	(a) Install mounting bolts of correct length. (b) Readjust wheel bearings to correct specifications.
FRONT BRAKES HEAT UP DURING DRIVING AND FAIL TO RELEASE	(a) Operator riding brake pedal. (b) Stop light switch improperly adjusted. (c) Sticking pedal linkage. (d) Frozen or seized piston. (e) Residual pressure valve in master cylinder. (f) Power brake malfunction.	(a) Instruct owner how to drive with disc brakes. (b) Adjust stop light to allow full return of pedal. (c) Free up sticking pedal linkage. (d) Disassemble caliper and free up piston. (e) Remove valve. (f) Replace.
LEAKY WHEEL CYLINDER	(a) Damaged or worn caliper piston seal. (b) Scores or corrosion on surface of cylinder bore.	(a) Disassembly caliper and install new seat. (b) Disassemble caliper and hone cylinder bore. Install new seal.
GRABBING OR UNEVEN BRAKE ACTION	(a) Causes listed under "Pull" (b) Power brake malfunction.	(a) Corrections listed under "Pull". (b) Replace.
BRAKE PEDAL CAN BE DEPRESSED WITHOUT BRAKING EFFECT	(a) Air in hydraulic system or improper bleeding procedure. (b) Leak past primary cup in master cylinder. (c) Leak in system. (d) Rear brakes out of adjustment. (e) Bleeder screw open.	(a) Bleed system. (b) Recondition master cylinder. (c) Check for leak and repair as required. (d) Adjust rear brakes. (e) Close bleeder screw and bleed entire system.
EXCESSIVE PEDAL TRAVEL	(a) Air, leak, or insufficient fluid in system or caliper. (b) Warped or excessively tapered shoe and lining assembly. (c) Excessive disc runout. (d) Rear brake adjustment required. (e) Loose wheel bearing adjustment. (f) Damaged caliper piston seal. (g) Improper brake fluid (boil). (h) Power brake malfunction.	(a) Check system for leaks and bleed. (b) Install new shoe and linings. (c) Check disc for runout with dial indicator. Install new or refinished disc. (d) Check and adjust rear brakes. (e) Readjust wheel bearing to specified torque. (f) Install new piston seal. (g) Drain and install correct fluid. (h) Replace.
BRAKE ROUGHNESS OR CHATTER (PEDAL PUMPING)	(a) Excessive thickness variation of braking disc. (b) Excessive lateral runout of braking disc. (c) Rear brake drums out-of-round. (d) Excessive front bearing clearance.	(a) Check disc for thickness variation using a micrometer. (b) Check disc for lateral runout with dial indicator. Install new or refinished disc. (c) Reface rear drums and check for out-of-round. (d) Readjust wheel bearings to specified torque.
EXCESSIVE PEDAL EFFORT	(a) Brake fluid, oil or grease on linings. (b) Incorrect lining. (c) Frozen or seized pistons. (d) Power brake malfunction.	(a) Install new shoe linings as required. (b) Remove lining and install correct lining. (c) Disassemble caliper and free up pistons. (d) Replace.
PULL	(a) Brake fluid, oil or grease on linings. (b) Unmatched linings. (c) Distorted brake shoes. (d) Frozen or seized pistons. (e) Incorrect tire pressure. (f) Front end out of alignment. (g) Broken rear spring. (h) Rear brake pistons sticking. (i) Restricted hose or line. (j) Caliper not in proper alignment to braking disc.	(a) Install new shoe and linings. (b) Install correct lining. (c) Install new brake shoes. (d) Disassemble caliper and free up pistons. (e) Inflate tires to recommended pressures. (f) Align front end and check. (g) Install new rear spring. (h) Free up rear brake pistons. (i) Check hoses and lines and correct as necessary. (j) Remove caliper and reinstall. Check alignment.

BRAKES
Domestic Vehicles

Domestic Vehicles

HYDRAULIC BRAKE SERVICE

Federal law requires cars to be equipped with two separate brake systems, so that if one system should fail, the other will provide enough braking power to safely stop the car. The standard approach has been to use a tandem master cylinder and separate hydraulic circuits for the front and rear brakes, or a diagonally split system separating opposite front and rear wheels. A tandem master cylinder actually uses two piston-and-seal assemblies in-line in a single bore. The dual system includes a red warning lamp on the instrument panel and, to activate it, a "Pressure Differential" valve which is connected to both sides of the system. The valve is sensitive to any loss of hydraulic pressure which results from a braking failure on either side of the system and alerts the driver by switching on the lamp. The lamp is connected to the ignition switch. With the switch in "start" position, the lamp is lit, furnishing a bulb check, but in "running" position, it will light only if a brake failure occurs. Although usual stops occur at moderate hydraulic pressures, during a "panic stop" the master cylinder develops pressure higher than 800 psi (pounds per square inch). Caliper disc brakes, being nonenergized, require more applying force than comparable energized-shoe/drum brakes. Caliper pistons, comparatively, are quite large and generally a higher pressure range is provided by power braking. Front disc-rear

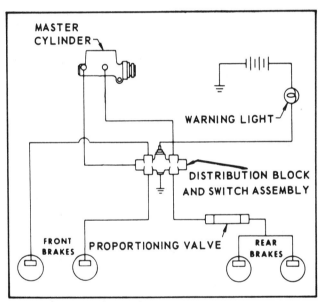

Brake system schematic

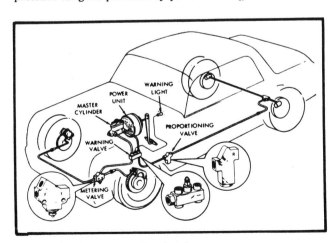

Typical dual system disc brake

drum brake vehicles are provided with pressure regulating units. Pressure-regulating units refine the braking balance, by changing the ratio of front-to-rear pressure, regulating it for moderate or severe stops as required to lessen skidding and diving. The "pressure metering" valve inhibits pressure to front disc brakes during easy, rolling stops. The "proportioning" valve reduces pressure to rear drum brakes in severe stops. One or both types of valves are found in various systems.

NOTE: All brake valving should be considered as nonrepairable. Replacement should be an exact duplicate of the unit that was designed for the car.

The dual master cylinder has two pistons, located one behind the other. The primary piston is actuated directly by mechanical linkage from the brake pedal. The secondary piston is actuated by fluid trapped between the two pistons. If a leak develops in front of the secondary piston, it moves forward until it

BRAKES
Domestic Vehicles

light switch incorporated in the master cylinder body. The piston is accessible by removing the large plug at the front of the master cylinder body. Only remove the plug when overhauling the cylinder, as brake fluid will escape.

Overhaul procedures on these new type master cylinders are basically the same as those on conventional master cylinder. bottoms against the front of the master cylinder. The fluid trapped between the pistons will operate one side of the split system. If the other side of the system develops a leak, the primary piston will move forward until direct contact with the secondary piston takes place, and it will force the secondary piston to actuate the other side of the split system. In either case the brake pedal drops closer to the floor board and less braking power is available.

HYDRAULIC BRAKE ACTUATING SYSTEM

Master Cylinders

The master cylinder unit is a highly calibrated unit specifically designed for the car it is on. Although the cylinders may look alike there are many differences in calibration. If replacement is necessary, make sure the replacement unit is the one specified for the car.

Some 1983 and later G.M. cars are equipped with "Quick Take-Up" master cylinders which provide a large volume of fluid to the wheel brakes at low pressure when the brake pedal is initially applied. This large volume of fluid is needed because of the new self retracting piston seals at the front disc brake calipers which pull the pistons into the calipers after the brakes are released, thereby preventing the brake pads from causing a drag on the rotors.

The master cylinder used on G.M. "X", "A" and "J" body front wheel drive cars has a hydraulically operated brake warning

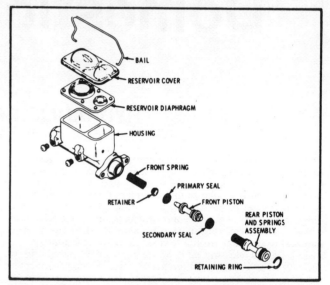

General Motors diesel engine type master cylinder

CONVENTIONAL TANDEM MASTER CYLINDER

Overhaul

1. Remove the secondary piston stop bolt from the bottom or inside the reservoir if so equipped.
2. Depress the primary piston and remove snap ring from retaining groove at the rear of the master cylinder bore.
3. Remove push rod and primary piston assembly from the master cylinder bore. Do not remove the screw that secures the primary return spring retainer, return spring, primary cup and protector on the primary piston. This assembly is factory pre-adjusted and should not be disassembled.
4. Remove the secondary piston assembly. Do not remove the outlet tube seats from the master cylinder body.
5. Inspect the parts for chipping, excessive wear or damage. When using a master cylinder repair kit, install all the parts supplied.
6. Be sure that all recesses, openings and internal passages are open and clean.
7. Inspect the master cylinder bore for signs of etching, pitting, scoring or rust. If necessary to hone the master cylinder bore to repair damage, do not exceed allowable home specifications, .003 thousands.
8. To assemble: dip all parts except the master cylinder body in clean brake fluid.
9. Carefully insert the complete secondary piston and return spring assembly in the master cylinder bore.
10. Install the primary piston assembly in the master cylinder bore.
11. Depress the primary piston and install the snap ring in the cylinder bore groove.
12. Install the push rod, boot and retainer on the push rod, if so equipped. Install the push rod assembly into the primary piston. Make sure the retainer is properly seated and holding the push rod securely.
13. Position the inner end of the push rod boot (if so equipped) in the master cylinder body retaining groove.
14. Install the secondary piston stop bolt, if used, with an O-ring if screw is on bottom outside of master cylinder casting. Pre-bleed the master cylinder before reinstalling in the car.
15. Install the cover and gasket on the master cylinder and secure the cover into position with the retainer.

QUICK TAKE UP & DIAGONAL SPLIT MASTER CYLINDER

Overhaul

NOTE: Plastic reservoirs need to be removed only for the following reasons.

1. Reservoir is damaged or the rubber grommet(s) between the reservoir and bore is leaking.
2. Removal of stop pin from Chrysler style plastic reservoir master cylinder to allow removal of pistons. Pin is located underneath front reservoir nipple.
3. If the G.M. quick take-up valve is defective, the entire master cylinder body must be replaced. The plastic reservoir may be reused on the new master cylinder body. The reservoir should be removed by first clamping the cylinder flange in a vise. Next remove the reservoir for the Chrysler style. Grasp the reservoir base on one end and pull away from the body. GM reservoirs must be removed by prying between the reservoir and casting with a pry bar. Grommets can be reused if they are in good condition. Whether or not the reservoir is removed, it and the cover or caps should be thoroughly cleaned.

BRAKES
Domestic Vehicles

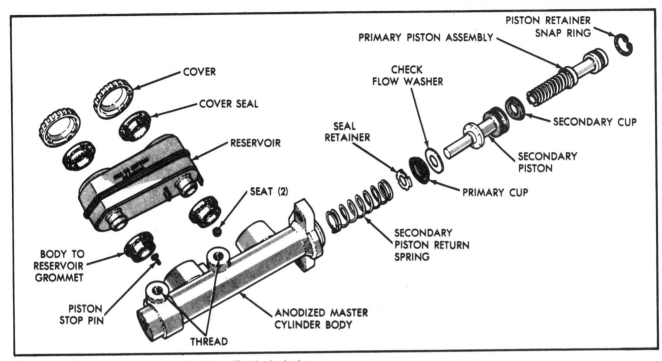

Chrysler Aluminum master cylinder — Exploded view

4. Remove the cylinder from the car and drain the brake fluid.
5. Mount the cylinder in a vise so that the outlets are up and remove the seal from the hub.
6. Remove the stop pin or screw from the bottom of the front reservoir, if present.
7. Remove the snap-ring from the front of the bore and remove the primary piston assembly.
8. Remove the secondary piston assembly using compressed air or a piece of wire.
9. Clan metal parts in brake fluid and discard rubber parts. Inspect the bore for damage or wear, and check pistons for damage and proper clearance in the bore. Aluminum cylinder bores cannot be honed. The cylinder must be replaced if the bore is scored.
10. If the bore is only slightly scored or pitted it may be honed. Always use hones that are in good condition and completely clean the cylinder with brake fluid when honing is completed. If any sign of wear or corrosion is apparent on "Quick Take-Up" master cylinder bores, the master cylinder must be replaced; it cannot be honed. If any evidence of contamination exists in the master cylinder the entire hydraulic system should be flushed and refilled with clean brake fluid. Blow out passages with compressed air.
11. Install new secondary seals in the two grooves in the flat end of the front piston. The lips of the seals will be facing away from each other.
12. Install a new primary seal and the seal protector on opposite end of the front piston with the lips of the seal facing outward.
13. Coat the seals with brake fluid. Install the spring on the front piston with the spring retainer in the primary seal.
14. Insert the piston assembly, spring end first, into the bore and use a wooden rod to seat it.
15. Coat the rear piston seals with brake fluid and install them into the piston grooves with the lips facing the spring end.
16. Assemble the spring onto the piston and install the as-

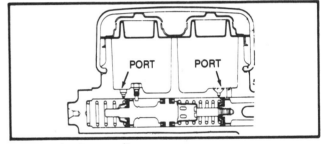

Feed and return ports

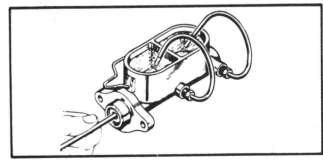

Pre-bleeding master cylinder

sembly into the bore spring first. Install the snapring.
17. Hold the piston at the bottom of the bore and install the stop screw.
18. On G.M. models with the hydraulic brake warning light switch, (Quick Take-Up Units) remove the allen head plug and remove the switch assembly with needle nose pliers. Remove the O-rings and retainers from the piston. Install new O-rings

1-15

BRAKES
Domestic Vehicles

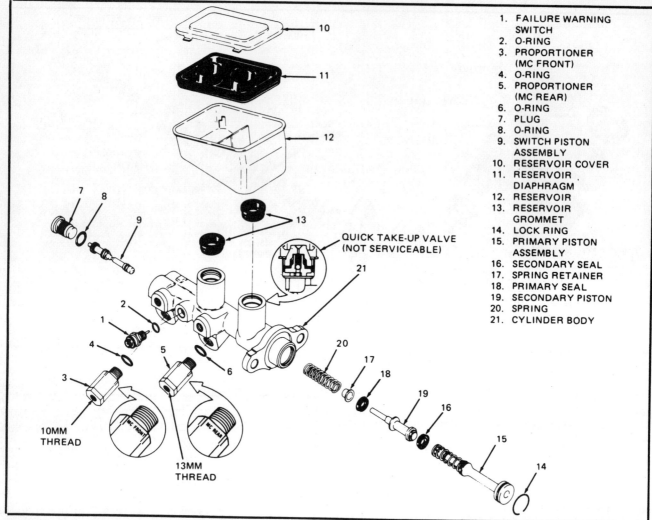

Typical diagonal split master cylinder

and retainers, fit the piston back into the master cylinder after lubricating with brake fluid. If any corrosion is present in the switch piston bore the master cylinder must be replaced; do not attempt to hone the bore.

19. Fit a new O-ring on the allen head plug and install the plug and tighten.

20. On all master cylinder, install a new seal in the hub, if equipped, then either bench bleed or bleed the cylinder on the car. Some master cylinders have bleed screws on the outlet flanges and may be bled without disturbing the wheel cylinders and calipers.

MASTER CYLINDER PUSH ROD

Adjustable

After assembly of the master cylinder to the power section or firewall, the piston cup in the hydraulic cylinder should just clear the compensating port hole when the brake pedal is fully released. If the push rod is too long, it will hold the piston over the port.

A push rod that is too short, will give too much loose travel (excessive pedal play). Apply the brakes and release the pedal all the way observing the brake fluid flow back into the master cylinder.

A full flow indicates the piston is coming back far enough to release the fluid. A slow return of the fluid indicates the piston is not coming back far enough to clear the ports. The push rod adjustment is too tight, and should be adjusted.

Non Adjustable

When installing a non-adjustable type push rod, make sure that the push rod is fully seated in the master cylinder.

Disc Brake Calipers

Caliper disc brakes can be dived into three types: the four-piston, fixed-caliper type; the single-piston, floating-caliper type, and the single-piston sliding-caliper type. Refer to the Brake Specifications Chart for applications.

In the four piston type (two in each side of the caliper) braking effect is achieved by hydraulically pushing both shoes against the disc sides. With the single piston floating-caliper type, the inboard shoe is pushed hydraulically into contact with the disc, while the reaction force thus generated is used to

BRAKES
Domestic Vehicles

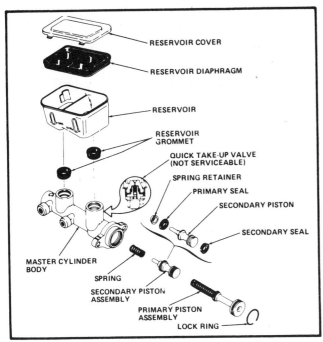

1981 and later General Motors Quick Take-Up master cylinder (not used on diesel engines)

pull the outboard shoe into frictional contact (made possible by letting the caliper move slightly along the axle centerline).

In the sliding caliper (single piston) type, the caliper assembly slides along the machined surfaces of the anchor plate. A steel key located between the machined surfaces of the caliper and the machined surfaces of the anchor plate is held in place with either a retaining screw or two cotter pins. The caliper is held in place against the anchor plate with one or two support springs.

SINGLE BORE

Overhaul

1. Raise the vehicle and safely support it. Remove the front wheels.
2. Working on one side at a time only, disconnect the hydraulic inlet line from the caliper and plug the end. Remove the caliper mounting bolts or pins, and shims (if used) and slide the caliper off the disc.
3. Remove the disc pads from the caliper or mounting adapter. If the old ones are to be reused, mark them so that they can be reinstalled in their original positions.
4. Open the caliper bleed screw and drain the fluid. Clean the outside of the caliper and mount it in a vise with padded jaws.

NOTE: When cleaning any brake components, use only brake fluid or denatured (Isopropyl) alcohol. Never use a mineral-based solvent, such as gasoline or paint thinner, since it will swell and quickly deteriorate rubber parts.

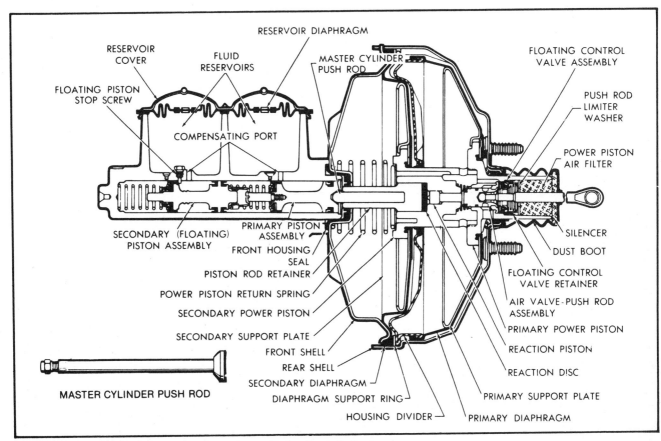

Typical Power brake booster and master cylinder installation

1-17

SECTION 1

BRAKES
Domestic Vehicles

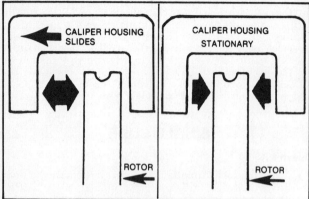
Floating (or sliding) caliper type Fixed caliper type

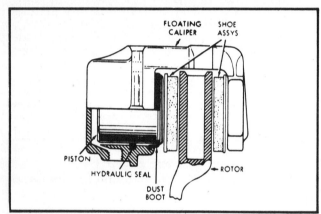
Floating caliper disc brake

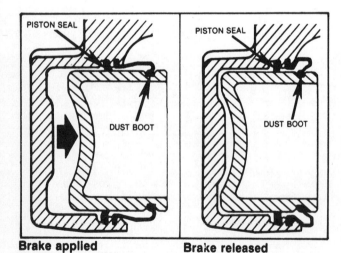
Brake applied Brake released

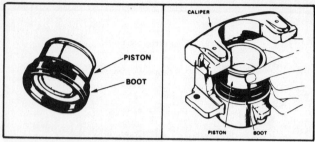
Assembling boot on piston Installing piston

piston in brake fluid. Blow out the caliper passages with an air hose.

7. Inspect the cylinder bore for scoring, pitting, or corrosion. Corrosion is a pitted or rough condition not to be confused with staining. Light rough spots may be removed by rotating crocus cloth, using finger pressure, in the bores. Do not polish with an in and out motion or use any other abrasive.

8. If the pistons are pitted, scored, or worn, they must be replaced. A corroded or deeply scored caliper should also be replaced.

9. Check the clearance of the piston in the bores using a feeler gauge. Clearance should be 0.002 to 0.006 inch. If there is excessive clearance the caliper must be replaced.

10. Replace all rubber parts and lubricate with brake fluid. Install the seals (or square cut rings) and boots in the grooves in each piston. The seal should be installed in the groove closest to the closed end of the piston with the seal lips facing the closed end. The lip on the boot should be facing the seal.

11. Lubricate the piston and bore with brake fluid. Position the piston return spring, large coil first, in the piston bore.

12. Install the piston in the bore, taking great care to avoid damaging the seal lip as it passes the edge of the cylinder bore.

13. Compress the lip on the dust boot into the groove in the caliper. Be sure the boot is fully seated in the groove, as poor sealing will allow contaminants to ruin the bore.

14. Install the disc pads in the caliper and remount the caliper on the hub. Connect the brake line to the caliper and bleed the brakes. Replace the wheels. Recheck the brake fluid level, check the brake pedal travel, and road test the vehicle.

FOUR BORE

Overhaul

1. Pull cotter pin from end of shoe assembly retaining pin. Remove the pin and shoe assembly from the caliper. Identify the inboard and outboard shoes if they are to be reused. Remove the end of brake hose at bracket by removing U-shaped retainer from the hose fitting and withdrawing the hose from bracket.

2. Separate the caliper halves by removing the two large bolts. Remove the two small O-rings from the cavities around the fluid transfer holes in the two ends of the caliper halves.

3. To free the piston boots so that the pistons may be removed, push the piston down into the caliper as far as it will go. Insert a suitable tool under the inner edge of the steel ring in the boot, and using the piston as a fulcrum, pry the boot from its seat in the caliper half. Use care not to puncture seal when removing pistons from caliper.

4. Remove the pistons and piston springs from the caliper half. Remove the pistons and piston springs from the caliper half. Remove the boots and seals from their grooves in the piston.

NOTE: Always use clean brake fluid to clean caliper parts. Never use mineral-base cleaning solvents such as

5. Pry the lip on (each) piston dust boot from its groove and remove the piston assemblies and spring from the bore. If necessary, air pressure may be used to force the piston out of the bore, using care to prevent the piston from popping out of control.

6. Remove the boot and seal from the piston and clean the

BRAKES
Domestic Vehicles

gasoline, kerosene, carbon-tetrachloride, acetone, paint thinner or units of like nature as these solvents deteriorate rubber parts, causing them to become soft and swollen in an extremely short time.

5. To install, clean all metal parts using clean brake fluid. Remove all traces of dirt and grease. After cleaning, wipe all fluid from boot counter bore in caliper and from boot groove in piston. These surfaces must be clean and dry at assembly to permit RTV to properly adhere.

6. Using an air hose, blow out all fluid passages in the caliper halves, making sure that there is no dirt or foreign material blocking any of these passages.

7. Discard all rubber parts. Boots, seals, and O-rings should be replaced with new service kit parts.

8. Carefully inspect the piston bores in the caliper halves. They must be frees of scores and pits. A scored or otherwise damaged bore will cause leaks and unsatisfactory brake operation. The bore surface should be restored by polishing with a very fine crocus cloth. If the bore surface cannot be restored using a very fine crocus cloth, it may be lightly honed. Replace the caliper half if either bore is damaged to the extent that light honing will not restore it.

9. Check the fit of the piston in the bore using a feeler gauge. Clearance should be, 0.0045 to 0.010 inch for the 1⅞ inch bore and 0.0035 to .009 inch for the 1⅜ inch bore. If the bore is not damaged, and the clearance exceeds either of the upper limits, a new piston that does meet the clearance specified will be required.

10. Assemble the seal in the groove in the piston which is closest to the flat end of the piston. The lip on the seal must face toward the large end of the piston.

11. Install the piston assembly in the bore using Piston Ring Compressor Tool J-22629 or J-22639 or equivalent. Use care not to damage the seal lip as piston is pressed past the edge of the bore.

12. The boot groove on the piston is the groove closest to the concave end of the piston. Insert a bead of silastic sealant GM1052366 or equivalent into the boot groove in the piston and assemble the boot in the groove. The fold in the boot must face toward the end of the piston with the seal on it.

13. Depress the pistons and check that they slide smoothly into the bore until the end of the piston is flush with the end of the bore. If not, recheck piston assembly and location of the piston spring and the seal.

14. Position Boot Seal Installer Tool J-22628 or J-22638 or equivalent over the piston and seat the steel boot retaining ring evenly in the counterbore as shown. The boot retaining ring must be flush or below the machined face of the caliper. Any distortion of uneven seating could allow contaminating and corrosive elements to enter the bore.

15. Depress pistons and while holding in a depressed position place a bead of silastic sealant GM1052366 or equivalent on outer diameter of the boot retaining ring forming a seal between the boot retainer ring and the housing.

16. Position the O-rings in the small cavities around the brake fluid transfer hole in both ends of the outboard caliper halves. Lubricate the hex head bolts with Delco Brake Lube (or equivalent) or dip in clean brake fluid. Fit caliper halves together and secure with bolts.

17. Carefully mount the assembled caliper over the edge of the disc. Use two screwdrivers to depress pistons so that the caliper can be lowered into position on the disc. Use care to prevent damage to boots on the edge of the disc as the caliper is mounted.

18. Secure the caliper to the mounting bracket with two hex head bolts. Refer to torque specifications in rear of manual for correct torque values. If replacing old shoe assemblies, be sure to install the shoes in the same position from which they were removed.

19. Install the shoe and lining assemblies as outlined in this section.

20. Place a new copper gasket on the male end of the front wheel brake hose. Install brake hose in the calipers. With the wheels straight ahead, pass the female end of the brake hose through the support bracket.

21. Make sure the tube seat is clean and connect the brake line tube nut to the caliper. Tighten securely.

22. Allowing the hose to seek a normal position without twist, insert hex of the hose fitting into the hole in the support bracket and secure it in place with the U-shaped retainer (maximum allowed twist is ±1 notch). Turn the steering geometry from lock to lock while observing the hose. Check that the hose does not touch other parts at any time during suspension or geometry travel. If contact does occur, remove the U-shaped retainer and rotate the end of the hose in the support bracket in a direction which will eliminate hose contact. Reinstall the retainer and recheck for hose contact. If it is satisfactory, place the steel tube connector in the hose fitting and tighten securely.

23. If rear brake caliper is being serviced, connect brake line to caliper.

24. Bleed brakes as outlined in this section.

25. Install wheels and lower vehicle. Do not move car until a firm pedal is obtained.

SLIDING OR FLOATING CALIPER, FROZEN CALIPER PISTON

Hydraulic Removal

1. Remove the caliper assembly from the rotor.
2. Remove brake pads and dust seal. With brake flexible line connected and bleed screw closed apply enough pedal pressure to move the piston most of the way out of the bore (brake fluid will begin to ooze past the piston inner seal).

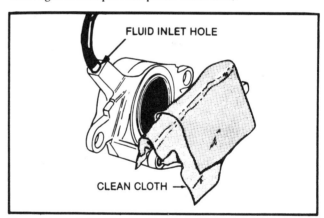

Removing piston hydraulically

Pneumatic Removal

1. Remove the caliper from the car.
2. With the bleed screw closed apply air pressure to force the piston out. Hydraulic and pneumatic methods of piston removal should be done carefully to prevent personal injury or piston damage.

FIXED CALIPER, FROZEN CALIPER PISTONS

NOTE: The hydraulic or pneumatic methods which apply to the single piston type caliper will not work on the multiple piston type brake caliper.

1-19

BRAKES
Domestic Vehicles

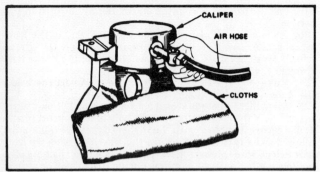
Removing piston pneumatically

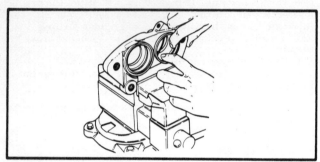
Installing fixed position rectangular ring seal (seal lip toward pressure slide)

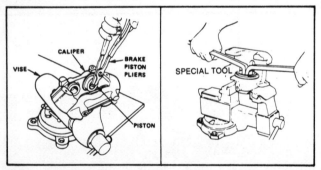
Removing pistons   Removing hollow and piston

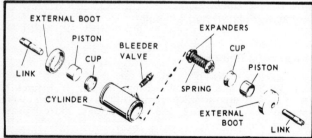
Wheel cylinder components

1. Remove the caliper from the car with the two halves separated.
2. Mount in a vise and use a piston puller (many types available) to remove the pistons.

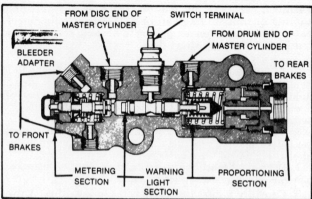
Push valve in when pressure bleeding— not necessary when using pedal bleed method

CALIPER CLEANING AND HONING

Castings may be cleaned with any type cleaning fluid after all rubber seals have been removed. It is important that all traces of cleaning fluid be completely removed from the caliper casting. Rubber components are compatible with alcohol and/or brake fluid.

Use a lint free wiping cloth to clean the caliper and parts. Black stains on pistons or walls, caused by the seals, will not do harm; however, extreme cleanliness is essential. Blow out passages with compressed air. A fine grade of crocus cloth may be used to correct minor imperfections in the cylinder bore. Slide crocus cloth with finger pressure in a circular rather than a lengthwise motion. Do not use any form of abrasive on a plated piston. Discard a piston which is pitted or has signs of plating wear.

If a fine stone honing of a caliper bore is necessary it should be done with skill and caution. Some cars can develop 800 psi. hydraulic pressure on severe application so the honing must never exceed .003 in. Also the dust seal groove must be free of rush or nicks so that a perfect mating surface is possible on piston and casting.

Wheel Cylinder

Overhaul

1. Raise the car and support it safely. Remove the wheel and drum from the side to be serviced.
2. Remove the brake shoes and clean the backing plate and wheel cylinder. Rebuilding can be done on the car, depending on the design of the brake backing plate. If the backing plate is recessed to the point that it is impossible to get a hone into the cylinder, the cylinder has to be removed.
3. To remove the cylinder: disconnect the brake line from the rear of the cylinder, remove the mounting bolts or retainers and remove the cylinder.

NOTE: On some models, in order to remove the rear wheel cylinders you must remove the wheel cylinder retainer. Insert two awls into the access slots and bend both tabs at the same time thereby releasing the cylinder. You must use a new retainer when reinstalling the wheel cylinder. The new retainer can be driven on using an 1 1/8 inch socket with an extension bar.

4. Remove the rubber boots (dust covers) from the ends of the cylinder. Remove the pistons, piston cups (expanders, if

BRAKES
Domestic Vehicles

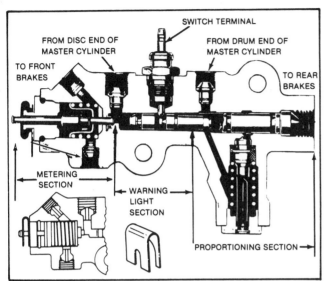

Hold valve out .060 in pressure bleed only – not necessary when using pedal bleed method

equipped) and spring from the inside of the cylinder. Remove the bleeder screw and make sure it is not clogged.

5. Discard all of the parts that the rebuilding kit will replace.
6. Examine the inside of the cylinder. If it is severely rusted, pitted or scratched install a new or rebuilt cylinder.
7. If the condition of the cylinder indicates that it can be rebuilt, hone the bore. Light honing will provide a new surface on the inside of the cylinder which promotes better cup sealing.
8. Wash out the cylinder with brake fluid after honing. Reassemble the cylinder using the new parts provided in the kit. When assembling the cylinder dip all parts in brake fluid.
9. Install the cylinder on the car. Reinstall the brakes, drum and wheel. Bleed the brake system.

Hydraulic Control Valves

PRESSURE DIFFERENTIAL VALVE

The pressure differential valve activates a dash panel warning light if pressure loss in the brake system occurs. If pressure loss occurs in one half of the split system the other system's normal pressure causes the piston in the switch to compress a spring until it touches an electrical contact. This turns the warning lamp on the dash panel to light, thus warning the driver of possible brake failure.

On some cars the spring balance piston automatically recenters as the brake pedal is released warning the driver only upon brake application. On other cars, the light remains on until manually cancelled. Valves may be located separately or as part of a combination valve. On GM front wheel drive cars, the valve and switch are usually incorporated into the master cylinder.

Resetting Valves

On some cars, the valve piston(s) remain off-center after failure until necessary repairs are made. The valve will automatically reset itself (after repairs) when pressure is equal on both sides of the system.

If the light does not go out, bleed the brake system that is opposite the failed system. If front brakes failed, bleed the rear brakes; this should force the light control piston toward center. If this fails, remove the terminal switch. If brake fluid is present in the electrical area, the seals are gone, replace the complete valve assembly.

METERING VALVE

The metering valve's function is to improve braking balance between the front disc and rear drum brakes, especially during light brake application. The metering valve prevents application of the front disc brakes until the rear brakes overcome the return spring pressure. Thus, when the front disc pads contact the rotor, the rear shoes will contact the brake drum at the same time.

Inspect the metering valve each time the brakes are serviced. A slight amount of moisture inside the boot does not indicate a defective valve, however, fluid leakage indicates a damaged or worn valve. If fluid leakage is present the valve must be replaced.

The metering valve can be checked very simply. With the car stopped, gently apply the brakes. At about an inch of travel a very small change in pedal effort (like a small bump) will be felt if the valve is operating properly. Metering valves are not serviceable, and must be replaced if defective.

PROPORTIONING VALVE

The proportioning (pressure control) valve is used, on some cars, to reduce the hydraulic pressure to the rear wheels to prevent skid during heavy brake application and to provide better brake balance. It is usually mounted in line to the rear wheels.

Whenever the brakes are serviced the valve should be inspected for leakage. Premature rear brake application during light braking can mean a bad proportioning valve. Repair is by replacement of the valve. Make sure the valve port marked "R" is connected toward the rear wheels.

On GM Quick Take-Up master cylinders, the proportioning valve(s) are screwed into the master cylinder. Since these cars have a diagonally split brake system, two valves are required. One rear brake line screws into each valve. The early type valves (GM front wheel drive) were steel and silver colored, an occasional "clunking" noise was encountered on some early models, but does not affect brake efficiency. Replacement valves are now made of aluminum. Never mix an aluminum valve with a steel valve, always use two aluminum valves.

COMBINATION VALVE

The combination valve may perform two or three functions. They are: metering, proportioning and brake failure warning. Variations of the two-way combination valve are: proportioning and brake failure warning or metering and brake failure warning. A three-way combination valve directs the brake fluid to the appropriate wheel, performs necessary valving and contains a brake failure warning. The combination valve is usually mounted under the hood close to the master cylinder, where the brake lines can easily be connected and routed to the front or rear wheels. The combination valve is non-serviceable and must be replaced if malfunctioning.

Brake Bleeding

The hydraulic brake system must be free of air to operate properly. Air can enter the system when hydraulic parts are disconnected for servicing or replacement, or when the fluid level in the master cylinder reservoirs is very low. Air in the system will give the brake pedal a spongy feeling upon application.

The quickest and easiest of the two ways for system bleeding is the pressure method, but special equipment is needed to ex-

BRAKES
Domestic Vehicles

ternally pressurize the hydraulic system. The other, more commonly used method of brake bleeding is done manually.

Bleeding Sequence

Bleeding may be required at only one or two wheels or at the master cylinder, depending upon what point the system was opened to air. If after bleeding the cylinder/caliper that was rebuilt or replaced and the pedal still has a spongy feeling upon application, it will be necessary to bleed the entire system.

Procedure

1. Master Cylinder: If the cylinder is not equipped with bleeder screws, open the brake line(s) to the wheels slightly while pressure is applied to the brake pedal. Be sure to tighten the line before the brake pedal is released. The procedure for bench bleeding the master cylinder is in the following section.
2. Power Brake Booster: If the unit is equipped with bleeder screws, it should be bled after the master cylinder. The car engine should be off and the brake pedal applied several times to exhaust any vacuum in the booster. If the unit is equipped with two bleeder screws, always bleed the higher located one first.
3. Combination Valve: If equipped with a bleeder screw.
4. Front/Back Split Systems: Start with the wheel farthest away from the master cylinder, usually the right rear wheel. Bleed the other rear wheel then the left front and right front.

NOTE: If you are unsuccessful in bleeding the front wheels, it may be necessary to deactivate the metering valve. This is accomplished by either pushing in, or pulling out a button or stem on the valve. The valve may be held by hand, with a special tool or taped; it should remain deactivated while the front brakes are bled.

5. Diagonally Split System: Start with the right rear then the left front. The left rear then the right front (refer to the following "GM Quick Take-Up Master Cylinder" section.
6. Rear Disc Brakes: If the car is equipped with rear disc brakes and the calipers have two bleeder screws, bleed the inner first then the outer. Do not allow brake fluid to spill on the car's finish, it will remove the paint. Flush the area with water.

Manual Bleeding

1. Clean the bleed screw at each wheel.
2. Start with the wheel farthest from the master cylinder (right rear).
3. Attach a small rubber hose to the bleed screw and place the end in a clear container of brake fluid.
4. Fill the master cylinder with brake fluid. (Check often during bleeding). Have an assistant slowly pump up the brake pedal and hold pressure.
5. Open the bleed screw about one-quarter turn, press the brake pedal to the floor, close the bleed screw and slowly release the pedal. Continue until no more air bubbles are forced from the cylinder on application of the brake pedal.
6. Repeat procedure on remaining wheel cylinders and calipers still working from cylinder/caliper farthest away from the master cylinder.
7. Master cylinders equipped with bleed screws may be bled independently. When bleeding the Bendix-type dual master cylinder it is necessary to solidly cap one reservoir section while bleeding the other to prevent pressure loss through the cap vent hole.
8. The disc should be rotated to make sure that the piston has returned to the unapplied position when bleeding is completed and the bleed screw closed.
9. The bleeder valve at the wheel cylinder must be closed at the end of each stroke, and before the brake pedal is released, to insure that no air can enter the system. It is also important that the brake pedal be returned to the full up position so the piston in the master cylinder moves back enough to clear the bypass outlets.

Pressure Bleeding Disc Brakes

Pressure bleeding disc brakes will close the metering valve and the front brakes will not bleed. For this reason it is necessary to manually hold the metering valve open during pressure bleeding. Never use a block or clamp to hold the valve open, and never force the valve stem beyond its normal position. Two different types of valves are used. The most common type requires the valve stem to be held in while bleeding the brakes, while the second type requires the valve stem to be held out (0.060 inch minimum travel). Determine the type of visual inspection.

Special adapters are required when pressure bleeding cylinders with plastic reservoirs. Pressure bleeding equipment should be diaphragm type; placing a diaphragm between the pressurized air supply and the brake fluid. This prevents moisture and other contaminants from entering the hydraulic system.

Front disc/rear drum equipped vehicles use a metering valve which closes off pressure to the front brakes under certain conditions. These systems contain manual release actuators which must be engaged to pressure bleed the front brakes.

1. Connect the tank hydraulic hose and adapter to the master cylinder.
2. Close hydraulic valve on the bleeder equipment.
3. Apply air pressure to the bleeder equipment. Follow equipment manufacturer's recommendations for correct air pressure.
4. Open the valve to bleed air out of the pressure hose to the master cylinder. Never bleed this system using the secondary piston stopscrew on the bottom of many master cylinders.
5. Open the hydraulic valve and bleed each wheel cylinder and caliper. Bleed rear brake system first when bleeding both front and rear systems.

BRAKES BLEEDING SEQUENCE CHART

Make	Model	System Split	Special Procedures	Bleeding Sequence
AMC	All except Alliance	Front-to-Rear	①	1. Passenger rear 2. Driver rear 3. Passenger front 4. Driver front
Chrysler	All	Front-to-Rear And Diagonal	①	1. Passenger rear 2. Driver rear 3. Passenger front 4. Driver front

1-22

BRAKES
Domestic Vehicles

BRAKES BLEEDING SEQUENCE CHART

Make	Model	System Split	Special Procedures	Bleeding Sequence
General Motors	Rear Drive and E,K Body ⑤	Front-to-rear	①	1. Passenger rear 2. Driver rear 3. Passenger front 4. Driver front
	Front Drive Except E,K Body ⑤	Diagonal	②③	1. Passenger rear 2. Drive front 3. Driver rear 4. Passenger front
Corvette		Front-to-Rear	①④	1. Driver rear-inner 2. Driver rear-outer 3. Passenger rear-inner 4. Passenger rear-outer 5. Driver front 6. Passenger front
Ford	Rear Drive-Front-to-Rear		①	1. Passenger rear 2. Driver rear 3. Passenger front 4. Driver front
	Front Drive	Diagonal		1. Passenger rear 2. Driver front 3. Driver rear 4. Passenger front

① It may be necessary to push on or pull out a button or rod on the metering valve to bleed front disc brakes, particularly when pressure bleeding.
② Use SLOW strokes only when manually bleeding the quick take-up system.
③ Always bleed the master cylinder first in the quick take-up system. Bleed the left front caliper fitting first and then the right hand fitting.
④ Raise the front of the vehicle slightly when bleeding the rear brakes.
⑤ E = Riviera, Eldorado and Toronado
K = Seville

DISC BRAKE APPLICATION CHART & SPECIFICATIONS

Make/Model	Text Reference Type	Caliper Style	Manufacturer	Anchor Bolt (ft lbs)	Bridge, Pin or Key Bolts (ft lbs)	Wheel Lugs (ft lbs)	Minimum Thickness Normal Standard	Minimum Thickness Machine To	Minimum Thickness Discard At	Rotor Parallel Variation	Max. Run-out
American Motors											
Eagle	1	Sliding	Bendix	100	30	75	.880	.815	.810	.0005	.003
Concord, Spirit	1	Sliding	Bendix	85	30	75	.880	.815	.810	.0005	.003
Chrysler Corporation (Front Wheel Drive)											
Aries, Reliant LeBaron, Dodge 400, 600-H.D. Brakes	6	Floating	K/H	70-100	18-22	80	9.35	.912	.882	.0005	.004
WO/H.D. Brakes	7	Floating	ATE	70-100	18-22	80	9.35	.912	.882	.0005	.004
E Class, New Yorker, Town & Country, Lazer	6	Floating	ATE or K/H	70-100	ATE 18-22 K/H 25-35	80	.935	.912	.882	.0005	.004
Omni, Horizon, Charger, Turismo	6	Floating	K/H	70-100	25-40	80	.500	.461	.431	.0005	.004
Aries, Reliant, LeBaron, Dodge 400	7	Floating	ATE	70-100	18-22	85	.935	.912	.882	.0005	.004
Omni, Horizon	6	Floating	K/H	70-100	25-40	85	.500	.461	.431	.0005	.004
Chrysler Corporation (Rear Wheel Drive) Cordoba, Diplomat Gran Fury, Mirada, New Yorker, Imperial	3	Sliding	Chrysler	95-125	15-20	85	1.010	.955	.940	.0005	.004

BRAKES
Domestic Vehicles

DISC BRAKE APPLICATION CHART & SPECIFICATIONS

Make/Model	Text Reference Type	Caliper Style	Manufacturer	Anchor Bolt (ft lbs)	Bridge, Pin or Key Bolts (ft lbs)	Wheel Lugs (ft lbs)	Normal Standard	Machine To	Discard At	Rotor Parallel Variation	Max. Run-out
Ford Motor Company (Front Wheel Drive) Escort, Lynx, LN7 EXP, Tempo, Topaz	5	Sliding	Ford	—	18–25	80–105	.945	—	.882	.0005	.003
Ford Motor Company (Rear Wheel Drive) Linc, Continental, Mark VII-Front	8	Sliding	Ford	—	40–60	80–105	1.030	—	.972	.0005	.003
Rear	6	Sliding	K/H	85–115	15–20	80–105	.945	—	.895	.0004	.004
Linc, Town Car, Crown Victoria, Grand Marquis	8	Sliding	Ford	—	40–60	80–105	1.030	—	.972	.0005	.003
Ford, Mercury-Front	8	Sliding	Ford	—	40–60	80–105	1.030	—	.972	.0005	.003
All models exc. noted	8	Sliding	Ford	—	30–40	80–105	.870	—	.810	.0005	.003
LTD, Cougar, XR7, Country Squire	1	Sliding	K/H	90–120	12–16	80–105	1.180	—	1.120	.0005	.003
Disc Brakes-Rear	6	Sliding	K/H	81–115	15–20	80–105	.945	—	.895	.0004	.0003
General Motors—Buick Electra Limited, Park Ave (Front Whl Drive)	2	Floating	Delco	—	35	70	1.043	.972	.957	.0005	.002
Electra, Estate Wagon (Rear Whl. Drive)	2	Floating	Delco	—	35	80 ①	1.037	.980	.965	.0005	.002
Riviera Front	2	Floating	Delco	—	35	100	1.037	.980	.965	.0005	.004
Rear	10	Floating	Delco	35	30	100	—	.980	.965	.0005	.004
Century W/H.D.	2	Floating	Delco	—	28	100	—	.972	.957	.0005	.002
Exc. H.D.	2	Floating	Delco	—	28	100	—	.830	.815	.0005	.002
Skyhawk-W/Vented Disc.	2	Floating	Delco	—	28	100	—	.830	.815	.0005	.002
W/Solid Disc.	2	Floating	Delco	—	28	100	—	.444	.429	.0005	.002
Regal, LeSabre	2	Floating	Delco	—	35	70–80	—	.980	.965	.0005	.004
Skylark	2	Floating	Delco	—	21–35	102	.885	.830	.815	.0005	.003
Century, Regal, LeSabre	2	Floating	Delco	—	35	80	—	.980	.965	.0005	.004
General Motors—Cadillac Cimarron	2	Floating	Delco	—	28	100	.885	.830	.815	.0005	.004
DeVille, Fleetwood (Rear Whl. Drive) Front	2	Floating	Delco	—	30	100	1.037	.980	.965	.0005	.004
Rear	10	Floating	Delco	35	30	100	.974	.910	.905	.0005	.003
CC Limousine	2	Floating	Delco	—	30	100	1.250	1.230	1.215	.0005	.004
Eldorado, Seville Front	2	Floating	Delco	—	28	100	1.000	.980	.965	.0005	.004
Rear	10	Floating	Delco	35	30	100	.974	.910	.905	.0005	.004
General Motors—Chevrolet Caprice, Impala	2	Floating	Delco	—	35	80 ②	1.030	.980	.965	.0005	.004
Malibu, Monte Carlo	2	Floating	Delco	—	35	80 ③	1.030	.980	.965	.0005	.004
Camaro-Front	2	Floating	Delco	—	21–35	80	1.030	.980	.965	.0005	.004
Rear	2	Floating	Delco	—	30–45	80	1.030	.980	.965	.0005	.004
Corvette-Front	9	Fixed	Delco	70	130	70 ④	1.285	1.230	1.215	.0005	.004
Rear	9	Fixed	Delco	70	60	70 ④	1.285	1.230	1.215	.0005	.004
Corvette-Front	12	Floating	Girlock	70	24	100	—	.724	—	.0005	.006
Rear	12	Floating	Girlock	44	24	100	—	.724	—	.0005	.006
Celebrity, Cavalier	2	Floating	Delco	—	28	100	Vented .885 / Solid .490	Vented .830 / Solid .444	Vented .815 / Solid .429	.0005	.004

BRAKES
Domestic Vehicles

DISC BRAKE APPLICATION CHART & SPECIFICATIONS

Make/Model	Text Reference Type	Caliper Style	Manufacturer	Anchor Bolt (ft lbs)	Bridge, Pin or Key Bolts (ft lbs)	Wheel Lugs (ft lbs)	Minimum Thickness Normal Standard	Minimum Thickness Machine To	Discard At	Rotor Parallel Variation	Max. Run-out
Citation	2	Floating	Delco	—	28	102	.885	.830	.815	.0005	.003
Chevette	4	Floating	Delco	70	28	70	.440	.390	.374	.0005	.005
	2	Floating	Girlock	—	21-25	70	—	.390	.374	.0005	.005
General Motors— Oldsmobile 98, Regency, Brougham (Front Whl. Drive)	2	Floating	Delco	—	35	70	1.043	.972	.957	.0005	.002
Full Size	2	Floating	Delco	—	35	80 ①	1.040	.980	.965	.0005	.004
Toronado—Front	2	Floating	Delco	—	35	100	1.000	.980	.965	.0005	.004
Rear	10	Floating	Delco	32	30	100	1.000	.980	.965	.0005	.004
Cutlass, Cutlass Supreme	2	Floating	Delco	—	35	80	1.040	.980	.965	.0005	.004

Flushing Hydraulic Brake Systems

Hydraulic brake systems must be totally flushed if the fluid becomes contaminated with water, dirt or other corrosive chemicals. To flush, simply bleed the entire system until all fluid has been replaced with the correct type of new fluid.

Bench Bleeding Master Cylinder

1. Connect two short pieces of brake line to the outlet fittings, bend them until the free end is below the fluid level in the master cylinder reservoirs.
2. Fill the reservoirs with fresh brake fluid. Pump the piston until no more air bubbles appear in the reservoir(s).
3. Disconnect the two short lines, refill the master cylinder and securely install the cylinder cap(s).
4. Install the master cylinder on the car. Attach the lines but do not completely tighten them. Force any air that might have been trapped in the connection by slowly depressing the brake pedal. Tighten the lines before releasing the brake pedal.

GM Quick Take Up System Bleeding

Bleed the master cylinder as follows. Disconnect the left front brake line at the master cylinder. Fill the cylinder with fluid until it flows from the opened port. Connect the line and tighten the fitting. Apply the brake pedal slowly one time and keep it applied. Loosen the same brake line fitting to allow any air to escape. Retighten the fitting and release the brake pedal slowly. Wait 15 seconds and repeat the procedure until all of the air is expelled. Bleed the right front connection in the same manner. Bleed the cylinders and calipers after you are sure all the air is out of the master cylinder. Rapid pumping will move the secondary piston down the bore and make it difficult to bleed the system. Always apply slow pedal pressure.

Powermaster Power Brake Unit

DESCRIPTION

The Powermaster unit is a complete, integral power brake apply system, consisting of an electro-hydraulic pump, fluid accumulator, pressure switch, fluid reservoir and a hydraulic booster, with an integral dual master cylinder. The nitrogen charged accumulator stores fluid at 510-675 psi for the hydraulic booster operation. The electro-hydraulic operates between pressure limits with the ignition switch on. When the pressure switch senses accumulator pressure is below 510 psi, the 12 volt pump operates to increase the accumulator fluid pressure to 675 psi. When the brake pedal is depressed, fluid from the accumulator acts on the booster power piston to apply the master cylinder which functions in the same manner as the conventional dual master cylinder.

Because of the excessively high hydraulic pressure, the system must be depressurized before any service operations are performed on the system. Failure to depressurize could result in personal injury and/or damage to the vehicle's painted surfaces.

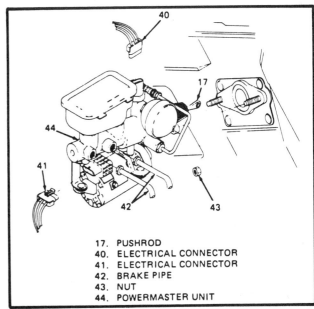

17. PUSHROD
40. ELECTRICAL CONNECTOR
41. ELECTRICAL CONNECTOR
42. BRAKE PIPE
43. NUT
44. POWERMASTER UNIT

Powermaster unit removal

DEPRESSURIZING THE POWERMASTER SYSTEM

1. With the ignition switch in the OFF position, apply and release the brake pedal a minimum of ten (10) times, using approximately 50 pounds of force on the brake pedal.

BRAKES
Domestic Vehicles

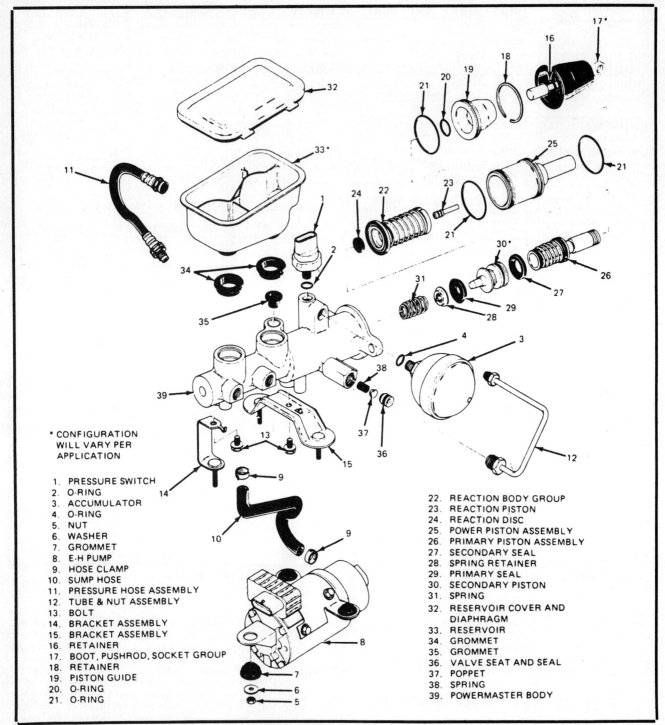

Exploded view of the powermaster power brake assembly

2. When loosening hoses or pipe fittings, wrap shop towels close to the fittings to prevent spraying of residual pressurized fluid.

Removal

1. Disconnect the power lead from the pressure switch.
2. Disconnect the electrical connector from the electro-hydraulic pump.
3. Disconnect the brake tubing fittings from the Powermaster unit.
4. Remove the two retaining nuts for the unit to dash panel.
5. Remove the brake pedal pushrod.

BRAKES
Domestic Vehicles

6. Remove the Powermaster unit from the vehicle.

Installation

1. Install the Powermaster unit, the brake pedal pushrod and install the two retaining nuts. Torque 22 to 30 ft. lbs.
2. Install the brake pipes to the unit.
3. Install the electrical connections to the unit.

BLEEDING OF UNIT

The brake system is bled in the conventional manner, either manually or by pressure. It must be remembered not to have the ignition switch on during the bleeding operation.

Powermaster Fluid Filling

1. Fill both sides of reservoir to the full marks on the inside of the reservoir. Use only clean new brake fluid meeting DOT specifications shown on reservoir cover.
2. Turn ignition "On". With the pump running, the brake fluid level in the booster side of the reservoir should decrease as brake fluid is moved to the accumulator.
3. If the booster side of the reservoir begins to run dry, add brake fluid to just cover the reservoir pump port until the pump stops.

NOTE: Pump must be shut off within 20 seconds. Turn ignition off after 20 seconds have elapsed. Check for leaks or flow back into reservoir from booster return port.

4. Properly install reservoir cover assembly to reservoir.
5. Turn ignition "OFF" and apply and release brake pedal 10 times, Remove reservoir cover and adjust booster fluid level to full mark.
6. Turn ignition "On". Pump will run and refill accumulator. Make sure that pump does not run longer than 20 seconds and that fluid level remains above pump sump port in reservoir.
7. Properly install reservoir cover. With ignition on, apply and release brake pedal to cycle pump on and off 10 to 15 cycles and remove air from booster section. Do not allow pump to run more than 20 seconds for each cycle.
8. Recheck high and low reservoir fluid levels per Steps 4 and 5. Check power master diagnosis if fluid levels do not stabilize high and low levels or if pump runs more than 20 seconds. Pump should not cycle without brake applications.

ELECTRO HYDRAULIC PUMP

Removal and Installation

1. Relieve the pressure from the powermaster unit and remove the reservoir cover and diaphragm.
2. Remove the end of the sump hose connected to the electrohydraulic pump and drain the reservoir sump pump.
3. Disconnect the electrical connector from the pump and disconnect the pressure hose assembly from the tube and nut assembly.
4. Disconnect the other end of the hose assembly from the pump.
5. Remove the three pump retaining bolts and remove the pump.
6. Installation is the reverse order of the removal procedure. Torque the pump retaining bolts to 23 to 35 ft. lbs. and the pressure hose assembly to 10 to 15 ft. lbs.

PRESSURE SWITCH

Removal and Installation

1. Relieve the pressure from the power master as previously outlined.

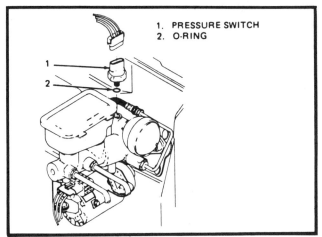

Pressure switch removal

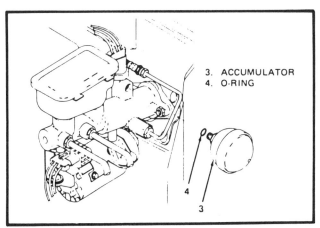

Accumulator removal

2. Disconnect the electrical connector from the pressure switch and remove the pressure switch with the O-ring.
3. Installation is the reverse order of the removal procedure. Be sure to use a new O-ring on the pressure switch and torque the pressure switch to 15 to 20 ft. lbs.

ACCUMULATOR

Removal and Installation

1. Relieve the pressure from the powermaster and remove the accumulator by unscrewing it from the powermaster unit.
2. Installation is the reverse order of the removal procedure. Be sure to install a new O-ring on the accumulator and torque the accumulator to 15 to 20 ft. lbs.

POWERMASTER OVERHAUL

Disassembly

1. Remove the powermaster from the vehicle as previously outlined.
2. Remove the reservoir cover with the diaphragm and empty the brake fluid from the reservoir.
3. Remove the following components, pressure switch, accu-

BRAKES
Domestic Vehicles

mulator, electro-hydraulic pump, pressure hose assembly, sump hose, clamps, tube and nut assembly and all brackets.

4. Remove the retainer from the groove in the powermaster body.

5. Remove the boot, retainer, pushrod, power piston group, remove the power piston group by pulling on the pushrod.

6. Disassemble the retainer and boot, pushrod, socket assembly and the piston guide from the power piston assembly.

7. Remove the O-ring from the piston guide and the O-rings from the power piston assembly and piston guide.

8. Remove the reaction body assembly from the power piston assembly and the reaction piston and disc from the reaction body assembly.

NOTE: The reaction body assembly and the power piston assembly have been disassembled as far as they can be. If there is any major problem with either of these assemblies, they must be replaced as a complete assembly.

9. Disassemble the primary piston from the secondary piston assembly, by blowing a small amount of compressed air into an outlet port at the blind end of the body (the other outlet port is plugged).

10. Remove the secondary seal, spring retainer and primary seal from the secondary piston. Remove the spring from the body core.

11. Place the powermaster body in a vise, be sure not to clamp across the powermaster body.

12. Using a small pry bar or equivalent remove the reservoir and the reservoir grommets.

13. Remove the valve seat and seal (do not reuse). It may be necessary to use an easy out to remove the valve seat.

14. Remove the poppet and spring and discard them. Before re-assembling the powermaster unit, be sure to clean all parts in denatured alcohol, except for the pressure switch and the electro-hydraulic pump.

Assembly

1. Use the clean fresh brake fluid to lubricate all parts before assembling the unit. Be sure to lubricate the new O-rings also. With the powermaster body still in the vise, install the new spring, poppet, valve seat and seal.

2. Bottom out the valve seat and seal by threading the nut of the tube and nut assembly in the powermaster body port.

3. Remove the powermaster body from the vise and install the grommets and the reservoir.

4. Install the spring into the powermaster body, along with the secondary seal, primary seal and spring retainer on the secondary piston.

5. Install the secondary piston assembly into the powermaster body.

6. Install the primary piston assembly into the powermaster body.

7. Assemble the reaction piston and disc into the reaction body assembly.

8. Install the two O-rings on the power piston assembly and install the reaction body assembly into the power piston assembly.

9. Install the power piston assembly into the powermaster body.

10. Install the O-ring on the piston guide and the O-ring in the piston guide.

11. Install the piston guide over the power piston in the powermaster body.

12. While depressing the piston guide install the retainer and the power piston.

13. Install the boot, pushrod, socket assembly, socket into the end of the power piston assembly and secure it with the retainer.

14. Install the brackets, sump hose, clamps, tube and nut assembly, electro-hydraulic pump and pressure hose assembly, accumulator and pressure switch.

15. Install the reservoir cover and diaphragm on the reservoir, and bench bleed the powermaster unit.

16. Install the powermaster unit on the vehicle and bleed the system as previously outlined. Also follow the instruction previously outlined on how to fill the powermaster unit with brake fluid.

Powermaster Diagnostic Procedure

PRELIMINARY PROCEDURE

1. Complete the fluid filling and bleeding procedures per powermaster bleed and fill instructions. Assure that pump cycle time and reservoir fluid levels are maintained within prescribed limits. Brake fluid temperature at 60 degrees to 80 degrees F. Warm fluid to 60 degrees F minimum by cycling pump.

2. Fully discharge accumulator by making 10 medium brake applications with ignition off.

3. Inspect for fluid leakage at brake pedal push rod, reservoir cover, hose and pipe connections, reservoir attaching points, pressure switch and accumulator.

4. Remove pressure switch from powermaster and install J-35126 test gauge adapter or equivalent. Reinstall pressure switch in test adapter. Attach pressure switch electrical connector. Close bleed valve.

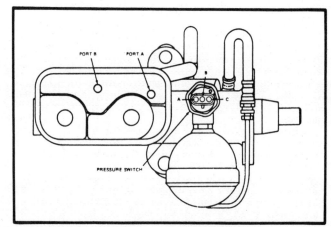

Powermaster vacuum ports and electrical terminals

FUNCTIONAL CHECK SEQUENCE

Test

1. Turn the ignition on. The electro-hydraulic pump will run and then shut off. Do not allow pump to run more than 20 seconds.

2. Observe the pump stops at 635 to 735 psi reading on the test gauge.

3. Slowly bleed off accumulator pressure with bleed valve return fluid to pump reservoir. Observe that the pump turns on again at 490 to 530 psi.

4. Have an assistant slowly apply brake pedal, reservoir cover off, and hold at steady medium force. Observe pressure gauge for indication of continuous pressure drop. Pressure drop rate should not cause pump to recycle within 30 seconds of first apply.

5. Turn the ignition off and remove the pressure switch electrical connector. With ohmmeter connected to the switch terminals B and C, and ignition off, slowly bleed off accumulator

BRAKES
Domestic Vehicles

pressure. Terminals B and C should close at 355 to 435 psi. This is the low pressure warning signal.

6. Continue to bleed off accumulator pressure and note sudden drop off in gauge pressure reading at 200 to 330 psi. This sudden drop is at the accumulator pre-charge pressure.

7. Assure that pump sump fluid level is at the full mark on the inside of the reservoir when accumulator is fully depressurized.

8. Turn ignition on and cycle the pump several times to remove air by opening and closing bleed valve. Pump on time should now be less than 10 seconds each cycle.

9. During pump on/off cycles, note sump reservoir fluid level. It normally will be just covering the sump hose port when pump is off and 1/2 full when pump comes on.

FINAL EVALUATIONS

1. Depressurize accumulator and remove J-35126 or equivalent test gauge adapter. Reinstall pressure switch and electrical connector and the reservoir cover.

2. With powermaster functioning normally, apply the brake pedal and note pedal travel. Pedal should not creep at steady pressure. Brake warning light should not indicate pressure differential between pressure circuits.

3. Observe running motor and pump sound from driver's seat. Compare to a good unit.

4. Consult the powermaster diagnosis for conditions and performance values which differ from normal.

POWERMASTER DIAGNOSIS

1. Symptom brake warning light on after engine start.

2. Parking brake applied. Temporarily release parking brake observe light. Reapply if light remains on.

3. Partial failure in brake hydraulic pressure circuit. Evaluate for excessive brake pedal travel. Evaluate for hard brake pedal force to stop. Evaluate for excessive stopping distances and early wheel lock-up tendency. Repair as necessary.

4. Low pressure in powermaster accumulator. Electrical failure check ignition, 30 amp fuse, pressure switch A/C terminals motor relay, connectors, wiring. Low fluid in reservoir.

5. Faulty warning light pressure switch. Check warning switch actuation pressure at Terminals B and C.

6. Symptom pump motor will not run. Electrical failure check ignition, 30 amp fuse, pressure switch terminals A/C closed, connector terminals, motor/relay and wiring.

7. Symptom pump motor runs does not shut off in 20 seconds. Turn ignition off after 20 seconds. Check reservoir fluid level. Check reservoir port A for backflow then replace power piston.

8. Check pump pressure. If pressure is low, check pressure line for obstruction, then replace pump and motor. If pressure is high, replace switch if higher than normal. Replace pump if lower than normal cut-off.

9. Symptom pump self cycles without brake applied. Check for accumulator precharge pressure, replace accumulator if low. Recheck self cycle.

10. Symptom pump self cycles while applying steady brake pressure. Does not self cycle without brake pressure. Check for accumulator precharge pressure, replace accumulator if low then recheck self cycle. Check for fluid backflow at reservoir port and replace power piston.

11. Symptom fluid level in pump reservoir does not cycle between full and nearly empty when accumulator is fully charged and fully depressurized. Check for air in fluid cycle 5 to 10 pump cycles to remove the air. Check accumulator precharge pressure and replace if low.

12. Symptom fluid level in pump reservoir does not cycle between half full and nearly empty at pressure switch limits. Check for reservoir full at fully depressurized accumulator. Check for accumulator precharge pressure if reservoir level after pump cycle is not nearly empty.

13. Symptom pump and motor noisy. Check for grounded tube and motor. Check for reservoir fluid level. Replace motor mount grommets.

14. Symptom fluid leakage. Check pump reservoir for excess fluid fill with accumulator fully depressurized. Check for tight reservoir cover and diaphragm. Wipe dry and identify source of leakage, then overhaul as necessary.

15. Symptom pump cycle time at pressure switch limit exceeds 10 seconds. Check for air in system recycle 5 to 10 pump cycles to remove air. Check for normal pressure switch points. Check for obstructed pump inlet and outlet fluid circuits. Check for faulty pump.

FRONT DISC BRAKES

ROTORS

Resurfacing Rotors

Manufacturers differ widely on permissible runout, but too much can sometimes be felt as a pulsation at the brake pedal. A wobble pump effect is created when a rotor is not perfectly smooth and the pad hits the high spots forcing fluid back into the master cylinder. This alternating pressure causes a pulsating feeling which can be felt at the pedal when the brakes are applied. This excessive runout also causes the brakes to be out of adjustment because disc brakes are self-adjusting; they are designed so that the pads drag on the rotor at all times and therefore automatically compensate for wear.

To check the actual runout of the rotor, first tighten the wheel spindle nut to a snug bearing adjustment, end-play removed. Fasten a dial indicator on the suspension at a convenient place so that the indicator stylus contacts the rotor face approximately one inch from its outer edge. Set the dial at zero. Check the total indicator reading while turning the rotor one full revolution. If the rotor is warped beyond the runout specification, it is unlikely that it can be successfully remachined.

Lateral Runout: A wobbly movement of the rotor from side to side as it rotates. Excessive lateral runout causes the rotor faces to knock back the disc pads and can result in chatter, excessive pedal travel, pumping or fighting pedal and vibration during the breaking action.

Parallelism (lack of): Refers to the amount of variation in the thickness of the rotor. Excessive variation can cause pedal vibration or fight, front end vibrations and possible "grab" during the braking action; a condition comparable to an "out-of-round brake drum." Check parallelism with a micrometer. "Mike" the thickness at eight or more equally spaced points, equally distant from the outer edge of the rotor, preferably at mid-points of the braking surface. Parallelism then is the amount of variation between maximum and minimum measurements.

Surface or Micro-inch finish, flatness, smoothness: Different from parallelism, these terms refer to the degree of perfection

SECTION 1

BRAKES
Domestic Vehicles

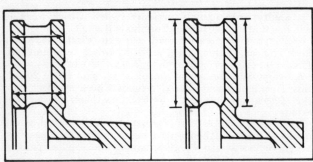

Taper variation not to exceed .003 in.

These surfaces to be flat and within .002 in.

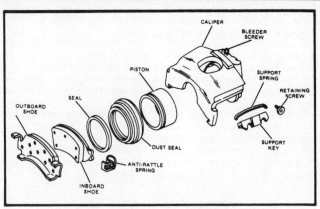

Type one Bendix caliper disc brakes (single piston)

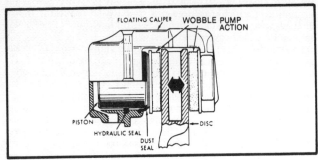

Wobble pump action

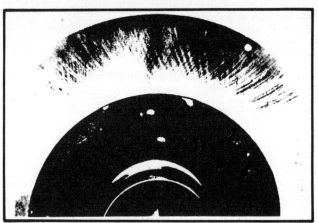

Ideal rotor surface condition

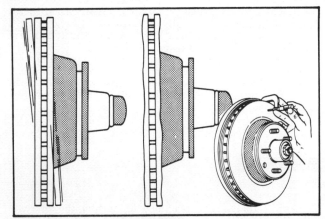

Excessive runout parallelism

of the flat surface on each side of the rotor; that is, the minute hills, valleys and swirls inherent in machining the surface. In a visual inspection, the remachined surface should have a fine ground polish with, at most, only a faint trace of non-directional swirls.

Removal and Installation

1. Raise the vehicle on a hoist or safely support on jackstands and remove the wheel.
2. Remove the caliper mounting bolts. Slide the caliper away from the disc and suspend it using a wire loop. On some cars, it is advisable to install a cardboard spacer between the pads to prevent the piston from coming out of its cylinder.
3. Remove the wheel bearing nut from the spindle and remove the outer wheel bearing roller assembly from the hub.
4. On Ford sliding caliper brakes, remove the wheel bearing adjusting nut and pull the hub and disc assembly outward enough to loosen the washer and outer wheel bearing. Push the assembly back onto the spindle and remove the washer and outer wheel bearing from the spindle.
5. Remove the hub and disc assembly from the spindle. Installation of hub and disc is in reverse order of removal. The disc is removable from the hub on the El Dorado, Toronado, and Corvette (rear only).
6. To separate the rear disc and hub on a Corvette the three hub-to-disc attaching rivets must be drilled out. This can be done with the hub and rotor mounted on the car. It is not necessary to install new rivets when the disc is installed.

CALIPER AND PAD

Inspection

Disc pads (lining and shoe assemblies) should be replaced in axle sets (both wheels) when the lining on any pad is worn to 1/16 inch at any point. If lining is allowed to wear past 1/16 inch minimum thickness, severe damage to disc may result. State inspection specifications take precedence over these general recommendations.

Note that disc pads in floating caliper type brakes may wear at an angle, and measurement should be made at the narrow end of the taper. Tapered linings should be replaced if the taper

BRAKES
Domestic Vehicles

TROUBLESHOOTING DISC BRAKES

Cause \ Symptom	Excessive Brake Pedal Travel	Brake Pedal Travel Gradually Increases	Excessive Brake Pedal Effort	Excessive Braking Action	Brakes Slow to Respond	Brakes Slow to Release	Brakes Drag	Uneven Braking Action (Side to Side)	Uneven Braking Action (Front to Rear)	Scraping Noise from Brakes	Brakes Squeak During Application	Brakes Squeak During Stop	Brakes Chatter (Roughness)	Brakes Groan at End of Stop	Brakes Tell-Tale Glows
Leaking Brake Line or Connection	X	XX	X					X							XX
Leaking Wheel Cylinder or Piston Seal	X	XX	X	X				X							X
Leaking Master Cylinder	X	XX	X												X
Air in Brake System	XX		X							X					XX
Contaminated or Improper Brake Fluid	X				X	X	X								X
Leaking Vacuum System			XX	X											
Restricted Air Passage in Power Head		X	X		XX	X									
Damaged Power Head			X	X	X	X	XX								
Worn Out Brake Lining			X	X				X	X	X	X	X		X	
Uneven Brake Lining Wear - Replace	X			X				X	X	X	X	XX		X	X
Glazed Brake Lining - Sand			XX		X			X	X		X	X			
Incorrect Lining Material - Replace			X	X				X	X			X		X	
Contaminated Brake Lining - Replace				XX				XX	XX	X	X	X		X	
Linings Damaged by Abusive Use - Replace			X	XX				X	X	X	X	X		X	
Excessive Brake Lining Dust - Remove with Air			X	XX				XX	XX		X	XX		X	
Heat Spotted or Scored Brake Drums or Discs				X				X	X		X	X	XX	X	
Out-of-Round or Vibrating Brake Drums												X	XX		
Out-of-Parallel Brake Discs	X												XX		
Excessive Disc Run-Out	X												X		
Faulty Automatic Adjusters	X							X	X						X
Incorrect Wheel Cylinder Sizes			X	X				X	X						
Weak or Incorrect Brake Shoe Retention Springs			X		X	X	XX	X	X	XX	X	XX			
Brake Assembly Attachments - Missing or Loose	X						X	X	X	X		X	X	X	
Insufficient Brake Shoe Guide Lubricant							X	X	X	XX	XX				
Restricted Brake Fluid Passage or Sticking Wheel Cylinder Piston		X	X		X	X	X	X	X						X
Improperly adjusted Stoplight Switch or Cruise Control Vacuum Dump						X									
Faulty Metering Valve	X		X	X	X	X	X		X						X
Faulty Proportioning Valve			X	X	X	X	X		X						
Brake Pedal Linkage Interference or Binding			X		X	XX	XX								
Improperly Adjusted Parking Brake							X								
Improperly Adjusted Master Cylinder Push Rod	X					X	XX								X
Incorrect Front End Alignment								XX							
Incorrect Tire Pressure								X	X						
Incorrect Wheel Bearing Adjustment	X											X	X		
Loose Front Suspension Attachments								X	X			XX	X	X	
Out-of-Balance Wheel Assemblies													XX		
Incorrect Body Mount Torque													X		
Need to Slightly Increase or Decrease Pedal Effort														XX	
Operator Riding Brake Pedal			X				X		X					X	
Sticking Caliper or Wheel Cylinder Pistons							XX								

XX – Indicates more probable cause(s)
X – Indicates other causes

BRAKES
Domestic Vehicles

exceeds 1/8 inch from end to end (the difference between the thickest and thinnest points).

To prevent costly paint damage, remove some brake fluid (don't re-use) from the reservoir and install the reservoir cover before replacing the disc pads. When replacing the pads, the piston is depressed and fluid is forced back through the lines to squirt out of the fluid reservoir.

When the caliper is unbolted from the hub do not let it dangle by the brake hose, it can be rested on a suspension member or wired onto the frame. All disc brake systems are self-adjusting and have no provision for manual adjustment.

Type One

KELSEY HAYES OR BENDIX SLIDING CALIPER DISC BRAKES (SINGLE PISTON)

Pad Removal

1. Remove half of the brake fluid from the master cylinder.
2. Remove the retaining screw holding the caliper support key.
3. Use a hammer and drift to drive the caliper retaining key and support spring out of the anchor plate.
4. Lift the caliper off of the rotor.
5. Support caliper so it doesn't hang by the brake hose.
6. Use a large C-clamp to force the piston back into its bore, being careful not to scratch the piston or bore, and being careful not to cut or tear the dust boot.
7. Remove the inboard pad and anti-rattle spring from the caliper support adapter.
8. Remove the outboard pad from the caliper. Check the condition of the rotor. If rotor run out exceeds manufacturer's specifications or has deep scratches, re-machine the rotor.
9. Clean all sliding surfaces on the adapter and caliper.

Pad Installation

1. Position the inboard brake pad and anti-rattle spring in the caliper support adapter.
2. Position the outboard brake pad in the caliper. Bend ears if necessary to provide slight interference fit in caliper.
3. Position the caliper over the rotor, take care not to damage the caliper piston dust boot.
4. Position the caliper support spring and support key into the slot and drive them into the opening between the lower end of the caliper and the lower anchor plate abutment.
5. Install and tighten the key retaining screw.
6. Fill the master cylinder with brake fluid. Bleed the system if necessary.

Type Two

DELCO FLOATING CALIPER (SINGLE PISTON)

Pad Removal

1. Remove half of the brake fluid from the master cylinder.
2. Position a large C-clamp over the caliper with the screw end against the outboard brake pad. Tighten the clamp until the caliper is pushed out enough to bottom the piston.
3. Remove the C-clamp. Remove the two caliper guide pin mounts and lift the caliper off of the rotor.
4. Support the caliper so there is no strain on the brake hose.
5. Press the inboard pad outward, then lift from the caliper.
6. Press the inboard pad outward, then lift from the caliper.
7. Remove and discard the four O-ring bushings and steel

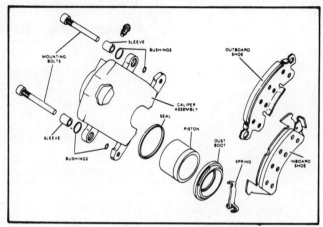

Type two Delco flating disc brakes (single piston)

sleeves if new ones are to be installed. Check the condition of the rotor. If rotor run out exceeds manufacturer's specifications or has deep scratches, re-machine the rotor.

Pad Installation

1. Lubricate and install the four O-ring bushings, install the sleeves pressing them through the O-rings until the sleeve end on the pad side is flush with caliper ear. Position the inboard pad so the pad contacts the piston and the two support spring ends. Note that the inboard and outboard pads are similar but not interchangeable.
2. Press down on the ears at the top of the inboard pad until the pad lies flat and the spring ends are just inside the lower edge of the pad.
3. Position the outboard pad with the ears toward the positioning pin holes and the tab on the inner edge of the pad resting in the notch in the edge of the caliper. Bend ears if necessary to provide slight interference fit in caliper.
4. Press the outboard pad tightly into position and use a pair of pliers to clinch the ears of the outboard pad over the outboard caliper half.
5. Position the caliper over the rotor.
6. Install the caliper mounting bolts and tighten to specification.
7. Fill the master cylinder with brake fluid.

Type Three

KELSEY HAYES CHRYSLER SLIDING CALIPER

Pad Removal

1. Remove half of the brake fluid from the master cylinder.
2. Remove caliper retaining clips and anti-rattle springs.
3. Lift the caliper off of the rotor.
4. Support the caliper so there is no strain on the brake hose.
5. Use a large C-clamp to force the piston back into its bore, being careful not to scratch the piston or bore, and being careful not to cut or tear the dust boot.
6. Pry the outboard pad from caliper.
7. Remove inboard pad from the adapter.
8. Check the condition of the rotor. If rotor run out exceeds manufacturer's specifications or has deep scratches, re-machine the rotor.

BRAKES
Domestic Vehicles

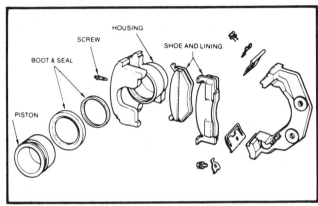

Type three—Kelsey Hayes sliding caliper

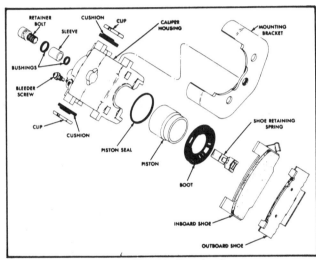

Type four—Chevette, T1000 disc brake

Pad Installation

1. Adjust ears of outboard pad to provide tight fit and install pad in caliper recess.
2. Install inboard pad with flanges inserted in adapter "ways."
3. Position the caliper on the rotor with the caliper engaging the adapter "ways."
4. Install anti-rattle springs and caliper retaining clips and torque retaining screws to 180 inch lbs.
5. Fill the master cylinder with brake fluid.

Type Four

CHEVETTE, T1000 DISC BRAKE

Pad Removal

1. Remove half of the brake fluid from the master cylinder.
2. Use a large C-clamp to force the piston back into its bore, being careful not to scratch the piston or bore, and being careful not to cut or tear the dust boot.
3. Remove the two hex head bolts that attach the caliper mounting bracket to the steering knuckle.

4. Support the caliper so there is no strain on the brake hose. Do not remove the socket head retainer bolt.
5. Remove the old shoe and lining assemblies. If the retaining spring does not come out with the inboard shoe, remove the spring from the piston.
6. Check the condition of the rotor. If rotor run out exceeds manufacturer's specifications or has deep scratches, re-machine the rotor.

Pad Installation

1. Before installing the inboard shoe, make sure that the shoe retaining spring is properly installed. Push the tab on the single-leg end of the spring down into the shoe hole, then snap the other two legs over the edge of the shoe notch.
2. Position the caliper over the rotor, lining up the bracket mounting holes. Install the mounting bolts.
3. Clinch the outboard shoe to the caliper. After clinching, radial and end play of the outboard shoe should be 0 to 0.005 inch.

Type Five

FORD FRONT DRIVE DISC BRAKE

Pad Removal

1. Remove master cylinder cap and check fluid level in reservoirs. Remove brake fluid until each reservoir is half full. Discard the removed fluid.

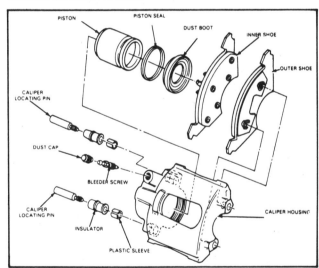

Type five—Ford front drive caliper

2. Remove wheel and tire assembly from rotor mounting face. Use care to avoid damage or interference with the caliper splash shield or bleeder screw fitting.
3. Remove brake caliper anti-rattle spring by applying upward pressure to center portion of spring until the spring tabs are free of the caliper holes.
4. Back out the caliper locating pins. Do not remove pins completely unless new bushings are to be installed. Reinstalling pins after complete removal can be difficult.
5. Lift caliper assembly from integral knuckle and anchor plate and rotor. Remove outer shoe and lining assembly from caliper assembly.
6. Remove inner shoe and lining assembly and inspect both

BRAKES
Domestic Vehicles

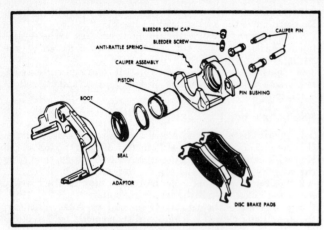

Type six—Kelsey Hayes floating caliper

rotor braking surfaces. Minor scoring or build-up of lining material does not require machining or replacement of the rotor.

7. Suspend caliper inside the fender housing. Use care not to damage caliper or stretch the brake hose.

Pad Installation

1. Use a 4 inch C-clamp and a block of wood 2 3/4 in. x 1 in. and approximately 3/4 inch thick to seat the caliper hydraulic piston in its bore. This must be done to provide clearance for the caliper assembly to fit over the rotor during installation. Extra care must be taken during this procedure to prevent damage to the aluminum piston. Metal or sharp objects cannot come into direct contact with the piston surface or damage will result.

2. Install the correct inner shoe and lining assembly in caliper piston(s). Do not bend shoe clips during installation in the piston or distortion and rattle can occur.

3. Install the correct outer shoe and lining assembly making sure clips are properly seated. Replace caliper anti-rattle spring. Refill master cylinder to at least 1/4 inch from the top in both reservoirs.

4. Install the wheel and tire assembly. Tighten wheel nuts 80 to 105 ft. lbs.

5. Pump the brake pedal prior to moving the vehicle to position brake linings.

6. Road test the vehicle.

Type Six

KELSEY-HAYES FLOATING CALIPER

Pad Removal

1. Remove half of the brake fluid from each master cylinder reservoir.
2. Remove the caliper guide pins, positioners and anti-rattle spring.
3. Lift the caliper from the rotor and support to prevent strain on the brake hose.
4. Pry the caliper piston back into the bore. Use a C-clamp if necessary.
5. Remove the brake pads from the caliper adaptor. Remove and discard the four bushings if they are to be replaced.

Pad Installation

1. Clean and lubricate the caliper guide pins and guide mounting surfaces. Install new guide bushings.

2. Position the new brake pads in the caliper adaptor.
3. Carefully lower the caliper over the adaptor. Install the guide pins and anti-rattle spring. The anti-rattle spring is installed with the end loop inboard on the caliper lug.
4. Fill the master cylinder with new fluid. Bleed the brakes if necessary.

Type Seven

ATE FLOATING CALIPER

Pad Removal

1. Remove the guide pin(s) and anti-rattle clips or springs.
2. Remove the caliper from the rotor by slowly sliding it up and away. Support the caliper so there is no strain on the brake hose. Late model calipers may be pivoted on the anchor bolt.
3. Remove the pads from the adaptor or caliper. In some cases the rotor must be removed to replace the inboard pad.
4. Push the caliper piston back into its bore.

Pad Installation

1. Install the pads and hardware into the adaptor or caliper.
2. Position the caliper over the rotor and install the guide pin(s), anti-rattle springs or clips. Fill the master cylinder with new brake fluid. Bleed the brake system if necessary.

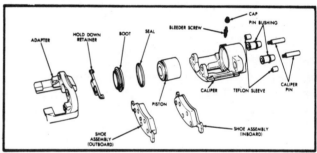

Type seven—ATE floating caliper

Type Eight

FORD FLOATING CALIPER

Pad Removal
EXCEPT MARK VII CONTINENTAL

1. Remove half of the brake fluid from the master cylinder reservoirs.
2. Remove the caliper guide pins.
3. Lift the caliper assembly from the rotor. Support the caliper so there is no strain on the brake hose.
4. Remove the outboard pad from the caliper. Remove the inboard pad from the piston. Step six can now be accomplished by using a C-clamp against the inboard pad.
5. Remove the insulators and inserts from the guide pin holes if they are to be replaced.
6. Push the caliper piston back into its bore.

Pad Installation
EXCEPT MARK VII CONTINENTAL

1. Install new guide bushings and insulators if they are to be replaced.
2. Install the inboard pad into the piston. Install the out-

BRAKES
Domestic Vehicles

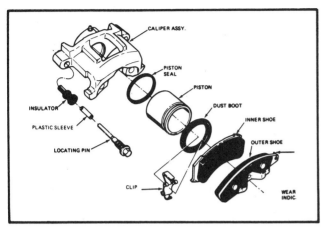

Type eight—Ford floating caliper

board making sure the buttons are seated into the caliper body. The wear indicator faces toward the front of the car.

3. Lower the caliper assembly onto the anchor plate and slide the guide pins through the holes in the caliper. When the guide pins reach the rubber insulators, they will require more pressure. After the pins bottom thread them into the hole. Take care not cross thread the guide pins.

4. Refill the master cylinder with new brake fluid. Bleed the brake system if necessary.

Pad Removal
MARK VII CONTINENTAL AND MUSTANG SVO

1. Raise the vehicle, and install safety stands. Block both front wheels.
2. Remove the wheel assemblies.
3. Disconnect the parking brake cable from the lever and bracket. Use care to avoid kinking or cutting the cable or return spring.
4. Remove the caliper locating pins.
5. Lift the caliper assembly away from the anchor plate by pushing the caliper upward toward the anchor plate, and then rotate the lower end out of the anchor plate.
6. If insufficient clearance between the caliper and shoe and lining assemblies prevents removal of the caliper, it is necessary to loosen the caliper end retainer 1/2 turn, maximum, to allow the piston to be forced back into its bore. To loosen the end retainer, remove the parking brake lever, then mark or scribe the end retainer and caliper housing to be sure that the end retainer is not loosened more than 1/2 turn. Force the piston back in its bore, and then remove the caliper.
7. If the retainer must be loosened more than 1/2 turn, the seal between the thrust screw and the housing may be broken, and brake fluid may leak into the parking brake mechanism chamber. In this case, the end retainer must be removed, and the internal parts cleaned and lubricated.
8. Remove the outer shoe and lining assembly from the anchor plate. Remove the two rotor retainer nuts and the rotor from the axle shaft.
9. Remove the inner brake shoe and lining assembly from the anchor plate. Remove the anti-rattle clip from anchor plate. If no further service than pad replacement is required, brake hose removal is not necessary. Do not support the caliper by the brake hose.
10. Remove the flexible hose from the caliper by removing the hollow retaining bolt that connects the hose fitting to the caliper.
11. Clean the caliper, anchor plate, and rotor assemblies and inspect for signs of brake fluid leakage, excessive wear, or damage. The caliper must be inspected for leakage both in the piston boot area and at the operating shaft seal area.
12. Lightly and or wire brush any rust or corrosion from the caliper and anchor plate sliding surfaces as well as the outer and inner brake shoe abutment surfaces. Inspect the brake shoes for wear. If either lining is worn to within 1/8 in. of the shoe surface, both shoe and lining assemblies must be replaced using the shoe and lining removal procedures.

Pad Installation
MARK VII CONTINENTAL AND MUSTANG SVO

1. If the end retainer has been loosened only 1/2 turn, reinstall the caliper in the anchor plate without shoe and lining assemblies. Tighten the end retainer 75 to 96 ft. lbs.
2. Install the parking brake lever on its keyed spline. The lever arm must point down and rearward. The parking brake cable will then pass freely under the axle. Tighten the retainer screw 16 to 22 ft. lbs. The parking brake lever must rotate freely after tightening the retainer screw. Remove the caliper from the anchor plate.
3. If new shoe and lining assemblies are to be installed, the piston must be screwed back into the caliper bore, using Tool T75P-2588-B or equivalent to provide installation clearance. Remove the rotor, and install the caliper, less shoe and lining assemblies, in the anchor plate. While holding the shaft, rotate the tool handle counterclockwise until the tool is seated firmly against the piston.
4. Now, loosen the handle about 1/4 turn. While holding the handle, rotate the tool shaft clockwise until the piston is fully bottomed in its bore; the piston will continue to turn even after it becomes bottomed. When there is no further inward movement of the piston and the tool handle is rotated until there is firm seating force, the piston is bottomed. Remove the tool and the caliper from the anchor plate.
5. Lubricate anchor plate sliding ways with D7AE-019590-A or equivalent grease. Use only specified grease because a lower temperature type of lubricant may melt and contaminate the brake pads. Use care to prevent any lubricant from getting on the braking surface. Install the anti-rattle clip on the lower rail of the anchor plate.
6. Install inner brake shoe and lining assembly on the anchor plate with the lining toward the rotor. Be sure shoes are installed in their original positions as marked for identification before removal. Install rotor and two retainer nuts.
7. Install the correct hand outer brake shoe and lining assembly on the anchor plate with the lining toward the rotor and wear indicator toward the upper portion of the brake.
8. Install the flexible hose by placing a new washer on each side of the fitting outlet and inserting the attaching bolt through the washers and fitting. Tighten 20 to 30 ft. lbs.
9. Rotate the caliper housing until it is completely over the rotor. Use care so that the piston dust boot is not damaged.
10. Piston Position Adjustment: Pull the caliper outboard until the inner shoe and lining is firmly seated against the rotor, and measure the clearance between the outer shoe and caliper. The clearance must be 1/32 to 3/32 inch. If it is not, remove the caliper, then readjust the piston to obtain required gap. Follow the procedure given in Step 3, and rotate the shaft counterclockwise to narrow gap and clockwise to widen gap (1/4 turn of the piston moves it approximately 1/16 inch.
11. A clearance greater than 3/32 inch may allow the adjuster to be pulled out of the piston when the service brake is applied. This will cause the parking brake mechanism to fail to adjust. It is then necessary to replace the piston/adjuster assembly.
12. Lubricate locating pins and inside of insulator with D7AZ-19A331-A or equivalent silicone grease. Add one drop of Loctite EOAC-19554-A or equivalent to locating pin threads.
13. Install the locating pins through caliper insulators and into the anchor plate; the pins must be hand inserted and hand started. Tighten to 29-37 ft. lbs.

SECTION 1

BRAKES
Domestic Vehicles

14. Connect the parking brake cable to the bracket and the lever on the caliper.
15. Bleed the brake system. Replace rubber bleed screw cap after bleeding. Fill master cylinder as required to within 1/8 inch of the top of the reservoir.
16. Caliper Adjustment: With the engine running, pump the service brake lightly (approximately 14 lbs. pedal effort) about 40 times. Allow at least one second between pedal applications. As an alternative, with the engine Off, pump the service brake lightly (approximately 87 lbs. pedal effort) about 30 times. Now check the parking brake for excessive travel or very light effort. In either case, repeat pumping the service brake, or if necessary, check the parking brake cable for proper tension. The caliper levers must turn to the Off position when the parking brake is released.
17. Install the wheel and tire assembly. Tighten the wheel lug nuts to specification. Install the wheel cover. Remove the safety stands, and lower the vehicle.
18. Be sure a firm brake pedal application is obtained, and then road test for proper brake operation, including parking brakes.

FORD FOUR WHEEL ANTI-LOCK BRAKE SYSTEM

General Description

The 1985 Mark VII and Continental models powered by the 5.0 liter V8 engine, are now equipped with an electronic anti-lock brake system. This new brake system is designed to prevent braking-induced wheel lock on any road surface and on high traction surfaces such as dry pavement, the system is seldom activated. On a surface where the traction is reduced such as ice on the road, the system will operate to provide the best braking possible, and will help to retain the driver's ability to steer the vehicle around a potential accident situation while maintaining the maximum braking power. The anti-lock system contains four sensors, one in each wheel hub. The function of each sensor is to constantly send signals to the central microprocessor which compares the speeds of all four wheels. There are three hydraulic brake circuits being employed in this system, one is for the left front wheel, one is for the right front wheel and the other one controls both rear wheel brakes simultaneously as an axle set. During a braking situation, if one wheel slows down faster than the other three, this would indicate imminent wheel lock (as in a skidding situation) and the microprocessor would signal the appropriate hydraulic braking circuit to pulse the braking pressure to that wheel or axle set. To aid the four wheel sensors, twin microprocessors are used in the system. Each computer checks itself against the other. If the computers do not agree with one another, the anti-lock feature of the brakes will be bypassed and braking will go back to the normal mode. A big difference between the systems with and without the anti-lock system is the power source for the brakes. The conventional hydro-boost actuation system for the Mark VII and the Continental is powered from the power steering pump. On the electronic anti-lock system, the power assist for brakes is generated as required, on demand, by an electric pump charging a nitrogen filled accumulator. There will be no noticeable difference between the two systems in the braking feel or efficiency during normal stopping.

MASTER CYLINDER AND HYDRAULIC BOOSTER

The master cylinder and the brake booster are arranged in the basic fore and aft position with the booster behind the master cylinder. The booster control valve is located in a parallel bore above the master cylinder centerline and is operated by a lever connected to the brake pedal pushrod.

ELECTRIC PUMP AND ACCUMULATOR

The electric pump is a high pressure pump design that runs at frequent intervals for a short period to charge the hydraulic accumulator that supplies the brake system. The accumulator is

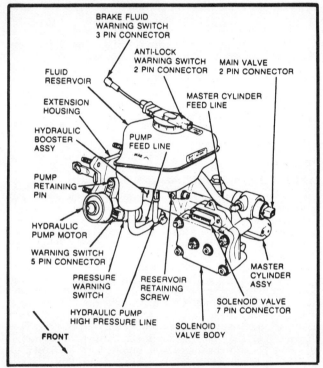

Master cylinder and booster assembly

a gas filled pressure chamber that is part of the pump and motor assembly. The electric motor, pump and accumulator assembly is shock mounted to the master cylinder/booster assembly.

VALVE BODY ASSEMBLY

The valve body assembly incorporates three pairs of solenoid valves, one pair for each front wheel, and a third pair for both the rear wheels combined. These solenoid valves are inlet-outlet valves with the inlet valve normally open and the outlet valve normally closed. The valve body itself is bolted to the inboard side of the master cylinder/booster assembly.

FLUID LEVEL WARNING SWITCHES

These two integral fluid level switches are incorporated in the

BRAKES
Domestic Vehicles

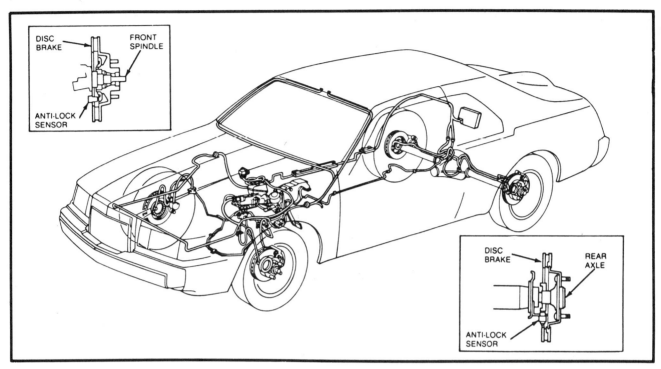

Ford anti-lock brake system

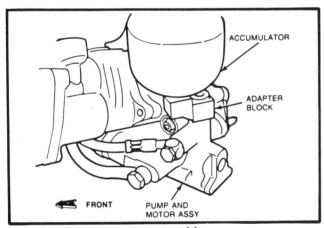

Accumulator and pump assembly

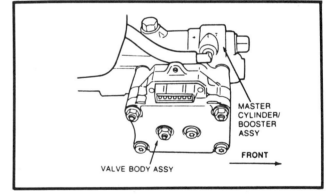

Valve body assembly

brake fluid reservoir cap assembly with two electrical connectors, one for each end of the cap, for wire harness connections.

WHEEL SENSORS

The sensors are four variable reluctance electronic sensor assemblies, each with a 104 tooth ring in the anti-lock system. The sensors are connected to an electronic controller through a wiring harness. The front sensors are bolted to brackets which are bolted to the front spindles. The front toothed sensor rings are pressed onto the inside of the front rotors. The rear sensors are bolted to brackets, that in turn are bolted to the rear disc brake axle adapters. The toothed rear sensor rings are pressed to the axle shafts, inboard of the axle shaft flange.

ELECTRONIC CONTROLLER

The controller is a self-contained non-serviceable unit, which consists of two microprocessors and the necessary circuitry for their operation. The function of the controller is to monitor the system operations during normal driving and during anti-lock braking. Any malfunction of the anti-lock brake system will cause the controller to shut off and bypass the anti-lock brake system. When the anti-lock brake system is bypassed, the normal power assisted braking will still remain.

CHECK ANTI-LOCK BRAKE WARNING LIGHT

The four wheel anti-lock system is self-monitoring. When the

BRAKES
Domestic Vehicles

ignition switch is in the run position, the electronic controller will perform a preliminary self-check on the anti-lock electrical system, this is indicated by 3-4 second energizing of the amber "Check Anti-Lock Brakes" lamp in the overhead console. This light will go out after the 3-4 second interval, unless there is a malfunction in the anti-lock brake system. If there is a malfunction the "Check Anti-Lock Brake" light and/or the brake lamp will stay lit, and diagnostic tests will then pinpoint the exact component needing service.

DIAGNOSIS AND TESTING

The diagnosis procedures were not available at the time of this publication, but there is a partial self-diagnostic capability incorporated in the EEC-IV system, and total capability is in order for the future model years. Ford also has taken the EEC-IV diagnostic box and set up an adapter so equipment already on the market can be used. There is also a comprehensive diagnostic guide available by the manufacturer.

BLEEDING THE BRAKE SYSTEM

The front brakes can be bled in the conventional manner, with or without the accumulator being charged. When bleeding the rear brakes the accumulator must be fully charged or the system has to be pressure bled as previously outlined in this section.

BLEEDING THE BRAKE SYSTEM WITH A CHARGED ACCUMULATOR

Be careful when opening the rear caliper bleeder screws, due to the high pressure in the system from a fully charged accumulator at the bleeder screws.
1. With the accumulator fully charged, have someone hold the brake pedal in the applied position and place the ignition switch in the run position and open the rear brake caliper bleed screws for 10 seconds at a time.
2. Repeat this procedure until the air is cleared from the brake fluid and close the brake caliper bleed screws.
3. Do this to all of the brake calipers and after the bleed screws are closed, pump the brake pedal a couple of times to complete the bleed procedure.
4. Adjust the brake fluid level in the reservoir to the max level with a fully charged accumulator.

MASTER CYLINDER

Removal and Installation

NOTE: The hydraulic pressure must be discharged from the brake system before removing the master cylinder. To discharge the system, turn the ignition key to the off position and pump the brake pedal at least 20 times until an increase in pedal force is clearly felt.

1. Disconnect the negative battery cable and the electrical connectors from the master cylinder reservoir cap, main valve, solenoid valve body, pressure warning switch, the hydraulic pump motor, and ground connector from the master cylinder.
2. Disconnect the brake lines from the solenoid valve body and plug the line openings in the valve body to prevent fluid loss. Do not allow the brake fluid to leak or spill onto any of the electrical connectors.
3. From inside the vehicle, disconnect the hydraulic booster pushrod from the brake pedal in the following order. Disconnect the stop light switch wires at the connector on the brake pedal. Remove the hairpin clip at the stop light switch on the brake pedal and move the switch off of the pedal pin far enough for the switch outer hole to clear the pin. Using a twisting mo-

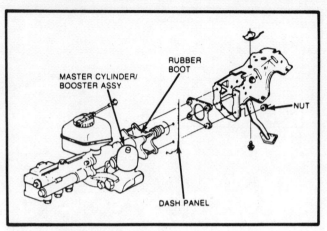

Master cylinder removal

tion, remove the switch, but be careful not to damage the switch during its removal. Remove the four retaining nuts at the dash panel and from inside the engine compartment remove the booster from the dash panel.
4. Installation is the reverse order of the removal procedure. After the unit has been installed, bleed the brake system as previously outlined in this section.

HYDRAULIC ACCUMULATOR

Removal and Installation

1. Discharge the pressure in the brake system and disconnect the electrical connection at the hydraulic pump motor.
2. Using an 8mm hex wrench or equivalent, unscrew the accumulator. Be sure that no dirt falls into the open port.
3. Using the same hex wrench or equivalent, remove the accumulator adapter block bolt and remove the block, if necessary.
4. Installation is the reverse order of the removal procedure, be sure to observe the following. Install new O-rings on the accumulator and adapter block. Torque the adapter block bolt 25 to 34 ft. lbs. and the accumulator 30 to 34 ft. lbs.
5. After installation, place the ignition switch in the on position and check to see if the "Check Anti-Lock Brake" light goes out after a maximum of one minute. Top off the master cylinder reservoir to the max mark with brake fluid.

HYDRAULIC PUMP MOTOR

Removal and Installation

1. Discharge the pressure in the brake system and disconnect the negative battery cable.
2. Disconnect the electrical connections at the hydraulic pump motor and pressure warning switch.
3. Remove the accumulator as previously outlined.
4. Remove the suction line between the reservoir and the pump at the reservoir by twisting the hose and pulling on it lightly. To prevent fluid loss, a large vacuum nipple can be slipped over the reservoir opening as the hose is removed.
5. Remove the retaining bolt on the pump high pressure line to the hydraulic booster housing at the housing. Make sure to save the two O-rings on both sides of the retaining bolt.
6. Remove the Allen head bolt that holds the pump and motor assembly to the extension housing located directly under the accumulator. Make sure to save the thick spacer between the extension housing and the shock mount.

BRAKES
Domestic Vehicles

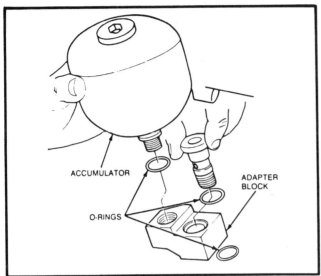

Accumulator removal and installation

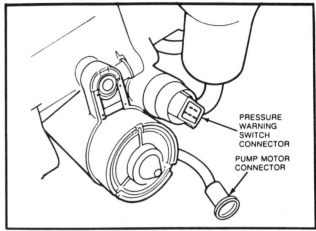

Hydraulic pump motor connectors

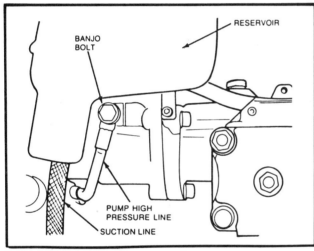

Hydraulic pump motor removal

7. Slide the pump assembly inboard to remove the assembly from the retainer pin located on the inboard side of the extension housing.

8. Installation is the reverse order of the removal procedure. Be sure to bleed the brake system and check to see if the "Check Anti-Lock Brake" light goes out after a maximum of one minute.

ELECTRONIC CONTROLLER

Removal and Installation

1. Disconnect the negative battery cable and the 35-pin connector from the electronic controller in the trunk of the vehicle in front of the forward trim panel.

2. Remove the three retaining screws holding the electronic controller to the seat back brace and remove the controller.

3. Installation is the reverse order of the removal procedure.

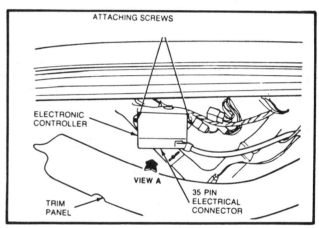

Electronic controller location

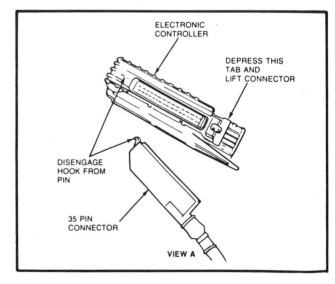

Electronic controller 35 pin connector

BRAKES
Domestic Vehicles

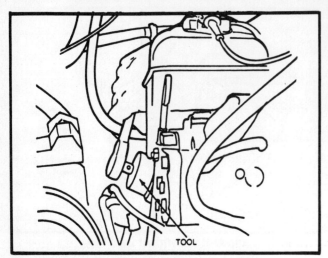

Removing or installing the pressure switch

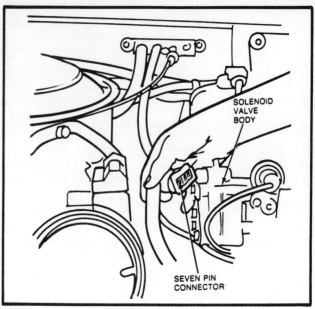

Location of the pressure switch

PRESSURE SWITCH

Removal and Installation

1. Discharge the pressure from the brake system and disconnect the negative battery cable.
2. Disconnect the solenoid valve body seven-pin connector and remove the pressure switch using special tool #T85P-20215-B or equivalent, a 1/2 to 3/8 inch adapter and a 3/8 inch drive ratchet.
3. Installation is the reverse of the removal procedure. Replace the O-ring on the switch. Torque the switch 15 to 25 ft. lbs.

FRONT WHEEL SENSOR

Removal

1. Disconnect the sensor electrical connector for the right or left front sensor from inside the engine compartment.
2. Raise and support the vehicle safely and disengage the wire grommet at the right or left hand shock tower.
3. Pull the sensor cable connector through the hole and be careful not to damage the connector.
4. Remove the sensor wire from the bracket on the shock tower and the side rail.
5. Loosen the set screw holding the sensor to the sensor bracket post. Remove the sensor through the hole in the disc splash shield.
6. Remove the sensor bracket or the sensor bracket post, if either has been damaged by removing the caliper, hub and rotor assembly (as previously outlined). Remove the two brake splash shield attaching bolts which hold the sensor bracket.

Installation

1. Install the sensor bracket and bracket post, if it has been removed. Torque the post retaining bolt 40 to 60 inch lbs. and the splash shield attaching bolts 10 to 15 ft. lbs. Install the hub and rotor assembly as previously outlined.

NOTE: If a sensor is going to be reused, the pole face must be cleaned of all dirt and grease and the pole face has to be scraped with a dull knife or equivalent so that the sensor slides freely on the post. Also glue a new front spacer paper (gasket) on the pole face.

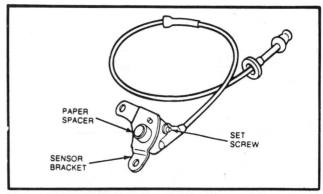

Paper spacer location

2. Install the sensor through the hole in the brake shield onto the sensor bracket post. Make sure the paper spacer does not come off during installation.
3. Push the sensor toward the toothed sensor ring until the new paper sensor contacts the ring. Hold the sensor against the sensor ring and torque the set screw to 21-26 inch lbs.
4. Insert the cable into the bracket on the shock strut, rail bracket and then through the inner fender apron to the engine compartment and the seat grommet.
5. Lower the vehicle and reconnect the sensor electrical connection.
6. Check the function of the sensor by road testing the vehicle and making sure the "Check Anti-Lock Brake" light does not stay on.

REAR WHEEL SENSOR

Removal

1. Disconnect the wheel sensor electrical connector located

BRAKES
Domestic Vehicles

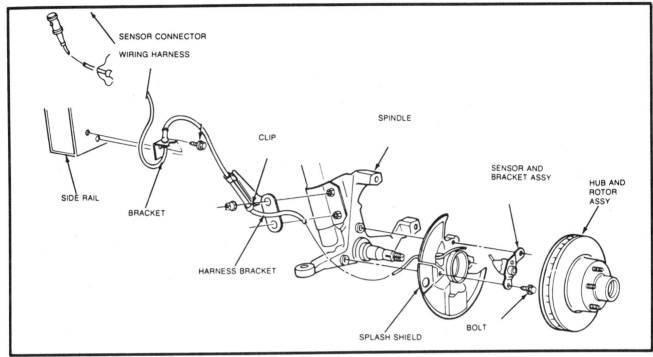

Removing the front wheel sensor

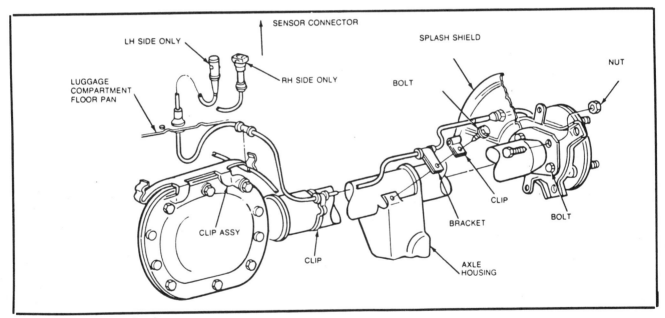

Removing the rear wheel sensor

behind the forward luggage compartment trim panel in the trunk.

2. Lift the carpet in the trunk and push the sensor wire grommet through the hole in the floor of the trunk.

3. Raise and support the vehicle safely and remove the appropriate wheel.

4. Remove the wheel sensor wiring from the axle shaft housing. The wiring harness has three different types of retainers, they are the following. The inboard retainer clip is located on the top of the differential housing. Just bend the clip far enough to remove the wiring harness. The second retainer clip is a C-clip and is located in the center of the axle shaft housing. Pull rearward on the clip to disengage the clip from the axle housing. The third clip is at the connection between the rear housing wheel brake tube and flexible hose. Remove the hold down bolt and open the clip to remove the wiring harness.

1-41

BRAKES
Domestic Vehicles

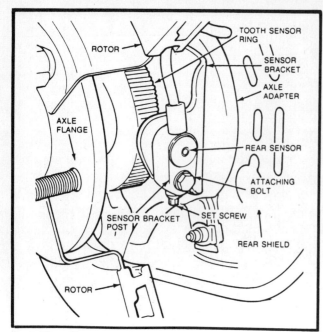

Installing the rear wheel sensor

NOTE: Be careful not to bend the C-clip open beyond the amount needed to remove the clip from the axle housing, because it could break.

5. Remove the rear wheel caliper and rotor assemblies as previously outlined in this section.
6. Remove the wheel sensor retaining bolt, slip the grommet out of the rear brake splash shield and pull the sensor wire outward through the hole.
7. If the sensor bracket is damaged, remove the bracket attaching screws and replace the bracket.

Installation

1. Install the sensor bracket if it was removed and torque the screws 11 to 15 ft. lbs.
2. If the sensor is going to be used again, it has to be cleaned just as the front wheel sensors were cleaned.
3. Insert the sensor into the large hole in the sensor bracket and install the retaining bolt into the sensor bracket post. Torque the retaining bolt 40 to 60 inch lbs.
4. Push the sensor toward the toothed ring until the new paper sensor touches the sensor ring, hold the sensor against the toothed ring and torque the set screw 21 to 26 inch lbs.
5. Install the caliper and rotor assemblies.
6. Push the wire and connector through the splash shield hole and engage the grommet into the shield eyelet. Install the sensor wire in the retainers along the axle housing.
7. Push the connector through the hole in the trunk and seat the grommet in the trunk floor pan.
8. Reconnect the cable electrical connector, and re-install the carpet in the trunk. Check the function of the sensor by road testing the vehicle and checking to see if the "Check Anti-Lock Brake" light goes out.

NOTE: If the toothed ring sensor is found to be malfunctioning on either the front or rear wheel, the rotor assembly has to be removed and the toothed ring sensor has to be pressed out and the new one pressed in. Brake pad removal and installation is covered in the Ford Floating Caliper section.

REAR DISC BRAKES

Rotors

Refer to front rotors for removal, installation or service procedures.

Calipers and Pads

Type Nine

DELCO FIXED CALIPER (FOUR PISTONS)

Pad Removal and Installation

1. Remove half of the brake fluid from the master cylinder.
2. Remove the brake pad retaining pins.
3. Pry the pistons back into their bore (being careful to pry both pistons at once so as not to force one out of its bore) and lift out one pad by tipping it down at the rear and up at the front.
4. Hold the rear piston in and slide the rear end of the new pad into place, being careful not to force out the front piston.
5. Check the condition of the rotor. If rotor run out exceeds .003 thousands, or has deep scratches, re-machine the rotor.
6. Now push the front piston back in to its bore and slide the front of the new pad into position.
7. Change the other pad in the same manner.

8. Reinstall the retaining pin through the caliper holes and through the holes in the pads.
9. Fill the master cylinder with fresh brake fluid.

Type Ten

GM REAR DISC BRAKE

Pad Removal and Installation

1. The calipers must be removed to replace linings. Remove two-thirds of the fluid in the front master cylinder.
2. Remove wheel and tire assembly, and reinstall one wheel mounting nut, flat side toward rotor, to prevent rotor from falling when caliper is removed.
3. Loosen tension on the parking brake cable at equalizer, and remove the cable from the parking brake lever at the caliper.
4. Remove return spring, lock nut, lever, lever seal and anti-friction washer (lever must be held in place while removing nut).
5. Using a C-clamp with the solid end of the lever stop and the screw end of the back of the outboard lining assembly, tighten clamp until piston bottoms in the caliper. Do not position C-clamp on actuator screw.

BRAKES
Domestic Vehicles

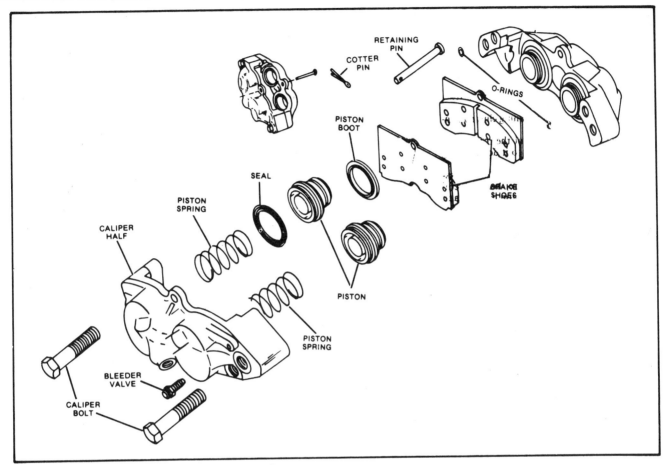

Type nine—Delco fixed caliper (four pistons)

6. Before removing the clamp lube the caliper housing surface (under the lever seal) with silicone.

7. Install new anti-friction washer, new lever seal and lever. Be certain to install lever on hex with arm pointing downward.

8. Rotate lever toward front of car and while holding in this position install nut and torque to 25 ft. lbs. Rotate lever back to stop.

9. Install lever return spring, and remove C-clamp. Springs are color coded, red for right side caliper and black for left.

10. Remove brake line from caliper and plug openings to retain fluid. If brake line nut is seized, brass bolt and block on caliper can be removed with brake line attached by removing bolt and block copper washers after removing caliper mounting bolts.

11. Remove caliper mounting bolts and remove caliper and brake shoes.

12. Remove two caliper mounting sleeves and four bushings, and install new parts using a silicone lubricant. (Sleeves are installed in inner bushings.)

13. Position new inboard shoe assembly on piston. D shaped in the indentation provided in the piston.

14. Install new outboard shoe assembly.

15. To reinstall caliper replace any corroded caliper mounting bolts with new parts. Wire bushing or sanding will damage the bolt plating.

16. If brass bolt and block was removed with brass pipe, unplug fittings and install bolt and block using two new copper gaskets. Torque to 30 ft. lbs. Be sure that all sleeves, bushings

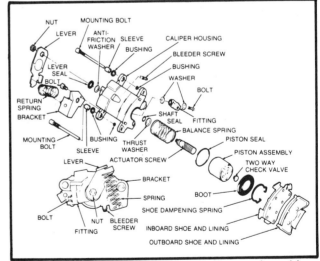

Type ten—General Motors rear disc and parking brake

and pins are well lubricated with silicone (mounting bolt should go under inboard shoe ears).

17. Install brake line tube nut into caliper and pump brake pedal to seat lining against rotor.

SECTION 1

BRAKES
Domestic Vehicles

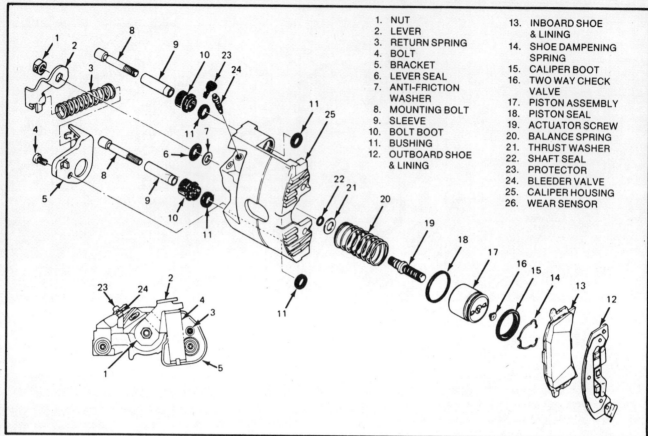

Type eleven — Delco floating rear brake caliper

18. Clinch upper ear of outboard shoe by positioning pliers with one jaw on top of upper ear and other jaw in notch or bottom of shoe, opposite upper ear. After clinching there should be no radial clearance between the shoe ears and caliper housing. Repeat clinching procedure if necessary.
19. Connect and adjust parking rake cables and bleed rear brake system.
20. Install wheel and tire assembly. Torque steel mounting nuts to 130 ft. lbs.

CALIPER PARKING BRAKE MECHANISM

When the parking brake is applied, the lever turn the actuator screw which is threaded into a nut in the piston assembly. This causes the piston to move outward and the caliper to slide inward mechanically, forcing the linings against the rotor. The piston assembly contains a self-adjusting mechanism for the parking brake.

Type Eleven

DELCO FLATING REAR BRAKE CALIPER

Pad Removal

1. Remove 2/3 of the brake fluid from the master cylinder.
2. Loosen the rear wheel lugs. Raise and support the rear of the vehicle on jackstands.

3. Mark the wheel and axle lug for same reinstallment location and remove the wheel assemblies.
4. Reinstall two lug nuts to retain the brake rotor.
5. Loosen the tension on the parking brake cable by backing off the equalizer.
6. After cable tension has been released, remove the cable end from the apply lever at the caliper.
7. Hold the apply lever in position and remove the retaining nut.
8. Remove the lever, lever seal and anti-fiction washer.

NOTE: If the parking brake levers are not disconnected from the caliper during pad removal and installation, damage to the piston assembly will occur when it is moved back in the caliper bore.

9. Position a C-clamp over the caliper and tighten until the piston bottoms in the caliper bore. Take care not to allow the C-clamp to contact the actuator screw on the caliper. Reinstall anti-friction washer, seal and lever.
10. If caliper service is required, disconnect the brake line. Plug all openings.
11. Remove the caliper mounting bolts using a 3/8 allen head socket or wrench.
12. Remove the caliper by lifting up and off the rotor. Do not permit the caliper to be suspended by the brake hose.
13. Remove the pads from the caliper. A suitable tool is required to pry the outboard pad from the caliper since it is retained by a spring button.
14. Remove the pin bushings and sleeves from the caliper ears.

Pad Installation

1. Install new sleeves and bushings after lubricating them. Insure that the sleeve is flush with the pad side of the caliper ear.
2. Install the inboard pad. Make sure that the D shaped retainer on the pad engages the D shaped slot in the caliper piston. Turn piston if necessary for correct alignment.
3. Be sure that the wear indicator is mounted on the leading edge of the pad for forward rotation of the wheel.
4. Slide the edge of the metal shoe under the ends of the dampening spring and snap the pad into position flat against the caliper piston.
5. Mount the outboard pad in position. Be sue it snaps into the caliper recess.
6. Install the caliper over the disc rotor in the reverse order of removal. Apply the brakes several times to seat the linings, after filing the master cylinder. Bleed brakes if necessary.

Type Twelve

GIRLOCK FLOATING FRONT AND REAR CALIPERS

Pad Removal

1. Remove 2/3 of the brake fluid from the master cylinder.
2. Loosen the rear wheel lugs. Raise and support the rear of the vehicle on jackstands.
3. Mark the wheel and axle lug for same reinstallment location and remove the wheel assemblies.
4. Reinstall two lug nuts to retain the brake rotor.
5. Loosen the tension on the parking brake cable by backing off the equalizer.
6. After cable tension has been released, remove the cable end from the apply lever at the caliper.
7. Hold the apply lever in position and remove the retaining nut.
8. Remove the lever, lever seal and anti-friction washer.

NOTE: If the parking brake levers are not disconnected from the caliper during pad removal and installation, damage to the piston assembly will occur when it is moved back in the caliper bore.

9. Position a C-clamp over the caliper and tighten until the piston bottoms in the caliper bore. Take care not to allow the C-clamp to contact the actuator screw on the caliper. Reinstall anti-friction washer, seal and lever.
10. If caliper service is required, disconnect the brake line. Plug all openings.
11. Remove the caliper mounting bolts using a 3/8 allen head socket or wrench.
12. Remove the caliper by lifting up and off the rotor. Do not permit the caliper to be suspended by the brake hose.
13. Remove the pads from the caliper. A suitable tool is required to pry the outboard pad from the caliper since it is retained by a spring button.
14. Remove the pin bushings and sleeves from the caliper ears.

Pad Installation

1. Install new sleeves and bushings after lubricating them. Insure that the sleeve is flush with the pad side of the caliper ear.
2. Install the inboard pad. Make sure that the D shaped retainer on the pad engages the D shaped slot in the caliper piston. Turn piston if necessary for correct alignment.
3. Be sure that the wear indicator is mounted on the leading edge of the pad for forward rotation of the wheel.

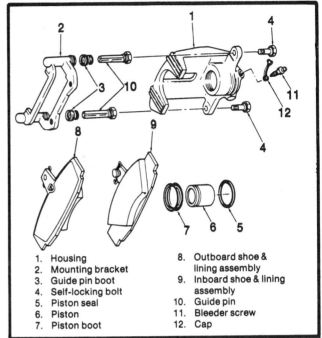

1. Housing
2. Mounting bracket
3. Guide pin boot
4. Self-locking bolt
5. Piston seal
6. Piston
7. Piston boot
8. Outboard shoe & lining assembly
9. Inboard shoe & lining assembly
10. Guide pin
11. Bleeder screw
12. Cap

Type twelve – Girlock front and rear brake caliper

4. Slide the edge of the metal shoe under the ends of the dampening spring and snap the pad into position flat against the caliper piston.
5. Mount the outboard pad in position. Be sure it snaps into the caliper recess.
6. Install the caliper over the disc rotor in the reverse order of removal. Apply the brakes several times to seat the linings, after filling the master cylinder. Bleed brakes if necessary.

Type Twelve

GIRLOCK FLOATING FRONT AND REAR CALIPERS

Pad Removal

1. Remove 2/3 of the brake fluid from the master cylinder.
2. Loosen the rear lugs and raise the rear of the vehicle. Support the vehicle on jackstands.
3. Remove the wheel assemblies. Install two lug nuts to hold the brake disc rotor in position.
4. Position a C-clamp over the caliper, one end on the outboard pad, the other on the inlet fitting bolt head.
5. Tighten the clamp to push the caliper piston until it bottoms in the bore.
6. Remove and discard the upper caliper self-locking bolt. Rotate the caliper on the lower bolt to expose the brake pads.
7. Remove the inner and outer pads from the caliper.
8. Clean the pad mounting frame on the caliper. Inspect the caliper for signs of fluid leakage. Remove and service caliper if necessary.

Pad Installation

1. Install the new inner and outer pad in position on the caliper.

1–45

BRAKES
Domestic Vehicles

2. Rotate the caliper back into position over the disc brake rotor.
3. Install a new self-locking bolt and tighten 22 to 25 ft. lbs.
4. Install wheel assemblies and lower vehicle.
5. Fill the master cylinder and pump the brake pedal several times to seat the pads. Bleed the brakes if necessary.

Parking Brake Adjustment

1. Raise the rear of the vehicle and support it safely, remove the rear wheels and place two lug nuts opposite of each other to ensure proper disc/drum position.
2. Loosen the parking brake cable so as to release any tension on the parking brake shoes. Rotate the disc so that the hole in the disc/drum face will be aligned with the star adjuster.
3. Using a brake spoon or equivalent, adjust the parking brake by inserting the tool into the hole in the disc face on the drivers side and move the handle of the tool toward the top of the fender skirt to adjust the shoes out and toward the ground to adjust the shoes in. Reverse this procedure on the passenger side.
4. Adjust the shoes until the disc/drum won't rotate and back the star adjuster off 5 to 7 notches on both sides.
5. Reinstall the rear wheels, lower the vehicle and road test.

NOTE: The asbestos dust thrown off from the brake linings or disc pads may be dangerous to your health if inhaled, Never use compressed air or your own breath to blow the dust from the brake assembly. Use a damp rag or a well filtered vacuum cleaner. Dispose of the rag or cleaner bag properly.

Drums

Conditions and Resurfacing

The condition of the brake drum surface is just as important as the surface to the brake lining. All drum surfaces should be clean, smooth, free from hard spots, heat checks, score marks and foreign matter imbedded in the drum surface. They should not be out of round, bell-mouthed or barrel shaped. It is recommended that all drums be first checked with a drum micrometer to see if they are within over-size limits. If drum is within safe limits, even though the drum surface appears smooth, it should be turned not only to assure a true drum surface but also to remove any possible contamination in the surface from previous brake linings, road dusts, etc. Too much metal removed from a drum is unsafe and may result in the following.

1. Brake to fade due to the thin drum being unable to absorb the heat generated.
2. Poor and erratic brake due to distortion of drums.
3. Noise due to possible vibration caused by thin drums.
4. A cracked or broken drum on a severe or very hard brake application. Brake drum run-out should not exceed 0.005 inch Drums turned to more than 0.060 inch oversize are unsafe and should be replaced with new drums, except for some heavy ribbed drums which have an 0.080 inch limit. It is recommended that the diameters of the left and right drums on any one axle be within when replacing drums. It is always good to replace the drums on both wheels at the same time.
5. If the drums are true, smooth up any slight scores by polishing with fine emery cloth. If deep scores or grooves are present, which cannot be removed by this method, then the drum must be turned.

DRUM BRAKE SPECIFICATIONS

Vehicle Make and Model	Brake Shoe ① Minimum Lining Thickness	Brake Drum Diameter		Wheel Lugs or Nuts Torque (ft-lbs)
		Standard Size	Machine To	
American Motors				
All exc. 6 cyl. Concord Wagon and Eagle	.030	9.000	9.060	75
6 cyl. Concord Wagon and Eagle	.030	10.000	10.060	75
Chrysler Corp.—Chrysler—Dodge—Plymouth				
Dodge 600, New Yorker	.030	8.661	—	85
Aries, Reliant, LeBaron Dodge 400	.030	7.870	7.900	85
Cordoba, Diplomat, Grand Fury, Mirada, New Yorker, Imperial				
w/10" rear brake	.030	10.000	10.060	85
w/11" rear brake	.030	11.000	11.060	85
Omni, Horizon	.030	7.870	7.900	85
Aries, Reliant	.030	7.870	7.900	85
Ford Motor Co.—Ford—Mercury—Lincoln				
Lincoln Continental				
Front	—	—	—	80-105
Rear	—	—	—	80-105
Thunderbird, XR-7				
w/9" rear brake	.030	9.000	9.060	80-105
w/10" rear brake	.030	10.000	10.060	80-105

BRAKES
Domestic Vehicles

DRUM BRAKE SPECIFICATIONS

Vehicle Make and Model		Brake Shoe ① Minimum Lining Thickness	Brake Drum Diameter Standard Size	Brake Drum Diameter Machine To	Wheel Lugs or Nuts Torque (ft-lbs)
'83	Cougar, Granada				
	w/9" rear brake	.030	9.000	9.060	80-105
	w/10" rear brake	.030	10.000	10.060	80-105
'82-'86	Escort, Lynx, EXP, LN7				
	w/7" rear brake	.030	7.000	7.060	80-105
	w/8" rear brake	.030	8.000	8.060	80-105
'83-'85	Mustang, Capri, Fairmount, Zephyr				
	w/9" rear brake	.030	9.000	9.060	80-105
	w/10" rear brake	.030	10.000	10.060	80-105
	Lincoln Town Car, Mark VI, LTD, Marquis				
	w/10" rear brakes	.030	10.000	10.060	80-105
	w/11" rear brakes	.030	11.030	11.090	80-105
General Motors Corp					
Buick					
	Century, Skyhawk	①	7.880	7.899	100
	Regal, LeSabre	①	9.500	9.560	80 ⑤
	Riviera				
	w/o rear disc brakes	①	9.500	9.560	100
	w/rear disc brakes	—	—	—	100
	Electra, Estate Wagon	①	11.000	11.060	100
	Skylark	①	7.880	7.899	103
Cadillac					
	Cimarron	.030	.880	7.899	100
	Fleetwood	.030	11.000	11.060	100
	Eldorado, Seville				
	Front	—	—	—	100
	Rear	—	—	—	100
	Fleetwood Limo, Commercial Chassis	①	12.000	12.060	100
Chevrolet					
	Celebrity, Cavalier	①	7.880	7.899	100
	Camaro				
	w/rear drum brakes	①	9.500	9.560	80†
	w/rear disc brakes	—	—	—	80†
	Malibu, Monte Carlo, El Camino	①	9.500	9.560	80†
	Citation	①	7.880	7.899	103
	Impala, Caprice				
	w/9½" rear brakes	①	9.500	9.560	80
	w/11" rear brakes	①	11.000	11.060	100
	Chevette	①	7.874	7.899	70
	Corvette				
	Front	—	—	—	70 ②
	Rear	—	—	—	70 ②
Oldsmobile					
	Ciera, Firenza	①	7.880	7.899	100
	Cutlass Supreme, 88	①	9.500	9.560	100 ③
'83-'84	Omega	①	7.880	7.899	103
	Toronado				
	w/o rear disc	①	9.500	9.560	100
	w/rear disc	—	—	—	100

1-47

SECTION 1

BRAKES
Domestic Vehicles

DRUM BRAKE SPECIFICATIONS

Vehicle Make and Model		Brake Shoe ① Minimum Lining Thickness	Brake Drum Diameter		Wheel Lugs or Nuts Torque (ft-lbs)
			Standard Size	Machine To	
'83-'84	Custom Cruiser, 88 (w/403), 98				
	w/9.5" rear brake	①	9.500	9.560	100
	w/11: rear brake	①	11.000	11.060	100
Pontiac					
	6000, J2000	①	7.880	7.899	100
	Firebird				
	w/rear drum brakes	①	9.500	9.560	80†
	w/rear disc brakes	—	—	—	80†
	T1000	①	7.874	7.899	70†
'83-'84	Phoenix (F.W.D.)	—	7.880	7.899	103
	Bonneville, Catalina, LeMans, Grand Prix, Grand Am, Safari				
	w/9.5" rear brakes	①	9.500	9.560	80
	w/11" rear brakes	①	11.000	11.060	80 ④

NOTE: State and local inspection regulations will take precedence over manufacturer's standards.
① .030 in. over rivet head, if bonded lining use .062 in.
② Aluminum whls; Corvette 80, Camaro 105, others 90.
③ 88 w/7/16 in. stud; 80 ft.-lbs.
④ 1/2 in. stud 100 ft.-lbs.
⑤ W/aluminum wheels, LeSabre 90 ft.-lbs., Regal 100 ft.-lbs.

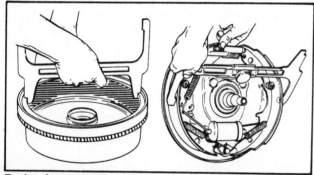

Brake drum gauge

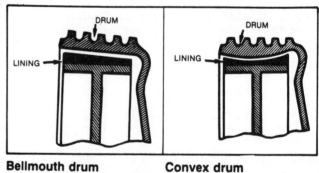

Bellmouth drum   Convex drum

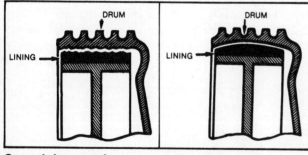

Scored drum surface   Concave drum

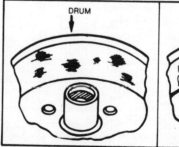

Hard or chill spots

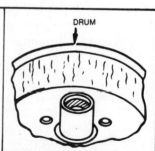

Heat checks

1-48

BRAKES
Domestic Vehicles

DRUM BRAKES

 TROUBLESHOOTING DRUM BRAKES

Trouble Symptoms

Possible Causes of Trouble Symptoms	One Brake Drags	All Brakes Drag	Hard Pedal	Spongy Pedal	Car Pulls to One Side	One Wheel Locks	Brakes Chatter	Excessive Pedal Travel	Pedal Gradually Goes to Floor	Brakes Uneven	Shoe Click Release	Noisy or Grabbing Brakes	Brakes Do Not Apply
Mechanical Resistance at Pedal or Shoes Damaged Linkage		X	X										
Brake Line Restricted	X	X	X		X								
Leaks or Insufficient Fluid				X				X	X				X
Improper Tire Pressure					X						X		
Improperly Adjusted or Worn Wheel Bearing	X				X								
Distorted or Improperly Adjusted Brake Shoe	X	X	X		X	X		X				X	
Faulty Retracting Spring	X				X								
Drum Out of Round	X				X		X						
Linings Glazed or Worn				X	X	X	X	X				X	X
Oil or Grease In Lining				X	X	X	X				X	X	X
Loose Carrier Plate	X					X	X						
Loose Lining					X	X							
Scored Drum											X	X	
Dirt on Drum-Lining Surface												X	
Faulty Wheel Cylinder	X				X	X						X	
Dirty Brake Fluid	X	X									X		X
Faulty Master Cylinder		X						X	X				X
Air in Hydraulic System	X			X				X					X
Self Adjusters Not Operating					X			X			X		
Insufficient Shoe-to-Carrier Plate Lubrication	X										X		
Tire Tread Worn						X							
Poor Lining to Drum Contact							X						
Loose Front Suspension							X						
"Threads" Left by Drum Turning Tool Pull Shoes Sideways											X		
Cracked Drum								X					
Sticking Booster Control Valve		X										X	

1–49

BRAKES
Domestic Vehicles

DRUM BRAKE APPLICATION CHART

Car and Years	Brake Type	Self-Adjuster Type
American Motors	Duo-Servo	Star & Screw
Chrysler Corp. 1982-86 all models ①	Duo-Servo	Star & Screw
Ford Motor Co. 1982-86 all models except below	Duo-Servo	Star & Screw
1982-86 Escort, Lynx, EXP, LN7	Non-Servo	Star & Screw (8 in. brake) Strut & Pin (7 in. brake)
General Motors Corp. 1982-86 all models	Duo-Servo	Star & Screw

①The rear drum brakes on Chrysler front wheel drive cars through 1982, are not automatically adjusted.

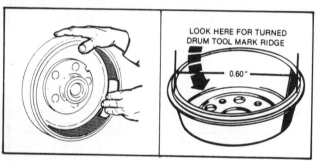

Sanding brake drums   Oversize drum

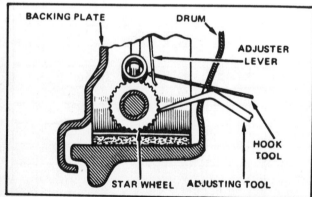

Release of adjusting lever with adjusting slot in brake drum

Duo-Servo Brake

Refer to the Drum Brake Application Chart for adjuster applications. In the Duo-Servo design, the force which the wheel cylinder applies to the shoes is supplemented by the tendency of the shoes to wrap or twist into the drum during braking. Thus two braking forces are applied at each drum every time the brakes are activated.

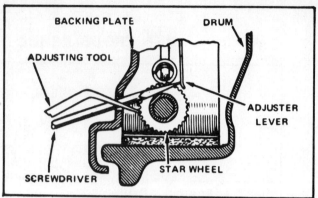

Release of adjusting lever with adjusting slot in backing plate

STAR AND SCREW ADJUSTER

The duo-servo brake, with star and screw type self-adjusters, is used on most late-model American cars. The same basic brake unit has been used on all cars. General Motors cars use a rod-operated lever to turn the star wheel, while all others use a cable-operated lever. This is the only difference, other than size, among units used on different models.

Brake Shoe Adjustment

The drum brakes used on today's cars are normally self-adjusting. They require manual adjustment only when the shoes have been replaced or when the star and screw adjuster has been disturbed.

NOTE: The drum brakes on most cars can be initially adjusted by removing the brake drum, measuring its internal diameter, then adjusting the shoes to that measurement and installing the drum. Use a vernier gauge to make the measurements. This method can be used on all models, and may be preferable to punching out the covering over the access hole in the backing plate or brake drum edge.

1. Remove the access slot plug from the backing plate or front of drum. On some cars, there is no access slot in the backing plate or in the front of the drums. Some have been filled in and must be punched out to gain access to the adjuster. Complete the adjustment and cover the hole with a plug to prevent entrance of dirt and water.

2. Using a brake adjusting spoon or screwdriver, pry downward on the end of the tool (starwheel teeth moving up) to tighten the brakes, or upward on the end of the tool (starwheel teeth moving down) to loosen the brakes.

NOTE: It may be necessary to use a small rod or suitable tool to hold the adjusting lever away from the star wheel. Be careful not to bend the adjusting lever.

3. When the brakes are tight almost to the point of being locked, back off on the starwheel until the wheel is able to rotate freely. The starwheel on each set of brakes (front or rear) must be backed off the same number of turns to prevent brake pull from side to side.

4. When all four brakes are adjusted, check brake pedal travel and then make several stops, while backing the car up, to equalize all the wheels.

Testing Adjuster

1. Raise the vehicle on a hoist, with a helper in the car, to apply the brakes.

BRAKES
Domestic Vehicles

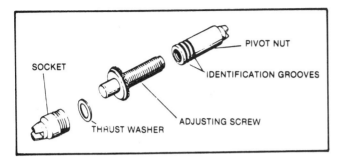

Adjusting screw assembly

2. On models with access plugs in the backing plate, loosen the brakes by holding the adjuster lever away from the starwheel and backing off the starwheel approximately 30 notches. On models without access plugs in the backing plate, remove wheel and drum, loosen the adjuster, then reinstall the drum and wheel.
3. Spin the wheel and brake drum in reverse and apply the brakes. The movement of the secondary shoe should pull the adjuster lever up, and when the brakes are released the lever should snap down and turn the starwheel.
4. If the automatic adjuster doesn't work, the drum must be removed and the adjuster components inspected carefully for breakage, wear, or improper installation.

BRAKE SHOES

Removal

1. Remove the brake drum.
2. Place the hollow end of a brake spring service tool on the brake shoe anchor pin and twist it to disengage one of the brake shoe retaining springs. Repeat this operation to remove the other spring. Be careful that the springs do not slip off the tool during removal, as the spring could break loose and cause personal injury.
3. Reach behind the brake backing plate and place a finger on the end of one of the brake holddown mounting pins. Using a pair of pliers, or special brake pin retainer tool, grasp the

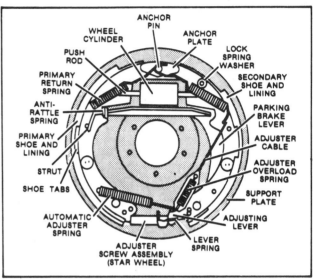

Chrsyler type rear drum brake assembly

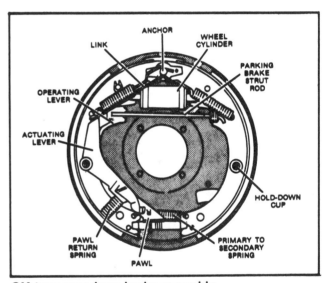

GM type rear drum brake assembly

washer on the top of the hold-down spring that corresponds to the pin that you are holding. Push down on the pliers and turn them 90° to align the slot in the washer with the head on the spring mounting pin. Remove the spring and washer and repeat this operation on the holddown spring of the other brake shoe.
4. On Ford and American Motors cars, place the tip of a suitable tool on the top of the brake adjusting screw and move the screwdriver upward to lift up on the brake adjusting lever. When there is enough slack in the automatic adjuster cable, disconnect the loop on the top of the cable from the anchor. Back off the adjusting screw while holding the adjustment lever away from the screw. Grasp the top of each brake shoe and move them outward to disengage from the wheel cylinder and parking brake link (if working on rear wheels). When the brake shoes are clear, lift them from the backing plate. Twist the shoes slightly and the automatic adjuster assembly will disassemble itself.

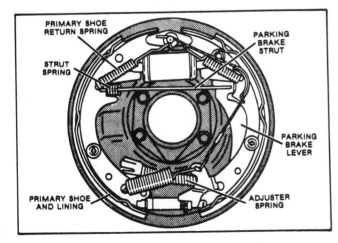

AMC, Ford type rear drum brake assembly

1-51

BRAKES
Domestic Vehicles

5. On GM cars, remove the automatic adjuster link. Remove the automatic adjuster lever, pivot, and override spring from the secondary spring as an assembly. Move the top of each brake shoe outward to clear the wheel cylinder pins and parking brake link (rear brakes). Lift the brakes from the backing plate and remove the adjusting screw.

6. On Chrysler cars, except some front wheel drive models, slide the automatic adjuster cable from the anchor pin and disengage it from the adjusting lever. Remove the cable, overload spring, and cable guide. Disconnect the automatic lever return spring and remove the spring and lever. Move the top of the brake shoes outward to clear the wheel cylinder pins and parking brake link (rear brakes). Lift the brakes from the backing plate and remove the adjusting screw.

7. Grasp the end of the brake cable spring with a pair of pliers and, using the brake lever as a fulcrum, pull the end of the spring away from the lever. Disengage the cable from the brake lever.

Installation

1. The brake cable must be connected to the secondary brake shoe before the shoe is installed on the backing plate. To do this, transfer the parking brake lever from the old secondary shoe to the new one. This is accomplished by spreading the bottom of the horseshoe clip and disengaging the lever. Position the lever on the new secondary shoe and install the spring washer and the horseshoe clip. Close the bottom of the clip after installing it. Grasp the metal tip of the parking brake cable with a pair of pliers. Position a pair of side cutters on the end of the cable coil spring and, using the pliers as a fulcrum, pull the coil spring back with the side cutters. Position the cable in the parking brake lever.

2. Apply a light coating of high-temperature grease to the brake shoe contact points on the backing plate. Position the primary brake shoe on the front of the backing plate and install the hold-down spring and washer over the mounting pin. Install the secondary shoe on the rear of the backing plate.

3. If working on rear brakes, install the parking brake link between the primary brake shoe and the secondary brake shoe.

4. On Ford and American Motors cars, install the automatic adjuster cable loop end on the anchor pin. Make sure that the crimped side of the loop faces the backing plate.

5. On GM cars, assemble the automatic adjuster lever, pivot, and override spring and install to the secondary spring as an assembly.

6. On Chrysler, install the automatic adjuster lever and return spring. Install the adjuster overload spring and cable. One end of the cable engages with the adjusting lever while the other slips over the anchor pin underneath the primary and secondary return springs.

7. Install the return spring in the primary brake shoe and, using the tapered end of a brake spring service tool, slide the top of the spring onto the anchor pin. Be careful to make sure that the spring does not slip off the tool during installation, as the spring could break loose and cause personal injury.

8. Install the automatic adjuster cable guide in the secondary brake shoe, making sure that the flared hole in the cable guide is inside the hole in the brake shoe. Fit the cable into the groove in the top of the cable guide.

9. Install the secondary shoe return spring through the hole in the cable guide and the brake shoe. Using the brake spring tool, slide the top of the spring onto the anchor pin.

10. Clean the threads on the adjusting screw and apply a light coating of high-temperature grease to the threads. Screw the adjuster closed, then open it one-half turn.

11. Install the adjusting screw between the brake shoes with the star wheel nearest to the secondary shoe. Make sure that the star wheel is in a position that is accessible from the adjusting slot in the backing plate.

12. Install the short, hooked end of the automatic adjuster spring in the proper hole in the primary brake shoe.

13. Connect the hooked end of the automatic adjuster cable and the free end of the automatic adjuster spring in the slot in the top of the automatic adjuster lever.

14. Pull the automatic adjuster lever (the lever will pull the cable and spring with it) downward and to the left, and engage the pivot hook of the lever in the hole in the secondary brake shoe.

15. Check the entire brake assembly to make sure everything is installed properly. Make sure that the shoes engage the wheel cylinder properly and are flush on the anchor pin. Make sure that the automatic adjuster cable is flush on the anchor pin and in the slot on the back of cable guide. Make sure that the adjusting lever rests on the adjusting screw star wheel. Pull upward on the adjusting cable until the adjusting lever is free of the star wheel, then release the cable. The adjusting lever should snap back into place on the adjusting screw star wheel and turn the wheel one tooth.

16. Expand the brake adjusting screw until the brake drum will just fit over the brake shoes.

17. Install the wheel and drum and adjust the brakes.

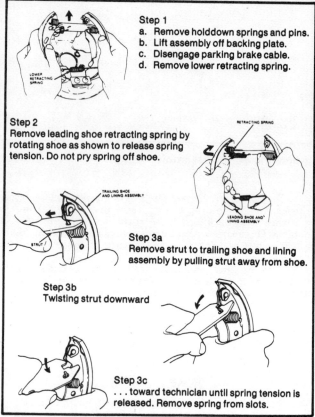

Step 1
a. Remove holddown springs and pins.
b. Lift assembly off backing plate.
c. Disengage parking brake cable.
d. Remove lower retracting spring.

Step 2
Remove leading shoe retracting spring by rotating shoe as shown to release spring tension. Do not pry spring off shoe.

Step 3a
Remove strut to trailing shoe and lining assembly by pulling strut away from shoe.

Step 3b
Twisting strut downward

Step 3c
...toward technician until spring tension is released. Remove spring from slots.

Non-servo – 7 in. Ford rear brakes

FORD NON-SERVO BRAKES

The star and screw adjuster is used on models with 8 in. diameter brake drums while the strut and pin adjuster is used on 7 inch diameter drums.

Adjustments

Normal shoe adjustments are automatic, however, when the

BRAKES
Domestic Vehicles

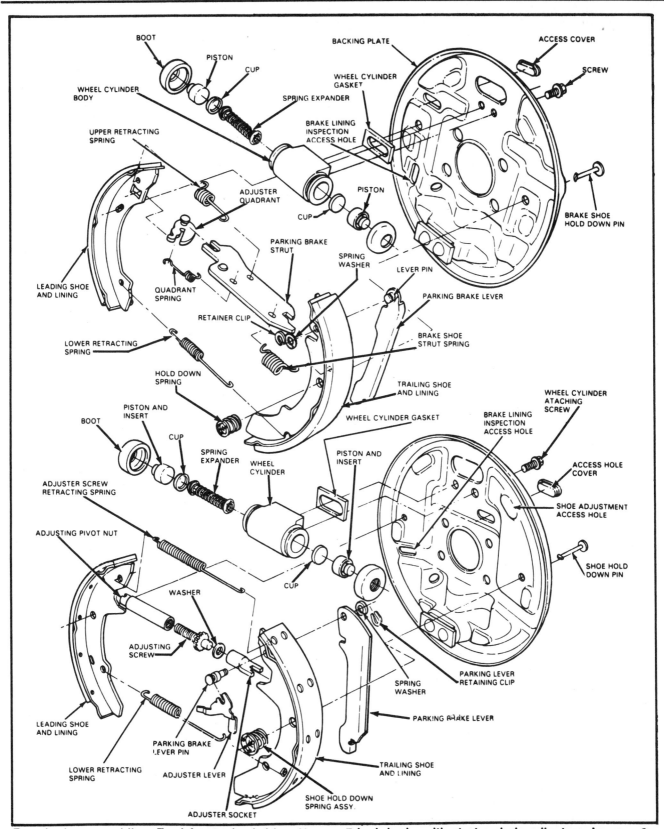

Rear brake assemblies, Ford front wheel drive. Upper—7 inch brake with strut and pin adjusters. Lower—8 inch brake with star and screw adjusters

BRAKES
Domestic Vehicles

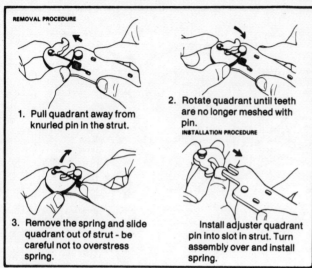

Quadrant removal and installation—7 in. Ford non-servo brakes

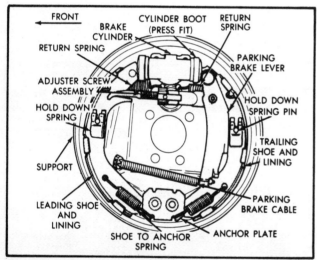

Chrysler non-servo rear brakes (non self adjusting)

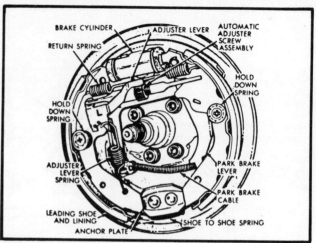

Rear wheel brake—1983 Chrysler front wheel drive vehicles

Removal and Installation

1. Remove the wheel and hub. Adjusters can be backed off through the back of the brake backing plate with a suitable tool if the drum will not come off.
2. Remove the holddown springs and pins. Lift the assembly off the brake backing plate and disengage the parking brake cable.
3. On 7 inch drums, remove the lower retracting spring. On 8 inch drums, remove all retracting springs and the adjuster lever.
4. The following removal procedures are for 7 inch drums only. Remove the leading shoe retracting spring by rotating shoe to release spring tension. Do not pry the spring off the shoe. Remove the strut to trailing shoe assembly by pulling the strut away from the shoe and twisting the shoe downward until spring tension is released. Remove the spring from the slots.
5. Installation is the reverse of removal. See adjustment procedure, above, for special information on initial adjustment techniques. Wheel bearings on 8 inch drums are adjusted in the same manner as 7 inch drums. See Step 2 of "Adjustment" procedure.

CHRYSLER NON-SERVO BRAKES

Adjustments

1. Remove the access slot plug from the upper part of the backing plate.
2. Using a thin brake adjusting spoon pry downward (left side) or upward (right side) on the end of the tool (starwheel teeth moving up) to tighten the brakes. The opposite applies to loosen the brakes.
3. When the brakes are tight almost to the point of being locked, back off on the starwheel 10 clicks. The starwheel on each side must be backed off the same number of turns to provide for even braking.

Removal and Installation

1. Remove the brake drum.
2. Unhook the parking brake cable from the secondary (trailing) shoe.
3. Remove the shoe-to-anchor retracting spring(s) and the upper spring (if equipped).
4. Remove the shoe hold down springs; compress them slightly and slide them off of the hold down pins or push in and twist them from the mount pin.

shoes have been replaced or the adjuster has been disturbed, the shoes should be initially adjusted by hand.
1. Raise the rear of the car and remove the wheels and drums. Drums are removed by releasing the parking brake, removing the dust cap, cotter pin, adjusting nut and wheel bearing, then pulling off the drum.
2. On 7 inch drums with strut and pin adjuster, pivot the adjuster quadrant until it meshes with the knurled pin and is in the third or fourth notch of the outboard end of the quadrant. Install the brake drum and wheel and adjust the wheel bearings by tightening the adjusting nut to 17-15 ft. lbs. while rotating the drum, then back off the adjusting nut about 100° and install the nut retainer and cotter pin.
3. 8 inch drums are adjusted in the same manner as the star and screw adjuster drums described under "Duo-Servo" brakes, above. See that section for procedure.
4. Complete adjustment by applying the brakes several times.

BRAKES
Domestic Vehicles

5. Remove the adjuster screw assembly by spreading the shoes apart. Disconnect the adjuster spring from the trailing shoes on self-adjuster models. The adjuster nut must be fully backed off.
6. Raise the parking brake lever. Pull the secondary (trailing) shoe away from the backing plate so pull-back spring tension is released.
7. Remove the secondary (trailing) shoe and disengage the spring end from the backing plate.
8. Raise the primary (leading) shoe to release spring tension. Remove the shoe and disengage the spring end from the backing plate.
9. Inspect the brakes linings.
10. Lubricate the six shoe contact areas on the brake backing plate and the web end of the brake shoe which contacts the anchor plate. Use a multi-purpose lubricant or a high temperature brake grease made for the purpose.
11. Chrysler recommends that the rear wheel bearings be cleaned and repacked whenever the brakes are renewed. Be sure to install a new bearing seal.
12. With the leading shoe return spring in position on the shoe, install the shoe at the same time as you engage the return spring in the end support.
13. Position the end of the shoe under the anchor.
14. With the trailing shoe return spring in position, install the shoe at the same time as you engage the spring in the support (backing plate).
15. Position the end of the shoe under the anchor.
16. Spread the shoes and install the adjuster screw assembly making sure that the forked end that enters the shoe is curved down.
17. Insert the shoe hold down spring pins and install the hold down springs.
18. Install the shoe-to-anchor springs and adjuster spring (if equipped).
19. Install the parking brake cable onto the parking brake lever.
20. Replace the brake drum and tighten the nut 240 to 300 inch lbs. while rotating the wheel.
21. Back off the nut enough to release the bearing preload and position the locknut with one pair of slots aligned with the cotter pin hole.
22. Install the cotter pin. The end play should be 0.001 to 0.003 inch.
23. Install the grease cap.

POWER BRAKES

Vacuum Operated Booster

Power brakes operate just as standard brake systems except in the actuation of the master cylinder pistons. A vacuum diaphragm is located on the front of the master cylinder and assists the driver in applying the brakes, reducing both the effort and travel he must put into moving the brake pedal.

The vacuum diaphragm housing is connected to the intake manifold by a vacuum hose. A check valve is placed at the point where the hose enters the diaphragm housing, so that during periods of low manifold vacuum brake assist vacuum will not be lost.

Depressing the brake pedal closes off the vacuum source and allows atmospheric pressure to enter on one side of the diaphragm. This causes the master cylinder pistons to move and apply the brakes. When the brake pedal is released, vacuum is applied to both sides of the diaphragm, and return springs re-

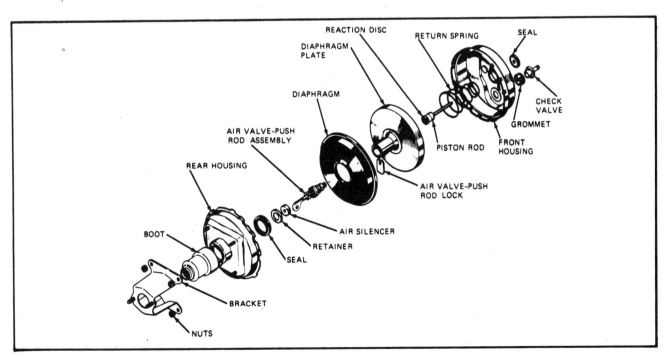

General Motors—Bendix—single diaphragm brake booster

BRAKES
Domestic Vehicles

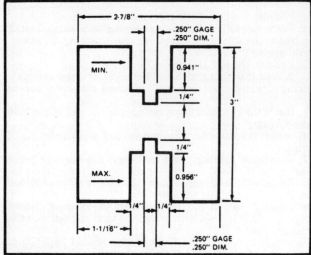

Ford brake booster to master cylinder push-rod gauge dimensions

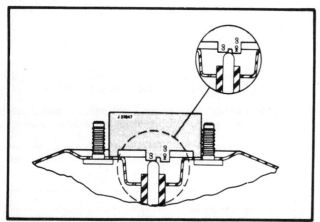

Using General Motors special tool J-22647 or equivalent to measure the brake booster to master cylinder push-rod

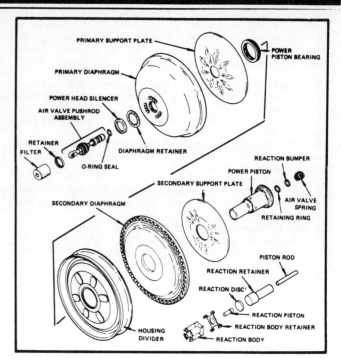

General Motors—Tandem brake booster, with an exploded view of the power piston group

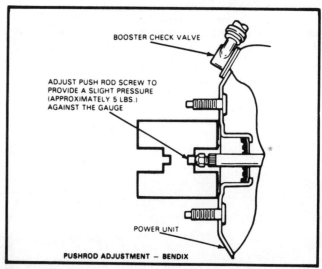

Measuring the Ford brake booster to master cylinder push-rod

turn the diaphragm and master cylinder pistons to the released position. If the vacuum fails, the brake pedal rod will butt against the end of the master cylinder actuating rod, and direct mechanical application will occur as the pedal is depressed.

The hydraulic and mechanical problems that apply to conventional brake systems also apply to power brakes, and should be checked or if the tests and chart below do not reveal the problem.

1. Operate the engine at idle with the transmission in Neutral without touching the brake pedal for at least one minute.
2. Turn off the engine, and wait one minute.
3. Test for the presence of assist vacuum by depressing the brake pedal and releasing it several times. Light application will produce less and less pedal travel, if vacuum was present. If there is no vacuum, air is leaking into the system somewhere.
4. Pump the brake pedal (with engine off) until the supply vacuum is entirely gone.
5. Put a light, steady pressure on the pedal.
6. Start the engine, and operate it at idle with the transmission in Neutral. If the system is operating, the brake pedal should fall toward the floor if constant pressure is maintained on the pedal.
7. Power brake systems may be tested for hydraulic leaks just as ordinary systems are tested, except that the engine should be idling with the transmission in Neutral throughout the test.

POWER BRAKE BOOSTER TROUBLESHOOTING

The following items are in addition to those listed in the General Troubleshooting Section. Check those items first.

BRAKES
Domestic Vehicles

HARD PEDAL

1. Faulty vacuum check valve.
2. Vacuum hose kinked, collapsed, plugged, leaky, or improperly connected.
3. Internal leak in unit.
4. Damaged vacuum cylinder.
5. Damaged valve plunger.
6. Broken or faulty springs.
7. Broken plunger stem.

GRABBING BRAKES

1. Damaged vacuum cylinder.
2. Faulty vacuum check valve.
3. Vacuum hose leaky or improperly connected.
4. Broken plunger stem.

PEDAL GOES TO FLOOR

Generally, when this problem occurs, it is not caused by the power brake booster. In rare cases, a broken plunger stem may be at fault.

Service and/or Overhaul

Most power brake boosters are serviced by replacement only. In many cases, repair parts are not available. A good many special tools are required for rebuilding these units. For these reasons, it would be most practical to replace a failed booster with a new or remanufactured unit.

AMC

If diagnosis indicates an internal malfunction in the power brake unit, service the unit as an assembly only. Do not attempt to disassemble, repair or adjust any power brake unit. If a unit must be replaced, use the master cylinder push rod supplied with the replacement unit. This push rod has been preset and gauged for use with the replacement unit.

CHRYSLER

Do not attempt to disassemble power brake unit as this booster is serviced as a complete assembly only. A properly installed power brake unit, with vacuum at the unit and pressure applied to the pedal, will vent the master cylinder (force a jet of fluid up through the front chamber vent port).

FORD

Adjustment of the pushrod and replacement of the check valve are the only services permitted on the brake booster. If any brake booster is damaged or inoperative, replace it with a new booster. The brake booster (excluding the check valve) is serviced only as an assembly.

GENERAL MOTORS

Use special tool number J22647 to gauge the power brake unit-to-master cylinder push rod length. The push rod length must fall within the go-no-go limits of the gauge.

Vacuum Pump

General Motors "J" body cars all use a vacuum pump designed to aid the 4 cylinder engine in maintaining a proper level of vacuum for the power brake system. The vacuum pump is equipped with an internal off-on switch in the controller. The switch will activate the pump when power brake vacuum falls

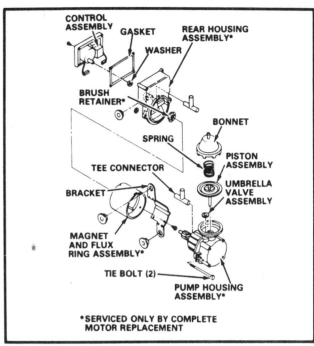

General Motors brake booster vacuum pump disassembled

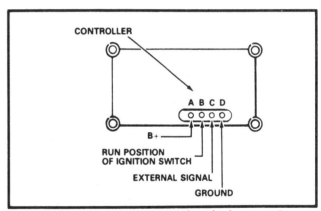

Brake booster vacuum pump electrical connector

below a designated level. The pump can only be activated by the controller when the ignition switch is in the run position.

Removal and Installation

1. Raise vehicle and remove splash shield from the driver's side of car.
2. Disconnect hoses at vacuum pump.
3. Disconnect electrical connector at vacuum pump.
4. Remove three nuts securing pump assembly to mounting bracket and remove pump.
5. Reverse procedure for installation or vacuum pump.

CONTROLLER

Removal and Installation

1. Remove tie bolts holding the body, pump housing, and rear housing together.

BRAKES
Domestic Vehicles

2. Remove rear housing. Do not damage tee connector.

3. Remove self tapping screws attaching control assembly to the rear housing. Detach brushes from brush holder and remove control assembly.

4. Remove old gasket and replace control assembly, using new gasket and washer. Ensure that both brushes are located in the brush holder cavity of the rear housing.

5. Set brush springs in preload state by locking them in provided slots above spring access slots. Place brushes in brush channels, ensuring that shunts are routed properly.

6. Place brush retainer in proper preassembly position and return spring to load condition.

7. Complete reassembly by reversing Steps 1 and 2.

PISTON ASSEMBLY

Removal and Installation

1. Uncrimp tabs on bonnet. Pump housing and bonnet must be held together during disassembly, due to spring load on bonnet.

2. Remove bonnet and spring.

3. Lift piston assembly out of pump housing assembly.

4. Remove and replace umbrella valve assembly.

5. Replace piston assembly and spring. Place bonnet on spring and compress. Crimp tabs allowing no tolerance for movement between bonnet and pump housing assembly.

Testing

At no time during testing of the vacuum pump should the outlet nozzle by blocked. Permanent damage to the diaphragm can result from a blocked outlet nozzle.

The pump should not draw over 8 amps and should automatically turn off within 5 to 10 seconds after the vacuum reaches a level between 10 and 15 in. Hg. After the pump shuts off, it should not leak down more than 2 in. Hg. in one minute. The brake warning light will come on to indicate trouble with the vacuum system. The brake light may or may not come on if the vacuum pump is running continuously.

ON THE VEHICLE TESTS

1. With the pump on the vehicle attach a vacuum gauge to the "T" connector between the controller and pump housing. Observe that the pump turns off within 5 to 10 seconds after the vacuum reaches 10 to 15 in. Hg. Check for less than 2 in. per minute leakage after the pump turns off.

2. If the pump runs continuously and no vacuum leaks can be found, replace the controller.

OFF THE VEHICLE TESTING

1. Remove the pump from the vehicle. Remove the T-fitting connector from between the pump housing and the controller. Attach a hand operated vacuum pump to the controller inlet and apply 20 in. Hg. of vacuum to the controller. If the switch leaks more than 2 in. Hg. per minute, then replace the switch.

2. Attach a hand vacuum pump to the pump housing inlet fitting and draw 20 in. Hg. of vacuum. If the pump leaks down more than 2 in. Hg. per minute, then replace the umbrella valve.

3. Plug the pump housing vacuum inlet, attach a hand vacuum pump to pump outlet port, plug the pump housing inlet and draw 20 in. Hg. of vacuum. If the pump leaks down more than 2 in. Hg. of vacuum per minute, then replace the pump piston assembly. Make sure the bonnet is tight and not leaking, before condemning the piston assembly.

Hydro-Boost II

Hydro-Boost II differs from conventional power brake systems, in that it operates from power steering pump fluid pressure, rather than intake manifold vacuum. The Hydro-Boost II unit contains a spool valve with an open center which controls the strength of pump pressure when braking occurs. A lever assembly controls the valve's position. A boost piston provides the force necessary to operate the conventional master cylinder on the front of the booster.

A reserve of at least two assisted brake applications is supplied by an accumulator which is spring loaded on earlier models and pneumatic on later models. The brakes can be applied manually if the reserve system is depleted.

All vehicles with Hydro-Boost II, on which the accumulator is an integral part of the Hydro-Boost II unit. All system checks, tests and troubleshooting procedures are the same for the two systems.

HYDRO-BOOST II SYSTEM CHECKS

1. A defective Hydro-Boost cannot cause any of the following conditions. Noisy brakes, fading pedal or pulling brakes. If any of these occur, check elsewhere in the brake system.

2. Check the fluid level in the master cylinder. It should be within 1/4 in. of the top. If it isn't, add only DOT3 or DOT4 brake fluid until the correct level is reached.

3. Check the fluid level in the power steering pump. The engine should be at normal running temperature and stopped. The level should register on the pump dipstick. Add power steering fluid to bring the reservoir level up to the correct level. Low fluid level will result in both poor steering and stopping ability.

NOTE: The brake hydraulic system uses brake fluid only, while the power steering and Hydro-Boost systems use power steering fluid only. Don't mix the two.

4. Check the power steering pump belt tension, and inspect all of the power steering/Hydro-Boost hoses for kinks or leaks.

5. Check and adjust the engine idle speed, as necessary.

6. Check the power steering pump fluid for bubbles. If air bubbles are present in the fluid, bleed the system. Fill the power steering pump reservoir to specifications the engine at normal operating temperature. With the engine running rotate the steering wheel through its normal travel 3 or 4 times, without holding the wheel against the stops. Check the fluid level again.

7. If the problem still exists, go on the Hydro-Boost test sections and troubleshooting chart.

Functional Test

1. Check the brake system for leaks or low fluid level. Correct as necessary.

2. Place the transmission in Neutral and stop the engine. Apply the brakes 4 or 5 times to empty the accumulator.

3. Keep the pedal depressed with moderate (25 to 40 lbs.) pressure and start the engine.

4. The brake pedal should fall slightly and then push back up against your foot. If not movement is felt, the Hydro-Boost system is not working.

Accumulator Leak Test

1. Run the engine at normal idle. Turn the steering wheel against one of the stops, hold it there for no longer than 5 seconds. Center the steering wheel and stop the engine.

2. There should be a minimum of 2 power (1-Hydro-Boost II) assisted brake applications when pedal pressure of 20 to 25 lbs. is applied.

3. Start the engine and allow it to idle. Rotate the steering

BRAKES
Domestic Vehicles

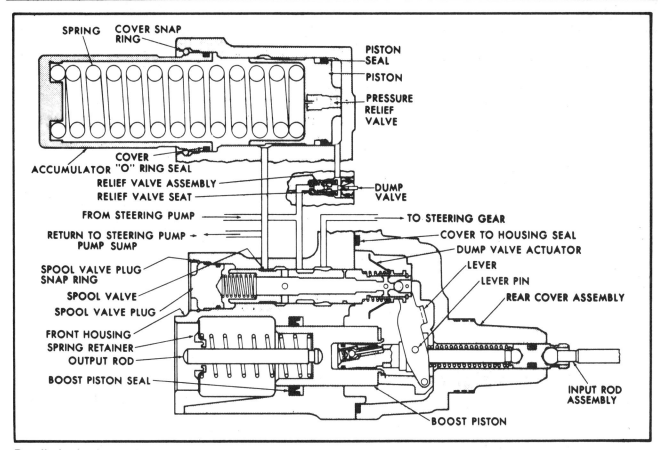

Bendix hydro-boost (spring accumulator type)

wheel against the stop. Listen for a light "hissing" sound, this is the accumulator being charged. Center the steering wheel and stop the engine.

4. Wait one hour and apply the brakes without starting the engine. As in Step 2, there should be at least 2 (1-Hydro-Boost II) stops with power assist. If not, the accumulator is defective and must be replaced.

Hydro-Boost System Bleeding

1. The system should be bled whenever the booster is removed and installed. Fill the power steering pump until the fluid level is at the base of the pump reservoir neck. Disconnect the battery lead from the distributor.

NOTE: On diesel engines remove the electrical lead to the fuel solenoid terminal on the injection pump before cranking the engine.

2. Jack up the front of the car, turn the wheels all the way to the left, and crank the engine for a few seconds.

3. Check steering pump fluid level. If necessary, add fluid to the "Add" mark on the dipstick.

4. Lower the car, connect the battery lead, and start the engine. Check fluid level and add fluid to the "Add" mark if necessary. With the engine running, turn the wheels from side to side to bleed air from the system. Make sure that the fluid level stays above the internal pump casting.

5. The Hydro-Boost system should now be fully bled. If the fluid is foaming after bleeding, stop the engine, let the system set for one hour, then repeat the second part of Step 4.

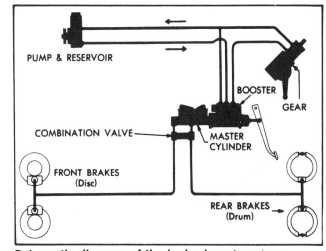

Schematic diagram of the hydro-boost system

6. The preceding procedures should be effective in removing excess air from the system, however sometimes air may still remain trapped. When this happens the booster may make a "gulping" noise when the brake is applied. Lightly pumping the brake pedal with the engine running should cause this noise to disappear. After the noise stops, check the pump fluid level and add as necessary.

BRAKES
Domestic Vehicles

TROUBLESHOOTING HYDRO-BOOST BRAKE BOOSTER

CONDITION	CAUSE	CORRECTION
Excessive Brake Pedal Effort	Loose or broken power steering pump belt.	Tighten or replace the belt.
	No fluid in power steering reservoir.	Fill reservoir and check for external leaks.
	Leaks in power steering, booster or accumulator hoses.	Replace faulty parts.
	Leaks at tube fittings, power steering, booster or accumulator connections.	Tighten fittings or replace tube seats, if faulty.
	External leakage at accumulator.	Replace "O" ring and retainer.
	Faulty booster piston seal causing leakage at booster flange vent.	Overhaul with new seal kit.
	Faulty booster input rod seal with leakage at input rod end.	Replace booster.
	Faulty booster cover seal with leakage between housing and cover.	Overhaul with new seal kit.
	Faulty booster spool plug seal.	Overhaul with spool plug seal kit.
Slow Brake Pedal Return	Excessive seal friction in booster.	Overhaul with new seal kit.
	Faulty spool action.	Flush steering system while pumping brake pedal.
	Broken piston return spring.	Replace spring.
	Restriction in return line from booster to pump reservoir.	Replace line.
	Broken spool return spring.	Replace spring.
Grabby Brakes	Broken spool return spring.	Replace spring.
	Faulty spool action caused by contamination in system.	Flush steering system while pumping brake pedal.
Booster Chatters – Pedal Vibrates	Power steering pump belt slips.	Tighten belt.
	Low fluid level in power steering pump reservoir.	Fill reservoir and check for external leaks.
	Faulty spool operation caused by contamination in system.	Flush steering system while pumping brake pedal.
Accumulator Leak Down – System does not hold charge	Contamination in steering hydro-boost system	Flush steering system while pumping brake pedal.
	Internal leakage in accumulator system	Overhaul unit using accumulator rebuild kit and seal kit.

HYDRO-BOOST TROUBLESHOOTING

HIGH PEDAL AND STEERING EFFORT (IDLE)

1. Loose/broken power steering pump belt.
2. Low power steering fluid level.
3. Leaking hoses or fittings.
4. Low idle speed.
5. Hose restriction.
6. Defective power steering pump.

HIGH PEDAL EFFORT (IDLE)

1. Binding pedal/linkage.
2. Fluid contamination.
3. Defective Hydro-Boost unit.

POOR PEDAL RETURN

1. Binding pedal linkage.
2. Restricted booster return line.
3. Internal return system restriction.

BRAKES
Domestic Vehicles

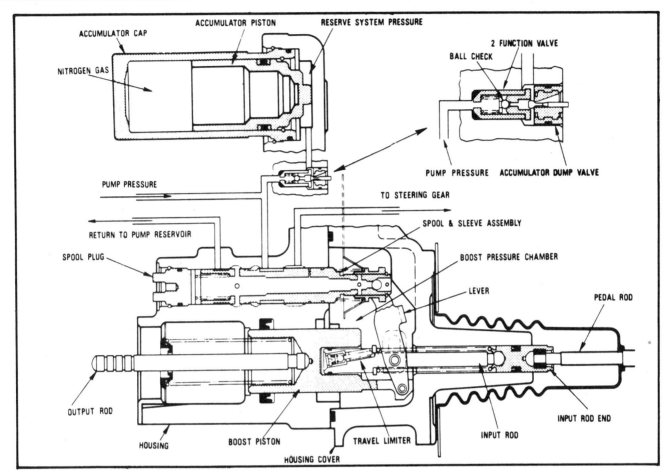

Bendix hydro-boost (nitrogen gas type)

PEDAL CHATTER/PULSATION

1. Power steering pump drivebelt slipping.
2. Low power steering fluid level.
3. Defective power steering pump.
4. Defective Hydro-Boost unit.

BRAKES OVERSENSITIVE

1. Binding pedal/linkage.
2. Defective Hydro-Boost unit.

NOISE

1. Low power steering fluid level.
2. Air in the power steering fluid.
3. Loose power steering pump drivebelt.

OVERHAUL

Ford Motor Company services the Hydro-Boost unit with a replacement new or rebuilt unit only. No provisions are made for overhaul of the unit. GM Hydro-Boost units may be overhauled. Do not attempt to interchange parts between Hydro-Boost units of different makes of cars, because of pressure differentials and differences of the tolerances of the internal parts. Pressure could exceed the normal accumulator release pressure of 1,400 psi, and injury or damage could result.

Disassembly

note: Have a drain pan ready to catch and discard leaking fluid during disassembly. Do not apply heat to the pneumatic accumulator. Do not attempt to repair an inoperative accumulator. Always replace with a new assembly. Before disposing of an inoperative accumulator drill a $1^{1}/_{16}$ inch diameter hole through the end of the accumulator pan. Do not drill through the piston end.

1. Remove the booster assembly from the car.
2. Secure the booster in a vise on a mounting bracket, if possible with the pedal rod down. Pump the pedal rod 4 to 5 times, assuring that accumulator pressure is depleted. Cut the strap securing the accumulator cap.
3. Depress the accumulator spring cap with a 12 in. C-clamp and unseat the retaining ring with a small punch and remove the ring.
4. Release the C-clamp slowly to relieve spring tension and remove the cap and spring.
5. To remove the piston, pressurize the booster thru the inlet port with air pressure, while the gear and return ports are plugged, and the piston will move out of its bore and can be removed.
6. If air is not available, form a hook from stiff wire and engage the piston in the piston fluid inlet hole. Wrap the wire around a suitable tool and pry against the housing to remove piston. Discard the piston.
7. Remove the accumulator plunger seat and guide assem-

SECTION 1

BRAKES
Domestic Vehicles

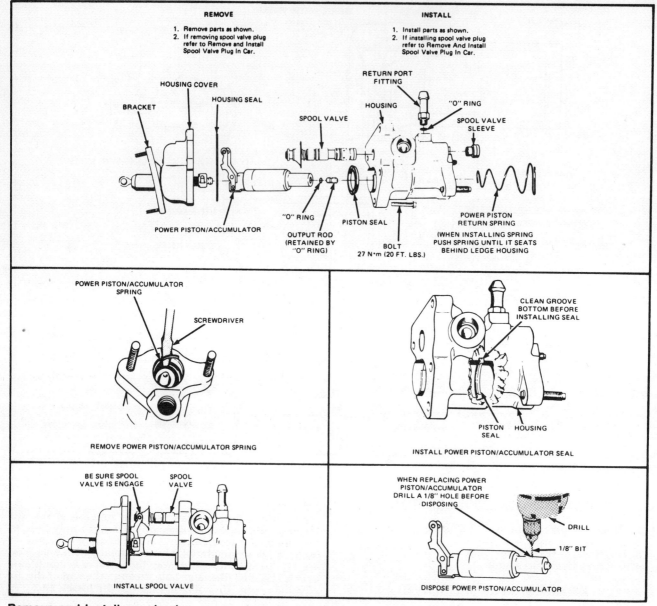

Remove and install spool valve, power piston/accumulator and seal

bly, and with a wire hook, remove the spacer-charging orifice and ball assembly and discard.

8. Loosen and remove five special bolts while holding the front housing and carefully lift off the front housing. A Torz® socket is required. The spool valve and power piston assembly will remain with the rear cover.

9. Remove the output rod and piston return spring from the power piston assembly and the spool valve spring from the valve. Remove the output rod retainer assembly from the housing.

10. Remove the spool valve and examine for scratches and wear marks. Reuse or replace as necessary.

11. Inspect the power piston for scratches and worn areas. Replace or reuse as necessary. If replacement of the power piston is necessary, snip off the staked end of the connecting pin and remove the pin with a small punch.

12. Clean and flush all parts with clean power steering fluid.

Assembly

1. Lower the new spacer-charging orifice and ball assembly into the accumulator valve bore on the front of the housing.

2. Mount a new O-ring onto the new accumulator plunger seat and guide the assembly and insert into the valve bore.

3. If a new power piston was needed, install a new pin in the hole to engage the piston connecting bracket to the small yoke in the lever and mushroom the end of the pin to avoid loss.

4. Install a new figure eight seal on the mating face of the rear housing and a new power piston seal in the front housing.

5. Insert the spool valve and spring into the bore while pulling up on the power piston and extending the lever to accept the sleeve on the spool valve. With the lever extended, put the

BRAKES
Domestic Vehicles

front housing over the rear housing and slide the lever pins into the slot in the sleeve of the spool valve.

6. Lower the front housing down into the rear cover while centering the power piston in the bore. If a seal protector is not available, extreme care must be exercised in seating the piston to the seal so that the seal lip is not damaged.

7. Install the five special bolts and torque to 20 ft. lbs. A Torx* socket is required.

8. Install the output rod, spring and new spring retainer, securing the retainer by taping it into place with a 7/8 in. deep well socket and a hammer.

9. Install the new accumulator piston assembly and install the new O-ring to the accumulator cap. With the 12 inch C-clamp, depress the cap and spring and install the retaining clip in the bore of the front housing.

10. Install the new strap from the drip pan to the accumulator cap.

11. Install the unit on the car and bleed the system.

Tandem Diaphragm Power Brake Booster

OVERHAUL

Disassembly

1. Remove the power booster from the vehicle, remove the push rod boot, silencer, front housing seal, grommet and vacuum check valve.

2. Scribe a line on the front and rear housing for alignment purposes when reassembling the unit.

3. Install the front housing into a suitable holding fixture, on Cadillac models, press down on the holding tool and turn counterclockwise to unlock the housing.

4. On all other models place the special spanner wrench (J-9504) or equivalent over the rear housing studs, press down and turn counterclockwise to unlock the housing.

5. Remove the power piston group, power piston return spring, and power piston bearing.

6. Remove the piston rod, reaction retainer and power head silencer.

7. Hold the assembly at the outside edge of the divider and diaphragms. Holding the push rod down against a hard surface, apply a light impact to dislodge the diaphragm retainer.

8. Remove the primary diaphragm, primary support plate, secondary support plate and diaphragm and power piston assembly.

NOTE: Before reassembling the unit, be sure to clean all plastic, metal and rubber parts in denatured alcohol. Air dry all parts and do not reinstall any rubber parts with cuts, nicks or distortion.

Assembly

NOTE: Before installing any of the rubber, plastic, and metal friction parts, lubricate them with silicone lube.

1. Assembly is the reverse order of the disassembly procedure.

2. After the housing have been installed align the scribe marks and press down with the holding fixture handle on Cadillac models or the spanner wrench on the other models, clockwise to lock the two housing together. Stake the two housing tabs into sockets at two new locations 180° apart.

3. Install the assembly on the vehicle, bleed the brake system and road test the vehicle.

HYDRO-BOOST II SPOOL VALVE PLUG

Removal and Installation

1. Turn engine off and pump brake pedal 4 or 5 times to deplete accumulator.

2. Separate the master cylinder from the booster with brake lines attached.

3. Push the spool valve plug in and use a small screwdriver to remove retaining ring. Remove the spool valve plug and O-ring.

4. Install the reverse order of removal. Bleed the Hydro-Boost system.

ANTI-LOCK BRAKING SYSTEM (ABS)

OPERATION

The Anti-Lock Braking System (ABS) is essentially a brake system enhancement. The purpose of ABS is to increase the driver's control over a vehicle during braking--especially steering control. When a vehicle equipped with a conventional brake system must brake suddenly, one or more wheels may lock up offering little or no steering control to avoid hazards. ABS is designed to prevent braked wheels from locking. The advantages of the system are considerable. For instance, during a high-speed stop while entering a curve, ABS is designed to allow the driver to steer through the curve while decelerating. Additionally, ABS is designed to enhance the braking action of each front wheel independently and the two rear wheels independent of the front wheels. This allows controlled braking even if one or more wheels encounters a slippery surface. In this situation, ABS will automatically sense the initial loss of adhesion in any one wheel and reduce or prevent further hydraulic pressure on that wheel's brake caliper, or if the rear wheels (both calipers) until adhesion is regained.

COMPONENTS

ABS is essentially the familiar split circuit hydraulic four wheel disc brake system to which a sophisticated electronic and mechanical override system has been carefully mated. Wheel speed sensors, an electronic control unit and a hydraulic unit that incorporates solenoid operated brake line valves are the major components of the system. The sensors monitor the rotation speed of the wheels and provide data about wheel acceleration and deceleration over very small intervals of time. The signals from the sensors are transmitted to the control unit. The control unit monitors the signals and compares them to a contained program. If one of the sensors suddenly shows a deceleration rate that exceeds the threshold values of the programmed system, indicating that a wheel is about to lock and skid, the computer activates the hydraulic control unit to maintain the optimum brake pressure in that wheel, or both rear wheels to prevent lock-up. If, for any reason, the ABS should malfunction the brakes will operate as a normal system without ABS and a warning light will go on indicating service is required.

SECTION 1

BRAKES
Domestic Vehicles

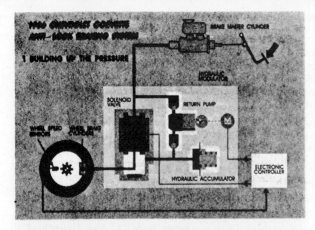

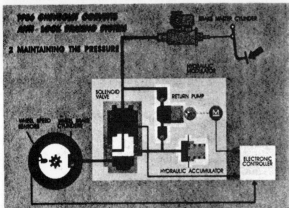

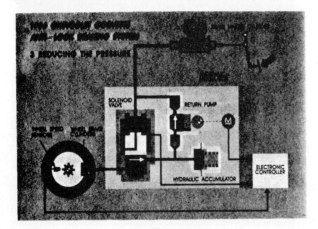

SYSTEMS

Ford and General Motors, with the exception of the Corvette use the Teves ABS system. Corvette is equipped with a Bosch ABS II system. Both systems achieve the same results to provide steering control during aggressive braking.

MASTER CYLINDER AND HYDRAULIC BOOSTER

The master cylinder and the brake booster are arranged in the basic fore and aft position with the booster behind the master cylinder. The booster control valve is located in a parallel bore above the master cylinder centerline and is operated by a lever connecting the brake pedal pushrod.

ELECTRIC PUMP AND ACCUMULATOR

The electric pump is a high pressure pump design that runs at frequent intervals for a short period to charge the hydraulic accumulator that supplies the brake system. The accumulator is a gas filled pressure chamber that is part of the pump and motor assembly. The electric motor, pump and accumulator assembly is shock mounted to the master cylinder booster assembly.

VALVE BODY ASSEMBLY

The valve body assembly incorporates three pairs of solenoid valves. One pair for each front wheel, and a third pair for both the rear wheels combined. These solenoid valves are inlet-outlet valves with inlet valve normally open and the outlet valve normally closed. The valve body itself is bolted to the inboard side of the master cylinder booster assembly.

FLUID LEVEL WARNING SWITCHES

These two integral fluid level switches are incorporated in the brake fluid reservoir cap assembly with two electrical connectors, one for each end of the cap, for wire harness connections.

WHEEL SENSORS

The sensors are four variable reluctance electronic sensor assemblies, each with a 104 tooth ring in the anti-lock system. The sensors are connected to the electronic controller through a wiring harness. The front sensors are bolted to brackets which are bolted to the front spindles. The front toothed sensor rings are pressed onto the inside of the front rotors. The rear sensors are bolted to brackets, that in turn are bolted to the rear disc brake axle adapters. The toothed rear sensor rings are pressed to the axle shafts, inboard of the axle shaft flange.

ELECTRONIC CONTROLLER

The controller is a self contained non-serviceable unit, which consists of two microprocessors and the necessary circuitry for their operation. The function of the controller is to monitor the system operations during normal driving and during anti-lock braking. Any malfunction of the anti-lock brake system will cause the controller to shut off and bypass the anti-lock brake system. When the anti-lock brake system is bypassed the normal power assisted braking will still remain.

CHECK ANTI-LOCK BRAKE WARNING LIGHT

The four wheel anti-lock system is self-monitoring. When the ignition switch is in the run position, the electronic controller will perform a preliminary self check on the anti-lock electrical system. This is indicated by a three to four second energizing of the amber. Check Anti-Lock Brakes lamp in the overhead console. This light will go out after the three to four second interval, unless there is a malfunction in the anti-lock brake system. If there is a malfunction the check anti-lock brake and/or the brake lamp will stay lit, and diagnostic tests will then pinpoint the exact component needing service.

BLEEDING THE BRAKE SYSTEM

The front brakes can be bled in the conventional manner, with

BRAKES
Domestic Vehicles

or without the accumulator being changed. When bleeding the rear brakes the accumulator must be fully charged or the system has to be pressure bled as previously outlined in this section.

Bleeding The Brake System With a Charged Accumulator

NOTE: **Be careful when opening the rear caliper bleeder screws, due to the high pressure in the system from a fully charged accumulator at the bleeder screws.**

1. With the accumulator fully charged have someone hold the rake pedal in the applied position and place the ignition switch in the run position and open the rear brake caliper bleed screws for 10 seconds at a time.
2. Repeat this procedure until the air is cleared from the brake fluid and close the brake caliper bleed screws.
3. Do this to all of the brake calipers and after the bleed screws are closed. Pump the brake pedal a couple of times to complete the bleeding procedure.
4. Adjust the brake fluid level in the reservoir to the max level with a fully charged accumulator.

MASTER CYLINDER

Removal and Installation

NOTE: **The hydraulic pressure must be discharged from the brake system before removing the master cylinder. To discharge the system, turn the ignition key to the off position and pump the brake pedal at least 20 times until an increase in pedal force is clearly felt.**

1. Disconnect the negative battery cable and the electrical connectors from the master cylinder reservoir cap, main valve, solenoid valve body, pressure warning switch, the hydraulic pump motor, and ground connector from the master cylinder.
2. Disconnect the brake lines from the solenoid valve body and plug the line openings in the valve body to prevent fluid loss. Do not allow the brake fluid to leak or spill onto any of the electrical connectors.
3. From inside the vehicle, disconnect the hydraulic booster pushrod from the brake pedal in the following order: Disconnect the stop light switch wires at the connector on the brake pedal. Remove the hairpin clip at the stop light switch on the brake pedal and move the switch off of the pedal pin far enough for the switch outer hole to clear the pin. Using a twisting motion, remove the switch, but be careful not to damage the switch during its removal. Remove the four retaining nuts at the dash panel and from inside the engine compartment. Remove the booster from the dash panel.
4. Installation is the reverse order of the removal procedure. Also after the unit has been installed, bleed the brake system as previously outlined in this section.

HYDRAULIC ACCUMULATOR

Removal and Installation

1. Discharge the pressure in the brake system and disconnect the electrical connection at the hydraulic pump motor.
2. Using an 8 mm hex wrench or equivalent, unscrew the accumulator. Be sure that no dirt falls into the open port.
3. Using the same hex wrench or equivalent, remove the accumulator adapter block bolt and remove the block, if necessary.
4. Installation is the reverse order of the removal procedure and be sure to observe the following: Install new O-rings on the accumulator and adapter block. Torque the adapter block bolt to 25 to 34 ft. lbs. and the accumulator to 30 to 34 ft. lbs. After installation, place the ignition switch in the ON position and check to see if the check Anti-Lock Brake light goes out after a maximum of one minute. Top off the master cylinder reservoir to the max mark with brake fluid.

HYDRAULIC PUMP MOTOR

Removal and Installation

1. Discharge the pressure in the brake system and disconnect the negative battery cable.
2. Disconnect the electrical connections at the hydraulic pump motor and pressure warning switch.
3. Remove the accumulator.
4. Remove the suction line between the reservoir and the pump at the reservoir by twisting the hose and pulling on it lightly. To prevent fluid loss, a large vacuum nipple can be slipped over the reservoir opening as the hose is removed.
5. Remove the retaining bolt on the pump high pressure line to the hydraulic booster housing at the housing. Make sure to save the two O-rings, that are on both sides of the retaining bolt.
6. Remove the Allen head bolt that holds the pump and motor assembly to the extension housing located directly under the accumulator. Make sure to save the thick spacer between the extension housing and the shock mount.
7. Slide the pump assembly inboard to remove the assembly from the retainer pin located on the inboard side of the extension housing.
8. Installation is the reverse order of the removal procedure. Be sure to bleed the brake system and check to see if the Check Anti-Lock Brake light goes out after a maximum of one minute.

ELECTRONIC CONTROLLER

Removal and Installation

1. Disconnect the negative battery cable. Disconnect the 35 pin connector from the electronic controller, which is located in the trunk of the vehicle in front of the forward trim panel.
2. Remove the three retaining screws holding the electronic controller to the seat back brace and remove the controller.
3. Installation is the reverse order of the removal procedure.

PRESSURE SWITCH

Removal and Installation

1. Discharge the pressure from the brake system and disconnect the negative battery cable.
2. Disconnect the solenoid valve body seven pin connector and remove the pressure switch using special tool #T85P-20215-B or equivalent, a 1/2 inch to 3/8 inch adapter and a 3/8 inch drive ratchet.
3. Installation is the reverse of the removal procedure and replace the O-ring on the switch. Torque the switch to 15 to 25 ft. lbs.

FRONT WHEEL SENSOR

Removal and Installation

1. Disconnect the sensor electrical connector for the right or left front sensor from inside the engine compartment.
2. Raise and support vehicle safely and disengage the wire grommet at the right or left hand shock tower.
3. Pull the sensor cable connector through the hole and be careful not to damage the connector.
4. Remove the sensor wire from the bracket on the shock tower and the side rail.
5. Loosen the set screw holding the sensor to the sensor

SECTION 1

BRAKES
Domestic Vehicles

1986 FRONT-WHEEL-DRIVE DE VILLE AND FLEETWOOD AVAILABLE ELECTRONIC BRAKING SYSTEM

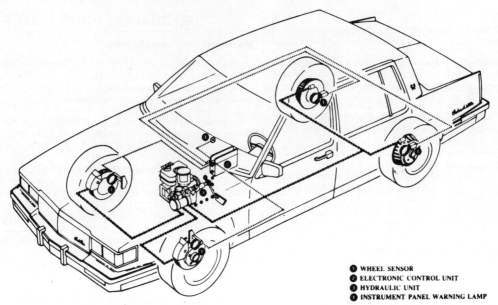

1. WHEEL SENSOR
2. ELECTRONIC CONTROL UNIT
3. HYDRAULIC UNIT
4. INSTRUMENT PANEL WARNING LAMP

bracket post. Remove the sensor through the hole in the disc brake splash shield.

6. Remove the sensor bracket or the sensor bracket post, if it has been damaged by removing the caliper, hub and rotor assembly (as previously outlined). Remove the two brake splash shield attaching bolts which hold the sensor bracket.

7. Install the sensor bracket with the sensor bracket post, if it has been removed. Torque the post retaining bolt 40 to 60 inch lbs. and the splash shield attaching bolts 10 to 15 ft. lbs. Install the hub and rotor assembly.

NOTE: If a sensor is going to be reused, the pole face must be cleaned of all dirt and grease and the pole face has to be scraped with a dull knife or equivalent so that the sensor slides freely on the post. Also glue a new front spacer paper on the pole face.

8. Install a new or old sensor through the hole in the brake shield onto the sensor bracket post. Make sure the paper spacer does not come off during installation.

9. Push the sensor toward the toothed sensor ring until the new paper sensor contacts the ring. Hold the sensor against the sensor ring and torque the set screw to 21 to 26 inch lbs.

10. Insert the cable into the bracket on the shock strut, rail bracket and then through the inner fender apron to engine compartment and the seat grommet.

11. Lower the vehicle and reconnect the sensor electrical connection.

12. Check the function of the sensor by road testing the vehicle and making sure the Check Anti-Lock Brake light does not stay on.

REAR WHEEL SENSOR

Removal and Installation

1. Disconnect the wheel sensor electrical connector located behind the forward luggage compartment trim panel in the trunk.

2. Lift the carpet in the trunk and push the sensor wire grommet through the hole in the floor of the trunk.

3. Raise and support the vehicle safely and remove the appropriate wheel.

4. Remove the wheel sensor wiring from the axle shaft housing. The wiring harness has three different types of retainers.

They are the following: The inboard retainer clip is located on the top of the differential housing. Just bend the clip far enough to remove the wiring harness. The second retainer clip is a C-clip and is located in the center of the axle shaft housing. Pull rearward on the clip to disengage the clip from the axle housing. The third clip is at the connection between the rear housing wheel brake tube and flexible hose. Remove the holddown bolt and open the clip to remove the wiring harness.

NOTE: Be careful not to bend the C-clip open beyond the amount needed to remove the clip from the axle housing, because it could break.

5. Remove the rear wheel caliper and rotor assemblies as previously outlined in this section.

6. Remove the wheel sensor retaining bolt, slip the grommet out of the rear brake splash shield and pull the sensor wire outward through the hole.

7. If the sensor bracket is damaged, remove the bracket attaching screws and replace the bracket.

8. Install the sensor bracket, if it was removed and torque the screws 11 to 15 ft. lbs.

9. If the sensor is going to be used again it has to be cleaned just as the front wheel sensors were cleaned.

10. Insert the sensor into the large hole in the sensor bracket and install the retaining bolt into the sensor bracket and torque the retaining bolt 40 to 60 in. lbs.

11. Push the sensor toward the toothed ring until the new paper sensor touches the sensor ring, hold the sensor against the toothed ring and torque the set screw 21 to 26 in. lbs.

12. Install the caliper and rotor assemblies.

13. Push the wire and connector through the splash shield hole and engage the grommet into the shield eyelet. Install the sensor wire in the retainers along the axle housing.

14. Push the connector through the hole in the trunk and seat the grommet in the trunk floor pan.

15. Reconnect the cable electrical connector, and install the carpet back down in the trunk. Check the function of the sensor by road testing the vehicle and checking to see if the Check Anti-Lock Brakelight goes out.

16. If the toothed ring sensor is found to be malfunctioning on either the front wheel or rear wheel, the rotor assembly has to be removed and the toothed ring sensor has to be pressed out and the new one pressed in.

BRAKES
Import Vehicles

Import Vehicles

BRAKE SYSTEM

Understanding the Brakes

HYDRAULIC SYSTEM

Basic Operating Principles

Hydraulic systems are used to actuate the brakes of all modern automobiles. The system transports the power required to force the frictional surfaces of the braking system together from the pedal to the individual brake units at each wheel. A hydraulic system is used for two reasons. First, fluid under pressure can be carried to all parts of an automobile by small hoses—some of which are flexible—without taking up a significant amount of room or posing routing problems. Second, a great mechanical advantage can be given to the brake pedal end of the system, and the foot pressure required to actuate the brakes can be reduced by making the surface area of the master cylinder pistons smaller than that of any of the pistons in the wheel cylinders or calipers.

The master cylinder consists of a fluid reservoir and a double cylinder and piston assembly. Double type master cylinders are designed to separate two two-wheel braking systems hydraulically in case of a leak. The standard approach has been to utilize two separate two-wheel circuits; one for the front wheels and another for the rear wheels.

Most newer models now use a diagonally split system; i.e. one front wheel and the opposite side rear wheel make up one braking circuit, while the remaining circuit consists of the other front wheel and its opposite side rear wheel.

Steel lines carry the brake fluid to a point on the vehicle's frame near each of the vehicle's wheels. The fluid is then carried to the wheel cylinders and/or calipers by flex-

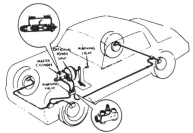
Dual braking system—front-to-rear split

Dual braking system—diagonally split

ible tubes in order to allow for suspension and steering movements.

The hydraulic system operates as follows: When at rest, the entire system, from the piston(s) in the master cylinder to those in the wheel cylinders or calipers, is full of brake fluid. Upon application of the brake pedal, fluid trapped in front of the master cylinder piston(s) is forced through the lines to the slave cylinders (wheel cylinders or calipers). Here, it forces the pistons outward, in the case of drum brakes, and inward toward the disc, in the case of disc brakes. The motion of the pistons is opposed by return springs mounted outside the cylinders in drum brakes, and by internal springs or spring seals, in disc brakes.

Upon release of the brake pedal, a spring located inside the master cylinder immediately returns the master cylinder pistons to the normal position. The pistons contain check valves and the master cylinder has compensating ports drilled in to it. These are uncovered as the pistons reach their normal position. The piston check valves allow fluid to flow toward the wheel cylinders or calipers as the pistons withdraw. Then, as the return springs force the shoes into the released position, the excess fluid flows back to the reservoir through the compensating ports. It is during the time the pedal is in the released position that any fluid that has leaked out of the system will be replaced through the compensating ports.

Dual circuit master cylinders employ two pistons, located one behind the other, in the same cylinder. The primary piston is actuated directly by mechanical linkage from the brake pedal. The secondary piston is actuated by fluid trapped between the two pistons. If a leak develops in front of the secondary piston, it moves forward until it bottoms against the front of the master cylinder, and the fluid trapped between the pistons will operate one side of the split system. If the other side of the system develops a leak, the primary piston will move forward until direct contact with the secondary piston takes place, and it will force the secondary piston to actuate the other side of the split system. In either case, the brake pedal moves farther when the brakes are applied, and less braking power is available.

All dual-circuit systems use a distributor switch to warn the driver when only half of the brake system is operational. This switch is located in a valve body which is mounted on the firewall or the frame below the master cylinder. A hydraulic piston receives pressure from both circuits, each circuit's pressure being applied to one end of the piston. When the pressures are in balance, the piston remains stationary. When one circuit has a leak, however, the greater pressure in that circuit during application of the brakes will push the piston to one

1-67

BRAKES
Import Vehicles

side, closing the distributor switch and activating the brake warning light.

In disc brake systems, this valve body also contains a metering valve and, in some cases, a proportioning valve (or valves). The metering valve keeps pressure from traveling to the disc brakes on the front wheels until the brake shoes or pads on the rear wheels have contacted the drums or rotors, ensuring that the front brakes will never be used alone. The proportioning valve throttles the pressure to the rear brakes so as to avoid rear wheel lock-up during very hard braking.

These valves may be tested by removing the lines to the front and rear brake systems and installing special brake pressure testing gauges. Front and rear system pressures are then compared as the pedal is gradually depressed. Specifications vary with the manufacturer and design of the brake system.

Brake system warning lights may be tested by depressing the brake pedal and holding it while opening one of the wheel cylinder bleeder screws. If this does not cause the light to go on, substitute a new lamp, make continuity checks, and, finally, replace the switch as necessary.

The hydraulic system may be checked for leaks by applying pressure to the pedal gradually and steadily. If the pedal sinks very slowly to the floor, the system has a leak. This is not to be confused with a springy or spongy feel due to the compression of air within the lines. If the system leaks, there will be a gradual change in the position of the pedal with a constant pressure.

Check for leaks along all lines and at wheel cylinders. If no external leaks are apparent, the problem is inside the master cylinder.

DISC BRAKES
Basic Operating Principles

Instead of the traditional expanding brakes that press outward against a circular drum, disc brake systems utilize a cast iron disc with brake pads positioned on either side of it. Braking effect is achieved in a manner similar to the way you would squeeze a spinning phonograph record between your fingers. The disc (rotor) is a one-piece casting which may be equipped with cooling fins between the two braking surfaces. The fins (if equipped) enable air to circulate between the braking surfaces making them less sensitive to heat buildup and more resistant to fade. Dirt and water do not affect braking action since contaminanats are thrown off by the centrifugal action of the rotor or scraped off by the pads. Also, the equal clamping action of the two brake pads tends to ensure uniform, straightline stops. All disc brakes are inherently self-adjusting.

There are three general types of disc brake:
1. A fixed caliper, two or four-piston type.
2. A floating caliper, single piston or double piston back-to-back type.
3. A sliding caliper, single piston or

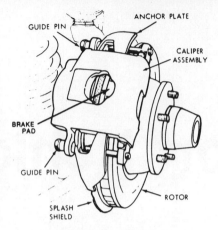

Typical disc brake assembly

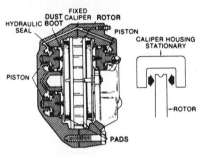

Typical fixed caliper disc brake (four piston shown)

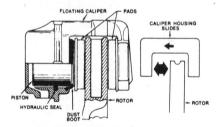

Typical floating caliper disc brake (sliding caliper similar)

double piston back-to-back type.

The fixed caliper design uses one or two pistons mounted on either side of the rotor (in each side of the caliper). The caliper is mounted rigidly and does not move.

The sliding and floating designs are quite similar. In fact, these two types are often lumped together. In both designs, the pad on the inside of the rotor is moved into contact with the rotor by hydraulic force. The caliper, which is not held in a fixed position, moves slightly, bringing the outside pad into contact with the rotor. There are various methods of attaching floating calipers. Some pivot at the bottom or top, and some slide on mounting bolts. In any event, the end result is the same.

DRUM BRAKES
Basic Operating Principles

Drum brakes employ two brake shoes mounted on a stationary backing plate. These shoes are positioned inside a circular cast iron (or aluminum) drum which rotates with the wheel assembly. The shoes are held in place by springs; this allows them to slide toward the drums (when they are applied) while keeping the linings and drums in alignment. The shoes are actuated by a wheel cylinder which is mounted at the top of the backing plate. When the brakes are applied, hydraulic pressure forces the wheel cylinder's two actuating links outward. Since these links bear directly against the top of the brake shoes, the tops of the shoes are then forced outward against the inner side of the drum. This action forces the bottoms of the two shoes to contact the brake drum by rotating the entire assembly slightly (known as servo action). When pressure within the wheel cylinder is relaxed, return springs pull the shoes back away from the drum.

Most modern drum brakes are designed to self-adjust themselves during application when the vehicle is moving in reverse. This motion causes both shoes to rotate very slightly with the drum, rocking an adjusting lever, thereby causing rotation of the adjusting screw by means of a star wheel.

POWER BRAKE BOOSTERS

Power brakes operate just as standard brake systems except in the actuation of the master cylinder pistons. A vacuum diaphragm is located on the front of the master cylinder and assists the driver in applying the brakes, reducing both the effort and travel he must put into moving the brake pedal.

The vacuum diaphragm housing is connected to the intake manifold by a vacuum hose. A check valve is placed at the point where the hose enters the diaphragm housing, so that during periods of low manifold vacuum, brake assist vacuum will not be lost.

Depressing the brake pedal closes off the vacuum source and allows atmospheric pressure to enter on one side of the diaphragm. This causes the master cylinder pistons to move and apply the brakes. When the brake pedal is released, vacuum is applied to both sides of the diaphragm, and return springs return the diaphragm and master cylinder pistons to the released position. If the vacuum fails, the brake pedal rod will butt against the end of the master cylinder actuating rod, and direct mechanical application will occur as the pedal is depressed.

HYDRAULIC CYLINDERS AND VALVES

Master Cylinders

The master cylinder is a type of hydraulic

BRAKES
Import Vehicles

pump that is operated by a push rod attached to the brake pedal or by a push rod that is part of the power brake booster. The cylinder provides a means of converting mechanical force into hydraulic pressure.

DUAL MASTER CYLINDER

In this type there are two separate hydraulic pressure systems. One of the hydraulic systems may be connected to the front brakes, and the other to the rear brakes, or the systems may connect diagonal wheels. If one system fails, the other system remains operational, thus providing an additional safety measure. There are two distinct fluid reservoirs and each has a vent and replenishing port that leads into the cylinder bore. These ports have been called compensating and inlet ports or bypass ports, and the terms have been used inconsistently causing confusion. The terms "vent" and "replenishing" ports are now standardized S.A.E. terms. An airtight seal for the reservoir is provided in the form of a rubber diaphragm, which is held in place by a metal cover. A bail type retainer or a bolt usually holds the cover on the reservoirs. The cover is vented to permit atmospheric pressure to enter above the diaphragm. The diaphragm prevents moisture and debris from contaminating the fluid. The cylinder bore contains the return springs, two pistons, and the seals. The piston stop bolt (if present) may be assembled in a threaded hole in the bottom of the cylinder.

Some master cylinders have the piston stop bolt assembled in a threaded hole in the side of the bore or in the bottom of the front reservoir, and others do not have stop bolts at all. Do not install a stop bolt in the reservoir of a master cylinder if one was not originally there. Some cylinders have a tapped hole, but no bolt was ever installed in production. *This was done on purpose, and is not an error.*

A retaining ring fits into a groove near the end of the bore and holds the piston assemblies in the cylinder bore.

Dual System—Applied

When the brake pedal is depressed, the push rod moves the primary piston forward in the cylinder bore. The primary vent port is sealed off by the lip of the primary cup. As a result, a solid column of fluid is created between the primary and secondary pistons.

With the help of the primary piston return spring, this column moves the secondary piston forward in the cylinder bore. This closes the secondary vent port. When both ports are closed, any further movement of the pushrod and pistons serves to increase the hydraulic pressure in the area ahead of each piston. This pressure is then transmitted through the two hydraulic brake systems to the brakes at each wheel.

Dual System—Released

When the brake pedal is released, the piston

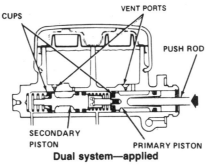
Dual system—applied

return springs move both pistons to their normal released position. The piston may move faster than the fluid can return from the wheel cylinders, creating a low pressure ahead of the piston.

To allow rapid pedal return, this low pressure must be relieved. Fluid flows from the reservoir through the replenishing port. It then flows around the outside of the piston and cup lips to the area ahead of the piston. Another design has this flow pass through small drilled holes in the piston face, then around the cup lips. This flow relieves the low pressure area ahead of the piston.

Due to this action, the area in the front of the pistons is kept full of brake fluid at all times. Any excess fluid is returned to the master cylinder reservoirs through the vent ports after the pistons reach their fully released positions. Tandem master cylinders on cars equipped with four wheel drum brakes may contain two residual check valves, one in each outlet port. Those on cars with front disc/rear drum brakes may contain one in the rear (drum) brake outlet port.

Partial System Failure

If a failure occurs in the hydraulic system served by the primary piston, this piston will move forward but will not develop pressure. The piston extension contacts the secondary piston and pedal effort is transmitted directly to that piston to build hydraulic pressure to operate the brakes in the secondary system.

If the secondary system suffers a leak or failure, both pistons move forward until the secondary piston bottoms out at the end of the master cylinder bore. Then the primary

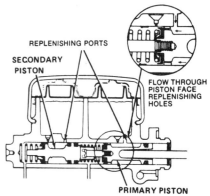
Dual system—released

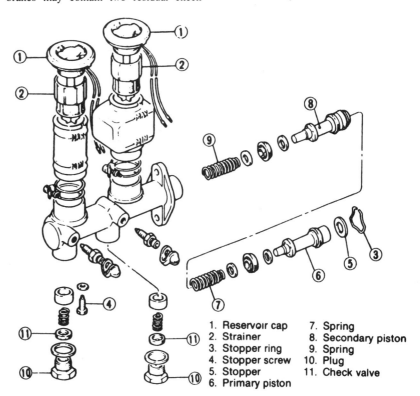

1. Reservoir cap
2. Strainer
3. Stopper ring
4. Stopper screw
5. Stopper
6. Primary piston
7. Spring
8. Secondary piston
9. Spring
10. Plug
11. Check valve

Exploded view of a dual system master cylinder

1-69

BRAKES
Import Vehicles

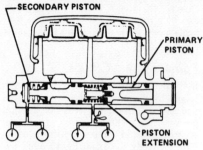

Primary system failure

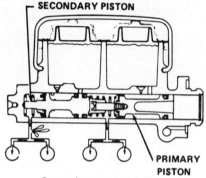

Secondary system failure

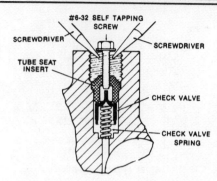

Removing the inserts from the master cylinder ports

piston develops hydraulic pressure to operate the brakes in the primary system.

The loss of about half the pedal stroke is usually experienced when a half system failure occurs.

MASTER CYLINDER SERVICING

Just like any other brake parts, master cylinders require periodic service. The usual reason for a master cylinder failure is that the cups don't seal any more. Fluid leaks past cups internally, and sometimes shows up as an external leak as well. A common symptom is a "spongy" brake pedal that goes all the way to the floor when all other brake system components are in good shape. The rubber parts wear with usage or may deteriorate with age or fluid contamination. Corrosion or deposits formed in the bore due to moisture or dirt in the hydraulic system may result in wear of the cylinder bore or the parts therein. Also, the fluid level in the reservoirs should be checked periodically. Whenever needed, clean brake fluid should be added to maintain the fluid level ¼ in. to ½ in. (6 to 13mm) from the top of the reservoir.

Removal and Disassembly

1. Clean the area around the master cylinder to prevent dirt and grease from contaminating the cylinder or the hydraulic lines. Disconnect the tubes, remove nuts or bolts that secure the master cylinder to the fire wall or power brake, and remove the master cylinder from the car (for further details, refer to appropriate car section).

2. Remove the reservoir cover, and drain the brake fluid from the reservoir. Then remove the piston stop bolt, if present, from the master cylinder. Remove the boot and snap ring, then slide the primary piston assembly out of the master cylinder. Next, remove the secondary piston assembly by tapping the master cylinder, or by using needle nose pliers to pull it from its bore, or by carefully using compressed air. Disassemble the secondary piston assembly.

3. Clamp the master cylinder in a vise with the outlet ports facing up. Test for the presence of a check valve by probing with wire through the hole in the tube seats. Replace tube seat(s) and check valve(s) only if a check valve is present and supplied in the rebuild kit. Remove the tube seat inserts, if required, by partially threading a self-tapping screw into each tube seat and using two screwdrivers to pry each seat out of the master cylinder. Remove the residual check valve and the spring from the outlet(s) (if present).

Plastic Reservoir Cleaning and Removal

Plastic reservoirs need to be removed only for the following reasons:

a. Reservoir is damaged or the rubber grommet(s) between the reservoir and bore is leaking.

b. Removal of the stop pin from Chrysler style plastic reservoir master cylinders to allow for the removal of pistons. Pin is located underneath front reservoir nipple.

The reservoir should be removed by first clamping the master cylinder flange in a vice. Next remove the reservoir. Grasp the reservoir base on one end and pull away from the body. Some must be removed by prying between the reservoir and casting with a pry bar. Grommets can be reused if they are in good condition. Whether or not the reservoir is removed, it and the cover or caps should be thoroughly cleaned.

Cleaning and Inspection

Thoroughly clean the master cylinder and any other parts to be reused in clean alcohol. DO NOT USE PETROLEUM PRODUCTS FOR CLEANING. If the bore is not badly scored, rusted or corroded, it is possible to rebuild the master cylinder in some cases. A slight bit of honing is permissible to clean cups are facing.

CAUTION
Aluminum cylinder bores cannot be honed. The cylinder MUST be replaced if the bore is scored.

Lubricate all new rubber parts with brake fluid or brake system assembly lubricant.

CAST IRON BORE CLEAN-UP

Crocus cloth or an approved cylinder hone should be used to remove lightly pitted, scored, or corroded areas from the bore.

CAUTION
If an aluminum master cylinder has pits or scratches in the bore, it must be replaced.

Brake fluid can be used as a lubricant while honing lightly. The master cylinder should be replaced if it cannot be cleaned up readily. After using the crocus cloth or a hone, the master cylinder should be thoroughly washed in clean alcohol or brake fluid to remove all dust and grit. If alcohol is used, dry parts thoroughly before reinstalling.

CAUTION
Other solvents should not be used.

Then the clearance between the bore wall and the piston (primary piston of a dual system master cylinder) should be checked. If a narrow (⅛ in. (3.2mm) to ¼ in. (6.4mm) wide) 0.006 in. (0.15mm) feeler gauge can be inserted between the wall and a new piston, the clearance is excessive, and the master cylinder should be replaced. The maximum clearance allowed for units containing pistons without replenishing holes is .009 in. (0.23mm).

ALUMINUM BORE CLEAN-UP

Inspect the bore for scoring, corrosion and pitting. If the bore is scored or badly pitted and corroded the assembly should be replaced. *Under no conditions should the bore be cleaned with an abrasive material.* This will remove the wear and corrosion resistant anodized surface. Clean the bore with a clean piece of cloth around a wooden dowel and wash thoroughly with alcohol. Do not confuse bore discoloration or staining with corrosion.

Reassembly and Installation

1. Carefully install the new cups or seals in the same positions and in reverse order of removal.

2. Use brake fluid or assembly fluid very generously to keep from damaging the seals.

3. Placing the small end of the pressure spring into the secondary piston retainer, slide the assembly into the cylinder bore, taking care not to nick or gouge any rubber part.

4. Place the spring retainer of the primary piston assembly over the secondary piston shoulder and push both assemblies into the bore.

5. Install and tighten the piston retaining screw and gasket, while holding the pistons in their seated positions. At the same time, reinstall any piston snap rings.

6. Install the residual check valve and spring in the proper master cylinder outlet (or both outlets, if originally present). If the tube seat inserts were removed, install new seats in both fluid outlets making sure that they are securely seated.

Bleeding and Checking

1. Bleed the hydraulic system as described later in this section.

NOTE: Be sure to bench bleed a rebuilt or new master cylinder before installation.

2. Check master cylinder vent port clearance by watching for a spurt of brake fluid in both reservoir vent holes when the brake pedal is slightly depressed, indicating proper port clearance.

Master Cylinder Push Rod Adjustment

After assembly of the master cylinder to the power section, the piston cup in the hydraulic cylinder should just clear the compensating port hole when the brake pedal is fully released. If the push rod is too long, it will hold the piston over the port.

A push rod that is too short, will give too much loose travel (excessive pedal play).

Apply the brakes and release the pedal all the way observing the brake fluid flow back into the master cylinder.

A full flow indicates the piston is coming back far enough to release the fluid.

A slow return of the fluid indicates the piston is not coming back far enough to clear the ports. The push rod adjustment is too tight, and should be shortened.

Wheel Cylinders

DRUM BRAKE WHEEL CYLINDER

The wheel cylinder performs in response to the master cylinder. It receives fluid from the hydraulic hose through its inlet port. As the pressure increases, the wheel cylinder cups and pistons are forced apart. As a result, the hydraulic pressure is converted into mechanical force acting on the brake shoes. The wheel cylinder size may vary from front to rear. The variation in wheel cylinder size (diameter) is one of the factors controlling the distribution of braking force in a vehicle.

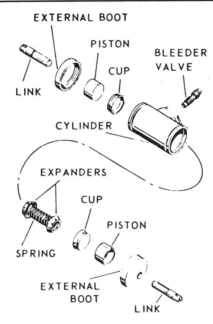

Typical wheel cylinder components

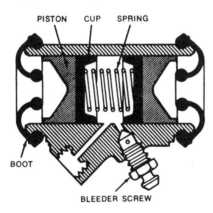

Double piston wheel cylinder

WHEEL CYLINDER OPERATION

The space between the cups in the cylinder bore must remain filled with fluid at all times. After depressing the brake pedal, additional brake fluid is forced into the cylinder bore. As a result of this, cups and pistons move outward in the cylinder bore pushing the shoe links and the brake shoes outward to contact the drum and apply the brakes.

On some designs, the end of the shoe web bears directly against the pistons and therefore, shoe links are not used.

SERVICE PROCEDURES

Wheel cylinders may need reconditioning or replacement whenever the brake shoes are replaced or when required to correct a leak condition. On many designs, the wheel cylinders can be disassembled without removing them from the backing plate. On some designs, however, the cylinder is mounted in an indention in the backing plate or a cylinder piston stop is welded to the backing plate. When servicing brakes of this type, the cylinder must be removed from the backing plate before being disassembled.

Diagnostic Inspection and Cleaning

Leaks which coat the boot and the cylinder with fluid, or result in a dropped reservoir fluid level, or dampen and stain the brake linings are dangerous. Such leaks can cause the brakes to "grab" or fail and should be immediately corrected. A leakage, not immediately apparent, can be detected by pulling back the cylinder boot. A small amount of fluid seepage dampening the interior of the boot is normal; a dripping boot is not. Unless other conditions causing a brake to pull, grab, or drag becomes obvious, the wheel cylinder is a suspect and should be included in general reconditioning.

Cylinder binding may be caused by rust, deposits, grime, or swollen cups due to fluid contamination, or by a cup wedged into an excessive piston clearance. If the clearance between the pistons and the bore wall exceeds allowable values, a condition called "heel drag" may exist. It can result in rapid cup wear and can cause the pistons to retract very slowly when the brakes are released.

A typical example of a scored, pitted or corroded cylinder bore is shown in the accompanying illustration. A ring of a hard, crystal-like substance is sometimes noticed in the cylinder bore where the piston stops after the brakes are released.

Light roughness or deposits can be removed with crocus cloth or an approved cylinder hone. While honing lightly, brake fluid can be used as a lubricant. If the bore cannot be cleaned up readily, the cylinder must be replaced.

NOTE: Aluminum wheel cylinders must not be honed.

―――――― CAUTION ――――――
Hydraulic system parts should not be allowed to come in contact with oil or grease, neither should those be handled with greasy hands. Even a trace of any petroleum based product is sufficient to cause damage to the rubber parts.
―――――――――――――――――

Reconditioning Wheel Cylinders

It is a common practice to recondition a wheel cylinder without dismounting it, however some brakes are equipped with external piston stops which prevent disassembly unless the cylinder is removed. In order to dismount, remove the shoe springs and spread the shoes apart, disconnect the brake line, remove the mounting bolts or retaining clips, and pull the cylinder free.

Pull the protective dust boots off the cyl-

BRAKES
Import Vehicles

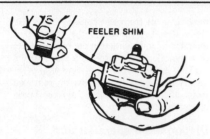

Checking the maximum piston clearance

inder. Internal parts should slide out, or be picked out easily. Parts can be driven out with a wooden dowel, or blown out at low pressure by applying compressed air to the fluid inlet port. Parts which cannot be removed easily indicate they are damaged beyond repair and the cylinder should be replaced.

Clean the cylinder and the parts in alcohol and/or brake fluid (do NOT use gasoline or other petroleum based products). Use only lint-free wiping cloths. Crocus cloth can be used to clean minute scratches, signs of rust, corrosion or discoloration from the cylinder bore and pistons. Slide the cloth in a circular rather than a lengthwise motion. A clean-up hone may be used. After a cylinder has been honed, inspect it for excessive piston clearance and remove any burrs formed on the edge of fluid intake or bleeder screw ports.

— CAUTION —
Do not rebuild aluminun cylinders.

To check the maximum piston clearance, place a ¼″ (6 mm) wide strip of feeler shim lengthwise in the cylinder bore.

If the piston can be inserted with the shim in place, the cylinder is oversize, and should be discarded. Depending upon the cylinder bore diameter, the shim (or the feeler gauge) thickness can vary as follows:

Cylinder Bore	Shim
¾ in.–1⁹⁄₁₆ in. (19–30mm)	.006″ (.15mm)
1¼ in.–1⁷⁄₁₆ in. (32–37mm)	.007 in. (.18mm)
1½ in. up (38mm)	.008 in. (.2mm)

Assemble the cylinder with the internal parts, making sure that the cylinder wall is wet with brake fluid. Insert the cups and pistons from each end of a double-end cylinder; do not slide them through the cylinder. Cup lips should always face inward.

Hydraulic Control Valves

PRESSURE DIFFERENTIAL VALVE

The pressure differential valve activates a dash panel warning light if pressure loss in the brake system occurs. If pressure loss occurs in one half of the split system, the other system's normal pressure causes the piston in the switch to compress a spring until it touches an electrical contact. This causes the warning lamp on the dash panel to light, thus warning the driver of possible brake failure.

On some cars the spring balance piston automatically recenters as the brake pedal is released, warning the driver only upon brake application. On other cars, the light remains on until manually cancelled.

Valves may be located separately or as part of a combination valve. On certain front wheel drive cars, the valve and switch are usually incorporated into the master cylinder.

Re-setting Valves

On some cars, the valve piston(s) remain off center after failure, until necessary repairs are made. The valve will automatically reset itself (after repairs) when pressure is equal on both sides of the system.

If the light does not go out, bleed the brake system that is opposite the failed system. If front brakes failed, bleed the rear brakes, this should force the light control piston toward center.

If this fails, remove the terminal switch. If brake fluid is present in the electrical area, the seals are gone, replace the complete valve assembly.

METERING VALVE

The metering valve's function is to improve braking balance between the front and rear brakes, especially during light brake application.

The metering valve prevents application of the front disc brakes until the rear brakes overcome the return spring pressure. Thus, when the front disc pads contact the rotor, the rear shoes will contact the brake drum at the same time.

Inspect the metering valve each time the brakes are serviced. A slight amount of moisture inside the boot does not indicate a defective valve, however, fluid leakage indicates a damaged or worn valve. If fluid

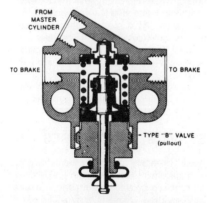

Typical pressure metering valve

leakage is present, the valve must be replaced.

The metering valve can be checked very simply. With the car stopped, gently apply the brakes. At about an inch of travel, a very small change in pedal effort (like a small bump) will be felt if the valve is operating properly. Metering valves are not serviceable, and must be replaced if defective.

PROPORTIONING VALVE

The proportioning (pressure control) valve is used, on some cars, to reduce the hydraulic pressure to the rear wheels to prevent skid during heavy brake application and to provide better brake balance. It is usually mounted in line to the rear wheels.

Whenever the brakes are serviced, the valve should be inspected for leakage. Premature rear brake application during light braking can mean a bad proportioning valve. Repair is by replacement of the valve. Make sure the valve port marked "R" is connected toward the rear wheels.

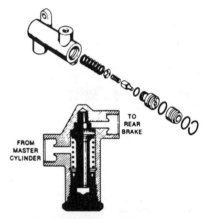

Typical proportioning valve

On some front wheel drive cars, the proportioning valve(s) is (are) screwed into the master cylinder. Since these cars usually have a diagonally split brake system, two valves are required. One rear brake line screws into each valve. The early type valves were steel, an occasional "clunking" noise was encountered on some early models, but does not affect brake efficiency. Replacement valves are now made of aluminum. Never mix an aluminum valve with a steel valve, always use two aluminum valves.

COMBINATION VALVE

The combination valve may perform two or three functions. They are; metering, proportioning and brake failure warning.

Variations of the two-way combination valve are; proportioning and brake failure warning or metering and brake failure warning.

A three-way combination valve directs the brake fluid to the appropriate wheel,

BRAKES
Import Vehicles

Two-way combination valve (metering and brake warning light switch)

3-way combination valve

performs necessary valving and contains a brake failure warning.

The combination valve is usually mounted under the hood close to the master cylinder, where the brake lines can easily be connected and routed to the front or rear wheels.

The combination valve is nonserviceable and must be replaced if malfunctioning.

Brake Bleeding

The hydraulic brake system must be free of air to operate properly. Air can enter the system when hydraulic parts are disconnected for servicing or replacement, or when the fluid level in the master cylinder reservoir(s) is very low. Air in the system will give the brake pedal a spongy feeling upon application.

The quickest and easiest of the two ways for system bleeding is the pressure method, but special equipment is needed to externally pressurize the hydraulic system. The other, more commonly used method of brake bleeding is done manually.

BLEEDING SEQUENCE

Bleeding may be required at only one or two wheels or at the master cylinder, depending upon what point the system was opened to air. If after bleeding the cylinder/caliper that was rebuilt or replaced, the pedal still has a spongy feeling upon application, it will be necessary to bleed the entire system.

Bleed the system in the following order:

1. **Master cylinder:** If the cylinder is not equipped with bleeder screws, open the brake line(s) to the wheels slightly while pressure is applied to the brake pedal. Be sure to tighten the line before the brake pedal is released. The procedure for bench bleeding the master cylinder is in the following section.

2. **Power Brake Booster:** If the unit is equipped with bleeder screws, it should be bled after the master cylinder. The car engine should be off and the brake pedal applied several times to exhaust any vacuum in the booster. If the unit is equipped with two bleeder screws, always bleed the higher one first.

3. **Combination Valve:** If equipped with a bleeder screw.

4. **Front/Back Split Systems:** Start with the wheel farthest away from the master cylinder, usually the right rear wheel. Bleed the other rear wheel, right front and then the left front.

NOTE: If you are unsuccessful in bleeding the front wheels, it may be necessary to deactivate the metering valve. This is accomplished by either pushing in, or pulling out a button or stem on the valve. The valve may be held by hand, with a special tool or taped, it should remain deactivated while the front brakes are bled.

5. **Diagonally Split System:** Start with the right rear then the left front. The left rear than the right front.

6. **Rear Disc Brakes:** If the car is equipped with rear disc brakes and the calipers have two bleeder screws, bleed the inner first and then the outer.

— CAUTION —
Do not allow brake fluid to spill on the car's finish, it will remove the paint. Flush the area with water.

MANUAL BLEEDING

1. Clean the bleeder screw at each wheel.
2. Start with the wheel farthest from the master cylinder (right rear).
3. Attach a small rubber hose to the bleeder screw and place the end in a clear container of brake fluid.
4. Fill the master cylinder with brake fluid (check often during bleeding). Have an assistant slowly pump up the brake pedal and hold pressure.
5. Open the bleed screw about one-quarter turn, press the brake pedal to the floor, close the bleed screw and slowly release the pedal. Continue until no more air bubbles are forced from the cylinder on application of the brake pedal.
6. Repeat the procedure on all remaining wheel cylinders and calipers.

Master cylinders equipped with bleed screws may be bled independently. When bleeding the Bendix-type dual master cylinder it is necessary to solidly cap one reservoir section while bleeding the other to prevent pressure loss through the cap vent hole.

NOTE: **The disc should be rotated to make sure that the piston has returned to the unapplied position when bleeding is completed and the bleed screw closed.**

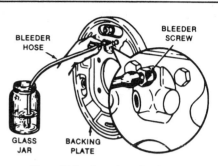

Manual bleeding drum brakes

— CAUTION —
The bleeder valve at the wheel cylinder must be closed at the end of each stroke, and before the brake pedal is released, to insure that no air can enter the system. It is also important that the brake pedal be returned to the full up position so the piston in the master cylinder moves back enough to clear the bypass outlets.

PRESSURE BLEEDING DISC BRAKES

Pressure bleeding disc brakes will close the metering valve and the front brakes will not bleed. For this reason it is necessary to manually hold the metering valve open during pressure bleeding. Never use a block or clamp to hold the valve open, and never force the valve stem beyond its normal position. Two different types of valves are used. The most common type requires the valve stem to be held in while bleeding the brakes, while the second type requires the valve stem to be held out (.060 in. minimum travel). Determine the type of valve by visual inspection.

— CAUTION —
Special adapters are required when pressure bleeding cylinders with plastic reservoirs.

Pressure bleeding equipment should be diaphragm type; placing a diaphragm between the pressurized air supply and the brake fluid. This prevents moisture and other contaminants from entering the hydraulic system.

NOTE: **Front disc/rear drum equipped vehicles use a metering valve which closes off pressure to the front brakes under certain conditions. These systems contain manual release actuators which must be engaged to pressure bleed the front brakes.**

1. Connect the tank hydraulic hose and adapter to the master cylinder.
2. Close the hydraulic valve on the bleeder equipment.
3. Apply air pressure to the bleeder equipment.

1-73

BRAKES
Import Vehicles

CAUTION
Follow the equipment manufacturer's recommendations for correct air pressure.

4. Open the valve to bleed air out of the pressure hose to the master cylinder.

NOTE: **Never bleed this system using the secondary piston stopscrew on the bottom of many master cylinders.**

5. Open the hydraulic valve and bleed each wheel cylinder and caliper. Bleed the rear brake system first when bleeding both front and rear systems.

FLUSHING HYDRAULIC BRAKE SYSTEMS

Hydraulic brake systems must be totally flushed if the fluid becomes contaminated with water, dirt or other corrosive chemicals. To flush, simply bleed the entire system until *all* fluid has been replaced with the correct type of new fluid.

BENCH BLEEDING MASTER CYLINDER

Bench bleeding the master cylinder before installing it on the car reduces the possibility of air getting into the lines.

1. Connect two short pieces of brake line to the outlet fittings, bend them until the free end is below the fluid level in the master cylinder reservoir(s).
2. Fill the reservoirs with fresh brake fluid. Pump the piston until no more air bubbles appear in the reservoir(s).
3. Disconnect the two short lines, refill the master cylinder and securely install the cylinder cap(s).
4. Install the master cylinder on the car. Attach the lines but do not completely tighten them. Force any air that might have been trapped in the connection by slowly depressing the brake pedal. Tighten the lines before releasing the brake pedal.

POWER BRAKES

Vacuum Operated Booster

Power brakes operate just as standard brake systems except in the actuation of the master cylinder pistons. A vacuum diaphragm is located on the front of the master cylinder and assists the driver in applying the brakes, reducing both the effort and travel he must put into moving the brake pedal.

The vacuum diaphragm housing is connected to the intake manifold by a vacuum hose. A check valve is placed at the point where the hose enters the diaphragm housing, so that during periods of low manifold vacuum brake assist vacuum will not be lost.

Depressing the brake pedal closes off the vacuum source and allows atmospheric pressure to enter on one side of the diaphragm. This causes the master cylinder pistons to move and apply the brakes. When the brake pedal is released, vacuum is applied to both sides of the diaphragm, and return springs return the diaphragm and master cylinder pistons to the released position. If the vacuum fails, the brake pedal rod will butt against the end of the master cylinder actuating rod, and direct mechanical application will occur as the pedal is depressed.

The hydraulic and mechanical problems that apply to conventional brake systems also apply to power brakes, and should be checked for if the tests and chart below do not reveal the problem.

Tests for a system vacuum leak as described below:

1. Operate the engine at idle with the transmission in Neutral without touching the brake pedal for at least one minute.
2. Turn off the engine, and wait one minute.
3. Test for the presence of assist vacuum by depressing the brake pedal and releasing it several times. Light application will produce less and less pedal travel, if vacuum was present. If there is no vacuum, air is leaking into the system somewhere. Test for system operation as follows:
1. Pump the brake pedal (with engine off) until the supply vacuum is entirely gone.
2. Put a light, steady pressure on the pedal.
3. Start the engine, and operate it at idle with the transmission in Neutral. If the system is operating, the brake pedal should fall toward the floor if constant pressure is maintained on the pedal.

Power brake systems may be tested for hydraulic leaks just as ordinary systems are tested, except that the engine should be idling with the transmission in Neutral throughout the test.

POWER BRAKE BOOSTER TROUBLESHOOTING CHART

The following items are in addition to those listed in the General Troubleshooting Section. Check those items first.

Hard Pedal

1. Faulty vacuum check valve.
2. Vacuum hose kinked, collapsed, plugged, leaky, or improperly connected.
3. Internal leak in unit.
4. Damaged vacuum cylinder.
5. Damaged valve plunger.
6. Broken or faulty springs.
7. Broken plunger stem.

Grabbing Brakes

1. Damaged vacuum cylinder.
2. Faulty vacuum check valve.
3. Vacuum hose leaky or improperly connected.
4. Broken plunger stem.

Pedal Goes to Floor

Generally, when this problem occurs, it is not caused by the power brake booster. In rare cases, a broken plunger stem may be at fault.

Overhaul

Most power brake boosters are serviced by replacement only. In many cases, repair parts are not available. A good many special tools are required for rebuilding these units. For these reasons, it would be most practical to replace a failed booster with a new or remanufactured unit.

ANTI-LOCK BRAKING SYSTEM (ABS)

OPERATION

The Anti-lock Braking System (ABS) is essentially a brake system enhancement. The purpose of ABS is to increase the driver's control over a vehicle during braking-especially steering control. When a vehicle equipped with a conventional brake system must brake suddenly, one or more wheels may lock up offering little or no steering control to avoid hazards. ABS is designed to prevent braked wheels from locking. The advantages of the system are considerable. For instance, during a high-speed stop while entering a curve, ABS is designed to allow the driver to steer through the curve while decelerating. Additionally, ABS is designed to enhance the braking action of each front wheel independently and the two rear wheels independent of the front wheels. This allows controlled braking even if one or more wheels encounters a slippery surface. In this situation, ABS will automatically sense the initial loss of adhesion in any one wheel and reduce or prevent further hydraulic pressure on that wheel's brake caliper, or if the rear wheels-both calipers until adhesion is regained.

COMPONENTS

ABS is essentially the familiar split circuit hydraulic four wheel disc brake system is which a sophisticated electronic and mechanical override system has been carefully mated. Three or four wheel speed sensors (depending on vehicle system design), an

BRAKES
Import Vehicles

electronic control unit and a hydraulic unit that incorporates solenoid operated brake line valves are the major components of the system. The sensors monitor the rotation speed of the wheels and provide data about wheel acceleration and deceleration over very small intervals of time. The signals from the sensors are transmitted to the control unit. The control unit monitors the signals and compares them to a contained program. If one of the sensors suddenly shows a deceleration rate that exceeds the threshold values of the programmed system-(indicating that a wheel is about to lock and skid)-the computor activates the hydraulic control unit to maintain the optimum brake pressure in that wheel, or both rear wheels to prevent lock-up. If, for any reason, the ABS should malfunction the brakes will operate as a normal system without ABS and a warning light will go on indicating service is required.

SERVICING DISC BRAKES

Disc Brake Caliper

An integral part of the caliper, the caliper bore(s) contains the piston(s) that direct thrust against the brake pads supported within the caliper. Since all braking forces (pad application force) are applied on each side of the rotor with no self energization, the cylinder and piston are large in comparison to a drum brake wheel cylinder.

Fixed-Type

A fixed type caliper is mounted solidly to the spindle bracket.

Pistons are located on both sides of the rotor, in inboard and outboard caliper halves. Fluid passes between caliper halves through an external crossover tube or through internal passages. A bleeder screw is located in the inboard caliper half. A dust boot protecting each cylinder fits in a circumferential groove on the piston.

Floating or Sliding-Type

Floating or sliding calipers are free to move in a fixed bracket or support.

The piston(s) is located only on the inboard side of the caliper housing, which straddles the rotor. The cylinder piston(s) applies the inboard brake shoe directly, and simultaneously hydraulic pressure slides the caliper in a clamping action which forces the caliper to apply the outboard brake shoe.

The actual applying movement is small. The unit merely grips during application, relaxes upon release, and the shoes do not retract an appreciable distance from the rotor. The fluid inlet port and the bleeder

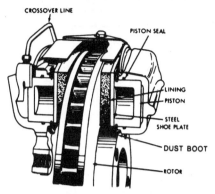

Fixed caliper disc brake

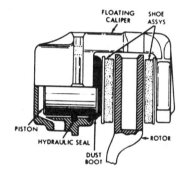

Floating caliper disc brake (sliding caliper similar)

screw are located on the inboard side of the caliper. A dust boot is fitted into a circumferential groove on the piston and into a recess at or near the outer end of the cylinder bore.

A scratched piston, nicked seal, or a sludge or varnish deposit which limits the sealing edge away from the piston will cause a fluid leak. A serious leak could develop if calipers are not reconditioned when new pads are installed. Then dust and road grime, gradually accumulating behind the dust boot, could be carried into the seal when the piston is shoved inward to accommodate new thick linings. Old seals may have taken a "set," thus preventing proper seating in the retainer groove and on the piston. Therefore, when reconditioning calipers, new seals should be installed.

OVERHAUL PROCEDURES

Before servicing, syphon or syringe about ⅔ of the fluid from the master cylinder reservoir; do not, however, lower the fluid level below the cylinder intake port.

1. To prevent a gravity loss of fluid, plug the brake line after disconnecting it from the caliper.
2. To overhaul, remove the caliper from the vehicle, allow the unit to drain, and remove the brake shoes.
3. For benchwork, clamp the caliper housing in a soft jawed vice.
4. On fixed-caliper types, remove the bridge bolts and separate the caliper into halves. Remove the sealing O-rings at crossover points, if the unit has internal fluid passages across the halves.
5. Whenever required, use special tools to remove pistons, dust boots, and seals. If compressed air is used, apply it gradually, gently ease the pistons from the cylinders, and trap them in a clean cloth; do not allow them to pop out. *Take care to avoid pinching hands or fingers.*
6. While removing stroking type seals and boots, work the lip of the boot from the groove in the caliper. After the boot is free, pull the piston, and strip the seal and boot from the piston.
7. While removing fixed position (rectangular ring) seals and boots, pull the piston through the boot. *Do not use a metal tool which would scratch the piston.* Use a small pointed wooden or plastic tool to lift the boots and seals from the grooves in the cylinder bore.

Cleaning, Inspection, and Installation

Use only alcohol and/or brake fluid and a lint free wiping cloth to clean the caliper and parts.

— **CAUTION** —
Other solvents should not be used. Blow out passages with compressed air. Always wear eye protection when using compressed air or cleaning calipers.

1. To correct minor inperfections in the cylinder bore, polish with a fine grade of crocus cloth working in a circular rather than a lengthwise motion. Do not use any form of abrasive on a plated piston. Discard a piston which is pitted or has signs of plating wear.
2. Inspect the new seal. It should lie flat and be round. If it has suffered a distorted "set" during its shelf life, do not use it. Lubricate the cylinder wall and parts with brake fluid.
3. While installing stroking type seals and boots, stretch the boot and the seal over the piston and seat them in position.
4. Use special alignment tools for inserting lip culp seals.
5. Install the fixed position (rectangu-

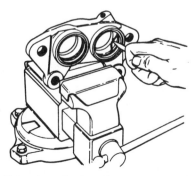

Removing a fixed position rectangular ring seal

1–75

SECTION 1

BRAKES
Import Vehicles

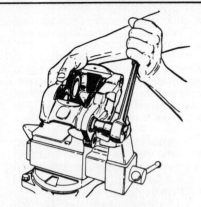

Removing the fixed caliper bridge bolts

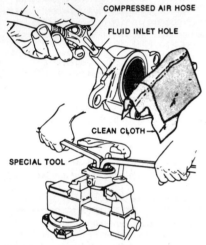

Removing a hollow-end piston with compressed air (top) or the special tool (bottom)

Checking maximum piston clearance

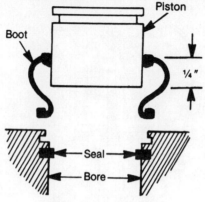

Typical boot installation

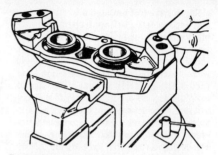

Replacing the O-rings in the internal cross-over passages

lar ring) seals and be sure the ring does not twist or roll in the groove.

6. Where the boot lip is retained inside the cylinder bore, the following method works well:

   a. Lubricate the bottom inside edge of the piston and brake seal in the caliper with brake fluid.

   b. Pull the boot over the bottom end of the piston so that the boot is positioned on the bottom of the piston with the lip about ¼ inch up from bottom end.

   c. Hold the piston suspended over bore.

   d. Insert the back boot lip into the groove in the caliper.

   e. Tuck the sides of the boot into the groove and work forward until only one bulge remains.

   f. Tuck the final bulge into the front of the groove.

   g. Push the piston carefully through the seal and boot to the bottom of the bore. The inside of the boot should slide on the piston and come to rest in the boot groove.

If the boot lip is retained outside the cylinder bore, first stretch the boot over the piston and seat it in its groove, then press the piston through the seal.

Fully depress the piston. You'll need 50 to 100 of pounds force to fasten the boot lip in place. On some designs, it is necessary to use a wooden drift or a special tool to seat the metal boot in the caliper counterbore below the face of the caliper.

Installing Fixed Caliper Bridge Bolts

If the caliper contains internal fluid cross-over passages, be sure to install the new O-ring seals at the joints.

Install high tensile strength bridge bolts on the mated caliper halves.

Never replace the bridge bolts with ordinary standard hardware bolts; order the bolts by part numbers only. Tighten the bridge bolts, using a specified torque wrench as follows specified by the manufacturer.

OVERHAUL NOTES

Field reports indicate that two factors determine whether to replace or rebuild calipers:

1. Can the piston or pistons be removed?
2. Will the bleed screw break off when removal is attempted? (Rebuilders will not accept a caliper with a broken bleed screw.) Since there is no way to predict how a bleed screw will react, follow this porcedure to attempt removal.

1. Insert a drill shank into the bleed screw hole (snug fit).
2. Tap the screw on all sides.
3. With a six point wrench apply pressure gently while working the drill up and down slightly.
4. If the drill starts to bind, the screw is beginning to collapse and cannot be removed intact.

Heating the caliper is another successful, but time consuming, bleed screw removal technique.

1. Remove the caliper from the car.
2. Heat the caliper.
3. Shrink the bleed screw by applying dry ice, and attempt removal.

BLEEDER SCREW REPLACEMENT

1. Using the existing hole in the bleeder screw for a pilot, drill ¼ in. hole completely through the existing bleeder.
2. Increase the hole to ⁷⁄₁₆ in.
3. Tap hole using a ¼ in. (18-National pipe tap) ½ in. deep-(full thread.)
4. Install bleeder repair kit.
5. Test for leaks and full brake pedal pressure.

FROZEN PISTONS

Sliding or Floating Caliper

1. **Hydraulic Removal:** remove the caliper assembly from the rotor.
2. Remove the brake pads and dust seal. With flexible brake line connected and the bleed screw closed, apply enough pedal pressure to move the piston most of the way out of the bore. (Brake fluid will begin to ooze past the piston inner seal.)

1. **Pneumatic Removal:** remove the caliper from the car.
2. With the bleed screw closed, apply air pressure to force the piston out.

--- **CAUTION** ---

Hydraulic and pneumatic methods of piston removal should be done carefully to prevent personal injury or piston damage.

Fixed Caliper

NOTE: The hydraulic or pneumatic methods which apply to the single piston type caliper will not work on the multiple type brake caliper.

BRAKES
Import Vehicles

Replacing a disc brake bleeder screw

1. Remove the caliper from the car with the two halves separated.
2. Mount in a vise and use a piston puller (many types available) to remove the pistons.

Brake Disc (Rotor)

ROTOR RUNOUT

Manufacturers differ widely on permissible runout, but too much can sometimes be felt as a pulsation at the brake pedal. A wobble pump effect is created when a rotor is not perfectly smooth and the pad hits the high spots forcing fluid back into the master cylinder. This alternating pressure causes a pulsating feeling which can be felt at the pedal when the brakes are applied. This

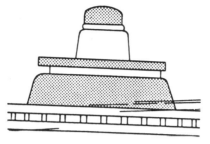

Excessive runout

excessive runout also causes the brakes to be out of adjustment because disc brakes are self-adjusting; they are designed so that the pads drag on the rotor at all times and therefore automatically compensate for wear.

To check the actual runout of the rotor, first tighten the wheel spindle nut to a snug bearing adjustment, end-play removed. Fasten a dial indicator on the suspension at a convenient place so that the indicator stylus contacts the rotor face approximately one inch from its outer edge. Set the dial at zero. Check the total indicator reading while turning the rotor one full revolution. If the rotor is warped beyond the runout specification, it is likely that it can be successfully remachined.

Lateral Runout: A wobbly movement of the rotor from side to side as it rotates. Excessive lateral runout causes the rotor faces to knock back the disc pads and can result in chatter, excessive pedal travel, pumping or fighting pedal and vibration during the breaking action.

Parallelism (lack of): Refers to the amount of variation in the thickness of the rotor. Excessive variation can cause pedal vibration or fight, front end vibrations and

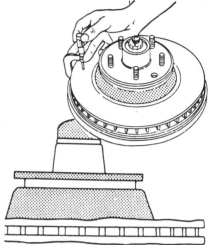

Parallelism

possible "grab" during the braking action; a condition comparable to an "out-of-round brake drum." Check parallelism with a micrometer. "mike" the thickness at eight or more equally spaced points, equally distant from the outer edge of the rotor, preferably at mid-points of the braking surface. Parallelism then is the amount of variation between maximum and minimum measurements.

Surface or Micro-inch finish, flatness, smoothness: Different from parallelism, these terms refer to the degree of perfection of the flat surface on each side of the rotor; that is, the minute hills, valleys and swirls inherent in machining the surface. In a visual inspection, the remachined surface should have a fine ground polish with, at most, only a faint trace of nondirectional swirls.

DISC BRAKE APPLICATION CHART

Make	Year/Model		Type No.
Audi	'80–'84 4000 Coupe		4
	'85–'87 4000S Coupe GT		4
	'84 4000S Quattro	Front	4
		Rear	4
	'85–'87 4000S Quattro	Front	4
		Rear	4
	'80 5000 exc Turbo		6
	'81–'87 5000 Turbo	Front	4
		Rear	4
	'83–'87 Quattro Turbo Coupe	Front	4
		Rear	4
BMW	'84–'87 318i		4
	'80–'81 320i		2
	'82–'83 320i	ATE system	2
		Girling system	4
	'84–'87 325e	Front	4
		Rear	4
	'82–'87 528e	Front	4
		Rear	4

BRAKES
Import Vehicles

DISC BRAKE APPLICATION CHART

Make	Year/Model		Type No.
BMW	'80–'81 528i	Front	2
		Rear	2
	'83–'87 533i, 535i, 633CSi, 635CSi	Front	4
		Rear	4
	'80–'82 633CSi, 733i	Front	2
		Rear	2
	'83–'87 733i, 735i	Front	2
		Rear	4
Chrysler Corp.	'80 Arrow	Front	5
		Rear	9
	'80 Colt exc Wagon		3
	'80 Colt Wagon	Front	5
		Rear	9
	'80–'84 Colt, Champ (to '82) FWD exc Turbo		3
	'85–'87 Colt exc Turbo		4
	'84–'87 Colt Turbo		4
	'80–'83 Challenger, Sapporo	Front	5
		Rear	9
	'84–'87 Conquest	Front	4
		Rear	9
	'84–'87 Vista		4
Honda	'80–'81 Accord		5
	'82–'86 Accord		4
	'80–'83 Civic Wagon		5
	'84–'86 Civic Wagon		4
	'80–'87 Civic		4
	'80–'83 Prelude		4
	'84–'87 Prelude	Front	4
		Rear	4
Hyundai	'86–'87 Excel		4
Isuzu	'81–'87 I-Mark		2
	'83–'87 Impulse	Front	4
		Rear	4
Mazda	'80–'84 GLC		5
	'85–'87 GLC, 323		4
	'80–'87 RX 7		4
	'80–'82 626		4
	'83–'87 626		4
Mercedes-Benz	'80–'87 Exc 190D, 190E, 300SD, 380SE, 380SEC, 380SEL, 500SEC, 500SEL	Front	2
		Rear	2
	'82–'87 190D, 190E, 300SD, 380SE, 380SEC, 380SEL, 500SEC, 500SEL	Front	4
		Rear	2
Merkur	'85–'87 XR4Ti		1
Mitsubishi	'83 Cordia, Tredia		4
	'84–'87 Cordia, Tredia		4
	'85–'87 Galant	Front	4
		Rear	9
	'85–'87 Mirage		4
	'83–'87 Starion	Front	4
		Rear	9

BRAKES
Import Vehicles

DISC BRAKE APPLICATION CHART

Make	Year/Model		Type No.
Nissan/Datsun	'80–'81 200 SX	Front	7
		Rear	11
	'82–'87 200 SX	Front	7
		Rear	4
	'80–'82 210		7
	'80–'81 280ZX	Front	4
		Rear	11
	'82–'83 280ZX	Front	4
		Rear	4
	'84–'87 300ZX	Front	4
		Rear	4
	'80–'82 310		7
	'80–'81 510		7
	'80 810		7
	'81 810	Front	4
		Rear	4
	'82–'87 Maxima	Front	4
		Rear	4
	'83–'87 Pulsar		4
	'83–'87 Sentra	Gas	4
		Diesel	4
	'82–'87 Stanza		4
Porsche	'80–'87 911	Front	2
		Rear	2
	'80–'87 924, 928, 944	Front	6
		Rear	6
Renault	'83–'87 Alliance, Encore		4
	'80–'87 Fuego		8
	'80–'84 LeCar		5
	'80–'83 18i		8
	'84–'87 18i Sportwagon		4
SAAB	'80–'87 900, 9000 Series	Front	7
		Rear	2
Subaru	'80–'87 All Models	Front	9
Toyota	'83–'87 Camry		4
	'80–'81 Celica		5
	'82–'87 Celica	Front	4
		Rear	4
	'80–'83 Corolla		3
	'84–'87 Corolla exc Coupe		4
	'84–'85 Corolla Coupe	Front	4
		Rear	4
	'80–'82 Corona		2
	'80 Cressida		2
	'81–'87 Cressida	Front	4
		Rear	3
		('85) Rear	4
	'81–'84 Starlet		4
	'80–'81 Supra	Front	5
		Rear	5
	'82–'87 Supra	Front	4
		Rear	4
	'80–'82 Tercel		4

1–79

SECTION 1

BRAKES
Import Vehicles

DISC BRAKE APPLICATION CHART

Make	Year/Model		Type No.
Toyota	'83–'87 Tercel		4
	'84–'87 VanWagon		4
VW (FWD)	'85–'87 All	Front	4
		Rear	4
	'80–'84 All w/KH calipers		1
	Models from 2/84		4
	'80–'81 All w/Girling calipers		7
VW (RWD)	'80–'87 Vanagon		2
Volvo	'80–'87 240, 260 Series	Front	2
		Rear	2
	'80–'87 740, 760 Series	Front	4
		Rear	2
Yugo	'86–'87 GV		5

TYPE 1

Kelsey-Hayes Floating Caliper

This unit is a single piston, one-piece caliper which floats on two guide pins screwed into the adapter (anchor plate). The adaptor, in turn, is held to the steering knuckle with two bolts. As the brake pads wear, the caliper floats along the adapter and guide pins during braking.

PAD REPLACEMENT

1. Raise the front of the vehicle and support it with jackstands. Remove the wheel.
2. Siphon some brake fluid from the master cylinder reservoir to prevent its overflowing when the piston is retracted into the cylinder bore.
3. Disconnect the brake pad warning indicator if so equipped.
4. Remove the anti-rattle springs.
5. Remove the guide pins that attach the caliper to the anchor plate.
6. Lift off the caliper and position it out of the way with some wire—you need not remove the brake lines.

— CAUTION —
Never allow the caliper to hang by its brake lines.

7. Slide the outer pad out of the anchor plate. Slide the rotor of the hub and remove the inner pad. Check the rotor as detailed in the appropriate section. Check the caliper for fluid leaks or cracked boots. If any damage is found, the caliper will require overhauling or replacement.
8. Carefully clean the anchor plate with a wire brush or some other abrasive material. Install the inner pad, rotor and outer pad, new brake pads into position on the

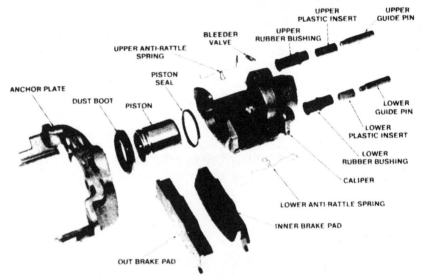

Exploded view of a Kelsey–Hayes floating caliper

Kelsey–Hayes floating caliper disc brake assembly

anchor plate. The inner pad usually has chamfered edges.

NOTE: When replacing brake pads, always replace both pads on both sides of the vehicle. Mixed pads will cause uneven braking.

9. Slowly and carefully push the piston into its bore until it's bottomed and then position the caliper onto the anchor plate. Install the guide pins by pushing them carefully into the bushings and threading them into the adapter.

NOTE: The upper guide pin is usually longer than the lower one.

10. Install the anti-rattle springs between the anchor plate and brake pads ears. The loops on the springs should be positioned inboard.
11. Fill the reservoir with brake fluid and pump the brake pedal several times to set the piston. It should not be necessary to bleed the system; however, if a firm pedal cannot be obtained, the system must be bled (see "Bleeding the Brakes" in this section).
12. Install the wheel and lower the vehicle.

BRAKES
Import Vehicles

TYPE 2

ATE, Girling, Sumitomo, Teves, etc. Fixed Caliper

These units are either two or four piston, two-piece calipers that are fixed directly to the steering knuckle or spindle.

Brake pads may be changed without removing the caliper on all of these models. There may be some differences in retainers or anti-rattle springs from the illustrations, but all versions are basically the same. Before removing any parts, carefully note the position of any springs, retainers or clips. Change pads on one wheel at a time and use the other as a reference.

All pads on all models are held in position by either retaining pins or retainer plates. The retainer plates are bolted to the caliper housing and need only be loosened and rotated out of the way for pad removal.

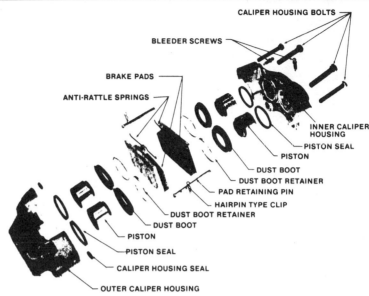

Exploded view of a four-piston fixed caliper

PAD REPLACEMENT

1. Raise the front (or rear) of the vehicle and support it with jackstands. Remove the wheel.

2. Siphon a sufficient quantity of brake fluid from the master cylinder reservoir to prevent the brake fluid from overflowing the master cylinder when removing or installing new pads. This is necessary as the pistons must be forced into the cylinder bore to provide sufficient clearance to remove the pads.

3. Some models may use a cover plate over the access hole for the pads, if so, remove it. Disconnect the brake pad lining wear indicator wire on models so equipped.

4. Carefully clean the exterior of the caliper with a wire brush and note the position of any dampening shims or anti-rattle springs.

5. Remove the pad retaining pins and any retaining clips holding them. Remove the anti-rattle springs if so equipped. Some pads may be held in position by a plate with a retaining bolt. If so, loosen the bolt and swing the plate away. Lift out the spreader spring if so equipped.

NOTE: **It is a good idea to remove one retaining spring or plate and then remove the anti-rattle springs or spreader spring. Remove the second retaining pin or plate last.**

6. Force the old pads away from the rotor for easy withdrawal and remove the pads from the caliper.

7. If so equipped, remove the lower anti-rattle springs and dampening shims using needlenose pliers.

8. Check the brake disc (rotor) as detailed in the appropriate section.

9. Examine the dust boot for cracks or damage and push the pistons back into the cylinder bores. If the pistons are frozen or if the caliper is leaking hydraulic fluid, it must be overhauled.

10. Install the anti-rattle spring or damping shims and slip the new pads into the caliper. If damping shims are used, be sure that the directional arrow on the shims face the forward rotation of the rotor.

11. Install one pad retaining pin and hairpin clip. Position the anti-rattle springs and/or spreader spring and then install the other pad retaining pin and clip.

12. Refill the master cylinder to the correct level with the proper brake fluid.

13. Replace the wheel and lower the vehicle. Pump the brake pedal several times to bring the pads into correct adjustment. Road test the vehicle.

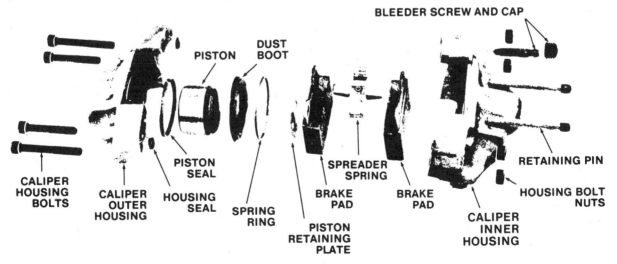

Exploded view of a two-piston fixed caliper

1-81

SECTION 1

BRAKES
Import Vehicles

Typical fixed caliper disc brake assembly

NOTE: If a firm pedal cannot be obtained, the system will require bleeding (see "Bleeding the Brakes" in this section).

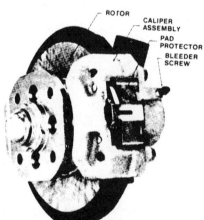

Exploded view of a Sumitomo Torque Plate floating caliper

TYPE 3

Sumitomo Torque Plate Floating Caliper

This unit is a single piston, two-piece caliper which floats on torque plate pins. The torque plate itself is bolted to the steering knuckle. The outer caliper half may be separated from the inner half although the caliper need not be separated or removed for pad replacement. There may be differences in the shape of the brake pads, retaining springs, etc. on various models using this system. Complete one side at a time using the other for reference.

PAD REPLACEMENT

1. Raise the front of the vehicle and support it on jackstands. Remove the wheel.
2. Siphon a sufficient quantity of brake fluid from the master cylinder reservoir to prevent the brake fluid from overflowing the master cylinder when removing or installing new pads. This is necessary as the piston must be forced into the cylinder bore to provide sufficient clearance to remove the pads.
3. Use a small prybar or other suitable tool and pry the pad protector off of the retaining pins.
4. Remove the center of the 'M' clip from the hole in the outboard pad and its ends from the retaining pins.

NOTE: To facilitate the reassembly operation later on, note how the 'M' clip and the 'K' spring are positioned in the caliper.

5. Pull out the retaining pins and remove the 'K' spring from the inboard pad.

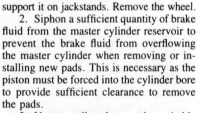

Typical Sumitomo Torque Plate floating caliper disc brake assembly

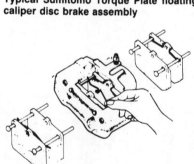

Spring and clip installation

6. Remove the inner and outer pads.
7. Check the brake disc (rotor) as detailed in the appropriate section.
8. Examine the dust boot for cracks or damage and then push the piston back into the cylinder bore. Use a C-clamp or other suitable tool to bottom the piston. If the piston is frozen, or if the caliper is leaking hydraulic fluid, the caliper must be overhauled or replaced.
9. Install new pads into the caliper.
10. Install one retaining pin.
11. Install the inboard pad 'K' spring. Hook one end of the 'K' spring under the retaining pin and the center of the 'K' spring over the top of the inboard pad. Insert the other retaining pin through the outboard pad, over the 'K' spring and through the inboard pad.
12. Insert the ends of the 'M' clip into the holes in the retaining pins and press the center of the spring into the hole in the outboard pad.
13. Install the pad protector.
14. Refill the master cylinder with fresh brake fluid.
15. Install the tire and wheel assembly and then pump the brake pedal several times to bring the pads into adjustment. Road test the vehicle.

NOTE: If a firm pedal cannot be obtained, bleed the system as detailed in "Bleeding the Brakes".

TYPE 4

ATE, Girling, etc. Floating Caliper

Although similar in many respects to a sliding caliper, this single piston unit floats on guide pins and bushings which are threaded into a mounting bracket. The mounting bracket is bolted to the steering knuckle.

Variations in pad retainers, shims, anti-rattle and retaining springs will be encountered but the service procedures are all ba-

BRAKES
Import Vehicles

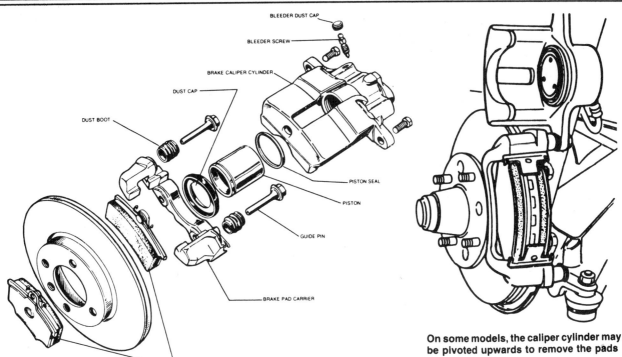

Exploded view of a typical Type 4 floating caliper

On some models, the caliper cylinder may be pivoted upwards to remove the pads

sically the same. Note the position of all springs, clips or shims when removing the pads. Work on one side at a time and use the other for reference.

PAD REPLACEMENT

1. Raise and support the front (or rear) of the vehicle on jackstands. Remove the wheel.
2. Siphon a sufficient quantity of brake fluid from the master cylinder reservoir to prevent the brake fluid from overflowing the master cylinder when removing or installing new pads. This is necessary as the piston must be forced into the cylinder bore to provide sufficient clearance to remove the pads.
3. Grasp the caliper from behind and pull it toward you. This will push the piston back into the cylinder bore.
4. Disconnect the brake pad lining wear indicator if so equipped. Remove any anti-rattle springs or clips if so equipped.

NOTE: Depending on the model and year of the particular caliper, you may not have to remove it entirely to get at the brake pads. If the caliper is the "swing" type, remove the upper or lower guide bolt, pivot the caliper on the other lossened bolt and swing it upward exposing the brake pads. If this method is employed, skip to Step 7.

5. Remove the caliper guide pins.
6. Remove the caliper from the rotor by slowly sliding it out and away from the rotor. Position the caliper out of the way

and support it with wire so that it doesn't hang by the brake line.
7. Slide the outboard pad out of the adapter.
8. Remove the inboard pad. Remove any shims or shields behind the pads and note their positions.
9. Install the anti-rattle hardware and then the pads (in their proper positions!).
10. If equipped with the parking brake, use a suitable tool/allen wrench to rotate the caliper piston back into the caliper bore.

If not equipped with the parking brake use a C-clamp to push the caliper piston into the bore.
11. Install any pad shims or heat shields.
12. Reposition the caliper and install the guide pins.

NOTE: If the caliper is the "swing" type, you need only pivot it back into position and install the lower guide pin. On 280ZX front calipers, insert a lever into the opening in the cylinder body as shown in the accompanying illustration and push the piston in by catching the torque member.

REMOVAL PROCEDURE

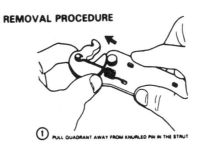

① PULL QUADRANT AWAY FROM KNURLED PIN IN THE STRUT

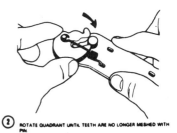

② ROTATE QUADRANT UNTIL TEETH ARE NO LONGER MESHED WITH PIN

INSTALLATION PROCEDURE

③ REMOVE THE SPRING AND SLIDE QUADRANT OUT OF STRUT – BE CAREFUL NOT TO OVERSTRESS SPRING.

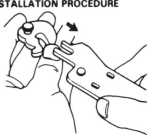

INSTALL ADJUSTER QUADRANT PIN INTO SLOT IN STRUT. TURN ASSEMBLY OVER AND INSTALL SPRING

SECTION 1

BRAKES
Import Vehicles

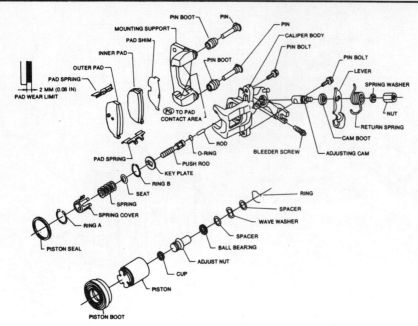

Exploded view of a Type 4 floating caliper with parking brake (Datsun shown, others similar)

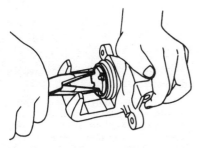

Using needle-nosed pliers to turn the piston before pushing it into the cylinder bore

13. Refill the master cylinder with fresh brake fluid.

14. Install the tire and wheel assembly and then pump the brake pedal several times to bring the pads into adjustment. Road test the vehicle.

NOTE: After installing new pads on models where caliper is mounted as a rear disc brake depress the pedal firmly (about 40 times-engine off) to set proper adjustment. Check the parking brake operation, adjust cable if necessary.

NOTE: If a firm pedal cannot be obtained, bleed the system as detailed in "Bleeding the Brakes".

TYPE 5

Akebono, Girling, etc. Sliding Caliper

A single piston sliding caliper system. The caliper is held to a mounting plate or adapter by guides or keys. Support plates are used under the pads to prevent rattling. Variations occur depending on model, however servicing is similar. Work on one side at a time using the other for reference.

NOTE: If the caliper is equipped with the parking brake system, refer to TYPE 7 for procedure.

PAD REPLACEMENT

1. Raise and support the vehicle on jackstands. Remove the wheel. Remove a sufficient amount of brake fluid from the master cylinder to allow for expansion when

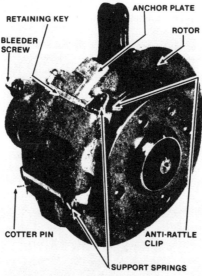

Typical sliding caliper disc brake assembly

fluid is forced back into the master cylinder when the caliper pistons are retracted.

2. Remove the clips or pins that retain the caliper guide(s) in position. Tap out the guide(s) or key(s). Make note of position for reinstallation.

3. Rock the caliper to retach the pads slightly and lift the caliper from the mounting bracket. Secure the caliper out of the way, do not permit the caliper to hang by the brake hose.

4. Remove the brake pads from the mounting bracket. Take note of the position of the pad support springs. They are not interchangeable and must be installed correctly.

5. Push the caliper piston back into the caliper bore with a suitable C-clamp.

6. Clean the metal contact points on the mounting bracket and caliper. Place the brake pads and support springs on the mounting bracket and reinstall the caliper.

7. Apply the brakes several times to position the pads. Fill the master cylinder and check for a firm pedal. Bleed the brake system if necessary.

TYPE 6

ATE, etc. Sliding Yoke Caliper

This unit may have one piston, or two in a single cylinder. It has a fixed mounting frame which is bolted to the steering knuckle. The pads are retained in the fixed frame. A floating frame, or yoke, slides on the fixed frame. The cylinder attaches to this yoke, creating a caliper. Braking pressure forces the piston against the inner pad. The reaction causes the yoke to move in the opposite direction, applying pressure to the outer pad. The yoke does not have to be removed to replace the pads.

PAD REPLACEMENT

1. Raise the front (or rear) of the vehicle and support it with jackstands. Remove the wheel.

2. Siphon a sufficient quantity of brake fluid from the master cylinder reservoir to prevent the brake fluid from overflowing the master cylinder when removing or installing new pads. This is necessary as the piston must be forced into the cylinder bore to provide sufficient clearance to remove the pads.

3. Disconnect the wire connector leading to the brake pad wear indicator.

4. Remove the brake pad retaining clips on the inside of the caliper and then drive out the retaining pins. Don't lose the pad positioner (spreader) that is held down by the pins.

BRAKES
Import Vehicles

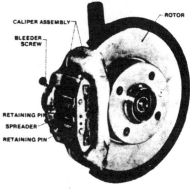

Typical ATE sliding yoke disc brake assembly

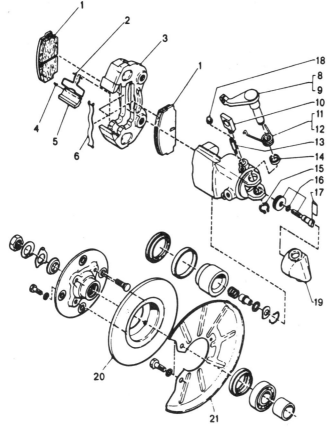

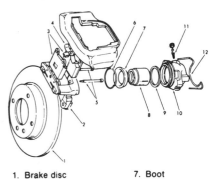

1. Brake disc
2. Caliper mounting frame
3. Pads
4. Cross spring
5. Retaining pins
6. Clamp ring
7. Boot
8. Piston
9. Seal
10. Cylinder
11. Bleeder nipple
12. Guide spring

Exploded view of an ATE sliding yoke caliper

1. Pad (disc brake F)
2. Spring (caliper)
3. Bracket (mounting)
4. Pin (caliper)
5. Stopper (plug)
6. Spring (pad)
7. Body caliper ass'y
8. Lever & spindle ass'y (LH)
9. Lever & spindle ass'y (RH)
10. Bracket (hand brake)
11. Spring (hand brake lever return LH)
12. Spring (hand brake lever return RH)
13. Bleeder screw (wheel cylinder)
14. Bushing (hand brake)
15. Retaining spring
16. Spindle ass'y
17. Connecting link
18. Cap (air bleeder)
19. Cap (lever)
20. Brake disc (F)
21. Cover (disc)

Exploded view of a Type 5 sliding caliper with parking brake (Subaru shown, others similar)

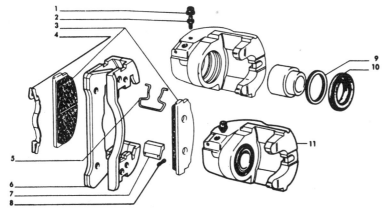

1. Bleeder screw cap
2. Bleeder screw
3. Brake pads
4. Spring
5. Caliper fastener spring
6. Caliper support bracket
7. Caliper locking block
8. Cotter pin
9. Piston seal
10. Piston dust boot
11. Assembled caliper

Exploded view of a Type 5 sliding caliper

5. Pry out the inner brake pad with a suitable tool inserted. Through the brake pad retaining pin holes.
6. The outer pads are secured by a notch at the top of the pad. Grasp the caliper assembly from the inside and pull it toward yourself. Remove the pad in the same manner as the inner and detach the wear indicator.
7. Check the brake disc (rotor) as detailed in the appropriate section.
8. Inspect the caliper and piston assembly for breaks, cracks or other damage. Overhaul or replace the caliper as necessary.
9. Use a C-clamp and press the piston back into the cylinder bore. If the caliper is equipped with a piston retaining plate, piston rotation may be necessary for proper plate location.
10. Install the wear indicator on the outer pad and then install both pads.
11. Installation of the remaining components is the reverse order or removal.
12. Top off the master cylinder with fresh brake fluid.
13. Pump the brake pedal several times to bring the pads into adjustment. Road test the vehicle. If a firm pedal cannot be obtained, bleed the brakes as detailed in "Bleeding the Brakes".

1-85

SECTION 1

BRAKES
Import Vehicles

TYPE 7

Girling/Annette Sliding Yoke Caliper

This unit is a double piston, one-piece caliper. The cylinder body contains two pistons, back-to-back, in a thru-bore. The cylinder body is bolted to the steering knuckle, with both pistons inboard of the rotor. A yoke, which slides on the cylinder body, is installed over the rotor and the caliper.

When the brakes are applied, hydraulic pressure forces the pistons apart in the double ended bore. The piston closest to the rotor applies force directly to the inboard pad. The other piston applies force to the yoke, which transmitts the force to the outer pad, creating a friction force on each side of the rotor.

One variation has a yoke that floats on guide pins screwed into the cylinder body.

The yokes do not have to be removed to replace the brake pads.

PAD REPLACEMENT

1. Raise and support the front (or rear) of the vehicle on jackstands. Remove the wheel.

2. Siphon a sufficient quantity of brake fluid from the master cylinder reservoir to prevent the brake fluid from overflowing the master cylinder when removing or installing new pads. This is necessary as the piston must be forced into the cylinder bore to provide sufficient clearance to remove the pads.

3. Disconnect the brake pad lining wear indicator if so equipped.

4. Remove the dust cover and/or anti-rattle (damper) clip if so equipped.

5. Lift off the wire clip(s) which hold the guide pins or retaining pin in place.

6. Remove the upper guide pin and the two hanger springs. Carefully tap out the lower guide pin.

— CAUTION —

The lower guide pin usually contains an anti-rattle coil spring—be careful not to lose this spring. If a retaining pin is used, pull the pin out and remove the two hanger springs.

7. Slide the yoke outward and remove the outer brake pad and the anti-noise shim (if so equipped).

8. Slide the yoke inward and remove the inner pad and anti-rattle shim.

9. Check the rotor as detailed in the appropriate section.

10. Inspect the caliper and piston assembly for breaks, cracks or other damage. Overhaul or replace the caliper as necessary.

11. Push the piston next to the rotor back into the cylinder bore until the end of the piston is flush with the boot retaining ring.

— CAUTION —

If the piston is pushed further than this, the seal will be damaged and the caliper assembly will have to be overhauled.

12. Retract the piston farthest from the rotor by pulling the yoke toward the outside of the vehicle.

13. Install the outboard pad. Anti-noise shims (if so equipped) must be located on the plate side of the pad with the triangular cutout pointing toward the top of the caliper.

14. Install the inboard pad with the shims (if so equipped) in the correct position.

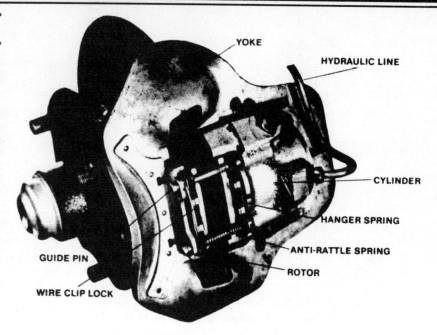

Typical Type 7 sliding yoke caliper disc brake assembly

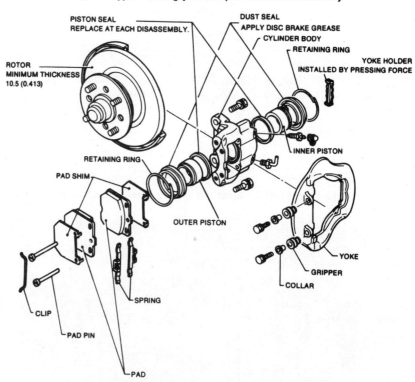

Exploded view of a Type 7 sliding yoke caliper

BRAKES
Import Vehicles

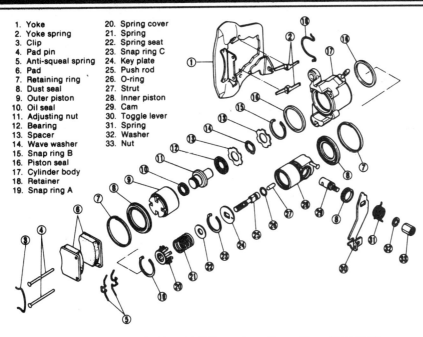

1. Yoke
2. Yoke spring
3. Clip
4. Pad pin
5. Anti-squeal spring
6. Pad
7. Retaining ring
8. Dust seal
9. Outer piston
10. Oil seal
11. Adjusting nut
12. Bearing
13. Spacer
14. Wave washer
15. Snap ring B
16. Piston seal
17. Cylinder body
18. Retainer
19. Snap ring A
20. Spring cover
21. Spring
22. Spring seat
23. Snap ring C
24. Key plate
25. Push rod
26. O-ring
27. Strut
28. Inner piston
29. Cam
30. Toggle lever
31. Spring
32. Washer
33. Nut

Exploded view of a Type 7 sliding yoke caliper with parking brake

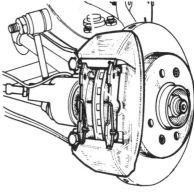

Typical Bendix floating caliper disc brake assembly

15. Replace the lower guide pin and the anti-rattle coil spring.
16. Hook the hanger springs under the pin and over the brake pads.
17. Install the upper guide pin over the ends of the hanger springs.

NOTE: If a single two-sided retaining pin is used, install the pin and then install the hanger springs as in Steps 16–17.

18. Insert the wire clip locks into the holes in the guide pins or retaining pin.
19. Refill the master cylinder with fresh brake fluid.
20. Install the tire and wheel assembly. Pump the brake pedal several times to bring the pads into adjustment. Road test the vehicle. If a firm pedal cannot be obtained, refer to "Bleeding the Brakes".

TYPE 8

Bendix Floating Caliper

This is a single piston unit that floats on guide pins and bellows bushings which are threaded into a mounting bracket. The mounting bracket is bolted to the stub axle carrier. This caliper is unique in that it is mounted on the leading edge of the brake disc, where most are mounted on the trailing edge.

The caliper does not have to be removed when replacing the brake pads.

Exploded view of the Bendix floating caliper

PAD REPLACEMENT

1. Raise and support the front of the vehicle on jackstands. Remove the wheel.
2. Siphon a sufficient quantity of brake fluid from the master cylinder reservoir to prevent the brake fluid from overflowing the master cylinder when removing or installing new pads. This is necessary as the piston must be forced into the cylinder bore to provide sufficient clearance to remove the pads.
3. Grasp the cylinder from behind and carefully pull it toward you. This will push the piston back into the cylinder bore.
4. Disconnect the brake pad lining wear indicator wires. Remove the anti-rattle springs.
5. Remove the retaining key clip on the upper side of the caliper. Remove the retaining key.
6. Lift out the brake pads.
7. Inspect the brake disc (rotor) as detailed in the appropriate section.
8. Inspect the caliper and piston assembly for breaks, cracks or other damage. Overhaul or replace the caliper as necessary.
9. Push the piston all the way back into its bore (a C-clamp may be necessary for this operation).
10. Slide the new pads into their original positions.
11. Slide the retaining key into position and replace the clip.
12. Reinstall the anti-rattle springs and the wear indicator.
13. Refill the master cylinder with fresh brake fluid.
14. Install the tire and wheel assembly and then pump the brake pedal several times to bring the pads into adjustment. Road test the vehicle.

NOTE: If a firm pedal cannot be obtained, bleed the system as detailed in "Bleeding the Brakes".

TYPE 9

Akebono Floating Caliper W/Parking Brake

This is a single piston unit that floats on guide pins and bushings which are threaded into a mounting bracket. The mounting bracket is bolted to the steering knuckle. This unit also incorporates a parking brake into the caliper.

When the parking is applied, the caliper lever rotates a cam against a pawl which pushes a threaded thrust screw in the caliper piston causing the piston to move out and apply the brakes. The thrust screw also moves on normal brake application maintaining correct adjustment.

The caliper does not have to be removed completely in order to remove the brake pads.

1-87

SECTION 1

BRAKES
Import Vehicles

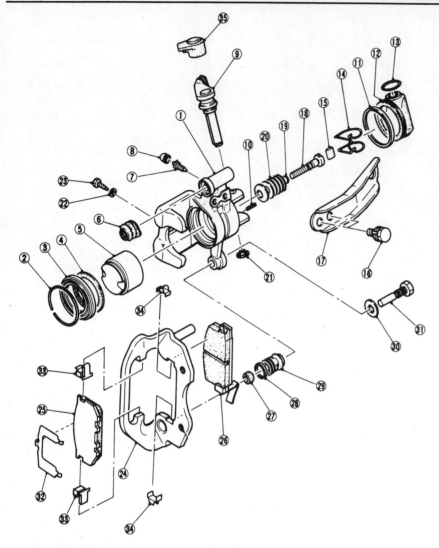

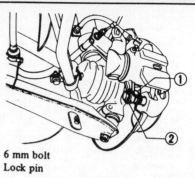

6 mm bolt
Lock pin

Removing the lock pin

8. Inspect the caliper and piston assembly for breaks, cracks or other damage. Overhaul or replace the caliper as necessary.

9. Turn the caliper piston clockwise into the cylinder bore and align the notches. Make sure the boot is not twisted or pinched.

NOTE: Do not force the piston into the cylinder bore. The piston is mounted on a threaded spindle which will bend under pressure.

10. Insert the new pads making sure that all shims and clips are in their original positions.
11. Swing the caliper down into position and install the lock pin and the 6 mm bolt.
12. Reconnect the parking brake.
13. Refill the master cylinder with fresh brake fluid.
14. Install the tire and wheel assembly and then pump the brake pedal several times to bring the pads into adjustment. Road test the vehicle.

NOTE: If a firm pedal cannot be obtained, bleed the system as detailed in "Bleeding the Brakes".

1 Caliper body	13 Gutter spring	25 Outer pad
2 Boot ring	14 Return spring	26 Inner pad
3 Piston boot	15 Connecting link	27 Rubber bushing
4 Piston seal	16 Bolt assembly	28 Retainer
5 Piston	17 Bracket	29 Lock pin boot
6 Guide pin boot	18 Spindle	30 Cone spring
7 Air bleeder screw	19 O-ring	31 Lock pin
8 Air bleeder cap	20 Cone spring	32 Shim
9 Lever & spindle	21 Snap ring	33 Outer pad clip
10 Spring pin	22 Spring washer	34 Inner pad clip
11 Cap ring	23 Bolt	35 Lever cap (upper)
12 Lever cap	24 Support	

Exploded view of the Type 9 floating caliper

PAD REPLACEMENT

1. Raise and support the front of the vehicle on jackstands. Remove the wheel.
2. Siphon a sufficient quantity of brake fluid from the master cylinder reservoir to prevent the brake fluid from overflowing the master cylinder when removing or installing new pads. This is necessary as the piston must be forced into the cylinder bore to provide sufficient clearance to remove the pads.
3. Release the parking brake and disconnect the cable from the caliper lever.
4. Remove the 6 mm lock pin bolt (lower front of the caliper). Loosen and remove the lock pin.
5. The caliper will now pivot on its upper support, swing it up and out of the way.
6. Remove the pads. Note the positions of the pad shims and the inner and outer pad clips.
7. Inspect the brake disc (rotor) as detailed in the appropriate section.

TYPE 10

Sumitomo, etc. Sliding Yoke Caliper W/ Integral Parking Brake

A single cylinder, dual piston caliper similar to the Girling sliding yoke caliper featuring a integral parking brake mechanism. Work on one side at a time using the other side as reference.

PAD REPLACEMENT

1. Raise and support the vehicle on jackstands. Remove the wheel.
2. Remove a sufficient amount of brake fluid from the master cylinder to allow for level expansion when the caliper pistons are retracted.

BRAKES
Import Vehicles

3. Release the parking brake and disconnect the cable from the caliper lever.

4. Unbolt and remove the caliper. Do not permit the caliper to be suspended by the brake hose or damage to the hose may occur.

5. Remove the pads making note of the location of the various spings, clips and pins.

6. Open the bleeder valve slightly to relieve pressure to the pistons and screw the outboard piston clockwise into the bore. Take care not to allow the inboard piston to come out of its bore.

7. Install the brake pads. Rotate the inboard piston so that the tab on the pad will align with the slot in the piston.

8. Install the caliper. Connect the parking brake cable. Bleed the system filling the master cylinder as necessary to maintain the proper level.

9. Install the wheel and lower the vehicle. Check the operation of the parking brake and test drive.

SERVICING DRUM BRAKES

A typical drum brake assembly includes a backing or support plate, with one or two wheel cylinders attached to it. Mounted on the backing plate are two lined brake shoes with shoe return springs and hold-down parts, and a means of adjusting the shoes to compensate for lining wear. A brake drum encloses these parts. The drum brakes on the rear of most vehicles also normally include the parts required for parking brakes. All of the drum brakes used on modern vehicles have these components but there is a variety of configurations for each.

Drum brakes are designed to be either "servo" or "non-servo" acting.

Servo Type Brakes

In these brakes the shoes are assembled to form a compound, "primary" and "secondary shoe unit joined at one end by an adjustable floating link. The drag of a normal (forward) drum rotation causes the primary shoe to leave its anchor and holds the secondary shoe anchored.

All of the forces applying and anchoring the primary shoe are transmitted through the shoe link, in a servo action, and also apply the secondary shoe, thus compounding its braking effect. When the drum is rotated backward, this compounding action of the shoes is reversed. When equipped with a double-end wheel cylinder (two opposed pistons), brake effectiveness can be substantially the same with either forward or reverse movement of the vehicle. With a single-end wheel cylinder (one piston), the brake is energized in only one direction. Since the secondary shoe performs more of the work in forward movement, it shows more lining wear. A longer or thicker lining is often used to offset this wear.

Non-Servo Type Brakes

In these brakes each shoe is separately anchored and their action is not compounded. On single cylinder brakes a "forward" or "leading" shoe is self-energized by the usual (forward) drum rotation while a "reverse" or "trailing" shoe is de-energized. When the drum is rotated backward, this action reverses, thus energizing the reverse shoe and de-energizing the forward shoe. The lining wear is unbalanced because the shoes perform different amounts of work; the wear is more rapid on the forward acting shoe during a forward stop.

Large two cylinder non-servo brakes, found on certain models, make use of two double-end wheel cylinders which enable the shoes to be anchored or actuated at either end. This arrangement is non-directional in effectiveness. With two-cylinder brakes, lining wear is balanced on both shoes.

MECHANICAL COMPONENTS

To be sure of restoring the brake components correctly after servicing, closely observe the arrangement of shoe hook-up parts as the brake is disassembled. These arrangements may vary on different models. Usually the brake shoes are held in a sliding fit by spring tensions, at rest upon their anchor by the return springs, and against support pads by spring or clip type hold-downs. Opposite the anchor, a star wheel adjuster links the shoe webs and provides a threaded adjustment which permits the shoes to be expanded or contracted. Some

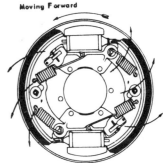

NON-DIRECTIONAL ACTING
Moving Forward

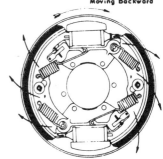

Moving Backward

Two cylinder, non-servo brake

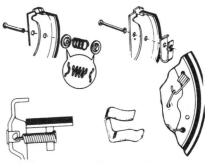

Different types of shoe hold-downs

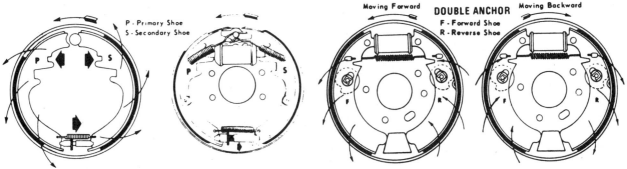

Servo brakes — Non-servo brakes

BRAKES
Import Vehicles

rear brakes have adjustable links. The shoes are held against the adjuster by a spring.

Shoe Hold-Downs

Various shoe hold-downs are shown in the illustration.

To unlock or lock the straight pin hold-downs, depress the locking cup and coil spring, or the spring clip, and rotate the pin or lock 90 degrees. On certain lever type adjusters, the inner (bottom) cup has a sleeve which aligns the adjuster lever.

Shoe Anchors

As shown in the illustration, there are various types of anchors such as the fixed non-adjustable type, or self-centering shoe sliding type, or, on some earlier models, adjustable fixed type providing either an eccentric or a slotted adjustment.

On adjustable anchors, when necessary to re-center the shoes in the drum or drum gauge, loosen the locknut enough to permit the anchor to slip out, but not so much that it can tilt.

On the eccentric type anchors, tighten the star wheel to heavy brake drag. Rotate the eccentric anchor in the direction which frees the brake until drag cannot be relieved. Tighten the anchor nut. Back off the star wheel to a normal manual adjustment.

On the slotted type anchor, tighten the star wheel to heavy drag. Tap the support plate until the anchor slips and frees the brake. Repeat this sequence until drag cannot be relieved. Tighten the locknut to the proper torque. Back off the star wheel to a normal manual adjustment.

Brake Shoes

In the same brake sizes, there can be differences in web thickness, shape of web cut-outs and positions of any reinforcements. Some vehicles require shoes made of higher tensile strength steels. Higher strength shoes usually are coded with a letter symbol stamped on the shoe web. Shoes with extra web holes or table nibs or tabs which do not cause interference generally are considered interchangeable with other shoes.

Stops

An eccentric stop under the primary or secondary shoe web on tilted front brakes prevents the shoes from bumping against the drum. Before adjusting the star wheel, loosen the lock nut on the support plate and rotate the eccentric in the forward direction until the shoe drags. Back-off until drag is relieved and tighten the lock nut.

PISTON STOPS

If the brake is equipped with piston stops, the wheel cylinder must be dismounted for reconditioning.

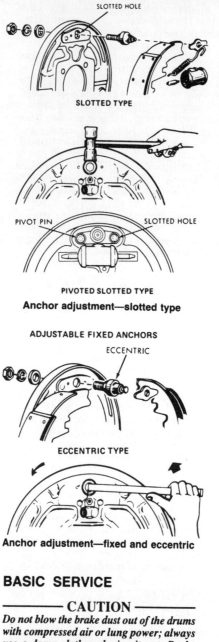

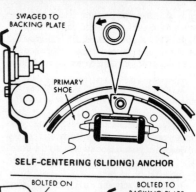

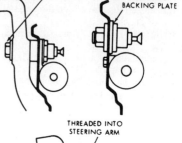

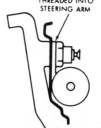

Different types of shoe anchors

BASIC SERVICE

CAUTION
Do not blow the brake dust out of the drums with compressed air or lung power; always use a damp cloth and wipe it out. Brake linings contain asbestos, a known cancer causing substance. Dispose of the cloth after use.

NOTE: Never work on a car supported only on a jack. Use a hydraulic lift or jack stands to support the vehicle while working.

Raising both front or rear wheels at once and supporting them on jack stands also allows comparison of the brake being serviced to the brake on the opposite side.

Check for Leaks

Press the brake pedal to ensure that there are no leaks in the hydraulic system. If the pedal does not remain hard, and drops to

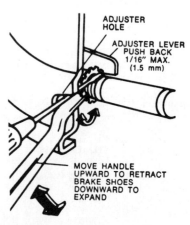

Some drums can be removed by backing off the self-adjuster

the floor, it is an indication of a leak in the master cylinder, hoses, wheel cylinders, or disc brake calipers. When performing this test, the engine should be running if the car is equipped with power brakes. With power brakes it is normal for the pedal to drop slightly when the engine starts. If it continues to drop, start looking for a leak.

Drum Removal

Safely support the car and release the park-

BRAKES
Import Vehicles

ing brake if working on the rear axle. Remove the lug nuts, the wheel/tire assembly and then pull off the drums. If the brake shoes have expanded too tightly against the drum, or have cut into the friction surface of the brake drum, the drums may be too tight for removal. In such a case, adjust the shoes inward before the brake drum is removed. On cars with self-adjusting mechanisms, reach through the adjusting slot with a very small prybar (or similar tool) and carefully push the self-adjusting lever away from the star wheel by a maximum of 1/16 in. (1.5mm). While holding the lever back, insert a brake adjusting tool into the slot and turn the star wheel in the proper direction until the brake drum can be removed. On cars with manual adjusting mechanisms, try lightly tapping the drum with a rubber mallet. If this does not work, simply reverse the manual adjustment procedures given later in this section until the drum can be removed.

Drum Inspection

Check the drums for any cracks, scores, grooves, or an out-of-round condition. Replace if cracked. Slight scores can be removed with fine emery cloth while extensive scoring requires turning the drum on a lathe.

If the friction surface of the brake drum appears scored or otherwise damaged beyond repair, it will require reconditioning. After machining, the drum diameter must not exceed the diameter cast on the drum or 0.060 in. (1.5mm) over the original nominal diameter. Carefully look for signs of grease or oil at the center of the assembly. If any leak is noticed, the seal should be replaced.

Rebuild the Cylinders

It is *always* a good idea to rebuild or replace the wheel cylinders when relining the brakes. This will help assure a properly operating brake system.

Remove Brake Shoes

It is convenient to disassemble one wheel at a time so the opposite side serves as a reference. Carefully note the colors and locations of different springs and parts. This is necessary to distinguish different springs that appear to be the same but have different tensions. If there are extra unused holes close to the ones in which the springs are located, use a dab of paint or other marking on the new shoes to identify the holes to be used. Replace any discolored springs and other parts found corroded or distorted. Use special tools whenever necessary. Examine the springs for signs of stretching or other defects and replace if their condition is at all questionable. Examine the flexible brake hoses and replace any that show signs of cracking or other damage.

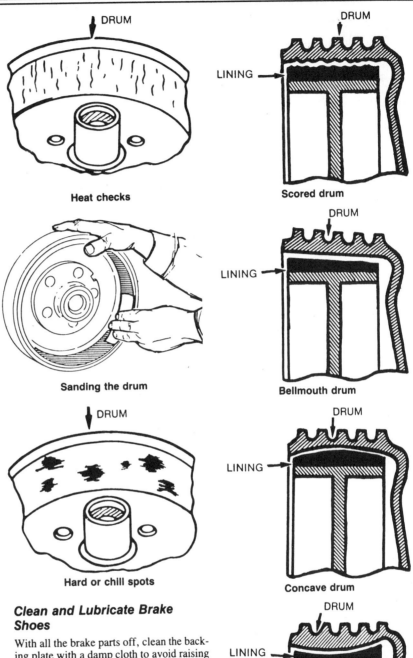

Clean and Lubricate Brake Shoes

With all the brake parts off, clean the backing plate with a damp cloth to avoid raising any asbestos dust, and dispose of the rag after use. Clean any rust with a wire brush. File smooth any ridges or rough edges on the contact points on the backing plate, and lubricate with approved brake lubricant. Clean and lightly lubricate the adjuster threads, and screw the adjuster all the way together to facilitate reassembly later on. Wash the wheel bearings with solvent and repack them with the proper grease. Check backing plate bolts to make sure they are tight.

Reassemble and Install Brake Shoes

Reassemble the brakes in the reverse order of disassembly. Make sure all parts are in their proper locations and that both brake shoes are properly positioned in either end of the adjuster. Also, both brake shoes should correctly engage the wheel cylinder push rods and parking brake links, and should be centered on the backing plate. Parking brake links and levers should be in place

BRAKES
Import Vehicles

on the rear brakes. With all the parts in place, try the fit of the brake drum over the new shoes. If not slightly snug, pull it off and turn the star wheel until a slight drag is felt when sliding the drum on. The use of a brake preset gauge will make this job easy. This makes final brake adjustment simpler. Then install the brake drum, wheel bearings, spindle nuts, cotter pins, dust caps, and wheel/tire assemblies, and make final brake adjustments as specified. Torque the spindle and lug nuts to specifications.

Bleed and Road-Test

Bleed the brakes to make sure of a high, hard brake pedal, and road-test the car. Most self-adjusting mechanisms are activated only during the rearward motion of a car. So, whenever servicing self-adjusting brakes, make sure that the road test includes enough stops, traveling in reverse, to allow the self-adjusters to perform the proper match-up of all wheels. Or, operate the parking brake several times if that activates the automatic adjuster.

CHILTON TIPS

- The primary brake shoe is the one toward the front of the car, and its lining is usually shorter than that on the secondary (rearward) shoe.
- Self-adjusting mechanisms are usually mounted on the secondary shoe.
- The star wheel part of an adjuster usually (but not always) goes toward the rear of the car.
- Different color springs belong in different locations.
- Self-adjusters and related parts are not interchangeable from one side of a car to the other since the direction of adjuster rotation varies from one side of a car to the other. Most adjusters on one side of a car have right hand threads, and the adjusters on the other side have left hand threads.
- Never press the brake pedal when one or more brake drums are off, or a wheel cylinder will pop apart.

DRUM BRAKE APPLICATION CHART

Make	Year/Model	Reference Number①	Adjuster Type
Audi	'80–'86 4000	12	Automatic
	'80–'84 5000		
BMW	'80–'83 320i	5	Manual (cam type)
Chrysler	'80–'84 Challenger/Sapporo, Arrow, Colt (rear wheel drive)	9	Automatic
	'80–'87 Colt/Champ (front wheel drive)	14	Automatic
Honda	'80–'87 Civic	11	Automatic
	'80–'81 Accord	8	Manual (bolt type)
	'82–'87 Accord	3	Semi-Automatic
	'80–'82 Prelude	11	Automatic
	'83 Prelude	3	Semi-Automatic
Hyundai	'86–'87 Excel	14	Automatic
Isuzu	'81–'87 All	9	Automatic
Mazda	'80–'83 RX-7	13	Automatic
	'80 GLC	1	Manual
	'81–'87 GLC, 323	15	Automatic
	'82 626	13	Automatic
	'83–'87 626	11	Automatic
Merkur	'85–'87 XR4Ti	2	Automatic
Mitsubishi	'83–'85 Cordia, Tredia	9	Automatic
Nissan/Datsun	'82–'87 Sentra, Pulsar	14	Automatic
	'80–'87 All Others	15	Automatic
Porsche	'80–'82 924	1	Manual
Renault	'81–'85 18i, Fuego, Sportwagon	9	Automatic
	'80–'83 Le Car	5	Manual (cam type)
	'82–'87 Alliance, Encore	3	Semi-Automatic
Subaru	'80–'84 All	8	Manual (bolt type)
Toyota	'80–'87 Celica/Supra	3	Semi-Automatic
	'80–'87 All Others	3, 4, 5, 6, 14②	Semi-Automatic, Manual (star wheel), Manual (bolt type), Automatic, Automatic
Volkswagen	'80–'87 Vanagon	1	Manual
	'80–'87 Rabbit, Scirocco, Jetta, Dasher, Quantum, Golf	10, 12②	Manual (star wheel), Automatic

BRAKES
Import Vehicles

DRUM BRAKE APPLICATION CHART

Make	Year/Model	Reference Number①	Adjuster Type
Yugo	'86–'87	5,6②	Manual (cam type), Automatic

①Due to the large number of different models covered in each section, slight differences in brake hardware and/or appearance may occur. The basic procedure as given, though, will apply to all models covered.

②When more than one reference number is given, compare the brake being serviced with those illustrated in each section until the correct one is found.

TYPE 1

Lockheed Non-Servo—Manual Adjuster

This brake consists of non-servo forward and reverse shoes with a double-end type wheel cylinder. The shoes anchor upon the slotted adjusting screws which permit them a sliding self centering action. Brakes are mounted with the cylinder at the top and the adjuster at the bottom.

REMOVAL & INSTALLATION

1. Raise the front/rear of the vehicle and support it on jackstands. Remove the tire and wheel.
2. Remove the drums (some vehicles may require special pullers).
3. Detach both retracting springs.
4. Remove the hold-down springs and lift the brake shoes from the backing plate. On the rear wheels, unhook the parking brake cable from parking brake lever before shoe removal.
5. Clean and lubricate the backing plate as detailed earlier.
6. Check the wheel cylinder for frozen pistons or fluid leaks. If any are found, rebuild or replace the cylinder. Disassemble the adjusters and clean and lubricate them.
7. Install the parking brake lever on a new reverse shoe (only on rear wheel brakes).
8. Place new brake shoes on the backing plate and attach the hold-down springs.

NOTE: Slots in the adjusting screws must be slanted toward the center of the assembly.

The ends of the shoes should engage the wheel cylinder piston slots, and the adjuster slots. If the adjuster screw ends have a slot with a bevel on one side. Make sure the bevel lines up with the bevel on the shoe web.

The end of the shoe with a slot for the parking strut should be installed near the wheel cylinder.

9. For rear wheels, hook the parking brake lever on the parking brake cable and then install the parking brake strut.
10. Install the heavier retracting spring between the toe or cylinder ends of the brake shoes.
11. Attach the lighter retracting spring to the heel or anchor ends of the shoes.
12. Replace the drums, bleed and adjust the assembly and road test the car.

ADJUSTMENT

Insert an adjusting spoon or a small screwdriver through the adjusting hole in the backing plate and expand the shoe assembly by revolving the notched adjusting wheel in a clockwise direction when facing the end of the wheel cylinder. Adjust the shoe until a heavy drag is felt when turning the wheel and drum; then, back off the adjustment until the wheel spins freely. Adjust one shoe at a time and repeat this procedure at all brake shoes.

TYPE 2

Fixed Anchor-Self-Adjusting

REMOVAL & INSTALLATION

1. Raise and support the rear of the vehicle on jackstands. Remove the wheel and brake drum.
2. Remove the holddown pins and springs by pushing down on and rotating the outer washer 90 degrees. It may be necessary to hold the back of the pin (behind the backing plate) while pressing down and turning the washer.
3. After the holddown pins and springs have been removed from both brake shoes, remove both shoes and the adjuster assembly by lifting up and away from the bottom anchor plate and shoe guide. Take care not to damage the wheel cylinder boots when removing the shoes from the wheel cylinder.
4. Remove the parking brake cable from the brake lever to allow the removal of the shoes and adjuster assembly.
5. Remove the lower shoe to shoe spring by rotating the leading brake shoe to release the spring tension. Do not pry the spring from the shoe.
6. Remove the adjuster strut from the trailing shoe by pulling the strut away from the shoe and twisting it downward toward

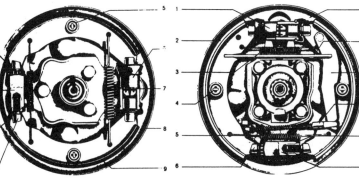

FRONT
1. Adjusting screw
2. Anchor block
3. Front return spring
4. Adjusting nut
5. Guide spring with cup and pin
6. Cylinder
7. Rear return spring

8. Back plate
9. Brake shoe with lining

REAR
1. Cylinder
2. Brake shoe with lining
3. Upper return spring
4. Spring with cup and pin

5. Lower return spring
6. Adjusting screw
7. Back plate
8. Connecting link
9. Lever
10. Brake cable
11. Adjusting nut
12. Anchor block

Typical Lockhead non-servo drum brake assembly (left—front brakes; right—rear brakes)

SECTION 1

BRAKES
Import Vehicles

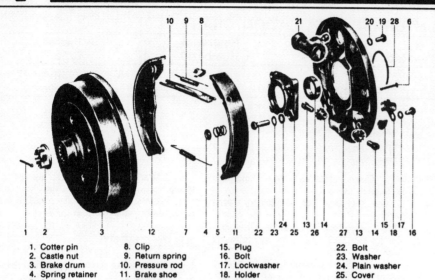

1. Cotter pin
2. Castle nut
3. Brake drum
4. Spring retainer
5. Spring
6. Pin
7. Return spring
8. Clip
9. Return spring
10. Pressure rod
11. Brake shoe
12. Brake lever
13. Adjusting screw
14. Adjusting nut
15. Plug
16. Bolt
17. Lockwasher
18. Holder
19. Bolt
20. Lockwasher
21. Wheel brake cylinder
22. Bolt
23. Washer
24. Plain washer
25. Cover
26. Spacer
27. Brake carrier
28. Seal

Exploded view of a Type 1 drum brake

19. Install the holddown pins, springs and washers.
20. Install the remaining components in the reverse order of removal.

ADJUSTMENT

1. Remove the brake drum.
2. Pivot the adjuster quadrant until the third or fourth notch from the outer end meshes with the knurled pin on the adjuster strut.
3. Reinstall the brake drum.

TYPE 3

Non–Servo—Semi–Automatic Adjuster

This brake consists of non–servo forward and reverse shoes with a double–ended type wheel cylinder. The brake shoes are adjusted automatically whenever the parking brake is applied.

yourself until the spring tension is released. Remove the spring from the slot.

7. Remove the parking brake lever from the shoe by disconnecting the horseshoe clip and spring washer and pulling the lever from the shoe.
8. If for any reason the adjuster assembly must be taken apart, do the following: pull the adjuster quadrant (U-shaped lever) away from the knurled pin on the adjuster strut by rotating the quadrant in either direction until the teeth are no longer engaged with the pin. Remove the spring and slide the quadrant out of the slot on the end of the adjuster strut. Do not put too much stress on the spring during disassembly.
9. Clean the brake backing (mounting) plate with a soft paint brush or vacuum cleaner.

— **CAUTION** —
Never inhale the dust from the brake linings. Asbestos dust when inhaled can be injurious to your health. Use a vacuum cleaner. Do not blow off the dust with air pressure.

10. Apply a thin film of high temperature grease at the points on the backing plate where the brake shoes make contact.
11. Apply a thin film of multi-purpose grease to the adjuster strut at the point between the quadrant and strut.
12. If the adjuster has been disassembled; install the quadrant mounting pin into the slot on the adjuster strut and install the adjuster spring.
13. Assemble the parking brake lever to the trailing shoe. Install the spring washer and a new horseshoe clip, squeeze the clip with pliers until the lever is secured on the shoe.
14. Install the adjuster strut attaching spring on to the trailing shoe. Attach the

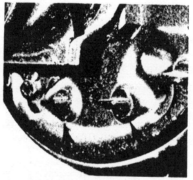

Rear drum brake shoe adjusters are found behind the two lower rubber plugs on the backing plate. The two higher plugs (if so equipped) are removed to check brake shoe wear

adjusting strut by fastening the spring in the slot and pivoting the strut into position. This will tension the spring. Make sure the end of the spring where the hook is parallel to the center line of the spring coils is hooked into the web of the brake shoe. The installed spring should be flat against the web and parallel to the adjuster strut.

15. Install the shoe to shoe spring with the longest hook attached to the trailing shoe.
16. Install the leading shoe to adjuster strut spring by installing the spring to both parts and pivoting the leading shoe over the quadrant and into position, this will tension the spring.
17. Place the shoes and adjuster assembly onto the backing plate. Spread the shoes slightly and position them into the wheel cylinder piston inserts and anchor plate. Take care not to damage the wheel cylinder boots.
18. Attach the parking brake cable to the parking brake lever.

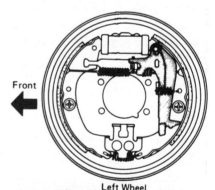

Typical Type 3 drum brake assembly

REMOVAL & INSTALLATION

1. Raise the rear of the vehicle and support it with jackstands. Remove the tire and wheel.
2. Remove the brake drum. Tap the drum lightly with a mallet in order to free it. If the drum cannot be removed easily, insert a screwdriver into the hole in the backing plate and hold the automatic adjusting lever away from the adjusting bolt. Using another screwdriver, relieve the brake shoe tension by turning the adjusting bolt clockwise. If the drum still will not come off, use a puller; but first make sure that the parking brake is released.

— **CAUTION** —
Do not depress the brake pedal once the brake drum has been removed.

3. Unhook the shoe tension springs from

BRAKES
Import Vehicles

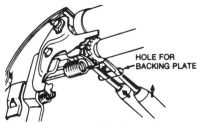

Backing off the brake shoes to remove the brake drum—Type 3

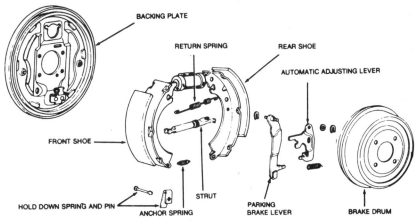

Exploded view of a Type 3 drum brake

the shoes with the aid of a brake spring removing tool.

4. Remove the brake shoe securing springs.

5. Disconnect the parking brake cable at the parking brake shoe lever.

6. Withdraw the shoes, complete with the parking brake shoe lever.

7. Unfasten the C-clip and remove the adjuster assembly from the shoes.

8. Inspect the shoes for wear and scoring.

9. Check the wheel cylinder for frozen pistons or fluid leaks. If any are found, rebuild or replace the wheel cylinder.

10. Clean and inspect all parts. Lubricate the backing plate bosses and anchor plate.

11. Check the tension springs to see if they are weak, distorted or rusted.

12. Inspect the teeth on the automatic adjuster wheel for chipping or other damage.

Installation is performed in the following order:

NOTE: Grease the point of the shoe which slides against the backing plate. Do not get grease on the linings.

1. Attach the parking brake shoe lever and the automatic adjuster lever to the rear side of the shoe.

2. Fasten the parking brake cable to the lever on the brake shoe.

3. Install the automatic adjuster and fit the tension spring on the adjuster lever.

4. Install the securing spring on the *rear* shoe and then install the securing spring on the *front* shoe.

NOTE: The tension spring should be installed on the anchor, before performing Step 4.

5. Hook one end of the tension spring over the rear shoe with the tool used during removal; hook the other end over the front shoe.

——— **CAUTION** ———
Be sure that the wheel cylinder boots are not being pinched in the ends of the shoes.

6. Test the automatic adjuster by operating the parking brake shoe lever.

7. Install the drum and adjust the brakes.

ADJUSTMENT

These brakes are equipped with automatic adjusters actuated by the parking brake mechanism. No periodic adjustment of the drum brakes is necessary if this mechanism is working properly. If the brake shoe to drum clearance is incorrect, and applying and releasing the parking brake a few times does not adjust it properly, the parts will have to be disassembled for repair.

TYPE 4

Non–Servo—Manual Adjuster

This brake is a non–servo unit with two single-piston wheel cylinders that act upon an individual shoe.

REMOVAL & INSTALLATION

1. Raise the rear of the vehicle and support it with jackstands. Remove the tire and wheel.

2. Remove the brake drum as previously detailed.

3. Remove the retracting springs and hold down spring clips.

4. Lift the shoes off of the backing plate.

5. Check the wheel cylinders for frozen pistons or leaks. If any are found, the wheel cylinder must be rebuilt or replaced.

6. Clean the backing plate and lubricate the bosses and bearing ends of the new shoes.

7. Place the new shoes in the slots of the wheel cylinders and the adjusters. Install the hold down spring clips.

8. Install the retracting springs. The blue, lighter weight spring is installed to the piston side of the shoe; the black, heav-

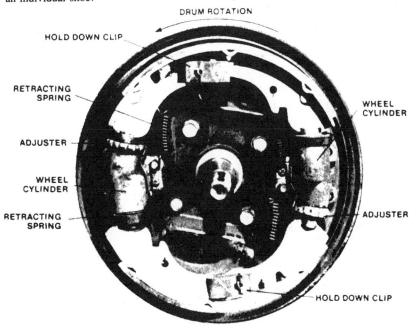

Typical Type 4 drum brake assembly

1-95

BRAKES
Import Vehicles

ier weight spring to the adjuster side of shoe.

9. The shoes must slide freely in the slots.

10. Replace the drum and wheel; bleed and adjust the brake.

11. Road test the car.

ADJUSTMENT

Adjust each shoe individually by rotating the notched adjuster (backing plate side) until a heavy drag is felt when turning the wheel and drum in a forward direction. Back off the adjuster until the wheel spins freely.

TYPE 5

Bendix Non–Servo, Self-Centering— Manual Adjuster

This brake consists of non–servo forward and reverse shoes with a double–ended type wheel cylinder. The brake shoes are self-centering by means of an anchor block and two adjusting cams.

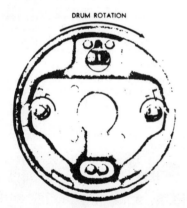

Adjustment of the Type 5 brake shoes is made by means of adjusting cams found on the backing plate

REMOVAL & INSTALLATION

1. Raise the rear of the vehicle and support it with jackstands. Remove the tire and wheel.

2. Remove the brake drums as previously detailed.

3. Remove the retracting spring (and hold-down springs, if used) and lift the shoe assembly from the backing plate. On the rear wheels, unhook the parking brake cable from the parking brake lever.

4. Separate the shoes by removing the connecting spring.

5. Check the wheel cylinder for frozen pistons or leaks. If any are found, rebuild or replace the cylinder.

6. Clean and lubricate the backing plate bosses and hold down clips.

7. Install the parking brake lever on the new reverse shoe.

8. Attach the connecting spring to the heel ends of the shoes.

9. Mount the shoes on the backing plate. For the rear wheels, hook the parking brake lever on the parking brake cable and install the parking brake strut.

10. Attach the retracting spring to the toe ends of the shoes.

11. Replace the drums. Center the shoes in the drums by depressing the pedal.

12. Bleed and adjust the brakes.

13. Road test the car.

ADJUSTMENT

Because of the self-centering feature of this brake, only minor adjustments to compensate for lining wear are necessary.

1. Jack up the car and support it with jackstands.

2. See that the parking brake lever is in the fully released position. Check the rear brake shoes to make certain they have not been moved away from the adjusting cams (partially applied) by improper adjustment of the cable (or cables) or by a sticking cable. If the shoes are not resting against the adjusting cams, back off or disconnect the parking brake cable.

3. Expand the brake shoes by turning the adjusting cams. If the adjusting cams have lock nuts, loosen the lock nut. Spin the wheel, while turning the adjusting cam in the proper direction, until a heavy drag is reached and then back the cam off gradually in the opposite direction until the wheel spins freely. When adjusting the forward shoe, spin the wheel in the forward direction. When adjusting the reverse shoe, spin the wheel in the reverse direction.

4. Apply the brake pedal firmly a few times and check all wheels to be sure that they spin freely. If a brake drag is noticed at this point, readjust the brake in accordance with Step 3.

TYPE 6

Bendix Non–Servo— Automatic Adjuster

This brake consists of non-servo forward and reverse shoes with a double–ended type wheel cylinder. The shoes anchor to the slotted anchor plate permitting them a sliding self-centering action.

REMOVAL & INSTALLATION

1. Raise the rear of the vehicle and support it with jackstands, Remove the tire and wheel.

2. Before removing the brake drum, remove the rubber grommet from the adjustment release hole in the backing plate. Insert a small prybar and push down on the adjustment latch. This will allow the shoes to retract and will eliminate possible interference between the brake shoes and brake drum.

3. Remove the brake drum and disengage the automatic adjuster spring from the strut and the reverse shoe. Remove the anti-noise spring.

4. Remove the upper shoe-to-shoe spring.

5. Remove the strut and disengage the shoes from the hold down springs. Check the springs for cracks or fatigue.

6. Unhook the lower shoe-to-shoe spring. Remove the spring and forward shoe assembly.

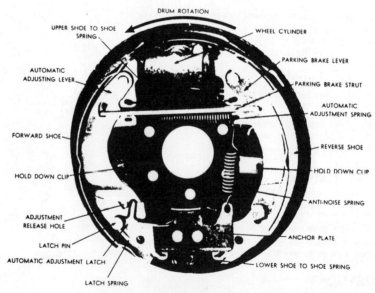

Typical Type 6 drum brake assembly

BRAKES
Import Vehicles

7. Disconnect the parking brake cable and remove the parking brake lever from the reverse shoe.

8. Remove the adjusting latch and the automatic adjusting lever from the forward shoe. Check for worn or damaged teeth.

To install:

1. Clean and inspect all parts. Replace if necessary.

2. Lubricate the shoe guide pads on the backing plate and the curved edges of the anchor plate.

3. Lubricate the latch pin (long pin) and attach the latch to the outer surface of the forward shoe web.

4. Lubricate the adjustment lever pin (short pin) and attach the adjustment lever to the web of the forward shoe.

5. Lubricate the parking brake lever pin and attach the parking brake lever to the web of the reverse shoe. The reverse shoe has a short lining.

6. Attach the parking brake cable.

7. Attach the lower shoe-to-shoe spring to the forward and reverse shoes.

8. Place the forward and reverse shoes in position on the backing plate and install the parking brake strut. Engage the tab on the strut in the slot of the adjustment lever and place the upper ends of the shoe webs against the wheel cylinder pistons.

9. Install the upper shoe-to-shoe spring.

10. Install the automatic adjustment spring. The spring must hold the strut tight against the parking brake lever.

11. Lubricate the inner surfaces of the shoe hold down springs and install the springs over the shoe webs.

12. Install the latch spring over the latch pivot pin. With the adjustment lever against the shoe rim, hook the spring over the latch.

13. Attach the anti-noise spring between the anchor plate and reverse shoe.

14. Install the drum and wheel. Bleed the system if necessary and then road test the vehicle.

ADJUSTMENT

An initial adjustment can be made after the brake drum has been removed and replaced by applying the handbrake several times. Aside from this initial adjustment, the brakes are self-adjusting and no further adjustment should ever be required.

TYPE 7

Non–Servo—Manual Adjuster

This brake consists of non-servo forward and reverse shoes with a double-ended wheel cylinder. The shoes are held in position by an adjuster/anchor plate and are manually adjusted.

REMOVAL & INSTALLATION

1. Raise the rear of the vehicle and support it with jackstands. Remove the tire and wheel assembly.

2. Remove the brake drum. If it is necessary to retract the shoes in order to remove a worn drum, remove the dust cover and back off on the adjusting bolt located on the inboard side of the backing plate.

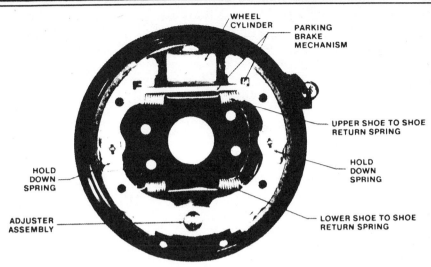

Typical Type 7 drum brake assembly

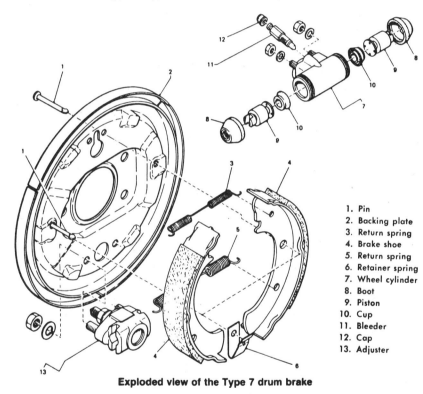

Exploded view of the Type 7 drum brake

1. Pin
2. Backing plate
3. Return spring
4. Brake shoe
5. Return spring
6. Retainer spring
7. Wheel cylinder
8. Boot
9. Piston
10. Cup
11. Bleeder
12. Cap
13. Adjuster

3. Remove the upper and lower shoe-to-shoe return springs.

4. Remove the brake shoe hold-down springs and lift the shoes from the backing plate.

5. Check the wheel cylinder for frozen pistons or fluid leaks. If any are found, rebuild or replace the wheel cylinder.

6. Clean and inspect all parts. Lubricate the backing plate bosses and the adjuster assembly.

7. Apply a thin coat of grease to the adjuster.

8. Back off all adjustment on the ad-

1-97

BRAKES
Import Vehicles

justing bolt located behind the backing plate.

9. Mount the new shoes to the backing plate and install the brake shoe hold-down springs. Be sure that the webs of the shoes are properly engaged in the parking brake mechanism, adjuster assembly and wheel cylinder.

10. Install the upper and lower shoe-to-shoe return springs.

NOTE: The upper and lower shoe-to-shoe return springs are not interchangeable and must be installed in the proper direction. Use the other wheel for reference if necessary when installing these springs.

11. Install the drum and then install the wheel and tire assembly.

NOTE: The drum may not fit if the adjustment bolt has not been backed off sufficiently or if the shoes are not centered properly on the backing plate.

12. Bleed the system if necessary and then road test the vehicle.

ADJUSTMENT

1. Block the front wheels, release the parking brake, raise the rear of the vehicle and support it with jackstands.

2. Depress the brake pedal several times and then release it.

3. Locate the adjuster on the inboard side of the brake backing plate and turn it clockwise until the wheel will no longer spin.

4. Back off the adjuster (counterclockwise) approximately two (2) turns or until the wheel just begins to turn freely.

TYPE 8

Non–Servo–Automatic Adjuster

This brake consists of non–servo forward and reverse shoes with a double–ended wheel cylinder. The shoes are held in position by an anchor plate and are automatically adjusted.

REMOVAL & INSTALLATION

1. Raise the rear of the vehicle and support it with jackstands. Remove the tire and wheel assembly.

2. Remove the drum. If it is necessary to retract the shoes in order to remove a worn drum, insert a small prybar through the hole in the backing plate and press down on the adjusting latch.

3. Remove the brake shoe hold-down springs.

4. Remove the upper strut-to-shoe spring and the upper shoe-to-shoe return spring.

5. Remove the reverse shoe along with the lower shoe-to-shoe return spring.

6. Holding the adjusting latch on the forward shoe downward, pull the adjusting lever toward the center of the brake and then remove the leading shoe assembly.

7. Check the wheel cylinder for frozen pistons or leaks. If any are found, rebuild or replace the cylinder.

8. Inspect old springs. If old springs are damaged or have been overheated, they should be replaced. Indications of overheated springs are paint discoloration and distortion.

9. Inspect the drum and recondition or replace if necessary.

10. Check the adjusting lever and adjusting latch for wear and damage. Replace damaged parts.

11. Clean and lubricate the backing plate bosses.

12. Remove the adjusting lever and latch from the forward shoe and install them on a new forward shoe.

13. Before mounting the forward shoe, rotate the adjusting lever outward away from the rim of the new shoe. Engage the adjusting lever with the parking brake strut. Rotate the adjusting lever inward until it touches the rim of the shoe and place the shoe against the backing plate.

14. Install the brake shoe hold-down spring to the forward shoe. Be sure that the web of the shoe is engaged with the slot in the wheel cylinder piston.

15. Hook one end of the lower shoe-to-shoe spring to the forward shoe and the other end to a new reverse shoe. Place the lower part of the reverse shoe on the anchor. Using the lower anchor as a pivot, rotate the reverse shoe upward toward the wheel cylinder and secure it to the backing plate with the hold-down spring.

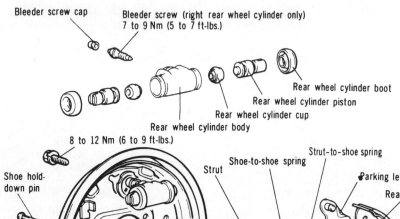

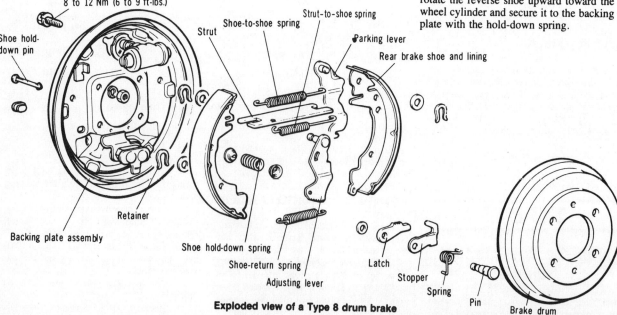

Exploded view of a Type 8 drum brake

BRAKES
Import Vehicles

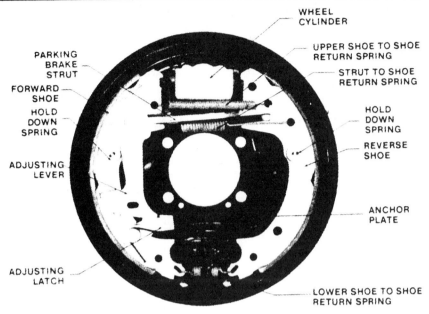

Typical Type 8 drum brake assembly

upper return spring. Disengage the shoe and remove the spring. To facilitate the reassembly operation, note how the upper return spring is positioned on the shoe and how it is connected to the hole in the anchor plate.

9. Remove the forward shoe in the same manner as above.

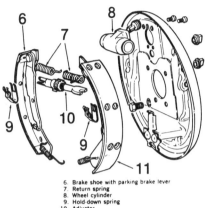

6. Brake shoe with parking brake lever
7. Return spring
8. Wheel cylinder
9. Hold-down spring
10. Adjuster
11. Brake shoe

Exploded view of a Type 9 drum brake

16. Install the upper shoe-to-shoe spring and the upper strut-to-shoe spring. Be sure that the web of the shoes is properly engaged with the slots in the wheel cylinder piston and parking brake lever.

17. Install the drum and wheel.

NOTE: The drum may not fit if the shoes are partially adjusted outward or if they are not centered properly on the backing plate.

18. Apply the brake pedal a few times to bring the brake shoes into adjustment.

19. Bleed the system and road test the vehicle.

ADJUSTMENT

This brake is automatically adjusted; no adjustment is either necessary or possible.

TYPE 9

Non–Servo—Manual Adjuster

This brake consists of non–servo forward and reverse shoes with a double–ended wheel cylinder. The shoes are held in position by an anchor plate and are manually adjusted.

REMOVAL & INSTALLATION

1. Raise the rear of the vehicle and support it with jackstands. Remove the tire and wheel assembly.
2. Remove the plug from the brake adjusting hole. Using a small prybar or other suitable tool, release the brake shoes by rotating the shoe adjuster downward on the right side of the vehicle and upward on the left side of the vehicle.
3. Remove the brake drum.
4. Remove the parking brake cable from the parking brake lever by compressing the cable return spring.
5. Remove the shoe-to-anchor springs located at the bottom.
6. Remove the brake shoe hold-down clips and pins.
7. Remove the adjuster screw assembly by spreading the shoes apart making sure that the adjuster screw is fully backed off.
8. Pull the reverse shoe away from the anchor plate to release the tension on the

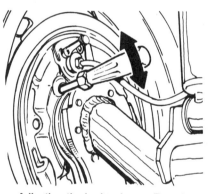

Adjusting the brake shoes—Type 9

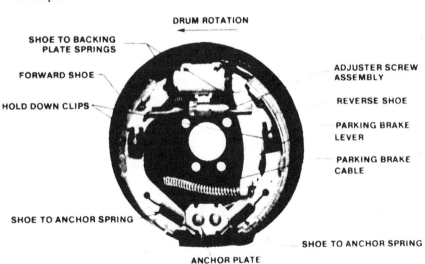

Typical Type 9 drum brake assembly

SECTION 1

BRAKES
Import Vehicles

10. Inspect the wheel cylinder and recondition or replace if necessary.

11. Clean and inspect the adjuster screw assembly. Apply a thin coat of lubricant to the adjuster threads.

12. Inspect the old springs. If old springs are damaged or have been overheated, they should be replaced. Indications of overheated springs are paint discoloration or distortion.

13. Lubricate the bosses on the anchor plate which make contact with the brake shoe tabs.

To install:

1. Remove the parking brake lever and attach the parking brake lever to the web of a new reverse shoe.

2. Position the upper return spring on the forward shoe and hook the other end of the spring into the hole in the backing plate.

3. Rotate the shoe outward with the upper part of the shoe against the wheel cylinder piston and insert the bottom part of the shoe under the anchor plate.

4. Repeat the above procedure for the reverse shoe.

5. With the adjuster screw fully retracted, position the straight forked end of the adjuster screw assembly on the parking brake lever. Make sure that the spring lock on the adjuster screw is on the outside and away from the adjusting hole.

6. Rotate the bottom of the forward shoe off the anchor plate and insert the curved fork end of the adjuster screw assembly into the web of the forward shoe.

NOTE: **Make sure that the curved portion of the forked end is facing downward and that the spring lock is on the outside and away from the adjusting hole.**

7. Insert the pins for the hold-down clips through the backing plate and web of the shoes. Install the hold-down clips.

8. Install the shoe-to-anchor springs.

9. Compress the brake cable return spring and attach the cable to the bottom of the parking brake lever.

10. Install the brake drum.

11. Install the wheel and tire assembly.

12. Adjust the brakes.

13. Bleed the system and road test the car.

ADJUSTMENT

Adjust the brakes through the adjusting hole located in the backing plate. Adjustment is made manually by spreading the adjuster screw assembly which is located directly under the wheel cylinder. Insert a small prybar or other suitable tool through the hole in the backing plate and rotate the adjuster wheel clockwise until the brakes drag as you turn the wheel in a forward direction. Turn the adjuster in the opposite direction until you just pass the point of drag. Repeat the procedure on the other wheel.

TYPE 10

Non-Servo-Automatic Adjuster

These brakes are a leading-trailing shoe design with a ratchet type self-adjusting mechanism. The shoes are held against the anchors at the top by a shoe to shoe spring. At the bottom the shoe webs are held against the wheel cylinder piston ends by a return spring.

The self-adjusting mechanism consists of a spacer strut and a pair of toothed ratchets attached to the primary brake shoe. The parking brake actuating lever is pivoted on the spacer strut.

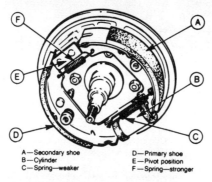

A—Secondary shoe
B—Cylinder
C—Spring—weaker
D—Primary shoe
E—Pivot position
F—Spring—stronger

Typical Type 10 drum brake assembly

The self-adjusting mechanism automatically senses the correct lining to drum clearance. As the linings wear, the clearance is adjusted by increasing the effective length of the spacer strut. This strut has projections to engage the inner edge of the secondary shoe via the handbrake lever and the inner edge of the large ratchet on the primary shoe. As wear on the linings increases, the movement of the shoes to bring them in contact with the drums becomes greater than the gap. The spacer strut, bearing on the shoe web, is moved together with the secondary shoe to close the gap. Further movement causes the large ratchet, behind the primary shoe, to rotate inwards against the spring-loaded small ratchet, and the serrations on the mating edges maintain this new setting until further wear on the shoes results in another readjustment. On releasing brake pedal pressure, the return springs cause the shoes to move into contact with the shoulders of the spacer strut/handbrake actuating lever, thus restoring the clearance between the linings and the drum proportionate to the gap shown.

REMOVAL & INSTALLATION

1. Raise the rear of the vehicle and support it with jackstands. Remove the wheel and tire assembly.

2. Disconnect the brake cable from the operating lever at the back of the backing plate. Remove the dust boot from the operating lever.

3. Remove the brake drum. If it is necessary to retract the shoes in order to remove a worn drum, insert a small prybar through the hole in the backing plate and lift the small ratchet lever on the adjuster assembly.

4. Remove the hold-down spring from the reverse shoe by depressing and rotating the washer. Remove the washer, spring and hold-down pin.

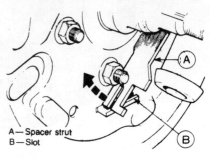

A—Spacer strut
B—Slot

Removing the spacer strut from the carrier plate

5. Twist the reverse shoe outwards and upwards away from the backing plate, taking care not to damage the wheel cylinder dust boot.

6. Unfasten the upper and lower shoe-to-shoe return springs and remove the shoe and springs.

7. Remove the hold-down spring from the forward shoe.

8. Lift the forward shoe along with the parking brake lever and spacer strut assembly away from the anchor plate. The operating lever should slide out of the hole in the backing plate.

NOTE: **To facilitate the reassembly operation later on, note how the parking brake lever is attached to the forward shoe.**

9. Disengage the parking brake lever and spacer strut assembly from the forward shoe by twisting inward to remove the tension on the spring.

10. Disassemble the ratchet assembly from the reverse shoe by first removing the retaining washers. Note how the ratchet levers are assembled so that you can put them back on a new shoe in the same way.

11. Rotate the large ratchet lever outward from under the tension spring and remove the ratchet lever from the shoe.

12. Remove the pressure spring and the small ratchet lever.

13. Check the wheel cylinder for frozen pistons or leaks. If any are found, rebuild or replace the wheel cylinder.

14. Inspect the old springs. If springs are damaged or have been overheated, they should be replaced. Indications of overheated springs are paint discoloration and distortion.

BRAKES
Import Vehicles

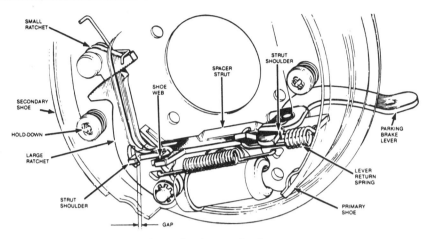

Self-adjusting mechanism

15. Inspect the drum and recondition or replace if necessary.
16. Check the adjusting lever and adjusting latch for wear and damage. Replace damaged parts.
17. Clean and lubricate the backing plate bosses.
18. Install the large ratchet on a new reverse shoe and secure it with a new retaining washer.
19. Install the small ratchet and pressure spring on the pivot of the new reverse shoe and secure with a new retaining washer. Be sure that the ratchet rotates and returns freely with spring pressure.
20. Pull back the small ratchet and rotate the large ratchet inward toward the rim of the shoe. Release the small ratchet and slowly rotate the large ratchet lever outward until the hole in the brake shoe web for the hold-down spring becomes completely exposed.
21. Attach the parking brake lever and spacer strut assembly to a new forward shoe. Hook the short side of the lever and strut tension spring to the slotted hole in the shoe. Hook the long end of the spring to the spacer strut part of the brake lever assembly and rotate the lever assembly until it is attached to the new shoe.
22. Place the forward shoe on the backing plate by inserting the operating lever through the hole in the backing plate. Insert the forked end of the spacer strut into the slotted carrier plate. Rest the upper part of the shoe against the anchor plate and the lower part against the wheel cylinder piston.
23. Insert the hold-down pin through the backing plate and web of the shoe. Install the hold-down spring and washer.
24. Before mounting the reverse shoe, note the slot in the long ratchet lever. The spacer strut must be engaged in the slotted hole in the long ratchet lever when the shoe is mounted to the backing plate.
25. Hook the stronger (thicker) shoe-to-shoe spring through the hole at the top of the already installed forward shoe.
26. Hook the reverse shoe to the other end of the shoe spring. Place the upper part of the shoe against the anchor plate and

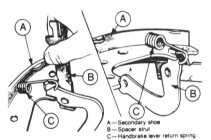

Assembling the secondary shoe components

using the anchor plate as a pivot, rotate the bottom part of the shoe outward. Position the lower part of the shoe against the wheel cylinder piston making sure that the spacer strut is engaged in the hole in the ratchet lever.
27. Insert the hold-down pin through the backing plate. Install the hold-down spring and washer.
28. Install the lower shoe-to-shoe spring (the weaker of the two springs) using a pair of pliers or other suitable tool.

29. Before installing the brake drum, lift the small ratchet lever upwards against the spring. This will allow the long ratchet lever to rotate inward toward the rim of the shoe and provide the clearance needed to install the drum.
30. Replace the dust boot over the operating lever behind the backing plate.
31. Reconnect the brake cable to the operating lever.
32. Install the brake drum and wheel.

NOTE: The drum may not fit if the shoes are partially adjusted outward or if they are not centered properly on the backing plate.

33. Apply the brake pedal a few times to bring the brake shoe into adjustment.
34. Bleed the system and road test the vehicle.

TYPE 11

Non–Servo—Automatic Adjuster

This brake consists of non–servo forward and reverse shoes with a double–ended wheel cylinder. The shoes are mounted by means of hold-down springs and an anchor plate. Adjustment is performed automatically.

REMOVAL & INSTALLATION

1. Raise the rear of the vehicle and support it with jackstands. Remove the tire and wheel assembly.
2. Remove the brake drum.

NOTE: If it is necessary to retract the shoes in order to remove a worn drum, insert a small prybar through one of the

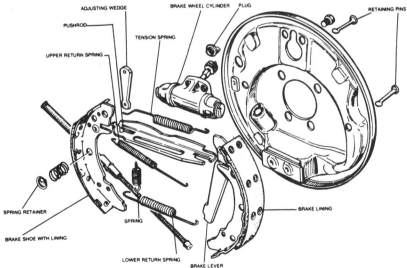

Exploded view of the Type 11 drum brake assembly

BRAKES
Import Vehicles

stud holes in the brake drum. Retract the adjuster wedge upward by pressing down on the prybar.

3. Remove the adjusting wedge spring.
4. Remove the upper and lower return springs.
5. Remove the hold-down springs.
6. Lift the shoes from the backing plate and disconnect the parking brake cable from the lever.
7. Disconnect the rear shoe from the push bar.
8. Clamp the push bar in a vise and remove the tensioning spring and adjusting wedge.
9. Check the wheel cylinder for frozen pistons or leaks. If any are found, rebuild or replace the wheel cylinder.
10. Inspect old springs. If old springs are damaged or have been overheated, they should be replaced. Indications of overheated springs are paint discoloration and distortion.
11. Inspect the brake drum and recondition or replace if necessary.
12. Clean and lubricate all contact points on the backing plate.
13. Attach the push bar and tensioning spring to the new front shoe.
14. Insert the adjusting wedge so that its lug is pointing toward the backing plate.
15. Remove the parking brake lever from the old rear shoe and attach it onto the new rear brake shoe.
16. Install the push bar onto the rear brake shoe and parking brake lever assembly.
17. Connect the parking brake cable to the lever and place the whole assembly onto the backing plate.
18. Install the hold-down springs.
19. Install the upper and lower return springs.
20. Install the adjusting wedge spring.
21. Center the brake shoes on the backing plate making sure the adjusting wedge is fully released before installing the drum.
22. Install the drum and wheel assembly and torque the wheel lugs to the manufacturer's specifications.
23. Apply the brake pedal several times to bring the brake shoes into adjustment.
24. Bleed the system and road test the vehicle.

ADJUSTMENT

This brake adjusts itself automatically; aside from the initial adjustments given in the previous section, no adjustments are either necessary or possible.

TYPE 12

Non–Servo—Manual Adjuster

This brake consists of non–servo forward

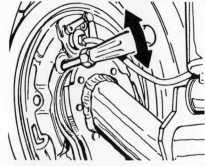

Adjusting the brake shoes—Type 12

Press the proportioning valve lever in the direction of the rear axle (Audi shown)

and reverse shoes with a double–ended wheel cylinder. The brake shoes are retained in position by an anchor plate and retaining springs.

REMOVAL & INSTALLATION

1. Raise the rear of the car and support it with jackstands.
2. Remove the wheels and then remove the brake drum.

NOTE: If the drum does not come off easily, the brakes will have to be backed off. First, push the lever on the proportioning valve toward the rear axle to relieve the residual brake pressure. Next remove the plug on the backing plate and turn the adjusting wheel to back off the brake shoes.

3. Use brake pliers to remove the upper and lower return springs.
4. Turn and remove the washers to release the brake shoe retaining springs.
5. Disconnect the parking brake cable by pressing the spring toward the front of the car and unhooking the cable from the brake lever.
6. Lift out the brake shoes. Make sure you take note of how the adjuster mechanism fits into the brake shoe web.
7. Check the wheel cylinder for frozen pistons or leaks. If any are found, rebuild or replace the wheel cylinder.
8. Inspect old springs. If old springs are damaged or have been overheated, they should be replaced. Indications of overheated springs are paint discoloration and distortion.
9. Inspect the brake drum and recondition or replace if necessary.
10. Clean and lubricate all contact points on the backing plate.
11. Fit the adjuster mechanism into the brake shoe web and then position the shoes onto the brake backing plate.
12. Push the retaining pin springs in and turn the washers onto the retaining pins.
13. Squeeze the spring and hook the parking brake cable to the parking brake lever.
14. Install the upper and lower retaining springs.
15. Install the brake drums and wheels and then lower the vehicle.
16. Bleed (if necessary) and adjust the brakes. Road test the vehicle.

ADJUSTMENT

1. Press the lever of the proportioning valve in the direction of the rear axle to relieve the residual brake pressure.
2. Remove the rubber plug from the brake backing plate.
3. Insert a small prybar into the hole and turn the adjusting wheel until the brake linings are just touching the drum.
4. Back off the adjusting wheel 6–8 teeth and replace the rubber plug.

TYPE 13

Non–Servo— Automatic Adjuster

This brake consists of non–servo forward and reverse shoes with a double–ended type wheel cylinder. The brake shoes are held in position by an anchor plate and two hold down springs. The brakes are self–adjusting.

REMOVAL & INSTALLATION

1. Raise the rear of the vehicle and support it with jackstands. Remove the wheel and tire assembly.
2. Remove the brake drums.
3. Remove the lower pressed metal spring clip, the shoe return spring (the large one piece spring between the two shoes), and the two shoe hold-down springs.
4. Remove the shoes and adjuster as an assembly, Disconnect the parking brake cable from the lever, remove the spring between the shoes and the lever from the rear (trailing) shoe. Disconnect the adjuster retaining spring and remove the adjuster, turn the star wheel in to the adjuster body after cleaning and lubricating the threads.
5. The wheel cylinder may be removed for service or replacement, if necessary.
6. Clean the backing plate with a wire brush. Install the wheel cylinder if it was

BRAKES
Import Vehicles

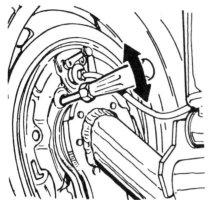

Adjusting the brake shoes—Type 13

removed. Lubricate all contact points on the backing plate, anchor plate, wheel cylinder-to-shoe contact and parking brake strut joints and contacts. Installation of the brake shoes, from this point, is the reverse of removal after the lever has been transferred to the new rear (trailing) shoe.

7. Pre-adjustment of the brake shoe can be made by turning the adjuster star wheel out until the drum will just slide on over the brake shoes. Before installing the drum, make sure the parking brake is not adjusted too tightly, if it is, loosen, or the adjustment of the rear brakes will not be correct.

8. If the wheel cylinders were serviced, bleed the brake system. The brake shoes are then adjusted by pumping the brake pedal and applying and releasing the parking brake. Adjust the parking brake stroke. Road test the car.

ADJUSTMENT

The brakes are self-adjusting. Aside from the initial adjustments given in the "Removal and Installation" section, no adjustments are either necessary or possible.

TYPE 14

Non-Servo—Automatic Adjuster

This brake consists of non-servo forward and reverse shoes with a double-ended type wheel cylinder. The shoes are held in position by an anchor plate and anti-rattle springs. The wheel cylinder is located at the top or bottom of the brake backing plate depending upon the particular application. These brakes are self-adjusting.

REMOVAL & INSTALLATION

1. Raise the rear of the vehicle and sup-

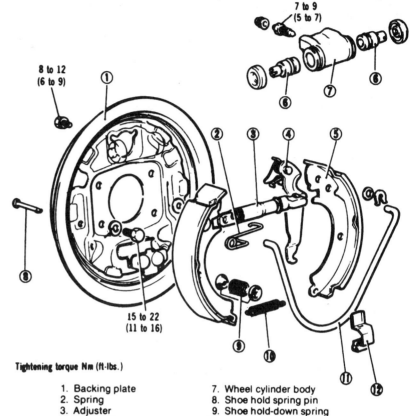

Tightening torque Nm (ft-lbs.)

1. Backing plate
2. Spring
3. Adjuster
4. Parking lever
5. Shoe and lining assembly
6. Piston
7. Wheel cylinder body
8. Shoe hold spring pin
9. Shoe hold-down spring
10. Shoe to shoe spring
11. Shoe return spring
12. Clip spring

Exploded view of the Type 14 drum brake

port it with jackstands. Remove the tire and wheel assembly.

2. Engage the parking brake. Pull the pin out and then remove the stopper from the toggle lever. Release the parking brake.

3. Remove the brake drum.

NOTE: If the brake drum cannot be easily removed, install two (2) bolts (8 mm) in the holes and drive it out.

4. Remove the return springs.
5. Push the anti-rattle spring retainers in and turn them so they can be removed from the pins.
6. Remove the brake shoes.
7. Clean the brake backing plate and check the wheel cylinder for leaks or other damage; replace as necessary.
8. Hook the return springs into the new shoes. The springs should be between the shoes and the backing plate. The longer return spring must be adjacent to the wheel cylinder. A very thin film of grease may be applied to the pivot points at the ends of the brake shoes. Grease the shoe locating buttons on the backing plate, also. Be careful not to get grease on the linings or drums.
9. Place one shoe in the adjuster and piston slots, and pry the other shoe into position.

10. Press and turn the anti-rattle spring retainers onto the pins.
11. Replace the drums and wheels. Adjust the brakes. Bleed the hydraulic system if the brake lines were disconnected.
12. Reconnect the handbrake, making sure that it does not cause the shoes to drag when it is released.

ADJUSTMENT

These brakes adjust themselves automatically with each application of the brake pedal or the parking brake; other than this, no adjustments are either necessary or possible.

TYPE 15

Non-Servo—Automatic Adjuster

This brake consists of non-servo forward and reverse shoes with a double-ended type wheel cylinder. The shoes are held in position by an anchor plate and anti-rattle

BRAKES
Import Vehicles

springs. The wheel cylinder is located at the top or bottom of the brake backing plate depending upon the particular application. These brakes are self-adjusting.

REMOVAL & INSTALLATION

1. Raise the rear of the vehicle and support it with jackstands. Remove the tire and wheel assembly.
2. Engage the parking brake. Pull the pin out and then remove the stopper from the toggle lever. Release the parking brake.
3. Remove the brake drum.

NOTE: If the brake drum cannot be easily removed, install two (2) bolts (8 mm) in the holes and drive it out.

4. Remove the return springs.
5. Push the anti-rattle spring retainers in and turn them so they can be removed from the pins.
6. Remove the brake shoes.
7. Clean the brake backing plate and check the wheel cylinder for leaks or other damage; replace as necessary.
8. Hook the return springs into the new shoes. The springs should be between the shoes and the backing plate. The longer return spring must be adjacent to the wheel cylinder. A very thin film of grease may be applied to the pivot points at the ends of the brake shoes. Grease the shoe locating buttons on the backing plate, also. Be careful not to get grease on the linings or drums.
9. Place one shoe in the adjuster and piston slots, and pry the other shoe into position.
10. Press and turn the anti-rattle spring retainers onto the pins.
11. Replace the drums and wheels. Adjust the brakes. Bleed the hydraulic system if the brake lines were disconnected.
12. Reconnect the handbrake, making sure that it does not cause the shoes to drag when it is released.

ADJUSTMENT

These brakes adjust themselves automatically with each application of the brake pedal or the parking brake; other than this, no adjustments are either necessary or possible.

BRAKES
Light Trucks & Vans

Light Trucks & Vans

HYDRAULIC SYSTEM SERVICE

Basic Hydraulic System

The hydraulic system controls the braking operation and consists of a master cylinder, hydraulic lines and hoses, control valves and calipers and/or wheel cylinders. When the brake pedal is depressed, the master cylinder forces brake fluid to the calipers and/or cylinders, via lines and hoses. Sliding rubber seals contain the fluid and prevent leakage.

Return springs in the master cylinder help the brake pedal return to the original unapplied position. Check valves (in most cases) regulate the return flow of the fluid to the master cylinder. Other valves, such as the metering valve, proportioning valve, or combination valve, regulate the flow of fluid to the caliper/wheel cylinder, to achieve efficient braking.

Single Braking Systems

On single brake systems, the master cylinder has only one piston which operates all of the wheel cylinders. The single brake system is confined to over the road vehicles above 10000 lbs. GVW, industrial and construction equipment.

Dual Braking Systems

The "dual" system differs from the "single" system by employing a "tandem" master cylinder, essentially two master cylinders (usually) formed by aligning two separate pistons and fluid reservoirs into one cylinder bore. Dual brake lines "split" the calipers and/or wheel cylinders into two groups, each actuated by a separate master cylinder piston. In event of failure of one of the "dual" systems, the other should provide enough braking power to safely stop the vehicle. The dual system usually includes a red warning light on the instrument panel which is activated by a pressure differential valve. The valve is sensitive to any loss of hydraulic pressure that might result from a braking failure on either side of the system.

Light trucks are equipped with either a front/rear wheel "split" or a diagonally "split" system. On front/rear systems, the front wheels are connected to one circuit while the rear wheels are connected to the other circuit. Diagonally split systems have diagonally opposite wheels connected to each circuit. Medium and heavy trucks may use the front/rear split or, if equipped with two wheel cylinders per wheel, each circuit will operate one cylinder per wheel.

General Information

Servicing the hydraulic brake system is chiefly a matter of adjustments, replacement of worn or damaged parts and correcting the damage caused by grit, dirt or contaminated brake fluid. Always make sure the brake system is clean and tightly sealed when a brake job is completed and that only approved heavy duty brake fluid is used.

The approved heavy duty type brake fluid retains the correct consistency throughout the widest range of temperature variation, will not affect rubber cups, helps protect the metal parts of the brake system against failure and assures long trouble free brake operation.

Never use brake fluid from a container that has been used for any other liquid. Mineral oil, alcohol, antifreeze, or cleaning solvents, even in very small quantities, will contaminate brake fluid. Contaminated brake fluid will cause piston cups and the valve(s) in the master cylinder to swell or deteriorate.

Brake adjustment is required after installation of new or relined brake shoes. Adjustment is also necessary whenever excessive travel of pedal is needed to start braking action.

LOW PEDAL

Normal brake lining wear reduces pedal reserve. Low pedal reserve may also be caused by the lack of brake fluid in the master cylinder. The wear condition may be compensated for by a

HYDRAULIC BRAKE SYSTEM TROUBLE DIAGNOSIS

Condition	Possible Cause	Correction
Dragging brakes	8. Obstruction in brake line. 9. Swollen cups in wheel cylinder or master cylinder. 10. Master cylinder linkage improperly adjusted.	8. Clean or replace brake line. 9. Recondition wheel or master cylinder. 10. Correctly adjust master cylinder linkage.
Hard pedal	1. Incorrect brake lining. 2. Incorrect brake adjustment. 3. Frozen brake pedal linkage. 4. Restricted brake line or hose.	1. Install matched brake lining. 2. Adjust brakes and check fluid. 3. Free up and lubricate brake linkage. 4. Clean out or replace brake line hose.
Brakes fade (high speed)	1. Improper brake adjustment. 2. Distorted or out of round brake drums. 3. Overheated brake drums. 4. Incorrect brake fluid (low boiling temperature). 5. Saturated brake lining.	1. Adjust brakes and check fluid. 2. Grind or replace the drums. 3. Inspect for dragging brakes. 4. Drain flush and refill and bleed the hydraulic brake system. 5. Reline brakes as necessary.

BRAKES
Light Trucks & Vans

HYDRAULIC BRAKE SYSTEM TROUBLE DIAGNOSIS

Condition	Possible Cause	Correction
Insufficient brakes	1. Improper brake adjustment. 2. Worn lining. 3. Sticking brakes. 4. Brake valve pressure low. 5. Slack adjuster to diaphragm rod not adjusted properly. 6. Master cylinder low on brake fluid.	1. Adjust brakes. 2. Replace brake lining and adjust brakes. 3. Lubricate brake pivots and support platforms. 4. Inspect for leaks and obstructed brake lines. 5. Adjust slack adjuster. 6. Fill master cylinder and inspect for leaks.
Brakes apply slowly	1. Improper brake adjustment or lack of lubrication. 2. Low air pressure. 3. Brake valve delivery pressure low. 4. Excessive leakage with brakes applied. 5. Restriction in brake line or hose.	1. Adjust brakes and lubricate linkage. 2. Check belt tension and compressor for output. Adjust as necessary. 3. Check valve pressure and clean or replace as necessary. 4. Inspect all fittings and lines for leaks and repair as necessary. 5. Clean or replace brake line or hose.
Spongy pedal	1. Air in hydraulic system. 2. Swollen rubber parts due to contaminated brake fluid. 3. Improper brake shoe adjustment. 4. Brake fluid with low boiling point. 5. Brake drums ground excessively.	1. Fill and bleed hydraulic system. 2. Clean hydraulic system and recondition wheel cylinders and master cylinder. 3. Adjust brakes. 4. Flush hydraulic system and refill with proper brake fluid. 5. Replace brake drums.
Erratic brakes	1. Linings soaked with grease or brake fluid. 2. Primary and secondary shoes mounted in wrong position.	1. Correct the leak and replace brake lining. 2. Match the primary and secondary shoes and mount in proper position.
Chattering brakes	1. Improper adjustment of brake shoes. 2. Loose front wheel bearings. 3. Hard spots in brake drums. 4. Out-of-round brake drums. 5. Grease or brake fluid on lining.	1. Adjust brakes. 2. Clean, pack and adjust wheel bearings. 3. Grind or replace brake drums. 4. Grind or replace brake drums. 5. Correct leak and replace brake lining.
Squealing brakes	1. Incorrect lining. 2. Distorted brakedrum. 3. Bent brake support plate. 4. Bent brake shoes. 5. Foreign material embedded in brake lining. 6. Dust or dirt in brake drum. 7. Shoes dragging on support plate. 8. Loose support plate. 9. Loose anchor bolts. 10. Loose lining on brake shoes or improperly ground lining.	1. Install correct lining. 2. Grind or replace brake drum. 3. Replace brake support plate. 4. Replace brake shoes. 5. Replace brake shoes. 6. Use compressed air and blow out drums and support plate and shoes. 7. Sand support plate platforms and lubricate. 8. Tighten support plate attaching nuts. 9. Tighten anchor bolts. 10. Replace brake shoes and cam-grind lining.
Brakes fading	1. Improper brake adjustment. 2. Improper brake lining. 3. Improper type of brake fluid. 4. Brake drums ground excessively.	1. Adjust brakes correctly. 2. Replace brake lining. 3. Drain, flush and refill hydraulic system. 4. Replace brake drums.
Dragging brakes	1. Improper brake adjustment. 2. Distorted cylinder cups. 3. Brake shoe seized on anchor bolt. 4. Broken brake shoe return spring. 5. Loose anchor bolt. 6. Distorted brake shoe. 7. Loose wheel bearings.	1. Correct adjust brakes. 2. Recondition or replace cylinder. 3. Clean and lubricate anchor bolt. 4. Replace brake shoe return spring. 5. Adjust and tighten anchor bolt. 6. Replace defective brake shoes. 7. Lubricate and adjust wheel bearings.

BRAKES
Light Trucks & Vans

minor brake adjustment. Check fluid level in master cylinder and add as required.

FLUID LOSS

If the master cylinder requires constant addition of hydraulic fluid, fluid may be leaking past the piston cups in the master cylinder or brake cylinders, the hydraulic lines; hoses or connections may be loose or broken. Loose connections should be tightened, or other necessary repairs or parts replacement made and the hydraulic brake system bled.

FLUID CONTAMINATION

To determine if contamination exists in the brake fluid, as indicated by swollen, deteriorated rubber cups, the following tests can be made.

Place a small amount of the drained brake fluid into a small clear glass bottle. Separation of the fluid into distinct layers will indicate mineral oil content. Be safe and discard old brake fluid that has been bled from the system. Fluid drained from the bleeding operation may contain dirt particles or other contamination and should not be reused.

BRAKE ADJUSTMENT

Self adjusting brakes usually do not require manual adjustment but in the event of a brake reline it may be advisable to make the initial adjustment manually to speed up adjusting time.

AUTOMATIC ADJUSTER CHECK

Raise and safely support the vehicle, have a helper in the driver's seat to apply brakes. Remove the plug from the adjustment slot to observe adjuster star wheel. Then, to exclude possibility of maximum adjustment which is, the adjuster refuses to operate because the closest possible adjustment has been reached; the star wheel should be backed off approximately 30 notches. It will be necessary to hold adjuster lever away from star wheel to allow backing off of the adjustment.

Spin the wheel and brake drum in reverse direction and apply brakes vigorously. This will provide the necessary inertia to cause the secondary brake shoe to leave the anchor. The wrap up effect will move the secondary shoe, and a cable or link will pull the adjuster lever away from the starwheel teeth. Upon release of brake pedal, the lever should snap back in position, turning star wheel. Thus, a definite rotation of adjuster star wheel can be observed if automatic adjuster is working properly. If by the described procedure one or more automatic adjusters do not function properly, the respective drum must be removed for adjuster servicing.

HYDRAULIC LINE REPAIR

Steel tubing is used in the hydraulic lines between the master cylinder and the front brake tube connector, and between the rear brake tube connector and the rear brake cylinders. Flexible hoses connect the brake tube to the front brake cylinders or calipers and to the rear brake tube connector.

When replacing hydraulic brake tubing, hoses, or connectors, tighten all connections securely. After replacement, bleed the brake system at the wheel cylinders or calipers and at the booster, if equipped with a bleeder screw.

BRAKE TUBE

If a section of the brake tube becomes damaged, the entire section should be replaced with tubing of the same type, size, shape, and length. Copper tubing should not be used in the hydraulic system. When bending brake tubing to fit the frame or rearaxle contours, be careful not to kink or crack the tube.

All brake tubing should be double flared to provide good leak proof connections. Always clean the inside of a new brake tube with clean isopropyl alcohol.

BRAKE HOSE

A flexible brake hose should be replaced if it shows signs of softening, cracking, or other damage.

When installing a new brake hose, position the hose to avoid contact with other vehicle components.

Hydraulic Control Valves

PRESSURE DIFFERENTIAL VALVE

Also known as a "warning valve", "dash-lamp valve" or "system effectiveness indicator". The valve activates a panel warning lamp in event of pressure loss failure. As pressure fails in one "split" system, the other system's normal pressure causes a piston in the switch to compress a spring and move until an electrical circuit is completed lighting the dash lamp. On some vehicles the spring balanced piston automatically recenters when the brake pedal is released, thus flashing the warning lamp only during brake application. On other vehicles the lamp will stay on until the cause of pressure loss is corrected.

Valves (pressure differential, metering or proportioning) may be located separately, but are usually part of a combination valve. On some brake systems the valve and switch are part of the master cylinder.

Resetting Valves

The pressure differential valve on many vehicles (equipped with a combination valve) will re-center automatically upon brake application after repairs to the system are completed. Other systems require manual resetting. Repair system as required, open a bleeder screw in the half of the system that did not fail. Turn on the ignition to light the warning lamp and slowly depress the brake pedal until the lamp goes out. If too much pressure is applied the piston will go to the other side and the procedure will have to be reversed by opening a bleeder screw in the opposite half of the system.

METERING VALVE

Often used on vehicles equipped with front disc and rear drum brakes, the metering valve improves braking balance during light brake applications by preventing application of the front disc brakes until pressure is built-up in the hydraulic system. The built up hydraulic pressure overcomes the tension of the rear brake shoe return springs. Thus, when the front brake pads contact the rotor the rear brakes shoes move outward to contact the brake drum at the same time.

The metering valve should be inspected whenever the brakes are serviced. A slight amount of moisture inside the boot does not indicate a defective valve, however a great deal of fluid indicates a worn valve and replacement is indicated. Make sure to install the brake lines in the correct ports when installing a new valve, crossed lines will cause the rear brakes to drag.

If a pressure bleeder is used to bleed a hydraulic system that includes a metering valve, the valve stem (inside the boot on some valves) must either be pushed in or pulled out, depending upon the type of valve. Never apply excessive pressure that might damage the valve. Never use a solid block or clamp to force the valve open. If the valve must be blocked, rig the stem with a yieldable spring load and take care not to exert more than normal pressure.

BRAKES
Light Trucks & Vans

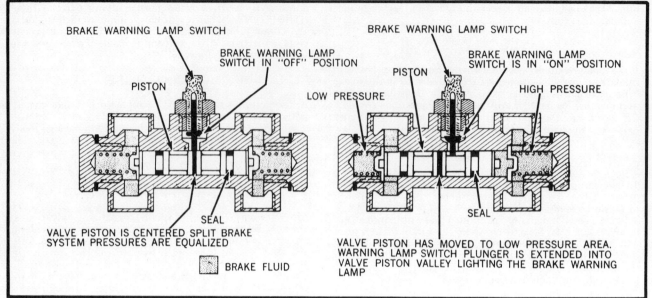
Differential valve system with split hydraulic brakes (© Ford Motor Co.)

If the brakes are to be bled manually using the brake pedal, the pressure developed is sufficient to overcome the metering valve and the stem need not be pushed in or pulled out.

PROPORTIONING VALVE

Used on vehicles equipped with front disc and rear drum brakes, the proportioning valve is installed in the line(s) to the rear drum brakes, and in a split system, below the pressure differential valve. By reducing pressure to the rear drum brakes, the valve helps to prevent premature lock-up during severe brake application and provides better braking balance.

Whenever the brakes are serviced, the valve should be inspected. To check valve operation, install hydraulic gauges ahead and behind the valve and determine that it has an operative transition point above which rear brake pressure is proportioned. If the valve is leaking replacement is required. Make sure the valve port marked "R" is connected to the rear brake line(s).

COMBINATION VALVE

A valve combining two or three functions (metering, proportioning, and/or brake warning) may be used. The combination valve is usually mounted under the hood close to the master cylinder, where the brake lines can be easily routed to the front and rear wheels. The combination valve is a non-serviceable unit, and if found to be malfunctioning, must be replaced as a unit.

Master Cylinder Service

CLEANING AND INSPECTION

Thoroughly clean the master cylinder and any other parts to be reused in clean alcohol. DO NOT USE PETROLEUM PRODUCTS FOR CLEANING. If the bore is not badly scored, rusted or corroded, it is possible to rebuild the master cylinder in some cases. A slight bit of honing is permissible to clean up and smooth out the bore. A master cylinder rebuilding kit and fresh fluid should be used. If the cylinder bore is badly pitted or corroded, or if it has been rebuilt before, the master cylinder should be replaced with a new one. Do not hone or repair a scratched or pitted bore of an aluminum master cylinder. Replace the master cylinder. Be sure to note the relative positions of all the parts, paying particular attention to the way the rubber cups are facing. Lubricate all new rubber parts with brake fluid or brake system assembly lubricant.

Cast Iron Bore Cleanup

Crocus cloth or an approved cylinder hone should be used to remove lightly pitted, scored, or corroded areas from the bore. Brake fluid can be used as a lubricant while honing lightly. The master cylinder should be replaced if it cannot be cleaned up readily. After using the crocus cloth or a hone, the master cylinder should be thoroughly washed in clean alcohol or brake fluid to remove all dust and grit. If alcohol is used, dry parts thoroughly before reinstalling. Other solvents should not be used. Check the clearance between the bore wall and the piston (primary piston of a dual system master cylinder) it should be as follows. If a narrow $1/8$ in. to $1/4$ in. wide. If a 0.006 in. feeler gauge can be inserted between the wall and a new piston, the clearance is excessive, and the master cylinder should be replaced. The maximum clearance allowed for units containing pistons without replenishing holes is 0.009 in.

Aluminum Bore Cleanup

Inspect the bore for scoring, corrosion and pitting. If the bore is scored or badly pitted and corroded the assembly should be replaced. Under no conditions should the bore be cleaned with an abrasive material. This will remove the wear and corrosion resistant anodized surface. Clean the bore with a clean piece of cloth around a wooden dowel and wash thoroughly with alcohol. Do not confuse bore discoloration or staining with corrosion.

Quick Take Up (GMC)

Disassembly and Assembly

1. Depress the primary piston and remove the snapring.
2. Remove the primary and secondary pistons and return springs from the cylinder bore.
3. Disassemble the secondary piston.

BRAKES
Light Trucks & Vans

4. Inspect the master cylinder bore. If it is corroded, replace the master cylinder. Never use abrasives on the bore.

NOTE: Always lubricate parts with clean, fresh brake fluid before assembly.

5. Install new seals on the secondary piston.
6. Install the spring and secondary piston into the cylinder.
7. Install the primary piston, depress and install the snapring.

Bendix Mini—Master

Disassembly and Assembly

1. Remove the reservoir cover and diaphragm, and drain the fluid from the reservoir.
2. Remove the four bolts that secure the body to the reservoir using special tool J–25085 or equivalent.

NOTE: Do not remove the two small filters from the inside of the reservoir unless they are damaged and are to be replaced.

3. Remove the small O-ring and the two compensating valve seals from the recessed areas on the bottom side of the reservoir.
4. Depress the primary piston using a tool with a smooth round end. Then remove the compensating valve poppets and the compensating valve springs from the compensating valve ports in the master cylinder body.
5. Remove the snapring at the end of the master cylinder bore. Then release the piston and remove the primary and secondary piston assemblies from the cylinder bore. It may be necessary to plug the front compensating valve port to remove the secondary piston assembly.
6. Lubricate the secondary piston assembly and the master cylinder bore with clean brake fluid.
7. Assemble the secondary spring (shorter of the two springs) in the open end of the secondary piston actuator, and assemble the piston return spring (longer spring) on the projection at the rear of the secondary piston.
8. Insert the secondary piston assembly, actuator end first, into the master cylinder bore and press the assembly to the bottom of the bore.
9. Lubricate the primary piston assembly with clean brake fluid. Insert the primary piston assembly, actuator end first, into the bore.
10. Place the snapring over a smooth round ended tool and depress the pistons in the bore.
11. Assemble the retaining ring in the groove in the cylinder bore.
12. Assemble the compensating valve seals and the small O-ring seal in the recesses on the bottom of the reservoir. Be sure that all seals are fully seated.
13. While holding the pistons depressed, assemble the compensating valve springs and the compensating valve poppets in the compensating valve ports.
14. Holding the pistons compressed, position the reservoir on the master cylinder body and secure it with the four mounting bolts. Torque the bolts 12 to 15 ft. lbs.

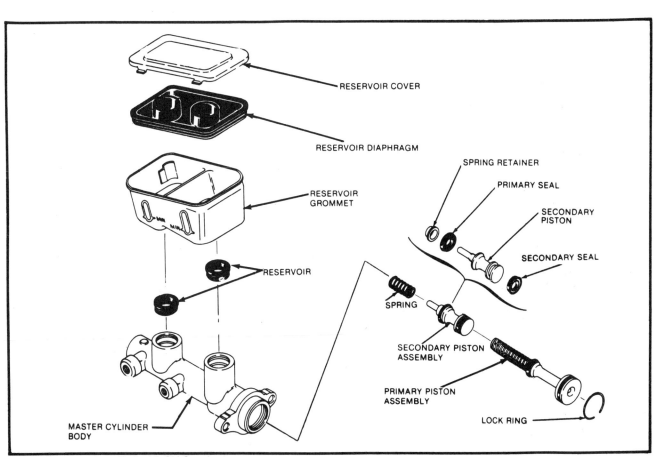

Quick take up master cylinder (© GM Corp.)

SECTION 1

BRAKES
Light Trucks & Vans

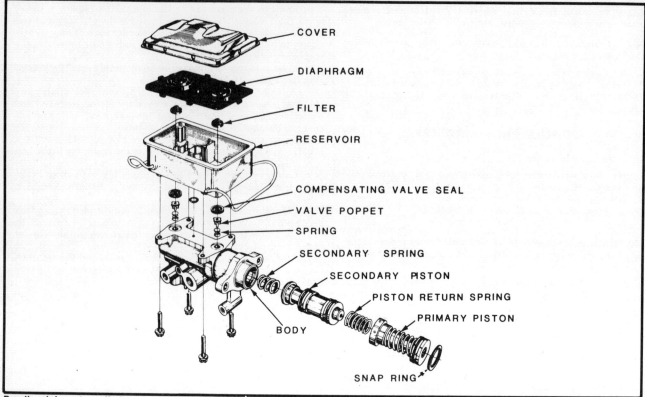
Bendix mini master cylinder—exploded view (© GM Corp.)

Bendix Tandem

Disassembly

1. Clean the outside of the master cylinder assembly. Remove the residual pressure valves.
2. Remove the tube seats by installing easy outs firmly into the seats. Tap lightly with a hammer to loosen, remove seats.
3. Slide clamp off master cylinder cover and remove the cover and its gasket. Drain the brake fluid from the master cylinder.
4. Remove the snapring from the open end of the cylinder with snap ring pliers. Remove the washer from cylinder bore.
5. Remove the front piston retaining screw. Carefully remove the rear piston assembly.
6. Remove the front piston assembly.

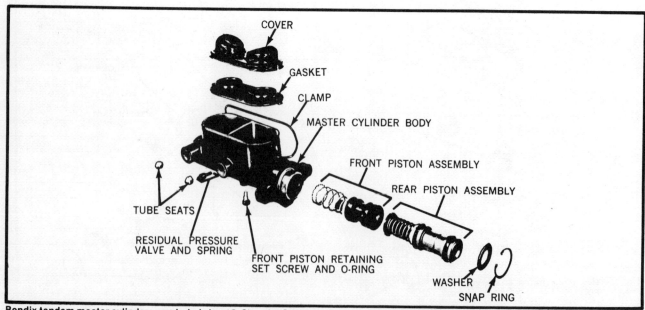
Bendix tandem master cylinder—exploded view (© Chrysler Corp.)

1-110

BRAKES
Light Trucks & Vans

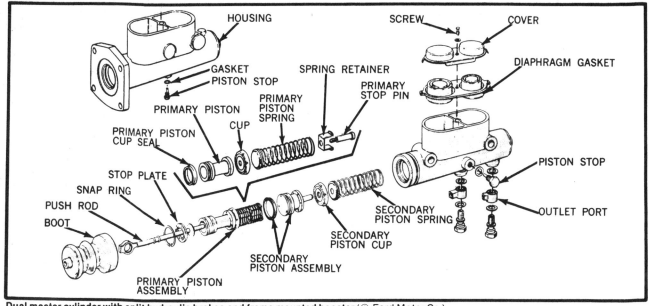

Dual master cylinder with split hydraulic brakes and frame mounted booster (© Ford Motor Co.)

Midland–Ross Tandem (Removable Reservoirs Type)

Disassembly and Assembly

1. Clean the outside of the cylinder and remove the filler cap and gasket (diaphragm). Pour out any brake fluid that may remain the reservoir. Stroke the push rod serveral times to remove fluid from the cylinder bore.
2. Remove the reservoir retainers, washers, and reservoir from the master cylinder body.
3. Remove the two rubber washers from the reservoir and the two O-rings from the reservoir retainers.
4. Remove the snapring, spring retainer and push rod spring.
5. Unscrew the retainer bushing counterclockwise and remove the push rod, retainer bushing, seal retainer and primary piston from the master cylinder.
6. Remove the primary piston from the push rod and discard it.
7. Remove the seal retainer, and retainer bushing from the push rod. Remove the two lip seals and two O-rings from the retainer bushing.
8. Unscrew the end cap counterclockwise and remove the end cap and secondary piston assembly from the master cylinder.
9. Remove the snapring from the secondary piston and remove the piston and return spring from the end cap and stop rod assembly.
10. Remove the two lip seals from the piston.

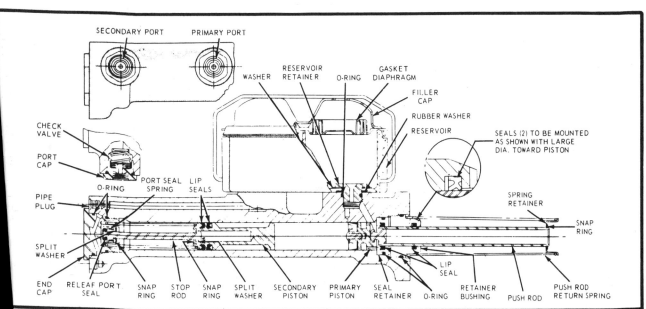

master cylinder with Midland Ross dash mounted booster (© Ford Motor Co.)

BRAKES
Light Trucks & Vans

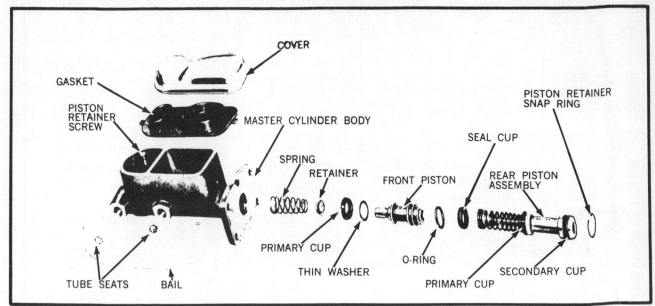

Chrysler tandem master cylinder—exploded view (© Chrysler Corp.)

Cleaning and Inspection

1. Clean all parts with a suitable solvent and dry with filtered compressed air. Wash cylinder bore with clean brake fluid and check for damage or wear.
2. If cylinder bore is lightly scratched or shows slight corrosion it can be cleaned with crocus cloth. Heavier scratches or corrosion can be removed by honing, providing that diameter of cylinder bore is not increased by more than 0.002 in. If master cylinder bore does not clean up at 0.002 in. when honed, the master cylinder should be replaced.
3. If master cylinder pistons are badly scored or corroded, replace them with new ones. All caps and seals should be replaced when rebuilding a master cylinder.

Assembly

NOTE: Before assembly of master cylinder, dip all parts in clean brake fluid and place on clean paper. Assembling master cylinder dry can damage rubber seals.

1. Coat master cylinder bore with brake fluid and carefully slide the front piston into cylinder body.
2. Slide the rear piston assembly into the cylinder bore. Compress pistons and install the front piston retaining screw.
3. Position washer in cylinder bore and secure with snapring.
4. Install the residual pressure valve and spring in the outlet port and install tube seats firmly.

Wagner Tandem

Disassembly

1. Clean the outside of the master cylinder. Remove the cylinder cover screw or spring retaining clip. Lift off the cover and the diaphragm gasket and pour off excess brake fluid. Use the push rod to stroke the cylinder forcing fluid from the cylinder through the outlet ports.
2. Loosen and remove the piston stop screw and gasket from the right hand side of the cylinder.
3. Pull back the push rod boot and remove the snap-ring from the groove in the end of the cylinder bore.
4. Remove the push rod and stop plate from the internal parts from the master cylinder. Remove the internal parts from the master cylinder. If the parts will not slide out apply air pressure at the secondary outlet port. If after applying air pressure, parts still do not move easily, check bore carefully for extensive damage which may eliminate possibility of rebuilding master cylinder.

Inspection and Repair

1. Clean all parts in clean brake fluid. Inspect the parts for chipping, excessive wear or damage. Replace them as required When using a master cylinder repair kit, install all the part supplied.
2. Check all recesses, openings and internal passages to sure they are open and free of foreign matter. Passages may probed with soft copper wire, 0.020 in. OD, or smaller.
3. Minor scratches or blemishes in the cylinder bore can removed with crocus cloth or a clean up hone. Do not overs the bore more than 0.007 in.

Assembly

1. Dip all parts except the master cylinder in clean hy lic brake fluid of the specified type.
2. Install the rear rubber cup on the secondary pisto the cup lip facing the rear. All other cups face the f closed end of the cylinder.
3. Assemble and install the secondary piston sprin cup, and the secondary piston.
4. Install the piston stop screw and gasket, making screw enters the cylinder behind the rear of the s piston.
5. Assemble and install the primary piston and parts.
6. Locate the stop plate in the seat in the bore a the snap ring into the groove at the rear of the cyl
7. Install the push rod boot onto the push rod and of the cylinder housing.
8. Bleed the master cylinder.

11. Remove the snapring from the end cap and remove the secondary piston stop rod, relief port seal spring, the two snaprings and the two split washers from the end cap.
12. Remove the relief port seal from the secondary piston stop rod.
13. Remove the O-rings from the end cap.
14. Remove the primary and secondary port caps and discard.
15. Remove the check valves and springs from the ports.
16. Remove the pipe plug from the end of the master cylinder.
17. Wash all metal parts in clean brake fluid before assembly. Dip all parts except the master cylinder body in clean hydraulic brake fluid of the specified type. When using a master cylinder repair kit, install all of the parts supplied.
18. Install the pipe plug in the end of the master cylinder.
19. Install a new primary piston into the front end of the master cylinder bore. Push the piston through the bore until it is flush with the retainer bushing recess. Use a nonmetallic object which will not scratch the bore.
20. Assemble the O-rings and the two lip seals on the retainer bushing. Be sure the slip seals fit into the undercuts in the center of the bushing with their large diameters toward the piston end.
21. Install the retainer bushing onto the closed end of the push rod and push it onto the push rod approximately half way. Be sure the lip seal at the piston end of the retainer bushing remains in the undercut portion of the retainer bushing.
22. Install the seal retainer onto the closed end of the push rod with the raised lip toward the retainer bushing.
23. Insert the push rod into the master cylinder bore and hook the push rod onto the primary piston.
24. Slide the seal retainer into the recess in the master cylinder bore.
25. Screw the retainer bushing into the master cylinder body and tighten 15 to 20 ft. lbs.
26. Install the push spring with the large end toward the master cylinder and install the spring retainer and snapring.
27. Install the O-rings on the end can.
28. Install the relief port seal on the secondary piston stop rod.
29. Place the port seal spring, split washer (largest of two), and snapring (largest of two) on the piston stop rod.
30. Slide the assembly into the end cap and engage the snapring into its groove.
31. Install the lip seals on the secondary piston with the large diameters facing outward.
32. Place the secondary piston return spring on the end cap assembly.
33. Compress the spring and place the remaining snapring and split washer on the piston stop rod.
34. Slide the piston stop rod into the secondary piston and engage the snap-ring in its groove.
35. Slide the end cap and piston assembly into the master cylinder bore and screw the end cap into the master cylinder body. Tighten the cap 15 to 20 ft. lbs.
36. Install washers on the reservoir retainer and place the retainers in the mounting holes of the reservoir.
37. Place the rubber washers and O-rings on the retainers.
38. Place the reservoir and retainer assembly on the master cylinder body and tighten the retainers 15 to 20 ft. lbs.
39. Replace the springs and check valves in the output ports of the cylinder.
40. Replace the primary and secondary port caps. Tighten 15 to 20 ft. lbs.
41. Install the mounting seal on the flange of the master cylinder. Install the filler cap and gasket (diaphragm).

Single and Double Barrel Master Cylinders (GMC)

Disassembly and Assembly

1. Clean the outside of the master cylinder.
2. Remove the snapring from the groove in the cylinder bore.
3. Remove the washer (stop plate) from the clutch bore.
4. Remove the piston assembly, primary cup, return spring and retainer assembly, check valve, and the check valve seat from the brake cylinder bore.
5. Remove the piston assembly, primary cup, and return spring and retainer assembly from the clutch cylinder bore.
6. Remove the cover and the bleeder screw valve from the housing.
7. Thoroughly clean all parts with brake fluid.
8. Check the clearance between the piston and the cylinder wall. It should be within 0.001 in. to 0.005 in.
9. Coat all internal parts with brake fluid.

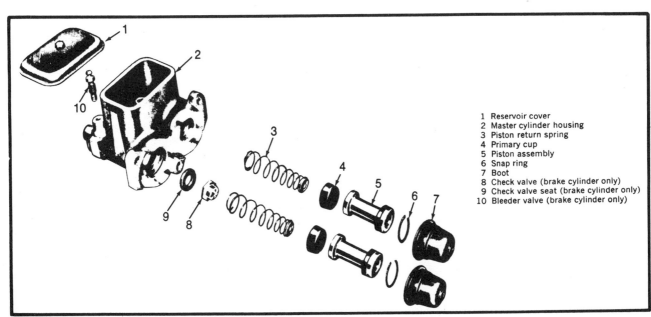

Dual reservoir master cylinder (© GM Corp.)

1 Reservoir cover
2 Master cylinder housing
3 Piston return spring
4 Primary cup
5 Piston assembly
6 Snap ring
7 Boot
8 Check valve (brake cylinder only)
9 Check valve seat (brake cylinder only)
10 Bleeder valve (brake cylinder only)

SECTION 1
BRAKES
Light Trucks & Vans

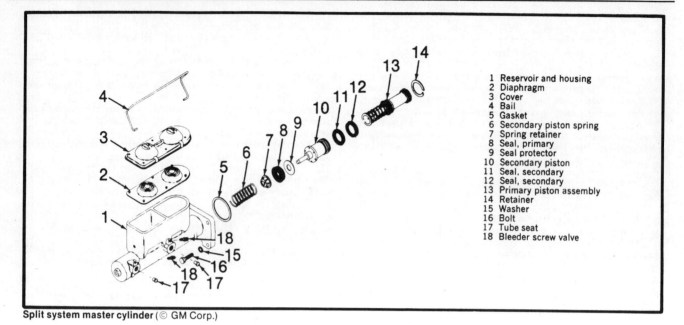

Split system master cylinder (© GM Corp.)

10. Install the parts in the brake cylinder bore. Install the check valve seat and then the check valve in the cylinder bore. Install the short return spring in the bore with the large diameter end of the spring over the check valve.

11. Install the primary cup in the cylinder bore with the lip of the cup toward the outlet end. Make sure the end of the return spring seats inside the cup.

12. Insert the piston and secondary cup assembly into the cylinder bore with the open end of the piston toward the open end of the cylinder bore.

14. Press all parts into the cylinder bore and install the washer (stop plate) if used and the snapring.

15. Install the parts in the clutch cylinder bore. Install the long return spring with the large diameter end first in the cylinder bore. Install the primary cup with the lip of the cup toward the outlet end.

16. Insert the piston and secondary cup into the cylinder bore, with the open end of the piston toward the open end of the cylinder bore. Press all parts into the cylinder bore and install the washer (stop plate) if used and the snapring.

17. Install the cover and the bleeder screw.

Split System Tandem (GMC)

Disassembly and Assembly

1. Remove the cover and reservoir seal.
2. Remove the retaining ring from the groove in the end of the cylinder of the cylinder bore.
3. Remove all parts from the cylinder bore.
4. Remove the bleeder screw valves.
5. Clean all parts in clean brake fluid.
6. Leave a coating of brake fluid on all internal parts and install parts in the cylinder bore using new rubber seals.
7. Install retainer ring and bleeder screws.

Bleeding Brakes

BENCH BLEEDING PROCEDURES

Bench bleed the master cylinder before installation. In order to expel air trapped in the cylinder, tandem master cylinders must be bench bled before they are installed on the vehicle.

Bench bleeding reduces the possibility of air getting in the brake lines. Follow this simple procedure for bench bleeding:

1. Route two shortened brake lines from the outlet connection(s) into the fluid reservoir(s), below the normal fluid level.
2. Fill the reservoir(s) with fresh brake fluid and pump the cylinder until air bubbles no longer appear in the reservoir. If the cylinder does not have a check valve at the outlet port, use a clean piece of rubber or plastic, or the end of your finger to close off the end of the tubing during the back stroke. Otherwise, the fluid will merely pump back and forth in the tubing.
3. When all air has been purged from the master cylinder, bend the tubes up out of the fluid, and remove them. Refill the cylinder and securely install the master cylinder cap.
4. Install the master cylinder on the vehicle. Attach the lines, but do not tighten the tube connection.

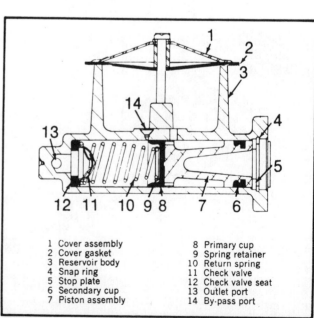

Single reservoir master cylinder (© GM Corp.)

BRAKES
Light Trucks & Vans

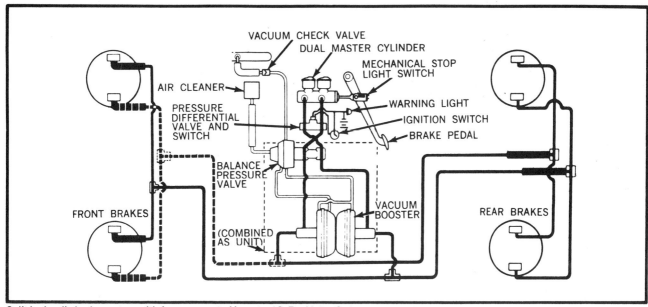

Split hydraulic brake system with frame mounted booster (© For Motor Co.)

5. Force out any air that might have been trapped in the connection by slowly depressing the pedal several times. Tighten the nut slightly before releasing pedal, and loosen before depressing each time. Catch the fluid in a rag to avoid damaging car finish. DO NOT BOTTOM THE PISTON. Tighten the connections when air bubbles are no longer present in the fluid. Make sure the master cylinder is adequately filled with brake fluid.

MANUAL BLEEDING

NOTE: See below for GM "Quick Take-Up" cylinder bleeding sequence

Bleed the longest line first on the individual system (i.e. front/rear split or diagonally front wheel, opposite side rear wheel split. If a single system, the right rear is usually the longest.) being serviced. During the complete bleeding operation, do not allow the reservoir to run dry. Keep the master cylinder reservoirs filled with the specified brake fluid. Never use brake fluid that has been drained from the hydraulic system.

1. Bleed the master cylinder at the outlet port side of the system being serviced.

NOTE: On a master cylinder without bleed screws, loosen the master cylinder to hydraulic line nut. Operate the brake pedal slowly until the brake fluid at the outlet connection is free of bubbles, then tighten the tube nut to the specified torque. Do not use the secondary piston stop screw located on the bottom of the master cylinder to bleed the brake system. Loosening or removing this screw could result in damage to the secondary piston or stop screw. Operate the brake pedal slowly until the brake fluid at the outlet connection is free of air bubbles, then tighten the bleed screw.

2. Position a suitable size (usually $3/8$ in.) box wrench on the bleeder fitting on the cylinder or caliper to be bled. Attach a rubber drain tube to the bleeder fitting. The end of the tube should fit snugly around the bleeder fitting.

3. Submerge the free end of the tube in a container partially filled with clean brake fluid, and loosen the bleeder fitting approximately $3/4$ turn.

4. Push the brake pedal down slowly thru its full travel. Close the bleeder fitting, then return the pedal to the full released position. Repeat this operation until air bubbles cease to appear at the submerged end of the bleeder tube.

5. When the fluid is completely free of air bubbles, close the bleeder fitting and remove the bleeder tube.

6. Repeat this procedure at the brake cylinder or caliper on the other side of the split system. Refill the master cylinder reservoir after each cylinder or caliper is bled. When the bleeding is complete, the master cylinder fluid level should be filled to within $1/4$ in. from the top of the reservoirs.

7. Centralize the pressure differential valve.

GM QUICK TAKE UP MASTER CYLINDER

Special procedures are required to manually bleed the quick take-up brake system used on some General Motors vehicles. Bleed the master cylinder first. Disconnect the left front brake line at the master cylinder, and fill the master cylinder until fluid flows from the port. Catch fluid in a rag and don't allow fluid or rag to contact car finish. Connect the line and tighten fitting.

Depress the brake pedal one time slowly and hold. Loosen same brake line fitting to purge air from the system. Retighten the fitting and release the brake pedal slowly. Wait 15 seconds. Then repeat the sequence, including the 15 second wait until all air is removed. Next bleed the right front connection in the same way as the left front.

Bleed the wheel cylinders and calipers only after you are sure that all the air has been removed from the master cylinder. Follow the specified RR, LF, LR, RF sequence and depress the brake pedal slowly one time before opening bleeder screw to release air. Tighten screw, slowly release pedal, and wait 15 seconds. Repeat all steps, including the 15 second delay until all air has been removed from the system. Rapid pumping of this system moves the secondary master cylinder piston down the bore in a manner that makes it difficult to bleed the left front/right rear part of the system.

SURGE BLEEDING

This method includes both manual and pressure bleeding, and

1–115

BRAKES
Light Trucks & Vans

deliberately creates a churning (higher pressure) turbulence in wheel cylinders so that any remaining air can be drawn off in the form of aerated fluid. It is important to remove all possible air before surging, this method is never used unless the routine manual or pressure bleeding method proves inadequate.

1. Bleed the brakes at all wheels in a usual manner.
2. At each wheel cylinder, in turn, open the bleeder screw and press the brake pedal down sharply several times. Close the bleeder screw. The action creates a turbulence in each cylinder, forcing out practically all of the remaining trapped air.

NOTE: After bleeding the brake system, road test to insure proper operation of the braking system.

BLEEDING THE POWER BRAKE UNIT

On power booster equipped vehicles, the engine should be turned off and the power system purged of vacuum or compressed air by depressing the brake pedal several times. After bleeding the master cylinder, bleed the power brake unit (if equipped with a bleeder screw).

Pressure multiplying type power units often have bleeder screws to remove the air trapped within the unit. If the unit has more that one bleeder screw, bleed the one at the pressure (main) cylinder first and the control valve second. When bleeding, manually close the bleeder screw before the pedal is allowed to back stroke each time.

Wheel Cylinders and Calipers

DRUM BRAKE WHEEL CYLINDER

The wheel cylinder performs in response to the master cylinder. It receives fluid from the hydraulic hose through its inlet port. As the pressure increases the wheel cylinder cups and pistons are forced apart. As a result, the hydraulic pressure is converted into mechanical force acting on the brake shoes. The wheel cylinder size may vary from front to rear. The variation in wheel cylinder size (diameter) is one of the factors controlling the distribution of braking force in a vehicle. Larger diameter wheel cylinders are normally specified for the front brakes of front engine passenger cars equipped with drum brakes. Bleeder screws are provided to remove air or vapor trapped in the system.

Three types of wheel cylinders are normally used with drum brakes.

Single Piston or "Single–end" Type

A single piston wheel cylinder has only one cup, piston, and dust boot and spring. It may also contain a cup filler or cup expander.

Double Piston or "Double–end" Straight Bore Type

The double piston, straight bore type is most commonly used. This type carries two opposed pistons, two cups and two boots.

Double Piston or "Double–end" Step Bore Type

This type is used on some of the non-servo brakes and has the same components as the straight bore type. Two different sized dust boots, cups, and pistons are used. Opposed pistons of different diameters exert different amounts of force.

SERVICE PROCEDURES

Wheel cylinders may need reconditioning or replacement whenever the brake shoes are replaced or when required to correct a leak condition. On many designs, the wheel cylinders can be diassembled without removing them from the backing plate. On some designs, however, the cylinder is mounted in an indention in the backing plate or a cylinder piston stop is welded to the backing plate. When servicing brakes of this type, the cylinder must be removed from the backing plate before being disassembled.

Diagnostic Inspection and Cleaning

Leaks which coat the boot and the cylinder with fluid, or result in a dropped reservoir fluid level, or dampen and stain the brake linings are dangerous. Such leaks can cause the brakes to "grab" or fail and should be immediately corrected. A leakage, not immediately apparent, can be detected by pulling back the cylinder boot. A small amount of fluid seepage dampening the interior of the boot is normal, however a dripping boot is not. Unless other conditions causing a brake to pull, grab, or drag becomes obvious, the wheel cylinder is a suspect and should be included in general reconditioning.

Cylinder binding may be caused by rust, deposits, grime, or swollen cups due to fluid contamination, or by a cup wedged into an excessive piston clearance. If the clearance between the pistons and the bore wall exceeds allowable values, a condition called "heel drag" may exist. It can result in rapid cup wear and can cause the pistons to retract very slowly when the brakes are released.

A ring of a hard, crystal like substance is sometimes noticed in the cylinder bore where the piston stops after the brakes are released.

Some front wheel cylinders have a baffle located between the opposed pistons. The baffle contains a small hole which causes the cylinder to act as a fluid shock absorber damping servo brake shoes as they become energized. These cylinders cannot be honed and should be replaced if the bore is pitted or corroded.

Hydraulic system parts should not be allowed to come in contact with oil or grease, neither should those be handled with greasy hands. Even a trace of any petroleum based product is sufficient to cause damage to the rubber parts.

RECONDITIONING DRUM BRAKE WHEEL CYLINDERS

It is a common practice to recondition a drum brake wheel cylinder without dismounting it, however some brakes are equipped with external piston stops which prevent disassembly unless the cylinder is removed. In order to dismount, remove the shoe springs and spread the shoes apart, disconnect the brake line, remove the mounting bolts or retaining clips, and pull the cylinder free.

Most wheel cylinders are attached to the backing plate with bolts and are easily removed for service or replacement. In recent years, some GM vehicles use a retaining clip for this purpose. To remove this type cylinder, use a special service tool, or insert $1/8$ in. diameter or less awls or pins into the slots between wheel cylinder pilot and retainer locking tabs. Bend both tabs away at the same time until tabs spring over the shoulder, releasing cylinder. Discard the old retainer.

To replace the wheel cylinder, use a new retainer and the following procedure.

1. Hold wheel cylinder against backing plate by inserting a block between the wheel cylinder and axle shaft flange.
2. Position wheel cylinder retainer clip so the tabs will be away from and in horizontal position with the backing plate when installing.
3. Press new retaining clip over wheel cylinder abutment and into position using $1 \frac{1}{8}$ in. 12 point socket. The retainer is in place when the tabs are snapped under the retainer abutment. Examine closely to be sure both retainer tabs are properly engaged.

Another variation of retainer clip is used on some imported vehicles. The retainer usually consists of two or three separate pieces which when slid together will lock themselves and the

BRAKES
Light Trucks & Vans

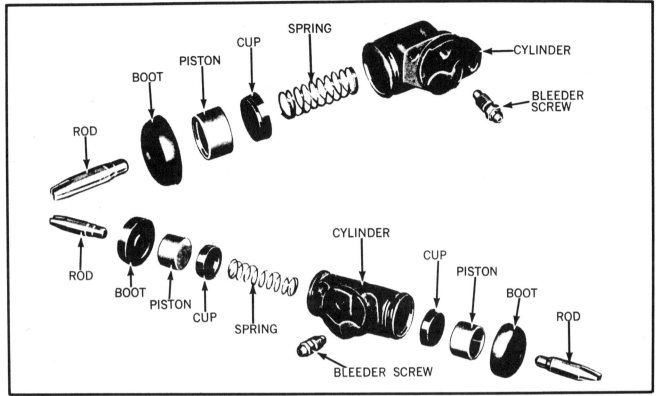
Typical wheel cylinder—exploded view (© Chrysler Corp.)

wheel cylinder in place. The retainers can be carefully removed without incurring damage which allows them to be reused. If they are damaged or corroded, however, they must be replaced.

Pull the protective dust boots off the cylinder. Internal parts should slide out, or be picked out easily. Parts can be driven out with a wooden dowel, or blown out at low pressure by applying compressed air to the fluid inlet port. Parts which cannot be removed easily indicate they are damaged beyond repair and the cylinder should be replaced.

Clean the cylinder and the parts in alcohol and/or brake fluid. (Do not use gasoline or other petroleum based products.) Use only lint free wiping cloths. Crocus cloth can be used to clean minute scratches, signs of rust, corrosion or discoloration from the cylinder bore and pistons. Slide the cloth in a circular rather than a lengthwise motion. A clean up hone may be used. After a cylinder has been honed, inspect it for excessive piston clearance and remove any burrs formed on the edge of fluid intake or bleeder screw ports.

NOTE: Do not rebuild aluminum cylinders. A cylinder that does not clean up at 0.002 in. should be discarded and a new cylinder installed. (Black stains on the cylinder walls are caused by the piston cups and will do no harm.)

Assemble the cylinder with the internal parts, making sure that the cylinder wall is wet with brake fluid. Insert the cups and pistons from each end of a double end cylinder; do not slide them through the cylinder. Cup lips should always face inward.

Disc Brake Caliper

An integral part of the caliper, the caliper bore(s) contains the piston(s) that direct thrust against the brake pads supported within the caliper. Since all braking forces (pad application force) are applied on each side of the rotor with no self energization, the cylinder and piston are large in comparison to a drum brake wheel cylinder.

FIXED CALIPER TYPE

A fixed type caliper is mounted solidly to the spindle bracket.

Pistons are located on both sides of the rotor, in inboard and outboard caliper halves. Fluid passes between caliper halves through an external crossover tube or through internal passages. A bleeder screw is located in the inboard caliper half. A dust boot protecting each cylinder fits in a circumferential groove on the piston.

FLOATING CALIPER TYPE

Floating or sliding calipers are free to move in a fixed bracket or support.

The piston is located only on the inboard side of the caliper housing, which straddles the rotor. The cylinder piston applies the inboard brake shoe directly, and simultaneously hydraulic pressure slides the caliper in a clamping action which forces the caliper to apply the outboard brake shoe.

The actual applying movement is small. The unit merely grips during application, relaxes upon release, and the shoes do not retract an appreciable distance from the rotor. The fluid inlet port and the bleeder screw are located n the inboard side of the caliper. A dust boot is fitted into a circumferential groove on the piston and into a recess at or near the outer end of the cylinder bore.

HYDRAULIC SEAL ARRANGEMENTS

Seal arrangements at the caliper pistons vary depending upon the brake manufacturer. Three makes of fixed caliper brakes,

BRAKES
Light Trucks & Vans

Bendix, Budd, and Delco-Moraine, use a ring seal which fits in a circumferential groove on the piston.

A fixed seal is now commonly used in brake calipers. During the very small applying movement of the piston, the elasticity of the fixed seal permits some deflection in the cylinder groove. The seal deflects as the brakes are applied and relaxes as the brakes are released, retracting the piston a small amount. Some GM types have a rolling seal that retracts the piston slightly further to reduce pad rubbing friction.

A scratched piston, nicked seal, or a sludge or varnish deposit which lifts the sealing edge away from the piston will cause a fluid leak. A serious leak could develop if calipers are not reconditioned when new pads are installed. Then dust and road grime, gradually accumulating behind the dust boot, could be carried into the seal when the piston is shoved inward to accommodate new thick linings. Old seals may have taken a "set," thus preventing proper seating in the retainer groove and on the piston. Therefore, when reconditioning calipers, new seals should be installed.

Service Procedures

Before servicing, syphon or syringe about ⅔ of the fluid from the master cylinder reservoir do not allow the, fluid level to fall below the cylinder intake port. To prevent a gravity loss of fluid, plug the brake line after disconnecting from the caliper. To recondition, remove the caliper from the vehicle, allow the unit to drain, and remove the brake shoes. For benchwork, clamp the caliper housing in a soft jaw vice. On fixed-caliper types, remove the bridge bolts and separate the caliper into halves. Remove the sealing O-rings at cross-over points, if the unit has internal fluid passages across the halves.

Whenever required, use special tools to remove pistons, dust boots, and seals. If compressed air is used, apply it gradually, gently ease the pistons from the cylinders, and trap them in a clean cloth; do not allow them to pop out. Take care to avoid pinching hands or fingers.

While removing stroking type seals and boots, work lip of boot from the groove in the caliper. After the boot is free, pull the piston, and strip the seal and boot from the piston.

While removing fixed position (rectangular ring) seals and boots, pull the piston through the boot. Do not use a metal tool which would scratch the piston. Use a small pointed wooden or plastic tool to lift the boots and seals from the grooves in the cylinder bore.

Cleaning, Inspection, and Installation

Use only alcohol and/or brake fluid and a lint free wiping cloth to clean the caliper and parts. Other solvents should not be used. Blow out passages with compressed air. Always wear eye protection when using compressed air or cleaning calipers.

To correct minor imperfections in the cylinder bore, polish with a fine grade of crocus cloth working in a circular rather than a lengthwise motion. Do not use any form of abrasive on a plated piston. Discard a piston which is pitted or has signs of plating wear.

Inspect the new seal. It should lie flat and be round. If it has suffered a distorted "set" during its shelf life, do not use it. Lubricate the cylinder wall and parts with brake fluid.

While installing stroking type seals and boots, stretch the boot and the seal over the piston and seat them in position.

Use special alignment tools for inserting lip cup seals. Be sure the seal does not twist or roll.

Where the boot lip is retained inside the cylinder bore the following method works well.

1. Lubricate bottom inside edge of piston and brake seal in caliper with brake fluid.
2. Pull boot over bottom end of piston so that boot is positioned on bottom of piston with lip about ¼ in. up from bottom end.
3. Hold piston suspended over bore.
4. Insert back boot lip into groove in caliper.
5. Then tuck the sides of boot into groove and work forward until only one bulge remains.
6. Tuck the final bulge into front of the groove.
7. Then push the piston carefully through the seal and boot to the bottom of the bore. The inside of the boot should slide on the piston and come to rest in the boot groove.
8. If the boot lip is retained outside the cylinder bore, first stretch boot over the piston and seat it in its groove, then press the piston through the seal. Fully depress the piston. You'll need 50 to 100 pounds force to fasten the boot lip in place. On some designs, it is necessary to use a wooden drift or a special tool to seat the metal boot in the caliper counterbore below the face of the caliper.

INSTALLING FIXED CALIPER BRIDGE BOLTS

If the caliper contains internal fluid cross-over passages, be sure to install the new O-ring seals at joints. Install high tensile strength bridge bolts on the mated caliper halves. Never replace the bridge bolts with ordinary standard hardware bolts.

Brake Disc (Rotor)

ROTOR RUNOUT

Manufacturers differ widely on permissible runout, but too much can sometimes be felt as a pulsation at the brake pedal. A wobble pump effect is created when a rotor is not perfectly smooth and the pad hits the high spots forcing fluid back into the master cylinder. This alternating pressure causes a pulsating feeling which can be felt at the pedal when the brakes are applied. This excessive runout also causes the brakes to be out of adjustment because disc brakes are self adjusting, they are designed so that the pads drag on the rotor at all times and therefore automatically compensate for wear. To check the actual runout of the rotor, first tighten the wheel spindle nut to a snug bearing adjustment, end play removed. Fasten a dial indicator on the suspension at a convenient place so that the indicator stylus contacts the rotor face approximately one in. from its outer edge. Set the dial at zero. Check the total indictor reading while turning the rotor one full revolution. If the rotor is warped beyond the runout specification, it is likely that it can be successfully remachined.

Lateral Runout: A wobbly movement of the rotor from side to side at it rotates. Excessive lateral runout causes the rotor faces to knock bac the disc pads and can result in chatter, excessive pedal travel, pumping or fighting pedal and vibration during the breaking action.

Parallelism (lack of): Refers to the amount of variation in the thickness of the rotor. Excessive variation can cause pedal vibration or fight, front end vibrations and possible "grab" during the braking action; a condition comparable to an "out-of-round brake drum." Check parallelism with a micrometer, "mike" the thickness at eight or more equally spaced points, equally distant from the outer edge of the rotor, preferably at mid-points of the braking surface. Parallelism then is the amount of variation between maximum and minimum measurements.

Surface of Micro-inch finish, flatness, smoothness: Different from parallelism, these terms refer to the degree of perfection of the flat surface on each side of the rotor, that is, the minute hills, valleys and swirls inherent in machining the surface. In a visual inspection, the remachined surface should have a fine ground polish with, at most, only a faint trace of nondirectional swirls.

BRAKES
Light Trucks & Vans

Disc Brake Surface Refinishing

To meet mandated brake system performance requirements, semi-metallic brake linings have been used for several years in some vehicle applications. In order to maintain the proper performance, it is important to correctly service these semi-metallic brake components as outlined in the following procedures.

Service Recommendations

1. Semi-metallic linings should be replaced with semi-metallic service linings, equal to the original equipment specifications.
2. Routine replacement of the disc pads does not require rotor refinishing, unless damage or extreme wear to the rotor has occurred.
3. Rotor refinishing should only be required if non-parallelism, excessive runout, rotor damage or scoring of the rotor surface has occurred.
4. If refinishing is necessary, the semi-metallic brake pads require a micro- inch surface refinish like new vehicle rotor specifications (10 to 50 micro-inches with non-directional swirl patterns).
5. The recommended procedure for obtaining this finish is outlined in the following chart.

ROTOR REFINISHING

Procedure	Rough Cut	Finish Cut
Spindle Speed	150 RPM	150 RPM
Depth of Cut Per Side	.005"	.002"
Tool Cross Feed Per Rev.	.006".010"	.002" Max.
Vibration Dampener	Yes	Yes
Swirl Pattern-120 GRIT	No	Yes

6. When refinishing brake rotors for semi-metallic linings, the following is important;
   a. The brake lathe must be in good working order and have the capability to produce the intended surface finish.
   b. Use the correct tool feed and arbor speeds. Too fast a speed or too deep a cut can result in a rough finish.
   c. Cutting tools must be sharp.
   d. Adapters must be clean and free of nicks.
   e. Lathe finish cuts should be further improved and made non-directional by dressing the rotor surface with a sanding disc power tool, such as AMMCO model 8350 Safe Swirl Disc Rotor Grinder or its equivalaent.
   f. Rotor surfaces are to be refinished to 10 to 50 micro-inches.
7. To become familiar with the required surface finish, drag the fingernail over the surface of a new rotor from parts stock or on a new vehicle. If your brake equipment cannot produce this smooth-a-finish when correctly used, contact the equipment manufacturer for corrective instructions.
8. When installing new rotors from service stock, do not refinish the surface as these parts are to the recommended finish. It also is not required to refinish a rotor on a vehicle which has a smooth finish.

Drum Brake Service

Basic Service

--- **CAUTION** ---

Do not blow the brake dust out of the drums with compressed air or lung power; always use a damp cloth, a vacuum unit and soft brush to gather the dust particles into a container for disposal. Use a nose/mouth protective cover Brake linings contain asbestos, a known cancer causing substance. Dispose of the residue safely.

NOTE: Never work on a vehicle supported only by a jack. Use a hydraulic lift and/or jack stands to support the vehicle safely.

Check For Leaks

Press the brake pedal to ensure that there are no leaks in the hydraulic system. If the pedal does not remain hard and drops to the end of its pedal travel, an internal or external fluid leakage is indicated in the master cylinder, hoses, wheel cylinders, or brake calipers. When performing this test, the engine should be running, if equipped with power brakes. With power brakes, it is normal for the pedal to drop slightly when the engine starts. If the pedal continues to drop, a leak in the system is indicated.

Drum Inspection

Check the drums for any cracks, scores, grooves, or out-of-round conditions. Slight scores can be removed with fine emery cloth, while extensive scoring requires machining the drum on a suitable drum lathe.

If the friction surface of the brake drum is scored or otherwise damaged beyond the allowable machining specification, it will require replacement. After machining, the drum diameter must not exceed the diameter specification cast on the drum or 0.060 in. (1.5mm) over the original nominal diameter. Carefully look for signs of grease, oil or brake fluid on the drum assembly and repair as required.

Rebuild the Wheel Cylinders

It is always a good practice to rebuild or replace the wheel cylinder when relining the brakes. This helps to assure a properly operating brake system and to prevent premature leakage of brake fluid past the cups and piston seals.

Clean and Lubricate

With the brake parts off, clean the backing plate with a damp cloth to avoid raising any asbestos dust and dispose of the rag after use. Clean any rust with a wire brush. File smooth any ridges or rough edges on the contact points of the backing plate. Lubricate the contact points with an approved brake lubricant. Clean and lightly lubricate the adjuster threads and screw the adjuster all the way together to facilitate reassembly of the brake components. If the wheel bearings are available, wash in solvent and repack with lubricant. Check the backing plate retaining bolts for tightness.

Reassemble And Install The Brake Shoes

Reassemble the brake shoes in the reverse order of their removal. Make sure all parts are in their proper position and that both brake shoes are properly positioned at either end of the adjuster assembly. Also, both brake shoes should correctly engage the wheel cylinder push rod and parking brake links, if equipped. With the brake shoes and components in position, measure the inside of the drum diameter and adjust the brake shoes to match the diameter with a brake shoe pre-set measuring tool. Install the brake drum and make final brake adjustment, as required. Install the remaining components and torque to specifications.

BLEED AND ROAD TEST

Bleed the air from the hydraulic system to insure a high, hard pedal and road test the vehicle. Self-adjusting mechanisms are activated by the application of the brake pedal when the vehicle is driven in reverse, driven forward or when the parking brake is applied. Be sure the road test course includes enough stops, enough traveling in reverse, and the use of the parking brake assembly, to allow the self adjusters to perform the proper adjustment on all wheels.

SECTION 1

BRAKES
Light Trucks & Vans

DRUM BRAKE SERVICE

Wagner

TWIN ACTION TYPE

Twin-action brake is a four-anchor type. Brake shoes are self-centering in operation, and both shoes are self-energizing in both forward and reverse.

Two wheel cylinders are mounted on opposite sides of the backing plate. One brake shoe is mounted above wheel cylinders and one below. Sliding pivot type anchor is used at front end of upper shoe and at rear end of lower shoe. Adjustable anchor is used at front end of lower shoe and at rear end of upper shoe. Four shoe return springs hold shoe ends firmly against anchors when brakes are released.

Anchor brackets are steel forgings, attached to flange on axle housing in conjunction with the backing plate. At adjustable anchor end of each shoe, shoe web bears against flat head of adjusting screw which threads into anchor bracket. The adjusting screw heads are notched and are rotated for brake adjustment through access holes in backing plate. A lock spring which fits over anchor bracket holds adjusting screw in position.

The brake backing plate has six machined bearing surfaces, three for each shoe, against which the inner edge of each shoe bears. Two brake shoe guide bolts are riveted to backing plate and extend through holes in center of brake shoe web. Shoes are retained on guide bolts by flat washers, nuts, and cotter pins.

Wheel cylinder push rods make contact between wheel cylinder pistons and brake shoes.

Inner edge of brake drum has a groove which fits over a flange on the edge of backing plate, forming a seal against the entrance of dirt and mud.

TWIN-ACTION TYPE REAR BRAKE

Brake Shoe Removal

1. Remove brake drums.

NOTE: If brake drums are worn severely, it may be necessary to retract the adjusting screws.

2. Remove the brake shoe pull back springs.

NOTE: Since wheel cylinder piston stops are incorporated in the anchor brackets, it is not necessary to install wheel cylinder clamps when the brake shoes are removed.

3. Loosen the adjusting lever cam cap screw, and while holding the star wheel end of the adjusting lever past the star wheel, remove the cap screw and cam.
4. Remove the brake shoe hold down springs and pins by compressing the spring and, at the same time, pushing the pin back through the flange plate toward the tool. Then, keeping the spring compressed, remove the lock (C-washer) from the pin with a magnet.
5. Lift off the brake shoe and self-adjuster lever as an assembly.
6. The self-adjuster lever can now be removed from the brake shoe by removing the hold-down spring and pin. Remove lever return spring also.

NOTE: The adjusting lever, override spring and pivot are an assembly. It is not recommended that they be disassembled for service purposes unless they are broken. It is much easier to assemble and disassemble the brake leaving them intact.

7. Thread the adjusting screw out of the brake shoe anchor and remove and discard the friction spring.
8. Clean all dirt out of brake drum. Inspect drums for roughness, scoring or out-of-round. Replace or recondition drums as necessary.
9. Carefully pull lower edges of wheel cylinder boots away from cylinders. If brake fluid flows out, overhaul of the wheel cylinders is necessary.

NOTE: A slight amount of fluid is nearly always present and acts as a lubricant for the piston.

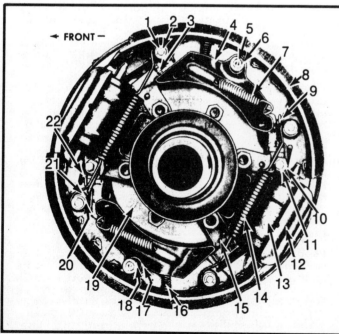

1. Hold-down pin spring lock
2. Hold-down pin
3. Adjusting screw
4. Adjusting lever
5. Adjusting lever pin spring
6. Hold-down spring cup
7. Lever override spring
8. Brake shoe and lining
9. Adjusting lever pivot
10. Adjusting lever cam
11. Adjusting lever bolt
12. Wheel cylinder shield
13. Wheel cylinder
14. Brake shoe return spring
15. Brake shoe anchor
16. Lever return spring
17. Adjusting lever pin sleeve
18. Hold-down spring
19. Brake backing plate
20. Hold-down pin retainer
21. Hold-down pin spring
22. Adjusting lever link

Twin action self adjusting brakes (© GM Corp.)

BRAKES
Light Trucks & Vans

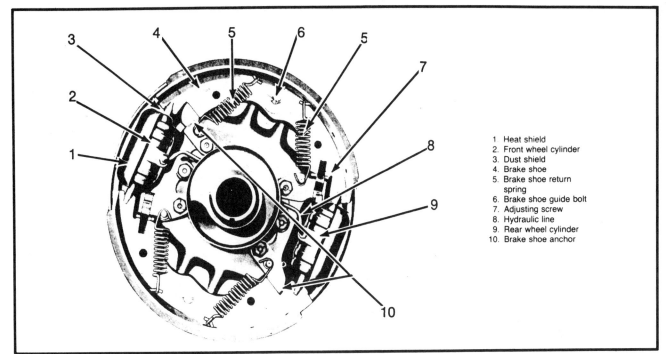

1. Heat shield
2. Front wheel cylinder
3. Dust shield
4. Brake shoe
5. Brake shoe return spring
6. Brake shoe guide bolt
7. Adjusting screw
8. Hydraulic line
9. Rear wheel cylinder
10. Brake shoe anchor

Twin action type brake installed (© GM Corp.)

10. Inspect flange plate for oil leakage past axle shaft oil seals. Install seals if necessary.

Brake Shoe Installation

1. Put a light film of lubricant on shoe bearing surfaces of brake flange plate and on threads of adjusting screw.
2. Thread adjusting screw completely into anchor without friction spring to be sure threads are clean and screw turns easily. Then remove screws, position a new friction spring on screw and reinstall in anchor.
3. Assemble self-adjuster assembly and lever return spring to brake shoe and position adjusting lever link on adjusting lever pivot.
4. Position hold-down pins in flange plate.
5. Install brake shoe and self-adjuster assemblies onto hold down pins. Insert ends of shoes in wheel cylinder push rods and legs of friction springs.

NOTE: Make sure the toe of the shoe is against the adjusting screw.

6. Install cup, spring and retainer on end of hold-down pin. With spring compressed, push the hold-down pin back through the flange plate toward the tool and install the lock on the pin.
7. Install brake shoe return springs.
8. Holding the star wheel end of the adjusting lever as far as possible past the star wheel, position the adjusting lever cam into the adjusting lever link and assemble with cap screw.
9. Check the brake shoes for being centered by measuring the distance from the lining surface to the edge of the flange plate. To center the shoes, tap the upper or lower end of the shoes with a plastic mallet until the distances at each end become equal.
10. Locate the adjusting lever 0.020 to 0.039 in. above the outside diameter of the adjusting screw thread by loosening the cap screw and turning the adjusting cam.

NOTE: To determine 0.020 to 0.039 in., turn the adjusting screw 2 full turns out from the fully retracted position. Hold a 0.060 in. wire gauge at a 90° angle with the star wheel edge of the adjusting lever. Turn the adjusting cam until the adjusting lever and threaded area on the adjusting screw just touch the wire.

11. Secure the adjusting cam cap screw and retract the adjusting screw.
12. Install brake drums and wheels.
13. Adjust the brakes by making several forward and reverse stops until a satisfactory brake pedal height results.

Wagner

TYPE F

Two identical brake shoes are arranged on backing plate so that their toes are diagonally opposite. Two single-end wheel cylinders are arranged so that each cylinder is mounted between the toe of one shoe and the heel of the other. The two wheel cylinder pistons apply an equal amount of force to the toe of each shoe. Each cylinder casting is shaped to provide an anchor block for the brake shoe heel.

Each shoe is adjusted by means of an eccentric cam which contacts a pin pressed into brake shoe web. Each cam is attached to the backing plate by a cam and shoe guide stud which protrudes through a slot in the shoe web and, in conjunction with flat washers and C-washers, also serves as a shoe hold-down. Two return springs are connected between the shoes, one at each toe and heel.

With vehicle moving forward, both shoes are forward acting (primary shoes), self-energizing in forward direction of drum rotation. With vehicle in reverse, both shoes are reverse acting since neither is self-energized in the reverse direction of drum rotation.

BRAKES
Light Trucks & Vans

Brake Shoe Removal

1. Remove both brake shoe return springs, using brake spring pliers.
2. Remove C-washer and flat washer from each adjusting cam and hold-down stud. Lift shoes off backing plate.

Cleaning and Inspection

1. Clean all dirt out of brake drum. Inspect drum for roughness, scoring, or out-of-round. Replace or recondition brake drum as necessary.
2. Inspect wheel bearings and oil seals.
3. Check backing plate attaching bolts to make sure they are tight. Clean all dirt off backing plate.
4. Inspect brake shoe return springs. If broken, cracked, or weakened, replace with new springs.
5. Check cam and shoe guide stud and friction spring on backing plate for corrosion or binding. Cam stud should turn easily with a wrench but should not be loose. If frozen, lubricate with kerosene or penetrating oil and work free.
6. Examine brake shoe linings for wear. Lining should be replaced if worn down close to rivet heads.

Brake Shoe Installation

1. Install anti-rattle spring washer on each cam and shoe guide stud, pronged side facing adjusting cam.
2. Place shoe assembly on backing plate with cam and shoe guide stud inserted through hole in shoe web; locate shoe toe in wheel cylinder piston shoe guide and position shoe heel in slot in anchor block.
3. Install flat washer and C-washer on cam and shoe guide stud. Crimp ends of C-washer together.
4. After installing both shoes, install brake shoe return springs. To install each spring, place spring end with short hook in toe of shoe, then using brake spring pliers, stretch spring and secure long hook end in heel of opposite shoe.
5. Install hub and brake drum assembly.
6. Adjust brake.
7. After checking pedal operation, road test vehicle.

Wagner Type FA
BRAKE SHOES

Removal

1. Block brake pedal in up position. Raise vehicle off ground and support.
2. Remove the brake drums. Disconnect the shoe retaining springs and hold down clips and lift off shoes.
3. Unhook the wedge actuating coil spring from the wedge.
4. Unhook the lever actuating spring from the shoe web, work the spring coil off the lever pivot pin and slide the spring "U" hook off the contact plug-lever pin.

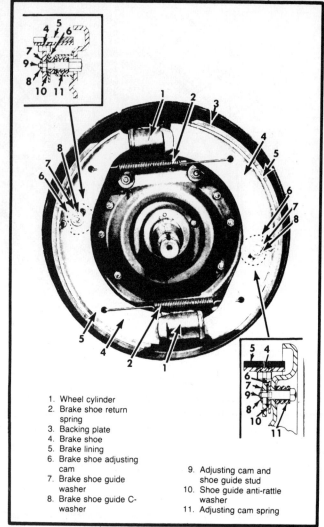

1. Wheel cylinder
2. Brake shoe return spring
3. Backing plate
4. Brake shoe
5. Brake lining
6. Brake shoe adjusting cam
7. Brake shoe guide washer
8. Brake shoe guide C-washer
9. Adjusting cam and shoe guide stud
10. Shoe guide anti-rattle washer
11. Adjusting cam spring

Type F brake assembly (© GM Corp.)

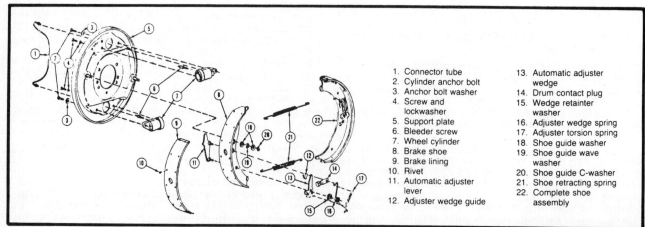

1. Connector tube
2. Cylinder anchor bolt
3. Anchor bolt washer
4. Screw and lockwasher
5. Support plate
6. Bleeder screw
7. Wheel cylinder
8. Brake shoe
9. Brake lining
10. Rivet
11. Automatic adjuster lever
12. Adjuster wedge guide
13. Automatic adjuster wedge
14. Drum contact plug
15. Wedge retainter washer
16. Adjuster wedge spring
17. Adjuster torsion spring
18. Shoe guide washer
19. Shoe guide wave washer
20. Shoe guide C-washer
21. Shoe retracting spring
22. Complete shoe assembly

Exploded view of Wagner type FA brake assembly (© Chrysler Corp.)

BRAKES
Light Trucks & Vans

5. Pull the adjuster lever from the opposite side of the shoe web, the contact plug through the shoe table and lift off the wedge washer, wedge and the wedge guide.

6. Clean all parts with the exception of the brake shoe linings in a suitable solvent. Inspect all components for wear or damage. Replace all parts that are in questionable condition.

Installation

1. Install the automatic adjuster, contact plug flush with the lining surface.

2. Position the wedge guide on the back side of the shoe with serrations facing away from the shoe table.

3. Position the wedge on the shoe with the serrations against matching serrations on the wedge guide with the slot aligned on the lever pivot pin hole.

4. Working from the drum side of the shoe, insert the contact plug, with the guide shank through the hole in the shoe table and over wedge guide and wedge.

5. Insert the adjuster lever pins through the shoe web from the opposite side, guiding actuating (center) pin into the mating hole of the contact plug shank.

6. Install the wedge washer over the shoulder of the pivot pin. Slide the U-hook of adjuster spring on the pin over the contact plug shank.

7. Attach the end of the wedge actuating spring to the U-hook of the adjusting spring. Position the coil of the adjuster torsion spring over the pivot pin and pull spring hook over the edge of shoe web.

8. Connect the wedge actuating spring on the raised hook of wedge fork.

9. Fully retract the wedge against the lever pivot pin, pressing upon contact plug to permit this movement. If the plug protrudes more than 0.005 in. above lining, clamp shoe in vise so that jaws press against adjuster lever. With a file, press down on the plug until it is even with the brake lining. Exercise caution when filing so as not to create a flat spot on the brake lining. If the fully extended plug is more than 0.005 in. below the surface of the lining, replace with a new contact plug.

10. Locate the shoe on hold-downs. Install the retracting springs, long ends of springs are at the ends of shoes.

Initial Adjustment

1. Fully release the manual cams.

2. Center each shoe by sliding up or down on its anchor slot until the leading and trailing edges of the lining are equal distant from the inner curl of the support plate.

3. Install the wheel and drum.

4. Rotate the manual adjuster cam in the direction of forward drum rotation, while rotating the drum in the same direction, until the shoe slightly drags on the brake drum. Back off adjuster until drag is just relieved. Use only sufficient adjustment torque to obtain drag that will just allow turning the wheel by hand (approximately 120 to 130 inch lbs.) as excessive torque may damage the adjuster mechanism.

5. Adjust other manual adjuster in the same manner, forward to tighten, and reverse to relieve drag.

6. Lower vehicle and road test. Automatic adjusters should operate from this point and additional manual adjustment should not be necessary.

Wagner Type FR-3

Each brake is equipped with two double-end wheel cylinders which apply hydraulic pressure to both the toe and the heel of two identical, self-centering shoes. The shoes anchor at either toe or heel, depending upon the direction of rotation. Each adjusting screw is threaded into or out of its support by means of an adjusting wheel. Adjusting wheels are accessible through adjusting slots in the backing plate.

BRAKE SHOES

Removal

1. Remove hub and brake drum assembly.
2. Install wheel cylinder clamps to hold pistons in cylinders.
3. Remove brake shoe return springs.
4. Remove lock wires, nuts, and washers from brake shoe guide bolts, then remove brake shoe assemblies.
5. Remove screws attaching adjusting wheel lock springs to anchor supports. Thread each adjusting screw from the shoe side of its anchor support by turning adjusting wheels, then lift adjusting wheels out of slots in anchor supports.

Installation

1. Install adjusting screws and wheels in anchor supports dry; use no lubricant. Insert each adjusting wheel in slot in anchor support, insert threaded end of adjusting screw in anchor support, then turn adjusting wheel to thread adjusting screw into anchor support. Insert anchor pins into holes in anchor supports, with slots in pins facing slots in supports.

2. Install brake shoes with cutaway end of shoe web next to adjusting screw and with ends of shoes engaging slots in wheel cylinder push rods and anchor pins. Install flat washer and nut on each brake shoe guide bolt. Tighten nuts finger-tight, then back off nuts only far enough to allow movement of shoes without binding.

3. Install brake shoe return springs, hooking one end of each spring in brake shoe web, then hook other end over anchor pins.

4. Remove wheel cylinder clamps.
5. Install hub and brake drum assembly.
6. Adjust brakes.

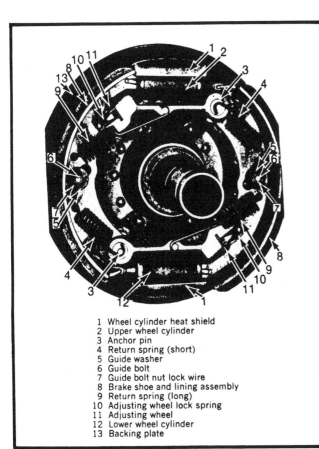

1 Wheel cylinder heat shield
2 Upper wheel cylinder
3 Anchor pin
4 Return spring (short)
5 Guide washer
6 Guide bolt
7 Guide bolt nut lock wire
8 Brake shoe and lining assembly
9 Return spring (long)
10 Adjusting wheel lock spring
11 Adjusting wheel
12 Lower wheel cylinder
13 Backing plate

Type FR3 brake assembly (© GM Corp.)

1-123

BRAKES
Light Trucks & Vans

7. After checking pedal operation, road test vehicle.

Wagner Type FR-3A and FR-5A

The FR-3A brake system is basically the same as the FR-3 except in that the FR-3A employs a self-adjuster assembly and the FR-3 does not. Each shoe is individually adjusted to compensate for wear by a link-crank system. This serves to maintain a high, firm brake pedal at all times.

BRAKE SHOES

Removal

1. Remove the hub and brake drum assemblies.
2. Remove the two springs for the automatic adjusters.
3. Remove the two long crank links from the adjuster assemblies by turning back the star wheel cranks until their slots align with the crank link U-hooks. Lift out the links and then slide the S-hooks out of the adjuster cranks.
4. Remove the short crank links by rotating the adjuster cranks until the link U-hooks clear the eccentrics on the brake shoe webs, then remove the small U-hooks from the adjuster cranks.
5. Spread the C-washers for the adjuster cranks and remove the cranks.
6. Remove the hold down bolt which holds the star wheel crank to the anchor support and remove the crank.
7. Remove the adjuster eccentric screw and eccentric from the brake shoe.
8. Remove the two long and the two short brake shoe return springs.

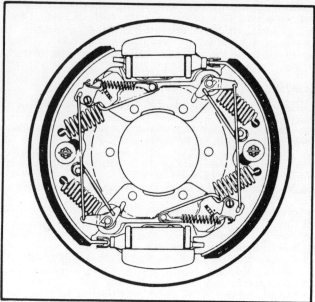

Wagner type FR-3A and FR-5A brake assembly (© Chrysler Corp.)

9. Remove the lock wires, hold-down nuts and washers from the hold down bolt and remove the brake shoes from the backing plate.
10. Thread each star wheel screw out of the anchor support

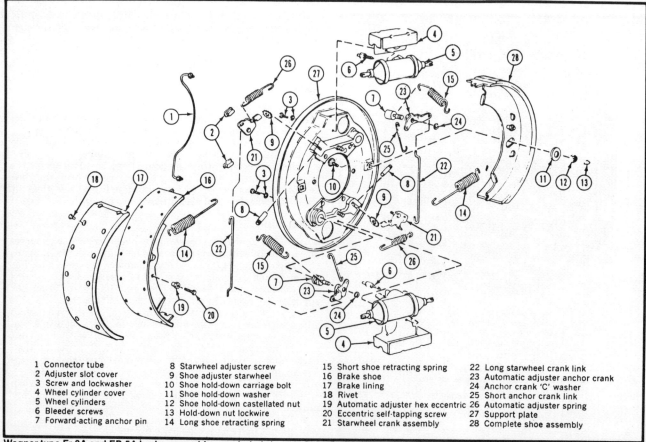

1 Connector tube
2 Adjuster slot cover
3 Screw and lockwasher
4 Wheel cylinder cover
5 Wheel cylinders
6 Bleeder screws
7 Forward-acting anchor pin
8 Starwheel adjuster screw
9 Shoe adjuster starwheel
10 Shoe hold-down carriage bolt
11 Shoe hold-down washer
12 Shoe hold-down castellated nut
13 Hold-down nut lockwire
14 Long shoe retracting spring
15 Short shoe retracting spring
16 Brake shoe
17 Brake lining
18 Rivet
19 Automatic adjuster hex eccentric
20 Eccentric self-tapping screw
21 Starwheel crank assembly
22 Long starwheel crank link
23 Automatic adjuster anchor crank
24 Anchor crank 'C' washer
25 Short anchor crank link
26 Automatic adjuster spring
27 Support plate
28 Complete shoe assembly

Wagner type Fr-3A and FR-5A brake assembly—exploded view (© Chrysler Corp.)

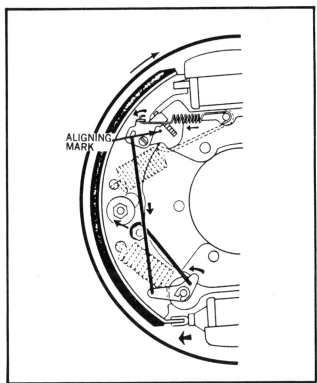

Wagner type FR-3A and FR-5A alignment marks (© Chrysler Corp.)

from the shoe side of the support and lift the star wheels from the support slots.

NOTE: There is a friction ring on each star wheel. DO NOT attempt to remove the friction ring from the star wheel screw. If necessary replace the star wheel and friction as an assembly.

Installation

1. Install the star wheels into the anchor support slots and thread the star wheel screws in from the shoe side with the friction ring end towards the shoe.

NOTE: Do not put any lubricant on the star wheel screws.

2. Position one brake shoe with the "toe" (cut away portion of web) in the adjuster slot and the "heel" in the anchor pin slot of the anchor supports. Install the brake shoe hold down bolt, washer and hold down nut. Tighten the nut finger tight and back off one turn and insert the lock wire in the nut.
3. Install the other brake shoe in the same manner as described in Step 2.
4. Install the long return springs in the shoe and hook them over the anchor pins. Then install the short springs in the same manner.
5. Install the adjuster eccentrics on the brake shoes and fasten them with a self-tapping screw, only make the screw finger tight to allow for final adjustment.
6. Install the adjuster cranks on the anchor pins with the long arm towards the shoe and the bushing towards the backing plate so that they rotate easily while resting against the return springs hooks.
7. Install the C-washer for the adjuster crank and crimp in place.
8. On the anchor support place the star wheel and fasten with the crank bolt.

9. On each adjuster crank assembly install the short links small hook into the short arm of the crank from the lower side and hook the other end of the link around the eccentric on the shoe web.
10. Install the long link S-hook into the long arm of the adjuster crank from the upper side. Then rotate the star wheel crank so the slot lines up with the U-hook on the long link. Insert the hook and rotate the crank back to its adjusting position.
11. Install the adjuster springs with the short hooks on the star wheel crank fingers and the long hook on the outer groove of the anchor pin on the wheel cylinder side.
12. Adjust the brakes in the following manner:
   a. Center the shoes on the backing plate.
   b. On the shoe web, rotate the eccentrics until the linkage aligns the star wheel crank pawl with the center line of the star wheel screw.
   c. When they are aligned, tighten the self-tapping screw to 19 ft. lbs. torque.
   d. Install hub and drum assembly and remove the plugs from the slots in the backing plate.
   e. Adjust the brakes in the normal manner to achieve the required amount of drag on each wheel.
13. Road test the vehicle to check for proper braking action.

Bendix Duo Servo Type

BRAKE SHOES

Removal and Installation

1. With the vehicle raised and supported safely, loosen the parking brake equlizer nut, if working on the rear wheel brakes.
2. Remove the drums and the brake shoe return springs, while noting the position of the secondary spring overlapping the primary spring.
3. Remove the brake shoe return retainers, springs and nails.
4. Slide the eye of the automatic adjuster cable off the top anchor and then unhook the cable from the adjusting lever. Remove the cable, cable guide and the anchor plate.
5. Disconnect the lever spring from the lever and disengage it from the shoe web. Remove the spring and lever.
6. Spread the anchor ends of the primary and secondary shoes and remove the parking brake strut and spring, if working on the rear wheels.
7. Disengage the parking brake cable from the parking brake lever and remove the brake assembly, if working on the rear wheels.
8. Remove the brake assembly and adjusting wheel assembly from the backing plate. Install a wheel cylinder piston retaining spring over the wheel cylinder.
9. Inspect the backing plate platforms for nick, burrs or extreme wear. After cleaning, apply a thin coat of lubricant to the suypport platforms.
10. If working on the rear wheels, attach the parking brake lever to the secondary shoe and retain in place with the attaching clip.
11. Position the primary and secondary shoes on a flat surface.
12. Lubricate the threads of the adjusting screw and install it between the primary and secondary shoes with the star wheel next to the secondary shoe. The star wheels are marked "R"(right side) and "L" (left side) and indicate their location on the vehicle.
13. Overlap anchor ends of the primary and secondary brake shoes and install the adjusting spring and lever.
14. Hold the brake shoes in their relative position on the backing plate and if working with the rear brake shoes, engage the parking brake cable into the parking brake lever.

BRAKES
Light Trucks & Vans

15. Retain the brake shoes with the retainer nails, springs and retainer, while installing the rear wheels parking brake strut and spring in position between the two shoes, if working on the rear wheels.
16. Complete the installation of the shoes to the backing plate and install the anchor pin plate.
17. Install the eye of the adjusting cable over the anchor pin and install the return spring between the primary shoe and the anchor spring.
18. Install the cable guide in the secondary shoe. Then install the secondary return spring, being sure the the secondary spring overlaps the primary spring.
19. Place the adjusting cable in the groove of the cable guide and engage the hook of the cable into the adjusting lever.
20. Be sure the adjuster operates satisfactorily and adjust the brake shoes to match the drum diameter.
21. Install the brake drum and retaining clips, make final brake shoe adjustments and prepare for road test.

Two—Piston Single Cylinder Hydraulically Actuated Type

Both shoes pivot on anchor pins at the bottom of the support plate. The shoes are actuated by one wheel cylinder which is of the double piston type. Specifications for heel and toe clearance of the shoes should be strictly followed to obtain efficient brake operation.

BRAKE SHOES

Removal And Installation

1. Back off the adjusting cam and remove the wheel and drum assembly.
2. Remove the brake shoe return spring
3. Install wheel cylinder piston clamp to prevent the pistons from being forced from the cylinder.
4. Remove the C-washer and retainer, guide spring retainer and guide spring from the anchor bolts to remove the brake shoes.
5. To install the brake shoes, reverse the removal procedure.

ADJUSTMENTS

Since tapered brake lining is thicker at the center than at the ends, the adjustment procedures outlined must be performed in order to assure maximum braking efficiency.

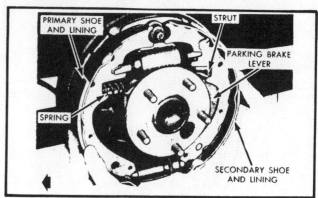
Installing brake shoes (© Chrysler Corp.)

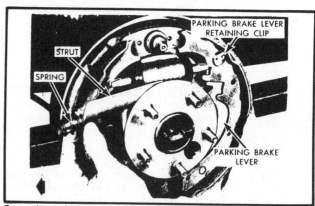
Removing or installing parking brake strut and spring—rear (© Chrysler Corp.)

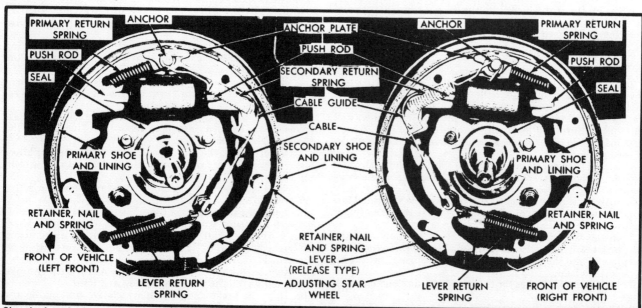
Chrysler front brake assembly (© Chrysler Corp.)

BRAKES
Light Trucks & Vans

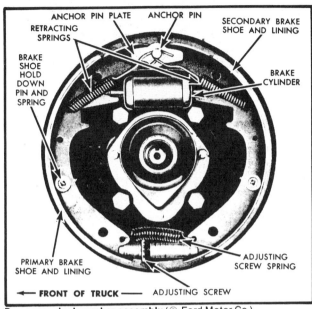

Duo-servo single anchor assembly (© Ford Motor Co.)

Minor Adjustment

1. Raise the vehicle and support safely so that the wheels of the brakes to be adjusted can be rotated freely.
2. While rotating the wheel forward and backward, adjust the shoe out to the drum with the adjusting cam until a light drag is obtained.
3. Back off the adjustment until the wheel is free to turn.
4. Repeat this operation on the other shoe. Continue to adjust the other brake shoes in a like manner.

Major Adjustment

1. Be sure the fluid level in the master cylinder is $3/8-1/2$ in. from the top of the reservoir.
2. Loosen the locknuts and turn the brake shoe anchor bolts to the fully released position.

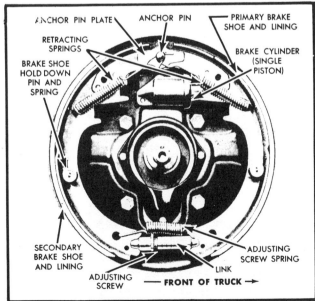

Uni-servo single anchor brake assembly (© Ford Motor Co.)

3. Adjust the anchor bolt and cam, the minor anchor bolt and the minor adjustment cam at the top of the brake shoe to give equal clearance at the toe and heel of the brake shoes. Make sure that sufficient center contact is maintained to produce a slight drag.
4. Lock the anchor adjusting nut. After adjusting the clearance on one shoe, repeat the procedure on the other shoe, then apply the brakes a couple of times to make sure the adjustment is to specifications.

NOTE: **Whenever cams are adjusted, check the brakes by applying pressure on the brake pedal several times so as to make sure wheel drag has not increased, since the spring loaded cams may cause shoe adjustment to change by shifting position. Wheels should have a slight drag at room temperature.**

Bendix Single Anchor Brakes

BRAKE SHOES

Removal

1. Remove the wheel and drum. Do not push down the brake pedal after the brake drum has been removed. On a truck equipped with a vacuum or air booster, be sure the engine is stopped and there is no vacuum or air pressure in the system before disconnecting the hydraulic lines.
2. Clamp the brake cylinder boots against the ends of the cylinder, and remove the brake shoe retracting springs from both shoes.
3. Remove the anchor pin plate.
4. Remove the hold-down spring cups and springs from the shoes, and remove the shoes and the adjusting screw parts from the carrier plate. Do not let oil or grease touch the brake linings. If the shoes on a rear brake assembly are being removed, remove the parking brake lever, link, and spring with the shoes. Unhook the parking brake cable from the lever as the shoes are being removed.
5. If the shoes are from a rear brake assembly, remove the parking brake lever from the secondary shoe.

Installation

1. Coat all points of contact between the brake shoes and the other brake assembly parts with Lubriplate® or a similar lubricant. Lubricate the adjusting screw threads.
2. Place the adjusting screw, socket, and nut on the brake shoes so that the star wheel on the screw is opposite the adjusting hole in the carrier plate. Then install the adjusting screw spring.
3. Position the brake shoes and the adjusting screw parts on the carrier plate, and install the hold-down spring pins, springs, and cups. When assemblying a rear brake, connect the parking brake lever to the secondary shoe, and install the link and spring with the shoes. Be sure to hook the parking brake cable to the lever.
4. Install the anchor pin plate on the pin.
5. Install the brake shoe retracting springs on both shoes. The primary shoe spring must be installed first.
6. Remove the clamp from the brake cylinder boots.
7. Install the wheel and drum.
8. Bleed the system and adjust the brakes. Check the brake pedal operation after bleeding the system.

Bendix Double Anchor Brakes

BRAKE SHOES

Removal and Installation

1. Remove the wheel and drum. Do not push down the brake pedal after the brake drum has been removed. On trucks

1–127

BRAKES
Light Trucks & Vans

equipped with vacuum boosters, be sure the engine is stopped and there is no vacuum in the system before disconnecting the hydraulic lines.

2. Clamp the brake cylinder boots against the ends of the cylinder, and remove the brake shoe retracting springs from both shoes.

3. At each shoe, remove the 2 brake shoe retainers and washers from the hold-down pins and remove the spring and pin from the carrier plate. Remove the anchor pin retainers and remove the shoes from the anchor pins. Do not allow grease or oil to touch the linings.

4. Clean all brake assembly parts. If the adjusting cams do not operate freely apply a small quantity of lubricating oil to points where the shaft of the cam enters the carrier plate. Wipe dirt and corrosion off the plate.

5. Clean the ledges on the carrier plate with sandpaper. Coat all points of contact between the brake shoes and the other brake assembly parts with high temperature grease.

6. Position the brake shoes on the carrier plate with the heel (lower) end of the shoes over the anchor pins and the toe (upper) end of the shoes engaged in the brake cylinder link. Install the hold-down spring pins, spring, washers and retainers.

7. Install the anchor pin retainers and then install the brake shoe return spring.

8. Turn the brake shoe adjusting cams to obtain maximum clearance for brake drum installation.

9. Install the wheel and drum assembly.

10. Bleed the brake system and adjust the brakes.

11. Check brake pedal operation and road test.

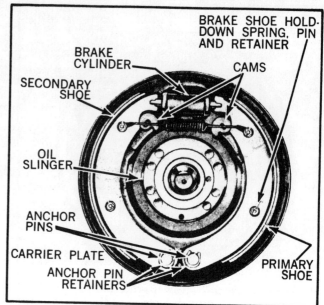

Double anchor brake assembly (© Ford Motor Co.)

Bendix Two Cylinder Brakes

BRAKE SHOE

Removal

1. Remove the wheel, and then remove the drum or the hub and drum assembly. Mark the hub and drum to aid assembly in the same position. On trucks equipped with vacuum or air boosters, be sure the engine is stopped and there is no vacuum or air pressure in the system before disconnecting the hydraulic lines.

2. Clamp the brake cylinder boots against the ends of the cylinder and remove the four brake shoe retracting springs.

3. Remove the brake shoe guide bolt cotter pin, nut, washer, and bolt from both shoes, and remove the shoes from the carrier plate.

4. Remove the clamp-type adjusting wheel lock from the anchor pin support, and unthread the adjusting screw and wheel assembly from the anchor pin support.

Installation

1. Clean the carrier plate ledges with sandpaper. Coat all points of contact between the brake shoes and other brake assembly parts with high temperature grease.

2. Thread the adjusting screw and wheel assembly into the anchor pin support and install the clamp-type adjusting wheel lock. Thread the adjusting wheel into the support so that the brake shoe will rest against the adjusting wheel end.

3. Place the brake shoe over the two brake shoe anchor pins, insert the ends in the brake cylinder links, and install the shoe guide bolt, washer, and nut. Finger tighten the nut, then back off one full turn, and install the cotter pin.

4. Install the four retracting springs.

5. Remove the cylinder clamps, install the drum or the hub and drum assembly, then install the wheel assembly. Align the marks on the hub and drum during installation.

6. Bleed and adjust the brakes.

7. Check pedal operation and road test.

BENDIX BRAKE SHOES

ADJUSTMENT

The brake drums should be at normal room temperature, when the brake shoes are adjusted. If the shoes are adjusted when the shoes are hot and expanded, the shoes may drag as the drums cool and contract.

A minor brake adjustment re-establishes the brake lining-to-drum clearance and compensates for normal lining wear.

A major brake adjustment includes the adjustment of the brake shoe anchor pins as well as the brake shoes. Adjustment of the anchor pin permits the centering of the brake shoes in the drum.

Adjustment procedures for each type of brake assembly are given under the applicable heading.

Minor Adjustment

The brake shoe adjustment procedures for the uniservo single anchor brake assembly are the same as those for the duo-servo single anchor type.

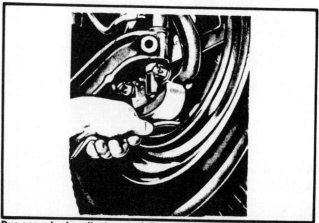

Duo-servo brake adjustment (© Ford Motor Co.)

BRAKES
Light Trucks & Vans

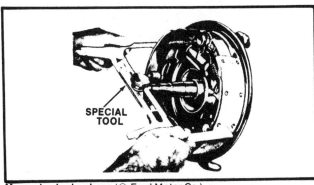

Measuring brake shoes (© Ford Motor Co.)

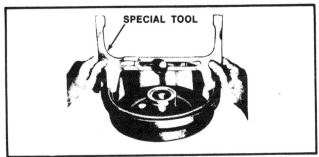

Measuring brake drum (© Ford Motor Co.)

A major brake adjustment should be performed when dragging brakes are not corrected by a minor adjustment, when brake shoes are relined or replaced, or when brake drums are machined.

Duo—Servo Single Anchor Brake

The duo-servo single-anchor brake is adjusted by turning an adjusting screw located between the lower ends of the shoes.
 1. Raise the truck until the wheels clear the floor.
 2. Remove the cover from the adjusting hole at the bottom of the brake carrier plate, and turn the adjusting screw inside the hole to expand the brake shoes until they drag against the brake drum.
 3. When the shoes are against the drum, back off the adjusting screw 10 or 12 notches so that the drum rotates freely without drag.
 4. Install the adjusting hole cover on the brake carrier plate.
 5. Check and adjust the other three brake assemblies. When adjusting the rear brake shoes, check the parking brake cables for proper adjustment. Make sure that there is clearance between the ends of the parking brake link and the shoes.
 6. Apply the brakes. If the pedal travels more than halfway down between the released position and the floor, too much clearance exists between the brake shoes and the drums. Repeat Steps 2 and 3 above. Internal inspection and/or bleeding may be necessary.
 7. When all brake shoes have been properly adjusted, road test the truck and check the operation of the brakes. Perform the road test only when the brakes will apply and the truck can be safely stopped.

SINGLE ANCHOR PIN

Major Adjustment

A major brake adjustment should be made when dragging brakes are not corrected by a minor adjustment, when brake shoes are relined or replaced, or when brake drums are machined.

 1. Raise the truck until the wheel clears the floor.
 2. Rotate the drum until the feeler slot is opposite the lower end of the secondary (rear) brake shoe.
 3. Insert a 0.010 in. feeler gauge through the slot in the drum. Move the feeler up along the secondary shoe unit it is wedged between the secondary shoe and the drum.
 4. Turn the adjusting screw (star wheel) to expand the brake shoes until a heavy drag is felt against the drum. Back off the adjusting screw just enough to establish a clearance of 0.010 in. between the shoe and the drum at a point $1\frac{1}{2}$ in. from each end of the secondary shoe. This adjustment will provide correct operating clearance for both the primary and secondary shoes. If the 0.010″ clearance cannot be obtained at both ends of the secondary shoe, the anchor pin must be adjusted.
 5. To adjust the anchor pin setting, loosen the anchor pin nut just enough to permit moving the pin up or down by tapping the nut with a soft hammer. Do not back the nut off too far or the shoes will move out of position when the nut is tightened. Tap the anchor pin in a direction that will allow the shoes to center in the drum and provide an operating clearance of 0.010 in.. Torque the anchor pin nut to 80–100 ft. lbs. Recheck the secondary shoe clearance at both the heel and toe ends of the shoe.
 6. When all brake shoes and anchor pins have been properly adjusted, road test the truck and check the operation of the brakes. Perform the road test only when the brakes will apply and the truck can be safely stopped.

Double Anchor Pin

Major Adjustment

 1. Raise the truck until the wheels clear the floor.
 2. Rotate the drum until the feeler slot is opposite the lower (heel) end of the secondary (rear) brake shoe.
 3. Insert a 0.007 in. feeler gauge through the slot in the drum. Move the feeler up along the secondary shoe until it is wedged between the shoe and the drum.
 4. Loosen the secondary shoe anchor pin nut. Turn the secondary shoe anchor pin until the brake shoe-to-drum clearance at a point $1\frac{1}{2}$ in. from the heel end of the shoe is 0.007 in. Remove the feeler gauge.
 5. Rotate the drum until the feeler slot is opposite the upper (toe) end of the secondary brake shoe.
 6. Insert a 0.010 in. feeler gauge through the slot in the drum. Move the feeler gauge down along the secondary shoe until it is wedged between the shoe and the drum. Turn the adjusting cam, to expand the brake shoe, until a heavy drag is felt against the drum.
 7. Turn the anchor pin until the brake shoe-to-drum clearance at a point $1\frac{1}{2}$ in. from the toe end of the shoe is 0.010 in.. Remove the feeler gauge.
 8. Torque the anchor pin nut to 80–100 ft. lbs. Recheck the heel and toe clearances.
 9. Using the preceding secondary brake shoe adjustment procedure as a guide, adjust the primary brake shoe-to-drum clearance.
 10. Road test the truck and check the operation of the brakes.

NOTE: Perform the road test only when the brakes will apply and the truck can be safely stopped.

Kelsey Hayes

FRONT BRAKE SHOES

Removal

 1. Raise the vehicle until the wheel clears the floor. Remove the wheel, drum and hub assembly.
 2. Clamp the wheel cylinder boots against the ends of the cylinder.

BRAKES
Light Trucks & Vans

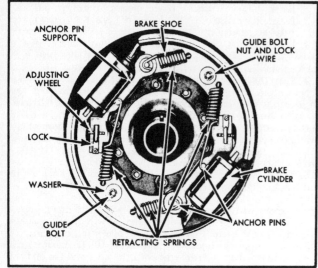

Two cylinder brake—equal length springs (© Ford Motor Co.)

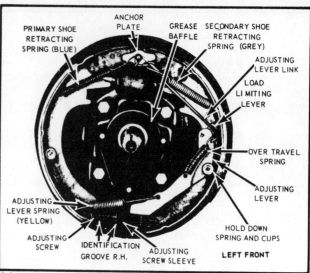

Kelsey-Hayes front brake assembly (© Ford Motor Co.)

3. Remove the brake shoe retracting springs from both shoes.
4. Remove the adjusting lever link, anchor plate and the adjusting lever spring.
5. Remove the hold down spring cups, springs and the adjusting lever.
6. Remove the brake shoes and adjuster screw assembly from the backing plate.

Installation

1. Clean all brake dust from the brake assembly parts with a clean dry rag.
2. Coat all points of contact between the shoes and other brake parts with high temperature grease.

3. Coat the adjuster screw with high temperature grease before assembly. Thread the adjuster screw into the adjuster screw sleeve.
4. Position the brake shoes on the backing plate and install the adjusting lever, hold down pins, springs and cups.
5. Position the adjuster screw assembly on the brake shoes so that the star wheel is opposite the adjusting slot in the backing plate. Install the adjusting lever spring.
6. Install the anchor plate and adjusting lever link.
7. Install the secondary brake shoe retracting spring.
8. Install the primary brake shoe retracting spring.
9. Remove the clamp from the wheel cylinder boots.
10. Install the wheel, drum and hub assembly.
11. Adjust the brakes. Subsequent adjustment will be automatic.

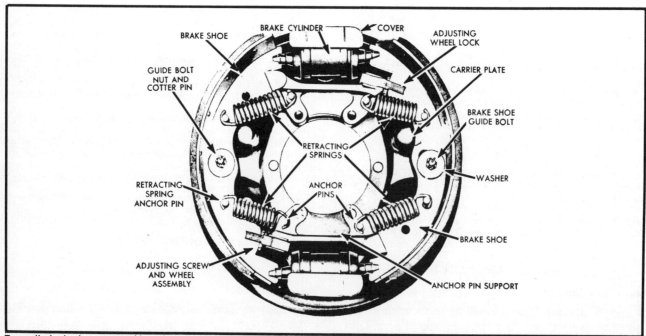

Two cylinder brake—unequal length springs (© Ford Motor Co.)

BRAKES
Light Trucks & Vans

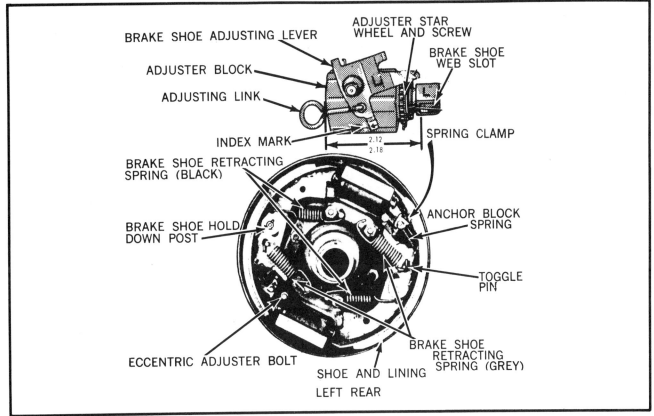

Kelsey-Hayes rear brake assembly (© Ford Motor Co.)

REAR BRAKE SHOES

Removal

1. Raise the truck until the wheel clears the floor.
2. Remove the wheel, hub and drum assembly.
3. Clamp the brake cylinder boots against the ends of the cylinder with brake piston clamps.
4. The two different types of brake shoe retracting springs and remove the springs.
5. Remove the brake shoe hold down post cotter key, nut, and shoe hold down washer.
6. Loosen and remove the eccentric adjuster bolt, lock washer, eccentric and adjusting link.
7. Remove the shoe and lining assembly from the backing plate.
8. Remove the anchor block spring and slide the adjuster assembly from the shoe web.
9. Remove the adjuster star wheel and screw from the adjuster block. Unthread the star wheel from the adjuster screw.

Installation

1. Wipe all brake dust from the brake assembly parts with a clean dry rag. Coat all points of contact between brake shoes and other parts with high temperature grease.
2. Coat the adjuster screw and the inside of the adjuster block with high temperature grease.
3. Thread the adjuster screw onto the star wheel and insert the adjuster screw assembly into the adjuster block. Maintain a 2.12–2.18 in. dimension from the end of the adjuster block to the adjuster screw web slot.
4. Install the adjuster assembly onto the shoe web and attach the anchor block spring.
5. Place the brake shoe over the retracting spring toggle pin and insert the ends of the shoe in the wheel cylinder links.
6. Install the shoe hold down washer and nut. Do not install the cotter pin.
7. Install the four brake shoe retracting springs. Make sure the retracting springs are installed. On 15 × 5 in. brakes the inner hook ends face the wheel cylinders. On 15 × 4 in. brakes the inner hook ends face the center of the axle.
8. Install the adjusting link, eccentric, lockwasher and adjuster bolt. Do not tighten.
9. Remove the brake piston clamps.
10. Tighten the shoe hold down nut until there is 0.015–0.025 in. clearance between the shoe and hold down washer with the shoe held against the backing plate. Install the cotter pin.
11. Center the shoes on the backing plate. Using a ½ in. wrench, rotate the adjuster eccentric until the adjusting lever is at the index mark. Tighten the eccentric adjuster bolt to specification.
12. Install the wheel, hub and drum assembly.
13. Adjust the brake to obtain a slight drag. Subsequent adjustments will be automatic.

BRAKE SHOES

ADJUSTMENT

The brake drums should be at normal room temperature, when the brake shoes are adjusted. If the shoes are adjusted when the shoes are hot and expanded, the shoes may drag as the drums cool and contract.

The brake shoes are automatically adjusted when the vehicle is driven in reverse and the brakes applied. A manual adjustment is required only after the brake shoes have been relined or replaced. The manual adjustment is performed while

BRAKES
Light Trucks & Vans

the drums are removed, using the tool and the procedure detailed below.

When adjusting the rear brake shoes, check the parking brake cables for proper adjustment. Make sure that the equalizer operates freely.

Adjustment of Brake Shoes:

1. Use special drum diameter to lining gauge and adjust the lining to the inside diameter of the drum braking surface.
2. Reverse the gauge and adjust the brake shoes to touch the gauge. The gauge contact points on the shoes must be parallel to the vehicle with the center line through the center of the axle. Hold the automatic adjusting lever out of engagement while rotating the adjusting screw, to prevent burring the screw slots. Make sure the adjusting screw rotates freely.
3. Apply a small quantity of high temperature grease to the points where the shoes contact the carrier plate, being careful not to get the lubricant on the linings.
4. Install the drums. Install the retaining nuts and tighten securely.
5. Install the wheels on the drums and tighten the mounting nuts to specification.
6. Complete the adjustment by applying the brakes several times while backing the vehicle.
7. After the brake shoes have been properly adjusted, check the operation of the brakes by making several stops while operating in a forward direction.

KELSEY HAYES SELF ADJUSTING BRAKES

Adjustment

TWO CYLINDER FRONT BRAKES

Two cylinder front brakes are adjusted by means of exposed, hex-head, self-locking cam adjusters. The brakes are to be manually adjusted initially. Subsequent adjustment is automatic.

Adjustment of Brakes

1. Raise the vehicle and check the front brakes for drag by rotating the wheels.
2. Adjust one shoe by rotating the wheel backward and forward while turning the cam hex-head with a wrench. Bring the shoe out to the drum until a light drag is felt. Do not apply excessive force on the hex head cam, as automatic adjuster parts can be damaged. Back off the adjustment until the wheel turns freely. Adjust the other cam on the same wheel in the same manner.
3. Adjust the other front wheel brake using the procedure above.
4. Apply the brakes and recheck the adjustment.

KELSEY HAYES SELF ADJUSTING BRAKES

Adjustment

Rear Brakes

The brake shoes are automatically adjusted when the vehicle is driven in reverse and the brakes applied. A manual adjustment is required only after the brake shoes have been relined or replaced.

The two-cylinder brake assembly brake shoes are adjusted by turning adjusting wheels reached through slots in the backing plate.

Two types of two-cylinder brake assemblies are used on truck rear wheels. The assemblies differ primarily in the retracting spring hookup, and in the design of the adjusting screws and locks. However, the service procedures are the same for both assemblies.

The brake adjustment is made with the vehicle raised. Check the brake drag by rotating the drum in the direction of forward rotation as the adjustment is made.

1. Remove the adjusting slot covers from the backing plate.
2. Turn the rear (secondary shoe) adjusting screw inside the hole to expand the brake shoe until a slight drag is felt against the brake drum.
3. Repeat the above procedure on the front (primary) brake shoe.
4. Replace the adjusting hole covers.
5. Complete the adjustment by applying the brakes several times while backing the vehicle.
6. After the brake shoes have been properly adjusted, check the operation of the brakes by making several stops while operating in a forward direction.

Rear Brake Assembly Used With Hydro-Max Power Booster

1984 AND LATER

The rear drum brakes are a completely new simplified design incorporating many air brake type features. Shoe and lining removal requires only removal of the two shoe retractor springs. The automatic adjusting mechanisms are part of the wheel cylinder pistons and are submerged in the brake fluid protected from any road contaminants. Automatic adjustment will take place in either forward or reverse direction. The lining blocks, four per wheel, are tapered from $3/4$ in. thick in the center to $5/8$ in. thick at the ends. There are two lining inspection holes in the backing plate with removable rubber plugs. The lining blocks have a wear limit groove in the edge of the material visible through the inspection holes. The shoes can be backed off manually through two 'adj' access holes in the backing plate for drum removal.

MANUAL ADJUSTMENT AND DE-ADJUSTMENT

Manual adjustment should not be considered as an alternative to the auto-adjuster. It has two functions:

1. To initially set shoe to drum clearance with automatic adjusting brakes, it is important to set the shoe to drum clearance prior to driving the vehicle. To adjust, using the backing plate as a fulcrum, turn the manual override wheel in a counterclockwise direction as viewed when looking down the piston.
2. To de-adjust the brake shoes (where there is a lipped drum condition), remove the plugs from the adjustment holes marked 'ADJ' in the backing plate. Insert a brake adjustment tool or a flat-bladed screwdriver until it engages a slot in the manual override wheel. Use the backing plate as a fulcrum to turn the manual override wheel in a clockwise direction (as viewed when looking down the piston) until the lining clears the drum. Repeat with the other shoe and remove the drum.

BRAKE SHOE AND LINING

Removal

NOTE: Always replace brake shoes and linings in axle sets. When replacing shoes, always replace the return springs to insure proper operation of the auto-adjuster.

1. Raise the vehicle and install safety stands. Remove the wheel and tire assembly.
2. Remove the adjuster plugs from the slots in the backing plate marked 'ADJ'. Remove the adjuster sight hole plugs.
3. Insert a brake adjustment tool or a flat-bladed screwdriver into the adjustment slot until the blade engages the slots in the adjustment wheel. Turn the wheel in a clockwise direction (when viewed looking down the piston) until the shoe and lining clears the drum. Remove the drum.

BRAKES
Light Trucks & Vans

4. In order for de-adjustment to take place, there must be a load applied to the wheel cylinder pistons. For this reason, fully de-adjust both wheel cylinders before removing the shoes.

NOTE: To avoid locking the auto-adjust mechanism in the fully de-adjusted position, wind out each wheel cylinder piston one complete turn after fully de-adjusting the pistons.

5. Remove the springs, with a removal tool.
6. Insert the removal tool in the loop on the return spring. Rest the fulcrum of the tool against the wheel cylinder body. Unhook the spring from the shoe web. Support the lower shoe and repeat procedure for other spring. Remove the shoes. Remove and discard the springs.

NOTE: Make sure the fulcrum of the tool does not rest on the adjusting wheel or the dust boot.

―――― CAUTION ――――
Do not use an air gun to remove dust from the backing plate. Remove dust with Brake Service Vacuum. Dust may also be removed with a damp rag.

7. Inspect the wheel cylinder/adjuster for leaks by removing the tappet head assembly. Lift the dust boot from the piston. If fluid escapes, rebuild or replace the wheel cylinder/adjuster.
8. Inspect the wheel cylinder/parking brake expander by removing the dust boot and inspecting for leaks. If leaks are present, rebuild or replace the wheel cylinder/expander.
9. If leakage is not evident, install the dust boot and tappet head assembly on the wheel cylinder. Temporarily place an elastic band around the cylinders to keep the pistons in place. Use a wire brush to remove any corrosion from the backing plate, taking care not to damage the wheel cylinder and boots.

Installation

1. Remove the elastic band from the wheel cylinders.
2. Lightly smear the abutment ends of the new shoes and the tips of the steady posts with high temperature grease. Keep the grease away from all hydraulic components and the shoe linings. The replacement brake shoes must be installed correctly to the brake, i.e. the linings are symmetrical on platform, although there is a taper on the shoe, the only way to correct installation is via the web profile.
3. Use the correct color of new shoe return springs. The RED colored springs with one coil is used on wheel cylinder/parking brake expander side of the backing plate on the right and left side. The GREEN colored shoe return spring with two coils is used on the wheel cylinder/adjuster side of the LEFT brake assembly. The YELLOW colored spring with two coils is used on the wheel cylinder/adjuster on the RIGHT brake assembly.

NOTE: After a spring is removed, it must be discarded and replaced with a new spring to insure proper operation of the auto-adjuster.

4. Install the springs on the shoes into position. Insert the end of the spring opposite the loop in the shoe. Use the spring removal-installation tool by placing it in the loop of the spring. Rest the fulcrum of the tool on the wheel cylinder body. Use the tool as a lever to lift up the spring and insert it into the hole in the shoe.

―――― CAUTION ――――
Make sure the fulcrum of the tool does not rest on the manual override wheel or dust boot of the wheel cylinder.

5. Install the drum or spider/drum assembly.
6. Manually adjust the wheel cylinders through the backing plate adjusting holes until the shoe to drum clearance is less than 1.8 mm (0.070 in.). Apply the brake pedal to centralize the shoes and release the brake pedal.
7. Check the shoe to drum clearance by using a feeler gauge placed through the lining inspection holes in the backing plate. If the reading is not 0.51–0.76mm (0.020–0.030 in.), manually adjust the wheel cylinder through the manual override wheel until the specified dimensions are obtained.
8. Bleed the brakes.
9. Install the wheel and tire. Install the lug nuts and tighten.
10. Insert the plugs in the inspection, adjuster slot and adjustment slots in the backing plate.
11. Check the brake fluid level in the master cylinder.

NOTE: The service brake system uses Ford Heavy Duty Brake Fluid, C6AZ-19542-A or -B (ESA-M6C25-A) or equivalent and is filled at the Hydro-Max master cylinder. The parking brake system uses Ford Automatic Transmission Fluid, ESP-M2C138-CJ, DEXRON® II or equivalent and is filled at the brake pump reservoir. DO NOT MIX FLUIDS.

12. Road test vehicle and check brake operation.

WHEEL CYLINDER/PARKING BRAKE EXPANDER

Removal

NOTE: If both the wheel cylinder/adjuster and the wheel cylinder/parking brake expander are to be serviced, the complete backing plate assembly may be removed by disconnecting the fluid lines (service and parking) and removing the bolts retaining the wheel cylinder assemblies and backing plate to the axle flange. Removal of the assemblies can then take place on the bench. IF THE BACKING PLATE ASSEMBLY IS TO BE REMOVED, PRESSURE MUST BE RELEASED BY CAGING THE SPRING IN THE PARKING BRAKE CHAMBER.

1. Attach a tube to the bleeder screw on the wheel cylinder/parking brake expander. Place the other end of the tube in a container. Unscrew the bleeder screw one-half turn. Pump the brake pedal to remove all brake fluid from the rear brake system. Discard the brake fluid.
2. Remove the shoe and lining.
3. Remove the parking brake chamber from the wheel cylinder/parking brake expander.
4. Remove the bolt retaining the bridge tube retainer to the backing plate and remove the retainer. Disconnect the bridge tube from the cylinders and remove the bridge tube.
5. Insert an Allen-head wrench into the bolts in the rear of the backing plate and remove the bolts retaining the wheel cylinder/parking brake expander to the backing plate.
6. Remove the three 9/16–18 in. bolts, nuts and washers retaining the wheel cylinder/parking brake expander and backing plate to the axle flange. Remove the wheel cylinder/parking brake expander and gasket from the backing plate.
7. Remove any rust or corrosion from the backing plate with a wire brush.

Installation

1. Install a new gasket on the wheel cylinder/parking brake expander. Position the assembly on the backing plate.
2. Install the three 9/16–18 in. bolts, nuts and washers retaining wheel cylinder/parking brake expander and backing plate to the axle flange. Tighten the bolts.
3. Install the Allen-head head bolts in the rear of the backing plate retaining the assembly to the plate.
4. Install the parking brake chamber.
5. Connect the bridge tube to the wheel cylinders. Install the bridge tube retainer and tighten the bolt.

BRAKES
Light Trucks & Vans

6. Install the shoe and lining and adjust the lining-to-drum clearance.

7. Fill the Hydro-Max master cylinder reservoir with clean Heavy Duty Brake Fluid. Bleed the service brakes.

8. If required, bleed the parking brakes. Check the brake pump reservoir with the parking brake applied and if required, fill to the specified level with DEXRON® II or equivalent.

NOTE: The service brake system uses Ford Heavy Duty Brake Fluid, C6AZ–19542–A or –B (ESA–M6C25–A) or equivalent and is filled at the Hydro-Max master cylinder. The parking brake system uses Ford Automatic Transmission Fluid, ESP–M2C138–CJ, DEXRON® II or equivalent and is filled at the brake pump reservoir. DO NOT MIX THE FLUIDS.

9. Apply the brakes and check for leaks. Road test the vehicle and check for proper operation.

WHEEL CYLINDER/ADJUSTER

Removal

NOTE: If both the wheel cylinder/adjuster and the wheel cylinder/parking brake expander are to be serviced, the complete backing plate assembly may be removed by disconnecting the fluid lines (service and parking) and removing the bolts retaining the wheel cylinder assemblies and backing plate to the axle flange. Removal of the assemblies can then take place on the bench. IF THE BACKING PLATE ASSEMBLY IS TO BE REMOVED, PRESSURE MUST BE RELEASED BY CAGING THE SPRING IN THE PARKING BRAKE CHAMBER.

1. Attach a tube to the bleeder screw on the wheel cylinder/parking brake expander. Place the other end of the tube in a container. Unscrew the bleeder screw one half turn. Pump the brake pedal to remove all brake fluid from the rear brake system. Discard the brake fluid.
2. Remove the brake shoe and lining.
3. Disconnect the hydraulic brake line from the fitting on the wheel cylinder/adjuster.
4. Remove the bolt retaining the bridge tube retainer to the backing plate and remove the retainer. Disconnect the bridge tube from the wheel cylinders and remove the bridge tube.
5. Insert an Allen-head wrench into the bolts in the rear of the backing plate and remove the bolts retaining the wheel cylinder/adjuster to the backing plate.
6. Remove the three 9/16–18 in. bolts, nuts and washers retaining the wheel cylinder/adjuster and backing plate to the axle flange. Remove the wheel cylinder/adjuster and gasket from the backing plate.
7. Remove any rust or corrosion from the backing plate with a wire brush.

Installation

1. Install a new gasket on the wheel cylinder/adjuster. Position the assembly on the backing plate.
2. Install the three 9/16–18 in. bolts, nuts and washers retaining the wheel cylinder/adjuster and backing plate to the axle flange. Tighten the bolts.
3. Install and tighten the Allen-head bolts in the rear of the backing plate that retain the assembly to the plate.
4. Connect the bridge tube to the wheel cylinders. Install the bridge tube retainer and tighten the bolt.
5. Connect the hydraulic brake line to the wheel cylinder/adjuster.
6. Install the shoe and lining and adjust the lining-to-drum clearance. Tighten the bleeder screw.
7. Fill the Hydro-Max master cylinder reservoir to the specified level with clean Heavy Duty Brake Fluid. Bleed the service brake system.

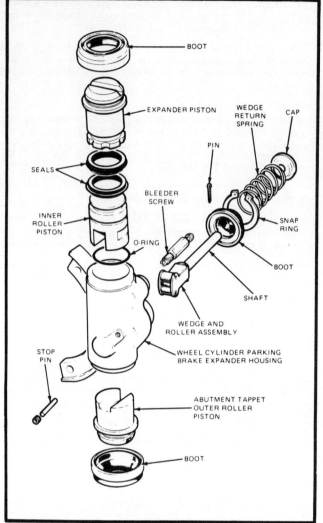

Exploded view wheel cylinder/parking brake expander

NOTE: The service brake system uses Ford Heavy Duty Brake Fluid, C6AZ–19542–A or –B (ESA–M6C25–A) or equivalent and is filled at the Hydro-Max master cylinder. The parking brake system uses Ford Automatic Transmission Fluid, ESP–M2C138–CJ, DEXRON® II or equivalent and is filled at the brake pump reservoir. DO NOT MIX THESE FLUIDS.

8. Apply the brakes and check for leaks. Road test the vehicle and check for proper operation.

WHEEL CYLINDER/ADJUSTER

Disassembly

NOTE: Use only clean Heavy Duty Brake Fluid to clean the parts. The same brake fluid is also required for lubrication purposes upon assembly.

1. Make sure cylinder, work bench, tools and hands are clean before proceeding with disassembly.
2. Remove the tappet head and manual adjuster assembly. Slip the dust boot off the housing.
3. Use air pressure to remove both piston and adjuster assemblies from housing. Use care when applying air pressure and removing the piston.

BRAKES
Light Trucks & Vans

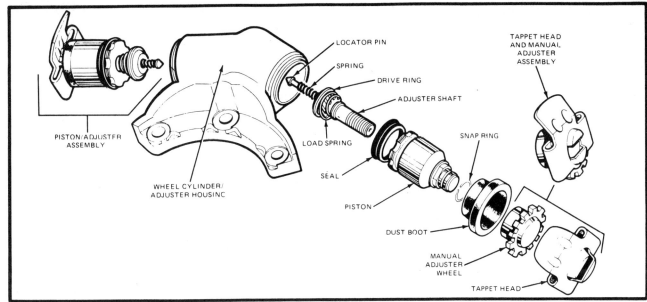

Exploded view wheel cylinder/adjuster

4. Remove the locator pin and spring.
5. Do not remove the adjuster shaft from the piston. Do not remove the drive ring and load spring from the adjuster shaft.
6. Use a round-tipped screwdriver to remove the seal from the piston so as not to score the seal groove. Discard the seal.
7. Wash all parts with clean Heavy Duty Brake Fluid.
8. Examine the cylinder bore in the housing for corrosion, ridges or score marks. Replace the housing if required. Make sure all parts are in good working condition.

NOTE: Some discoloration of the bore surface near the mouth of the wheel cylinder (alternatively 'expander'), and of the piston diameter may be apparent in service. This is the natural result of brake actuation over a peroid of operation.

Assembly

1. Install a new seal in the groove on the piston.
2. Turn the adjuster shaft until it bottoms in the piston and then unscrew the shaft one full turn.

NOTE: It is essential that the adjuster shaft rotates freely in the piston to allow for the initial self-adjuster movement to take place. If the shaft does not rotate freely in the piston or binds, replace the piston and adjuster shaft.

3. Install the spring and locator pin in the adjuster shaft.
4. Lubricate the piston with clean Heavy Duty Brake Fluid. Install the piston assembly in the housing cylinder bore.
5. Install the dust boot onto the wheel cylinder.
6. Install the tappet head and manual adjuster assembly.

NOTE: Do not turn the manual adjuster wheel.

WHEEL CYLINDER/PARKING BRAKE EXPANDER

Disassembly

1. Place the wheel cylinder/parking brake expander in a soft jawed vise.
2. Press down on the shaft and return spring and remove the pin from the shaft. Slowly release wedge return, spring and remove the cap and return spring from the shaft.
3. Remove the snap ring that retains the shaft and boot in the housing. Remove and discard the dust boot.
4. Remove stop pin from the housing.
5. Remove and discard the dust boots from the housing.
6. Remove the abutment tappet outer roller piston.
7. Remove the wedge and roller assembly.
8. Remove the bleed screw.
9. Push the expander piston and the inner roller piston out of the opposite side of the housing.
10. Remove and discard the O-rings and wiper seals from the pistons.
11. Wash all parts in clean Heavy Duty Brake Fluid.
12. Inspect the cylinder bore and pistons for evidence of corrosion, ridges or scoring. Replace, if required.

Assembly

— CAUTION —

Care must be taken when assembling the wheel cylinder/parking brake expander. Both Ford Heavy Duty Brake Fluid, C6AZ-19542-A or -B (ESA-M6C25-A) or equivalent and a synthetic based grease are used within the housing (BATCO 5-86-B Grease or equivalent or Castrol G 148 or equivalent) because they are compatible with each other. THE USE OF ANY MINERAL BASED GREASE WILL CAUSE CONTAMINATION OF THE SEALS AND MAY RESULT IN BRAKE FAILURE. Use only the specified lubricants.

1. Thoroughly smear the wedge, cage and roller with the grease supplied. Fit the wedge seal to the stem as close to the cage as possible. Insert the assembly into the housing and retain with a new circlip.
2. Coat the outside diameter and the ramp on the abutment tappet outer roller piston with BATCO S-86-B grease or equivalent or Castrol G 148 or equivalent. Install the outer roller piston in the bore so the ramp engages the wedge and roller assembly. Install a new dust boot over the outer roller piston on the housing.
3. Coat the inner roller piston ramp and O-ring with or equivalent and install the O-ring on the piston.

1-135

BRAKES
Light Trucks & Vans

4. Coat the wiper seals with clean Heavy Duty Brake Fluid, or equivalent. Install the wiper seals on the piston so seal lips face the wedge and roller assembly.
5. Install the inner roller piston so the ramp engages the wedge and roller.
6. Install the expander piston in the bore. Install a new dust boot over the expander piston on the housing.
7. Push down on each tappet until the cage and roller is engaged and install the stop pin. Tighten to 21 Nm (16 ft. lbs).
8. Install the return spring and cap over the shaft. Compress the spring with the cap and install the cotter pin. Slowly release the spring and cap into position.

NOTE: If the parking brake cable is replaced, prestretch it by applying the parking brake hard about three times before attempting adjustment.

DISC BRAKE SERVICE

Floating Caliper Disc Brakes

This disc brake is a floating caliper design with one or two pistons on one side of the rotor. It is a two piece unit consisting of the caliper and cylinder housing. The caliper is mounted to the anchor plate on two mounting pins which travel in bushings in the anchor plate. The bushings and pins are protected by toot type seals.

Two brake shoe and lining assemblies are used in each caliper, one on each side of the rotor. The shoes are identical and are attached to the caliper with two mounting pins.

The cylinder housing contains the two pistons. The pistons are fitted with an insulator on the front and a seal on the back lip. A friction ring is attached to the back of the piston with a shouldered cap screw. The pistons and cylinder bores are protected by boot seals which are fitted to a groove in the piston and attached to the cylinder housing with retainers. The cylinder assembly is attached to the caliper with two cap screws and washers.

The anchor plate is bolted directly to the spindle. It positions the caliper assembly over the rotor forward of the spindle.

DISC BRAKE SHOE ADJUSTMENT

The front disc brake assembly is designed so that it is inherently self-adjusting and requires no manual adjustment.

DISC BRAKES—TROUBLE DIAGNOSIS

Cause	Correction
1. Master cylinder fluid level low.	1. Fill to proper level with approved fluid. (Fluid level drops as disc brake linings wear.)
2. Poor quality brake fluid (low boiling point) in system.	2. Drain hydraulic system and fill with approved.
3. Air in hydraulic system.	3. Bleed hydraulic system and refill with approved fluid.
4. Hoses soft or weak (expanding under pressure).	4. Replace defective hoses. Combination valve and all cups and seals in complete brakes.
1. Power brake malfunctioning.	1. Check and repair power unit.
2. Linings soiled with brake fluid, oil or grease.	2. Replace shoes and linings.
3. Lines, hoses or connections dented, kinked, collapsed, clogged or disconnected.	3. Repair or replace defective parts.
4. Master cylinder cups swollen.	4. Drain hydraulic system, flush system with brake fluid and replace combination valve and all cups and seals in complete brake system.
5. Master cylinder bore corroded or rough.	5. Repair or replace master cylinder.
6. Caliper pistons frozen or seized.	6. Disassemble caliper and free pistons (replace if necessary).
7. Caliper cylinder bores corroded or rough.	7. Disassemble caliper and remove corrosion or roughness, or replace caliper.
8. Pedal push rod and linkage binding.	8. Free and lubricate.
9. Metering valve not working.	9. Replace combination valve.
GRABBING OR PULLING (Severe Reaction To Pedal Pressure and Out of Line Stops)	
1. Linings soiled with brake fluid, oil or grease.	1. Replace shoes and linings.
2. Caliper loose.	2. Tighten caliper mounting bolts to specified torque.
3. Lines, hoses or connection dented, kinked, collapsed or clogged.	3. Repair or replace defective parts.
4. Master cylinder bore corroded or rough.	4. Repair or replace master cylinder.
5. Caliper pistons frozen or seized.	5. Disassemble caliper and free pistons (replace if necessary).
6. Caliper cylinder seals soft or swollen.	6. Drain hydraulic system, flush system with brake fluid and replace all cups and seals in complete brake system.
7. Caliper cylinder bores corroded or rough.	7. Disassemble caliper and remove corrosion or roughness, or replace caliper.

BRAKES
Light Trucks & Vans

DISC BRAKES—TROUBLE DIAGNOSIS

Cause	Correction
8. Pedal linkage binding (and suddenly releasing).	8. Free and lubricate linkage.
9. Metering valve not functioning properly.	9. Replace combination valve.
FADING PEDAL (Pedal Falling Away Under Steady Pressure)	
1. Poor quality brake fluid (low boiling point) in system.	1. Drain hydraulic system and fill with approved fluid.
2. Hydraulic connections loose; lines or hoses ruptured (causing leakage).	2. Tighten or replace defective parts.
3. Master cylinder cup worn or damaged. (primary, secondary or both).	3. Repair master cylinder.
4. Master cylinder bore corroded, worn or scored.	4. Repair or replace master cylinder.
5. Caliper cylinder seals worn or damaged.	5. Replace seals.
6. Caliper cylinder bores corroded, worn or scored.	6. Disassemble caliper and remove corrosion or scoring, or replace caliper.
7. Bleed screw open.	7. Close bleed screw and bleed hydraulic system.
NOISE AND CHATTER (May Be Accompanied By Brake Roughness and Pedal Pumping)	
1. Disc has excessive lateral runout.	1. Replace or machine disc.
2. Disc has excessive thickness variations (out of parallel).	2. Replace or machine disc.
3. Disc has casting imperfections.	3. Replace disc.
4. Car creeping or moving slowly with brakes applied (may produce groan or crunching noise).	4. Increase or decrease pedal effort slightly.
5. Squeal, during application.	5. A small amount of high-pitched squeal is inherent in disc brake design and must be considered normal. Some relief may be obtained with service package backing.
DRAGGING BRAKES (Slow or Incomplete Release of Brakes)	
1. Lines, hoses or connections dented, kinked, collapsed or clogged.	1. Repair or replace defective parts.
2. Master cylinder compensating port restricted by swollen primary cup.	2. Drain hydraulic system, flush system with brake fluid and replace combination valve and all cups and seals in complete brake system.
3. Residual pressure check valve in lines to front wheels.	3. Remove check valve.
4. Caliper pistons frozen or seized.	4. Disassemble caliper and free pistons (replace if necessary).
5. Caliper cylinder seals swollen.	5. Drain hydraulic system, flush system with clean brake fluid and replace combination valve and all cups and seals in complete brake system.
6. Caliper cylinder bores corroded or rough.	6. Disassemble caliper and remove corrosion or roughness, or replace caliper.
7. Hydraulic push rod on power brake out of adjustment or binding (causing primary cup to restrict master cylinder compensating port).	7. Adjust or free and lubricate.

Automatic adjustment for lining wear is achieved by the piston and friction ring sliding outward in the cylinder bore. The piston assumes a new position in the cylinder and maintains the correct adjustment.

FRONT DISC BRAKE SHOE AND LINING

NOTE: Refer to following section for caliper and pad service for the Ford Ranger and Bronco II.

Replace shoe and lining assemblies when lining is worn to a minimum of $1/16''$ in thickness (combined thickness of shoe and lining $1/4$ in. minimum).

Removal

1. Remove the shoe and lining mounting pins, anti-rattle springs and old shoe and lining assemblies.

Installation

1. Remove the master cylinder cover.
2. Loosen the piston housing-to-caliper mounting bolts sufficiently to permit the installation of new shoe and lining assemblies. Do not move pistons.
3. Install new shoe and lining assemblies. Install the brake shoe mounting pins and anti-rattle springs. Be sure that the spring tangs are located in the holes provided in the shoe plates.
4. Torque the brake shoe mounting pins to 17–23 ft. lbs.
5. Reset the pistons to the correct location in the cylinders by placing shims or feeler gauges of 0.023–0.035 in. thickness between the shoe plate of the outboard shoe and lining assembly and the caliper; then, retighten the piston housing-to-caliper mounting bolts. Keep the cylinder housing square with the caliper.

BRAKES
Light Trucks & Vans

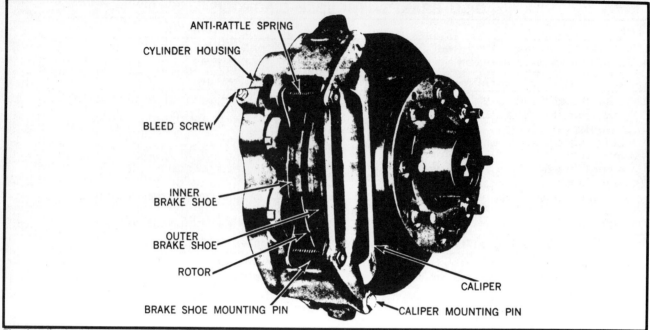

Floating disc brake caliper (© Ford Motor Co.)

6. Loosen the piston housing-to-caliper mounting bolts and remove the shims.
7. Torque the piston housing-to-caliper mounting bolts to 155–185 ft. lbs.
8. Check the master cylinder reservoirs.
9. Install the master cylinder cover.

DISC BRAKE CALIPER

Removal

1. Remove the wheel and tire assembly.
2. Remove the pins and nuts retaining the caliper assembly to the anchor plate.
3. Disconnect the brake hose from the caliper and remove the caliper.

Installation

1. Connect the brake hose to the caliper.
2. Position the caliper assembly to the anchor plate and install the retaining pins and nuts. Torque the nuts to specifications.
3. Install drum and wheel and bleed brake system.

NOTE: If the caliper assembly is leaking, the piston assemblies must be removed from the piston housing and replaced. If the cylinder bores are scored, corroded or ex-

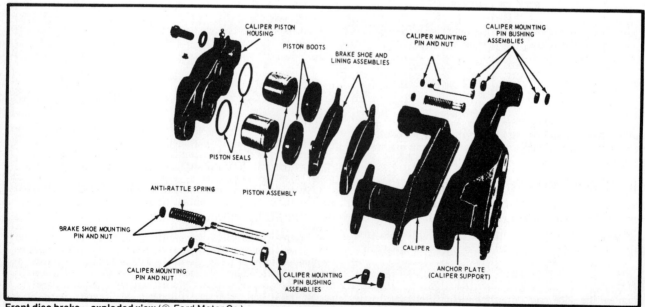

Front disc brake—exploded view (© Ford Motor Co.)

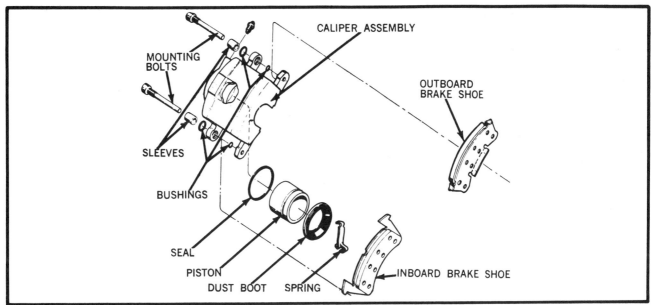
Bolt mounted caliper—exploded view (© GM Corp.)

cessive wear is evident, the piston housing must be replaced. Do not hone the cylinder bores. Piston assemblies are not available for oversize bores. The piston housing must be removed from the caliper for replacement.

Disassembly

1. Remove the two pins and nuts retaining the caliper to the support. Disconnect the flexible brake hose and plug the end to prevent brake fluid leakage.
2. Remove the boot retainers and remove the dust boots from the pistons and cylinder housing.
3. Position the caliper assembly in a vise.
4. Place a block of wood between the caliper and the cylinders, and apply low pressure air to the brake hose inlet. One piston will be forced out.
5. Reverse the piston and install it by hand pressure back into the cylinder bore far enough to form a seal. Block the reversed piston from moving out of the bore and place the wooden block between the remaining piston and the caliper.
6. Force out the second piston with low pressure air. Care should be taken as the piston is forced out of the bore.
7. Remove the two bolts and separate the caliper from the cylinder housing.

Assembly

The piston assembly and dust boots are not to be reused. A new set is to be used each time the caliper is assembled.

1. Apply a film of clean brake fluid in the cylinder bores and on the piston assemblies. Do not apply brake fluid on the insulators.
2. Start the piston assemblies into the cylinder bores using firm hand pressure. Exercise care to avoid cocking the piston in the bore.
3. Lightly tapping with a rawhide mallet, seat each piston assembly until the friction ring bottoms out in the cylinder bore.
4. Install the piston dust boots and retainers.
5. Position the piston housing on the caliper and install the piston housing-to-caliper mounting bolts and washers. Torque the piston housing-to-caliper mounting bolts to 155–185 ft. lbs.
6. Install the flexible brake hose.
7. Bleed the brake system and centralize the pressure differential valve.

NOTE: Do not move the vehicle after working on the disc brakes until a firm brake pedal is obtained.

Sliding Caliper Disc Brakes—Single Piston

CHRYSLER CORP. AND GENERAL MOTORS CORP. TYPES

This caliper is a one piece type with a single piston on the inboard side. The piston is made of steel and is plated to resist wear and corrosion. The piston has a square cut seal which provides for a seal between the piston and the caliper cylinder wall. A rubber dust boot located in a groove in the cylinder helps keep contamination from the piston and cylinder wall.

The caliper is mounted on an adapter which is mounted on the steering knuckle.

DISC BRAKE ADJUSTMENT

No adjustment is required on this unit other than applying the pedal several times after the unit has been worked on. This is to seat the shoes and after this is done the hydraulic pressure maintains the proper clearance between the brake shoes and the rotor.

BRAKE DISC PADS

Replace the disc pads when the linings are worn within $\frac{1}{16}$ in. of the shoe or the rivets.

Removal

1. Remove the master cylinder cover and if the cylinder is more than $\frac{1}{3}$ full remove the fluid necessary to make the cylinder only $\frac{1}{3}$ full. This is done to prevent any overflow from the cylinder when the piston is pushed into the bore of the caliper.
2. Raise vehicle on hoist and remove the front wheels.
3. Compress the piston back into the bore by using a large C-clamp and compressing the unit until the piston bottoms in the bore.

1–139

BRAKES
Light Trucks & Vans

4. Remove the two retaining bolts that hold the caliper into the support. If the caliper has retaining clips remove the retaining clips and anti-rattle springs. If the caliper has key type retainers, remove the key retaining screws, and using a hammer and drift, punch drive the key out of the caliper.

5. Slide the caliper off the rotor disc. Be careful not to damage the dust boot on the piston when removing the caliper.

NOTE: Do not let the caliper hang with the brake hose supporting the weight. This can cause damage to the hose which could result in a loss of brakes. Set the caliper on the front suspension arm or tie rod.

6. Remove the outer shoe from the caliper. It may be necessary to tap the shoe to loosen it from the caliper. Remove the inner shoe from the caliper or spindle assembly depending on where the shoe stays.

7. Remove the shoe support spring from the piston.

Cleaning and Inspection

Clean the sliding surfaces of the caliper and clean any dirt from the mounting bolts, clips or keys.

Inspect the boot on the piston for signs of cracks, cuts or other damage. Check to see if there is signs of fluid leaking around the seal on the piston. This will show up in the boot. If there is an indication of a fluid leak, the entire caliper has to be disassembled and the seal replaced.

Installation

1. Make sure that the piston is fully bottomed in the cylinder bore and install the outboard shoe in the recess of the caliper.

NOTE: On shoes with anti-rattle springs be sure to install the spring before installing the shoe in the caliper.

2. Place the outer shoe on the caliper and press it into place with finger pressure.

3. Position the caliper on the rotor and carefully slide it down into position over the rotor.

4. Install the caliper mounting bolts and torque them to 35 ft. lbs. On models with retaining clips install the anti-rattle springs and the retaining clips and torque the retaining screws to 200 inch lbs. On models with key type retainers press down the caliper and install the key in its slot and drive it in place with a hammer and drift. Install the retaining screw and torque to 12–18 ft. lbs.

5. Install the wheels and lower the vehicle. Check the master cylinder fluid level and add any fluid necessary to bring it up to the proper level.

6. Pump the brake pedal several times until a firm brake pedal is established. Road test the vehicle to check for proper operation.

DISC CALIPER

Removal

1. Remove the cover on the master cylinder and check if the fluid level is $1/3$ full. If it is more than $1/3$ full remove the necessary amount to bring the level down. This step is necessary to avoid overflow from the master cylinder when the piston is compressed into the cylinder bore.

2. Raise the vehicle and remove the wheel.

3. Compress the piston into the caliper bore and remove the brake hose from the caliper. Tape the end of the hose to prevent dirt from entering the line.

4. Remove the caliper retaining bolts, clips or wedges and remove the caliper from the vehicle.

Disassembly

1. Clean the outside of the caliper with clean brake fluid and drain any fluid from the caliper.

2. Remove the piston from the caliper by connecting the hydraulic line to the caliper and gently stroking the brake pedal. This will push the piston from the caliper bore.

3. With care remove the boot from the caliper piston bore.

4. Remove the piston seal from the caliper bore using a piece of wood or plastic.

NOTE: DO NOT use a metal tool to remove the seal. This can damage the bore or burr the edges of the seal groove.

5. Remove the bleeder valve.

Cleaning and Inspection

1. Clean all the parts with clean brake fluid and blow out all the passages in the caliper.

NOTE: When ever the caliper is disassembled, discard the boot and piston seal. These parts must not be reused.

2. Inspect the outside of the piston for signs of wear, corrosion, scores or any other defects. If any defects are detected replace the piston.

3. Check the caliper bore for the same defects as the piston. However, the bore can be cleaned up to a point with crocus cloth. If there are any marks that will not clean up with the cloth the caliper must be replaced.

Assembly and Installation

1. Lube the caliper bore and the piston with clean brake fluid and position the seal for the piston in the cylinder bore groove.

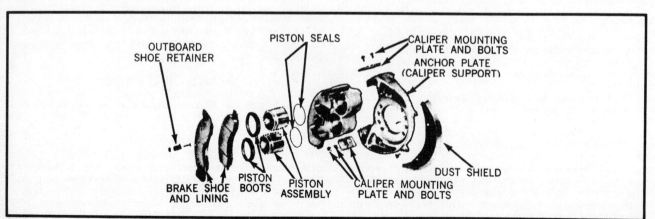

Sliding caliper disc brakes—bouble piston (© Ford Motor Co.)

BRAKES
Light Trucks & Vans

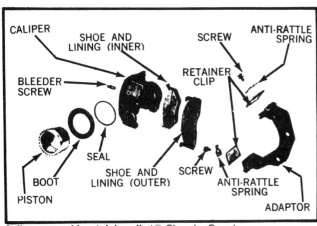

Caliper assembly retaining clip (© Chrysler Corp.)

5. Install the bleeder screw in the caliper and install the caliper back on the vehicle.
6. Connect the brake hoses and bleed the calipers of air. When bleeding is done pump the pedal several times to develop a firm brake pedal.

Sliding Caliper Disc Brakes (Double Piston)

BRAKE SHOE AND CALIPER

Removal

1. Drain about $2/3$ of the total brake fluid from the reservoir.
2. Jack up the vehicle and remove the front wheels.
3. Remove the four screws and remove the caliper hold-down assembly.
4. Lift the caliper off the hub and rotor. If the caliper is to be removed, disconnect the hydraulic line; if not, lay the caliper on the suspension or support with a length of wire.
5. Remove the inner and outer shoe and lining.

Disassembly

1. Drain the brake fluid from the caliper and clean the exterior with clean brake fluid.
2. Place a small block of wood under the caliper pistons and place a protective pad over the exterior. Remove the pistons by directing compressed air into the caliper fluid outlet.
3. Remove and discard piston boots.
4. Remove the piston seals from the groove in the caliper bore.

Assembly

1. Clean all parts in clean brake fluid and blow dry.
2. Dip the new piston seal in clean brake fluid and install it into the cylinder groove.

NOTE: Be sure that the seal is not rolled or twisted in the groove.

3. Install the dust boot in the cylinder groove.
4. Coat the outside diameter of the piston with clean brake fluid. Use something plastic or wood and gradually work the dust boot around the piston.
5. Press the piston straight into the caliper bore until it bottoms. Position the boot in the piston groove.

2. Install the dust boot into the groove in the piston with the fold faces toward the open end of the piston.
3. Install the piston in the bore being careful not to unseat the piston seal in the bore.
4. With the piston bottomed in the cylinder position the boot in the groove in the caliper. Make sure that the retaining ring in the seal is pressed down evenly around the cylinder.

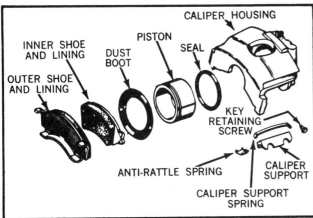

Disc brake caliper retainer—key type (© Ford Motor Co.)

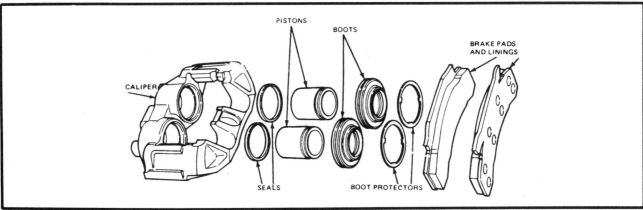

Rail slider caliper—6000 and 7000 pound front axle

1-141

BRAKES
Light Trucks & Vans

Installation

1. Install a new shoe and lining assembly into the anchor plate.
2. Push the pistons to the bottom of the piston bore. Place a small block of wood over both pistons and boots. Push the pistons to the bottom of the bores with a C-clamp.
3. Install the outer shoe and lining onto the caliper and install the shoe hold-down spring and pin.
4. Install the caliper assembly over the hub, rotor and inner shoe, and position into the inner grooves in the anchor plate.
5. Install the caliper hold-down parts and tighten to 40 ft. lbs.
6. Add extra heavy duty brake fluid to bring the level to $\frac{1}{4}$ in. from the top of the reservoir.
7. Bleed the system and add fluid as necessary.

DISC BRAKES

Ford Ranger and Bronco II

INSPECTION

Replace the front pads when the pad thickness is at the minimum thickness recommended by Ford Motor Co. ($\frac{1}{32}$ in.), or at the minimum allowed by the applicable state or local motor vehicle inspection code. Pad thickness may be checked by removing the wheel and looking through the inspection port in the caliper assembly.

FRONT CALIPER AND DISC BRAKE PADS

Removal & Installation

NOTE: Always replace all disc pad assemblies on an axle. Never service one wheel only.

1. To avoid fluid overflow when the caliper piston is pressed into the caliper cylinder bores, siphon or dip part of the brake fluid out of the larger master cylinder reservoir (connected to the front disc brakes). Discard the removed fluid.
2. Raise the vehicle and install jack stands. Remove a front wheel and tire assembly.
3. Place an eight in. C-clamp on the caliper and tighten the clamp to bottom the caliper piston in the cylinder bore. Remove the clamp.

NOTE: Do not use a screwdriver or similar tool to pry piston away from the rotor.

4. There are three types of caliper pins used: a single tang type, a double tang type and a split-shell type. The pin removal process is dependent upon how the pin is installed (bolt head direction). Remove the upper caliper pin first.

NOTE: On some applications, the pin may be retained by a nut and torx-head bolt (except the split-shell type).

5. If the bolt head is on the outside of the caliper, use the following procedure:
    a. From the inner side of the caliper, tap the bolt within the caliper pin until the bolt head on the outer side of the caliper shows a separation between the bolt head and the caliper pin.
    b. Using a hacksaw or bolt cutter, remove the bolt head from the bolt.
    c. Depress the tab on the bolt head end of the upper caliper pin with a screwdriver, while tapping on the pin with a hammer. Continue tapping until the tab is depressed by the v-slot.
    d. Place one end of a punch ($\frac{1}{2}$ in. or smaller) against the end of the caliper pin and drive the caliper pin out of the caliper toward the inside of the vehicle. Do not use a screwdriver or other edged tool to help drive out the caliper pin as the v-grooves may be damaged.

---- CAUTION ----
Never reuse caliper pins. Always install new pins whenever a caliper is removed.

6. If the nut end of the bolt is on the outside of the caliper, use the following procedure:
    a. Remove the nut from the bolt.
    b. Depress the lead tang on the end of the upper caliper pin with a screwdriver while tapping on the pin with a hammer. Continue tapping until the lead tang is depressed by the v-slot.
    c. Place one end of a punch ($\frac{1}{2}$ in. or smaller) against the end of the caliper pin and drive the caliper pin out of the caliper toward the inside of the vehicle. Do not use a screwdriver or other edged tool to help drive out the caliper pin as the v-grooves may be damaged.
7. Repeat the procedure in Step 4 for the lower caliper pin.
8. Remove the caliper from the rotor. If the caliper is to be removed for service, remove the brake hose from the caliper.
9. Remove the outer pad. Remove the anti-rattle clips and remove the inner pad.
10. To install, place a new anti-rattle clip on the lower end of the inner pad. Be sure the tabs on the clip are positioned properly and the clip is fully seated.
11. Position the inner pad and anti-rattle clip in the pad abutment with the anti-rattle clip tab against the pad abutment and the loop-type spring away from the rotor. Compress the anti-rattle clip and slide the upper end of the pad in position.
12. Install the outer pad, making sure the torque buttons on the pad spring clip are seated solidly in the matching holes in the caliper.
13. Install the caliper on the spindle, making sure the mounting surfaces are free of dirt and lubricate the caliper grooves with Disc Brake Caliper Grease. Install new caliper pins, making sure the pins are installed with the tang in position.
14. The pin must be installed with the lead tang in first, the bolt head facing outward (if equipped) and the pin properly positioned. Position the lead tang in the v-slot mounting surface and drive in the caliper until the drive tang is flush with the caliper assembly. Install the nut (if equipped) and tighten to 32–47 inch lbs.

---- CAUTION ----
Never reuse caliper pins. Always install new pins whenever a caliper is removed.

15. If removed, install the brake hose to the caliper.
16. Bleed the brakes as described earlier in this chapter.
17. Install the wheel and tire assembly. Torque the lug nuts to 85–115 ft. lbs.
18. Remove the jack stands and lower the vehicle. Check the brake fluid level and fill as necessary. Check the brakes for proper operation.

BRAKES
Light Trucks & Vans

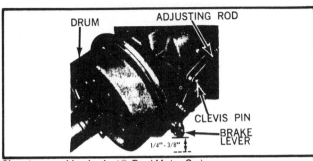

Shoe type parking brake (© Ford Motor Co.)

Parking Brakes

INTERNAL SHOE TYPE

Adjustment

NINE INCH DIAMETER DRUM

1. Release the parking brake lever in the cab.
2. From under the truck, remove the cotter pin from the parking brake linkage adjusting clevis pin. Remove the clevis pin.
3. Lengthen the parking brake adjusting link by turning the clevis. Continue to lengthen the adjusting link until the shoes seat against the drum when the clevis pin is installed.
4. Remove the clevis pin and shorten the linkage adjustment until there is 0.010 in. clearance between the shoes and the drum. The measurement should be taken at all points around the drum with the clevis pin installed.
5. Install a new cotter pin in the clevis retaining pin and check the brake operation.

Twelve Inch Diameter Drum

There is no internal adjustment on this brake. Adjustment is made on the linkage. Remove the clevis pin, loosen the nuts on the adjusting rod, and turn the clevis on the rod until a $1/4$–$3/8$ in. free play is obtained at the brake lever. Tighten the nuts, and connect the clevis to the bellcrank with the clevis pin.

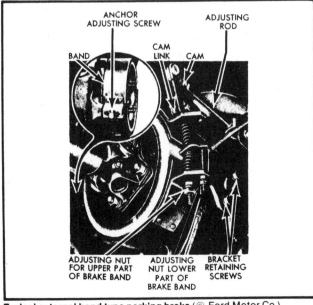

Typical external band type parking brake (© Ford Motor Co.)

EXTERNAL BAND TYPE

Adjustment

1. On cable-controlled parking brakes, move the parking brake lever to the fully released position. On a vehicle with a rod-type linkage, set the lever at the first notch.
2. Check the position of the cam to make sure the flat portion is resting on the brake band bracket. If the cam is not flat with the bracket, remove the clevis pin from the upper part of the cam, and adjust the clevis rod to allow the flat portion of the cam to rest on the brake band bracket. Install the clevis pin and cotter pin.
3. Remove the lock wire from the anchor adjusting screw, and turn the adjusting screw clockwise until a clearance of 0.010 in. is established between the brake lining and the brake drum at the anchor bracket. Install the lock wire in the anchor adjusting screw.
4. Loosen the lock nut on the adjusting screw for the lower half of the brake band, and adjust the screw to establish a 0.010 in. clearance between the lining and the brake drum at the lower half of the brake band. Tighten the lock nut.
5. Turn the upper band adjusting rod nut until a 0.010 in. clearance is established between the upper half of the band and the drum.
6. Apply and release brake several times to insure full release.

PARKING BRAKE INCLUDED WITH REAR BRAKES

Before attempting parking brake adjustment, make sure that the rear brakes are fully adjusted.
1. Raise and support the rear axle. Release the parking brake.
2. Apply the pedal or handle one to four clicks.
3. Adjust the cable equalizer nut under the truck until a moderate drag can be felt when the rear wheels are turned forward.
4. Release the parking brake and check that there is no drag when the wheels are turned forward.

NOTE: If the parking brake cable is replaced, prestretch it by applying the parking brake hard about three times before attempting adjustment.

Parking Brake Assembly Used With Hydro–Max Power Booster

1984 AND LATER

The parking brake is actuated by a spring/ramp assembly. The spring is compressed by Hydro-Max pump hydraulic pressure of 110 PSI minimum. The parking brakes are set by the release of this hydraulic pressure. The unit is sealed and full of fluid to prevent road contamination. As long as there is pump pressure, the parking brake can be released with a parking brake control in the cab. If the parking brake is set and no pump pressure is available, the spring can be caged by turning the nut, pinned to the stud, clockwise until the distance between the nut and spring can is $2 \pm 1/4$ in.. This cages the spring in the spring chamber, releasing the parking brake.

WHEEL CYLINDER/PARKING BRAKE EXPANDER

Removal

NOTE: If both the wheel cylinder/adjuster and wheel cylinder/parking brake expander are to be serviced, the complete backing plate assembly may be removed by disconnecting the fluid lines (service and parking) and re-

SECTION 1
BRAKES
Light Trucks & Vans

moving the bolts retaining the wheel cylinder assemblies and backing plate to the axle flange. Removal of the wheel cylinder assemblies can then take place on the bench. IF THE BACKING PLATE ASSEMBLY IS TO BE REMOVED, PRESSURE MUST BE RELEASED BY CAGING THE SPRING IN THE PARKING BRAKE CHAMBER.

1. Attach a tube to the bleeder screw on the wheel cylinder/parking brake expander. Place the other end of the tube in a container. Unscrew the bleeder screw one-half turn. Pump the brake pedal to remove all brake fluid from the rear brake system. Discard the brake fluid.
2. Remove the shoe and lining.
3. Remove the parking brake chamber from the wheel cylinder/parking brake expander.
4. Remove the bolt retaining the bridge tube retainer to the backing plate and remove the retainer. Disconnect the bridge tube from the cylinders and remove the bridge tube.
5. Insert an Allen-head wrench into the bolts in the rear of the backing plate and remove the bolts retaining the wheel cylinder/parking brake expander to the backing plate.
6. Remove the three 9/16–18 in. bolts, nuts and washers retaining the wheel cylinder/parking brake expander and backing plate to the axle flange. Remove the wheel cylinder/parking brake expander and gasket from the backing plate.
7. Remove any rust or corrosion from the backing plate with a wire brush.

Installation

1. Install a new gasket on the wheel cylinder/parking brake expander. Position the assembly on the backing plate.
2. Install the three 9/16–18 in. bolts, nuts and washers retaining wheel cylinder/parking brake expander and backing plate to the axle flange. Tighten the bolts.
3. Install the Allen-head head bolts in the rear of the backing plate retaining the assembly to the plate.

Parking brake chamber caging spring

4. Install the parking brake chamber.
5. Connect the bridge tube to the wheel cylinders. Install the bridge tube retainer and tighten the bolt.
6. Install the shoe and lining and adjust the lining-to-drum clearance.
7. Fill the Hydro-Max master cylinder reservoir with clean Heavy Duty Brake Fluid. Bleed the service brakes.
8. If required, bleed the parking brakes. Check the brake pump reservoir with the parking brake applied and if required, fill to the specified level with DEXRON® II or equivalent.

NOTE: The service brake system uses Ford Heavy Duty Brake Fluid, C6AZ–19542–A or –B (ESA–M6C25–A) or equivalent and is filled at the Hydro-Max master cylinder. The parking brake system uses Ford Automatic Transmission Fluid, ESP-M2C138–CJ, DEXRON® II or equivalent and is filled at the brake pump reservoir. DO NOT MIX THE FLUIDS.

9. Apply the brakes and check for leaks. Road test the vehicle and check for proper operation.

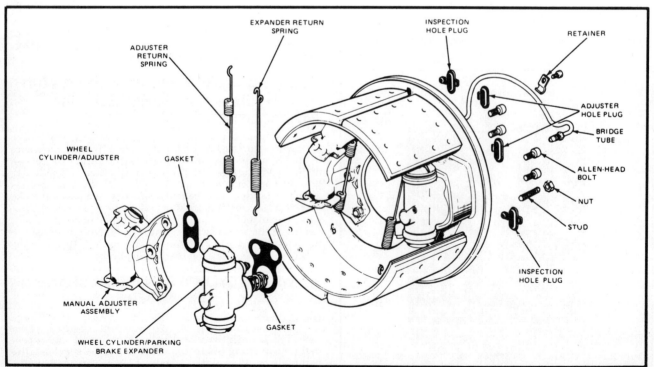

Drum brake system and related components

BRAKES
Light Trucks & Vans

PARKING BRAKE CHAMBER

Removal

NOTE: Droplets of fluid on the cap screws retaining the end cap to the housing of the parking brake chamber indicate a normal operating condition. It does not indicate a leak in the chamber.

1. Place chocks under the wheels to prevent vehicle movement.
2. Cage the spring in the chamber by turning the release bolt and nut counterclockwise. The spring is caged when the release bolt extends 50.80–57.15mm (2.00–2.25 in.) from the filter retainer to the bottom of the release nut.

CAUTION

Do not use an impact wrench to cage the spring. Air tools can damage the piston and prevent proper spring caging.

---

3. Remove the system pressure by opening the bleeder valve.
4. Disconnect the fluid line from the fitting and cap the line. Remove the inlet fitting and bleeder screw from the chamber.
5. Loosen the jam nut that retains the chamber to the wheel cylinder/parking brake expander. Rotate the chamber counterclockwise until the fluid inlet port points downward and the fluid drains out. Discard the fluid.
6. Rotate the chamber counterclockwise and remove the chamber from the wheel cylinder/parking brake expander. Cap the fluid inlet fitting to prevent the entry of contaminants into the chamber.

Installation

1. Prior to installing the chamber, make sure the spring is caged. The spring is caged when the release bolt extends 50.80 ± 6.35mm (2.0 ± 1/4 in.) from the filter retainer to the bottom of the release nut.
2. If removed, install the jam nut on the mounting tube until the nut reaches the end of the threads.
3. Position the chamber on the wheel cylinder/parking brake expander. Rotate the chamber until it stops in the cylinder. Make sure the jam nut does not interfere with the chamber.
4. Rotate the chamber in the opposite direction the bleeder screw is facing straight up and the fluid inlet port faces away from the axle.
5. Tighten the jam nut to 203 Nm (150 ft. lbs.), making sure the chamber does not rotate out of position. Install the bleeder screw and fluid inlet fitting in the chamber.
6. Tighten the bleeder fitting. Uncap the fluid line and the chamber fluid inlet port. Connect the fluid line to the chamber.
7. Check the fluid level in the brake pump reservoir. If required, fill the reservoir to the specified level with DEXRON® II or equivalent.
8. Bleed the parking brake chambers.
9. Check for leaks with the parking brakes in the applied position and in the released position. Repair as required.
10. Uncage the spring by turning the nut on the release bolt clockwise until the nut is snug against the filter retainer. Tighten the nut to 28–40 Nm (20–30 ft. lbs.).

PARKING BRAKE CHAMBER

Disassembly

1. Remove the parking brake chamber.
2. Uncage the spring by rotating the release bolt and nut clockwise until the nut extends 3.17mm (1/8 in.) above the filter retainer.

NOTE: To facilitate the spring uncaging procedure, apply 120 psi of air pressure to the fluid inlet port. Remove air pressure when the spring is uncaged.

CAUTION

Do not use an impact wrench to uncage the spring. The piston may be damaged and prevent proper spring caging.

---

3. Drive the roll pin from the release bolt and nut. Unscrew the nut and remove the filter retainer and filter.
4. In the end of the release bolt, cut a slot 1.587mm (1/16 in.) wide and 3.175mm (1/8 in.) deep. Insert a screwdrier in the slot and rotate the release bolt clockwise until the bolt is completely beneath the end cap and all threads are disengaged.
5. Place the chamber assembly in a press with the mounting tube supported by tube-type press stock and the end cap assembly facing the press ram. Bring the press ram down firmly on the end cap and make sure the press ram is centered on the end cap.

NOTE: The press must have a minimum capacity of 2500 lbs. and a minimum press ram travel of 6 in.

6. Remove the six cap screws retaining the end cap to the chamber assembly.
7. Slowly relax the pressure on the press ram and allow the spring to expand until it raises the end cap six in. above the chamber housing. Remove the spring and end cap from the chamber housing.
8. Drive the piston from the chamber housing with a brass drift and mallet.
9. With the piston supported, drive out the release bolt and socket by striking the release bolt with a hammer. Remove and discard all the O-rings from the rod socket. Discard the release bolt.
10. Remove the retaining ring from inside the chamber housing.
11. Position a brass drift through the mounting tube so it rests against the insert and drive out the insert. Remove and discard all the O-rings on the insert.
12. Inspect all parts for wear or damage and replace as required. If the spring shows signs of rust or corrosion, clean with a wire brush and recoat the spring with a rust inhibitor.

Assembly

CAUTION

All parts must be clean before assembly. Foreign particles such as dirt or metal chips in the chamber may result in premature O-ring replacement, leakage and reduced service life.

---

1. Coat the three new O-rings for the insert with DEXRON® II or equivalent. Install the O-rings on the insert.
2. Coat all interior surfaces of the chamber housing with DEXRON® II or equivalent.
3. Place the insert in position in the mounting tube and install the retaining ring.
4. Coat the piston and a new O-ring (large) and piston bearing with DEXRON® II or equivalent. Install the O-ring and bearing assembly on the piston.
5. Place the piston assembly in the chamber housing. Compress the O-ring and bearing assembly into the piston groove with a suitable tool (such as plastic tie wrap).
6. Carefully line up the piston rod with the hole in the insert. Gently tap the piston assembly into the chamber housing. The compressor tool should be loose. Remove the tool. Continue to tap the piston into the chamber until completely seated.
7. Position the chamber housing assembly in a press so the mounting tube portion is supported by an appropriate size tube type press stock and the end cap portion of housing is facing upward. Insert the spring in the chamber and place the end cap on top of the spring. Make sure the cap screw holes in the end cap are aligned with the holes in the housing. Slowly and carefully

BRAKES
Light Trucks & Vans

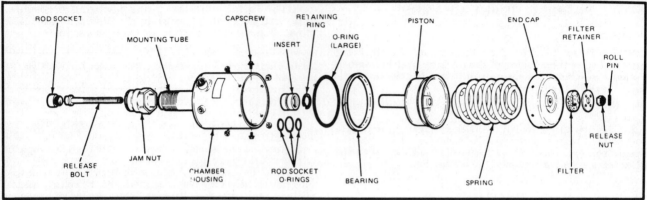

Parking brake chamber—exploded view

bring the press ram down to compress the end cap and spring. Make sure the press ram is centered on the end cap and the end cap is in alignment with the housing. Compress the spring until the holes in the end cap and housing are aligned. DO NOT RELEASE PRESSURE.

―――――――― **CAUTION** ――――――――
The press must have a minimum capacity of 2500 lbs. and a minimum travel of six in.

8. Install the six cap screws retaining the end cap to the chamber. Tighten the cap screws alternately and evenly to 55–81 Nm (40–60 ft. lbs.).
9. Install a new release bolt through the mounting tube and piston. Insert a $5/16$ in. hex driver and rotate the bolt through the end cap. Tighten the bolt by hand until the bolt protrudes through the end cap and is snug.
10. Install a new filter and the filter retainer over the release bolt. Screw on the release nut until the nut slots align with bolt hole. Install the roll pin.
11. Cage the spring by rotating the release bolt and nut counterclockwise until the bolt and nut projects 50.80–57.15mm (2.0–2.25 in.) from the bottom of the nut to the top of the filter retainer.

NOTE: To facilitate spring caging, apply 120 psi of air pressure to the fluid inlet. Release air pressure after the spring is caged.

―――――――― **CAUTION** ――――――――
Do use an impact wrench to cage the spring. Air tools may damage the piston and prevent proper spring caging.

12. Place the chamber assembly in a press so the end cap rests on the press bed. Position the rod socket into the end of the piston rod.
13. Press the rod socket securely in place.
14. Check that the rod socket is properly installed by measuring the distance between the top of the rod socket and the top of the mounting tube. The distance should be 81.28 ± 1.52mm (3.20 ± 0.06 in.).
15. Install the chamber.

BLEEDING PARKING BRAKES

1. Check the fluid level in the brake pump reservoir with the parking brakes applied. If required, fill the reservoir to the specified level with Ford Automatic Transmission Fluid, ESP-M2C138-CJ, DEXRON®II or equivalent.
2. Place chocks under the wheels to prevent vehicle movement.

3. Start the engine and run at normal curb idle speeds. Run the engine for at least two minutes to purge all air from the lines in the system before the Parking Brake Unitized Valve.
4. Place a wrench on the parking brake chamber bleeder fitting of the right rear chamber. (On tandem axles, start bleeding on right rear/rear axle chamber.) Attach a rubber drain tube to the bleeder fitting making sure the end of the tube fits snugly around the fitting.
5. Submerge the free end of the tube in a container partially filled with Ford Automatic Transmission Fluid, ESP-M2C138-CJ, DEXRON®II or equivalent. Loosen the bleeder fitting three-quarters of a turn.
6. Apply the parking brake. Close the bleeder fitting and release the parking brake. Repeat this operation until air bubbles no longer appear in the submerged end of the bleeder tube.
7. When the fluid is completely free of air bubbles, tighten the bleeder fitting and remove the bleeder tube.
8. Repeat this procedure on the left rear parking brake chamber. (On tandem axles, bleed the chamber on the left side of the rear/rear axle, then bleed the chamber on the right side of the forward/rear axle and finally, bleed the chamber on the left side of the forward/rear axle).
9. Check Fluid level and refill as required.

CAGING PARKING BRAKE CHAMBER SPRING

Cage the spring by rotating the release bolt and nut counterclockwise until the bolt and nut projects 50.80–57.15mm (2.0–2.25 in.) from the bottom of the nut to the top of the filter retainer.

―――――――― **CAUTION** ――――――――
Do use an impact wrench to cage the spring. Air tools may damage the piston and prevent proper spring caging.

UNCAGING PARKING BRAKE CHAMBER SPRING

Uncage the spring by rotating the release bolt and nut clockwise until the nut extends 3.17mm ($1/8$ in.) above the filter retainer.

―――――――― **CAUTION** ――――――――
Do not use an impact wrench to uncage the spring. The piston may be damaged and prevent proper spring caging.

Brakes Specifications

INDEX

NOTE: Should state inspection regulations exceed manufacturer's specifications for lining or rotor/drum reserve, the state inspection regulation specification **must** be used.

DOMESTIC CARS

American Motors	1-148
Chrysler Corporation	1-149
Ford Motor Company	1-150
General Motors	
Buick	1-152
Cadillac	1-153
Chevrolet	1-153
Oldsmobile	1-154
Pontiac	1-155

DOMESTIC LIGHT TRUCKS

American Motors — Jeep	1-156
Chrysler Corporation	1-157
Ford Motor Company	1-159
General Motors	
Chevrolet	1-160
GMC	1-160
International	1-161

IMPORT CARS

Audi	1-161
BMW	1-161
Buick Opel	1-162
Chevrolet	1-162
Datsun	1-163
Dodge	1-164
Fiat	1-164
Ford	1-165
Honda	1-165
Isuzu	1-166
Luv	1-166
Mazda	1-166
Mercedes-Benz	1-167
Mitsubishi	1-168
Plymouth	1-168
Renault	1-168
SAAB	1-169
Subaru	1-169
Toyota	1-169
Volkswagen	1-170
Volvo	1-171

DESIGN ILLUSTRATIONS 1-173

ABBREVIATIONS

B	Large
ft. lbs.	Foot pounds
inch lbs.	Inch pounds
L	Lower
S	Small
T.I.R.	Total indiactor Reading
U	Upper

SECTION 1

BRAKES
Specifications

DOMESTIC PASSENGER CARS

NOTICE: Should state inspection regulations exceed manufacturer's specifications for lining or rotor/drum reserve, the state inspection regulation specification must be used.

VEHICLE	BRAKE SHOE	BRAKE DRUM		BRAKE PAD	BRAKE ROTOR					CALIPER		WHEEL	WHEEL BEARING SETTING		
		DIAMETER			MIN. THICKNESS										
YEAR, MAKE, MODEL	Minimum Lining Thickness	Standard Size	Machine To	Minimum Lining Thickness	Machine To	Discard At	Variation From Parallelism	Runout T.I.R.	DESIGN	Mounting Bolts Torque (ft-lbs)	Bridge, Pin or Key Bolts Torque (ft-lbs)	Lugs or Nuts Torque (ft-lbs)	STEP 1 Tighten Spindle Nut (ft-lbs)	STEP 2 Back Off Retorque (in-lbs)	STEP 3 Lock, or Back Off and Lock
AMERICAN MOTORS—ALLIANCE, ENCORE (SEE RENAULT)															
86-82 Eagle	.030	10.000	10.060	.030*	.815	.810	.0005	.004	1	100	30	75	25	6	Step 2
83-82 Concord exc. Wagon; Spirit	.030	9.000	9.060	.030*	.815	.810	.0005	.003	1	85	30	75	25	6	Step 2
83-82 Concord Wagon	.030	10.000	10.060	.030*	.815	.810	.0005	.003	1	85	30	75	25	6	Step 2
81 All exc. 6 cyl. Concord Wagon and Eagle	.030	9.000	9.060	.030	.815	.810	.0005	.003	1	80	15	75	25	6	Step 2
81 6 cyl. Concord Wagon and Eagle	.030	10.000	10.060	.030*	.815	.810	.0005	.003∎	1	80	15	75	25	6	Step 2
80-78 Spirit exc. 4 cyl., Concord, Gremlin, exc. 4 cyl; Pacer, Eagle, AMX	.030	10.000	10.060	.062	.815	.810	.0005	.003∎	1	80	15	75	25	6	Step 2
80-78 Spirit & Gremlin w/4 cyl.	.030	9.000	9.060	.062	.815	.810	.0005	.003	1	80	15	75	25	6	Step 2
78 Matador	.030	10.000	10.060	.062	—	1.120	.0005	.003	1	80	15	75	25	6	Step 2
77 Gremlin, Hornet, Pacer	.030	10.000	10.060	.062	.815	.810	.0005	.003	1	80-90	15-18	75-90	25	6	Step 2
77 Matador	.030	10.000	10.060	—	—	1.120	.0005	.003	1	80-90	15-18	75-90	25	6	Step 2
76-75 Pacer w/drum brakes, Front	.030	9.000	9.060	—	—	—	—	—	—	—	—	60-90	22	—	—
Rear	.030	9.000	9.060	—	—	1.120	.0005	.003	1	80	15	75	22	6	Step 2
76-75 Pacer w/disc brakes	.030	9.000	9.060	.062	1.130	1.120	.0005	.003	1	80	15	60-90	22	6	Step 2
76-75 Hornet & Gremlin w/6 cyl	.030	10.000	10.060	.062	1.130	—	—	—	1	80	15	75	22	6	Step 2
76-75 Hornet & Gremlin w/V8 and All Matador	.030	9.000	9.060	—	—	—	.0005	.003	1	80	15	75	20	6	Step 2
74-72 Hornet, Gremlin & Javelin 6 cyl. w/drum brakes	.030	10.000	10.060	—	—	—	—	—	—	—	—	75	20	6☐	Step 2
74-72 Hornet, Gremlin & Javelin V8 & Horner V8, Gremlin V8 w/6 cyl. w/disc brakes	.030	9.000	9.060	.062	—	.940	.0005	.005	2	80	—	75	20	6	Step 2
74-73 Hornet V8, Gremlin V8 Javelin V8, All Matador & Ambassador w/disc brakes	.030	10.000	10.060	.062	—	.940	.0005	.005	2	80	—	75	20	6☐	Step 2
72 Hornet, Gremlin & Javelin w/6 cyl	.030	10.000	10.060	.062	—	.940	.0005	.005	2	80	—	75	20	12	Step 2

∎ Eagle-.004   ☐ 1972-12 in/lbs.   * .030" over rivet head; if bonded lining, use .062"

1-148

BRAKES
Specifications

CHRYSLER CORP.—CHRYSLER-DODGE-PLYMOUTH

Model															
86-83 Aries, Reliant, Lancer, GTS LeBaron, Dodge 400, 600															
w/7⅞" rear brake	.030	7.874	7.904	.030	.912	.882	.0005	.004	☐	70-100‡	††	80•	20-25	Handtight	Step 2
w/8²¹⁄₃₂" rear brake	.030	8.661	8.691	.030	.912	.882	.0005	.004	☐	70-100‡	††	80•	20-25	Handtight	Step 2
86-83 E Class New Yorker, Town & Country Wagon, Daytona, Laser, Caravelle	.030	8.661	8.691	.030	.912	.882	.0005	.004	☐	70-100‡	††	80•	20-25	Handtight	Step 2
86-83 Omni, Horizon, Charger, Turismo	.030	7.874	7.904	.030	.461	.431	.0005	.004	8	70-100‡	††	80•	20-25	Handtight	Step 2
86-82 Cordoba, Diplomat, Gran Fury, Mirada, Fifth Ave., Newport, New Yorker, Imperial															
w/10" rear brake	.030	10.000	10.060	.125	.970	.940	.0005	.004	5	95-125	15-20	85	20-25	Handtight	Step 2
w/11" rear brake	.030	11.000	11.060	.125	.970	.940	.0005	.004	5	95-125	15-20	85	20-25	Handtight	Step 2
82 Aries, Reliant, LeBaron Dodge 400	.030	7.870	7.900	.030	.912	.882	.0005	.004	6	70-100	18-22	85	20-25	Handtight	Step 2
82-78 Omni, Horizon	.030	7.870	7.900	.030	.461	.431	.0005	.005 ■	8	70-100	25-40	85	20-25	Handtight	Step 2
81 Aries, Reliant	.030	7.870	7.900	.030	.912	.882	.0005	.004	6	70-100	18-22	85	20-25	Handtight	Step 2
81-78 Aspen, Volare, LeBaron, Diplomat, St. Regis, Cordoba, Gran Fury, Magnum, Mirada Newport, New Yorker, Imperial															
w/10" rear brakes	.030	10.000	10.060	.030	.970	.940	.0005	.005 ■	5	95-125	15-20	85	20-25	Handtight	Step 2
w/11" rear brakes	.030	11.000	11.060	.030	.970	.940	.0005	.005 ■	5	95-125	15-20	85	20-25	Handtight	Step 2
78 Fury, Monaco w/10" rear brakes	.030	10.000	10.060	.030	.970	.940	.0005	.004	5	95-125	25-40	85	20-25	Handtight	Step 2
w/11" rear brakes	.030	11.000	11.060	.030	.970	.940	.0005	.004	5	95-125	25-40	85	20-25	Handtight	Step 2
77-75 Chrysler, Gran Fury, Royal Monaco	.030	11.000	11.060	.030	1.195	1.180	.0005	.004	5	95-125	25-35	85†	20-25	Handtight	Step 2
77-76 Volare & Aspen, exc. wagon	.030	10.000	10.060	.030	.970	.940	.0005	.004	5	95-125	15	85	20-25	Handtight	Step 2
77-76 Volare & Aspen Wagon	.030	11.000	11.060	.030	1.195	1.180	.0005	.004	5	95-125	15	85	20-25	Handtight	Step 2
77-76 Fury, Coronet, Cordoba, Monaco, Charger SE	.030	11.000	11.060	.030	.970	.940	.0005	.004	★	95-125	25-40	85	20-25	Handtight	Step 2
76-75 Valiant, Dart w/ft. drum:															
Front	.030	10.000	10.060	—	—	—	—	—	—	—	—	70	—	Handtight	—
Rear	.030	9.000	9.060	—	—	—	—	—	—	—	—	70	20-25	Handtight	Step 2
76-75 Valiant, Dart w/ft. disc	.030	10.000	10.060	.030	.970	.940	.0005	.004	5	95-125	15	70	20-25	Handtight	Step 2
75 Imperial: Front	—	—	—	.030	1.195	1.180	.0005	.004	5	95-125	25-35	85	20-25	Handtight	—
Rear	—	—	—	.030	.970	.940	.0005	.004	5	95-125	25-35	85	20-25	Handtight	—
75 Fury, Cornet, Charger, Cordoba Rear	.030	10.000	10.060	.030	.970	.940	.0005	.004	2	95-125	25-35	70	20-25	Handtight	Step 2
75 Fury & Coronet Wagon	.030	11.000	11.060	.030	.970	.940	.0005	.004	2	95-125	25-35	70	20-25	Handtight	Step 2
74 Imperial Front	—	—	—	.030	1.195	1.180	.0005	.004	5	75-100	30-35	65	8	Skip	Step 2
Rear	—	—	—	.030	.970	.940	.0005	.004	5	95-125	25-35	65	8	Skip	1 Slot
74-73 Chrysler, Dodge, Plymouth Full Size	.030	11.000	11.060	.030	.970	.940	.0005	.004	5	75-100	25-35	65	8	Skip	1 Slot
74-73 Charger, Coronet, Challenger, Belvedere, Satellite, Barracuda, Valiant & Dart w/ft. disc	.030	10.000	10.060	.030	.970	.940	.0005	.004	★	75-100	25-35	65	8	Skip	1 Slot
74-73 Valiant, Dart w/ft. drum:															
Front	.030	10.000	10.060	—	—	—	—	—	—	—	—	65	—	—	—
Rear	.030	9.000	9.060	—	—	—	—	—	—	—	—	65	—	—	—

* .030" over rivet head; if bonded lining, use .062"   ■ 1982-80-.004   • 1986-84-.95 ft/lbs.   ★ 2 or 5   ☐ 6 or 8   † 1975-65 ft/lbs.   †† K.H. 25-35, ATE 18-22   ‡ 1986-84 130-190 ft/lbs.

1–149

SECTION 1

BRAKES
Specifications

VEHICLE	BRAKE SHOE	BRAKE DRUM DIAMETER		BRAKE PAD	BRAKE ROTOR MIN. THICKNESS				DESIGN	CALIPER		WHEEL	WHEEL BEARING SETTING		
YEAR, MAKE, MODEL	Minimum Lining Thickness *	Standard Size	Machine To	Minimum Lining Thickness	Machine To	Discard At	Variation From Parallelism	Runout T.I.R.		Mounting Bolts Torque (ft-lbs)	Bridge, Pin or Key Bolts Torque (ft-lbs)	Lugs or Nuts Torque (ft-lbs)	STEP 1 Tighten Spindle Nut (ft-lbs)	STEP 2 Back Off Retorque (in-lbs)	STEP 3 Lock, or Back Off and Lock
73-72 Imperial	.030	11.000	11.060	.030	1.195	1.180	.0005	.0025	5	75-100	30-35	65	8	Skip	1 Slot
72 Chrysler, Monaco, Polara, Fury, GT, Suburban S.W.V.I.P.	.030	11.000	11.060	.030	1.195	1.180	.0005	.0025	5	75-100	30-35	65	8	Skip	1 Slot
72 Charger, Coronet, Crestwood, SE, Super Bee, Belvedere, Satellite, GTX, Regent, Road Runner, Sebring, Barracuda, Challenger w/discs	.030	10.000	10.060	.030	.970	.940	.0005	.0025	5	75-100	30-35	65	8	Skip	1 Slot
72 Charger, Coronet, Crestwood, SE, Super Bee, Belvedere, Satellite, GTX, Regent, Road Runner, Sebring, Barracuda, Challenger w/10" drum	.030	10.000	10.060	—	—	—	—	—	—	—	—	65	8	Skip	1 Slot
w/11" drum	.030	11.000	11.060	—	—	—	—	—	—	—	—	65	8	Skip	1 Slot
72 Dart, Demon, GT, GTS, Swinger, Valiant, Duster, Scamp, Signet, V100, V200															
w/10" drums	.030	10.000	10.060	—	—	—	—	—	—	—	—	55	6	Skip	1 Slot
w/9" drums	.030	9.000	9.060	—	—	—	—	—	—	—	—	55	6	Skip	1 Slot
w/ft. disc	.030	10.000	10.060	.030	.790	.780	.0005	.0025	9	50-80	70-80	55	6	Skip	1 Slot
FORD MOTOR CO.—FORD—MERCURY—LINCOLN															
86 Taurus, Sable sedan	.030	8.858	8.799	.125	—	.896	.0005	.003	44	—	18-25	80-105	17-25	10-15	Step 2
Wagon	.030	9.843	9.783	.125	—	.896	.0005	.003	44	—	18-25	80-105	17-25	10-15	Step 2
86-85 Merkur	.040	10.000	10.060	.060	.927	.897	.0006	.002	44	43-44	18-23	55-70	217	—	—
86-84 Lincoln Continental, Mark VII, SVO (86-85) Front	—	—	—	.125	—	.972	.0005	.003	10	—	40-60	80-105	17-25	10-15	Step 2
Rear	—	—	—	.125	—	.895	.0005	.004	64	80-110	29-37	80-105	—	—	—
86-83 Lincoln Town Car, Mark VI, Crown Victoria, Grand Marquis															
w/10" rear brake	.030	10.000	10.060	.125	—	.972	.0005†	.003	10	—	40-60	80-105	17-25	10-15	Step 2
w/11" rear brake	.030	11.030	11.090	.125	—	.972	.0005†	.003	10	—	40-60	80-105	17-25	10-15	Step 2
86-83 LTD, Marquis, Fairmont Futura, Zephyr															
w/9" rear brake	.030	9.000	9.060	.125	—	.810	.0005†	.003	10	—	30-40	80-105	17-25	10-15	Step 2
w/10" rear brake	.030	10.000	10.060	.125	—	.810	.0005†	.003	10	—	30-40	80-105	17-25	10-15	Step 2
86-83 Thunderbird, XR-7, Cougar, Mustang (exc. 86-85 SVO), Capri															
w/9" rear brake	.030	9.000	9.060	.125	—	.810	.0005†	.003	10	—	30-40	80-105	17-25	10-15	Step 2
w/10" rear brake	.030	10.000	10.060	.125	—	.810	.0005†	.003	10	—	30-40	80-105	17-25	10-15	Step 2
83-82 Lincoln Continental Front	—	—	—	.125	—	.972	.0005	.003	10	—	40-60	80-105	17-25	10-15	Step 2
Rear	—	—	—	.125	—	.895	.0005	.004	11	80-110	29-37	80-105	—	—	—

* .030" over rivet head; if bonded lining, use .062". ■ Lincoln-.00025 ● 1986-84 Alum. whls. 1985-84-LTD, Marquis .0003 ● 85-84 Escort, Lynx, EXP: .0004 ††1985-84 Escort, Lynx, EXP: .002
■ 1982-80-.004 ● 1986-84-95 ft/lbs. ★ 2 or 5 □ 6 or 8 † 1975-65 ft/lbs. ‡K.H. 25-35, ATE 18-22 ‡ 86-84 130-190 ft/lbs. † 1986-.0003 with Alum. whls.

1-150

BRAKES
Specifications

Model	Lining Thickness Front	Drum Diameter Original	Drum Diameter Max	Lining Thickness Rear	Rotor Diameter	Rotor Runout	Rotor Parallelism	Rotor Min Thickness	Caliper Bolt	Wheel Bearing	Lug Nut	Master Cylinder	Notes
86-81 Escort, Lynx, EXP, LN7, Tempo, Topaz													
w/7" rear brake	.030*	7.090	7.149	.125	—	.882	.0005*	.003††	44	—	18-25	80-105	17-25	10-15	Step 2
w/8" rear brake	.030*	8.000	8.060	.125	—	.882	.0005*	.003††	44	—	18-25	80-105	17-25	10-15	Step 2
82-81 Thunderbird, XR-7													
w/9" rear brake	.030	9.000	9.060	.125	—	.810	.0005	.003	10	—	30-40	80-105	17-25	10-15	Step 2
w/10" rear brake	.030	10.000	10.060	.125	—	.810	.0005	.003	10	—	30-40	80-105	17-25	10-15	Step 2
82-81 Cougar, Granada													
w/9" rear brake	.030	9.000	9.060	.125	—	.810	.0005	.003	10	—	30-40	80-105	17-25	10-15	Step 2
w/10" rear brake	.030	10.000	10.060	.125	—	.810	.0005	.003	10	—	30-40	80-105	17-25	10-15	Step 2
82-79 Mustang, Capri, Fairmont, Zephyr w/9" rear brake	.030	9.000	9.060	.125	—	.810	.0005	.003	10	—	30-40	80-105	17-25	10-15	Step 2
w/10" rear brake	.030	10.000	10.060	.125	—	.810	.0005	.003	10	—	30-40	80-105	17-25	10-15	Step 2
82-79 Lincoln Town Car, Mark VI, LTD, Marquis, w/10" rear brakes	.030	10.000	10.060	.125	—	.972	.0005	.003	10	—	40-60	80-105	17-25	10-15	Step 2
w/11" rear brakes	.030	11.030	11.090	.125	—	.972	.0005	.003	10	—	40-60	80-105	17-25	10-15	Step 2
80 Thunderbird, Cougar	.030	9.000	9.060	.125	—	.810	.0005	.003	10	—	30-40	80-105	17-25	10-15	Step 2
80-79 Granada, Monarch, Versailles w/o rear disc brakes	.030	10.000	10.060	.125	—	.810	.0005	.004	1	U105-L65	12-16	80-105	17-25	10-15	Step 2
w/rear disc brakes	—	—	—	—	—	.895	.0005	.004	11	90-120	12-16	80-105	—	10-15	—
80-77 Pinto, Bobcat, Mustang II	.030	9.000	9.060	.030	—	.810	.0005	.003	1	U105-L65	12-16	80-105	17-25	10-15	Step 2
79 LTD II, Thunderbird, Cougar	.030	11.030	11.090	.125	—	1.120	.0005	.003	1	90-120	12-16	80-105	17-25	10-15	Step 2
79-77 Mark V w/o rear disc brakes	.030	11.030	11.090	.125	—	1.120	.00025	.003	1	90-120	12-16	80-105	17-25	10-15	Step 2
w/rear disc brakes	—	—	—	—	—	.895	.0004	.004	11	90-120	12-16	80-105	17-25	10-15	—
78 Ford, Lincoln, Mercury, Custom 500, LTD II, Cougar, Ranchero II, Squire, Thunderbird	.030	11.030	11.090	.125	—	1.120	.0005 ■	.003	1	90-125	12-16	70-115	17-25	10-15	Step 2
78 Fairmont, Zephyr exc. Wagon	.030	9.000	9.060	.125	—	.810	.0005	.003	10	—	30-40	70-115	17-25	10-15	Step 2
Wagon	.030	10.000	10.060	.125	—	.810	.0005	.003	10	—	30-40	70-115	17-25	10-15	Step 2
78-77 Maverick, Comet, Granada, Monarch, Versailles	.030	10.000	10.060	.125	—	.810	.0005	.003	1	U105-L65	12-16	70-115	17-25	10-15	Step 2
78-77 LTD II, Thunderbird, Cougar	.030	11.030	11.090	.125	—	1.120	.0005	.003	1	90-120	12-16	70-115	17-25	10-15	Step 2
77-73 Lincoln, Mark IV w/o rear disc brakes	.030	11.030	11.090	.125	—	1.120	.00025	.003	1	90-120	12-16	70-115	17-25	10-15	Step 2
w/rear disc brakes	—	—	—	—	—	.895	.0004	.004	11	90-120	12-16	70-115	—	10-15	—
77-74 Ford, Mercury, Meteor w/o rear disc brakes	.030	11.030	11.090	.125	—	1.120	.0005	.003	1	90-120	12-16	70-115	17-25	10-15	Step 2
76-75 Granada, Monarch, Maverick, Comet w/o rear disc brakes	.030	10.000	10.060	.125	—	.810	.0005	.003	1	90-120	12-16	70-115	17-25	10-15	Step 2
w/rear disc brakes	—	—	—	.125	—	.895	.005	.003	11	U105-L65	12-16	70-115	—	10-15	—
76-75 Pinto, Bobcat, Mustang	.030	9.000	9.060	.030	—	.810	.0005	.003	1	90-120	12-16	70-115	17-25	10-15	Step 2
76-74 Torino, Montego, Cougar w/10" rear brakes	.030	10.000	10.060	.030	—	1.120	.0005	.003	1	90-120	12-16	70-115	17-25	10-15	Step 2
w/11" rear brakes	.030	11.030	11.090	.030	—	1.120	.005	.003	1	90-120	12-16	70-115	17-25	10-15	Step 2
76-73 Thunderbird	.030	11.030	11.090	.125	—	1.120	.00025	.003	1	U105-L65	12-16	70-115	17-25	10-15	Step 2
74 Maverick, Comet	.030	10.000	10.060	.125	—	.810	.0005	.003	1	90-120	12-16	70-115	17-25	10-15	Step 2
74 Pinto, Mustang	.030	9.000	9.060	.030	—	.875	.0005	.002	1	U105-L65	12-16	—	17-25	10-15	Step 2
73 Ford, Mercury, Meteor, Lincoln, Continental	.030	11.030	11.090	.125	—	1.120	.0005	.003	1	90-120	—	70-115	17-25	10-15	Step 2

* .030" over rivet head; if bonded lining, use .062"   ■ Lincoln-.00025   † 1986-.0003   †† 1986-84 Escort, Lynx, EXP: .002
• 1986-84 Escort, Lynx, EXP: .0004   1985-84 LTD, Marquis .0003

BRAKES
Specifications

VEHICLE	BRAKE SHOE	BRAKE DRUM		BRAKE PAD	BRAKE ROTOR						CALIPER			WHEEL	WHEEL BEARING SETTING		
		DIAMETER			MIN. THICKNESS												
YEAR, MAKE, MODEL	Minimum Lining Thickness	Standard Size	Machine To	Minimum Lining Thickness	Machine To	Discard At	Variation From Parallelism	Runout T.I.R.	DESIGN	Mounting Bolts Torque (ft-lbs)	Bridge, Pin or Key Bolts Torque (ft-lbs)	Lugs or Nuts Torque (ft-lbs)	STEP 1 Tighten Spindle Nut (ft-lbs)	STEP 2 Back Off Retorque (in-lbs)	STEP 3 Lock, or Back Off and Lock
73 Torino, Montego	.030	10.000	10.060	.030	—	1.120	.0005	.003	1	90-120	25-35	70-115	17-25	10-15	Step 2
73-72 Pinto	.030	9.000	9.060	.030	.700	.685	.0007	.003	21	U105-L65	—	70-115	17-25	10-15	Step 2
73-72 Cougar, Mustang	.030	10.000	10.060	.030	.890	.875	.0007	.002	23	U125-L65	—	70-115	17-25	10-15	Step 2
73-72 Maverick, Comet w/9" drum	.030	9.000	9.060	—	—	—	—	—	—	—	—	70-115	17-25	10-15	Step 2
w/10" drum	.030	10.000	10.060	.030	—	—	—	—	1	90-120	25-35	70-115	17-25	10-15	Step 2
72 Torino, Montego	.030	10.000	10.060	—	—	—	—	—	—	—	—	70-115	17-25	10-15	Step 2
72 Ford, Mercury, Meteor, Thunderbird, Lincoln, Mark IV	.030	11.030	11.090	.030	1.135	1.120	.0007	.003	24	U125-L105	25-35	70-115	17-25	10-15	Step 2
GENERAL MOTORS CORP—BUICK															
86 Riviera: Front	—	—	—	.030	.971	.956	.0005	.004	71	83	63	100	NOT ADJUSTABLE		
Rear	*	—	—	.030	.444	.429	.0005	.003	72	83	63	100	NOT ADJUSTABLE		
86-83 Regal, LeSabre (RWD)	*	9.500	9.560	.030	.980	.965	.0005	.004	4	—	28□	80†	12	Handtight	1/2 Flat
85-83 Riviera w/o rear disc brakes	*	9.500	9.560	.030	.980	.965	.0005	.004	41	32	28□	100	12	Handtight	1/2 Flat
w/rear disc brakes	—	—	—	.030	.980	.965	.0005	.004	4	—	30	100	—	—	—
86-83 Electra, Estate Wagon (RWD)	*	11.000	11.060	.030	.980	.965	.0005	.004	4	—	28□	100	12	Handtight	1/2 Flat
86-83 Century w/H.D., Electra, LeSabre (FWD)	—	—	—	.030	.972	.957	.0005	.002•	4‡	—	—	100	—	—	—
86-82 Century exc. H.D.	*	8.858	8.880	.030	.830	.815	.0005	.002•	40	—	—	100	NOT ADJUSTABLE		
86-82 Skyhawk	*	8.858	8.880	.030*	.830	.815	.005	.002•	40‡	—	28□	100	NOT ADJUSTABLE		
w/vented disc (86-82)	*	7.880	7.899	.030*	.444	.429	.0005	.002	40	—	28	100	NOT ADJUSTABLE		
w/solid disc (82)	*	7.880	7.899	.030	.980	.965	.0005	.004	4	—	35	80†	12	Handtight	1/2 Flat
82 Regal, LeSabre	*	9.500	9.560	.030	.980	.965	.0005	.004	4	—	35	100	—	—	—
82-79 Riviera w/o rear disc brakes	*	9.500	9.560	.030	.980	.965	.0005	.004	41	32	30	100	12	Handtight	1/2 Flat
w/rear disc brakes	—	—	—	.030	.980	.965	.0005	.004	4	—	35	100	—	—	—
82-79 Electra, Estate Wagon	*	11.000	11.060	.030	.980	.965	.0005	.002•	4	—	35	103	12	Handtight	1/2 Flat
86-80 Skylark, Regal (FWD)	*	7.880	7.899	.030	.830	.815	.0005	.004	40‡	—	28□	100	—	—	—
81-79 Century, Regal, LeSabre	*	9.500	9.560	.030	.980	.965	.0005	.004	4	—	35	80†	12	Handtight	1/2 Flat
80-76 Skyhawk	*	9.500	9.560	.030	.830	.815	.0005	.004	15	—	—	80	12	Handtight	1/2 Flat
79 Skylark	*	9.500	9.560	.030	.980	.965	.0005	.005	4	—	35	80	12	Handtight	1/2 Flat
78-77 Electra, Estate Wagon, Riviera	*	11.000	11.060	.030	.980	.965	.0005	.004	4	—	35	80	12	Handtight	1/2 Flat
78-77 Century, LeSabre, Regal, Skylark	*	9.500	9.560	.030	.980	.965	.0005	.004	4	—	35	80■	12	Handtight	1/2 Flat
76-75 Riviera, Electra, Estate Wagon, LeSabre, Custom	*	12.000	12.060	.030	1.230	1.215	.0005	.005	4	—	30-40	90	19	132	1/6 Turn
76-75 Apollo, Skylark	*	9.500	9.560	.030	.980	.965	.0005	.005	4	—	35	70	19	132	1/6 Turn
75 Century, Regal	*	9.500	9.560	.030	.980	.965	.0005	.005	4	—	35	70	19	132	1/6 Turn
75 Skyhawk	*	9.000	9.060	.030	.455	.440	.0005	.005	15	—	35	70	19	132	1/6 Turn
74-72 Electra, Custom, LeSabre, Centurion, Riviera, Wildcat	*	11.000	11.060	.030	1.230	1.215	.0005	.005	4	—	30-40	75	12	Handtight	1/2 Flat
74-73 Estate Wagon	*	12.000	12.060	.030	1.230	1.215	.0005	.005	4	—	30-40	75	12	Handtight	1/2 Flat
74-73 Century, Regal	*	9.500	9.560	.030	.980	.965	.0005	.005	4	—	35	65-70	19	132	1/6 Turn

‡86-85 use 62 □1986-85 38 ft/lbs. "030" over rivet head, if bonded lining use .062" ■ w/½" stud 100 ft/lbs. † w/Aluminum whls. LeSabre 90 ft/lbs., Regal 100 ft/lbs. •1986-84 .004

BRAKES
Specifications

						.980	.965	.0005	.004	4		35	65-70	19	22	1/6 Turn
72 GS, Gran Sport, Skylark, Sports Wagon	•	—	—	.030												
‡86-85 use 62 □1986-85 38 ft/lbs. "030" over rivet head, if bonded lining use .062" ■w½" stud 100 ft/lbs. †w/Aluminum whls. LeSabre 90 ft/lbs. Regal 100 ft/lbs. •1986-84 .004																
GENERAL MOTORS CORP.—CADILLAC																
86 Eldorado, Seville Front	•	—	—	.030	.971	.956	.0005	.004	71	83		63	100	—	NOT ADJUSTABLE	
Rear	.030	—	8.917	.030	.444	.429	.0005	.003	72	83		63	100	—	NOT ADJUSTABLE	
86 Fleetwood, DeVille (FWD)	.030	8.863	7.899	.030	.972	.965	.0005	.002•	62	—		38	100	—	NOT ADJUSTABLE	
86-82 Cimarron	—	8.858	8.880	.030	.830	.815	.0005	.004	40‡	—		28 □	100	—	NOT ADJUSTABLE	
86-82 Fleetwood, DeVille (FWD)	.030	11.000	11.060	.030	.972	.957	.0005	.004	62	—		35 □	100	12	Handtight	Step 2
85-82 Eldorado, Seville					.980	.965	.0005	.004	4	—		28 □	100	—	NOT ADJUSTABLE	
Front	—	—	—	.030	.980	.965	.0005	.004	4	—						
Rear	—	—	—	.030					41	35		38	100	—	NOT ADJUSTABLE	
85-77 Fleetwood Limo, Commercial Chassis	•	12.000	12.060	.062	1.230	1.215	.0005	.004	4	—		30	100	12	Handtight	Step 2
81-79 Fleetwood, Brougham, DeVille, Seville, (RWD) w/o rear disc brakes (1979)	•	11.000	11.060	.062	.980	.965	.0005	.004	4	—		30	100	12	Handtight	Step 2
				.062	.980	.965	.0005	.004	41	34		30	100			
81-79 Eldorado, Seville (FWD) Front Disc	•	—	—	.062	.980	.965	.0005	.004	4	—		30	100	12	Handtight	Step 2
Rear Disc	—	—	—	.062	.980	.965	.0005	.004	41	35		30	100	12	Handtight	Step 2
78-77 DeVille	•	11.000	11.060	.062	.980	.965	.0005	.004	4	—		30	100	12	Handtight	Step 2
78-77 Brougham, Seville: Front	—	—	—	.062	.980	.965	.0005	.003	4	—		30	100			
Rear	—	—	—	.062	.910	.905	.0005	.008	41	35		30	130			
78-77 Seville	•	11.000	11.060	.062	—	1.170	.0005	.008	4	—		30	130	15	Handtight	Step 2
76-74 Calais, Brougham, DeVille, Fleetwood 75, Commercial Chassis	•	12.000	12.060	.062	1.220	1.215	.0005	.005	4	—		30	80			
76 Eldorado: Front	•	—	—	.062	1.205	1.190	.0005	.005	4	U60-L80		30-40	100†	15	Handtight	Step 2
Rear	•	—	—	.062	1.205	1.190	.0005	.008	41	35		30	130			
75-72 Eldorado	•	11.000	11.060	.062	1.205	1.190	.0005	.008	4	35		30	130			
73-72 Calais, Fleetwood 60, 75 DeVille, Commercial Chassis	•	12.000	12.060	.062	1.220	1.215	.0007	.005	4	U60-L80		30	130	15	Handtight	Step 2

".030" over rivet head, if bonded lining use .062" ‡1975-74 130 ft/lbs. •1986-84 .004 □1986-84 .004 †Aluminum whls; Corvette 80, Camaro 105, others 90. ‡1986-85 38 ft/lbs.

						.980	.965	.0005	.004	4		35	65-70	19	22	
GENERAL MOTORS CORP.—CHEVROLET																
86 Camaro w/rear drum brakes	•	9.500	9.560	.030	.980	.965	.0005	.005	4	—		30-45	80	12	Handtight	½ Flat
w/rear disc brakes	•	—	—	.030	.986	.956	.0005	.005	41	—		30-45	80	12	Handtight	½ Flat
86 Chevette	•	7.880	7.899	.030	.404	.374	.0005	.005	50	70		30-45	80	12	Handtight	½ Flat
86-84 Corvette: Front	—	—	—	.062	.724	—	.0005	.006	61	70††		24	100†	—	NOT ADJUSTABLE	
Rear	—	—	—	.062	.724	—	.0005	.006	61	44††		24	100†	—	NOT ADJUSTABLE	
85-84 Chevette	•	7.874	7.899	.030	.430	.374	.0005	.005	50	—		21-25 □	70	12	Handtight	½ Flat
83 Chevette	•	7.874	7.899	.030	.390	.374	.0005	.005	50	—		21-25 □	70	12	Handtight	½ Flat
86-83 Celebrity w/H.D.	•	8.858	8.880	.030	.972	.957	.0005	.002•	4‡	—		28 □	100	—	NOT ADJUSTABLE	
86-82 Celebrity exc. H.D.	•	8.858	8.880	.030	.830	.815	.0005	.002•	40‡	—		28 □	100	—	NOT ADJUSTABLE	
86-82 Cavalier																
w/vented disc (86-82)	•	7.880	7.899	.030	.830	.815	.0005	.002•	40‡	—		28 □	100	—	NOT ADJUSTABLE	
w/solid disc (82)	•	7.880	7.899	.030	.444	.429	.0005	.002	40	—		28	100	—	NOT ADJUSTABLE	

‡1986-85 use 62 •86-84 .004 □1986-85 38 ft/lbs. ".030" over rivet head, if bonded lining use .062" +1986 80 ft/lbs. †Aluminum whls; Corvette 80, Camaro 105, others 90. ††1986-85 front 133 ft/lbs. rear 70 ft/lbs.

1-153

BRAKES
Specifications

VEHICLE	BRAKE SHOE	BRAKE DRUM		BRAKE PAD	BRAKE ROTOR					CALIPER			WHEEL	WHEEL BEARING SETTING		
		DIAMETER			MIN. THICKNESS											
YEAR, MAKE, MODEL	Minimum Lining Thickness	Standard Size	Machine To	Minimum Lining Thickness	Machine To	Discard At	Variation From Parallelism	Runout T.I.R.	DESIGN	Mounting Bolts Torque (ft-lbs)	Bridge, Pin or Key Bolts Torque (ft-lbs)	Lugs or Nuts Torque (ft-lbs)	STEP 1 Tighten Spindle Nut (ft-lbs)	STEP 2 Back Off Retorque (in-lbs)	STEP 3 Lock, or Back Off and Lock
85-82 Camaro w/rear drum brakes	*	9.500	9.560	.030	.980	.965	.0005	.004	4	—	21-35	80†	12	Handtight	½ Flat
w/rear disc brakes	—	—	—	.030	.980	.965	.0005	.004	41	—	30-45	80†	—	—	—
86-82 Malibu, Monte Carlo, El Camino	*	9.500	9.560	.030	.980	.965	.0005	.004	4	—	35	80†	12	Handtight	½ Flat
85-80 Citation	*	7.880	7.899	.030	.830	.815	.0005	.002●	40‡	—	28	103	12	NOT ADJUSTABLE	
86-79 Impala, Caprice w/9½" rear brakes	*	9.500	9.560	.030	.980	.965	.0005	.004	4	—	35	80	12	Handtight	½ Flat
w/11" rear brakes	*	11.000	11.060	.030	.980	.965	.0005	.004	4	—	35	100	12	Handtight	½ Flat
82-78 Chevette	*	7.874	7.899	.030	.390	.374	.0005	.005	20	—	28	70	12	Handtight	½ Flat
82-77 Corvette: Front	—	—	—	.030*	1.230	1.215	.0005	.005	13	70	130	70†	12	Handtight	½ Flat
Rear	—	—	—	.030*	1.230	1.215	.0005	.005	13	70	60	70†	12	Handtight	½ Flat
81-79 Malibu, Camaro, Nova, Monte Carlo, El Camino	*	9.500	9.560	.030	.980	.965	.0005	.004	4	—	35	80†	12	Handtight	½ Flat
80-76 Monza	*	9.500	9.560	.030	.830	.815	.0005	.005	15	—	—	80†	12	Handtight	½ Flat
78 Caprice, Camaro, Impala, Nova exc. Wagon	*	9.500	9.560	.030	.980	.965	.0005	.004	4	—	35	80†	12	Handtight	½ Flat
Wagon	*	11.000	11.060	.030	.980	.965	.0005	.005	4	—	35	100	12	Handtight	½ Flat
78 Malibu, Monte Carlo, El Camino	*	11.000	11.060	.030	.980	.965	.0005	.004	4	—	35	80	12	Handtight	½ Flat
77 Chevette	*	7.880	7.899	.030	.456	.441	.0005	.005	20	—	28	70	12	Handtight	½ Flat
77 Impala, Caprice exc. Wagon	*	9.500	9.560	.030*	.980	.965	.0005	.005	4	—	35	80	12	Handtight	½ Flat
Wagon	*	11.000	11.060	.030	.980	.965	.0005	.005	4	—	35	100	12	Handtight	½ Flat
77 Chevelle, Monte Carlo, El Camino, Malibu	*	11.000	11.060	.030	.980	.965	.0005	.005	4	—	35	80	12	Handtight	½ Flat
77 Camaro, Nova	*	9.500	9.560	.030	.980	.965	.0005	.005	4	—	35	80	12	Handtight	½ Flat
77-76 Vega	*	9.500	9.560	.030	.455	.440	.0005	.005	15	—	—	80†	12	Handtight	½ Flat
76 Chevette	*	7.870	7.899	.030	.448	.433	.0005	.005	20	—	28	70	12	Handtight	½ Flat
76-72 Corvette: Front	—	—	—	.030*	1.230	1.215	.0005	.005	13	70	130	75	12	Skip	1-1½ Slot
Rear	—	—	—	.030*	1.230	1.215	.0005	.005	13	75	50	75	12	Skip	1-1½ Slot
76-72 Bel Air, Impala, Caprice	*	11.000	11.060	.030*	1.230	1.215	.0005	.005	4	—	35	80	12	Skip	1-1½ Slot
76-72 Nova	*	9.500	9.560	.030*	.980	.965	.0005	.005	4	—	35	70	12	Skip	1-1½ Slot
76-73 Chevelle, Camaro, Monte Carlo, El Camino	*	9.500	9.560	.030*	.980	.965	.0005	.005	4	—	35	70	12	Skip	1-1½ Slot
75 Monza	*	9.000	9.060	.030	.455	.440	.0005	.005	15	—	—	70	12	Skip	1 Flat
75-72 Vega	*	9.000	9.060	.030	.455	.440	.0005	.005	15	—	—	65	12	Skip	1 Flat
72 Chevelle, Camaro, El Camino	*	9.500	9.560	.030*	.980	.965	.0005	.005	4	—	35	70	12	Skip	1-1½ Slot

●1986-84 .004   □1986-85 38 ft/lbs.   *.030" over rivet head. if bonded lining use .062"   †Aluminum whls; Corvette 80, Camaro 105, other 90.   ††1986-85 front 133 ft/lbs. rear 70 ft/lbs.

GENERAL MOTORS CORP.—OLDSMOBILE																
86 Toronado: Front	—	—	—	.030	.971	.956	.0005	.004	71	83	63	100	—	NOT ADJUSTABLE		
Rear	—	—	—	.030	.444	.429	.0005	.003	72	83	63	100	—	—	—
86-83 Ciera w/H.D. 88, 98 (FWD)	*	8.880	8.880	.030	.972	.957	.0005	.002●	4‡	—	28 □	100	—	NOT ADJUSTABLE		
86-82 Ciera exc. H.D.	*	8.858	8.880	.030	.830	.815	.0005	.002●	40‡	—	28 □	100	—	NOT ADJUSTABLE		

*.030" over rivet head, if bonded lining use .062"   ●1986-84 .004   □1986-84 28 ft/lbs.   ‡88 w/7/16" stud, 1983 Cutlass; 80 ft/lbs.

1-154

BRAKES
Specifications

Model		Drum Diameter Original	Max. Machine O/S	Lining Min. Thickness	Disc Original Thickness	Disc Min. Thickness	Disc Max Runout	Disc Max Parallelism	Wheel Nuts ft/lbs	Cal. Mtg. Bolts ft/lbs	Master Cyl. Bolts ft/lbs	Push Rod Adjustment	Pedal Free Play
86-82 Firenza	*	7.880	7.899	.030	.830	.815	.0005	.002•	40‡	—	28□	100	NOT ADJUSTABLE	
w/vented disc (86-82)	*	7.880	7.899	.030	.444	.429	.0005	.002	40	—	28	100	NOT ADJUSTABLE	
w/solid disc (82)	*	9.500	9.560	.030	.980	.965	.0005	.004	4	—	35□	100†	Handtight	½ Flat
86-82 Cutlass Supreme, 88 (RWD)	*	7.880	7.899	.030	.830	.815	.0005	.002□	40‡	—	28	103	NOT ADJUSTABLE	
86-80 Omega, Calais	*	9.500	9.560	.030	.980	.965	.0005	.004	4	32	35	100	NOT ADJUSTABLE	
85-79 Toronado w/o rear disc	—	—	—	—	.980	.965	.0005	.004	41		30	100		
w/rear disc	*	11.000	11.060	.030	.980	.965	.0005	.004	4	—	35□	100	Handtight	½ Flat
86-78 Custom Cruiser, 88 (w/403), 98 (RWD)	*	9.500	9.560	.030	.980	.965	.0005	.004	4	—	35	80	Handtight	½ Flat
81-78 Cutlass, 88 (w/o 403)	*	9.500	9.560	.030	.830	.815	.0005	.004	15	—	—	80	Handtight	½ Flat
80-76 Starfire	*	9.500	9.560	.030	.930	.965	.0005	.005	4	—	35	80	Handtight	½ Flat
79-78 Omega w/5 Speed	*	9.500	9.560	.030	.980	.965	.0005	.004	4	—	35	80	Handtight	½ Flat
79-78 Omega w/o 5 Speed	*	11.000	11.060	.030	.980	.965	.0005	.002	4	—	35-40	130	NOT ADJUSTABLE	
78-72 Toronado	—	—	—	.062	1.185	1.170	.0005	.004						
77 Customer Cruiser, 88 (w/403), 98	*	11.000	11.060	.030	.980	.965	.0005	.004	4	—	40	100	Handtight	Step 2
77 Omega w/o 5 Speed	*	9.500	9.560	.030	.980	.965	.0005	.004	4	—	40	80	Handtight	Step 2
77 88 (w/o 403)	*	9.500	9.560	.030	.980	.965	.0005	.005	4	—	40	80	Handtight	Step 2
77-76 Omega w/5 Speed, Cutlass	*	11.000	11.060	.030	.980	.965	.0005	.005	4	—	40	80	Handtight	Step 2
76 Omega, w/o 5 Speed	*	9.500	9.560	.030	.980	.965	.0005	.004	4	—	40	80	Handtight	Step 2
76 Cutlass	*	11.000	11.060	.062	.980	.965	.0005	.002	4	—	35-40	130	Handtight	Step 2
76-72 88, 98 exc. Wagon & H.D. Pkg	*	11.000	11.060	.062	1.230	1.215	.0005	.005	4	—	40	80	Handtight	Step 2
76-72 88 Wagon & H.D. Pkg	*	12.000	12.060	.062	1.230	1.215	.0005	.005	4	—	40	80	Handtight	Step 2
75 Starfire	*	9.000	9.060	.062	.455	.440	.0005	.005	15	—	40	80	Handtight	Step 2
75-72 Omega, Cutlass exc. Vista Cruiser	*	9.500	9.560	.030	.980	.965	.0005	.004	4	—	40	80	Handtight	Step 2
75-72 Vista Cruiser	*	11.000	11.060	.030	.980	.965	.0005	.004	4	—	40	80	Handtight	Step 2

".030" over rivet head, if bonded lining use .062" †88 w/7/16" stud, 1983 Cutlass; 80 ft/lbs; •1986-84 .004 □1986-85 38 ft/lbs

GENERAL MOTORS CORP.—PONTIAC

Model		Drum Orig.	Drum Max.	Lining Min.	Disc Orig.	Disc Min.	Runout	Parallelism	Cal. Bolts	Caliper	Wheel Nuts	Master Cyl.	Push Rod	Pedal
86 T1000	*	7.880	7.899	.030	.404	.374	.0005	.005	50	70	21-35	80	Handtight	½ Flat
86 Firebird w/rear drum brakes	*	9.500	9.560	.030	.980	.965	.0005	.005	4	—	38	80†	Handtight	½ Flat
w/rear disc brakes	—	—	—	.030	.986	.956	.0005	.005	41	—	38	80†	—	—
86-84 Fiero: Front	—	—	—	.062	.444	.390	.0005	.004	62	35	—	81†	—	—
Rear	—	—	—	.062	.444	.390	.0005	.005	63	35	21-25	81†	—	½ Flat
85-83 T1000	*	7.874	7.899	.030	.390	.374	.0005	.005	50	—	28■	70	Handtight	—
86-83 6000 w/H.D.	*	8.858	8.880	.062	.972	.957	.0005	.002●	4‡	—	28■	100	NOT ADJUSTABLE	
86-82 6000 exc. H.D.	*	8.858	8.880	.030	.830	.815	.0005	.002●	40‡	—	28	100	NOT ADJUSTABLE	
86-82 J2000	*	7.880	7.899	.030	.830	.815	.0005	.002●	40‡	—	28	100	NOT ADJUSTABLE	
w/vented disc (86-82)	*	7.880	7.899	.030	.444	.429	.0005	.002	40	—	28	100	NOT ADJUSTABLE	12
w/solid disc (82)	*	9.500	9.560	.030	.980	.965	.0005	.004	4	—	21-35	80†	Handtight	½ Flat
85-82 Firebird w/rear drum brakes	—	—	—	.062	.830	.815	.0005	.002●	41	—	30-45	80†	Handtight	½ Flat
w/rear disc brakes	—	7.880	7.899	.030	.830	.815	—	—	40‡	—	35	103	—	—
86-80 Phoenix, Grand Am (FWD)														
86-77 Bonneville, Catalina, LeMans, Grand Prix, Grand Am (RWD), Parisienne, Safari	*	9.500	9.560	.030	.980	.965	.0005	.004	4	—	35	80	Handtight	½ Flat
w/9.5" rear brakes	*	11.000	11.060	.030	.980	.965	.0005	.004	4	—	35	80□	Handtight	½ Flat
w/11" rear brakes														

‡1986-85 use 62 *.030" over rivet head, if bonded lining use .062" □½" stud 100 ft/lbs. †w/Aluminum whls. 105 ft/lbs. •1986-84 .004 ■1986-85 38 ft/lbs.

SECTION 1 BRAKES
Specifications

VEHICLE YEAR, MAKE, MODEL	BRAKE SHOE Minimum Lining Thickness *	BRAKE DRUM DIAMETER Standard Size	BRAKE DRUM Machine To	BRAKE PAD Minimum Lining Thickness	BRAKE ROTOR MIN. THICKNESS Machine To	BRAKE ROTOR Discard At	BRAKE ROTOR Variation From Parallelism	BRAKE ROTOR Runout T.I.R.	DESIGN	CALIPER Mounting Bolts Torque (ft-lbs)	CALIPER Bridge, Pin or Key Bolts Torque (ft-lbs)	WHEEL Lugs or Nuts Torque (ft-lbs)	WHEEL BEARING SETTING STEP 1 Tighten Spindle Nut (ft-lbs)	STEP 2 Back Off Retorque (in-lbs)	STEP 3 Lock, or Back Off and Lock
82-81 T1000	•	7.874	7.899	.030	.390	.374	.0005	.005	20	70	28	70	12	Handtight	½ Flat
81-77 Firebird, Ventura, Phoenix (R.W.D.)	•	9.500	9.560	.030	.980	.965	.0005	.004	4	—	35	80	12	Handtight	½ Flat
80-76 Sunbird, Astre w/rear disc (1981-79)	—	—	—	.030	.921	.905	.0005	.004	41	—	30	80	12	—	½ Flat
76-72 Catalina, Bonneville, Gran Ville exc. Wagon	—	9.500	9.560	.030	.830	.815	.0005	.004	15	—	35	80	12	Handtight	½ Flat
Wagon, Grand Safari	.125	11.000	11.060	.125	1.230	1.215	.0005	.004	4	—	35	75	12	Handtight	½ Flat
76-74 LeMans, Firebird, Grand Prix, Ventura exc. Wagon	.125	12.000	12.060	.125	1.230	1.215	.0005	.004	4	—	35	75	12	Handtight	½ Flat
Wagon	.125	9.500	9.560	.125	.980	.965	.0005	.004	4	—	35	70	12	Handtight	½ Flat
73-72 Ventura	.125	11.000	11.060	.125	.980	.965	.0005	.004	4	—	35	70	12	Handtight	½ Flat
73-72 LeMans, Firebird, Grand Prix	.125	9.500	9.560	.125	—	.980	.0005	.004	4	—	35	70	Snug	Handtight	Step 2
	.125	9.500	9.560	.125	.980	.965	.0005	.004	4	—	35	70	Snug	Handtight	Step 2

‡1986-85 use 62 * .030" over rivet head, if bonded lining use .062" □½" stud 100 ft/lbs. •1986-84 .004 †w/Aluminum whls. 105 ft/lbs. ■1986-85 38 ft/lbs.

DOMESTIC LIGHT TRUCKS

NOTICE: Should state inspection regulations exceed manufacturer's specifications for lining or rotor/drum reserve, the state inspection regulation specification must be used.

VEHICLE YEAR, MAKE, MODEL	BRAKE SHOE Minimum Lining Thickness *	BRAKE DRUM DIAMETER Standard Size	BRAKE DRUM Machine To	BRAKE PAD Minimum Lining Thickness	BRAKE ROTOR MIN. THICKNESS Machine To	BRAKE ROTOR Discard At	BRAKE ROTOR Variation From Parallelism	BRAKE ROTOR Runout T.I.R.	DESIGN	CALIPER Mounting Bolts Torque (ft-lbs)	CALIPER Bridge, Pin or Key Bolts Torque (ft-lbs)	WHEEL Lugs or Nuts Torque (ft-lbs)	WHEEL BEARING SETTING STEP 1 Tighten Spindle Nut (ft-lbs)	STEP 2 Back Off Retorque (in-lbs)	STEP 3 Lock, or Back Off and Lock
AMERICAN MOTORS—JEEP															
86-84 Cherokee, Wagoneer, Comanche, (Sportwagon)	.030	10.000	10.060	.030	—	.815	—	.004	1	77	25-35	75	•	•	•
86-82 CJ, Scrambler	.030*	10.000	10.060	.062	—	.815	.001	.005	1	—	30	75	50	Skip	1/6 Turn
86-82 Cherokee, Wagoneer, J-10	.030*	11.000	11.060	.062	—	1.215	.001	.005	4	35	30	75	50	Skip	1/6 Turn
86-82 J-20	.030*	12.000	12.060	.062	—	1.215	.001	.005	4	35	30	75 ■	50	Skip	1/6 Turn
81-79 CJ	.030*	10.000	10.060	.062	—	.815	.001	.005	1	—	15	75	50	Skip	1/3 Turn
81-78 Cherokee, Wagoneer, J-10	.030*	11.000	11.060	.062	—	1.215	.001	.005	4	35	—	75	50	420	1/3 Turn
81-78 J-20	.030*	12.000	12.060	.062	—	1.215	.001	.005	4	35	15	75 ■	50	420	1/3 Turn
78-77 CJ	.030	11.000	11.060	.062	—	1.125	.001†	.005	1	—	—	75	50	Skip	1/3 Turn

*.030" over rivet head, if bonded lining use .062" †1977-.0005
•torque hub bolts to 75 ft/lbs., torque hub nut to 175 ft/lbs. □while rotating whl., tighten nut until whl. binds.
•torque hub bolts to 75 ft/lbs. ■J20 (8400 GVW) 130 ft/lbs.

BRAKES
Specifications

Model	Col2	Col3	Col4	Col5	Col6	Col7	Col8	Col9	Col10	Col11	Col12
77-76 Cherokee, Wagoneer, J-10	.030*	11.000	11.060	—	—	.0005	.005	4	—	75	50	Skip	1/4 Turn
77-76 J-20	.030*	12.000	12.060	—	—	.0005	.005	4	—	75	50	Skip	1/4 Turn
76-72 CJ Front & Rear	.030	11.000	11.060	—	—	.003	—	—	—	75 ■	—	Skip	1/6 Turn
75-74 Cherokee, Wagoneer, J-10	.030	11.000	11.060	—	—	.003	.005	4	—	75	50	Skip	1/4 Turn
75-74 J-20	.030	12.000	12.060	—	—	.003	.005	4	—	75 ■	—	Skip	1/6 Turn
73-72 all w/11" brakes	.030	11.000	11.060	—	—	—	—	—	—	75	—	Skip	1/6 Turn
73-72 all w/12" brakes	.030	12.000	12.060	—	—	—	—	—	—	75	—	Skip	1/6 Turn
73-72 8,000 GVW Camper	.030	12.125	12.185	—	—	—	—	—	—	75	—	Skip	1/3 Turn

*.030" over rivet head, if bonded lining use .062" †1977-.0005 □while rotating whl., tighten nut until whl. binds. ■J20 (8400 GVW) 130 ft/lbs.
•torque hub bolts to 75 ft/lbs., torque hub nut to 175 ft/lbs.

CHRYSLER CORP.—DODGE, PLYMOUTH

Model	Col2	Col3	Col4	Col5	Col6	Col7	Col8	Col9	Col10	Col11	Col12	Col13
86-84 Caravan, Mini Ram Van, Voyager	.030	9.000	9.060	.030*	.833	.0005	.005	8	130-190	25-35	95	20-25	Handtight	Step 2
84-83 Rampage, Scamp	.030	7.874	7.904	.030	.461	.0005	.004	8	130-190	25-35	80††	20-25	Handtight	Step 2
82 Rampage	.030	7.780	7.900	.030	.461	.0005	.004	8	70-100	25-40	85	20-25	Handtight	Step 2
86-84 Ramcharger 4 × 2	.030	11.000	11.060	.030	1.220	.0005	.004	5	110	17	105	8	360-480	1/3 Turn
4 × 4	.030	11.000	11.060	.030	1.220	.001	.005	5	150	17	105	50	360-480	1/3 Turn
83-79 Ramcharger, Trail Duster														
4 × 2	.030	10.000	10.060	.030	1.220	.0005	.004	5	110	17	105	8	Handtight	Step 2
4 × 4	.030	10.000	10.060	.030	1.220	.001	.005	5	150	17	105	50	360-480	1/3 Turn
78-74 Ramcharger, Trail Duster														
4 × 2	.030	11.000	11.060	.030	1.220	.0005	.004	5	95	17	85-125	8	Handtight	Step 2
4 × 4	.030	11.000	11.060	.030	1.220	.0005	.004	5	95	17	85-125	100	Skip	1 Slot
86-84 D150	.030	10.000	10.060	.030	1.220	.0005	.004	5	110	17	105	8	Handtight	Step 2
83-75 D100/150 exc. w/9¼" R. axle	.030	10.000	10.060	.030	1.220	.0005	.004	5	110★	17	105	8	Handtight	Step 2
76-75 D100 w/9¼" rear axle	.030	10.000	10.060	.030	1.220	.0005	.004	5	100	17	65-85	20-25	Handtight	Step 2
74-73 D100 exc. w/11" front brakes	.030	11.000	11.060	.030	1.220	.0005	.004	5	100	17	65-85	20-25	Handtight	Step 2
73-72 D100 w/11" front brakes	.030	10.000	10.060	—	—	—	—	—	—	30-35	70-90	20-25	Handtight	Step 2
72 D100	.030	11.000	11.060	.030	1.180	.0005	.0025	2	75-100	17	105	8	Handtight	Step 2
86-81 D200/250 w/3300 lb. F.A.	.030	12.000	12.060	.030	1.190	.001	.005	5	110	17	105	8	Handtight	Step 2
86-79 D200/250 w/4000 lb. F.A., D300/350	.030	12.000	12.060	.030	1.220	.001	.005	5	160	17	105•	8	Handtight	Step 2
80-79 D200 w/3300 lb. F.A.	.030	12.000	12.060	.030	1.160	.001	.005	5	100	17	105	8	Handtight	Step 2
80-79 D200 w/3300 lb. F.A. GVW over 6200 lb.	.030	12.000	12.060	.030	1.220	.001	.005	5	100	17	105	8	Handtight	Step 2
78-75 D200 w/6600 lb. GVW	.030	12.120	12.180	.030	1.160	.0005	.004	5	160	17	105•	8	Handtight	Step 2
78-75 D200 over 6000 GVW, D300	.030	12.120	12.180	.030	1.160	.0005	.004	5	160	17	65-85	20-25	Handtight	Step 2
74 D200 w/6000 lb. GVW	.030	12.000	12.060	.030	1.160	.0005	.004	5	160	17	68-85•	20-25	Handtight	Step 2
74 D200 over 6000 GVW, D300	.030	12.000	12.060	.030	—	.0005	.004	5	100	17	70-90	20-25	Handtight	Step 2
73 D200 w/6000 lb. GVW	.030	12.000	12.060	.030	1.125	.0005	.004	5	100	17	70-90	20-25	Handtight	Step 2
73 D200 over 6000 GVW	.030	12.000	12.060	.030	1.180	.0005	.004	5	100	17	70-90	20-25	Handtight	Step 2
73 D200 w/12.57" rotor diameter	.030	12.000	12.060	.030	1.125	.0005	.004	5	100	17	70-90•	20-25	Handtight	Step 2
73-72 D300	.030	12.120	12.180	.030	—	.0005	.0025	2	75-100	30-35	79-90	20-25	Handtight	Step 2
72 D200 w/6000 lb GVW	.030	12.000	12.060	.030	1.125	.0005	.0025	2	75-100	30-35	70-90	20-25	Handtight	Step 2
72 D200 w/12.57 rotor diameter	.030	12.000	12.060	.030	1.180	.0005	.0025	2	75-100	30-35	70-90	20-25	Handtight	Step 2
72 D200 w/12.82 rotor diameter	.030	12.120	12.180	.030	1.125	.0005	.0025	2	75-100	17	105	20-25	Handtight	Step 2
86-84 W150	.030	11.000	11.060	.030	1.220	.001	.005	2	150	17	105	50	360-480	1/3 Turn

•⅝" whl. stud (86-75) 200 ft/lbs.; (74) 175-225 ft/lbs.; (73-72) 125-175 ft/lbs. w/dual rear whls.   ■⅝" bolt 140-180 ft/lbs.   ★ (1980-75) 100 ft/lbs.
□w/dual rear whls. (78-76) 300-350 ft/lbs. (75-72) F125-175 ft/lbs. R300-350 ft/lbs.   †2 or 5   +w/cone type nut shown; w/flanged 300-350 ft/lbs.
*.030" over rivet head; if bonded lining, use .062".   ††1984—95 ft/lbs.

BRAKES
Specifications

VEHICLE	BRAKE SHOE	BRAKE DRUM		BRAKE PAD	BRAKE ROTOR						CALIPER		WHEEL	WHEEL BEARING SETTING		
		DIAMETER			MIN. THICKNESS											
YEAR, MAKE, MODEL	Minimum Lining Thickness *	Standard Size	Machine To	Minimum Lining Thickness	Machine To	Discard At	Variation From Parallelism	Runout T.I.R.	DESIGN	Mounting Bolts Torque (ft-lbs)	Bridge, Pin or Key Bolts Torque (ft-lbs)	Lugs or Nuts Torque (ft-lbs)	STEP 1 Tighten Spindle Nut (ft-lbs)	STEP 2 Back Off Retorque (in-lbs)	STEP 3 Lock, or Back Off and Lock
83-79 W150/W100	.030	10.000	10.060	.030	1.220	1.190	.001	.005	5	150	17	105	50	360-480	1/3 Turn
78-77 W100	.030	11.000	11.060	.030	1.220	1.190	.0005	.004	5	100	17	105	50	360-480	1/3 Turn
76-75 W100	.030	11.000	11.060	.030	1.220	1.190	.0005	—	5	100	17	105	30-40	Handtight	1/3 Turn
74-72 W100	.030	11.000	11.060	—	—	—	—	—	—	—	—	70-90	—	Skip	Step 2
86-79 W200/250 Exc. w/Spicer 60	.030	12.000	12.060	.030	1.160	1.130	.001	.005	5	150	17	105	50	360-480	1/3 Turn
86-79 W200/250 w/Spicer 60, W300/350	.030	12.000	12.060	.030	1.160	1.130	.001	.005	21	160	15	105●	50	360-480	1/3 Turn
78-75 W200 exc. w/Spicer 60	.030	12.000	12.060	.030	1.160	1.130	.0005	.004	5	100	17	105	50	360-480	1/3 Turn
78-75 W200 w/Spicer 60, W300	.030	12.000	12.060	.030	1.160	1.130	.0005	.004	21	160	15	105	50	360-480	1/3 Turn
74-72 W200 exc. w/4500 lb. F.A. Front	.030	11.000	11.060	—	—	—	—	—	—	—	—	—	—	—	—
Rear	.030	12.120	12.180	—	—	—	—	—	—	—	—	—	50	Skip	1/3 Turn
74-72 W200 w/4500 lb. F.A., W300	.030	12.000	12.060	—	—	—	—	—	—	—	—	—	50	Skip	1/3 Turn
86 B150, B250	.030	12.000	12.060	—	—	—	—	—	—	—	—	—	50	Skip	1/3 Turn
85-76 B100/150, PB100/150	.030	11.000	11.060	.125	1.210	1.180	.0005	.004	5	95-125	15	85-110	30-40	Handtight	Step 2
75-73 B100, PB100	.030	10.000	10.060	.030	1.220	1.190	.0005	.004	5	95-125	14-22	85-125	30-40	Handtight	Step 2
w/11" rear brakes	.030	11.000	11.060	.030	1.220	1.190	.0005	.004	5	95-125	14-22	85-125	30-40	Handtight	Step 2
72 B100 w/10" rear brakes	.030	10.000	10.060	.030	1.220	1.190	.0005	.004	†	75-100	17	65-85	20-30	Handtight	Step 2
w/11" rear brakes	.030	11.000	11.060	.030	1.220	1.190	.0005	.004	†	75-100	17	65-85	20-30	Handtight	Step 2
85-84 B250, PB250	.030	11.000	11.060	.030	1.220	1.190	.0005	.004	5	75-100	17	65-85	20-30	Handtight	Step 2
83-78 B200/250, PB200/250	.030	11.000	11.060	.030	1.220	1.190	.0005	.004	5	95-125	14-22	85-125	30-40	Handtight	Step 2
77-76 B200, PB200	.030	11.000	11.060	.030	1.220	1.190	.0005	.004	5	75-100	17	65-85	20-30	Handtight	Step 2
75-73 B200, PB200	.030	11.000	11.060	.030	1.220	1.190	.0005	.004	5	95-125	14-22	85-125	30-40	Handtight	Step 2
72 B200 w/10" brakes	.030	10.000	10.060	.030	1.220	1.190	.0005	.004	†	75-100	17	65-85	20-30	Handtight	Step 2
w/11" brakes	.030	11.000	11.060	.030	1.220	1.190	.0005	.004	†	75-100	17	65-85	20-30	Handtight	Step 2
86 B350 w/3600 lb. F.A.	.030	12.000	12.060	.125	1.210	1.180	.0005	.004	5	95-125	15	85-110	30-40	Handtight	Step 2
w/4000 lb. F.A.	.030	12.000	12.060	.125	1.155	1.125	.0005	.004	5	95-125	15	175-225+	30-40	Handtight	Step 2
85-79 B300/350, CB300/350, PB300/350	.030	12.000	12.060	.030	1.220	1.190	.0005	.004	4	95-125	14-22	85-125	30-40	Handtight	Step 2
w/3600 lb. F.A.	.030	12.000	12.060	.030	1.160	1.130	.0005	.004	5	95-125	14-22	175-225	30-40	Handtight	Step 2
w/4000 lb. F.A.	.030	12.000	12.060	.030	1.160	1.130	.0005	.004	5	95-125	14-22	65-85☐	30-40	Handtight	Step 2
78-76 B300, CB300, PB300	.030	12.000	12.060	.030	1.160	1.130	.0005	.004	5	140-180	17	65-85	20-30	Handtight	Step 2
75-73 B300, CB300, PB300	.030	12.000	12.060	.030	1.160	1.130	.0005	.004	†	140-180	17	65-85	20-30	Handtight	Step 2
72 B300, CB300															

● ⅝" whl. stud (86-75) 200 ft/lbs.; (74) 175-225 ft/lbs. (73-72) 125-175 ft/lbs. w/dual rear whls. 325 ft/lbs. ★ (1980-75) 100 ft/lbs. ■ ⅝" bolt 140-180 ft/lbs.
☐ w/dual rear whls. (78-76) 300-350 ft/lbs.; (75-72) F125-175 ft/lbs. R300-350 ft/lbs. † 2 or 5 + w/cone type nut shown; w/flanged 300-350 ft/lbs.
* .030" over rivet head; if bonded lining, use .062". ††1984—95 ft/lbs.

FORD MOTOR CO.															
86-83 Ranger 4x2	.030	9.000	9.060	.030	—	.810	.001	.003	**59**	—	32-47	85-115	17-25	10-15	Step 2
86-84 Bronco II 4x2	.030	9.000	9.060	.030	—	.810	.001	.003	**10**	—	32-47	85-115	17-25	10-15	Step 2
86-83 Bronco II 4x4, Ranger 4x4	.030	9.000	9.060	.030	—	.810	.001	.003	**10**	—	32-47	85-115	35	90°	16 in/lbs.
86-81 Bronco	.030	11.031	11.091	.030	—	1.120	.0007	.003	**1**	74-102	12-20	90	50†	Skip	45°
80-76 Bronco	.030	11.031	11.091	.030	—	1.120	.0007	.003	**1**	50-60	12-20	90	50	Skip	90°
75-72 Bronco w/10" brakes	.030	10.000	10.060	—	—	—	—	—	—	—	—	90	50	Skip	90°
w/11" brakes	.030	11.000	11.060	—	—	—	—	—	—	—	—	90	50	Skip	90°
83-80 F100 w/46-4900 GVW	.030	10.000	10.060	.030	—	.810	.0005	.003	**1**	—	12-20	90	22-25	Skip	1/8 Turn
86-77 F100, F150, E100, E150	.030	11.031	11.091	.030	—	1.120	.0007	.003	**1**	—	12-20	90	22-25	Skip	1/8 Turn
76-75 F100, F150, E100, E150	.030	11.031	11.091	.030	—	1.120	.0007	.003	**1**	—	12-20	90	17-25	Skip	2 Slots
74 E100	.030	10.000	10.060	—	—	—	—	—	—	—	—	90	17-25	Skip	2 Slots
74-73 F100	.030	11.031	11.091	.030	—	1.120	.0003	.003	**1**	—	12-20	90	17-25	Skip	2 Slots
72 E100	.030	11.031	11.091	—	—	—	—	—	—	—	—	90	17-25	Skip	2 Slots
86-77 F250 (6900 GVW std.)	.030	12.000	12.060	.030	—	1.120	.0007	.003	**1**	—	12-20	90‡	22-25	Skip	1/8 Turn
76-75 F250 (6900 GVW std.)															
w/12" rear brakes	.030	12.000	12.060	.030	—	1.120	.0007	.003	**1**	—	12-20	90	17-25	Skip	2 Slots
w/12⅛" rear brakes	.030	12.125	12.185	.030	—	1.120	.0007	.003	**1**	—	12-20	90	17-25	Skip	2 Slots
74-73 F250 (6900 GVW std.)															
w/12" rear brakes	.030	12.000	12.060	.030	—	1.120	.0007	.003	**1**	—	12-20	90	17-25	Skip	2 Slots
w/12⅛" rear brakes	.030	12.125	12.185	.030	—	1.120	.0007	.003	**1**	—	12-20	90	17-25	Skip	2 Slots
72 F250 (Std.) w/12" brakes	.030	12.000	12.060	—	—	—	—	—	—	—	—	90	17-25	Skip	2 Slots
w/12⅛" brakes	.030	12.125	12.185	—	—	—	—	—	—	—	—	90	17-25	Skip	2 Slots
86-78 F250 (6900 GVW H.D.), F350, E250, E350	.030	12.000	12.060	.030	—	1.180	.0007	.003	**19**	74-102	12-20★	90■	22-25	Skip	1/8 Turn
77 F250, (6900 GVW H.D.), F350, E250, E350	.030	12.000	12.060	.030	—	1.214	.0007	.003	**19**	74-102	12-20★	90■	22-25	Skip	1/8 Turn
76 F250 (6900 H.D.), F350, E250, E350															
w/12" rear brakes	.030	12.000	12.060	.030	—	1.180	.0007	.003	**19**	74-102	12-20★	90■	17-25	Skip	2 Slots
w/12⅛" rear brakes	.030	12.125	12.185	.030	—	1.180	.0007	.003	**19**	74-102	12-20★	90■	17-25	Skip	2 Slots
75-72 F250 (6900 GVW H.D.), F350															
w/12" rear brakes	.030	12.000	12.060	.030	—	.940	.001	.003	**23**	55-72	17-23★	90■	17-25	Skip	2 Slots
w/12⅛" rear brakes	.030	12.125	12.185	.030	—	.940	.001	.003	**23**	55-72	17-23★	90■	17-25	Skip	2 Slots
75 E250, E350	.030	12.000	12.060	.030	—	.940	.001	.003	**23**	55-72	17-23★	90■	17-25	Skip	2 Slots
74-72 E250	.030	11.031	11.091	—	—	—	—	—	—	—	—	90	17-25	Skip	2 Slots
74-72 E350	.030	12.000	12.060	—	—	—	—	—	—	—	—	135■	17-25	Skip	2 Slots
86-81 F100, F150, F250 w/Dana 44IFS Front Axle (4x4)	.030	11.031	11.091	.030	—	1.120	.0007	.003	**1**	74-102	12-20	90	50†	Skip	45°
80-76 F100, F150 (4x4)	.030	11.031	11.091	.030	—	1.120	.0007	.003	**1**	50-60	12-20	90	50	Skip	90°
75-72 F100 (4x4) Front	.030	11.000	11.060	—	—	—	—	—	—	—	—	90	50	Skip	90°
Rear	.030	11.031	11.091	—	—	—	—	—	—	—	—	90	50	Skip	90°
86-83 F250 Exc. Dana 44IFS Front Axle (4x4), F350 (4x4)	.030	12.000	12.060	.030	—	1.180	.0007	.003	**19**	74-102	12-20★	100■	50	372-468	135°/150°
82-81 F250 (4x4)	.030	12.000	12.060	.030	—	1.180	.0007	.003	**19**	74-102	12-20★	90	50	Skip	90°
82-81 F350 (4x4)	.030	12.000	12.060	.030	—	1.180	.0007	.003	**19**	74-102	12-20★	90■	50	372-468	135°/150°
80-76 F250 (4x4), F350 (4x4)	.030	12.000	12.060	.030	—	1.180	.0007	.003	**19**	74-102	12-20★	90■	50	Skip	90°
75-72 F250 (4x4) Front	.030	12.125	12.185	—	—	—	—	—	—	—	—	90	50	Skip	90°
Rear	.030	12.000	12.060	—	—	—	—	—	—	—	—	90	50	Skip	90°
75-72 F350 (4x4)	.030	12.000	12.060	.030	—	.940	.001	.003	**23**	55-72	17-23★	135■	50	Skip	90°

★Caliper bridge bolt 155-185 ft/lbs. †1986 70 ft/lbs. ‡ 1986 140 ft/lbs.
■1986-85 140 ft/lbs. 350 w/single rear whls. (84-78) 145 ft/lbs., (77-72) 135 ft/lbs., w/dual rear whls. (84-78) 220 ft/lbs. (77-72) 210 ft/lbs.

SECTION 1 BRAKES
Specifications

VEHICLE YEAR, MAKE, MODEL	BRAKE SHOE Minimum Lining Thickness	BRAKE DRUM DIAMETER Standard Size	BRAKE DRUM DIAMETER Machine To	BRAKE PAD Minimum Lining Thickness	BRAKE ROTOR MIN. THICKNESS Machine To	BRAKE ROTOR MIN. THICKNESS Discard At	BRAKE ROTOR Variation From Parallelism	BRAKE ROTOR Runout T.I.R.	DESIGN	CALIPER Mounting Bolts Torque (ft-lbs)	CALIPER Bridge, Pin or Key Bolts Torque (ft-lbs)	WHEEL Lugs or Nuts Torque (ft-lbs)	WHEEL BEARING SETTING STEP 1 Tighten Spindle Nut (ft-lbs)	WHEEL BEARING SETTING STEP 2 Back Off Retorque (in-lbs)	WHEEL BEARING SETTING STEP 3 Lock, or Back Off and Lock
GENERAL MOTORS CORP.															
86-85 Astro Van	*	9.500	9.560	.030	.980	.965	.0005	.004		—	30-45	90	12	Handtight	1/2 Flat
86-82 S, T 10/15	.030	9.500	9.560	.030	.980	.965	.0005	.004	4	—	21-35	80†	12★	Handtight	1/2 Flat
86-83 Blazer, Jimmy (S, T Series)	.062	9.500	9.560	.030	.980	.965	.0005	.004	4	—	21-35	80†	12★	Handtight	1/2 Flat
86-83 Blazer, Jimmy (K Series)	.062	11.150	11.210	.030	1.230	1.215	.0005	.004	4	—	35	70-90	50	420	3/8 Turn
82-76 Blazer, Jimmy (4x2)	.062	11.150	11.210	.030	1.230	1.215	.0005	.004	4	—	35	75-100	12	Handtight	1/2 Flat
(4x4)	.062	11.000	11.060	.030	1.230	1.215	.0005	.004	4	—	35	70-90	50	420	3/8 Turn
75-74 Blazer, Jimmy (4x2)	.030	11.000	11.060	.030	1.230	1.215	.0005	.004	4	—	35	75-100	12	Handtight	1/2 Flat
w/11" rear brakes (4x4)	.030	11.000	11.060	.030	1.230	1.215	.0005	.005	4	—	35	70-90	50	420	1/4 Turn
w/11⅛" rear brakes (4x4)	.030	11.150	11.210	.030	1.230	1.215	.0005	.005	4	—	35	70-90	50	Skip	1/4 Turn
73-72 Blazer, Jimmy (4x2)	.030	11.000	11.060	.030	1.230	1.215	.0005	.005	4	—	35	65-90	50	420	1/4 Turn
(4x4)	.030	11.000	11.060	.030	1.230	1.215	.0005	.005	4	—	35	55-75	Snug	420	1/4 Turn
86-76 Suburban, C,G 10/15 w/11.0" rotor.															
w/11" brakes	.062	11.000	11.060	.030	.980	.965	.0005	.004	4	—	35	75-100	12	Handtight	1/2 Flat
w/11⅛" rear brakes	.062	11.150	11.210	.030	.980	.965	.0005	.004	4	—	35	75-100	12	Handtight	1/2 Flat
w/1.25" rotor															
w/11" rear brakes	.062	11.000	11.060	.030	1.230	1.215	.0005	.004	4	—	35	75-100	12	Handtight	1/2 Flat
75-74 Suburban, C,G 10/15	.062	11.150	11.210	.030	1.230	1.215	.0005	.004	4	—	35	75-100	12	Handtight	1/2 Flat
w/11" rear brakes	.030	11.000	11.060	.030	1.230	1.215	.0005	.005	4	—	35	75-100	12	Handtight	1/2 Flat
73-72 Suburban, C,G 10/15	.030	11.150	11.210	.030	1.230	1.215	.0005	.005	4	—	35	75-100	12	Handtight	1/2 Flat
w/11⅛" rear brakes	.030	11.000	11.060	.030	1.230	1.215	.0005	.005	4	—	35	65-90	Snug	Skip	1/4 Turn
86-76 Suburban, C,G, 20/25, 30/35 under 8600 GVW															
w/11⅛" rear brakes	.062	11.150	11.210	.030	1.230	1.215	.0005	.004	4	—	35	90-120☐	12	Handtight	1/2 Flat
w/13" rear brakes	.062	13.000	13.060	.030	1.230	1.215	.0005	.004	4	—	35	90-120☐	12	Handtight	1/2 Flat
86-76 C,G, 30/35 over 8600 GVW	.062	13.000	13.060	.030	1.480	1.465	.0005	.004	1	—	15	90-120●	12	Handtight	1/2 Flat
75-74 Suburban, C,G 20/25, 30/35 under 8600 GVW															
w/11⅛" rear brakes	.030	11.150	11.210	.030	1.230	1.215	.0005	.005	4	—	35	90-120☐	12	Handtight	1/2 Flat
w/13" rear brakes	.030	13.000	13.060	.030	1.230	1.215	.0005	.005	4	—	35	90-120	12	Handtight	1/2 Flat
75-74 C,G 30/35 w/11⅛" rear brakes	.030	11.150	11.210	.030	1.480	1.465	.0005	.005	1	—	15	90-120	12	Handtight	1/2 Flat
73-72 C 20/25 w/11⅛" rear brakes	.030	12.000	12.060	.030	1.230	1.215	.0005	.005	4	—	35	90-120	Snug	Skip	1/4 Turn
w/12" rear brakes	.030	13.000	13.060	.030	1.230	1.215	.0005	.005	4	—	35	90-120	Snug	Skip	1/4 Turn
w/13" rear brakes	.030	11.000	11.060	.030	1.230	1.215	.0005	.005	4	—	35	75-100	Snug	Skip	1/4 Turn
73-72 G 20/25	.030	13.000	13.060	.030	1.230	1.215	.0005	.005	4	—	35	90-120	Snug	Skip	1/4 Turn
73-72 C,G 30/35 w/13" rear brakes	.030	15.000	15.060	.030	1.230	1.215	.0005	.005	4	—	35	90-120	Snug	Skip	1/4 Turn
w/15" rear brakes	.030	15.000	15.060	.030	1.230	1.215	.0005	.005	4	—	35	90-120	Snug	Skip	1/4 Turn

☐ G 20/25, K 10/15 is 75-100 ft/lbs. ● w/8 bolt whl. 110-140 ft/lbs., w/10 bolt whl. 130-200 ft/lbs.
■ K20/25 is 90-120 ft/lbs. † 4x2 std. shown, 4x2 & 4x4 w/½" bolt; 100 ft/lbs. ★ 4x2 only, 4x4 not adjustable

BRAKES
Specifications

Model												
86-76 K 10/15, 20/25	.062	11.150	11.210	.030	1.230	1.215	.0005	.004	4	—	35	70-90	50	420	3/8 Turn
w/11 11/16" rear brakes	.062	13.000	13.060	.030	1.230	1.215	.0005	.004	4	—	35	70-90■	50	420	3/8 Turn
75-74 K 10/15, 20/25	.030	11.000	11.060	.030	1.230	1.215	.0005	.005	4	—	35	90-120□	50	420	1/4 Turn
w/11" rear brakes	.030	11.150	11.210	.030	1.230	1.215	.0005	.005	4	—	35	90-120□	50	420	1/4 Turn
w/13" rear brakes	.030	13.000	13.060	.030	1.230	1.215	.0005	.005	4	—	35	90-120□	50	420	1/4 Turn
73-72 K10/15, 20/25	.030	11.000	11.060	.030	1.230	1.215	.0005	.005	4	—	35	55-75	50	420	1/4 Turn
w/11" rear brakes	.030	11.150	11.210	.030	1.230	1.215	.0005	.005	4	—	35	55-75	50	420	1/4 Turn
w/13" rear brakes	.030	13.000	13.060	.030	1.230	1.215	.0005	.005	4	—	35	55-75	50	420	1/4 Turn
86-77 K 30/35	.062	13.000	13.060	.030	1.480	1.465	.0005	.004	4	—	35	90-120●	50	420	3/8 Turn

□ G 20/25, K 10/15 is 75-100 ft/lbs. ● w/8 bolt whl. 110-140 ft/lbs., w/10 bolt whl. 130-200 ft/lbs.
■ K20/25 is 90-120 ft/lbs. † 4x2 std. shown, 4x2 & 4x4 w/1/2" bolt; 100 ft/lbs. ★ 4x2 only, 4x4 non adjustable

INTERNATIONAL HARVESTER CORP.

80-72 Scout II	.030	11.000	11.060	.125	—	1.120	.0005	.005	1	—	12-18	—	30	Skip	1/4 Turn
75-74 Travel All (½ Ton) 100/150	.030	11.000	11.060	.125	—	1.120	.0005	.005	1	—	12-18	—	To Drag	Skip	To Slot
75-74 Travel All (¾ Ton) 200	.030	12.000	12.060	.125	—	1.120	.0005	.005	1	—	12-18	—	To Drag	Skip	To Slot

IMPORTED CARS AND LIGHT TRUCKS

NOTICE: Should state inspection regulations exceed manufacturer's specifications for lining or rotor/drum reserve, the state inspection regulation specification must be used.

AUDI														
86-85 4000S Quattro Front	—	—	—	.079	.728	.709	.0003	.0011	61	52	25	80	■	—	—
Rear	—	—	—	.079	.807	.787	.0008	.002	33	47	25	80	—	—	—
86-85 4000S Coupe GT	.097	—	7.894	.079	.728	.709	.0003	.0011	31	52	25	80	—	—	—
85-83 Quattro Turbo Coupe Front	—	—	—	.079	.335	.315	.0008	.002	31*	83	25	80	■	—	—
Rear	—	—	—	.079	.807	.787	.0008	.002	33	47	25	80	—	—	—
84 4000S Quattro Front	—	—	—	.079	.807	.787	.0008	.002	31	83	25	80	■	—	—
Rear	—	—	—	.079	.413	.394	—	.002	33	47	25	80	—	—	—
84-80 4000 Series, Coupe	.097	—	7.894	.079	.807	.787	.0008	.002	33	50□	25	65	■	NOT ADJUSTABLE	—
86-81 5000 Series exc. Turbo, Quattro	.098	9.055	9.094	.079	.807	.787	.0008	.002	37	83	25	80	■	—	—
80-78 5000 Series exc. Turbo	.098	9.055	9.094	.079	.807	.787	.0008	.004	39	83	18	80	■	NOT ADJUSTABLE	—
86-80 5000 Turbo, Quattro Front	—	—	—	.078	.807	.787	.0008	.002	31*	83	—	80	■	—	—
Rear	—	—	—	.079	.335	.315	—	.002	33	47	25	80	■	—	—
79-78 Fox	.097	7.870	7.900	.078	.413	.393	.0008	.002	37	43	25	65	■	—	—
77-73 Fox	.097	7.870	7.900	.078	.413	.393	—	.004	39	43	—	65	■	—	—
BMW														
86-84 325, 325es Front	—	—	—	.080	.921	—	.0008	.008	6	—	22-25	65-79	—	NOT ADJUSTABLE	—
Rear	—	—	—	.080	.331	—	—	.008	6	—	22-25	65-79	—	—	—
85-84 318i	.060	—	9.035	.080	.594	—	.0008	.008	6	—	22-25	72-80	—	NOT ADJUSTABLE	—

■ Seat brg. while turning wheel, back off nut until thrust washer can be moved slightly by screwdriver w/finger pressure, lock. (Quattro models not adjustable)
□ w/standard bolt, 1980-81 & early 1982; 36 ft/lbs. *1986-85 use ill. 61

SECTION 1
BRAKES
Specifications

VEHICLE	BRAKE SHOE	BRAKE DRUM DIAMETER		BRAKE PAD	BRAKE ROTOR MIN. THICKNESS					CALIPER			WHEEL	WHEEL BEARING SETTING		
YEAR, MAKE, MODEL	Minimum Lining Thickness *	Standard Size	Machine To	Minimum Lining Thickness	Machine To	Discard At	Variation From Parallelism	Runout T.I.R.	DESIGN	Mounting Bolts Torque (ft-lbs)	Bridge, Pin or Key Bolts Torque (ft-lbs)	Lugs or Nuts Torque (ft-lbs)	STEP 1 Tighten Spindle Nut (ft-lbs)	STEP 2 Back Off Retorque (in-lbs)	STEP 3 Lock, or Back Off and Lock
86-83 524td, 528e, 533i, 535i, 633 csi, 635 csi Front	—	—	—	.080	—	□	—	.008	6	80-89	14-18	72-80	NOT ADJUSTABLE		
Rear	—	—	—	.080	.331	□	—	.008	6	43-48	14-18	72-80			
86-83 733i, 735i Front	—	—	—	.080	.921	.827	.0008	.008	16	58-69	—	72-80	22-24	12●	Step 2
Rear	—	—	—	.080	.331	.335	.0008	.008	6	43-48	14-18	72-80	—	—	—
82 528e Front	—	—	—	.138	.803	—	.0008	.008	6	89	14-18	72-80	NOT ADJUSTABLE		
Rear	—	—	—	.138	.331	—	.0008	.008	6	43-48	14-18	72-80			
81-79 528i Front	—	—	—	.080	.846	.827	.0008	.008	16	—	—	59-65	22-24†	12●	Step 2
Rear	—	—	—	.080	.354	.335	.0008	.008	17	—	—	59-65	—	—	—
82-79 633CSi Front	—	—	—	.080	.846	.827	.0008	.008	16	—	—	59-65	22-24†	12●	Step 2
Rear	—	—	—	.080	.728	.709	.0008	.008	17	—	—	59-65	—	—	—
82-78 733i Front	—	—	—	.080	.846	.827	.0008	.008	16	—	—	60-66	22-24†	12●	Step 2
Rear	—	—	—	.080	.374	.354	.0008	.006	17	59-70	—	60-66	—	—	—
83-82 320i w/ATE	—	9.842	9.882	.118	.846	.827	.0008	.006	16	44-49	—	59-65	22-24†	12●	Step 2
w/Girling	—	9.842	9.882	.118	.480	.461	.0008	.008	17	58-69	—	59-65	22-24†	12●	Step 2
81-77 320i	—	9.842	9.882	.118	.846	.827	.0008	.008	14	58-69	—	59-65	22-24†	12●	Step 2
78-77 530i Front	—	—	—	.080	.480	.461	.0008	.008	17	58-69	33-40	59-65	22-24	12	Step 2
Rear	—	—	—	.080	.354	.335	.0008	.008	16	44-48	16-19	59-65	—	—	—
79-77 630CSi Front	—	—	—	.080	.846	.827	.0008	.008	17	58-69	—	59-65	22-24	12	Step 2
Rear	—	—	—	.080	.728	.709	.0008	.008	16	44-48	16-19	59-65	—	—	—
76-72 2002	.120	9.060	9.100	.080	.354	—	—	.008	16	58-69	16-19	59-65	22-24	12	Step 2
74-72 2002tii	.120	9.060	9.100	.080	.459	—	—	.008	16	58-69	33-40	59-65	22-24	12	Step 2

† Keep nut stationary and tighten bearing cover two full turns. ● Loosen nut and retighten a maximum of 24 ft/lbs.
□ Minimum safe thickness stamped on rotor surface.

BUICK OPEL																
79-76	.040	9.000	9.040	.067	—	.339	—	.006	14	36	—	50	21	Handtight	Step 2
75	.040	9.000	9.040	.067	—	.465	—	.004	17	72	—	65	18	Skip	1/4 Turn
74-72	*	9.060	9.090	.067	.404	.394	.0006	.006	17	72	—	65	18	Skip	1/4 Turn

* .030" over rivet head; if bonded lining use .062".

CHEVROLET IMPORTS																
86-85 Nova	.039	7.874	7.913	.039	—	.492	—	.006	65	65	18	76	NOT ADJUSTABLE		
86-85 Spectrum	.039	7.090	7.140	.390	—	.378	—	.0059	57	40	36	65	NOT ADJUSTABLE		
86-85 Sprint	.110†	7.090	7.160	.315†	—	.315	—	.0028	66	17-26	—	29-50	58-86	Skip	Skip

† measurement of shoe and lining.

BRAKES
Specifications

Model													
DATSUN/NISSAN													
86-85 Sentra w/gas eng.	.059	8.000	8.050	.079	—	.394	.0012	.0028	30	40-47	16-23	58-72	29-33	Skip	1/4 Turn
w/diesel eng.	.059	8.000	8.050	.079	—	.630	.0012	.0028	45	40-47	23-30	58-72	29-33	Skip	1/4 Turn
84-83 Sentra w/gas eng.	.059	7.090	7.130	.080	—	.394	.0012	.0028	30	40-47	16-23	58-72	29-33	Skip	1/4 Turn
w/diesel eng.	.059	8.000	.8050	.080	—	.630	.0012	.0028	45	53-72	23-30	58-72	29-33	Skip	1/4 Turn
86 Pulsar NX	.059	8.000	8.050	.079	—	.433	.0012	.0028	30	40-47	23-30	58-72	29-33	Skip	1/4 Turn
85 Pulsar NX	.059	8.000	8.050	.080	—	.394	.0012	.0028	30	40-47	16-23	58-72	29-33	Skip	1/4 Turn
84-83 Pulsar, NX	.059	7.090	7.130	.079	—	.394	.0012	.0028	30	40-47	16-23	58-72	29-33	Skip	1/4 Turn
86 Stanza Wagon 4x2	.059	9.000	9.060	.080	—	.787	.0012	.0028	37	53-72	16-23	58-72	18-25	78-104	Step 2
4x4	.059	10.240	10.300	.079	—	.787	.0012	.0028	37	53-72	16-23	58-72	—	NOT ADJUSTABLE	
86-84 Stanza exc. wagon	.059	8.000	8.050	.080	—	.630	.0012	.0028	45	53-72	23-30	58-72	29-33	Skip	1/4 Turn
83-82 Stanza	.059	8.000	8.050	.063	—	.630	.0008	.006☐	38	53-72	23-30	58-72	22-25	Skip	1/4 Turn
82-79 210	.059	8.000	8.050	.063	—	.331	.0012	.005	38	53-72	—	58-72	18-22	Skip	1/4 Turn
78-74 B210	.059	8.000	8.051	.063	—	.331	.0012	.005	38	33-41	—	60	22-25	Skip	1/4 Turn
73-72 1200	.059	8.000	8.050	.079	—	.331	.0015	.005	38	40-47	—	58-65	29-33	Skip	Step 2
82-79 310	.059	8.000	8.051	.063	—	.339	.0012	.006	38	40-47	—	58-65	18-22	Handtight	1/6 Turn
78-76 F10	.039	8.000	8.050	.063	—	.331	—	.006	36	53-72	12-15	58-65	18-22	Skip	1/6 Turn
81-78 510	.059	9.000	9.055	.080	—	.331	.0012	.005	26	53-72	—	58-65	18-22	Skip	1/6 Turn
73-72 510	.059	9.000	9.055	.040	—	.331	.0012	.005	38	53-72	—	58-65	18-22	Skip	1/6 Turn
76-75 610	.059	9.000	9.055	.063	—	.331	—	.006	26	53-72	12-15	58-65	18-22	Skip	1/6 Turn
74-73 610	.059	9.000	9.055	.040	—	.331	.0012	.006	36	53-72	—	58-65	18-22	Skip	1/6 Turn
77-76 710	.059	9.000	9.055	.063	—	.331	—	.006	26	53-72	12-15	58-65	22-25	Skip	1/6 Turn
75-73 710	.059	9.000	9.055	.063	.341	.787	.0012	.0028	30	53-72	16-23	58-72	—	—	—
86-85 810 Maxima front disc	—	—	—	.079	—	.630	—	.0028	46	28-38	16-23	58-72	—	—	—
rear disc	.059	—	—	.079	—	.354	.0012	.0028	30	53-72	12-15	58-72	18-22	Skip	1/6 Turn
84-83 810 Maxima w/rear drum	—	9.000	9.055	.079	—	.630	—	.0028	46	28-38	16-23	58-72	—	—	—
w/rear disc	.059	—	—	.079	—	.354	.0012	.006	30	53-72	12-15	58-72	18-22	Skip	1/6 Turn
82 810 Maxima w/rear drum	—	9.000	9.055	.079	—	.630	—	.006	46	28-38	16-23	58-72	—	—	—
w/rear disc	.059	—	—	.079	—	.339	.0012	.006	30	53-72	12-15	58-72	18-22	Skip	1/6 Turn
82 810 Maxima w/rear drum	—	9.000	9.055	.079	—	.630	—	.006	46	28-38	12-15	58-72	—	—	—
81 810 w/rear drum	—	9.000	9.055	.063	—	.339	.0028	.0028	42	53-72	—	58-72	—	—	—
81 w/rear disc	.059	—	—	.079	—	.413	.0012	.0028	36	53-72	12-15	58-65	18-22	Skip	1/6 Turn
80-77 810	—	9.000	9.055	.080	—	.630	.0028	.0028	42	53-72	23-30	58-72	—	—	—
86-84 200SX w/rear drum	—	9.000	9.055	.079	—	.354	.0012	.0028	36	28-38	16-23	58-72	18-22	Skip	1/6 Turn
w/rear disc	—	—	—	.079	—	.413	.0012	.0024	46	53-72	16-23	58-72	18-22	Skip	1/6 Turn
83 200SX front disc	—	—	—	.079	—	.354	.0012	.0028	36	28-38	12-15	58-72	18-22	Skip	1/6 Turn
82 200SX front disc	—	—	—	.063	—	.413	.0012	.005	46	53-72	16-23	58-72	—	—	—
82 200SX rear disc	—	—	—	.079	—	.339	.0012	.006	36	28-38	16-23	58-72	18-22	Skip	1/6 Turn
81-80 200SX w/rear drum	—	—	—	.063	—	.413	.0028	.006	36	53-72	12-15	58-72	—	—	—
81 200SX rear disc	—	—	—	.079	—	.339	.0028	.006	42	28-38	—	58-72	18-22	Skip	1/6 Turn
80 200SX rear disc	.059	9.000	9.055	.080	—	.630	.0012	.005	38	53-72	23-30	58-65	18-22	Skip	1/6 Turn
79 200SX	—	9.000	9.055	.079	—	.331	—	.005	38	53-72	16-23	58-72	18-22	Skip	1/6 Turn
78-77 200SX	—	—	—	.060	—	.331	—	.0028	37	53-72†	16-23	58-72	—	—	—
86-84 300ZX front disc	—	—	—	.080	—	.787	—	.0028	46	28-38	16-23	58-72	18-22	Skip	1/6 Turn
rear disc	.059	—	—	.080	—	.354	.0012	.004☐	37	53-72††	16-23	58-72	18-22	Skip	1/6 Turn
83-82 280ZX front disc	—	—	—	.080	—	.709	—	.0028	46	53-72	16-23	58-72	—	—	—
rear disc	—	—	—	.080	—	.339	.0012	.006	46	28-38	16-23	58-72	18-22	Skip	1/6 Turn

★ 1986-85 .0008" †† 1985-28-38 ft/lbs. †2 WD only; 4WD 108-145 ft/lbs. ☐ 1983-.0028" • 2 WD only; 4WD 58-72 ft/lbs.—loosen—+15°30°.
■ Front shown, 84 rear; 159-188 ft/lbs. 1986-85 exc. Alum. wheel 166-203 ft/lbs., Alum. wheel 58-72 ft/lbs.

SECTION 1 BRAKES Specifications

VEHICLE YEAR, MAKE, MODEL	BRAKE SHOE Minimum Lining Thickness	BRAKE DRUM DIAMETER Standard Size	BRAKE DRUM Machine To	BRAKE PAD Minimum Lining Thickness	BRAKE ROTOR Machine To	BRAKE ROTOR Discard At	BRAKE ROTOR Variation From Parallelism	BRAKE ROTOR Runout T.I.R.	DESIGN	CALIPER Mounting Bolts Torque (ft-lbs)	CALIPER Bridge, Pin or Key Bolts Torque (ft-lbs)	WHEEL Lugs or Nuts Torque (ft-lbs)	WHEEL BEARING SETTING STEP 1 Tighten Spindle Nut (ft-lbs)	WHEEL BEARING SETTING STEP 2 Back Off Retorque (in-lbs)	WHEEL BEARING SETTING STEP 3 Lock, or Back Off and Lock
81-79 280 ZX front disc	—	—	—	.080	—	.709	.0012	.004	37	53-72	16-23	58-72	18-22	Skip	1/6 Turn
rear disc	—	—	—	.080	—	.339	.0012	.006	42	28-38	—	58-72	—	—	—
78-72 280Z, 260Z, 240Z	.059	9.000	9.055	.080	.423	.413	.0015	.004	14	53-72	—	58-65	18-22	Skip	1/6 Turn
86-84 720 Pickup single whl.	.059	10.000	10.060	.080	—	.787	.0012★	.0028	37	53-72	16-23	87-108	25-29•	Skip	1/8 Turn
dual whl.	.059	8.660	8.720	.080	—	.787	.0012★	.0028	37	53-72	16-23	87-108	25-29•	Skip	1/8 Turn
83-82 720 Pickup	.059	10.000	10.055	.080	—	.413	.0012	.006	36	53-72	12-15	87-108	25-29†	Skip	1/6 Turn
81 720 Pickup	.059	10.000	10.055	.080	—	.413	.0028	.006	36	53-72	12-15	87-108	25-29†	Skip	1/6 Turn
80 720 Pickup	.059	10.000	10.055	.080	—	.413	.0028	.006	36	53-72	12-15	58-72	25-29†	Skip	1/6 Turn
79-78 620 Pickup	.059	10.000	10.055	.080	—	.413	—	.006	36	53-72	12-15	58-72	25-29†	Skip	1/6 Turn
77-72 620 Pickup	.059	10.000	10.055	.080	—	.413	.0015	.006	36	53-72	12-15	58-65	22-25†	Skip	1/6 Turn

★1986-85-.0008"   ††1985-28-38 ft/lbs.   †2 WD only; 4 WD 108-145 ft/lbs.   •2WD only; 4WD 58-72 ft/lbs.   —loosen—+15°30°
■front shown, 84 rear; 159-188 ft/lbs. 1986-85. Alum. wheel 166-203 ft/lbs., Alum. wheel 58-72 ft/lbs.

DODGE—COLT, COLT VISTA, CONQUEST, CHALLENGER, D-50, RAM 50 PICKUP

VEHICLE	Min Lining	Std Size	Machine To	Min Lining	Machine To	Discard At	Variation	Runout	DESIGN	Caliper Mtg	Bridge/Pin	Lugs	Step 1	Step 2	Step 3
86-84 Conquest Front	—	—	—	.040	—	.880	—	.006	57	58-72	61-69	50-57•	14	48	Step 2
Rear	—	—	—	.040	—	.650	—	.006	56	29-36	36-43	50-57•	—	—	—
86-84 Colt Vista 4x2	.040	8.000	8.150	.040	—	.650	—	.006	69	58-72	16-23	50-57•	14	48	Step 2
4x4	.040	9.000	9.078	.040	—	.880	—	.004	69	58-72	16-23	50-57•	—	—	—
86-84 Colt Turbo	.040	7.100	7.150	.040	—	.645	—	.006	69	43-58‡	16-23	51-58■	14	48	Step 2
86-85 Colt (FWD) exc. Turbo	.040	7.100	7.150	.040	—	.450	—	.006	69	58-72	16-23	50-57•	—	—	—
84-79 Colt (FWD) exc. Turbo	.040	7.100	7.150	.040	—	.450	—	.006	18	43-58	58-69	51-58■	14	18	Step 2
80-78 Colt exc. Wagon	.040	9.000	9.050	.080	—	.450	—	.006	18	51-65□	—	51-58■	14.5	43	Step 2
80-78 Colt Wagon															
w/rear drum brakes	.040	9.000	9.050	.040	—	.330	—	.006	34	51-65	—	51-58■	14.5	43	Step 2
w/rear disc brakes	—	—	—	.040	—	.330	—	.006	27	29-36	—	51-58■	—	—	—
77-76 Colt	.040	9.000	9.050	.040	—	.450	—	.006	18	51-65□	—	51-58■	14.5	43	Step 2
75-73 Colt	.040	9.000	9.050	.040	—	.450	—	.006	18	51-65□	—	51-58■	14.5	43	Step 2
72 Colt	.040	9.000	9.050	.040	—	.374	—	.006	14	30-36	—	51-58■	14.5	43	Step 2
83-78 Challenger															
w/rear drum brakes	.040	9.000	9.050	.040	—	.430	—	.006	34	51-65	—	51-58■	14.5	43	Step 2
w/rear disc brakes	—	—	—	.040	—	.330	—	.006	27	29-36	—	51-58■	—	—	—
86-83 Ram-50															
w/9½" rear brake	.040	9.500	9.570	.040	—	.720	—	.006	34	51-65	—	51-57	21.7††	69.6	Step 2
w/10" rear brake	.040	10.000	10.070	.040	—	.720	—	.006	34	51-65★	—	51-57	94-145†	216	30°
82-79 D-50, Ram-50 Pickup	.040	9.500	9.550	.040	—	.720	—	.006	34	51-65	—	51-58	21.7	48	Step 2

□Caliper to adapter shown, adapter to caliper 29-36 ft/lbs.   ★1986-85 4WD 58-72 ft/lbs.   ■w/aluminum wheels; 58-72 ft/lbs.   ‡1986-84 58-72 ft/lbs.
•w/aluminum wheels; 65-80 ft/lbs.   ††4WD   ††2WD

FIAT

VEHICLE															
82-74 X1/9 front	—	—	—	.080	—	.354	.002	.006	43	36	—	51	\multicolumn{2}{c}{NOT ADJUSTABLE}	
Rear	—	—	—	.080	—	.354	.002	.006	27	36	—	51	—	—	—
82-75 2000 & 124 Front	—	—	—	.080	—	.354	—	.006	22	59	—	65	14.5	60	1/3 Turn
Rear	—	—	—	.080	—	.372	—	.006	27	55	—	65	—	—	—

BRAKES
Specifications

Model												
74-72 124 Front	—	—	—	.368	—	—	22	36	—	51	14.5	1/3 Turn
Rear	.060	7,293	.080	.372	—	.006	27	40	—	51	—	1/3 Turn
82-79 Strada	.181	9,000	.080	.368	—	.006	33	—	—	65	NOT ADJUSTABLE	1/2 Flat
81-79 Brava	.181	9,000	.060	—	.002	.006	22	50	—	65	14.5	60
78-75 131 Series	.060	7,336	.060	.368	.002	.006	22	50	—	65	14.5	60
79-73 128	—	7,300	.080	.354	.002	.006	22	36	—	51	—	1/3 Turn
FORD—CAPRI, COURIER, FIESTA												
77-76 Capri	.030	9,000	.100	.460	.0004	.002	14	45-50	*	50-55	17-25	1/2 Turn
74-72 Capri	.030	9,000	.100	—	—	.0035	14	40-50	*	50-55	17-25	1/2 Turn
82-79 Courier	.039	10,236	.276■	.433	.0005	.004	35	—	*	58-65	17-25	Skip
78-77 Courier	.039	10,244	.315■	.433	.0005	.004	35	—	*	58-65□	17-25	Skip
76-74 Courier	.039	10,236	—	—	—	.006	—	—	—	85	17-25	1/2 Turn
73-72 Courier	.039	10,244	—	—	—	—	—	—	—	—	15-18	180°
80-78 Fiesta	.060	7,000	.060	.340	—	.006	22	38-45	—	63-85	—	Handtight
■ Measurement of shoe & lining ★ Machining not recommended □ 74 is 85 ft/lbs.												
HONDA												
86-84 Civic CRX w/1300, HF	.080	7,090	.120	.350	.0006	.004	37	56	*	80	NOT ADJUSTABLE	
w/1500 exc. HF	.080	7,090	.120	.590	.0006	.004	30	56	*	80	NOT ADJUSTABLE	
86-84 Civic Hatchback w/1300	.080	7,090	.120	.390	.0006	.004	37	56	*	80	NOT ADJUSTABLE	
w/1500	.080	7,090	.120	.590	.0006	.004	37	56	*	80	NOT ADJUSTABLE	
83 Civic Hatchback	.079	7,090	.063	.350	.0006	.006	37	56	20	58	18	48 Step 2
w/1300 4 SPD	.079	7,090	.063	.390	.0006	.006	37	56	20	58	18	48 Step 2
w/1300 5 SPD	.079	7,090	.063	.590	.0006	.006	37	56	20	58	18	48 Step 2
w/1500	.079	7,090	.063	—	.0006	.004	—	—	—	—	—	—
82 Civic Hatchback	.079	7,090	.063	.350	.0006	.006	37	56	20	58	18	48 Step 2
w/1300 4 SPD	.079	7,090	.063	.390	.0006	.006	37	56	20	58	18	48 Step 2
exc. 1300 4 SPD	.079	7,090	.063	.590	.0006	.006	37	56	20	58	18	48 Step 2
81-80 Civic CVCC Hatchback	.079	7,080	.063	.354	.0006	.006	38	36-43	—	51-65	NOT ADJUSTABLE	
79-75 Civic, CVCC	.079	7,080	.063	.354	.0006	.006	38	36-43	*	51-65	NOT ADJUSTABLE	
74-73 Civic	.079	7,080	.063	—	.0006	.004	37	56	20	80	NOT ADJUSTABLE	
86-84 Civic Sedan	.080	7,090	.120	.590	.0006	.006	37	56	20	58	18	48 Step 2
83 Civic Sedan	.079	7,090	.063	.390	.0006	.006	37	56	*	58	18	48 Step 2
82-81 Civic Sedan	.080	7,870	.120	.590	.0006	.004	30	56	—	80	NOT ADJUSTABLE	
86-84 Civic Wagon	.079	7,910	.063	.390	.0006	.006	22	56	63	58	18	48 Step 2
83-80 Civic Wagon	.079	7,910	.120	.590	—	.006	18	63	33†	51-65	18	36 Step 2
79-76 Civic Wagon	.079	7,910	.120	.449	.437	.006	73	53	20	80	18	36 Step 2
86 Accord	.080	7,870	.120	.670	—	.006	30	56	20	80	18	48 Step 2
85-84 Accord	.080	7,870	.059	.670	—	.006	37	56	—	58	18	48 Step 2
83-82 Accord	.079	7,870	.063	.600	—	.006	22	56	*	58	18	36 Step 2
81-79 Accord	.079	7,080	.063	.413	—	.006	22	58-65	20	50-65	18	36 Step 2
78-76 Accord	.079	7,080	.063	.449	.437	.006	22	56	*	79	18	48 Step 2
86 Prelude exc. Fuel Inj. Front	—	—	.118	.670	—	.004	30	56	22	79	18	48 Step 2
Rear	—	—	.063	.310	—	.006	46	28	33†	79	—	—
w/Fuel Inj. Front	—	—	.120	.670	—	.006	73	53	17	79	18	48 Step 2
Rear	—	—	.063	.310	—	.006	46	28	*	79	—	—
85-84 Prelude Front	—	—	.120	.670	—	.004	30	56	22	79	18	48 Step 2
Rear	—	—	.063	.310	—	.006	33	28	*	80	—	—
83 Prelude	.080	7,870	.060	.670	—	.004	30	56	22	80	18	36 Step 2
82-79 Prelude	.079	7,080	.063	.413	—	.006	37	56	13	80	18	— Step 2

* Upper Pin-14 ft/lbs., Lower Pin-13 ft/lbs. †Lower bolt only.

SECTION 1
BRAKES
Specifications

VEHICLE YEAR, MAKE, MODEL	BRAKE SHOE Minimum Lining Thickness*	BRAKE DRUM DIAMETER Standard Size	BRAKE DRUM Machine To	BRAKE PAD Minimum Lining Thickness	BRAKE ROTOR MIN. THICKNESS Machine To	BRAKE ROTOR MIN. THICKNESS Discard At	BRAKE ROTOR Variation From Parallelism	BRAKE ROTOR Runout T.I.R.	DESIGN	CALIPER Mounting Bolts Torque (ft-lbs)	CALIPER Bridge, Pin or Key Bolts Torque (ft-lbs)	WHEEL Lugs or Nuts Torque (ft-lbs)	WHEEL BEARING SETTING STEP 1 Tighten Spindle Nut (ft-lbs)	WHEEL BEARING SETTING STEP 2 Back Off Retorque (in-lbs)	WHEEL BEARING SETTING STEP 3 Lock, or Back Off and Lock
ISUZU															
86-85 Impulse Front	—	—	—	.040	—	.654	.0006	.005	47	36	27	87	22	Handtight	1.1-3.3☐
Rear	—	—	—	.040	—	.654	.0006	.005	47	36	15	87	—	—	—
86-85 I-Mark	.039	7.090	7.140	.039	—	.378	—	.006	54	40	36	60†	22	Handtight	Step 2
84-83 Impulse Front	—	—	—	.125	.706	.654	.0012	.005	47	35-38	25-28	80-94	22	Handtight	1.3-3.3☐
Rear	—	—	—	.125	.706	.654	.0012	.006	47	35-38	13-16	80-94	—	—	—
84-81 I-Mark	.039	9.000	9.040	.067	.354	.338	—	.005	49	36	—	50†	22	Handtight	Step 2
86-84 Pickup, Trooper II	.039	10.000	10.039	.039	.668	.654	.0012	.005	47	62-65	22-25	58-80†	22	Handtight	2.2☐■
83-81 Pickup	.039	10.000	10.039	.039	.453	.437	.003	.005	47	64	15	65	22	Handtight	1.8-2.6■

† w/aluminum whls. 90 ft/lbs.   ☐ Radial pull in lbs. at lug nuts.   ■ 4x2 shown, 4x4-3.3.

LUV—CHEVROLET															
82-81	.059	10.000	10.059	.236	.668	.653	.003	.005	47	64	15	65	22■	Handtight	Step 2
80-76	.059	10.000	10.059	.236	.668	.653	.003	.005	22	64	—	65	22■	Handtight	Step 2
75-72	.059	10.000	10.059	—	—	—	—	—	—	—	—	65	22	Handtight	Step 2

■Wheel bearing adjustment for 2 WD only

MAZDA															
86 323 w/rear drum brake	.040	7.870	7.910	.118	—	.630	—	.003	55	••	—	65-87	18-22	.6-1.9	Step 2
w/rear disc brake	—	—	—	.040	—	.350	—	.003	70	••	—	65-87	18-22	.9-2.2	Step 2
85 GLC	.040	7.090	7.130	.118	—	.390	—	.004	57	33-40	—	65-80	16-19	.9-2.2	Step 2
84-81 GLC exc. Wagon	.040	7.090	7.130	.040	—	.393	—	.003	35	33-40	—	65-80	16-19	.9-2.2	Step 2
83-81 GLC Wagon	.040	7.874	7.913	.040	—	.472	—	.003	35	33-40	—	65-80	14-18	.9-1.4	Step 2
80-77 GLC	.040	7.874	7.914	.040	—	.472	—	.004	35	33-40	—	65-87	14-18	1.0-1.3☐	Step 2
86 626 w/rear drum brake	.040	7.870	7.910	.118	—	.710	—	.004	58	••	—	65-87	18-22	.9-2.2	Step 2
w/rear disc brake	—	—	—	.040	—	.350	—	.004	70	••	—	65-87	18-22	.9-2.2	Step 2
85 626 w/gas eng.	.040	7.870	7.910	.118	—	.470	—	.004	58	—	—	65-87	18-22	.9-2.2	Step 2
w/diesel eng.	—	—	—	.118	—	.710	—	.004	58	—	—	65-87	18-22	.9-2.2	Step 2
84-83 626	.040	7.874	7.913	.040	—	.490	—	.004	58	—	—	65-87	18-22	.9-2.2	Step 2
82-79 626	.040	9.000	9.040	.040	—	.472	—	.004	37	33-40	—	65-80	14-18	.8-1.9	Step 2
86 RX7 Front	—	—	—	.040	—	.790	—	.004	32	58-72	23-30	65-87	14-22	.9-2.2	Step 2
w/14" Wheels	—	—	—	.118	—	.790	—	.004	16	58-72	—	65-87	14-22	.9-2.2	Step 2
w/15" Wheels Rear	—	—	—	.040	—	.310	—	.004	70	33-40	22-30	65-87	—	—	—
w/14" Wheels	—	—	—	.118	—	.670	—	.004	70	33-40	22-30	65-87	—	—	—
w/15" Wheels	.040	7.874	7.914	.040	—	.354	—	.004	32	—	—	65-80	18-22	1.0-1.4☐	Step 2
85-80 RX7 w/front disc	—	—	—	.040	—	.787	—	.004	32	—	22-30	65-80	—	—	—
w/solid rear disc	—	—	—	.040	—	—	—	—	32	—	22-30	65-80	—	—	—
w/vented rear disc	—	—	—	.040	—	—	—	—	32	—	22-30	65-80	—	—	—

■ Measurement of shoe & lining   ☐ Radial pull in lbs. at lug nuts.   *Upper bolt: 11-18 ft/lbs., Lower bolt: 15-22 ft/lbs.

BRAKES
Specifications

Model												
79 RX7	.040	7.874	—	.040	—	.670	—	35	33-40	—	65-80	SEAT	1.0-1.4 □	Step 2
78-76 Cosmo Front	—	7.874	—	.276 ■	—	.669	—	35	40-47	—	65-72	SEAT	.9-2.2 □	Step 2
Rear	.040	—	—	.276 ■	—	.354	—	35	40-47	—	65-72	SEAT	.9-2.2 □	Step 2
77-74 808	.040	7.874	7.914	.256 ■	—	.394	—	18	—	—	65-72	SEAT	9-2.2 □	Step 2
73-72 808	.040	7.874	7.914	.256 ■	—	.394	—	18	—	—	65	SEAT	9-2.2 □	Step 2
78-74 RX4	.040	9.000	9.040	.276 ■	—	.433	—	35	36-40	—	65-72	SEAT	9-2.2 □	Step 2
77-75 RX3	.040	7.874	7.914	.256 ■	—	.394	—	18	—	—	65	SEAT	9-2.2 □	Step 2
73-72 RX3	.040	7.874	7.914	.276 ■	—	.433	—	18	—	—	65	SEAT	9-2.2 □	Step 2
73-72 RX2	.040	10.240	10.310	.118	—	.710	—	37	65-80	23-30	87-108	14-22	Handtight	1.3-2.5 □
86 B2000 Pickup	.040	10.236	—	.040	—	.748	—	35	40-47	—	58-65	18-22	Handtight	1.3-2.5 □
84 B2200 Pickup	.040	10.236	—	.040	—	.433	—	35	40-47	—	58-65	18-22	Handtight	1.3-2.5 □
84 B2000 Pickup	.040	10.236	—	.276 ■	—	.748	—	35	40-47	—	58-65	SEAT	1.3-2.4 □	Step 2
83-82 B2200 Pickup	.040	10.236	—	.276 ■	—	.433	—	18	40-47	—	58-65	SEAT	1.3-2.4 □	Step 2
83-77 B1800/B2000 Pickup	.040	10.236	—	—	—	—	—	—	—	—	—	—	—	—
76-72 B 1600 Pickup	.040	10.236	—	.276 ■	—	.433	—	35	—	—	58-65	SEAT	1.3-2.4 □	Step 2
77-74 Rotary Pickup	.040	10.275	—	.276 ■	—	.433	—	35	—	—	58-65	SEAT	1.3-2.4 □	Step 2

■ Measurement of shoe & lining   □ Radial pull in lbs. at lug nuts.   *Upper bolt: 11-18 ft/lbs., Lower bolt: 15-22 lbs.

MERCEDES BENZ														
85-84 190D,E Front	—	—	—	.138	—	.354	—	—	—	—	81	Snug Up	Skip	1/3 Turn
Rear	—	—	—	.079	—	.295	—	—	—	—	81	—	—	—
85-81 300SD, 380SE, SEC, SEL, 500 SEC, SEL Front	—	—	—	.138*	.787	.764	.005	—	85	26	81	Snug Up	Skip	1/3 Turn
Rear	—	—	—	.138*	.339	.327	.005	—	66	26	81	—	—	—
85-77 230, 240D, 300D, CD, TD 280E, CE Front	—	—	—	.079	—	.417	.005	14	85	—	81	Snug Up	Skip	1/3 Turn
Rear	—	—	—	.079	—	.327	.005	14	66	—	81	—	—	—
85-73 380 SL, SLC, 450 SL, SLC Front	—	—	—	.079	—	.811†	.005	14	85	—	81	Snug Up	Skip	1/3 Turn
Rear	—	—	—	.079	—	.327	.005	14	66	—	81	—	—	—
80 280 SE, 300 SD, 450 SEL Front	—	—	—	.079	—	.764	.005	14	85	—	81	Snug Up	Skip	1/3 Turn
Rear	—	—	—	.079	—	.327	.005	14	66	—	81	—	—	—
79-73 280S, SE, 300SD, 450 SE, SEL, 6.9 Front	—	—	—	.079	—	.787	.005	14	85	—	81	Snug Up	Skip	1/3 Turn
Rear	—	—	—	.079	—	.327	.005	14	85	—	81	—	—	—
76-72 220, D, 220/8, 230, 240D, 250, C, 280, C, 300D Front	—	—	—	.079	—	.435●	.005	14	85	—	81	Snug Up	Skip	1/3 Turn
Rear	—	—	—	.079	—	.327	.005	14	66	—	81	—	—	—
73-72 280 SE, SEL, 300 SEL Front	—	—	—	.079	—	.435††	.005	14	85	—	81	Snug Up	Skip	1/3 Turn
Rear	—	—	—	.079	—	.327	—	14	66	—	81	Snug Up	Skip	—
72 600 Front	—	—	—	.354	—	.725	—	14	—	—	125	—	Skip	1/3 Turn
Rear	—	—	—	.079	—	.570	—	14	—	—	125	—	—	—

* w/floating caliper shown; w/fixed caliper .079   † w/57mm caliper piston shown; w/60mm caliper piston up to 3/80 .787 from 3/80 .763
● w/57mm caliper piston shown; w/60mm caliper piston .417   †† 1st version shown; 2nd version .700

1-167

SECTION 1: BRAKES Specifications

VEHICLE YEAR, MAKE, MODEL	BRAKE SHOE Minimum Lining Thickness*	BRAKE DRUM DIAMETER Standard Size	BRAKE DRUM DIAMETER Machine To	BRAKE PAD Minimum Lining Thickness	BRAKE ROTOR MIN. THICKNESS Machine To	BRAKE ROTOR Discard At	BRAKE ROTOR Variation From Parallelism	BRAKE ROTOR Runout T.I.R.	DESIGN	CALIPER Mounting Bolts Torque (ft-lbs)	CALIPER Bridge, Pin or Key Bolts Torque (ft-lbs)	WHEEL Lugs or Nuts Torque (ft-lbs)	WHEEL BEARING SETTING STEP 1 Tighten Spindle Nut (ft-lbs)	WHEEL BEARING SETTING STEP 2 Back Off Retorque (in-lbs)	WHEEL BEARING SETTING STEP 3 Lock, or Back Off and Lock
MITSUBISHI															
86–85 Galant															
w/rear disc brakes	.040	8.000	8.050	.040	—	.650	—	.006	69	58–72	16–23	50–57+	14	48	Step 2
86–85 Mirage exc. Turbo	—	7.100	7.150	.040	—	.330	—	.006	70	36–43	16–23	50–57+	14	48	Step 2
w/Turbo	.040	7.100	7.150	.040	—	.450	—	.006	69	58–72	16–23	50–57+	14	48	Step 2
86–84 Tredia, Cordia	.040	8.000	8.050	.040	—	.650	—	.006	69	58–72	16–23	50–57+	14	48	Step 2
86–83 Pickup	.040	—	—	.040	—	.650	—	.006	60	58–72	16–23	50–57+	14	48	Step 2
w/9½" rear brakes	.040	9.500	9.550	.040	—	.720	—	.006	34	51–65	—	51–57	21.7 ■	69.6	Step 2
w/10" rear brakes	.040	10.000	10.070	.040	—	.720	—	.006	34	51–65 □	—	51–57	94–145•	216	30°
86–83 Montero	.040	10.000	10.040	.040	—	.720	—	.006	34	51–65 □	—	51–57	95–145	216	30°
83 Tredia, Cordia	.040	8.000	8.050	.040	—	.450	—	.006	58	58–72	16–23	72–87*	14	48	Step 2
86–83 Starion Front	—	—	—	.040	—	.880	—	.006	57	58–72	61–69	50–57††	14	48	Step 2
Rear	—	—	—	.040	—	.650	—	.006	56	29–36	36–43	50–57††	14	48	Step 2

† w/aluminum wheels; 57–72 ft/lbs. †† w/aluminum wheels; (1983) 57–72 ft/lbs. (1984) 65–80 ft/lbs.
+ w/aluminum wheels; 66–81 ft/lbs. ■ 2WD • 4WD □ 1986–85 *1983; 50–57 ft/lbs.

VEHICLE YEAR, MAKE, MODEL	BRAKE SHOE Minimum Lining Thickness*	BRAKE DRUM Standard Size	BRAKE DRUM Machine To	BRAKE PAD Minimum Lining Thickness	BRAKE ROTOR Machine To	BRAKE ROTOR Discard At	BRAKE ROTOR Variation From Parallelism	BRAKE ROTOR Runout T.I.R.	DESIGN	CALIPER Mounting Bolts	CALIPER Bridge Pin/Key Bolts	WHEEL Lugs/Nuts	STEP 1	STEP 2	STEP 3
PLYMOUTH—ARROW, CHAMP, COLT, COLT VISTA, CONQUEST, SAPPORO & ARROW PICK UP															
80–76 Arrow w/rear drum brakes	.040	9.000	9.050	.040	—	.450	—	.006	34	51–65	—	51–58 ■	14.5	43	Step 2
w/rear disc brakes	—	—	—	.040	—	.330	—	.006	27	29–36	—	51–58	—	—	—
82–79 Arrow Pickup	.040	9.500	9.550	.040	—	.720	—	.006	34	51–65	—	51–58	21.7	48	Step 2
82–79 Champ (FWD)	.040	7.100	7.150	.040	—	.450	—	.006	18	43–58	—	51–58	14	48	Step 2
83–78 Sapporo w/rear drum brakes	.040	9.000	9.050	.040	—	.430	—	.006	34	51–65	—	51–58	14.5	43	Step 2
w/rear disc brakes	—	—	—	.040	—	.330	—	.006	27	29–36	—	51–58•	14	48	Step 2
86–84 Conquest Front	—	—	—	.040	—	.880	—	.006	57	58–72	61–69	50–57•	14	48	Step 2
Rear	—	—	—	.040	—	.650	—	.006	56	29–36	36–43	50–57•	14	48	Step 2
86–84 Colt Vista 4x2	.040	8.000	8.150	.040	—	.880	—	.006	69	58–72	16–23	50–57•	14	48	Step 2
4x4	.040	9.000	9.078	.040	—	.640	—	.004	69	43–58	16–23	50–57•	14	48	Step 2
86–84 Colt Turbo	.040	7.100	7.150	.040	—	.450	—	.006	69	58–72	16–23	50–57•	14	48	Step 2
86–85 Colt exc. Turbo	.040	7.100	7.150	.040	—	.450	—	.006	69	43–58	16–23	50–57•	14	48	Step 2
84–83 Colt exc. Turbo	.040	—	—	.040	—	.450	—	.006	18	43–58	58–69	51–58■	14	48	Step 2

■ w/aluminum wheels; 58–72 ft/lbs. • w/aluminum wheels; 65–80 ft/lbs.

VEHICLE YEAR, MAKE, MODEL	BRAKE SHOE Minimum Lining Thickness*	BRAKE DRUM Standard Size	BRAKE DRUM Machine To	BRAKE PAD Minimum Lining Thickness	BRAKE ROTOR Machine To	BRAKE ROTOR Discard At	BRAKE ROTOR Variation From Parallelism	BRAKE ROTOR Runout T.I.R.	DESIGN	CALIPER Mounting Bolts	CALIPER Bridge Pin/Key Bolts	WHEEL Lugs/Nuts	STEP 1	STEP 2	STEP 3
RENAULT															
86–84 18i Sportwagon	.188 ■	9.000	9.040	.359 ■	—	.709	—	.003	31	48	26	59	22	1/6 Turn	□□
83–80 18i	.020	8.996	9.035	.276 ■	—	.433	—	.003	29	74	42	59	22	1/6 Turn	□□
86–83 Alliance, Encore	.020	8.000	8.060	.276 ■	—	.433	—	.003	51	74	26	63	NOT ADJUSTABLE		
84–82 Fuego Turbo	.020	8.996	9.035	.276 ■	—	.709	—	.003	29	74	42	59	NOT ADJUSTABLE		
85–80 Fuego exc. Turbo	.020	8.996	9.035	.276 ■	—	.433	—	.003	29	74	42	45–60	NOT ADJUSTABLE		
80–74 Gordini	.203 ■	9.000	9.035	.276 ■	—	.354	.0004	.008	43	—	—	45–60	20	1/4 Turn	□□□□
84–76 LeCar/5 Series	.203 ■	7.096	7.136	.275 ■	—	.354	—	.004	43	50	—	40–45	20	1/4 Turn	□□□□

■ Measurement of shoe & lining □ .001"–.002" end play

1–168

BRAKES
Specifications

SAAB										
86 9000 Front	—	—	—	.039	.0006	.003	61	52-82	—	76-90	NOT ADJUSTABLE
Rear	—	—	—	.039	.0006	.003	74	52-67	—	76-90	NOT ADJUSTABLE
78-75 99 Front	—	—	—	.080	.0006	.004	28	—	—	65-80	NOT ADJUSTABLE
74-70 Front	—	—	—	.080	.0006	.004	17	—	—	65-80	NOT ADJUSTABLE
78-70 99 Rear	—	—	—	.040	.0006	.004	17	—	—	65-80	NOT ADJUSTABLE
86-79 900 Front	—	—	—	.040	.0006	.004	28	82-88	—	65-80	NOT ADJUSTABLE
Rear	—	—	—	.040	.0006	.004	17	52-66	—	65-80	NOT ADJUSTABLE
SUBARU										
86-85 All Models w/front disc brake	.060	7.090	7.170	.295 ■	—	—	48	36-51	†	58-72	36 — 1/8 Turn — 1.9-3.2 □
w/rear disc brake	—	—	—	.256 ■	—	—	61	34-43	16-23	58-72	36 — 1/8 Turn — 1.9-3.2 □
84-83 All Models										
w/solid disc (83)	.060	7.090	7.120	.295 ■	—	—	48	36-51	12-17	58-72	36 — 1/8 Turn — 1.9-3.2 □
w/vented disc (84-83)	.060	7.090	7.120	.295 ■	—	—	48	36-51	12-17	58-72	36 — 1/8 Turn — 1.9-3.2 □
82-80 All Models	.060	7.090	7.120	.295 ■	—	—	48	36-51	33-54	58-72	36 — 1/8 Turn — 1.9-3.2 □
79-75 All Models	.060	9.000	9.040	.060	—	—	27	36-51	—	40-54	36 — 1/8 Turn — 1.9-3.2 □
74-72 Sedan, Wagon Front	.060	7.090	7.120	—	—	—	—	—	—	40-54	36 — 1/8 Turn — 2.2-3.4 □
Rear	.060	7.090	7.120	.060	—	—	27	36-51	—	40-54	36 — 1/8 Turn — 2.2-3.4 □
74-72 Coupe										

■ Measurement of shoe & lining   □ Radial pull in lbs. at lug nuts.   † upper bolt 33-40 ft/lbs., lower bolt 23-30 ft/lbs.

TOYOTA										
86-85 MR2 Front	—	—	—	.040	—	.006	32	65	18	76	NOT ADJUSTABLE —
Rear	.040	7.874	7.913	.040	—	.006	70	43	14	76	— — — —
86-83 Camry	.040	7.874	7.913	.040	—	.006	65	65	18	76	NOT ADJUSTABLE —
86-84 Corolla exc. Coupe	.040	9.000	9.079	.040	—	.006	65	47	18	76	NOT ADJUSTABLE —
86-84 Corolla Cpe. w/frt. disc brake	—	—	—	.040	—	.006	66	47	14	76	21 Handtight 0-2.3 ■
w/rear disc brake	—	—	—	.040	—	.006	70	47	14	76	— — — —
83-80 Corolla	.040	9.000	9.040	.040	—	.006	18	40-54	58-68	66-86	21 Handtight .7-1.5 ■
79-75 Corolla 1600	.040	9.000	9.040	.040	—	.006	22	40-54	—	65-86	21 Handtight .7-1.5 ■
74-72 Corolla 1600	.040	9.000	9.040	.040	—	.006	22	48	—	70-75	21 Handtight .7-1.5 ■
79-75 Corolla 1200	.040	7.874	7.914	.040	—	.006	18	30-40	—	58-69	21 Handtight .7-1.5 ■
74-72 Corolla 1200	.040	7.874	7.914	.040	.360	.006	22	40-54	18	65-86	21 Handtight .7-1.5 ■
86-85 Tercel Sedan	.040	7.087	7.126	.040	—	.006	65	65	18	66-86	22 Handtight 9-2.2 ■
Wagon	.040	7.087	7.126	.040	—	.006	65	65	18	66-86	— — — —
84-83 Tercel 4 x 2	.040	7.874	7.913	.040	—	.006	65	65	18	66-86	22 NOT ADJUSTABLE .9-2.2 ■
4 x 4	.040	7.874	7.913	.040	—	.006	7	33-39	11-15	66-86	22 Handtight 8-1.9 ■
82-80 Tercel	.040	7.087	7.913	.040	—	.006	65	69	18	76	— — — —
86 Celica w/rear drum brakes	—	—	—	.040	—	.006	32	34	14	76	NOT ADJUSTABLE —
w/rear disc brakes	—	—	—	.040	—	.006	47	59-75	12-17	66-86	22 Handtight .8-1.9 ■
85-82 Celica w/rear drum brakes	.040	9.000	9.040	.118	—	.006	7	29-39	—	66-86	— — — —
w/rear disc brakes	—	—	—	.118	—	—	—	.750	—	.670	— — — —
81 Celica	.040	9.000	9.040	.040	—	.006	22	40-54	—	65-86	21 Handtight .7-1.5 ■
80-77 Celica	.040	9.000	9.040	.040	—	.006	22	40-54	—	65-86	21 Handtight .7-1.5 ■

■ Radial pull in lbs. at lug nuts   • torque locknut to 33 ft/lbs., brg. preload: 2.2-8.6 lbs.

SECTION 1
BRAKES
Specifications

VEHICLE	BRAKE SHOE	BRAKE DRUM DIAMETER		BRAKE PAD	BRAKE ROTOR				DESIGN	CALIPER		WHEEL	WHEEL BEARING SETTING		
					MIN. THICKNESS										
YEAR, MAKE, MODEL	* Minimum Lining Thickness	Standard Size	Machine To	Minimum Lining Thickness	Machine To	Discard At	Variation From Parallelism	Runout T.I.R.		Mounting Bolts Torque (ft-lbs)	Bridge, Pin or Key Bolts Torque (ft-lbs)	Lugs or Nuts Torque (ft-lbs)	STEP 1 Tighten Spindle Nut (ft-lbs)	STEP 2 Back Off Retorque (in-lbs)	STEP 3 Lock, or Back Off and Lock
76 Celica	.040	9.000	9.040	—	—	.350	—	.006	22	48	—	65-86	21	Handtight	.7-1.5 ■
75-72 Celica	.040	9.000	9.040	—	—	.350	—	.006	22	48	—	70-75	21	Handtight	.7-1.5 ■
86-82 Supra Front	—	—	—	.118	—	.750	—	.006	47	59-75	12-17	66-86	22	Handtight	8-1.9 ■
Rear	—	—	—	.118	—	.669	—	.006	7	30-39	12-17	66-86	—	—	—
81-79 Supra Front	—	—	—	.118	—	.450	—	.006	22	40-54	—	65-86	21	Handtight	.7-1.5 ■
Rear	.040	7.870	7.910	—	—	.354	—	.006	27	29-39	—	65-86	—	—	—
84-81 Starlet	.040	9.000	9.040	—	—	.450	—	.006	7	29-54	11-15	65-86	22	Handtight	8-1.9 ■
82-75 Corona	.040	9.000	9.040	—	—	.450	—	.006	14	67-87 □	—	65-86	21	Handtight	.7-1.5 ■
74 Corona deluxe	.040	9.000	9.040	—	—	.450	—	.006	14	67-87 □	—	65-86	21	Handtight	.7-1.5 ■
74 Std. Corona	.040	9.010	9.050	—	—	.350	—	.006	22	53	—	65-86	21	Handtight	.7-1.5 ■
73-72 Corona	.040	9.000	9.040	—	—	.350	—	.006	22	53	—	66-86	22	Handtight	8-1.9 ■
86-83 Cressida w/rear drum brake	.040	—	—	.040	—	.830	—	.006	54	68-86	62-68	66-86	—	—	—
86-85 Cressida rear disc brake	—	—	—	.040	—	.670	—	.006	54	34	18	76	—	—	—
84-83 Cressida rear disc brake	—	—	—	.040	—	.669	—	.006	55	29-39	37-43	65-86	21	Handtight	8-2.2 ■
82-81 Cressida	.040	9.000	9.040	—	—	.450	—	.006	54	68-86	62-68	65-86	21	Handtight	8-2.2 ■
80-78 Cressida	.040	9.000	9.040	—	—	.450	—	.006	14	67-87	—	65-86	21	Handtight	.7-1.5 ■
76-73 Corona Mark II	.040	9.000	9.040	—	—	.450	—	.006	14	67-87	—	65-94	21	Handtight	.7-1.5 ■
73-72 Corona Mark II	.060	11.610	11.650	—	—	.453	—	.006	14	67-87	—	65-86	22	Handtight	.7-1.5 ■
74-75 Land Cruiser	.060	11.400	11.440	—	—	.748	—	.005	16	52-76	—	66-86	43	35-60	6.2-12.6 ■
74-72 Land Cruiser	.040	—	—	.040	—	—	—	—	—	—	—	65-87	43	35-60	3.9-5.0 ■
86-85 Pickup 2 W.D. (4x2) ½ ton	.040	10.000	10.060	.040	—	.827	—	.006	54	80	65	76	25	Handtight	1.3-4.0 ■
1 ton, Heavy	.040	10.000	10.060	.040	—	.945	—	.006	65	80	29	76	25	Handtight	1.3-4.0 ■
84 Pickup 2 W.D. (4x2)	.040	10.000	10.060	.040	—	.827	—	.006	54	73-86	62-68	76	25	Handtight	1.3-4.0 ■
83-75 Pickup 2 W.D. (4x2)	.040	10.000	10.060	.040	—	.453	—	.006	14	68-86	—	66-86	22	Handtight	1.3-3.8 ■
74-72 Pickup 2 W.D. (4x2)	.040	10.000	10.060	.040	—	—	—	—	—	—	—	65-87	22	Handtight	1.3-3.8 ■
72 Pickup 2 W.D. (4x2) w/9" brake	.040	9.000	9.060	.040	—	—	—	—	—	—	—	65-87	22	Handtight	1.3-3.8 ■
86 Pickup 4x4	.060	11.614	11.654	.040	—	.748	—	.006	16	90	—	76	43	Handtight	6.2-12.6 ■
85-84 Pickup 4x4	.040	10.000	10.060	.040	—	.453	—	.006	16	55-75	—	76	43	Handtight	18•
83-79 4 x 4 Pickup	.040	10.000	10.060	.040	—	.453	—	.006	16	55-75	—	66-86	43	35-60	6.2-12.6 ■
83-79 4 x 2 Cab & Chassis	.040	10.000	10.060	.040	—	.748	—	.006	5	80-126	29-39	66-86	22	Handtight	1.3-3.8 ■
86-84 VanWagon	.040	10.000	10.060	.040	—	.748	—	.006	66	61	14	76	21	Handtight	8-1.9 ■

■ Radial pull in lbs. at lug nuts. □ 14" whl. shown, w/13" whl. 51-65 ft/lbs. • torque locknut to 33 ft/lbs., brg. preload; 2.2-8.6 lbs.

■ Seat brg. while turning wheel, back off nut until thrust washer can be moved slightly by screwdriver w/finger pressure, lock.
* .098 riveted; .059 bonded • Design 38 or 39 † Measurement of shoe & lining †† 1986-83; 73-87 ft/lbs. + Use ill. 67 from 2/84.

VOLKSWAGEN—F.W.D. VEHICLES EXCEPT VANAGON

VEHICLE	BRAKE SHOE Minimum Lining Thickness	BRAKE DRUM Standard Size	Machine To	BRAKE PAD Minimum Lining Thickness	Machine To	Discard At	Variation	Runout	DESIGN	Mounting Bolts	Bridge Bolts	Lugs	STEP 1	STEP 2	STEP 3
86-85 Golf, GTI, Jetta															
w/solid disc	.098	7.087	7.106	.276†	.413	.393	—	.002	67	—	18	81	■	—	—
w/vented disc	.098	7.087	7.106	.276†	.728	.708	—	.002	67	—	18	81	■	—	—
w/rear disc brakes	—	—	—	.276†	—	.315	—	—	68	65	26	81	■	—	—

1-170

BRAKES
Specifications

Model													
86-85 Scirocco, Rabbit Conv.	.098	7.086		.250†	.413	.393	—	.002	67	50	30	73-87	—	—
84-83 Rabbit GTI	.098	7.086	7.150	.375†	.728	.709	—	.002	8	50	30	73-87	—	—
84-82 Rabbit, Jetta, Scirocco	.098	7.086	7.105	.250†	.413	.393	—	.002	8+	50	30	65††	—	—
84-82 Rabbit Pickup	.098	7.874	7.894	.250†	.413	.393	—	.002	8	50	30	65††	—	—
81-80 Rabbit, Jetta, Pickup w/Kelsey-Hayes Caliper	∗	7.086	7.105	.080	.413	.393	—	.004	8	36☐	30	65	—	—
81-80 Rabbit, Jetta, Scirocco w/Girling Caliper	∗	7.086	7.105	.080	.413	.393	—	.004	38	36	—	65	—	—
79-77 Rabbit w/drum brakes, Front	.040	9.059	9.079	—	—	—	—	—	—	—	—	65	—	—
Rear	∗	7.086	7.105	—	—	—	—	—	—	—	—	65	—	—
79-75 Rabbit, Scirocco w/Girling Caliper	∗	7.086	7.105	.080	.452	.413	—	.004	38	36	—	65	—	—
79-75 Rabbit, Scirocco w/ate Caliper	.098	7.086	7.105	.080	.452	.413	—	.004	39	43	—	65	—	—
86-82 Quantum	∗	7.874	7.894	.276†	.413	.393	—	.002	8+	50	30	65††	—	—
81-73 Dasher	∗	7.850	7.900	.080	.413	.393	—	.004	37	43☐	25	65	—	—
77 Dasher	∗	7.850	7.900	.080	.413	.393	—	.004	•	43	—	65	—	—
76-74 Dasher	∗	7.850	7.900	.080	.433	.413	.0008	.004	•	43	—	65	—	—

■ Seat brg. while turning wheel, back off nut until thrust washer can be moved slightly by screwdriver w/finger pressure, lock.
∗ .098 riveted; .059 bonded • Design 38 or 39 ☐ w/self locking bolt 50 ft/lbs. †† 1986-83; 73-87 ft/lbs. + Use ill. 67 from 2/84
† Measurement of shoe & lining

VOLKSWAGEN—REAR DRIVE VEHICLES AND VANAGON

Model													
TYPE I—Beetle, Super Beetle, Karman Ghia													
79-72 Super Beetle, Front	.100	9.768	9.803	—	—	—	—	—	—	—	—	87-94	Snug-Up	Handtight	Step 2
Rear	.100	9.055	9.094	—	—	—	—	—	—	—	—	87-94	—	—	—
78-72 Beetle, Front	.100	9.059	9.068	—	—	—	—	—	—	—	—	87-94	Snug-Up	Handtight	Step 2
Rear	.100	9.055	9.094	—	—	—	—	—	—	—	—	87-94	—	—	—
74-73 Karman Ghia	.100	9.055	9.094	.080	—	.335	.0008	.004	17	58-65	—	94	Snug-Up	Handtight	Step 2
72 Karman Ghia	.100	9.055	9.094	.080	—	.335	.0008	.008	17	58-65	—	108	Snug-Up	Handtight	Step 2
TYPE II—Van, Bus, Wagon, Vanagon, Transporter, Kombi													
86 Vanagon	.098	9.921	9.921	.079	—	.512	—	.004	61	26	—	132	Snug-Up	Handtight	Step 2
85-80 Vanagon	.098	9.921	9.921	.078	.452	.433	—	.004	■	115	—	123	Snug-Up	Handtight	Step 2
79-77 Van, Bus, Wagon, Transporter, Kombi	.098	9.921	9.921	.080	.492	.453	.0008	.004	■	123	—	94	Snug-Up	Handtight	Step 2
76-73 Van, Bus, Kombi, Wagon, Transporter	.100	9.921	9.921	.080	—	.472	.0008	.004	17	116	—	94	7	Handtight	Step 2
72 Van, Bus, Kombi, Wagon, Transporter	.100	9.921	9.921	.080	.482	.472	.0008	.004	17	72	—	94	7	Handtight	Step 2
TYPE III—Square Back/Fastback 73-72 All	.100	9.768	9.803	.080	—	.393	.0008	.008	17	56-65	—	94	Snug-Up	Handtight	Step 2
TYPE IV—411, 412 Sedan 74-72 All	.100	9.768	9.803	.080	—	.393	—	.004	17	58-65	—	94	—	Handtight	Step 2

■ Design 14 or 17

VOLVO

Model													
86-83 740, 760 GLE Front w/solid disc	—	—	—	.120	.433	—	—	.003	52	74	†	63	42	12	Step 2
w/vented disc	—	—	—	.112	.790	—	—	.003	52	74	†	63	42	12	Step 2
Rear	—	—	—	.078	.330	—	—	.004	53	43	—	63	—	—	—

† upper pin 19 ft/lbs.; lower pin 25 ft/lbs.

BRAKES
Specifications

VEHICLE	BRAKE SHOE Minimum Lining Thickness*	BRAKE DRUM DIAMETER Standard Size	BRAKE DRUM Machine To	BRAKE PAD Minimum Lining Thickness	BRAKE ROTOR MIN. THICKNESS Machine To	BRAKE ROTOR MIN. THICKNESS Discard At	BRAKE ROTOR Variation From Parallelism	BRAKE ROTOR Runout T.I.R.	DESIGN	CALIPER Mounting Bolts Torque (ft-lbs)	CALIPER Bridge, Pin or Key Bolts Torque (ft-lbs)	WHEEL Lugs or Nuts Torque (ft-lbs)	WHEEL BEARING SETTING STEP 1 Tighten Spindle Nut (ft-lbs)	WHEEL BEARING SETTING STEP 2 Back Off Retorque (in-lbs)	WHEEL BEARING SETTING STEP 3 Lock, or Back Off and Lock
YEAR, MAKE, MODEL															
86-75 240 Series, Front w/ATE; vented	—	—	—	.062	.897	—	.0008	.004	16	65-70	—	70-95	50	Skip	1/3 Turn to slot
Solid	—	—	—	.062	.519	—	.0008	.004	16	65-70	—	70-95	50	Skip	1/3 Turn to slot
w/Girling: vented	—	—	—	.062	.818	—	.0008	.004	16	65-70	—	70-95	50	Skip	1/3 Turn to slot
Solid	—	—	—	.062	.519	—	.0008	.004	16	65-70	—	70-95	50	Skip	1/3 Turn to slot
Rear w/ATE	—	—	—	.062	.331	—	.0008	.004	14	38-46	—	70-95	—	—	—
w/Girling	—	—	—	.062	.331	—	.0008	.004	17	38-46	—	70-95	—	—	—
74-72 140 Series, Front Disc	—	—	—	.062	.457	—	.0012	.004	16	65-70	—	70-100	50	Skip	1/3 Turn to slot
Rear w/Girling	—	—	—	.062	.331	—	.0012	.006	14	45-50	—	70-100	—	—	—
Rear w/ATE	—	—	—	.062	.331	—	.0012	.006	17	45-50	—	70-100	—	—	—
82-76 260 Series, Front w/vented disc	—	—	—	.062	.818	—	.0008	.004	16	65-72	—	72-94	50	Skip	1/3 Turn to slot
w/solid disc	—	—	—	.062	.519	—	.0008	.004	16	65-72	—	72-94	50	Skip	1/3 Turn to slot
Rear w/ATE	—	—	—	.062	.331	—	.0008	.004	14	38-46	—	72-94	—	—	—
w/Girling	—	—	—	.062	.331	—	.0008	.004	17	38-46	—	72-94	—	—	—
75-72 160 Series, Front	—	—	—	.062	.900	—	.0012	.004	16	65-70	—	70-95	50	Skip	1/3 Turn to slot
Rear w/Girling	—	—	—	.062	.331	—	.0012	.006	14	38-47	—	70-95	—	—	—
Rear w/ATE	—	—	—	.062	.331	—	.0012	.006	17	38-47	—	70-95	—	—	—
73-72 1800 Sport Coupe, Front	—	—	—	.062	—	.520	.0012	.006	16	65-70	—	70-100	50	Skip	1/3 Turn to slot
Rear	—	—	—	.062	—	.331	.0012	.006	17	45-80	—	70-100	—	—	—

† Upper pin 19 ft/lbs; Lower pin 25 ft/lbs.

CALIPER DESIGN ILLUSTRATIONS

1–172

BRAKES
Specifications

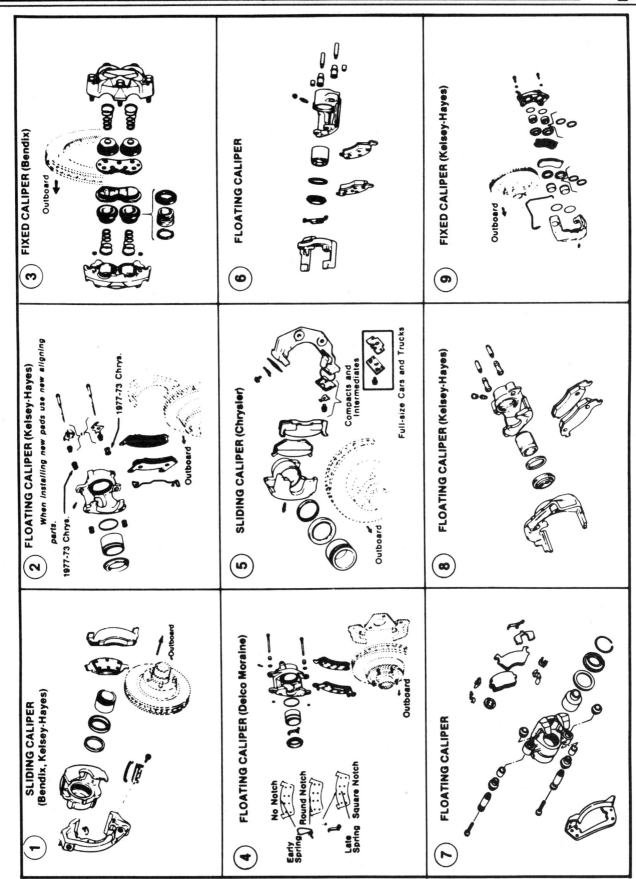

BRAKES
Specifications

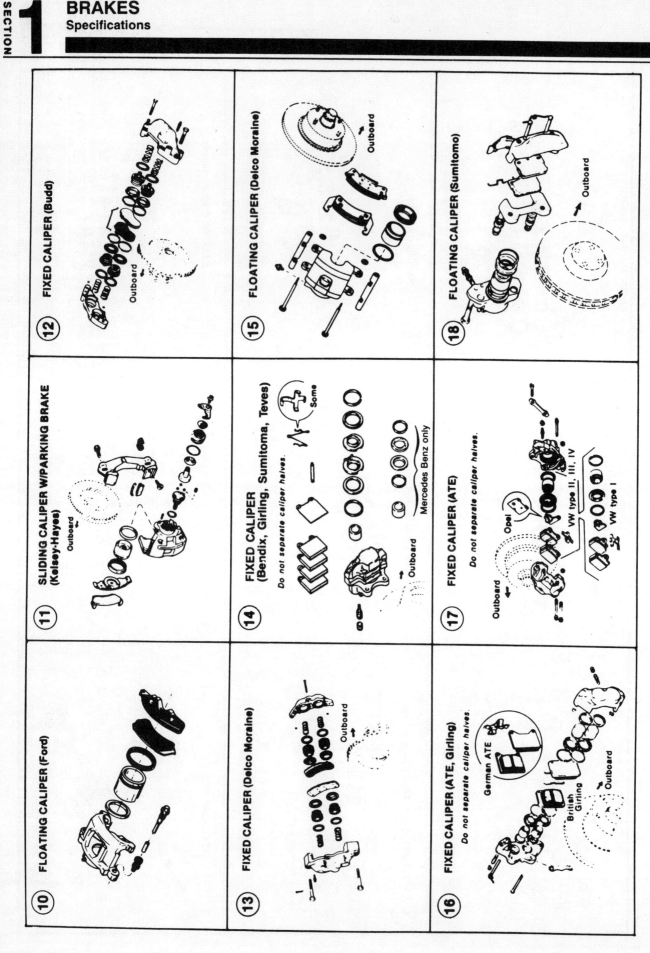

1-174

BRAKES
Specifications

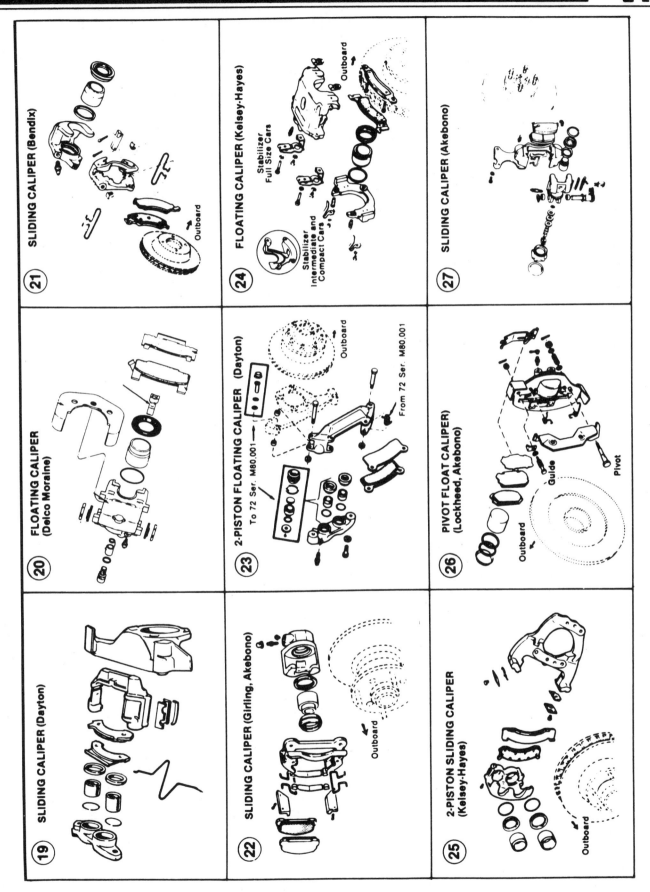

BRAKES
Specifications

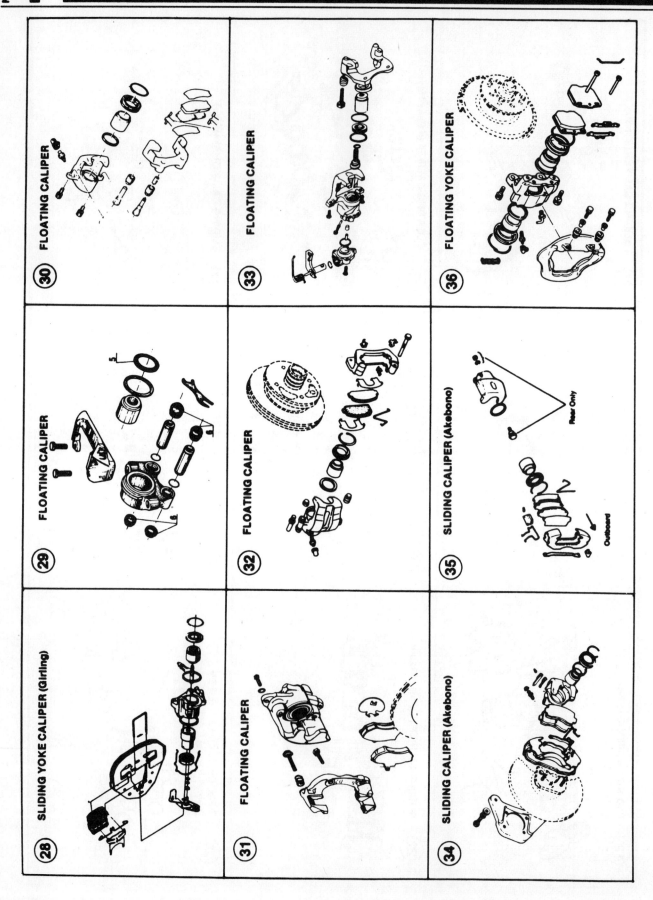

1-176

BRAKES
Specifications

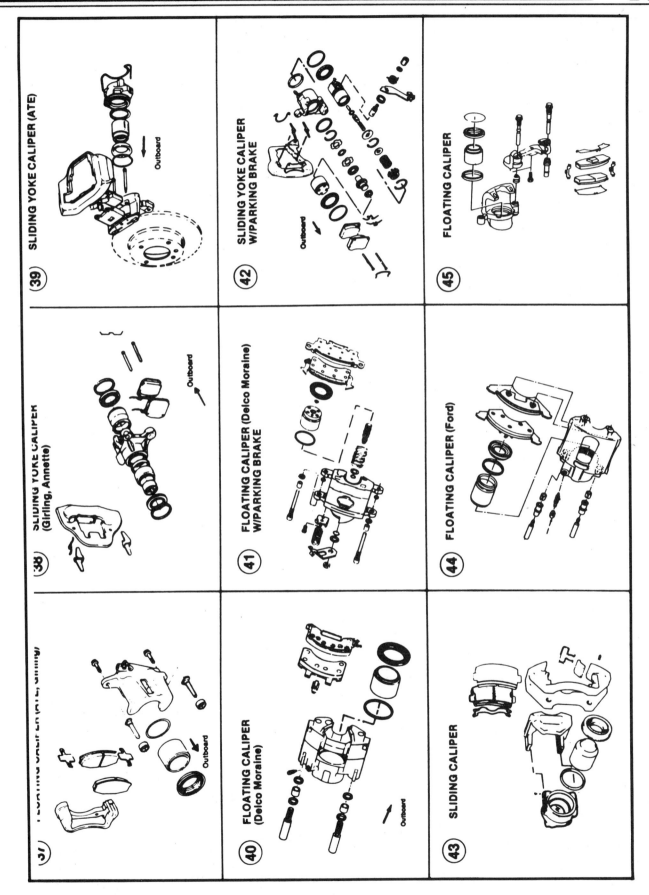

BRAKES
Specifications

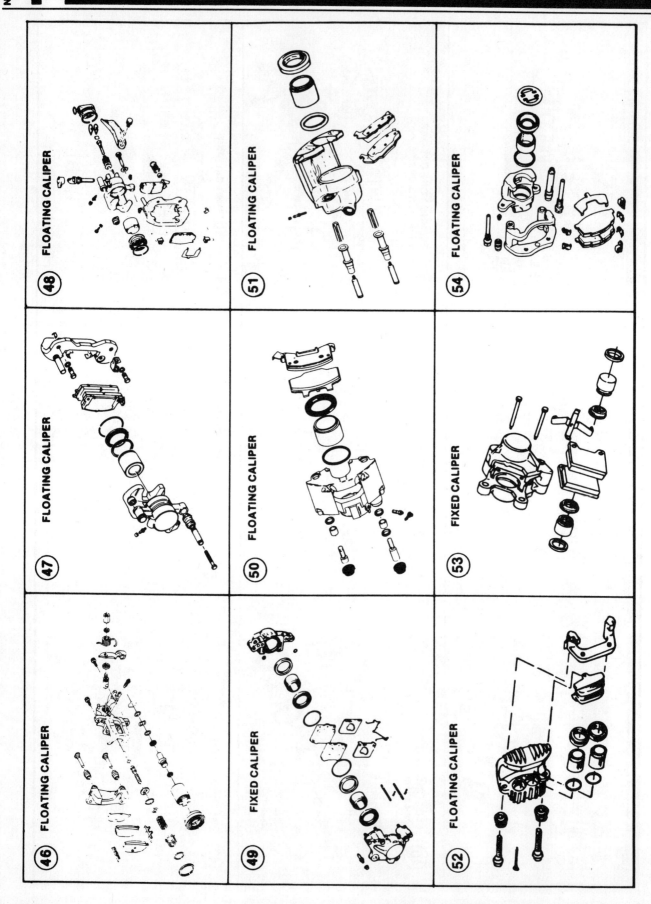

BRAKES
Specifications

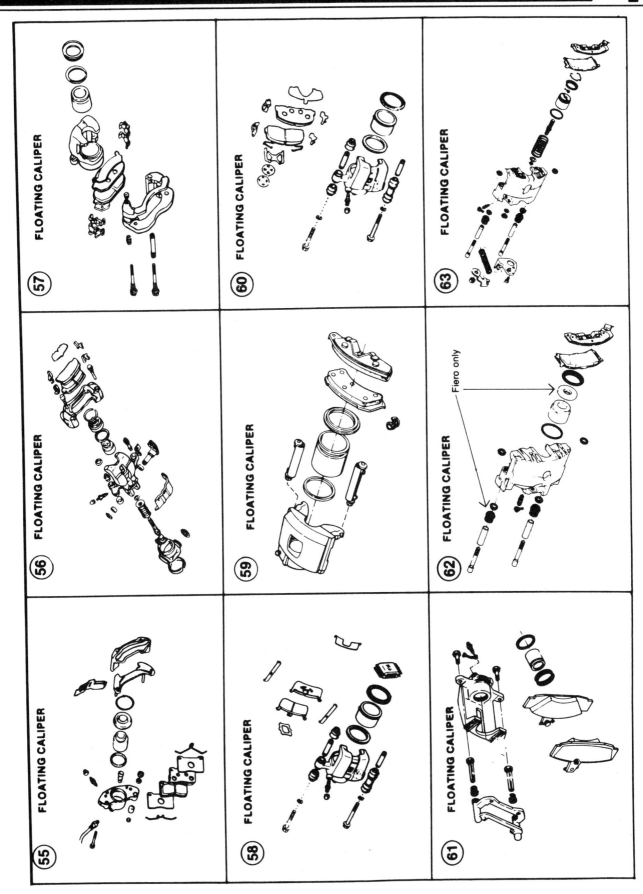

1-179

SECTION 2: SUSPENSION
General Information

INDEX

DOMESTIC CARS

GENERAL INFORMATION..	2-3 — 2-36	
TROUBLE DIAGNOSIS	2-37	

SUSPENSION SERVICE

AMERICAN MOTORS
- Front 2-40
- Rear 2-46

CHRYSLER CORPORATION
- Front Wheel Drive
  - Front 2-51
  - Rear 2-57
- Rear Wheel Drive
  - Front 2-63
  - Rear 2-73
- Electronic Level Controls............. 2-74

FORD MOTOR COMPANY
- Front Wheel Drive
  - Front 2-82
  - Rear 2-88
- Rear Wheel Drive
  - Front 2-91
  - Rear 2-112
- Air Suspension 2-98

GENERAL MOTORS
- Body Style Identification 2-115
- Front Wheel Drive
  - "A & X" Body
    - Front 2-116
    - Rear 2-118
  - "C" Body
    - Front 2-120
    - Rear 2-123
  - "F" Body
    - Front 2-125
    - Rear 2-128
  - "E & H"
    - Front 2-129
  - Rear 2-135
  - "J" Body
    - Front 2-138
    - Rear 2-143
  - "P" Body
    - Front 2-146
    - Rear 2-149
  - "S" Body (Nova)
    - Front 2-152
    - Rear 2-154
- Rear Wheel Drive
  - Exc. "F, T, P" Body & Corvette
    - Front 2-156
    - Rear 2-164
  - "T" Body
    - Front 2-168
    - Rear 2-172
  - Corvette
    - Front 2-173
    - Rear 2-176

IMPORT CARS

SUSPENSION SERVICE

AUDI
- Front 2-185
- Rear 2-187

BMW
- Front 2-188
- Rear 2-191

CHRYSLER CORPORATION
- Front 2-193
- Rear 2-196

HONDA
- Front 2-200
- Rear 2-204

HYUNDAI
- Front 2-207
- Rear 2-209

ISUZU
- Front 2-210
- Rear 2-214

MAZDA
- Front 2-215
- Rear 2-217

MERCEDES-BENZ
- Front 2-219
- Rear 2-226

MERKUR
- Front 2-229
- Rear 2-231

MITSUBISHI
- Front 2-231
- Rear 2-234

NISSAN/DATSUN
- Front 2-236
- Rear 2-238

PORSCHE 911, TURBO
- Front 2-243
- Rear 2-245

PORSCHE 924, 928, 944
- Front 2-246
- Rear 2-248

RENAULT
- Front 2-250
- Rear 2-251

SUBARU
- Front 2-254
- Rear 2-256

TOYOTA
- Front 2-257
- Rear 2-264

VOLKSWAGEN
- Front Wheel Drive
  - Front 2-270
  - Rear 2-274
- Rear Wheel Drive
  - Front 2-277
  - Rear 2-278

VOLVO
- Front 2-279
- Rear 2-282

LIGHT TRUCKS & VANS

CHEVROLET/GMC
- Front 2-283
- Rear 2-302

DODGE/PLYMOUTH
- Front 2-304
- Rear 2-321

FORD
- Front 2-323
- Rear 2-342

NAVISTAR/INTERNATIONAL
- Front 2-345
- Rear 2-355

JEEP VEHICLES
- Front 2-358
- Rear 2-363

SUSPENSION SPECIFICATIONS INDEX 2-365

SUSPENSION
General Information

Suspension

SUSPENSION AND STEERING

Theory and Principles

EVOLUTION OF AUTOMOBILE SUSPENSION SYSTEMS

Decades ago, the front suspension of a carriage consisted of a single solid axle that pivoted in the center and was attached to, and turned by, a horse, which supplied the motive and directional power. The evolution of the horseless carriage brought higher speeds and new problems: a need for driver control, better braking, and directional stability. The single pivot front axle was difficult to control. It prohibited the use of front brakes, as the slightest imbalance of side-to-side braking effort would cause the vehicle to veer from one side to the other.

The single pivot axle was soon replaced by the Ackerman axle, which provided a fixed position axle that could not be turned for steering. Steering was through two pivot points, one on each end of the axle, and connected to it with vertical kingpins. Ackerman axles gave way to a later design known as the Elliot axle which provided a tilted kingpin with a double yoke on the spindle and a short axle to steer the wheel. As speeds increased and front brakes were needed, the Reverse Elliot axle with kingpin inclination (KPI) was developed, giving greater stability and permitting the introduc-

Single pivot front axle

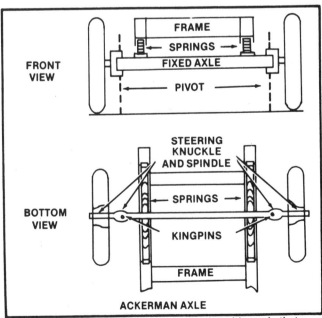

The Ackerman axle design provided a fixed position axle that could not be turned for steering. It provided two pivot points, one on each end of the axle, with vertical kingpins

2–3

SECTION 2

SUSPENSION
General Information

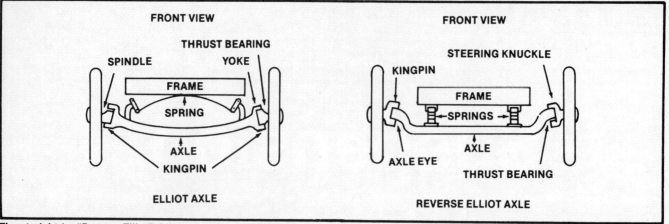

The straight (or "Reverse Elliot") front axle improved stability with a four point support instead of three, allowed a shorter turning radius and permitted the automobile to be set closer to the road

tion of front brakes. The Reverse Elliot, or basic straight front axle, improved stability with a four point support instead of three, allowed a shorter turning radius, and permitted the automobile to be set closer to the road.

For many years, every automobile used a straight front axle. However, ride control engineers soon learned that a bump hit on one side transferred shock to the other side and jolted the whole automobile. They also learned that the greater amount of unsprung weight or weight not supported by springs (wheels, tires, brakes, axles, suspension pieces, etc.), the harsher the ride. Straight axles had to be heavy to be strong, and contributed significantly to unsprung weight. Other drawbacks were evident. If in the process of turning one wheel gripped the road and gave good sta-

bility while the other wheel hit a rough portion of the road, the shock was transferred to the stable wheel causing it to lose some of its grip. Because of the harsh ride and poor handling characteristics of the straight axle, new designs incorporating independent suspension were developed.

INDEPENDENT SUSPENSIONS

There are several types of independent suspensions, but all share a common feature: Each wheel is supported independently of the other, so that movement of one wheel, up or down, will not necessarily cause movement of the other. Independent sus-

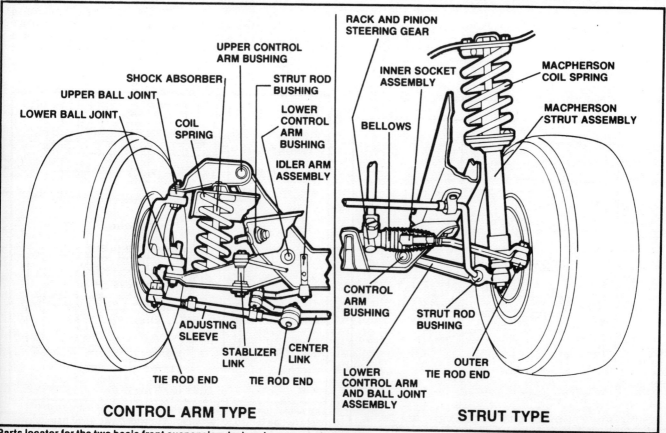

Parts locator for the two basic front suspension designs in use on today's cars

2-4

SUSPENSION
General Information

pensions are much lighter than solid axles, reducing unsprung weight and greatly improving riding and handling characteristics.

Early independent suspensions were not without problems, the greatest of which was poor and uneven tire wear. In most of these designs, the upper and lower control arms were of equal length. If the control arms are of equal length, the wheel will remain vertical. Because of the up and down motion, however, the arms become relatively shorter and longer with respect to the horizontal plane of the road surface. This creates side-to-side scuffing and significant tire wear. For example, as a wheel moves up and down it will be brought closer to the chassis. Since it remains vertical, it will be forced to scuff at the road surface, and scuff again in the opposite direction when the wheel returns to the normal position. Equal length control arms also caused the wheels to toe inward while turning, creating directional instability.

By using a short upper control arm and a long lower control arm, it allows the wheel to pivot at the road surface and more in and out at the top only. As the wheel moves up and down, the wheel stays in place at the road surface with respect to lateral travel, but moves in and out at the top. One might think that constantly tilting the wheel in and out at the top would create excessive tire wear, but the movement is slight; indeed, locating the pivot point at the road surface actually reduces tire wear.

By the late 1960's, engineers had learned that by positioning the arms at slightly different angles, the wheel would move out away from the frame at the top as the spring was initially compressed, which improved handling and stability in a crosswind. By further changing the position of the arms, front end dipping could be somewhat relieved when the brakes were applied.

Although an automobile may be properly aligned to factory specifications, there are other factors involved that directly affect ride, handling, and tire wear. One of these factors is dampening. Softer springing to improve ride characteristics created a need for better dampening of road shock. The early shock dampeners (shock absorbers) were indirect acting. The later shock absorbers (called direct acting or airplane type) were mounted directly to the suspension and frame and controlled the amount of road shock transmitted to the vehicle. Shock absorbers that check compression only are known as single acting; shock absorbers that check rebound as well are known as double acting. The shock absorber action partially controls the spring action, thus protecting the spring and the suspension. This action also keeps wheel bounce to a minimum and gives the tire greater contact with the road.

Despite the use of shock absorbers, there were still some forces that could not be absorbed, such as large bumps and deep holes. These hazards were severe enough to damage the suspension components. The use of rubber bumpers attached to frame members and control arms reduced suspension damage by absorbing the harder shocks.

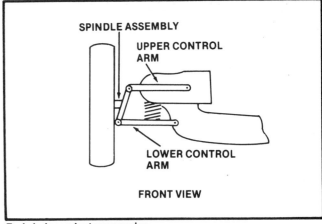

Early independent suspension

Other problems encountered with independent suspensions involved excessive lean on turns, oscillation of the front of the car on a straight road, and side-to-side roll. Although the automobile had proper alignment in the static position, it showed signs of misalignment while in motion and exhibited excessive tire wear. Wheels had excessive tilt while cornering, which caused the outer edges of the tires to wear prematurely. To solve this problem, the stabilizer bar or "sway bar" was introduced. It was first installed in the rear and later added to the front. The stabilizer bar is a round metal bar mounted in rubber blocks and secured by brackets to the front portion of the frame. The ends of the bar are attached to the outer ends of the lower control arms. The twisting or torsional action of the bar permits the vehicle to corner at much higher speeds and remain reasonably flat.

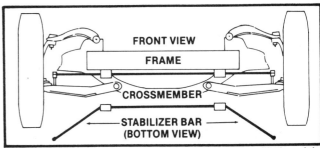

The stabilizer bar reduced excessive lean on turns, oscillation of the front of the car on straight road, and side-to-side roll

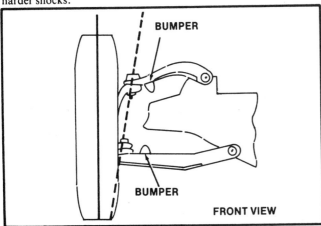

Rubber bumpers absorbed the harder shocks

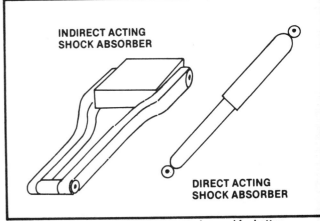

Softer springing to improve riding created a need for better dampening of road shock

2-5

SUSPENSION
General Information

A-ARM SUSPENSIONS

Early independent suspension systems used kingpins and bushings to connect the steering knuckle with the axle or control arms. The bushings above the kingpins were eccentric and also served to adjust the alignment of the suspension. However, failure to lubricate the suspension bushings meant that the alignment adjustments became frozen and expensive repairs followed.

The control arms were shaped as an "A" (hence the term A-arm). The wider the "A" was spread at its base, the greater the support and stability. In these early designs, the upper control arms also acted as indirect shock absorbers in many suspension systems. When balljoints replaced kingpins and bushings, the alignment adjustments were moved to other locations, usually where the A-arm mounted to the frame.

A further refinement was the addition of the brake or strut rod. The lower A-arm was redesigned as a single straight arm and the strut or brake rod was added to it. The strut rod connected the outer end of the lower arm and the front of the frame, creating a still wider "A" effect for greater stability. The strut rod also provided strength and prevented the single lower arm from buckling under or being torn off. This system permits the strut rod to be mounted in rubber where it connects to the front of the frame. When the wheel hits a rough portion of the road it can "give" slightly, reducing road shock transmitted to the frame and yielding a better ride.

Dual arm suspensions used a coil spring between the upper and lower control arms, to control suspension travel and reduce shock. Development of unitized construction permitted the use of a coil spring above the upper arm with both spring and shock absorber mounted on the upper arm instead of on the lower arm. This system is a short-and-long-arm design (the lower arm is longer than the upper control arm) but requires a very strong inner fender panel (spring tower) to absorb the spring that shock absorber forces. The advantage of the higher and wider mounting of the springs is greater resistance to the roll forces of the automobile. Coupled with a strut rod-type lower arm, this system provides excellent ride and handling characteristics.

MACPHERSON STRUTS

The single control arm suspension is better known as the "MacPherson strut." This strut is a one-piece design incorporating a shock absorber and concentric coil spring, giving excellent anti-roll characteristics due to the high placement of the spring. Spindle load and road shocks are transferred directly to the spring without going through a control arm, producing a much smoother ride. In this type of suspension the stabilizer bar may sometimes serve as the strut rod. As with the coil spring on the upper arm, a strong all-welded inner fender is required.

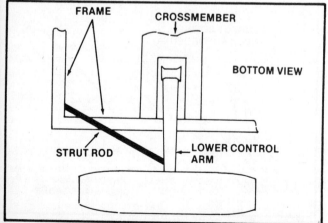
The strut rod provided greater stability and strength, and prevented damage to the single lower control arm

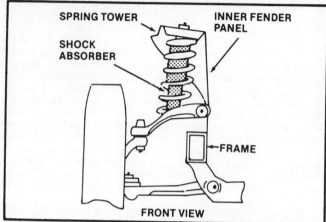
The advantage of the higher and wider mounting of the springs is greater resistance to the roll forces of the automobile. The darkened area shows typical shock absorber location

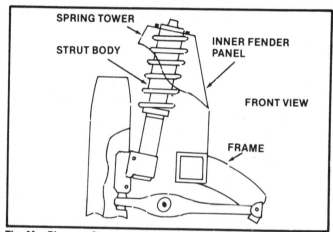
The MacPherson Strut provides for the spindle load and road shock to be transferred directly to the spring without going through a control arm. The shock absorber is located inside the strut body

Weight has become a very important factor in the design of modern automobiles. The strut suspension, combined with the unibody, permits the elimination of the heavy frame and the upper control arm. The unibody supports the upper end of the coil spring, eliminating the need for a frame horn or extension of the crossmember for coil spring support.

Space and accessibility to the serviceable areas of the automobile are also very important considerations in the manufacturing of smaller, lighter automobiles. When all available options are added (power steering, power brakes, air conditioning, and so on), available space becomes extremely limited. After an automobile has been sold and delivered to the customer, it must be readily serviceable. In the past, many small automobiles without strut-type suspensions posed problems in normal service procedures because of extremely cramped working conditions. The introduction of transversely mounted engines further complicated space problems. Because the strut suspension eliminates so many previously necessary parts, it occupies much less space while allowing the automobile to be "downsized" for even greater weight reduction, better economy, and greater ease of service.

The strut suspension, much simpler than previous mechanisms, generally is easier and less expensive to maintain because it contains fewer moving parts. There are many design variations in strut suspensions, but most have a coil spring at the top of the strut and a shock absorber within the strut body. Others have a coil spring

SUSPENSION
General Information

supported by the control arm and still others have a shock absorber mounted independently of the strut body. Some types use a single control arm in combination with a longitudinal strut rod that attaches to the chassis, while others may use a wide double "A" type control arm with no strut rod. All modern strut suspensions eliminate the upper control arm, and all are mounted in rubber to minimize noise and road shocks, resulting in a smoother ride. Because the strut suspension reduces the number of pivot points, total free play in the moving parts has been reduced. Consequently, there is less loss of motion in the steering and suspension, resulting in more responsive handling characteristics. Another consequence, however, is greater sensitivity to any imbalanced or out-of-round conditions in the tires. Strut suspensions using a coil spring mounted at the top of the strut are often regarded as having superior ride and handling characteristics because of the wider and higher placement of the spring, but strut suspensions with the coil spring mounted at the control arm, independent of the strut, will ride and handle just as well under normal driving conditions.

GEOMETRIC PRINCIPLES

The suspension and steering systems are a group of connected levers pivoting within controlled and limited arcs. The laws of geometry are dependable and when the levers are properly positioned, all angles on the wheels at the road surface are controlled.

The suspension and steering systems are actually two separate systems. the suspension system supports or suspends the automobile. All of its adjustments are referenced to the horizontal. Because of the effects each system has on the other, the correction of the angles within these systems are done at the same time.

Of the many different types of suspensions in use today, all have one thing in common: they all use the same principles of geometry. If the relevant geometric angles are thoroughly understood, the type of design is of little consequence to the mechanic. If the angles to be considered - caster, camber, steering axis inclination, included angle, (the angles referenced to the vertical) toe, toe-out-on-turns, thrustline, (the angles referenced to the horizontal and "squaring" of the automobile - are understood, making the adjustments is relatively easy.

For years it has been known that any adjustments made in the suspension system (caster, camber, etc. or any adjustments referenced to the vertical) would change the steering or toe adjustments (those referenced to the horizontal). This is still true. However, it was also thought that toe adjustments were made last because these adjustments would not affect the suspension adjustments. *This is not true.* You will see why changing a toe adjustment on *one* rear wheel can change both *camber and toe* of each front wheel when the automobile is driven.

As suspension and steering systems evolved, it became more evident that all angles were closely related. It is very certain that every angle that is considered in the total alignment procedure is important, and is there for a specific purpose. Because of the very close relationship of all angles, one might think that all the adjustments must be made simultaneously to attain a perfect total alignment. This is mentioned to stress the importance of following a specific proper procedure and why you must use a machine that has the logic to reference to geometric centerline throughout the *entire* total alignment procedure. The geometric centerline is more commonly known as the thrustline. *The thrustline is the base reference from which the total alignment procedure must begin and continue to be referenced to throughout the total alignment procedure.* When a four wheel alignment is being done, alignment of the rear wheels is done *first*. This *establishes* the thrustline. The front wheels are then aligned referencing *all* angles in both systems to the thrustline. This ensures a proper total alignment. In the case of a two wheel only alignment, that is when the rear axle is in a fixed non-adjustable position, the thrustline must be able to be found in *all* angles in the front wheels aligned to this reference. You will see how a vehicle requiring a two front wheel only alignment, with high caster setting, *could* be misaligned if the thrustline is not used as a reference throughout the entire total alignment.

We will examine each of the angles individually and see why they are used and how they are related. We will examine other factors involved in proper total alignment. We will see how some of these factors may affect an automobile even after alignment adjustments have been completed.

When the angles are in harmony with one another, an automobile will follow the desired course without wander, scuffing, dragging, or slipping. The vehicle will be free-rolling. Steering will tend to return to the straight position and will be controllable with minimal effort.

PURPOSES OF DESIGN AND REASONS FOR ALIGNMENT

The engineers that designed the various suspension and steering systems have also done extensive research to determine the best possible alignment specifications for each type of system used on a particular vehicle. The purpose of this research is to obtain maximum stability, handling and tire life.

Gravity plays a very important part in the design of suspension systems. Because of these design features, the front of the vehicle is raised when the wheels are steering. The vehicle is constantly seeking to be at its lowest point and when alignment is correct, the lowest point will be when the wheels are pointed dead straight ahead. These design features greatly influence the stability and driveability of the vehicle.

Each tire has a footprint at the road surface. The goal is to have the maximum amount of tire footprint, at the road surface, under all drive conditions and road speeds. Because of the infinite number of variables involved, this goal is rarely obtained. However, the engineers have determined the best alignment settings possible for all conditions the average driver will experience. It may sometimes appear to the novice, that specifications seem incorrect while the vehicle is in the static position, nonetheless, when the vehicle is given motion, the alignment settings will be as correct as possible. As you study the individual angles you will see how this is accomplished. The altering of alignment specifications is not a recommended practice. These specifications generally provide the maximum tire life and stability possible.

Although the suspension and steering systems may be considered by some to be one of the weaker parts of an automobile, in most cases, the engineers have done an excellent job. The suspension and steering systems must be able to withstand brutal forces. These systems must be able to stay intact while the wheels are locked by the brakes, hitting holes and bumps and at the same time support tremendous bouncing and shifting weight. After all these severe shocks and stresses, the systems must return to the *exact* position from which they started. While an automobile is being drive, these systems are under constant stresses, strains, and shocks. They must be able to withstand this for many thousands of miles. Given the durability of these systems today, this is truly outstanding engineering!

Suspension and steering system parts cannot be made absolutely rigid. If they were, they could break upon receiving severe shocks. There is a certain amount of built in "give" in these parts. In addition to this, as anything that has constant movement, there is a certain amount of wear at the joints; a little here and a little there. Torsion bars or springs will sag just a bit over the miles. Combined, this "give", sag, and wear is probably not enough to justify parts replacement, however, any or all of the above are good reasons to have the wheels periodically realigned. Proper wheel alignment is the "fine tuning" of these efficiently designed systems and ensures maximum stability, handling, and tire life.

2-7

SECTION 2 SUSPENSION
General Information

Suspension Designs

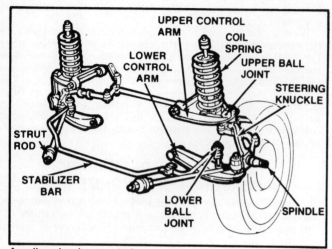

A coil spring is mounted on top of the control arm with the shock absorber in the center of the coil spring. Only the upper control arm is A-arm design. This design is used primarily on Ford

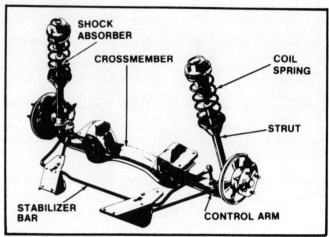

MacPherson strut type front suspension

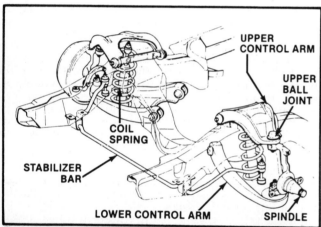

Typical unequal length of A-arm suspension used on many American sedans. The shock absorber and coil spring are positioned between the upper and lower control arms. Note that the control arms (A-arms) are not the same length.

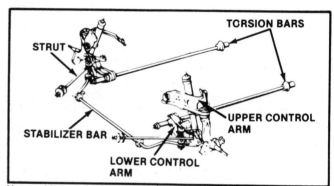

Unequal length A-arm suspension with torsion bars.

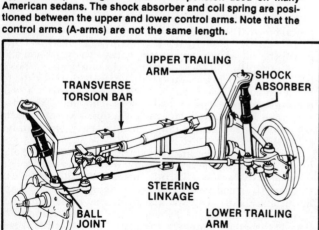

Transverse torsion bar front suspension used primarily by VW on rear engine cars.

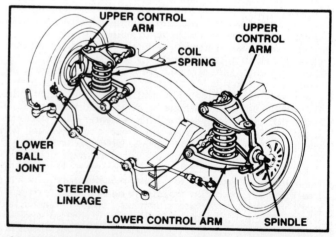

A truck's independent front suspension is very similar to the unequal length A-arm suspension used on passenger cars. It functions in the same manner, but the components are often beefier to handle the extra stress.

2-8

SUSPENSION
General Information

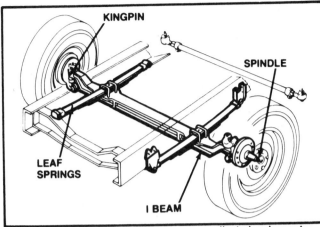

A straight I beam front suspension is uncomplicated and meant to handle heavy loads.

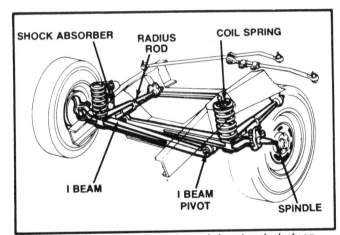

The twin I beam front suspension is used almost exclusively on Ford trucks. The coil spring is mounted between the frame and the I beam that carries each wheel. The I beam pivots at the other end, and a radius rod locates the fore-and-aft position of each beam.

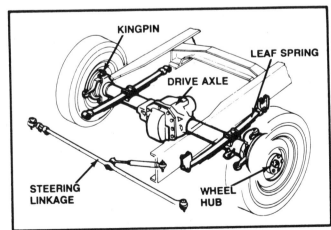

A four wheel drive front suspension is basically the same as an I beam suspension except that a front drive axle takes the place of an I beam.

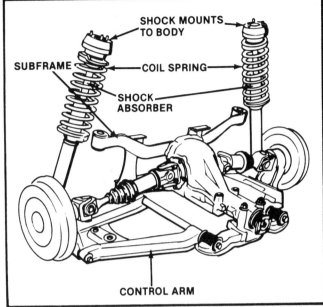

The rear strut suspension combines the coil spring and shock absorber in one unit, attached to the body and the wheel spindle. The lower control arm, strut and rear axle are usually mounted on some sort of subframe, which is attached to the car body.

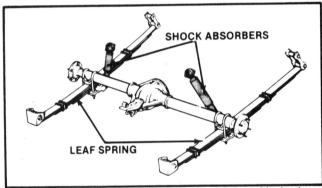

A basic, uncomplicated leaf spring rear suspension with shock absorbers to control vibration and vertical movement of the axle.

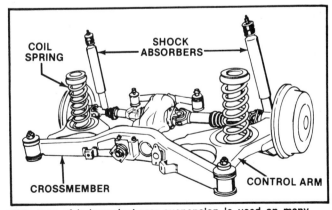

This type of independent rear suspension is used on many sports cars. Coil springs are used between the control arm and the vehicle body. The control arms pivot on a crossmember and are attached to the spindle at the other end. A shock absorber is attached to the spindle or the control arm.

2-9

SECTION 2

SUSPENSION
General Information

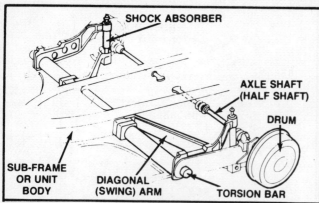

Swing arm rear suspensions are typical of rear-engined VWs. The rear wheels are independently sprung on diagonal arms or swing arms. The torsion bars are anchored at either end by splines which can also be used to adjust suspension height.

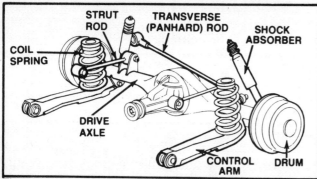

These non-independent rear suspensions differ from similar designs in that coil springs replace leaf springs and strut rods and control arms serve to position the rear axle.

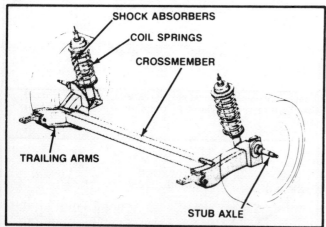

An independent (dead) rear axle is mostly used on front wheel drive cars. Trailing arms holding the wheel and shock absorber are attached to a rigid crossmember.

Wheel Alignment

PREALIGNMENT

Pre-Drive

A thorough pre-drive is recommended. It will enable you to determine if there are other problems with the automobile. For example: A customer may detect a vibration in the automobile and decide for himself that it needs a wheel alignment. It may very well need a total wheel alignment, however, the best alignment possible will not correct an out of balance wheel condition or a bad universal joint. If the customer leaves your shop and still has the vibration, it is quite possible for him to think you did not do a proper job. He may never return! A knee on the bumper will not yield a proper shock absorber check. A pre-drive will, and at the same time will allow you to note any driving irregularities or peculiarities the automobile may have. A pre-drive increases your business volume and customer satisfaction as well.

Tire Check

Before an automobile is aligned, it should be given a thorough prealignment check. Proper tire pressure is very important, so pressure must be checked on each tire before an alignment is started. Incorrect tire pressure has a drastic effect on alignment

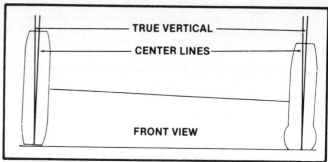

Incorrect tire pressure can result in misalignment

settings. For example, assume there are two underinflated tires on the left side of a car with normal pressure in the tires on the right side. The car would be lower in curb height on the left side, increasing the camber on that side and decreasing camber on the right side. No matter how painstakingly done, if the alignment were adjusted with the tires in this condition, camber would be altered when the tires were brought to proper pressure. Experiments have shown this can change camber as much as one-half degree, an excessive amount of change.

There are a number of other factors concerning tires that should be examined and noted while tire pressures are being checked. Look for uneven wear patterns caused by misalignment. Badly worn tires can also be indicative of other suspension problems, such as worn ball joints, bent suspension components, worn shocks or other components that can affect vehicle performance and/or alignment. See the section on Diagnosis and Troubleshooting.

Determine if tires are matched. If there are two different sizes, they should differ only front to rear. If there are two new tires and two good but older tires, the new tires should both be on the front or both be on the rear. Conventional tires should not be used with radial tires. A common misconception is that radial tires may be used with conventional tires by installing the first pair of radial tires in the rear. Mixing radial and bias ply tires is not a recommended practice. Different size footprints can create a safety hazard, especially when braking on wet, muddy, or gravelly surfaces. The best policy is to follow the tire or vehicle manufacturer's recommendations.

Suspension Check

In addition to checking tires, learn to do a thorough parts check. Loose wheel bearings, for example, not only distort the alignment readings, but are hazardous and can lead to expensive repairs. Loose steering parts greatly affect "running toe" as do other suspension parts. Loose balljoints or bushings not only change "running toe," but caster and camber as well. Bad stabilizer bar bushings or links can cause excessive vibration, especially on a rough corner. A bad strut rod bushing can cause an automobile to pull when the brakes are applied. The brakes may be fine, but they erroneously become suspect if the bad strut rod bushing remains undetected. The list of cause and effect is long. If loose parts are discovered during a prealignment check, alignment should not be adjusted until the necessary repairs are made. There is no way to set the static alignment correctly for the running alignment, given the uncertain tolerances created by loose parts. Performing an alignment on an automobile with faulty steering or suspension parts will also compromise the performance of the vehicle and the safety of the occupants.

Thrustline and Centerline

Every vehicle has a true centerline and if the vehicle is built geometrically perfect, the true centerline will be pointed in the exact direction of travel of the vehicle.

All vehicles are not build geometrically perfect, and whether the vehicle is built as a frame type, unibody, or unitized body, all vehicles are built within certain tolerance limits. These tolerances apply to the installation of the front and rear suspension systems as well. Since tolerance means "an acceptable margin of deviation from a given norm," all vehicles are not exactly alike. Some vehicles are manufactured and assembled on the outer limits of all tolerances of the manufacturer. Consequently, the thrustline of the vehicle may not coincide with the true centerline.

All angles of the suspension and steering systems are important, however, if there is a "most important" angle, it would have to be the thrustline. You will see as you study about this invisible line called the thrustline, that it is the true direction of travel of the vehicle. You will see that for a proper total alignment, you must use an alignment machine that has the logic to find this invisible line.

As you can see in the illustration, the thrustline may *not* be in the exact position as the true centerline. Nonetheless, the thrustline may be well within the tolerance specifications of the manufacturer. However, the front wheels must be steered in the exact same direction as the heading of the rear wheels for the vehicle to go straight. As you study the balance of the individual angles, you will see why it is important that the front wheels be aligned for the *running alignment* which is the true direction of travel of the vehicle. While they may still be well within tolerance, this condition can and does exist in many automobiles, regardless of their type of construction.

FRAME-TYPE VEHICLES

Thrustline and Centerline

Light trucks and some large automobiles are still being built with frame type construction. There are many frame type vehicles still being driven, therefore, the alignment mechanic must have knowledge of frame type vehicles still being driven. Proper frame alignment is very important to proper alignment of all four wheels and while frame straightening is a special operation, the alignment mechanic should know how to check frame alignment. A misaligned frame can greatly affect the overall alignment.

In most frame-type automobiles, adjustments provide compensation for the geometric imbalances. Most of these automobiles provide for front suspension adjustments, but few provide for rear suspension adjustments. Frame alignment is checked by referring to the frame centerline. However, if all the adjustments can be made to specification, and all other geometric angles are within tolerance, checking frame alignment is not necessary. If one or more angles cannot be brought to specification, a further check must be made to find the cause of the problem.

When an automobile is driven straight ahead, all four wheels should run parallel to the frame centerline. Front and rear axles should be at, or nearly at, 90° angles to the frame centerline.

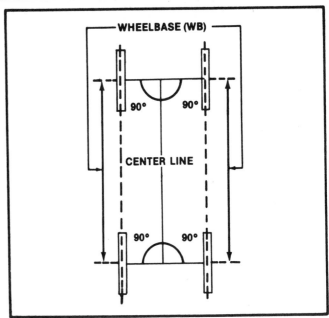

All four wheels should be parallel to the frame centerline

The wheelbase is the distance from the center of the front axle to the center of the rear axle. Any difference in side-to-side wheelbase measurement can affect alignment. Some common causes for deviations in wheelbase measurement are:
1. Front wheel set back
2. Rear axle shift
3. Front wheel set back and rear axle shift
4. A swayed frame
5. A diamond frame.

In a "set back" frame, one front wheel is set back closer to its corresponding rear wheel. Thus, the wheelbase on one side is shorter. If the condition is caused by a bent suspension and is severe enough, a proper alignment probably cannot be obtained. If

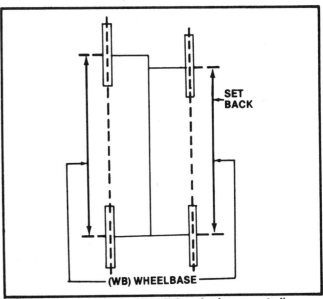

Set back damage. Although parallel to the frame centerline, the one wheel is set back

2-11

SECTION 2 SUSPENSION
General Information

the set back is caused by a bent frame, a proper alignment is not possible. And if it is severe, the automobile likely will pull toward the shorter wheelbase.

"Rear axle shift" is a condition in which one side of the rear axle has shifted toward the front or the rear. This is the most common of all vehicles geometric imbalances. Rear axle shift can be determined by measuring from a common point on the frame to the rear wheel on each side of the vehicle and comparing measurements. A modern alignment will find and measure rear axle shift in minutes. The machine *must* find and measure the amount of shift to be able to use the thrustline as a reference. Regardless of the direction of shift, or the cause, the thrustline has been changed to a different angle relative to the frame centerline, changing the direction in which the vehicle will tend to travel. The thrustline is the true direction of travel of the vehicle. If the front wheels remain parallel to the frame centerline, but the thrustline has changed, the vehicle will no longer tend to go straight. For the vehicle to go straight ahead, the front wheels must be steered in the exact direction of the thrustline, resulting in "dog tracking or crabbing" under the conditions described here. Many vehicles have an acceptable amount of deviation in thrustline from its true position. This can be compensated for, and a straight steering wheel obtained, through normal alignment procedures.

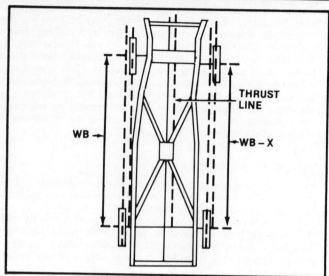
Swayed frame. The automobile has been hit in the left front area changing the wheelbase and it no longer tracks properly

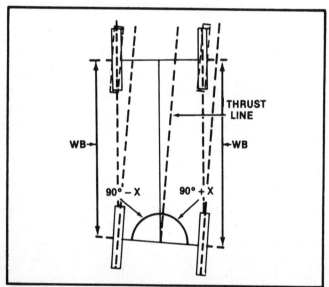
Axle shift. The left rear wheel has shifted toward the front of the automobile changing the thrust line which also changes the direction of travel

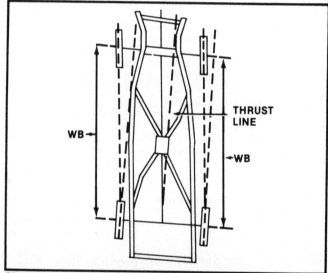
Diamond frame. The automobile is no longer rectangular and the thrust line has been changed

A swayed frame occurs when the automobile has been hit at an angle in the front, resulting in a shorter opposite side wheelbase. The rear wheels will not track properly with respect to the front wheels. Such an automobile cannot be made to drive properly or safely until the frame alignment has been corrected.

A diamond frame means that one side frame rail has been driven back such that the frame is no longer rectangular. Any automobile with this problem must be sent to a frame or body shop before an alignment can be completed. Faulty driving and tracking symptoms caused by a diamond frame could be similar to those caused by rear axle shift, so identifying the difference between the two problems is important.

Rear axle offset does not affect side-to-side wheelbase measurement, but while the wheels remain parallel to each other and to the frame centerline, the rear wheels will not track properly as compared to the front wheels. However, the direction of the thrustline remains parallel to the centerline. The human eye cannot easily detect problems like this, but modern alignment machines are capable of distinguishing rear axle offset from rear axle shift.

HOW TO DETECT FRAME PROBLEMS

Proper frame alignment is the foundation of properly functioning steering and suspension systems. Calibrated frame gauges quickly determine the extent and location of any bend in the frame. Even the slightest bend of any type will be detected. Frame gauges are of reasonable cost, can be installed in minutes, and are not difficult to learn to use and read.

The best alignment mechanics know all of the angles of suspension and steering systems and how to use and read frame gauges. The alignment machine will indicate to the mechanic if the vehicle is in "square". A vehicle that is in "square" is actually rectangular in shape, but the term "square" indicates that: (1) the thrustline has been referenced to, (2) all four wheels are parallel to each other and the thrustline, (3) front and rear axles will be at right angles to the thrustline, (4) front and rear

SUSPENSION
General Information

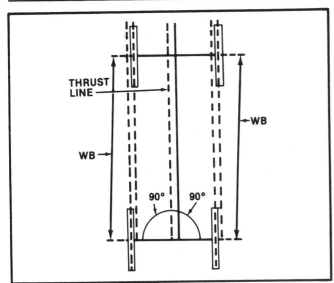

You must be able to determine if rear axle offset is within tolerance

axle midpoint will be at the vehicle thrustline. With this knowledge and equipment, you can quickly and accurately analyze any alignment problems.

In the past, obtaining a straight steering wheel has been one of the more difficult tasks in the alignment procedure. Allowable tolerances of the manufacturing process have their greatest influence in the steering system adjustments. Since the thrustline is the true direction of travel of the vehicle, steering adjustments are set with reference to the thrustline, which also assures a straight steering wheel.

UNITIZED BODY

In unitized body construction, there is a short boxframe toward the front and toward the rear to support the suspension and other components. The body itself is the main structural member. These automobiles are aligned in the same manner as frame type vehicles.

UNIBODY

Because of the need to reduce weight and increase fuel efficiency, the heavier frame-type automobile is being replaced by the lighter, more fuel efficient "unibody" chassis/body units. The body is bolted to the frame of a frame type automobile and all references begin at the frame centerline, with no references made to the body. Alignment of unibody automobiles begins with all references being made to the geometric centerline of the body. Unibody construction is a single, completely spot-welded, fully integrated body frame assembly with inner fender panels braced within the structure to support the engine, driveline, suspension, and other components. This type of construction has replaced the independent frame assembly. With unibody construction, all angles that are referenced to the vertical pose no problems; however, these and all other angles must be referenced to the geometric centerline of the body. This change of reference has created new challenges in the total alignment procedure. Alignment machines that do not find and use the geometric centerline of the unibody as a reference, are obsolete for unibody alignment work.

Frame gauges may be used to find the true body centerline, however, this could be the wrong reference and there is no way to feed this information into an alignment machine. There is no accurate or quick way to measure from the benchmarks on the underside of the body for the misalignment. Therefore, the alignment machine must be able to find the geometric centerline and use it as reference to bring the automobile into "square".

Let us examine what we do when we bring an automobile into "square". Since the body itself is the frame in a unibody vehicle, the true body centerline is the base reference from which the automobile was *built*. As was mentioned before, there are allowable tolerances within the building processes; it is possible that the wheels of the automobile were *not* aligned referencing to the true body centerline. However, when the automobile was built, the wheels are aligned to the *geometric centerline*. The geometric centerline is as folows: all angles to the rear wheels are aligned first and are referenced to the *midpoint* between the two front wheels. Then, all angles of the two front wheels are aligned to the *midpoint* between the two rear wheels. This establishes a straightline between the midpoints of both the front and rear wheels. This line is theoretically called the geometric centerline. When the wheels are aligned referencing to the geometric centerline it assures that: the rear wheels are tracking properly behind the front wheels, and that all four wheels are pointed in the exact same direction. The geometric centerline and the true body centerline may or may not be the exact same line; *it is of no concern if the automobile was built within the limits of all tolerances*. These same geometric principles apply whether the automobile is front wheel or rear wheel drive.

When the vehicle is in a static position, or motionless, the line between the midpoints of both the front and rear wheels is termed the geometric centerline. However, when the power of the engine is transferred to the drive wheels, (front or rear) the vehicle is "thrust" into motion. If the vehicle was properly aligned to the geometric centerline in the static position, the "thrust" of the vehicle will be in a straight line. Factually, he vehicle was properly aligned in the static position so that when it is "thrust" into motion it has a proper *running alignment*. The primary concern is what the alignment of all four wheels will be when the vehicle is being driven. When the vehicle was aligned, we either had to establish a correct geometric centerline or reference to an existing correct geometric centerline so that when the vehicle is "thrust" into motion, the vehicle will go straight. Since the *running alignment* is our primary concern, when the vehicle is "thrust" into motion, the geometric centerline becomes the *thrustline*, and that is what this invisible line is now commonly called.

When a vehicle has been damaged, because of an accident, the repairs must be made referencing to the *true body centerline*. Since the body of a unibody vehicle *is* the frame, corrections for serious body damage must be made to this reference. In order to find the true body centerline of this vehicle, you would have to put the vehicle on a machine that is dedicated to just such a purpose. Normally, this is quite time consuming and generally, these machines are used only in body and frame shops. It would seem to some, that a four wheel alignment that references to the true body centerline would be ideal. While a "perfect" alignment with reference made to the true body centerline would be ideal, it really is not economically sensible, and it would *not* be to any actual advantage. Any minor corrections that could be made would have to be done by heavy duty pullers designed for this specific purpose. This could not be done in the alignment shop nor does it *need* to be done.

Unibody vehicles that have been damaged and repaired using these dedicated systems, that refer to the true body centerline, are repaired using the same tolerance limitations that were used when the vehicle was built. After the body (frame) damage has been corrected, the vehicle is usually aligned referencing to the geometric centerline because it is the true motion or running alignment. If a unibody vehicle can easily be aligned, it is reasonably safe to assume that it is within the safe limits of all tolerances.

An automobile with unibody design, front wheel drive, and a

2-13

SECTION 2
SUSPENSION
General Information

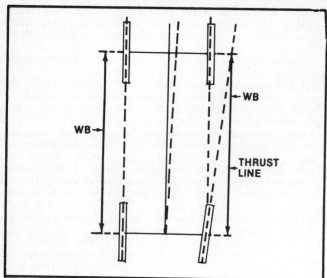

The body centerline must be known before the toe can be properly set and the thrustline established

solid rear axle may have any of the problems described in this section. These problems can only be found using an alignment machine with advanced capabilities.

A unibody automobile with independent rear suspension may also have some additional problems not previously discussed. For example, if one wheel were toed out, where would the thrustline be? The thrustline would be at the midpoint of the angle between the centerline of the properly aligned wheel and the direction of the misaligned wheel. The same theory would apply if one wheel were toed in.

Misaligned unibody toe in the rear wheels may cause the thrustline to deviate from the centerline. How is the variance measured? Deviation of the thrustline relative to the centerline, measured in degrees, equals one-half of the net difference in the toe of the two wheels. Three possibilities exist: (1) one rear wheel toed in or out with the other having zero toe, (2) both wheels toed in or out, and (3) one wheel toed in with the other toed out. The thrustline deviation can be stated in somewhat more specific terms. The toe of one rear wheel minus the toe of the other rear wheel equals the net toe of the rear wheel assembly. Dividing this net toe by two gives the average net toe of the rear wheel assembly and is the degree of deviation of the thrustline from the centerline. For example, if the left rear wheel is toed out by 3° and the right rear wheel is toed out by 1°, the net toe of the two rear wheels is 2° (3° − 1° = 2°) or an average of 1° toe out for the two wheels of the rear wheel assembly (2° ÷ 2° = 1°).

If the right rear wheel were toed in the same direction as the left rear wheel (that is, left rear wheel toed out at 3° and right rear wheel toed in at 1°), the two angles would be added together and then divided by two to find the thrustline deviation. In this case, the net toe would be 4° (3° + 1° = 4°) and the deviation of the thrustline would be 2° (4° ÷ 2° = 2°). In the latter example, the entire rear assembly would tend to steer to the left, and as with rear steering vehicles (lift trucks) the automobile would steer to the right. The driver would have to correct by steering the front wheels toward the left the same number of degrees as the thrustline deviation.

An automobile with faulty rear wheel alignment will tend to steer in a direction opposite to the direction of the net toe. For example, if the net toe is toward the left (left rear wheel toed out or right rear wheel toed in) the automobile will steer toward the right. If the automobile is steered in a direction opposite to the net toe, and the deviated thrustline, the automobile will tend to oversteer and the rear of the automobile may tend to spin outward. If the automobile is steered to the same direction of the net toe, and the deviated thrustline, the automobile will tend to understeer. The above conditions become more evident if the automobile is driven on wet, loose, or slippery surfaces.

If both rear wheels are toed in or toed out by exactly the same degree, thrustline deviation will be zero and the automobile will steer straight. However, the rear tires will be subjected to excessive scuffing wear, and the automobile will display instability in turns and on uneven road surfaces.

A rear wheel that is set back but is parallel to the body centerline may or may not materially affect the way the automobile drives, depending upon the severity of this condition. If the condition is pronounced, the automobile may tend to pull to the side of the shorter wheelbase. Nonetheless, the automobile is indeed out of square.

There are numerous ways in which the wheels of an independent rear suspension might be misaligned, and proper equipment is essential to the diagnostic process.

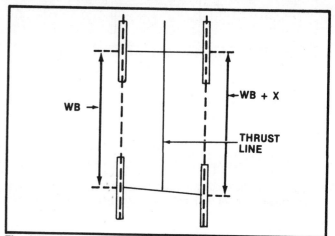

The right rear wheel is set back but is still parallel to the body centerline

Many automobiles with unibody designs provide few suspension adjustments. In the past, these automobiles were difficult to align. Now, however, many automobile and parts manufacturers have made available offset bushing and various parts kits to aid in performing normal alignment adjustments. But if the alignment procedure reveals that a unibody automobile has a bend in the structure, the automobile must be sent to a frame or body shop with proper equipment. Older frame-rack machines simply cannot detect body misalignments with any degree of accuracy because they were never designed to find the proper references on unibodies. Machines designed especially for unitized and unibody collision repair work, however, are available and can help restore the structure to a condition equal to the original specifications.

Today, there is no reason to not align a vehicle properly. Most new alignment machines will immediately recognize the thrustline. These machines have the logic build in to know how to align the rear wheels to the midpoint of the front suspension and then can use this as the reference for the balance of the alignment. Before you invest in alignment equipment, you should be sure it has the capability of aligning *all* angles referencing to the thrustline.

The Suspension System
CAMBER
Dual Arm Suspension

Camber is the tilt of the wheel in or out at the top as viewed

2-14

SUSPENSION
General Information

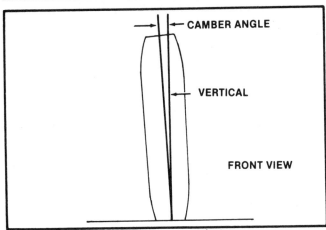

Camber is the tilt of the wheel in or out at the top

from the front and is measured in degrees from true vertical. When set at zero degrees of camber, the centerline of a wheel is at a 90° angle to the road surface. A wheel that is tilted out at the top is said to have positive camber. A wheel that is tilted in at the top is said to have negative camber. Camber has a great effect on the amount of tire footprint on the road surface. Properly adjusted camber will maximize the footprint on the road surface when the vehicle is in motion under the most conditions. Misaligned camber is a cause of uneven tire wear.

Camber does not remain the same in a steered position as it is when the wheels are set for straight ahead. Many vehicles use high caster settings, which causes a change in camber when the wheels are steered. (Caster is the forward or rearward angle formed between true vertical and a line drawn through the steering pivot points to the road surface.) As a vehicle is steered, positive caster creates positive camber on the inside wheel and negative camber on the outside wheel. Negative caster settings have the opposite effect. The sharper the vehicle is steered, the greater the effect caste will have on camber.

Steering axis inclination also has an effect on camber when the wheels are steered. (Steering axis inclination, or SAI, is the inward tilt of the upper part of the suspension and a line drawn through the steering pivot points to the road surface.) Some front wheel drive vehicles have a very high degree of SAI to reduce torque pull in the steering upon acceleration. The greater the degree of SAI, the greater will be the effect of SAI on camber when the wheels are steered.

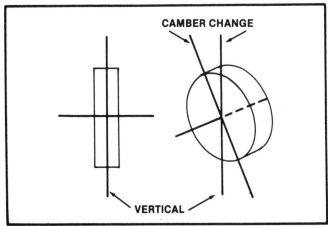

Positive or negative caster causes a change in camber when the wheels are steered

With some model vehicles (front and rear wheel drive), because of the effects of caster and SAI, on camber, just the slightest amount of wheel steer will greatly affect camber. This condition is called *camber roll* and is the reason why camber *must* be aligned referencing to the motion angle, the thrustline. Unless the alignment machine has the logic to align camber referencing to the thrustline, it would be possible to have uneven tire wear that the alignment machine could not detect.

The lengths and positions of the upper and lower control arms cause camber to become negative as the spring is compressed. As the spring is expanded, camber changes to zero or slightly negative. Because of inertia, the body of an automobile will lean to the outside of a curve, thus compressing the outside spring. This is called weight transfer. Consequently, the forces that create side stress problems are also the forces that reasonably solve them. Changes in outside camber counteract side stress problems created in turns.

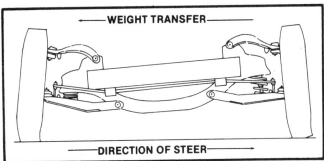

In hard cornering, because of inertia, weight shifts to the outside wheel causing that wheel to have negative camber. The negative camber on the outside wheel partially counteracts the side stress forces

Many alignment mechanics have not considered camber to be a directional control angle, but it is; a simple illustration shows why this is so. A cone shaped object will tend to roll in a circle, and so will a tire if forced to roll in a conical position. Thus an automobile with negative camber on the right side and positive camber on the left side will tend to pull to the left if this condition is severe enough, assuming the road is flat and all other alignment angles are set properly.

Another simple example demonstrates how seriously camber affects directional control. The left sides of the tires in the illustration of misaligned camber are smaller in diameter and are also buckled, thus creating a greater rolling resistance on the left. The right sides of the tires are larger in diameter and are more free-rolling. This forced conical position of the tires creates a pulling condition that would become more severe if the brakes were

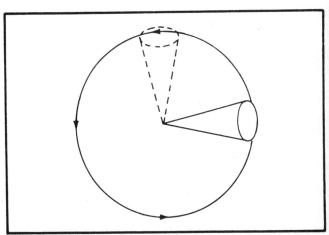

A cone will roll in a circle

SUSPENSION
General Information

applied. Thus, misaligned camber not only causes uneven tire wear—it can also be a serious safety hazard. Uneven tire wear, of course, is expensive, but an automobile that loses stability when the brakes are applied is dangerous.

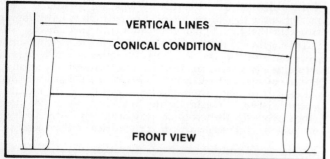

A badly misaligned camber can create a conical condition

Each vehicle manufacturer has established the proper static camber setting for maximum tire wear; this specification considers vehicle loads. Many specifications also call for less camber on the right side, to counter the effect of road crown. Due to the different tolerance specifications used in the manufacturing of different automobiles, camber specifications will vary from one automobile to another. There is no single camber specification that can be used successfully on all automobiles. For a proper camber adjustment, each wheel must be brought to the exact position of the thrust line before camber corrections or adjustments are made, and it probably will be necessary to reset the wheel to the thrust line position if a great amount of camber adjustment is made during the procedure. Altering camber specifications is not recommended.

STEERING AXIS INCLINATION

Dual Arm Construction

The next angle to be examined is steering axis inclination (SAI), also referred to as kingpin inclination (KPI) or balljoint inclination. Steering axis inclination is viewed from the front and is found by drawing a line (also called the pivot line) through the centers of the upper and lower balljoints and extending the line to the road surface. The degrees of difference from the true vertical to the pivot line is the SAI. Steering axis inclination is created by moving the upper balljoint closer to the center of the automobile relative to the lower balljoint. (No consideration is given at this point to the position of this balljoint relative to the front or rear of the automobile; this is called caster, and is covered in the next section.)

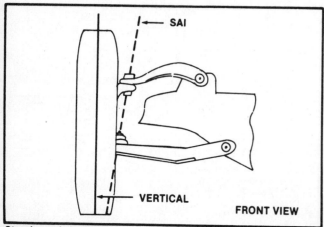

Steering axis inclination

When an automobile is steered, SAI produces an equal downward force on both front wheels. To see why this is so, look at the tip of the wheel spindle in the illustration. Because of the tilt of the steering knuckle and spindle, the spindle moves through an arc.

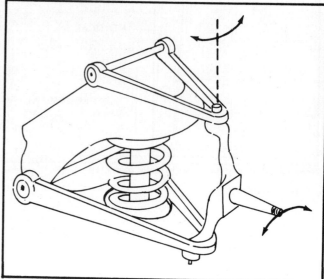

When an automobile is steered in either direction, SAI produces an equal downward force on the wheels

The highest point of the arc is when the spindle is centered—that is, when the wheels are straight ahead. If the spindle is turned from this position, it moves through its arc, resulting in a downward force on the wheel attached to it. Since the wheels cannot be forced into the road, the car is raised; it returns to its original lower position when the wheels are straightened out. The force of gravity causes the automobile to go straight ahead unless forced or steered in another direction. When the steering wheel is released after a turn, the weight of the car will make the automobile return to the straight ahead position.

If the inclination angle is less on one side than on the other, the automobile will pull to the lesser side. The automobile will steer easier to the side of least inclination, and if the condition is severe enough, the steering wheel may have to be helped to the straight ahead position.

On some earlier independent suspension systems, a large amount of SAI was used and hard steering resulted. Steering also returned to straight ahead position quite readily. This was offset by the use of negative caster. Thus, it is possible to set an automobile with negative caster, and have it return to straight ahead without wandering. Ideally, a properly engineered SAI angle will give the wheels a tendency to return to the straight ahead position, giving stability and directional control, but without creating hard steering.

The introduction of SAI not only aided directional stability of the automobile—it also permitted the use of four wheel brakes. If the upper balljoint were directly over the lower balljoint, the pivot to the road surface would be vertical. If the automobile were in a static position and the wheels were steered, they could rotate around the pivot only if they were free to roll. If brakes were applied, the wheels could not roll, making it impossible for the wheels to rotate around the pivot. The same principle would apply if the automobile were in motion. The harder the braking action, the harder the automobile would steer. Further, irregular road conditions or unequal braking would cause serious pulling conditions.

Earlier engineers tilted the top of the kingpin inward so that the centerline of the pivot projected to the contact point of the wheel on the road surface. This new angle was termed "kingpin inclination," and it eliminated the need for the wheel to be free to roll as it was steered.

SUSPENSION
General Information

Another component of the SAI angle is scrub radius. Scrub radius is the distance between the pivot line and the centerline of the wheel at the tire's contact point with the road. Only one pivot point exists in the tire footprint. The rest of the tire footprint must rotate around this point as the wheels are steered. At first, engineers believed this point should be as close to the centerline of the tire as possible. Later, however, they discovered that a pivot point in the exact center of the tire caused a scrubbing action in opposite directions within the same tire footprint. That is, when the wheels were steered, tire surfaces were forced to scrub one direction on one side of the pivot and in the opposite direction on the other side. This scrubbing action is called squirm and it causes a slight instability in turns.

The remedy for squirm is to bring the pivot point to the inside edge of the tire; this causes the scrubbing action to be in one direction only and relieves tire squirm. The result is better stability while turning with no appreciable difference in tire wear. Front wheel drive vehicles have used a different principle in scrub radius. Be sure to read the "MacPherson Strut Suspension" section and see why this principle has changed.

As we have seen, camber is the inward or outward tilt of the wheel at the top, and steering axis inclination is the inward tilt of the steering pivot line. The arithmetic sum of camber and SAI is

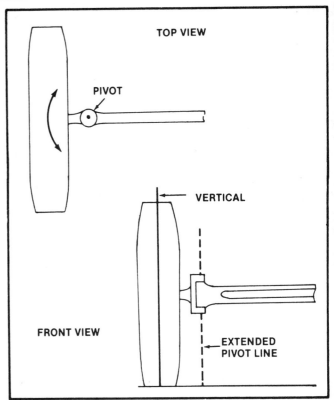

Without SAI the wheel is forced to rotate around the pivot as it is steered. The extended pivot line does not point to the contact point of the wheel on the road surface

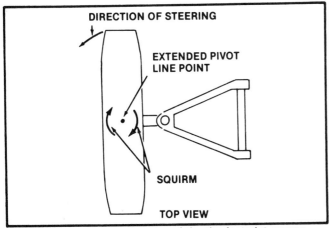

A pivot point in the exact center of the tire footprint causes scrubbing action in opposite directions when the wheels are steered. This is called squirm

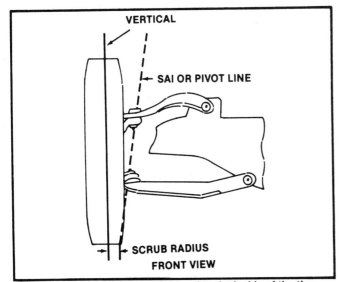

When the extended pivot line is moved to the inside of the tire footprint, the distance from the center line is the scrub radius

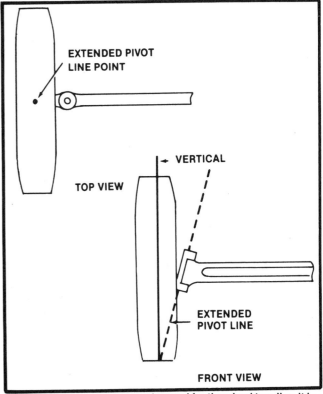

Kingpin inclination eliminates the need for the wheel to roll as it is steered. The wheel can be steered in place

2–17

SUSPENSION
General Information

called the included angle. Included angle, like caster and SAI, is referenced from the true vertical. It is a nonadjustable angle that cannot change unless the spindle assembly has been damaged. By determining the camber and SAI, the mechanic may then use the included angle to determine if suspension parts have been bent.

In the example shown here, the SAI is at 8° and camber at 0°, with an included angle of 8° (8° plus 0° equals 8°). If the camber angle were adjusted to +2° (positive), the SAI angle would become 6°. It makes no difference how this adjustment is made. The top balljoint must be moved out 2° or the bottom balljoint in 2° to change the camber to +2°. Since SAI as well as camber and the included angle are referenced from the true vertical, the SAI has been changed to 6°. The included angle did not change (6° plus 2° equals 8°). Camber was the intended adjustment, but in the process, the SAI angle also changed.

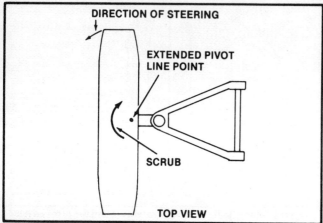
Scrubbing in one direction only relieves tire squirm

There has been some confusion as to whether the SAI angle is adjustable or not, as a result of the common misconception that the angle of the spindle assembly is the reference for the SAI. It is not. (There is an exception to this, please be sure to read the section "MacPherson Strut Suspension".) The part of the spindle that connects the upper and lower balljoints is referenced to the axle angle. This angle, the included angle, is manufactured into the spindle assembly and is not adjustable. If the included angle is referenced to the true vertical, the SAI and camber angles can be found. The line that is at a 90° angle to the axle is the centerline of the wheel. This line is measured against the true vertical to determine the degree of camber.

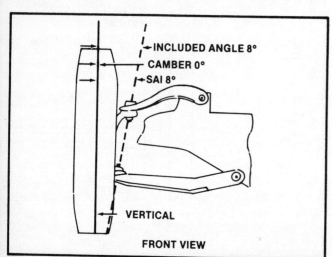
The arithmetic sum of camber and SAI equals the included angle

The line drawn through the portion of the spindle that connects the balljoints (the pivot line) is referenced to the true vertical and is the SAI. Camber plus SAI equals the nonadjustable angle - the included angle. (When camber is negative, it is subtracted from the SAI angle to find the included angle.) The SAI angle *does* change when camber is adjusted. Therefore, the SAI angle should be read only after the camber adjustment has been made.

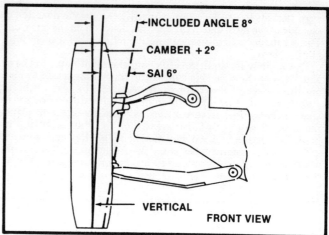
When the camber is adjusted, SAI also changes

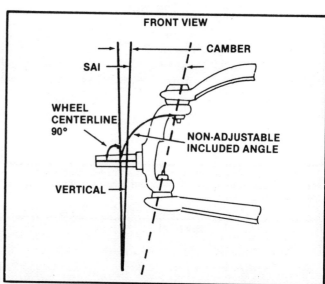
Camber plus SAI equals the non-adjustable included angle

Knowing and understanding SAI, camber, and included angle can help the mechanic to quickly determine if there is a bent part and where it is located. For example, the factory specification requires an SAI of 6° and camber of +2°, the included angle will be 8°. If these specifications can be met through normal adjustment procedures, the spindle assembly, upper control arm and lower control arm are not bent. However, let's examine some cases where there are bent or damaged parts using the above specifications.

Problem One: Assume the SAI is 6° and the camber is 0°. SAI of 6° plus camber of 0° equal an included angle of 6°. Using the above specification of 8°, this is incorrect. If camber were adjusted out to +2°, the SAI would be reduced to 4°. SAI of 4° plus camber of +2° equal an included angle of 6°—still incorrect. If the included angle specification cannot be met, the spindle is bent.

SUSPENSION
General Information

REFERENCE CHART

Angles			Problem
SAI	Camber	Included angle	
Correct	Less than specification	Less than specification	Bent spindle
Less than specification	Greater than specification	Correct	Bent lower control arm
Greater than specification	Less than specification	Correct	Bent upper control arm
Less than specification	Greater than specification	Greater than specification	Bent lower control arm and spindle

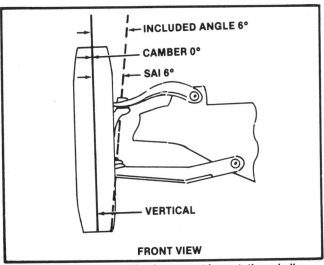

If the included angle specification cannot be met, the spindle is bent

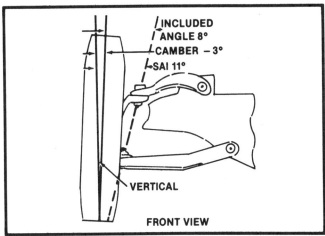

If the camber is negative, it is subtracted from the SAI angle to find the included angle

Problem Two: Assume an SAI of 4° and a camber of +4°. The included angle is correct, indicating that the spindle is not bent, but SAI and camber are incorrect. Suppose that camber cannot be adjusted by more than 1° toward negative. The SAI would then be 5° and the camber + 3°. Included angle remains correct, but SAI and camber are still incorrect. If the SAI is less than the specification and the included angle is correct, the lower control arm is bent.

Problem Three: Suppose another automobile has an SAI of 8° and camber of 0° and camber can be adjusted to only + 1°. SAI would then be 7°. The included angle is correct, but SAI and camber are incorrect. If the SAI is greater than the specification and the included angle is correct, the upper control arm is bent.

If the included angle is incorrect, the spindle is bent. If the SAI is also incorrect, both the spindle and another part are bent. Parts usually do not stretch when bent, so something is probably shorter.

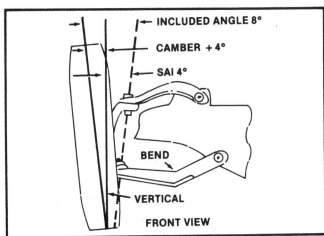

If, after maximum adjustment, camber is greater than the specification, SAI is less than the specification and the included angle is correct, the lower control arm is bent

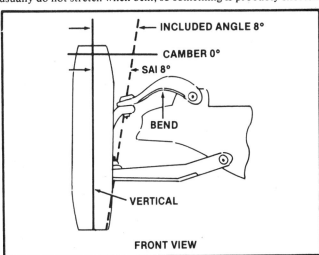

If, after a maximum adjustment, camber is less than the specification, SAI is greater than the specification, and the included angle is correct, the upper control arm is bent

SECTION 2
SUSPENSION
General Information

An SAI greater than or less than the manufacturer's specification indicates where to look for damage. This does not take into account any bent frame rails or crossmembers.

CASTER

Dual Arm Construction

Caster is referenced from the side and is the angle formed by drawing a line through the upper and lower balljoints and extending this line to the road surface. Caster angle is measured in degrees from true vertical. If the upper balljoint is located more toward the rear of the automobile than the lower balljoint, caster is positive. If the upper balljoint is ahead of the lower balljoint, caster is negative.

Caster is not considered to be a wear angle, but is used to aid directional stability, determine the amount of effort required to turn the wheels from the straight ahead position, and assist in the returning action of the wheels to the straight ahead position.

To better understand the effect of caster angle on the steering of an automobile, consider the action of a caster wheel on a tool cabinet. If the caster wheel is permitted to pivot freely, the contact point is always behind the pivot point. By reversing the direction of the cabinet, the caster will swing around behind the pivot point.

In the accompanying illustration, line A has been drawn through the pivot to the floor. Another line, B, has been drawn through the axle of the wheel to the point where the pivot line touches the floor. Measuring in degrees from line B to the true vertical indicates the degree of positive caster.

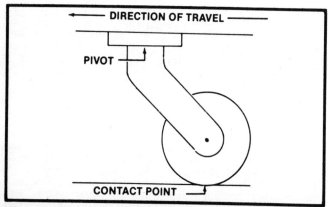
When permitted, the caster will always swing to the rear of the pivot

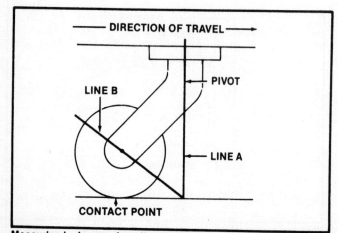
Measuring in degrees from line B to line A indicates the degree of caster

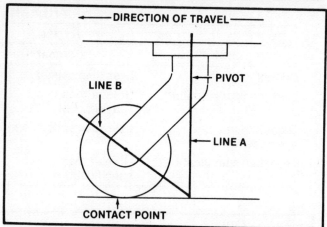
Locking the caster and changing the movement of the cabinet would create instability

If the caster were locked so that it could not swing and the direction of movement of the cabinet were again reversed, the cabinet would not be directionally stable. Although the caster could not swing, it would want to do so. This illustrates negative caster and suggests the possible effect of negative caster when applied to an automobile.

If the caster fork were replaced by balljoints, line A would be drawn through the balljoints to find the angle of the pivot. In this case, the contact point is behind the pivot point showing positive caster. To determine the degree, measure from the true vertical.

If you could see down through the top of a tire, you would see the tire footprint on the road surface and also the part of the footprint to which the pivot line is pointing. In the case illustrated, the pivot point is exactly at the center of the tire footprint, indicating zero (0) degrees caster. This means the footprint of the tire on the road surface is divided evenly. The wheel would have no tendency to turn in either direction, because zero caster contributes nothing to directional stability.

When caster is positive, the extended pivot point is ahead of most of the tire footprint. The portion of the tire footprint to the inside of the pivot point equals the portion to the outside. The remainder of the footprint (the dark area in the figure) must "drag" while being steered. Most of the drag or friction is behind the pivot point and is the tire force that makes the wheel return to the straight ahead position.

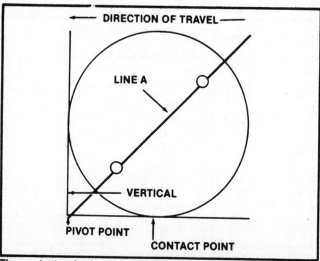
The contact point is behind the pivot point showing positive caster

2-20

SUSPENSION
General Information

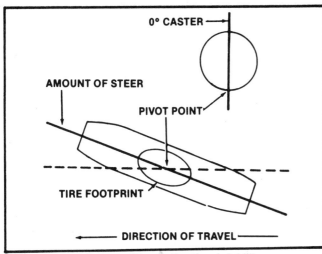

Zero caster contributes nothing to directional stability

Negative caster is illustrated when the pivot point is to the rear of the footprint. Again, the portion of the inside of the footprint with respect to the pivot point is equal to the portion to the outside. In the case of negative caster, most of the drag is ahead of the pivot point. Thus, when the automobile is steered, friction tends to turn the wheel farther from the straight ahead position.

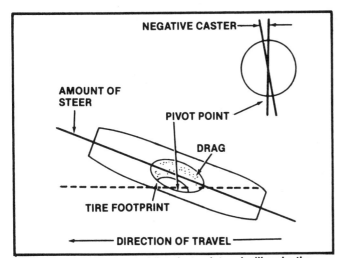

Most of the drag is ahead of the pivot point and will make the wheel tend to steer farther

If the tire forces were the only forces involved in caster and directional stability, any error in caster balance from side-to-side would be revealed while turning. However, other forces are involved. When the automobile is steered, positive caster forces the spindle, or axle, down on the inside wheel, which raises that side of the automobile. On the outside wheel the spindle is raised, which lowers that side of the automobile. When the steering wheel is released, and the caster is equal on both sides, gravity will tend to equalize the weight of the automobile on the front wheels, forcing the wheels to return to the straight ahead position.

If an automobile is set with negative caster, the inside spindle will rise and permit that side of the automobile to lower. The outside spindle will lower and raise that side of the automobile. With a negative caster setting, the "drag" of the tire footprint is ahead of the pivot. If this condition were severe enough, the "drag" forces of the tire footprint could overcome the weight equalization and SAI forces, and the wheels would not tend to return to the straight ahead position. This would be an extreme condition but helps to illustrate the next point. If an automobile had positive caster on

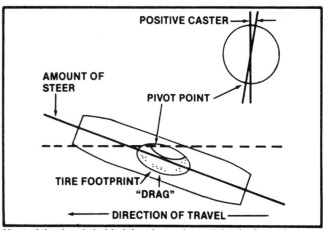

Most of the drag is behind the pivot point and the tire force that makes the wheel return to the straight ahead position

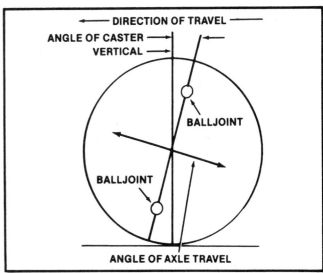

Gravity will tend to equalize the weight of the automobile forcing the front wheels to return to the straight ahead position

one side and negative caster on the other, a pulling condition would exist. If the automobile were steered to the side of positive caster, both sides of the automobile would raise, as shown earlier. Because of gravity, the automobile would seek its lowest point. Both sides of the automobile would be lowered if the wheels were steered to the side with negative caster. The "drag" force of the negative side footprint (which is mostly ahead of the pivot point) would be stronger than the drag force on the positive side footprint (which is mostly behind the pivot point). If the size of the drag in the negative side footprint were equal in area to the size of the drag in the footprint on the positive side, the amount of drag on the negative side being pushed would be greater. The uneven forces, coupled with the tendency of the automobile to seek the lowest possible point, would cause the automobile to continue to pull.

A pulling condition may exist even without positive caster on one side and negative caster on the other. Assuming all other angles are set to proper specifications and the road surface is flat, an automobile will pull to the side that is set with the least positive caster. With an imbalance in caster settings, pulling will be constant and not greatly affected by braking.

High pressure tires were used on very early automobiles, resulting in a small tire footprint. There was very little tire drag, and these automobiles were very unstable at higher speeds. Large amounts of positive caster increased stability, but also created hard steering and increased road shock at the steering wheel (road shock

2–21

SUSPENSION
General Information

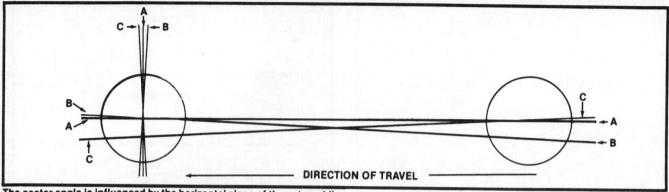
The caster angle is influenced by the horizontal plane of the automobile

caused by positive caster results from the pivot line being pointed at irregularities of the road). A considerable amount of shock was transferred directly through the steering mechanism to the steering wheel. High pressure tires were replaced by "balloon" tires that had much larger footprints. The increase in drag permitted a reduction in positive caster settings, resulting in easier steering, equal or better stability, and much less road shock.

The development of radial tires has provided a much larger footprint, which increases steering drag and stability, but since more road shock is absorbed by the flexible sidewalls of the radial tire, caster can be increased to more positive settings. Hydraulic power steering systems cushion road shock and have overcome hard steering associated with high positive caster, but at the same time, hydraulic power steering systems resist sudden reversals of direction. This offsets the increased force with which the wheels return to the straight ahead position as a result of increased positive caster and explains why some automobile models have very high positive caster settings with power steering but less positive caster without power steering.

Caster angle is also influenced by the horizontal plane of the automobile. Rear springs that sag will increase the caster angle toward positive. Front springs that sag will decrease caster toward negative.

Altering the height of the rear of an automobile by extending rear shackle lengths or adding rear air shock absorbers pumped all the way up can change the intended front suspension angles to the point of making the automobile dangerous to drive. Raising the rear of the automobile not only upsets the front end geometry, it also increases rear end sway and weakens every aspect of safety designed into the automobile. Automobiles used in competitive racing that have had suspension and steering systems altered have been totally redesigned by experts. Complete systems have been changed in harmony and are geometrically designed to handle safely at high speeds under specific conditions. These automobiles would not drive or ride well at normal speeds under all conditions.

Caster specifications are not normally altered and sometimes are not adjustable. However, the results are predictable if the angles are altered even slightly. If the caster settings are equally decreased to the negative on an automobile without power steering, the automobile will tend to steer easier but will also tend to wander at high speeds, and consequently will not be as stable. The automobile will be more affected by crosswinds. An older person who drives only in town might find the automobile easier to steer. Nonetheless, the reduction in caster is not recommended. Increased positive caster can improve high speed stability but it also can create some adverse conditions. For example, road shock increases and a greater work load is placed on the power steering system, potentially shortening its life. It is for these reasons that changes in caster settings from the factory specifications are not recommended.

The Steering System

TOE

Toe, viewed from the top, is the difference of the distance between the extreme fronts of the wheels and the distance between the extreme rears. If the distance is less between the wheels at the front than at the rear, the wheels are toed in. If the distance is greater, the wheels are toed out.

As far as tire wear is concerned, toe is the most critical of alignment adjustments. A wheel suffering from an incorrect toe setting will scuff along the road surface as it is forced to continue in the direction of travel. Any other adjustment made will affect toe, and thus toe is always the last alignment adjustment made.

In rear wheel drive automobiles, toe is set slightly inward while the car is in a stationary position, because toe changes as the automobile is driven. Rolling resistance of the tires, coupled with the rearward movement of the suspension, changes the setting of the steering linkage (and thus the toe) as the automobile moves forward. If the steering linkage is in the front of the suspension, it is expanded or pushed open. If the steering linkage is in the rear of the suspension, it is compressed. Either way, the toe opens when the automobile moves forward. The "running toe" (toe setting when the car is moving forward) will be zero when all the normal steering tolerances are taken up.

Front wheel drive automobiles have the same travel, but in the opposite way. Because the front wheels are pulling the automobile, they tend to pull inward toward each other. The specifications of these automobiles commonly call for toe-out in the static position.

The slightest amount of looseness in the steering linkage or suspension parts can create a very unfavorable effect on the "running

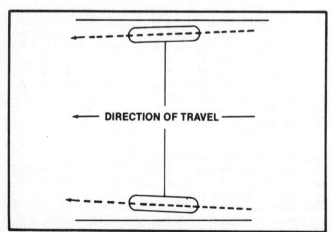
Toe is the distance between the extreme fronts of the wheels compared to the distance between the extreme rears

SUSPENSION
General Information

toe." Looseness will cause running toe-out even if a proper static setting has been made. Although too much running toe-in or running toe-out will have a serious effect on tire wear, even a small amount of running toe-out can cause the car to wander and will seriously affect directional stability while braking.

Early model automobiles used a great amount of positive camber. For many years it was thought that positive camber would give the wheels a tendency to turn out due to the forced conical position of the tire, so additional toe-in was specified. However, when the "balloon" tire was introduced and higher road speeds were common, tires began to wear unevenly, showing greater wear on the outsides. Research and tests proved that there is no relationship between the angles of camber and toe. They are entirely independent.

In older cars equipped with a single beam axle, the steering linkage was a simple affair that attached one end of a tie-rod to the arm that emerged from the steering box (the pitman arm) and the other end to a steering arm at one wheel. The other wheel was connected by means of another tie-rod that ran from its steering arm to the other tie-rod. As the automobile was steered, the pitman arm pushed or pulled the tie-rod steering the one wheel. Simultaneously, the other tie-rod was pushed or pulled by the first so that both wheels were steered. The system included an adjustment to set toe. As the wheels were forced up and down by road irregularities, the entire system went up and down together. This was very simple and worked well, but when independent suspensions appeared, steering geometry changed.

Early independent suspension designs exhibited a tendency for the front wheels to close as the frame of the car dropped due to road irregularities. As the automobile moved upward from going over a rise, the wheels opened. Any additional weight in the automobile would lower the frame, lengthen the tie-rods, and spread the wheels at the rear.

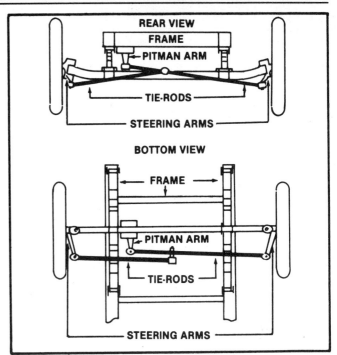

Single beam axles had a very simple steering system

In the modern automobile, tie-rods are nearly the same length as the lower control arm, and are positioned in the automobile at approximately the same angle as the lower control arms. If the frame is raised or lowered from any cause, the tie-rods assume the same position and angle as the lower control arms. Consequently, they do not change lengths (or the toe setting of the wheels). The inner ends of the tie-rods are attached to a center link (or, in a rack and pinion system, the steering rack), and only the center link (or rack) moves up and down with the frame.

Early independent front suspension experienced excessive changes in toe as the suspension was raised or lowered from load changes and road conditions

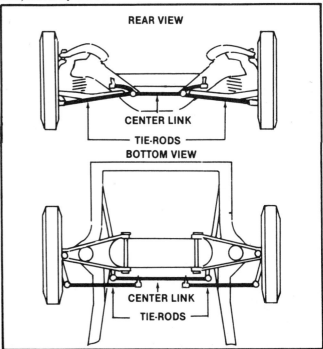

In modern automobiles, the tie rods are nearly the same length as the lower control arms, and are positioned at approximately the same angles as the lower control arms

2-23

SECTION 2: SUSPENSION
General Information

Although there are many different types of steering systems in use today, they all conform to the same geometric principles. Some steering linkages are in front of the suspensions and some are behind. In most cases, the location of the steering linkage is determined by the available space in the design of the car. Regardless of the location of the steering system, it will be built in harmony with the suspension system.

Some suspension and steering systems are intentionally built with greater tolerances. Consequently, no single static toe setting can be used on every automobile. It is extremely important that static toe be set exactly to the manufacturer's specification for the specific automobile being aligned. It is also important that toe be set using the thrust line as the reference. Any deviation in the thrust line from the centerline will cause toe-out on turns and affect the running toe.

Toe-Out On Turns

Zero toe is the ideal toe when driving straight ahead, but what happens to the toe when the automobile is steered? The wheels must toe-out as they are steered, and the farther the wheels are steered, the greater the toe-out must become. This change is called toe-out on turns. That is, when viewed from the top, the distance between the fronts of the two wheels is greater than the distance between the rears while the wheels are turned. The more sharply the wheels are turned, the greater the difference in the two distances. Toe-out is designed to prevent the tires from scuffing and to add stability in turns.

Although it might seem logical that zero toe would be the ideal setting under all conditions, a closer look at steering geometry demonstrates that this is not so. When an automobile is steered, both front wheels revolve around the same center. Regardless of the direction steered, the outside wheel steers in a larger arc than the inside wheel. The inside wheel is also ahead of the outside wheel and must steer more sharply to remain perpendicular to the radius. When the steering is returned to the straight ahead position, the wheels must return to the parallel position.

This is similar to the principle of a lever moving in a circle. In the illustration, the lever pivots at point A. If the other end moves from point B to point C, the lever has travelled half the distance along the horizontal line but not half the distance of the arc. If the lever is moved to point D, it has moved the other half of the horizontal line, but a much greater distance of the arc.

When we apply this principle to the automobile steering mechanism, we see (in the next figure) that both levers (steering arms) have travelled the same distance on the horizontal line, but because the levers are on different segments of their arc, the connecting lines (wheels) have turned at different angles. The centerline of the automobile and the mid-point of the rear axle determine the *theoretical* angle of the steering arms, which are not parallel to the

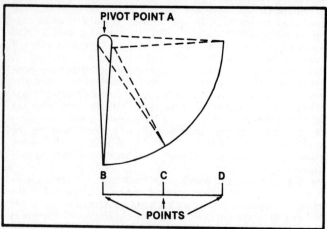
If the lever is moved to point C, the lever has travelled half the distance of the horizontal line, but not half the distance of the arc

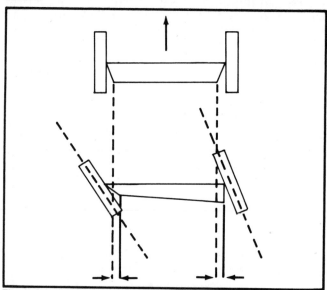
The levers (steering arms) are in different segments of their arcs demonstrating different turning angles

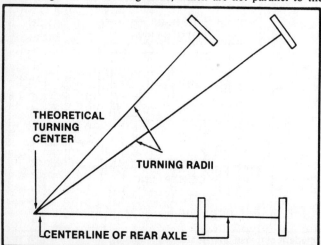
Toe out on turns prevents the tires from scuffing and adds stability when the automobile is steered

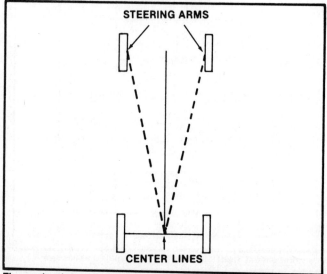
The angle of the steering arms is theoretically determined by the wheelbase of the automobile

SUSPENSION
General Information

front wheels. Each is angled inward, pointing directly to the center of the rear axle, and toe-out on turns becomes automatic as a result of this design feature.

Another consideration in determining the angle of the steering arms is the slip angle. As the speed of the automobile increases, centrifugal force acts upon the automobile and causes all the tires to slip somewhat. The tires are not actually slipping, but rather are twisting at the point of contact. Because of this twisting condition, the actual turning center is considerably ahead of the theoretical turning center, and the difference between centers is referred to as the slip angle.

If the front slip angle is greater than the rear slip angle, understeer exists. With an understeer condition, the steering wheel must be progressively turned inward to maintain the same turning radius. The greater the speed, the worse the condition. If the rear slip angle is greater than the front slip angle, oversteer exists. With an oversteer condition, the rear of the automobile will tend to slide to the outside of the turn at high speeds, and the steering wheel will have to be turned to the outside of the turn to prevent the automobile from spinning around. Equal front and rear slip angles create neutral steer, which while desirable, is rarely attained. Varying speeds, varying amounts of fuel in the tank, and changes in number and location of passengers in the automobile all contribute to the changes in slip angle. An automobile decelerating into a turn may have understeer that changes to neutral steer as power is lightly applied through the turn. Steering characteristics may further change to oversteer as the automobile comes out of the turn with still more power applied. Modern automobiles with normal alignment settings, proper tire sizes, and proper tire pressures may display no perceptible symptoms of understeer or oversteer at normal road speeds. Nevertheless, varying degrees of understeer and oversteer will occur, depending on variations in speed, weight distribution, and road conditions. An automobile with two underinflated front tires or two overinflated rear tires would produce an understeer condition, perhaps noticeable at normal road speeds. If the unequal tire inflation were reversed, the automobile would experience oversteer. Thus, proper tire size and inflation are critical to the handling ability originally engineered into the automobile.

Many other factors are involved in determining slip angles—engine size (power-to-weight ratio), tire size, tire pressures, suspension geometry, firmness of the suspension, stabilizer bars, weight distribution, weight transfer, center of gravity and the roll center are all considered in determining proper angles for the steering arms.

Different automobiles have different slip angles. Consequently, different automobiles with the same length wheelbase may have various toe-out-on-turns specifications. Use of the specification set forth by the manufacturer for the specific automobile being checked is important.

Toe-out on turns is a nonadjustable angle. The steering arms are installed with preset manufactured angles. If the toe-out on turns specification cannot be met, one or both steering arms is bent.

This is a critical wear angle which should be checked in every alignment procedure. An incorrect toe-out on turns will adversely affect tire wear and stability while steering.

MacPherson Strut Suspension

CAMBER

Under normal driving conditions, the wheels of the strut type suspension will remain to the desired camber settings. When the spring is greatly compressed, camber will go slightly negative; when the spring is greatly expanded, the camber will go slightly positive. The longer the control arm, the less camber change there will be. These camber changes occur only under extreme conditions and actually improve handling and driving characteristics. Most strut suspensions respond in this manner.

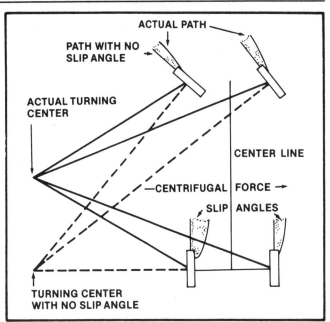

The actual turning center is considerably ahead of the theoretical turning center

Some vehicles provide for camber adjustments and some do not. With those that do provide adjustments, you will find the maximum adjustment that can be made is slight; usually 1° or less. This is due to the long length of the strut. If you were to pivot a one foot lever on one end, the lever would become the radius of a two foot circle. A three foot lever would become the radius of a six foot circle. If you were to move the outer end of the one foot lever ½" laterally, the movement in degrees within the two foot circle would be greater than moving two foot lever ½" within the six foot circle. This demonstrates why the strut suspension adjustments respond to a lesser degree than those made in a dual arm suspension. The length of the spindle in a dual arm suspension is approximately $1/3$ the length of the strut in a strut suspension. A $1/4$" of adjustment in a dual arm suspension will yield a greater degree change than a $1/4$" adjustment in a strut suspension.

On vehicles that do not provide adjustments, there are offset bushing and shim kits available. Again, these kits will give only a slight adjustment and for the same reason as above. You must always remember, *the unibody is the frame.* If the faulty alignment requires more adjustment than either of these adjustments can provide, look for something to be either bent or badly worn.

STEERING AXIS INCLINATION

Steering axis inclination is the major directional control angle. This angle becomes more important in unibody vehicles with MacPherson strut suspensions and front wheel drive. It gives these lighter vehicles more stability and is the primary angle to reduce torque steer. Torque steer is the tendency of a vehicle to involuntarily steer to one side or the other upon acceleration of the vehicle. (There are several causes of torque steer; these problems will be addressed later in this section.) As we saw in the SAI section of dual arm suspensions, SAI causes both sides of the front of the vehicle to raise when the wheels are steered. The greater the degree of SAI, the higher the front of the vehicle will be raised when the wheels are steered. Thus, gravity aids in stability and helps to resist torque steer.

2-25

SUSPENSION
General Information

MACPHERSON STRUT SUSPENSIONS

The basic geometric angles used in the strut suspension—caster, camber, steering axis inclination, and included angle—are the same

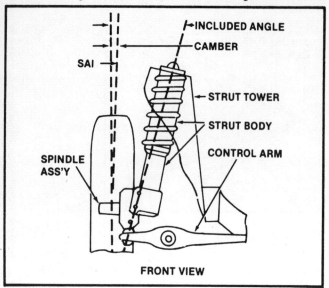

The same basic geometric angles are used in the strut suspension

The illustration of this strut type suspension shows that the angles of camber, SAI, and the included angle are the same as those in almost all other types of suspension. There are only three points that an adjustment can be made—the top of the strut, the balljoint, or the control arm bushing(s). Any camber adjustment made would change the camber, SAI, but *not the included angle*. This general rule has been an aid in diagnosis of and location of bent parts in suspension systems. Later in this section, you will see how this rule is applied.

The above illustration demonstrates a different type of strut suspension. It may appear the same in the illustration and in the vehicle, but it *is* different. It has an adjustment for camber within the strut. This change of adjustment location greatly affects the angles that will change when a camber adjustment is made. When camber is adjusted in this suspension, there is a change in camber, the included angle, *but not the SAI angle*. In this suspension, it is the SAI angle that is non-adjustable and is permanently fixed. From the introduction of the SAI angle in the Elliot axle until this type of suspension, the included angle was the permanently fixed angle. This suspension system voids the general rule! Since the distance from the adjustment to the pivot (the balljoint) is short, just a slight amount of adjustment will effect a great amount of camber changes. However, if you need to make a considerable amount of camber adjustment, look for something bent or badly worn. Generally, it is accepted that side-to-side difference in SAI should not exceed 1°. While a camber adjustment of more than 1° can be made with this type of suspension, you must be sure the side-to-side difference in SAI does not exceed the specification and that *both* sides are within the specification.

CASTER

Caster in the strut suspension is the same as it is in any other suspension. Where it is adjusted may be different. But the purposes and reasons for it are the same. Correct caster adjustment is somewhat more critical in front wheel drive vehicles because of torque steer. There are several causes of torque steer and these will be discussed in the torque steer part of this section.

The Steering System

The two angles—toe and toe-out on turns—are very critical tire wear angles. They become more important in front wheel drive strut suspension vehicles. The front wheels *steer* the vehicle and they *propel* the vehicle as well. Consequently, the tires on the front wheels are subjected to greater wear. The slightest amount of misalignment in the entire steering system can result in a great amount of uneven tire wear. It is extremely important that the complete steering system be aligned using the thrustline method. Toe-out on turns must be checked with every alignment. Misaligned toe in the rear wheels can show peculiar wear patterns in the tires of front wheel drive vehicles. Because of the lighter weight in the rear of these vehicles, and if the misalignment is great, the wear will appear as 45° scalloping across the entire tread of the tire in evenly spaced intervals. The vehicle is actually hopping. If the misalignment is severe enough, the driver will feel a sensation similar to that of a rear wheel being out of balance. We saw in the thrustline section that the thrustline is the direction of travel. In some front wheel drive vehicles, the rear of the vehicle is too light to control direction. The rear wheels are being pulled in one direction while having a constant tendency to roll in another. This constant hopping and tendency to roll in another direction, causes severe instability and is downright dangerous on wet or slick road.

TORQUE STEER

The basic definition of torque steer is the action of the driving wheels affecting steering performance during acceleration, causing a lateral deviation of direction. Stated in more simple terms: The vehicle will involuntarily pull to one side or the other when accelerated. A certain amount of torque steer may *normally* be experienced during heavy throttle application on some front wheel drive vehicles. This is due to one axle (half shaft) being longer than the other which causes a difference in the axle angles and this difference creates more torque toe-in effect to one front wheel, especially on manual transmission equipped vehicles.

Some front wheel drive vehicles have an excessive amount of torque steer. The most common of all causes of torque steer can be found in the tire and wheel assemblies. Tire and wheel assemblies have the most significant effect on torque steer. A slightly smaller diameter tire will cause a pull to that side. Inspect the front tires for differences in size, band, or construction. If you suspect the cause to be tires, change the front wheels and tires *side to side* and retest the vehicle. If the vehicle then goes straight or pulls to the other direction, you have isolated the problem. Be sure to check the front tires for equal pressure.

In addition to tire and wheel assemblies, there are three basic causes of excessive torque steer: 1. mechanical conditions. 2. collision damage. 3. improper collision damage repair.

Mechanical conditions could be any of the following: Looseness in the control arm bushing(s), strut rod bushing or mounting, (if so equipped) steering gear mounting, rack & pinion bushings, anything that will permit one wheel to pull forward or that will otherwise cause a change in toe upon acceleration. A loose suspension or steering part can also cause a pull to the opposite side of deceleration. A binding or tight drive axle joint can cause a pull. If new springs have been installed, a temporarily high suspension can change the drive axle angles and cause a pull. High suspension can also cause a wobble at low speeds, however, these symptoms will disappear as mileage is increased and new springs settled in. Incorrect, worn or loose engine mounts can change drive axle angles. Look for anything that can cause a change in drive axle angles or toe to change, either can have an adverse effect on the steering geometry. As you have seen, failure to align the front wheels referenced to

SUSPENSION
General Information

the thrustline, can cause a change in both camber and toe when the car is given motion. This can cause torque steer.

Collision damage is obvious, however, the damage may be visually slight and thought to be of no consequence. The damage may be hidden. Check for a bent or misaligned cradle (the assembly that supports the engine and driveline components between the side rails). Are the side rails themselves bent or out of alignment? Is any part of the suspension and steering system bent? A difference in side-to-side caster can cause a constant pull but is accentuated upon acceleration. Front wheel set back can cause a pulling condition, however, some vehicles have a built in setback condition. If caster is equal on both sides, and the sheet metal is fine, the setback may have been built in to help resist torque steer. If caster is unequal and setback is found, there is something shifted or bent. The cradle may be back on the side with the least amount of caster or a suspension part is bent. Measure side-to-side wheelbase, compare these measurements to the specifications; this can help you to determine the problem. Determine if the wheels must be steered a great amount to be able to be aligned to the thrustline. If they do, something is twisted in the front. Can the rear wheels be easily aligned to the midpoint of the front suspension? If not, something is twisted or bent in the rear. Minor caster adjustments can be made to correct slight caster differences. You can slot the top of the spring tower if no other adjustment can be made for a minor caster change. The point is: if you can make minor adjustments to correct for damage it may be fine, however, if major changes must be made to correct for torque steer, the vehicle should be sent to a shop equipped to make such corrections.

Improper collision repair is basically checked the same as collision damage. One of the major problems could be that all alignment angles are correct, but the vehicle has a bad torque steer condition. One of the more common errors is that the cradle is still misaligned, but the outer end of the control arm was brought ahead, by whatever means, to compensate for the short wheelbase on that side. This could also bring the caster back to correct, *but,* the axle angles are incorrect which is a cause of torque steer. Look for a misaligned rack and pinion and engine mounts. Misaligned side rail supports can cause problems. Check for wrong length halfshafts. Many repair shops will use substandard parts. Universal "fits-all" type halfshafts may fit, but are not the exact length they should be; this can cause torque steer.

SCRUB RADIUS

Some front wheel drive vehicles, with four wheel strut suspension systems, have the scrub radius entirely to the inside of the tire. The extended pivot line points to the extreme outside edge of the tire. The reason for this is the completely different way these vehicles were built and the problems that followed. Because of torque steer, the SAI angle had to be increased considerably. To do this, the top of the strut towers had to be brought in toward the center of the vehicle which cramped the engine compartment, especially with transverse mounted engines. By extending the pivot line to the outside edge of the front tires, it permitted the distance between the strut towers to be greater thus somewhat relieving the cramping of the engine area.

Bringing the scrub radius to the inside solved more than one problem. A more important one is stability while braking in the event of loss of $1/2$ of the brake system. Frame type automobiles are built to have approximately 60% of their weight on the front wheels and 40% on the rear wheels. The safety type brake system divides the master cylinder and uses one half to activate the front brakes and the other half to activate the rear brakes. In the event of a brake failure in one half of the system, either the front or the rear brakes could stop the vehicle with a minimum amount of instability.

Front wheel drive automobiles brought new problems. Approximately 85% of the weight of the automobile is on the front wheels and only 15% on the rear. If the rear half of the system failed, the front bakes could safely stop the automobile, however, if the front half of the system failed, the rear brakes could not safely stop the automobile. The cure for this problem was to split the brake system another way. One half of the brake system goes to the left front wheel and to the right rear while the other half of the system goes to the opposite two wheels. Engineers soon learned that having the scrub radius entirely to the outside, dangerous instability was present while braking with only one front wheel. The opposite side rear brake, with only 15% of the vehicle weight, could not offset the braking of the one front wheel and the imbalance. By bringing the scrub radius more toward the center of the automobile and to the inside of the extended pivot point, stability was greatly increased.

LOCATING BENT PARTS

Finding bent parts in the strut suspension may be a bit more difficult, however, using the alignment machine may be very helpful. The following information applies only to those strut suspensions that *do not* have a camber adjustment within the strut body itself.

If the SAI angle is greater than the specification, you can be reasonably certain that the strut tower and body is bent in this vehicle. If the SAI angle is less than the specification, it will be a bent control arm or whatever it is mounted to. You can determine if the strut is all right if the included angle specification can be met. It would not matter hat the SAI and camber readings would be, if an adjustment can be made so that the included angle specification can be met, the strut is not bent.

Steering axis inclination (SAI) is the major directional control angle. Unibody automobiles with MacPherson strut suspension have increased SAI for greater stability, which also relieves torque steer in front wheel drive automobiles with MacPherson struts. Steering axis inclination must be measured in every alignment procedure. If the SAI is correct but the camber is incorrect, then the included angle will also be incorrect, indicating that the spindle and/or strut body is bent. The unibody structure probably has not been damaged, however. When the SAI specification is correct, the mechanic can be reasonably certain that the unibody strut tower is undamaged. If the SAI is less than the specification and the included angle is correct, either the control arm is bent or the strut tower is out at the top. A bent control arm can be found easily. However, a shifted strut tower can be nearly impossible to find without the proper equipment. Relying on hood and fender alignment to determine if the strut tower is bent is not accurate. Even if a bend in the unibody structure is detected, the bend cannot be accurately corrected without a machine designed with this capability.

If the SAI is less than specification and the camber is correct, but the included angle is incorrect, any or all of these three possibilities exist: (1) a bent spindle and/or a bent strut body, (2) a bent control arm, or (3) a strut tower out of location at the top.

If the SAI is greater than the manufacturer's specification, the wheel likely has been hit high from the side, pushing the top of the strut tower inward. The included angle would be used to determine if the strut body or spindle has been bent.

The included angle is used to determine if there is a bent component in the suspension. The SAI is used to determine if the unibody is bent or twisted. A bent control arm will change the SAI, but is usually readily detectable. If the SAI remains incorrect after the control arm has been replaced, the strut tower probably has shifted. Worn parts also affect the readings.

Caster or camber must never be corrected by bending any part of the strut assembly! Bending may crack or structurally weaken the strut and may cause it to break at any time, potentially resulting in a serious accident. Any bent strut assembly must be replaced. If the misalignment is somewhat more severe (that is, affecting more than simply caster or camber), the alignment machine will help in locat-

2–27

SECTION 2: SUSPENSION
General Information

ing the bent part or parts. If the included angle specification cannot be met, a damaged spindle may be the culprit. However, in this case, the strut body may be bent as well. A bent strut body or bent spindle will change the included angle. In most cases, if the spindle is bent, the strut body will also be bent, and the entire strut unit must be replaced.

Locating bent parts in the strut suspension that *does* have a camber adjustment within the strut body itself can be done in the same manner as above, however, the key to this is the SAI angle. Since the SAI angle is permanently fixed, there is no need to adjust camber to determine if it is within specification. If the SAI angle is correct you can be reasonably sure that the strut tower and body are right. If the camber cannot be adjusted to within specification, the included angle specification could not be met. This would tell you that the strut and/or spindle is bent.

KNOW WHAT YOU ARE DOING

Unibody vehicles with front wheel drive and all strut suspension do have improved handling characteristics; they are more responsive in all respects. However, they are also less forgiving of any errors in repairs. Everything must be nearly perfect to retain the fine responsiveness originally built into the vehicle. The alignment mechanic should not be satisfied with anything less than preferred settings, or at the least, very close to them. Damage or mechanical problems should be corrected only by qualified technicians. Because the body is the frame, all metal in unibodies are tempered to a very precise degree, therefore, *any* welding done on these units *must* be done only by technicians that have had the proper training for just such welding. It is essential that the *problem* be corrected and not just compensated for. With these units, the alignment mechanic should not go beyond his limitations. To do so could result in some startling and serious consequences.

Rear Suspension

In the past, rear suspension systems were generally ignored. Few domestic cars had any means of adjustments. Unless a vehicle had a particular bothersome problem, the rear suspension was taken for granted as being all right. However, any suspension can have worn parts or be out of alignment, whether it has provisions for alignment adjustments or not. Both solid axle and independent rear suspensions can have worn or bent parts, resulting in misalignment.

New equipment makes parts checking and alignment of rear suspensions much easier and more accurate. Do not overlook the opportunity to increase your profits and customer satisfaction. The rear wheels are not subjected to steering stresses, nonetheless, rear wheel drive vehicles are subjected to tremendous torque and various other forces. Balljoints are not common in rear suspensions, but bushings, control arms and many other parts are. Independent rear suspensions with worn or bent parts will drive in a very strange and unpredictable manner. Torque steer is common in rear drive cars with faulty rear suspension parts.

Incorrect camber or toe is often found in solid rear axle front drive vehicles. Every part of any suspension is designed to have "give". Occasionally, after an encounter with a large bump or hole in the road, you will find misalignment. There are shim kits for this problem. The point is, don't assume that because it is a solid axle that it cannot get out of alignment; it can, and so can driving solid rear axles! Bent solid rear driving axles must be replaced.

Loose or bent suspension parts, in the rear of any vehicle, can cause misalignment and reduce the size of the footprint at the road surface on one or both rear wheels. This not only results in uneven tire wear, but more importantly, it jeopardizes the safety of the occupants. If one wheel has a reduced tire contact patch it will tend to lock up or hop under strong braking. It will skid more readily under other adverse conditions. Unless the loose parts are found, this wheel may show as being in perfect alignment when it is in a static condition. The primary purpose of alignment is to ensure that all four wheels will be in proper alignment when the vehicle has motion.

MONO AXLE AND FOUR WHEEL DRIVE SUSPENSIONS

Mono axle front suspensions are used on many four wheel drive vehicles and heavy trucks. Four wheel drive vehicles equipped with

MACPHERSON STRUT SUSPENSION REFERENCE CHART

Angles			Problem
SAI	Camber	Included Angle	
Correct	Less than specification	Less than specification	Bent spindle and/or bent strut body
Correct	Greater than specification	Greater than specification	Bent spindle and/or bent strut body
Less than specification	Greater than specification	Correct	Bent control arm or strut tower out at top
Greater than specification	Less than specification	Correct	Strut tower in at top
Greater than specification	Greater than specification	Greater than specification	Strut tower in at top and spindle and/or strut body bent
Less than specification	Greater than specification	Greater than specification	Bent control arm or strut tower out at top plus bent spindle and/or strut body
Less than specification	Less than specification	Less than specification	Strut tower out at top and spindle and/or strut body bent or bent control arm

SUSPENSION
General Information

mono front axles have greater road clearance because the spring is mounted above the axle. These units are subjected to many off-road uses and require added strength and higher frame and body placement. Because the mono axle is rigid, there is no camber change as the spring is compressed or expanded. The wheels remain in the pre-set camber position at all times. There are no caster/

Mono axle alignment can be accomplished in two ways, bending and shimming. If caster requires adjustment of 1° or less, inserting a wedge shim between the spring and the axle will usually correct the misalignment. There are also special bending kits available that can twist the outer end of the axle to correct caster. These kits will also bend to correct camber. *Under no circumstances should heat be used to make any corrective bends!*

Many four wheel drive vehicles are driven under adverse conditions and therefore require a periodic alignment and parts check. Alignment procedures should follow the manufacturer's recommendations. A serious alignment problem, such as a bent axle housing or spindle assembly, will necessitate replacement of the damaged parts. Inner and outer axle housing bearing seats are perfectly aligned so that a straight axle will turn freely. The slightest bend in the axle housing will misalign the bearings and force the axle to bend as it rotates. This places stress on the axle bearings which greatly shortens axle bearing life and requires greater power to turn the axle. The added friction creates heat which may cause multiple problems that could be quite costly. *A bent axle housing cannot be properly straightened; it must be replaced.* There is no way the bearing seats can be perfectly realigned by attempting to straighten the axle housing, even if the desired suspension alignment settings have been achieved. Any bent suspension or steering parts should be replaced. The accompanying table shows how to find bent parts.

Some four wheel drive vehicles have independent front suspension. If the vehicle has upper and lower control arms, the reference chart under "Steering Axis Inclination" may be used to locate bent parts. If the suspension is twin I beam type, the diagnostic portion of the section dealing with twin I beams will explain how to locate bent parts. Geometric principles apply equally to front wheel, rear wheel, and four wheel drive vehicles.

Twin I Beam Suspension

The "Twin I Beam" front suspension was developed to combine independent front wheel action with the strength and dependability of the mono beam axle. Twin I beam axles allow each front wheel to absorb bumps and road irregularities independently, while providing sturdy, simple construction. The outer ends of the I beams are attached to the spindle and to the radius arms. The inner ends are attached to a pivot bracket fastened to the frame near the opposite side of the vehicle. This type of construction provides good anti-roll characteristics. The use of a progressive coil spring (a spring that becomes increasingly stiff as it is compressed) provides good riding qualities on normal roads and sturdiness off the road.

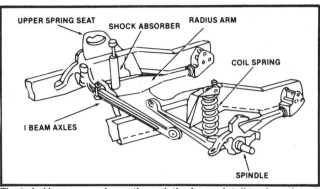

The twin I beams are shown through the frame detail to show the mounting details

The radius arms permit the I beam to move up and down, stabilize any front to rear movement of the I beam, and help maintain the proper caster setting. The spindle is mounted to the I beam by a spindle bolt (kingpin). There are no ball joints.

The action of the I beam produces good handling characteristics. As the spring is compressed, the camber increasingly moves toward negative. The exact opposite occurs as the spring is expanded. The more the spring is expanded, the more positive the camber becomes. In cornering, body roll of the vehicle increases as speed and sharpness of the turn increase. Up to the critical point where the vehicle will either roll over or lose traction, these cornering forces compress the outside front spring (creating negative camber in that wheel) and expand the inside spring (creating positive camber in the inside wheel). These changes tend to increase the cornering ability of the vehicle. Connection of the I beam to the opposite side of the vehicle provides changes in camber identical to those described with respect to roll forces, giving good resistance to crosswind forces.

Vehicle height is most important for the proper alignment of camber. Many twin I beam suspensions have been needlessly bent and re-aligned to achieve proper camber when the problem may have been caused by nothing more than a weak spring. As the spring weakens and sags it becomes more compressed, lowering vehicle height, which causes camber to go toward negative. In most cases, all that is required is that the weakened springs be replaced (they must be replaced in pairs). When the vehicle is restored to the proper height, the camber will usually be correct. Excessive vehicle height will cause positive camber. An incorrect set of springs will result in incorrect camber settings. In the case of *new* springs, however, vehicle height may be temporarily too high. A few hundred miles of driving will lower the height to normal. Camber should be correct after new springs have had a chance to settle in.

If the vehicle height is proper and a camber misalignment still exists, determine where the problem is before any corrective steps are taken. Specifically, check the SAI. If the SAI is correct and

MONO AXLE REFERENCE CHART

	Angles		Problem
SAI	Camber	Included Angle	
Correct	Greater than specification	Greater than specification	Bent spindle assembly
Greater than specification	Less than specification	Correct	Bent axle housing
Less than specification	Greater than specification	Correct	Bent axle housing
Less than specification	Greater than specification	Greater than specification	Bent spindle assembly plus bent axle housing

2-29

SECTION 2
SUSPENSION
General Information

camber is not, the included angle is incorrect. As previously discussed, if the included angle is incorrect, the spindle assembly is bent. Bending an I beam to correct a camber problem makes no sense if the included angle is incorrect—bending does not correct the problem, but rather adds another. If the vehicle height and SAI are correct and the included angle is incorrect, the solution is to replace the spindle assembly. If the vehicle height is correct and the included angle is correct, but the SAI and camber are incorrect, the problem is a bent I beam. In the following reference table, vehicle height is assumed to be absolutely correct. *The relationships listed in the table will not be true if the vehicle height is incorrect.*

The twin I beam suspension is a simple but very strong suspension. It is not easily bent out of alignment. Nonetheless, many "corrective bends" have been made because of improper vehicle height. A simple way to determine vehicle height is to physically measure it as it is prescribed by the manufacturer. The key to proper twin I beam alignment is proper vehicle height and the included angle. If vehicle height is correct and the included angle is correct, the vehicle is in proper alignment. When springs have sagged, replace them with new springs. Front spring shim kits are available, however, raising the vehicle height by shimming the springs will comprise the riding and handling characteristics. As was previously mentioned, the springs are progressive springs, and when they have weakened, they should be replaced. If only one spring has weakened, you must replace *both* springs. *Do not shim just one spring,* it could possibly upset the geometry of the entire system and create hazardous handling and driving conditions.

Occasionally, a bent I beam may be isolated as an alignment problem. At this point, and this point only, a corrective bend should be considered unless the I-Beam suspension is equipped with ball joints. The manufacturer does not recommend any type of corrective bending. However, when I beams have actually become bent by accident, corrective bends have been performed with satisfactory results. If the error of misalignment in the I beam is one degree or less, a *cold bend* may be performed, but only with the proper tools. *Under no circumstances whatsoever should heat be used to make this corrective bend!* If the error in misalignment caused by the I beam is more than one degree, replace the I beam.

If an error in caster is found, something is bent. Two possibilities exist: (1) a bent radius arm, or (2) a twisted I beam somewhere between where the radius arm is attached and the spindle assembly. In both cases, replace the damaged part. Do not attempt to straighten either part.

If an error in caster is found, something is bent. Two possibilities exist: (1) bent radius arm, or (2) a twisted I beam somewhere between where the radius arm is attached and the spindle assembly. In both cases, *replace the damaged* part. Do not attempt to straighten either part. When caster readings are taken, you must be sure that the vehicle height is correct, both front and rear. Many trucks will have oversize rear tires or rear helper springs that would raise the rear of the vehicle. Proper caster readings are taken with reference being made to the degree angle of the frame. Having the correct tools and following the manufacturers recommendations is the only way to properly take caster readings and correct caster on trucks.

A particular problem often encountered with twin I beam suspension is tire cupping, which is sometimes called scalloping. Frequently, alignment is blamed for this condition. Altering a good alignment to a misalignment will not correct tire cupping problems. Three conditions usually cause tire cupping: (1) loose parts, (2) out-of-round or out-of-balance tires and wheels, and (3) bad or weak shock absorbers. These problems may occur in any combination. If there are no loose parts, the tires and wheels are in balance and are not out-of-round, the alignment is at specification, and the shock absorbers *appear* to be good, it is usually the shock absorbers that are at fault, regardless of their appearance. The I beam is nothing more than a very long control arm. The longer the control arm, the greater the leverage and the amount of unsprung weight. This increases the dampening required. If the shock absorber strength is less than it should be and a nearly undetectable out-of-balance condition exists, cupping will occur. Loose parts, even those within tolerance, together with a slightly weak shock absorber, can cause a cupping condition. These problems can be nearly impossible to solve unless the real culprit is identified—the weak shock absorber. The solution is to replace the shock absorbers with heavy duty shock absorbers. This will not greatly affect the riding qualities, and if everything else is correct, the cupping condition will be eliminated. Replacing standard size wheels and tires with larger ones increases the amount of unsprung weight and the need for dampening. The solution is to install a second pair of front shock absorbers giving a total of four front shock absorbers. Many twin I beam truck suspensions have provisions for the installation of the second pair.

Type one: The I beam is constructed of solid metal and is longer than types two and three. This is the only model that has kingpins and is the only one on which a "corrective bend" can be made.

Type two: This model has ball joints and a camber adjustment is provided. *Under no circumstances should a "corrective bend" be made on this suspension.* It is of different construction (box type and not solid metal), and a bend of any kind will weaken the structure, creating a serious problem later on. An adjustment is provided and if the alignment requires more than the camber adjustment can provide, you have a problem that bending would not correct. Follow the twin I beam diagnosis in this section. The problem can be isolated, and when the faulty part or parts are replaced, alignment will be proper.

Type three: Essentially the same as type two with this exception; the I beam is made of solid metal. However, do not make a "corrective bend" on this I beam. Consider it as a control arm the same as you would on any other suspension. Use the twin I beam diagnosis chart and correct the actual problem. Later model twin I beam suspension systems have caster correction kits available. If caster requires a major adjustment isolate the problem and replace the bent part or parts.

Early model twin I beam suspensions provided only one tie rod adjustment, making the correction of off-center steering wheel more difficult. If the error is not great, perhaps the manufacturing tolerances are on the outer limits, but still within the specifications. if the error is great the steering wheel may have been removed and improperly replaced or a steering arm is seriously bent. The most probable is that it was not aligned referencing to the thrustline, nonetheless, do not overlook bent parts. The procedure to correct an off-center steering wheel on a twin I beam steering system is simple if you have the proper equipment and nearly impossible if you do not. With adequate alignment equipment, the cause of the off-center steering wheel can be located. The steering wheel will be dead center after proper alignment procedures have been completed with references to the thrustline.

TWIN I BEAM REFERENCE CHART

	Angles		Problem
SAI	Camber	Included Angle	
Correct	Greater than specification	Greater than specification	Bent spindle assembly
Greater than specification	Less than specification	Correct	Bent I beam
Less than specification	Greater than specification	Correct	Bent I beam
Less than specification	Greater than specification	Greater than specification	Bent I beam plus bent spindle assembly

SUSPENSION
General Information

FRONT SUSPENSION AND STEERING LINKAGE TROUBLESHOOTING—REAR WHEEL DRIVE

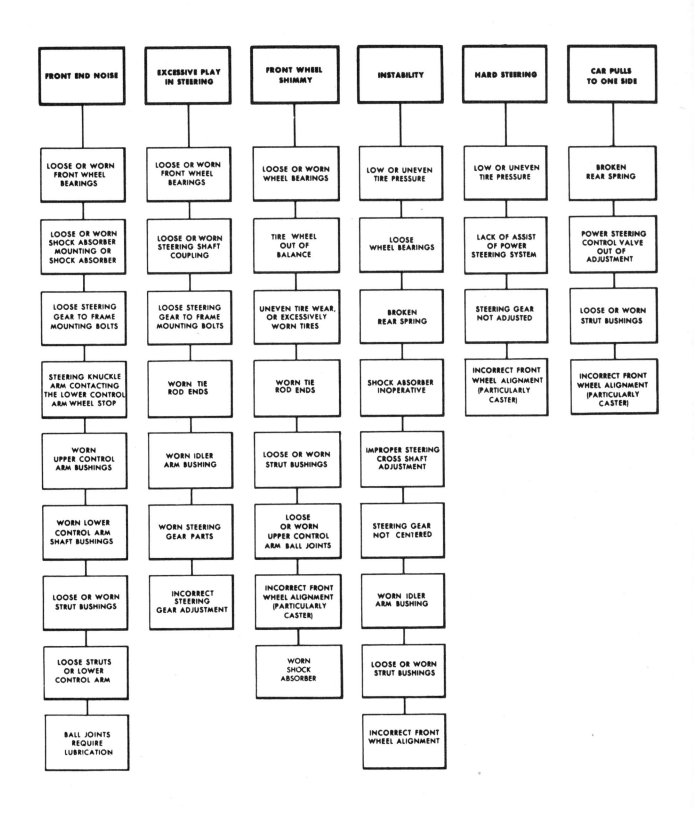

SECTION 2
SUSPENSION
General Information

FRONT SUSPENSION AND STEERING LINKAGE TROUBLESHOOTING—FRONT WHEEL DRIVE

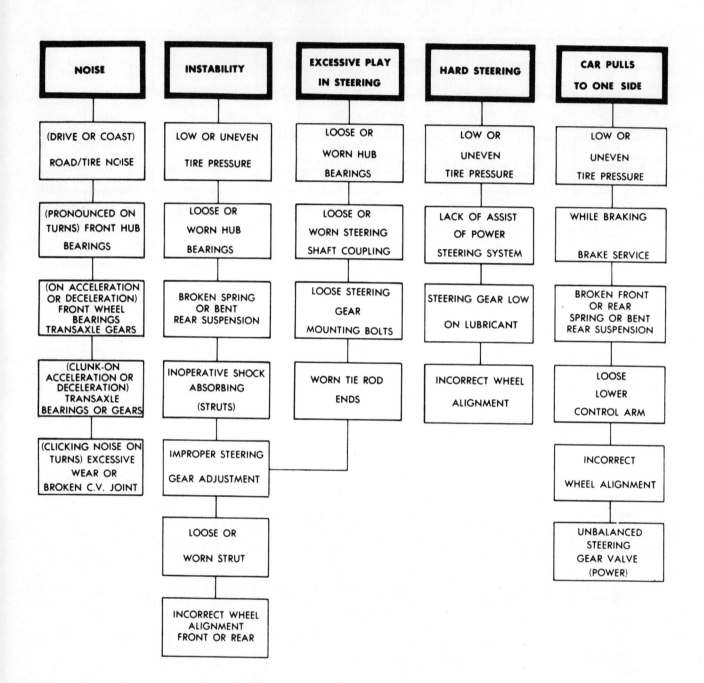

2-32

SUSPENSION
General Information

TAPERED WHEEL BEARING TROUBLESHOOTING

CONSIDER THE FOLLOWING FACTORS WHEN DIAGNOSING BEARING CONDITION:

1. GENERAL CONDITION OF ALL PARTS DURING DISASSEMBLY AND INSPECTION.
2. CLASSIFY THE FAILURE WITH THE AID OF THE ILLUSTRATIONS.
3. DETERMINE THE CAUSE.
4. MAKE ALL REPAIRS FOLLOWING RECOMMENDED PROCEDURES.

GOOD BEARING

BENT CAGE

CAGE DAMAGE DUE TO IMPROPER HANDLING OR TOOL USAGE.

REPLACE BEARING.

BENT CAGE

CAGE DAMAGE DUE TO IMPROPER HANDLING OR TOOL USAGE.

REPLACE BEARING.

GALLING

METAL SMEARS ON ROLLER ENDS DUE TO OVERHEAT, LUBRICANT FAILURE OR OVERLOAD.

REPLACE BEARING – CHECK SEALS AND CHECK FOR PROPER LUBRICATION.

ABRASIVE STEP WEAR

PATTERN ON ROLLER ENDS CAUSED BY FINE ABRASIVES.

CLEAN ALL PARTS AND HOUSINGS, CHECK SEALS AND BEARINGS AND REPLACE IF LEAKING, ROUGH OR NOISY.

ETCHING

BEARING SURFACES APPEAR GRAY OR GRAYISH BLACK IN COLOR WITH RELATED ETCHING AWAY OF MATERIAL USUALLY AT ROLLER SPACING.

REPLACE BEARINGS – CHECK SEALS AND CHECK FOR PROPER LUBRICATION.

MISALIGNMENT

OUTER RACE MISALIGNMENT DUE TO FOREIGN OBJECT.

CLEAN RELATED PARTS AND REPLACE BEARING. MAKE SURE RACES ARE PROPERLY SEATED.

INDENTATIONS

SURFACE DEPRESSIONS ON RACE AND ROLLERS CAUSED BY HARD PARTICLES OF FOREIGN MATERIAL.

CLEAN ALL PARTS AND HOUSINGS, CHECK SEALS AND REPLACE BEARINGS IF ROUGH OR NOISY.

FATIGUE SPALLING

FLAKING OF SURFACE METAL RESULTING FROM FATIGUE.

REPLACE BEARING – CLEAN ALL RELATED PARTS.

SECTION 2 SUSPENSION
General Information

TAPERED WHEEL BEARING TROUBLESHOOTING

BRINELLING

SURFACE INDENTATIONS IN RACEWAY CAUSED BY ROLLERS EITHER UNDER IMPACT LOADING OR VIBRATION WHILE THE BEARING IS NOT ROTATING.

REPLACE BEARING IF ROUGH OR NOISY.

CAGE WEAR

WEAR AROUND OUTSIDE DIAMETER OF CAGE AND ROLLER POCKETS CAUSED BY ABRASIVE MATERIAL AND INEFFICIENT LUBRICATION. CHECK SEALS AND REPLACE BEARINGS.

ABRASIVE ROLLER WEAR

PATTERN ON RACES AND ROLLERS CAUSED BY FINE ABRASIVES.

CLEAN ALL PARTS AND HOUSINGS, CHECK SEALS AND BEARINGS AND REPLACE IF LEAKING, ROUGH OR NOISY.

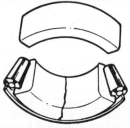

CRACKED INNER RACE

RACE CRACKED DUE TO IMPROPER FIT, COCKING, OR POOR BEARING SEATS.

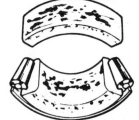

SMEARS

SMEARING OF METAL DUE TO SLIPPAGE, SLIPPAGE CAN BE CAUSED BY POOR FITS, LUBRICATION, OVERHEATING, OVERLOADS OR HANDLING DAMAGE.

REPLACE BEARINGS, CLEAN RELATED PARTS AND CHECK FOR PROPER FIT AND LUBRICATION.

REPLACE SHAFT IF DAMAGED.

FRETTAGE

CORROSION SET UP BY SMALL RELATIVE MOVEMENT OF PARTS WITH NO LUBRICATION.

REPLACE BEARING. CLEAN RELATED PARTS. CHECK SEALS AND CHECK FOR PROPER LUBRICATION.

HEAT DISCOLORATION

HEAT DISCOLORATION CAN RANGE FROM FAINT YELLOW TO DARK BLUE RESULTING FROM OVERLOAD OR INCORRECT LUBRICANT.

EXCESSIVE HEAT CAN CAUSE SOFTENING OF RACES OR ROLLERS.

TO CHECK FOR LOSS OF TEMPER ON RACES OR ROLLERS A SIMPLE FILE TEST MAY BE MADE. A FILE DRAWN OVER A TEMPERED PART WILL GRAB AND CUT METAL, WHEREAS, A FILE DRAWN OVER A HARD PART WILL GLIDE READILY WITH NO METAL CUTTING.

REPLACE BEARINGS IF OVER HEATING DAMAGE IS INDICATED. CHECK SEALS AND OTHER PARTS.

STAIN DISCOLORATION

DISCOLORATION CAN RANGE FROM LIGHT BROWN TO BLACK CAUSED BY INCORRECT LUBRICANT OR MOISTURE.

RE-USE BEARINGS IF STAINS CAN BE REMOVED BY LIGHT POLISHING OR IF NO EVIDENCE OF OVERHEATING IS OBSERVED.

CHECK SEALS AND RELATED PARTS FOR DAMAGE.

SUSPENSION
General Information

HOW TO READ TIRE WEAR

The way your tires wear is a good indicator of other parts of the suspension. Abnormal wear patterns are often caused by the need for simple tire maintenance, or for front end alignment.

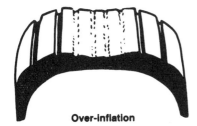

Over-inflation

Excessive wear at the center of the tread indicates that the air pressure in the tire is consistently too high. The tire is riding on the center of the tread and wearing it prematurely. Occasionally, this wear pattern can result from outrageously wide tires on narrow rims. The cure for this is to replace either the tires or the wheels.

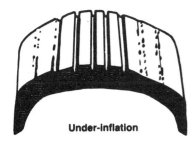

Under-inflation

This type of wear usually results from consistent under-inflation. When a tire is under-inflated, there is too much contact with the road by the outer treads, which wear prematurely. When this type of wear occurs, and the tire pressure is known to be consistently correct, a bent or worn steering component or the need for wheel alignment could be indicated.

Feathering

Feathering is a condition when the edge of each tread rib develops a slightly rounded edge on one side and a sharp edge on the other. By running your hand over the tire, you can usually feel the sharper edges before you'll be able to see them. The most common causes of feathering are incorrect toe-in setting or deteriorated bushings in the front suspension.

SECTION 2 SUSPENSION
General Information

One side wear

When an inner or outer rib wears faster than the rest of the tire, the need for wheel alignment is indicated. There is excessive camber in the front suspension, causing the wheel to lean too much putting excessive load on one side of the tire. Misalignment could also be due to sagging springs, worn ball joints, or worn control arm bushings. Be sure the vehicle is loaded the way it's normally driven when you have the wheels aligned.

Cupping

Cups or scalloped dips appearing around the edge of the tread almost always indicate worn (sometimes bent) suspension parts. Adjustment of wheel alignment alone will seldom cure the problem. Any worn component that connects the wheel to the suspension can cause this type of wear. Occasionally, wheels that are out of balance will wear like this, but wheel imbalance usually shows up as bald spots between the outside edges and center of the tread.

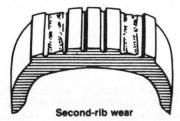

Second-rib wear

Second-rib wear is usually found only in radial tires, and appears where the steel belts end in relation to the tread. It can be kept to a minimum by paying careful attention to tire pressure and frequently rotating the tires. This is often considered normal wear but excessive amounts indicate that the tires are too wide for the wheels.

SUSPENSION
Domestic Models

Domestic Vehicles
PRINCIPLES OF WHEEL ALIGNMENT

Introduction

The term wheel alignment is well known, but it might be better understood as motion balance, because correct vehicle alignment involves balancing all of the forces created by friction, gravity, centrifugal force and momentum, while a vehicle is in motion.

A correct alignment job will make a vehicle run smoother, have better road-handling characteristics, have better steering ability and operate with more stability while running in a straight line and around curves. It also eliminates unnecessary road friction, which causes abnormal tire wear and decreases fuel mileage.

Wheel alignment is more than just the simple tracking of front and rear wheels to insure that they roll freely on a straight-ahead course. Front end alignment must also be maintained during turns and other steering maneuvers even thought the road surface and other irregularities can cause the wheels to move up and down almost constantly. In addition to steering control, front end alignment also provides directional stability which helps the driver hold a straight course without making continuous steering corrections. Ideally, only very light pressure on the steering wheel should be enough to keep the vehicle headed on a straight course. However, a slight resistance (light turning load) is needed to help stabilize steering control. In effect, this resistance helps give the driver something to turn against, thus reducing the tendency to over-control.

In the past, caster, camber and toe were all adjustable. However, with the introduction of the MacPherson type of front suspension and its modified versions, caster adjustment can only be accomplished by bending the suspension. Camber is adjustable on some types, but not on others. This does not mean that these angles are not important. Proper setting of both caster and camber are absolutely essential to good stability, easy steering control and maximum tire life.

SUSPENSION DIAGNOSIS

SYMPTOM	PROBABLE CAUSE
Excessive tire wear on outside shoulder	Excessive positive camber
Excessive tire wear on inside shoulder	Excessive negative camber
Excessive tire wear on both shoulders	Rounding curves at high speeds Underinflated tires
Saw-tooth tire wear	Too much toe-in or toe-out
One tire wears more than the other	Improper camber Defective brakes Defective shock absorber
Tire treads cupped or dished	Out-of-round tires Out-of-balance condition Defective shock absorber
Front wheels shimmy	Defective idler arm bushing Out-of-round tires Out-of-balance condition Excessive positive caster Uneven caster
Vehicle vibrates	Defective tires One or more of all 4 tires out-of-round One or more of all 4 tires out-of-balance Driveshaft bent Driveshaft sprayed with undercoating
Car tends to wander either to the right or left	Improper toe setting Looseness in steering system or ball joints Uneven caster Tire pull
Vehicle swerves or pulls to side when applying brakes	Uneven caster Brakes need adjustment Out-of-round brake drum Defective brakes Underinflated tire
Car tends to pull either to the right or left when taking hands off steering wheel	Improper camber Unequal caster Tires worn unevenly Tire pressure unequal

SECTION 2

SUSPENSION
Domestic Models

SUSPENSION DIAGNOSIS

SYMPTOM	PROBABLE CAUSE
Car is hard to steer	Tires underinflated
	Power steering defective
	Too much positive caster
	Steering system too tight or binding
Steering has excessive play or looseness	Loose wheel bearings
	Loose ball joints
	Loose bushings
	Loose idler arm
	Loose steering gear assembly
	Worn steering gear or steering gear bearings

With the introduction of small front drive cars, the rear suspension now becomes a service item. Some rear axles have alignment adjustments, some do not. See each type axle in their respective sections.

What Is Toe

Toe-in is the amount, in fractions of an inch, that the wheels are closer together at the extreme front of the tire than they are at the rear. If the wheels are farther apart at the front, they toe-out.

ZERO RUNNING TOE

When a vehicle is moving, zero toe provides parallel rolling of the two front wheels. This stabilizes steering and reduces side-slipping and tire wear to a minimum. To obtain a running toe of zero for average driving, it is usually necessary to provide a small amount of toe-in when the vehicle is at rest. This offsets small deflections due to rolling resistance and brake applications which tend to toe the wheels outward. When a vehicle is rolling forward a force is set up which compresses the tie-rod and the tie-rod ends. This lets the wheels spread outward at the front. Very little looseness in the steering linkage will allow wheels set at recommended toe-in actually to toe-out under running conditions.

It has been common to talk about toe-in. However, many front drive cars require toe-out settings.

The foregoing explanation of the forces at work trying to change toe when the car is rolling helps explain why it is important to set to to the preferred specification. It also explains the futility of trying to obtain zero running toe on a card having loose steering tie-rod.

--- CAUTION ---
Tie rod ball ends are designed to allow complete freedom of movement without binding. However, because each rod has two tie-rod ends, both of them must be centered exactly to avoid binding and interference when the wheels move up and down or are turned.

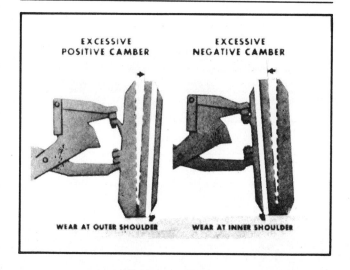

EXCESSIVE POSITIVE CAMBER — WEAR AT OUTER SHOULDER
EXCESSIVE NEGATIVE CAMBER — WEAR AT INNER SHOULDER

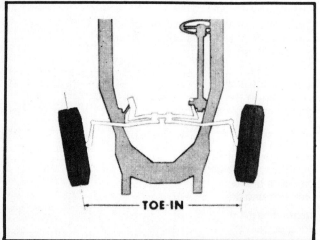

TOE-IN

What Is Caster

Caster is defined as the angle, when viewed from the side, between the steering axis and the vertical. It is simply a new name that describes what used to be known as the "kingpin axis." Caster is positive when the top of the axis is inclined rearward and negative when it is inclined forward.

EFFECT OF CASTER ON STEERING

The mechanic's creeper illustrates the principle and effect of caster on steering geometry. When the force is applied to the creeper, the caster turn on their pivots until the caster wheels are in line with the force applied and the wheels are then trailing behind the pivot point. The same principle applies to the front wheel of a car. If the steering axis pivot is tilted backward at the top of the projected axis contacts the road ahead of the point of tire contact. This produces the same effect as the creep-

SUSPENSION
Domestic Models

er's caster. The pivot axis pulls the wheel behind it, adding to the car's directional stability.

As the positive caster angle is increased, the effort required to turn the car from a straight-ahead course is increased. The tendency of the front wheels to straighten out rapidly when leaving a turn is also increased.

NOTE: Although the caster angle tends to cause the car body and frame to lift when the wheels are turned, this is a very minor factor in the stability and returnability of present-day cars. There are two reasons for this: first, caster angles are so small that the lift is negligible; second, the higher rolling resistance of today's low-pressure tires produces far more directional stability from the caster effect that did smaller high pressure tires used in the past.

On a power steering car too much positive caster indirectly can cause low-speed shimmy, increased road shock and high-speed wander. Unequal caster-either positive or negative-between the wheels is also undesirable. If all other factors are equal, a car will lead or drift toward the side with the least amount of caster. The worst possible situation exists when there is no negative caster at one wheel and positive caster on the other. Under these conditions, the wheel with negative caster tries to turn outward and the car pulls toward the side with negative caster when brakes are applied.

What Is Camber

Camber is the amount, in degrees, that the wheel is inclined from the vertical as viewed from the front of the car. If the top of the wheel leans away from the car, the camber is positive. If the top of the wheel leans toward the car, the camber is negative.

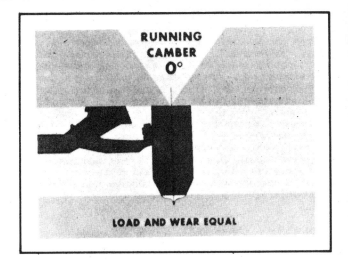

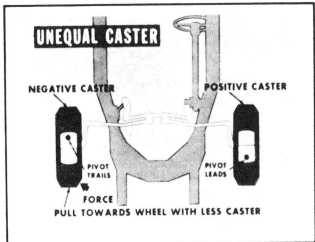

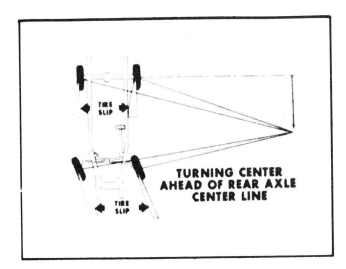

EFFECTS OF INCORRECT CAMBER

Too much positive camber causes tire wear at the outer shoulder, too much negative camber causes tire wear at the inner shoulder. However, you should also remember that rounding turns at high speeds also causes the tires to show more wear at the shoulders. Therefore, such tires wear may not always be caused by incorrect camber.

Similarly, continuous high-speed driving on curves to the right and left will produce wear at the inner and outer shoulders of both tires. This often results in a wear pattern that looks very much like under inflation wear. Again, such wear may not be caused by incorrect camber and will not be corrected by adjusting the camber.

Under all driving conditions, the best average running camber is obtained when camber is adjusted to specifications. Deviation from specifications can cause hard steering, unstable steering and wander. Unequal camber can contribute to a low-speed shimmy condition.

CAMBER ANGLES CHANGE WITH VERTICAL WHEEL MOVEMENT

When a front wheel moves upward (or the car body and frame moves downward), the top of the wheel moves inward, producing negative camber. The upward movement of the wheel is usually called jounce motion of jounce travel.

ZERO RUNNING CAMBER

Maximum tire life is obtained when the average running camber is zero. That is, when the wheel and the tire are vertical, the tire tread in contact with the road is uniform from side-to-side. Therefore, the load and wear are distributed equally over the entire tread.

SUSPENSION
Domestic Models

CAMBER ON TURNS

When a car goes into a turn at high speeds, centrifugal force causes a significant weight shift toward the outside wheel. The body and frame moves downward, producing negative camber at the outside wheel. At the same time, weight is reduce at the inside wheel and the body and frame tries to lift. This movement produces little camber change or a small amount of positive camber at the inside wheel.

The negative camber at the outside wheel has the very desirable effect of bracing the tire tread against sideslip. The combination of negative camber and increased tracting force (more weight on outside wheel) minimizes tire slip and increases cornering stability. Similarly, zero camber or a small amount of positive camber at the inside wheel helps minimize tire slip at this point.

Toe Out On Turns

Toe-out on turns is an important non-adjustable measurement that should be checked after all other steering geometry angles are measured and adjusted to their straight-ahead specifications. When the wheels are turned, a fifth angle becomes important. This is commonly referred to as toe-out on turns.

In theory, all four car wheels should turn about the same center to minimize sideslipping of the tires. Since both rear wheels are connected by a solid axle, the front wheels should turn in circles whose centerlines intersect the centerline of the rear axle. Actually, all four tires slip because of centrifugal force when rounding a corner at any speed greater than a brisk walk. As a result, the real turning center is considerably ahead of the true centerline of the rear axle.

FRONT SUSPENSION AMERICAN MOTORS

Except for two types of steering knuckles, service procedures for both two wheel drive and four wheel drive vehicles are very similar. On two wheel drive cars, ball joints can be replaced in a conventional manner. On four wheel drive models, the complete control arm must be replaced.

FRONT WHEEL ALIGNMENT

Front wheel alignment, or steering geometry, refers to the various angles assumed by the components which form the front wheel turning mechanism. There are three adjustable, alignment angles which are caster, camber and toe-in.

Caster describes the forward or rearward tilt (from vertical) of the steering knuckle. Tilting the top of the knuckle rearward provides positive caster. Tiling the top of the knuckle forward provides negative caster. Caster is directional stability angle which enables the front wheels to return to a straight-ahead position after turns.

Adjust caster by loosening the strut rod jamnut and turning the rod adjusting nuts in or out to move the lower control arm forward or rearward to obtain the desired caster angle. Tighten adjusting nuts to 65 ft. lbs. (88 Nm) torque and jamnut to 75 ft. lbs. (102 Nm) torque when adjustment is completed.

Camber describes the inward and outward tilt of the wheel relative to the center of the automobile. An inward tilt of the top of the wheel produces negative camber. An outward tilt produces positive camber. Camber greatly affects tire wear. Incorrect camber will cause abnormal wear on the tire outside or inside edge.

Adjust camber by turning lower control arm inner pivot bolt eccentric. Tighten pivot bolt locknuts to 110 ft. lbs. (149 Nm) torque after completing camber adjustment.

Toe-in is a condition that exists when the measured distance at the front of each tire is less than the distance at the rear of the tires. When the distance at the front is less than the rear, the tires are toed-in. toe-in compensates for normal steering play and causes the tires to roll in a straight-ahead manner. Incorrect toe-in will ear the tires to a feathered edge.

Adjust toe-in by turning tie rod adjuster tubes in or out to shorten or lengthen tie rods to obtain desired toe-in. Place front wheels in straight-ahead position and center steering wheel and gear. Turn tie rod adjusting tubes equally in opposite directions to obtain desired toe-in setting. If steering wheel spoke position was disturbed during toe-in adjustment, correct spoke position by turning tie rod tubes equally in the same direction until desired position is obtained.

Front Wheel Bearings

TWO WHEEL DRIVE

When repacking and adjusting front wheel bearings, use an EP-type, lithium base wheel bearing lubricant. Pack the bearings with a generous amount of lubricant and place extra lubricant in the rotor hub cavity between the bearings. Always use a new grease seal during assembly.

When inspecting, replacing, or repacking bearings, be sure the inner cones of the bearings are free to creep on the spindle. The bearings are designed to creep to allow a constantly changing load contact between the cones and the rollers. Polishing and applying lubricant to the spindle will permit this movement and prevent rust formation.

Two wheel drive suspensions (© American Motors Corp.)

SUSPENSION
Domestic Models

Adjustment

1. Raise and support front of the automobile.
2. Remove hub cap, grease cap and O-ring, cotter pin and nutlock.
3. On automobiles with styled wheels, remove wheel and hub cap. Install wheel.
4. Tighten spindle nut to 25 ft. lbs. (34 Nm) torque while rotating wheel to seat bearings.
5. Loosen spindle nut $\frac{1}{3}$ turn. While rotating wheel, tighten spindle nut to 6 inch lbs. (0.7 Nm) torque.
6. Install nutlock to spindle nut so cotter pin holes in nutlock and spindle are aligned.
7. Install replacement cotter pin, grease cap and hub cap.
8. On automobiles with styled wheels, remove wheel, install hub cap and install wheel.

FOUR WHEEL DRIVE

Adjustment

Four wheel drive models have a unique front axle hub and bearing assembly. The assembly is sealed and does not require lubrication, periodic maintenance, or adjustment. The hub has ball bearings which seat in races machined directly into the hub. There are darkened areas surrounding the bearing race areas of the hub. These darkened areas are from a heat treatment process, are normal, and should not be mistaken for a problem condition.

Removal

1. Raise and support the vehicle safely.
2. Remove the wheel, caliper and rotor.
3. Remove the bolts attaching the axle shaft flange to the halfshaft.
4. Remove the cotter pine, lock nut and axle hub nut from the assembly.
5. Remove the halfshaft.
6. Remove the steering arm from the steering knuckle.
7. Remove the caliper anchor plate from the steering knuckle.
8. Remove the Torx© head bolts retaining the hub assembly.
9. Remove the hub assembly from the steering knuckle pine.
10. Remove any grease remaining in the steering knuckle cavity.

NOTE: Remove and save the front hub spacer. During removal, it may be lodged on the end of the halfshaft or on the hub shaft. If a replacement hub is to be used, the hub spacer must be installed on the new hub assembly.

Installation

1. Partially fill the hub cavity of the steering knuckle with chassis lubricant and install the hub assembly. Make sure that the hub spacer is installed on the hub shaft.
2. Install the inner seal in the steering knuckle pin.
3. Install the splash seal on the hub and bearing carrier. Install the O-ring on the hub and bearing carrier.
4. Install the carrier attaching bolts. Tighten to 75 ft. lbs. (102 Nm) torque.
5. Install the caliper anchor plat and plate retaining bolts. Tighten the retaining bolts to 100 ft. lbs. (136 Nm) torque.
6. Install the steering arm bolts. Tighten to 100 ft. lbs. (136 Nm) torque.
7. Install the halfshaft.
8. Install the axle flange to shaft bolts. Tighten to 45 ft. lbs. (61 Nm) torque.
9. Install the hub washer and nut. Tighten the hub nut to 175 ft. lbs. (237 Nm) torque.
10. Install the lock nut and new cotter pin. Install the rotor, caliper and wheel.

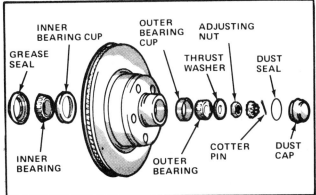

Front wheel bearing – two wheel drive (© American Motors Corp.)

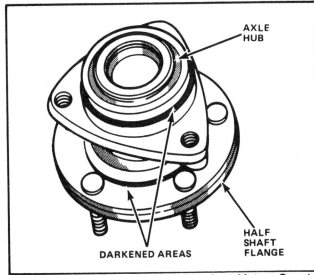

Axle hub assembly – Eagle (© American Motors Corp.)

SHOCK ABSORBER

Removal
ALL MODELS

1. Remove the lower retaining nuts, washer and grommets.
2. Remove upper mounting bracket bolts/nuts from wheelhouse panel.
3. Remove upper bracket and shock absorber from wheelhouse panel.
4. Remove upper retaining nut from shock absorber and remove upper bracket from shock absorber.

Installation

1. Install washers, grommets, upper mounting bracket and nut on shock absorber if removed. Tighten nut to 8 ft. lbs. (11 Nm) torque.
2. Extend shock absorber piston fully.
3. Install grommets on lower mounting studs and position shock absorber in wheelhouse panel.
4. Insert lower mounting studs into lower springs seat and install lower grommets, flat washers and nuts. Tighten nuts to 15 ft. lbs. (20 Nm) torque.

2-41

SUSPENSION
Domestic Models

5. Install and tighten upper mounting bracket attaching nuts/bolts to 20 ft. lbs. (27 Nm) torque.

ADJUSTABLE SHOCKS

To adjust the shock, compress the piston completely. Holding the upper part of the shock, turn the shock until the lower arrow is aligned with the desired setting. A click will be heard when the desired setting is reached. Install the shock as follows:

1. Fit the grommets, washers, upper mounting bracket and nut on the shock, in the reverse order of removal. Tighten the nut to 8 ft. lbs.
2. Fully extend the shock and install two grommets onto the lower mounting studs.
3. Lower the shock through the hole in the wheelwell. Fit the lower attachment studs through the lower spring seat.
4. Install the grommets, washers, and nuts. Tighten the nuts to 15 ft. lbs.
5. Secure the upper mounting bracket with its attachment nuts and bolts. Tighten to 20 ft. lbs.

NOTE: When installing new shock absorbers, purge them of air by extending them in their normal position and compressing them while inverted. Do this several times. It is normal for new shock absorbers to be more resistant to extension than to compression.

Upper Ball Joint

TWO WHEEL DRIVE

Inspection

1. Remove upper ball joint lubrication plug and install a dial indicator gauge through the lubrication hole so that you can measure the up and down movement of the ball joint socket.
2. Place a pry bar under tire to load ball joint and raise tire several times to seat gauge tool pin.
3. Pry tire upward to load ball joint and record gauge reading; then release tire to unload ball joint and record gauge reading. Perform this operation several times to ensure accuracy.
4. The difference between load-no-load readings represents ball joint clearance. If clearance is more than 0.080 in. (2.3mm), ball joint should be replaced.

Replacement

1. Install 2 X 4 X 5 in. wood block on frame side sill and under upper control arm.
2. Raise and support front of car.
3. Remove wheel, caliper and rotor.
4. Remove ball stud cotter pin and retaining nut.
5. Install ball joint remover and loosen ball stud in steering knuckle. Do not remove tool at this time.
6. Place support stand under lower control arm.
7. Remove heads from ball joint attaching rivets using chisel or grinding tool.
8. Drive rivets out of ball joint and control arm using hammer and punch.
9. Disengage ball joint from control arm.
10. Remove tool from ball joint stud and remove ball joint from steering knuckle.
11. Position replacement ball joint in control arm and align bolt holes.
12. Install ball joint attaching bolts (supplied in ball joint replacement kits) and tighten nuts to 25 ft. lbs. (34 Nm) torque.
13. Install steering knuckle and retaining nut on ball joint stud. Tighten nut to 75 ft. lbs. (102 Nm) torque and install a new cotter pin.
14. Install rotor, caliper and wheel.

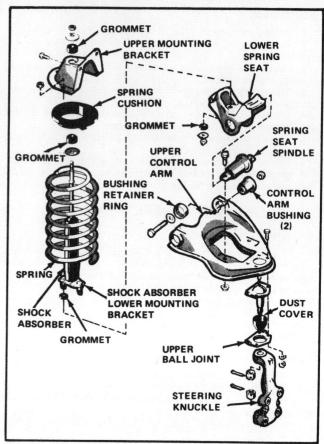

Shock absorber assembly (© American Motors Corp.)

FOUR WHEEL DRIVE

Inspection

Inspection procedures are the same as for two wheel drive models.

Replacement

If a ball joint is worn (upper and lower) the complete arm assembly must be replaced. Do not attempt to service the ball joint separately.

Lower Ball Joint

TWO WHEEL DRIVE

Inspection

1. Raise and support front of automobile.
2. Move lower portion of wheel and tire alternately toward and away from center of automobile. Perform this operation several times.
3. Lower ball join is spring-equipped and preloaded in its socket at all times to minimize looseness and compensate for wear. If lower joint exhibits any lateral movement (shake), ball joint should be replaced.

Replacement

1. Install 2 X 4 X 5 in. wood block on frame side sill and under upper control arm.

SUSPENSION
Domestic Models

2. Raise and support the front of automobile safely.
3. Remove wheel, caliper and rotor.
4. Disconnect stut rod at lower control arm.
5. Disconnect steering arm from steering knuckle.
6. Remove ball stud cotter pin and retaining nut.
7. Install ball joint removal tool and loosen ball stud in steering knuckle. Do not remove tool at this time.
8. Place support stand under lower control arm.
9. Remove heads from ball joint attaching rivets using chisel or grinding tool.
10. Drive rivets out of ball joint and control arm using punch.
11. Disengage ball joint from control arm.
12. Remove ball joint from control arm.
13. Position replacement ball joint on control arm and align bolt holes.
14. Install but to not tighten attaching bolts supplied in replacement ball joint kit.
15. Attach strut rod to lower control arm. Tighten rod attaching bolts to 75 ft. lbs. (102 Nm) torque.
16. Tighten ball joint attaching bolts to 25 ft. lbs. (34 Nm) torque.
17. Apply chassis grease to steering stops.
18. Install ball joint stud in steering knuckle.
19. Install retaining nut on ball stud. Tighten nut to 75 ft. lbs. (102 Nm) torque and install replacement cotter pine.
20. Install steering arm on steering knuckle.
21. Install rotor, caliper and wheel.

FOUR WHEEL DRIVE

Inspection and Replacement

See the upper ball joint inspection and replacement procedures for four wheel drive models.

Coil Spring
ALL MODELS

Identification

A plastic identification tag which has the spring part number printed on it, is attached to each coil spring. Whenever a spring must be replaced, refer to this part number when ordering a replacement spring.

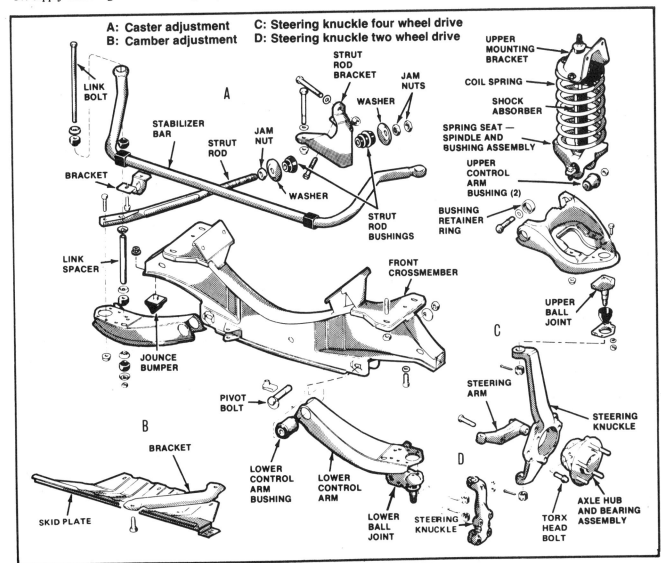

Four wheel drive suspension (© American Motors Corp.)

SUSPENSION
Domestic Models

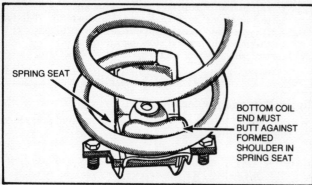

Lower spring seal position (© American Motors Corp.)

REMOVAL

1. Remove shock absorbers and mounting brackets.
2. Install spring compressor and compress spring approximately 1 in. (25.4mm).
3. Remove lower spring seat pivot bolt retaining nuts.
4. Raise front of automobile until control arms are free of lower spring seat.
5. Remove wheel.
6. Pull lower spring seat away from automobile, and guide lower spring seat out and over upper control arm.
7. Remove spring compressor tool and remove lower retainer, spring seat and spring.

— **CAUTION** —

Do not use impact wrench to turn the compression nut. An impact wrench will place unnecessary stress on the compressor tool bolt threads which could result in thread damage or bolt breakage.

Installation

1. Install spring compressor tool.

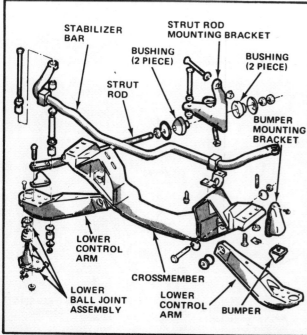

Lower control arm components – Spirit/Concord (© American Motors Corp.)

2. Install spring upper cushion on top coil of spring. Tape cushion in place to retain it.
3. Install spring in lower spring seat.

NOTE: **One side of the lower spring seat has a formed shoulder to help locate the spring properly. Position the spring on the seat so the cut-off end of the spring bottom coil seats against this shoulder. If the spring seat was removed for service, be sure the shouldered end of the spring seat and cut-off end of the spring bottom coil are installed so they face the engine compartment.**

4. Position spring in upper seat.
5. Align lower spring seat pivot so that retaining studs will enter upper control arm when spring is in position. Be sure spring lower coil end is properly positioned on seat.
6. Compress spring until lower spring seat pivot studs can be aligned with holes in upper control arm.
7. Turn compression nu counterclockwise and guide spring seat pivot studs into control arm.
8. Install wheel.
9. Remove supports and lower automobile.
10. Install and tighten lower spring seat pivot retaining nuts to 35 ft. lbs. (47 Nm) torque.
11. Remove spring compressor tool.
12. Install shock absorber and mounting bracket.

Upper Control Arm
ALL MODELS

Removal

1. Remove shock absorber and mounting bracket. Install spring compressor tool.
2. Remove lower spring seat pivot retaining nuts and turn compressor tool until spring is compressed approximately 2 inches (5.03 cm).
3. Raise and support front of automobile.
4. Remove wheel.
5. Remove upper ball joint stud cotter pin and retaining nut.
6. Remove upper ball joint stud from steering knuckle.
7. Remove control arm inner pivot bolts and control arm from wheelhouse panel.

Installation

1. Position control arm in wheelhouse panel and install inner pivot bolts.

— **CAUTION** —

Do not tighten the pivot bolts until the automobile is resting on the wheels as ride height may be affected.

2. Install steering knuckle and retaining nut on ball join stud. Tighten nut to 75 ft. lbs. (102 Nm) torque and install a new cotter pin.
3. Turn spring compressor tool compression nut and guide spring seat pivot studs into control arm.

Lower Control Arm
TWO WHEEL DRIVE

Removal

1. Raise and support front of automobile.
2. Remove wheel, caliper and rotor.
3. Disconnect steering arm from steering knuckle.
4. Remove lower ball joint stud cotter pin and retaining nut.
5. Remove ball stud from steering knuckle.
6. Disconnect stabilizer bar from control arm, if equipped.
7. Disconnect strut rod from control arm.

SUSPENSION
Domestic Models

8. Remove inner pivot bolt and remove control arm from crossmember.

Installation

1. Position control arm in crossmember and install inner pivot bolt.

--- CAUTION ---
Do not tighten the inner pivot bolt until the automobile weight is supported by the wheels as ride height may be affected.

2. Install steering knuckle and retaining nut on ball joint stud. Tighten nut to 75 ft. lbs. (102 Nm) torque and install replacement cotter pin.
3. Connect strut rod to control arm. Tighten bolts to 75 ft. lbs. (102 Nm) torque.
4. Connect stabilizer bar to control arm, if equipped. Tighten bolts to 7 ft. lbs. (9 Nm) torque.
5. Connect steering arm to steering knuckle.
6. Install rotor, caliper and wheel.
7. Place a jack under control arm. Raise jack to compress spring slightly and tighten control arm inner bolt to 110 ft lbs. (149 Nm) torque.

FOUR WHEEL DRIVE

Removal

1. Raise and support the vehicle safely. Remove cotter pin, nut lock and hub nut.
2. Raise and support front of automobile.
3. Remove wheel, caliper and rotor.
4. Remove lower ball joint cotter pin and retaining nut.
5. Remove ball stud from steering knuckle.
6. Remove halfshaft flange bolts.
7. Remove halfshaft.
8. Remove bolts attaching strut rod to control arm.
9. Disconnect stabilizer bar from control arm.
10. Remove inner pivot bolt and remove control arm.

Installation

1. Position control arm in crossmember and install inner pivot bolt.

--- CAUTION ---
Do not tighten the inner pivot bolt until the automobile weight is supported by the wheels as ride height may be affected.

2. Insert ball stud in steering knuckle and install retaining nut on ball joint stud. Tighten nut to 75 ft. lbs. (102 Nm) torque and install replacement cotter pin.
3. Connect stabilizer bar to control arm. Tighten lock nut to 7 ft. lbs. (9 Nm) torque.
4. Connect strut rod to control arm. Tighten bolts to 75 ft. lbs. (102 Nm) torque.
5. Install halfshaft-to-axle flange bolts.
6. Tighten flange bolts to 45 ft. lbs. (61 Nm) torque.

NOTE: *If control arm is worn or bushing is not tight when installed, control arm must be replaced.*

Steering Knuckle And Spindle

TWO WHEEL DRIVE

Removal

1. Raise and support front of automobile.
2. Remove wheel, caliper, and rotor.
3. Remove caliper, anchor plate, adapter, steering spindle, and steering arm from knuckle.
4. Remove upper and lower ball joint stud cotter pins and retaining nuts.
5. Remove all joint studs from steering knuckle.

Installation

1. Install steering knuckle and retaining nuts on ball joint studs. Tighten nuts to 75 ft. lbs. (102 Nm) torque and install new cotter pins.
2. Install steering arm, spindle, caliper anchor plate, and adapter. Tighten bolts to 55 ft. lbs. (75 Nm) torque.
3. Install rotor, caliper and wheel.

FOUR WHEEL DRIVE

Removal

1. Raise and support the vehicle. Remove cotter pin, nut lock and hub nut.
2. Remove wheel, caliper and rotor.
3. Remove halfshaft-to-axle flange bolts.
4. Remove halfshaft.
5. Remove steering arm from steering knuckle.
6. Remove caliper anchor plate from steering knuckle.
7. Remove three Torx® head attaching bolts remaining front wheel hub assembly.
8. Remove hub assembly from knuckle.
9. Remove rear hub seal from steering knuckle using small pry bar.
10. Remove upper and lower ball joint stud cotter pins and retaining nuts.
11. Remove ball joint studs from steering knuckle using a strike tool to loosen and remove studs from knuckle.

Installation

1. Install steering knuckle and ball joint retaining nuts on ball joint studs. Tighten nuts to 75 ft. lbs. (102 Nm) torque and install new cotter pins.
2. Install hub rear seal.
3. Partially fill hub cavity of steering knuckle with chassis lubricant and install hub assembly in knuckle.
4. Tighten hub Torx® head bolts to 75 ft. lbs. (102 Nm) torque.

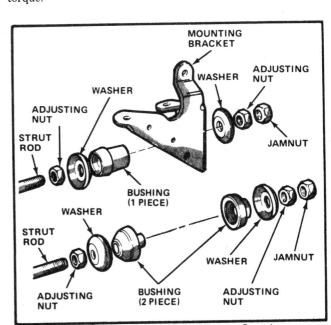

Strut rod bushings (© American Motors Corp.)

SUSPENSION
Domestic Models

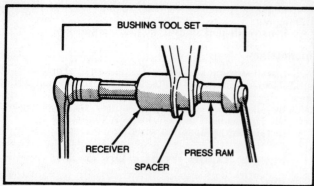

Control arm bushing replacements (© American Motors Corp.)

5. Install caliper anchor plate and retaining bolts.
6. Tighten caliper anchor plate bolts to 100 ft. lbs. (136 Nm) torque.
7. Install steering arm and bolts.
8. Tighten steering arm bolts to 100 ft. lbs. (136 Nm) torque.
9. Install halfshaft and shaft-to-axle flange bolts.
10. Tighten halfshaft-to-axle flange to 45 ft. lbs. (61 Nm) torque.
11. Install rotor, caliper and hub nut.
12. Install wheel.
13. Lower automobile.
14. Tighten hub nut to 180 ft. lbs. (244 Nm) torque. Install nut lock and a new cotter pin.

Strut Rod And Bushing
ALL MODELS
Replacement

1. Raise and support front of automobile.
2. Remove the jamnut and caster adjustment nut from strut rod.
3. Disconnect strut rod from lower control arm and remove rod, bushings and washers.
4. On automobiles with one-piece bushing, lubricate the bushing with soapy water and install.

NOTE: A special tool is required to press the one-piece bushing in and out of the mounting bracket.

REAR SUSPENSION AMERICAN MOTORS

Shock Absorber
Removal

NOTE: When installing new shock absorbers, it is first necessary to purge them of air by repeatedly extending them in their normal position and compressing them while inverted. It is normal for new shock absorbers to be more resistant to extension than to compression.

1. Raise and support the rear of the vehicle. Support the axle assembly safely.

Rear suspension components (© American Motors Corp.)

NOTE: When removing air adjustable shock absorbers, it is first necessary to remove the air line from the shock absorber. Tag the line for ease of assembly. Support the suspension in such a way that no strain is on shock absorber.

2. Remove locknut, retainer, and grommet which attach shock absorber lower mounting stud to spring plate.
3. Compress shock absorber by hand and disengage lower mounting stud from spring plate.
4. Remove bolts and lockwashers attaching shock absorber upper mounting bracket to underbody panel and remove shock absorber.
5. Remove locknut, retainer, and grommet which attach mounting bracket to shock absorber upper mounting stud and remove bracket.
6. Remove remaining grommets and retainers from shock absorber upper and lower mounting studs.

Installation

1. Install retainer and grommet on shock absorber mounting stud. Be sure locating shoulder on grommet faces end of mounting stud.
2. Install mounting bracket on shock absorber upper mounting stud with flat side of bracket facing underbody panel.
3. Install second grommet on mounting stud and install retainer and locknut to 8 ft. lbs. (11 Nm) torque.

NOTE: Be sure the locating shoulders on the grommets are centered in the mounting bracket hole before tightening the locknut.

4. Position assembled mounting bracket and shock absorber on mounting studs in underbody panel. Install lockwashers and bolts. Tighten bolts to 28 ft. lbs. (38 Nm) torque on Pacers and 15 ft. lbs. (20 Nm) torque on all other models.

NOTE: If an adjustable shock absorber is being installed, adjust the ride control setting as necessary before connecting the shock to the spring clip plate.

5. Engage shock absorber lower mounting stud in spring clip plate.
6. Install second grommet with shoulder of grommet facing spring tie plate and install retainer and locknut. Tighten locknut to 8 ft. lbs. (11 Nm) torque.

Leaf Spring

Removal

1. Raise and support rear of the vehicle. Support axle assembly with hydraulic jack.
2. Remove shock absorber lower mounting locknut, retainer and grommet.
3. Remove U-bolts, spring clamps, and clamp bracket.
4. Remove pivot bolt and nut from spring front eye.
5. Remove shackle nuts, shackle plate, and shackle at rear spring eye. Remove spring.

BUSHING REPLACEMENT

1. Remove bushings from spring eyes using arbor press and suitable size socket or section of pipe.
2. Install replacement bushings in spring eyes using arbor press and suitable size socket or section of pipe. Be sure bushings are centered in spring eyes.

Installation

1. Insert shackle pins into spring rear eye and rear hanger.
2. Position front spring eye in front hanger and install pivot bolt and pivot bolt locknut. Tighten locknut to 110 ft. lbs. (149 Nm) torque.
3. Install shackle plate and locknuts on shackle pins. Tighten locknuts to 30 ft. lbs. (41 Nm) torque.
4. Install clamp bracket, spring isoclamps, spring plate and U-bolts.
5. Engage shock lower mounting stud in spring plate and install grommet, retainer and locknut. Tighten nut to 8 ft. lbs. (11 Nm) torque.

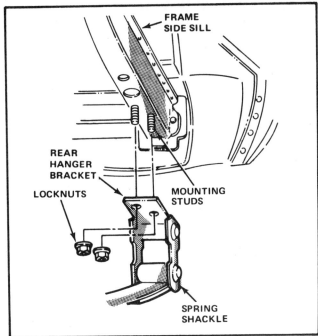

Rear spring, rear mounting (© American Motors Corp.)

Stabilizer Bar

Removal

1. Raise and support the vehicle safely.
2. Remove the nuts and grommets attaching the stabilizer bar to the connecting lings.
3. Remove the bolts attaching the stabilizer bar mounting clamps to the spring clip plates.

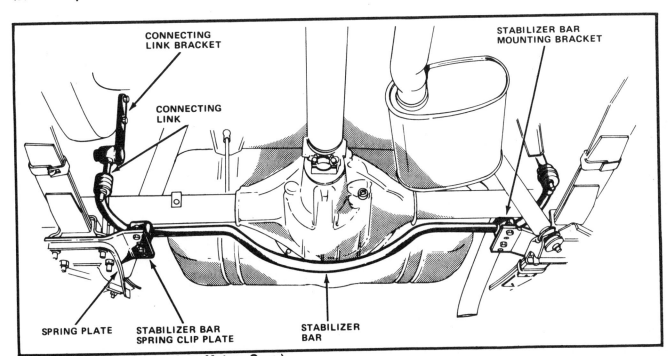

Rear stabilizer bar (© American Motors Corp.)

SUSPENSION
Domestic Models

4. Remove the stabilizer bar from the vehicle.

Installation

1. Position stabilizer bar and mounting clamps on spring clip plates and install clamp bolts finger tight.
2. Install stabilizer bar on connecting links and install grommets and locknuts.
3. Tighten connecting link locknuts to 7 ft. lbs. (9 Nm) torque and tighten stabilizer mounting clamp bolts to 25 ft. lbs. (34 Nm) torque.

Automatic Load Leveling System

Operation

The load leveling system automatically adjusts the rear height with changes in the vehicle loading. The system consists of a compressor assembly, exhaust solenoid, air dryer, compressor relay, air adjustable shock absorbers, height sensors, air tubing, wiring and a pressure limiter.

Adjustment is achieved through the on-board compressor. The electrically operated compressor is mounted in the engine compartment. Although the compressor is controlled automatically, it can be switched to manual by means of a three position switch located in the compressor mounting bracket.

When the three position switch is in the automatic mode, the compressor is operated by the height sensor and compressor relay.

An auxiliary air hose is also included with the system. This hose can be connected to an auxiliary air valve on the compressor. It can be used to inflate air mattresses, tires, etc.

COMPRESSOR

The compressor assembly is a positive displacement single piston air pump powered by a 12 volt permanent magnet motor. The compressor head casting contains piston, intake and exhaust valves plus a solenoid operated exhaust valve which releases air from the system when energized.

NOTE: The compressor is not a serviceable item. If diagnosis indicates the compressor has malfunctioned, replace the compressor as an assembly only. Do not attempt to repair it.

AIR DRYER

The air dryer, which is attached to the compressor output, provides a dual function. It contains a dry chemical which absorbs moisture from the sir before it is delivered to the shocks and returns the moisture to the air when it is exhausted. The air dryer also contains a valve arrangement which maintains a minimum air pressure in the system between 7-14 psi.

EXHAUST SOLENOID

The exhaust solenoid, which provides a dual function, is located in the compressor assembly. When energized, it exhausts air from the system. This operation is controlled by the height sensor. The solenoid also acts as a relief valve to limit maximum output pressure of the compressor.

COMPRESSOR RELAY

The relay is located on the compressor bracket. It is a single pole/single throw type switch. When energized, it completes the 12 volt circuit to the compressor motor. This operation is controlled by the height sensor.

HEIGHT SENSOR

The height sensor, which is an electronic device, controls the compressor relay and exhaust solenoid ground circuit.

To prevent falsely activating the exhaust solenoid circuit or the compressor relay, during normal driving motions, the height sensor provides a 7-15 second delay before either circuit is completed.

The sensor also will limit the time the compressor or exhaust solenoid is energized to a maximum time of $3\frac{1}{2}$ minutes. The time limit is designed to prevent unnecessary running time of the compressor in the event of a system leak or exhaust solenoid malfunction. Turning the ignition "off and on" will reset the electronic timer circuit to the $3\frac{1}{2}$ minute maximum run time.

The electronic timer circuit is also reset for each change in exhaust and compressor signals from the height sensor. The height sensor, on most vehicles, is located on the side sill in the rear of the vehicle with the sensor actuator arm attached to the rear axle housing by means of a short link.

ADJUSTABLE AIR SHOCKS

The adjustable air shock is essentially a conventional shock absorber encased in an air chamber. The shocks are constructed with a rubber sleeve attached to the shock reservoir and dust tube. This creates a flexible chamber which will extend the shock absorber when air pressure is increased in the air chamber. When air pressure is released, the weight of the vehicle collapses the shock absorber.

RAISING THE AUTOMOBILE

When weight is added to the rear of the car, the body is forced downward which causes the height sensor actuating arm to rotate upward. This action causes the height sensor to electrically start the internal time delay circuit. When the time delay (7-15 seconds) has occurred, the sensor then completes the compressor relay circuit to ground. With the relay energized, the 12 V (+) circuit to the compressor is complete and the compressor runs, sending air to the air adjustable shock absorbers through air lines. When the body reaches its original trim height ($\pm \frac{3}{4}$ inch) the sensor opens the compressor relay circuit, shutting of the compressor.

LOWERING THE AUTOMOBILE

A high body condition has the effect of rotating the height sensor actuating arm downward. The height sensor then senses the high condition and starts the time delay circuit. When the time delay (7-15 seconds) has elapsed, the sensor completes the exhaust solenoid circuit to ground. With the exhaust solenoid energized, air escapes from the shocks exiting through the air dryer and exhaust solenoid valve.

As the automobile body lowers, the height sensor actuating arm is rotated toward its original position. When the automobile body reaches its original height ($\pm \frac{3}{4}$ inch), the sensor opens the exhaust valve solenoid circuit, which prevents more air from escaping.

A minimum air pressure of 7-14 psi is maintained on the automobile. The minimum pressure provides improved ride characteristics when the automobile has a minimum load. The compressor relief valve is designed to operate at 120-150 psi.

SYSTEM OPERATION

Testing

1. With the vehicle on a level surface, measure the distance between the bumper and the floor.

SUSPENSION
Domestic Models

2. Turn the ignition on.
3. Apply approximately 200 pounds to the trunk of the vehicle. There should be a 7–15 second delay before the compressor is activated and the vehicle starts to raise. The vehicle should raise to within $3/4$ inch of the measurement made in Step 1.
4. Failure of the vehicle to return to $3/4$ inch of the original position could be due to improper adjustment of the height sensor.

COMPRESSOR

Testing

The following test can be performed with the compressor either on the vehicle or on the bench.

1. Disconnect the wiring from the compressor motor and the exhaust solenoid terminals.
2. Disconnect the existing pressure line from the dryer and attach pressure gauge J-22124A or equivalent to the dryer fitting.
3. Connect a 12 volt (+) power supply to the compressor through an ammeter and note the following:
   a. Current draw should not exceed 14 amps.
   b. When the gauge reads approximately 100 psi minimum, shut off the compressor and note if pressure is maintained or it leaks down (allow pressure to stabilize).
   c. If the compressor is permitted to run until it reaches maximum output pressure of 120–150 psi, the solenoid exhaust valve will act as a pressure relief valve. The resulting leak down when the compressor is shut off will indicate a false leak.

HEIGHT SENSOR

Testing

1. Turn the ignition off then on. This will reset the height sensor timer circuitry to the 3 $1/2$ minute maximum run time.
2. Raise and support the vehicle safely.

NOTE: Be sure that the rear wheels or the axle housing are supported as close as possible to the trim height dimension. Inspect the wiring for proper connections along with the harness ground.

3. Disconnect the link from the height sensor arm.
4. Move the sensor arm up. There should be a 7–15 second delay before the compressor begins operating and the air shocks inflate. As soon as the air shock begin to fill, stop the compressor by moving the sensor arm downward.
5. Move the sensor arm down below the position where the compressor stopped. There should be a 7–15 second delay before the shocks are able to be deflated.
6. Reconnect the link to the height sensor arm before making any adjustments.

TRIM HEIGHT

Adjustment

NOTE: The link must be attached to the metal arm when making the adjustment.

1. Loosen the locknut securing the metal arm to the height sensor plastic arm.
2. To increase the vehicle trim height, move the black plastic actuator arm upward to the top of slot and tighten the locknut.
3. To lower the vehicle trim height, follow Step 1 and move the plastic arm downward to the bottom of the slot.

NOTE: If all the adjustments are used, inspect the vehicle for proper trim height.

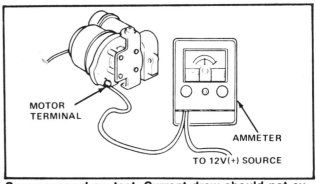

Compressor draw test. Current draw should not exceed 14 amps (© American Motors Corp.)

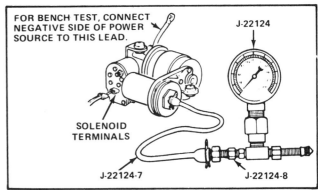

Compressor pressure test (© American Motors Corp.)

SYSTEM LEAK

Testing

1. Connect pressure gauge J-22124-A or equivalent into the system between the dryer and the system air line. Make sure that the shut off valve is on the compressor side of the gauge.
2. With the shut off valve open, apply air pressure through the service valve on the gauge until a reading of 100–120 is obtained.

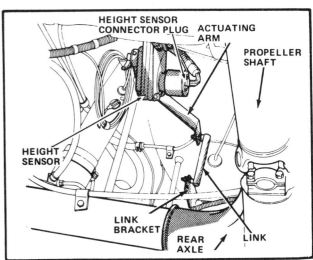

Height sensor (© American Motors Corp.)

2-49

SUSPENSION
Domestic Models

3. If a leak is indicated, close the shut off valve and continue to observe for a pressure drop.
4. If the gauge pressure continues to drop, the leak is external to the compressor.
5. If the gauge does not drop any further, the leak is internal to the compressor.

Component Replacement
COMPRESSOR

Removal and Installation

1. Bleed air from the system.

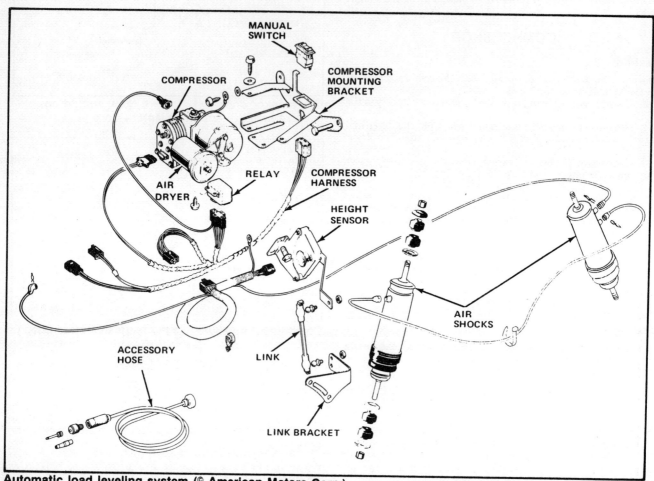

Automatic load leveling system (© American Motors Corp.)

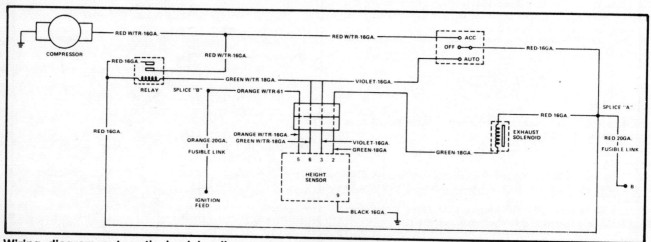

Wiring diagram—automatic load leveling system (© American Motors Corp.)

SUSPENSION
Domestic Models

2. Disconnect the air line at the compressor air dryer.
3. Tag and disconnect electrical connections from the compressor.
4. Remove the compressor mounting bolts. Remove the compressor assembly from the vehicle.
5. Installation is the reverse of the removal procedure. Check compressor for proper operation.

HEIGHT SENSOR

Removal

1. Disconnect connector plug from sensor.

2. Disconnect link from sensor actuating arm.
3. Remove bolts attaching height sensor to underbody and remove sensor.

Installation

1. Position height sensor on underbody and install sensor attaching bolts.
2. Connect sensor actuating arm to link.
3. Connect wiring harness connector plug to height sensor.

NOTE: Due to the diagnostic complexity of the Automatic Load Levelling System, detailed diagnostic procedures can be found in Chilton's Chassis Electronic Service Manual.

FRONT SUSPENSION – CHRYLSER FRONT DRIVE CARS

Alignment

PRE-ALIGNMENT CHECK

There are six factors which are the foundation to front wheel alignment: Height, caster, camber, toe-in, steering axis inclination and toe-out on turns. Of these six basic factors, only camber and toe are mechanically adjustable.

1. Before any attempt is made to change or correct the wheel alignment, inspection and necessary corrections must be made on those parts which influence the steering of the vehicle.
2. Check and inflate tires to recommended pressures.
3. Check front wheel and tire assembly for radial runout.
4. Check struts (shock absorbers for extra-stiff, notchy or spongy operation.
5. Front suspension should be checked only after vehicle has the recommended tire pressures, full tank of fuel, no passenger or luggage compartment load and is on a level floor or alignment rack.
6. To obtain accurate reading, vehicle should be jounced in the following manner just prior to taking measurement. Grasp bumpers at center (rear bumper first) and jounce up and down several times. Always release bumpers at bottom of down cycle after jouncing both rear and front ends an equal number of times.

Camber Adjustment

1. Loosen cam and through bolts (each side).
2. Rotate upper cam bolt to move upper (knuckle and) wheel in or out to specified camber.
3. Tighten bolts to 85 ft. lbs. (115 Nm).

Toe Adjustment

1. Center steering wheel and hold with steering wheel clamp.
2. Loosen tie rod locknuts. Rotate rods to align toe to specifications.

Front Wheel Bearings

The vehicle is equipped with permanently sealed front wheel bearings. There is no periodic lubrication, maintenance, or adjustment recommended for these units.
Service repair or replacement of front drive bearing, hub, brake dust shield or knuckle will require assembly removal from the vehicle.

Strut Damper Assembly

Removal

1. Raise the vehicle and support safely. Remove wheel and tire assembly.
2. Remove cam adjusting bolt, through bolt and brake hose-to-damper bracket retaining screw.
3. Remove strut damper-to-fender shield mounting nut washer assemblies.

Disassembly

1. Compress coil with spring compressor tool.
2. Hold strut rod while loosening strut rod nut. Remove nut.
3. Remove retainers and bushings.
4. Remove coil spring.

NOTE: Mark spring for replacement in original position.

5. Check retainers for cracks or distortion.
6. Check bearings for binding. Check that they contain an adequate supply of lubricant.

Front suspension (© Chrysler Corp.)

SECTION 2
SUSPENSION
Domestic Models

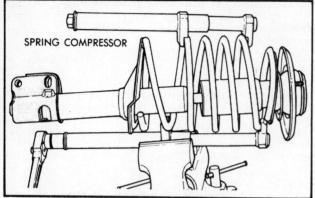
Compressor coil springs (© Chrysler Corp.)

Loosening strut rod nut (© Chrysler Corp.)

Assembly

1. Reassemble and hold strut rod while tightening rod nut to 55 ft. lbs. (81 Nm).

NOTE: Perform Step 1 before releasing spring compressor tool.

2. Release spring compressor tool.

NOTE: Springs are rated separately for each side of vehicle depending on optional equipment and type of service.

During assembly of spring to strut damper, ensure that coil end is seated in strut damper spring seat recess.

Installation

1. Install unit into fender reinforcement and install retaining nut and washer assemblies. Torque to 27 ft. lbs. (20 Nm).
2. Position knuckle leg into strut and install upper (cam) and lower through bolts.
3. Attach brake hose retainer to damper; tighten screw to 10 ft. lbs. (12 Nm).
4. Index cam bolt to original mark and tighten bolts to 85 ft. lbs. (110 Nm) torque.
5. Install wheel and tire assembly. Tighten wheel nuts to 80 ft. lbs. (108 Nm).

Lower Ball Joint

Inspection

The lower ball joint is checked at the lube fitting. Try to turn the lube fitting. If it turns or wobbles, the ball joint is worn and should be replaced.

Removal and Installation

NOTE: On some models, the front ball joints are welded to the control arms and are not to be pressed out. Those that are welded must be serviced by complete replacement of the control arm and ball joint assembly.

1. Pry off seal.
2. Position a receiving cup tool C-4699-2 or equivalent to support the lower control arm.
3. Install a 1 $\frac{1}{16}$ in. deep socket over the stud and against the joint upper housing.
4. Press the joint assembly from the arm.
5. To install, position the ball joint housing into the control arm cavity.
6. Position the assembly in a press with installer tool C-4699-1 or equivalent supporting the control arm.
7. Align the ball joint assembly then press it until the housing ledge stops against the control arm cavity down flange.
8. To install a new seal, support the ball joint housing with installing tool C-4699-2 or equivalent and position a new seal over the stud against the housing.
9. With a 1 $\frac{1}{2}$ socket, press the seal onto the joint housing with the seat against the control arm.

Lower Control Arm

Removal

1. Remove front inner pivot through bolt.
2. Remove rear stub strut nut, retainer and bushing.
3. Remove ball joint-to-steering knuckle clamp bolt.
4. Seperate ball joint stud from steering knuckle.

Pivot Bushing Replacement

1. Position support tool between flanges of lower control arm and around bushing to prevent control arm distortion.
2. Install $\frac{1}{2}$ X 2 $\frac{1}{2}$ in. bolt into bushing.
3. With receiving cup on press base, position control arm inner flange against cup wall to support flange while receiving bushing.

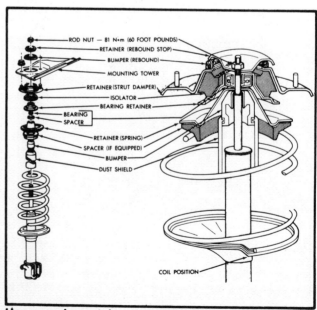
Upper spring retainer assembly (© Chrysler Corp.)

SUSPENSION
Domestic Models

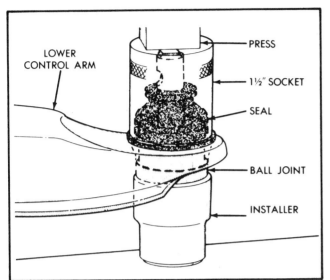
Ball joint seal installation (© Chrysler Corp.)

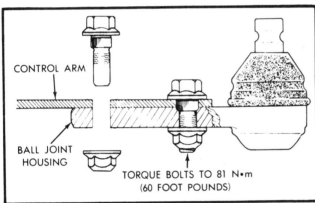
Ball joint bolted to lower control arm (© Chrysler Corp.)

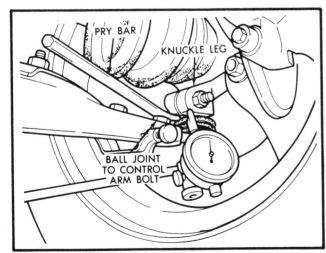
Checking ball joint for excessive clearance, using a dial indicator (© Chrysler Corp.)

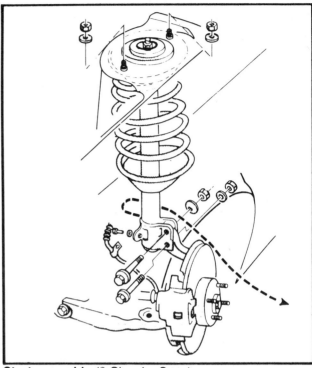
Strut assembly (© Chrysler Corp.)

4. Remove bushing by pressing against bolt head.
5. To install, position support tool between flanges of control arm.
6. Install bushing inner sleeve and insulator into installer tool C-4699-1 or equivalent cavity with the bushing outer shell flange against the tool wall. Position assembly onto press base and align control arm to receive bushing.

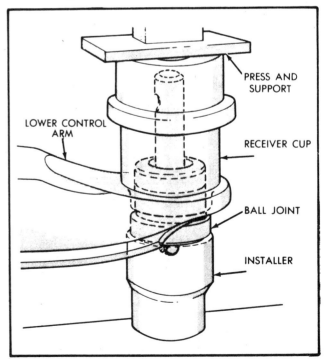
Ball joint assembly (© Chrysler Corp.)

2-53

SECTION 2: SUSPENSION
Domestic Models

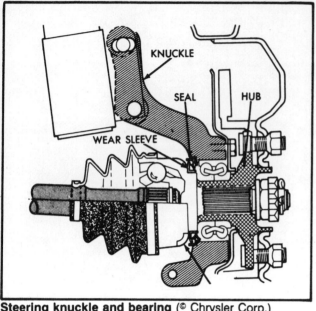
Steering knuckle and bearing (© Chrysler Corp.)

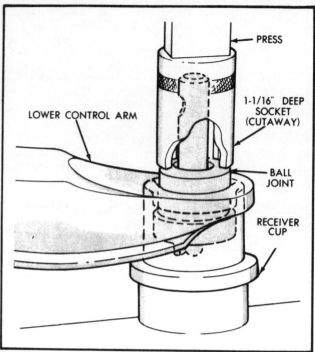
Ball joint removal (© Chrysler Corp.)

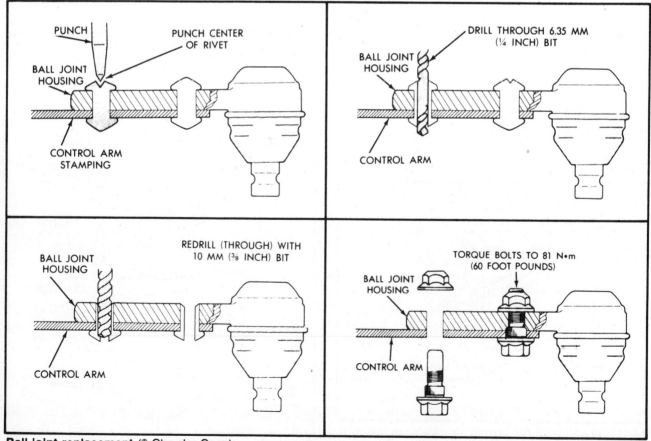
Ball joint replacement (© Chrysler Corp.)

2-54

SUSPENSION
Domestic Models

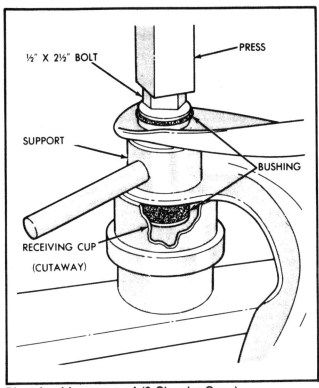

Pivot bushing removal (© Chrysler Corp.)

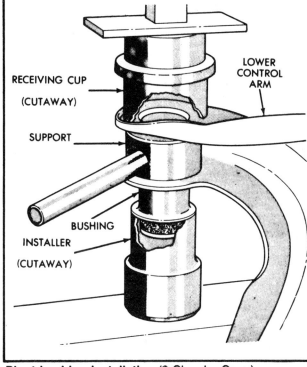

Pivot bushing installation (© Chrysler Corp.)

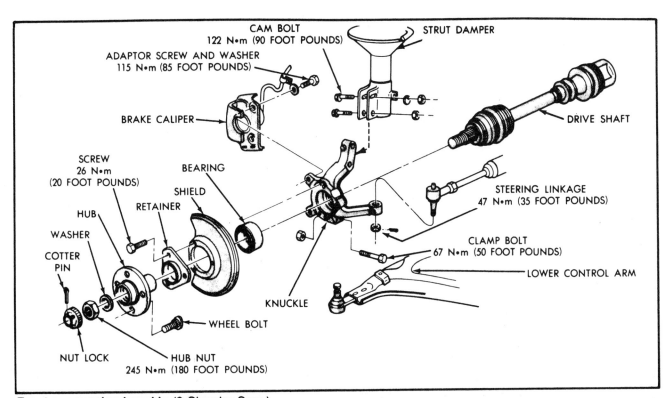

Front suspension knuckle (© Chrysler Corp.)

SUSPENSION
Domestic Models

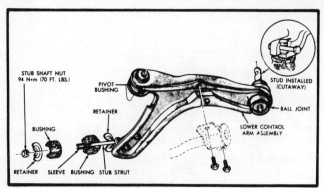
Lower control arm (© Chrysler Corp.)

7. Position receiving cup tool to support control arm outer flange while receiving bushing.
8. Press bushing into control arm until bushing flange seats against control arm.

Installation

1. Install retainer, bushing and sleeve on stub strut.
2. Positon control arm over sway bar and install rear stub strut and front pivot into crossmember.
3. Install front pivot bolt; install nut but do not tighten yet.
4. Install stub strut bushing and retainer and loosely assemble nut.
5. Install ball joint stud into steering knuckle and install clamp bolt. Tighten clamp bolt to 50 ft. lbs. (67 Nm).
6. Position sway bar end bushing retainer to control arm. Install retainer bolts and tighten nuts to 22 ft. lbs. (30 Nm).
7. Lower vehicle. With suspension support vehicle (control arm at design height) tighten front pivot bolt to 100 ft. lbs. (135 Nm) and stub strut nut to 70 ft. lbs. (94 Nm) torque.

Knuckle

The front suspension knuckle provides for steering, braking and alignment and also supports the front (driving) hub (and axle) assembly.
Service repair or replacement of front drive bearing, hub, brake dust shield or knuckle will require removal of the assembly from the vehicle.

Removal and Installation

1. Remove cotter pin and lock.
2. Loosen hub nut while vehicle is on the floor and brakes applied.

NOTE: The hub and driveshaft are splined together through the knuckle (bearing) and retained by the hub nut.

3. Remove wheel and tire assembly.
4. Remove hub nut.

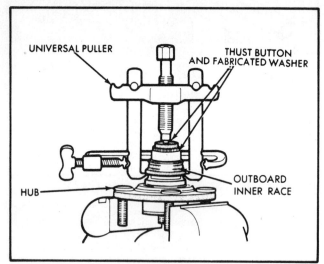
Outboard inner race removal (© Chrysler Corp.)

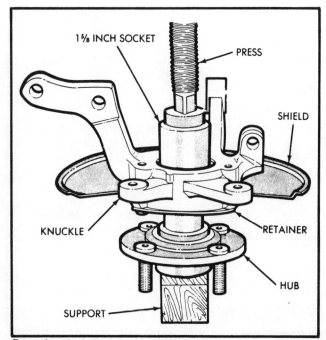
Pressing hub into knuckle bearing (© Chrysler Corp.)

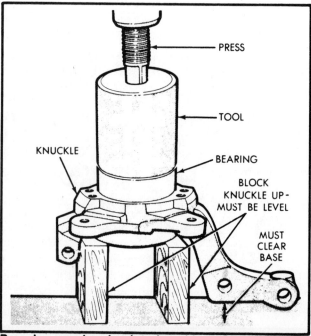
Pressing new bearing into knuckle (© Chrysler Corp.)

SUSPENSION
Domestic Models

CAUTION

Ensure that splined driveshaft is "free" to separate from spline in hub during knuckle removal from vehicle. A pulling force on the shaft can separate the inner universal joint. Tap lightly with soft brass punch if required.

5. Disconnect tie rod end from steering arm.
6. Disconnect brake hose retainer from strut damper.
7. Remove clamp bolt securing ball joint stud into steering knuckle. Remove brake caliper adaptor screw and washer assemblies.
8. Support caliper with wire hook. Do not allow assembly to hang by brake hose.
9. Remove braking disc (rotor).
10. Mark camber position on upper cam adjusting bolt. Loosen both bolts.
11. To remove assembly from vehicle, support knuckle and remove cam adjusting and through bolts, then move upper knuckle "leg" out of strut damper bracket and lift knuckle off of ball joint stud.

CAUTION

Support driveshaft during knuckle removal. Do not allow driveshaft to "hang" after separating steering knuckle from vehicle (severe angles will damage inboard universal joint boot).

12. Installation is the reverse of the removal operation.

Hub And Bearing

Removal and Installation

NOTE: Do not reuse bearing.

1. Remove hub (out of bearing) with hub remover tool and fabricated washer.
   a. Place washer and thrust button on hub.
   b. Back out one retainer screw-to-hub, as far as it will go.
   c. Position tool and install two screws firmly into tapped brake adaption extensions and one screw into retaining screw threads.
   d. Tighten press screw to remove hub through bearing.

NOTE: Bearing inner races will separate; outboard race will stay on hub.

2. Remove bearing outer race from hub with thrust button from tool and universal puller.
3. Remove brake dust shield and bearing retainer.
4. Installation is the reverse of the removal procedure.

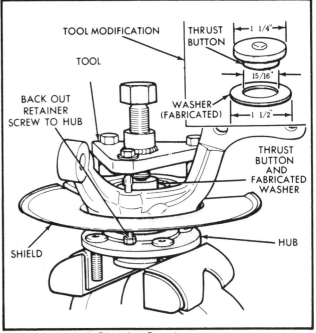

Hub removal (© Chrysler Corp.)

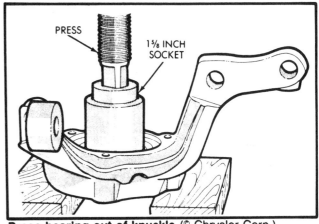

Press bearing out of knuckle (© Chrysler Corp.)

REAR SUSPENSION—CHRYSLER FRONT DRIVE CARS

General Information

Chrysler front drive cars have two types of rear suspension systems. A semi-independent rear suspension which is used on "L" body cars (Omni, Horizon) and a trailing arm solid axle design used on "K" cars (Aries, Reliant), "E" cars (Caravelle, 600, New Yorker), "G" cards (Daytona, Laser, and "H" cars (Lancer, LeBaron GTS). The suspensions are similar, although certain service procedures are varied.

Alignment
ALL MODELS

Because of the trailing arm rear suspension of the vehicle, and the incorporation of stub axles or wheel spindles, it is possible to align both the camber and toe of the rear wheels. Alignment is controlled by adding shim stock of 0.010 in. thickness between the spindle mounting surface and spindle mounting plate.

If rear wheel alignment is required, place vehicle on alignment rack and check alignment specifications. Follow equipment manufacturer's recommendations for their equipment. Maintain rear alignment within Chrylser corporation recommendations.

Installation of Rear Alignment Shims

1. Block front tires so vehicle will not move.
2. Release parking brake.
3. Hoist vehicle so that rear suspension is in full rebound and tires are off the ground.

2–57

SECTION 2
SUSPENSION
Domestic Models

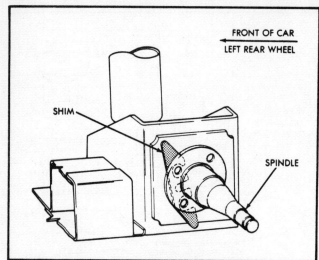

Shim installation for toe out (© Chrysler Corp.)

4. Remove wheel and tire assembly.
5. Pry off grease cap.
6. Remove cotter pin and castle lock.
7. Remove adjusting nut.
8. Remove brake drum.
9. Loosen the brake assembly and spindle mounting bolts enough to allow clearance for shim installation.

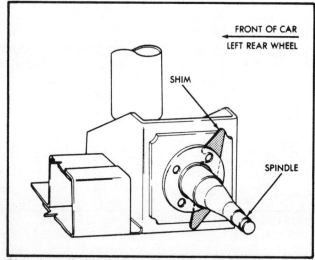

Shim installation for toe in (© Chrysler Corp.)

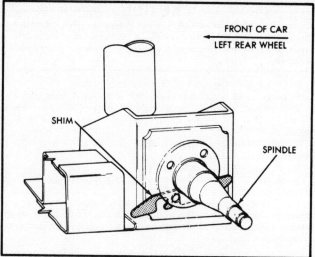

Shim installation for negative camber (© Chrysler Corp.)

NOTE: Do not remove mounting bolts.

10. Install shims for desired wheel change.

NOTE: Wheel alignment changes by 0.3° per shim.

11. Tighten four brake assembly and spindle mounting bolts. Tighten to 45 ft. lbs. (60 Nm) torque.
12. Install brake drum.
13. Install washer and nut. Tighten adjusting nut to 20–25 ft. lbs. (27–34 Nm) while rotating wheel. Back off adjusting nut with wrench to completely release bearing preload. Finger tighten the adjusting nut.
14. Position nut lock with one pair of slots in line with cotter pin hole. Install cotter pin. The end play should be 0.001–0.003 in. (0.025–0.076 mm). Clean and install grease cap.
15. Install wheel and tire assembly. Tighten wheel nuts to 80 ft. lbs. (108 Nm) torque.
16. Lower vehicle.
17. Recheck alignment specifications.

Wheel Bearings
"L" BODY CARS

Lubrication

The lubricant in the wheel bearings should be inspected whenever the drums are removed to inspect or service the brake system, but at least every 22,500 miles (36,000 km). The bearings should be cleaned and repacked with a higher temperature

CHRYSLER CORP.

	CV	E	G	H	K	KC	L	M	S
PLYMOUTH		Caravelle			Reliant		Horizon Turismo	Gran Fury	Voyager
DODGE	600 Convertible	600	Daytona	Lancer	Aries		Omni Charger	Diplomat	Caravan
CHRYSLER	LeBaron	New Yorker	Laser	LeBaron GTS		Limo		Fifth Avenue	

SUSPENSION
Domestic Models

REAR WHEEL ALIGNMENT

Model	Acceptable Alignment Range	Preferred Setting
Camber		
Horizon-Turismo, Omni-Charger	−1.25° to −.25° (−1¼ to 1/4)	−.75° ± .5° (1/2)
Reliant-Aries, Caravelle-600-New Yorker, Daytona-Laser, Lancer-LeBaron	−1.0° to 0° (−1° to 0°)	−.5° ± .5° (1/2°)
Toe*		
Horizon-Turismo, Omni-Charger	specified in inches 5/32″ OUT to 11/32″ IN	3/32″ IN
	specified in degrees 0.3° OUT to 0.7° IN	0.2° IN
Reliant-Aries, Caravelle-600-New Yorker, Daytona-Laser, Lancer-LeBaron	specified in inches 3/16″ OUT to 3/16″ IN	0″ ± 1/8″
	specified in degrees 0.50° OUT to 0.50° IN	0° ± .25°

*Toe Out when backed on alignment racks is Toe In when driving

multipurpose E.P. grease whenever the brake drums are resurfaced.

NOTE: Do not add grease to the wheel bearing. Relubricate completely.

Discard the old seal. Thoroughly clean the old lubricant from the bearings and from the hub cavity. Inspect the rollers for signs of pitting or other surface distress. Light bearing discolorations should be considered normal. Bearings must be replaced if any defects exist. Repack the bearings with a high temperature multipurpose E.P. grease. The use of a bearing packet is recommended. A small amount of new grease should also be added to the hub cavity.

Adjustment

1. Install hub assembly on spindle.
2. Install outer bearing, thrust washer and nut.

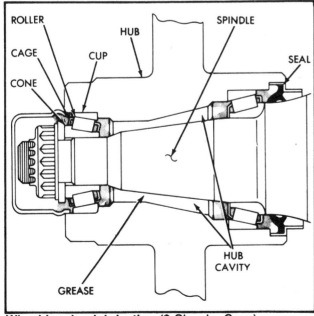

Wheel bearing lubrication (© Chrysler Corp.)

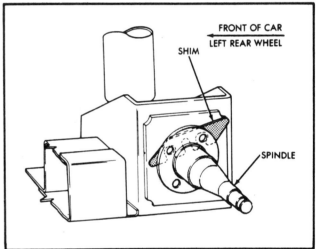

Shim installation for positive camber (© Chrysler Corp.)

3. Tighten wheel bearing adjusting nut to 20–25 ft. lbs. (27–34 Nm) while rotating hub.
4. Back off adjusting nut to release all pre-load, then tighten adjusting nut finer-tight.
5. Position lock on nut with one pair of slots in line with cotter pin hole. Install cotter pin.
6. Install grease cap and wheel and tire assemblies.

"K", "E", "G", "H" BODY CARS

Lubrication

The lubricant in the rear wheel bearings should be inspected whenever the drums are removed to inspect or service the brake system, or at least every 48,000 kilometers (30,000 miles). Bearings should be cleaned and repacked with a high temperature multipurpose E.P. grease whenever the brake drums are resurfaced.

NOTE: Do not add grease to the wheel bearings. Relubricate completely.

Discard the old seal. Thoroughly clean the old lubricant from the bearings and from the hub cavity.

SUSPENSION
Domestic Models

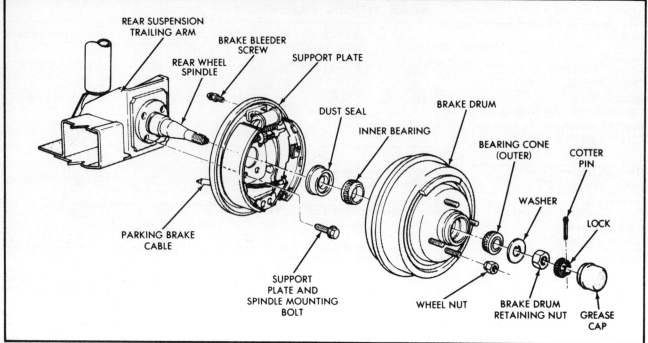

Rear wheel bearing (© Chrysler Corp.)

Inspect the rollers for signs of pitting or other surface distress. Light bearing discoloration should be considered normal. Bearings must be replaced if any defects exist. Repack the bearings with a high temperature multipurpose E.P. grease. Use of a bearing packer is also recommended. A small amount of new grease should also be added to the hub cavity.

Adjustment

1. Tighten adjusting nut to 20–25 ft. lbs. (27–34 Nm) torque, while rotating wheel.
2. Stop rotations and back off adjusting nut with wrench to completely release bearing preload.
3. Finger tighten adjusting nut.
4. Position nut lock with one pair of slots in line with cotter pin hole.
5. Install cotter pin.
6. The end-play should be 0.0012–0.003 in (0.025–0.076mm).

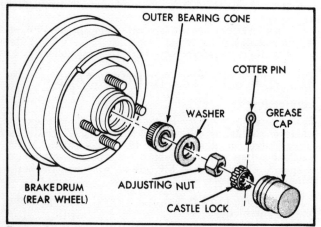

Rear wheel components (© Chrysler Corp.)

7. Clean and install grease cap.
8. Install wheel and tire assembly.

Shock Absorber And Coil Spring Assembly

"L" BODY CARS

Removal

1. Locate upper shock absorber mounting nut protective cap inside of vehicle at upper rear wheel well area (on two-door models, remove lower rear quarter trim panel).
2. Unsnap cap. Use care to retain sound insulation material inside cap.
3. Remove upper shock absorber mounting nut, isolator retainer, and upper isolater. Remove lower shock absorber mounting bolt.
4. Remove shock absorber and coil spring assembly from trailing arm bracket. The shock absorber and coil spring assembly should now be free of vehicle.
5. Place coil spring compressor tool on coil spring and place in vise.

--- **CAUTION** ---

Always grip 4 or 5 coils of spring in retaining nut. If coil spring is not compressed enough, serious injury could occur when retaining nut is loosened.

8. Remove lower isolator, shock rod sleeve, and upper spring seat.
9. Carefully remove shock absorber from coil spring.

Installation

1. Install lower spring seat on shock absorber. Orient seat recess to centerline of lower bushing.
2. Install dust shield and jounce bumper on shock absorber.
3. Carefully slip the unit inside the coil spring. Install upper

SUSPENSION
Domestic Models

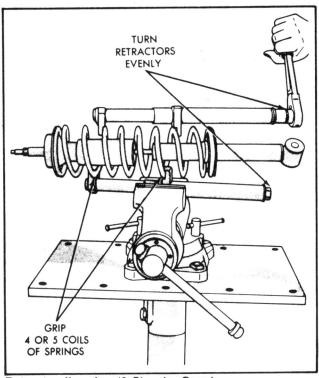

Retract coil spring (© Chrysler Corp.)

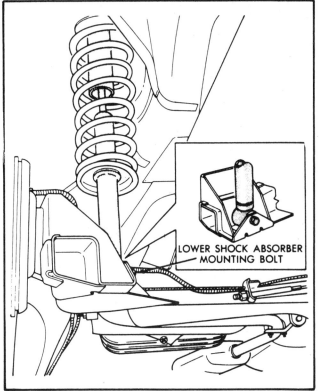

Lower shock absorber mounting bolt (© Chrysler Corp.)

spring seat. Make sure that the leveled surface on both spring seats are in position against the ends of the coil spring.

4. Install sleeve on shock rod. Install retaining nut on end of shock rod. Tighten retaining nut to 20 ft. lbs. (27 Nm) torque.

5. Carefully loosen both coil spring retractors evenly and remove retractors from unit.

6. Install lower end of unit in trailing arm bracket. Insert bolt. Finger tighten only. Make sure that upper end of unit is in proper hole at top of wheel well.

7. Tighten lower shock absorber bolt to 40 ft. lbs. (55 Nm) torque.

8. Install upper isolator, isolator retainer, and upper mounting nut. Hold shock absorber rod end and tighten nut to 20 ft. lbs. (27 Nm) torque.

9. Install sound insulation material and snap protective cap on securely.

Shock Absorbers

"K", "E", "G", "H" BODY CARS

Removal

1. Support axle and remove wheel and tire assembly.
2. Remove upper and lower shock absorber fasteners and remove sock absorbers.

Installation

1. Position shock absorber and install fasteners; loosely assembly lower fastener. Tighten upper fastener to 40 ft. lbs. (54 Nm).
2. Install wheel and tire assembly, tighten wheel stud and nuts to 80 ft. lbs. (108 Nm). Lower vehicle to ground.
3. With suspension supporting vehicle, tighten lower shock absorber fastener to 40 ft. lbs. (54 Nm).

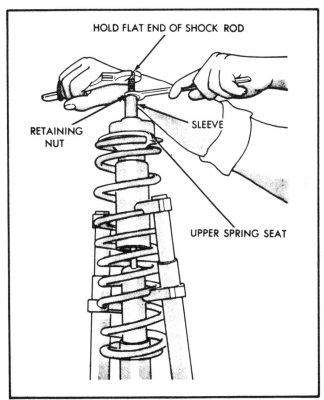

Loosen retaining nut (© Chrysler Corp.)

2-61

SUSPENSION
Domestic Models

Coil Springs And Jounce Bumper

"K", "E", "G", "H" BODY CARS

Removal

1. Raise the vehicle and support safely.
2. Support axle assembly and remove both lower shock absorber attaching bolts. 3. Lower axle assembly until spring upper isolator can be removed (do not stretch brake hose.)
4. Remove two screws hold jounce bumper assembly to rail. Remove jounce bumper assembly.

Installation

1. Position jounce bumper to rail. Install and tighten attaching screws to 5 ft. lbs. (7 Nm).
2. Install isolator over jounce bumper and install spring.
3. Raise axle and loosely assembly both shock absorber-to-axle screws. Remove axle support and lower vehicle.
4. With suspension supporting vehicle, tighten both shock absorber attaching screws to 40 ft. lbs. (54 Nm).

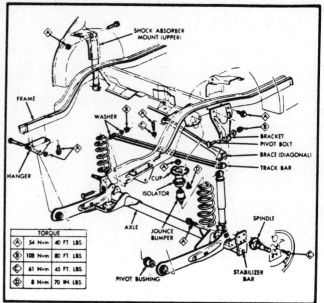

K car flex arm rear suspension (© Chrysler Corp.)

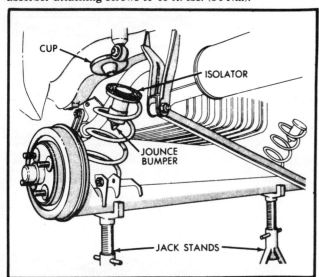

K car spring and jounce bumper (© Chrysler Corp.)

Rear Axle Assembly

"L" BODY CARS

Removal

NOTE: Support the car on the rear crossmember; let the axle hang down.

1. Remove wheel and tire assembly.
2. Remove brake fittings and retaining clips holding flexible brake line.
3. Remove parking brake cable adjusting connection nut.
4. Release both parking brake cables from brackets by slipping ball-end of cables through brake connectors. Pull parking brake cable through bracket.
5. Pry off grease cap.
6. Remove cotter pin and castle lock.
7. Remove adjusting nut. Remove brake drum.
8. Remove four (4) brake assembly and spindle retaining bolts.
9. Set spindle aside and using a piece of wire, hang brake assembly out of way.
10. Remove shock absorber mounting brackets.

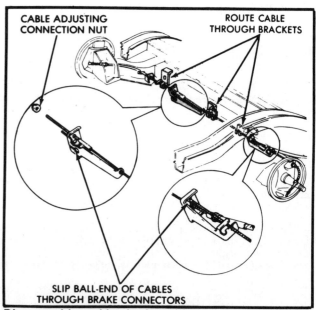

Disassemble parking brake cable and rear crossmember (© Chrysler Corp.)

11. Remove trailing arm-to-hanger bracket mounting bolt.
12. Remove axle assembly.

Installation

1. Using jacks, raise rear axle assembly into position under vehicle.
2. Install trailing arm-to-hanger mounting bracket; finger tighten bolts only.
3. Install shock absorber mounting bolts loosely. Remove jacks.
4. Place spindle and brake assembly in position. Install four (4) retaining bolts finger-tight.
5. Tighten the four retaining bolts to 45 ft. lbs. (60 Nm) torque.

SUSPENSION
Domestic Models

6. Install brake drum.
7. Install washer and nut. Tighten adjusting nut to 20–25 ft. lbs. (27–34 Nm) while rotating wheel. Back off adjusting nut with wrench to completely release bearing pre-load. Finger-tighten adjusting nut.
8. Position nut lock with one pair of slots in line with cotter pin hole. Install cotter pin. The end-play should be 0.001–0.003 in. (0.025–0.076mm). Clean and install grease cap.
9. Put parking brake cable through the bracket.
10. Slip ball-end of parking brake cables through brake connectors on parking brake bracket.
11. Install both retaining clips.
12. Install parking brake cable adjusting connection nut. Tighten until all slcak is removed from cables.
13. Install retaining clips and brake tube fittings. Tighten fittings to 9 ft. lbs. (12 Nm).
14. Bleed rear brake system and readjust brakes.
15. Install wheel and tire assembly. Tighten wheel nuts to 80 ft. lbs. (108 Nm) torque.
16. With vehicle on ground, tighten trailing arm-to-hanger bracket mounting bolts to 40 ft. lbs. (55 Nm) torque.
17. Tighten shock absorber mounting bolts to 40 ft. lbs. (55 Nm) torque.

"K", "E", "G", "H" BODY CARS

Removal

1. Raise the vehicle and support safely. Support the rear axle with adjustable jack stands. Remove the wheel assemblies.
2. Separate the parking brake cable at the connector and cable housing at the floor pan racket.
3. Separate the brake tube assembly from the brake hose at the training arm bracket and remove the lock.
4. Remove the lower shock absorber through bolts and the track bar to axle through bolts. Support the track bar end with wire to keep out of the way.
5. Lower the axle until the spring and isolator assemblies can be removed.
6. Support the pivot bushing end of the trailing arms and remove the pivot bushing hanger bracket to frame screws. Carefully lower the axle assembly and remove from under the vehicle.

Installation

1. Raise and support the axle assembly on adjustable jack stands.
2. Attach the pivot bushing hanger brackets to the frame rail and tighten the attaching bolts.
3. Install the springs and isolators and carefully raise the axle assembly.
4. Install the shock absorber and track bar through bolts. Do not tighten.
5. Position the spindle and brake support to the axle while routing the parking brake cable through the trailing arm opening and the brake tube over the trailing arm. Install the four retaining bolts loosely, then tighten to 45 ft. lbs. torque.

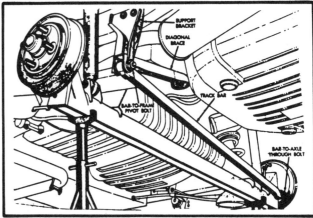
Flex arm rear suspension (© Chrysler Corp.)

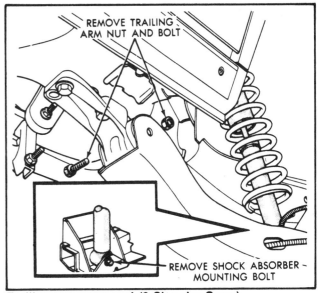
Trailing arms removal (© Chrysler Corp.)

6. Install hub and drum, if previously removed.
7. Route the parking brake cable through the fingers in the retaining bracket and lock housing end into the floor pan bracket. Install the cable end into the intermediate connector.
8. Install the brake hose end fitting into the bracket and install the lock. Tighten as required.
9. Install wheel assemblies and lower vehicle to floor. Tighten the lower shock absorber bolts to 40 ft. lbs. and the track bar bolt to 80 ft. lbs. torque.
10. Bleed the brake system as required.

FRONT SUSPENSION—CHRYSLER REAR DRIVE CARS

Alignment

There are six basic factor which are the foundation to front wheel alignment: height, caster, camber, toe-in, steering axis inclination and toe-out on turns. All are mechanically adjustable except steering axis inclination and toe-out on turns. The latter two are valuable in determining if parts are bent or damaged, particularly when the camber and caster adjustments cannot be brought within the recommended specifications.

2–63

SECTION 2

SUSPENSION
Domestic Models

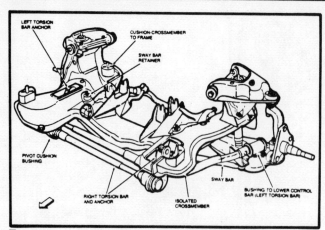

Transverse torsion bar suspension (© Chrysler Corp.)

All adjustments and checks should be made in the following sequence:
1. Front suspension height.
2. Caster and camber.
3. Toe-in.
4. Steering axis inclination (not adjustable).
5. Toe-out on turns (not adjustable).

HEIGHT

Front suspension heights must be measured with the recommended tire pressures and with no passenger or luggage compartment load. The car should have a full tank of gasoline or equivalent weight compensation. It must be on a level surface.

Procedure

Rock the vehicle at the center of the front and rear bumpers at least six times to eliminate friction effects before making the vehicle height measurements. Allow the vehicle to settle on its own weight.

For Gran Fury and Newport/New Yorker, measure from the bottom of the front frame rail, between the radiator yoke and the forward edge of the front suspension crossmember, to the ground. For all other torsion bar front suspension models, measure from the head of the front suspension crossmember front isolator bolt to the ground.

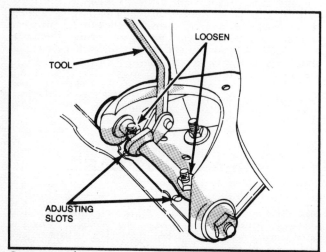

Caster/camber adjustment (© Chrysler Corp.)

On all models, check the measurements against those given in the Front End Alignment Specifications. Adjust the height by turning the torsion bar adjusting bolt clockwise to raise or counterclockwise to lower. The height should not vary more than $1/8$ inch from side to side.

CAMBER AND CASTER

1. Prepare vehicle for measuring wheel alignment.
2. Determine initial camber and caster readings to confirm variance to specifications before loosening pivot bar bolts.
3. Remove foreign material from exposed threads of pivot bar bolts.
4. Loosen nuts slightly hold pivot (caster/camber) bar. Slightly loosening the pivot bar nuts will allow the upper control arm to be repositioned without slipping to end of adjustment slots.
5. Position claw of tool on pivot bar and pin of tool into holes provided in tower or bracket. Make adjustments by moving pivot bar in or out. Adjust as follows:

Camber: Move both ends of upper control arm in or out exactly equal amounts. Camber settings should be held as close as possible to "preferred" settings.

Caster: Moving one end of the bar will change caster (and camber). To preserve camber while adjusting caster, move each end of the upper control arm pivot bar exactly equal amounts in opposite directions. For example, to increase positive caster, move front of pivot bar away from engine, then move rear of pivot bar towards engine in an equal amount. Caster should be nearly equal as possible on both wheels.

TOE

The toe setting should be in the final operation of the front wheel alignment adjustments. In all cases, follow equipment manufacturers procedure.
1. Secure steering wheel in "straight ahead" position. On vehicles equipped with power steering, start engine before centering wheel. (Engine should be kept running while adjusting toe).
2. Loosen tie rod clamp bolts.
3. Adjust toe by turning tie rod sleeves.

NOTE: To avoid a binding condition in either tie rod assembly, rotate both tie rod ends in direction of sleeve travel during adjustment. This will ensure that both ends will be in the center of their travel when tightening sleeve clamps.

4. Shut off engine.
5. Position sleeve clamps so ends do not locate in the sleeve slot, then tighten clamp bolts as specified. Be sure the clamp bolts are indexed at or near bottom to avoid possible interference with torsion bars when vehicle is in full jounce.

Upon completion of alignment operations, it is essential that the splash shields, if removed, be correctly reinstalled with all holding clips in place.

Front Wheel Bearings

LUBRICATION

Under normal service, the lubricant in front wheel bearings should be inspected whenever brake drums or disc brake rotors are removed to inspect or service the brake system, but at least every 30,000 miles (48,000 kilometers).

For severe service vehicles (such as taxi and police vehicles involving frequent brake application), wheel bearings should be inspected whenever the rotors are removed to inspect or service the brake system, or at least eery 9,000 miles (14,000 kilometers), whichever occurs first.

SUSPENSION
Domestic Models

Check lubricant to see that it is adequate in quantity and quality. If grease is low in quantity, contains dirt, appears dry or has been contaminated with water to produce a milky appearance, bearings should be cleaned and completely repacked. Never add grease to wheel bearings.

When lubrication is required, discard old seal. Thoroughly clean old lubricant from bearings and from hub cavity. Inspect rollers for signs of pitting or other surface distress. Light bearing discolorations should be considered normal. Bearings must be replaced if any defects exist. For all service, repack the bearings with a high temperature wheel bearing grease. Use of a bearing packer is recommended. A small amount of new grease should also be added to hub cavity.

Adjustment

1. Tighten adjusting nut to 20–25 ft. lbs. (27–34 Nm) while rotating wheel. Stop rotation and back off adjusting nut with wrench to completely release bearing pre-load. Next, finger-tighten adjusting nut. Position nut lock with one pair of slots in line with cotter pin hole. Install cotter pin. The resulting adjustment should be 0.0001–0.0003 in. end play.
2. Clean and install grease cap. Install wheel and tire assembly.

Removal

1. In the event the bearing cap in found defective during inspection, remove grease cap, cotter pin, nut lock and bearing adjusting nut.
2. Remove the disc brake sliding caliper retaining clips and anti-rattle springs.
3. Slowly slide caliper housing assembly up and away from brake disc and support caliper housing or steering knuckle arm. Do not let caliper housing hang by brake hose, as possible brake hose damage may result.
4. Remove thrust washer and outer bearing cone.
5. Slide wheel hub and disc assembly off the spindle.
6. Carefully drive out inner seal and remove bearing cone with $\frac{3}{4}$ in. diameter non-metallic rod.

Installation

1. Using a bearing drive tool, install new cone. Care must be taken to fully seat new cup against shoulder of hub.
2. Force lubricant between all bearing cone rollers or repack using a suitable bearing packer. A small amount of grease should be added to hub cavity.
3. Install inner cone and a new seal with lip of seal facing inward. Position seal flush with end of hub. The seal flange may be damaged if proper tool is not used.
4. Clean spindle and apply a light coating of wheel bearing lubricant over polished surfaces.
5. Install hub and braking disc assembly on spindle and install outer bearing cone, thrust washer and adjusting nut. Refer to bearing adjustment procedure.
6. Slowly slide caliper housing assembly down on brake disc assembly into position on adaptor. Install caliper retaining clips and anti-rattle springs. Tighten to 15 ft. lbs. (20 Nm).
7. Install tire and wheel.

Shock Absorbers

Removal

NOTE: To remove the front shock absorbers on all models, you may find it more convenient to remove the wheel assembly and perform the removal from under the fender.

1. Loosen and remove nut and retainer from upper end of shock absorber piston rod.
2. Raise vehicles so wheels are clear of floor and remove lower attachment.

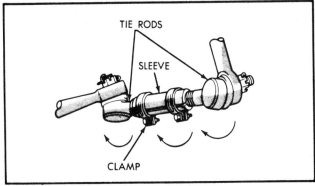

Tie rod adjustment (© Chrysler Corp.)

3. Compress shock absorber completely by pushing upward. Remove from vehicle by pulling down and out of upper shock absorber mounting bushing.
4. Check appearance of upper shock absorber mounting bushing.

If it appears worn, damaged or deteriorated, remove bushing by first pressing out inner sleeve with a suitable tool then prying out or cutting out the rubber bushing. (This bushing will take some set after it has been in service and must be replaced once it has been removed).

Installation

1. To install upper rubber bushing, remove inner seal sleeve and immerse bushing in water (do not use oil) and with a twisting motion, start bushing into hole of upper mounting bracket.

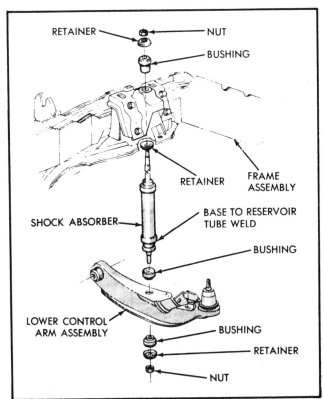

Front shock absorbers—transverse torsion bar suspension (© Chrysler Corp.)

2–65

SUSPENSION
Domestic Models

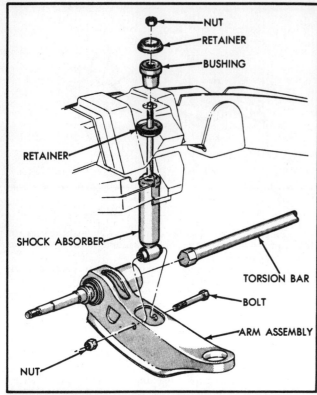

Front shock absorber – conventional torsion bar suspension (© Chrysler Corp.)

Tap into position with a hammer. Reinstall steel inner sleeve in bushing.

2. Test and expel air from shock absorber. Compress to its shortest length. Install upper bushing lower retainer and insert rod through upper bushing. Install upper retainer and nut; tighten to 25 ft. lbs. (34 Nm).

NOTE: In each case, install all retainers with the concave side in contact with the rubber.

3. Position and align lower eye of shock absorber with that of lower control arm mounting holes. Install shock absorber and tighten nut to 50 ft. lbs. (68 Nm) on bolt-and-nut-type. On suspensions with removal bushings, tighten retainer nut to 35 ft. lbs. (47 Nm) with full weight of vehicle on the wheels.

NOTE: When tightening retaining nut, be sure to grip shock absorber at the base area below the weld to avoid reservoir damage.

Upper Ball Joint

Inspection

1. Position jack under the lower control arm and raise wheel clear of floor. Remove wheel cover, grease cap and cotter pin.
2. Tighten bearing adjusting nut enough to remove all play between hub, bearings and spindle.
3. Lower jack to allow tire to lightly contact floor (most of vehicle weight relieved from the tire). It is important that the tire have contact with the floor.
4. Grasp the top of the tire and apply force inward and outward. While this force is being applied, have an observer check for any movement at the ball joints between the upper control arm and the knuckle.
5. If any lateral movement is evident, replace the ball joint.

Removal

1. Place ignition switch in Off or Unlocked position.
2. Raise front of vehicle with hand jack and place jack stand under lower control arm. Position jack stand as close to wheel as possible. Be sure jack stand is not in contact with brake splash shield. Rubber rebound bumper must not contact frame.

— CAUTION —
Torsion bar will remain in loaded position.

3. Remove wheel cover, wheel and tire assembly.
4. Remove cotter pins and nuts from upper and lower ball joints to facilitate use of ball joint removal tool.
5. Slide tool on lower ball joint stud allowing tool to rest on knuckle arm. Set tool securely against upper stud.
6. Tighten tool to apply pressure to upper stud and strike knuckle sharply with hammer to loosen stud. Do not attempt to force stud out only with tool.
7. After removing tool, disengage upper ball joint from knuckle. Support knuckle and brake assembly to prevent damage to brake hose or lower ball joint.
8. Remove upper ball joint by turning counterclockwise from upper control arm.

Installation

1. Screw ball joint squarely into control arm as far as possible by hand. Make certain ball joint threads engage those of control arm correctly if original arm is used. Seals should always be replaced once they have been removed.
2. Tighten ball joint until it bottoms on housing. Tighten to 125 ft. lbs. (180 Nm).
3. Position new seal over balljoint stud and install using tool adapter. Make sure seal is seated on ball joint housing.
4. Postion upper ball join stud in steering knuckle and install nut. Tighten nut to 100 ft. lbs. (136 Nm).
5. Install lower ball joint stud nut and tighten to 100 ft. lbs. (136 Nm). Install cotter pin and lubricate upper ball joint.
6. Torsion bar will remain in loaded position.
7. Install wheel and tire assembly with wheel cover.

Lower Ball Joint

Inspection

1. Raise the front of the vehicle and install safety floor stands under both lower control arm as far outboard as possible. The upper control arms must not contact the rubber rebound bumpers.
2. With the weight of vehicle on the control arm, install dial indicator and clamp assembly to lower control arm.
3. Position dial indicator plunger tip against knuckle arm and zero dial indicator.
4. Measure axial travel of the knuckle arm with respect to the control arm, by raising the lowering the wheel using a pry bar under the center of the tire.
5. If during measurement you find the axial travel of the control arm is 0.030 in. (0.76 mm) or more, relative to the knuckle arm, the ball joint should be replaced.

Removal

1. Place ignition switch in Off or Unlocked position.
2. Raise vehicle on hoist to place front suspension in rebound. Place jack stands under front frame for additional support.
3. Remove wheel cover, wheel and tire assembly.
4. Remove brake caliper and support with wire hook. Do not hang caliper by brake hose alone.
5. Remove hub and rotor assembly and splash shield. Disconnect shock absorber at lower control arm.

6. Unwind torsion bar.
7. Remove upper and lower ball joint stud cotter pins and nuts. Slide tool over upper stud until tool rests on steering knuckle.
8. Turn threaded portion of tool locking it securely against lower stud. Tighten tool enough to place lower ball joint stud under pressure, then strike steering knuckle arm sharply with a hammer to loosen stud. Do not attempt to force stud out of knuckle with tool alone.
9. Use tool to press ball joint out of lower control arm.

Installation

1. Press new ball joint into lower control arm assembly.
2. Place a new seal over ball joint with adapter tool. Press retainer portion of seal down on ball joint housing until it is securely locked in position.
3. Insert ball joint stud into opening in knuckle arm and install stud retaining nuts; tighten as specified. Install cotter pins and lubricate ball joint.
4. Place a load on torsion bar by turning adjusting bolt clockwise.
5. Install wheel, tire and brake assembly and adjust front wheel bearing.
6. Lower vehicle to floor. Adjust front suspension heights.

Torsion Bars (Longitudinal Type)

Longitudinal torsion bars have a hex formed on each end. One hex end is installed in the lower control arm anchor, the opposite end is anchored in the frame or body crossmember.

Torsion bars are identified for use by length and thickness (depending on carline, body, engine, etc.) and are not interchangeable side for side. The bars are marked on either right or left by the letter "R" or "L" stamped on one end of the bar.

Removal

1. Lift vehicle on hoist to place front suspension in rebound.
2. Release load from torsion both bars by turning the anchor adjusting bolt in lower control arm counterclockwise.
3. Remove lock ring from anchor at rear of bar. Install drivetool to remove torsion bar. (If necessary, remove transmission torque shaft to provide clearance). Place tool toward rear of bar to allow sufficient room for striking pad of tool. Do not apply heat to torsion bar, front anchor or rear anchor.
4. Remove tool and slide bar out through rear anchor. Do not damage balloon seal when removing bar.

Inspection

1. Inspect torsion bar and seal for damage; replace if damaged.
2. Remove all foreign matter from hex opening(s) in anchors and from hex end(s) of torsion bar.
3. Inspect torsion bar adjusting bolt and swivel and replace if there is corrosion or other damage. Lubricate for easy installation.

Installation

1. Insert torsion bar through rear anchor.
2. Lubricate inside surface of balloon seal and slide seal over torsion bar (cupped end toward read of bar).
3. Coat both hex ends of torsion bar with lubricant.
4. Slide torsion bar in hex opening of lower control arm.

NOTE: If torsion bar hex opening does not index with lower control arm hex opening, loosen lower control arm pivot shaft nut, rotate pivot shaft to index with torsion bar. Install torsion bar. Do not tighten pivot shaft nut while suspension is in rebound.

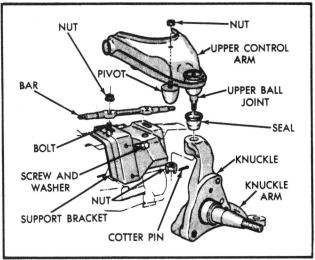

Knuckle control arm and ball joint (© Chrysler Corp.)

5. Install lock ring, making sure it is seated in its groove.
6. Pack rear anchor openings at lock ring area and area under seals with lubricant. Position lip of seal in groove of anchor.
7. Turn adjusting bolt clockwise to place a load on torsion bar.
8. Lower vehicle to floor and tighten pivot shaft nut to 145 ft. lbs. (197 Nm).
9. Adjust front suspension height.

Torsion Bars (Transverse–Type)

Torsion bars are formed with an angle for transverse mounting. Each bar is hex shaped on the anchor end with a replaceable torsion bar-to-lower control arm bushing on the opposite end and a pivot cushion bushing (permanently attached) midway on the bar creating right and left hand assemblies.

The hex end of the bar is anchored in the crossmember (opposite the affected wheel), extends parallel to the front crossmember, through the pivot cushion bushing (also attached to the crossmember), turns, and attaches to the lower control arm through the torsion bar to lower control arm bushing.

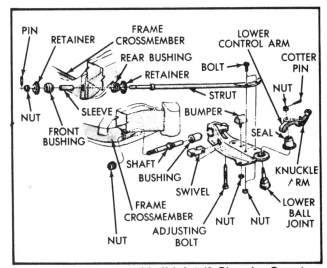

Lower control arm and ball joint (© Chrysler Corp.)

SUSPENSION
Domestic Models

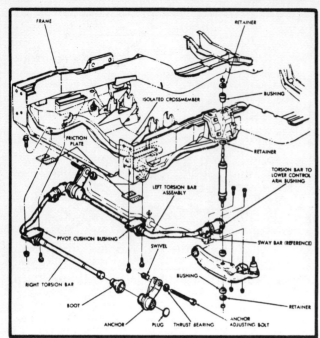

Transverse torsion bar front suspension (© Chrysler Corp.)

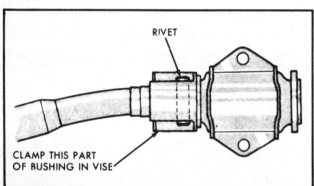

Bushing removal, torsion bar to lower control arm (© Chrysler Corp.)

Removal

1. Raise car on hoist and support vehicle so that front suspension is in full rebound position.
2. Release load on both torsion bars by turning anchor adjusting bolts in frame crossmember counterclockwise. Remove anchor adjusting bolt on torsion bar to be removed.
3. Raise lower control arms until clearance between crossmember ledge (at jounce bumper) and torsion bar end bushing is 2 7/8 in. (63mm). Support lower control arms at this design height (equal to three passenger position with vehicle on ground). This is necessary to align sway bar and lower control arm attaching points for disassembly and component re-alignment and attachment during reassembly.
4. Remove sway bar-to-control arm attaching bolt and retainers.
5. Remove two bolts attaching torsion bar end bushing to lower control arm.
6. Remove two bolts attaching torsion bar pivot cushion bushing to crossmember, and remove torsion bar and anchor assembly from crossmember.

7. Carefully separate anchor from torsion bar.

TORSION BAR – TO – LOWER CONTROL ARM BUSHING REPLACEMENT

Service replacement bars include pivot cushion bushing and torsion bar to lower control arm bushing.

1. Clamp assembly in vise with rivet head up (hex end of bar down).

CAUTION

Never clamp the bar in a vise unless soft vise jaw inserts (brass, aluminum, etc.) are used.

2. Centerpunch the rivet head and drill a 3/8 in. (9.5 mm) diameter hole approximately 1/2 in. (12.5 mm) deep. A short length of 5/16 in. (8 mm) rod can be used to remove the rivet. It may be necessary to remove flange of rivet head before driving rivet out.

CAUTION

Do not enlarge the 7/16 in (11 mm) diameter hole in the bar.

3. Remove bushing from bar.
4. Install new bushing. Rough area under bushing may be cleaned with sandpaper if necessary for easy assembly. New bushing should go on by hand.
5. Install bushing retaining bolt and tighten nut to 50 ft. lbs. (68 Nm).

Inspection

1. Inspect seal for damage, replace if necessary.
2. Inspect bushing-to-lower control arm and pivot cushion bushing.

NOTE: Inspect seals on cushion bushing for cuts, tears or severe deterioration that may allow moisture under cushion. If corrosion is evident, the torsion bar assembly should be replaced.

3. Remove all foreign matter from hex opening(s) in anchors and from hex end(s) of torsion bar.
4. Inspect torsion bar adjusting bolt and swivel and replace if there is any sign of corrosion or other damage. Lubricate for easy installation.

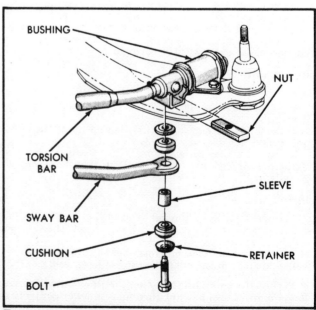

Transverse bar lower control arm mounting (© Chrysler Corp.)

SUSPENSION
Domestic Models

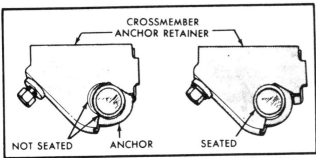
Torsion bar anchor assembly (© Chrysler Corp.)

Installation

1. Carefully slide balloon seal over end of torsion bar (cupped end toward hex).
2. Coat hex end of torsion bar with lubricant.
3. Install torsion bar hex end into anchor bracket. With torsion bar in a horizontal position, the ears of the anchor bracket should be positioned nearly straight up. Position swivel into anchor bracket ears.
4. Place bushing end of bar into position on top of lower control arm. Then, install anchor bracket assembly into crossmember anchor retainer and install anchor adjusting bearing and bolt.
5. Attach pivot cushion bushing to crossmember with two bolt and washer assemblies. Leave bolt and washer assemblies loose enough to install friction plates.
6. With lower control arms at "design height", install two bolt and nut assemblies attaching torsion bar bushing to lower control arm. Tighten to 70 ft. lbs. (95 Nm).
7. Ensure that torsion bar anchor bracket is fully seated in crossmember. Install friction plates between crossmember and pivot cushion bushing with open end of slot to rear and bottomed out on mounting bolt. Tighten cushion bushing bolts to 85 ft. lbs. (115 Nm).

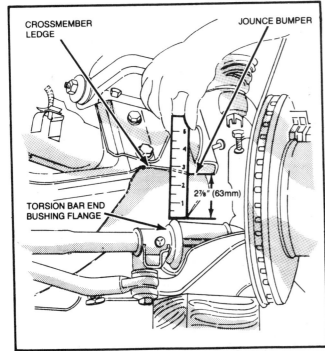
Measuring design height (© Chrysler Corp.)

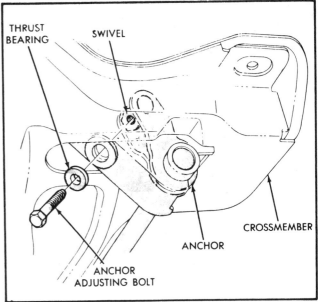
Transverse torsion bar anchor bolt (© Chrysler Corp.)

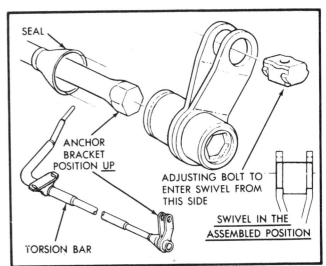
Correct anchor and swivel installation (© Chrysler Corp.)

8. Position balloon seal over anchor bracket.
9. Reinstall bolt, through sway bar, retainer cushions and sleeve, and attach to lower control arm end bushing. Tighten bolt to 50 ft. lbs. (68 Nm).
10. Load torsion bar by turning anchor adjusting bolt clockwise.
11. Lower vehicle and adjust torsion bar height to specifications.

Upper Control Arm

Removal

1. Place ignition switch in Off or Unlocked position.
2. Raise front of vehicle with hand jack and remove wheel cover, wheel and tire assembly.
3. Position short jack stand under lower control arm near splash shield and lower hand jack. Observe that jack stand

SECTION 2
SUSPENSION
Domestic Models

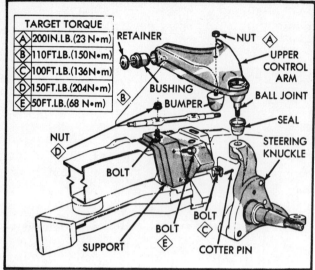

Upper control arm and knuckle assembly (© Chrysler Corp.)

does not contact shield and rebound bumpers are under no load.

4. On some models, remove brake caliper and set aside to provide clearance for ball joint remover tool.
5. Remove cotter pin and nut from upper and lower ball joints to facilitate use of tool to free ball joint.
6. Slide spreader tool over lower ball joint stud to allow tool to rest on steering knuckle arm. Tigthen tool to apply pressure to upper ball joint stud and strike steering knuckle boss sharply with hammer to loosen stud. Do not attempt to force stud out of knuckle with tool alone.
7. After removing tool, support brake and knuckle assembly to prevent damage to brake hose or lower ball joint, then disengage upper ball joint from knuckle.

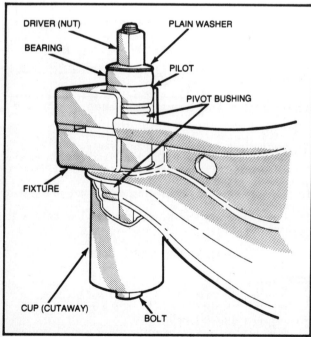

Lower bushing removal tool (© Chrysler Corp.)

8. From under hood, remove engine splash shield to expose upper control arm pivot bar.
9. Scribe a line on support bracket along inboard edge of pivot bar (to re-establish supsension alignment during reassembly).
10. Remove pivot bolts or nuts and lift upper control arm with ball joint and pivot bar assembly from bracket.

Disassembly (Bushings)

1. Place upper control arm in vise and remove pivot bar nuts and bushing retainers.
2. Bolt support tool C-4253-1 to pivot bar.
3. Place puller tool C-4253-2 over end of pivot bar and reinstall nut. Snug bolts against arm.
4. Screw bolts equally until bushing is free in arm and remove tool and bushing.

Assembly (Bushings)

1. With control arm in vise, put pivot bar in arm and attach support bracket spacer tool.
2. Slip bushings over each end of pivot bar and pilot into holes in arm.
3. Install bushing cups over both bushings and press bushings together until both bushings are fully seated in arm. Pound bushings in place at the same time or use an arbor press. Bushing flange must be bottomed on control arm extrusion.
4. Install retainers and nuts on pivot bar. Snug nuts against retainers.

NOTE: Pivot bar bushing retainer nuts are to be tightened to specifications AFTER suspension (upper control arm) is at design height.

Installation

1. Place upper control arm with ball joint and pivot bar on bracket. Install and snug attaching bolts against arm.
2. Seat inboard edge of pivot bar on mounting bracket. Tighten bolts to 150 ft. lbs. (204 Nm).
3. Replace engine splash shield.
4. Install ball joint stud through steering knuckle and install upper and lower ball joints nuts, tighten to specifications and install cotter pins.
5. With vehicle at design height tighten pivot bar nuts to 110 ft. lbs. (150 Nm).

LOWER CONTROL ARM (LONGITUDINAL-TYPE TORSION BARS)

Removal

1. Place ignition switch in Off or Unlocked position.
2. Remove rebound bumper.
3. Raise vehicle on hoist to place front suspension in rebound. Place jack stands under front frame for additional support.
4. Remove wheel cover, wheel and tire assembly.
5. Remove brake caliper and set aside. Do not hang caliper by brake hose alone. Disconnect shock absorber lower bolt.
6. Remove hub and rotor assembly, splash shield and lower shock mounting nut. Remove bolt and nut.
7. Remove two (2) strut bar attaching bolts from lower control arm.
8. Remove automatic transmission gear shift torque shaft assembly if required for tool clearance.
9. Measure torsion bar anchor bolt depth into lower control arm before unwinding torsion bar. Unwind bar.
10. Remove torsion bar.
11. Separate lower ball joint from knuckle arm.

12. Remove lower control arm shaft nut from control arm shaft and push out shaft from frame crossmember. Strike threaded end of shaft with soft hammer to loosen if necessary. Remove lower control arm and shaft as an assembly from the vehicle.

13. In the event the shaft bushing indicates wear and deterioration, replacement is recommended.

Disassembly (Bushings)

1. Place lower control arm in vise and remove torsion bar adjusting bolt and swivel.
2. Place lower control arm assembly in arbor press with torsion bar hex opening up and with a support around anchor on bottom end.
3. Place a brass drift into hex opening and press shaft out of lower control arm. The bushing inner shell will remain on shaft.
4. Cut and remove rubber portion of bushing from control arm or shaft. Remove bushing outer shell by cutting with a chisel. Use care not to cut into control arm.
5. Remove bushing inner shell with pivot shaft.

Assembly (Bushings)

1. Position new bushing on shaft (flange end of bushing first). Press shaft into inner sleeve until bushing seats on shoulder of shaft.
2. Press shaft and bushing assembly into lower control arm using an arbor press.
3. Install torsion bar adjusting bolt and swivel.

Installation

1. Position lower control arm with shaft in crossmember. Install lower control arm shaft nut and finger-tighten nut.
2. Position lower ball joint stud into knuckle arm and tighten nut to 100 ft. lbs. (136 Nm).
3. Install torsion bar into lower control arm.
4. Replace transmission gear shaft torque shaft if removed.
5. Position strut bar with two attaching bolts to lower control arm. Tighten to 100 ft. lbs. (136 Nm).
6. Attach brake splash shield and secure lower shock mounting bolt to lower control arm.
7. Attach hub and rotor and install brake caliper.
8. Install wheel and tire assembly.
9. Lower vehicle to floor and adjust front suspension heights. Tighten lower control arm pivot shaft nut to 145 ft. lbs. (197 Nm). Install rebound bumper and tighten to 17 ft. lbs. (23 Nm).
10. Adjust wheel alignment.

LOWER CONTROL ARM (TRANSVERSE–TYPE TORSION BARS)

Removal

1. Raise car on hoist and remove wheel and tire assembly.
2. Remove brake caliper retaining screws, clips and anti-rattle springs and remove caliper from adaptor and support caliper assembly on wire hook. (Do not hang caliper by brake hose.).
3. Remove hub and rotor assembly and splash shield.
4. Remove shock absorber lower nut, retainer and insulator.
5. Release load on both torsion bars by turning anchor adjusting bolts counter clockwise. Releasing both torsion bars is required because of sway bar reaction from the opposite torsion bar.
6. Raise lower control arm until clearance between crossmember ledge (at jounce bumper) and torsion bar to lower control arm bushing is 2 7/8 inches (73 mm). Support control arm at this "design height" and remove two bolts attaching torsion bar end bushing to lower control arm.
7. Separate lower ball joint from knuckle arm.
8. Remove lower control arm pivot bolt and lower control arm.

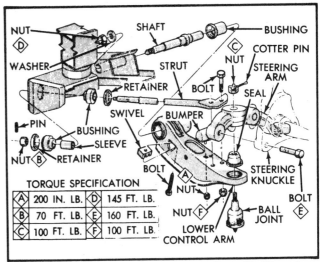

Longitudinal torsion bar suspension (© Chrysler Corp.)

Disassembly (Bushings)

1. Place lower control arm in vise and install bushing removal tool.
2. Place support fixture between flanges of control arm and around bushing. Proper fixture position is required to prevent control arm distortion during bushing removal.
3. Position cup over flanged bushing end with bolt through cup and bushing.

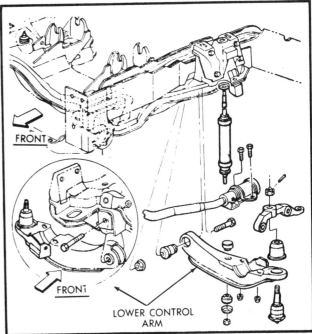

Lower cobtrol arm with transverse torsion bar suspension (© Chrysler Corp.)

2–71

SUSPENSION
Domestic Models

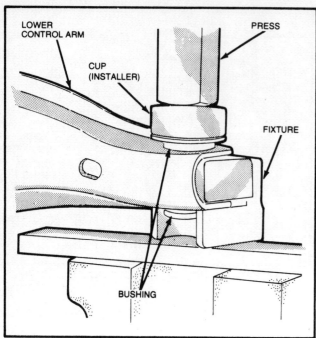

Lower control arm bushing installation (© Chrysler Corp.)

4. Install pilot, thrust washer, plain washer and nut on through bolt.
5. Press bushing out of lower control arm by hold bolt on cup end while turning nut of pilot end.

Assembly (Bushings)

1. Place support fixture on lower control arm flanges and position assembly on base of suitable press. Proper fixture position is required to prevent control arm distortion during bushing installation.
2. Position flange end of new bushing into cup squarely and press bushing into control arm until bushing flange seats on arm.

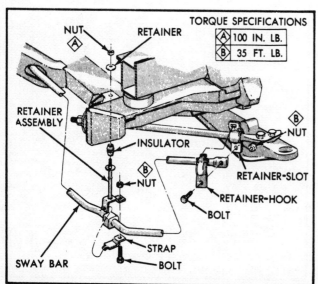

Longitudinal type torsion bar (© Chrysler Corp.)

Installation

1. Position lower control arm in crossmember, install pivot bolt and finger-tighten flanged nut.
2. Position lower ball joint stud into steering knuckle arm and tighten nu to 100 ft. lbs. (136 Nm). Insert cotter pin.
3. Install torsion bar into lower control arm.
4. Load torsion bar by returning torsion bar adjusting bolt depth to original position before removal.
5. Position strut bar with two attaching bolts to lower control arm. tighten to 100 ft. lbs. (136 Nm).
6. Attach brake splash shield and secure lower shock mounting bolt to lower control arm.
7. Attach hub and rotor and install brake caliper.
8. Install wheel and tire assembly.
9. Lower vehicle to floor and adjust front suspension heights. Tighten lower control arm pivot shaft nut to 145 ft. lbs. (197 Nm). Install rebound bumper and tighten to 17 ft. lbs. (23 Nm).
10. Adjust wheel alignment.

Steering Knuckle Arm

Removal

1. Place ignition switch in Off or Unlocked position.
2. Remove rebound bumper.
3. Raise vehicle on hoist to place front suspension in rebound. Use jack stands under front frame to additional support.
4. Remove wheel cover, wheel and tire assembly.
5. Remove brake caliper and hang out of way with wire hook during this operation to prevent damage to brake hose.
6. Remove hub and brake disc assembly.
7. Remove brake splash shield from steering knuckle.
8. Unload torsion bars, by turning anchor adjusting bolt counterclockwise.
9. Disconnect tie rod from steering knuckle arm. Use care not to damage seals.
10. Remove lower ball joint stud from knuckle arm.
11. Separate knuckle arm from steering knuckle by removing two (2) nuts and two (2) attaching bolts.
12. Remove steering knuckle arm.

Installation

1. Attach steering knuckle arm to knuckle and install two bolts and nuts. Tighten to 160 ft. lbs. (217 Nm).
2. Attach lower ball joint stud to knuckle arm. Tighten nut to 100 ft. lbs. (136 Nm) and install cotter key.
3. Attach tie rod end to steering knuckle arm and inside nut. Tighten to 40 ft. lbs. (54 Nm) and install cotter pin.
4. Load torsion bar by turning adjusting bolt on lower control arm counterclockwise.
5. Install brake splash shield onto steering knuckle.
6. Install hub and disc assembly. Adjust wheel bearings. Install caliper.
7. Install wheel and tire assembly and attach wheel cover.
8. Lower vehicle to floor, adjust front suspension heights and wheel alignment as necessary.

Sway Bar (Longitudinal Torsion Bar)

Removal

1. Place ignition switch in Off or Unlocked position.
2. Raise car on hoist to place front suspension in rebound.
3. Remove wheel cover, wheel and tire assembly.
4. Remove nut and bolt on each end of bar attaching sway bar to strut clamp. Remove nut and bolt from both sway bar link straps to free sway bar from lines.
5. Remove sway bar by pulling unit out through frame cross-

SUSPENSION
Domestic Models

member openings in direction of area where wheel has been removed.

6. In the event strut cushions and sway bar bushings show excessive wear or deterioration of rubber, replacement is recommended.

Installation

1. On side where wheel assembly has been removed, install sway bar with center offset in downward position (color code on bar is always on driver's side).

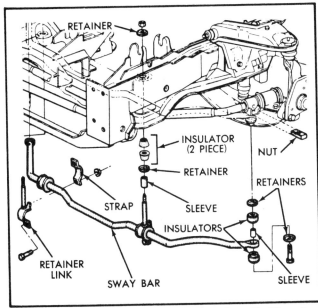

Sway bar—transverse type torsion bar (© Chrysler Corp.)

2. Attach sway bar with bolt and nut to strut retainer clamp on each side of bar and tighten to 35 ft. lbs. (47 Nm).

3. Lower vehicle to floor and attach both sway bar frame link straps. Tighten to 35 ft. lbs. (47 Nm).

Sway Bar (Transverse Torsion Bar)

Removal

1. Raise vehicle and support safely.

NOTE: Sway bar-to-lower control arm attaching points are aligned ONLY when lower control arms are at "design height" (equal to three passenger position with vehicle on ground). If frame contact or twin post hoist is used, release load on torsion bar by turning adjuster bolts counterclockwise. Raise lower control arms until clearance between crossmember ledge (at jounce bumper) and torsion bar to lower control arm bushing is 2 7/8 in. (73 mm). Support lower control arms with jack stand during sway bar removal and installation.

2. With lower control arms supported as described in note above, remove sway bar-to-torsion bar bushing attaching bolts, retainers, cushions and sleeves.

3. Remove retainer assembly strap bolts and retainer straps. Remove sway bar.

4. Inspect cushions and bushings for excessive wear or deterioration and replace if required.

Installation

1. Position sway bar bushings against retainers and install retainer straps. Loosely assemble retainer bolts.

2. Reinstall bolt through sway bar retainer, cushions and sleeve and attach to torsion bar lower control arm bushings. Tighten bolt to 50 ft. lbs. (68 Nm) torque.

3. Tighten sway bar retainer and strap bolts to 30 ft. lbs. (41 nm).

4. Load torsion bar by turning anchor adjusting bolt in crossmember clockwise.

5. Lower vehicle and adjust torsion bar height to specifications.

REAR SUSPENSION—CHRYLSER REAR DRIVE CARS

Shock Absorber

Removal

1. Raise the vehicle and support safely.

2. Using floor stands or equivalent, raise the axle to relieve the load on the shock absorbers.

3. Remove the shock absorber lower end as follows: Loosen and remove the nut, retainer and bushing from the spring plate.

NOTE: When loosening the retaining nut, grip the shock absorber at the base (below the base to reservoir tube weld) to avoid reservoir damage.

4. Loosen and remove the nut and bolt from the upper shock absorber mounting. Remove the shock absorber.

Installation

1. Expel air from new shock absorber.

2. Position and align upper eye of shock absorber with mounting holes in crossmember and install bolt and nut. Do not fully tighten.

3. Install shock absorber lower end, as follows: Install upper bushing on shock absorber stud and pull stud through spring plate mounting hole. Install lower bushing, cupped washer and nut. Tighten as specified.

4. Lower vehicle until full weight of vehicle is on the wheels. Tighten upper nut 70 ft. lbs. (95 Nm). Tighten lower nut to 35 ft. lbs. (47 Nm).

Springs

MEASURING SPRING HEIGHT

When measuring rear spring heights, place vehicle on a level floor, have correct front suspension height on both sides, correct tire pressures and no passenger or luggage compartment load and a full tank of fuel.

1. Jounce car several times (from bumper first). Release bumpers at same point in each cycle.

2. Measure shortest distance from top of axle housing to the rail at side of rear axle bumper strap (at rear of bumper)

3. Measure both right and left sides.

If these measurements vary by more than 3/4 in. (side to side), it is an indication that one of the rear springs may need replacing.

SECTION 2

SUSPENSION
Domestic Models

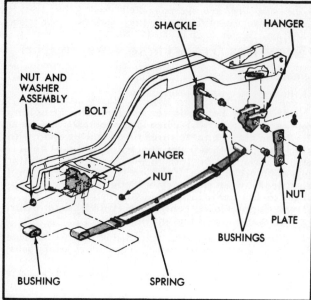

Rear suspension cars with transverse front suspension (© Chrysler Corp.)

Removal

1. Using floor stands under axle assembly, raise axle assembly to relieve weight on rear springs.
2. Disconnect rear shock absorber at spring plate. Lower axle assembly, permitting rear springs to hang free (on vehicles so equipped, disconnect rear sway bar links).
3. Loosen and remove U-bolt nuts and remove U-bolts and spring plate.
4. Loosen and remove the nuts holding from spring hanger to body mounting bracket.
5. Loosen and remove rear spring hanger bolts and let spring drop far enough to pull front springer hanger bolts of body mounting bracket holes.
6. Loosen and remove from pivot bolt from front spring hanger.
7. Loosen and remove shackle nuts and remove shackle from rear spring.

BUSHING REPLACEMENT

It is recommended that the spring assembly be removed from the vehicle for bushing replacement on the bench.
1. Bend two locking tabs away from spring eye on opposite side and remove bushing.
2. Press old bushing out.
3. Press new bushing in.

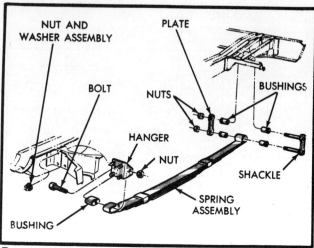

Rear suspension models with longitudinal torsion bar suspension (© Chrysler Corp.)

Installation

1. Assemble shackle to spring. Do not fully tighten bolt nut.
2. Install front spring hanger and insert pivot bolt and nut; do not fully tighten.
3. Install rear spring hanger-to-body bracket.
4. Raise and insert spring hanger mounting bracket bolts. Tighten to 30–35 ft. lbs. (42–46 Nm).
5. Align axle assembly with spring center bolt. Position center bolt over lower spring plate. Insert U-bolt and nut. tighten to 45 ft. lbs. (60 Nm).
6. Connect shock absorbers.
7. Lower car. Tighten pivot bolts to 105 ft. lbs. (142 Nm). Tighten shackle nuts to 35 ft. lbs. (46 Nm).

CHRYSLER AUTOMATIC AIR LOAD LEVELING SYSTEM

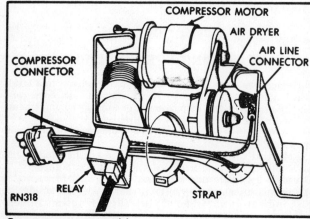

Compressor assembly

General Information

The automatic air load leveling system includes the following: compressor assembly, height sensor assembly, wiring harness, air lines, compressor relay, air shock absorbers and air dryer. The system is used to assist the standard suspension system on such equipped vehicles.

System Components

COMPRESSOR ASSEMBLY

The compressor assembly is driven by an electric motor and supplies air pressure between 120–200 psi. A solenoid operated exhaust valve, located in the compressor head assembly, releases air when energized. This valve limits maximum blow off pressure to 200 psi and will maintain a minimum system pressure of 100 psi.

2-74

SUSPENSION
Domestic Models

HEIGHT SENSOR ASSEMBLY

The height sensor assembly is an electronic device that controls the ground circuits for the compressor relay and the exhaust valve solenoid. An electronic timer within the unit controls the run time from one minute and 45 seconds to three minutes and 30 seconds. This prevents damage to the leveling system.

Also included in the system, is an air replenishment cycle that is controlled by the height sensor assembly. When the ignition switch is turned to the "ON" position, after a 30–60 second delay, the compressor will run 3–5 seconds.

In order to prevent excessive cycling between the compressor and the exhaust solenoid circuits during normal ride conditions a 13–27 second delay is incorporated into the electronic timer. The height sensor is mounted on the right rear frame rail. A link from the sensor actuating arm is attached to the rear of the track bar. When the arm moves up the compressor relay is energized, when the arm moves down the exhaust solenoid is energized.

AIR LINES AND FITTINGS

The air lines are equipped with snap on fittings with two "O" rings located on the male fittings. A retainer spring locks the male fitting into a groove on the female fitting.

NOTE: When replacing air lines and/or fittings, do not kink air lines or route new components near moving components of the vehicle.

COMPRESSOR RELAY

The relay is mounted to a bracket on the compressor assembly. When the compressor is energized it allows the compressor to operate. The unit is controlled by the height sensor.

AIR ADJUSTABLE SHOCK ABSORBERS

Air shock absorbers are hydraulic shock absorbers with a neoprene bladder sealing the upper and lower sections together, forming an air cylinder.

AIR DRYER

The air dryer is attached to the compressor and is not serviceable. This component serves two purposes; it absorbs moisture from the atmosphere before it enters the system and through internal valving, maintains a residual pressure of 14–21 psi.

System Operation

RAISING VEHICLE HEIGHT

When weight is added to the rear suspension, the body of the vehicle is lowered causing the height sensor actuating arm to rotate upward. This action causes the internal time delay circuit to activate. After a time delay of 13–27 seconds, the sensor grounds pin number 3 and completes the ground circuit to the compressor relay.

When the relay is energized, the compressor motor runs and air is sent through the system. As the shock absorbers inflate, the body moves upward to the corrected position. When the body reaches the corrected level, the sensor stops the compressor operation.

LOWERING VEHICLE HEIGHT

When the weight is removed from the vehicle the body moves

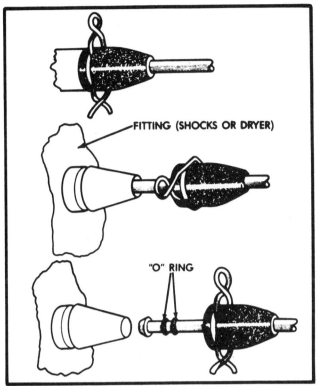

Air line fittings

upward, causing the height sensor actuating arm to rotate downward and activate the internal time delay circuit.

After a time delay of 13–27 seconds, the sensor activates the exhaust solenoid circuit. Air is exhausted from the shock absorbers through the air dryer and exhaust solenoid to the atmosphere.

As the body lowers, the height sensor actuating arm rotates towards its original position. When the body reaches the original vehicle height the sensor opens the exhaust solenoid valve circuit.

Troubleshooting

COMPRESSOR

Performance Test

This test can be performed on the vehicle in order to evaluate the compressor current draw, leak down and pressure output.

1. Tag and disconnect the compressor motor wiring harness connector.
2. Disconnect the compressor air line at the air line "T" connector.
3. Connect an air pressure gauge into the air line between the compressor and the air hose tee.
4. Connect an ammeter in series between the orange wire in the wiring harness and the dark green wire to the compressor motor
5. Connect a ground wire from the black wire on the compressor motor to a known ground located on the frame.
6. If the current draw to the compressor motor exceeds 14 amps, replace the compressor assembly.
7. When the air pressure stabilizes at 120 psi, disconnect the positive (+) lead from the connector.

SUSPENSION
Domestic Models

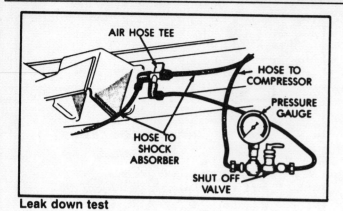

Leak down test

CAUTION

If any of the following conditions exist, replace the compressor assembly.

---

1. Air pressure leaks down below 90 psi before it remains steady.
2. Output pressure builds up to less than 110 psi psi when it stabilizes.

If the compressor is allowed to run during this test until it reaches its maximum output pressure (200 psi), the solenoid exhaust valve will act as a pressure relief valve. The resulting leak down after the compressor is shut off will indicate a false leak.

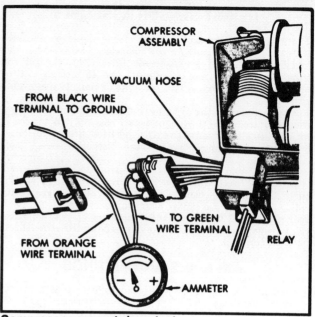

Compressor current draw test

HEIGHT SENSOR

Performance Test

1. Cycle the ignition off then on. This resets the height sensor timer circuit.

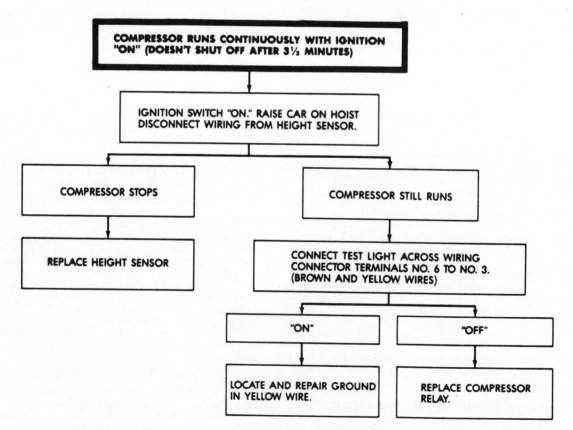

SUSPENSION
Domestic Models

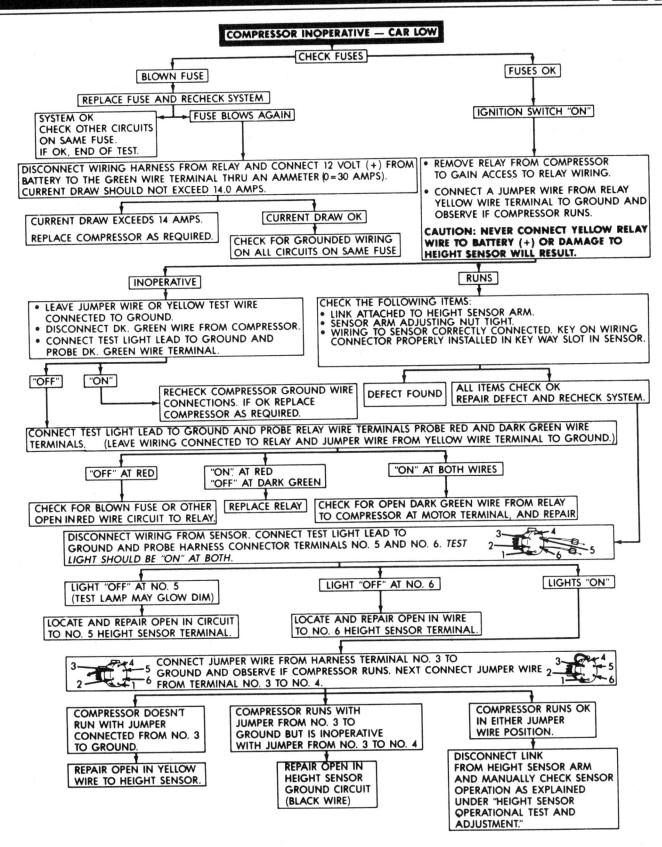

2-77

SECTION 2: SUSPENSION
Domestic Models

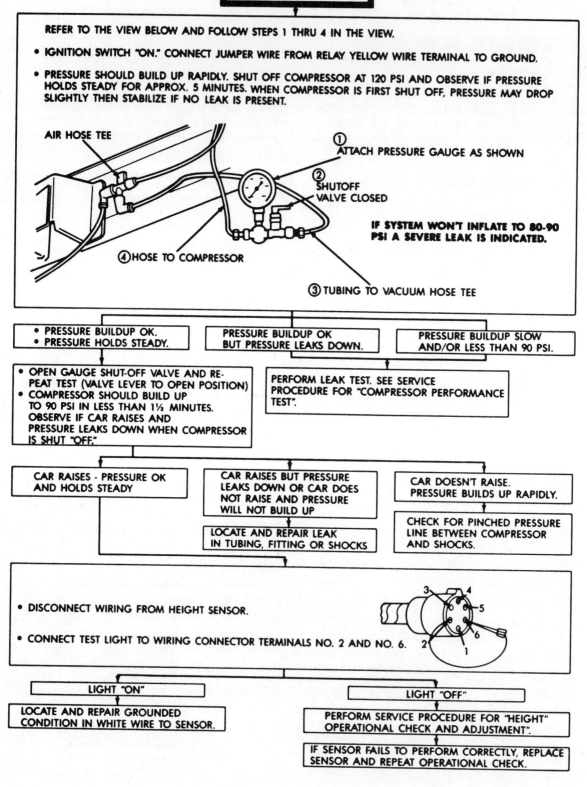

LOSS OF AIR PRESSURE

REFER TO THE VIEW BELOW AND FOLLOW STEPS 1 THRU 4 IN THE VIEW.

- IGNITION SWITCH "ON." CONNECT JUMPER WIRE FROM RELAY YELLOW WIRE TERMINAL TO GROUND.
- PRESSURE SHOULD BUILD UP RAPIDLY. SHUT OFF COMPRESSOR AT 120 PSI AND OBSERVE IF PRESSURE HOLDS STEADY FOR APPROX. 5 MINUTES. WHEN COMPRESSOR IS FIRST SHUT OFF, PRESSURE MAY DROP SLIGHTLY THEN STABILIZE IF NO LEAK IS PRESENT.

AIR HOSE TEE

① ATTACH PRESSURE GAUGE AS SHOWN
② SHUTOFF VALVE CLOSED
③ TUBING TO VACUUM HOSE TEE
④ HOSE TO COMPRESSOR

IF SYSTEM WON'T INFLATE TO 80-90 PSI A SEVERE LEAK IS INDICATED.

- PRESSURE BUILDUP OK.
- PRESSURE HOLDS STEADY.

- PRESSURE BUILDUP OK BUT PRESSURE LEAKS DOWN.

- PRESSURE BUILDUP SLOW AND/OR LESS THAN 90 PSI.

- OPEN GAUGE SHUT-OFF VALVE AND REPEAT TEST (VALVE LEVER TO OPEN POSITION)
- COMPRESSOR SHOULD BUILD UP TO 90 PSI IN LESS THAN 1½ MINUTES. OBSERVE IF CAR RAISES AND PRESSURE LEAKS DOWN WHEN COMPRESSOR IS SHUT "OFF."

PERFORM LEAK TEST. SEE SERVICE PROCEDURE FOR "COMPRESSOR PERFORMANCE TEST".

CAR RAISES - PRESSURE OK AND HOLDS STEADY

CAR RAISES BUT PRESSURE LEAKS DOWN OR CAR DOES NOT RAISE AND PRESSURE WILL NOT BUILD UP

CAR DOESN'T RAISE. PRESSURE BUILDS UP RAPIDLY.

LOCATE AND REPAIR LEAK IN TUBING, FITTING OR SHOCKS

CHECK FOR PINCHED PRESSURE LINE BETWEEN COMPRESSOR AND SHOCKS.

- DISCONNECT WIRING FROM HEIGHT SENSOR.
- CONNECT TEST LIGHT TO WIRING CONNECTOR TERMINALS NO. 2 AND NO. 6.

LIGHT "ON"

LOCATE AND REPAIR GROUNDED CONDITION IN WHITE WIRE TO SENSOR.

LIGHT "OFF"

PERFORM SERVICE PROCEDURE FOR "HEIGHT OPERATIONAL CHECK AND ADJUSTMENT".

IF SENSOR FAILS TO PERFORM CORRECTLY, REPLACE SENSOR AND REPEAT OPERATIONAL CHECK.

SUSPENSION
Domestic Models

SECTION 2

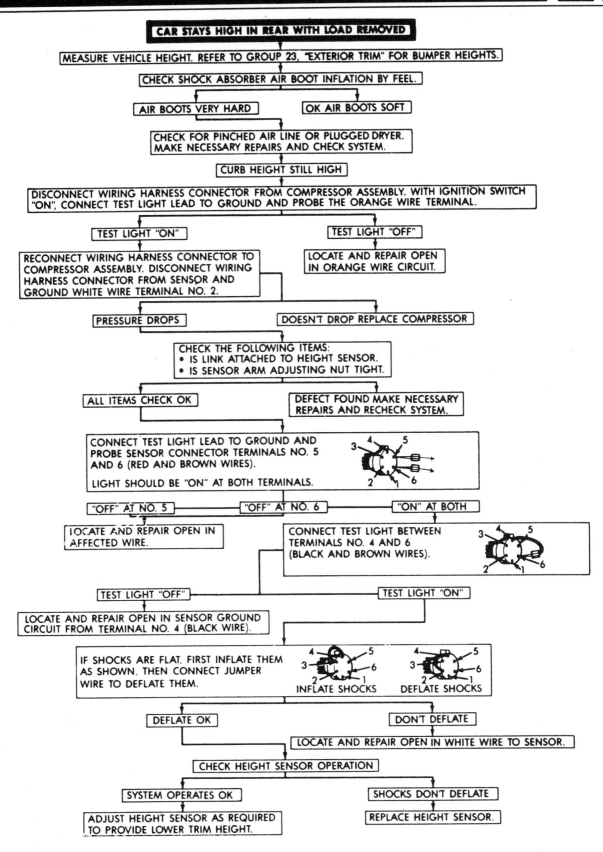

CAR STAYS HIGH IN REAR WITH LOAD REMOVED

MEASURE VEHICLE HEIGHT. REFER TO GROUP 23, "EXTERIOR TRIM" FOR BUMPER HEIGHTS.

CHECK SHOCK ABSORBER AIR BOOT INFLATION BY FEEL.

- AIR BOOTS VERY HARD
- OK AIR BOOTS SOFT

CHECK FOR PINCHED AIR LINE OR PLUGGED DRYER. MAKE NECESSARY REPAIRS AND CHECK SYSTEM.

CURB HEIGHT STILL HIGH

DISCONNECT WIRING HARNESS CONNECTOR FROM COMPRESSOR ASSEMBLY. WITH IGNITION SWITCH "ON", CONNECT TEST LIGHT LEAD TO GROUND AND PROBE THE ORANGE WIRE TERMINAL.

- TEST LIGHT "ON"
- TEST LIGHT "OFF" → LOCATE AND REPAIR OPEN IN ORANGE WIRE CIRCUIT.

RECONNECT WIRING HARNESS CONNECTOR TO COMPRESSOR ASSEMBLY. DISCONNECT WIRING HARNESS CONNECTOR FROM SENSOR AND GROUND WHITE WIRE TERMINAL NO. 2.

- PRESSURE DROPS
- DOESN'T DROP REPLACE COMPRESSOR

CHECK THE FOLLOWING ITEMS:
- IS LINK ATTACHED TO HEIGHT SENSOR.
- IS SENSOR ARM ADJUSTING NUT TIGHT.

- ALL ITEMS CHECK OK
- DEFECT FOUND MAKE NECESSARY REPAIRS AND RECHECK SYSTEM.

CONNECT TEST LIGHT LEAD TO GROUND AND PROBE SENSOR CONNECTOR TERMINALS NO. 5 AND 6 (RED AND BROWN WIRES).
LIGHT SHOULD BE "ON" AT BOTH TERMINALS.

- "OFF" AT NO. 5
- "OFF" AT NO. 6
- "ON" AT BOTH

LOCATE AND REPAIR OPEN IN AFFECTED WIRE.

CONNECT TEST LIGHT BETWEEN TERMINALS NO. 4 AND 6 (BLACK AND BROWN WIRES).

- TEST LIGHT "OFF" → LOCATE AND REPAIR OPEN IN SENSOR GROUND CIRCUIT FROM TERMINAL NO. 4 (BLACK WIRE).
- TEST LIGHT "ON"

IF SHOCKS ARE FLAT, FIRST INFLATE THEM AS SHOWN. THEN CONNECT JUMPER WIRE TO DEFLATE THEM.

INFLATE SHOCKS | DEFLATE SHOCKS

- DEFLATE OK
- DON'T DEFLATE → LOCATE AND REPAIR OPEN IN WHITE WIRE TO SENSOR.

CHECK HEIGHT SENSOR OPERATION

- SYSTEM OPERATES OK → ADJUST HEIGHT SENSOR AS REQUIRED TO PROVIDE LOWER TRIM HEIGHT.
- SHOCKS DON'T DEFLATE → REPLACE HEIGHT SENSOR.

SUSPENSION
Domestic Models

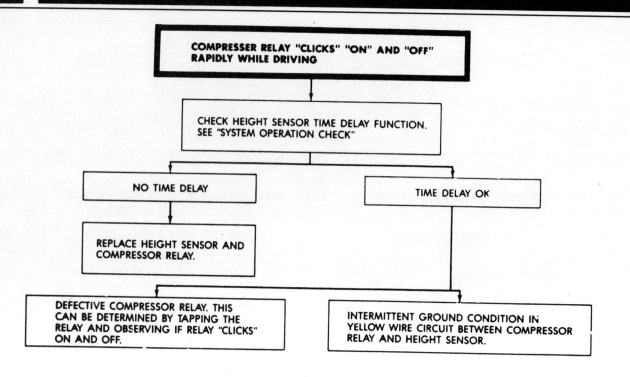

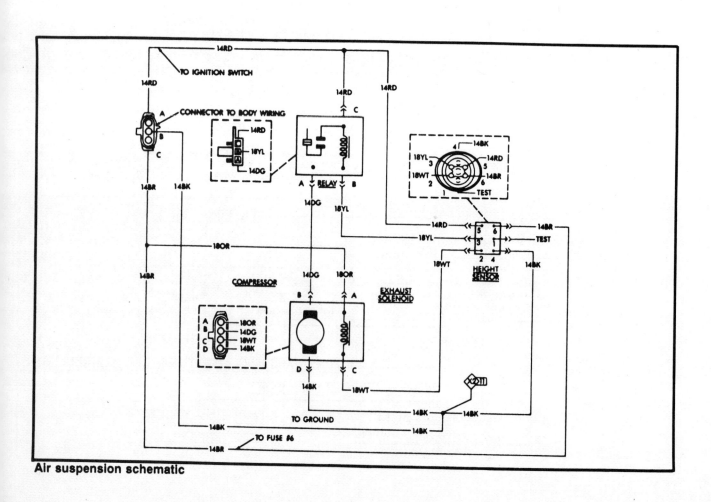

Air suspension schematic

SUSPENSION
Domestic Models

2. Raise the vehicle and support safely.
3. Inspect the wiring to see that it is properly secured to the height sensor.
4. Disconnect the link from the height sensor arm and move up. There should be a 13–27 second delay before the compressor turns on and the shocks begin to inflate.

— **CAUTION** —

As soon as the shock absorber bladder fills, stop the compressor by moving the sensor arm down as damage to the air bladder can result.

5. There should be a 13–27 second delay after the sensor arm is moved down before the shocks begin to deflate.

Component Replacement

COMPRESSOR

Removal

1. Disconnect the negative battery cable.
2. Raise the vehicle and support safely.
3. Remove cover from the compressor assembly and discharge the air system.
4. Remove the air hose from the electrical connectors.
5. Remove the compressor assembly mounting bolts. Lower the assembly from the vehicle.
6. Remove the mounting bracket bolts and slide the mounting bracket away from the compressor.

Installation

1. Install the mounting bracket on the compressor and install the bolts and tighten to 70 inch pounds (8 Nm).
2. Install the compressor assembly to the frame rail and tighten the bolts to 70 inch pounds (8 Nm).
3. Connect the air hose and electrical connectors to the compressor.

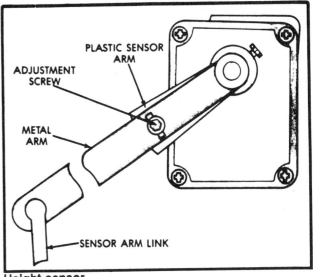

Height sensor

4. Install the compressor cover and tighten the bolts to 70 inch pounds (8 Nm).
5. Lower the vehicle and connect the negative battery cable.
6. Turn the ignition switch to the "ON" position and then back to "OFF" in order to reset the height sensor timing circuits.
7. Check the operation of system.

HEIGHT SENSOR

Removal

1. Disconnect the negative battery cable.

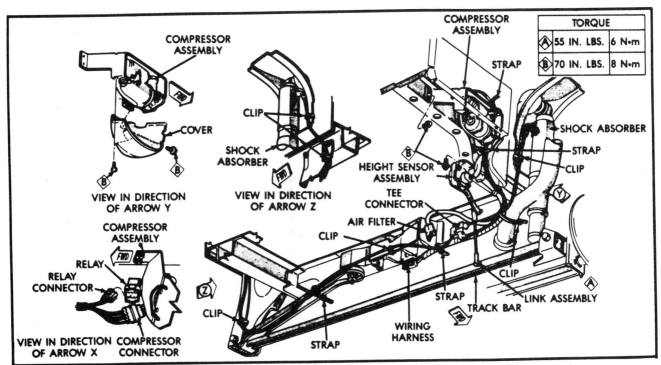

Automatic air load levelling system

SUSPENSION
Domestic Models

2. Raise the vehicle and support safely.
3. Tag and disconnect the electrical connector and link from the sensor arm.
4. Remove the mounting bolts and remove the height sensor assembly from the frame rail.

Installation

1. Install height sensor assembly and tighten mounting bolts to 70 inch pounds (8 Nm).
2. Connect the link and electrical connector to sensor.
3. Lower the vehicle and connect the negative battery cable.
4. Check system operation.

Adjustment

1. Loosen the lock nut on the sensor arm.
2. To increase vehicle height, move the sensor arm upward and tighten.
3. To decrease vehicle height, move the sensor arm downward and tighten.
4. Check for proper system operation and vehicle height.

NOTE: Due to the diagnostic complexity of the Automatic Load Levelling System, detailed diagnostic procedures can be found in Chilton's Chassis Electronic Service Manual.

FRONT SUSPENSION—FORD FRONT WHEEL DRIVE CARS

Description

The front wheel drive front suspension is a MacPherson strut design with cast steering knuckles. The shock absorber strut assembly includes a rubber top mount and a coil spring insulator, mounted on the shock strut.

The entire strut assembly is attached at the top by two bolts, which retain the top mount of the strut to the body side apron. The lower end of the assembly is attached to the steering knuckle. A forged lower arm assembly is attached to the underbody side apron and steering knuckle. A stabilizer bar connects the outer end of lower arm to the engine mount bracket. The drive shaft outer stub shaft and wheel hub are attached inside the steering knuckle hub by a pressed fit of mating splines. The assembly is secured by a staked nut on the end of the stub shaft. The hub rotates on two non-adjustable tapered roller bearings which seat against cups in the steering knuckle.

Taurus and Sable shock absorbers are gas pressurized which will result in the struts being fully extended when not restrained. If a strut does not fully extend when inspected, it is damaged and should be replaced.

Wheel Alignment

TOE

Toe is the difference in distance between the front and the rear of the front wheels.

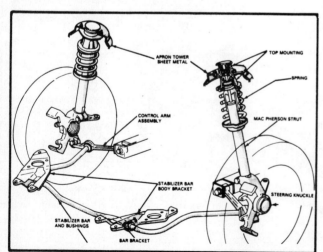

Front wheel drive suspension (© Ford Motor Co.)

1. Start the engine (power steering only) and move the steering wheel back and forth several times until it is in the straight ahead or centered position.
2. Turn the engine off (power steering only) and lock the steering wheel in place using a steering wheel holder. Loosen and slide off small outer clamp front boot prior to starting toe adjustment to prevent boot from twisting.
3. Adjust left and right tie rods until each wheel has one-half of the desired total toe specification.

NOTE: When jam nuts are loosened for toe adjustment, the nuts must be tightened to specifications. Attach boot clamp after setting is completed and make sure boot is not twisted.

CASTER AND CAMBER

Caster and camber angles of this suspension system are preset at the factory and cannot be adjusted. Measurement procedures are for diagnostic purposes.

NOTE: Caster measurements must be made on the left side by turning the left wheel through the prescribed angle of sweep and on the right side by turning the right wheel through the prescribed angle of sweep.

FRONT WHEEL TURNING ANGLE

When the inside wheel is turned 20 degrees, turning angle of outside wheel should be specified. The turning angle cannot be adjusted directly, because it is a result of the combination of caster, camber and toe adjustments and should, therefore, be measured only after the toe adjustment has been made.

NOTE: If the turning angle does not measure to specifications, check the knuckle or other suspension or steering parts for a bent condition.

Wheel Bearings

Front wheel bearings are located in the front knuckle, not the rotor. The bearings are protected by inner and outer grease seals and an additional inner grease shield immediately inboard of the inner grease seal. The wheel hub is installed with an interference fit to the constant velocity universal joint outer race shaft. The hub nut and washer are installed and tightened to 180–200 ft. lbs. (240–270 Nm). The rotor fits loosely on the hub assembly and is secured when the wheel and wheel nuts are installed.

Adjustment

The front wheel bearings have a set-right design that requires

SUSPENSION
Domestic Models

no scheduled maintenance. The bearing design relies on component stack-up and deformation/torque at assembly to determine bearing setting. Therefore, bearings cannot be adjusted. In addition to the maintaining bearing adjustment, the hub nut torque of 180–200 ft. lbs. (240–270 Nm) restricts bearing/hub relative movement and maintains axial position of the hub. Due to the importance of the hub nut torque/tension relationship, certain precautions must be taken during service.

1. The hub nut must be replaced with a new nut whenever the nut is backed off or removed after the nut has been staked. Never re-use the nut.
2. The hub nut must not be backed off after reaching the required torque of 180–200 ft. lbs. (240–270 Nm) during installation.
3. The hub nut collar must be staked into the outboard constant velocity joint slot with the proper tool to make sure the required torque is maintained during vehicle operation. The nut collar must not split or crack when staked. If the collar splits or cracks, the nut must be replaced.
4. Impact type tools must not be used to tighten the hub nut or bearing damage will result.
5. The hub and constant velocity joint splines have an interference fit requiring special tools for removal and assembly. The hub nut must not be used to accomplish assembly.
6. To remove the hub nut, apply sufficient torque to the nut to overcome the prevailing torque feature of the crimp in the nut collar. Do not use tools such as a screwdriver or chisel to remove the crimp.

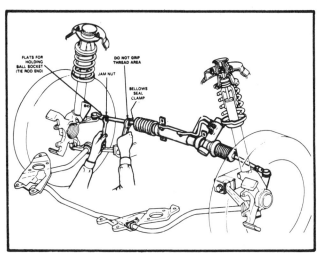

Tie rod adjustment (© Ford Motor Co.)

2. Raise the vehicle and support safely. Remove brake caliper. Do not remove caliper pins from the caliper assembly. Lift caliper off the rotor and hang it free of the rotor. Do not allow caliper assembly to hang from the brake hose. Support caliper assembly.
3. Remove rotor from hub by pulling if off the hub bolts.
4. Install hub remover/install tool, T81P-1104-A with T81P-1104-C adaptors T81P-B or equivalent, and remove the hub. If outer bearing is seized on the hub, use a puller to remove the bearing. Be careful not to damage bearing if it is being re-used and not to raise burrs on the hub journal diameter. If bearings are being re-used, carefully inspect both bearing cone and rollers, bearing cups and lubrication for any signs of damage or contamination. If damage or contamination exists, replace all bearing components including cups and seals. In the event the bearings are acceptable, clean and repack bearing components. Inner and outer grease retainers and hub nut must be replaced whenever bearings are inspected.
5. Remove front suspension knuckle.

Installation

1. Place front knuckle and bearing assembly in a vise so that the inner knuckle bore faces upward (to prevent inner bearing from falling out of the knuckle). Start hub into outer knuckle bore and push the hub by hand through outer and inner wheel bearings as far as possible.

── CAUTION ──

Prior to assembly, remove burrs, nicks, score marks, foreign material (rust, dirt, etc.) from hub bearing journal. Due to the very close tolerance 0.0005 in. (0.012 mm) between the wheel bearing inside diameter and the hub assembly, it is important to install hub completely through inner and outer wheel bearings. Hand pressure only is essential to this procedure. Forcing or jamming bearing race (cone) on the hub barrel will cause burrs that can prevent proper installation. Do not strike hub with any type of tool.

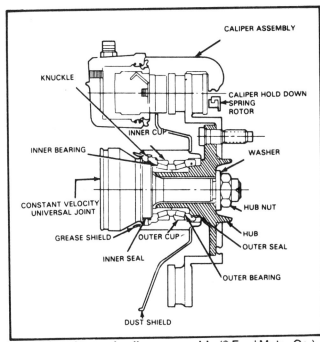

Wheel bearing and caliper assembly (© Ford Motor Co.)

Front Hub

Removal

1. Remove hub retaining nut and washer by applying sufficient torque to the nut to overcome prevailing torque features of the crimp in the nut collar. Do not use tools such as a screwdriver or chisel to remove the crimp or use an impact-type tool to remove the hub nut. The hub nut must be discarded after removal.

NOTE: Crocus cloth may be used to remove burrs, score marks and rust from the hub barrel.

2. With the hub fully seated in the bearings, position hub and knuckle assembly to front strut. Attach the knuckle to the strut.
3. Lightly lubricate the constant velocity joint stub shaft splines using S.A.E. 30 motor oil.
4. Using hand pressure only, insert splines of the constant velocity joint stub shaft into knuckle/hub assembly as far as

SECTION 2

SUSPENSION
Domestic Models

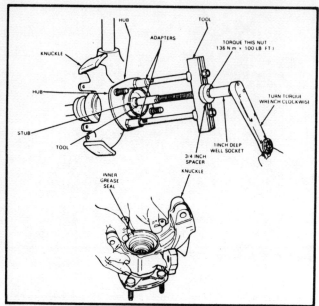

Step 1. Install hub to knuckle after bearing installation. Knuckle must be positioned as shown and hub must be inserted through bearings using hand pressure only. **Step 2.** Install hub tro constant velocity universal joint splined stub shaft. Tighten tool nut to 150 Nm (110 ft. lbs.) using torque wrench to seat hub (© Ford Motor Co.)

possible. Install hub installer tool T81P-1104-C-B-A to the hub and stub shaft.

——— **CAUTION** ———
Care must be taken during installation to prevent hub from backing out of bearing assembly. Otherwise, it will be necessary to reassemble hub through bearings.

5. Tighten hub installer tool to 120 ft. lbs. (163 Nm) torque to make sure that the hub is fully seated. Remove tool and install washer and hub nut. Tighten the hub nut finger tight.
6. Install disc brake rotor and brake caliper.
7. Lower vehicle and block wheels to prevent rolling.
8. Tighten wheel nuts to 80–105 ft. lbs. (109–142 Nm) torque.

Strut, Spring And Upper Mount

Removal

1. Raise front of vehicle and place jack stands under frame jack pads, rearward of the wheels. Do not raise the Taurus or Sable by the lower control arms.
2. Remove tire and wheel assembly.
3. Remove brake line flex hose clip from strut. If necessary, remove the brake rotor in order to gain working clearance.
4. Place a floor jack under lower control arm and raise strut as far as possible without lifting vehicle.
5. Install spring compressor tool by placing top jaw on second coil from top and bottom jaw so as to grip a total of five coils. Compress spring until there is about ⅛ in. between any two coils.

——— **CAUTION** ———
Use hand wrenches (no impact wrenches)

6. Using a pry bar, slightly spread knuckle-to-strut pinch joint.
7. Place a piece of wood, 2 in. by 4 in., and 7 ½ in. long, against shoulder on the knuckle. Using a short pry bar between wood block and lower spring seat, separate the strut from the knuckle.
8. Remove two top mounting nuts.
9. Remove strut, spring and top mount assembly from vehicle.
10. Place an 18 mm deep socket on strut shaft nut. Insert a 6 mm allen wrench into shaft end and the clamp mount into a vise. Remove top shaft mounting nut from shaft while holding allen wrench with vise grips or a suitable extension.
11. Remove strut top mount components and spring.

Installation

1. Position compressed spring in lower spring seat. Be sure that:
   a. Pigtail of spring is indexed to seat.
   b. Spring compressor tool is positioned 90 degrees from metal tab on lower part of strut.
2. Using a new nut, assemble top mount components to strut.

——— **CAUTION** ———
Be sure that the correct assembly sequence and proper positioning of bearing and seal assembly is followed. If bearing and seal assembly is out of position, damage to bearing will result.

3. Tighten shaft nut to torque of 48–62.5 ft. lbs. (65–85 Nm).
4. Install strut, spring, upper mount and spring compressor into the vehicle as an assembly.
5. Position two top mounts attacking studs through holes in apron and start two new nuts. Do not tighten nuts at this time.
6. Install strut into steering knuckle pinch point.
7. Install a new pinch bolt in the steering knuckle and tighten to torque of 68–81 ft. lbs. (90–110 Nm).

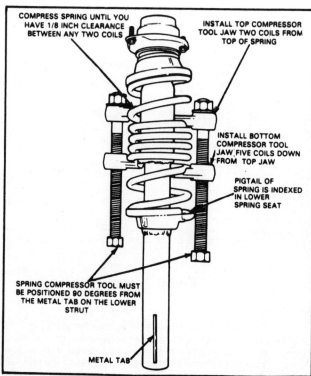

Spring comprssor tool (© Ford Motor Co.)

SUSPENSION
Domestic Models

8. Tighten two upper mount attaching nuts to torque of 22–29 ft. lbs. (30–40 Nm).
9. Remove spring compressor from the vehicle. As the compressor is loosened, be sure the spring ends are indexed in upper and lower spring seats.
10. Install brake line flex hose clip to strut.
11. Install tire and wheel assembly.
12. Remove jack stands and lower vehicle.

Steering Knuckle

Removal

1. Raise vehicle on a hoist.
2. Remove tire and wheel assembly.
3. Remove cotter pin from the tie rod end strut and remove fitted nut.
4. Remove tie rod end from knuckle.
5. Remove the brake caliper and rotor. Wire off to the side in order to gain working clearance.
6. Remove the hub from the driveshaft.
7. Remove lower arm-to-steering knuckle pinch bolt and nut. (A drift punch may be used to remove production bolt.) Using a screwdriver, slightly spread the knuckle-to-lower arm pinch joint and remove lower arm from steering knuckle.

NOTE: Be sure steering column is in unlocked position and do not use a hammer to separate ball joint from knuckle.

8. Remove shock absorber strut-to-steering knuckle pinch bolt. Using a pry bar, slightly spread knuckle-to-strut pinch joint.
9. Remove steering knuckle from the shock absorber strut.
10. Place assembly on a bench and remove the seals are bearings.
11. Remove rotor splash shield from knuckle. Remove the hub, retainer ring and bearing

Installation

1. Install a rotor splash shield.

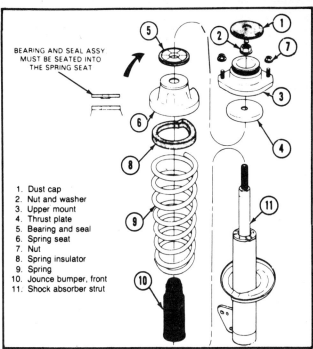

Top mount components (© Ford Motor Co.)

1. Dust cap
2. Nut and washer
3. Upper mount
4. Thrust plate
5. Bearing and seal
6. Spring seat
7. Nut
8. Spring insulator
9. Spring
10. Jounce bumper, front
11. Shock absorber strut

2. Install bearings and seals.
3. Install steering knuckle onto shock absorber strut and install a new pinch bolt in knuckle to retain strut. Tighten nut to torque of 66–81 ft. lbs. (99–110 Nm).
4. Install hub on the driveshaft.
5. Install lower control arm to knuckle, ensuring that ball stud groove is properly positioned. Install a new nut and bolt. Tighten to torque of 37–44 ft. lbs. (50–60 Nm).
6. Install the brake caliper.
7. Position tie rod end into knuckle, install a new slotted nut and tighten to torque of 25–35 ft. lbs. (31–47 Nm). If necessary, advance nut to align slot and install a new cotter pin.
8. Install tire and wheel assembly.

Lower Ball Joint Check

1. Raise vehicle on a frame contact hoist or by floor jacks placed beneath the underbody until wheel fall to the full down position.
2. Ask an assistant to grasp lower edge of the tire and move wheel and tire assembly in and out.
3. As wheel is being moved in and out, observe the lower end of the knuckle and the lower control arm. Any movement between the lower end of knuckle and lower arm indicates abnormal ball joint wear.
4. If any movement is observed, install a new lower control arm assembly.

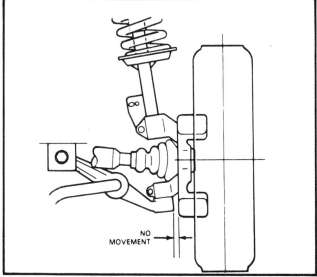

As wheel is being moved in and out, observe the lower end of the knuckle and the lower control arm. Any movement between lower end of knuckle and the lower arm indicates abnormal ball joint wear (© Ford Motor Co.)

Lower Control Arm

Removal

1. Raise vehicle and support safely. Remove the wheel and tire assembly.
2. Remove nut from the stabilizer bar. Pull off large dished water.
3. Remove lower control arm inner pivot bolt and nut.
4. Remove lower control arm ball joint pinch bolt. Slightly

2-85

SUSPENSION
Domestic Models

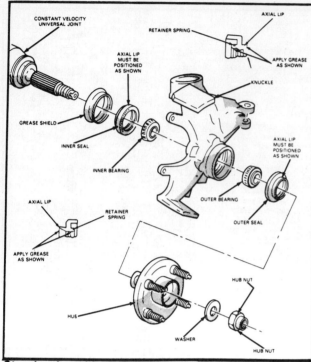

Steering knuckle (© Ford Motor Co.)

spread the knuckle pinch joint and separate control arm from the steering knuckle. A drift punch may be used to remove the bolt.

NOTE: Be sure steering column is in unlocked position, and DO NOT use a hammer to separate ball joint from knuckle.

5. Remove the lower control arm pivot bolt and nut. Remove the lower control arm assembly from the tension strut.

Installation

1. Assemble lower control arm ball joint stud to the steering knuckle, insuring that the ball stud groove is properly positioned.

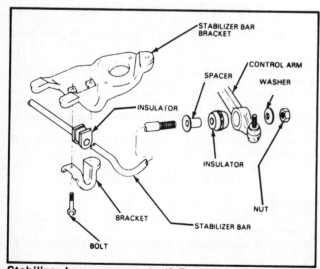

Stabilizer bar components (© Ford Motor Co.)

2. Insert a new pinch bolt and nut. Tighten nut to torque of 37–44 ft. lbs. (50–60 Nm).
3. Position lower control arm onto stabilizer bar and then position lower control arm to the inner underbody mounting. Install a new nut and bolt. Tighten bolt to torque of 44–55 ft. lbs. (60–75 Nm).
4. Assemble stabilizer bar, dished washer and a new nut to stabilizer bar. Tighten nut to torque of 59–73 ft. lbs. (80–110 Nm).

Stabilizer Bar and Insulators

Removal

1. Raise the vehicle and support safely.
2. Remove nut from the stabilizer bar from each lower control arm and remove the large dished washer.
3. Remove the stabilizer bar insulator mounting bracket bolts and remove the stabilizer bar assembly.
4. Remove the worn insulators from the bar.

Installation

1. Coat the bar and insulators using petroleum jelly. Slide the new insulators onto the bar and position in the approximate final position.
2. Install the washer spacers onto the bar ends and push the mounting brackets over the insulators.
3. Insert the ends of the bar into the lower control arms. Using new bolts, attach the bar and insulator mounting brackets to body. Tighten to 80–92 Nm. (59–68 ft. lbs.).
4. Using the new nuts and original dished washers, attach the bar to the lower control arms. Tighten the nuts to 133–156 Nm. (98–115 ft. lbs.).

Stabilizer Bar – To – Control Arm Insulator

Removal

1. Raise vehicle on a hoist.
2. Remove stabilizer bar-to-control arm nut and dished washer.
3. Remove control arm inner pivot nut and bolt and pull arm down from the underbody and away from the stabilizer bar.
4. Using Tool T81P–5493–A and T74P–3044–A1 or equivalent, remove old insulator bushing from the control arm.

Installation

1. Saturate new insulator bushing and lower arm with vegetable oil.

NOTE: Use only vegetable oil. Any mineral or petroleum based oil or brake fluid will deteriorate the rubber bushing.

2. Using Tool T81P–5493–A and T74P–3044–A1 or equivalent, install new insulator bushing in lower control arm by tightening the C-clamp very slowly until bushing pops in place.
3. Position the control arm onto the stabilizer bar. Washer end of spacer must seat against the stabilizer bar machined shoulder.
4. Install the control arm to on the underbody using a new nut and bolt. Tighten to 48–55 ft. lbs. (65–75 Nm).
5. Install a new nut and the original dished washer on stabilizer bar. Tighten to 98–115 ft. lbs. (133–156 Nm).

Lower Arm Inner Pivot Bushing

Removal

1. Raise vehicle on a hoist.

SUSPENSION
Domestic Models

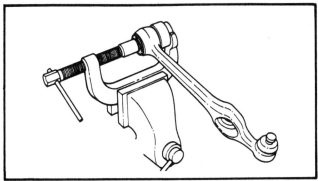

Inner pivot bushing Installation (© Ford Motor Co.)

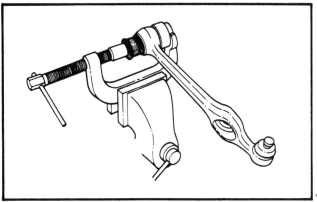

Inner pivot bushing removal (© Ford Motor Co.)

2. Remove stabilizer bar to control arm nut and dished washer.
3. Remove control arm inner pivot nut and bolt and pull arm down from underbody and away from stabilizer bar.
4. Using a sharp knife, carefully cut away retaining lip of bushing prior to its removal.
5. Using Tool T81P-5493-B and T74P-3044-1 or equivalent, remove old bushing from control arm.

NOTE: This operation can be done in vehicle without removing arm from knuckle.

Installation

1. Saturate new bushing and lower control arm with vegetable oil or equivalent.

NOTE: Use only vegetable oil. Any mineral or petroleum based oil or brake fluid will deteriorate the rubber bushing.

2. Using Tool T81P-5493-B and T74P-3044-A1 or equivalent, install new bushing in lower control arm.
3. Position control arm onto stabilizer bar. Be sure washer is in place.
4. Tighten control to underbody and install a new nut and bolt. Tighten torque of 44–55 ft. lbs. (60–75 Nm).
5. Install a new nut and the original dished washer on stabilizer bar. Tighten nut to torque of 59–73 ft. lbs. (80–100 Nm).

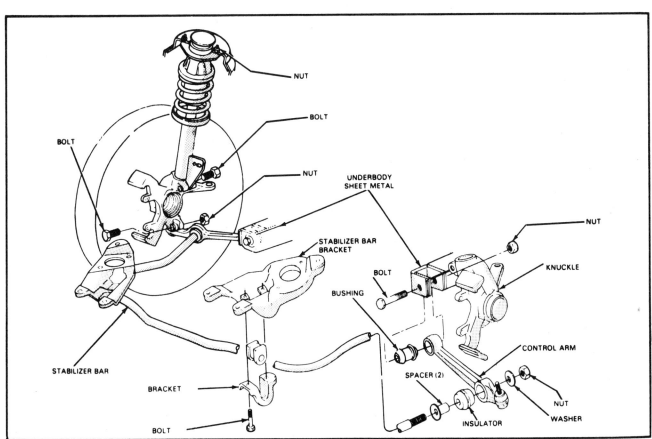

Front suspension fasteners. Bolts must be installed in direction shown (© Ford Motor Co.)

SUSPENSION
Domestic Models

REAR SUSPENSION – FORD FRONT WHEEL DRIVE CARS

Wheel Bearings

Adjustment

Tighten adjusting nut "A" to 17–25 ft. lbs. (23–24 Nm) while rotating hub and drum assembly. Back off adjusting nut approximately 100 degrees. Position nut retainer "B" over adjusting nut so slots are in line with cotter pin hole without rotating adjusting nut. Install cotter pin.

NOTE: The spindle has a prevailing torque feature that prevents adjusting the nut by hand.

Component Replacement

The following applies regarding components that are replaced individually or as an assembly.
1. The shock absorber strut upper mounting is separately servicable.
2. The shock absorber strut is not repairable and must be replaced as an assembly.
3. Lower control arm bushings are not servicable. They must be replaced with a lower control arm and bushing assembly.
4. Tie rod bushings can be serviced separately at both the forward and rearward locations. However, if the tie rod requires replacement, new bushings must be installed in the spindle at the same time.
5. Coil springs are servicable. If a rear coil spring is replaced, the upper spring insulator must also be replaced.

Shock Absorber Strut

Removal

1. Remove rear component access panels. Four-door model requires removal of quarter trim panel.
2. Loosen but do not remove top shock absorber attaching nut.

NOTE: If the shock absorber is to be re-used, do not grip the shock absorber shaft with pliers or vise grips, as this will damage the shaft surface finish.

3. Raise vehicle and support safely. Remove the tire and wheel assembly.

NOTE: If a frame contact hoist is used, support the lower control arm with a floor jack. If a twin post hoist is used, support the body with floor jacks on lifting pads forward of the tie-rod body bracket.

4. Remove clip retaining the brake flexible hose to the rear shock and carefully move hose aside.
5. Loosen two nuts and bolts retaining shock to the spindle. DO NOT remove bolts at this time.
6. Remove top mounting nut, washer and rubber insulator.
7. Remove two bottom mounting bolts and remove the shock from the vehicle.

Installation

1. Extend shock absorber to its maximum length.
2. Install a new lower washer and insulator assembly, using tire lubricant to ease insertion into the quarter panel shock tower. (Use of a lubricant other than ESA-M1B6-B or soapy water is not recommended as it may damage the rubber insulator).
3. Position upper part of the shock absorber shaft into shock tower opening in the body and push slowly on lower part of the shock until mounting holes are lined up with mounting holes in the spindle.
4. Install a new lower mounting bolts and nuts. DO NOT tighten at this time.

NOTE: The heads of both bolts must be to the rear of the vehicle.

5. Place a new upper insulator and washer assembly and nut on the upper shock absorber shaft. Tighten nut to torque of 60–70 ft. lbs. (81–95 Nm), using the 18 X 18 X 43 mm deep socket with external hex (Tool D81P-18045-A3) while holding the

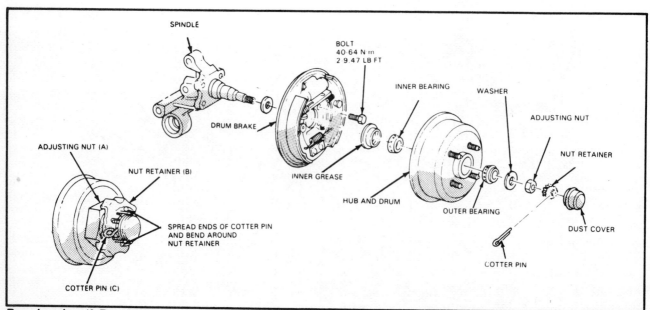

Rear bearing (© Ford Motor Co.)

SUSPENSION
Domestic Models

strut shaft with a 6 mm allen wrench. Do not grip the shaft with pliers or vise grips.

6. Tighten two lower mounting bolts to torque of 90–100 ft. lbs. (122–135 Nm).
7. Install brake flex hose and retaining clip.
8. Install wheel and tire assembly.
9. Install quarter trim panels on four-door models and access panels on other models.

Lower Control Arm

Removal

NOTE: **The lower control arm is replaced as a unit. The bushing is not serviceable.**

1. Raise the vehicle and support safely. Remove tire an wheel assembly.
2. On Taurus and Sable sedans, remove the brake proportioning valve from the left side of the front arm. Disconnect the parking brake cable.
3. Place a floor jack under the lower control arm between spindle and spindle end mounting.
4. Remove nuts from control arm-to-body mounting and control arm-to-spindle mounting. Do not remove bolts at this time.
5. Remove spindle end mounting bolt. Slowly lower floor jack until spring and spring insulator can be removed.
6. Remove bolt from the body end and remove control arm from the vehicle.

Installation

1. Attach lower control arm to body bracket using a new bolt and a new nut. DO NOT tighten at this time. Install this bolt with bolt head to the front of the vehicle.

2. Place spring in spring pocket in lower control arm. Be sure the spring pigtail is in the proper index in lower control arm and the insulator is at the top of the spring, properly seated and indexed. Insulator must be replaced on the spring before spring is placed in position.
3. Using a floor jack, raise lower control arm until it comes in line with mounting hole in the spindle.
4. Install lower control arm to spindle using a new nut, new bolt and new washers. DO NOT tighten at this time. Install this bolt with the bolt head to the front of the vehicle.
5. Using a floor jack, raise lower control arm to its curb height.
6. Tighten control arm-to-spindle bolt to torque of 90–100 ft. lbs. (122–135 Nm).

Strut And Spring

Removal and Installation

TEMPO/TOPAZ

1. Raise jack only enough to contact body.
2. Open truck lid and loosen, but do not remove two nuts retaining the upper strut mount to body.
3. Raise vehicle. Remove wheel and tire.
4. Place a jack stand under the control arms to support the suspension.

—————— CAUTION ——————
Care should be taken when removing the strut that the rear brake flex hose is not stretched or the steel brake tube is not bent.

5. Remove the bolt attaching the brake hose bracket to the strut and carefully move it out of the way.

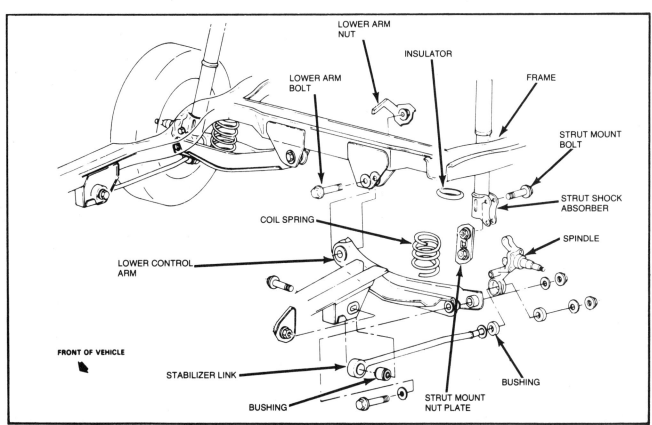

All bolts must be insatalled in direction shown (© Ford Motor Co.)

SUSPENSION
Domestic Models

6. Remove two bolts retaining the jounce bumper bracket strut to spindle.
7. Remove jounce bumper bracket from the vehicle.
8. Remove sock strut from the spindle.
9. Remove two upper mount-to-body nuts.
10. Remove strut from vehicle.
11. Install in reverse order. Torque top mount to body bolts 20–30 ft. lbs. and strut to spindle bolts 70–96 ft. lbs. Always install new strut to spindle bolts.

Tension Strut

Removal and Installation
TAURUS/SABLE – STATION WAGON

1. Raise the vehicle and support safely. Place a floor jack under the rerar lower suspension arm and raise the arm to normal curb height.
2. Remove the wheel and tire assembly.
3. Remove the nut and bolt retaining the tension strut to the lower suspension arm.
4. Remove the tension strut to body bracket retaining nut. Remove the strut assembly.
5. Installation is the reverse of the removal procedure. Tighten tension strut to body bracket bolt 40–55 ft. lbs.

TAURUS/SABLE – SEDAN

1. From inside the trunk, loosen, but do not remove the three nuts retaining the upper shock strut mount to the body.
2. Raise the vehicle and support safely. Remove the wheel and tire assembly.
3. Remove the tension strut to spindle retaining nut.
4. Remove the tension strut to body retaining nut.
5. Move the spindle rearward enough to gain working room in order to remove tension strut. Remove tension strut.
6. Installation is the reverse of the removal procedure using new bushings and washers on both ends of the new tension strut.

Control Arm

Removal and Installation
TEMPO/TOPAZ

1. Raise and safely support vehicle.
2. Remove tire and wheel assembly.
3. Remove arm-to-spindle bolt and nut.
4. Remove center mounting bolt and nut.
5. Remove arm from vehicle.
6. Install in reverse order. Torque arm to body bolt 40–55 ft. lbs. and arm spindle bolt 60–86 ft. lbs.

TAURUS/SABLE

1. Raise vehicle and support safely. Do not raise vehicle by tension strut.
2. Disconnect the brake proportioning valve from the left side front arm.
3. Disconnect the parking brake cable.
4. Remove the arm-to-spindle bolt, washer and nut. Remove the arm-to-body retaining nut.
5. Remove the control arm from the vehicle.
6. Installation is the reverse of the removal procedure.

NOTE: When installing new control arms, the offset on all arms must face up. The arms are stamped bottom on the lower edge. The flange edge of the right side rear arm stamping must face the front of the vehicle.

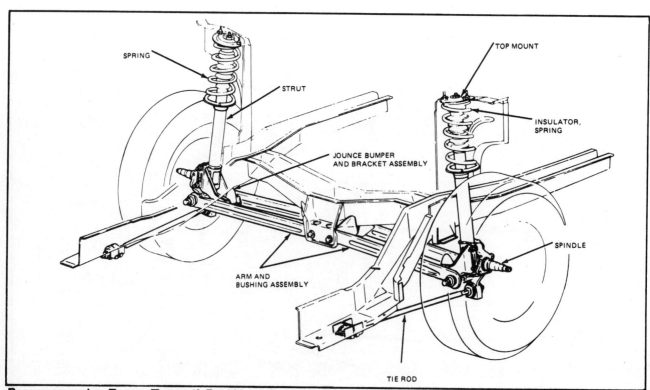

Rear suspension Tempo/Topaz (© Ford Motor Co.)

Tie Rod

Removal and Installation
TEMPO/TOPAZ

1. Raise car only enough to contact the body. From inside the trunk loosen, but DO NOT remove the two strut top mount-to-body nuts.
2. Raise vehicle and place a jack stand under the suspension to support it. Remove wheel and tire assembly.
3. Remove two top mount studs.
4. Remove nut retaining tie rod to the spindle. Remove nut retaining tie rod to body.
5. Lower the jack stand enough so that the upper strut mount studs are out of the holes in the body.
6. Move spindle rearward enough so that the tie rod can be removed.
7. Place new washers and bushings on both ends of new tie rod. Bushings at front and rear of the tie rod are different. The rear bushings have indentations in them.
8. Insert one end into the body bracket and install a new bushing washer and nut. DO NOT tighten at this time.
9. Pull back on the spindle enough so that the tie rod end can be installed in the spindle.
10. Install a new bushing, washer and nut. DO NOT tighten at this time.
11. Raise the jack stand enough to hold the two strut mounting studs in place.
12. Install two new strut-to-body mount nuts. Tighten to 20–30 ft. lbs.
13. Raise the suspension to curb height and tighten the two tie rod nuts to 52–74 ft. lbs.
14. Remove jack stand. Install tire and wheel assembly. Lower vehicle.

FRONT SUSPENSION—FORD, LINCOLN, MERCURY REAR DRIVE CARS, SINGLE ARM DESIGN

Description

The design utilizes shock struts with coil springs mounted between the lower arm and a spring pocket in the crossmember. The shock struts are non-repairable, and they must be replaced as a unit. The ball joints and lower suspension arm bushings are not separately serviced, and they also must be replaced as a suspension arm, bushing, and ball joint assembly.

Wheel Alignment

Caster and camber angles of this suspension are set at the factory, and cannot be adjusted in the field. Toe is adjustable.

TOE

Start the engine and move the steering wheel back and forth several times until it is in the straight ahead or centered position. Turn the engine off, and lock the steering wheel in place using a steering wheel holder. Adjust the left and right spindle rod sleeves until each wheel has one-half the desired total toe specification.

NOTE: For all vehicles, whenever the jam nuts are loosened for toe adjustment, the nuts must be tightened to 33–50 ft. lbs. (48–67 Nm).

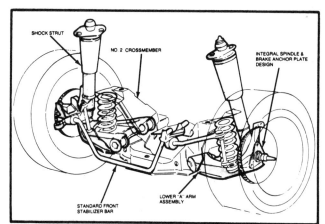

Single arm front suspension Thunderbird/XR7, Fairmont/Zephyr, Granada/Cougar, Mustang/Capri (© Ford Motor Co.)

Wheel Bearings

Replacement and Lubrication

1. Raise the vehicle until the tire clears the floor, and remove wheel from hub and rotor.
2. Remove the caliper from the spindle, and wire it to the underbody to prevent damage to the brake hose.
3. Remove the grease cap from the hub. Remove the cotter pin, nut lock, adjusting nut, and flatwasher from the spindle. Remove the outer bearing cone and roller assembly.
4. Pull the hub and rotor assembly off the spindle.
5. Using tool 1175–AC or equivalent, remove and discard the grease retainer. Remove the inner bearing cone and roller assembly from the hub.
6. Clean the inner and outer cups with solvent. Inspect the cups for scratches, pits, excessive wear, and other damage. If the cups are worn or damaged, replace.

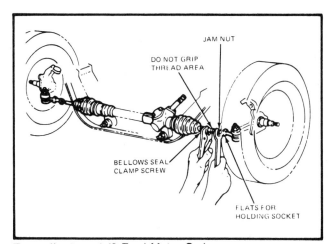

Toe adjustment (© Ford Motor Co.)

SUSPENSION
Domestic Models

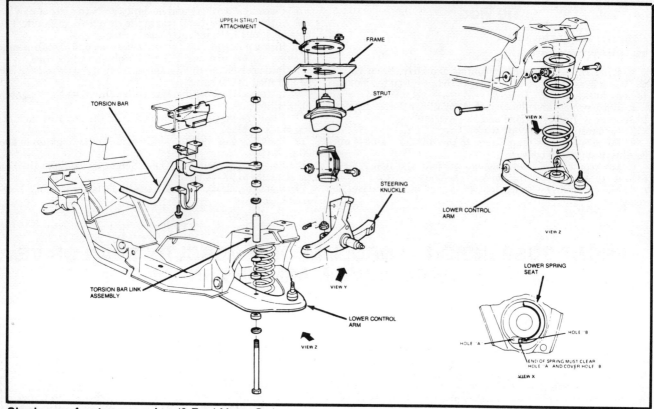

Single arm front suspension (© Ford Motor Co.)

Adjustment

If the wheel is loose on the spindle or does not rotate freely, adjust the front wheel bearings as follows:
1. Raise the vehicle until the tire clears the floor.
2. Remove the wheel cover. Remove the grease cap from the hub.
3. Wipe the excess grease from the end of the spindle. Remove the cotter pin and nut lock.
4. Loosen the adjusting nut three turns. Rock the wheel, hub and rotor assembly in and out several times to push the shoe and linings away from the rotor.
5. While rotating the wheel, hub and rotor assembly, tighten the adjusting nut to 17–25 ft. lbs. (24–33 Nm) to seat the bearings.
6. Loosen the adjusting nut one-half turn, then retighten 10–15 inch lbs. (1.2–1.6 Nm), using a torque wrench.
7. Place the nut lock on the adjusting nut, so the castellations on the lock are in line with the cotter pin hole in the spindle.
8. Install a new cotter pin, an bend the ends around the castellated flange of the nut lock.
9. Check the front wheel rotation. If the wheel rotates properly, reinstall the grease cap and wheel cover. If rotation is noisy or rough, follow the inspection, lubrication, and replacement procedures.
10. Before driving the vehicle, pump the brake pedal several times to restore normal brake pedal travel.

Lower Ball Joint

Inspection

1. Support the vehicle in normal driving position with both ball joints loaded.
2. Wipe the wear indicator and ball joint cover checking surface, so they are free of dirt and grease.
3. The checking surface should project outside the cover. If the checking surface is inside the cover, replace the lower arm assembly.

Shock/Strut

Removal

1. Place the ignition key in the unlocked position to permit free movement of front wheels.
2. From the engine compartment, remove the one strut to upper mount attaching nut. A screwdriver in the slot will hold the rod stationary while removing the nut.
3. Raise the front of the vehicle by the lower control arms, and position safety stands under the frame jacking pads, rearward of the wheels.
4. Remove the tire and wheel assembly.
5. Remove brake caliper and rotate out of position.
6. Remove the two lower nuts and bolts attaching the strut to the spindle.
7. Lift the strut up from the spindle to compress the rod, then pull down and remove the strut.

Installation

1. With the rod half extended, place the rod through the upper mount, and hand start with a new 16 mm nut, engaging as many nut threads as possible.
2. Extend the strut, and position into the spindle.
3. Install two new lower mounting bolts, and hand start nuts.

SUSPENSION
Domestic Models

4. Tighten the new 16 mm strut to upper mount attaching nut, inside the engine compartment to 60–75 ft. lbs. (81–102 Nm). A screwdriver in the slot will hold the rod stationary while the nut is being tightened.
5. Remove the suspension load from the lower control arms by lowering the hoist, and tighten the lower mounting nuts to 150–180 ft. lbs. (203–244 Nm).

Lower Suspension Arm

Removal

1. Raise the vehicle and support safely. Allow the control arms to hang free.
2. Remove the wheel and tire assembly.
3. Remove the brake caliper. Wire if off to the side to obtain working room. Remove the brake rotor and dust shield.

NOTE: 1983 Lincoln Continentals do not require the removal of the caliper, rotor or dust shield.

4. Disconnect the tie rod assembly from the steering spindle. Remove the steering gear bolts and position the gear so that the suspension arm bolt may be removed.
5. Disconnect the stabilizer bar link from the lower arm.
6. Remove the cotter pin from the ball joint stud nut, and loosen the ball joint nut one or two turns.
7. Tap the spindle boss to relive the stud pressure.
8. Install a spring compressor tool into the lower arm spring pocket hole, through coil spring into upper plate.
9. Tighten the nut on the compressor tool until a drag on the nut is felt.
10. Remove the ball joint nut and raise the entire strut and spindle assembly. wire it off to the side to obtain working room.
11. Remove the suspension arm-to-crossmember bolts. The compressor tool forcing nut may have to be tightened or loosened for ease of bolt removal.
12. Loosen the compression rod forcing nut until the spring tension is relived. Remove the forcing nut, lower suspension arm and coil spring.

Installation

1. Place the insulator on the top of the spring. Position the spring into the lower arm spring pocket.
2. Position the spring and lower arm onto the compression tool.
3. Tighten the nut on the compression tool. Position the lower arm into the crossmember. Install new lower arm-to-crossmember bolts and nuts. Do not tighten at this time.
4. Unwire the strut and spindle assembly and attach to the ball joint stud. Install a new ball joint stud nut. Do not tighten at this time.
5. Using a suitable jack, raise the suspension arm to a normal position. Remove the compressor tool.
6. While the jack is still in place, tighten the lower arm-to-crossmember attaching nuts to 150–180 ft. lbs. (203–244 Nm) for 1983 and later vehicles.
7. Tighten the ball joint stud nut to 100–200 ft. lbs. (136–163 Nm). Install new cotter pin.
8. Install the brake shield, rotor and brake caliper if removed.
9. Install the steering gear-to-crossmembr bolts and nuts if removed. Tighten to 90–100 ft. lbs. (122–135 Nm).
10. Install the tie rod assembly into the steering spindle. Install the retaining nut. Tighten to 35 ft. lbs. (47 Nm). Install a new cotter pin.
11. Connect the stabilizer bar link to the lower suspension arm. Tighten to 9–12 ft. lbs. (12–16 Nm).
12. Install the wheel and tire assembly. Check and adjust the front wheel alignment if necessary.

Spring

Removal

1. Raise the front of the vehicle and position safety stands under both sides at the jack pads just back of the lower arms. Remove the wheel and tire assembly.
2. Disconnect the stabilizer link bar from the lower bar.
3. Remove the steering gear bolts, and move the steering gear out of the way.
4. Disconnect the tie rod from the steering spindle.
5. Using the spring compressor tool D78P–5310–A or equivalent, install one plate with the pivot ball seat down into the coils of the spring. Rotate the plate so that it is fully seated into the lower suspension arm spring seat.
6. Install the other plate with the pivot ball seat up into the coils of the spring. Insert the ball nut through the coils of the spring, so it rests in the upper plate.
7. Insert compression rod into the opening in the lower arm through the lower and upper plate. Install the ball nut on the rod, and return the securing pin.

NOTE: This pin can only be inserted one way into the upper ball nut because of a stepped hole design.

8. With the upper ball nut secured, turn the upper plate, so it walks up the coil until it contacts the upper spring seat.
9. Install the lower ball nut, thrust bearing and forcing nut on the compression rod.
10. Rotate the nut until the spring is compressed enough, so it is free in its seat.
11. Remove the two lower control arm pivot bolts and nuts to disengage the lower arm from the frame crossmember. Remove the spring assembly.
12. If a new spring is to be installed, mark the position of the upper and lower plates on the spring with chalk. Measure the compressed length of the spring as well as the amount of spring curvature to assist in the compression and installation of a new spring.
13. Loosen the nut to relieve spring tension, and remove the tools from the spring.

Installation

1. Assemble the spring compressor tool and locate tool through spring.
2. Before compression the coil spring, be sure the upper ball nut securing pin is inserted properly.
3. Compress the coil spring.
4. Position the coil spring assembly into the lower arm.

NOTE: Be sure that the lower end (pigtail) of the coil spring is properly positioned between the two holes in the lower arm spring pocket depression.

5. Install coil spring. Reverse removal procedures.

Spindle Assembly

Removal

1. Raise the front of the vehicle, and position safety stands under both sides at the jacking pads just behind the lower arms.
2. Remove the wheel and tire assembly.
3. Remove the brake caliper, rotor, and dust shield.
4. Remove the stabilizer link from the lower arm assembly.
5. Remove the tie rod end from the spindle.
6. Remove the cotter pin from the ball joint stud nut, and loosen the ball joint stud nut one or two turns.

— CAUTION —
Do not remove the nut from the ball joint stud at this time.

SECTION 2
SUSPENSION
Domestic Models

7. Tap the spindle boss to relieve the stud pressure.
8. Place a floor jack under the lower arm, compress the coil spring and remove the stud nut.
9. Remove the two bolts and nuts attaching the spindle to the shock strut. Compress the shock strut until working clearance is obtained.
10. Remove the spindle assembly.

Installation

1. Place the spindle on the ball joint stud, and install the stud nut. Do not tighten at this time.
2. Lower the shock strut until the attaching holes are in line with the holes in the spindle. Install two new bolts and nuts.
3. Tighten the ball joint stud nut to 80–120 ft. lbs. (108–163 Nm) and install a new cotter pine.
4. Tighten the shock strut-to-spindle attaching nuts to 150–180 ft. lbs. (203–244 Nm).
5. Lower the floor jack from under the suspension arm, and remove the jack.
6. Intall the stabilizer link and tighten the attaching bolt and nut to 8–12 ft. lbs. (11–16 Nm).
7. Attach the tie-rod end, and tighten the retaining nut to 35–47 ft. lbs. (47–64 Nm).
8. Install the disc brake dust shield, rotor and caliper.
9. Install the wheel and tire assembly. Check the front wheel alignment and adjust if necessary.

Stabilizer Bar Link Insulators

Removal

To replace the link insulators on each stabilizer link, use the following procedure:
1. Raise the vehicle on a hoist.
2. Remove the nut, washer and insulator from the upper end of the stabilizer bar attaching link bolt.
3. Remove the bolt and the remaining washers, insulators and spacer.

Installation

1. Install the stabilizer bar link insulators by reversing the above steps.
2. Tighten the attaching nuts to 8–12 ft. lbs. (11–16Nm).

Stabilizer Bar And Insulator

Removal

1. Raise the vehicle on a hoist.
2. Disconnect the stabilizer from each stabilizer link and both stabilizer insulator attaching clamps. Remove the stabilizer bar assembly.
3. Cut the worn insulators and plastic sleeves from the stabilizer bar.

Installation

1. Coat the necessary parts of the stabilizer bar with Ford Suspension Rubber Insulator Lubricant or an equivalent lubricant; install new plastic sleeves with the flange inboard, and slide insulators onto the stabilizer bar and over sleeves. Be sure the insulator is fully seated against the flange.
2. Using a new nut and bolt, secure each end of the stabilizer bar to the lower suspension arm. Tighten these nuts to 8–12 ft. lbs. (11–16 Nm).
3. Using new fasteners, clamp the stabilizer bar to the attaching brackets on the side rail. Tighten these bolts to 20–25 ft. lbs. (27–33 Nm).

FRONT SUSPENSION – FORD, LINCOLN, MERCURY REAR DRIVE CARS, SPRING ON LOWER ARM DESIGN

Wheel Alignment

ADJUSTMENT

Caster and Camber

1. Check suspension with the front wheels in the straight ahead position. Run the engine so the power steering control valve will be in the center (neutral) position (if equipped).
2. Check caster and camber and record the readings.
3. Compare camber and caster readings with specifications to determine if adjustment is required to bring vehicle to nominal setting.
4. If adjustment is required, insert alignment tools into frame holes and "snug" the tool hooks finger-tight against the upper arm inner shaft. Then tighten hex nut of each alignment tool 1 additional "hex flat."
5. Loosen upper arm inner shaft-to-frame attaching bolts so that the lockwashers on bolts are unloaded. Then firmly tap bolt heads to assure loosening of the lower assemblies.
6. Adjust camber and caster on each wheel.
7. Torque upper arm inner shaft-to-frame attaching bolts to 100–140 ft. lbs. (136–190 Nm). It is not necessary to recheck caster and camber after this adjustment procedure is performed.
8. Check toe-in and steering wheel spoke position and adjust both (as required) at the same time.

Toe and Steering Wheel Spoke Position

After adjusting caster and camber, check the steering wheel spoke position with the front wheels in straight-ahead position. If the spokes are not in their normal position, they can be properly adjusted while toe is being adjusted.
1. Loosen the two clamp bolts on each spindle connecting rod sleeve.
2. Adjust toe. If the steering wheel spokes are in their normal position, lengthen or shorten both rods equally to obtain correct toe.
3. If the steering wheel spokes are not in their normal position, make the necessary rod adjustments to obtain correct toe and steering wheel spoke alignment.
4. When toe and the steering wheel spoke position are both correct, lubricate clamp, bolts and nuts and tighten the clamp bolts on both connecting rod sleeves and specification. The sleeve position should not be changed when the clamp bolts are tightened for proper clamp bolt orientation.

Wheel Bearings

Adjustment and Inspection

If the wheel is loose on the spindle or does not rotate freely, adjust the front wheel bearings as follows:
1. Raise the vehicle until the tire clears the floor.

2-94

SUSPENSION
Domestic Models

2. Remove the wheel cover. Remove the grease cap from the hub.
3. Wipe the excess grease from the end of the spindle. Remove the cotter pin and nut lock.
4. Loosen the adjusting nut three turns and rock the wheel, hub and rotor assembly in and out several times to push the shoe and linings away from the rotor.
5. While rotating the wheel, hub and rotor assembly, torque the adjusting nut to 17–25 ft. lbs. (23–24 Nm) to seat the bearings.
6. Loosen the adjusting nut one-half turn, then retighten to 10–15 inch lbs. (1.2–1.6 Nm) using a torque wrench.
7. Place the nut lock on the adjusting nut so the castellations on the lock are in line with the cotter pine hole in the spindle.
8. Install a new cotter pin and bend the ends around the castellated flange of the nut lock.
9. Check front wheel rotation. If the wheel rotates properly, reinstall the grease cap and wheel cover. If rotation is noisy or rough, remove wheel hub and check for bearing problems.
10. Before driving the vehicle, pump the brake pedal several times to restore normal brake pedal travel.

Replacement and Lubrication

1. Raise the vehicle until the tire clears the floor. Remove wheel from hub and rotor.
2. Remove the caliper from the spindle and wire it to the underbody to prevent damage to the brake hose.
3. Remove the grease cap from the hub. Remove the cotter pin, nut lock, adjusting nut and flatwasher from the spindle. Remove the outer bearing cone and roller assembly.
4. Pull the hub and rotor assembly off the spindle.
5. Using tool 1175–AC or equivalent, remove and discard the grease retainer. Remove the inner bearing cone and roller assembly from the hub.
6. Clean the inner and outer bearing cups with solvent. Inspect the cups for scratches, pits, excessive wear and other damage. If the cups are worn or damaged, remove them with tools D80L–927–A and T77F–1102–A or equivalent and replace them.
7. Remove the old lubricant from the spindle and inside of the hub assembly.
8. If the inner and outer bearing cups were removed, install replacement cups.

NOTE: When installing replacement cups in the bearings, support the rotor on the hub barrel with a wooden block in order to avoid damage to the wheel studs.

9. Clean the old grease from all the surrounding surfaces.
10. Using a bearing packer, pack the bearing cone with C1AZ–19590–B or equivalent grease. Grease the bearing cup surfaces.
11. Install the inner bearing cone and roller assemblies in the inner cup. Coat the new grease retainer with grease and install the retainer. Be sure the retainer is properly seated.
12. Install the hub and rotor assembly on the spindle.
13. Install the outer bearing cone and roller assembly and flatwasher on the spindle. Install the adjusting nut. Adjust the wheel bearings as outline earlier. Install the grease cap.
14. Install the caliper on the spindle if removed earlier.
15. Install the wheel and tire assembly. Lower the vehicle.

NOTE: Before driving the vehicle, pump the brake pedal several times to restore normal brake pedal travel.

Shock Absorber

Removal

1. Remove the nut, washer and bushing from the shock absorber upper end. Raise the vehicle and support safely.

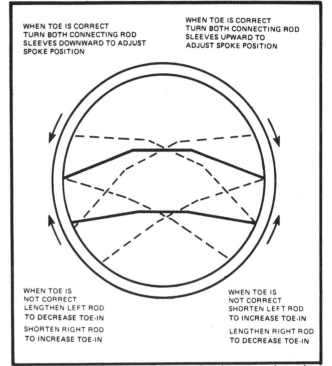

Adjusrt both rods equally to maintain normal spoke position (© Ford Motor Co.)

2. Remove the two thread-cutting screws attaching the shock absorber to the lower arm and remove the shock absorber. Lightly wire brush the shock studs to free of rust, oil or corrosion.
3. Remove the shock absorber.

Installation

1. Place a washer and bushing on the shock absorber top stud and position the shock absorber inside the front spring. Install the thread-cutting screws and torque to specifications. If the threads in the lower arm become stripped or damaged, the removed thread cutting screws should be re-used, along with $5/16$–18 lock nuts. Torque to the same specifications as when thread cutting screws are secured directly to the lower arm.
2. Remove the safety stands and lower the vehicle.
3. Place a bushing and washer on the shock absorber top stud and install nut.

Spring

Removal

1. Raise the vehicle. Remove the wheel and tire assembly.
2. Disconnect the stabilizer bar link from the lower arm.
3. Remove the two bolts attaching the shock absorber to the lower arm assembly.
4. Remove the upper nut, retainer and grommet from the shock absorber and remove the shock.
5. Remove the steering center link from the pitman arm.
6. Support the vehicle with safety stands under the jacking pads and lower the hoist for working room.
7. Using the spring compressor tool, install one plate with the pivot ball seat facing downward into the coils of the spring. Rotate the plate so that it is flush with the upper surface of the lower arm.
8. Install the other plate with the pivot ball seat facing upward into the coils of the spring. Insert the upper ball nut

SUSPENSION
Domestic Models

through the coils of the spring so the nut rests in the upper plate.

9. Insert the compression rod into the opening in the lower arm through the upper and lower plate and upper ball nut. Insert the securing pin through the upper ball nut and compression rod.

NOTE: This pin can only be inserted one way in the upper ball nut because of a stepped hole design.

10. With the upper ball arm secured, turn the upper plate so that it walks up the coil until it contacts the upper spring seat, then back off one half turn.
11. Install the lower ball nut and thrust washer on the compression rod, and screw on the forcing nut.
12. Tighten the forcing nut until the spring is compressed enough so that it is free in its seat.
13. Remove the two lower arm pivot bolts and disengage the lower arm from the frame crossmember; remove the spring assembly.
14. If a new spring is to be installed, mark the position of the upper and lower plates on the spring with chalk, and measure the compressed length of the spring and amount of spring curvature to assist in compression and installation of a new spring.
15. Loosen the forcing nut to relieve spring tension and remove the tools from the spring.

Installation

1. Assemble the spring compressor and locate in the same position as indicated in Step 14 of Spring Removal.
2. Before compressing the coil spring, be sure that the upper ball nut securing pin in inserted properly.
3. Compress the coil spring until the spring height reaches the dimension obtained in Step 14 of Spring Removal.
4. Position the coil spring assembly into the lower arm.
5. To install coil spring, reverse removal procedure.

Ball Joints

Inspection
The checking surface should project outside the cover. If the checking surface is inside the cover, replace the lower arm assembly, or install ball joint kit.

Replacement
The manufacturer recommends replacing the complete arm if the ball joint is worn. However, aftermarket suppliers have ball joint kits available to replace worn ball joints without replacing the complete arm.

Upper Arm

Removal

1. Raise the front of the vehicle and position safety stands under both sides of the frame just back of the lower arm.
2. Remove the wheel and tire.
3. Remove the cotter pin from the upper ball joint stud nut.
4. Loosen the upper ball joint stud nut one or two turns.

--- CAUTION ---
Do not remove the nut from the stud at this time.

5. Insert ball joint press tool between the upper and lower ball joint studs with the adapter screw on top.
6. With a wrench, turn the adapter screw until the tool places the stud under compression. Tap the spindle near the upper stud with a hammer to loosen the stud in the spindle.

NOTE: Do not loosen the stud from the spindle with tool pressure only. Do not contract the boot seal with hammer.

7. Remove the tool from between the joint studs and place a floor jack under the lower arm.
8. Remove the upper arm attaching bolts, and remove the upper arm assembly.

Installation

1. Transfer the bumper from the old arm to the new arm.
2. Position the upper arm in new shaft to the frame bracket, and install the two attaching bolts and washers to a snug fit.
3. Connect the upper ball joint stud to the spindle and install the attaching nut. Torque the nut to specifications and continue to tighten the nut until the cotter pin hole in the stud is in line with the nut slots, then install the cotter pine.
4. Install the wheel and tire adjust the wheel bearing.
5. Remove the safety stands and lower the front of the vehicle.
6. Adjust caster, camber and toe-in to specifications.

Upper Arm Bushings

Replacement (With Arm Removed)

1. Remove the nuts and washer from both ends of the upper arm shaft.

NOTE: Use the existing C-clamp tool part number T74P-3044-A-1 or equivalent and adapters to remove the bushings.

2. Position the shaft and new bushings to the upper arm and install the bushings and shaft to the upper arm.

NOTE: The front bushing is a larger diameter than the rear, requiring that adapter part number T79P-3044-A2 or equivalent, is used when installing rear bushings.

3. Make certain that the inner shaft is positioned so that the serrated side contacts the frame.
4. Install an inner washer, rear bushing only, and two outer washers and new nuts on each side of the inner shaft.

Lower Arm

Removal

1. Raise the front of the vehicle and position safety stands under both sides of the frame just back of the lower arms.
2. Remove the wheel and tire.

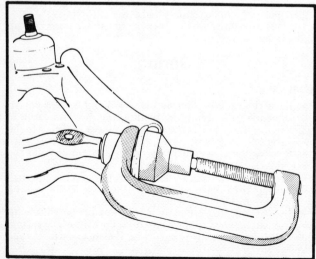
Upper arm bushing installation (© Ford Motor Co.)

SUSPENSION
Domestic Models

3. Remove the brake caliper and rotor and dust shield.
4. Remove the shock absorber.
5. Disconnect the stabilizer bar link from the lower arm.
6. Disconnect the steering center link from the pitman arm.
7. Remove the cotter pin from the lower ball joint nut.
8. Loosen the lower ball joint stud nut one or two turns.
9. Install ball joint press tool between the upper and lower ball joints.
10. Install the coil spring compression tools and remove the spring.
11. Remove the ball joint nut, and remove arm assembly.
12. With a wrench, turn the adapter screw until the tool places the stud under compression. Tap the spindle near the lower stud with a hammer to loosen the stud in the spindle.
13. Remove the ball joint press tool.
14. Place a floor jack under the lower arm.
15. Gently lower the arm until all tension is relieved.
16. Remove lower arm center bolt and remove arm.

Installation

1. Position the arm assembly ball joint stud into the spindle and install the nut. Torque to specifications and install a new cotter pin.
2. Position the coil spring into the upper spring pocket; raise the lower arm and align the holes in the arm with the holes in the crossmember. Install bolts and nuts. Do not tighten at this time.

NOTE: Be sure that the pigtail of the lower coil of the spring is in the proper location of the seat of the lower arm, between the two holes.

3. Remove the spring compressor tool.
4. Connect the steering center link at the pitman arm, install the nut and tighten to specifications. Install a new cotter pin.
5. Install the shock absorber and torque fasteners to specifications.
6. Install the jounce bumper and torque nut to specifications.
7. Install the dust shield, rotor and caliper.
8. Position the stabilizer link to the lower arm and install the bolt and attaching nut.
9. Install the wheel and tire.
10. Remove the safety stands and lower the vehicle. After the vehicle has been lowered to floor and at curb height, torque the lower pivot bolts to 100–140 ft. lbs. (136–189 Nm).

Wheel Spindle

Removal

1. Raise the front of the vehicle and position safety stands under both sides of the frame just back of the lower arm.
2. Remove the wheel and tire.
3. Remove the brake rotor, caliper and dust shield.
4. Disconnect the tie rod end from the spindle.
5. Remove the cotter pins from both ball joint stud nuts and loosen the nuts one or two turns.

NOTE: Do not remove the nuts at this time.

6. Position the ball joint remover tool between the upper and lower ball joint studs.
7. Turn the tool with a wrench until the tool places the studs under compression. With a hammer, sharply hit the spindle near the studs to break it loose in the spindle.
8. Position a floor jack under the lower arm at the lower ball joint area.
9. Remove the upper and lower ball joint stud nuts, lower the jack carefully to remove the spindle.

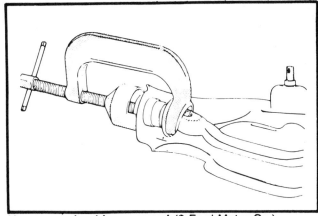

Upper arm bushing removal (© Ford Motor Co.)

Installation

1. Position the spindle on the lower ball joint stud and install the stud nut. Torque the nut to specification and install a new cotter pin.
2. Raise the lower arm and guide the upper ball joint stud into the spindle. Install the stud nut.
3. Torque the nut to specifications and install a new cotter pin. Remove the floor jack.
4. Disconnect the tie rod to the spindle. Install the nut and torque to specifications. Install a new cotter pine.
5. Install the brake dust shield, caliper and rotor.
6. Install the wheel and tire assembly.
7. Remove the safety stands and lower the vehicle.
8. Check caster, camber and toe-in and adjust as required.

Stabilizer Bar End Bushing

Replacement

1. Raise the vehicle on a hoist.
2. Remove the nut, washer and insulator from the lower end of the stabilizer bar attaching bolt.

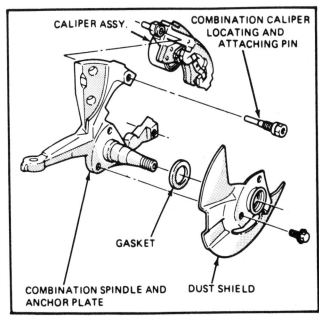

Spindle mounting (© Ford Motor Co.)

SECTION 2
SUSPENSION
Domestic Models

3. Remove the bolt and remaining washers, insulators, and the spacer.
4. Assemble a cup washer and new insulator on the bolt.
5. Insert the bolt through the stabilizer bar, then install new insulator and cup washer.
6. Install the spacer, cup washer and another new insulator on the bolt.
7. Insert the bolt through the lower arm and install a new insulator and cup washer. Install the attaching nut.

Stabilizer Bar And/Or Insulator

Removal

1. Raise the vehicle on a hoist, and place jack stands under the lower arm.
2. Disconnect the stabilizer from each stabilizer link and both stabilizer insulator clamps. Remove the stabilizer bar assembly.
3. Cut the worn insulators from the stabilizer bar.

Installation

1. Coat the necessary parts of the stabilizer bar with Ford Suspension Rubber Insulator Lubricant or an equivalent lubricant; slide insulators onto the stabilizer bar.
2. Using a new nut and bolt, secure each end of the stabilizer bar to the lower suspension arm, making sure the bolt head is at the bottom. Tighten nuts to 6–12 ft. lbs. (9–16 Nm).
3. Using new fasteners, clamp the stabilizer bar to the attaching brackets on the side rail. Tighten bolts to 14–26 ft. lbs. (19–35 Nm).

FRONT AND REAR SUSPENSION – 1984 AND LATER MARK VII AND CONTINENTAL 1985 AND LATER FORD CROWN VICTORIA AND MERCURY GRAND MARQUIS

DESCRIPTION AND OPERATION

Description

Air suspension is an air-operated, microcompressor controlled, suspension systems which replaces the conventional coil spring suspension and provides automatic front and rear load leveling.

Four air springs, made of rubber and plastic, support the vehicle load at the front and rear wheels.

The front air springs are mounted to a spring pocket in the crossmember and on the lower suspension arms similar to the conventional spring system. The rear air springs are mounted ahead of the rear axle outside the body sub-frame side members and on the lower rear suspension arms similar to the conventional rear spring system.

A single cylinder piston-type electrically operated air compressor, mounted on the left fender apron, supplies the air pressure for operating the system. A regenerative-type dryer is attached to the compressor manifold. All air flow during compression or venting passes through the dryer. A vent solenoid, located on the compressor manifold, controls air exhaustion.

The air flow to the entire system is controlled by the interaction of the air compressor, solenoids, height sensors and control module. All the air operated parts of the system are connected by nylon tubing.

Operation

The air suspension leveling system operated by adding or removing air in the springs to maintain the level at a pre-determined front and rear suspension height. The predetermined distance is known as the vehicle trim height. Trim height is controlled by the height sensors. Distance of the body to ground will change with tire size and inflation pressure.

The height sensors are attached to the body and the suspension arms, and will lengthen or shorten with suspension travel. Three height sensors are used: One at the left front wheel, one at the right front wheel, and one for the rear suspension.

The system works in the following manner: As weight is added to the vehicle, the body will settle under the load. As the body lowers, the heights sensors shorten (low out-of-trim), generating a signal to the control module which activates the air compressor (through a relay) and opens the air spring solenoid valves. As the body rises, the height sensors lengthen. When the pre-set trim height is reached, the air compressor is turned off and the solenoid valves are closed by the control module.

A similar action takes place whenever weight is removed from the vehicle. As weight is removed, the body will rise, which causes the height sensors to lengthen (high out-of-trim), generating a signal to the control module which opens the air compressor vent solenoid and opens the air spring solenoid valves. As the body lowers, the height sensors shorten and when the pre-set trim height is reached, the air compressor vent solenoid is closed and the spring solenoid valves are closed by the control module.

Air required for levelling the vehicle is distributed from the air compressor to each air spring by four nylon air lines which start at the compressor dryer and terminate at the individual air springs. The dryer is a common pressure manifold for all four air lines so orientation of these lines at the compressor is not required. However, the air lines are color coded to identify to which air spring they are attached. The dryer contains a desiccant (silica gel) which dries the compressed air before delivering the air to the springs. during venting of any air spring, the previously dried air passes through the dryer to remove moisture from the desiccant (regeneration). Air required for compression and vent air enter and exit through a common port on the compressor head. Vented air is also controlled by a solenoid valve in the compressor head.

CAUTION

The compressor relay, compressor vent solenoid and all air spring solenoids have internal diodes for electrical noise suppression and therefore are polarity sensitive. Care must be taken when servicing these components not to switch the battery feed and ground circuits or component damage will result.

A microcomputer based module controls the air compressor motor (through a relay), vent solenoid and the four air spring solenoids to provide the air requirements of the springs. The module also provides power and ground to the three digital height sensors and continuously monitors input from the three height sensors and the ignition Run/Brake On/Door Open circuits. These inputs are used by the module to make vehicle levelling decisions which are then carried out by the air system components controlled by the module. For service, the module provides a series of diagnostic tests, a routine for filling the springs and operates a warning lamp.

The control logic for operating the system is given below:

IGNITION OFF

1. Operates for 30 minutes after the ignition switch is

SUSPENSION
Domestic Models

turned from Run to Off, then the system is inoperable through the module.

2. WIll service down requests (lower vehicle) as required during the 30 minutes EXCEPT if any height sensor was reading a high (vehicle) when the ignition was turned from Run to Off. The vent time is limited to 10 seconds for the rear springs and 3 seconds for the front..

3. At one hour after ignition is turned to Off, the system will correct for a low vehicle if required. Compressor run time is limited to 15 seconds for the rear springs and 30 seconds for the front springs.

IGNITION IN RUN

1. Ignition in the Run position for less than 45 seconds.
   a. Will service first rear or front up requests (raise vehicle) immediately if required.
   b. Will not service down requests (lower vehicle).
2. Ignition in the Run position for more than 45 seconds.
   a. If a door(s) is open with the brake not engaged, up requests (raise vehicle will be serviced immediately and down request (lower vehicle) will be serviced after the door(s) is closed.
   b. If the doors are not open and the brake is not engaged, service all (up or down) requests by a 45 second averaging method.
   c. If the brake is engaged and a door(s) is open, service up requests (raise vehicle) immediately but do not service down (lower vehicle) requests.
   d. If the brake is engaged and the doors are closed, all requests (up or down) will not be serviced except that if a rear up request (raise vehicle) is in progress it will be completed.

GENERAL

1. Down requests (lower vehicle) will not be serviced if any door is open.
2. Requests are serviced in the following order: rear up, front up, rear down, front down.
3. During Ignition In Run, if any up or down request cannot be serviced within three minutes, the warning lamp will come On and stay On for that ignition cycle. However, only the request which was being serviced is affected. that is, if a time-out failure occurred during a left front up correction, the module would continue to service future left front down requests and all right front and rear requests.
4. The rear spring solenoids are always operated in tandem, but the front spring solenoids may operate independently.
5. Front and rear requests (up or down) will never be serviced simultaneously.
6. Turning the ignition from Run to Off clears all memory in the module and therefore the warning lamp may not immediately indicate a failure when the ignition is returned to the Run position.

--- **CAUTION** ---
When charging the battery, the ignition switch must be in the OFF position if the air suspension switch is On or damage to the air compressor relay or motor may occur.

However, use of a battery charge while performing the diagnostic test or air spring fill options is acceptable. Set to a rate to maintain, but not damage the vehicle battery.

ADJUSTMENTS

This adjustment procedure must be used prior to alignment, pinion angle or ride height checking. This method causes the system to perform a vent to trim.

NOTE: *If vehicle is significantly colder or warmer than alignment area (20 degrees or greater differential), time*

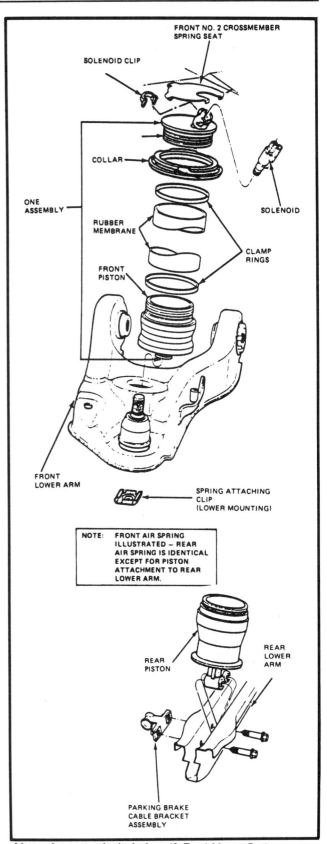

Air spring — exploded view (© Ford Motor Co.)

2-99

SECTION 2

SUSPENSION
Domestic Models

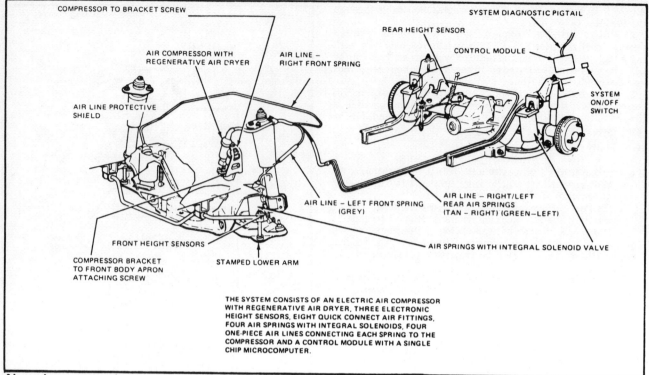

Air spring suspension system (© Ford Motor Co.)

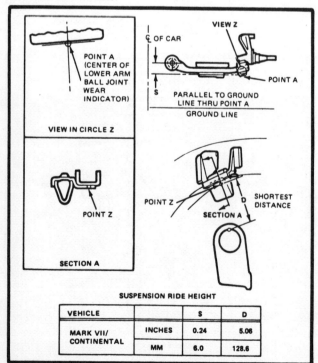

Check dimensions S and D (© Ford Motor Co.)

must be allowed for the vehicle to warm or cool to the temperature of the alignment area prior to Steps 1, 2 and 3.

1. Drive onto alignment rack, position vehicle, turn ignition OFF and exit vehicle.
2. Level rack as required, re-enter vehicle and turn ignition to Run position (do not start).
3. Allow one minute for vehicle to level, then push trunk release, turn ignition to the Off position and exit vehicle.
4. Allow 20 seconds for vehicle to vent to trim height (all doors must be closed) then turn air suspension system switch to the Off position (in trunk on LH side).
5. Check alignment (pinion angle) per specified procedures.

Ride Height

The FRONT SUSPENSION ride height of S dimension is adjusted by moving the front left and/or right sensor attaching stud (there are three adjustment positions provided on the bracket). Loosen the attaching screw and adjust up or down as required. A one position change to the sensor attachment point will yield approx. 0.5 in. (12.7 mm) change (up or down) to the S dimension.

The REAR SUSPENSION ride height D dimension is adjusted by moving the rear sensor attaching bracket up or down relative to the right rear upper arm (a slot adjustment is provided on the bracket). Loosen the attaching nut and adjust up or down as required. A one index mark change to the sensor attachment point will yield approximately 0.25 in. (6.35 mm) change (up or down) to the D dimension.

DIAGNOSIS

Leak Checks

If the air spring system is suspected of leakage, the standard soap solution check procedure is acceptable.

Warning Lamp (Check Suspension)

The air suspension warning lamp, located in the overhead console, has three main functions:

1. During normal operation with the ignition in the Run po-

SUSPENSION
Domestic Models

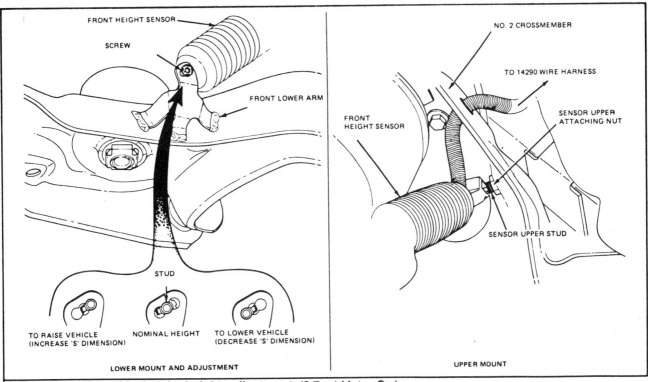

Front suspension ride height height adjustment (© Ford Motor Co.)

sition, the lamp glowing continuously indicates a possible air suspension system problem.

2. During diagnostic testing the lamp blinks at a rate of 1.8 blinks per second to show that diagnostic routine (in the module) has been entered and then blinks the test number that is being run during the test sequence.

3. During the air spring fill routine, the lamp blinks at a rate of 1 blink every two seconds to show that the air fill routine (in the module) has been entered.

Observation of the warning lamp during normal operation with the ignition switch On, can aid in detecting some system problems.

1. On a vehicle operating normally, the warning lamp will glow for approximately one second and then go out when the

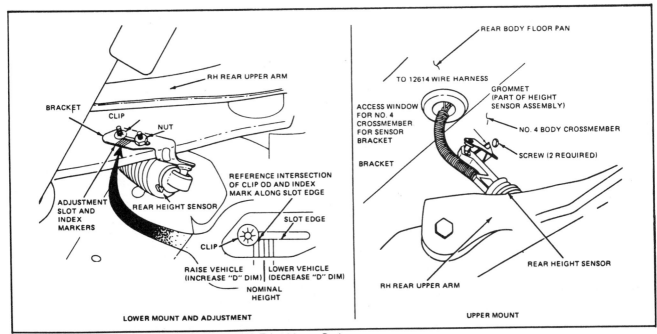

Rear suspension ride height adjustment (© Ford Motor Co.)

2-101

SUSPENSION
Domestic Models

ignition is turned from the Off to Run position. The lamp does not operate when the ignition is in either the Off or Start position.

2. If lamps does not go out after turning the ignition from the Off to the Run position, it indicates no battery power to the module.

3. If lamp glows for approximately ½ second, goes out, and then glows continuously after 5–8 seconds, when the ignition is turned to the Off to Run position, a height sensor or harness problem is indicated.

4. After ignition is turned from the Off to the Run position, if the lamp comes on and glows continuously at any time after 8 seconds, a system problem is indicated.

5. Once the warning lamp comes on during an ignition On cycle, it will glow continuously for that ignition On cycle.

6. Erratic operation of the warning lamp (blinking or occasional flashing) during an ignition On cycle, indicates a system problem.

DIAGNOSTIC AND AIR FILL INSTRUCTIONS

The control module has the capability of performing either a series of diagnostic tests on the air suspension system or to selectively fill the front and/or rear air springs. Specific instructions for using the air fill capability are in Removal and Installation. Instructions for entering diagnostics and test descriptions follow.

Entering Diagnostics

1. Turn On the air suspension switch. Diagnostic pigtail is to be ungrounded.

2. Install battery charger to reduce battery drain.

3. Cycle the ignition from the Off to the Run position, hold in the Run position for a minimum of five seconds, then return to the Off position. Driver's door is open with all other doors shut.

4. Change the diagnostic pigtail from an ungrounded state to a grounded state by attaching a lead from the diagnostic pigtail to vehicle ground. The pigtail must remain grounded during the spring fill sequence.

5. Turn the ignition switch to the Run position. (Do not start vehicle) The warning lamp will blink continuously at a rate of 1.8 blinks per second to indicate diagnostics has been entered and is ready.

WARNING LAMP FUNCTION

During diagnostics, the warning lamp continuously blinks either the "ready" status or the current test number.

DOOR FUNCTION

Each successive transition from DOOR CLOSED TO DOOR OPEN will cause the module to advance to the next step in the test sequence.

TERMINATION DIAGNOSTICS

Diagnostics may be terminated and the module returned to the normal operational mode at any time by cycling the ignition, actuating the brake or ungrounding the diagnostic pigtail.

Test Steps

The following tests will be run during Diagnostics.

For Tests 1, 2 and 3, PASS/FAIL will be determined by the module at the conclusion of Step A, B or C.

For Tests 4 through 10, PASS/FAIL will be determined by the technician observing the operation of the specific component.

TEST 1

Rear Suspension.

TEST 2

Right Front Suspension.

TEST 3

Left Front Suspension.

The following steps occur in each of the first three tests:

1. Raise the (rear, right front, left front) of the vehicle for 15 seconds. Continue raising the vehicle for an additional 15 seconds (30 seconds total maximum) or until a 'Vehicle High' signal or an illegal sensor read is received from the (rear, right front, left front) sensor.

2. Lower the (rear, right front, left front) of vehicle for 30 seconds or until a "Vehicle Low signal, or an illegal sensor read is received from the (rear, right front, left front) sensor.

3. Raise the (rear, right front, left front) of vehicle 30 seconds or until a Vehicle Trim signal, or an illegal sensor read is received from the (rear, right front, left front) sensor.

If the expected signal is not received within the 30 second limit, the test will stop and the warning lamp will turn on continuously. Also, if an illegal sensor read is received, the test will stop and the warning lamp will flash rapidly.

NOTE: The failed test may then be repeated by closing/opening the door or the next test may be initiated by closing/opening the door twice within 15 seconds.

TEST 4

NOTE: Hz (Hertz) = one cycle per second.

Compressor is cycled On/Off at 0.25 Hz. The compressor is limited to cycling a maximum of 50 times.

TEST 5

Vent solenoid is cycled open/closed at 1 Hz.

TEST 6

Left front solenoid is cycled open/closed at 1 Hz. and the vent solenoid is opened. Left front corner of the vehicle will drop slowly at test progresses.

TEST 7

Right front solenoid is cycled open/closed at 1 Hz. and the vent solenoid is opened. Right front corner of the vehicle will slowly drop so test progresses.

TEST 8

Right rear solenoid is cycled open/closed at 1 Hz. and the vent solenoid is opened. Right rear corner of the vehicle will drop slowly at test progresses.

TEST 9

Left rear solenoid is cycled open/closed at 1 Hz. and the vent solenoid is opened. Left rear corner of the vehicle will drop slowly as test progresses.

TEST 10

Actuating the brake, turning the ignition switch to Off, or disconnecting the diagnostic lead returns the module from diagnostics to the normal operating mode.

DIAGNOSTIC TESTING INDEX

1. QUICK TEST: Perform System Self-Diagnostic or Quick Test.
2. PINPOINT TEST B: Cannot Enter/Sequence or Exit Self-Test Diagnostic Quick Test.
3. PINPOINT TEST C: Diagnose Sensor Related Problem.
4. PINPOINT TEST D: Diagnose Vehicle Rear Problem.
5. PINPOINT TEST E: Diagnose Vehicle Right Front Problem.
6. PINPOINT TEST F: Diagnose Vehicle Left Front Problem.
7. PINPOINT TEST G: Diagnose Compressor Motor Electrical Problem.
8. PINPOINT TEST H: Diagnose Compressor Vent Solenoid Electrical Problem.

SUSPENSION
Domestic Models

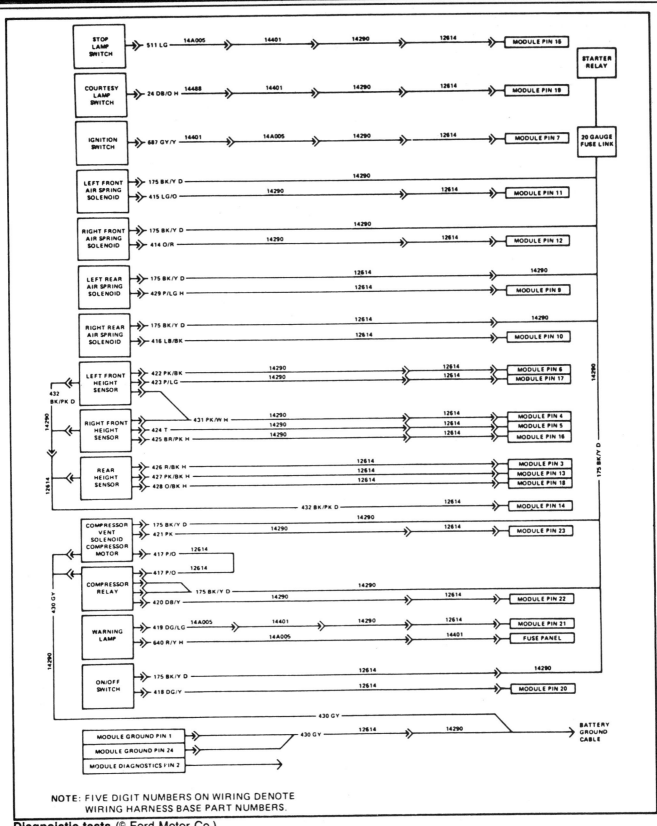

NOTE: FIVE DIGIT NUMBERS ON WIRING DENOTE WIRING HARNESS BASE PART NUMBERS.

Diagnoistic tests (© Ford Motor Co.)

SUSPENSION
Domestic Models

Air Spring System Components

CAUTION

Do not remove an air spring under any circumstances when there is pressure in the air spring. Do not remove any components supporting an air spring without either exhausting the air or providing support for the air spring.

Suspension Fasteners

Suspension fasteners are important attaching parts in that they could affect performance of vital components and systems and/or could result in major service expense. They must be replaced with fasteners of the same part number, or with an equivalent part, if replacement becomes necessary. DO NOT use a replacement part of lesser quality or substitute design.

Torque values must be used, as specified, during assembly to assure proper retention of parts. New fasteners must be used whenever old fasteners are loosened or removed or when new component parts are installed.

Air Spring Suspension

Air compressor (less dryer), regenerative dryer, O-ring, mounting bracket and the isolator mounts are all serviced as separate components.

Height sensors and modules are replaceable.

Air springs are replaceable as assemblies (includes solenoid valve).

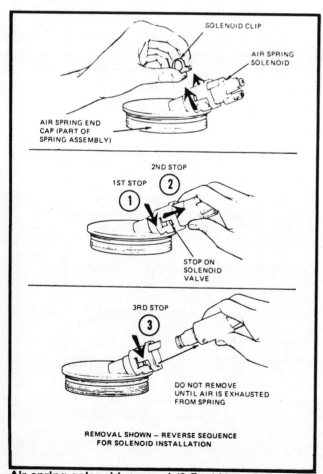

Air spring solenoid removal (© Ford Motor Co.)

Air spring solenoid valves an external O-rings are replaceable.

Air lines are replaceable, however quick connect unions and bulk tubing are available to mend a damaged air line.

Collect and O-ring of the quick connect fitting is replaceable.

Suspension, Front

Gas filled shock absorber struts must be replaced as assemblies. they are not serviceable. Replace only the damaged shock absorber strut. It is not necessary to replace in matched pairs.

Strut upper mounts may be replaced individually.

Air springs are replaced as assemblies. It is not necessary to replace in pairs.

Lower control arm is replaceable an assembly with the ball joint and bushings included.

Spindle is replaceable.

Stabilizer bar is replaceable with stabilizer bar-to-body insulators included.

Stabilizer bar-to-body bushing is replaceable.

Suspension, Rear

The following rear suspension components may be replaced individually:

Gas filled shock absorbers must be replaced as assemblies.

They are not serviceable. Replace only the damaged shock absorber. It is not necessary to replaced in matched pairs.

Air springs are replaced as assemblies. It is not necessary to replace in pairs.

Lower control arms, including both end bushings, are replaceable as assemblies (Must be replaced in pairs).

Upper control arm axle and bushings are replaceable as assemblies (Must be replaced in pairs).

Upper control arm axle and bushings are replaceable individually. (Must be replaced in pairs).

Stabilizer bar is replaceable with stabilizer bar-to-axle insulator included.

Stabilizer bar-to-body bushings are replaceable.

HOIST LIFTING, JACKING, TOWING RESTRICTIONS

CAUTION

The electrical power supply to the air suspension system must be shut off prior to hoisting, jacking or towing an air suspension vehicle. This can be accomplished by disconnecting the battery or turning off the power switch located in the trunk on the LH side. Failure to do so may result in unexpected inflation or deflation of the air springs which may result in shifting of the vehicle during these operations.

HOISTING AND BODY SUPPORT PROCEDURES

1. Position vehicle over hoist and then turn ignition Off and shut off the air suspension power switch located in the trunk on the left side.

CAUTION

The following Hoist Restrictions must be observed:

---

2. A body hoist is the recommended method for vehicle hoisting. When this hoist is used, raise the vehicle using the standard support procedures. The suspension will be supported in rebound by the front struts and the rear shock absorbers after the vehicle is lifted. Also support vehicle at four corners with jack stands as a safety precautions. Do not use suspension hoists.

3. If a body hoist is not available, an alternate method approved for vehicle hoisting is to use a standard hydraulic floor

SUSTENSION
Domestic Models

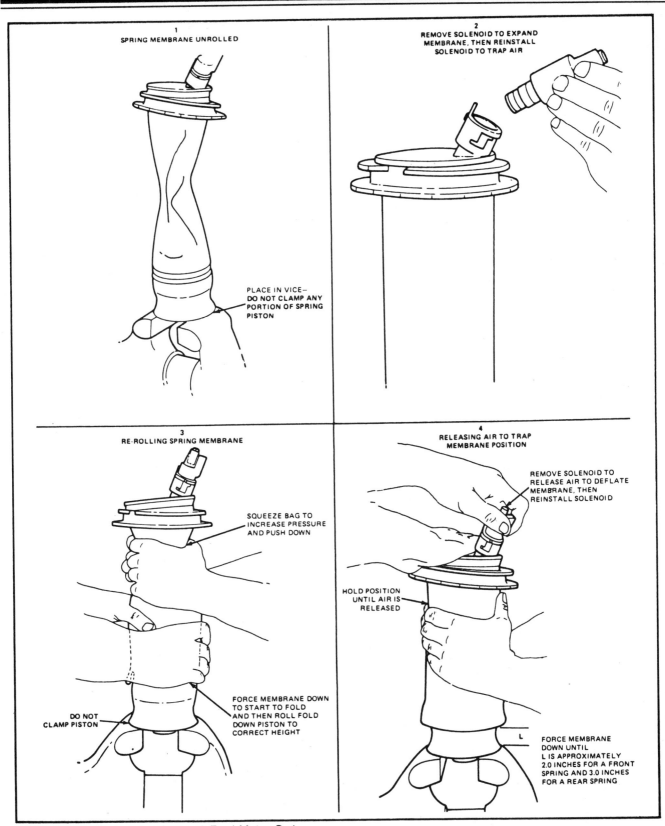

Air spring folding procedures (© Ford Motor Co.)

2–105

SECTION 2
SUSPENSION
Domestic Models

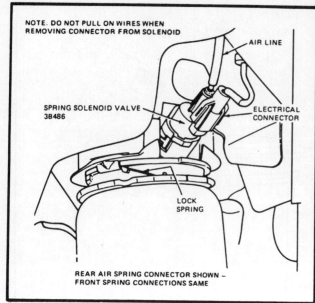

Air spring solenoid connector (© Ford Motor Co.)

jack. Raise the front of the vehicle at the No. 2 crossmember until the tires are above the floor. Support vehicle body with jack stands at each front corner and then lower floor jack so that the front suspension is in full rebound. Repeat this procedure or the rear suspension except raise the body at the rear jacking location.

— CAUTION —
Power to the air system must be shut off by turning the air suspension switch (in luggage compartment) Off or by disconnecting the battery when servicing any air suspension components.

1. Do not attempt to install or inflate any air spring that has become unfolded.
2. Any spring which has unfolded must be refolded prior to being installed in a vehicle.
3. Do not attempt to inflate any air spring which has been collapsed while uninflated from the rebound hanging position to the jounce stop.
4. After inflating an air spring in hanging position, it must be inspected for proper shape.
5. Failure to follow the above procedures may result in a sudden failure of the air spring or suspension system.

AIR SPRING SOLENOID

Removal
The air spring solenoid valve has a two stage solenoid pressure relief fitting similar to a radiator cap. A clip is firs removed, and rotation of the solenoid out of the spring will release air from the assembly before the solenoid can be removed.
1. Turn the air suspension switch Off.
2. Hoist vehicle and support safely. Suspension must be at full rebound.
3. Remove wheel and tire assembly.
4. Disconnect electrical connector and then disconnect the air line.
5. Remove solenoid clip.
6. Rotate solenoid counterclockwise to the first stop.
7. Pull solenoid straight out slowly to the second stop to bleed air from the system.

— CAUTION —
Do not fully release solenoid until air is completely bled from the air spring.

8. After air is fully bled from the system, rotate counterclockwise to the third stop, and remove solenoid from the air spring assembly.

Installation
1. Check solenoid O-ring for abrasion or cuts. Replace O-ring as required. Lightly grease O-ring area of solenoid with silicone dielectric compound or equivalent.
2. Insert solenoid into air spring end cap and rotate clockwise to the third stop, push into the second stop, then rotate clockwise to the first stop.
3. Install solenoid clip.
4. Connect the air line and the electrical connector.
5. Refill the air spring(s) as outlined.
6. Install the wheel and tire assembly.

Air Spring Fill
1. Turn On the air suspension switch. Diagnostic pigtail is to be ungrounded.

NOTE: Lower hoist as required, but do not apply a load to the suspension.

2. Install battery charger to reduce battery drain.
3. Cycle the ignition from the Off to the Run position, hold in the Run position for a minimum of five seconds, then return to the Off position. Driver's door is open with all other doors shut.
4. Change the diagnostic pigtail from an ungrounded state to a grounded state by attaching a lead from the diagnostic pigtail to vehicle ground. The pigtail must remain grounded during the spring fill sequence.
5. While applying the brakes, turn the ignition switch to the Run position. (The door must be open and do not start vehicle). The warning lamp will blink continuously once every two seconds to indicate the spring pump sequence has been entered.
6. To fill a rear spring(s), close and open door once. After a 6 second delay, the rear spring will be filled for 60 seconds.
7. To fill a front spring(s), close and open the door twice. After a 6 second delay, the front spring will be filled for 60 seconds.
8. To fill rear and front springs, fill the rear springs first. When the rear fill has finished, close and open the door once to initiate the front spring fill.
9. Terminate the air spring fill by turning the ignition switch to Off, actuating the brake, or ungrounding the diagnostic pigtail. The diagnostic pigtail must be ungrounded at the end of the spring fill.
10. Lower hoist completely and start vehicle and allow vehicle to level with doors closed.

AIR SPRING—FRONT OR REAR

Removal
1. Turn the air suspension switch Off.
2. Hoist vehicle and support safely. Suspension must be at full rebound.
3. Remove tire and wheel assembly.
4. Remove air spring solenoid.
5. Remove spring to lower arm fasteners. Remove clip for front spring and/or remove bolts for rear spring.
6. Push down or spring clip on the collar of the air springs and rotate collar counterclockwise to release the spring from the body spring seat.
7. Remove air spring.

SUSPENSION
Domestic Models

Installation

1. Install air spring solenoid as outlined.
2. Correctly position the solenoid. For LH illustration (front or rear spring), the notch on the collar is to be in line with the centerline of the solenoid. For RH installation (front or rear), the flat on the collar is to be in line with the centerline of the solenoid.
3. Install the air spring into the body spring seat, taking care to keep the solenoid air and electrical connections clean and free of damage. Rotate the air spring collar until the spring clips snaps into place. Be sure that the air spring collar is retained by the three rolled tabs on the body spring seat.
4. Attach air line and electrical connector to the solenoid assembly.
5. Align and secure lower arm to spring attachment with suspension at full rebound and supported by shock absorbers.

— CAUTION —
The air springs may be damaged if suspension is allowed to compress before spring is inflated.

6. Replace tire and wheel assembly.
7. Remove floor jacks and lower vehicle until tire and wheel assembly are 1–3 in. above floor.
8. Refill the air spring(s) as outlined.

AIR COMPRESSOR AND DRYER ASSEMBLY

Removal

1. Turn the air suspension switch Off.
2. Disconnect the electrical connector located on the compressor.
3. Remove the air line protector cap from the dryer by releasing the two latching pins located on the bottom of the cap 180 degrees apart.
4. Disconnect the four air lines from dryer.
5. Remove the three screws retaining the air compressor to mounting brackets.

Installation

1. Position air compressor and dryer assembly to mounting bracket and install the three mounting screws.
2. Connect all four air lines into the dryer.
3. Connect the electrical connection.
4. Install the air line protector cap onto the dryer.
5. Turn air suspension switch On.

DRYER, AIR COMPRESSOR

Removal

1. Turn the air suspension switch Off.
2. Remove the air line protector cap from the dryer by releasing the two latching pins located on the bottom of the cap 180 degrees apart.
3. Disconnect the four air lines from the dryer.
4. Remove the dryer retainer clip and screw.
5. Remove from the head assembly.

Installation

1. Check to ensure the old O-ring is not in the head assembly.
2. Check dryer end to ensure new O-ring is in proper position.
3. Insert dryer into head assembly and install retainer clip and screw.
4. Connect the four air lines into the dryer.
5. Install the air line protector cap onto the dryer.
6. Turn the air suspension switch On.

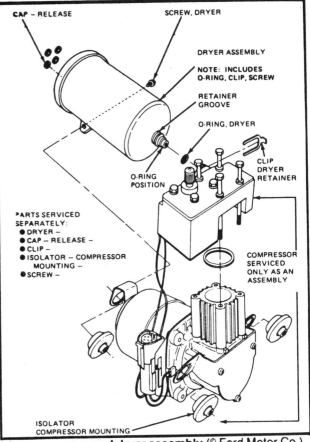

Air compressor and dryer assembly (© Ford Motor Co.)

HEIGHT SENSORS—FRONT

Removal

1. Turn the air suspension switch Off.
2. Disconnect sensor electrical connector. The front sensor connectors are located in the engine compartment behind the shock towers.
3. Push the front sensor connector through the access hole in the rear of the shock tower.
4. Hoist vehicle and support safely. Suspension must be at full rebound.
5. Disconnect the bottom and then the top end of the sensor from the attaching studs.
6. Disconnect the sensor wire harness from the plastic clips on the shock tower and remove sensor.

Installation

1. Connect the top and then the bottom end of the sensor to the attaching studs. Route the sensor electrical connector as required to connect the vehicle wire harness.
2. Lower vehicle.
3. Connect the sensor connector.
4. Turn air suspension switch On.

HEIGHT SENSOR—REAR

Removal

1. Turn the air suspension switch Off.
2. Disconnect the sensor electrical connector located in the

SECTION 2

SUSPENSION
Domestic Models

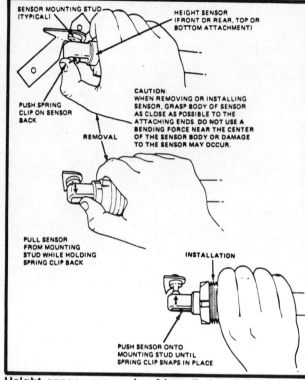

Height sensor removal and Installation (© Ford Motor Co.)

luggage compartment in front of the forward trim panel. Also pull the luggage compartment carpet back for access to the sensor sealing grommet located on the floor plan.
3. Hoist vehicle and support safely. Suspension must be at full rebound.
4. Disconnect the bottom and then the top end of the sensor from the attaching studs.
5. Push upwards on the sealing grommet to unseat and then push sensor through the floor pan hole into the luggage compartment.
6. Lower vehicle.

Installation

1. Connect the sensor connector and then push sensor through the floor pan hole being sure to seat the sealing grommet. Replace luggage compartment carpet.
2. Hoist vehicle and support safely.
3. Connect the top and then the bottom end of the sensor.
4. Lower vehicle.
5. Turn air suspension switch On.

CONTROL MODULE

Removal

1. Turn air suspension switch Off. Ignition switch is also to be Off.
2. Remove LH luggage compartment trim panel.
3. Disconnect wire harness from module.
4. Remove three attaching nuts.
5. Remove module.

Installation

1. Position module and secure with the attaching nuts.
2. Connect wire harness to module.

3. Attach LH luggage compartment trim panel.
4. Turn air suspension switch On.

NYLON AIR LINE

If a leak is detected in an air line, it can be serviced by carefully cutting the line with a sharp knife to ensure a good, clean straight cut. Then, install a service fitting.

QUICK CONNECT FITTINGS

If a leak is detected in any of the eight quick connect fittings, it can be serviced using a repair kit containing a new O-ring, collet, release ring, and O-Ring removal tool. The outer housing of the fitting cannot be serviced.

AIR SUSPENSION SWITCH

Removal and Installation

1. Disconnect electrical connector.
2. Depress retaining clips switch to brace, and remove switch.
3. Push switch into position in the brace, making sure retaining clips are fully seated.
4. Connect electrical connector.

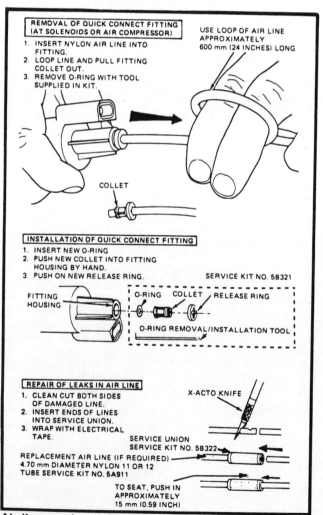

Air line service (© Ford Motor Co.)

2-108

SUSPENSION
Domestic Models

COMPRESSOR RELAY

Removal and Installation

1. Disconnect electrical connector.
2. Remove screw retaining relay to left front shock tower and remove relay.
3. Position relay on shock tower and install retaining screw.
4. Connect electrical connector.

Front Suspension Components

CAUTION

Power in the air system must be shut off by turning the air suspension switch (in luggage compartment) Off or by disconnecting the battery when servicing any suspension components.

STABILIZER BAR AND/OR BUSHING

Removal

1. Turn air suspension switch Off.
2. Raise vehicle on a hoist and support safely.
3. Disconnect stabilizer bar from each link and bushing U-clamps. Remove the stabilizer bar assembly.
4. Remove adapter brackets and U-clamps.
5. Cut the worn bushings from the stabilizer bar.

Installation

1. Coat the necessary parts of the stabilizer bar with Ford Rubber Suspension Insulator Lubricant or equivalent, and slide bushings onto the stabilizer bar. Reinstall U-clamps.
2. Reinstall adapter brackets on U-clamps.
3. Using a new nut and bolt, secure each end of the stabilizer bar to the lower suspension arm. Tighten nuts to specification.
4. Using new bolts, clamp the stabilizer bar to the attaching brackets on the side rail. Tighten bolts to specification.
5. Lower the vehicle.
6. Turn air suspension switch On.

SHOCK STRUT

Removal

1. Turn air suspension switch Off.
2. Turn the ignition key to the unlocked position to allow free movement of front wheels.
3. From the engine compartment, loosen but do not remove the strut-to-upper mount attaching nut.
4. Hoist vehicle and support safely, and position safety stands under the lower control arms as far outboard as possible being sure that the lower sensor mounting bracket is clear. Lower hoist until vehicle weight is supported by lower arms.
5. Remove tire and wheel assembly.
6. Remove brake caliper, then rotate out of position and wire securely.
7. Remove the strut-to-upper mount attaching nut and then the two lower nuts and bolts attaching the strut to the spindle.

NOTE: The strut should be held firmly during the removal of the last bolt since the gas pressure will cause the strut to fully extend when removed.

8. Lift the strut up from the spindle to compress the rod and then remove the strut.
9. Remove jounce bumper.

Installation

1. Prime new strut by extending and compressing strut rod five times.
2. Install jounce bumper.
3. Place strut rod through the upper mount, and hand start and secure a new 16 mm nut.
4. Compress strut, and position onto the spindle.
5. Install two new lower mounting bolts and hand start nuts.
6. Raise hoist to remove vehicle load from the lower control arms and tighten the lower mounting nuts.
7. Install brake caliper.
8. Install the tire and wheel assembly.
9. Remove safety stand and lower the vehicle to the ground.
10. Turn air suspension switch On.
11. Front wheel alignment should be checked and adjusted, if out of specification.

UPPER MOUNT ASSEMBLY

Removal

NOTE: Upper mounts are one piece units and cannot be disassembled.

1. Turn air suspension system Off.
2. Turn the ignition key to the unlocked position to allow free movement of the front wheels.
3. From the engine compartment, loosen but do not remove the three upper mount retaining nuts. Vehicle should be in place over a hoist and must not be driven with these nuts removed. Do not remove the pop rivet holding the camber plate in position.
4. Loosen strut rod nut at this time.
5. Hoist the vehicle and support safely, and position safety stands under the lower control arms as far outboard as possible being sure that the lower sensor mounting bracket is clear. Lower hoist until vehicle weight is supported by the lower arms.
6. Remove the tire and wheel assembly.
7. Remove brake caliper and rotate out of position and wire securely.
8. Remove the upper mount retaining nuts and the two lower nuts and bolts that attach the strut to the spindle.

NOTE: The strut should be held firmly during the removal of the last bolt since the gas pressure will cause the strut to fully extend when removed.

9. Lift the strut up from the spindle to compress the rod, and then remove the strut.
10. Remove upper mount from strut.

Installation

1. Install new upper mount on strut and hand start a new nut.
2. Position the upper mount studs into the body and start and secure three new nuts. Secure the strut rod nut.
3. Compress the strut and position onto the spindle.
4. Install two new lower mounting bolts, and hand strut nuts.
5. Raise hoist to remove vehicle load from the lower control arms and tighten the lower mounting nuts to 126–179 ft. lbs. (170–244 NNm).
6. Install the brake caliper.
7. Install the tire and wheel assembly.
8. Remove safety stands and lower vehicle to the ground.
9. Turn air suspension switch On.
10. Front wheel alignment should be checked and adjusted if out of specification.

SPINDLE ASSEMBLY

Removal

1. Turn air suspension switch Off.
2. Hoist vehicle and support safely.

SUSPENSION
Domestic Models

3. Remove the wheel and tire assembly.
4. Remove the brake caliper, rotor and dust shield.
5. Remove the stabilizer link from the lower arm assembly.
6. Remove the tie rod end from the spindle.
7. Remove the cotter pin from the ball joint stud nut and loosen the ball joint nut one or two turns.

CAUTION
DO NOT remove the nut from the ball joint stud at this time.

8. Tap spindle boss smartly to relive stud pressure.
9. Place a floor jack under the lower arm, compress the air spring and remove the stud nuts.
10. Remove the bolts and nuts attaching the spindle to the shock strut. Compress the shock strut until working clearance is obtained.
11. Remove the spindle assembly.

Installation

1. Place the spindle on the ball joint stud and install the new stud nut. DO NOT tighten at this time.
2. Lower the shock strut until the attaching holes are in line with the holes in the spindle. Install two new bolts and nuts.
3. Tighten ball joint stud nut and install cotter pin.
4. Lower floor jack from under the suspension arm, and remove jack.
5. Tighten the shock strut to spindle attaching nuts.
6. Install stabilizer bar link and tighten attaching nut.
7. Attach tie rod end and tighten the retaining nut.
8. Install the disc brake dust shield, rotor and caliper.
9. Install the wheel and tire assembly.
10. Remove the safety stands and lower the vehicle.
11. Turn air suspension switch On.
12. Front wheel alignment should be checked and adjusted if out of specification.

SUSPENSION CONTROL ARM

Removal

1. Turn air suspension switch Off.
2. Raise vehicle on a hoist and support safely so the control arms hang free (full rebound).
3. Remove the wheel and tire assembly.
4. Disconnect the tie rod assembly from the steering spindle.
5. Remove the steering gear bolts if necessary and position the gar so that the suspension are bolt may be removed.
6. Disconnect stabilizer bar link from the lower arm.
7. Disconnect lower end of the height sensor from the lower control arm sensor mounting stud. Remove sensor mounting stud and screw from lower arm, noting the position of stud on the lower arm bracket.
8. Remove the cotter pin from the ball joint stud nut and loosen the ball join nut one or two turns. DO NOT remove the nut at this time. Tap spindle boss smartly to relieve stud pressure.
9. Vent air spring(s) to atmospheric pressure. Refer to Air Spring Solenoid Removal. Then, reinstall solenoid. Refer to Air Spring Solenoid Installation.
10. Remove air spring to lower arm fastener clip.
11. Remove the ball joint nut, and raise the entire strut and spindle assembly (strut, rotor, caliper and spindle). Wire it out of the way to obtain working room.
12. Remove the suspension arm to crossmember nuts and bolts, and remove the arm from the spindle.

Installation

1. Position the arm into the crossmember and install a new arm to crossmember bolts and nuts. DO NOT tighten at this time.
2. Remove the wire from the strut and spindle assembly and attach to the ball joint stud. Install a new ball joint stud nut. DO NOT tighten at this time.
3. Position air spring in arm and install a new fastener.
4. Attach sensor mounting stud and screw to lower arm in the same position as on the replaced arm. Connect lower end of sensor to lower arm mounting stud.
5. With a suitable jack, raise the suspension arm to curbheight.
6. With the jack still in place, tighten the lower arm to crossmember attaching nut to 150–180 ft. lbs. (203–244 Nm).
7. Tighten ball join stud nut to 100–120 ft. lbs. (136–163 Nm) and install a new cotter pin. Remove jack.
8. Install the steering gear to crossmember bolts and nuts (if removed). Hold the bolts and tighten nuts to 90–100 ft. lbs. (122–135 Nm).
9. Position the tie rod assembly into the steering spindle, and install the retaining nut. Tighten the nut to 35 ft. lbs. (47 Nm) and continue tightening the nut to align the next castellation with cotter pin hole in the stud. Install a new cotter pin.
10. Connect the stabilizer bar link to the lower suspension arm and tighten the attaching nut to 9–12 ft. lbs. (12–16 Nm).
11. Install the wheel and tire assembly, and lower the vehicle but DO NOT allow tires to touch the ground.
12. Turn air suspension switch On.
13. Refill air spring(s) as outlined.
14. Front wheel alignment should be checked and adjusted if out of specification.

REAR SUSPENSION COMPONENTS

CAUTION
Power to the air system must be shut off by turning the air suspension switch (in luggage compartment) Off or by disconnecting the battery when servicing any suspension components.

SHOCK ABSORBER

Removal

1. Turn air suspension switch Off.
2. Open the luggage compartment and remove the inside trim panels to gain access to the upper shock stud.
3. Loosen but DO NOT REMOVE the shock rod attaching nut.
4. Hoist vehicle and support safely. Position two safety stands under the rear axle. Lower hoist until vehicle weight is supported by the rear axle.
5. Remove the upper attaching nut, washer and insulator and then remove the lower shock protective cover (right shock only) and lower shock absorber crossbolt and nut from the lower shock brackets.
6. From under the vehicle, compress the shock absorber to clear it from the hold in the upper shock tower.

CAUTION
Shock absorbers will extend unassisted. Do not apply heat or flame to the shock absorber during removal.

7. Remove shock absorber.

SUSPENSION
Domestic Models

Installation

1. Prime new shock absorber by extending and compressing shock absorber five times.
2. Place the inner washer and insulator on the upper attaching stud. Position stud through the shock tower mounting hole and position an insulator, washer or stud from the luggage compartment. Hand start the attaching nut and then secure.
3. Place shock absorber's lower mounting eye between the ears of the lower shock mounting bracket, compressing shock as required. Insert the bolt (bolt head must seat on the inboard side of the shock bracket), through the shock bracket and the shock absorber mounting eye. Hand start and then secure the original attaching nut.
4. Install the protective cover, to the RH shock absorber. This is done by inserting the bolt point and nut into the cover's open end, sliding the cover over the shock bracket and snapping the closed end of the cover over the bolt head. Properly installed, the cover will completely conceal the bolt point, nut, and bolt head. The rounded or closed end of the cover should be pointing inboard.
5. Raise hoist and remove safety stands from under axle then lower vehicle.
6. Reinstall the inside trim panels.
7. Turn air suspension switch On.

LOWER CONTROL ARM

Removal

NOTE: If one arm requires replacement, replace the other arm also.

1. Turn air suspension switch Off.
2. Hoist vehicle and support safely. Suspension will be at full rebound.
3. Remove tire and wheel assembly.
4. Vent air spring(s) at atmospheric pressure. Refer to Air Spring Solenoid Removal. Then, reinstall solenoid. Refer to Air Spring Solenoid Installation.
5. Remove the two air spring-to-lower arm bolts and remove the air spring from the lower arm.
6. Remove the frame-to-arm and the axle-to-arm bolts and remove the arm from the vehicle.

Installation

1. Position the lower arm assembly into the front arm brackets and insert a new, arm-to-frame pivot bolt and nut with nut facing outwards. DO NOT tighten at this time.
2. Position the rear bushing in the axle bracket and install a new arm-to-axle pivot bolt and nut with nut facing outwards. DO NOT tighten at this time.
3. Install two new air spring-to-arm bolts. DO NOT tighten at this time.
4. Using a suitable jack, raise the axle to curb height. Tighten the lower arm front bolt, the rear pivot bolt, and the air spring to arm bolt being sure that the air spring piston is flat on the lower arm. Remove the jack.
5. Replace tire and wheel assembly.
6. Lower the vehicle.
7. Turn air suspension switch On.
8. Refill air spring(s) as outlined.

UPPER CONTROL ARM AND AXLE BUSHING

Removal

NOTE: If one arm requires replacement, replace the other arm also.

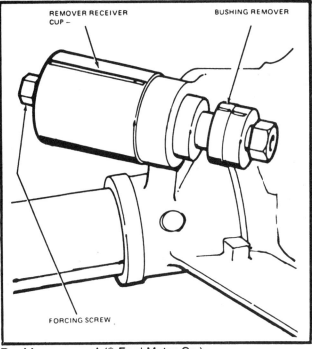

Bushing removal (© Ford Motor Co.)

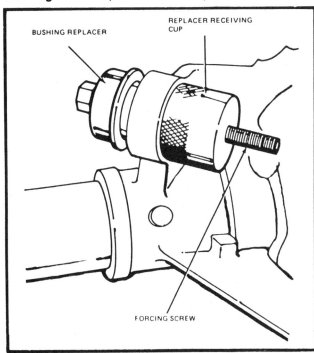

Bushing installation (© Ford Motor Co.)

1. Turn air suspension switch Off.
2. Hoist vehicle and support safely. Suspension will be at full rebound.
3. On the RH side, detach rear height sensor from side arm. Note position of the sensor adjustment bracket on the upper arm.
4. Remove upper arm-to-axle pivot nut and bolt.

2-111

SUSPENSION
Domestic Models

5. Remove upper arm-to-axle frame pivot bolt and nut. Remove upper arm from vehicle.

If upper arm axle bushing is to be replaced, use the following procedure:

6. Place the upper arm axle bushing remover tool in position, and remove the bushing assembly.

AIR SUSPENSION BOLT TORQUE SPECIFICATIONS

Front Suspension—Description	Nm	ft. lbs.
Lower arm to No. 2 crossmember—nut	203-244	150-180
Stabilizer bar mounting clamp to bracket—bolt	27-34	20-25
Stabilizer bar to lower arm—nut	12-16	9-12
Spindle to shock strut—nut	203-244	150-180
Shock strut to upper mount—nut	75-125	55-92
Ball joint to spindle—nut	136-163	100-120
Shock upper mount to body—nut	84-102	62-75
Steering gear to No. 2 crossmember—nut	122-136	90-100
Tie rod end to spindle—nut	47-64	35-47
Compressor bracket to frame—bolt	3-5	30-40*
Air compressor to compressor bracket—bolt	3-5	30-40*
Sensor upper attachment to frame—nut	35-46	26-34
Sensor lower attachment to arm—bolt	10-16	8-12

Rear Suspension—Description	Nm	Ft. Lbs.
Shock absorber to frame—nut	26-37	17-27
Upper arm to frame—bolt	135-142	100-105
Upper arm to axle—bolt	122-135	90-100
Lower arm to frame—bolt	135-142	100-105
Lower arm to axle—bolt	122-135	90-100
Shock absorber to clevis bracket bolt	61-81	45-60
Clevis bracket to axle—nut	75-90	55-70
Stabilizer bar to axle—bolt	41-48	30-35
Stabilizer bar to body—nut	17-24	13-18
Air spring to lower arm—bolt	41-48	30-35
Sensor upper bracket to frame—bolt	12-17	110-150*
Sensor lower bracket to arm—nut	8-14	7-10

7. Using the installer tool, install the bushing assembly into the bushing ear of the rear axle.

Installation

1. Place the upper arm into the bracket of the body side rail. Insert a new upper arm-to-frame pivot bolt and nut (nut facing outboard). DO NOT tighten at this time.
2. Align the upper arm-to-axle pivot hole with the hole in the axle bushing. If required, raise the axle using a suitable jack to align. Install a new pivot bolt and not (nut inboard). DO NOT tighten at this time.
3. On the RH side, reattach rear height sensor to the arm. Set the adjustment bracket to the same position as on the replaced arm and tighten nut.
4. Using a suitable jack, raise the axle to curb height and tighten the front upper arm bolt, and the rear upper arm bolt.
5. Remove the jack stands supporting the axle.
6. Lower vehicle.
7. Turn air suspension switch On.

STABILIZER BAR BUSHINGS

Removal

1. Turn air suspension switch Off.
2. Hoist vehicle and support safely.
3. Disconnect the stabilizer bar from each link and bushing U-clamp. Remove the stabilizer bar assembly.
4. Remove the U-clamps.
5. Cut the worn bushings from the stabilizer bar.

Installation

1. Coat the necessary parts of the stabilizer bar with Ford Rubber Suspension Insulator Lubricant or equivalent and slide new bushings onto the stabilizer bar. Reinstall U-clamps.
2. Using new bolts and nuts, attach stabilizer bar to the axle. Do not tighten bolts at this time.
3. Using new bolts and nuts, attach the link end of the stabilizer bar to the body. Tighten the link attaching nut and then the axle attaching bolts.
4. Lower vehicle.
5. Turn air suspension switch On.

NOTE: Due to the diagnostic complexity of the Automatic Load Levelling System, detailed diagnostic procedures can be found in Chilton's Chassis Electronic Service Manual.

REAR SUSPENSION—FORD, LINCOLN, MERCURY REAR DRIVE CARS, FOUR–BAR LINK DESIGN

SHOCK ABSORBER

Removal

1. Remove the attaching nut, washer and insulator from the shock absorber's upper stud.
2. Raise the vehicle on a hoist, and support the rear axle.
3. From underneath the vehicle, compress the shock absorber to clear it from the hole in the upper shock tower.
4. Remove the lower shock absorber bolt, washer and nut from the axle bracket.
5. Remove the shock absorber.

Installation

1. Expel all air from the new shock absorber.
2. Compress the shock absorber and position the shock's mounting eye to the axle bracket mounting hole. Place a new load bearing washer between the shock eye and axle bracket. Install a new Torx drive belt or equivalent through the shock eye, washer and axle bracket, then hand start the bolt into a new self-wrenching nut. Do not tighten at this time.
3. After compressing the shock absorber, place the absorber's lower mounting eye between the ears of the lower shock mounting bracket. Then insert the bolt. the bolt head must seat on the inboard side of the shock bracket, through the shock bracket and the shock absorber mounting eye. Install the prevailing torque attaching nut. Do not tighten the nut at this time.
4. Place the inner washer and insulator on the upper attaching stud.

SUSPENSION
Domestic Models

5. Extend the shock absorber's upper stud, and position it through the mounting hole in the shock tower.

6. Fairmont/Zephyr, Mustang/Capri/Cougar: While holding the shock absorber in position, tighten the lower attaching bolt to 70 ft. lbs. (94 Nm) using tool number D80P-2100-T55 or equivalent. Allow the self wrenching nut to rotate freely so that the wrenching tab seats on the outboard leg of the axle bracket. Do not restrain the nut using any other method.

7. Thunderbird/XR-7: While holding the shock absorber in position, tighten the lower shock cross bolt to 70 ft. lbs. (94.9 Nm).

8. Thunderbird/XR-7: Install the protective cover (only one is required) to the right hand shock absorber. This is done by inserting the bolt point and nut into the cover's open end, sliding the cover over the bolt head. Properly installed, the cover will completely conceal the bolt point, nut and bolt head. The rounded or closed end of the cover should be pointing inboard.

9. Lower the vehicle. Install the insulator, outer washer, and a new nut to the upper shock stud, and tighten. Install the rubber cap on the shock stud. Install the inside panel trim covers.

SPRING

Removal

NOTE: **If vehicle is equipped with a rear stabilizer bar, remove the bar.**

1. Raise the vehicle and support the body at the rear body crossmember.
2. Lower the hoist until the rear shocks are fully extended.

NOTE: **The axle must be supported by the hoist, a transmission jack or jack stands.**

3. Place a transmission jack under the lower arm axle pivot bolt, and remove the bolt and nut. Lower the transmission jack slowly until the coil spring load is relieved.
4. Remove the coil spring and insulators from the vehicle.

Installation

1. Place the upper spring insulator on top of the spring. Tape in place if necessary.
2. Place the lower spring insulator on the lower arm (except Mustang/Capri). Install the internal damper into the spring (except Thunderbird/XR-7).
3. Position the coil spring on the lower arm spring seat so that the pigtail on the lower arm is at the rear of the vehicle and pointing toward the left side of the vehicle. Slowly raise the transmission jack until the arm is in position. Insert a new rear pivot bolt and nut with the nut facing outwards. Do not tighten at this time.
4. Lower the transmission jack. Raise the axle to curb height. Tighten the lower arm pivot bolt to 100 ft. lbs. (135 Nm).
5. If vehicle was equipped with a rear stabilizer bar, install bar.
6. Remove crossmember supports and lower the vehicle.

LOWER ARM

Removal

1. Raise the vehicle and support body at the rear body crossmember.
2. Lower the hoist until the rear shocks are fully extended.

NOTE: **The axle must be supported by the hoist, a jack or stands.**

3. Place the transmission jack under the lower arm rear pivot bolt, and remove the bolt and nut.

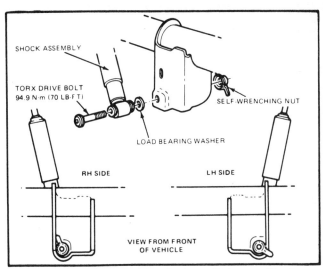

Rear shock lower installation—Fairmont/Zephyr, Mustang/Capri, Granada/Cougar (© Ford Motor Co.)

4. Lower the jack slowly until the coil spring can be removed.
5. Remove the lower arm assembly.

Installation

1. Position the lower arm assembly into the front arm bracket and insert a new front pivot bolt and nut with nut facing outward. Do not tighten at this time.
2. Install coil spring. Holding the spring in position, use the jack under the rear of the lower arm. Raise the jack until the arm is in position. Insert a new rear pivot bolt and nut with nut facing outwards. Do not tighten at this time.
3. Lower the jack. Raise the axle with the hoist to curb height. Tighten the lower arm front bolt and rear pivot bolt heads to 100 ft. lbs. (135 Nm).
4. If vehicle is equipped with a rear stabilizer bar, install bar.
5. Remove crossmember supports and lower vehicle.

UPPER ARM AND AXLE BUSHINGS

Replacement

1. Remove upper arm rear pivot bolt and nut.

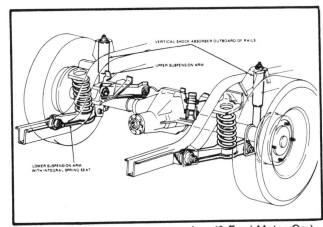

Four bar link coil rear suspension (© Ford Motor Co.)

2-113

SECTION 2
SUSPENSION
Domestic Models

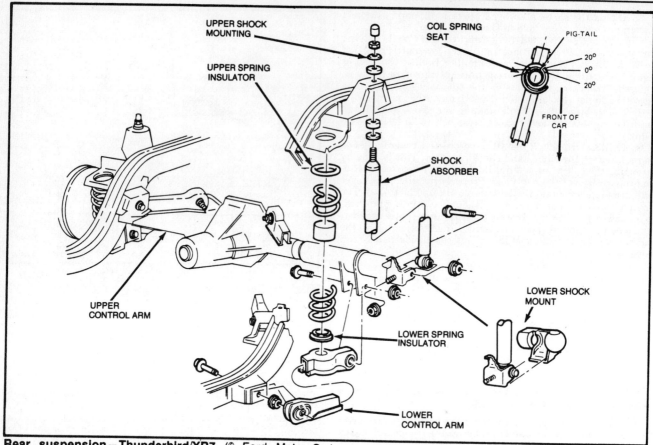

Rear suspension—Thunderbird/XR7 (© Ford Motor Co.)

2. Remove front pivot bolt and nut. Remover upper arm from vehicle.
3. Place the upper arm rear bushing remover tool in position, and remove the bushing assembly.
4. Using the installer tool, install the bushing assembly into the bushing end of the rear axle.

STABILIZER BAR

Removal

1. Raise vehicle on hoist.

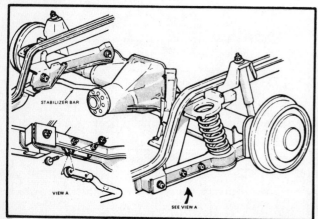

Rear stabilizer bar installation (© Ford Motor Co.)

2. Remove four bolts attaching stabilizer bar to brackets in lower arms.
3. Remove stabilizer bar from vehicle.

Installation

NOTE: Make sure bar is not installed upside down. A color code is provided on stabilizer bar (passenger side only) as an aid for proper orientation.

1. Align four holes in stabilizer bar with holes in brackets in lower arms.
2. Install four new bolts and tighten nuts to 20 ft. lbs. (27 Nm).
3. Visually inspect installation to insure adequate clearance between stabilizer bar and lower arm.

STABILIZER BAR BRACKETS

Removal

1. Raise vehicle on hoist and support body at rear crossmember.
2. Remove stabilizer bar.
3. Disconnect the shock absorbers at the lower shock bracket.
4. Slowly lower the suspension until the front bracket-to-arm bolt clears the body side rail.

NOTE: Do not stress the brake hose when lowering the suspension.

5. Remove both bracket to arm bolts and nuts, and remove bracket from the arm.

SUSPENSION
Domestic Models

GENERAL MOTORS BODY IDENTIFICATION

GENERAL MOTORS FRONT WHEEL DRIVE "A" & "X" BODY	"A" BODY—	Buick Century Custom • Century Limited • Chevrolet Celebrity
	"X" BODY—	Buick Skylark • Skylark Sport • Skylark Limited • Skylark Custom • Skylark "T" Type • Chevrolet Citation • Oldsmobile Omega • Omega Brougham • Pontiac Phoenix • Phoenix SJ • Phoenix LJ • Phoenix SE • Phoenix LE
"C" BODY		BUICK Electra Limited • Park Ave • "T" Type CADILLAC Fleetwood Brougham • DeVille OLDSMOBILE 98 Regency • 98 Regency Brougham
"E" & "K" BODY	"E" BODY—	Buick Riviera • Riviera Luxury • Riviera "T" Type • Riviera Convertible • Cadillac Eldorado • Olds Toronado Brougham
	"K" BODY—	Cadillac Seville
"J" BODY		BUICK Skyhawk • Skyhawk Custom • Skyhawk Limited • Skyhawk "T" Type CADILLAC Cimarron CHEVROLET Cavalier OLDSMOBILE Firenza • Firenza "S", "SX" & "LX" PONTIAC J2000 • J2000 LE • J2000 SE
"N" BODY		Buick Somerset Regal • Oldsmobile Calais • Pontiac Grand AM
GENERAL MOTORS REAR WHEEL DRIVE		
BUICK		1981 Century • Century Limited • Regal • Regal Limited • Regal Sport Coupe • Le Sabre • Le Sabre Custom • Le Sabre Limited • Electra • Estate Wagon
CADILLAC		DeVille • Fleetwood
CHEVROLET		Caprice Classic • Corvette • Malibu • Monte Carlo • Impala
"F" BODY		Camaro • Camaro Berlinetta • Firebird • Firebird Trans Am • SE
"P" BODY		Pontiac Fiero • Fiero SE
"T" BODY		Chevette • Chevette Scooter • Pontiac T1000 • 1000
OLDSMOBILE		Cutlass Supreme • Cutlass Salon • Cutlass Supreme Brougham • Cutlass Calais • Delta 88 Royal • Delta 88 Royale Brougham • Custom Cruiser • Olds 98 • Olds 98 Regency • Olds 98 Regency Brougham • Hurst Olds
PONTIAC		Bonneville • Bonneville LE • Bonneville Brougham • Catalina • Le Mans • Grand Prix • Grand Prix LE • Grand Prix Brougham • Safari SW • Parisienne • Parisienne Brougham • Parisienne Brougham Wagon

Installation

1. Insert the bracket into the arm and align holes in arm and bracket. Install new bolts and nuts. Tighten the nut to 70 ft. lbs. (94 Nm).
2. Raise the suspension and reassemble the rear shock absorber lower attachment using a new attaching nut.
3. Install stabilizer bar.
4. Remove crossmember supports and lower the vehicle.

SECTION 2

SUSPENSION
Domestic Models

FRONT SUSPENSION—GENERAL MOTORS A AND X BODY CARS—FRONT WHEEL DRIVE

Wheel Alignment

Front alignment consists of the camber adjustment and toe setting. The caster setting is built into the vehicle with no provisions for adjustment.

Two bolts clamp the lower end of the MacPherson strut assembly to the upper arm of the steering knuckle. The lower of the two bolts has an eccentric washer at the head providing the camber adjustment. These special high tensile bolts with the loose nuts are torqued to 210 ft. lbs. (270 Nm). The camber setting is plus .5 degrees with a .5 degree tolerance.

The toe adjustment is conventional, with adjusting sleeves at the tie rod ends held in place with locking jam nuts. The toe setting is plus .1 degree with a tolerance of ± 1 degree.

Wheel Bearings

The front wheel bearing is a double row ball design. It is a prelubricated sealed bearing and requires no regular maintenance. The bearing in a loose fit in the steering knuckle. The drive axle outer joint shaft is a splined fit through the bearing. The hub nut and washer are used to pre-load the bearing.

DIAGNOSIS

Check for proper drive axle nut torque, 185 ft. lbs. (250 Nm). Clean threads, remove nut, install new nut and torque to proper specifications. Free the shoes from the disc or remove calipers. Reinstall two wheel nuts to ensure disc to bearing. Mount dial indicator. Grasp disc and use a push-pull movement. Do not rock discs as this will give a false reading. If looseness exceeds 0.005 in. (0.508mm) replace bearing.

Removal

1. Loosen hub nut.
2. Raise car and remove wheel.
3. Remove hub nut and discard.
4. Remove brake caliper.
5. Remove three hub and bearing attaching bolts. If old bearing is being reused, mark attaching bolts and corresponding holes for installation.
6. Install tool J–28733 or equivalent and remove bearing. If excessive corrosion is present, make sure bearing is loose in knuckle before using tool.

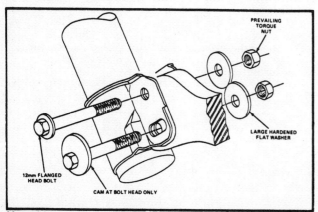

Knuckle strut mounting bolts for camber adjustment (© General Motors Corp.)

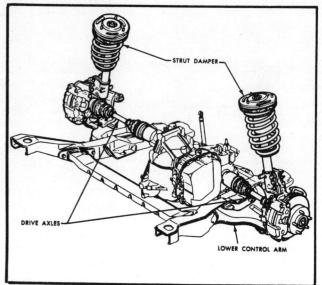

Front suspension components (© General Motors Corp.)

NOTE: A boot protector should be installed whenever servicing front suspension components to prevent damage to the drive axle boot.

Installation

1. Clean and inspect bearing mating surfaces and steering knuckle bore for dirt, nicks and burrs.
2. If installing steering knuckle seal, use tool J–28671 or equivalent. Apply grease to seal and knuckle bore.
3. Push bearing on axle shaft.
4. Torque new hub until bearing is seated.
5. Install brake caliper.
6. Lower car.
7. Apply final torque to hub nut.

TORQUES

Top strut nut: 68 ft. lbs. (90 Nm)
 Top mount nuts: 18 ft. lbs. (24 Nm)
Lower strut bolts: 140 ft. lbs. (190 Nm)
Hub nut: 170–192 ft. lbs. (230–260 Nm)

MAC PHERSON STRUT

Removal

1. Support the car so that there is no weight on the lower control arm.
2. Remove wheel.
3. Clean up and mark camber adjusting cam.
4. Remove the brake hose clip.
5. Remove top three bolts from the lower strut bolts.
6. Remove the strut assembly, and take a sample to work bench.

---- CAUTION ----
A reliable spring compressor tool is essential to disassemble and assemble strut bumper to avoid personal injury.

SUSPENSION
Domestic Models

7. Compress spring with compressor until there is not pressure on the upper spring seat.

CAUTION
Do not compress spring until it bottoms.

8. Remove the top nut from the strut shaft and remove the bumper shaft and the top mounting assembly.
9. Remove the spring from the strut assembly.

Installation

1. Install the new strut assembly into the spring and attach the mounting components on to the strut assembly.
2. Tighten the top strut nut 68 ft. lbs. (90 Nm) and remove the spring compressor.
3. Install the spring and strut assembly, first in the upper spring seat, then connect the lower end of the strut to the lower control arm.
4. Install brake caliper and wheel.

STEERING KNUCKLE

Removal

1. Remove wheel and wheel bearing.
2. Mark and remove the lower strut bolts.
3. Remove tie-rod end and ball joint.
4. Remove steering knuckle.

Installation

1. Install knuckle to ball joint and tighten.
2. Loosely install knuckle to strut
3. Install front wheel bearing.
4. Jack control arm into position and install tie-rod end.
5. Tighten cam bolts.
6. Reset steering camber and toe.
7. Install brake caliper and wheel.

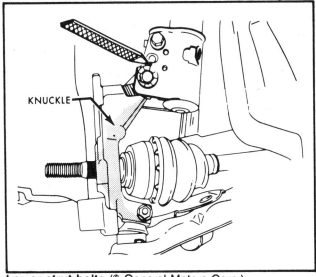

Lower strut bolts (© General Motors Corp.)

LOWER CONTROL ARM BALL JOINT

Inspection

1. Raise front suspension by placing jack or lift under the cradle.
2. Grasp the wheel at top and bottom and shake to of wheel in an in-and-out motion. Observe for any horizontal movement of the knuckle relative to the control arm. Replace ball joint if such movement is noted.
3. If the ball stud is disconnected from the knuckle and any looseness is detected, or if the ball stud can be twisted in its socket using finger pressure, replace the ball joint.

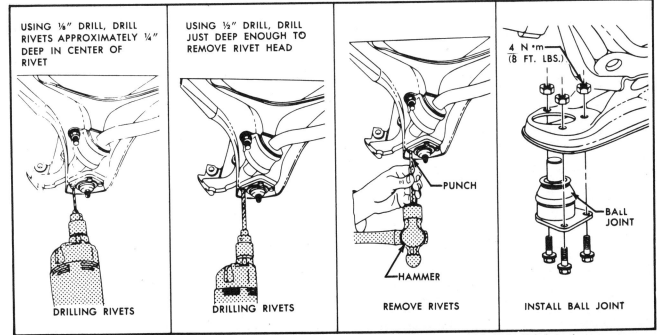

Ball joint inspection: vertical movement—.000 and horizontal movement—.000. If ball joint shows any movement, replace. It is not necessary to remove control arm to replace ball joint (© General Motors Corp.)

SECTION 2

SUSPENSION
Domestic Models

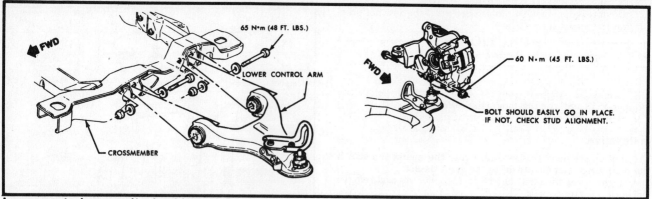

Lower control arm and/or bushings (© General Motors Corp.)

Removal

1. Raise car and remove wheel.
2. Remove parts.
3. Remove ball joint from knuckle.

Installation

1. Install ball join to knuckle.
2. Install parts.
3. Install wheel and lower car.
4. Check toe-in setting. Adjust as required.

REAR SUSPENSION – GENERAL MOTORS A AND X BODY CARS – FRONT WHEEL DRIVE

Wheel Bearings

The rear wheel bearing is a double row ball bearing. It is pre-lubricated and sealed at the factory. The bolt on the bearing should be replaced if the looseness exceeds recommendations.

Removal

1. Remove wheel and brake drum.

NOTE: Do not hammer on brake drum as damage to the bearing could result.

2. Remove four hub and bearing assembly-to-rear axle attaching bolts and remove hub and bearing assembly from axle.

NOTE: If studs must be removed from the hub, do not remove with a hammer as damage to bearing will result.

Installation

1. Install hub and bearing assembly to rear axle. Tighten bolts to 35 ft. lbs. (55 Nm).

Shock Absorber

Removal

1. Open deck lid, remove trim cover and remove upper shock attaching nut.
2. Raise car on hoist and support rear axle assembly.
3. Remove lower attaching bolt and nut and remove shock.

Installation

1. Install shock at lower attachment, feed bolt through holes and loosely install nut.
2. Lower car enough to guide upper stud through body opening and install nut loosely.
3. Torque lower nut to 34 ft. lbs. (47 Nm).
4. Lower car all the way and torque upper nut. Torque to 7 ft. lbs. (10 Nm).

Track Bar

Removal

1. Raise car on hoist and supper rear axle.
2. Remove nut and bolt at both the axle and body attachments and remove track bar.

Installation

1. Position track bar in left hand reinforcement and loosely

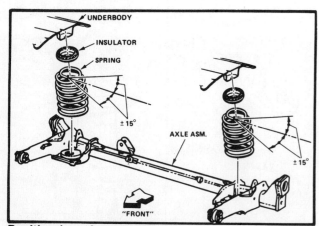

Position leg of upper coil on springs parallel to axle assembly and towards left-hand side of vehicle within limits shown. When removing rear springs, do not use a twin-post type hoist. The swing arc tendency of the rear axle assembly when certain fasteners are removed may cause it to slip from the hoist. Perform operation on floor if necessary (© General Motors Corp.)

SUSPENSION
Domestic Models

install bolt and nut. The open side of the bar and nut must face rearward.

2. Place other end of track bar in body reinforcement and install bolt and nut (nut must be at the rear of reinforcement of both attachments). Torque nut at axle bracket to 33 ft. lbs. (45 Nm) and torque nut at body reinforcement to 34 ft. lbs. (47 Nm).

Spring

Removal and Installation

NOTE: Do not use a twin-post type hoist. The swing are tendency of the rear axle when some fasteners are removed may cause it to slip from the hoist.

1. Raise the car. Support the rear axle while removing the brake line brackets, the track bar and the shock absorber lower mounts.
2. Lower rear axle and remove springs.
3. When installing, position springs correctly.

Control Arm Bushing

Removal

1. Raise car on hoist and support rear axle under front side of spring seat.
2. If removing right bushing, disconnect parking brake cable from hook guide.
3. Remove dual parking brake cables from bracket attachment and pull out of way.
4. Disconnect brake line bracket attachment from frame.
5. Remove shock lower attaching nut and bolt and pull string out of way.
6. Remove four control arm bracket-to-underbody attaching bolts and allow control arm to rotate downward.
7. Remove nut and bolt from bracket attachment and remove bracket.
8. Press bushing out of control arm.

Installation

1. Press bushing into control arm.

NOTE: Cut-outs on rubber portion of bushing must face front and rear.

2. Install bracket to control arm and torque nut to 34 ft. lbs. (47 Nm). Bracket must be at a 45 degree angle.
3. Raise control arm into position and install four control arm bracket-to-underbody attaching bolts. Torque to 20 ft. lbs. (27 Nm).
4. Replace spring and insulator and install shock lower attaching nut and bolt. Torque to 34 ft. lbs. (47 Nm).
5. Install brake line bracket to frame and torque screw to 18 ft. lbs. (11 Nm).
6. On right side only, reconnect brake cables to bracket, and reinstall brake cable to hook. Adjust cable as necessary.

REAR AXLE ASSEMBLY

Removal

NOTE: When removing rear axle assembly, do not use a twin post type hoist. The swing arc tendency of the rear axle assembly when certain fasteners re removed may cause it to slip from the hoist.

1. Remove the wheel and brake drum.

NOTE: Do not hammer on brake drum as damage to the bearing could result.

2. Disconnect parking brake cable from hook connection.
3. Remove brake line brackets from frame.
4. Remove shock lower attaching bolts and nuts at axle and disconnect shocks from axle.

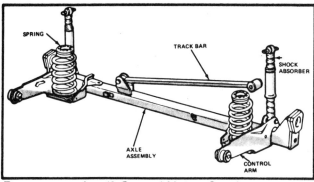

Rear suspension (© General Motors Corp.)

5. Remove track bar attaching nuts and bolt at axle and disconnect track bar.
6. Lower rear axle and remove coil springs and insulators.
7. Disconnect brake lines from control arm attachments.
8. Remove brake cable from rear axle attachments.
9. Remove hub attaching bolts and remove hub and bearing assembly. Move backing plate out of way.
10. Remove control arm bracket-to-underbody attaching bolts (four per side) and lower axle down to bench. This may require two people to steady axle.
11. Remove control arm brackets from control arms.

Installation

1. Install control arm brackets to control arms. Torque nut to 34 ft. lbs. (47 Nm). Brackets must be at a 45 degree angle.
2. Place axle assembly on transmission jack and raise into position. Attach control arms to underbody with four bolts per side. Torque bolts to 20 ft. lbs. (27 Nm).
3. Install backing plate and hub and bearing assembly to rear axle. Torque bolts to 35 ft. lbs. (55 Nm).
4. Install brake line connections to frame.
5. Attach brake cable to rear axle assembly.
6. Position coil springs and insulators in seats and raise rear axle. Leg of upper coil on springs must be parallel to axle assembly and face outboard on both sides.
7. Install shock absorbers to rear axle and torque nuts to 34 ft. lbs. (47 Nm).
8. Install track bar to rear axle and torque nut to 33 ft. lbs. (45 Nm).
9. Install brake line to control arm brackets and torque screws to 8 ft. lbs. (11 Nm).
10. Connect parking brake cable to guide hook and adjust as necessary.
11. Install brake drums and wheels. Torque lug nuts to 103 ft. lbs. (140 Nm).
12. Remove transmission support and lower car.
13. Bleed brake system and refill reservoir.

SUPERLIFT SHOCK ABSORBERS

The Superlift system is an assist type leveling device which the driver controls manually be varying air pressure in the system. The leveling unit is a combination of a pliable neoprene boot and air cylinder built around a hydraulic shock absorber.

PRECAUTIONS

To insure satisfactory functioning of the Superlift system, observe the following precautions:

1. Maintain a minimum of 10 psi (70 kPa) for best ride characteristics with an empty car.
2. Vary pressure up to a maximum of 90 psi (620 kPa) to level the car with loads.

SUSPENSION
Domestic Models

FRONT SUSPENSION—GENERAL MOTORS C BODY CARS—FRONT WHEEL DRIVE

General Description

The front suspension is a MacPherson Strut design. The control arm pivots from the frame mounted cradle. It is mounted in rubber bushings. The upper end of the strut assembly is isolated by a rubber mount which contains a non-serviceable bearing which allows for wheel turning. The lower end of the steering knuckle pivots on a ball joint which is riveted to the control arm. The ball joint is attached to the steering knuckle with a castellated nut and cotter pin.

Wheel Alignment

Before adjusting caster and camber angles, the front bumper should be raised and released three times to allow car to return to its normal standing height.

FRONT CASTER

Adjustment

1. Remove the to strut mounting nuts.
2. Separate the top strut mount from the inner wheelhouse.
3. Drill two $^{11}/_{32}$ in. holes at the front and rear of the round strut mounting hole. Remove the excess metal between the existing hole and the drilled hole.
4. Reinstall the strut mount in the holes. Install the washers and nuts.
5. Caster is set by moving the top of strut rearward or forward as required.
6. Tighten the strut mounting nuts.

FRONT CAMBER

Adjustment

1. Loosen both strut to knuckle nuts.
2. Install camber adjusting tool J-29862 or equivalent.
3. Set camber to specification.

FRONT TOE

Adjustment

1. Loosen the lock nut on both inner tie rods.
2. Set toe to specification by turning the inner tie rod accordingly.
3. Tighten the lock nuts on inner tie rods to 52 ft. lbs. (70 Nm).

Suspension Description

The front suspension is a MacPherson strut design. The control arm pivots from the cradle and is mounted in rubber bushings. The upper end of the strut is isolated by a rubber mount which contains a non-serviceable bearing which allows for wheel turning. The lower end of the steering knuckle pivots on a ball joint which is riveted to the control arm. The ball joint is fastened to the steering knuckle with a castellated nut and cotter pin.

Hub And Bearing Assembly

Removal and Installation

1. Raise the vehicle and support safely.

1. Nut 24 N·m (18 lbs. ft.)
2. Washer
3. Strut Assy.
4. Cover
5. Drill 8.731mm (11/32") holes
6. File here

Front caster adjustment (© General Motors Corp.)

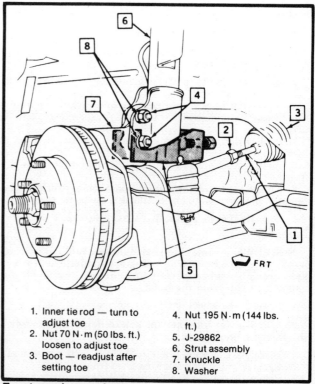

1. Inner tie rod — turn to adjust toe
2. Nut 70 N·m (50 lbs. ft.) loosen to adjust toe
3. Boot — readjust after setting toe
4. Nut 195 N·m (144 lbs. ft.)
5. J-29862
6. Strut assembly
7. Knuckle
8. Washer

Front camber and toe adjustmnet

SUSPENSION
Domestic Models

2. Remove the tire and wheel assembly.
3. Install drive axle boot seal protector J–28712 or equivalent on all outer constant velocity joints and tool number J–34754 or equivalent on all inner joints.
4. Remove the hub nut.
5. Remove the brake caliper and rotor. Wire off to the side to gain working clearance.
6. Attach tool J–28733 or equivalent and loosen splines between the hub and drive axle.
7. Remove the hub attaching bolts, splash shield, hub and bearing assembly.
8. Installation is the reverse of removal procedure. Lubricate hub bearing seal with grease prior to installation. Tighten the hub and bearing bolts to 70 ft. lbs. (95 Nm).

Strut Assembly

Removal and Installation

1. Remove the nuts attaching the top strut assembly to the vehicle body.
2. Raise the vehicle and support safely. Install jack stands under the vehicle cradle and lower the vehicle so the weight of the vehicle rests on the stands and not the control arm.
3. Install drive axle boot seal protectors. Care must be taken in order to prevent overextension of the inner Tri-Pot joints.
4. Remove the brake line retaining bracket from the strut assembly.
5. Remove the strut to steering knuckle bolts.
6. Remove the strut assembly from the vehicle.
7. Installation is the reverse of removal procedure. Note the following:
   a. Check wheel alignment.
   b. Tighten the strut to body bolts to 18 ft. lbs. (24 Nm).
   c. Tighten the strut to steering knuckle bolts to 144 ft. lbs. (195 Nm)

NOTE: Vehicle must now have a front wheel alignment performed.

Spring Assembly

Removal and Installation

1. Strut assembly must be removed from vehicle to perform this procedure. Refer to strut assembly removal outlined earlier in this section.

NOTE: A reliable spring compressor tool is essential to disassemble the strut to avoid personal injury.

2. Mount the spring assembly in a suitable holding fixture.
3. Install spring compressor until there is no pressure on the upper spring seat.

NOTE: Do not compress the spring until it bottoms.

5. Remove the top nut from the strut shaft and remove the shaft and top mounting assembly.
6. Remove the spring from the strut assembly.
7. Installation is the reverse of the removal procedure. Tighten the strut nut to 74 ft. lbs. (100 Nm).

Ball Joints

Inspection

1. Raise the front of the vehicle with lift placed under the engine cradle. The front wheels should be clear of the ground.
2. Grasp the wheel at the top and the bottom and shake the wheel in and out.
3. If any movement is seen on the steering knuckle, the ball joints are defective and must be replaced. Note the movement elsewhere may be due to loose wheel bearings or other defective components. Take note of the knuckle to control arm connection.
4. If the ball stud is disconnected from the steering and any looseness is noted, replace the ball joints.

Removal and Installation

1. Raise the front of the car and support it with jackstands underneath the engine cradle. Lower the car slightly so that the weight rest primarily on the jack stands.
2. Remove the wheel and tire assemblies.
3. Install drive axle covers to protect the drive axle boot seals.
4. Pull the cotter pin from the ball joint and install a ball joint separator tool. Turn the castellated nut counterclockwise to separate the ball joint from the steering knuckle.
5. Use a 1/8 in. drill bit to drill a hole approximately 1/4 in. deep in the center of each of the three ball joint rivets.
6. Use a 1/2 in. drill bit to drill off the rivet heads. Drill only enough to remove the rivet head.
7. Use a hammer and punch to remove the rivets. Drive them out from the bottom.
8. Loosen the stabilizer bar bushing assembly nut.
9. Pull down on the control arm and remove the ball joint from the steering knuckle and control arm.
10. Install the new ball joint in the steering knuckle and line up the holes with those in the control arm.
11. Install the three ball joint nuts facing down and tighten the nuts to 50 ft. lbs. (68 Nm).
12. Install the castellated nut and tighten to 81 ft. lbs. (110 Nm).

NOTE: Tightening the nut for cotter pin alignment is allowed, but do not loosen it once the torque has been reached.

13. Install the cotter pin.
14. Installation of the remaining components is in the reverse order of removal.

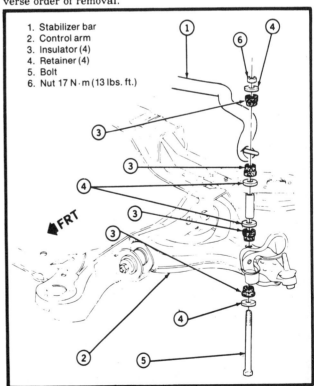

1. Stabilizer bar
2. Control arm
3. Insulator (4)
4. Retainer (4)
5. Bolt
6. Nut 17 N·m (13 lbs. ft.)

Stabilizer bar bushing assembly

SECTION 2
SUSPENSION
Domestic Models

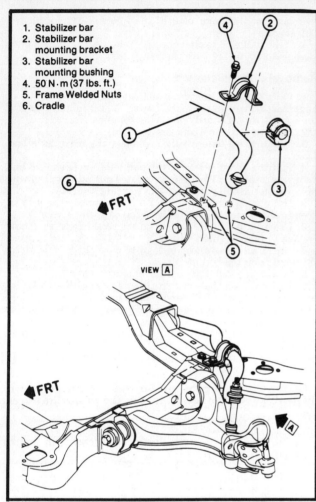

1. Stabilizer bar
2. Stabilizer bar mounting bracket
3. Stabilizer bar mounting bushing
4. 50 N·m (37 lbs. ft.)
5. Frame Welded Nuts
6. Cradle

Stabilizer bar installation (© General Motors Corp.)

STABILIZER BAR

Removal and Installation

1. Raise the vehicle and support safely. Place jack stands underneath the engine cradle. Lower the car slightly so that the weight rests on the jack stands.
2. Remove the wheel and tire assembly.
3. Install drive axle covers to protect the drive axle boot seals.

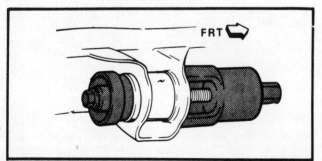

Bushing removal tool installation (© General Motors Corp.)

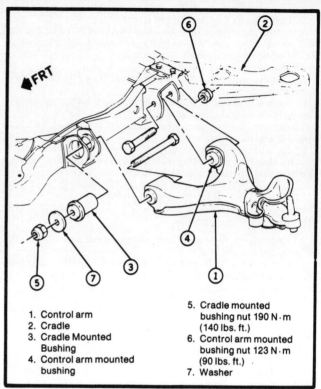

1. Control arm
2. Cradle
3. Cradle Mounted Bushing
4. Control arm mounted bushing
5. Cradle mounted bushing nut 190 N·m (140 lbs. ft.)
6. Control arm mounted bushing nut 123 N·m (90 lbs. ft.)
7. Washer

Control arm mounting – disassembled view (© General Motors Corp.)

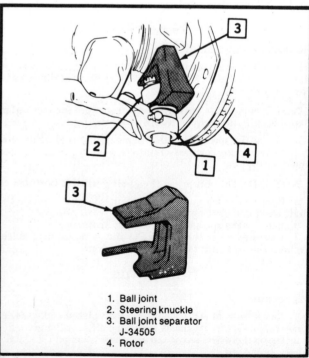

1. Ball joint
2. Steering knuckle
3. Ball joint separator J-34505
4. Rotor

Ball joint separator tool installation (© General Motors Corp.)

SUSPENSION
Domestic Models

4. Remove the bolts from both sides connecting the stabilizer bar bushings to the control arm.
5. Remove the stabilizer bar mounting bolts.
6. Remove the exhaust pipe between the exhaust manifold and the catalytic converter.
7. Remove the stabilizer bar from the vehicle by sliding it over the right steering knuckle.
8. Installation is the reverse of the removal procedure. Tighten the bolts to the following specifications:
   a. Stabilizer bar mounting bracket bolts-37 ft. lbs. (50 Nm).
   b. Bushing assembly nut-12 ft. lbs. (17 Nm).
   c. Tie rod end nut-35 ft. lbs. (47 Nm) 52 ft. lbs. (70 Nm) maximum torque for cotter pin alignment.

CONTROL ARM BUSHINGS

Removal and Installation

1. Raise the vehicle and support it with jack stands underneath the engine cradle. Lower the vehicle slightly so that the weight rests primarily on the jack stands.
2. Remove the tire and wheel assemblies.
3. Install drive axle covers to protect the drive axle boot seals.
4. Remove the stabilizer bar bushing to control arm bolt.
5. Remove the cotter pin from the ball and install the ball joint separator tool.
6. Remove the control arm mounting bolts. Remove the control arm from the vehicle.
7. Installation is the reverse of the removal procedure.

Cradle Mounted Bushing

Removal

1. Remove the lower control arm from the vehicle as outlined earlier in this section.
2. Assemble bushing tool according to the diagram with the small end facing the bushing.
3. Tighten the bolt until the bushing is driven out of the cradle assembly.

Installation

1. Assemble the bushing tool with the large end facing the bushing.
2. Install the bushing tool end nut.
3. Tighten the bushing tool nut until the bushing is firmly seated in the cradle.

REAR SUSPENSION—GENERAL MOTORS C BODY CARS—FRONT WHEEL DRIVE

General Description

These vehicles are equipped with independent coil spring rear suspension systems. Each suspension knuckle is supported with a lower control arm and an air adjustable Superlift® strut. A stabilizer bar minimizes body roll. Each control arm is equipped with a suspension adjustment link. This link allows for toe adjustment. It will also minimize alignment variation caused by suspension movement. The rear control arm is attached to the suspension knuckle by means of a ball joint which helps reduce friction. The entire suspension design allows for movement of one wheel without affecting the wheel on the opposite side of the vehicle.

The control arms are made of low carbon, mild steel. The hub and bearing is machined for precise contact.

The hub and wheel bearing is one assembly. This eliminates the need for wheel bearing adjustments. It does not require maintenance or adjustments.

Electronic Level Control is standard equipment on all models to maintain rear trim height under a wide range of operating conditions.

Ball Joints

Inspection

The lower ball joint is inspected for wear by visual observation alone. The vehicle must be supported by the wheels during inspection. Wear is indicated by retraction of the half-inch diameter nipple into the ball joint cover. Normal wear will be indicated by the nipple retracting slowly into the ball joint cover.

Ball stud tightness in the knuckle boss should also be checked when inspecting the ball joint. In order to accomplish this, shake the wheel and feel for movement of the stud end or castellated nut at the knuckle boss. A loose nut can also indicate a bent or damaged stud.

Removal and Installation

1. Raise and support the vehicle safely. Remove the tire and wheel assemblies from the rear of the vehicle.
2. Disconnect the Electronic Level Control (ELC) height sensor link (right control arm) and/or the parking brake cable retaining link (left control arm).
3. Remove the cotter pin and the castellated nut from the outer suspension adjustment link.
4. Separate the outer suspension link from the knuckle.
5. Support the control arm with a suitable jack. The lower control arm must be supported to prevent the coil spring from forcing the control arm downward.
6. Remove the ball and cotter pin.
7. Remove the castellated nut and then reinstall it with the flat side facing upward. Do not tighten.

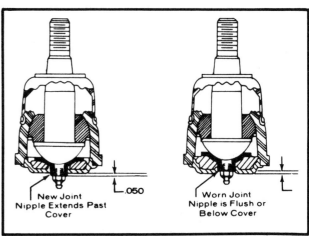

Lower ball joint inspection (© General Motors Corp.)

SECTION 2

SUSPENSION
Domestic Models

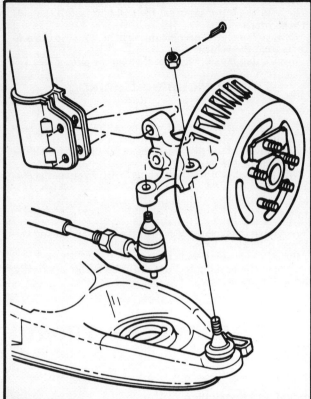

Rear lower ball joint – exploded view (© General Motors Corp.)

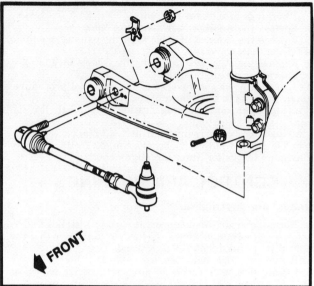

Suspension adjustment link (© General Motors Corp.)

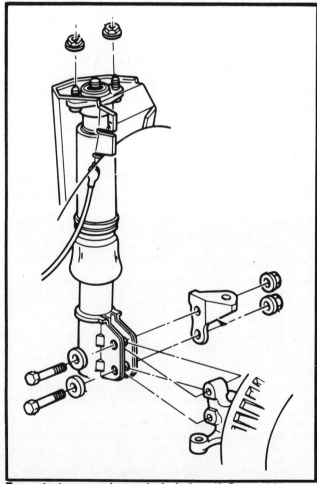

Rear strut removal – exploded view (© General Motors Corp.)

8. Install a ball joint separator tool and separate the ball joint from the knuckle.
9. Separate the ball joint from the control arm.
10. Installation is the reverse of the removal procedure. Note the following:
    a. Tighten the NEW castellated nut to 7.5 ft lbs. (10 Nm). Tighten the nut additional ⅔ of a turn.
    b. Align the slot in the nut to the cotter pin hole by tightening only. Do not loosen the nut in order to align the holes.

Superlift Strut®

Removal and Installation

1. Remove the inner side trunk cover. Remove the strut tower mounting nuts.
2. Raise the vehicle and support safely. Remove the rear wheel and tire assemblies.
3. Disconnect and plug the ELC air line.
4. Remove the strut anchor nuts, washers and bolts from the rear knuckle and knuckle bracket.
5. Remove the strut from the vehicle.
6. Installation is the reverse of the removal procedure. Tighten the strut tower mounting nuts to 19 ft. lbs. (25 Nm). Tighten the strut anchor nuts to 144 ft. lbs. (195 Nm). Check rear wheel alignment.

Coil Springs

Removal and Installation

1. Raise and support the vehicle safely.
2. Remove the rear wheels.

SUSPENSION
Domestic Models

3. Separate the rear stabilizer bar from the knuckle bracket and remove it.
4. Disconnect the ELC height sensor line and the parking brake cable retaining clip.
5. Position the special tool J–23028–01 or equivalent so as to cradle the control arm bushings.

NOTE: Special tool J–23028–01 should be secured to a suitable jack.

6. Raise the jack to remove the tension from the control arm pivot bolts.

NOTE: Secure a chain around the spring and through the control arm as a safety precaution.

7. Remove the rear control arm pivot bolt and nut.
8. Move the jack so as to relive any tension from the control arm pivot bolt. Remove the bolt and nut.
9. Lower the jack to allow the jack to pivot downward.
10. When all pressure has been removed from the coil spring, remove the safety chain, spring and insulators.
11. Installation is the reverse of the removal procedure. Control arm mounting nuts should not be tightened until the vehicle is unsupported and resting on its wheels at normal trim height.

Rear Stabilizer Bar And Bushings

Removal and Installation

1. Raise the vehicle and support safely. Remove the rear wheel and tire assemblies.
2. Remove the stabilizer bar bolts, nuts and bar retainers.
3. Remove the bushing clip bolt.
4. Bend the open end of the support assembly downward.
5. Remove the stabilizer bar and bushings.
6. Installation is the reverse of the removal procedure.

Suspension Adjustment Link

Removal and Installation

1. Raise and support the vehicle safely. Remove the rear wheel and tire assemblies.
2. Remove the cotter pin and castellated nut from the knuckle.

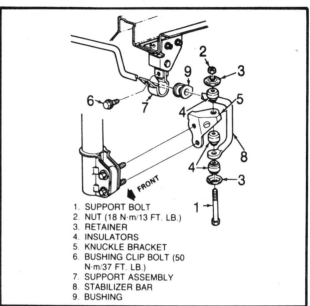

1. SUPPORT BOLT
2. NUT (18 N·m/13 FT. LB.)
3. RETAINER
4. INSULATORS
5. KNUCKLE BRACKET
6. BUSHING CLIP BOLT (50 N·m/37 FT. LB.)
7. SUPPORT ASSEMBLY
8. STABILIZER BAR
9. BUSHING

Rear stabilizer bar and bushging assembly (© General Motors Corp.)

3. Separate the outer suspension link from the knuckle.

NOTE: When separating the linkage joint, no attempt should be made to disengage the joint by driving a wedge between the joint and the attached part. Seal damage may result.

4. Remove the link assembly retaining nut and retainer. Remove the suspension adjustment link.
5. Installation is the reverse of the removal procedure. Check rear wheel alignmnent.

NOTE: Due to the diagnostic complexity of the Automatic Load Levelling System, detailed diagnostic procedures can be found in Chilton's Chassis Electronic Service Manual.

FRONT SUSPENSION – GENERAL MOTORS F BODY CARS – REAR WHEEL DRIVE

Wheel Alignment

CASTER AND CAMBER ADJUSTMENTS

Caster and camber can be adjusted by moving the position of the upper strut mount assembly. Moving the strut mount forward/rearward adjusts the caster while moving the strut mount inward and outward, adjusts the camber.

The position of the strut mount can be changed after loosening the three retaining nuts. The weight of the vehicle will normally cause the strut assembly to move to the full inboard position.

Install special tool J–29724 or its equivalent between the strut mount and a fender bolt and tighten the tool's turnbuckle until the proper camber reading is obtained. If an adjustment of caster is required, tap the strut mount either forward or rearward with a rubber mallet until the caster reading is obtained. Tighten the three mount screws to 20 ft. lbs. (28 Nm).

Remove the tool from the strut mount to fender bolt and reinstall the fender bolt in place.

Suspension Description

Each wheel is independently connected to the frame by a steering knuckle, strut assembly, ball joint, and lower control arm. The steering knuckles move in a prescribed three dimensional arc. The front wheels are held in proper relationship to each other by two tie rods which are connected to the steering knuckles and to a relay rod assembly.

Coil chassis springs are mounted between the spring housings on the front crossmember and the lower control arm. Ride control is provided by a double, direct acting strut assemblies. The upper portion of each strut assembly extends through the fender well and attaches to the upper mount assembly with a nut.

Side roll of the front suspension is controlled by a spring

2-125

SECTION 2

SUSPENSION
Domestic Models

steel stabilizer shaft. It is mounted in rubber bushings which are held to the frame side rails by brackets. The ends of the stabilizer are connected to the lower control arms by link bolts isolated by rubber grommets.

The inner ends of the lower control arm have pressed-in bushings. Bolts, passing through the bushings, attach the arm to the suspension crossmember. The lower ball joint assembly is a press fit in the arm and attaches to the steering knuckle with a torque prevailing nut.

WHEEL BEARINGS

The proper functioning of the front suspension cannot be maintained unless the front wheel tapered roller bearings are correctly adjusted. The bearings must be a slip fit on the spindle and the inside diameter of the bearings should be lubricated to insure proper operation. The spindle nut must be a free-running fit on the threads.

Adjustment

1. Remove dust cap from hub.
2. Remove cotter pin from spindle and spindle nut.
3. Tighten the spindle nut to 12 ft. lbs. (16 Nm) while turning the wheel assembly forward by hand to fully seat the bearings. This will remove any grease or burrs which could cause excessive wheel bearing play later.
4. Back off the nut to the "just loose" position.
5. Hand tighen the spindle nut. Loosen spindle nut until either hole in the spindle lines up with a slot in the nut. Not more than $1/2$ flat.
6. Install new cotter pin. Bend the ends of the cotter pin against nut, cut off extra length to ensure ends will not interfere with the dust cap.
7. Measure the looseness in the hub assembly. There will be from 0.001 to 0.005 in. (0.03–0.13 mm) end play when properly adjusted.
8. Install dust cap on hub.

STRUT ASSEMBLY

Removal and Installation

1. Raise vehicle.

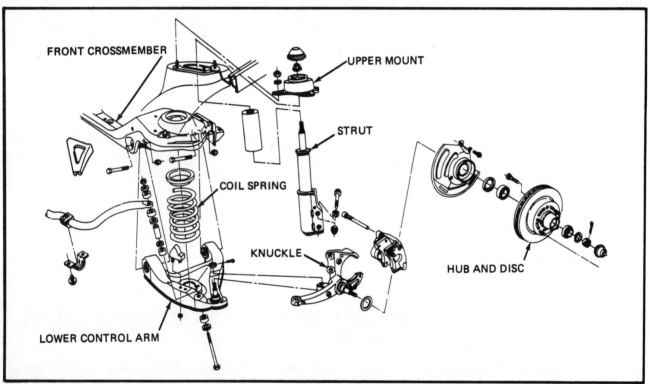

1982 and later F body front suspension—exploded view (© General Motors Corp.)

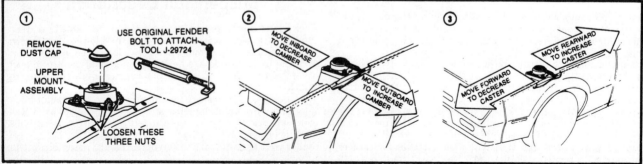

Front end alignment procedure (© General Motors Corp.)

2-126

SUSPENSION
Domestic Models

2. Remove wheel and tire.
3. Support lower control arm with jackstand.
4. Remove brake hose bracket.
5. Remove two strut-to-knuckle bolts.
6. Remove cover from upper mount assembly.
7. Remove nut from upper end of strut.
8. Remove strut and shield.
9. Reverse order of removal to replace strut.

LOWER BALL JOINT

Removal

1. Raise car, support with floor stands under frame.
2. Remove tire and wheel assembly.

---- **CAUTION** ----

Floor jack must remain under control arm spring seat during removal and installation to retain spring and control arm in position.

4. Remove cotter pin, and loosen castellated nut. Use tool J-24292A or equivalent to break ball joint loose from knuckle. Remove tool and separate joint from knuckle.
5. Guide lower control arm out of opening in splash shield with a putty knife or similar tool.
6. Remove grease fittings, and install tools as shown below. Press ball joint out of lower control arm.

Inspection

Inspect the tapered hole in the steering knuckle. Remove any dirt. If out-of-roundness, deformation or damage is noted, the knuckle MUST be replaced.

Installation

1. Position all joint into lower control arm and press in until it bottoms on the control arm, using tools as illustrated below. Grease purge on seal must be located facing inboard.
2. Place ball joint stud in steering knuckle.
3. Torque ball stud to 90 ft lbs. (120 Nm). Then tighten as additional amount enough to align slot in nut with hole in stud. Install cotter pin.
4. Install and lubricate all joint fitting until grease appears at the seal.
5. Install tire and wheel assembly.
6. Check front alignment.

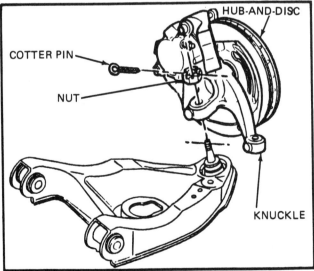

Knuckle, hub and disc assembly (© General Motors Corp.)

COIL SPRING/LOWER CONTROL ARM

Removal and Installation

1. Raise vehicle using a hoist.
2. Remove the wheel and tire.
3. Remove stabilizer link and bushings at lower control arm.
4. Remove pivot bolt nuts. DO NOT remove pivot bolts at this time.
5. Install tool J-23028 adaptor or equivalent, to jack and place into position with tool J-23028 or equivalent, supporting bushings.
6. Install jackstand under outside frame rail on opposite side of vehicle.
7. Raise tool J-23028 or equivalent, enough to remove both pivot bolts.
8. Lower tool J-23028 or equivalent, carefully, as shown below.
9. Remove spring and insulator tape insulator to spring.
10. Remove ball joint from knuckle using tool J-2492A or equivalent, as outlined earlier.
11. Replace bushings in lower control arm.
12. Install parts in reverse order of removal.

NOTE: After assembly, end of spring coil must cover all or part of one inspection drain hole. The other hole must be partly exposed or completely uncovered.

KNUCKLE, HUB AND DISC

Removal and Installation

1. Siphon master cylinder to avoid leakage.
2. Raise vehicle.
3. Remove wheel and tire.
4. Remove brake from stut.
5. Remove caliper support safely.
6. Remove hub and disc.
7. Remove splash shield.
8. Disconnect tie rod from knuckle.
9. Support lower contol arm.
10. Disconnect ball joint from knuckle, using tool J-2492A or equivalent.

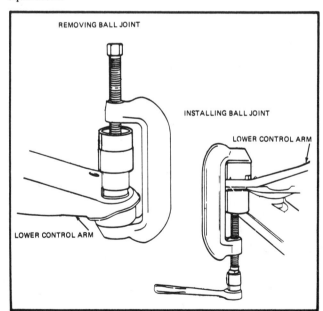

Removal and Installation of lower control arm ball joint (© General Motors Corp.)

2-127

SUSPENSION
Domestic Models

11. Remove two bolts attaching strut to knuckle and remove knuckle.
12. Reverse order of removal to install.

STABILIZER SHAFT

Removal and Installation
1. Raise vehicle on hoist.
2. Remove link bolt, nut, grommets, spacer and retainers.
3. Remove insulators and brackets.
4. Remove stabilizer shaft.
5. Install parts in reverse order of removal.
6. Hold stabilizer shaft at approximately 55.0 mm from bottom of side rail when tightening stabilizer shaft insulators.
7. Lower vehicle.

REAR SUSPENSION – GENERAL MOTORS F BODY CARS – REAR WHEEL DRIVE

Suspension Description

The rear axle assembly is attached to the vehicle through a link type suspension system. The axle housing is connected to the body by two lower control arms and a track bar. A single torque arm is used in place of upper control arms and is rigidly mounted to the rear axle housing at the rear and through a rubber bushing to the transmission at the front. Coil springs are used to support the weight of the car and ride control is provided by shock absorbers mounted to the rear of the axle housing. A stabilizer shaft is optional.

SHOCK ABSORBERS

Removal
1. Hoist car and support rear axle.
2. From above, pull back carpeting and remove shock absorber upper mounting nut.

NOTE: Axle assembly must be supported before removing upper shock absorber nut to avoid possible damage to brake lines, track bar and prop shaft.

3. Loosen and remove shock absorber lower mounting nut from shock absorber and remove shock.

Installation
1. Position shock absorber through body mounting hole and loosely install the lower shock absorber mounting nut.
2. From above, install the upper shock absorber retainer and nut and torque nut to 13 ft. lbs. (17 Nm).
3. Torque lower shock absorber nut to 70 ft. lbs. (95 Nm).
4. Remove rear axle support and lower car.

COIL SPRINGS AND INSULATORS

Removal
1. Hoist car on non-twin post-type joist and support rear axle assembly with an adjustable lifting device.
2. Remove track bar mounting bolt at axle assembly and loosen track bar bolt at body brace.
3. Disconnect rear brake hose clip at underbody to allow additional axle drop.
4. Remove right and left shock absorber lower attaching nuts.
5. Remove prop shaft on vehicles equipped with 4 cylinder engines.
6. Carefully lower rear axle and remove springs(s) and or insulator(s).

NOTE: DO NOT suspend rear axle by brake hose. Damage to hose could result.

Installation
1. Position springs and insulators in spring seats and raise rear axle until rear axle supports weight of vehicle at normal curb height position.
2. Install shocks to rear axle and torque nuts to 70 ft. lbs. (95 Nm).
3. Throughly clean track bar to axle assembly bolt and nut.
4. Reinstall track bar mounting bolt at axle and torque nut to 93 ft. lbs. (125 Nm). Torque track bar to body bracket nut to 58 ft. lbs. (78 Nm).
5. Install brake line clip to underbody.
6. Install prop shaft on 4 cylinder engine equipped cars.
7. Remove adjustable lifting device from beneath axle and lower car.

TRACK BAR

Removal
1. Hoist car and support rear axle, at curb height position.
2. Remove track bar mounting bolt and nut at rear axle and at body bracket.
3. Remove track bar.

Installation
1. Position track bar in body bracket and loosely install bolt and nut.

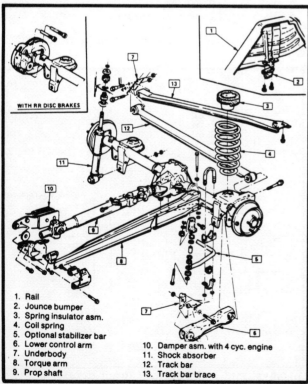

1. Rail
2. Jounce bumper
3. Spring insulator asm.
4. Coil spring
5. Optional stabilizer bar
6. Lower control arm
7. Underbody
8. Torque arm
9. Prop shaft
10. Damper asm. with 4 cyc. engine
11. Shock absorber
12. Track bar
13. Track bar brace

F body rear suspension-exploded view

SUSPENSION
Domestic Models

2. Thoroughly clean track bar to axle assembly bolt and nut.
3. Position track bar at axle and install bolt and nut, torque bolt to 93 ft. lbs. (125 Nm).
4. Torque track bar nut at body bracket to 58 ft. lbs. (78 Nm).
5. Remove rear axle support and lower car.

TRACK BAR BRACE

Removal

1. Hoist car and support rear axle.
2. Remove heat shield screws from track bar brace.
3. Remove three track bar brace to body brace screws.
4. Remove nut and bolt at body bracket and remove track bar brace.

Installation

1. Position track bar brace and loosely install nut and bolt at body bracket.
2. Position other end of track bar brace at body bracket and install three screws. Torque screws to 34 ft. lbs. (47 Nm).
3. Torque track bar nut at body brace to 58 ft. lbs. (78 Nm).
4. Install heat shield screws to track bar brace.
5. Remove rear axle support and lower car.

REAR LOWER CONTROL ARM

NOTE: If both control arms are being replaced, remove and replace one control arm at a time to prevent the axle from rolling or slipping sideways making replacement difficult.

Removal

1. Hoist car and support rear axle at curb height position.
2. Remove lower control arm to axle housing bolt and control arm to underbody bolt.
3. Remove control arm.

Installation

1. Position control arm and install front and rear nuts and bolts.
2. Torque front and rear bolts to 68 ft. lbs. (93 Nm).
3. Remove rear axle support and lower car.

BUSHING (REAR LOWER CONTROL ARM)

Removal

1. Remove control arm as specified in Rear Lower Control Arm Removal Procedure.
2. Place receiver J–25317–2 or equivalent over flanged side of bushing.
3. Use an arbor press to force the bushing out of the arm, using large O.D. of a driver such as J–21465–8 contacting O.D. of bushing outer sleeve.

Installation

To install the bushing, reverse the tool and push bushing into position. Connect the rear control arms as outlined in Rear Lower Control Arm Installation procedure.

TORQUE ARM

Removal

NOTE: Coil springs must be removed before removing torque arm to avoid rear axle forward twist which may cause vehicle damage.

1. Hoist car on a non-twin post-type hoist and support rear axle assembly with an adjustable lifting device.
2. Remove track bar mounting bolt at axle assembly and loosen track bar bolt at body brace.
3. Disconnect rear brake hose clip at underbody to allow additional axle drop.
4. Remove right and left shock absorber lower attaching nuts.
5. Remove prop shaft on vehicles equipped with 4 cylinder engines.
6. Carefully lower rear axle and remove coil springs.

NOTE: Do not suspend rear axle by brake hose. Damage to hose could result.

7. Remove torque arm rear attaching bolts.
8. Remove front torque arm outer bracket and remove torque arm.

Installation

1. Position torque arm and loosely install torque arm bolts.
2. Install front torque arm bracket and torque nuts to 20 ft. lbs. (27 Nm).
3. Torque rear torque arm nuts to 100 ft. lbs. (135 Nm).
4. Position springs and insulators in spring seats and raise rear axle until rear axle supports weight of axle of vehicle at normal curb height position.
5. Install shocks to rear axle and torque nuts to 70 ft. lbs. (95 Nm).
6. Thoroughly clean track bar to axle assembly bolt and nut.
7. Reinstall track bar mounting bolt at axle and torque nut to 93 ft. lbs. (125 Nm), torque track bar to body bracket nut to 58 ft. lbs. (78 Nm).
8. Install brake line clip to underbody.
9. Install prop shaft on 4 cylinder engine equipped cars.
10. Remove adjustable lifting device and lower car.

FRONT SUSPENSION—GENERAL MOTORS E AND K BODY CARS—FRONT WHEEL DRIVE

General Description

The front suspension consists of control arms, stabilizer bar, shock absorber and a right and left torsion bar. Torsion bars are used instead of the conventional coil springs. The front end of the torsion bar is attached to the lower control arm. The rear of the torsion bar is mounted into an adjustable arm at the torsion bar crossmember. The trim height of the car is controlled by this adjustment.

Wheel Alignment

HEIGHT ADJUSTMENT (TORSION BAR SUSPENSION MODELS)

The standing height must be checked and adjusted if necessary, before performing the front end alignment procedure. The standing height is controlled by the adjustment setting of

SECTION 2: SUSPENSION
Domestic Models

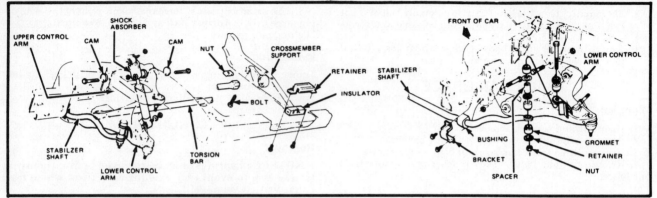

Front suspension (© General Motors Corp.)

the torsion bar adjusting bolt.

Clockwise rotation of the bolt increases the front height; counterclockwise decreases the front height.

Car must be on a level surface, gas tank full, or a compensating weight added, front seat all the way to the rear, and front and rear tires inflated to the proper pressures. Doors, hood and trunk must be closed and no passengers or additional weight should be in car or trunk.

These tolerances are production specifications on bumper height. If there is more than 1 inch (25mm) difference, side to side, at the wheel well opening, corrective measures may need to be implemented on a case by case basis. These are curb height dimensions which include a full tank of fuel.

CAMBER AND CASTER ADJUSTMENTS

These adjustments can be made either from under car or under hood, as desired. If under hood approach is used, however, be sure to recheck alignment after all operations are completed. Change in weight distribution caused by opened hood is sufficient to disturb final alignment settings.

1. Loosen nuts on upper suspension arm front and rear cam bolts.

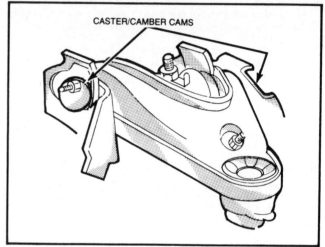

Caster/camber adjustment (© General Motors Corp.)

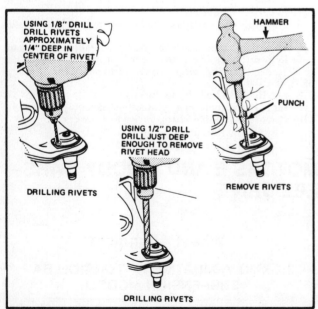

Upper ball joint removal and installation (© General Motors Corp.)

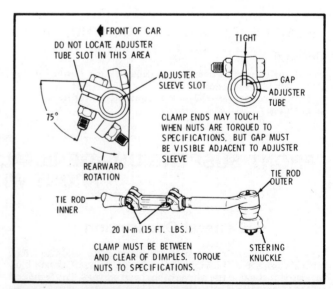

E series tie rod clamp and sleeve positioning. Bolts must be installed in direction shown (© General Motors Corp.)

SUSPENSION
Domestic Models

2. Note camber reading and rotate front bolt to correct for ½ of incorrect reading or as near as possible.

3. Rotate rear cam bolt to bring camber reading to 0°.

NOTE: Do not use a socket to adjust rear cam bolt on left side as brake pipes could be damaged. An offset box end wrench is recommended at this adjustment point.

4. Tighten front and rear bolts and check caster. If caster requires adjustment, proceed with Step 5; if not, move to Step 8.

5. Loosen front and rear cam bolt nuts.

6. Using camber scale on alignment equipment, rotate front bolt so that the camber changes an amount equal to ¼ of the desired caster change. (A caster change-to-camber change ratio of about 2 to 1 is inherent to the Eldorado and Seville suspension system. That is, when one cam is rotated sufficiently to change camber 1°, caster reading will change about 2°.).

If adjusting to correct for excessive negative caster, rotate

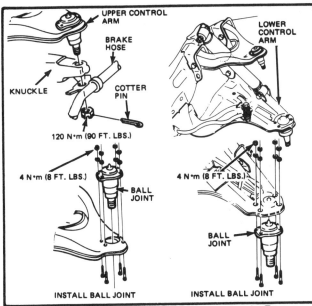

To remove: hoist car and remove wheel. Remove knuckle. Remove hub and bearing assembly, knuckle and knuckle seal. Remove parts as shown. To install: Install poarets as shown. Install steering knuckle. Install hub and bearingg assembly, knuckle and knuckle seal. Install wheel and lower hoist (© General Motors Corp.)

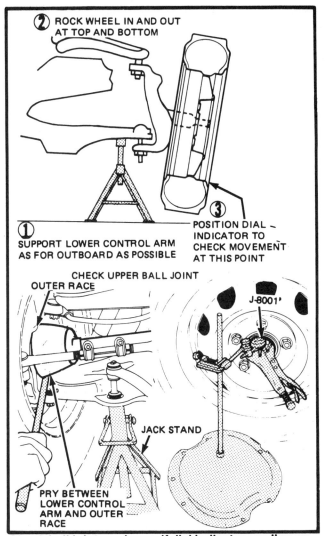

Check ball joint as shown. If dial indicator reading exceeds 3.2mm (.125 in.) or if ball stud is disconnected from knuckle and any looseness is detected or ball stud can be twisted in its socket with fingers, replace ball joint (© General Motors Corp.)

front bolt to increase positive camber. If adjusting to correct for excessive positive caster, rotate front bolt to increase negative camber.

7. Rotate the rear bolt until camber setting returns to its corrected position (Step 3).

8. Tighten upper suspension arm cam nuts to 95 ft. lbs. (130 Nm). Hold head of bolt securely; any movement of the cam will affect final setting and will require a recheck of the camber and caster adjustments.

TOE ADJUSTMENT

Before checking toe-in, make certain that the intermediate rod height is correct.

Toe-in is adjusted by turning the tie rod adjuster tubes at the outer ends of each tie rod after loosening clamp bolts. The readings should be taken only when the front wheels are in a straight ahead position so that the steering gear is on its high spot.

1. Center steering wheel, raise car, and check wheel runout.

2. Loosen tie rod adjuster nus and adjust tie rods to obtain proper toe setting.

3. Position tie rod adjuster clamps so that openings of clamps are facing up. Interference with front suspension components could occur while turning if clamps are facing down.

Wheel Bearings (Tapered Roller Bearings)

Lubrication

For normal application, clean and repack front wheel bearings with a high melting point wheel bearing lubricant at each front brake lining replacement or 30,000 miles (48,000 km),

SUSPENSION
Domestic Models

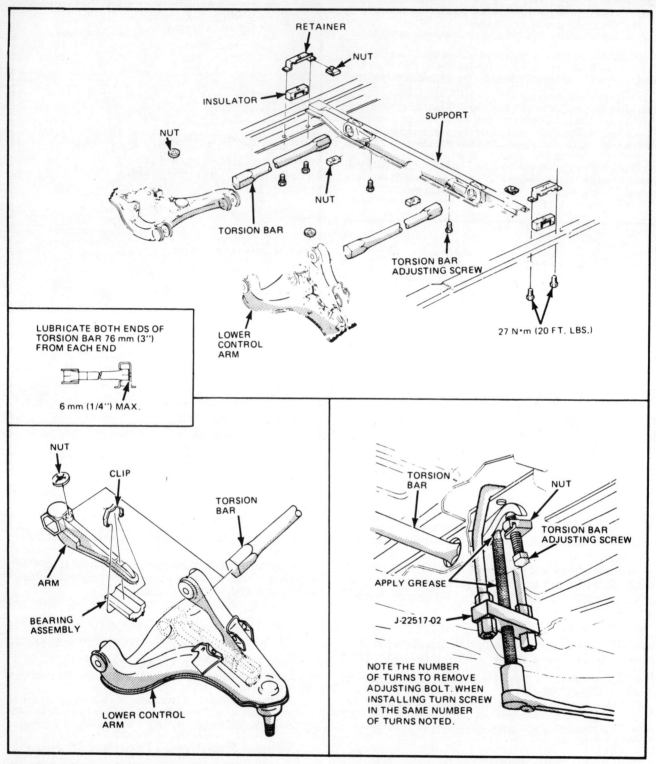

To remove: hoist car and remove parts as shown. To remove torsion bar (s) only: remove torsion bar adjusting screw as shown. Slide torsion bar forward in lower control arm until torsion bar clears support. Then pull down on bar and remove from control arm. To install: install parts as shown. Lower hoist and adjust trim height (© General Motors Corp.)

SUSPENSION
Domestic Models

Bolt on Sealed Wheel Bearing Diagnosis

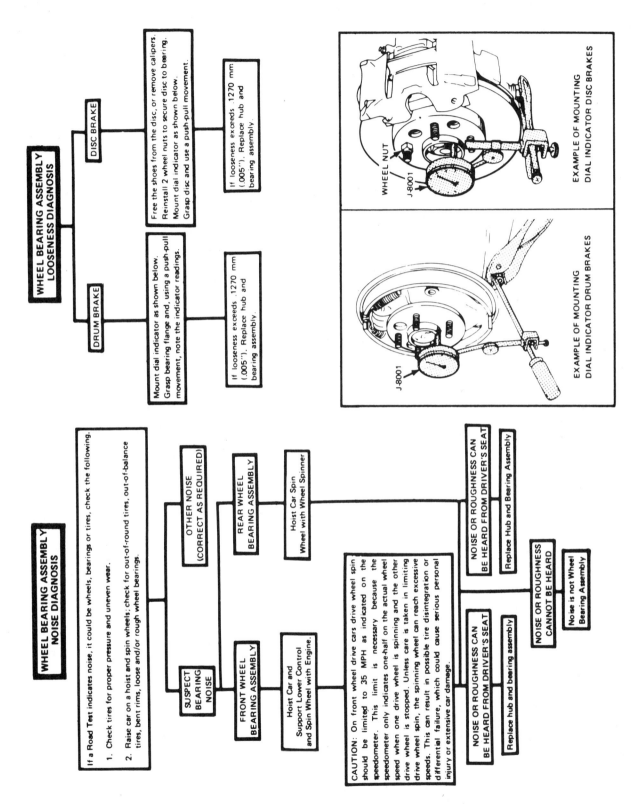

Bolt on sealed wheel bearing diagnosis (© General Motors Corp.)

2-133

Section 2

SUSPENSION
Domestic Models

whichever comes first. For heavy duty application, clean and repack front wheel bearings at each front brake lining replacement or 15,000 miles (24,000 km), whichever comes first. Use wheel bearing lubricant; "long fiber" or "viscous" type lubricant should not be used. Do not miss wheel bearing lubricants. Be sure to thoroughly clean bearings and hubs of all old lubricant before repacking.

NOTE: Tapered roller bearings have a slightly loose feel when properly adjusted. They must never be overtightened (pre-loaded) or severe bearing damage may result.

Adjustment

The proper functioning of the front suspension cannot be maintained unless the front wheel tapered roller bearings are correctly adjusted. Cones must be a slip fit on the spindle and the inside diameter of cones should be lubricated to insure that the cones will creep. Spindle nut must be a free-running fit on threads.

1. Remove cotter pin from spindle and spindle nut.
2. Tighten the spindle nut to 12 ft. lbs. (16 Nm) while turning the wheel assembly forward by hand to fully seat the bearings. This will remove any grease or burrs which could cause excessive wheel bearing play later.
3. Back off the nut to the "just loose" position.
4. Hand tighten the spindle nut. Loosen spindle nut until either hole in the spindle lines up with a slot in the nut (not more than ½ flat).
5. Install a new cotter pin. Bend the ends of the cotter pin against nut. Cut off extra length to ensure ends will not interfere with the dust cap.
6. Measure the looseness in the hub assembly. There will be from 0.001–0.005 in. (0.03–0.13mm) end play when properly adjusted.
7. Install dust cap on hub.

Wheel Bearings (Bolt On—Type Bearings)

The E and K Series have front and rear sealed wheel bearings. The bearings are pre-adjusted and require no lubrication maintenance or adjustment. There are darkened areas on the bearing assembly. These darkened areas are from a heat treatment process and do not require bearing replacement.

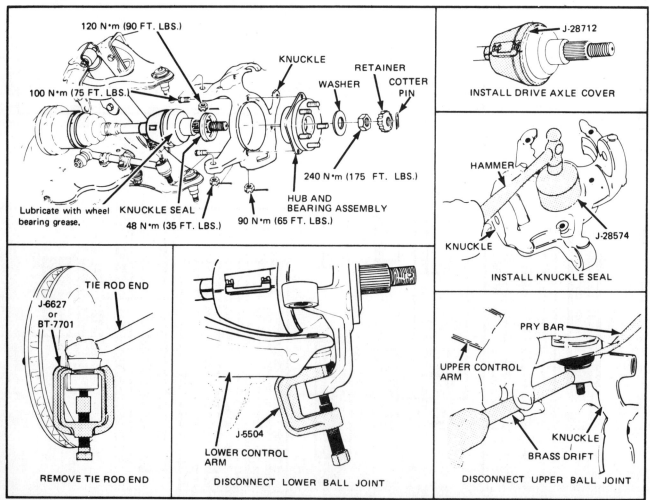

Hub and bearing assembly, knuckle and knuckle seal removal and installation. To remove: hoist car and remove wheel, remove disc, and remove parts as shown. To install: be sure that bearing surfaces are clean and free of burrs. Install parts as shown. Install disc, install wheel and lower hoist (© General Motors Corp.)

SUSPENSION
Domestic Models

REAR SUSPENSION—GENERAL MOTORS E AND K BODY CARS—FRONT WHEEL DRIVE

Description

The E and K Series have a semi-trailing arm-type rear suspension system with a relatively long control arm for a minimum camber change. The system consists of boxed control arms, coil springs, super-lift shock absorbers and a stabilizer bar.

The control arms are welded together. The hub and bearing attachment plane is machined for precise suspension alignment.

The hub and wheel bearing is a unit assembly which eliminates the need for wheel bearing adjustments and does not require periodic maintenance.

Operation

The left and right rear wheel suspensions, being independent of each other, permit the vertical movement of one wheel without affecting the wheel on the opposite side of the car.

This independent wheel movement is obtained by an A frame control arm. The control arm is hinged at the frame to provide the up and down movement of the wheel. The solid stabilizer bar forces the wheel to travel in through a controlled arc.

The control arm also carries the rear brake mounting bracket and hub and bearing assembly.

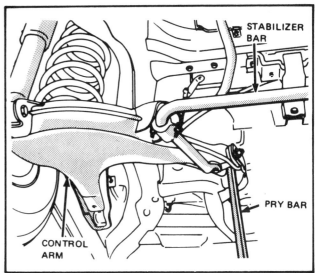

Rear wheel alignment diagnosis (© General Motors Corp.)

Alignment

Satisfactory operation may occur over a wide range of rear wheel alignment settings. Nevertheless, should settings vary beyond certain tolerances, readjustment of alignment is advisable. The specifications stated in column 1 of the charts should be used as guidelines.

These specifications provide an acceptable all-around operating range in that they prevent abnormal tire wire caused by wheel alignment.

In the event the actual settings are beyond the specifications set forth in column 1 of 2 (whichever is applicable), or whenever for other reasons the alignment is being reset, the factory recommends that the specifications given in column 3 of the charts be used.

Rear wheel alignment should be checked and adjusted as necessary in the following procedure.
1. Check front and rear trim heights.
2. Check electronic level control for proper operation.
3. Using an alignment machine, use one of the following procedures.

Preferred method:
  a. If machine does not have guide line, place tape on floor from wheel plate to rear of car to use as a guide for lining up car on machine.
  b. Back car onto alignment machine placing rear wheel on wheel plates.
  c. Place a straightedge at same rib of tire at front and rear and measure distance from inside edge of straight edge and edge guide line. The measurement from the guide line to the straight edge must be greater at the rear tire by $\frac{5}{8}$ in. (16mm) ± $\frac{1}{4}$ in. (6mm).

Alternate method:
  a. Place a one inch tape on the floor along the righthand side of the center line between the wheel plates of the alignment machine.
  b. Back car onto alignment machine making sure car is as straight as possible.

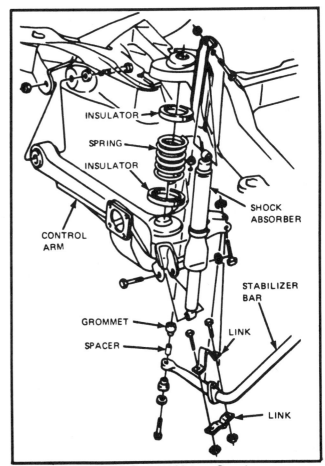

Rear suspension (© General Motors Corp.)

2-135

SECTION 2

SUSPENSION
Domestic Models

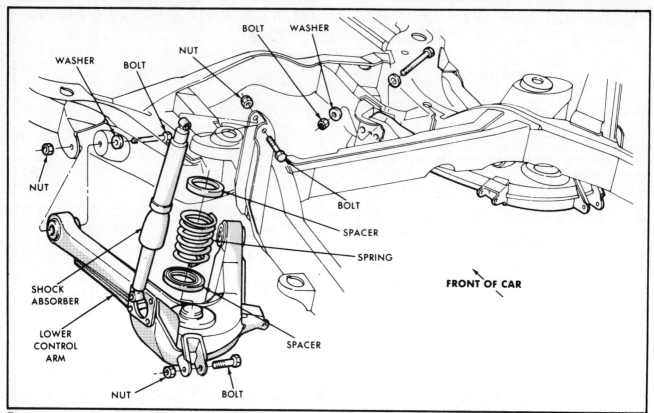

Rear suspension components (© General Motors Corp.)

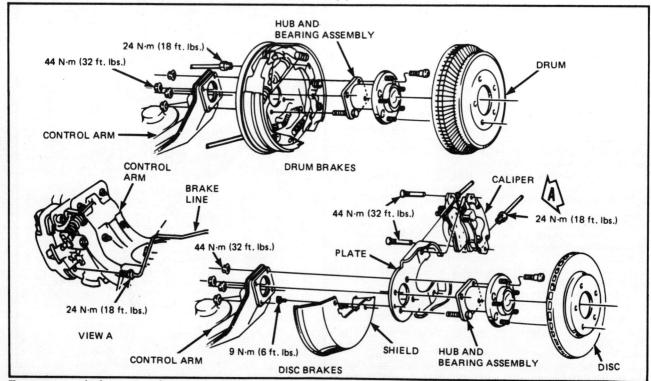

To remove: hoist car and remove wheel as shown. Remove parts as shown. To install: be sure that bearing surfaces are clean and free of burrs. Install parts as shown. If equipped with disc brakes, bleed brakes. Install wheel and lower hoist (© General Motors Corp.)

SUSPENSION
Domestic Models

REAR WHEEL ALIGNMENT DIAGNOSIS

PROBLEM	CAUSE	CORRECTION
Toe not adjustable within specifications	Lower control arm bent	Replace control arm
	Frame bent	Bring frame within specification
	Car not properly centered on alignment machine	Center car on alignment machine
	Bearing mounting flange bent	Replace bearing assembly
	Wheel bent	Replace wheel
Camber out of specification	Control arm bent	Replace control arm
	Frame bent	Bring frame within specifications
	Spindle-bearing	Properly mount

REAR WHEEL ALIGNMENT

	Specifications for Diagnosis for Warranty Repair or Customer Paid Service	Specifications for Periodic Motor Vehicle Inspection	Specification for Alignment Resetting
Camber (measure only)	$-1.3°$ to $+0.3°$	$-1.5°$ to $+0.5°$	Refer to Rear Suspension Diagnosis
Toe-in per wheel	$0.00°$ to $+0.30°$ ($0''$ to $+5/32''$)	$-0.25°$ to $+0.55°$ ($-1/8''$ to $+9/32''$)	$+0.15° \pm 0.15°$ ($+3/32'' \pm 3/32''$)
Toe-in both wheels	$0.00°$ to $+0.60°$ ($0''$ to $+5/16''$)	$-0.50°$ to $+1.10°$ ($-1/4''$ to $+9/16''$)	$+0.30° \pm 0.30°$ ($+5/32'' \pm 5/32''$)

NOTE: It is important that toe-in be measured per wheel. If equipment is not available to measure each wheel, measure toe-in both wheels. When resetting be sure that toe-in on each wheel is the same.

CAUTION: With car backed onto alignment machine, toe-in and toe-out are reversed. Toe-in will be read as toe-out. It is very important that the readings be made and understood properly.

TORQUE SPECIFICATIONS

	N·m	Ft.Lbs.
Stabilizer		
Stabilizer Link Nut	18	13
Stabilizer Bar Brkt. to Frame Bolts & Nuts	33	24
Shock Absorber		
Shock Absorber Upper Attaching Nut	130	95
Shock Absorber to Control Arm Bolts	100	75
Control Arms		
Upper Control Arm to Frame Attaching Nuts	95	70
Lower Control Arm to Frame Attaching Nuts	120	90

TORQUE SPECIFICATIONS

	N·m	Ft.Lbs.
Ball Joints		
Service Ball Joints to Upper Control Arm	11	8
Lower	90	65
Upper	120	90
Front Wheel Drive		
Drive Axle Nut	240	175
Hub and Bearing to Knuckle Bolts	100	75
Torsion Bar Crossmember Retainer Bolts	27	20
Drive Axle to Output Shaft Bolts	80	60
Tie Rod to Knuckle Nut	54	40

SECTION 2
SUSPENSION
Domestic Models

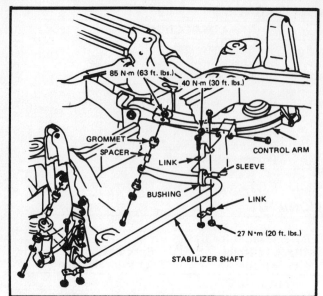

Stabilizer bar and/or bushing removal and installation. To remove: hoist vehicle and remove parts as shown. To: install: install parts as shown, and lower vehicle. (© General Motors Corp.)

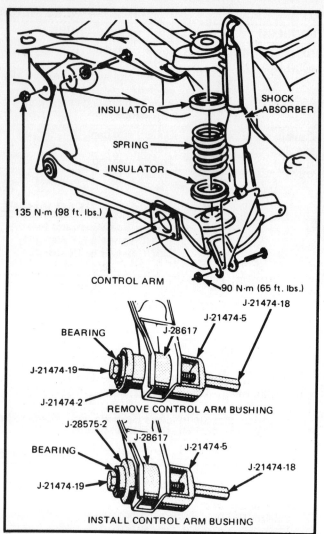

Rear control arm, spring and/or bushing removal and installation. To remove: hoist car and remove wheel. Remove hub and bearing assembly. Remove parts as shown. To install: install parts as shown. Install hub and bearing assembly. Bleed brakes. Install wheel and lower hoist (© General Motors Corp.)

  c. Hanging plumbs on the front and rear crossmembers at gage holes will give guide lines for near perfect centering of the car.
  4. Attach alignment mirrors to rear wheel and take toe and camber readings.

NOTE: With car backed onto alignment machine, toe-in and toe-out are reversed. Toe-in will be read as toe-out. It is very important that the readings be made and understood properly.

  5. Toe adjustments are made at the inner pivot bushings. Loosening the nut and bolt at the inner bushing will enable the toe to be moved in or out as necessary.
  6. Tighten bushing mounting nut to 97 ft. lbs. (135 Nm) and recheck toe for correct setting. It may be necessary to use a pry bar to move the control arm. Moving the control arm rearward increases toe-in; moving it forward increases toe-out.
  7. Check camber.

FRONT SUSPENSION – GENERAL MOTORS J BODY CARS – FRONT WHEEL DRIVE

Description

The front suspension is a MacPherson strut design. The lower control arms pivot from the lower side rails. Rubber bushings are used for the lower control arm pivots. The upper end of the strut is isolated by a rubber mount which contains a non-serviceable bearing for wheel turning. The tie rods connect to the steering arm on the strut, just below the spring seat. The lower end of the wheel steering knuckle pivots on a ball stud for wheel turning. The ball stud is riveted in the lower control arm and is fastened to the steering knuckle with a castellated nut and cotter pin.

Wheel Alignment
TOE

Toe is controlled by the tie rod position. To adjust toe setting, loosen the clamp bolts at the outer end of the tie rod. Rotate adjuster to align toe to specifications. Tighten bolts to 15 ft. lbs. (20 Nm).

2-138

SUSPENSION
Domestic Models

Adjustment

1. Loosen clamp bolts at the outer tie rod.
2. Square the vehicle.
3. Rotate adjuster to set toe to specifications.
4. Tighten clamp bolts.

CAMBER

In special circumstances when camber adjustment becomes necessary, refer to the following procedure for instructions on modifying the front suspension strut assembly.

Adjustment

1. Position the car on the alignment equipment. Follow the manufacturer's instructions to obtain the camber reading.
2. Use appropriate extensions to reach around both sides of the tire. Loosen both strut-to-knuckle bolts just enough to allow movement between the strut and the knuckle. Remove tools.
3. Grasp the top of the tire firmly, and move the tire inboard or outboard until the correct camber reading is obtained.
4. Carefully reach around the tire with extensions and tighten both bolts enough to hold the correct camber while the wheel and tire is removed to allow final torque.
5. With wheel and tire removed, torque both bolts to specifications.
6. Reinstall wheel and tire. Tighten nuts to specifications.

MacPherson Strut

Removal

1. Raise hood and disconnect upper strut-to-body nuts.
2. Hoist car, allowing front suspension to hang free.
3. Remove wheel and tire.
4. Install drive axle cover.
5. Disconnect tie rod from strut.
6. Remove both strut-to-knuckle bolts.
7. Remove strut.

Installation

1. Install strut by reversing removal Steps 1–6.
2. Place flats on both mounting bolts in a horizontal position.
3. Torque all fasteners to specifications.

STRUT MODIFICATION (ONLY FOR ADJUSTMENT OF CAMBER SETTING)

1. Place strut in vise. (It is not necessary to remove the strut from the car. Filing can be accomplished by disconnecting strut from knuckle).
2. File the holes in outer flanges to enlarge the bottom holes until they match the slots in the inner flanges.

STRUT DISASSEMBLY

1. Mount strut compressor in vise.
2. Place strut assembly in bottom adapter of compressor and install J26584-86 or equivalent (make sure adapter captures strut and that locating pins are engaged).
3. Rotate strut assembly to align top mounting assembly lip with strut compressor support notch.
4. Insert both J26584-88 top adapters between the top mounting assembly and the top spring seat. Position top adapters so that the split line is in the 9 o'clock–3 o'clock position.
5. Using a ratchet with 1 inch socket, turn compressor forcing screw clockwise until top support flange contact the J26584-88 top adapters. Continue turning the screw, compressing the strut spring approximately ½ inch (four complete turns).

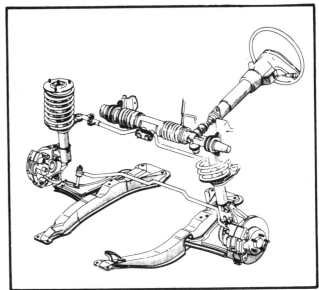

Front suspension (© General Motors Corp.)

NOTE: Never bottom spring or strut damper rod.

6. The top nut can now be removed from the strut damper shaft and the top mounting assembly (containing bearing) can be lifted off the strut assembly.
7. Turn strut compressor forcing screw counterclockwise until the strut spring tension is relieved. Remove top adapters, bottom adapter, then remove components.

STRUT ASSEMBLY

--- **CAUTION** ---

Never place a hard tool such as pliers or a screwdriver against the polished surface of the damper shaft. The shaft can be held up from the top end with your fingers, or with the extension, to prevent it from receding into the strut assembly, while the spring is being compressed.

1. Clamp strut compressor body J26584 in vise.
2. Place strut assembly in bottom adapter of compressor and install J26584–86 or equivalent (make sure adapter captures strut and locating pins are engaged).

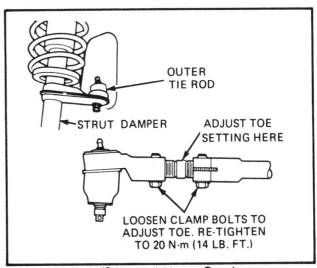

Toe adjustment (© General Motors Corp.)

SECTION 2
SUSPENSION
Domestic Models

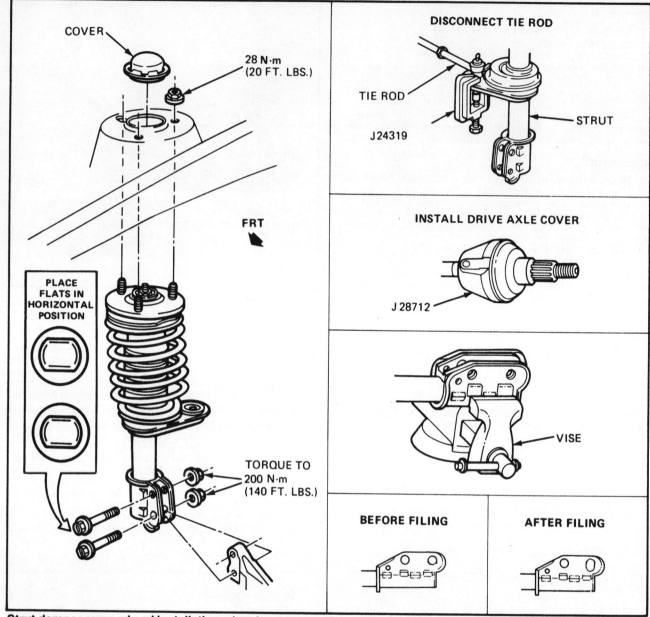

Strut damper removal and installation, showing modification procedure for camber adjustment
(© General Motors Corp.)

3. Rotate strut assembly until mounting flange is facing out, directly opposite the compressor forcing screw.
4. Position spring on strut making sure spring is properly seated on bottom spring plate.
5. Install strut spring seat assembly on top of spring.
6. Place both J26584-88 top adapters over spring seat assembly.
7. Turn compressor forcing screw until compressor top support just contacts top adapters (do not compress spring at this time).
8. Install a long extension with a socket to fit the hex on the damper shaft through the top spring seat. Use the extension to guide the components during reassembly.
9. Compress spring by turning screw clockwise until approximately 1 ½ inch of damper shaft extends through the top spring plate.

NOTE: Do not compress spring until it bottoms.

10. Remove extension and socket, position top mounting assembly over damper shaft and install nut.
11. Turn forcing screw counterclockwise to back off support, remove top adapters and bottom adapter, and remove strut assembly from compressor.

NOTE: Special tool J-26584 or equivalent must be used to disassemble and assemble strut damper, or damage could result.

SUSPENSION
Domestic Models

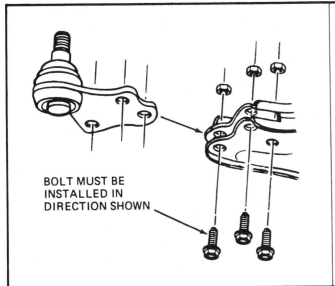
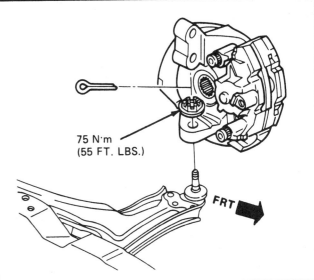

Install ball joint to control (© General Motors Corp.)

REPLACE STRUT CARTRIDGE

The internal piston rod, cylinder assembly, and fluid can be replaced utilizing a service cartridge and nut. Internal threads are located immediately below a cut line groove.

1. Mount strut in vise. Do not overclamp. Excessive clamping may damage tube and/or bracket.
2. Locate cut line groove. It is important to locate groove as accurately as possible because mislocation will result in thread damage. Cut a round groove with a pipe cutter until reservoir tube is completely cut through.
3. Remove and discard end cap, cylinder, and piston rod assembly. Remove strut from vise and discard fluid.
4. Reclamp strut in vise. A flaring cup tool is provided in service package to flare and debur cut edge of reservoir tube to accept service nut. Place flaring cup on open end of reservoir tube. Strike flaring cup with a mallet or hammer until flaring cup's flat outer surface rests on reservoir tube. Remove the flaring cup tool and discard. At this time, try nut to assure positive start and smooth threading into reservoir tube threads. Remove nut after this check. Flaring cup must be placed in contact with tube so there is not gap between cup and tube when struck.
5. Place strut cartridge in reservoir tube. Turn cartridge until it settles into indentations at base of tube so cartridge cannot be easily turned. Place nut over cartridge.
6. Using tool J-29778 or equivalent for 53 mm hex nut and a torque wrench, tighten to 140–170 ft. lbs. (190–230 Nm) in upright mounting position. Stroke the piston rod once or twice to check for proper operation.

Ball Joints

Inspection

Car must be supported by the wheels so weight of car will properly load the ball joints.

The lower ball joint is inspected for wear by visual observation alone. Wear is indicated by the protrusion of the ½ in. (12.7mm) diameter nipple into which the grease fitting is threaded. This round nipple projects 0.050 in. (1.27mm) beyond the surface of the ball joint cover on a new, unworn joint. Normal wear will result in the surface of this nipple retreating very slowly inward.

Removal

1. Raise the vehicle and support safely. Remove the wheel and tire assembly.
2. If no countersink is found on the lower side of the rivets, carefully locate the center of the rivet body and mark with a punch.
3. Use the proper sequence to drill out rivets.
4. Use tool J-29330 or equivalent to separate joint from knuckle.
5. Disconnect stabilizer from control arm.
6. Remove ball joint.

Installation

1. Install new ball joint using three bolts.
2. Reverse removal Steps 1–5 to install. Tighten all fasteners to specifications.
3. Check toe setting. Adjust as required.

Control Arm

Removal

1. Raise car and support safely. Remove wheel and tire.

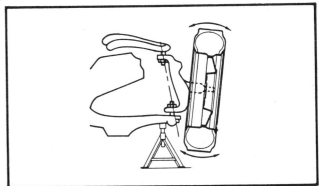

Support lower control arm as far outboard as possible. Position dial indicator to check movement at point shown. Rock wheel in and out at top and bottom (© General Motors Corp.)

2-141

SUSPENSION
Domestic Models

2. Disconnect stabilizer bar from control arm and/or support.
3. Separate knuckle from ball joint using tool J-29330 or equivalent.
4. Remove control arm/support.

Installation

1. Install control arm/support.
2. When installing support, install the center bolts first.
3. Install ball joint to knuckle.
4. Install wheel and tire.
5. Lower car. Check toe setting, adjust as required.

Hub and Bearing

Removal

NOTE: The car must not be moved while the driveshaft is out of the hub-and-bearing, nor until the hub nut is installed to final torque.

1. Loosen hub nut.
2. Raise car and support safely. Remove wheel and tire.
3. Install boot cover J-28712 or equivalent.
4. Remove hub nut.
5. Remove caliper and rotor.

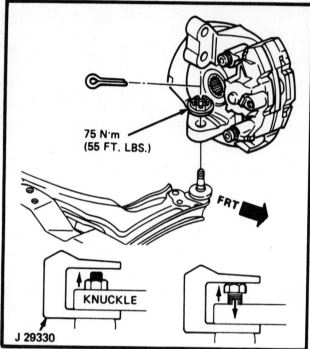

Place J29330 into position as shwon. Loosen nut and back off until the nut contacts the tool. Continue backing off the nut until the nut forces the ball stud out of the knuckle (© General Motors Corp.)

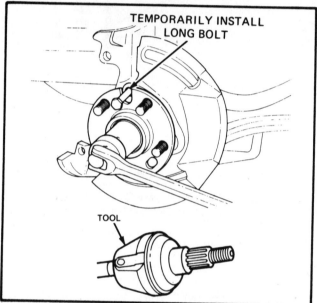

Installing drive axle cover (© General Motors Corp.)

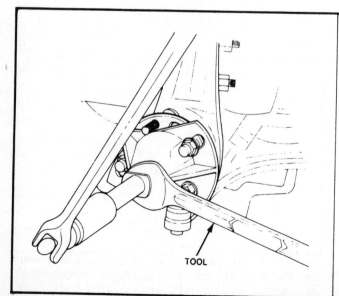

Hub and bearing assembly—removal and installation (© General Motors Corp.)

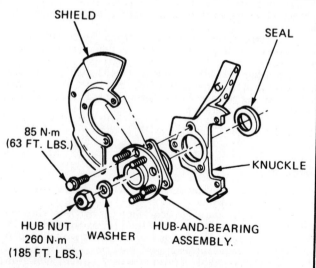

SUSPENSION
Domestic Models

6. Remove hub-and-bearing mounting bolts. Remove shield. If bearing assembly is to be re-used, mark attaching bolt and corresponding hole for installation in the same position.

7. Install tool J–28733 or equivalent, and turn bolt to press the hub-and-bearing assembly off of the drive shaft. If excessive corrosion is present, make sure the hub-and-bearing is loose in the knuckle before using the tool.

8. If installing a new bearing assembly, replace the steering knuckle seal, using tool J–22388.

Installation

1. Clean and inspect bearing mating surfaces and steering knuckle bore for dirt, nicks and burrs.

2. If installing knuckle seal, apply grease to seal and to bore of knuckle.

3. Replace parts in reverse order or removal. When attaching hub-and-bearing mounting bolts, use one long bolt to extend through cut-out. This will serve as a reaction point to allow enough torque on hub nut to seat axle shaft into bearing. After tightening hub nut to 70 ft. lbs. (100 Nm), remove long bolt and replace with normal bolt.

4. Lower car. Apply final torque to hub nut, 185 ft. lbs. (260 Nm).

Steering Knuckle

Removal

1. Raise vehicle and support safely. Remove wheel and tire.
2. Remove front wheel hub-and-bearing.
3. Disconnect ball joint from knuckle, using tool J–29330 or equivalent.

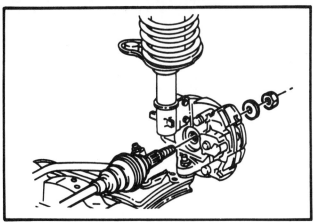

Removing axle stub from knuckle assembly (© General Motors Corp.)

4. Remove both strut-to-knuckle mounting bolts. Remove steering knuckle.

Installation

1. Install both strut-to-knuckle mounting bolts loosely.
2. Install knuckle to ball joint. Torque ball joint nut to 55 ft. lbs. (75 Nm). Install cotter pin.
3. Tighten mounting bolts to 140 ft. lbs. (200 Nm).
4. Install remaining components.

REAR SUSPENSION – GENERAL MOTORS J BODY CARS – FRONT WHEEL DRIVE

Description

This vehicle has a semi-independent rear suspension which consists of an axle with trailing arms and twisting cross beam, two coil springs, two shock absorbers, two upper spring insulators, and two spring compression bumpers. The axle assembly attaches to the underbody through a rubber bushing located at the front of each control arm. the brackets are integral with the underbody side rails. The axle structure itself maintains the geometrical relationship of the wheels relative to the body. A serviceable stabilizer bar is available as an option. It is attached to the inside of the axle beam and to the lower surface of the control arms as a subassembly of the axle.

The two coil springs support the weight of the car in the rear. Each spring is retained between a seat in the underbody and a seat welded to the top of the control arm. A rubber cushion is used to isolate the coil spring upper end from the underbody seat, while the lower end sits on a combination compression bumper and spring insulator.

The double-acting rear shock absorbers are filled with a calibrated amount of fluid, and sealed during production. They are non-adjustable, non-refillable, and cannot be disassembled. The only service the shock absorbers require is replacement if they have lost their resistance, are damaged, or are leaking fluid.

The lower ends of the shock absorbers are attached to the axle assembly, with bolts and paddle nuts. The upper ends are attached to the body in the wheelhouse area with conventional insulators, washers and nuts.

A single unit hub-and-bearing assembly is bolted to both ends of the rear axle assembly. This hub-and-bearing assembly is a sealed unit. The bearing is not replaceable as a separate unit.

Shock Absorber

Removal

---— **CAUTION** ———

Do not remove both shock absorbers at one time as suspending rear axle at full length could result in damage to brake lines and hoses.

1. Open deck lid, remove trim cover and remove upper shock attaching nut. Remove one shock at a time when both shocks are being replaced.

2. Raise vehicle on hoist and support rear axle assembly. When lifting vehicle with body hoist it will be necessary to support rear axle with adjustable jack stands. When lifting vehicle with suspension hoist care should be taken to align axle on the hoist prior to lifting.

3. Remove lower attaching bolt and nut and remove shock.

Installation

1. Install shock absorbers at lower attachment, feed bolt through holes and loosely install paddle nut.
2. Lower vehicle enough to guide upper stud through body opening and install nut loosely.
3. Torque lower bolt to 41 ft. lbs. (55 Nm).
4. Remove axle support and lower car all the way and torque upper nut. Torque to 13 ft. lbs. (17 Nm).

SECTION 2: SUSPENSION
Domestic Models

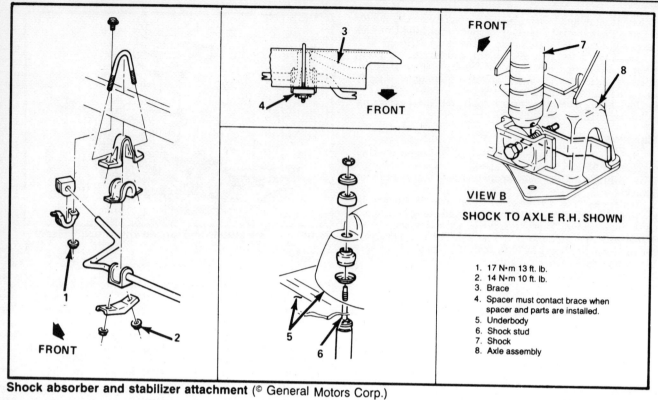

Shock absorber and stabilizer attachment (© General Motors Corp.)

5. Replace rear trim cover.

Stabilizer Bar

Removal

1. Raise vehicle on hoist and support body with jack stands.
2. Remove nuts and bolts at both the axle and control arm attachments and remove bracket, insulator and stabilizer bar.

Installation

1. Install U-bolts, upper clamp, spacer and insulator and trailing axle. Position stabilizer bar in insulators and loosely install lower clamp and nuts.
2. Attach the end of stabilizer bar to control arms and torque all nuts to 13 ft. lbs. (17 Nm).
3. Torque axle attaching nut to 10 ft. lbs. (14 Nm).
4. Lower vehicle and remove from hoist.

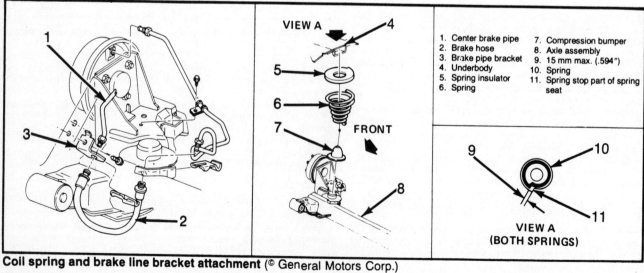

Coil spring and brake line bracket attachment (© General Motors Corp.)

2-144

SUSPENSION
Domestic Models

Springs and Insulators

Removal

CAUTION

When removing rear springs do not use a twin-post hoist. The swing arc tendency of the rear axle assembly when certain fasteners are removed may cause it to slip from the hoist. Perform operation on floor if necessary.

1. Raise vehicle using frame contact type hoist if possible and support rear control arms with jack stands. If necessary to lift vehicle with twin post hoist, lift by tires and support the control arms or body with jack stands.
2. Remove wheel and tire assembly.
3. Remove right and left brake line bracket attaching screws from body and allow brake line to hang free.
4. Remove right and left shock absorber lower attaching bolts.

CAUTION

Do not suspend rear axle by brake hoses. Damage to hoses could result.

5. Lower rear axle and remove spring(s) and/or insulators(s).

Installation

1. Position springs and insulators in seats and raise axle. The ends of the upper coil on the spring must be positioned in the seat of the body. Prior to installing spring it will be necessary to install upper insulators to the body with adhesive to keep it in position while raising the axle assembly and springs.
2. Connect shocks to rear axle and torque bolts to 41 ft. lbs. (55 Nm). It will be necessary to bring the axle assembly to standing height prior to torquing bolts on the shocks.
3. Install brake line brackets to body and torque screws to 8 ft. lbs. (11 Nm).
4. Install wheel and tire assembly. Torque lug nuts to 102 ft. lbs. (140 Nm).
5. Remove jack stands and lower vehicle.

Control Arm Bushing

Removal

1. Raise vehicle on hoist.
2. Remove wheel and tire assembly and support body with jack stands.
3. If removing right bushing, disconnect brake line from body. If left bushing is being removed, disconnect brake line bracket from body, and parking brake cable from hook guide on the body. Replace only one bushing at a time.
4. Remove nut, bolt and washer from the control arm and bracket attachment, and rotate control arm downward.
5. Remove bushing as follows:
   a. Install J29376-1 or equivalent on control arm over bushing and tighten attaching nuts until tool is securely in place.
   b. Install J21474-19 bolt through plate J29376-7 and install into J29376-1 receiver.
   c. Place J29376-6 remover into position on bushing and install nut J21474-18 onto J21474-19 bolt.
   d. Remove bushing from control arm by turning bolt.

Installation

1. Install bushing on bolt and position onto housing. Align bushing installer J29376-4 arrow with arrow on receiver for proper indexing of bushing.
2. Install nut J21474-18 onto bolt J21474-19.
3. Press bushing into control arm by turning bolt. When bushing is in proper position the end flange will be flush against the face of the control arm.
4. Use a screw type jack stand to position control arm into bracket and install bolt and nut. Do not torque bolt at this time. It will be necessary to torque the bolt of the control arm with vehicle at standing height.
5. Install brake line bracket to frame and torque screw to 8 ft. lbs. (11 Nm).
6. If left side was disconnected, reconnect brake cables to bracket, and reinstall brake cable to hook. Adjust cable as necessary.
7. While supporting vehicle at standing height, tighten control arm bolt to 67 ft. lbs. (90 Nm).
8. Remove jack stands and install wheel assembly and lower vehicle from hoist.

Hub and Bearing

Removal

1. Raise vehicle on hoist.
2. Remove wheel and tire assembly and brake drum.

CAUTION

Do not hammer on brake drum as damage to the assembly could result.

3. Remove hub and bearing assembly-to-rear axle attaching bolts and remove hub and bearing assembly from axle. The top rear attaching bolt will not clear the brake shoe when removing the hub and bearing assembly. Partially remove hub and bearing assembly prior to removing this bolt.

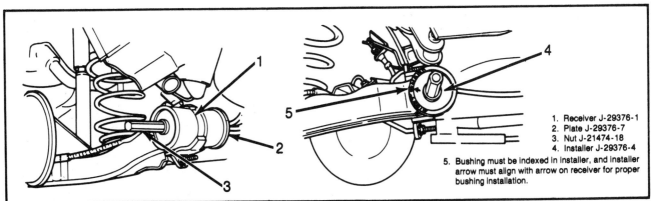

1. Receiver J-29376-1
2. Plate J-29376-7
3. Nut J-21474-18
4. Installer J-29376-4
5. Bushing must be indexed in installer, and installer arrow must align with arrow on receiver for proper bushing installation.

Control arm bushing Installation (© General Motors Corp.)

SECTION 2
SUSPENSION
Domestic Models

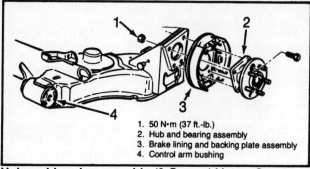

1. 50 N·m (37 ft.-lb.)
2. Hub and bearing assembly
3. Brake lining and backing plate assembly
4. Control arm bushing

Hub and bearing assmbly (© General Motors Corp.)

--- CAUTION ---
Do not hammer on brake drum as damage to the bearing could result.

4. Remove shock absorber lower attaching bolts and paddle nuts at axle and disconnect shocks from control arm.
5. Disconnect parking brake cable from the axle assembly.
6. To insure that axle assembly is not suspended by brake lines, disconnect brake line at the brackets from axle assembly.
7. Lower rear axle and remove coil springs and insulators.
8. Remove control arm bolts from underbody bracket and lower axle.
9. Remove hub attaching bolts and remove hub, bearing and packing plate assembly. Be careful not to drop hub and bearing assembly, as damage to the bearing could result.

Installation

1. Install backing plate, and hub-and-bearing assembly to rear axle. Hold nuts and torque bolts to 39 ft. lbs. (52 Nm).
2. Install stabilizer bar, if so equipped, by attaching nut and bolts to axle assembly and at the end to the control arms.
3. Place axle assembly on transmission jack and raise into position. Attach control arms to underbody bracket with bolts and nuts. Do not torque bolts at this time. It will be necessary to torque the bolt of the control arm at standing height.
4. Install brake line connections to axle assembly.
5. Attach brake cable to rear axle assembly.
6. Position coil springs and insulators in seats and raise rear axle. The end of upper coil on the springs must be parallel to axle assembly and seated in pocket.
7. Install shock absorber lower attachment bolts and paddle nuts to rear axle, torque bolt to 41 ft. lbs. (55 Nm).
8. Connect parking brake cable to guide hook and adjust as necessary.
9. Install brake drums, wheels and tire assembly. Torque lug nuts to 103 ft. lbs. (140 Nm).
10. Bleed brake system and refill reservoir.

Installation

1. Position top rear attaching bolt in hub-and-bearing assembly prior to the installation in the axle assembly.
2. Install remaining bolts and nuts. Torque bolts to 39 ft. lbs. (52 Nm).
3. Install brake drum, and wheel and tire assembly. Torque lug nuts to 103 ft. lbs. (140 Nm).
4. Lower vehicle and remove from hoist.

Rear Axle Assembly

Removal

1. Raise vehicle on hoist and support assembly with jack stands under the control arms.
2. Remove stabilizer bar from axle assembly, if so equipped.
3. Remove wheel and tire assembly and brake drum.

FRONT SUSPENSION – GENERAL MOTORS – P BODY CARS

Wheel Alignment

Caster Adjustment

Caster angle can be changed with a realignment of washers located between the legs of the upper control arm. For adjustment, a kit containing two washers, one of 3 mm thickness and one of 9 mm thickness, must be used. Install as shown to adjust caster. See Upper Control Arm Removal and Installation.

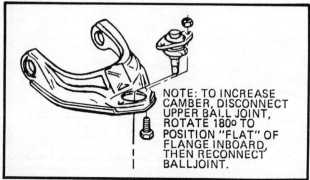

Upper ball joint/camber adjustment (© General Motors Corp.)

Camber Adjustment

Camber angle can be increased approximately 1° by removing the upper ball joint, rotating it one-half turn, and reinstalling it with the flat of the upper flange on the inboard side of the control arm.

Toe–In Adjustment

Toe-in is the turning in of the wheels. The actual amount of toe-in is normally only a fraction of a degree. The purpose of toe specifications is to ensure parallel rolling of the rear wheels. (Excessive toe-in or toe-out may increase tire wear). Toe-in also serves to offset the small deflections of the wheel support system which occurs when the car is rolling forward. In other words, even when the wheels are set slightly to toe-in when the car is standing still, they tend to roll parallel on the road when the car is moving.

Suspension Description

The front suspension system uses conventional long and short arm design and coil springs. the control arms attach to the vehicle with bolts and bushings at the inner pivot points, and to the steering knuckle/front wheel spindle assembly at the outer pivot points. Lower ball joints use the "wear indicator" feature.

SUSPENSION
Domestic Models

Front Wheel Bearings

NOTE: Tapered roller bearings are used on all series vehicles and they have a slightly loose feel when properly adjusted. A design feature of front wheel tapered roller bearings is that they must never be preloaded. Damage can result from preloading.

Adjustment

1. Raise vehicle and support safely.
2. Remove wheel.
3. Remove dust cap from hub.
4. Remove cotter pin from spindle and spindle nut.
5. Tighten the spindle nut to 12 ft. lbs. (16 Nm) while turning the wheel assembly forward by hand to fully seat the bearings. This will remove any grease or burrs which could cause excessive wheel bearing play later.
6. Back off the nut to the "just loose" position.
7. Hand tighten the spindle nut. Loosen spindle nut until either hole in the spindle lines up with a slot in the nut. (Not more than 1/2 flat.)
8. Install new cotter pin. Bend the ends of the cotter pin against nut, cut off extra length to ensure ends will not interfere with the dust cap.
9. Measure the looseness in the hub assembly. There will be from 0.001–0.005 in. (0.025–0.127 mm) end play when properly adjusted.
10. Install dust cap on hub.
11. Replace the wheel cover or hub cap.
12. Lower vehicle to floor.
13. Perform the same operation for each front wheel.

Ball Joints

Upper Ball Joint—Removal

1. Raise the vehicle on a hoist.
2. Remove the tire and wheel assembly.
3. Support the lower control arm with a floor jack.
4. Remove upper ball stud nut, then reinstall nut finger tight.
5. Install Tool J–26407 or equivalent with the cup end over the lower ball stud nut.
6. Turn the threaded end of tool until upper ball stud is free of steering knuckle.
7. Remove Tool and remove nut from ball stud.
8. Remove two nuts and bolts attaching ball joint to upper control arm. Note which way the flat of the ball joint is pointing before removing it. The direction of this flat on the ball

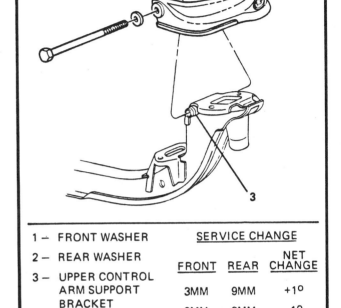

1 – FRONT WASHER	SERVICE CHANGE		
2 – REAR WASHER	FRONT	REAR	NET CHANGE
3 – UPPER CONTROL ARM SUPPORT BRACKET	3MM	9MM	+1°
	9MM	3MM	–1°

Caster adjustment (© General Motors Corp.)

joint flange should be in the same direction as the one removed unless a change in camber is desired.

9. Remove ball joint.

Upper Ball Joint—Installation

Inspect the tapered hole in the steering knuckle. Remove any dirt and if any out-of-roundness, deformation, or damage is noted, the knuckle MUST be replaced.

1. Install bolt and nuts attaching ball joint to upper control arm and torque to 28 ft. lbs. (39 Nm), then mate the upper control arm ball stud to the steering knuckle.
2. Install the ball stud nut and torque to 35 ft. lbs. (47 Nm). Then turn 1/6 of a turn to align cotter pin.
3. Install cotter pin.
4. Install the tire and wheel assembly.
5. Lower the vehicle to the floor.
6. Set toe.

Lower Ball Joint

Removal and Installation

The lower ball joint is welded to the lower control arm and cannot be serviced separately. Replacement of the entire lower control arm will be necessary if the lower ball joint requires replacement. See lower control arm removal.

Front Spring/Lower Control Arm

Removal

1. Raise vehicle on a hoist and support to vehicle on the crossmember.
2. Remove wheel and tire assembly.
3. Disconnect stabilizer bar from the lower control arm.
4. Disconnect the tie rod from the steering knuckle.

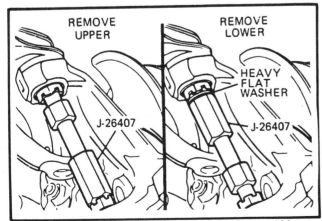

Removal of ball joints from knuckle (© General Motors Corp.)

SECTION 2

SUSPENSION
Domestic Models

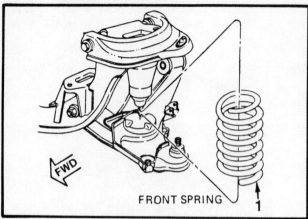

Front coil spring positioning (© General Motors Corp.)

5. Disconnect the shock absorber at the lower control arm.
6. Support the lower control arm with a jack.
7. Remove the nut from the lower ball joint, then use tool J-26407 or equivalent to press the ball joint out of the knuckle.
8. Swing the knuckle and hub out of the way.
9. Loosen the lower control arm pivot bolts.
10. Install a chain through the coil spring as a safety precaution.

CAUTION

The coil spring is under load and could result in personal injury if it were released too quickly. Be sure to install a chain and to slowly lower the jack.

---

11. Slowly lower the jack and remove the spring.
12. Remove the pivot bolts at the chassis and the crossmember and remove the lower control arm.

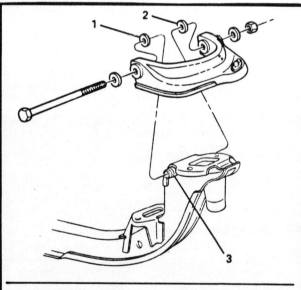

		SERVICE CHANGE		
1 — FRONT WASHER				
2 — REAR WASHER		FRONT	REAR	NET CHANGE
3 — UPPER CONTROL ARM SUPPORT BRACKET		3MM	9MM	+1°
		9MM	3MM	−1°

Front control arm shim arrangement (© General Motors Corp.)

13. Removal of the pivot bolt at the crossmember may require the loosening or removal of the steering assembly mounting bolts.

Installation

1. Install the lower control arm and pivot bolts at crossmember and body. Tighten slightly but do not torque.
2. Position the spring and install the spring into the upper pocket. Align spring bottom to lower control arm pocket.
3. Install spring lower end onto lower control arm. It may be necessary to have an assistant help you compress the spring far enough to slide it over the raised area of the lower control arm seat.
4. Use a jack to raise the lower control arm and compress the coil springs.
5. Install the ball joint through the lower control arm and into the steering knuckle. Install nut to ball joint stud and torque to 55 ft. lbs. (75 Nm). Install a new cotter pin.
6. Connect the stabilizer bar and torque the bolt to 16 ft. lbs. (22 Nm).
7. Connect the tie rod and torque to 29 ft. lbs. (39 Nm).
8. Install the shock absorber to the lower control arm and torque the bolt to 35 ft. lbs. (47 Nm).
9. If the bolts were removed or loosened at the steering assembly replace with new bolts and torque to 21 ft. lbs. (29 Nm).
10. With the suspension system in its normal standing height, torque the lower control arm to body bolt at 62 ft. lbs. (85 Nm) and the lower control arm to crossmember nut at 52 ft. lbs. (70 Nm).
11. Check and set alignment as necessary.

Upper Control Arm

Removal

1. Raise vehicle on a hoist.
2. Remove the tire and wheel assembly.
3. Remove rivet holding brake line clip to upper control arm.
4. Support the lower control arm with a floor jack.
5. Remove upper ball joint from steering knuckle, as described earlier.
6. Remove control arm pivot bolt and remove control arm from vehicle.
7. Transfer ball joint if not damaged or worn.

Installation

Washers and shims must be reinstalled as removed unless a change in geometry is desired.

1. Install upper control arm and pivot bolt to vehicle. The inner pivot bolt must be installed with the bolt head toward the front.
2. Install the pivot bolt nut.
3. Position the control arm in a horizontal plane and torque the nut to 66 ft. lbs. (90 Nm).

NOTE: Bolt may turn when torqued to minimum if nut is not backed up with a wrench. This does not mean the joint is loose.

4. Install ball joint to upper control arm and to steering knuckle, as described earlier. Install nut, torque to 35 ft. lbs. (47 Nm). Install a new cotter pin.
5. Install wheel and tire.
6. Lower vehicle to floor.

Steering Knuckle

Removal

1. Raise vehicle on a hoist and support the lower control arm with a jackstand.

SUSPENSION
Domestic Models

SECTION 2

CAUTION

This keeps the coil spring compressed. Use care to support adequately, or personal injury could result.

---

2. Remove the tire and wheel assembly.
3. Remove the disc brake caliper. Secure the caliper to the suspension using wire. Do not allow the caliper to hang by the brake hose. Insert a piece of wood between the shoes to hold the piston in the caliper bore. (The block of wood should be about the same thickness as the brake disc.).
4. Remove the hub and disc.
5. Remove the splash shield.
6. Remove both ball stud nuts (See Ball Joint Removal).
7. Remove the tie rod end from the steering knuckle.
8. Using Tool J–26407 or equivalent press the upper ball stud from the steering knuckle.
9. Reverse Tool to the other ball stud and press lower ball stud from the steering knuckle.
10. Remove ball stud nuts and remove the steering knuckle.

Installation

1. Place steering knuckle in position and insert the upper and lower ball studs into knuckle bosses.
2. Install ball stud nuts as tightened to specifications. For L.C.A., torque to 55 ft. lbs. (75 Nm). For U.C.A., torque to 35 ft. lbs. (47 Nm). Install new cotter pins.
3. Install splash shield to the steering knuckle. Torque to 7 ft. lbs. (10 Nm).
4. Install the tire rod end to the steering knuckle. Torque to 29 ft. lbs. (39 Nm), and install cotter pin.
5. Repack the wheel bearings, follow the Procedure as outlined above. Then install the hub and disc, bearings and nut. Torque to specifications as outlined above.
6. Install the brake caliper.
7. Install the tire wheel assembly.
8. Remove the jackstand and lower the vehicle to the floor.

REAR SUSPENSION – GENERAL MOTORS P BODY CARS

Suspension Description

The rear suspension is a MacPherson Strut design. This combination strut and shock adapts to the rear wheel drive. The lower control arms pivot from the engine cradle. The cradle has isolation mounts to the body and conventional rubber bushings are used for the lower control arm pivots. The upper end of the strut is isolated by a rubber mount.

Rear Wheel Camber

Adjustment

1. Position the vehicle on alignment equipment, and follow the manufacturers instructions to obtain a camber reading.
2. Use appropriate sockets and extensions to reach around both sides of the tire and LOOSEN both strut-to-knuckle bolts enough to allow movement between the strut and the knuckle. Remove the tools.
3. Grasp the top of the tire firmly, and move it inboard or outboard until the correct camber is obtained.
4. Again reach around the tire, as in Step 2, and tighten both bolts to 140 ft. lbs. (190 Nm).
5. If the accessibility to the bolts prevents applying complete torque, it will be necessary to apply only PARTIAL torque (just enough to hold the correct camber position), then to remove the wheel-and-tire in order to apply FINAL torque. After complete torquing, install the wheel-and-tire.
6. Repeat on other side.

Whenever adjusting casting, it is important to always use two washers totaling 12 mm thickness, with one washer at each end of locating tube.

Rear Wheel Bearings

Removal

STEEL WHEEL

1. Raise vehicle and remove wheel and tire.
2. Install drive axle boot protectors tool J–33162 or equivalent.
3. Remove hub nut, and discard.
4. Remove caliper and rotor.
5. Remove hub-and-bearing attaching bolts. If bearing assembly is being reused, mark attaching bolt and corresponding holes for installation.

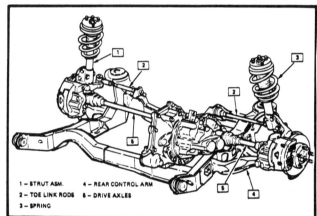

1 – STRUT ASM. 4 – REAR CONTROL ARM
2 – TOE LINK RODS 5 – DRIVE AXLES
3 – SPRING

MacPherson strut design rear suspension (© General Motors Corp.)

6. Install J–28733 or equivalent, and remove hub and bearing assembly. If excessive corrosion is present make sure hub-and-bearing is loose in knuckle before using tool.
7. If installing new bearing, replace knuckle seal. Car must be moved without hub nut installed to proper torque.

ALUMINUM WHEEL

1. Set parking brake.
2. Raise vehicle and support safely.
3. Remove wheel and tire assembly.
4. Remove hub nut.
5. Refer to steel wheel removal Step 2 through Step 7.

Installation

1. Clean and inspect bearing mating surfaces and knuckle bore for dirt, nicks and burrs.
2. If installing knuckle seal, use tool J–28671 or equivalent, apply grease to seal & knuckle bore.
3. Push hub-and-bearing on axle shaft.
4. Install parts as shown.
5. Apply PARTIAL torque to new hub nut, until hub-and-bearing assembly is seated 74 ft. lbs. (100 Nm).
6. Install rotor and caliper.

2–149

SECTION 2
SUSPENSION
Domestic Models

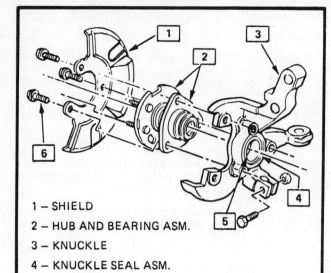

1 – SHIELD
2 – HUB AND BEARING ASM.
3 – KNUCKLE
4 – KNUCKLE SEAL ASM.
5 – FILL HUB BEARING CAVITY BETWEEN SEALING LIPS WITH .8 GRAMS OF CHASSIS LUBRICANT.
6 – BOLT 75-95 N·m (55-70 FT. LB.)

Rear wheel bearing arrangement (© General Motors Corp.)

7. Lower car.
8. Apply FINAL torque to hub nut. (200 ft. lbs. (270 Nm).

Rear Strut Damper Assembly

Removal

1. Remove the engine compartment lid.
2. Raise vehicle and support at rear control arm.
3. Remove upper strut nuts and washers.
4. Remove brake line clip.
5. Scribe mark strut and knuckle to assure proper assembly.
6. Remove strut mounting bolts and remove strut and spacer plate.

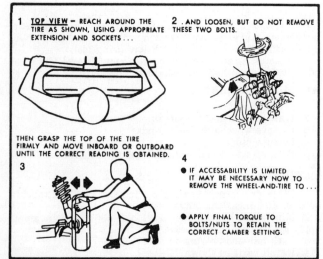

Rear wheel camber adjustment (© General Motors Corp.)

Installation

1. Set strut assembly and spacer plate into position and install mounting bolts and nuts.
2. Align scribe marks on strut and knuckle. Replace bolts in the same order in which they were removed.
3. Tighten knuckle nuts to 140 ft. lbs. (190 Nm).
4. Install brake line clip.
5. Lower vehicle and install upper strut nuts and washers. Tighten upper strut nuts to 18 ft. lbs. (24 Nm).
6. Install engine compartment lid.

Rear Strut Assembly/Spring Replacement

Disassembly

NOTE: Special tool J-26584 must be used to disassemble and assembly strut damper. Care must be used not to damage the special coating on the coil springs, or damage could occur to the coils.

1. Clamp J-26584 Strut Compressor in vise.
2. Place strut assembly in bottom adapter of compressor and install J-26584-89 (make sure adapter captures the strut and locating pins are engaged).
3. Rotate strut assembly to align top mounting assembly lip with strut compressor support notch.
4. Insert J-26584-430 top adapter on the top spring seal. Position top adapters so that the long stud is at high location to strut flange.
5. Using a ratchet with 1 in. socket, turn compressor forcing screw clockwise until top support flange contracts the J-26584-430 top adapter. Continue turning the screw compressing the strut spring.

SCRIBING PROCEDURE

1. USING A SHARP TOOL, SCRIBE THE KNUCKLE ALONG THE LOWER OUTBOARD STRUT RADIUS, AS IN VIEW A.
2. SCRIBE THE STRUT FLANGE ON THE INBOARD SIDE, ALONG THE CURVE OF THE KNUCKLE, AS IN VIEW B.
3. MAKE A CHISEL MARK ACROSS THE STRUT/KNUCKLE INTERFACE, AS IN VIEW C.
4. ON REASSEMBLY, CAREFULLY MATCH THE MARKS TO THE COMPONENTS.

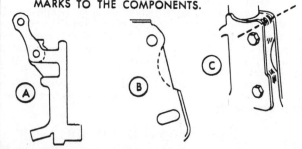

Strut and knuckle scribe marks (© General Motors Corp.)

6. Place J-26584-430 top adapter over spring seat assembly.
7. Turn strut compressor forcing screw counterclockwise until the strut spring tension is relieved. Remove top adapters, bottom adapter, then remove strut.

SUSPENSION
Domestic Models

Assembly

1. Clamp strut compressor body J-26584 in vise.
2. Place strut assembly in bottom adapter of compressor and install J-26584-89 (make sure adapter captures strut and locating pins are engaged).
3. Rotate strut assembly until mounting flange is facing out, directly opposite the compressor forcing screw.
4. Position spring and components on strut, as shown below. Make sure spring is properly seated on bottom spring plate.
5. Install strut spring seat assembly on top of spring. The long stud must be 180° from strut mounting flange.
6. Place J-26584-403 top adapter over spring seat assembly.
7. Turn compressor forcing screw until compressor top support just contacts top adapters (do not compress spring at ths time).
8. Install J-26584-27 Strut Alignment Rod through top spring seat and thread rod onto damper shaft, hand tight.
9. Compress spring by turning screw clockwise until enough of the damper shaft is exposed to where the nut can be threaded securely, and thread nut on damper shaft. DO NOT COMPRESS SPRING UNTIL IT BOTTOMS.

NOTE: Be sure that the damper shaft comes through the CENTER of the spring seat opening, or damage could occur.

10. Remove alignment rod, position strut mount over damper shaft and spring seat studs. Install washer and nut.
11. Turn forcing screw counterclockwise to back off support and remove strut assembly from compressor.

Control Arm Ball Joint

Removal

1. Raise car and remove wheel.
2. Remove clamp bolt from lower control arm ball stud.
3. Disconnect the ball joint from the knuckle. It may be necessary to tap the stud with a mallet.
4. Remove the rivets as shown.

Installation

1. Install ball joint to lower control arm.
2. Position knuckle over ball stud, to allow the clamp bolt to be installed. Torque to specs.
3. Install wheel and lower car.
4. Check toe-in setting. Adjust as required.

INSPECTION
BALL JOINT SEALS

Ball joint seals should be carefully inspected for cuts and tears. Whenever cuts or tears are found, the ball joint MUST be replaced.

KNUCKLE ASSEMBLY

Inspect the hole in the knuckle assembly clamp area. Remove any dirt. If out-of-roundness, deformation, or damage is noted, the knuckle MUST be replaced.

Lower Control Arm and/or Bushings

Removal

1. Raise car and remove wheel.
2. Remove ball joint clamping bolt.
3. Separate knuckle from ball joint.
4. Remove lower control arm pivot bolts at frame.
5. Remove control arm.

Installation

1. Install parts in reverse order of removal.
2. Install wheel; lower car.
3. Check toe-in and camber settings. Adjust as required.

Rear Knuckle

Removal

1. Refer to rear wheel bearing removal.
2. Remove toe-link rod at knuckle.
3. Remove clamp bolt. Disconnect knuckle from ball stud.

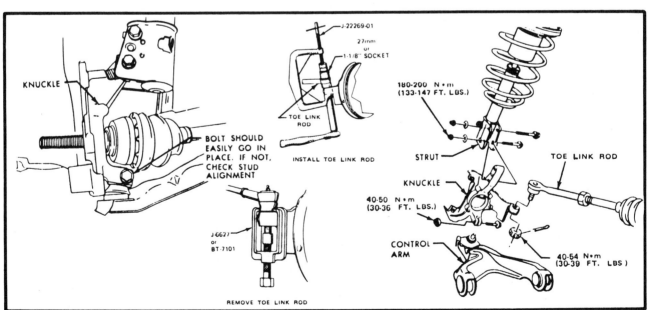

Remove/Install rear knuckle assembly (© General Motors Corp.)

SECTION 2
SUSPENSION
Domestic Models

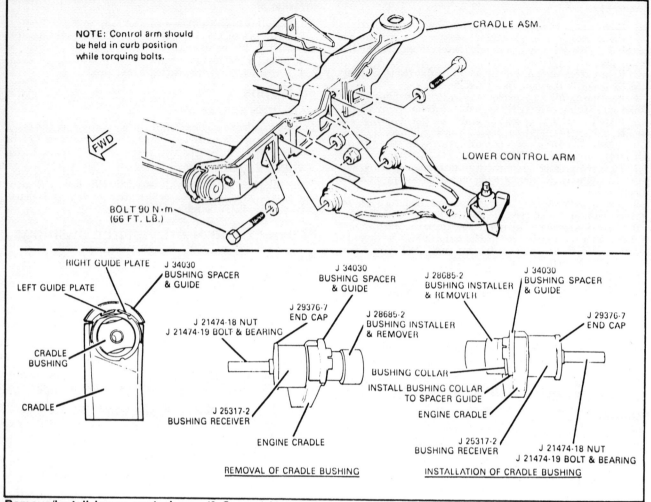

Remove/install lower control arm (© General Motors Corp.)

4. Remove bolt through bolts holding strut-to-knuckle. Remove knuckle.

NOTE: Whenever separating the ball joint from the knuckle, be careful not to cut or tear the ball joint seal, or damage to the ball joint could occur. If the seal is cut or torn, the ball joint MUST be replaced.

Installation

1. Install knuckle to ball joint.
2. Loosely install knuckle to strut.
3. Install toe-link rod to knuckle.
4. Refer to wheel bearing installation.
5. Set camber and toe to specifications.

FRONT SUSPENSION—S BODY (NOVA)

Description

The front suspension is a MacPherson strut independent suspension. The upper end of the strut is anchored to the vehicle body by a strut support. The strut and strut support are isolated by a rubber mount. The lower end of the strut is connected to the upper end of the steering knuckle. The lower end of the knuckle is attached to a ball joint, which is attached to the suspension control arm.

Movement of the steering wheel is transmitted to the tie rod end and then to the knuckle, which in turn moves the wheel and tire.

Front Wheel Toe Alignment

Toe is adjusted by changing the tie rod length. Loosen the boot clamps and slide out of the working area. Loosen the left and right tie rod end locknuts. Turn the left and right tie rods to specification.

NOTE: In this adjustment, the left and right tie rods must be equal in length.

After adjustment is completed, reinstall the boot clamps, tighten the nuts and check to insure that the rack boots are not twisted.

SUSPENSION
Domestic Models

Front Camber And Caster Adjustment

Camber is adjusted by loosening the upper and lower strut to knuckle bolts and nuts, then rotate the cam to the correct specification. After adjustment is completed, tighten the nuts and bolts to specification.

Caster is not adjustable. If the caster is found to be out of specification, inspect the vehicle for loose, bent or otherwise worn suspension components. Replace as necessary.

NOTE: In order to prevent an incorrect reading of camber or caster, jounce the vehicle three times before inspection.

FRONT STRUT ASSEMBLY

Removal and Installation

1. Disconnect the negative battery cable. Remove the strut to body attaching nuts.
2. Raise the vehicle and support safely. Remove the wheel and tire assembly. Loosen the axle shaft nut if the knuckle is to be removed.
3. Remove the brake hose clip at the strut bracket. Disconnect the brake flex hose at the brake pipe. Remove the brake hose clips.
4. Pull the brake hose through the opening in the strut bracket. Tape the end of the brake hose to prevent dirt contamination.
5. Remove the two brake caliper mounting bracket bolts and remove the caliper. Support the caliper so it will not hang by the brake hose. Do not disconnect the brake hose from the caliper.
6. Mark the adjusting cam and remove both strut to knuckle attaching bolts.
7. Remove the strut assembly from the vehicle. Remove the camber adjusting cam from the knuckle.
8. Inspect components for signs of wear, cracks or distortion. Replace as necessary.
9. Installation is the reverse of the removal procedure. Before installing any removed dust covers, pack all areas with grease. Bleed the brake system. Check wheel alignment and correct as required.

CONTROL ARM

Removal and Installation

1. Disconnect the negative battery cable. Raise the vehicle and support safely.
2. Remove the lower control arm to steering knuckle attaching bolts and nuts. Remove the control arm and inspect the arm and bushing for distortion or cracking. Repair or replace as needed.
3. To remove the rear lower control arm bushing, remove the nut retainer and bushing. Torque the bushing nut to 76 ft. lbs.
4. Installation is the reverse of the removal procedure. Always replace self locking nuts with new ones. Check wheel alignment and adjust if necessary.

LOWER BALL JOINT

Removal and Installation

1. Disconnect the negative battery cable. Raise the vehicle and support safely. Remove the wheel and tire assembly.
2. Using tool J-35413 or equivalent, separate the ball joint from the knuckle.
3. Remove the two nuts and bolts attaching the ball joint and control arm.

Front toe adjustment-Nova

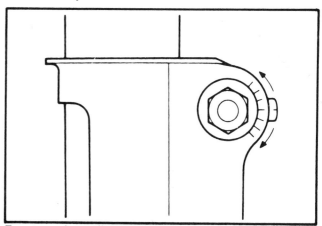
Front camber adjustment-Nova

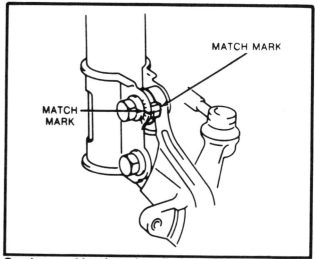
Camber marking for reinstallation

4. Remove the ball joint. Inspect the ball joint for excessive wear or damage to the boot seal.
5. Installation is the reverse of the removal procedure. Torque the knuckle to ball joint nut to 82 ft. lbs. (111 Nm). Torque the control arm attaching nuts to 47 ft. lbs. (64 Nm).

SECTION 2: SUSPENSION
Domestic Models

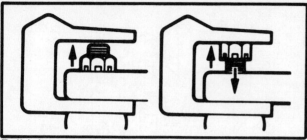
Ball joint removal

NOTE: Do not remove the hub assembly from the knuckle unless it is absolutely necessary. Should removal become necessary, the grease seals must be replaced with new ones.

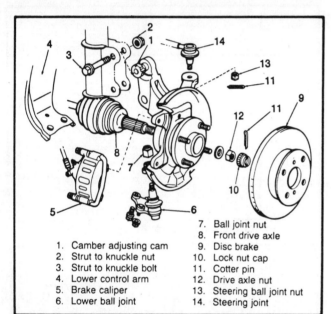

1. Camber adjusting cam
2. Strut to knuckle nut
3. Strut to knuckle bolt
4. Lower control arm
5. Brake caliper
6. Lower ball joint
7. Ball joint nut
8. Front drive axle
9. Disc brake
10. Lock nut cap
11. Cotter pin
12. Drive axle nut
13. Steering ball joint nut
14. Steering joint

Hub/knuckle-exploded view

— **CAUTION** —
Never reuse a self locking nut. Always replace a self locking nut with a new one.

KNUCKLE ASSEMBLY

Removal

1. Disconnect the negative battery cable. Raise the vehicle and support safely. Remove the tire and wheel assembly.
2. Remove the brake hose retaining clip at the strut. Disconnect the flex hose from the brake pipe.
3. Remove the caliper bracket to knuckle mounting bolts. Support the caliper. Remove the disc.
4. Remove the drive axle nut.
5. Using tool J-25287 or equivalent, push out the drive axle.
6. Remove the cotter pin. Remove the tie rod to knuckle attaching nut. Separate the tie rod using tool J-24319-01 or equivalent.
7. Remove the ball joint to control arm attaching nuts and bolts.
8. Matchmark the camber relationship and remove the strut to knuckle attaching bolts and nuts. Remove the knuckle.

Installation

1. Install the lower ball joint to the lower control arm. Torque the nuts and bolts to 47 ft. lbs. (64 Nm).
2. Install the ball joint to knuckle. Torque the nut to 14 ft. lbs. (20 Nm) and remove the nut. Install a NEW nut and torque to 82 ft. lbs. (111 Nm).
3. Install the camber adjusting cam to knuckle. Connect the steering knuckle to the strut lower bracket. Insert the bolts from the rear to the front and align the matchmarks of the camber adjusting cam and strut. Torque to 105 ft. lbs. (142 Nm).
4. Install the tie rod to knuckle attaching nuts. Torque to 38 ft. lbs. (56 Nm). Install the cotter pin.
5. Torque the lower control arm to ball joint to 47 ft. lbs. (64 Nm). Install the brake caliper. Torque to 65 ft. lbs. (88 Nm).
6. Install the brake hose connector to the brake pipe. Install the wheel and tire assembly. Lower vehicle.
7. Apply grease to the axle shaft threads. Install the attaching nut. Torque to 137 ft. lbs. (186 Nm). Install the cotter pin and cap.
8. Bleed the brake system. Check wheel alignment and adjust if necessary. Always install a new grease seal and self locking nut.

REAR SUSPENSION – P BODY (NOVA)

Rear Wheel Toe Alignment

Toe is adjusted by rotating the cam located on the rear lower control arm. Loosen the bolt and rotate the nut to get the correct specification.

1. Measure the distance and find the difference between the left and right disc wheels and the centerline of the adjustment cam. If the distance is not within specification, inspect each component for damage or excessive wear.
2. Turn each cam an equal amount in the opposite direction. The toe will change approximately 2 mm for each turn of the cam.

Rear Camber and Caster Adjustment

Caster and camber cannot be adjusted. Should the camber setting be found to be out of adjustment, inspect the suspension parts for damage or wear.

NOTE: In order to prevent incorrect readings of the camber setting, jounce the vehicle three times before inspection.

REAR STRUT ASSEMBLY

Removal and Installation

1. Disconnect the negative battery cable.
2. Working inside the vehicle, remove the shock absorber cover and package tray bracket.
3. Raise the vehicle and support safely. Remove the rear wheel and tire assemblies.
4. Disconnect the brake line at the wheel cylinder and plug it.

SUSPENSION
Domestic Models

5. Disconnect the flexible hose from the shock absorber.
6. Remove the retaining bolts holding the strut to the axle carrier. Disconnect the strut.
7. Remove the three upper strut mounting nuts and remove the strut assembly.
8. Installation is the reverse of the removal procedure. Bleed the brake system.

SUSPENSION ARM (REAR)

Removal and Installation

1. Raise the vehicle and support safely.
2. Remove the bolt and nut holding the rear suspension arm to the axle carrier.
3. Remove the cam and bolt holding the rear suspension arm to the body. Remove the suspension arm.
4. Installation is the reverse of the removal procedure.

NOTE: Remember where the cam plate mark is before disassembly.

SUSPENSION ARM (FRONT)

Removal and Installation

1. Raise the vehicle and support safely.
2. Remove the bolt and nut holding the front suspension arm to the axle carrier.
3. Remove the bolt and nut holding the front suspension arm to the body. Remove the front suspension arm.
4. Installation is the reverse of the removal procedure. Lower vehicle and check rear wheel alignment.

STRUT ROD

Removal and Installation

1. Raise vehicle and support safely.
2. Remove the strut rod to axle carrier retaining nut.
3. Remove the strut rod to body retaining nut. Remove the strut rod.
4. Installation is the reverse of the removal procedure.

REAR AXLE HUB

Removal and Installation

1. Raise the vehicle and support safely.
2. Remove the tire and wheel assembly.
3. Remove the brake drum and wire out of the way.
4. Remove the four bolts which hold the axle hub assembly to the axle carrier.
5. Remove the axle/hub bearing assembly. Remove the "O" ring.
6. Installation is the reverse of the removal procedure.

REAR AXLE CARRIER

Removal and Installation

1. Raise the vehicle and support safely.
2. Remove the rear axle hub as outlined above.
3. Disconnect the brake line from the wheel cylinder and plug the line.

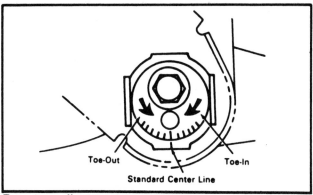

Rear toe adjustment

4. Remove the axle carrier to strut rod retaining nut. Remove the axle carrier to front suspension and rear suspension arm retaining nuts.
5. Remove the bolts and nuts holding the axle carrier to the strut. Remove the axle carrier.
6. Installation is the reverse of the removal procedure. Bleed the brake system and check rear wheel alignment. Adjust as necessary.

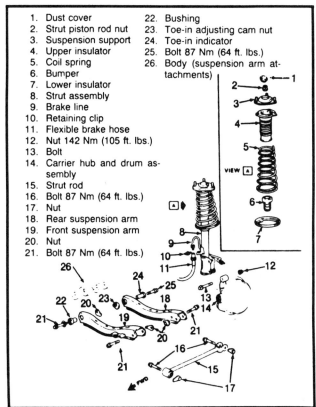

Rear suspension-exploded view

1. Dust cover
2. Strut piston rod nut
3. Suspension support
4. Upper insulator
5. Coil spring
6. Bumper
7. Lower insulator
8. Strut assembly
9. Brake line
10. Retaining clip
11. Flexible brake hose
12. Nut 142 Nm (105 ft. lbs.)
13. Bolt
14. Carrier hub and drum assembly
15. Strut rod
16. Bolt 87 Nm (64 ft. lbs.)
17. Nut
18. Rear suspension arm
19. Front suspension arm
20. Nut
21. Bolt 87 Nm (64 ft. lbs.)
22. Bushing
23. Toe-in adjusting cam nut
24. Toe-in indicator
25. Bolt 87 Nm (64 ft. lbs.)
26. Body (suspension arm attachments)

SECTION 2 SUSPENSION
Domestic Models

FRONT SUSPENSION—GENERAL MOTORS REAR DRIVE CARS— EXCEPT F, T, P BODY AND 1984 AND LATER CORVETTE

Wheel Alignment

Front wheel alignment factors are caster, camber, toe-in, toe-out, and trim height. Before any corrections are made, the car must be on a level surface with a full gas tank and the front seat to the rear. All doors must be closed with no passengers or excess weight in the car.

CASTER AND CAMBER ADJUSTMENTS

To adjust caster and camber, loosen the upper control arm shaft-to-frame nuts, add or subtract shims as required, and retorque nuts.

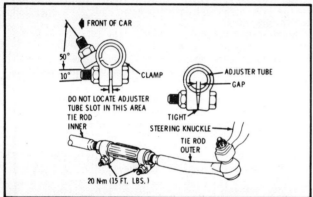

Bolts must be installed in direction shown. Rotate both inner and outer tie rod housings rearward to the limit of ball joint travel before tightening clamps. With this same rearward rotation all bolt centerlines must be between angles shown after tightening clamps. Clamp ends may touch when nut is torqued to specifications, but gap must be visible adjacent to adjuster sleeve. Clamp must be between and clear of dimples. Torque nuts to specifications (© General Motors Corp.)

A normal shim pack will leave at least two threads of the bolt exposed beyond the nut. The difference between front and rear shim packs must not exceed 40 inches.

If these requirements cannot be met in order to reach specifications, check for damaged control arms and related parts. Always tighten the nut on the thinner shim pack first, for improved shaft-to-frame clamping force and torque retention.

TOE—IN ADJUSTMENT

Toe-in can be increased or decreased by changing the length of the tie-rods. A threaded sleeve is provided for this purpose.

When the tie-rods are mounted ahead of the steering knuckle, they must be decreased in length in order to increase toe-in.

Loosen the clamp bolts at each end of the steering tie rod adjustable sleeves. With steering wheel set in straight ahead position, turn tie rod adjusting sleeves to obtain proper toe-in adjustment.

NOTE: Before locking clamp bolts on the rods, make sure that the tie rod ends are in alignment with their ball studs by rotating both tie rod ends in the same direction as far as they will go. Then tighten adjuster tube clamps to specified torque. Make certain that adjuster tubes and clamps are positioned correctly.

TOE—OUT

Toe-out on turns refers to the difference in angles between the front wheels and the car frame during turns. Toe-out on turns is non-adjustable.

TRIM HEIGHT ADJUSTMENT

When checking trim height, the car should be parked on a level surface, full tank of gas, front seat rearward, doors closed and the tire pressure as specified.

If there is more than 1 in. (24 mm) difference side to side at the wheel well opening, corrective measures should be taken to make the car level.
1. Check tire sizes.
2. Check tire wear.

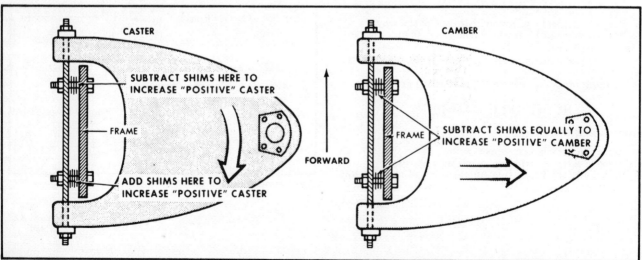

Caster and camber adjustment (© General Motors Corp.)

SUSPENSION
Domestic Models

3. Check coil spring height.
4. Check for worn suspension parts.

Wheel Bearings

For normal use, clean and repack front wheel bearings with a high melting point wheel bearing lubricant at each front brake lining replacement or 30,000 miles (48,000 km), whichever comes first. For heavy duty application such as police cars and taxi cabs, clean and repack front wheel bearings at each front brake lining replacement or 15,000 miles (24,000 km) whichever comes first.

"Long fiber" or "viscous" type lubricants should not be used. Do not mix wheel bearing lubricants. Be sure to thoroughly clean bearings and hubs of all old lubricant before repacking.

NOTE: Tapered roller bearings used in these cars have a slightly loose feel when properly adjusted. They must never be over-tightened (pre-loaded) or severe bearing damage may result.

Adjustment

The proper functioning of the front suspension cannot be maintained unless the front wheel taper roller bearings are correctly adjusted. Cones must be a slip fit on the spindle and the inside diameter of cones should be lubricated to insure that the cones will creep. Spindle nut must be a free running fit on threads.

1. Remove dust cap from hub.
2. Remove cotter pin from spindle and spindle nut.
3. Tighten the spindle nut to 12 ft. lbs. (16 Nm) while turning the wheel assembly forward by hand to fully seat the bearings. This will remove any grease or burrs which could cause excessive wheel bearing play later.
4. Back off the nut to the "just loose" position.
5. Hand-tighten the spindle nut. Loosen spindle nut until either hole in the spindle lines up with a slot in the nut (not more than $\frac{1}{2}$ flat).
6. Install new cotter pin. Bend the ends of the cotter pin against nut. Cut off extra length to ensure ends will not interfere with the dust cap.
7. Measure the looseness in the hub assembly. There will be from 0.001–0.005 in. (0.03–0.13mm) end-play when properly adjusted.
8. Install dust cap on hub.

Shock Absorbers

Removal

1. Raise car on hoist and with an open end wrench hold the shock absorber upper stem from turning. Remove the upper stem retaining nut, retainer and rubber grommet.
2. Remove the two bolts retaining the lower shock absorber pivot to the lower control arm and pull the shock absorber assembly out from the bottom.

Installation

1. With the lower retainer and rubber grommet in place over the stem, install the shock absorber (fully extended) up through the lower control arm and spring so that the upper stem passes through the mounting hole in the upper control arm frame bracket.
2. Install the upper rubber grommet, retainer and attaching nut over the shock absorber upper stem.
3. With an open end wrench, hold the upper stem from turning and tighten the retaining nut.
4. Install the retainers attaching the shock absorber lower pivot to the lower control arm, torque and lower car to floor.

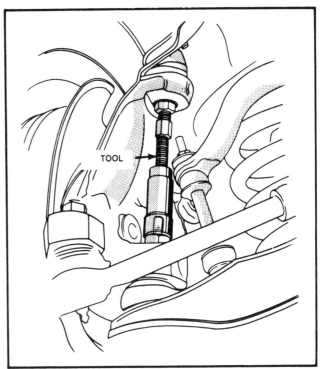

Disconnecting upper ball joint (© General Motors Corp.)

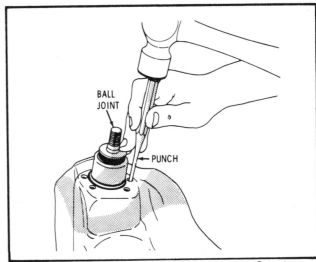

Upper ball joint removal (© General Motors Corp.)

Upper Ball Joint

Inspection

1. Raise the car and position floor stands under the left and right lower control arm as near as possible to each lower ball joint. Car must be stable and should not rock on the floor stands. Upper control arm bumper must not contact frame.
2. Position dial indicator against the wheel rim.
3. Grasp front wheel and push in on bottom of tire while pulling out at the top. Read gauge, then reverse the push-pull

2-157

SUSPENSION
Domestic Models

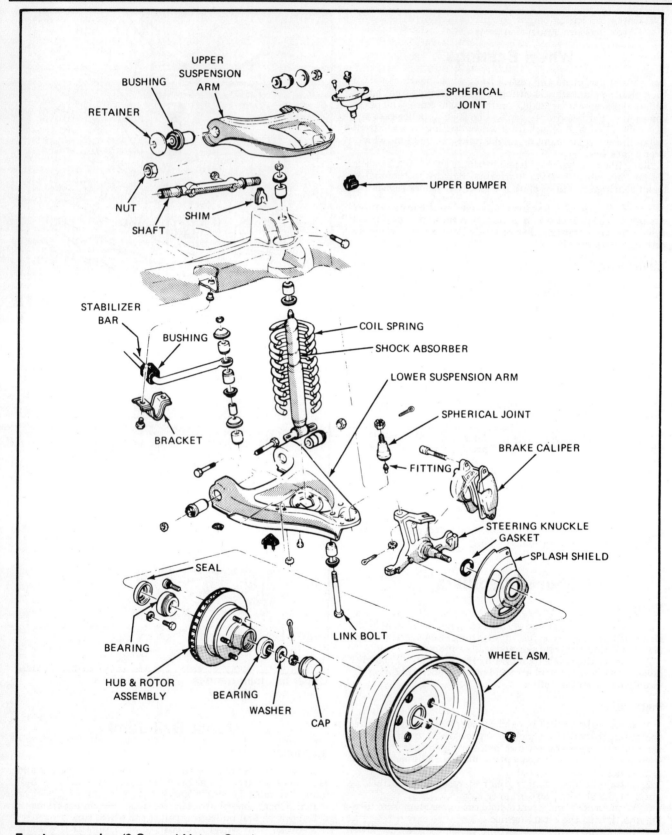

Front suspension (© General Motors Corp.)

SUSPENSION
Domestic Models

procedure. Horizontal deflection on dial indicator should not exceed 1.25 in. (3.18 mm).

4. If dial indicator reading exceeds 0.125 in. (3.18 mm), or if ball stud has been disconnected from knuckle assembly and any looseness is detected, or the stud can be twisted in its socket with your fingers, replace ball joint.

Removal

1. Raise front of car and support lower control arm with floor stands.

— CAUTION —
Floor jack or stand must remain under control arm spring seat during removal and installation to retain spring and control arm in position.

Since the weight of the car is used to relieve spring tension on the upper control arm, the floor stands must be positioned between the spring seats and ball joints of the lower control arms for maximum leverage.

2. Remove wheel, then loosen the upper ball joint from the steering knuckle as follows:
   a. Remove upper ball joint nut and install push tool.
   b. Apply pressure on stud by expanding the tool until the stud breaks loose.
3. Remove tool and upper ball joint nut, then pull stud from knuckle. Support the knuckle assembly to prevent weight of the assembly from damaging the brake hose.
4. With control arm in the raised position, drill four rivets $1/4$ in. deep using a $1/8$ in. diameter drill.
5. Drill off rivet heads using a $1/2$ in. diameter drill.
6. Punch out rivets using a small punch, and remove ball joint.

Installation

1. Position new ball joint in control arm and install four attaching bolts. Torque nuts to 8 ft. lbs. (11 Nm).
2. Connect ball joint to steering knuckle. Torque nut to 30 ft. lbs. (40 Nm).

Lower Ball Joint

Inspection

Car must be supported by the wheel so weight of car will properly load the ball joints.

The lower ball joint is inspected for wear by visual observation alone. Wear is indicated by protrusion of the $1/2$ in. (12.7 mm) diameter nipple into which the grease fitting is threaded. This round nipple projects 0.050 in. (1.27 mm) beyond the surface of the ball joint cover on a new, unworn joint. Normal wear will result in the surface of this nipple retreating slowly inward.

To inspect for wear, wipe grease fitting and nipple free of dirt and grease as for a grease job. Observe or scrape a scale, screwdriver or fingernail across the cover. If the round nipple is flush or inside the cover surface, replace the ball joint.

Removal

1. Raise the car, support with floor stands under frame.
2. Remove tire and wheel assembly.
3. Place floor jack under control arm spring seat.

— CAUTION —
Floor jack must remain under control arm spring seat during removal and installation to retain spring and control arm in position.

4. To disconnect the lower control arm ball joint from the steering knuckle. Remove the cotter pin from ball joint stud and remove stud nut. Tool J-8806 can be used to break the ball joint loose from knuckle after stud breaks loose.

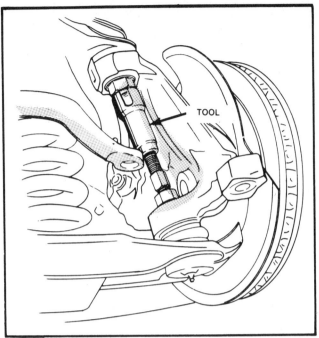

Disconnecting lower ball joint (© General Motors Corp.)

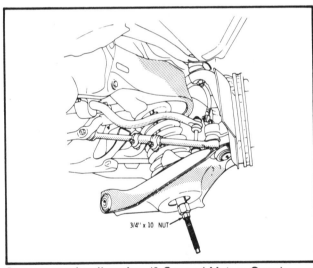

Compressed coil spring (© General Motors Corp.)

5. Guide lower control arm out of opening in splash shield with a putty knife or similar tool.
6. Block knuckle assembly out of the way by placing a wooden block between frame and upper control arm.
7. Remove ball joint seal by prying off retainer with a pry bar or driving off with a chisel.
8. Remove grease fittings and install special tool to remove lower ball joint from lower control arm.

Installation

1. Position the lower ball joint into the lower control arm and with special press tools, press the ball joint in until it bottoms in the control arm.

NOTE: *The grease purge on the seal must be located facing inboard.*

2-159

SUSPENSION
Domestic Models

2. Place the ball joint stud into the bottom hole of the steering knuckle. Force stud into tapered hole to a torque of 40 ft. lbs. (55 Nm).

3. Install the stud nut and torque to 90 ft. lbs. (120 Nm).

NOTE: Some replacement ball joints will use cotter pins. Be sure they are in place when required.

4. Install the grease fitting and lubricate the ball joint until the grease appears at the seal.

5. Install the wheel and lower the vehicle.

Coil Spring

Removal

1. Place transmission in neutral so steering wheel is unlocked.
2. Clean shock upper threads; oil, then remove nut, washer, and grommet.
3. Hoist car. Remove wheel and tire.
4. Remove stabilizer link nut, grommets washers, and bolt.
5. Support car with floor stands.
6. Remove shock.
7. Install coil spring tool. Make sure tool is fully seated into lower control arm spring seat.
8. Rotate nut until spring is compressed enough so that it is free in its seat.
9. Remove the two lower control arm pivot bolts and disengage lower control arm from frame. Rotate arm with spring rearward and remove spring from arm.

Installation

1. Install spring on bench.
2. Insert compressed spring into place.
3. Twist spring into proper position.

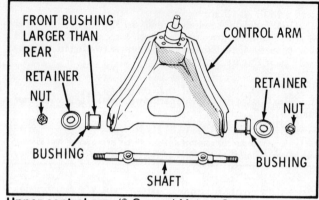

Upper control arm (© General Motors Corp.)

4. Carefully lift lower control arm and attach the lower control arm pivot bolts. Tighten nuts to 90 ft. lbs. (120 Nm).

Upper Control Arm

Removal

1. Raise front of car and support lower control arm with floor stands.

--- CAUTION ---

Floor jack must remain under control arm spring seat during removal and installation to retain spring and control arm in position.

NOTE: Since the weight of the car is used to relieve spring tension on the upper control arm, the floor stands must be positioned between the spring seats and ball joints of the lower control arms for maximum leverage.

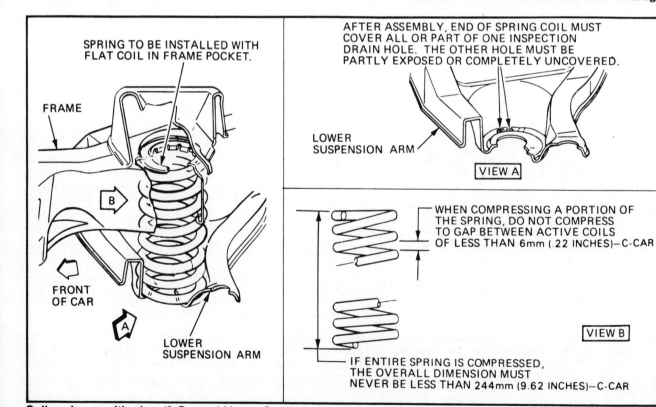

Coil spring positioning (© General Motors Corp.)

SUSPENSION
Domestic Models

2. Remove wheel, then loosen the upper ball joint from the steering knuckle as follows:
   a. Remove the upper ball joint nut.
   b. Apply pressure on stud by expanding the tool until the stud breaks loose.
3. Remove tool and upper ball joint nut, then pull stud free from knuckle. Support the knuckle assembly to prevent weight of the assembly from damaging the brake hose.
4. Remove the upper control arm attaching bolts to allow clearance to remove upper control arm assembly.
5. Remove upper control arm from the car.

PIVOT SHAFT BUSHING REPLACEMENT
1. Remove upper control arm assembly from the car.
2. Remove nuts from ends of pivot shaft.
3. Position control arm assembly and tools and push bushing out of control arm.
4. To install bushings, place pivot shaft in control arm and push new bushing into control arm and over end of pivot.

Installation
1. Position the upper control arm attaching bolts loosely in the frame and install the pivot shaft on the attaching bolts.
2. Install the inner pivot bolt with the heads to the front on the front bushing and to the rear on the rear bushing.
3. Install the alignment shims between the pivot shaft and frame on their respective bolts. Torque the nuts to 73 ft. lbs. on the B models and 45 ft. lbs. on the rear drive A body models.
4. Remove the temporary support from the hub assembly and connect the ball joint to the steering knuckle.
5. Install the wheel, check front end alignment and adjust as required.

Lower Control Arm

Removal
1. Place transmission in neutral so steering wheel is unlocked.
2. Clean chock upper threads; oil, then remove nut, bolt, washer and grommet.
3. Hoist car. Remove wheel and tire.
4. Remove stabilizer link nut, grommets washers, and bolt.
5. Support car with floor stands and lower hoist. Remove shock.
6. Loosen the lower ball joint nut and use tool J–8806 or equivalent. Apply pressure on stud by expanding the tool until the stud breaks loose.
7. Install spring compressor in through front spring. Compress spring until all tension is off lower control arm.
8. Remove pivot bolts and ball joint.
9. Remove complete control arm.

Installation
1. Install the ball joint stud into the lower hole of the steering knuckle. Install the nut loosely.
2. Position the spring on the lower arm and safety chain it to the lower arm to prevent personal injury.
3. Position a jack under the rear of the arm and raise the arm rear pivot bushings into position on the crossmember brackets. Install the pivot bolts.
4. Remove the jack and safety chain. Lower the vehicle to the ground and torque the bolts as noted in the front spring removal and installation section.

LOWER CONTROL ARM BUSHINGS

Removal and Installation
The removal and installation of the lower control arm bushings require the use of a press tool. A flare is found on the bushings

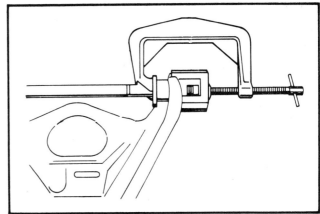
Upper control arm bushing removal (© General Motors Corp.)

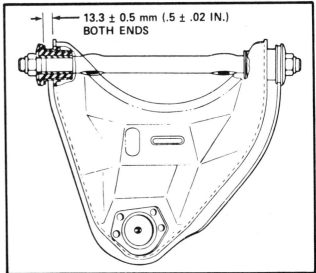
Upper control arm pivot shaft bushing installation clearance (© General Motors Corp.)

used except with the rear wheel drive A models. When the bushings are installed and the flare is required, the tool must be capable of flaring the metal flange on the bushing to its proper angle.

Steering Knuckle

It is recommended that the car be raised and supported so that the front coil spring remains compressed, yet the wheel and steering knuckle assembly remain accessible. If a frame hoist is used, support the lower suspension arm with an adjustable jack stand to retain spring in the curb height position.

Removal
1. Raise car on hoist and support lower suspension arm.
2. Remove wheel and tire assembly.
3. Remove tie-rod end from steering knuckle.
4. Remove brake caliper and hub and rotor assembly. Use a piece of wire to attach caliper to upper suspension arm.

NOTE: Never allow caliper to hang from brake hose, as hose may be damaged.

SUSPENSION
Domestic Models

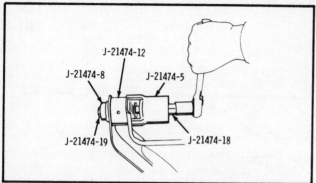

Bushing removal. Press out old bushing (© General Motors Corp.)

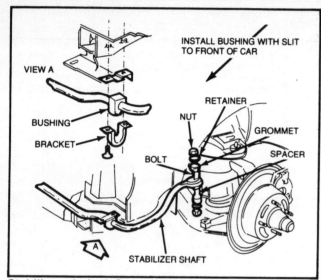

Stabilizer shaft (© General Motors Corp.)

5. Remove splash shield.
6. Remove upper and lower ball joint studs.
7. Remove studs from steering knuckle.

Installation

1. Place steering knuckle into position and insert upper and lower ball joint studs into knuckle bosses.
2. Install stud nuts and torque upper and lower nuts to 40 ft. lbs. (55 Nm).
3. Install splash shield. Torque screw to 7 ft. lbs. (9.5 Nm).
4. Install hub and rotor assembly.
5. Install outer bearing, spindle washer and nut. Adjust bearing.
6. Install brake caliper.
7. Install wheel and tire assembly. Tighten nuts to 100 ft. lbs. (140 Nm).
8. Lower car to floor.
9. Check front wheel alignment.

Stabilizer Shaft

Removal

1. Hoist car.
2. Disconnect each side of stabilizer linkage by removing nut from link bolt. Pull bolt from linkage and remove retainers, grommets and spacer.
3. Remove bracket-to-frame or body bolts and remove stabilizer shaft, rubber bushings and brackets. Some models require a special tool to remove stabilizer shaft bolt.

Installation

To replace, reverse sequence of operations, being sure to install with the identification forming on the right side of the car. The rubber bushings should be positioned squarely in the brackets with the slit in the bushings facing the front of car. Torque stabilizer link nut to 13 ft. lbs. (18 Nm) and bracket bolts to 24 ft. lbs. (33 Nm).

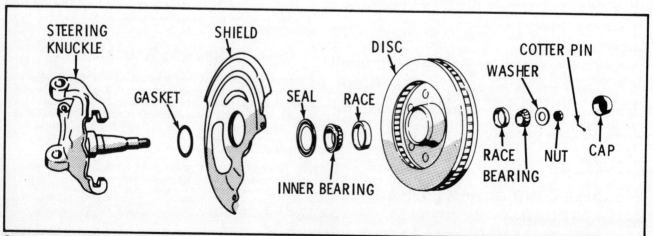

Steering knuckle and hub assembly (© General Motors Corp.)

SUSPENSION
Domestic Models

REAR SUSPENSION—GENERAL MOTORS REAR DRIVE SOLID AXLE EXCEPT F, T, P BODY AND '84 AND LATER CORVETTE

Description

COIL SPRING SYSTEM

The rear axle assembly is attached to the frame through a link type suspension system. Two rubber bushed lower control arms mounted between the axle assembly and the frame maintain fore and aft relationship of the axle assembly to the chassis. Two rubber bushed upper control arms, angularly mounted with respect to the centerline of the car, control driving and braking torque and sideways movement of the axle assembly. The rigid axle holds the rear wheels in proper alignment.

The upper control arms are shorter than the lower arms, causing the differential housing to "rock" or tilt forward on compression. This rocking or tilting lowers the rear propeller shaft to make possible the use of a lower tunnel in the rear floor pan area.

The rear upper controls arms control drive forces, side sway and pinion nose angle. Pinion angle adjustment can greatly affect car smoothness and must be maintained as specified.

The rear chassis springs are located between brackets on the axle tube and spring seats in the frame. The springs are held in the seat pilots by the weight of the car and by the shock absorbers which limit axle movement during rebound.

Ride control is provided by two identical direct double acting shock absorbers angle-mounted between brackets attached to the axle housing and the rear spring seats.

Shock Absorbers

Removal

Raise car and support rear axle to prevent stretching of brake hose. The lower end has a stud which is an integral part of the shock. Remove the nut and tap shock free from bracket. To disconnect the shock at the top, on all models, remove the two bolts, nuts and lockwashers.

Installation

Loosely attach shock at both ends. Tighten upper bolts and nuts to 20 ft. lbs. (26 Nm). Tighten lower nut to 65 ft. lbs. (90 Nm).

Coil Springs

Removal

1. Hoist rear of car on axle housing and support at frame rails with floor stands. Do not lower hoist at this time.

NOTE: Do not allow the rear brake hose to become kinked or stretched.

2. Disconnect brake line at axle housing.
3. Disconnect upper control arms at axle housing.
4. Remove shock at lower mount.
5. Lower hoist at rear axle.
6. Remove spring.

Installation

1. Install coil spring.
2. Hoist vehicle at rear axle.
3. Install shock at lower mount.
4. Install upper control arm bolts at axle housing.
5. Connect brake line at axle housing.
6. Remove jack stands and lower car.

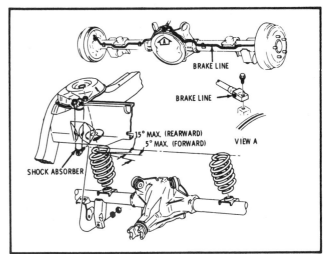

Coil spring mounting (© General Motors Corp.)

Upper Control Arm

CAUTION

If both control arms are to be replaced, remove and replace one control arm at a time to prevent the axle from rolling or slipping sideways. This might occur with both upper control arms removed, making replacement difficult.

Removal

1. Remove nut from rear arm to rear axle housing bolt and while rocking rear axle, remove the bolt. On some cars disconnecting lower shock absorber stud will provide clearance. Use support under rear axle nose to aid in bolt removal.
2. Remove from and rear arm attaching nuts and bolts.
3. Remove suspension arm and inspect bushings.

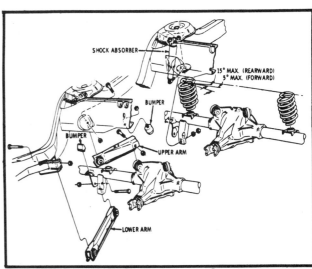

Rear suspension (© General Motors Corp.)

2-163

SUSPENSION
Domestic Models

Installation

To install, reverse removal procedure. Torque nuts with car resting at normal trim height.

Lower Control Arm

— CAUTION —

If both control arms are to be replaced, remove and replace one control arm at a time to prevent the axle from rolling or slipping sideways. This might occur with both lower control arms removed, making replacement difficult.

Removal

1. Raise car and support under axle housing.
2. Remove rear arm-to-axle housing bracket bolt.
3. Remove front arm-to-bracket bolts and remove lower control arm.

Installation

To replace arm, reverse the removal sequence of operations. Torque arm attaching nuts with the weight of the car on the rear springs.

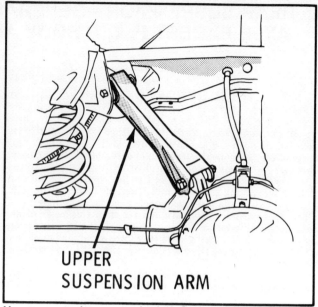

Upper control arm—typical (© General Motors Corp.)

REAR SUSPENSION—GENERAL MOTORS REAR DRIVE INDEPENDENT SUSPENSION EXCEPT F, T, P BODY AND '84–'85 CORVETTE

Description

The rear suspension features a transverse spring mounted on a fixed differential carrier. Each rear wheel is mounted by a three-link independent suspension. These three links are made up of wheel drive shaft, a camber control strut rod and a wheel spindle support arm.

Rear Wheel Alignment

To align the rear suspension, "back" the car onto the machine normally used to align front suspension. Camber will now be read in the normal manner. However, with the vehicle "backed" in, toe-in will now read as toe-out, while toe-out will be read as toe-in.

NOTE: Check condition of strut rods. They should be straight. Rear wheel alignment could be affected if they are bent.

CAMBER

Wheel camber angle is obtained by adjusting the eccentric cam and bolt assembly located at the inboard mounting of the strut rod. Place rear wheels on alignment machine and determine camber angle. To adjust, loosen cam bolt nut and rotate cam and bolt assembly until specified camber is reached. Tighten nut securely and torque to specifications.

TOE–IN

Wheel toe-in is adjusted by inserting shims of varying thickness inside the frame side member on both sides of the torque control arm pivot bushing. Shims are available in thickness of $\frac{1}{64}$ in. (40mm), $\frac{1}{32}$ in. (79mm), $\frac{1}{8}$ in. (3.18mm), and $\frac{1}{4}$ in. (6.35mm).

To adjust toe-in, loosen torque control arm pivot bolt. Remove cotter pin retaining shims and remove shims. Position torque control arm to obtain specified toe-in. Shim the gap toward vehicle centerline between torque control arm bushing and frame side inner wall. Do not use thicker shim than necessary, and do not use undue force when shimming inner side of torque control arm. To do so may cause toe setting to change.

Shim outboard gap as necessary to obtain solid stackup between torque control arm bushing and inner wall of frame side member. After correct shim stack has been selected, install cot-

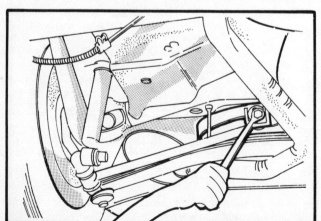

Rear wheel camber adjustment (© General Motors Corp.)

SUSPENSION
Domestic Models

ter pin (with loop outboard) through shims. Torque nut to specifications, and install cotter pin. If specified torque does not permit cotter pin insertion, tighten nut to next flat.

Wheel Bearing

END–PLAY CHECK

The tapered-roller spindle bearings should have end play of 0.001–0.008 in (0.003–0.20 mm). During inspection, check end play and, when necessary, adjust as outlined in this section.

1. Raise vehicle on hoist, being careful not to bend the strut rods.
2. Disengage bolt lock tabs and disconnect outboard end of axle drive shaft from wheel spindle flange.
3. Mark camber cam in relation to bracket. Loosen and turn camber bolt until strut rod forces torque control arm outward. Position loose end of axle-drive shaft to one side for access to spindle.
4. Remove wheel and tire assembly. Mount dial indicator on torque control arm adjacent surface and rest pointer on flange or spindle end.
5. Grasp brake disc and move axially (in and out) while reading movement on dial indicator. If end movement is within the 0.001–0.008 in. (0.003–0.20 mm) limit, bearings do not require adjustment. If not within 0.001–0.008 in. (0.003–0.20) limit, record reading for future reference and adjust bearings.

SPINDLE

Removal and Installation

1. Apply parking brake to prevent spindle from turning and remove cotter pin and nut from spindle.
2. Release parking brake and remove drive spindle flange from splined end of spindle. It may be necessary to use tool J–8614–01 or equivalent to remove flange from spindle.
3. Remove brake caliper.
4. Install thread protector J–21859–1 or equivalent over spindle threads. Remove drive spindle from spindle support, using tool J–22602.

When using tool J–22602 to remove drive spindle, make sure puller plate is positioned vertically in the torque control arm before applying pressure to the puller screw.

5. When the spindle is removed, the outer bearing will remain on the spindle. the inner bearing, tubular spacer, end-play adjustment shim and both outer races will remain in the spindle support.

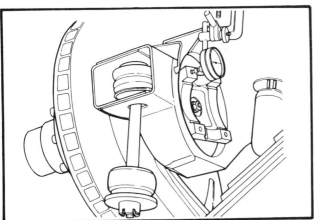

Spindle bearing end-play check (© General Motors Corp.)

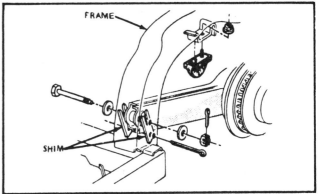

Toe in adjusting shim location (© General Motors Corp.)

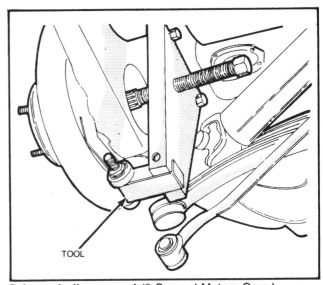

Drive spindle removal (© General Motors Corp.)

6. Remove bearing, spacer and shim. Record shim thickness for later use.

Bearing Replacement and Adjustment, and Spindle Installation

1. With the spindle assembly on a bench, place the two halves of J–24489–1 into position between the outer bearing and the oil seal.
2. Mount J–8433–1 to J–24489–1 and draw bearing off spindle.
3. Remove outer oil seal from spindle shaft and inspect for damage. Replace if necessary.
4. Remove the outer races from the spindle support and install new ones, using J–7817 for reinstallation.
5. Pack new bearings with EPB-2 bearing lubricant, or equivalent.
6. Check bearing end play as measured in Step 5 of Wheel Bearing End Plat Check. Use the same adjusting shim thickness as the original. If end play was not within limits, use the following steps to determine the proper shim thickness.

   a. If end play was greater than 0.008 in. (0.20 mm), it will be necessary to reduce shim thickness to bring end play within limits.
   b. For example, if end-play reading was 0.13 in. (0.33 mm), and the shim measured 0.144 in. (3.66 mm), you have to decrease the shim thickness. Reducing the shim by 0.010

2-165

SUSPENSION
Domestic Models

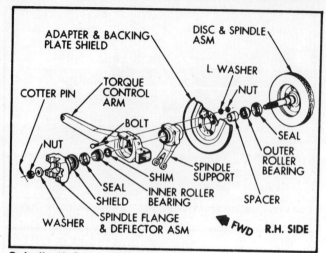

Spindle (© General Motors Corp.)

in. (0.25 mm), from 0.144–0.134 in. (3.40 mm), will also reduce end play by 0.010 in. (0.25 mm), from 0.013–0.003 in. (0.33–0.08 mm).

   c. If no end play was found on inspection, add 0.003 in. (0.08 mm) to the original shim as a starting point.

7. To check bearing end-play before final installation, use J-24626 or equivalent as follows:

   a. Mount the outer bearing onto the large shoulder, with the large end of the bearing against the flange.

   b. Place the tubular spacer, with the large end against the outer bearing, and the shim selected in Step 6 onto J-24626.

   c. Place the tool into position in the spindle support and install inner bearing, large washer and nut.

   d. Tighten nut to 100 ft. lbs. (140 Nm) to simulate actual installed conditions.

   e. Mount a dial indicator and check bearing end-play.

   f. After shim thickness as necessary to obtain end-play from 0.001–0.008 in. (0.03–0.20 mm). Shims are available in thicknesses from 0.097–0.145 in. (2.46–3.68 mm).

   g. Remove J-24626 from spindle support.

8. Install outer bearing into outer race. Install outer oil seal into bore of spindle support, making sure it is firmly seated.

9. Carefully install spindle assembly through the outer oil seal (being careful not to dislodge seal from the bore) and through the outer bearing.

10. Place the tubular spacer and the shim selected in Step 7 onto the spindle shaft.

11. Place the inner bearing onto the spindle shaft.

12. Thread too J-24490-1 onto the spindle shaft, then install sleeve J-24490-2, and washer and nut. Tighten nut against sleeve. Spindle shaft will now be drawn through the bearings to its final installed position.

13. Remove J-24490-1 and J-24490-2.

14. Position drive flange over spindle, making sure flange is aligned with spindle splines. Install washer and nut on spindle, then tighten nut to specifications and install cotter pin. If specified torque does not permit cotter pin insertion, tighten nut to next flat.

15. Install caliper onto disc.

16. Install axle drive shaft, wheel and tire assembly, adjust camber cam to original position and torque all components to specifications.

Spindle Support

Removal

1. Remove wheel spindle as outlined previously.
2. Disconnect parking brake cable from actuating lever.
3. Remove four nuts securing spindle support to torque control arm and withdraw brake backing plate. Position it out of the way.
4. Disconnect shock absorber lower eye from strut rod mounting shaft. It may be necessary to support spring outer end before disconnecting shock absorber, as shock absorber has internal rebound control.
5. Remove cotter pin and nut from strut rod mounting shaft, then pull shaft from support and strut rod.
6. Separate support from torque control arm.

Installation

1. Position support over torque arm bolts with strut rod fork toward center of vehicle and downward.
2. Place backing plate in position, install four nuts and torque to specifications.
3. Install strut rod and shock absorber mounting shaft onto support arm. Install shock absorber. Torque to specifications.
4. Connect parking brake cable to actuating lever.
5. Install drive spindle assembly.

Shock Absorber

Removal

1. Raise vehicle on hoist.
2. Disconnect shock absorber upper mounting bolt.
3. Remove lower mounting nut and locker washer.
4. Slide shock upper eye out of frame bracket and pull lower eye and rubber grommets off strut and mounting shaft.

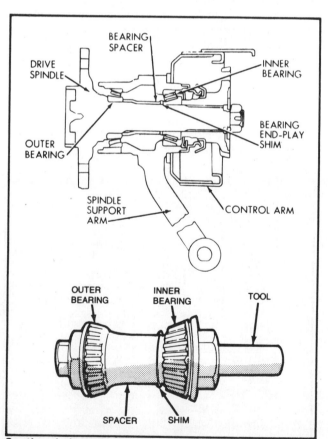

Sectional view of spindle assembly (© General Motors Corp.)

5. Inspect grommets and shock absorber upper eye for excessive wear.

Installation

1. Slide upper mounting eye into frame mounting bracket and install bolt, locker washer and nut.
2. Place rubber grommet, shock lower eye, inboard grommet, washer and nut over strut rod shaft. Install washer with curve pointing inboard (away from grommet).
3. Torque nuts to specifications.
4. Lower vehicle and remove from hoist.

Strut Rod and Bracket

Removal

1. Raise vehicle on hoist.
2. Disconnect shock absorber lower eye from strut rod shaft.

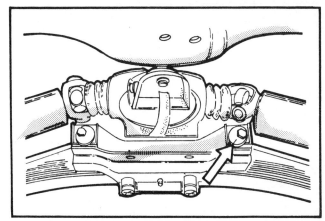

Marking camber cam and bracket (© General Motors Corp.)

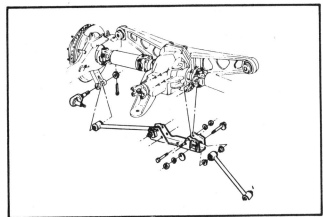

Strut rod mounting. Install strut with putboard end angled forward (© General Motors Corp.)

3. Remove strut rod shaft cotter pin and nut. Withdraw shaft by pulling toward front of vehicle.
4. Mark relative position of camber adjusting cam and bracket, so they may be reassembled in same location.
5. Loosen camber bolt and nut. Remove four bolts, lock washers and flat washers securing strut rod bracket to carrier and lower bracket.
6. Remove cam bolt nut and cam and bolt assembly. Pull strut down out of bracket and remove bushing caps.
7. Inspect strut rod bushings for wear and replace where necessary.

Installation

1. Place bushing caps over inboard bushing and slide rod into bracket. Install cam and bolt assembly and adjust cam to line up with mark of bracket. Tighten nut but do not torque at this point.
2. Raise bracket and assemble to carrier lower mounting surface. Be sure both flat washer and lock washer are between bolt and bracket. Torque bolts to specifications.
3. Raise outboard end of strut rod into fork so that flat on shaft lines up with corresponding flat in spindle fork. Install retaining nut, but do not torque.
4. Place shock absorber lower eye and bushing over strut shaft, install washer and nut and torque to specifications.
5. With weight on wheels torque camber cam nut and strut rod shaft nut to specifications. Then install cotter pin through rod bolt.
6. Check rear wheel camber and adjust where necessary.
7. Lower vehicle and remove from hoist.

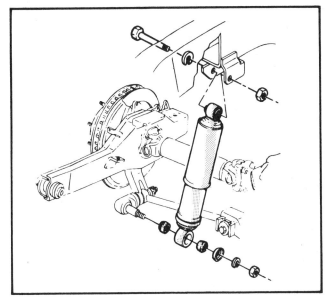

Shock absorber mounting (© General Motors Corp.)

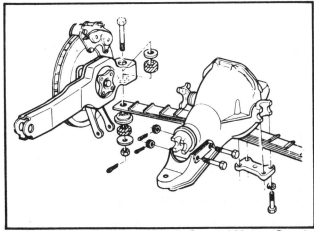

Transverse spring mounting (© General Motors Corp.)

SUSPENSION
Domestic Models

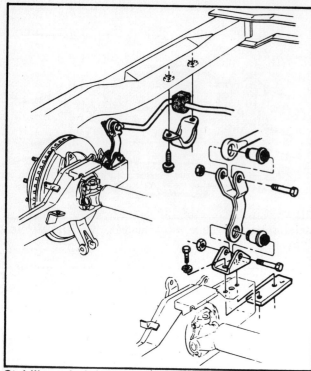

Stabilizer shaft Installation (© General Motors Corp.)

Transverse Spring

Removal

1. Raise vehicle on hoist allowing axle to hang. Remove wheels and tires.
2. Install a C-clamp on spring approximately 9 in. (23 cm.) from one end. Tighten securely.
3. Place adjustable lifting device under spring with lifting pad of jack inboard of link bolt near the C-clamp. Place a suitable piece of wood between jack pad and C-clamp screw. The C-clamp is merely acting as a stop so the jack will not slip when the spring is released. The wood block is used to protect the clamp threads from distortion due to contact with the jack pad.
4. Raise jack until all load is off link. Remove link cotter key and link nut. Remove cushion. Do not grip shank of spring link bolt with Vise Grips. Use new bolt if the bolt surface is scored or damaged.
5. Carefully lower jack until spring tension is released.
6. Repeat Steps 2–5 for other side.
7. Remove four bolts and washers securing spring center clamp plate.
8. Slide spring out from under vehicle.

Installation

1. Place spring on carrier cover mounting surface, indexing center bolt head with hole in cover.
2. Place center clamp plate in position and install bolts and washers. Snug bolts to position spring and torque to specifications.
3. Install C-clamp as in Step 2 of removal procedure.
4. Place adjustable lifting device inboard of link bolt near C-clamp. Add wooden block as in Step 3 of removal procedure.
5. Raise spring outer end until spring is nearly flat, aligning torque arm with spring end.
6. Install new attaching parts. Whenever servicing spring or removing spring attaching parts, always install new link bolts, rubber cushions, retainers, nuts and cotter pins.
7. Lower jack making sure cushions remain indexed in retainers. Remove C-clamp.
8. Remove jack and repeat for other side.
9. Place vehicle weight on wheels and torque center clamp bolts to specifications.

Torque Control Arm

Removal

1. Disconnect spring on side torque arm is to be removed. Follow Steps 1–5 of the spring removal procedure. If vehicle is so equipped, disconnect stabilizer shaft from torque arm.
2. Remove shock absorber lower eye from strut rod shaft.
3. Disconnect and remove strut rod shaft and swing strut rod down.
4. Remove four bolts securing axle drive shaft to spindle flange and disconnect driveshaft. It may be necessary to force torque arm outboard to provide clearance to lower driveshaft.
5. Disconnect brake line at caliper and from torque arm. Disconnect parking brake cable.
6. Remove torque arm pivot bolt and toe-in shims and pull torque arm out of frame. Tape shims together and identify for correction reinstallation.

Installation

1. Place torque arm in frame opening.
2. Install pivot bolt. Place toe-in shims in original position on both sides of torque arm. Install cotter pin retaining shims with loop of pin pointed outboard. Do not tighten pivot bolt nut at this time.
3. Raise axle driveshaft into position and install to drive flange. Torque bolts to specifications.
4. Raise strut rod into position and insert strut rod shaft so that flat lines up with flat in spindle support fork. Install nut and torque to specifications.
5. Install shock absorber lower eye and tighten nut to specifications.
6. Connect spring end as outlined under spring installation. Step 3–6. If vehicle is so equipped connect stabilizer shaft to torque arm.
7. Install brake line at caliper and torque arm. Bleed brakes.
8. Install wheel and tire. Torque the torque arm pivot bolt to specifications and install cotter pin with weight on wheels.

FRONT SUSPENSION—GENERAL MOTORS T BODY—REAR DRIVE CARS

Description

The front suspension system uses conventional long and short arm design and coil springs. The control arms attach to the vehicle with bolts and bushings at the inner pivot points, and to the steering knuckle/front wheel spindle assembly at the outer pivot points. The lower ball joints use the wear indicator feature used on other General Motor original equipment ball joints.

SUSSPENSION
Domestic Models

Wheel Alignment

CAMBER

Camber angle can be increased approximately one degree. Remove the upper ball joint, rotate it one-half turn and reinstall it with the flat of the upper flange on the inboard side of the control arm.

CASTER

Shims placed between the upper control arm and legs control caster. Always use two washers totalling 12 mm thickness, placing one washer at each end of the locating tube.

TOE

Adjust by changing tie-rod position Loosen the nuts at the steering knuckle end of the tie rod and the rubber cover at the lower end. Rotate tie-rod to change adjustment.

Wheel Bearings

NOTE: Tapered roller bearings are used on all series vehicles and they have a slightly loose feel when properly adjusted. A design feature of front wheel tapered roller bearings is that they must never be pre-loaded. Damage can result from pre-loading.

The proper functioning of the front suspension cannot be maintained unless the front wheel taper roller bearings are correctly adjusted. Cones must be a slip fit on the spindle and the inside diameter of cones should be lubricated to insure that the cones will creep. Spindle nut must be a free-running fit on threads.

Inspection

1. Raise vehicle and support at front lower control arm.

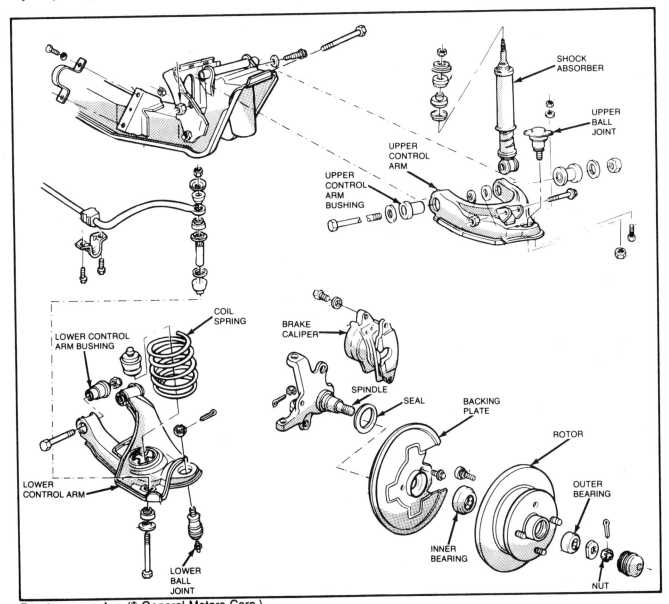

Front suspension (© General Motors Corp.)

SUSPENSION
Domestic Models

2. Spin wheel to check for unusual noise or roughness.
3. If bearings are noisy, tight, or excessively loose, they should be cleaned, inspected, and relubricated prior to final adjustment. If it is necessary to inspect bearings, movement should be from 0.001–0.005 in. (0.025–0.127 mm). If movement is not in this range, adjust bearings.

Adjustment

1. Remove hub cap or wheel disc from wheel.
2. Remove dust cap from hub.
3. Remove cotter pin from spindle and spindle nut.
4. Tighten the spindle nut to 12 ft. lbs (16 Nm) while turning the wheel assembly forward by hand to fully seat the bearings. This will remove any grease or burrs which could cause excessive wheel bearing play later.
5. Back off the nut to the "just loose" position.
6. Hand-tighten the spindle nut. Loosen spindle nut until either hole in the spindle lines up with a slot in the nut (not more than ½ flat).
7. Install new cotter pin. Bend the ends of the cotter pin against nut, cut off extra length to ensure ends will not interfere with the dust cap.
8. Measure the looseness in the hub assembly. There will be from 0.001–0.005 in. (0.025–0.127 mm) end-play when properly adjusted.
9. Install dust cap on hub.
10. Replace the wheel cover or hub cap.

Shock Absorbers

Removal

1. Hold the shock absorber upper stem and remove the nut, upper retainer and rubber grommet.
2. Raise vehicle on a hoist.
3. Remove the bolts from the lower end of the shock absorber.
4. Lower the shock absorber from the vehicle.

Installation

1. With the lower retainer and rubber grommet in position, extend the shock absorber stem and install the stem through the wheelhouse opening.
2. Install the lower bolts. Torque to 35–50 ft. lbs. (48–70 Nm).
3. Lower the vehicle to the floor.
4. Install the upper rubber grommet, retainer and nut to the shock absorber stem.
5. Hold the stem and tighten the nut to 60–120 inch lbs. (7–13 Nm). Torque is obtained by running nut to unthreaded portion of stud.

Upper Ball Joint

Removal

1. Raise the vehicle on a hoist.
2. Remove the tire and wheel assembly.
3. Support the lower control arm with a floor jack.
4. Remove upper ball stud nut. Reinstall nut finger-tight.
5. Install spreader tool and push stud loose from knuckle.
6. Remove tool and remove nut from ball stud.
7. Remove two nuts and bolts attaching ball joint to upper control arm, then remove ball joint.

Installation

Inspect the tapered hole in the steering knuckle. Remove any dirt and if any out-of-roundness, deformation, or damage is noted, the knuckle must be replace.

1. Install bolts and nuts attaching ball joint to upper control arm, then mate the upper control arm ball stud to the steering knuckle. The ball joint studs use a special nut which must be discarded whenever loosened and removed. On reassembly, use a standard nut to draw the ball joint into position on the knuckle. Torque the standard nut to 22 ft. lbs. (30 Nm), then remove that nut and install a new special nut for final installation.
2. Install the ball stud nut and torque to 29–36 ft. lbs (39–49 Nm).
3. Install the tire and wheel assembly.

Lower Ball Joint

Removal

1. Raise vehicle on hoist.
2. Remove the tire and wheel assembly.
3. Support the lower control arm with a hydraulic floor jack.
4. Remove lower ball stud nut, then reinstall nut finger-tight.
5. Install spreader tool and push the ball joint stud until it is free of the steering knuckle.
6. Remove tool and remove nut from ball stud.
7. Remove ball joint from lower control arm.

Installation

Inspect the tapered hole in the steering knuckle. Remove any dirt and if any out-of-roundness, deformation, or damage, is noted, the knuckle must be replaced.

1. Mate the ball stud through the lower control arm and into the steering knuckle. The ball joint studs use a special nut which must be discarded whenever loosened and removed. On reassembly, use a standard nut to draw the ball joint into position on the knuckle. Torque the standard nut then remove that nut and install a new special nut for final installation.
2. Install the ball stud nut and torque to 41–54 ft. lbs. (56–73 Nm).
3. Install the tire and wheel assembly.

Front Spring – Lower Control Arm

Removal

1. Remove wheel and tire assembly.

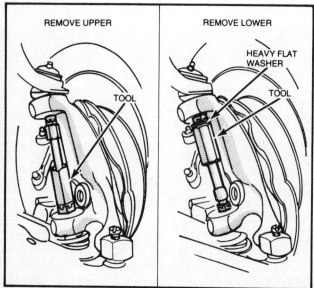

Removal of ball joint from knuckle (© General Motors Corp.)

SUSPENSION
Domestic Models

2. Disconnect stabilizer from lower control arm. Disconnect tie-rod from steering knuckle.
3. Support lower control arm with a jack.
4. Remove the nut from the lower ball joint. Install spreader tool and push the ball joint stud loose in the steering knuckle.
5. Swing the knuckle-and-hub out of the way, and attach securely with wire.
6. Loosen lower control arm pivot bolts.
7. Install chain through coil spring as a safety precaution.

CAUTION
The coil spring is under load. Be sure to install a chain and to slowly lower the jack.

---

8. Slowly lower the jack.
9. When the spring is extended as far as possible, use a prybar to carefully lift the spring over the lower control arm seat. Remove the spring.
10. Remove pivot bolts, and then remove lower control arm.

Installation

1. Install lower control arm and pivot bolts to underbody brackets.
2. Position spring and install spring into upper pocket. Use tape to hold insulator onto spring.
3. Install spring lower end onto lower control arm. It may be necessary to have an assistant help you compress the spring far enough to slide it over the raised area of the lower control arm seat.
4. Use a jack to raise the lower control arm and compress the coil spring.
5. The ball joint studs uses a special nut which must be discarded whenever loosened and removed. On reassembly, use a standard nut to draw the ball joint into position on the knuckle, then remove that nut and install a new special nut for final installation. Install the ball joint through the lower control arm and into the steering knuckle. Install nut to ball joint stud and torque to 41–53 ft. lbs. (56–73 Nm).
6. Connect stabilizer bar and tie rod. Install wheel and tire assembly. Torque to specifications.

Upper Control Arm

Removal

1. Raise vehicle on a hoist.
2. Remove the tire and wheel assembly.
3. Support the lower control arm with a floor jack.
4. Remove upper ball joint from steering knuckle.
5. Remove control arm pivot bolts and remove control arm from vehicle.

Installation

1. Install upper control arm and pivot bolt to vehicle. The inner pivot bolt must be installed with the bolt head toward the front.
2. Install the pivot bolt nut.
3. Position the control arm in a horizontal plane and torque the nut to 43–50 ft. lbs. (59–68 Nm).
4. The ball joint studs use a special nut which must be discarded whenever loosened and removed. On reassembly, use a standard nut to draw the ball joint into position on the knuckle, then remove that nut and install a new special nut for final installation. Install ball joint to upper control arm and to steering knuckle, as described earlier. Install nut; tighten to specifications.
5. Install wheel and tire; torque to specifications.
6. Lower vehicle to floor.

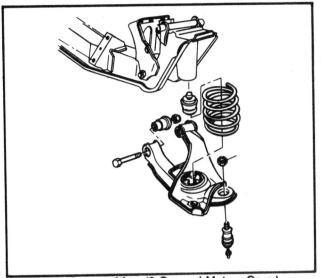

Correct spring position (© General Motors Corp.)

Steering Knuckle

Removal

1. Raise vehicle and support the lower control arm with a jackstand.

CAUTION
This keeps the coil spring compressed. Use care to support safely.

---

2. Remove the tire and wheel assembly.
3. Remove the disc brake caliper. Do not allow the caliper to hang by the brake hose. Insert a piece of wood between the shoes to hold the piston in the caliper bore. The block of wood should be about the same thickness as the brake disc.
4. Remove the hub and disc.
5. Remove the splash shield.
6. Remove the tie-rod end from the steering knuckle.
7. Loosen both ball stud nuts. Using a spreader tool, push both the upper and lower ball studs from the steering knuckle.
8. Remove ball stud nuts and remove the steering knuckle.

Installation

1. Place steering knuckle in position and insert the upper and lower ball studs into knuckle bosses.

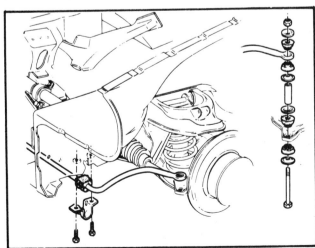

Stabilizer bar attachment (© General Motors Corp.)

2-171

SECTION 2

SUSPENSION
Domestic Models

2. The ball joint studs use a special nut which must be discarded whenever loosened and removed. On reassembly, use a standard nut to draw the ball joint into position on the knuckle. Torque the standard nut then remove that nut and install a new special nut for final installation. Install ball stud nuts and tighten to specifications.

Stabilizer Bar

Removal

1. Raise the vehicle on a hoist.

2. Remove the stabilizer bar nut and bolt from lower control arm.
3. Remove stabilizer bar bracket from body.

Installation

1. Hold stabilizer bar in place and install the body bushings and brackets.
2. Install the retainers, grommets and spacers to the lower control arm and install nuts.
3. Lower the vehicle to the floor.
4. Torque nut to 12–18 ft. lbs. (16–24 Nm). Torque is obtained by running nut to unthreaded portion of link bolt.

REAR SUSPENSION – GENERAL MOTORS T BODY – REAR DRIVE CARS

Description

The solid rear axle is attached to the body through two tubular lower control arms, a straight track rod, two shock absorbers and a bracket at the front end of the axle extension. Variable rate coil springs mount between the axle and body.

Two rubber bushed lower control arms mounted between the axle assembly and the frame maintain fore and aft relationship of the axle assembly to the chassis. The rigid axle holds the rear wheels in proper alignment.

The rear chassis springs are located between brackets on the axle tube and spring seats in the frame. The springs are held in the seat pilots by the weight of the car and by the shock absorbers which limit axle movement during rebound.

Ride control is provided by two identical direct double acting shock absorbers angle-mounted between brackets attached to the axle housing and the rear spring seats.

Shock Absorber

Removal

1. Support rear axle assembly.
2. Remove upper attaching bolts and lower attaching bolt and nut.

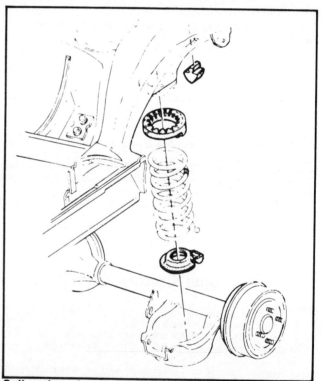

Coil spring placement (© General Motors Corp.)

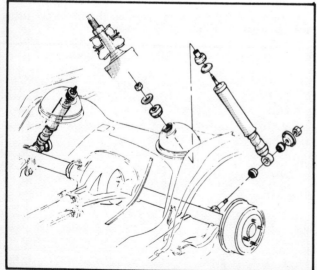

Rear shock absorber mounting (© General Motors Corp.)

3. Remove shock absorber.

Installation

1. Install retainer and rubber grommet onto shock.
2. Place shock into installed position and install upper retaining bolts. Torque to specifications.
3. Install bolt and nut onto lower shock attachment. Torque to specifications.
4. Lower vehicle and remove from hoist.

Coil Spring

Removal

1. Raise vehicle on hoist.

SUSPENSION
Domestic Models

2. Support rear axle with an adjustable lifting device.
3. Disconnect both shock absorbers from lower brackets.
4. Disconnect rear axle extension bracket.

> **CAUTION**
> Be sure to use caution when disconnecting extension assembly. Be sure to support assembly safely.

5. Lower axle and remove springs and spring insulators. One or both springs may be removed at this point.

> **CAUTION**
> When lowering axle, do not stretch brake hose running from frame to axle or damage to the brake line may result

Installation

1. Install insulators on top and bottom of springs then position spring between upper and lower seats.
2. Raise axle and reconnect shock absorbers. Torque nut to specifications.
3. Remove lifting device from axle.
4. Lower vehicle and remove from hoist.

Lower Control Arm and Tie Rod

Removal

> **CAUTION**
> If both control arms are to be replaced, removed and replace one control arm at a time to prevent the axle from rolling or slipping sideways.

1. Raise the car.
2. Support the rear axle.
3. Disconnect the stabilizer bar.
4. Remove the control arm front and rear attaching bolts and remove the control arm.
5. Remove the track rod attaching bolts and remove the track rod.

BUSHING REPLACEMENT

1. Use appropriate tools to press bushings out of control arm/tie rod.
2. Inspect for distortion, burrs, etc.
3. Press bushing into place.

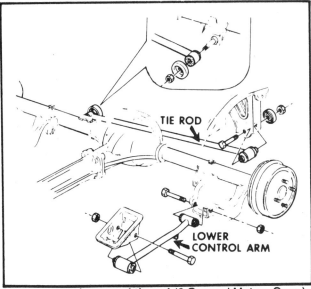

Lower control arm and tie rod (© General Motors Corp.)

Installation

1. Place control arm into position and install front and rear bolts. Torque to specifications.
2. Place tie-rod into position; torque bolts to specifications. Car must be at curb height when tightening pivot bolts. Tighten pivot bolts to 33 ft. lbs. (45 Nm).
3. Reattach stabilizer bar.

Stabilizer Bar

Removal

1. Raise vehicle on hoist.
2. Remove bolts securing brackets to body and link to axle and remove bar.

Installation

1. Place stabilizer into position. Install bolts and nuts. Torque to 15 ft. lbs. (20 Nm).

FRONT SUSPENSION – GENERAL MOTORS – 1984 AND LATER CORVETTE – REAR WHEEL DRIVE

Wheel Alignment

CASTER AND CAMBER

Before adjusting caster and camber angles, the front bumper should be raised and released twice to allow vehicle to return to its normal height.

Caster and camber adjustments are made by means of shims inserted between the upper control arm shaft and the frame bracket. Shims may be added, subtracted or transferred to change the readings as follows:

Caster

Transfer shims, front to rear or rear to front.

Camber

Front Suspension-Change shims at both the front and rear of the shaft.

Adding an equal number of shims at both front and rear of the cross shaft will increase positive camber.

Caster and camber can be adjusted in one operation. Toe-in must be checked after chaning camber or caster.

To adjust caster and camber, loosen the upper control arm shaft to frame nuts, add or subtract shims as required, per alignment correction charts, and retorque nuts.

A normal shim pack will leave at least two (2) threads of the bolt exposed beyond the nut. The difference between front and rear shim packs must not exceed 0.40 in. (10 mm).

If these requirements cannot be met in order to reach specifi-

2-173

SECTION 2
SUSPENSION
Domestic Models

FRONT WHEEL ALIGNMENT SPECS

To Align	Curb Service Checking	Curb Service Setting
Caster	3° ± 0.8° (C)	3° ± 0.5° (B)
Camber	0.8° ± 0.5° (C)	0.8° ± 0.5° (B)
Toe-In (Degrees per WHL)	.15° ± .15° (D)	.15° ± .10° (D)

NOTE Vehicle must be jounced 3 times before checking alignment to eliminate false geometry readings.

(A) Front suspension (Z) dimension and rear suspension (D) dimension are held as indicated in "trim heights" chart. Wheel alignment specifications to be as indicated on this chart.
(B) Left and right side to be equal within .50°.
(C) Left and right side to be equal within 1.0°.
(D) Toe-in left and right side to be set separately per wheel and steering wheel must be held in straight ahead position within ±2.5°.

Front wheel alignment (© General Motors Corp.)

cations, check for damaged cross member, bushings, control arms and related parts.

Always tighten the nut on the thinner shim pack first, for improved shaft to frame clamping force and torque retention.

Suspension Description

The front suspension uses aluminum for all its major components. These include forged aluminum arms and knuckles and replaces conventional coil springs with a lighter, more durable transverse fiberglass monoleaf spring.

Other elements of the front suspension system include long life stabilizer link bushings and the use of spindle offset. Spindle offset is achieved by moving the center of the wheel rearward from the conventional location on line through the ball joints. The displacement contributes to the directional sense. Combined with +3° caster, spindle offset gives an effect similar to higher caster without the poor responsiveness.

Knuckle Hub and Bearing

Removal

1. Hoist vehicle and remove wheel and tire.
2. Remove hub and bearing assembly.
3. Remove splash shield.
4. Disconnect upper and lower ball joint using tool J-33436-9 or equivalent.
5. Remove knuckle.

Installation

1. Reverse removal procedure.

SUSPENSION
Domestic Models

2. Torque nuts and screws to specifications.
3. When installing ball joints and tie rod end to knuckle, do not back off nut for cotter pin insertion.
4. Install wheel and tire and lower vehicle.

Upper Control Arm

Removal

1. Raise vehicle and remove wheel and tire.
2. Remove spring protector.
3. Using spring compressor J–33432 or equivalent compress and loosen spring.
4. Using tool J–33436 or equivalent disconnect upper ball joint from knuckle.
5. Remove upper control arm as shown.

Installation

1. Install control arm as shown.
2. Torque nuts to specifications.
3. Cotter pin at ball joint must be installed from rear to front. Do not back off nut for cotter pin insertion.
4. Check front alignment. Adjust if necessary.

Lower Control Arm

Removal

1. Raise vehicle and remove wheel and tire.
2. Remove spring protector.
3. Using tool J–33432 or equivalent compress spring.
4. Remove lower shock bracket.
5. Using tool J–33436 or equivalent disconnect lower ball joint.

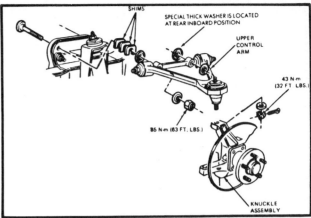

Upper control arm installation (© General Motors Corp.)

6. Remove lower control arm.

Installation

1. Install lower control arm to front crossmember.
2. Hold suspension at curb height and torque bolt to specification. Retain suspension at curb height. Do not allow it to move below rebound.
3. Install shock absorber bracket to control arm. Torque nuts to specifications.
4. Install ball joint to knuckle.
5. Loosen spring and remove tool.
6. Install spring protector.
7. Install wheel and tire and lower vehicle.

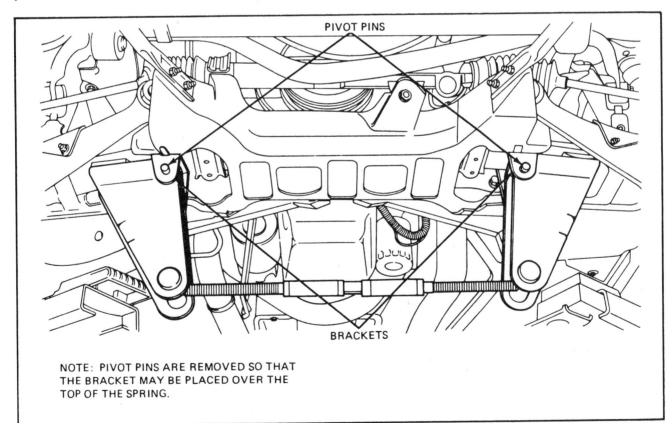

Transverse monoleaf spring installation (© General Motors Corp.)

SECTION 2
SUSPENSION
Domestic Models

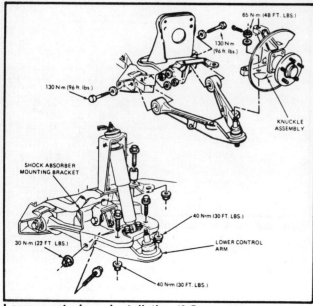

Lower control arm installation (© General Motors Corp.)

Front Transverse Spring

Removal

1. Raise vehicle on a hoist.
2. Remove wheel and tire assembly.
3. Remove both spring protectors.
4. Install spring compressing tool J–33432 or equivalent.
5. Disconnect lower ball joints.
6. Compress spring tool.
7. Remove shock mounting bracket to lower control arm attaching bolts.
8. Remove spring mounting bolts.
9. Release and remove spring compressing tool.

Installation

1. Install the spring by reversing the removal procedure.
2. Torque all nuts and bolts to specifications.
3. Check the vehicle's wheel alignment.

NOTE: During removal and installation of transverse spring take care to prevent damage.

REAR SUSPENSION – GENERAL MOTORS 1984 AND LATER CORVETTE REAR WHEEL DRIVE

Rear Wheel Alignment

To align the rear suspension, "back" the vehicle onto the alignment machine normally used to align front suspension. Camber will now be read in the normal manner. However, with the vehicle, "backed" in, toe-in will now read as toe-out, while toe-out, will be read as toe-in.

NOTE: Check condition of camber control support rods. They should be "straight." Rear wheel alignment could be affected if they are bent.

CAMBER

Wheel camber angle is obtained by adjusting the eccentric cam and bolt assembly located at the inboard mounting of the support rods. Place rear wheels on alignment machine and determine camber angle. To adjust, loosen cam bolt nut and rotate cam and bolt assembly until specified camber is reached. Tighten nut securely and torque to specification.

TOE–IN

Toe-in can be adjusted by loosening the lock nuts on the tie rod ends and turning the adjusting sleeves until the desired setting is obtained.

Suspension Description

The rear suspension features a light weight fiberglass transverse spring mounted to the fixed differential carrier cover beam. Light weight aluminum components such as the knuckles, upper and lower control arms, camber control support rods, differential carrier cover beam and the drive line support beam are used throughout the rear suspension. Each wheel is mounted by a five link independent suspension. The five links are identified as the wheel drive shaft, camber control support rod, upper and lower control arms and tie rod. The advantages of this suspension unit include a reduction of unsprung weight as well as an overall weight reduction. In addition, wheel tramp is eliminated, and handling is improved because of the independent action of each rear wheel.

The axle drive shafts and the camber control support rods act together in maintaining an almost constant camber change throughout the entire arc of wheel travel. Fore-aft motion of the wheel is controlled by the upper and lower control arms. Each rear wheel has a short spindle, hub and bearing assembly and knuckle contained at the rear of the upper and lower control arms. The knuckle also acts as a mount for the brake caliper mounting and parking brake backing plate assembly.

Several techniques can be employed to achieve this independent wheel movement. A five-link independent rear suspension is used on Corvettes. The five-link design can be compared to a right angle. The wheel is located at the right angle formed by the control arms and the "lateral links" (the camber control support rod and the rear wheel drive shaft). The points of the triangle are hinged to provide up-down wheel travel. The solid links thus force the wheel to travel through a controlled arc with fore-aft position determined by the control arms and lateral position held by the lateral links.

Aside from controlling wheel location, each portion of the suspension has additional functions. The controls arms and knuckle carries the brake caliper, thus, all brake torque and braking tractive forces are transmitted through the arms. the lateral links transmit side forces to the fixed differential, and through the rubber bushings in the cover beam to the frame. The upper link, or wheel drive shaft, transmits acceleration torque through the differential to the frame.

The final duty of the lateral links is to maintain the camber angle of the wheel throughout its travel. Since the camber control support rod and the wheel drive shaft are of different

SUSPENSION
Domestic Models

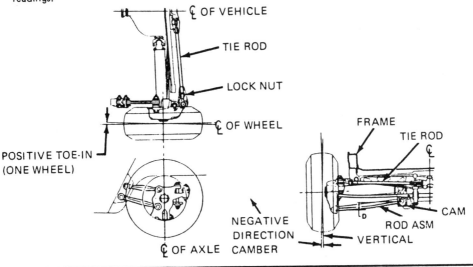

REAR WHEEL ALIGNMENT SPECS

To Align	Service Setting
Camber	0° ± .5°
Toe-In (Deg per WHL)	0.15° ± .06°

CAMBER

(1) Finger tighten lateral strut attaching parts at inboard end of strut asm.
(2) Rotate cam to obtain specified camber.
(3) Hold camber setting and torque nut with rear suspension set at "D" height of 71.0, As shown on trim height chart.

TOE IN

(1) Loosen Lock Nut
(2) With special tool in serrated area, rotate tie rod to obtain specified toe-in.
(3) Torque lock nut and adjust boot to cover lock nut & serrations.

CHECKING

NOTE: Vehicle must be jounced three times before checking alignment, to eliminate false geometry readings.

Rear wheel alignment (© General Motors Corp.)

lengths, a certain amount of camber change occurs through jounce and rebound. The overall result of the camber control support rod and wheel drive shaft geometry holds the wheel in a near vertical position at all times.

Direct double-acting shock absorbers are attached at the upper eye to a frame bracket and at the lower eye to the knuckle which has a threaded stud for the shock absorber lower eye.

The transversely mounted spring is clamp bolted at the center section to a lower mounting surface on the differential carrier cover beam. The outer ends of the spring are provided with a hole through which the spring is link bolted to the rear of the knuckle.

A stabilizer shaft is used which attaches to the section of the knuckle, and extends rearward where it is connected to the frame by two rubber bushings and mounting brackets.

A single unit hub and bearing assembly is bolted to each knuckle. The hub and bearing assembly supports the drive axle shaft and spindle allowing torque to be transferred from the differential carrier to the wheel and tire. This hub and bearing assembly is a sealed unit and no maintenance is required.

Rear Hub and Bearing

Removal

1. Raise vehicle and support safely.
2. Remove wheel and tire.
3. Remove brake caliper.
4. Remove brake rotor.
5. Disconnect tie rod end from the knuckle.
6. Disconnect transverse spring from the knuckle.

SECTION 2

SUSPENSION
Domestic Models

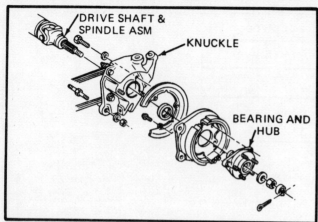

Rear hub and bearing attachment (© General Motors Corp.)

7. Scribe mark on cam bolt and mounting bracket so they can be realigned in the same position.
8. Remove cam bolt and separate spindle support rod from the mounting bracket.
9. Remove the trunnion straps at the side gear yoke shaft. Push out on the knuckle and separate axle shaft from the side gear yoke shaft. Remove the axle shaft from the vehicle.
10. Using J–34161 (Torx #45) or equivalent remove the hub and bearing mounting bolts.
11. Remove hub and bearing from the vehicle and support the parking brake backing plate.

Installation

To install, reverse the removal procedures and include the following:
1. Inspect the spindle seal and replace if necessary.
2. Torque all bolts to specification.
3. Check and adjust rear suspension alignment as necessary.

Rear Wheel Spindle

Removal

1. Raise vehicle and support safely.
2. Remove wheel and tire.
3. Disconnect tie rod end from the knuckle.
4. Disconnect transverse spring from the knuckle.
5. Scribe mark on cam bolt and mounting bracket so they can be realigned in the same position.
6. Remove cam bolt and separate spindle support rod from the mounting bracket.

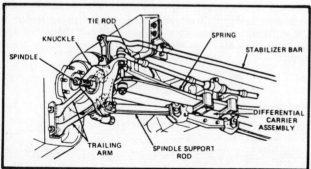

Corvette rear suspension (© General Motors Corp.)

7. Remove the trunnion straps at the spindle yoke. Push out on the knuckle and separate axle shaft from the spindle.
8. Remove the spindle from the hub and bearing.

Installation

To install, reverse the removal procedures and include the following:
1. Inspect the spindle seal and replace if necessary.
2. Torque all bolts to specifications.
3. Check and adjust rear suspension alignment as necessary.

Rear Knuckle

Removal

1. Raise vehicle and support safely.
2. Remove wheel and tire.
3. Remove brake caliper.
4. Remove brake rotor.
5. Disconnect tie rod end from the knuckle.
6. Disconnect transverse spring from the knuckle.
7. Disconnect stabilizer bar from the knuckle.
8. Disconnect parking brake cable from the backing plate.
9. Disconnect shock absorber from the knuckle. Use a back-up wrench on the mounting stud.
10. Disconnect spindle support rod from the knuckle.
11. Disconnect upper and lower control arms from the knuckle.
12. Lower the knuckle assembly and slide spindle out from the hub and bearing.
13. Using J–34161 (Torx #T55) or equivalent remove the hub and bearing bolts. Remove hub and bearing with parking brake backing plate from the knuckle.
14. Remove splash shield from the knuckle.

Installation

To install, reverse the removal procedures and include the following:
1. Install a new spindle seal.
2. Torque all bolts to specifications.
3. Check and adjust the rear suspension alignment as necessary.

Rear Transverse Spring

Removal

1. Raise vehicle and support safely.
2. Remove one rear wheel and tire.

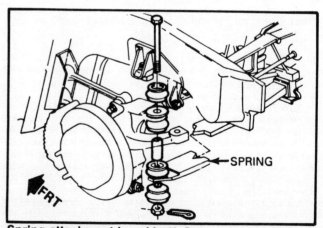

Spring attachment knuckle (© General Motors Corp.)

SUSPENSION
Domestic Models

3. Remove cotter pins, retaining nuts, bushings and link bolt retaining the spring to the knuckles.
4. Remove transverse spring attaching bolts, spacers, insulators and spring from the cover beam.

Installation

1. Position spring, insulators and spacers to the cover beam. Install attaching bolts and torque to specification.
2. Position spring to knuckles. Install bolts, insulators, spacers and nut. Tighten nut until slot in nut aligns with hole in bolt and then install cotter pin.
3. Install wheel and tire.
4. Lower vehicle.

Spindle Support Rod

Removal

1. Raise vehicle and support safely.
2. Scribe mark on cam bolt and mounting bracket so they can be realigned in the same position.
3. Remove cam bolt and separate spindle support rod from the mounting bracket.
4. Remove the spindle support rod bolt at the knuckle and remove rod.

Installation

To install, reverse the removal procedures and include the following:
1. Tighten all bolts to specification.
2. Check and adjust rear suspension alignment as necessary.

Upper/Lower Control Arms

Removal

1. Raise vehicle.
2. Remove shock absorber. Use a backup wrench on the mounting stud.
3. Remove control arm bolt at the knuckle.
4. Remove control arm bolt at the body bracket and remove the arm.

Installation

To install, reverse removal procedures and include the following:
1. Torque all bolts to specifications.

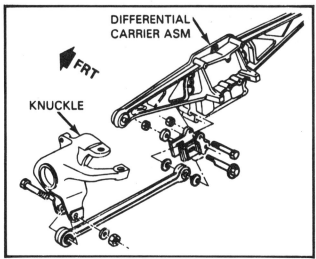
Spindle support rod (© General Motors Corp.)

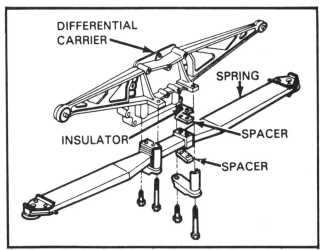
Spring attachment at cover beam (© General Motors Corp.)

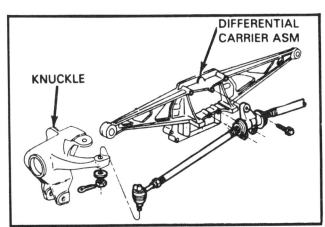
Tie rod assembly (© General Motors Corp.)

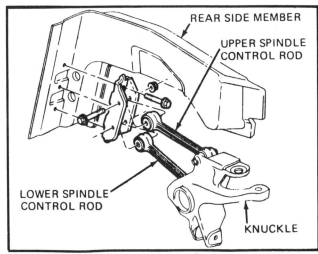
Control arms (© General Motors Corp.)

SECTION 2

SUSPENSION
Domestic Models

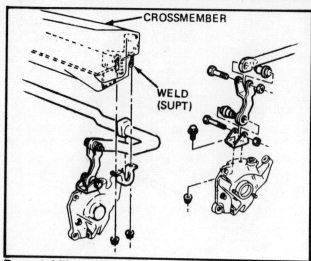

Rear stabilizer bar (© General Motors Corp.)

and nut and torque to specification. Use a backup wrench on the mounting stud.
3. Lower vehicle.

Rear Axle Tie Rod and/or Adjuster Sleeve

Removal

1. Raise vehicle.
2. Remove cotter pin and retaining nut from tie rod end at knuckle.
3. Loosen jam nut on tie rod end.
4. Using J-24319-01 or equivalent press tie rod end out of the knuckle.
5. Remove tie rod end from the adjusting sleeve.

Installation

To install, reverse the removal procedures and include the following:
1. Tighten all bolts to specification.
2. Check and adjust the rear suspension alignment as necessary.

Rear Shock Absorber

Removal

1. Raise vehicle.
2. Disconnect the shock absorber at the knuckle. Use a back-up wrench on the mounting stud.
3. Remove upper shock absorber attaching bolt.

Installation

1. Position shock absorber to body bracket. Install attaching bolt and torque to specifications.
2. Position shock absorber to the knuckle. Install washer

Stabilizer Bar

Removal

1. Raise vehicle.
2. Remove spare tire and tire carrier.
3. Disconnect stabilizer bar from knuckles.
4. Remove stabilizer bar bushing retainers, bushings and bar from the vehicle.

Installation

To install, reverse the removal procedures and include the following:
1. Torque all bolts to specifications.

GENERAL MOTORS ELECTRONIC LEVEL CONTROL

Description

The electronic level control (ELC) system automatically adjusts the rear height with varying car loads. The system is activated when weight is added to, or removed from the rear of the car.

Components

The electronic level control system consists of the following components:
1. Compressor
2. Air adjustable shock absorbers
3. Electronic height sensor
4. Compressor relay (two with E series)
5. Exhaust solenoid
6. Air dryer
7. Wiring and air tubing
8. Pressure regulator (E series only)

The E and K series front drive cars (torsion bar front suspension) have a pressure limiter value added to the system. This valve is located in the engine compartment in the pressure line which runs from the compressor to the shocks. The limiter allows a maximum of 85 psi (586kPa) ± 5 psi (34kPa) to reach the rear shocks.

COMPRESSOR RELAY

This relay is a single pole single throw type that completes the 12V(+) circuit to the compressor motor when energized. The compressor relay is located on the compressor mounting bracket.

COMPRESSOR

The basic compressor assembly is a positive displacement single piston air pump powered by a 12 volt DC permanent magnet motor. The compressor head casting contains piston intake and exhaust valves plus a solenoid.

AIR DRYER

The air dryer is attached externally to the compressor output and provides a dual function.
1. It contains a dry chemical that absorbs moisture from the air before it is delivered to the shocks and returns the moisture to the air when it is being exhausted. This action provides a long chemical life.
2. The air dryer also contains a valving arrangement that maintains 8–15 pounds minimum air pressure in the shock absorbers (except the E and K series which have 14–20 lb. retention.

SUSPENSION
Domestic Models

EXHAUST SOLENOID

The exhaust solenoid is located in the compressor head assembly and provides two functions.

1. It exhausts air from the system when energized. The height sensor controls this function.
2. It acts as a blow off valve to limit maximum pressure output of the compressor.

HEIGHT SENSOR

The height sensor is an electronic device that controls two basic circuits.

1. Compressor relay coil ground circuit.
2. Exhaust solenoid coil ground circuit.

To prevent falsely actuating the compressor relay or exhaust solenoid circuits during normal ride motions, the sensor circuitry provides an 8-14 second delay before either circuit can be completed.

In addition, the sensor electronically limits compressor run time or exhaust solenoid energized time to a maximum of 3 ½ minutes. This time limit function is necessary to prevent continuous compressor operation in case of a solenoid malfunction. Turning the ignition "off" and "on" resets the electronic timer circuit to renew the 3 ½ minute maximum run time. The height sensor is mounted to the frame crossmember in the rear. The sensor actuator arm is attached to the rear upper control arm by a link.

Air Lines and Fittings

NOTE: While the lines are flexible for easy routing and handling, care should be taken not to kink them and to keep them from coming in contact with the exhaust system.

When the air line is attached to the shock absorber fittings or compressor dryer fitting the retainer clip snaps into a groove in the fitting locking the air line in position. To remove the air line, spread the the retainer clip, release it from the groove and pull on the air line.

System Operation Check

NOTE: When certain tests require raising the car on a hoist, the hoist should support the rear wheels or axle housing. When a frame type hoist is used, two additional jack stands should be used to support the rear axle housing in its normal curb weight position.

1. Select a suitable location at rear wheelhouse opening and measure distance to floor.

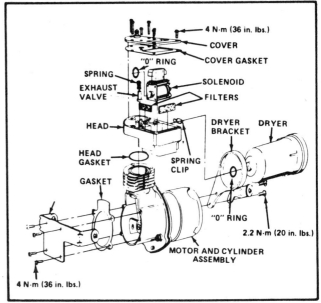
Compressor assembly (© General Motors Corp.)

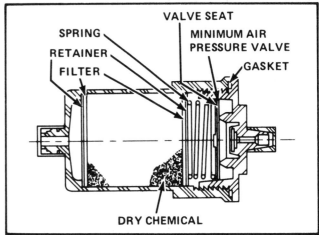
Air dryer (© General Motors Corp.)

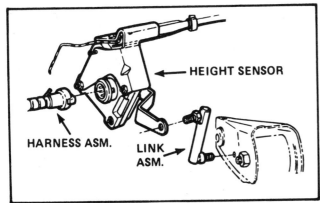
Height sensor (© General Motors Corp.)

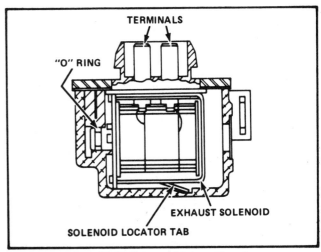
Exhaust solenoid (© General Motors Corp.)

SECTION 2 SUSPENSION
Domestic Models

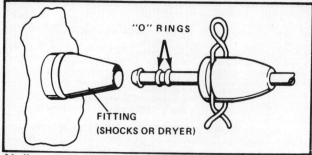

Air line retainer clip (© General Motors Corp.)

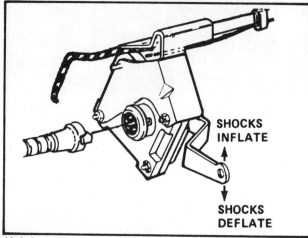

Height sensor opporational check (© General Motors Corp.)

2. Start engine momentarily. Leave switch "ON".
3. Apply load to rear of car (two people or approximately 300–350 pounds).
   a. There should be 8–14 second delay before compressor turns on and the car begins to raise.
   b. Car should raise to with $3/4$ in. (19 mm) of measurement made in Step 1 by the time the compressor shuts off. If car does not raise, refer to the diagnosis chart.

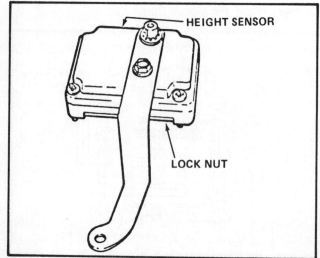

Height sensoe adjustment 1° = ¼ in. at bumper. Adjustment of 5° total (© General Motors Corp.)

NOTE: Failure of car to return to within $3/4$ in. (19 mm) of unloaded dimension can be caused by unusually heavy loading in the trunk which exceeds the capacity of the system. If this type of loading is encountered, remove it and repeat test.

4. Remove load applied in Step 3.
   a. There should be 8–14 second delay before car begins to lower.
   b. Car should lower to within $3/4$ in. (19 mm) of measurement made in Step 1 in less than $3 \frac{1}{2}$ minutes.

Compressor/Dryer Performance Test

COMPRESSOR CURRENT DRAW, PRESSURE OUTPUT AND LEAK DOWN TEST

1. Disconnect wiring from compressor motor and exhaust solenoid terminals.
2. Disconnect existing pressure line from dryer and attach pressure gauge to dryer fitting.
3. Connect ammeter to 12V source and to compressor.
   a. Current draw should NOT exceed 14 amp.
   b. When gauge read 110–120 psi SHUT COMPRESSOR OFF and observe if pressure leaks down.

NOTE: If compressor is permitted to run until it reaches its maximum output pressure, the solenoid exhaust valve will act as a relief valve. The resulting leak down when compressor is shut off will indicate a false leak.

   c. Leak down pressure should not drop below 90 psi when compressor is shut off.

RESIDUAL AIR CHECK

1. Remove air line from dryer fitting and attach it to gauge. Attach gauge air line to dryer fitting.
2. Turn ignition "ON" and perform system check, to inflate shocks.
3. Turn ignition "OFF" and deflate system through compressor service valve. Gauge should read 8–15 psi after system is deflated.

COMPRESSOR/DRYER DIAGNOSIS CHART

Malfunction	Correction
1. Current draw exceeds 14 amps.	1. Replace motor cylinder assembly.
2. Compressor Inoperative.	2. Replace motor cylinder assembly.
3. Pressure build up OK but leaks down below 90 psi before holding steady.	3. Replace solenoid exhaust valve assembly.
4. Compressor pressure leaks down to 0 psi.	4. Leak test compressor/dryer assembly.
5. Compressor output less than 110 psi and current draw normal.	5. Perform compressor/dryer leak test. If no leak is found, replace motor/cylinder assembly.

SUSPENSION
Domestic Models

Height Sensor Operational Check/Adjustment

OPERATIONAL CHECK

1. Turn ignition switch "ON" and raise car on hoist. If frame hoist is used, rear wheel or axle must be supported. Jacks should be adjusted upward until axle housing and/or wheels reach trim/curb weight position.

2. Compare neutral position of the height sensor metal arm with position of sensor arm being tested. (Shocks should have minimum air pressure.) If neutral position varies more than 3–4° check for correct sensor and/or link, sensor mounting bolts tight, sensor mounting bracket not bent. Make necessary corrections as required.

3. Disconnect link from height sensor arm.

4. Disconnect and reconnect wiring to height sensor to assure resetting the sensor time limit function. Failure to do this can result in erroneous diagnosis.

5. Move sensor metal arm upward approximately 1 ½–2 in. above neutral position. There should be 8–15 seconds delay before compressor turns "ON". As soon as shocks noticeably inflate move sensor arm down slowly and note arm position where compressor stops. This position should be very close to the neutral position.

6. Move arm down approximately 1 ½ in. below the point where the compressor stopped. There should be 8–15 seconds delay before shocks start to deflate. Allow shocks to deflate until only the retention pressure is left in the shocks (approximately 8–15 lbs.).

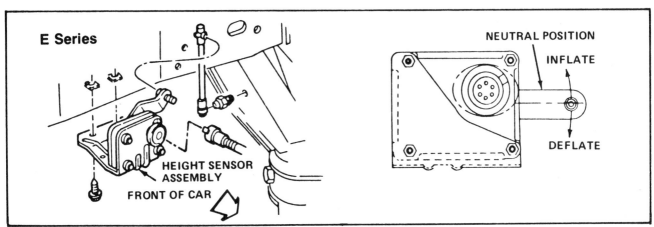

Front drive (© General Motors Corp.)

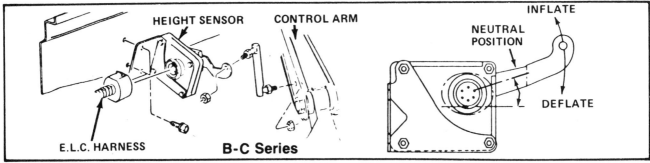

Rear drive (© General Motors Corp.)

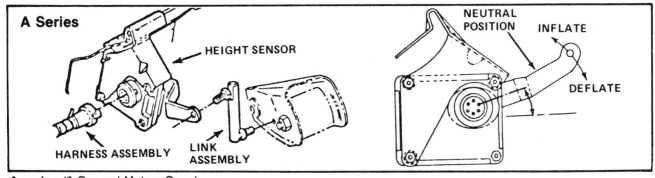

A series (© General Motors Corp.)

SUSPENSION
Domestic Models

Trim Adjustment

NOTE: Link should be attached to metal arm when making the adjustment.

1. Loosen lock nut that secures metal arm to height sensor plastic arm.
2. To increase car trim height move white plastic actuator arm upward and tighten lock nut.

NOTE: If all adjustment is used up, check trim height.

3. To lower car trim height, follow Step 1 and move plastic arm down.
4. If adjustment cannot be made, check for correct height sensor.

Compressor and Bracket

Removal

1. Remove negative battery cable.
2. Deflate system through service valve.
3. Disconnect high pressure line at air dryer by revolving spring clip 90° while holding connector end and removing tube assembly.
4. Remove two relay-to-compressor bracket screws and allow relay to hang to one side.
5. Remove support bracket screws.
6. Remove the radiator support to compressor bracket screws.
7. Disconnect solenoid and motor connectors.
8. Remove compressor and bracket assembly.
9. Remove three compressor mounting bracket screws then remove bracket.
10. If replacing compressor assembly remove dryer, dryer bracket, and compressor cylinder housing bracket and gasket.

Installation

1. If compressor was replaced install dryer and bracket and torque to 20 inch lbs. (2.2 Nm).
2. Install mounting brackets to compressor assembly and torque screws to 36 inch lbs. (4 Nm).
3. Connect solenoid and motor connectors.
4. Install two radiator support to compressor bracket screws and torque to 48 inch lbs. (6 Nm).
5. Install support bracket screws and torque to 7 ft. lbs. (10 Nm).
6. Install two compressor relay attaching screws.
7. Rotate clip on high pressure line until clip snaps in groove, then connect high pressure line at air dryer.
8. Cycle ignition switch and test for system operation and leaks at air dryer.

Air Dryer

Removal

1. Deflate system through service valve.
2. Disconnect high pressure line at air dryer by revolving spring clip and removing tube assembly.
3. Disconnect air dryer from compressor by revolving spring clip and sliding air dryer assembly away from compressor head through its bracket. Remove O-ring from compressor head.

Installation

Lubricate dryer O-ring with petroleum jelly or equivalent before installing dryer in head casting.

1. Reverse removal procedure.
2. Check for leaks.

Air Line Repair

The air lines used on the superlift shock absorbers and the electronic level control systems can be repaired by splicing in a coupling at the leaking area.

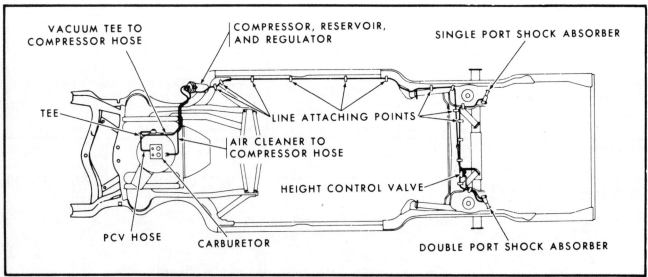

Automatic level control system component locations except Eldorado and Seville (© General Motors Corp.)

SUSPENSION
Import Models

Import Vehicles

FRONT SUSPENSION AUDI

MacPherson Strut

REMOVAL & INSTALLATION

4000 and 5000

1. With the car on the ground, remove the front axle nut and loosen the wheel bolts.
2. Raise and support the front of the car and remove the wheels.
3. Remove the brake caliper mounting bolts and the brake line bracket. Remove the brake caliper with the line still attached to it and wire it out of the way.
4. Remove the wheel bearing housing-to-ball joint clamp bolt.
5. Remove the retaining nut and press off the tie–rod end.
6. If equipped with a stabilizer bar, remove the retaining bolt and remove the stabilizer bar end clamps. Pivot the stabilizer bar downward. (4000 only).
7. Remove the two center stabilizer bar clamps and then unbolt it from the lower control arm. Remove the stabilizer bar (5000 only).
8. Pry the lower control arm down and remove the ball joint from the wheel hub.

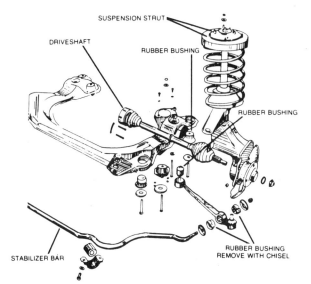

Front suspension and drive axle—4000 and 5000 models—5 cyl

1. Cotter pin
2. Tie-rod
3. Axle driveshaft
4. Circlip
5. Retainer nut
6. Brake caliper
7. Wheel bearing
8. Hub
9. Brake disc
10. Axle nut

4000 front suspension

9. Using the removal tool No. OTC 827B, press the halfshaft from the wheel hub.
10. While holding the shock absorber piston rod with an internal socket wrench, remove the retaining bolt and then remove the strut assembly (4000 only).
11. Remove the spring strut cover (Quattro and Turbo) and remove the three strut retaining nuts, then remove the strut assembly (5000 series only).
12. Installation is in the reverse order of removal. Note the following:
    a. When installing the stabilizer bar, the positioning is correct if the clamps are difficult to install in the rubber bushings. Attach the clamps loosely, take a short test drive to bring the bushings into the correct position and then tighten to 18 ft. lbs.
    b. Tighten the ball joint bolt to 36 ft. lbs. (4000) or 47 ft. lbs. (5000).
    c. Tighten the axle nut to 167 ft. lbs. on the 4000, 203 ft. lbs. on the Quattro and Turbo and 207 ft. lbs. on all other 5000 models.

OVERHAUL

For all spring and shock absorber removal and installation procedures and any other strut overhaul procedures, please refer to "Strut Overhaul" in the Unit Repair section.

Lower Control Arm and Ball Joint

REMOVAL & INSTALLATION

5000 Models

1. Remove the ball joint clamp nut.
2. Pry the control arm down and out of the clamp.
3. Remove the nut on the end of the stabilizer bar.
4. Loosen the control arm-to-subframe mounting bolts and then pull the control arm off of the end of the stabilizer bar.
5. Remove the bolts and remove the control arm.

NOTE: The ball joint and control arm are one unit and can only be replaced as a unit.

6. Installation is the reverse of removal. Torque the control arm-to-ball joint nut to 48 ft. lbs., the control arm-to-subframe bolt to 63 ft. lbs. and the stabilizer bar-to-control arm nut to 81 ft. lbs. Check the toe and camber adjustments.

Ball Joints

REMOVAL & INSTALLATION

4000 Models

1. Mark the position of the ball joint

SECTION 2

SUSPENSION
Import Models

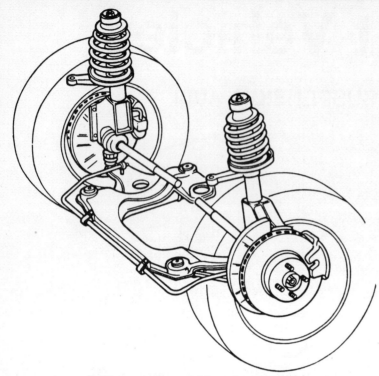

Front suspension and drive axle—5000 models

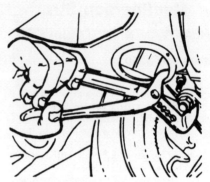

Adjusting the camber on the 4000 and Quattro without the special tool

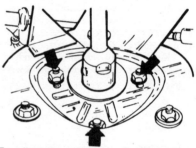

To adjust the camber on the 5000, move the strut assembly in the slots of the spring strut mounting plate

flange on the lower control arm.

2. Remove the ball joint retaining clamp nut and pull the ball joint/control arm down and out of the retaining clamp.

3. Unscrew the two ball flange retaining nuts and remove the ball joint.

4. Installation is the reverse of removal. Tighten the clamp nut to 47 ft. lbs. and tighten the ball joint flange nuts to 47 ft. lbs. (Turbo) or 36 ft. lbs. (non-Turbo).

Lower Control Arm

REMOVAL & INSTALLATION

4000 Models

1. Raise the front of the vehicle and support it with jackstands.

2. Remove the ball joint as detailed earlier.

3. If equipped with a stabilizer bar, disconnect the end of the stabilizer bar and pull it down.

4. Remove the two control arm-to-subframe bolts and the control arm.

5. Installation is in the reverse order of removal. Check control arm bushings for cracking or undue wear. Tighten the control arm-to-subframe bolts to 43 ft. lbs. and the stabilizer bar mounting bolts to 18 ft. lbs.

Front Wheel Bearings

ADJUSTMENT

There is no front wheel bearing adjustment.

The bearing is pressed into the steering knuckle. Axle nut torque is 167 ft. lbs. for the 4000, 203 ft. lbs. for the 5000 and 207 ft. lbs. for the Turbo and Quattro. The axle nut should be tightened only with the wheels resting on the ground.

REMOVAL & INSTALLATION

1. Raise the front of the car and support it with jackstands.
2. Remove the wheels.
3. Remove the caliper assembly.
4. Remove the brake disc.
5. Unscrew two retaining screws and remove the splash shield.
6. Using the removal tool No. OTC 827B, press the halfshaft from the wheel hub.
7. Press off the wheel hub.
8. Remove the inner and outer circlips and press out the wheel bearing.

NOTE: The wheel bearing will be damaged upon removal and must be replaced.

9. Installation is in the reverse order of removal.

Wheel Alignment

NOTE: A suitable alignment rack is necessary to check and adjust the camber and toe-in to specifications.

Before checking wheel alignment, tire pressures should be brought up to specifications and the front ride height checked. The vehicle should be bounced and settled before each alignment check or adjustment. Make camber adjustment first, then toe-in. Caster is set at the factory and is not adjustable other than the replacement of damaged suspension parts. Make all adjustments with the wheels straight ahead.

CAMBER ADJUSTMENT

4000 Models

Loosen both ball joint flange mounting bolts on the control arm. Using the special tool US 4490, or by just moving the ball joint by hand, adjust the camber to specifications.

5000 Models

Loosen the spring strut plate mounting bolts. Attach a socket wrench to the top piston rod nut and move the assembly in the slots until the camber is correct. Tighten all bolts.

TOE-IN ADJUSTMENT

Toe-in can be determined by measuring and comparing the distance between the center of the tire tread, front and rear, or by measuring and comparing the distance between the inside edges of the wheel rims, front and rear. If the wheel rims are used as the basis of measurement, the car should be rolled forward slightly and a second set of

SUSPENSION
Import Models

measurements taken. This avoids any error induced by bent wheels. If at all possible, a toe-in gauge should be used; it will give a much more accurate measurement. Toe-in is adjusted at the steering tie-rods. Turn both rods to lengthen or shorten them an equal amount. If the tie–rods are not adjusted equally, the steering wheel will be crooked and the turning arcs of the front wheels will be changed. If the steering wheel is crooked, it must be removed and repositioned. Tighten the clamps when adjustment is complete.

REAR SUSPENSION AUDI

Rear Struts

For all spring and shock absorber removal and installation procedures and any other strut overhaul procedures, refer to "Strut Overhaul"

REMOVAL & INSTALLATION

4000 (exc. Quattro)

NOTE: Always remove and install the suspension struts one at a time. Do not allow the rear axle to hang in place as this may cause damage to the brake lines.

1. With the car at ground level, open the trunk and remove the sheet metal trim from around the shock tower.
2. Remove the rubber cap.
3. Remove the strut mounting nut.
4. Raise the rear of the car and support it with jackstands.
5. Remove the lower strut mounting bolt from the axle beam and remove the strut.
6. Installation is the reverse of removal. Torque the upper strut mounting bolt to 14 ft. lbs. and the lower strut mounting bolt to 43 ft. lbs.

5000 (exc. Quattro)

NOTE: The struts must be removed with the weight of the vehicle on the rear wheels. If not, a spring compressor must be used on the rear springs.

1. If the vehicle is not on its wheels, install the spring compressor and compress the spring. Do Not attempt to remove the shock with the rear wheels raised without a compressor.
2. Remove the upper strut mounting nut.
3. Remove the lower strut mounting nut.
4. Remove the shock absorber.
5. Installation is the reverse of removal. Torque the lower mounts to 40 ft. lbs. (1980–83) or 66 ft. lbs. (1984–86) and the upper to 14 ft. lbs.

4000 Quattro

1. Loosen the lug nuts and remove the axle nut cover.
2. Remove the axle nut and then raise and support the rear of the vehicle. Remove the wheels.
3. Using a tie-rod end puller, remove the tie-rod end.
4. Remove the brake caliper mounting bolts. Disconnect the brake line from its bracket and then position the caliper out of the way.
5. Remove the brake disc and the ball joint clamp bolt and then pry the ball joint out of the hub.
6. Using a four-armed puller tool No. OTC 8278, press the halfshaft from the strut/wheel hub assembly.
7. Have a helper hold the strut assembly and remove the upper strut mounting nut. Remove the strut.

NOTE: The rear axle assembly must not be under tension while removing the strut.

8. Installation is in the reverse order of removal. Please note the following torque figures:
- Tie-rod nut—29 ft. lbs. (non-Turbo) or 43 ft. lbs. (Turbo).
- Axle nut—203 ft. lbs.
- Upper strut mounting nut—43 ft. lbs.
- Ball joint clamp bolt—54 ft. lbs. (non-Turbo) or 47 ft. lbs. (Turbo).

5000 Quattro

1. Raise and support the rear of the vehicle on jackstands. Remove the wheel.
2. Open the trunk and remove the shock absorber covers, then remove the shock absorber-to-body nuts/bolts.
3. Remove the shock absorber-to-rear wheel knuckle assembly. Remove the shock absorber from the vehicle.
4. To install, reverse the removal procedures. Torque the shock absorber-to-body nuts/bolts to 15 ft. lbs. and the shock absorber-to-rear wheel knuckle assembly bolt to 66 ft. lbs.

OVERHAUL

For all spring and shock absorber removal and installation procedures and any other strut overhaul procedures, refer to "Strut Overhaul"

Rear Control Arms

REMOVAL & INSTALLATION

4000 Quattro

1. Raise and support the rear of the vehicle on jackstands, under the frame and differential.
2. Using a scribing tool, mark the position of the ball joint carrier with the control arm.
3. Remove the ball joint carrier-to-control arm nuts and the lock plate, then separate the ball joint carrier from the control arm.
4. Remove the control arm-to-subframe bolts and the control arm from the vehicle.
5. To install, reverse the removal procedures. Torque the control arm-to-subframe bolts to 43 ft. lbs. Check the rear wheel alignment.

5000 Quattro

On this vehicle, the control arm is known as the Trapezoidal Arm. It is connected to the wheel bearing housing and to two separate cross-members.

1. Raise and support the rear of the vehicle on jackstands, placed under the frame and the differential.
2. Remove the wheel. Along the trapezoidal arm, remove the speed sensor wiring bracket nuts/bolts and the guide.
3. Remove the wheel bearing housing-to-trapezoidal arm front and rear bolts.
4. Remove the trapezoidal arm-to-rear cross-member bolt.
5. At the brake pressure regulator, disconnect the spring.
6. Remove the trapezoidal arm-to-front cross-member nut and the trapezoidal arm from the vehicle.
7. To install, reverse the removal procedures. Torque the trapezoidal arm-to-front cross-member nut to 44 ft. lbs., the trapezoidal arm-to-rear cross-member bolt to 63 ft. lbs., the trapezoidal arm-to-wheel bearing housing bolts to 125 ft.lbs. and the speed sensor guide nut/bolts to 7 ft. lbs. Adjust the rear wheel alignment.

NOTE: Before installing the trapezoidal arm, be sure to coat the fasteners with locking compound D-6.

Rear Wheel Bearings

ADJUSTMENT

All Models—Except Quattro

1. Raise and support the rear of the vehicle on jackstands.
2. Remove the grease cap.
3. Remove the cotter pin and the locking nut.
4. While turning the wheel (so that the wheel bearing does not jam), tighten the adjusting nut firmly.
5. Back the nut off slightly. Using a screwdriver, try to move the thrust washer with finger pressure; when the thrust washer can be move slightly, the correct adjust-

SECTION 2: SUSPENSION
Import Models

ment has been met.

6. To install, place new grease in the grease cap and reverse the removal procedures.

Rear Wheel Alignment

Only the 4000 and the 5000 Quattro's are provided with rear wheel camber and toe adjustments.

NOTE: It is advised to take the vehicle to a qualified alignment shop to have the alignment performed correctly.

FRONT SUSPENSION BMW

MacPherson Strut Assembly

For removal of springs from struts and all strut overhaul procedures, please refer to "Strut Overhaul" in the Unit Repair section.

─────── CAUTION ───────
Macepherson strut springs are under tremendous pressure and any attempt to remove them without proper tools could result in serious personal injury.

REMOVAL & INSTALLATION

318i and 325e

1. Remove the front wheel. Disconnect the brake pad wear indicator plug and ground wire. Pull the wires out of the holder on the strut.

Lock wire location at strut assembly

2. Unbolt the caliper and pull it away from the strut, suspending it with a piece of wire from the body. Do not disconnect the brake line.
3. Remove the attaching nut and then detach the push rod on the stabilizer bar at the strut.
4. Unscrew the attaching nut and press off the guide joint.
5. Unscrew the nut and press off the tie rod joint.
6. Press the bottom of the strut outward and push it over the guide joint pin. Support the bottom of the strut.
7. Unscrew the nuts at the top of the strut (from inside the engine compartment) and then remove the strut.

Install in reverse order, keeping the following points in mind:

a. Replace the self-locking nuts that fasten the top of the strut.
b. Tie rod and guide joints must have both pins and both bores clean for reassembly. Replace both self-locking nuts.
c. Torque the control arm to spring strut attaching nut to 43–51 ft. lb. Torque the spring strut to wheel well nuts to 16–17 ft. lb.

320i

1. Raise the vehicle and support safely. Remove the wheel.
2. Detach the bracket at the strut assembly.
3. Disconnect and suspend the brake caliper with a wire from the vehicle body. Do not disconnect the brake line.
4. Remove the cotter pin and castle nut. Press the tie-rod off the steering knuckle.
5. Remove the three retaining nuts and detach the strut assembly at the wheel house.
6. Installation is the reverse of removal.

524 td, 528e, 533i, 535i, 633CSi, and 635CSi

1. Raise the vehicle and support safely. Remove the wheel.
2. Disconnect the bracket at the strut assembly.
3. Disconnect the brake caliper and suspend from the vehicle body with wire. Do not remove the brake hose.
4. Remove the lock wire and disconnect the tie rod arm at the strut assembly.
5. Remove the three retaining nuts and detach the strut assembly at the wheelhouse.
6. Installation is the reverse of removal.

1980–82 733i

1. Raise the vehicle and support safely. Remove the wheel.
2. Disconnect the vibration strut from the control arm.
3. Disconnect the bracket and clamps from the strut assembly.
4. Disconnect the wire connection and press out the wire from the clamp on the spring strut tube.
5. Remove the brake caliper and suspend it from the vehicle body with a wire. Do not remove the brake hose.
6. Disconnect the tie-rod from the shock absorber.

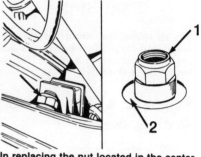

In replacing the nut located in the center of the control arm, use a replacement nut (1) and washer (2) of the type shown

7. Remove the three retaining nuts and disconnect the strut assembly from the wheelhouse.
8. Installation is the reverse of removal.

733i (1983–84) 735i

1. Raise the vehicle and support it safely by the body. Remove the front wheel. Detach the brake line bracket and clamp from the strut.
2. Pull off the rubber cover, and then use an Allen type wrench to unbolt the antilock sensor at the rear of the caliper. Remove it.
3. Pull the antilock sensor electrical connector out of the holder and unplug it. Disconnect the ground wire.
4. Detach the caliper and suspend it from the body without disconnecting the brake line.
5. Support the strut in a secure manner at the bottom. Remove the self-locking nut and through bolt connecting the lower end of the vibration strut where it connects to the control arm.
6. Then, remove the three self-locking nuts from the top of the strut housing in the engine compartment, and remove the strut.
7. Install the strut in reverse order, noting the following points:
a. Use new self-locking nuts on the top of the strut housing and at the connection to the control arm.
b. When reconnecting the lower strut to the track arm, clean the bolt threads and bolt holes and install the bolts with a special bolt tightener HWB No. 81 22 9 400 086 or equivalent.

SUSPENSION
Import Models

Control Arm

REMOVAL & INSTALLATION

524 td, 528e, 528i, 533i, 535i, 633CSi, and 635CSi

1. Raise the vehicle and support safely. Remove the wheel.
2. Disconnect the stabilizer at the control arm.
3. Remove the tension strut nut on the control arm.
4. Disconnect the control arm at the front axle support and remove it from the tension strut.
5. Remove the lock wire, remove the bolts and take the control arm off the spring strut.
6. Remove the cotter pin and nut.
7. Using special tool BMW 00-7-500

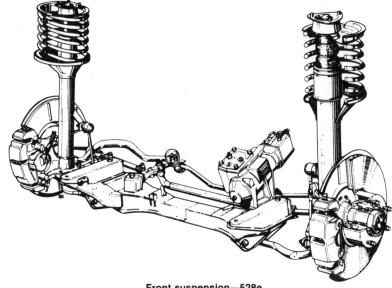

Front suspension—528e

Stabilizer-to-control arm attaching nut —530i, 630i

or equivalent, pull the guide joint from the tie-rod arm.

8. Installation is the reverse of removal.

318i and 325e

1. Remove the front wheel. Disconnect the rear control arm bracket where it connects to the body by removing the two bolts.
2. Remove the nut and disconnect the thrust rod on the front stabilizer bar where it connects to the center of the control arm.
3. Unscrew the nut which attaches the front of the stabilizer bar to the crossmember and remove the nut from above the crossmember. Then, use a plastic hammer to knock this support pin out of the crossmember.
4. Unscrew the nut and press off the guide joint where the control arm attaches to the lower end of the strut.
5. Reverse the procedure to install. Keep these points in mind:
   a. Replace the self-locking nut that fastens the guide joint to the control arm.
   b. Make sure the support pin and the bore in the crossmember are clean before inserting the pin through the crossmem-

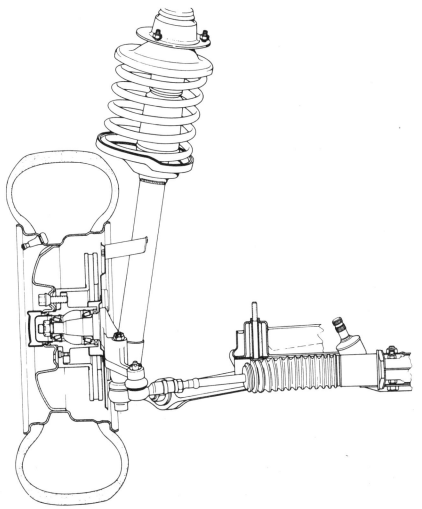

Front suspension—320i

2-189

SECTION 2: SUSPENSION
Import Models

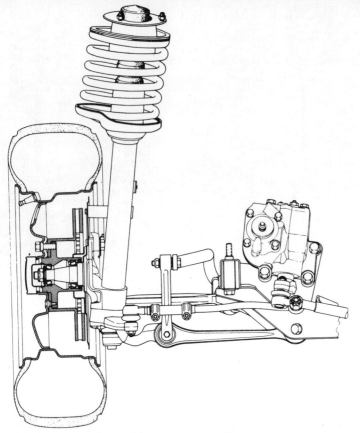

Front suspension—733i

ber. Replace the original nut with a replacement nut and washer equivalent to those shown in the illustration.

c. Torque the control arm-to-spring strut nut to 43–51 ft. lb. Torque the control arm support to crossmember nut to 29–34 ft. lb. Torque the push rod on the stabilizer bar to 29–34 ft. lb.

320i

1. Disconnect the stabilizer at the control arm.
2. Disconnect the control arm at the front axle support.
3. Remove the cotter pin and castle nut.
4. Press the control arm off the steering knuckle with special tool BMW 31-1-100 or equivalent.
5. Installation is the reverse of removal.

733i and 735i

1. Raise the vehicle and support safely. Remove the wheel.
2. Disconnect the vibration strut from the control arm.
3. Disconnect the control arm from the axle carrier.
4. Disconnect the tie rod arm from the front strut.
5. Remove the cotter pin and castle nut. Press off the control arm with special tool BMW 31-1-110 or equivalent.

Vibration strut attaching bolt—733i

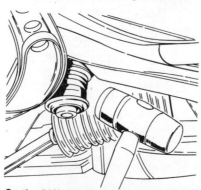

On the 318i, knock the pin loose with a soft hammer, as shown

6. Installation is the reverse of removal. Use new self-locking nuts on the connections at the vibration strut and axle carrier. On 1983 and later models, when reconnecting the arm to the tie rod arm, use a bolt tightener HWB No. 81 22 9 400 086 and make sure the threads and bolt-holes are clean.

Front Wheel Bearings

ADJUSTMENT

NOTE: The 318i and 325e wheel bearings cannot be adjusted.

320i, 533i, 535i, 633i, 635i, 733i, 735i

1. Raise the vehicle, support it and remove the front wheel.
2. Remove the end cap, and then straighten the cotter pin and remove it. Loosen the castellated nut.
3. While continuously spinning the brake disc, torque the castellated nut down to 22–24 ft. lbs. Keep turning the disc thru-out this and make sure it turns at least two turns after the nut is torqued and held.
4. Loosen the nut until there is end play and the hub rotates with the nut.
5. Torque the nut to no more than 2 ft. lbs. Finally, loosen slowly just until castellations and the nearest cotter pin hole line up and insert a new cotter pin.
6. Make sure the slotted washer is free to turn without noticeable resistance; otherwise, there is no end play and the bearings will wear excessively.

528i

1. Remove the wheel. Remove the locking cap from the hub by gripping it carefully on both sides with a pair of pliers.
2. Remove the cotter pin from the castellated nut, and loosen the nut.
3. Spin the disc constantly while torquing the nut to 7 ft lbs. Continue spinning the disc a couple of turns after the nut is torqued and held.
4. Loosen the castellated nut ¼–⅓ turn—until the slotted washer can be turned readily.
5. Fasten a dial indicator to the front suspension and rest the pin against the wheel hub. Preload the meter about 0.039 in. to remove any play.
6. Adjust the position of the castellated nut while reading the play on the indicator. Make the play as small as possible while backing off the castellated nut just until a new cotter pin can be inserted. The permissible range is 0.0008–0.004 in.
7. Install the new cotter pin, locking cap, and the wheel.

SUSPENSION
Import Models

524 td, 528e

1. Raise the vehicle and support it safely. Remove the front wheels.
2. Remove the bearing cap and the selflocking nut.
3. Install a new nut and tighten it to specified torque.
4. Lock the nut by hitting it with a round punch several times on the outer lip.
5. Install bearing cap and front wheels. Lower the vehicle.

REMOVAL & INSTALLATION (PACKING)

Except 318i, 325e, 733i, and 735i

1. Remove the wheel. Unbolt and remove the caliper. Hang it from the body. Do not disconnect or stress the hose. On models with a separate disc, remove the locking cap by gripping carefully on both sides with a pair of pliers, remove the cotter pin from the castellated nut, and remove the nut and, where equipped, the slotted washer. Then, remove the entire hub and bearing.
2. Remove the shaft sealing ring and take out the roller bearing.
3. On most models, the outer bearing race may be forced out through the recesses in the wheel hub. A BMW puller 00 8 550 or the equivalent may also be used. On the 733i, the recesses are not provided and a puller is necessary.
4. Clean all bearings and races and the interior of the hub with alcohol, and allow to air dry.

NOTE: Do not dry with compressed air as this can damage the bearings by rolling them over one another unlubricated or force one loose from the cage and injure you.

5. Replace *all* bearings and races if there is any sign of scoring or galling.
6. Press in the outer races with a suitable sleeve. Pack a new shaft seal with graphite grease and refill the hub with fresh grease.
7. Assemble in this order: outer race; inner race; outer race; inner race; shaft seal.
8. If necessary, adjust the wheel bearing play as described above.

318i and 325e

NOTE: The bearings on the 318i and 325e are only removed if they are worn. They cannot be removed without destroying them (due to side thrust created by the bearing puller). They are not periodically disassembled, repacked and adjusted.

1. Remove the front wheel and support the car. Remove the attaching bolts and remove and suspend the brake caliper, hanging it from the body so as to avoid putting stress on the brake line.
2. Pull off the brake disc and pry off the dust cover with a small prybar. Remove the setscrew.
3. Using a chisel, knock the tab on the collar nut away from the shaft. Unscrew and discard the nut.
4. Pull off the bearing with a puller and discard.
5. If the inside bearing inner race remains on the stub axle, unbolt and remove the dust guard. Bend back the inner dust guard and pull the inner race off with a special tool capable of getting under the

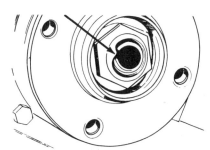

On 318i, unlock the collar nut as shown with a chisel, by applying force in the direction shown by the arrow

race (BMW 00 7 500 and 33 1 309 or equivalent. Reinstall the dust guard.
6. Then install a special tool (BMW 31 2 120 or equivalent) over the stub axle and screw it in for the entire length of the guide sleeve's threads. Press the bearing on.
7. Reverse the remaining removal procedures to install the disc and caliper. Torque the wheel hub collar nut to 188 ft. lb. Lock the collar nut by bending over the tab.

733i and 735i

1. Remove the front wheel. Detach the brake line clamp and bracket on the strut. On cars with ABS, remove the rubber boot and unbolt and remove the anti-lock sensor.
2. Detach the caliper without disconnecting the brake line and suspend it.
3. Use a tool such as 21 2 000 to remove the end cap. Remove the cotter pin and unscrew the castellated nut. Then, remove the stepped washer, brake disc, and wheel hub.
4. Use an Allen type wrench to unscrew the bolt and separate the disc from the wheel hub.
5. Lift out the shaft seal. Remove the tapered roller bearings. Knock out the outer races if they show scoring with a punch by tapping all around.
6. Press in new outer races with special tools such as 31 2 061 for the inside bearings and 31 2 062 for the outside bearings. Pack the two inner races with wheel bearing grease. The order of installation is: outer race, inner race, outer race, inner race, shaft seal. To install the seal, lubricate the sealing lip of the shaft seal with grease and install the seal with a tool such as 31 2 040. Repack all bearings thoroughly with wheel bearing grease before installation.
7. Insall in reverse order, adjusting the wheel bearings as described above.

Front Suspension Alignment

CASTER AND CAMBER

Caster and camber are not adjustable, except for replacement of bent or worn parts.
On the 318i and 325e, camber that is out of specification because of excessive tolerances can be corrected by installing eccentric mounts. This cannot be done to correct misalignment caused by a collision, however.

TOE-IN ADJUSTMENT

Toe-in is adjusted by changing the length of the tie rod and tie rod end assembly. When adjusting the tie rod ends, adjust each by equal amount (in the opposite direction) to increase or decrease the toe-in measurement.

REAR SUSPENSION BMW

MacPherson Strut Assembly

For all spring and shock absorber removal and installation procedures and any other strut overhaul procedures, please refer to "Strut Overhaul"

CAUTION
MacPherson strut springs are under tremendous pressure and any attempt to remove them without proper tools could result in serious personal injury.

REMOVAL & INSTALLATION

1. On model 1980–82 733i remove the rear seat and back rest. On 1983 and later 733i, 735i, remove the trim from over the wheel well in the trunk.
2. Jack up the car and support the control arms.

SECTION 2 SUSPENSION
Import Models

NOTE: On the 318i and 325e, the spring and shock absorber are separate. The control arm must be securely supported throughout this procedure.

CAUTION
The coil spring, shock absorber assembly acts as a strap so the control arm should always be supported.

---

3. Remove the lower shock retaining bolt.
4. Disconnect the upper strut retaining nuts at the wheel arch and remove the assembly.

NOTE: On the 318i and 325e, this is located behind the trim panel in the trunk. On these models only, the shock absorber, because it is separate from the spring, may now be replaced.

5. Using the appropriate spring compressor, compress the coil spring far enough to remove the centering cup, then release the coil spring and separate the spring, boot and shock absorber.
6. Install in reverse order, using new gaskets between the unit and the wheel arch, and new self-locking nuts on top of the strut. Torque the shock-to-body nuts to 16–17 ft. lb.; spring retainer-to-wheel house nuts (6 cyl.) to 16–17 ft. lb.; lower bolt to 52–63 ft. lb. (4 cyl.), 90–103 ft. lb. (6 cyl.). On '83 and later 733i, 735i, replace the gasket that goes between the top of the strut and the lower surface of the wheel well. Final torquing of the lower strut bolt should be done with the car in the normal riding position.

Rear Spring
REMOVAL & INSTALLATION
318i and 325e

1. Disconnect the rear portion of the exhaust system and hang it from the body.
2. Disconnect the final drive rubber mount, push it down, and hold it down with a wedge.
3. Remove the bolt that connects the rear stabilizer bar to the strut on the side you're working on. Be careful not to damage the brake line.

NOTE: Support the lower control arm securely with a jack or other device that will permit you to lower it gradually, while maintaining secure support.

4. Then, to prevent damage to the output shaft joints, lower the control arm *only* enough to slip the coil spring off the retainer.

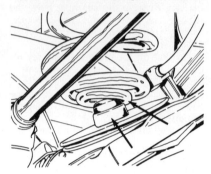

On the 318i, lower the trailing arm just enough to get the spring off the locating tang

5. Make sure, in replacing the spring, that the same part number, color code, and proper rubber ring are used. Reverse all removal procedures to install, making sure that the spring is in proper position, keeping the control arm securely supported until the shock bolt is replaced, and tightening stabilizer bar and lower shock mount bolts with the control arm in the normal ride position. Torque the stabilizer bolt to 22–24 ft. lb., and the shock bolt to 52–63 ft. lb.

Stabilizer
REMOVAL & INSTALLATION

1. Disconnect the stabilizer from the trailing arm by removing the connect bolt.
2. Disconnect the stabilizer on the crossmember.

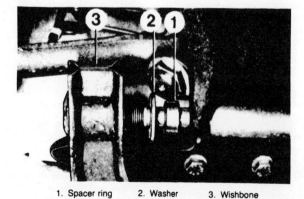

1. Spacer ring 2. Washer 3. Wishbone

Detaching the lower arm at front axle beam

3. Check the rubber bushings for wear and replace as necessary.

Rear Control Arm
REMOVAL & INSTALLATION
524td, 528e, and 528i

1. Remove the parking brake lever.
2. Plug the front hose to prevent loss of brake fluid in the reservoir.
3. Support the body.
4. Disconnect the brake line at the brake hose.
5. Disconnect the driveshaft.
6. Disconnect the stabilizer and coil spring at the control arm.
7. Disconnect the control arm at the axle carrier.
8. Installation is the reverse of removal. Bleed brake system.

733i and 735i

1. Remove the rear wheel.
2. Pull up the parking brake lever and disconnect the output shaft at the drive flange.
3. Remove the parking brake lever.
4. Remove the brake fluid from the reservoir.
5. Disconnect the brake line.
6. Disconnect the control arm from the rear axle carrier.
7. Disconnect the shock absorber and remove the control arm.
8. Installation is the reverse of removal. When reattaching the trailing arm, insert the bolt on the inner bracket first. Final tighten all mounting bolts with the car resting on its wheels. Bleed the brake system.

Trailing Arm
REMOVAL & INSTALLATION
533i, 535i, 633CSi and 635CSi

1. Remove the parking brake lever.
2. Remove the rear wheel.
3. Using vise grips, clamp the front hose to prevent loss of fluid.
4. Support the body.
5. Pull the parking brake cable out of the pipe.
6. Disconnect the stabilizer and spring strut at the trailing arm.
7. Disconnect the brake line at the brake hose.
8. Disconnect the trailing arm at the rear axle support.
9. Detach the output shaft.
10. Disconnect the brake pad wear indicator wire at the right trailing arm and take the wire out of the clamps.
11. Installation is the reverse of removal. Bleed the system.

SUSPENSION
Import Models

318i, 320i, and 325e

1. Raise the vehicle and remove the rear wheel. Apply the parking brake and disconnect the output shaft at the rear axle shaft. Then, on the 320i, disconnect the parking brake cable at the handbrake. On the 318i and 325e, remove the parking brake lever.
2. Remove the brake fluid from the master cylinder reservoir on the 318i and 325e. Disconnect the brake line connection on the rear control arm on both types of car. Plug the openings.
3. Support the control arm securely. Disconnect the shock absorber at the control arm. On 318i and 325e, lower the control arm slowly and remove the spring. On the 320i, the control arm need not be lowered slowly because the spring is integral with the strut.
4. Remove the nuts and then slide the bolts out of the mounts where the control arm is mounted to the axle carrier.
5. Install in reverse order. Torque the bolts holding the trailing arm to the axle carrier to 48–54 ft. lb. On the 318i and 325e, make sure the spring is positioned properly top and bottom. Torque the strut bolt to 52–63 ft. lb. Reinstall the handbrake or reconnect the cable and adjust. Then apply the brake and reconnect the output shaft. Reconnect the brake line, replenish with the proper brake fluid, and bleed the system.

FRONT SUSPENSION CHRYSLER CORP.

MacPherson Strut

REMOVAL & INSTALLATION

Rear Wheel Drive

ALL EXCEPT CONQUEST

1. Loosen the lug nuts, jack up the front of the car (after blocking the rear wheels) and support safely on jackstands.
2. Remove the brake caliper, hub and brake disc rotors. Disconnect the stabilizer link from the lower arm, remove the three steering knuckle-to-strut assembly bolts. Carefully force the lower control arm down and separate the strut assembly and the steering knuckle. Unscrew the three retaining nuts at the top and remove the strut assembly.

NOTE: On some models the lower splash shield may interfere with the strut removal.

4. To install the strut, position the strut assembly in the fender. Install the upper retaining nuts hand tight.
5. Apply sealer on the mounting flange and fasten the strut assembly to the steering knuckle. Connect the brake hose if removed.
6. Tighten the upper retaining nuts to 7–11 ft.lbs. Torque the knuckle bolts to 30–36 ft.lbs.
7. Assemble the stabilizer link and fasten to the lower control arm if removed.
8. Assemble the remaining parts in the opposite order of removal. If the brake line was disconnected during the strut removal, the brake system will have to be bled.

CONQUEST

1. Raise and support the front end on jackstands. Let the wheels hang.
2. Remove the caliper and suspend it out of the way.
3. Remove the hub and rotor assembly.
4. Remove the brake dust cover.
5. Unbolt the strut from the knuckle.
6. Remove the strut-to-fender liner nuts and lift out the strut. Installation is the reverse of removal. Torque the strut-to-fender liner nuts to 18–25 ft.lb.; the strut-to-knuckle arm bolts to 58–72 ft.lb.

Front Wheel Drive

1980–84 COLT AND CHAMP, 1985 AND LATER VISTA

1. Raise and support the front end.
2. Remove the front wheels.
3. Detach the brake hose from the clip on the strut.
4. Remove the three nut securing the strut to the fender housing.
5. Unbolt the strut from the knuckle.
6. Remove the strut from the car.
7. Installation is the reverse of removal. Torque the strut-to-knuckle bolts to 55–65 ft.lb. Torque the strut-to-fender housing nuts to 7–11 ft.lbs. on Colts and Champs; 18–25 ft.lb. on Vistas.

1985 AND LATER COLT

1. Raise and support the front end on jackstands.
2. Remove the front wheels.
3. Detach the brake hose from the clip on the strut.

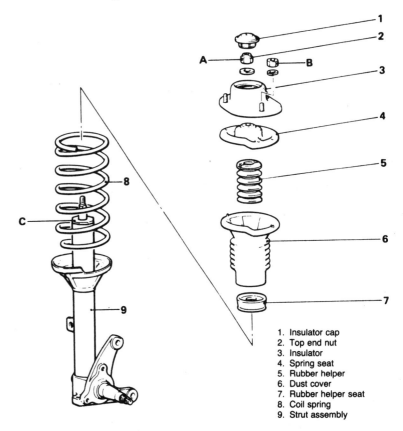

1. Insulator cap
2. Top end nut
3. Insulator
4. Spring seat
5. Rubber helper
6. Dust cover
7. Rubber helper seat
8. Coil spring
9. Strut assembly

Conquest front strut

SECTION 2

SUSPENSION
Import Models

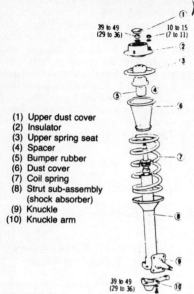

(1) Upper dust cover
(2) Insulator
(3) Upper spring seat
(4) Spacer
(5) Bumper rubber
(6) Dust cover
(7) Coil spring
(8) Strut sub-assembly (shock absorber)
(9) Knuckle
(10) Knuckle arm

Front strut on all rear wheel drive cars except Conquest

4. Unbolt the strut from the knuckle.
5. Remove the dust cover from the upper end of the strut.
6. Insert a ½ in. drive socket in the top of the strut to hold the gland nut.
7. Insert an allen wrench in the socket and into the retaining nut. Loosen the retaining nut while keeping the gland nut from turning. Remove the socket, allen wrench and retaining nut.
8. Remove the strut from the car.
9. Installation is the reverse of removal. Torque the strut-to-knuckle bolts to 55–65 ft.lb.; the upper retainer nut to 33–43 ft.lb.

OVERHAUL

For all spring and shock absorber REMOVAL and INSTALLATION procedures, and all strut overhaul procedures, please refer to "Strut Overhaul"

Ball Joint

REMOVAL & INSTALLATION

Rear Wheel Drive
1980

Remove the lower control arm. Remove the snap-ring retaining the ball joint to the control arm. Press out the ball joint using a ball joint remover or have your local automotive machine shop do the job for you. Press a new ball joint into the control arm. Make sure the ball joint does not cock when installing in control arm seat. Installation is in reverse order. Refer to control arm section for torque specifications.

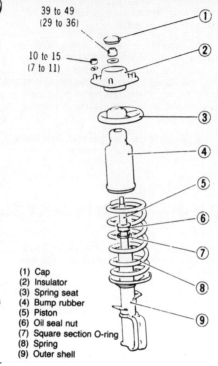

(1) Cap
(2) Insulator
(3) Spring seat
(4) Bump rubber
(5) Piston
(6) Oil seal nut
(7) Square section O-ring
(8) Spring
(9) Outer shell

Tightening torque Nm (ft-lbs.)

1980—84 front wheel drive Colt front strut

1981 AND LATER

Remove the tire and wheel assembly. Remove the brake caliper and support from the mounting adapter. Anchor assembly out of the way with wire to the strut spring. Remove the tie rod end nut and separate the tie rod end from the steering knuckle using a removing tool. Remove the bolts securing the strut assembly to the steering knuckle. Tap the connection with a plastic hammer to separate. Remove the ball joint to control arm mounting bolts and remove the ball joint with the knuckle arm attached. Remove the ball joint stud nut and separate the ball joint and knuckle arm. Installation is the reverse of removal. Torque specifications are; Ball joint stud nut; 43–52 ft.lbs. Strut to knuckle bolts; 58–78 ft.lbs. Ball joint to control arm bolts; 43–51 ft.lbs.

— **CAUTION** —
When self-locking nuts are removed, always replace with new self-locking nuts.

Front Wheel Drive
1980–84

Unbolt the ball joint from the control arm. Use a ball joint removing tool and separate the ball joint from the steering knuckle after removing the stud retaining nut. Install in the reverse order. Torque specifications are; Ball joint to control arm; 69–87 ft.lbs. Ball joint stud nut; 40–51 ft.lbs.

1985 AND LATER COLT

The ball joint is not replaceable. The ball joint and lower control arm must be replaced as an assembly.

1985 AND LATER VISTA

NOTE: This procedure requires a hydraulic press.

1. See the "Lower Control Arm" procedure and disconnect the lower arm.
2. Remove the ball joint dust cover.
3. Using snapring pliers, remove the snapring from the ball joint.
4. Using an adapter plate and driver, such as tool MB990800, press the ball joint from the arm.
5. Installation is the reverse of removal, inverting the tool in the press for installation. Coat the lip and interior of the dust cover with lithium based chassis lube.

Lower Control Arm

REMOVAL & INSTALLATION

Rear Wheel Drive
EXCEPT CONQUEST

1. Loosen the lug nuts and then raise the front end of the car.
2. Remove the caliper, hub, and disc.
3. Disconnect the stabilizer link and strut bar from the lower control arm. Depending on year, remove idler arm support from the chassis and move steering linkage to gain clearance.
4. Remove the three steering knuckle-to-strut assembly bolts.
5. Carefully force the lower control arm down and separate the strut assembly and the steering knuckle.
6. Unscrew the three retaining nuts at the top and withdraw the strut assembly, if necessary.
7. Using a puller, disconnect the steering knuckle arm and the tie-rod ball joint.
8. Again using a puller, disconnect the knuckle arm and the lower arm ball joint.
9. Remove the control arm-to-crossmember bolts and remove the control arm.
10. Install the lower arm on the crossmember. Tighten the bolts to 51–57 ft.lbs. (1980) and 58–69 ft. lbs. (1981 and later) the chamfered end of the nut should be facing the round surface of the bracket.
11. Tighten the steering knuckle arm-to-control arm ball joint nut to 30–40 ft. lbs. (1980) and 52–69 ft. lbs. (1981 and later).
12. Install the strut assembly into the fender. Tighten the top mounting nuts to 7–10 ft. lbs. (1980) and 18–25 ft. lbs. (1981 and later).
13. Apply sealer to the lower end of the strut. Install and tighten the strut-to-steering knuckle arm bolts to 30–36 ft. lbs. (1980) and 58–72 ft. lbs. (1981 and later).

SUSPENSION
Import Models

14. Assemble the stabilizer link and fasten it and strut bar to the lower control arm.
15. Install the backing plate, brake disc, hub, and caliper. Tighten strut bar to 18–25 ft. lbs.
16. Install the wheel and lower the car.
17. Jounce the car up and down a few times and then tighten the stabilizer bolt to 7–10 ft. lbs.

CONQUEST

NOTE: This procedure requires the use of a special tool.

1. Raise and support the front end on jackstands.
2. Remove the front wheels.
3. Remove the caliper and suspend it out of the way.
4. Remove the hub and rotor assembly.
5. Disconnect the stabilizer bar and strut bar from the lower arm.
6. Using a separator, remove the attaching nut and disconnect the tie rod from the knuckle arm.
7. Unbolt the strut from the knuckle arm.
8. Unbolt the lower control arm and knuckle arm assembly from the crossmember.
9. Using special tool MB990635, separate the knuckle arm from the lower control arm.
10. Installation is the reverse of removal. Apply sealant to the flange of the knuckle arm where it mates with the strut. Torque the lower control arm shaft bolt to 60–70 ft.lbs.; the ball joint-to-knuckle arm nut to 45–55 ft.lbs.

Front Wheel Drive Models

1980–84

1. Loosen front wheel lugs, block rear wheels, jack up the front of the car and support on jackstands.
2. Remove the front wheels. Remove the lower splash shield.
3. Disconnect the lower ball joint by unfastening the nuts and bolts mounting it to the control arm. It is not necessary to remove the ball joint from the knuckle.
4. Remove the strut bar and the control arm inner mounting nut and bolt. Remove the control arm.
5. Assembly is the reverse of removal.
Torque Specifications are:
- Inner mount bolt; 69–87 ft.lbs.
- Ball joint mount: 69–87 ft.lbs.
- Ball joint nut: 40–51 ft.lbs.

1985 AND LATER COLT

1. Raise and support the front end. Remove the wheels. Remove the splash shield.
2. Disconnect the stabilizer bar from the lower arm.
3. Using a ball joint separator, disconnect the ball joint from the knuckle.
4. Unbolt the lower arm from the body and remove it from the car.

NOTE: The ball joint cannot be separated from the control arm, but must be replaced as an assembly.

5. If the stabilizer bar is to be removed, disconnect the tie rod from the knuckle, and unbolt and remove the stabilizer.
6. Check all parts for wear and damage and replace any suspect part.
7. Using an in.lb. torque wrench, check the ball joint starting torque. Nominal starting effort should be 22–87 inch lbs. Replace it if otherwise.
8. Installation is the reverse of removal. Use a new dust cover, the lip and inside of which is coated with lithium based chassis lube. The dust cover should be hammered into place with a driver, such as tool MB990800. Install the stabilizer bar so that the serrations on the horizontal part protrude 6mm to the inside of the clamp and 23mm of threaded stud appear below the nut at the control arm. The washer on the lower arm shaft should be installed as shown in the accompanying illustration. The left side lower arm shaft has left-handed threads. The lower arm shaft nut must be torqued with the wheels hanging freely. Observe the following torques:
- Knuckle-to-strut: 54–65 ft.lbs.
- Lower arm shaft-to-body: 118–125 ft.lbs.
- Stabilizer bar-to-body: 12–20 ft.lbs.
- Ball joint-to-knuckle: 44–53 ft.lbs.
- Lower arm-to-shaft: 70–88 ft.lbs.

1985 AND LATER VISTA

1. Raise and support the front end.
2. Remove the wheels.
3. Disconnect the stabilizer bar and strut bar from the lower arm.
4. Remove the nut and disconnect the ball joint from the knuckle with a separator.
5. Unbolt the lower arm from the crossmember.
6. Check all parts for wear or damage and replace any suspect part.
7. Using an inch lbs. torque wrench, check the ball joint starting torque. Starting torque should be 20–86 inch lbs. If it is not within that range, replace the ball joint.

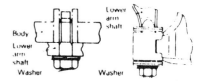

Installation of washer on lower shaft

8. Installation is the reverse of removal. Tighten all fasteners with the wheels hanging freely. Observe the following torques:
- Ball joint-to-knuckle: 44–53 ft.lb.

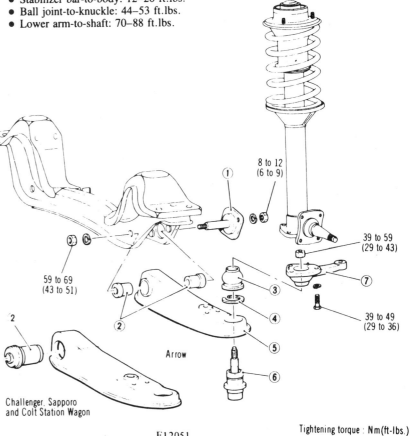

Tightening torque : Nm(ft-lbs.)

Lower control arm and ball joint used on all rear wheel drive cars, except Conquest

SECTION 2
SUSPENSION
Import Models

- Arm-to-crossmember:
  2wd: 90–111 ft.lb.
  4wd: 58–68 ft.lb.
  Stabilizer bar hanger brackets: 7–9 ft.lbs.

NOTE: When installing the stabilizer bar, the nut on the bar-to-crossmember bolts and the bar-to-lower arm bolts, are not torqued, but turned on until a certain length of thread is exposed above the nut:
  2wd stabilizer bar-
  to-crossmember: 0.31–0.39 in.
  2wd stabilizer bar-
  to-lower control arm: 0.31–0.39 in.
  4wd stabilizer bar-
  to-crossmember: 0.31–0.39 in.
  4wd stabilizer bar-
  to-lower arm: 0.51–0.59 in.

Front Wheel Bearings

NOTE: Please refer to the "Drive Axle" Section for FWD Models

REMOVAL & INSTALLATION

Rear Wheel Drive

1. Remove the caliper (pin type) or the caliper and support (sliding type).

NOTE: On sliding type calipers, remove the caliper and support as a unit by unfastening the bolts holding it to the adapter ("backing plate"). Support the caliper with wire, do not allow the weight to be supported by the brake hose.

2. Pry off the dust cap. Tap out and discard the cotter pin. Remove the locknut.
3. Being careful not to drop the outer bearing, pull off the brake disc and wheel hub.
4. Remove the grease inside the wheel hub.
5. Using a brass drift, carefully drive the outer bearing race out of the hub.
6. Rewmove the inner bearing seal and bearing.
7. Check the bearings for wear or damage and replace them if necessary.
8. Coat the inner surface of the hub with grease.
9. Grease the outer surface of the bearing race and drift it into place in the hub.
10. Pack the inner and outer wheel bearings with grease. (see repacking).

NOTE: If the brake disc has been removed and/or replaced; tighten the retaining bolts to 25–29 ft. lbs.

11. Install the inner bearing in the hub. Being careful not to distort it, install the oil seal with its lip facing the bearing. Drive the seal on until it outer edge is even with the edge of the hub.
12. Install the hub/disc assembly on the spindle, being careful not to damage the oil seal.
13. Install the outer bearing, washer, and spindle nut. Adjusting the bearing as follows.

ADJUSTMENT

Rear Wheel Drive

1. Remove the wheel and dust cover. Remove the cotter pin and lock cap from the nut.
2. Torque the wheel bearing nut to 14.5 ft. lbs. (19.6 Nm) and then loosen the nut. Retorque the nut to 3.6 ft. lbs. (4.9 Nm) and install the lock cap and cotter pin.
3. Install the dust cover and the wheel.

Front Wheel Alignment

CASTER AND CAMBER

Caster is preset at the factory. It requires adjustment only if the suspension and steering linkage components are damaged, in which case, repair is accomplished by replacing the damaged part. A slight caster adjustment can be made by moving the nuts on the front anchors of the strut bars.

TOE-IN

Toe-in is the difference in the distance between the front wheels, as measured at both the front and the rear of the front tires.

Toe-in is adjusted by turning the tie rod turnbuckles as necessary. The turnbuckles should always be tightened or loosened the same amount for both tie rods; the difference in length between the two tie rods should not exceed 0.2 in. On the Challenger and Sapporo, only the left tie rod is adjustable.

REAR SUSPENSION CHRYSLER CORP.

Leaf Springs

REMOVAL & INSTALLATION

Rear Wheel Drive Except Conquest and Station Wagons

1. Remove the hub cap or wheel cover. Loosen the lug nuts.
2. Raise the rear of the car. Install a stand at the exact point at which the two dimples locate the support point on the sill flange.

— CAUTION —
Damage to the unit body can result from installing a stand at any other location.

3. Disconnect the lower mounting nut of the shock absorber.
4. Remove the four U-bolt fastening nuts from the spring seat.

NOTE: It's not necessary to remove the shock absorber, leave the top connected.

5. Place a floor jack under the rear axle and raise it just enough to remove the load from the springs. Remove the spring pad and seat.
6. Remove the two rear shackle attaching nuts and remove the rear shackle.
7. Remove the front pin retaining nut. Remove the two pin retaining bolts and take off the pin.
8. Remove the spring.

NOTE: It is a good safety practice to replace used suspension fasteners with new parts.

9. Install the front spring eye bushings from both sides of the eye with the bushing flanges facing out.
10. Insert the spring pin assembly from the body side and fasten it with the bolts. Temporarily tighten the spring pin nut.
11. Install the rear eye bushings in the same manner as the front, insert the shackle pins from the outside of the car, and temporarily tighten the nut after installing the shackle plate.
12. Install the pads on both sides of the spring, aligning the pad center holes with the spring center bolt collar, and then install the spring seat with its center hole through the spring center collar.

13. Attach the assembled spring and spring seat to the axle housing with the axle housing spring center hole meeting with the spring center bolt and install the U-bolt nuts. Tighten the nuts to 33–36 ft. lbs.
14. Tighten the lower shock absorber nut to 12–15 ft. lbs. on all models.
15. Lower the car to the floor, jounce it a few times, and then tighten the spring pin and shackle pin nuts to 36–43 ft. lbs.

Coil Springs

REMOVAL & INSTALLATION

Front Wheel Drive and Station Wagons Except 4WD Vista

1. Raise and support the car safely allowing the rear axle to hang unsupported.
2. Place a jack under the rear axle, and remove the bottom bolts or nuts of the shock absorbers.
3. Lower the rear axle and remove the left and right coil springs.
4. Installation is the reverse of removal.

SUSPENSION
Import Models

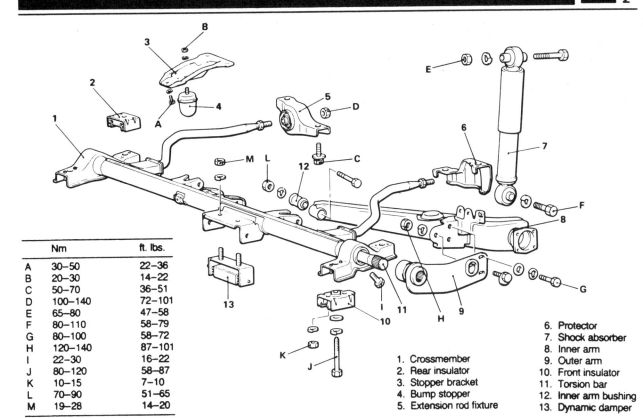

	Nm	ft. lbs.
A	30-50	22-36
B	20-30	14-22
C	50-70	36-51
D	100-140	72-101
E	65-80	47-58
F	80-110	58-79
G	80-100	58-72
H	120-140	87-101
I	22-30	16-22
J	80-120	58-87
K	10-15	7-10
L	70-90	51-65
M	19-28	14-20

1. Crossmember
2. Rear insulator
3. Stopper bracket
4. Bump stopper
5. Extension rod fixture
6. Protector
7. Shock absorber
8. Inner arm
9. Outer arm
10. Front insulator
11. Torsion bar
12. Inner arm bushing
13. Dynamic damper

4wd Vista rear suspension

NOTE: When installing the spring, pay attention to the difference in shape between the upper and lower spring seats.

Torsion Bar and Control Arms

Instead of springs, the 4 wd Vista uses transversely mounted torsion bars housed inside the rear crossmember, attached to which are inner and outer control arms. The conventional style shock absorbers are mounted on the inner arms.

REMOVAL & INSTALLATION

4wd Vista

1. Raise and support the car with jackstands under the frame.
2. Remove the differential as described above.
3. Remove the intermediate shafts and axleshafts.
4. Remove the rear brake assemblies.
5. Disconnect the brake lines and parking brake cables from the inner arms.
6. Remove the main muffler.
7. Raise the inner arms slightly with a floor jack and disconnect the shock absorbers.
8. Matchmark, precisely, the upper ends of the outer arms, the torsion bar ends and the top of the crossmember bracket and remove the inner and outer arm attaching bolts.
9. Remove the extension rods fixtures' attaching bolts.
10. Remove the crossmember attaching bolts and remove the rear suspension assembly from the car.
11. Unbolt and remove the damper from the crossmember.
12. Remove the front and rear insulators from both ends of the crossmember.
13. Loosen, but do not remove, the lockbolts securing the outer arm bushings at both ends of the crossmember.
14. Pull the outer arm from the crossmember. Many times, the torsion bar will slide out of the crossmember with the outer arm.
15. Remove the torsion bar from either the crossmember or outer arm.

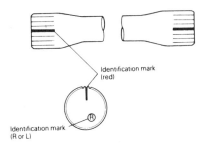

Torsion bar suspension identifying marks

16. Inspect all parts for wear or damage. Inspect the crossmember for bending or deformation.
17. Inner arm bushings may be replaced at this time using a press. The thicker end of the bushings goes on the inner side.
18. Prior to installation note that the torsion bars are marked with an L or R on the outer end, and are not interchangeable.
19. If the original torsion bars are being installed, align the identification marks on the torsion bar end, crossmember and outer arm, install the torsion bar and arm and tighten the lockbolts. Skip Step 20. If new torsion bars are being installed, procede to Step 20.
20. A special alignment jib must be fabricated. See the accompanying illustration for the dimension needed to make this jig. The jig is bolted to the rear insulator hole on the crossmember bracket as shown. Install the crossmember and inner arms. Insert the torsion bar into the outer arm, aligning the red identification mark on the torsion bar end with the matchmark made on the outer arm top side. Install the torsion bar and arm so that the center of the flanged bolt hole on the arm is 32mm below the lower marking line on the jig. Then, pull the outer arm off of the torsion bar, leaving the bar undisturbed in the crossmember. Reposition the arm on the torsion bar, one serration counterclockwise from its former position. This will make the previously measured dimension, 33mm above the lower line. In any event, when the outer arm and

2-197

SECTION 2: SUSPENSION
Import Models

torsion bar are properly positioned, the marking lines on the jig will run diagonally across the center of the toe-in adjustment hole as shown. When the adjustment is complete, tighten the lockbolts. The clearance between the outer arm and the crossmember bracket, at the torsion bar, should be 5.0–7.0mm.

21. The remainder of installation is the reverse of removal. Observe the following torques:
   Extension rod fixture bolts: 45–50 ft.lb.
   Extension rod-to-fixture nut: 95–100 ft.lbs.
   Shock absorber lower bolt: 75–80 ft.lbs.
   Outer arm attaching bolts: 65–70 ft.lbs.
   Toe-in bolt: 95–100 ft.lbs.
   Lockbolts: 20–22 ft.lbs.
   Crossmember attaching bolts: 80–85 ft.lbs.
   Front insulator nuts: 7–10 ft.lbs.
   Inner arm-to-crossmember bolts: 60–65 ft.lbs.
   Damper-to-crossmember nuts: 15–20 ft.lbs.

22. Lower the car to the ground and check the ride height. The ride height is checked on both sides and is determined by measuring the distance between the center line of the toe-in bolt hole on the outer arm, and the lower edge of the rebound bumper. The distance on each side should be 4.00–4.11 inches. If not, or if there is a significant distance between sides, the torsion bar(s) positioning is wrong.

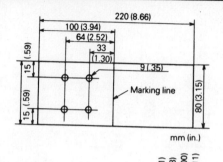

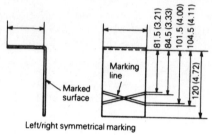

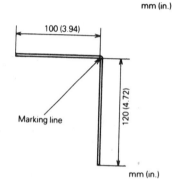

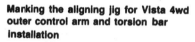

Marking the aligning jig for Vista 4wd outer control arm and torsion bar installation

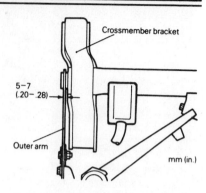

Outer arm-to-crossmember spacing

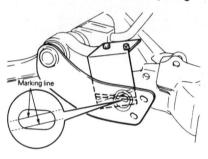

Final alignment of the outer control arm

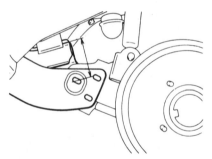

Ride height adjustment point on the 4wd Vista

Mac Pherson Strut

REMOVAL & INSTALLATION

Conquest

1. Raise and support the rear end on jackstands.
2. Remove the rear wheels.
3. Unclip the brake hose at the strut.
4. Unbolt the intermediate shaft from the companion flange.
5. Unbolt the strut assembly from the axleshaft housing. Remove the housing coupling bolt. Separate the strut from the housing by pushing the housing downward while prying open the coupling on the housing.
6. Remove the strut upper end attaching nuts, found under the side trim in the cargo area.
7. Lift out the strut.
8. Installation is the reverse of removal. Torque the upper end nuts to 20–25 ft.lb.; the strut-to-housing bolts to 50 ft.lb.; the coupling bolt to 50 ft.lbs.

OVERHAUL

For all spring and shock absorber removal and installation procedures, and all strut overhaul procedures, please refer to "Strut Overhaul"

Shock Absorbers

REMOVAL & INSTALLATION

Except 4wd Vista

1. Remove the hub cap or wheel cover. Loosen the lug nuts.
2. Raise the rear of the car. Support the car with jack stands.

NOTE: The body sill is marked with two dimples to locate the support position. Never place a stand anywhere but between these marks or you'll damage the body.

3. Remove the wheel. Remove the upper mounting bolt and nut.
4. While holding the bottom stud mount nut with one wrench, remove the locknut with another wrench.
5. Remove the shock absorber.
6. Check the shock for:
   a. Excessive oil leakage; some minor weeping is permissible;
   b. Bent center rod, damaged outer case, or other defects.
   c. Pump the shock absorber several times, if it offers even resistance on full strokes it may be considered serviceable.
7. Install the upper shock mounting nut and bolt. Hand tighten the nut.
8. Install the bottom eye of the shock over the spring stud. Tighten the lower nut to 12–15 ft. lbs. on rear wheel drive models; 47–58 ft.lbs. on front wheel drive models.
9. Finally, tighten the upper nut to 47–58 ft. lbs. on all models except station wagons, which are tightened to 12–15 ft. lbs.

4wd Vista

1. Raise and support the rear end on jackstands under the frame.
2. Remove the rear wheels.
3. Using a floor jack, raise the inner control arm slightly.
4. Unbolt the top, then the bottom of

SUSPENSION
Import Models

the shock absorber. Remove it from the car.

5. Installation is the reverse of removal. Torque the top nut to 55-58 ft.lb.; the bottom bolt to 75–80 ft.lb.

Lower Control Arm

REMOVAL & INSTALLATION

Rear Wheel Drive—Except Conquest

1. Support the vehicle body on safety stands. Use a jack under the rear axle to raise the rear axle assembly slightly.
2. Remove the wheel and the upper control arm rod.
3. Detach the parking brake rear cable from the lower arm.
4. Remove the lower arm from the rear axle housing and from the bracket attached to the body.
5. Temporarily install the lower arm (check for marking on left side arm) and torque the bolts to 94.0–108.0 ft. lbs. (127.0–147.0 Nm). Torque the assist link bushing bolt to 47.0–58.0 ft. lbs. (64.0–78.0 Nm).
6. With the special nut assembly placed securely against the rear axle housing bracket, install the upper control arm rod to the bracket. Torque the bolt to 94.0 ft. lbs. (127.0–147.0 Nm).

— CAUTION —
Always use new bolts.

Conquest

1. Raise and support the rear end on jackstands. Allow the wheels to hang freely.
2. Remove the rear wheels.
3. Disconnect the parking brake cable from the lower arm.
4. Disconnect the stabilizer bar.
5. Unbolt the lower control arm from the axleshaft housing.
6. Unbolt the lower control arm from the front support.
7. Unbolt the lower control arm from the crossmember and remove it.
8. Installation is the reverse of removal. Apply a thin coating of chassis lube to the cutout portion of the lower arm-to-axleshaft housing shaft. Do not allow the grease to touch the bushings. Insert the shaft with the mark on its head facing downward. When positioning the lower control arm on the crossmember, align the mark on the crossmember with the line on the plate. Torque the lower control arm-to-front support bolts to 108 ft.lb.; the arm-to-crossmember bolts to 108 ft.lb.; the arm to axleshaft housing bolts to 60 ft.lb.; the arm locking pin to 15 ft.lb. Have the rear wheel alignment checked.

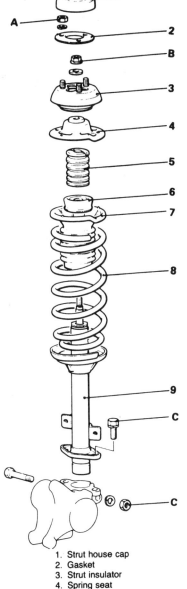

1. Strut house cap
2. Gasket
3. Strut insulator
4. Spring seat
5. Rubber helper
6. Rubber helper seat
7. Dust cover
8. Coil spring
9. Strut

Conquest rear strut

Trailing Arm

REMOVAL & INSTALLATION

Front Wheel Drive—Except 4wd Vista

1. Support the side frame on jack stands and remove the rear wheels. Remove the rear brake assembly.
2. Remove the muffler and jack the control arm just enough to raise it slightly.
3. Remove the shock absorber and lower the jack. Remove the coil spring and temporarily install the shock absorber to the control arm.
4. Disconnect the brake hoses at the rear suspension arms and remove the rear suspension from the body as an assembly.
5. Install the fixture-to-body bolts and torque to 36.0–51.0 ft. lbs. (49.0–69.0 Nm).
6. Install the coil springs and loosely install the shock absorbers. Tighten the shock absorber bolts to specification after the vehicle is lowered to the floor.
7. Install the rear brake assembly.
8. Lower the vehicle and tighten the suspension arm end nuts on all except Colt Vista: to 36.0–51.0 ft. lbs. (49.0–69.0 Nm)

Colt Vista: 94–108 ft. lbs. (130–150 Nm) and the shock bolts to 47.0–58.0 ft. lbs. (64.0–78.0 Nm), for all models.
9. Install the brake drums and wheels.
10. Bleed the brake system and adjust the rear brake shoe clearance.

Rear Wheel Bearings

REMOVAL, INSPECTION AND INSTALLATION

1980–84 Colt, 1985 and Later

1. Loosen the lug nuts, raise the rear of the car and support it on jackstands. Remove the wheel.
2. Remove the grease cap, cotter pin, nut and washer.
3. Remove the brake drum. While pulling the drum, the outer bearing will fall out. Position your hand to catch it.
4. Pry out the grease seal and discard it.
5. Remove the inner bearing.

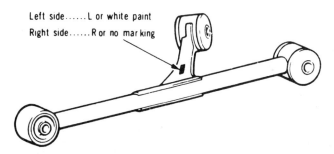

Rear lower control arm identifying marks for all rear wheel drive cars except Conquest

SUSPENSION
Import Models

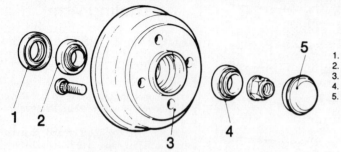

Rear wheel bearing assembly on 1985 and later Colt

1. Oil seal
2. Inner bearing
3. Brake drum
4. Outer bearing
5. Hub cap

6. Check the bearing races. If any scoring, heat checking or damage is noted, they should be replaced.

NOTE: When bearing or races need replacement, replace them as a set.

7. Inspect the bearings. If wear or looseness or heat checking is found, replace them.
6. If the bearings and races are to be replaced, drive out the races with a brass drift.
7. Before installing new races, coat them with lithium based wheel bearing grease. The races are most easily installed using a driver made for that purpose. They can, however, be driven into place with a brass drift. Make sure that they are fully seated.
8. Thoroughly pack the bearings with lithium based wheel bearing grease. Pack the hub with grease.
9. Install the inner bearing and coat the lip and rim of the grease seal with grease. Drive the seal into place with a seal driver.
10. Mount the drum onto the hub, slide the outer bearing into place, install the washer and thread the nut into place, finger-tightly.
11. Install a torque wrench on the nut. While turning the drum by hand, tighten the nut to 15 ft.lbs. Back off the nut until it is loose, then tighten it to 4 ft.lbs. If your torque wrench is not all that accurate below 10 ft.lb. (most aren't), use an in.lb. torque wrench and tighten the nut to 48 inch lbs.
12. Install the lock cap and insert a new cotter pin. If the lock cap and hole don't align, and repositioning the cap can't accomplish alignment, back off the nut no more than 15°. If that won't align the holes either, try the adjustment procedure over again.

1985 and later Colt

NOTE: Special tools are needed for this procedure.

1. Loosen the lug nuts. Raise the rear of the car and support it on jackstands.
2. Remove the wheel.
3. Remove the grease cap.
4. Remove the nut.
5. Pull the drum off. The outer bearing will fall out while the drum is coming off, so position your hand to catch it.
6. Pry out the oil seal. Discard it.
7. Remove the inner bearing.
8. Check the bearing races. If any scoring, heat checking or damage is noted, they should be replaced.

NOTE: When bearings or races need replacement, replace them as a set.

9. Inspect the bearings. If wear or looseness or heat checking is found, replace them.
10. If the bearings and races are to be replaced, drive out the race with a brass drift.
11. Before installing new races, coat them with lithium based wheel bearing grease. The races are most easily installed using a driver made for that purpose. They can, however, be driven into place with a brass drift. Make sure that they are fully seated.
12. Thoroughly pack the bearings with lithium based wheel bearing grease. Pack the hub with grease.
13. Install the inner bearing and coat the lip and rim of the grease seal with grease. Drive the seal into place with a seal driver.
14. Mount the drum on the axleshaft. Install the outer bearing. Don't install the nut at this point.
15. Using a pull scale attached to one of the lugs, measure the starting force necessary to get the drum to turn. Starting force should be 5 lbs. If the starting torque is greater than specific, replace the bearings.
16. Install the nut on the axleshaft. Thread the nut on, by hand, to a point at which the back face of the nut is 2–3mm from the shoulder of the shaft (where the threads end).
17. Using an inch lb. torque wrench, turn the nut counterclockwise 2 to 3 turns, noting the average force needed during the turning procedure. Turning torque for the nut should be about 48 inch lbs. If turning torque is not within 5 inch lb., either way, replace the nut.
18. Tighten the nut to 75–110 ft.lbs.
19. Using a stand-mounted gauge, check the axial play of the wheel bearings. Play should be less than 0.0079 in. If play cannot be brought within that figure, you probably have assembled the unit incorrectly.
20. Pack the grease cap with wheel bearing grease and install it.

FRONT SUSPENSION HONDA

All models except the 1983 and later Prelude, the 1984 and later Civic models and the 1986 and later Accord models use a MacPherson strut type front suspension. Each steering knuckle is suspended by a lower control arm at the bottom and a combined coil spring/shock absorber unit at the top. A front stabilizer bar, mounted between each lower control arm and the body, doubles as a locating rod for the suspension. Caster and camber are not adjustable and are fixed by the location of the strut assemblies in their respective sheet metal towers.

The 1983 and later Prelude and the 1986 and later Accord models use a completely redesigned front suspension. A double wishbone system, the lower wishbone consists of a forged transverse link with a locating stabilizer bar. The lower end of the shock absorber has a fork shape to allow the driveshaft to pass through it. The upper arm is located in the wheel well and is twist mounted, angled forward from its inner mount, to clear the shock absorber.

The 1984 and later Civic models also use a redesigned front suspension. This change was made to lower the hood line, thus making the car more aero-dynamic. The new suspension consists of two independent torsion bars and front shock absorbers similar to a front strut assembly, but without a spring. Both lower forged radius arms are connected with a stabilizer bar.

Shock Absorbers

REMOVAL & INSTALLATION

1983 and Later Prelude and 1986 and Later Accord

1. Raise the front of the car and support on jackstands. Remove the front wheels.
2. Remove the shock absorber locking bolt.
3. Remove the shock fork bolt and remove the shock fork.
4. Remove the shock absorber assembly.

NOTE: For spring and shock absorber disassembly procedures, please refer to "Strut Overhaul"

SUSPENSION
Import Models

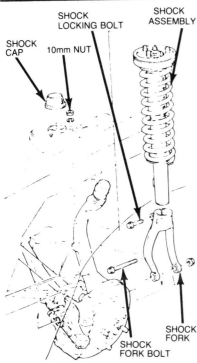

Front shock mounting—1983 and later Prelude, and 1986 and later Accord

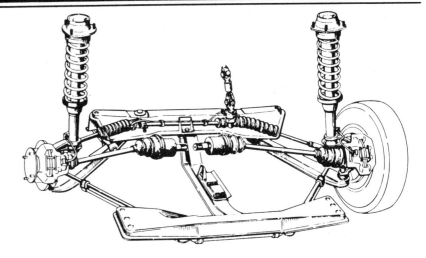

Typical front suspension and steering gear—1980–82 models

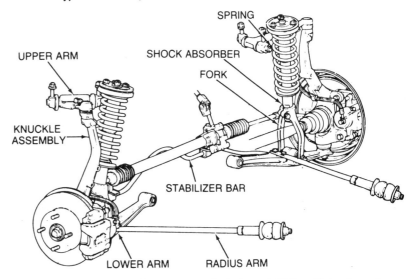

Front suspension—1983 and later Prelude, and 1986 and later Accord

5. Installation is the reverse of the removal procedure, taking note of the following:
   a. Align the shock absorber aligning tab with the slot in the shock absorber fork.
   b. The mounting base bolt should be tightened with the weight of the car placed on the shock.
   c. Torque the upper mounting bolts to 29 ft. lbs., the shock locking bolt to 32 ft. lbs. and the shock fork bolt to 47 ft. lbs.

1984 and Later Civic Models

1. Raise the front of the car and support on jackstands. Remove the front wheels.
2. Remove the brake hose clamp bolt.
3. Place a floor jack beneath the lower control arm to support it.
4. Remove the lower shock retaining bolt from the steering knuckle, then slowly lower the jack.

— **CAUTION** —
Be sure the jack is positioned securely beneath the lower control arm at the ball joint. Otherwise, the tension from the torsion bar may cause the lower control arm to suddenly "jump" away from the shock absorber as the pinch bolt is removed.

5. Compress the shock absorber by hand, then remove the two upper lock nuts and remove from the car.
6. Installation is the reverse of the removal procedure, taking note of the following:

   a. Use new self locking nuts on the top of the shock assembly and torque to 28 ft. lbs.
   b. Tighten the lower pinch bolt to 47 ft. lbs.
   c. Install and tighten the brake hose clamp to 16 ft. lbs.

Strut Assembly

For all spring and shock absorber removal and installation procedures and any other strut overhaul procedures, please refer to "Strut Overhaul"

INSPECTION

1. Check for wear or damage to bushings and needle bearings.
2. Check for oil leaks from the struts.
3. Check all rubber parts for wear or damage.
4. Bounce the car to check shock absorber effectiveness. The car should continue to bounce for no more than two cycles.

REMOVAL & INSTALLATION

1. Raise the front of the car and support it with safety stands. Remove the front wheels.
2. Disconnect the brake pipe at the strut and remove the brake hose retaining clip.
3. Remove the caliper and carefully hang from the undercarriage of the car with a piece of wire.
4. On 1980–85 Accord models, disconnect the stabilizer bar from the lower arm.
5. Loosen the bolt on the knuckle that retains the lower end of the shock absorber. Push down firmly while tapping it with a hammer until the knuckle is free of the strut.
6. Remove the three nuts retaining the upper end of the strut and remove the strut from the car.

2–201

SECTION 2
SUSPENSION
Import Models

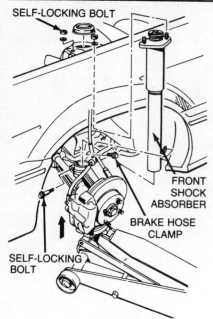

Front shock mounting—1984 and later Civic

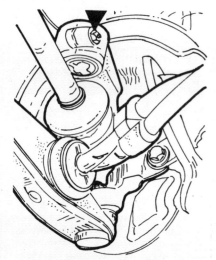

Lower strut retaining (pinch) bolt

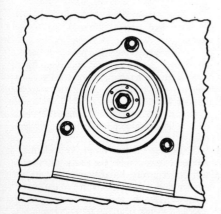

Typical strut upper mounting nuts

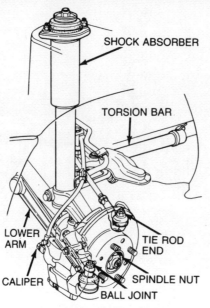

Front suspension—1984 and later Civic

7. To install, reverse the removal procedure. Be sure to properly match the mating surface of the strut and the knuckle notch. Tighten the knuckle bolt to 40 ft. lbs. (43–51 ft. lbs.—Accord).

Torsion Bar Assembly

REMOVAL & INSTALLATION

1984 and Later Civic Models

1. Raise the front of the car and support on jackstands.
2. Remove the height adjusting nut and the torque tube holder.
3. Remove the 33 mm circlip.
4. Remove the torsion bar cap, then remove the torsion bar clip by tapping the bar out of the torque tube.

NOTE: The torsion bar will slide easier if you move the lower arm up and down.

5. Tap the torsion bar backward, out of the torque tube and remove the torque tube.
6. Install a new seal onto the torque tube. Coat the torque tube seal and torque with grease, then install them on the rear beam.
7. Grease the ends of the torsion bar and insert into the torque tube from the back.
8. Align the projection on the torque tube splines with the cutout in the torsion bar splines and insert the torsion bar approximately (10 mm) 0.394 in.

NOTE: The torsion bar will slide easier if the lower arm is moved up and down.

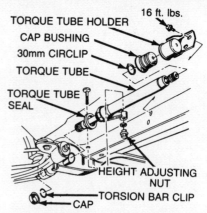

Torsion bar assembly—1984 and later Civic

9. Install the torsion bar clip and cap, then install the 30mm circlip and the torque tube cap.

NOTE: Push the torsion bar to the front so there is no clearance between the torque tube and the 30 mm circlip.

10. Coat the cap bushing with grease and install it on the torque tube. Install the torque tube holder.
11. Temporarily tighten the height adjusting nut.
12. Remove the jackstands and lower the car to the ground. Adjust the torsion bar spring height.

TORSION BAR ADJUSTMENT

1. Measure the torsion bar spring height between the ground and the highest point of the wheel arch:

   Coupe(CRX)   25.35 ± 0.20in.
   Hatchback   25.43 ± 0.20in.
   Sedan   25.63 ± 0.20in.
   Wagon   25.55 ± 0.20in.

2. If the spring height does not meet the specifications above, make the following adjustment.

   a. Raise the front wheels off the ground.
   b. Adjust the spring height by turning the height adjusting nut. Tightening the nut raises the height, and loosening the nut lowers the height.

NOTE: The height varies 0.20 in. per turn of the adjusting nut.

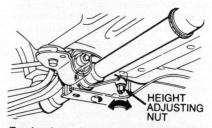

Torsion bar adjustment—1984 and later Civic

c. Lower the front wheels to the ground, then bounce the car up and down several times and recheck the spring height to see if it is within specifications.

Lower Ball Joints

INSPECTION

Check ball joint play as follows:
  a. Raise the front of the car and support it with safety stands.
  b. Clamp a dial indicator onto the lower control arm and place the indicator tip on the knuckle, near the ball joint;
  c. Place a pry bar between the lower control arm and the knuckle. Replace the lower control arm if the play exceeds 0.020 in.

REMOVAL & INSTALLATION

All Except the 1983 and later Prelude and 1986 and later Accord

If the ball joint play exceeds 0.020 in. the ball joint and lower control arm or lower radius arm (1984 and later Civics) must be replaced as an assembly.

1983 and later Prelude and 1986 and later Accord

NOTE: This procedure is performed after the removal of the steering knuckle and requires the use of the following special tools or their equivalent: Honda part no. 07965-SB00100 Ball Joint Remover/Installer, 07965-SB00200 Ball Joint Removal Base, 07965-SB00300 Ball Joint Installation Base, and 07974-SA50700 Clip Guide Tool.

1. Pry the snap-ring off and remove the boot.
2. Pry the snap-ring out of the groove in the ball joint.
3. Install the ball joint removal tool with the large end facing out and tighten the ball joint nut.
4. Position the ball joint removal tool base on the ball joint and set the assembly in a large vise. Press the ball joint out of the steering knuckle.
5. Position the new ball joint into the hole of the steering knuckle.
6. Install the ball joint installer tool with the small end facing out.
7. Position the ball joint installation base tool on the ball joint and set the assembly in a large vise. Press the ball joint into the steering knuckle.
8. Seat the snap-ring in the groove of the ball joint.
9. Install the boot and snap-ring using the clip guide tool.

Radius Arm

REMOVAL & INSTALLATION

1984 and later Civic Models Only

1. Raise the front of the car off the ground and support on jackstands. Remove the front wheels.
2. Place a floor jack beneath the lower control arm, then remove the ball joint cotter pin and nut.

--- CAUTION ---
Be sure to place the jack securely beneath the lower control arm at the ball joint. Otherwise, the tension from the torsion bar may cause the arm to suddenly "jump" away from the steering knuckle as the ball joint is removed.

3. Using a ball joint remover, remove the ball joint from the steering knuckle.
4. Remove the radius arm locking nuts and the stabilizer locking nut, then separate the radius arm from the stabilizer bar.
5. Remove the lower arm bolts and remove the radius arm by pulling it down and then forward.
6. Installation is the reverse of the re-

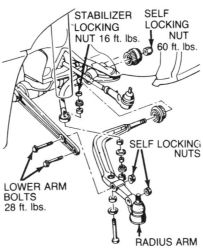

Radius arm—1984 and later Civic

moval procedure. Tighten all the rubber bushings and dampered parts only after the car is placed back on the ground.

Lower Control Arm

REMOVAL & INSTALLATION

All Except 1984 and Later Civic Models

1. Raise the front of the car and support it with safety stands. Remove the front wheels.
2. Disconnect the lower arm ball joint. Be careful not to damage the seal.

3. Remove the stabilizer bar retaining brackets, starting with the center brackets.
4. Remove the lower arm pivot bolt.
5. Disconnect the radius rod and remove the lower arm.
6. To install, reverse the removal procedure. Be sure to tighten the components to their proper torque.

Front End Alignment

CASTER & CAMBER ADJUSTMENT

Caster and camber cannot be adjusted on any Honda except the 1983 and later Prelude. If caster, camber or kingpin angle is incorrect or front end parts are damaged or worn, they must be replaced.

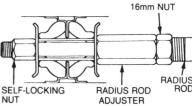

Caster adjustment—1983 and later Prelude, and 1986 and later Accord

1983 and Later Prelude and 1986 and Later Accord

NOTE: Wheel alignment adjustments must be performed in the following order: camber, caster and then toe-in.

The camber adjustment can be made by loosening the two nuts on the upper control arm and sliding the ball joint until the camber meets specifications. The caster adjustment can be made by loosening the 16mm nuts on the front beam radius rods and then turning the locknut to make the adjustment. Turning the nut clockwise decreases the caster and turning it counterclockwise increases the caster. After adjusting to specifications, hold the nylon locknut and lightly tighten the adjuster. Tighten the 16mm nut to 58 ft. lbs., then tighten the locknut to 32 ft. lbs. while holding the 16mm nut.

TOE-OUT ADJUSTMENT

Toe is the difference of the distance between the forward extremes of the front tires and the distance between the rearward extremes of the front tires. On Hondas, the fronts of the tires are further apart than the rear to counteract the pulling-together effect of front wheel drive.

Toe-out can be adjusted on all Hondas by loosening the locknuts at each end of the tie rods. To increase toe-out, turn the right tie rod in the direction of forward wheel rotation and turn the left tie rod in the opposite direction. Turn both tie rods an equal amount until toe-out meets specifications.

SECTION 2
SUSPENSION
Import Models

REAR SUSPENSION HONDA

All Civic sedan, hatchback and 1984 and later 2wd station wagon models and the Accord and Prelude utilize an independent MacPherson strut arrangement for each rear wheel. Each suspension unit consists of a combined oil spring/shock absorber strut, a lower control arm, and a radius rod or arm. The Civic 4wd Station Wagon uses coil springs which are mounted directly on the axle housing independent of the rear shock absorbers. The Accord and Prelude have adjustable rear suspension.

1980–83 station wagon models use a more conventional leaf spring rear suspension with a solid rear axle. The springs are three-leaf, semi-elliptic types located longitudinally with a pair of telescopic shock absorbers to control rebound. The solid axle and leaf springs allow for a greater load carrying capacity for the wagon.

Rear Strut Assembly

For all spring and shock absorber removal and installation procedures and any other strut overhaul procedures, please refer to "Strut Overhaul" in the Unit Repair section.

REMOVAL & INSTALLATION

1. Raise the rear of the car and support it with safety stands.
2. Remove the rear wheel.
3. Disconnect the brake line at the shock absorber (if so equipped). Remove the retaining clip and separate the brake hose from the shock absorber.
4. Disconnect the parking brake cable at the backing plate lever.
5. Remove the lower strut retaining bolt or pinch bolt and hub carrier pivot bolt. To remove the pivot bolt, you first have to remove the castle nut and its cotter pin.
6. Remove the upper strut retaining nuts and remove the strut from the car.
7. To install, reverse the removal procedure. Be sure to install the top of the strut in the body first. After installation, bleed the brake lines.

Lower Control Arm and Radius Arm

REMOVAL & INSTALLATION
Except 1980–83 Station Wagon

1. Raise the rear of the car and support it with safety stands.
2. Remove the rear wheels and brake drums.
3. Disconnect the hydraulic brake line and parking brake cable.
4. Remove the backing plate assembly.
5. Remove the radius arm nuts and bolts,

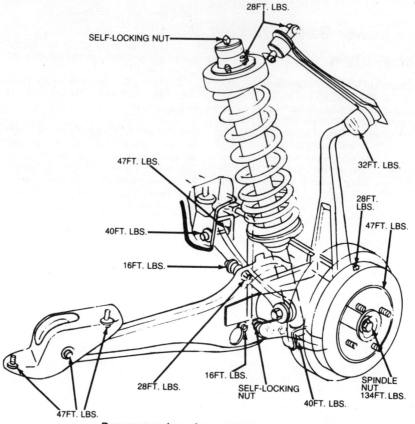

Rear suspension—Accord 1986 and later

and remove the radius arm. Unscrew the stabilizer bolt and remove the stabilizer bar (if so equipped).

6. Remove the shock absorber pinch bolt, then separate the hub carrier from the shock absorber.
7. Remove the lower control arm retaining bolts, then remove the arm.

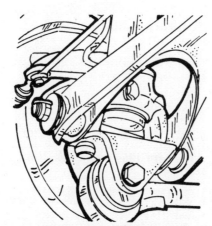

1983 Civic Sedan rear control arm

Leaf Spring

REMOVAL & INSTALLATION
1980–83 Station Wagon Only

1. Raise the rear of the car and support it on stands placed on the frame. Remove the wheels.
2. Remove the shock absorber lower mounting bolt.
3. Remove the nuts from the U-bolt and remove the U-bolts, bump rubber, and clamp bracket.
4. Unbolt the front and rear spring shackle bolts, remove the bolts, and remove the spring.
5. To install, first position the spring on the axle and install the front and rear shackle bolts. Apply a soapy water solution to the bushings to ease installation. Do not tighten the shackle nuts yet.
6. Install the U-bolts, spring clamp bracket and bump rubber loosely on the axle and spring.
7. Install the wheels and lower the car. Tighten the front and rear shackle bolts to 33 ft. lbs. Also tighten the U-bolt nuts to 33 ft. lbs., after the shackle bolts have been

SUSPENSION
Import Models

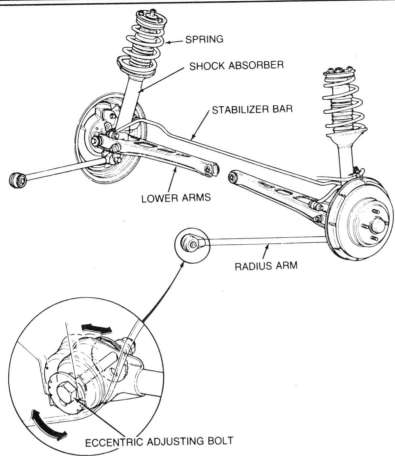

Rear suspension—Accord 1980-85 and Prelude 1983 and later

Shock Absorbers

REMOVAL & INSTALLATION

1980–83 Station Wagon and 4wd Models Only

1. It is not necessary to jack the car or remove the wheels unless you require working clearance. Unbolt the upper mounting nut and lower bolt and remove the shock absorber. Note the position of the washers and lock washers upon removal.
2. Installation is the reverse. Be sure the washers and lock washers are installed correctly. Tighten the upper mount to 44 ft. lbs. and the lower mount to 33 ft. lbs.

Rear Wheel Hub Bearings

REPLACEMENT

All exccept 1984 and later Civic and Prelude, and 1986 and later Accord.

1. Slightly loosen the rear lug nuts. Raise the car and support safely on jack stands.
2. Release the parking brake. Remove the rear wheels.
3. Remove the rear bearing hub cap and cotter pin and pin holder.
4. Remove the spindle nut, then pull the hub and drum off of the spindle.
5. Drive the outboard and inboard bearing races out of the hub. Punch in a criss-cross pattern to avoid cocking the bearing race in the bore.
6. Clean the bearing seats thoroughly before going on to the next step.
7. Using a bearing driver, drive the inboard bearing race into the hub.
8. Turn the hub over and drive the outboard bearing race in the same way.
9. Check to see that the bearing races are seated properly.
10. When fitting new bearings, you must pack them with wheel bearing grease. To

tightened.

8. Install the shock absorber to the lower mount. Tighten to 33 ft. lbs.

Rear Wheel Alignment

Caster and camber are fixed as on the front suspension. However, toe-out is adjustable (except 1984 and later Civics and 1986 and later Accords) by means of an eccentric adjusting bolt at the forward anchor of the radius rod.

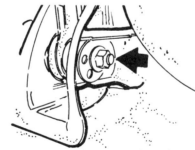

Rear toe adjustment location—Accord 1980-82

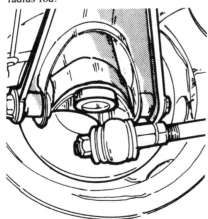

Rear toe adjustment location—Civic 1980-82

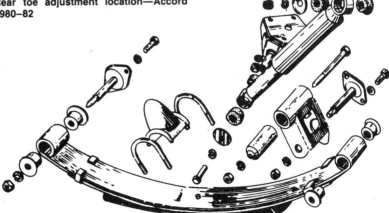

Rear suspension—Civic station Wagon 1980-83

2-205

SECTION 2: SUSPENSION
Import Models

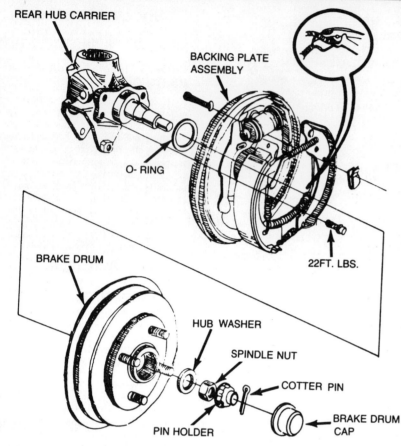

Rear hub and bearing—all except 1984 and later Civic and Prelude, and 1986 and later Accord

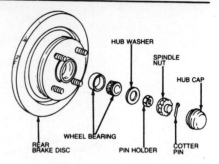

Rear hub and bearing—Prelude 1984 and later

do this, place a glob of grease in your left palm, then, holding one of the bearings in your right hand, drag the face of the bearing heavily through the grease. This must be done to work as much grease as possible through the ball bearings and the cage. Turn the bearing and continue to pull it through the grease, until the grease is thoroughly packed between the bearing balls and the cage, all around the bearing. Repeat this operation until all of the bearings are packed with grease.

11. Pack the inside of the hub with a moderate amount of grease. Do not overload the hub with grease.

12. Apply a small amount of grease to the spindle and to the lip of the inner seal before installing.

13. Place the inboard bearing into the hub.

14. Apply grease to the hub seal, and carefully tap into place. Tap in a criss-cross pattern to avoid cocking the seal in the bore.

15. Slip the hub and disc over the spindle, then insert the outboard bearing, hub washer, and spindle nut.

16. Follow the procedures below for adjustment.

do this, place a glob of grease in your left palm, then, holding one of the bearings in your right hand, drag the face of the bearing heavily through the grease. This must be done to work as much grease as possible through the ball bearings and the cage. Turn the bearing and continue to pull it through the grease, until the grease is thoroughly packed between the bearing balls and the cage, all around the bearing. Repeat this operation until all of the bearings are packed with grease.

11. Pack the inside of the hub with a moderate amount of grease. Do not overload the hub with grease.

12. Apply a small amount of grease to the spindle and to the lip of the inner seal before installing.

13. Place the inboard bearing into the hub.

14. Apply grease to the hub seal, and carefully tap into place. Tap in a criss-cross pattern to avoid cocking the seal in the bore.

15. Slip the hub and drum over the spindle, then insert the outboard bearing, hub washer, and spindle nut.

16. Follow the procedures below under "Wheel Bearing Adjustment".

1984 and later Prelude

1. Slightly loosen the rear lug nuts. Raise the car and support safely on jack stands.
2. Release the parking brake. Remove the rear wheels.
3. Remove the bolts retaining the brake caliper and remove the caliper from the knuckle. Do not let the caliper hang by the brake hose, support it with a length of wire.
4. Remove the rear bearing hub cap and cotter pin and pin holder. Remove the spindle nut, then pull the hub and disc off of the spindle.
5. Drive the outboard and inboard bearing races out of the disc. Punch in a criss-cross pattern to avoid cocking the bearing race in the bore.
6. Clean the bearing seats thoroughly before going on to the next step.
7. Using a bearing driver, drive the inboard bearing race into the disc.
8. Turn the disc over and drive the outboard bearing race in the same way.
9. Check to see that the bearing races are seated properly.
10. When fitting new bearings, you must pack them with wheel bearing grease. To

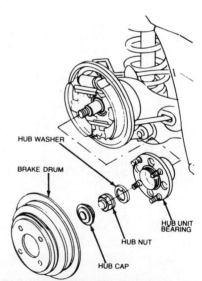

Rear hub and bearing—Civic 1984 and later, and Accord 1986 and later

SUSPENSION
Import Models

1984 and later Civic and 1986 and later Accord

1. Slightly loosen the rear lug nuts. Raise the car and support safely on jack stands.
2. Release the parking brake. Remove the rear wheel and the brake drum.
3. Remove the rear bearing hub cap and nut.
4. Pull the hub unit off of the spindle.
5. Installation is the reverse order of removal. Tighten the new spindle nut to 134 ft.lbs., then stake the nut.

ADJUSTMENT

1. Apply grease or oil on the spindle nut and spindle threads.
2. Install and tighten the spindle nut to 18 ft.lbs. and rotate the drum/disc 2–3 turns by hand, then retighten the spindle nut to 18 ft.lbs.
3. Repeat the above step until the spindle nut hold that torque.
4. Loosen the spindle nut to 0 ft.lbs.

NOTE: Loosen the nut until it just breaks free, but doesn't turn.

5. Retorque the spindle nut to 4 ft.lbs.
6. Set the pin holder so the slots will be as close as possible to the hole in the spindle.
7. Tighten the spindle nut just enough to align the slot and hole, then secure it with a new cotter pin.

FRONT SUSPENSION HYUNDAI

MacPherson Strut

REMOVAL & INSTALLATION

1. Raise the vehicle and support it by the body or crossmembers. Remove the front wheels. Detach the brake hose bracket at the strut.
2. Remove the four nuts securing the strut to the fender well.
3. Unbolt the strut lower end from the knuckle.
4. Remove the strut from the car.
5. Installation is the reverse of removal. Torque the strut-to-knuckle bolts to 55–65 ft. lbs.; the strut-to-fender well nuts to 7–11 ft. lbs.

OVERHAUL

For all spring and shock absorber removal and installation procedures, and all strut overhaul procedures, please refer to "Strut Overhaul"

Ball Joints

INSPECTION

Support the vehicle on axle stands. Disconnect the ball joint at the lower end of the strut as described below; you need not remove the lower control arm entirely. Install the nut back onto the ballstud. Then, with an in.lb. torque wrench, measure the torque required to start the ball joint rotating. The figures are 26–86 in. lbs. If the figures are within specification, the ball joint is satisfactory. If the figure is too high, the joint should be replaced. If the figure is too low, you can still reuse the joint, provided its rotation is smooth and even. If there is roughness, or play, it must be replaced.

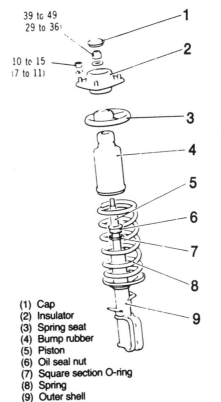

(1) Cap
(2) Insulator
(3) Spring seat
(4) Bump rubber
(5) Piston
(6) Oil seal nut
(7) Square section O-ring
(8) Spring
(9) Outer shell

Front strut assembly

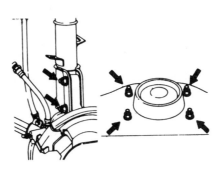

Strut attaching points

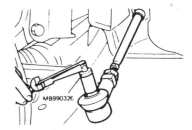

Check ball joint starting torque with an in. lb. torque wrench

REMOVAL & INSTALLATION

If the ball joint requires replacement, the entire lower control arm should be replaced, as described below.

Stabilizer Bar

REMOVAL & INSTALLATION

1. Raise and support the front of the car. Remove the front wheels. Remove the under cover.
2. Disconnect the stabilizer bar from the lower control arm by removing the nut and long bolt, and the various spacers, washers, and rubber bushings, keeping all parts in order.
3. Disconnect the tie rod ends from the steering knuckles as described below. In addition, remove the lower bolt from the rear engine roll stopper for access to the stabilizer bar mounts.
4. Remove the bolts at either end of the stabilizer bar retainer and remove the stabilizer bar from the crossmember.
5. Install the stabilizer bar in reverse order, torquing the stabilizer bar retainer bolts to 12–19 ft. lbs.. Make sure the bushings fit squarely inside the retainer brackets. Mount the bar so the distance between the inner end of the retaining bracket and the inner edge of the marked portion of the bar is 4–6mm. Torque the link

2-207

SECTION 2
SUSPENSION
Import Models

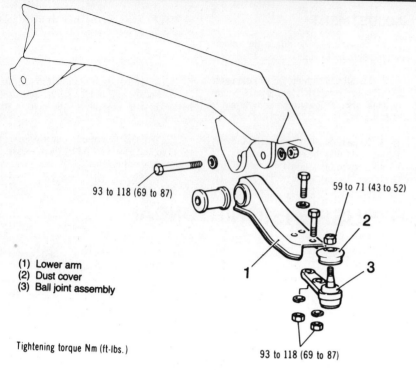

(1) Lower arm
(2) Dust cover
(3) Ball joint assembly

Tightening torque Nm (ft-lbs.)

Lower control arm and ball joint

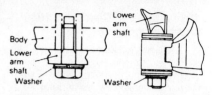

Lower control arm washer must be installed as shown

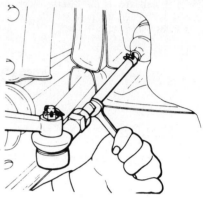

Adjusting the toe-in

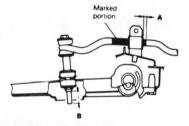

Mount the bar so the distance between the inner end of the retaining bracket and the inner edge of the marked portion is .16–.24 in. Torque the link at either end so the amount of threads exposed is .83–.91 in.

for the bar at either side until 21–23mm of threads are exposed.

Lower Control Arm

REMOVAL & INSTALLATION

1. Support the vehicle securely by the crossmember and remove the front wheel.
2. Remove the under cover.
3. Disconnect the stabilizer bar from the lower arm. Remove the nut from underneath the control arm and take off the washer and spacer.
4. Remove the ball joint stud nut and press the tool off with a tool such as MB991113.
5. Remove the bolts which retain the spacer at the rear and the nut and washers on the front of the lower arm shaft (at the front). Slide the arm forward, off the shaft and out of the bushing.
6. Replace the dust cover on the ball joint. The new cover must be greased on the lip and inside with #2 EP Multipurpose grease and pressed on with a tool such as MB990800 and a hammer until it is fully seated.
7. Installation is the reverse of removal. The nut on the stabilizer bar bolt must be torqued until the link shows 21–23mm of threads below the bottom of the nut. Also, the washer for the lower arm must be installed as shown. The left side arm shaft has a left hand thread. Finally, tighten the arm shaft to the lower arm with the weight of the car with no passengers or luggage on the front suspension. Torque the nut for the lower arm shaft to 69–87 ft. lbs.; the bolts for the spacer at the rear to 43–58 ft. lbs.; the ball stud nut to 43–52 ft. lbs.

Front Wheel Alignment

ADJUSTMENT

Both caster and camber on these vehicles are fixed by the design of the suspension. If alignment readings are outside specifications, bent or damaged parts must be replaced.

Toe-in is adjusted by unfastening the clips for the rubber boots on the inner ends of the tie rods, loosening the locknuts at the outer ends, and rotating the tie rod, using a wrench on the flats. The toe will move outward as the left side turnbuckle is turned toward the front of the vehicle (with the wrench underneath) and the right one is turned toward the rear. For each half turn of the tie rod, toe in will be adjusted by 6mm. Steering axis inclination should be checked after toe-in is adjusted to ascertain that the adjustment is correct. The turnbuckles should be turned the same distance in opposite directions.

Front Wheel Bearings

For all service on front wheel bearings, please refer to "Front Hub and Bearing".

SUSPENSION
Import Models

REAR SUSPENSION HYUNDAI

Coil Spring

REMOVAL & INSTALLATION

1. Support the car securely at the rear of the body at approved points and remove the rear wheels.
2. Support the rear suspension arm with a floorjack. Then, remove the lower shock absorber attaching bolt, nut, and lockwasher.
3. Slowly lower the jack just to the point where the spring can be removed and remove the spring. If the spring is being replaced, transfer the spring seat to the new spring.
4. Install the spring in reverse order of the removal procedure, making sure the smaller diameter is upward. Make sure the spring identification and load markings match up.
5. Torque the lower shock mounting nut/bolt to 65–80 ft. lbs.

Shock Absorber

REMOVAL & INSTALLATION

1. Follow Steps 1 and 2 of the coil spring procedure.
2. Remove the upper shock absorber bolt, nut, and lockwasher, and remove the shock.
3. Install in reverse order, torquing both upper and lower bolts to 65–80 ft. lbs.

Rear Suspension Alignment

All Hyundai models have independent rear suspension systems on which unusual wear can cause alignment problems. Alignment cannot be adjusted, but camber can be checked in case handling or tire wear problems should occur. Alignment problems can be corrected only by replacement of major suspension components that have become severely worn or bent through abuse. In all cases, camber must be corrected through parts replacement.

Rear Wheel Bearings

REMOVAL, INSPECTION, PACKING, INSTALLATION AND ADJUSTMENT

1. Loosen the lug nuts and raise and support the rear of the car.

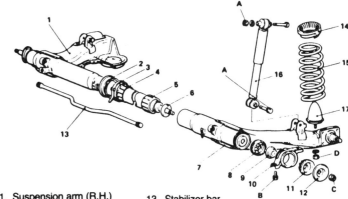

1. Suspension arm (R.H.)
2. Dust cover
3. Clamp
4. Bushing A
5. Bushing B
6. Rubber stopper
7. Suspension arm (L.H.)
8. rubber bushing (inner)
9. Rubber stopper
10. Fixture
11. Rubber bushing (outer)
12. Washer
13. Stabilizer bar
14. Spring seat
15. Coil spring
16. Shock absorber
17. Bump stopper

	Nm	ft. lbs.
A	65–80	46–56
B	50–70	36–51
C	80–100	56–70
D	18–25	13–18

Rear suspension

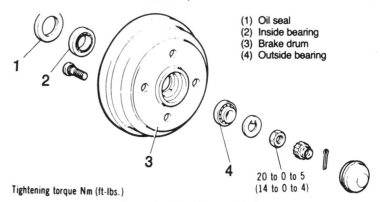

(1) Oil seal
(2) Inside bearing
(3) Brake drum
(4) Outside bearing

Tightening torque Nm (ft-lbs.) 20 to 0 to 5 (14 to 0 to 4)

Rear wheel bearings and hub

2. Remove the grease cap, cotter pin serrated nut cap, axle shaft nut and washer.
3. Pull outward on the brake drum, positioning your hand to catch the outer bearing when it fall out.
4. Pry the inner grease seal from the hub and discard it.
5. Remove the inner bearing. If the bearings are being replaced, drive the bearing races from the hub.
6. Clean all old grease from the hub and bearings. If the old bearing are beings reused, clean them in a safe solvent and inspect them carefully.
7. If the bearings are being replaced, coat the new races with EP lithium wheel bearing grease and drive them into the hub, making sure they are fully and squarely seated.
8. Pack the hub cavity with new EP lithium wheel bearing grease, until the cavity is full.
9. Pack the bearings completely.
10. Install the inner bearing and drive a new grease seal into place.
11. Install the drum on the spindle and install the outer bearing, washer and shaft nut. Tighten the nut to 15 ft. lbs. while turning the drum, to seat the bearings. Back off on the nut until it is loose, then torque it to 48 in.lb. (4 ft. lbs.).
12. Install the serrated nut cap and a new cotter pin. If the cotter pin holes have to be aligned, back off on the nut no more than 15 degrees. If that won't align the holes, repeat the adjustment procedure.

SECTION 2 SUSPENSION
Import Models

FRONT SUSPENSION ISUZU

Front Wheel Bearings

REPLACEMENT, REPACKING & ADJUSTMENT

I-Mark (RWD) and Impulse

1. Raise the car and safely support it with jackstands. Remove the front wheel.
2. Remove the grease cap. Remove and discard the cotter pin, then remove the spindle nut and washer.
3. Wiggle the hub and the wheel bearing will pop out enough to grab it. Wipe any dirt or old grease off of the spindle.

Removing spindle nut on front hub

NOTE: If the rear wheel bearing is to be inspected, the hub will have to be removed. Remove and tie up the brake caliper as outlined in the brake section and the hub will be free to be removed.

4. Inspect the wheel bearing for signs of wear, nicks or obvious damage. Clean the bearings in solvent and blow dry with compressed air.

— **CAUTION** —
Do not give in to the impulse of spinning the clean bearing with the air nozzle. Running a dry bearing at high rpm while holding it in your hand can damage the bearing.

5. Carefully repack the front wheel bearings with new high temperature wheel bearing grease. Put a glob of grease in your palm and force it into the bearing with a scraping motion until the grease comes out the top. Coat the packed bearing with a covering of grease and install it with the taper side in.
6. Clean and install the washer and spindle nut. Torque the spindle nut to 22 ft. lbs. while rotating the hub. This will seat the bearing.
7. Back off the spindle nut completely, then turn it back all the way using only your fingers. Once the spindle nut is snug, insert a new cotter pin.

NOTE: If the holes on the spindle nut and spindle do not align, tighten the nut only enough to align. A properly adjusted wheel bearing has a small amount of end-play and a slightly loose nut when adjusted in this manner.

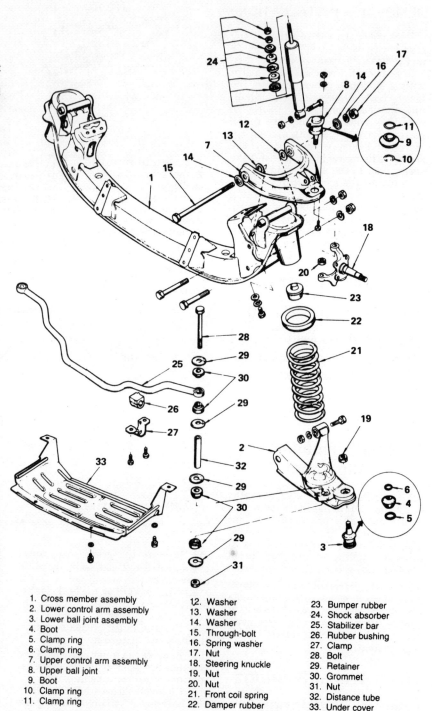

1. Cross member assembly
2. Lower control arm assembly
3. Lower ball joint assembly
4. Boot
5. Clamp ring
6. Clamp ring
7. Upper control arm assembly
8. Upper ball joint
9. Boot
10. Clamp ring
11. Clamp ring
12. Washer
13. Washer
14. Washer
15. Through-bolt
16. Spring washer
17. Nut
18. Steering knuckle
19. Nut
20. Nut
21. Front coil spring
22. Damper rubber
23. Bumper rubber
24. Shock absorber
25. Stabilizer bar
26. Rubber bushing
27. Clamp
28. Bolt
29. Retainer
30. Grommet
31. Nut
32. Distance tube
33. Under cover

Exploded view of front suspension system—all except Impulse

SUSPENSION
Import Models

Shock Absorbers

REMOVAL & INSTALLATION

1. Raise the car and support it safely. Remove the front wheel.
2. Disconnect the shock absorber from the upper control arm using two wrenches.
3. Remove the shock absorber nuts from the engine compartment.
4. Remove the shock absorber. Installation is the reverse of removal. Torque the control arm nut to specifications. Tighten the top nut to the end of the threads on the rod. Use lock nuts.

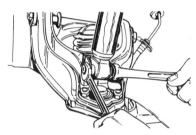

Disconnecting shock absorber from upper control arm—except Impulse

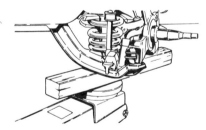

Shock absorber installation in engine compartment

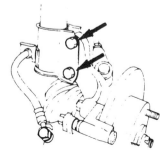

Lifting the control arm with a hydraulic jack—except Impulse

Removing the lower strut bolts

MacPherson Strut

REMOVAL & INSTALLATION

I-Mark (FWD)

1. Loosen the front wheel lug nuts, raise and support the front of the vehicle safely and remove the wheel and tire assembly.
2. Remove the brake hose clip-to-strut bolt (if so equipped). Install a drive axle cover, to protect the axle boot.
3. Remove the bolts attaching the strut to the steering knuckle.
4. Remove the two strut tower nuts and remove the strut assembly from the vehicle.
5. Installation is the reverse order of the removal procedure. The two strut tower nuts must be tightened before the lower strut bolts. The strut tower nuts should be torqued to 40 ft. lbs. and the lower strut bolts should be torqued to 80 ft. lbs.

OVERHAUL

For all spring and shock absorber removal and installation procedures, and all strut overhaul procedures, please refer to "Strut Overhaul"

Coil Springs

REMOVAL & INSTALLATION

All Models Except Impulse and I-Mark (FWD)

1. Raise the car and safely support it with jackstands. Remove the wheel.
2. Remove the tie rod end cotter pin and castle nut. Discard the cotter pin.

Brake caliper attaching bolts—except Impulse

3. Use a suspension fork to separate the tie rod end from the steering knuckle.
4. Remove the lower shock absorber bolt and push the shock up as far as possible.
5. Remove the stabilizer bar bolt and grommet assembly from the lower control arm.
6. Remove the upper brake caliper bolt and slide the hose retaining clip back about 1/2 in.
7. Place the lifting pad of a hydraulic floor jack under the outer extreme of the control arm and raise the lower control arm until it is level.

CAUTION
Secure a safety chain through one coil near the top of the spring and attach it to the upper control arm to prevent the spring from coming out unexpectedly. The coil spring will come out with a lethal force, so don't take any chances.

8. Loosen the lower ball joint lock nut until the top of the nut is flush with the top of the ball joint. Using tool J-26407 or equivalent, disconnect the lower ball joint from the steering knuckle.
9. Remove the hub assembly and steering knuckle from the lower ball joint and support with a wire or rope out of the way.
10. Pry the lower control arm down, using extreme caution so as not to injure yourself. Remove the spring.
11. Installation is the reverse of removal. Properly seat the spring and use the safety chain. Use the hydraulic jack to compress the new spring until the control arm is level.
12. Torque the ball joint lock nut to 58 ft. lbs. Torque the lower shock absorber mounting nuts and all other attaching hardware according to the specifications chart at the end of the section.

Front Stabilizer Bar

REMOVAL & INSTALLATION

I-Mark (RWD)

1. Raise and support safely the front of the vehicle.
2. Remove the engine dust cover from underneath the engine.
3. Remove the stabilizer bar bolt and grommet assemblies from the lower control arms.
4. Remove the clamps securing the stabilizer bar.
5. Installation is the reverse order of the removal procedure.

I-Mark (FWD)

1. Raise and support the front of the vehicle safely.
2. Remove the nuts, bolts and insulators retaining the stabilizer bar to the tension rod.
3. Remove the stabilizer bar.
4. To install, reverse the removal procedure. Align the front side of the insulator edge with the paint mark on the upper rear edge of the tension bar.

Tension Bar

REMOVAL & INSTALLATION

I-Mark (FWD)

1. Raise and support the vehicle on jackstands.

SECTION 2
SUSPENSION
Import Models

2. If equipped with a stabilizer bar, remove the nuts, bolts and insulators retaining it to the tension rod.
3. Remove the nut and washer retaining the tension rod to the body.
4. Remove the nuts and bolts retaining the tension rod to the control rod.
5. Remove the tension rod.
6. To install, reverse the removal procedure.

Ball Joints

INSPECTION

The maximum permissible axial play in the ball joint is 0.008 in. (0.2 mm). Replace any joint that exceeds this value.

NOTE: The lower ball joint is splined to the lower control arm. The upper ball joints are offset to allow for camber setting.

REMOVAL & INSTALLATION

I-Mark (RWD)
UPPER BALL JOINT

1. Raise the car and support it safely. Remove the wheel.
2. Remove the upper brake caliper bolt and slide the hose retaining clip back about ½ in.

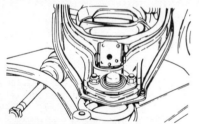

Installing the upper ball joint in the control arm—except Impulse

3. Remove the lower shock absorber nut and bolt and push the shock absorber up.
4. Place a hydraulic jack under the outer extreme of the lower control arm and raise until level.
5. Loosen the upper ball joint nut until the top of the nut is flush with the top of the ball joint.
6. Using special tool J-26407 or equivalent, disconnect the upper ball joint from the steering knuckle.
7. Remove the two bolts connecting the upper ball joint to the upper control arm. Remove the ball joint.
8. Installation is the reverse of removal. Install the new ball joint in the control arm so that the cut-off portion is facing outward. Torque the ball joint lock nut to 40 ft. lbs. Torque all attaching nuts and bolts. For specifications, see the torque chart at the end of the section.

NOTE: The car should be aligned whenever any suspension components are replaced.

LOWER BALL JOINT

1. Raise the car and support it safely. Remove the front wheel.
2. Remove the tie rod end cotter pin and castle nut. Discard the pin and separate the tie rod end with a suspension fork. Remove the tie rod from the steering knuckle.
3. Remove the stabilizer bar bolt and grommet assembly from the lower control arm.
4. Remove the upper brake caliper bolt and slide the brake hose retaining clip back about ½ in.
5. Remove the shock absorber lower bolt and push the shock up.

CAUTION
Secure a safety chain through the upper and lower control arms to prevent the possibility of the spring coming out and causing serious damage or injury. Allow enough room to get the ball joint out.

---

6. Place a hydraulic jack under the outer extremity of the lower control arm and raise it until level.
7. Loosen the ball joint lock nut until the top of the nut is flush with the top of the ball joint.
8. Using special tool J-26407 or equivalent, disconnect the lower ball joint from the steering knuckle.
9. Remove the hub assembly and steering knuckle from the lower ball joint and support with a wire or rope.
10. Remove the lower ball joint from the control arm using tool J-9519-03 or equivalent.
11. Installation is the reverse of removal. Do not strike the ball joint bottom. Torque the ball joint nut to 50 ft. lbs.
12. Torque all bolts to specifications as listed in the chart at the end of the section.

REMOVAL & INSTALLATION

I-Mark (FWD)

1. Loosen the wheel nuts.
2. Raise the vehicle and support it on jackstands.
3. Remove the wheel and tire assembly.
4. Remove the two nuts retaining the ball joint to the tension rod and control arm assembly.

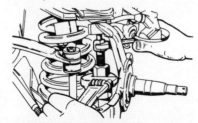

Disconnecting the lower ball joint with the special tool—except Impulse

5. Remove the pinch bolt retaining the ball joint to the steering knuckle.
6. Remove the ball joint.
7. To install, reverse the removal procedure.

Control Arms

REMOVAL & INSTALLATION

I-Mark (RWD)
UPPER

1. Raise the car and support it safely. Remove the front wheel.
2. Remove the upper brake caliper bolt and slide the brake hose retainer clip back about ½ in.

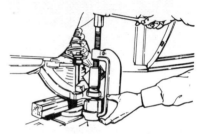

Removing the lower ball joint from the control arm—except Impulse

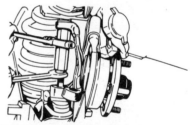

Removing the lower ball joint from the steering knuckle—except Impulse

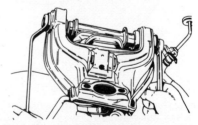

Removing the upper control arm from the crossmember—except Impulse

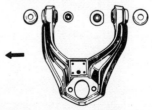

Installation of the upper control arm—except Impulse

SUSPENSION
Import Models

SECTION 2

3. Remove the lower shock bolt and push the shock absorber up.

4. Place a hydraulic jack under the control arm on the outer extreme and raise the control arm until it is level.

5. Loosen the upper ball joint lock nut until the top of the nut is flush with the top of the ball joint. Disconnect the upper ball joint from the steering knuckle using tool J-26407 or equivalent.

6. Disconnect and remove the through bolt connecting the upper control arm to the crossmember. Remove the upper control arm.

7. Installation is the reverse of removal. On installation, make sure the smaller washer is on the inner face of the front arm and the larger washer is on the inner face of the rear arm.

8. Torque all attaching hardware to the specifications in the chart at the end of the section. Align the front end.

NOTE: Always check the camber when working around the upper control arm area.

LOWER

1. Follow Steps 1–10 for removal of coil springs.

2. Remove the bolts connecting the lower control arm to the crossmember and the body.

3. Installation is the reverse of removal. Torque all bolts and nuts to specifications.

NOTE: When reinstalling front end components, it's best to snug all the bolts and nuts first, then lower the car so that there is weight on the suspension when final torque adjustments are made.

I-Mark (FWD)

1. Raise and support the front of the vehicle.

2. Remove the control arm to tension arm retaining nuts and bolts.

3. Remove the nut/bolt securing the control arm to the body.

4. Remove the control arm and check for cracking or distortion.

5. To install, reverse the removal procedure.

NOTE: Raise the control arm to a distance of 15.18 in. from the top of the wheel well to the center of the hub. Use 41 ft. lbs. of torque to fasten the control

Removing the lower control arm—except Impulse

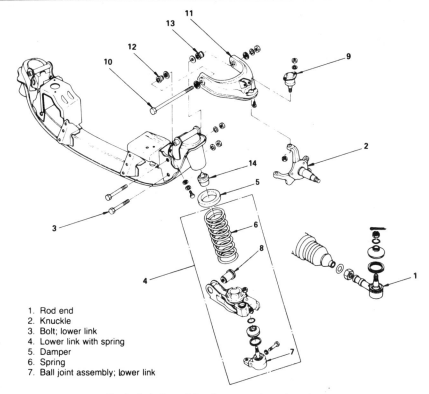

1. Rod end
2. Knuckle
3. Bolt; lower link
4. Lower link with spring
5. Damper
6. Spring
7. Ball joint assembly; lower link

Exploded view of front suspension—Impulse

arm to the body and 80 ft. lbs. to secure the control arm to the tension rod. This procedure aligns the bushing arm to the body.

Control Arms, Knuckles and Coil Springs

REMOVAL & INSTALLATION

Impulse

1. Raise the car and support it safely. Remove the front wheel.

2. Remove the tie rod end cotter pin and castle nut and remove the tie rod end using tool J-21687-02.

3. Using coil spring compressor J-33992, compress the coil spring.

— CAUTION —
Secure a safety chain through one coil near the top of the spring and attach it to the upper control arm to prevent the spring from coming out.

4. Compress the coil spring until its top end releases from the cushion rubber at the center of the upper link, then remove the coil spring.

5. Remove the lower link bolt, then remove the coil spring together with the lower link.

6. Remove the lower ball joint assembly.

7. Press out the lower link bushing.

8. Remove the upper ball joint assembly.

9. Press out the upper link bushings.

10. Installation is the reverse of removal with the following precautions:

a. When installing the upper control arm washers install the small washer on the inboard side of the rear end.

b. Leave the upper and lower control arm link bolts semi-tight as they are to be torqued to specifications after completion of installation with the wheels lowered to the floor. Upper—47 ft. lbs., lower—61 ft. lbs.

c. When installing the upper ball joint to the control arm the cutaway portion of the ball joint should be turned outward.

Front End Alignment

NOTE: Steering problems are not always the result of improper alignment. Before aligning the car, check the tire pressure and check all suspension components for damage or excessive wear.

CAMBER ADJUSTMENT

On all models except the Impulse, camber angle can be increased approximately 1° by removing the upper ball joint, rotating it ½ turn and reinstalling it with the cut-off portion of the upper flange on the inboard side of the control arm. On the Impulse, camber is not adjustable. Replace parts as necessary to correct alignment.

2–213

SECTION 2: SUSPENSION
Import Models

CASTER ADJUSTMENT

The caster angle is pre-set at the factory and cannot be adjusted.

TOE-IN ANGLE ADJUSTMENT

Toe-in is controlled by adjusting the tie-rod. To adjust the toe-in setting loosen the nuts at the steering knuckle end of the tie-rod. Rotate the rod as required to adjust the toe-in. Retighten the cover and locknuts, check that the rubber bellows is not twisted. For all specifications, see the Alignment Specs in the front of the section.

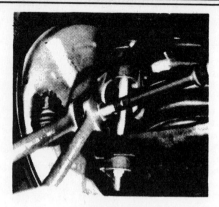

Toe-in adjustment

REAR SUSPENSION ISUZU

Shock Absorbers

REMOVAL & INSTALLATION

1. Raise the car and support it safely under the axle housing.
2. Disconnect the lower end of the shock absorber from the axle.
3. Remove the fuel tank cover from inside the trunk on all models except the Impulse and disconnect the upper end of the shock absorber.
4. Working from under the car, remove the shock absorber.
5. Installation is the reverse of removal. Use lock nuts at each end. Torque all shock absorber nuts to specifications.

Coil Springs

REMOVAL & INSTALLATION

1. Raise the rear of the car on the axle housing and support it at the jack side brackets with jackstands.
2. Position a hydraulic jack under the differential housing, but use a light contact pressure.
3. Disconnect the shock absorber lower mounting bolts.
4. Slowly lower or separate the axle assembly from the car body to the point where the spring becomes loose enough to allow removal.

——— CAUTION ———
Do not stress the brake hoses when lowering the axle.

5. Installation is the reverse of removal. Position the spring correctly. Make sure that the insulator is in position on top of the spring.
6. Torque the shock absorber lower bolts to 29 ft. lbs.

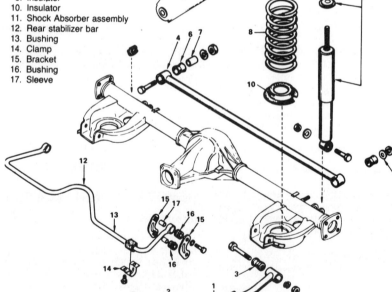

1. Control arm
2. Bushing
3. Bushing
4. Lateral rod
5. Bushing
6. Bushing
7. Sleeve
8. Spring
9. Insulator
10. Insulator
11. Shock Absorber assembly
12. Rear stabilizer bar
13. Bushing
14. Clamp
15. Bracket
16. Bushing
17. Sleeve

Exploded view of rear suspension system—all except Impulse

Rear Control Arm

REMOVAL & INSTALLATION

1. Raise the car and support it safely.
2. Remove the bolt connecting the control arm to the axle case.
3. Remove the bolt connecting the control arm to the body.
4. Remove the control arm assembly.
5. Installation is the reverse of removal. Torque the bolts to specifications.

NOTE: When reinstalling the control arm assembly, leave the bolts semi-tight.

SUSPENSION
Import Models

Lower the car before torquing any nuts. The vehicle weight should be on all suspension components when torquing the nuts.

Rear Axle Hub
REMOVAL & INSTALLATION

I-Mark (FWD)

1. Raise the rear end of the vehicle and support it on jackstands. Remove the rear wheels.
2. Remove the hub cap, cotter pin, hub nut, washer and outer bearing.

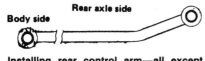

Installing rear control arm—all except Impulse

3. Remove the hub.
4. To install, reverse the removal procedure.

NOTE: If the cotter pin holes are out of alignment upon reassembly, use a wrench to tighten the nut until the hole in the shaft and a slot of the nut align.

Rear Wheel Bearing
REMOVAL & INSTALLATION

I-Mark (FWD)

1. Refer to the "Rear Axle Hub Removal and Installation" procedure in this section, then remove the hub.
2. Using a slide hammer puller and attachment, pull the oil seal from the hub. Remove the inner bearing.
3. Using a brass drift and a hammer, drive both bearing races from the hub.
4. Clean, inspect and/or replace all parts.
5. To install, pack the bearings with grease, coat the oil seal lips with grease and reverse the removal procedure.

ADJUSTMENT

I-Mark (FWD)

1. Torque the rear hub nut to 22 ft.lbs.
2. Rotate the hub two or three times, then untighten the nut completely.
3. Tighten the nut fully by hand and insert the cotter pin.
4. If the cotter pin holes are not aligned, tighten the nut just enough to align the holes.

FRONT SUSPENSION MAZDA

MacPherson Struts

REMOVAL & INSTALLATION

1. Remove the wheel cover and loosen the lug nuts.
2. Raise the front of the vehicle and support it with jackstands. Do not jack it or support it by any of the front suspension members. Remove the wheel.

—— CAUTION ——
Be sure the car is securely supported. Remember, you will be working underneath it.

3. Remove the brake caliper and disc on all models except the GLC and 626 with front wheel drive.
4. Unfasten the nuts which secure the upper shock mount to the top of the wheel arch.
5. Unfasten the two bolts that secure the lower end of the shock to the steering knuckle arm.
6. Remove the shock and coil spring as a complete assembly.
7. Installation is in the reverse order.

—— CAUTION ——
The coil springs are retained under considerable pressure. They can exert enough force to cause serious injury. Exercise extreme caution when servicing.

OVERHAUL

For all spring removal and installation procedures and any other strut overhaul procedures, please refer to "Strut Overhaul"

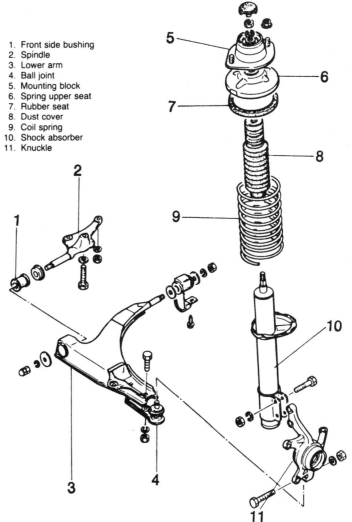

1. Front side bushing
2. Spindle
3. Lower arm
4. Ball joint
5. Mounting block
6. Spring upper seat
7. Rubber seat
8. Dust cover
9. Coil spring
10. Shock absorber
11. Knuckle

Typical front suspension assembly—front wheel drive models

SECTION 2
SUSPENSION
Import Models

Front Control Arm

REMOVAL & INSTALLATION

1980–85 Rear Wheel Drive

Remove the control arm and steering knuckle as an assembly.

1. Remove the wheel.
2. Remove the cotter pin and nut, which secure the tie-rod end, from the knuckle arm; then use a puller to separate them.
3. Unbolt the lower end of the shock absorber.
4. Remove the nut, then withdraw the rubber bushing and washer which secure the stabilizer bar to the control arm.
5. Unfasten the nut and bolt which secure the control arm to the frame member.
6. Push outward on the strut assembly while removing the end of the control arm from the frame member.
7. Remove the control arm and steering knuckle arm as an assembly.
8. Separate the knuckle arm from the control arm with a puller.
9. Installation of the control arm is the reverse of removal.

1986 and later RX-7

1. Raise the front of the car and support it safely with jackstands.
2. Remove the lower splash shield.
3. Disconnect the front stabilizer link from the control arm.
4. Remove the pinch bolt, then separate the lower ball joint from the steering knuckle.
5. Remove the front control arm mounting bolt.
6. Remove the control arm bushing bracket bolts and lower the control arm from the car.
7. Installation is the reverse of removal. Check the front end alignment.

Front Wheel Drive
GLC

1. Loosen the wheel lugs, raise the car and safely support it on jackstands. Remove the front wheel.
2. Remove the thru-bolt connecting the lower arm to the steering knuckle.
3. Remove the bolts and nuts mounting the control arm to the body (two inner and three outer).
4. Remove the lower control arm. The ball joint can be serviced at this time if necessary.
5. Installation is in the reverse order of removal.
Mounting Torque;
Ball Joint-to-Steering Knuckle: 32–40 ft. lbs.
Outer Bolts: 43–54 ft. lbs.
Inner Bolts: 69–86 ft. lbs.

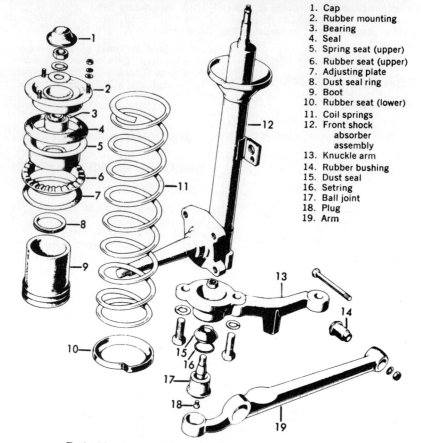

1. Cap
2. Rubber mounting
3. Bearing
4. Seal
5. Spring seat (upper)
6. Rubber seat (upper)
7. Adjusting plate
8. Dust seal ring
9. Boot
10. Rubber seat (lower)
11. Coil springs
12. Front shock absorber assembly
13. Knuckle arm
14. Rubber bushing
15. Dust seal
16. Setring
17. Ball joint
18. Plug
19. Arm

Typical front suspension assembly—rear wheel drive models

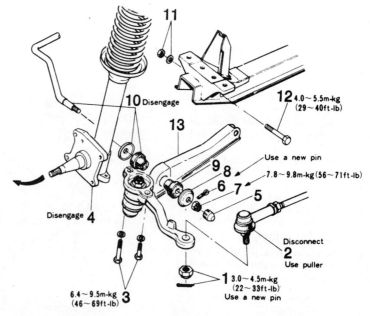

1. Castle nut and cotter pin
2. Tie-rod end
3. Strut to lower arm bolts
4. Strut and wheel spindle
5. Cap nut
6. Cotter pin
7. Castle nut
8. Washer
9. Bushing
10. Anti-roll bar and bushing
11. Nut and washer
12. Suspension arm bolt
13. Suspension arm

GLC rear wheel drive—strut and lower control arm

SUSPENSION
Import Models

1983 AND LATER 626 AND 323

1. Raise the front end and support it with safety stands.
2. Remove the wheel and splash shield.
3. Remove the stabilizer link from the control arm.
4. Remove the lower control arm-to-frame attaching bolts.
5. Remove the pinch bolt and separate the ball joint from the steering knuckle.
6. Remove the lower control arm.
7. Installation is the reverse of removal. Torque the lower arm mounting bolts to 69–86 ft. lbs.

Ball Joints

INSPECTION

All Exc. GLC

1. Perform Steps 1–5 of the "Control Arm Removal" procedure.
2. Check the ball joint dust boot.
3. Check the amount of pressure required to turn the ball stud, by hooking a pull scale into the tie rod hole in the knuckle arm. Pull the spring scale until the arm just begins to turn; this should require 4.4–7.7 lbs. on 626 models; 1–2.2 lbs. on 323 models.

If the reading is lower than 14 oz. on the 1980–85 RX-7, replace the ball joint and the suspension arm as a unit.

GLC

1. Check the dust boot for wear.
2. Raise the wheels off the ground and grip the tire at the top and bottom and alternately push and pull to check ball joint end-play. Wear limit is 0.04 in. If necessary, replace the ball joint and control arm assembly.

REMOVAL & INSTALLATION

1980–85 RX-7 and 1980–82 626

1. Remove the control arm.
2. Remove the set-ring and the dust boot.
3. Press the ball joint out of the control arm.
4. Clean the ball joint mounting bore and coat it with kerosene.
5. Press the ball joint into the control arm.

NOTE: If the pressure required to press the new ball joint into place is less than 3,300 lbs., the bore is worn and the control arm must be replaced.

GLC

The ball joint and control arm assembly is replaced as a unit on models through 1980. On 1981 and later models it may be removed during lower control arm removal. See "Control Arm Removal and Installation." Torque the ball joint nut to 43–51 ft. lbs. (thru 1980), check the revolving torque of the ball joint on 1981 and later models (63–109 ft. lbs).

1983 and Later 626

The ball joint and control arm assembly is replaced as a unit.

1986 and later 323 and RX-7

The ball joint can be unbolted from the control arm after separation from the steering knuckle.

Wheel Bearings— Front Discs

REMOVAL & INSTALLATION

All 1980–85 Rear Drive Models

1. Remove the brake disc/hub.
2. Drive the seal out and remove the inner bearing.
3. Drive the outer bearing races out.
4. Installation is the reverse of removal. Repack the bearings and the hub cavity with lithium grease. Adjust the bearing.

ADJUSTMENT

NOTE: This operation is performed with the wheel, grease cap, nut lock, and cotter pin removed.

1. To seat the bearings, back off on the adjusting nut three turns and then rotate the hub/disc assembly while tightening the adjusting nut.
2. Back off on the adjusting nut about ⅙ of a turn.
3. Hook a spring scale in one of the bolt holes on the hub.
4. Pull the spring scale squarely, until the hub just begins to rotate. The scale reading should be 0.9–2.2 lbs.—all passenger cars exc. GLC, 626 and RX-7. GLC: 0.33–1.32 lbs.; 626: 0.77–1.92 lbs.; RX-7: 0.99–1.43 lbs. Tighten the adjusting nut until the proper spring scale reading is obtained.
5. Place the castellated nut lock over the adjusting nut. Align one of the slots on the nut-lock with the hole in the spindle and fit the cotter pin into place.

Front End Alignment

CASTER AND CAMBER

Caster and camber are preset by the manufacturer. They require adjustment only if the suspension and steering linkage components are damaged. In this case, adjustment is accomplished by replacing the damaged part, except for the GLC and 626 (FWD).

On these models, the caster and camber may be changed by rotating the shock absorber support. If they can't be brought to within specifications, replace or repair suspension parts as necessary.

On 1983 and later 626 and 1986 and later 323 and RX-7 models the camber and caster can be adjusted by about 28' by changing the position of the mounting or support block.

1. Jack up the front end of the vehicle and support it with jack stands.
2. Loosen the four nuts that hold the mounting block to the strut tower. The standard strut mount position and the value for three optional positions are given in the following chart:

Optional position	Value at optional position	
	Camber angle	Caster angle
A	0°	28'
B	28'	28'
C	28'	0°

NOTE: 30' on 1986 and later RX-7

3. Tighten the four nuts to 17.0–22.0 ft. lbs. (23–30 Nm).

TOE-IN ADJUSTMENT

To adjust the toe-in, loosen the tie-rod locknuts and turn both tie rods an equal amount, until the proper specification is obtained.

REAR SUSPENSION MAZDA

Springs

REMOVAL & INSTALLATION

GLC (RWD)

1. Raise the rear end of the vehicle and place jack stands under the frame side rails.
2. Remove the rear wheel.
3. Remove the upper and lower shock absorber bolts and nuts and remove the shock absorber.
4. Place a jack under the lower arm to support it.
5. Remove the pivot bolt and nut that secures the rear end of the lower arm to the axle housing.
6. Slowly lower the jack to relieve the spring pressure on the lower arm, then remove the spring.
7. Install the spring in the reverse order of removal.

SECTION 2
SUSPENSION
Import Models

1980–85 RX-7 and 1980–82 626

1. Raise the rear of the vehicle and support it safely.
2. Remove the rear wheel.
3. Position an hydraulic floor jack under the rear axle housing.
4. Disconnect the shock absorber lower end and the lower link bolt (just to the front of the lower shock bolt). Remove the rear bolt from the upper link.

NOTE: These trailing control arm links run parallel from the front to rear on the vehicle, with the smaller watt links running side to side on the vehicle.

5. Disconnect the front ends of the stabilizer bar (if equipped).
6. Remove the right and left watt links at the rear axle housing.
7. Carefully lower the rear axle housing and remove the coil spring and the rubber seat.
8. The installation is the reverse of the removal.

Shock Absorbers

REMOVAL & INSTALLATION

1980–85 RX-7 and 1980–82 626

Remove the trim panel from the rear of the luggage compartment.

1. Unfasten the nuts, then remove the washers and rubber bushings from the upper shock absorber mounts.
2. Unfasten the nut and bolt which secure the end of the rear shock to the axle housing.
3. Remove the shock.
4. Installation is the reverse of removal. Torque the shock absorber bolts to 54 ft.lbs.

GLC (RWD)

1. Raise the rear end of the vehicle and place jack stands under the frame side rails.
2. Remove the rear wheel.
3. Remove the upper and lower shock absorber bolts and nuts and remove the shock absorber.
4. Install the shock absorber in the reverse order of removal.

MacPherson Strut

REMOVAL & INSTALLATION

1981 and Later GLC and 1983 and Later 626 and 323

1. Remove the side trim panels from inside the "trunk," or the rear seat and trim. Loosen and remove the top mounting nuts from the strut mounting block assembly.
2. Loosen the rear wheel lugs, raise the car and safely support it on jackstands.
3. Remove the rear wheels. Disconnect the flexible brake hose from the strut.
4. Disconnect the trailing arm from the lower side of the strut: Separate the lateral link and strut by removing the bolt assembly.
5. Remove the strut from the lower unit by removing the two through nuts and bolts,
6. Remove the strut and brake assembly.

— CAUTION —
The coil springs are retained under considerable pressure. They can exert enough force to cause serious injury. Exercise extreme caution when servicing.

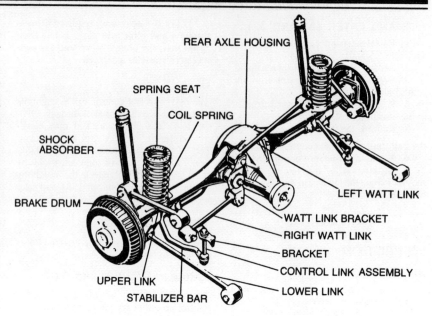

Rear suspension assembly—1980–85 RX-7

1. Crossmember
2. Rear stabilizer bar
3. Lateral link
4. Mounting block
5. Spring upper seat
6. Rubber seat
7. Dust cover
8. Coil spring
9. Shock absorber
10. Rear hub spindle
11. Trailing link

Rear suspension assembly—GLC front wheel drive

SUSPENSION
Import Models

1986 and later RX-7

1. Raise the rear of the car and support it safely with jackstands.
2. Remove the two top shock flange mounting bolts.
3. Remove the bottom shock absorber mounting bolt and remove the shock and spring assembly from the car.
4. Installation is the reverse of removal.

NOTE: If equipped with adjustable shock absorbers, disconnect the electrical connector from the top shock absorber tower before removal.

OVERHAUL

For all strut overhaul procedures, please refer to "Strut Overhaul"

Trailing Arm

REMOVAL & INSTALLATION

1986 and later RX-7

1. Raise the rear of the car and support it safely.
2. Remove the axle shaft locknut.
3. Remove the brake caliper and wire it out of the way. Do not let the caliper hang by its brake hose.
4. Remove the brake caliper mounting bracket and the brake rotor.
5. Remove the shock absorber.
6. Disconnect the rear stabilizer bar from the trailing arm.
7. Remove the drive axle shaft and rear hub assembly.
8. Disconnect the lateral link. Remove the inner trailing arm mounting bolt.
9. Remove the outer trailing arm bolt and remove the trailing arm from the car.
10. Installation is the reverse of removal.

Rear Wheel Bearings

REPLACEMENT

All front wheel drive models use a conventional roller bearing with the race pressed into the rotor or brake drum. To replace the bearing, remove the rear brake drum or rotor and lift out the bearing cage assembly. Use a blunt drift to knock the bearing race out, then press in a new race using a bench press and suitable manderl. When the drum or rotor is installed, torque the center nut to 18–22 ft. lbs. (25–29 Nm) to preload the bearing, then crimp the nut collar using a suitable drift to lock the nut in place.

FRONT SUSPENSION MERCEDES-BENZ

Springs

REMOVAL & INSTALLATION

190D, 190E and 300E

1. Raise the front of the vehicle and support it with jackstands. Remove the wheel.
2. Remove the engine compartment lining underneath the vehicle (if so equipped).
3. Install a spring compressor so that at least 7½ coils are engaged.
4. Support the lower control arm with a floor jack and then loosen the retaining nut at the upper end of the damper strut.

CAUTION
NEVER loosen the damper strut retaining nut unless the wheels are on the ground, the control arm is supported or the springs have been removed.

5. Lower the jack under the control arm slightly and then remove the spring toward the front.
6. On installation, position the spring between the control arm and the upper mount so that when the control arm is raised, the end of the lower coil will be seated in the impression in the control arm.
7. Use the jack and raise the control arm until the spring is held securely.
8. Using a new nut, tighten the upper end of the damper strut to 44 ft. lbs. (60 Nm).
9. Slowly ease the tension on the spring compressor until the spring is seated properly and then remove the compressor.
10. Installation of the remaining components is in the reverse order of removal.

1. Wishbone
2. Steering knuckle
3. Torsion bar
4. Damper strut
5. Front spring
6. Spring-rubber mount
7. Front end

Front spring—190D and 190E

380SL, 380SLC, 450SL, 450SLC and 560SL

NOTE: Be extremely careful when attempting to remove the front springs as they are compressed and under considerable load.

1. Raise the front of the vehicle and support it with jackstands. Remove the wheels.
2. Remove the front shock absorber and disconnect the sway bar.
3. First punchmark the position of the eccentric adjusters, then loosen the hex bolts.
4. Support the lower control arm with a jack.

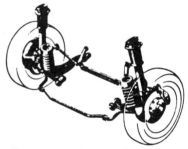

Front suspension—190D and 190E

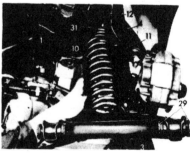

1. Front axle carrier
3. Lower control arm
4. Upper control arm
10. Front spring
11. Front shock absorber
12. Torsion bar
29. Rubber mounting
31. Rubber mounting for front spring

Front spring removal—380SL, 380SLC, 450SL, 450SLC and 560SL

5. Knock out the eccentric pins and gradually lower the arm until spring tension is relieved.
6. The spring can now be removed.

NOTE: Check caster and camber after installing a new spring.

2-219

SECTION 2: SUSPENSION
Import Models

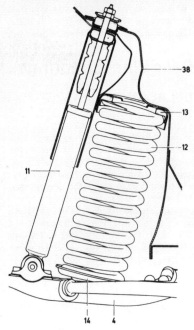

4. Lower control arm
11. Front shock absorber
12. Front spring
13. Rubber mount for front spring
14. Retainer for front spring
38. Front end

Front spring—all models except 190 series, 300 series (1986–87), 380SL, 380SLC, 450SL, 450SLC and 560SL

7. Installation is the reverse of removal.
8. For ease of installation, tape the rubber mounts to the springs.
9. If the eccentric adjusters were not matchmarked, install the eccentric bolts as illustrated under "Front End Alignment".

All Other Models

1. Raise and support the front of the car and support the lower control arm.
2. Remove the wheel. Unbolt the upper shock absorber mount.
3. Install a spring compressor and compress the spring.
4. Remove the front spring with the lower mount.
5. Installation is the reverse of removal. Tighten the upper shock absorber suspension.

Shock Absorbers

Shock absorbers are normally replaced only if leaking excessively (oil visible on outside cover) or if worn internally to a point where the car no longer rides smoothly and rebounds after hitting a bump. A good general test of shock absorber condition is made by bouncing the front of the car. If the car rebounds more than two or three times it can be assumed that the shock absorbers need replacement.

REMOVAL & INSTALLATION
380SL, 380SLC, 450SL, 450SLC and 560SL

1. For removal and installation of shock absorbers, it is best to jack up the front of the car until the weight is off of the wheels and support the car securely on jack stands.
2. When removing the shock absorbers, it is also wise to draw a simple diagram of the location of parts such as lockrings, rubber stops, locknuts and steel plates, since many shock absorbers require their own peculiar installation of these parts.
3. Raise the hood and locate the upper shock absorber mount.
4. Support the lower control arm with a jack.
5. Unbolt the mount for the shock absorber at the top. Remove the coolant expansion tank to allow access to the right front shock absorber.
6. Remove the nuts which secure the shock absorber to the lower control arm.
7. Push the shock absorber piston rod in, install the stirrup, and remove the shock absorber.
8. Remove the stirrup, since this must be installed on replacement shock absorbers.
9. Installation is the reverse of removal. Always use new bushings when installing replacement shock absorbers.

All Other Models (Except 190D, 190E and 300E)

1. Jack and support the car. Support the lower control arm.
2. Loosen the nuts on the upper shock absorber mount. Remove the plate and ring.
3. Place the shock absorber vertical to the lower control arm and remove the lower mounting bolts.
4. Remove the shock absorber. Be sure to disconnect and plug the pressure line on models with level control.
5. Installation is the reverse of removal. On Bilstein shocks, do not confuse the upper and lower plates.

NOTE: The 1981 and later 380SEL shock absorber uses a protective plastic sleeve that must be installed between the lower retainer and lower rubber ring. Also, a slot is provided for holding the piston rod, in place of the 2 flats used previously.

NOTE: 380SEC shock absorbers have the same part number regardless of manufacturer. However, the shock absorbers on these cars have a larger diameter and narrower lower mounting eye than other models.

Damper Strut

REMOVAL & INSTALLATION
190D, 190E and 300E

1. Raise the front of the vehicle and support it with jackstands. Remove the wheel.
2. Using a spring compressor, compress the spring until any load is removed from the lower control arm.

NOTE: When using a spring compressor, be sure that at least 7½ coils are engaged before applying tension.

3. Support the lower control arm with a floor jack. Loosen the retaining bolt for the upper end of the damper strut by holding the inner piston rod with an Allen wrench and then unscrewing the nut. *NEVER use an impact wrench on the retaining nut.*

--- CAUTION ---
Never unscrew the nut with the axle half at full rebound—the spring may fly out with considerable force, causing personal injury.

4. Unbolt the two screws and one nut and then disconnect the lower damper strut from the steering knuckle.
5. Remove the strut down and forward. Be sure to disconnect and plug the pressure line on models with level control. Secure the steering knuckle in position so that it won't tilt.
6. Installation is in the reverse order of removal. Please note the following:
   a. When attaching the lower end of the damper strut to the steering knuckle, first position all three screws; next tighten the two lower screws to 72 ft. lbs. (100 Nm); finally, tighten the nut on the upper clamping connection screw to 54 ft. lbs. (75 Nm).
   b. Tighten the retaining nut on the upper end of the damper strut to 44 ft. lbs. (60 Nm).

Steering Knuckle and Ball Joints

CHECKING BALL JOINTS

1. To check the steering knuckles or ball joints, jack up the car, placing a jack directly under the front spring plate. This unloads the front suspension to allow the maximum play to be observed.
2. Late model ball joints need to be replaced only if dried out with plainly visible wear and/or play.

REMOVAL & INSTALLATION
190D, 190E and 300E

1. Raise the front of the vehicle and support it with jackstands. Remove the wheel.
2. Install a spring compressor on the spring.
3. Remove the brake caliper and then

SUSPENSION
Import Models

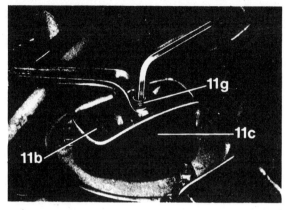

- 11b Rebound limiter
- 11c Rubber mount
- 11g Piston rod

Remove the upper damper strut retaining nut by locking the piston rod with an Allen wrench—190D and 190E

wire it out of the way. Be careful not to damage the brake line.

4. Remove the brake disc and wheel hub.

NOTE: On vehicles equipped with ABS, remove the speed sensor.

5. Unscrew the three socket-head bolts and then remove the brake backing plate from the steering knuckle.

6. Tighten the spring compressor until all tension and/or load has been removed from the lower control arm.

7. Disconnect the steering knuckle arm from the steering knuckle (this is the arm attached to the tie rod).

――― **CAUTION** ―――
There must be no tension on the lower control arm.

8. Unscrew the three bolts and disconnect the lower end of the damper strut from the steering knuckle.

9. Remove the hex-head clamp nut at the supporting joint (lower ball joint).

10. Remove the steering knuckle.

11. Installation is in the reverse order of removal. Please note the following:

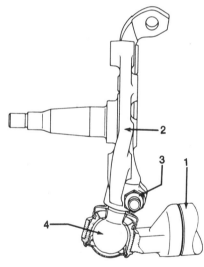

1. Wishbone
2. Steering knuckle
3. Bolt with nut
4. Ball joint

Steering knuckle/ball joint—190D and 190E

a. Tighten the supporting joint clamp nut to 70 ft. lbs. (125 Nm).

b. Refer to the "Damper Strut Removal and Installation" procedure when connecting the lower end of the damper strut to the steering knuckle.

380SL, 380SLC, 450SL, 450SLC and 560SL

1. This should only be done with the front shock absorber installed. If, however, the front shock absorber has been removed, the lower control arm should be supported

with a jack and the spring should be clamped with a spring tensioner. In this case, the hex nut on the guide joint should not be loosened without the spring tensioner installed.

2. Jack up the front of the car and support it on jack stands.
3. Remove the wheel.
4. Remove the brake caliper.
5. Unbolt the steering relay lever from the steering knuckle. For safety, install spring clamps on the front springs.
6. Remove the hex nuts from the upper and lower ball joints.
7. Remove the ball joints from the steering knuckle with the aid of a puller.
8. Remove the steering knuckle.
9. Installation is the reverse of removal. Be sure that the seats for the pins of the ball joints are free of grease.
10. Bleed the brakes.

All Other Models

1. Raise and support the car. For safety, it's a good idea to install some type of clamp on the front spring. Position jack stands at the outside front against the lower control arms.

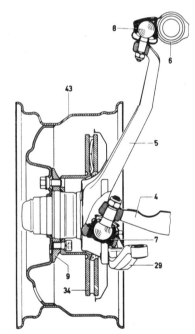

4. Lower control arm
5. Steering knuckle
6. Upper control arm
7. Support joint
8. Guide joint
9. Front wheel hub
29. Steering knuckle arm
34. Brake disc
43. Wheel

Steering knuckle/ball joint—all except 190 series, 300 series (1986–87), 380SL, 380SLC, 450SL, 450SLC and 560SL

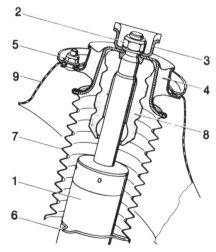

1. Damper strut
2. Hex. nut
3. Rebound stop
4. Rubber mount
5. Hex. nuts
6. Stop ring
7. Sleeve
8. Additional PU spring
9. Front end

Upper damper strut mounting—190D and 190E

2-221

SECTION 2 SUSPENSION
Import Models

2. Remove the wheel.
3. Remove the steering knuckle arm from the steering knuckle.
4. Remove and suspend the brake caliper.
5. Remove the front wheel hub.

NOTE: On vehicles equipped with ABS, disconnect the speed sensor.

6. Loosen the brake hose holder on the cover plate.
7. Loosen the nut on the guide joint and remove the joint from the steering knuckle.
8. Loosen the nut on the support joint.
9. Swivel the steering knuckle outward and force the ball joint from the lower control arm.
10. Remove the steering knuckle.
11. If necessary, remove the cover plate from the steering knuckle.
12. Installation is the reverse of removal. Use self-locking nuts and adjust the wheel bearings.

Upper Control Arm

NOTE: The 190D, 190E and 300E models have no upper control arm.

REMOVAL & INSTALLATION

All Models Except 380SL, 380SLC, 450SL, 450SLC and 560SL

1. Raise and support the car. Position jack stands at the outside front against the lower control arms.
2. Remove the wheel.
3. Loosen the nut on the guide joint.
4. Remove the guide joint from the steering knuckle.
5. Secure the steering knuckle with a hook on the upper control arm stop to prevent it from tilting.
6. Loosen the clamp screw and separate the upper control arm from the torsion bar.
7. Loosen the upper control arm bearing at the front and remove the upper control arm.
8. Installation is the reverse of removal. Use new self-locking nuts and check the front wheel alignment.

380SL, 380SLC, 450SL, 450SLC and 560SL

1. The front shock absorbers should remain installed. Never loosen the hex nuts of the ball joints with the shock absorber removed, unless a spring clamp is installed.
2. Jack the front of the car and remove the wheel.
3. Support the front end on jack stands.
4. Remove the steering arm from the steering knuckle.

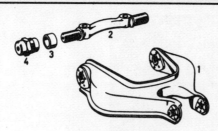

1. Upper control arm
2. Pivot pin
3. Rubber sealing ring
4. Threaded bushing

Upper control arm and pivot shaft

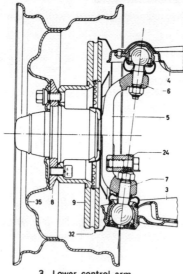

3. Lower control arm
4. Upper control arm
5. Steering knuckle
6. Guide joint
7. Supporting joint
8. Front wheel hub
9. Brake disc
24. Steering knuckle arm
32. Cover plate
35. Wheel

Steering knuckle/ball joint – 380SL, 380SLC, 450SL, 450SLC and 560SL

5. Separate the brake line and brake hose from each other and plug the openings.
6. Support the lower control arm and unscrew the nuts from the ball joints.
7. Remove the ball joints from the steering knuckle.
8. Loosen the bolts on the upper control arm and remove the upper control arm.
9. Installation is the reverse of removal.

CAUTION
Mount the front hex bolt from the rear in a forward direction, and the rear hex bolt from the front in a rearward direction.

10. Bleed the brakes.

Lower Control Arm

REMOVAL & INSTALLATION

All Models Except 190D, 190E, 300E, 380SL, 380SLC, 450SL, 450SLC and 560SL

The lower control arm is the same as the front axle half. For safety install a spring compressor on the coil spring.

1. Raise and support the front of the car and remove the wheels.
2. Remove the front shock absorber. Loosen the top mount first.
3. Remove the front springs.
4. Separate and plug the brake lines.
5. Remove the track rod from the steering knuckle arm.
6. Matchmark the position of the eccentric bolts on the bearing of the lower control arm in relation to the crossmember.
7. Remove the shield from the cross yoke.
8. Support the front axle half.
9. Loosen the eccentric bolt on the front and rear bearing of the lower control arm and knock them out.
10. Remove the bolt from the cross yoke bearing.

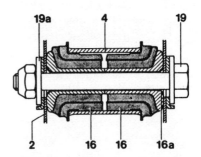

1. Frame cross member
2. Wishbone
3. Torsion rubber bushing
4. Clamping sleeve
5. Eccentric bolt (camber adjustment)
6. Eccentric washer

Cross section of the front lower control arm bushing on 190 models

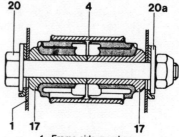

1. Frame side member
2. Wishbone
3. Torsion rubber bushing
4. Eccentric bolt (caster adjustment)
5. Eccentric washer

Cross section of the rear lower control arm bushing on 190 models

SUSPENSION
Import Models

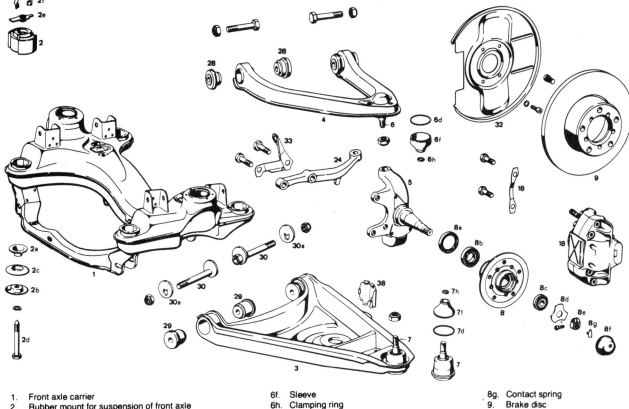

1. Front axle carrier
2. Rubber mount for suspension of front axle
2a. Stop buffer for inward deflection
2b. Stop plate
2c. Stop buffer for outward deflections
2d. Hex. bolt with snap-ring
2e. Fastening nut
2f. Nut holder
3. Lower control arm
4. Upper control arm
5. Steering knuckle
6. Guide joint
6d. Circlip
6f. Sleeve
6h. Clamping ring
7. Supporting joint
7d. Circlip
7f. Sleeve
7h. Clamping ring
8. Front wheel hub
8a. Radial sealing ring
8b. Inside tapered roller bearing
8c. Outside tapered roller bearing
8d. Washer
8e. Clamp nut
8f. Wheel cap
8g. Contact spring
9. Brake disc
18. Brake caliper
18a. Lockwasher
24. Steering knuckle arm
28. Rubber slide bearing
29. Rubber bearing (torsion bearing)
30. Cam bolt
30a. Cam washer
32. Cover plate
33. Holder for brake hose
38. Protective cap for steering lock

Lower control arm and pivot shaft

11. Loosen the screw at the opposite end of the cross yoke bearing.
12. Pull the cross yoke bearing down slightly.
13. Loosen the support of the upper control arm on the torsion bar. Remove the clamp screw from the clamp.

Loosen the bolts on the upper control arm and remove the upper control arm.

Mark the position of the inner eccentric pins, relative to the frame, on the bearing of the control arm.

14. Remove the upper control arm bearing on the front end.

15. Remove the front axle half.

16. Installation is the reverse of removal. Tighten the eccentric bolts of the lower control arm bearing with the car resting on the wheels. Bleed the brakes and check the front end alignment.

190D, 190E and 300E

1. Remove the engine compartment lining at the bottom of the vehicle (if so equipped).
2. Raise the front of the vehicle and support it with jackstands. Remove the wheel.
3. Support the lower control arm with jackstands and then disconnect the torsion bar bearing at the control arm.
4. Remove the spring as detailed earlier in this chapter.
5. Disconnect the tie rod at the steering knuckle and then press out the ball joint with the proper tool.
6. Remove the brake caliper and position it out of the way. Be sure that you do not damage the brake line.
7. Remove the brake disc/wheel hub assembly.
8. Disconnect the lower end of the

10. Unscrew and remove the pins.
11. Remove the jackstands and remove the lower control arm.
12. Installation is in the reverse order of removal. Please note the following:

   a. Tighten the eccentric bolts on the inner arm to 130 ft. lbs. (180 Nm).

   b. To facilitate torsion bar installation, raise the opposite side of the lower control arm with a jack.

   c. Tighten the clamp nut on the tie rod ball joint to 25 ft. lbs. (35 Nm).

   d. When installing the rear torsion bar bushing, on the 300E, the flats on the cone must be vertical.

380SL, 380SLC, 450SL, 450SLC and 560SL

1. Since the front shock absorber acts as a deflection stop for the front wheels, the lower shock absorber attaching point should not be loosened unless the vehicle

2-223

SUSPENSION
Import Models

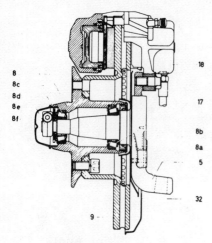
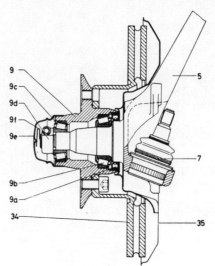
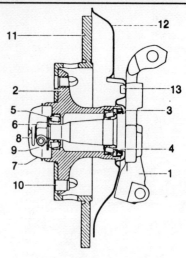

5. Steering knuckle
8. Wheel hub
8a. Radial sealing ring
8b. Tapered roller bearing, outside
8c. Tapered roller bearing, inside
8d. Washer
8e. Clamping nut
8f. Wheel cap
9. Brake disc
17. Brake hose
18. Brake caliper
32. Cover plate

Wheel bearing cutaway—380SL, 380SLC, 450SL, 450SLC and 560SL

5. Steering knuckle
7. Supporting joint
9. Front wheel hub
9a. Radial sealing ring
9b. Tapered roller bearing, inside
9c. Tapered roller bearing, outside
9d. Clamping nut
9e. Wheel cap
9f. Contact spring
34. Brake disc
35. Cover plate

Wheel bearing cutaway—all except 380SL, 380SLC, 450SL, 450SLC and 560SL

1. Steering knuckle
2. Front wheel hub
3. Radial seal ring
4. Tapered roller bearing, inner
5. Tapered roller bearing, outer
6. Clamping nut
7. Grease cap
8. Contact spring
9. Washer
10. Clamping sleeve
11. Brake disk
12. Brake backing plate
13. Allen screws

Wheel bearing cutaway—190D and 190E

is resting on the wheels or unless the lower control arm is supported.
2. Jack up the front of the vehicle and support it on jack stands.
3. Support the lower control arm.
4. Loosen the lower shock absorber attachment.
5. Unscrew the steering arm from the steering knuckle.
6. Separate the brake line and brake hose and plug the openings.
7. Remove the front spring.
8. Unscrew the hex nuts on the ball joints.
9. Remove the lower ball joint and remove the lower control arm.
10. Installation is the reverse of removal. Bleed the brakes and check the front end alignment.

Wheel Bearings

REMOVAL & INSTALLATION

If the wheel bearing play is being checked for correct setting only, it is not necessary to remove the caliper. It is only necessary to remove the brake pads.
1. Remove the brake caliper.
2. Pull the cap from the hub with a pair of channel-lock pliers. Remove the radio suppression spring, if equipped.

3. Loosen the socket screw of the clamp nut on the wheel spindle. Remove the clamp nut and washer.
4. Remove the front wheel hub and brake disc.
5. Remove the inner race with the roller cage of the outer bearing.
6. Using a brass or aluminum drift, carefully tap the outer race of the inner bearing until it can be removed with the inner race, bearing cage, and seal.
7. In the same manner, tap the outer race of the bearing out of the hub.
8. Separate the front hub from the brake disc.
9. To assemble, press the outer races into the front wheel hub.
10. Pack the bearing cage with bearing grease and insert the inner race with the bearing into the wheel hub.

Dial indicator set-up for checking wheel bearing play

11. Coat the sealing ring with sealant and press it into the hub.
12. Pack the front wheel hub with 45–55 grams of wheel bearing grease. The races of the tapered bearing should be well packed and also apply grease to the front faces of the rollers. Pack the front bearings with the specified amount of grease. Too much grease will cause overheating of the lubricant and it may lose its lubricity. Too little grease will not lubricate properly.
13. Coat the contact surface of the sealing ring on the wheel spindle with Molykote® paste.
14. Press the wheel hub onto the wheel spindle.
15. Install the inner race and cage of the outer bearing.
16. Install the steel washer and the clamp nut.

ADJUSTMENT

1. Tighten the clamp nut until the hub can just be turned.
2. Slacken the clamp nut and seat the bearings on the spindle by rapping the spindle sharply with a hammer.
3. Attach a dial indicator, with the pointer indexed, onto the wheel hub.
4. Check the end-play of the hub by pushing and pulling on the flange. The end-play should be approximately 0.0004–0.0008 in.

SUSPENSION
Import Models

5. Make an additional check by rotating the washer between the inner race of the outer bearing and the clamp nut. It should be able to be turned by hand.

6. Check the position of the suppressor pin in the wheel spindle and the contact spring in the dust cap.

7. Pack the dust cap with 20–25 grams of wheel bearing grease and install the cap.

8. Install the brake caliper and bleed the brakes.

Front End Alignment

Caster and camber are critical to proper handling and tire wear. Neither adjustment should be attempted without the specialized equipment to accurately measure the geometry of the front end.

CASTER/CAMBER ADJUSTMENT

All Models Except 380SL, 380SLC, 450SL, 450SLC and 560SL

The front axle provides for caster and camber adjustment, but both wheel adjustments can only be made together. Adjustments are made with cam bolts on the lower control arm bearings.

The front bearing cam bolt is used to set caster, while the rear bearing cam bolt is used for camber.

380SL, 380SLC, 450SL, 450SLC and 560SL

Caster and camber are dependent upon each other and cannot be adjusted independently. They can only be adjusted simultaneously.

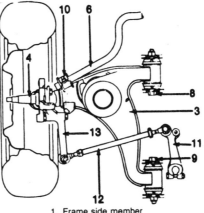

1. Frame side member
2. Frame cross member
3. Wishbone
4. Steering knuckle
5. Supporting ball joint
6. Torsion bar
7. Damper strut
8. Eccentric bolt of front bushing (camber adjustment)
9. Eccentric bolt of rear bushing (caster adjustment)
10. Torsion bar bushing on wishbone
11. Pitman arm
12. Tie rod
13. Steering knuckle arm

Caster and camber adjustment points—190D and 190E

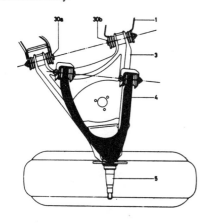

1. Front axle carrier
3. Lower control arm
4. Upper control arm
5. Steering knuckle
30a. Cam bolt front (caster)
30b. Cam bolt rear (camber)

Caster and camber adjusting points on the 380SL, 380SLC, 450SL, 450SLC and 560SL

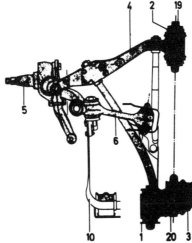

1. Frame side member
2. Frame cross member for front axle
3. Cross yoke
4. Lower control arm
5. Steering knuckle
6. Upper control arm
10. Torsion bar
19. Cam bolt of front bearing (camber adjustment)
20. Cam bolt of rear bearing (caster adjustment)

Caster and camber adjusting points on the 280SE, 1980 300SD and 450SEL

Caster is adjusted by turning the lower control arm around the front mounting, using the eccentric bolt.

Camber is adjusted by turning the lower control arm about the rear mounting, using

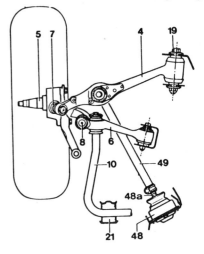

1. Frame side member
2. Frame cross member for front axle
4. Lower control arm
5. Steering knuckle
6. Upper control arm
7. Supporting joint
8. Guide joint
10. Torsion bar
19. Eccentric bolt (camber adjustment)
21. Torsion bar mounting on front end
48. Supporting joint
48a. Ball pin (caster adjustment)
49. Supporting tube

Caster and camber adjustment points on all other models

the eccentric bolt. Bear in mind that caster will be changed accordingly.

When camber is adjusted in a positive direction, caster is changed in a negative direction, and vice versa. Adjustment of camber by 0°15′ results in a caster change of approximately 0°20′. Adjustment of caster by 1° results in a camber change of approximately 0°7′.

TOE-IN ADJUSTMENT

Toe-in is the difference of the distance between the front edges of the wheel rims and the rear edges of the wheel rims.

To measure toe-in, the steering should be in the straight ahead position and the marks on the pitman arm and pitman shaft should be aligned.

Toe-in is adjusted by changing the length of the two tie rods or track rods with the wheels in the straight ahead position. Some older models have a hex nut locking arrangement rather than the newer clamp, but adjustment is the same.

NOTE: Install new tie rods so that the left-handed thread points toward the left-hand side of the car.

2-225

SECTION 2 SUSPENSION
Import Models

REAR SUSPENSION MERCEDES-BENZ

Springs

REMOVAL & INSTALLATION

190D, 190E and 300E

1. Raise the rear of the vehicle and support it with jackstands. Remove the wheel.
2. Disconnect the holding clamps for the spring link cover and then remove the cover.
3. Install a spring compressor and compress the spring until the spring link is free of all load.
4. Disconnect the lower end of the shock absorber.
5. Increase the tension on the spring compressor and remove the spring.
6. Installation is in the reverse order of removal. Please note the following:
   a. Position the spring so that the end of the lower coil is seated in the impression of the spring seat and the upper coil seats properly in the rubber mount in the frame floor.
   b. Do not release tension on the spring compressor until the lower end of the shock absorber is connected and tightened to 47 ft. lbs. (65 Nm).

380SL, 380SLC, 450SL, 450SLC and 560SL

1. Jack up the rear of the car.
2. Remove the rear shock absorber.
3. With a floor jack, raise the control arm to approximately a horizontal position. Install a spring compressor to aid in this operation.
4. Carefully lower the jack until the control arm contacts the stop on the rear axle support.
5. Remove the spring and spring compressor with great care.
6. Installation is the reverse of removal. For ease of installation, attach the rubber seats to the springs with masking tape.

All Others

1. Raise and support the rear of the car and the trailing arm.
2. Remove the rear shock absorber.
3. Be sure that the upper shock absorber attachment is released first.
4. Compress the spring with a spring compressor.
5. Remove the rear spring with the rubber mount.

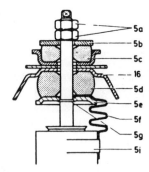

5a. Nut
5b. Washer
5c. Upper rubber ring
5d. Lower rubber ring
5e. Plate
5f. Dust protection
5g. Lockring
5i. Clamping strap
16. Dome on frame floor

Rear shock absorber lower mount—300SDL, all 380 models, 420SEL, 450SL, 450SLC, all 500 models and all 560 models

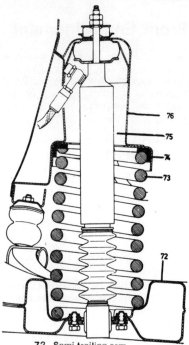

72. Semi-trailing arm
73. Rear spring
74. Rubber mounting
75. Shock absorber or spring strut
76. Dome on frame floor

Rear spring—all models except 190 and 300 (1986 and later) series, 380SL, 380SLC, 450SL, 450SLC and 560SL

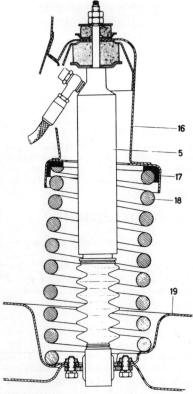

5. Spring strut
16. Dome on frame floor
17. Rubber mount
18. Rear spring
19. Semi-trailing arm

Rear spring—380SL, 380SLC, 450SL and 450SLC

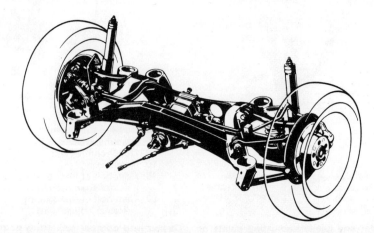

Rear suspension on the 190D and 190E. Five separate links keep the wheels in alignment under all conditions.

2-226

SUSPENSION
Import Models

6. Installation is the reverse of removal. When installing the shock absorber, tighten the lower mount first.

Shock Absorber

REMOVAL & INSTALLATION

190D, 190E, 300E, 380SL, 380SLC, 450SL, 450SLC and 560SL

1. Jack up the rear of the car and support the control arm.
2. From inside the trunk (sedans), remove the rubber cap, locknut, and hex nut from the upper mount of the shock absorber. On 380SL and 450SL, the upper mount of the rear shock absorber is accessible after removing the top, top flap, rear seat, backrest and lining. On the 380SLC and 450SLC, remove the rear seat, backrest and cover plate.
3. Unbolt the mounting for the rear shock absorber at the bottom and remove the shock absorber. Be sure to disconnect and plug the pressure line on the 190E-16.
4. Installation is the reverse of removal.

All Other Models

1. Remove the rear seat and backrest.
2. Remove the cover from the rear wall.
3. Raise and support the car and the trailing arm.

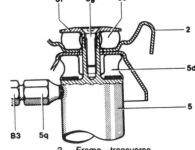

2. Frame—transverse member
5. Suspension strut
5c. Upper rubber mount
5d. Lower rubber mount
5g. Special screw
5i. Plate
5q. Stud
B3. Pressure line (pressure hose), pressure reservoir—suspension strut

Rear shock absorber upper mount—1983 and later 300TD. Retrofitting is not possible

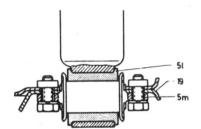

5l. Rubber mounting
5m. Fastening clip
19. Semi-trailing arm

Rear shock absorber upper mount—1981 and later 300SD, 300SDL, 380SE, 380SEC, 380SEL, 380SL, 380SLC, 420SEL, 450SL, 450SLC, 500SEC, 500SEL, 560SEC, 560SEL and 560SL (5f not used in U.S.)

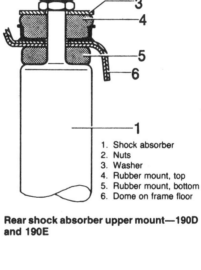

1. Shock absorber
2. Nuts
3. Washer
4. Rubber mount, top
5. Rubber mount, bottom
6. Dome on frame floor

Rear shock absorber upper mount—190D and 190E

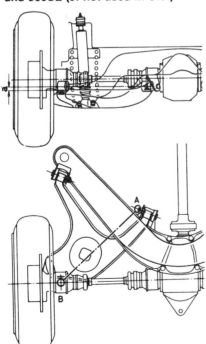

Rear wheel camber adjustment on all models except 190D, 190E, 380SL, 380SLC, 450SL, 450SLC and 560SL

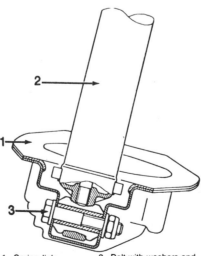

1. Spring link
2. Shock absorber
3. Bolt with washers and self-locking nut

Rear shock absorber lower mount—190D and 190E

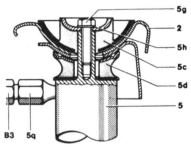

Rear shock absorber upper mount—1981-82 300TD

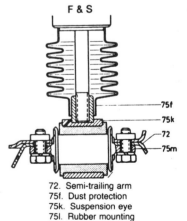

72. Semi-trailing arm
75f. Dust protection
75k. Suspension eye
75l. Rubber mounting
75m. Fastening stirrup

Rear shock absorber lower mount—240D, all 280 models, all 1980 300 models and the 450SEL

2-227

SUSPENSION
Import Models

4. Loosen the nuts on the upper mount. Remove the washer and rubber ring.
5. Loosen the lower mount and remove the shock absorber downward.
6. Installation is the reverse of removal. Tighten the upper mounting nut to the end of the threads.

Independent Rear Suspension Adjustments

Suspension adjustments should only be checked when the vehicle is resting on a level surface and is carrying the required fluids (full tank of gas, engine oil, etc.).

CAMBER

All Models

Rear wheel camber is determined by the position of the control arm. The difference in height (a) between the axis of the control arm mounting point on the rear axle subframe and lower edge of the cup on the constant velocity joint is directly translated in degrees of camber.

TOE-IN

Toe-in, on the rear wheels, is dependent on the camber of the rear wheels.

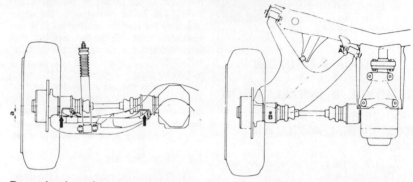

Rear wheel camber measurement on the 380SL, 380SLC, 450SL, 450SLC and 560SL. The control arm position (difference in height between the axis of the rear control arm mount (A) and the lower edge of the cup on the outer edge of the CV-joint)

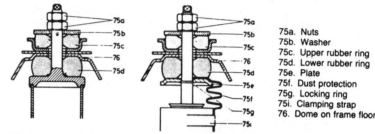

75a. Nuts
75b. Washer
75c. Upper rubber ring
75d. Lower rubber ring
75e. Plate
75f. Dust protection
75g. Locking ring
75i. Clamping strap
76. Dome on frame floor

Rear shock absorber upper mount—240D, 280E, 280CE, 1980 300 models and 450SEL (5f not used in U.S.)

Hydropneumatic Suspension

The hydropneumatic suspension is used on the 190E-16, 380SEL, 500SEL, 560SEC and 560SEL. Service of this system should be left to a Mercedes-Benz dealer or other qualified service establishment.

OPERATION

The system is a gas pressure system with a hydraulic level control. The car is supported by 4 suspension struts that also serve as shock absorbers. The suspension consists of a strut and pressure reservoir, connected by a line. The load is transmitted to the pressure reservoirs via the struts, resulting in an adjustment of the gas cushion in each pressure reservoir.

To regulate the level of the car, the oil level in the struts is increased or reduced by the hydraulic system, composed of an oil pump, pressure regulator, main pressure reservoir, and oil reservoir. The pressure regulator also contains a level selector valve as part of the unit.

The oil volume is controlled by a levelling valve at the front and rear axle and by the level selector valve. This allows adjustment of the vehicle level by using the level selector switch on the dashboard. When the engine is not running, the main pressure reservoir supplies the system.

A hydraulic oil pump, driven by the engine, pumps oil from the oil reservoir to the main pressure reservoir. When the maximum oil pressure in the main reservoir is reached, the pressure regulator in the reservoir valve unit reverses the flow of oil. If the pressure in the reservoir drops to a pre-set minimum (as a result of operation of the system) the pressure regulator again reverses the flow of oil, pumping oil into the pressure reservoir until maximum pressure is reached, when the flow will be reversed once more.

The oil in the pressure reservoir is connected to the level selector valve and to the individual levelling valves by pressure lines. If the car level drops, due to an increased load, the levelling valve opens the passage to the suspension struts, allowing the passage of oil until normal vehicle attitude is reached. If the level rises, due to a decreased load, the levelling valve opens and allows oil to flow from the suspension struts back to the oil reservoir, until the car resumes its normal level.

SUSPENSION
Import Models

FRONT SUSPENSION MERKUR

Alignment

NOTE: Caster and camber are not adjustable. Toe should be adjusted to 1/64" out to 3/16" in—preferred is 5/64" in.

MacPherson Strut

REMOVAL & INSTALLATION

For all strut overhaul procedures, please refer to "Strut Overhaul" in the Unit Repair section.

1. Raise and support the front of the vehicle on jackstands after loosening the front wheel lug nuts.
2. Remove the wheel and tire assembly. Position a floor jack under the lower control arm, raise the jack until it is slightly lower than the control arm.
3. Remove the pinch bolt that secures the strut to the lower control arm. Use a small pry bar to spread the mounting flange ears and push down on the lower control arm to separate the arm and strut. Lower the jack if necessary, but do not allow the brake hose to stretch. When separated, rest the control arm on the jack.
4. Hold the top of the strut by inserting a 6 mm hex wrench in the slot provided and remove the locknut.
5. Remove the strut assembly from the vehicle.
6. Install replacement strut in the reverse order of removal.

Front Wheel Bearing

REMOVAL & INSTALLATION

The front wheels on the Merkur are attached to spindle shafts which are supported on opposed, tapered roller bearings. The spindle hub casting forms a housing for the bearings and shaft while providing the necessary suspension and steering connection points. Self-setting wheel bearings are used that never require adjustment and are lubricated with high temperature, long life grease.

1. Raise and safely support the vehicle. Remove the front wheels and brake calipers. Suspend the calipers on wire to prevent brake hose damage.
2. Matchmark the rotor and wheel stud. The unit is balanced by the factory and must be installed in the same position to maintain balance.
3. Remove the cotter pin and the tie-rod end attaching nut. Remove the tie-rod from the spindle.

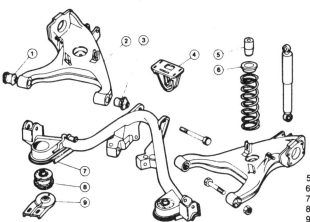

1. Outer Rubber Bush
2. Lower Suspension Arm
3. Inner Rubber Bush
4. Axle Housing Rubber Mounting
5. Bump Rubber
6. Spring Seat Pad
7. Rear Axle Cross Member
8. Mounting Rubber
9. Guide Plate

Front suspension

4. Remove the cotter pin and control arm attaching nut and remove the control arm from the knuckle.
5. Remove the strut mounting to knuckle pinch bolt. Spread the ears and remove the knuckle from the strut assembly.
6. Place the spindle and hub in a vise, wheel studs pointing downward and clamped between two pieces of wood and the vise jaws.
7. Remove the bearing plug from the rear of the knuckle using a flat drift.
8. Use a 27mm socket to remove the spindle bearing locknut.

NOTE: Spindles from the right side of the vehicle are equipped with left handed threads and are loosened by turning clockwise. Spindles from the left side of the vehicle are equipped with right handed threads which are loosened by turning counterclockwise. The spindles are marked with an R or L on the large hexagonal recess.

9. Lift the spindle carrier and inner bearing off the (hub) spindle shaft. Remove the inner bearing and splined washer. If the bearing is to be reused, tag for location identification.
10. Clamp the spindle carrier (knuckle) in a vise and remove the grease seal using a flat prybar. Remove the outer bearing and tag for location identification.
11. Remove bearing cups from the spindle, if necessary, using a bearing puller jaws on a slide hammer.
12. Clean and inspect all parts. Press new bearing cups into the spindle. Pack the wheel bearing with high temperature grease.
13. Install the outer bearing and grease seal in the spindle (knuckle). Install the

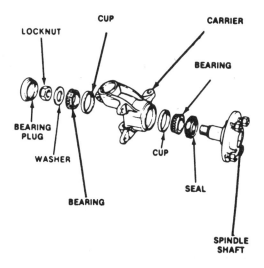

Merkur XR4Ti spindle/Knuckle exploded view

2-229

SECTION 2: SUSPENSION
Import Models

spindle shaft (hub).

14. Install the inner bearing and splined washer. Install the spindle bearing locknut and tighten to 202–232 ft. lbs. Install the bearing cover plug.

NOTE: Be sure the spindle is mounted secure in the vise but do not damage the studs. The amount of torque required for the locknut is extremely important. If a higher or lower torque is applied bearing failure is likely to occur.

15. Install the spindle (knuckle) in the reverse order of removal. Torque the attaching parts as follows:
- Strut to Spindle Pinch Bolt- 59–66 ft. lbs.
- Lower Control Arm Nut- 48–63 ft. lbs.
- Tie Rod End Nut- 15–23 ft. lbs.

Control Arm/Stabilizer Bar

REMOVAL & INSTALLATION

1. Raise and support the front of the vehicle. Remove the front wheels.
2. Remove the cotter pin and attaching nut and separate the control arm from the spindle carrier.
3. Remove the pivot bolt attaching the control arm to the crossmember.
4. Remove the nut attaching the stabilizer bar to the control arm. Remove the front washer/plastic cover from the end of the stabilizer bar.
5. Remove the control arm and bushings as an assembly.
6. Remove the rear washer/plastic cover from the end of the stabilizer bar. Service bushings as necessary, they are pressed into the bar.
7. Install in the reverse order of removal. Tighten all mounting bolts snugly and lower the vehicle so full weight is on the suspension. Tighten the bolts as follows: Control Arm Pivot Bolt- 11 ft. lbs. + 90° Stabilizer Nut- 52–81 ft. lbs.

Stabilizer Bar

REMOVAL & INSTALLATION

1. Raise and support the vehicle safely. Remove the attaching nuts and front washer covers from the ends of the stabilizer bar.
2. Remove the four bolts securing the two U-brackets and torque brace to the body.
3. Detach one control arm pivot bolt and pull the control arm out of the crossmember. Pull the stabilizer out of the lower control arms and remove it from the vehicle.
4. Service as required and install in the reverse order. Refer to Control Arm procedure for torque specifications.

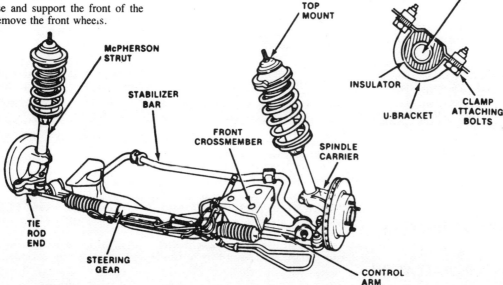

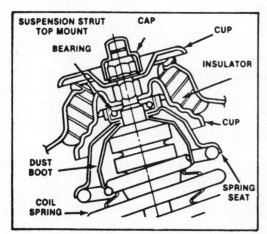

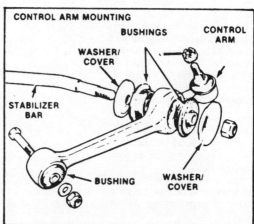

Merkur XR4Ti front suspension

SUSPENSION
Import Models

REAR SUSPENSION MERKUR

Coil Spring

REMOVAL & INSTALLATION

1. Loosen the rear wheel lug nuts. Raise and support the rear of the vehicle. Remove the tire and wheel assembly and brake drum.
2. Position a floor jack under the rear control arm and take a slight amount of weight off of the spring.
3. Disconnect the rear brake hose at the body bracket (rubber line from steel line).
4. Remove the bolts that attach the rear axle flange and brake backing plate to the control arm. Remove the axle shaft from the vehicle. Secure the backing plate back in position with two bolts to prevent damage to the steel brake line.
5. Remove the bolt that secures the lower mounting eye of the shock absorber to the control arm.

--- CAUTION ---
Make sure that the lower arm and spring tension is supported by a floor jack.

6. Remove the bolts and pin that mount the axle beam to the floor pan/crossmember mount.
7. Carefully lower the suspension arm on the jack and remove the coil spring and rubber spring seat.
8. Insert the rear spring with the rubber spring seat onto the lower arm. Raise the jack slowly and make sure the spring and seat are correctly located. Install in the reverse order of removal.

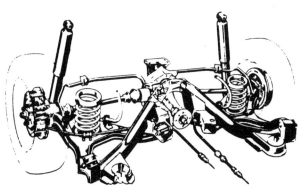

Rear suspension components

Shock Absorber

REMOVAL & INSTALLATION

1. Raise and support the rear of the vehicle on jackstands. Position a floor jack under the lower control arm of the side requiring shock replacement. Raise the jack until it contacts the control arm.
2. Remove the luggage compartment cover. Remove the upper shock mount trim cover.
3. Remove the upper shock mount through bolt and nut.
4. Remove the lower shock mount bolt and nut. Remove the shock absorber.
5. Install the replacement shock in the reverse order.

Stabilizer Bar

REMOVAL & INSTALLATION

1. Loosen the wheel lug nuts on one side of the vehicle.
2. Raise and support the rear of the vehicle.
3. Remove one wheel and tire assembly.
4. Use a small pry bar and unclip the connector from the end of the bar to lower control arm. Repeat on the other side of the vehicle.
5. Remove the bushing bracket retaining bolts from the floor pan and remove the stabilizer bar assembly.
6. Install in the reverse order.

FRONT SUSPENSION MITSUBISHI

MacPherson Strut

REMOVAL & INSTALLATION

For all spring and shock removal and installation procedures and any other strut overhaul procedures, please refer to "Strut Overhaul" in the Unit Repair section.

Cordia and Tredia

1. Jack up and support the front of the car.
2. Remove the front wheel. Remove the brake line from the strut.
3. Disconnect the strut assembly from the steering knuckle by removing the two bolts/nuts. Support the strut and remove the two nuts and washers fastening it to the wheel well. Remove the strut.

NOTE: When installing the strut, apply a non-hardening sealer to the mating surfaces of the strut and knuckle arm.

4. Installation is the reverse of removal. Torque the strut insulator-body to 18–25 ft. lb.; strut bar-to-crossmember to 98–115 ft. lb.; strut bar-to-lower arm to 43–50 ft lb.; strut to steering knuckle to 54–65 ft. lbs. Align the front end. Bleed the front brakes after installation.

Galant and Mirage

NOTE: To perform this procedure, you'll need a deep well socket with flats on the top for an open-end wrench, and an Allen wrench. They will have to fit the nut on top of the strut and the shaft at its center, respectively.

1. Raise the vehicle and support it by the body or crossmembers. Remove the front wheels. On the Mirage, detach the brake hose bracket at the strut.
2. Remove the two nuts, bolts, and lockwashers attaching the lower end of the strut to the steering knuckle.
3. Remove the dust cover from the top of the strut on the wheel well. Support the strut from underneath. Install the socket wrench on the nut at the top of the strut, and a box or open end wrench on the socket. Then, install the Allen wrench through the center of the socket, long part downward. Hold the Allen wrench in place, if necessary by using a small diameter pipe as a cheater. Turn the socket to loosen the nut. Remove the nut and the lower and remove the strut.
4. To install, reverse the removal procedure. Torque the bolts attaching the bot-

2–231

SUSPENSION
Import Models

tom of the strut to the knuckle to 53–63 ft. lbs. on the Mirage and 65–76 ft. lbs. on the Galant. The nut at the top of the strut must be torqued with the shaft of the shock held from turning with the Allen wrench, as during the loosening process. Since it is not usually possible to use a torque wrench on the flats of a socket, you'll have to estimate the torque you're applying. It should be 36–43 ft. lbs.

Starion

1. Remove the front wheel and caliper. Remove the front hub with disc and dust cover.
2. Disconnect the stabilizer linkage and the lower. Remove the strut assembly, knuckle arm and strut insulator retaining bolts and remove the strut assembly from the wheelhouse.
3. Installation is the reverse of removal. Observe the following torque:
Strut to wheelhouse attaching nuts— 18–25 ft. lbs.

NOTE: When installing the strut, apply a non-hardening sealer to the mating surfaces of the strut and knuckle arm.

OVERHAUL

For all spring and shock absorber removal and installation procedures, and all strut overhaul procedures, please refer to "Strut Overhaul" in the Unit Repair section.

Ball Joint

REMOVAL & INSTALLATION

Front Wheel Drive

1. Riase the support the front of the car. Remove the wheels.
2. Disconnect the stabilizer bar and strut from the lower arm.
3. Remove the ball joint mounting nut and separate the ball joint from the front knuckle. The ball joint stud must be pressed off. You can use special tool MB991113 or equivalent.

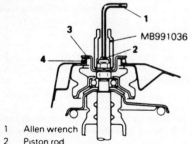

1. Allen wrench
2. Piston rod
3. Stopper
4. Stopper rubber

Removing the upper strut attaching nut with a socket wrench and Allen wrench on the Galant and Mirage

4. Remove the lower control arm by removing the bolt(s)/nut(s) attaching it to the crossmember.
5. Remove the dust cover from the ball joint. Remove the mounting snap ring.
6. Press the ball joint out of the lower control arm.
7. Press the new ball joint into place. Install a new snap ring with snap ring pliers.
8. Apply multipurpose grease to the lip and to the inside of the dust cover. Use a special tool (and hammer) such as MB990800-3-01 to drive on a new dust cover. It must go in and make contact with the snap ring.
9. Install the lower control arm to the crossmember, making sure there is not torque on it. Install and tighten the bolt and nut.

Starion

1. Raise and support the front of the car. Remove the wheels.
2. Remove the strut end from the steering knuckle.
3. Remove the ball joint-to-knuckle arm mounting nut. You'll need a tool which can be bolted to the holes in the knuckle arm and which will then press downward on the center of the Ball Stud—MB990241-01 or equivalent.

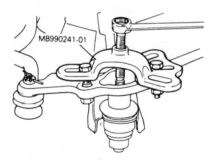

Pressing the ball joint off—Starion

4. Remove the ball joint-to-control arm nuts and bolts and remove the ball joint.
5. Install in reverse order. Torques in ft. lbs.: Ball joint mounting bolts—43–51; strut to knuckle arm bolts—58–72; ballstud nut—43–52.

Lower Control Arm

REMOVAL & INSTALLATION

Cordia and Tredia

1. Support the vehicle securely by the crossmember and remove the front wheel.
2. Disconnect the stabilizer bar and strut bar from the lower control arm by removing the one attaching bolt for the stabilizer bar and the two bolts for the strut bar.
3. Remove the ballstud nut and then press the ball joint stud out of the knuckle with a tool such as MB991113.

4. Remove the nut and bolt attaching the inner end of the stabilizer bar to the crossmember, and pull the stabilizer bar and bushing out of the crossmember.
5. Installation is the reverse of removal. Install all parts and tighten nuts and bolts just snug. Then, complete tightening, torquing the lower arm-to-crossmember attaching nut/bolt to 87–108 ft. lbs. and the ball joint stud nut to 43–52 ft. lbs. Torque the strut rod-to-stabilizer bar bolt/nut to 43–50 ft. lbs.

Starion

1. Support the vehicle securely by the crossmember and remove the front wheel.
2. Disconnect the stabilizer bar where the link bolts to the control arm by removing the nut underneath the arm. Remove the nut and bolt attaching the strut bar to the control arm.
3. Disconnect the tie rod at the knuckle arm. Use a fork-like tool such as MB990778-01, a standard type tool for pulling ball joint studs. First, loosen the stud nut until it is near the top of the threads and then hammer the tool between the ball joint of the tie rod end and the knuckle arm. When the ballstud comes loose, remove the nut and disconnect the stud.
4. Unbolt the MacPherson strut from the knuckle arm.
5. Unbolt the inner end of the ball joint assembly to disconnect it from the outer end of the control arm.
6. Remove the nut, bolt, and lockwasher, and pull the inner end of the control arm out of the crossmember.
7. Installation is the reverse of removal. Torque the bolt fastening the control arm to the crossmember to 58–69 ft. lbs.; the bolts attaching the ball joint to the outer end to 43–51 ft. lbs.; the ballstud nut to 43–52 ft. lbs.; and the strut attaching bolts to 58–72 ft. lbs. Tighten the nut for the stabilizer bar link until .59–.67 in. of thread shows below the bottom of the nut.

Mirage and Galant

1. Support the vehicle securely by the crossmember and remove the front wheel.
2. On the Mirage, remove the under cover.
3. Disconnect the stabilizer bar from the lower arm. On the Galant, remove the nut at the top and remove the washer and bushing, keeping them in order. On the Mirage, you can remove the nut from underneath the control arm and take off the washer and spacer.
4. On the Galant with electronically controlled suspension, if you're removing the right arm, disconnect the height sensor rod from the lower arm. Loosen the ball joint stud nut and then press the stud out of the control arm, using a fork-like tool (MB990778-01) and hammer on the Galant; on the Mirage, remove the stud nut and

SUSPENSION
Import Models

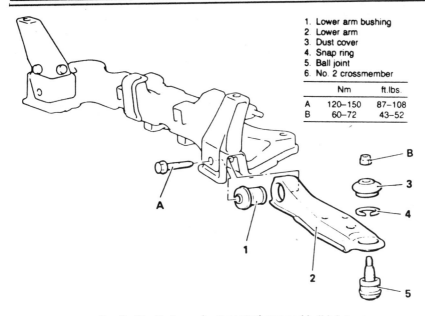

Cordia, Tredia lower front control arm and ball joint

1. Lower arm bushing
2. Lower arm
3. Dust cover
4. Snap ring
5. Ball joint
6. No. 2 crossmember

	Nm	ft.lbs.
A	120–150	87–108
B	60–72	43–52

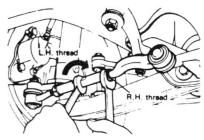

Increasing toe-in on the Starion

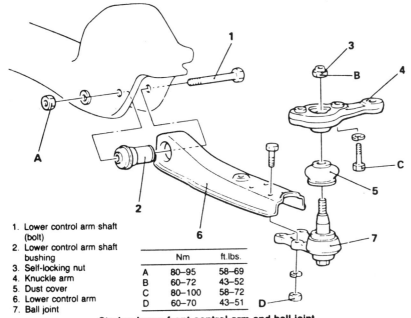

1. Lower control arm shaft (bolt)
2. Lower control arm shaft bushing
3. Self-locking nut
4. Knuckle arm
5. Dust cover
6. Lower control arm
7. Ball joint

	Nm	ft.lbs.
A	80–95	58–69
B	60–72	43–52
C	80–100	58–72
D	60–70	43–51

Starion lower front control arm and ball joint

the car with no passengers or luggage on the front suspension. On the Mirage, torque the nut for the lower arm shaft to 69–87 ft. lbs.; the bolts for the spacer at the rear to 43–58 ft. lbs. and the ballstud nut to 43–52 ft. lbs. On the Galant, torque the nut for the nut/bolt-to-crossmember to 69–87 ft. lbs., the ballstud to 42–50 ft. lbs., and the bolt for retaining the rear bushing to the body to 58–72 ft. lbs.

Front Wheel Bearing

REMOVAL & INSTALLATION

Rear Wheel Drive

1. Raise and support the front of the car. Remove wheel. Remove the caliper.
2. Pry off the dust cap. Tap out and discard the cotter pin. Remove the locknut.
3. Being careful not to drop the outer bearing, pull off the brake disc and wheel hub.
4. Remove the grease inside the wheel hub.
5. Using a brass drift, carefully drive the outer bearing race out of the hub.
6. Remove the inner bearing seal and bearing.
7. Check the bearings for wear or damage and replace them if necessary.
8. Coat the inner surface of the hub with grease.
9. Grease the outer surface of the bearing race and drift it into place in the hub.
10. Pack the inner and outer wheel bearings with grease. (see repacking.)

NOTE: If the brake disc has been removed and/or replaced, tighten the retaining bolts to 25–29 ft. lbs.

11. Install the inner bearing in the hub. Being careful not to distort it, install the oil seal with its lip facing the bearing. Drive the seal on until its outer edge is even with the edge of the hub.
12. Install the hub/disc assembly on the spindle, being careful not to damage the oil seal.
13. Install the outer bearing, washer, and spindle nut. Adjust the bearing.

press the tool off with a tool such as MB991113.

5. On the Galant, remove the nuts and bolts which retain the bushings to the crossmember at the front and which retain the bushing retainer to the crossmember at the rear and pull the arm out. On the Mirage, remove the bolts which retain the spacer at the rear and the nut and washers on the front of the lower arm shaft (at the front). Slide the arm forward, off the shaft and out of the bushing.

6. Replace the dust cover on the ball joint. The new cover must be greased on the lip and inside with #2 EP Multipurpose grease and pressed on with a tool such as MB990800 and a hammer until it is fully seated.

7. Installation is the reverse of removal. On the Galant, make sure the nut on the stabilizer bar bolt is torqued to give .63–.7 in. of thread exposed between the top of the nut and the end of the link. On the Mirage, the nut must be torqued until the link shows .83–.91 in. of threads below the bottom of the nut. Also on the Mirage, the washer for the lower arm must be installed as shown. The left side arm shaft has a left hand thread. Final tighten the arm shaft to the lower arm with the weight of

2-233

SECTION 2: SUSPENSION
Import Models

ADJUSTMENT

1. Remove the wheel and dust cover. Remove the cotter pin and lock cap from the nut.
2. Torque the wheel bearing nut to 14.5 ft. lbs. (19.6 Nm) and then loosen the nut. Retorque the nut to 3.6 ft. lbs. (4.9 Nm) and install the lock cap and cotter pin.
3. Install the dust cover and the wheel.

Front End Alignment

Camber is pre-set at the factory and cannot be adjusted. Caster should not require adjustment, although adjustment (to a certain extent) is possible by adjusting the length of the strut bar. Loosen both nuts and turn in or out as required. Toe adjustment is possible by adjusting both tie-rod end turnbuckles (the same amount) on the Cordia and Tredia, or the left tie-rod end turnbuckle on the Starion.

Before turning the turnbuckles on the Cordia and Tredia, unfasten the clips for the rubber boots on the inner ends of the turnbuckles. Using a wrench from below on the flats in the middle of the turnbuckle, the toe will move out as the left side turnbuckle is turned toward the front of the vehicle and the right toward the rear. For the Starion, see the illustration.

On the Mirage and Galant, toe-in is adjusted as for the Cordia and Tredia. Neither caster nor camber can be adjusted.

On the Starion, toe-in can usually be adjusted by turning the one turnbuckle. You should, however, check to make sure that the difference between the length of right and left tie rods is not greater than .2 in. If it is, you should remove the tight tie rod at the knuckle and bring the length within specifications; and then toe-in is brought to correct values also.

REAR SUSPENSION MITSUBISHI

Shock Absorbers

REMOVAL & INSTALLATION

Cordia, Tredia, and Mirage

1. Remove the hub cap or wheel cover. Loosen the lug nuts.
2. Raise the rear of the car. Support it safely.
3. Position a floor jack under the lower control arm. Remove the upper shock moujting bolt and nut.
4. Compress the shock slightly and remove the lower mounting bolt.
5. Remove the shock absorber.
6. Installation is the reverse of removal.

Coil Spring

REMOVAL & INSTALLATION

Cordia, Tredia, and Mirage

1. Raise and support the car safely allowing the rear axle to hang unsupported.
2. Place a jack under the work side control arm and remove the bottom bolts of the shock absorbers.
3. Lower the arm and remove the coil spring.
4. Installation is the reverse of removal.

MacPherson Strut

REMOVAL & INSTALLATION

Starion

1. Support the rear of car on jackstands at the frame rails. Position a floor jack under the lower control arm and raise it slightly.
2. Disconnect the rear brake hose from the strut assembly.
3. Disconnect the axle shaft from the wheel side flange.
4. Remove the strut assembly-to-axle housing mounting bolts. Separate the strut assembly from the axle housing. Lower the floor jack and push down on the housing while opening the coupling with a small pry bar.
5. Remove the upper strut mounting nuts from under the side trim in rear hatch.
6. Remove the strut assembly.
7. Install in reverse order. Tighten the top mounting nuts to 18–25 ft. lbs. and the lower mountings to 36–51 ft. lbs.

Galant

1. Support the vehicle with axle stands. Remove the rear wheels.
2. Place a floor jack under the axle/arm assembly and raise it slightly. Then, remove the forward trim from the trunk and remove the cap and strut mounting nuts and washers.
3. Remove the nut, pull the through-bolt out where the strut connects with the axle/arm assembly, and remove the strut assembly.
4. Installation is the reverse of removal. Torque the upper strut mounting nuts to 33–40 ft. lbs. and the lower through-bolt to 72–87 ft. lbs.

OVERHAUL

For all spring and shock removal and installation procedures and any other strut overhaul procedures please refer to "Strut Overhaul"

Rear Control Arm

REMOVAL & INSTALLATION

Cordia, Tredia, and Mirage

1. Raise and support the vehicle. Remove the rear wheels. Remove the rear brake assemblies. Remove the muffler.
2. Disconnect the parking brake cable from the suspension arm on both sides.

	Nm	ft.lbs.
A	65–80	47–58
B	70–90	51–65
C	80–100	58–72
D	18–25	13–18

1. Suspension arm (R.H.)
2. Dust cover
3. Clamp
4. Bushing A
5. Bushing B
6. Rubber stopper
7. Suspension arm (L.H.)
8. Fixture
9. Rubber bushing
10. Washer
11. Spring upper seat
12. Coil spring
13. Spring lower seat
14. Shock absorber
15. Bump stopper

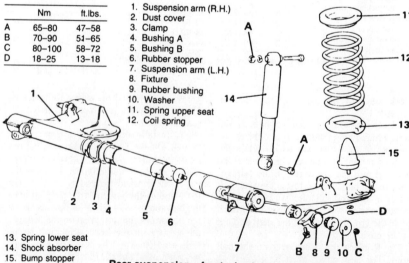

Rear suspension—front wheel drive models

SUSPENSION
Import Models

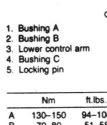

1. Bushing A
2. Bushing B
3. Lower control arm
4. Bushing C
5. Locking pin

	Nm	ft.lbs.
A	130–150	94–108
B	70–80	51–58
C	15–20	11–14

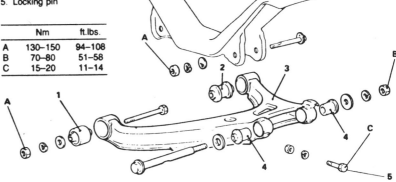

Rear lower control arm on Starion

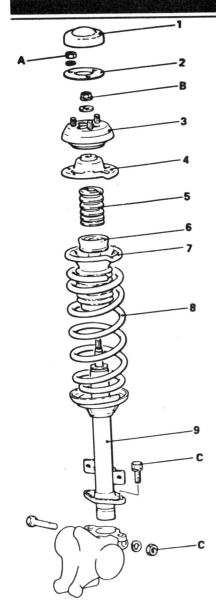

	Nm	ft.lbs.
A	25–35	18–25
B	70–90	51–65
C	50–70	36–51

1. Strut house cap
2. Gasket
3. Strut insulator
4. Spring seat
5. Rubber helper
6. Rubber helper seat
7. Dust cover
8. Coil spring
9. Strut

Starion rear strut assembly

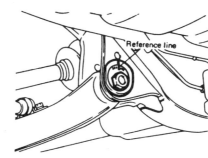

Matchmark crossmember and bushing alignment before moving the rear control arm on the Starion

3. Jack up the suspension arm on both sides just slightly. Then remove both lower shock absorber attaching bolts. Then, lower the jack carefully and, when it can be disengaged, remove the spring. Keep the spring in the position it was in when installed so it can be installed in the same direction.

4. Disconnect the brake hoses at the suspension arms. Then, support the rear suspension assembly while you remove the two mounting bolts on either side and remove the assembly.

5. Installation is the reverse of removal. Lower shock mounting bolts are torqued to 47–58 ft. lbs. Suspension assembly-to-body mounting bolts are torqued to 51–65 ft. lbs. on the Cordia and Tredia, and to 36–51 ft. lbs. on the Mirage.

Starion

1. Raise and support the rear of the car on jackstands.
2. Disconnect the parking brake from the control arm brackets. Disconnect the stabilizer bar.
3. Remove the nut and bolt connecting the lower control arm to the front support.
4. Matchmark the relationship between the crossmember and the eccentric bushing so alignment can be restored at assembly. Remove the nut and bolt connecting the lower control arm to the crossmember.
5. Remove the lower control arm from the car.
6. Install in reverse order.

Rear Wheel Bearings

REMOVAL & INSTALLATION

NOTE: Special tools are needed for this procedure:

1. Loosen the lug nuts. Raise the rear of the car and support it on jackstands.
2. Remove the wheel.
3. Remove the grease cap.
4. Remove the nut.
5. Pull the drum off. The outer bearing will fall out while the drum is coming off, so position your hand to catch it.
6. Pry out the oil seal. Discard it.
7. Remove the inner bearing.
8. Check the bearing races. If any scoring, heat checking or damage is noted, they should be replaced.

NOTE: When bearings or races need replacement, replace them as a set.

9. Inspect the bearings. If wear or looseness or heat checking is found, replace them.
10. If the bearings and races are to be replaced, drive out the race with a brass drift.
11. Before installing new races, coat them with lithium based wheel bearing grease. The races are most easily installed using a driver made for that purpose. They can, however, be driven into place with a brass drift. Make sure that they are fully seated.
12. Thoroughly pack the bearings with lithium based wheel bearing grease. Pack the hub with grease.
13. Install the inner bearing and coat the lip and rim of the grease seal with grease. Drive the seal into place with a seal driver.
14. Mount the drum on the axleshaft. Install the outer bearing. Don't install the nut at this point.

SECTION 2: SUSPENSION
Import Models

15. Using a pull scale attached to one of the lugs, measure the starting force necessary to get the drum to turn. Starting force should be 5 lbs. If the starting torque is greater than specified, replace the bearings.

16. Install the nut on the axleshaft. Thread the nut on, by hand, to a point at which the back face of the nut is 2–3mm from the shoulder of the shaft (where the threads end).

17. Using an inch lb. torque wrench, turn the nut counterclockwise 2 to 3 turns, noting the average force needed during the turning procedure. Turning torque for the nut should be about 48 inch lbs. If turning torque is not within 5 inch lbs., either way, replace the nut.

18. Tighten the nut to 75–110 ft. lbs.

19. Using a stand-mounted gauge, check the axial play of the wheel bearings. Play should be less than .0079 in. If play cannot be brought within that figure, you probably have assembled the unit incorrectly.

20. Pack the grease cap with wheel bearing grease and install it.

FRONT SUSPENSION NISSAN/DATSUN

The independent front suspension on all models covered here uses MacPherson struts. Each strut combines the function of coil spring and shock absorber. The spindle is mounted to the lower part of the strut which has a single ball joint. No upper suspension arm is required in this design. The spindle and lower suspension transverse link (control arm) are located fore and aft by the tension rods to the front part of the chassis on most models. A cross-chassis sway bar is used on all models.

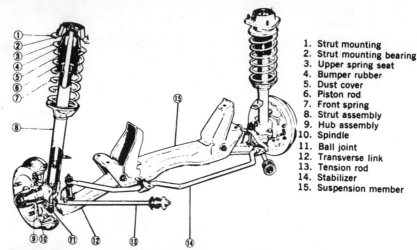

1. Strut mounting
2. Strut mounting bearing
3. Upper spring seat
4. Bumper rubber
5. Dust cover
6. Piston rod
7. Front spring
8. Strut assembly
9. Hub assembly
10. Spindle
11. Ball joint
12. Transverse link
13. Tension rod
14. Stabilizer
15. Suspension member

B210 strut-type front suspension. On the 280Z, the tension rods are replaced by compression rods running to the rear

MacPherson Strut

REMOVAL & INSTALLATION

All Models

1. Jack up the car and support it safely. Remove the wheel.
2. Disconnect and plug the brake hose.
3. Disconnect the tension rod (compression rod on the "Z" series) and stabilizer bar from the transverse link.
4. Unbolt the steering arm.
5. Place a jack under the bottom of the strut.
6. Open the hood and remove the nuts holding the top of the strut. If your 300ZX or 1985 and later Maxima is equipped with adjustable shocks, disconnect the electrical lead.
7. Lower the jack slowly and cautiously until the strut assembly can be removed.
8. Reverse the procedure to install. The self locking nuts holding the top of the strut must be replaced.

OVERHAUL

For all spring and shock absorber removal and installation procedures and any other strut overhaul procedures, please refer to "Strut Overhaul"

Ball Joint

INSPECTION

The lower ball joint should be replaced when play becomes excessive. Datsun does not publish specifications for this, giving instead a rotational torque figure for the ball joint. However, this requires removal for measurement. An effective way to determine play is to jack up the car until the wheel is clear of the ground. Do not place the jack under the ball joint; it must be unloaded. Place a long bar under the tire and move the wheel up and down. Keep one hand on top of the tire while doing this. If ¼ in. or more of play exists at the top of the tire, the ball joint should be replaced. Be sure the wheel bearings are properly adjusted before making this measurement. A double check can be made; while the tire is being moved up and down, observe the ball joint. If play is seen, replace the ball joint.

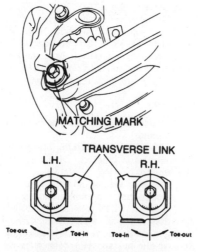

Adjusting the rear wheel toe—1986 Stanza wagon (4 × 4)

REMOVAL & INSTALLATION

Rear Wheel Drive Models

NOTE: On most late-model vehicles, the transverse link (lower control arm) must be removed and then the ball joint must be pressed out.

The ball joint should be greased every 30,000 miles. There is a plugged hole in the bottom of the joint for installation of a grease fitting.

1. Raise and support the car so the wheels hang free. Remove the wheel.
2. Unbolt the tension rod (compression rod on "Z" series) and stabilizer bar from transverse link.
3. Unbolt the strut from the steering arm.
4. Remove the cotter pin and ball joint stud nut. Separate the ball joint and steering arm.
5. Unbolt the ball joint from the transverse link.

SUSPENSION
Import Models

6. Reverse the procedure to install a new ball joint. Grease the joint after installation.

Front Wheel Drive Models

1. Jack up the car and support it on stands.
2. Remove the wheel.
3. Remove the halfshaft.
4. Separate the ball joint from the steering knuckle with a ball joint remover, being careful not to damage the ball joint dust cover if the ball joint is to be used again.
5. Remove the other ball joint from the transverse link and remove the ball joint.

Installation is the reverse of removal. Tighten the ball stud attaching nut (from ball joint–to–steering knuckle) to 22–29 ft. lbs., and the ball joint–to–transverse link bolts to 40–47 ft. lbs.

Lower Control Arm (Transverse Link) and Ball Joint

REMOVAL & INSTALLATION

You'll need a ball joint remover for this operation.

1. Jack up the vehicle and support it with jackstands; remove the wheel.
2. Remove the splash board, if so equipped.
3. Remove the cotter pin and castle nut from the side rod (steering arm) ball joint and separate the ball joint from the side rod. You'll need either a fork type or puller type ball joint remover.
4. Separate the steering knuckle arm from the MacPherson strut.
5. Remove the tension rod and stabilizer bar from the lower arm. Front wheel drive models do not have tension rods.
6. Remove the nuts or bolts connecting the lower control arm (transverse link) to the suspension crossmember on all models.
7. On the 810 and Maxima, to remove the transverse link (control arm) on the steering gear side, separate the gear arm from the sector shaft and lower steering linkage; to remove the transverse link on the idler arm side, detach the idler arm assembly from the body frame and lower steering linkage.
8. Remove the lower control arm (transverse link) with the suspension ball joint and knuckle arm still attached.

Installation is the reverse of removal with the following notes.

9. When installing the control arm, temporarily tighten the nuts and/or bolts securing the control arm to the suspension crossmember. Tighten them fully only after the car is sitting on its wheels.
10. Lubricate the ball joints after assembly.

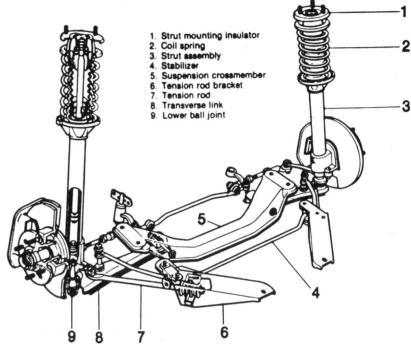

1. Strut mounting insulator
2. Coil spring
3. Strut assembly
4. Stabilizer
5. Suspension crossmember
6. Tension rod bracket
7. Tension rod
8. Transverse link
9. Lower ball joint

Strut-type front suspension, 200SX. 510 and early 810 similar; later 810, Maxima, Sentra, Pulsar, Stanza similar except rack and pinion steering

Front Wheel Bearing

ADJUSTMENT

NOTE: For wheel bearing adjustment on front wheel drive models, please refer to "Drive Axles".

1. Raise and support the vehicle safely, remove the front wheels and the brake shoes.
2. While rotating disc, torque wheel bearing lock nut to 18–22 ft. lbs. on all models.
3. Loosen lock nut approximately 60 degrees on all models. Install adjusting cap and align groove of nut with hole in spindle. If alignment cannot be obtained, change position of adjusting cap. Also, if alignment cannot be obtained, loosen lock nut slightly but not more than 15 degrees.
4. Install brake shoes.

Tension Rod And Stabilizer-Bar

REMOVAL & INSTALLATION

Rear Wheel Drive Models Only

1. Support, and set the load of the vehicle.
2. Remove the tension rod-to-frame lock nuts. Remove the to mounting bolts at the transverse link (lower control arm) and then slide out the tension rod.
3. Unbolt the stabilizer bar at each transverse link.
4. Remove the four stabilizer bar bracket bolts and remove the stabilizer bar.
5. Installation is in the reverse order of removal.

—Tighten the stabilizer bar-to-transverse link bolts to 12–16 ft. lbs. (16–22 Nm)
—Tighten the stabilizer bar bracket bolts to 22–29 ft. lbs. (29–39 Nm)
—Tighten the tension rod-to-transverse link nuts to 31–43 ft. lbs. (42–59 Nm)
—Tighten the tension rod-to-frame nut (bushing end) to 33–40 ft. lbs. (44–54 Nm)
—Never tighten any bolts or nuts to their final torque unless the care is resting, unsupported, on the wheels
—Be certain the tension rod bushings are installed as shown in the illustration.

Tension rod bushing positioning—rear wheel drive models

SUSPENSION
Import Models

Stabilizer Bar

REMOVAL & INSTALLATION

Sentra, Pulsar and Stanza Wagon

1. Disconnect the parking brake cable at the equalizer on the Stanza wagon.
2. Disconnect the front exhaust pipe at the manifold and position it out of the way.
3. Remove the stabilizer bar-to-transverse link (lower control arm) mounting bolts.
4. Remove the four stabilizer bar bracket mounting bolts and then pull the bar out, around the link and exhaust pipe.
5. Installation is in the reverse order of removal. Never tighten the mounting bolts unless the car is resting on the ground with normal weight upon the wheels.

Front End Alignment

The major front end adjustments for the Datsun are given here—caster/camber, toe, and steering angle.

CASTER AND CAMBER

Caster is the forward or rearward tilt of the upper end of the kingpin, or the upper ball joint, which results in a slight tilt of the steering axis forward or backward. Rearward tilt is referred to as a positive caster, while forward title is referred to as a negative caster.

Camber is the inward or outward tilt from the vertical, measured in degrees, of the front wheels at the top. An outward tilt gives the wheel positive camber. Proper camber is critical to assure even tire wear.

Since caster and chamber are adjusted traditionally by adding or subtracting shims behind the upper control arms, and the Datsuns covered in this guide have replaced the upper control arm with the MacPherson strut, the only way to adjust caster and camber is to replace bent or worn parts of the front suspension.

TOE

Toe is the amount, measured in a fraction of an inch, that the wheels are closer together at one end than the other. Toe-in means that the front wheels are closer together at the front than the rear, toe-out means the rears are closer than the front. Datsuns are adjusted to have a slight amount of toe-in. Toe-in is adjusted by turning the tie-rod, which has a right-hand threat on one and a left-hand threat on the other.

You can check your vehicle's toe-in yourself without special equipment if you make careful measurements. The wheels must be straight ahead.

1. Toe-in can be determined by measuring the distance between the center of the tire treads, at the front of the tire and at the rear. If the treat pattern of your car's tires make this impossible, you can measure between the edges of the wheel rims, but make sure to move the car forward and measure in a couple of places to void errors caused by bent rims or wheel runout.
2. If the measurement is not within specifications, loosen the locknuts at both ends of the tie-rod (the driver's side locknut is left-hand threaded).
3. Turn the top of the tie-rod toward the front of the car to reduce toe-in, or toward the rear to increase it. When the correct dimension is reached, tighten the locknuts and check the adjustment.

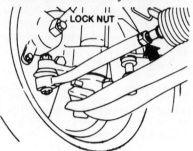

Toe adjustment is made at the tie rod

NOTE: The length of the tie-rods must always be equal to each other.

STEERING ANGLE ADJUSTMENT

The maximum steering angle is adjusted by stopper bolts on the steering arms. Loosen the locknut on the stopper bolt, turn the stopped bold in or out as required to obtain the proper maximum steering angle and retighten the locknut.

REAR SUSPENSION NISSAN/DATSUN

Springs

REMOVAL & INSTALLATION

Leaf Spring Type

510 AND 810 (1980–82) STATION WAGONS

1. Raise the rear of the car and support it with jackstands.
2. Remove the wheels and tires.
3. Disconnect the lower end of the shock absorber and remove the U-bolt nuts.
4. Place a jack under the rear axle.
5. Disconnect the spring shackle bolts at the front and rear of the spring.
6. Lower the jack slowly and remove the spring.
7. Installation is the reverse of removal.

Coil Spring Type

Four types of coil spring suspension are used:

Trailing Arm Type—310, Sentra, and Pulsar

Four Bar Link Type—210, 510, 810 and 1980–84 200SX

MacPherson Strut Type—810, Maxima, 280Z and ZX, 300ZX, Stanza and Stanza Wagon

I.R.S. Coil Spring Type—1984 and later 200SX with independent rear suspension

--- CAUTION ---
Coil springs are under considerable tension and can exert enough force to cause bodily injury. Exercise extreme caution when working with them.

TRAILING ARM TYPE

1. Raise the rear of the vehicle and support it on jack stands.
2. Remove the wheels.
3. Disconnect the handbrake linkage and return spring.
4. Unbolt the axleshaft flange at the wheel end.
5. Unbolt the rubber bumper inside the bottom of the coil spring.
6. Jack up the suspension arm and unbolt the shock absorber lower mounting.
7. Lower the jack slowly and cautiously. Remove the coil spring, spring seat, and rubber bumper.
8. Reverse the procedure to install, making sure that the flat face of the spring is at the top.

FOUR BAR LINK TYPE

1. Raise the car and support it with jackstands.
2. Support the center of the differential with a jack or other suitable tool.
3. Remove the rear wheels.
4. Remove the bolts securing the lower ends of the shock absorbers.
5. Lower the jack under the differential slowly and carefully and remove the coil springs after they are fully extended.
6. Installation is in the reverse order of removal.

I.R.S. Coil Spring Type

This suspension is similar to the I.R.S. MacPherson strut type, except this type uti-

SUSPENSION
Import Models

lizes separate coil springs and shock absorbers, instead of strut units.

1. Set a suitable spring compressor on the coil spring.
2. Jack up the rear end of the car.
3. Compress the coil spring until it is of sufficient length to be removed. Remove the spring.
4. When installing the spring, be sure the upper and lower spring seat rubbers are not twisted and have not slipped off when installing the coil spring.

MACPHERSON STRUT TYPE

For all strut overhaul procedures, please refer to "Strut Overhaul" in the Unit Repair section.

810 (Maxima), 280ZX, 300ZX and Stanza

The shock absorber and spring must be removed as a unit.

1. Block the front wheels. Raise the rear of the car until the suspension hangs free and support the car with stands under the frame members.
2. Remove the upper shock absorber mounting nuts (inside the trunk on the 810 and Maxima, or under covers in the cargo compartment in the ZX).

NOTE: On the 300ZX and the 1985 and later Maxima, be sure to disconnect the electrical lead if equipped with adjustable shock absorbers.

3. Disconnect the lower end of the shock and remove the assembly.
4. Installation is the reverse. Install the top end of the unit first.

Torsion Bar Type

1986 STANZA WAGON (2WD) ONLY

1. Raise the rear of the vehicle and support it with jackstands.

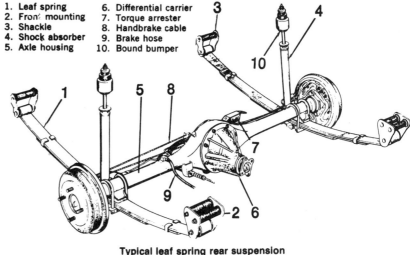

1. Leaf spring
2. Front mounting
3. Shackle
4. Shock absorber
5. Axle housing
6. Differential carrier
7. Torque arrester
8. Handbrake cable
9. Brake hose
10. Bound bumper

Typical leaf spring rear suspension

2. Remove the wheel and tire assembly. Release the parking brake.
3. Remove the inner hub cap, the cotter pin and the wheel bearing lock nut. Remove the brake drum.
4. Disconnect and plug the hydraulic brake line. Disconnect the parking brake cable.
5. Remove the four brake backing plate mounting bolts and then slide the backing plate along with the inner wheel bearing off of the rear axle.
6. Disconnect the rear stabilizer bar.
7. Unbolt the anchor arm bracket and then remove the inner bushing bracket mounting bolts. Remove the torsion bar.
8. Installation is in the reverse order of removal. Tighten the inner bushing and anchor arm mounting bolts to 36–43 ft. lbs. (49–59 Nm). Tighten the stabilizer bar bolts to 65–80 ft. lbs. (88–108 Nm).

Shock Absorber

REMOVAL & INSTALLATION

200SX, 210, 1986 Stanza Wagon (2WD) and 510 Sedans

1. Open the trunk and remove the cover panel (if necessary) to expose the shock mounts. Pry off the mount covers, if so equipped. On leaf spring models, jack up the rear of the vehicle and support the rear axle on stands.
2. Remove the two nuts holding the top of the shock absorber. Unbolt the bottom of the shock absorber.
3. Remove the shock absorber.
4. Installation is the reverse of removal.

510, 810 (Maxima) Station Wagons

1. Raise the rear of the car and support the axle on jack stands.
2. Remove the lower retaining nut on the shock absorber.
3. Remove the upper retaining bolt(s).
4. Remove the shock from under the car.

1. Raise and support the rear of the car.
2. Remove the wheels.
3. Support the rear arm with a jack at the lower end.
4. Remove the upper and lower shock mounting nuts.
5. Slowly and carefully lower the jack and remove the shock.
6. Installation is the reverse.

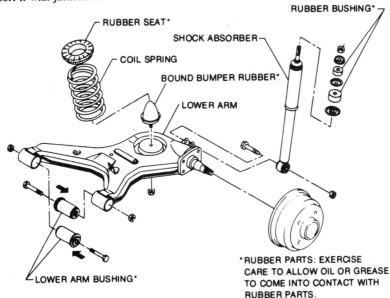

One side of a typical trailing arm rear suspension

SECTION 2: SUSPENSION
Import Models

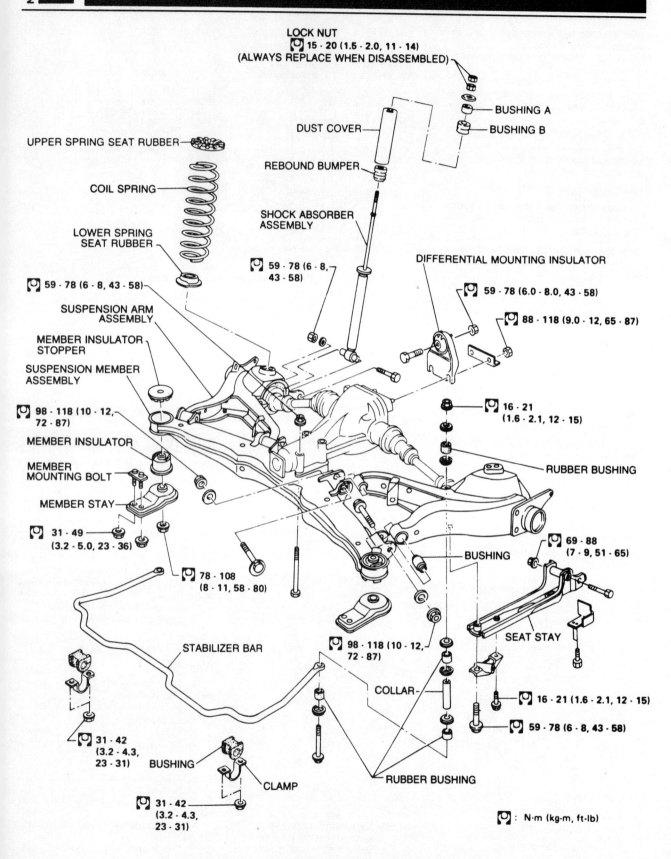

IRS coil spring rear suspension—200SX shown; others similar

2-240

SUSPENSION
Import Models

810, 280ZX, Maxima, 300ZX and Stanza

For all strut removal and installation procedures and all strut overhaul procedures, please refer to "Strut Overhaul" in the Unit Repair section.

Rear Wheel Bearings

ADJUSTMENT

310

The rear wheel bearings on the 310 are adjusted in the same manner as front wheel bearings. Use this procedure for all 310 models.

1. Raise the rear of the car.
2. Remove the wheel.
3. Remove and discard the cotter pin.
4. Tighten the wheel bearing lock nut to 29–33 ft. lbs.
5. Rotate the drum back and forth a few revolutions to snug down the bearing.
6. After turning the wheel, recheck the torque of the nut then loosen it 90° from its position.
7. Check the drum rotation. If it does not move freely, check for dragging brake shoes, or dirty bearings.

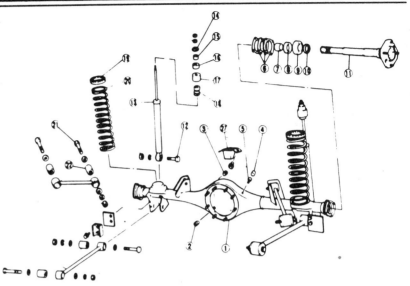

1. Rear axle case
2. Drain plug
3. Filler plug
4. Breather cap
5. Breather
6. Rear axle case end shim
7. Bearing collar
8. Oil seal
9. Rear axle bearing
10. Bearing spacer
11. Rear axle shaft
12. Shock absorber lower end bolt
13. Shock absorber assembly
14. Special washer
15. Shock absorber mounting bushing
16. Shock absorber mounting bushing
17. Bound bumper cover
18. Bound bumper rubber
19. Shock absorber mounting insulator
20. Coil spring
21. Upper link bushing bolt
22. Upper link bushing

Four link rear suspension

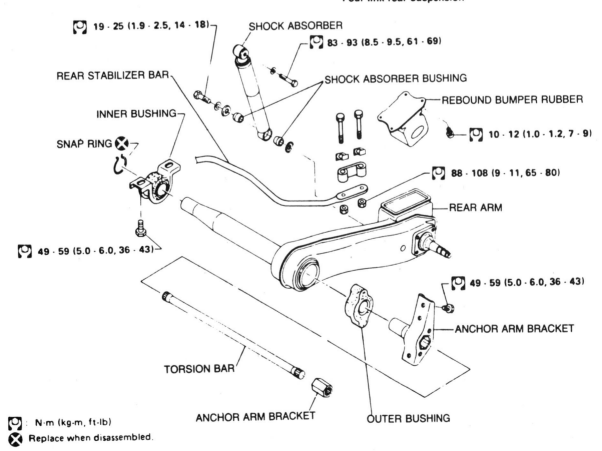

Torsion bar type rear suspension—1986 Stanza wagon (2 wheel drive)

2-241

SECTION 2: SUSPENSION
Import Models

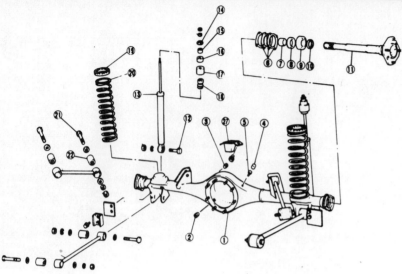

1. Rear axle case
2. Drain plug
3. Filler plug
4. Breather cap
5. Breather
6. Rear axle case end shim
7. Bearing collar
8. Oil seal
9. Rear axle bearing
10. Bearing spacer
11. Rear axle shaft
12. Shock absorber lower end bolt
13. Shock absorber assembly
14. Special washer
15. Shock absorber mounting bushing
16. Shock absorber mounting bushing
17. Bound bumper cover
18. Bound bumper rubber
19. Shock absorber mounting insulator
20. Coil spring
21. Upper link bushing bolt
22. Upper link bushing

Four link rear suspension

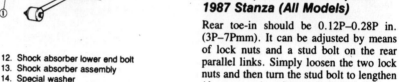

Adjusting the rear wheel toe—200SX shown

8. Align the cotter pin hole in the spindle with that in the locknut, and install a new cotter pin.
9. Tighten the nut no more than 15° to align the holes.

Sentra/Pulsar, Stanza and Maxima (1985 and Later)

1. Apply multi-purpose grease to the following parts:
   a. threaded portion of the wheel spindle
   b. mating surfaces of the lock washer and outer wheel bearing
   c. inner hub cap
   d. grease seal lip.
2. Tighten the wheel bearing nut to 18–25 ft. lbs. (25–34 Nm).
3. Turn the wheel several times in both directions to seat the bearing correctly.
4. Loosen the wheel bearing nut until there is no preload and then tighten it to 6.5–8.7 ft. lbs. (9–12 Nm). Turn the wheel several times again and then retighten it to the same torque again.
5. Install the adjusting cap and align any of its slots with the hole in the spindle.

NOTE: If necessary, tighten the lock nut as much as 15 degrees in order to align the spindle hole with one in the adjusting cap.

6. Rotate the hub in both directions several times while measuring its starting torque and axil play. They should be as follows:

Axial play: 0
Starting torque:
  W/grease seal—6.9 inch lbs. (0.8 Nm) or less.
  When measured at wheel hub bolt—3.1 lbs. (13.7 N).

7. Correctly measure the rotation from the starting force toward the tangential direction against the hub bolt. The above figures do not allow for any "dragging" resistance. When measuring starting torque, confirm that no "dragging" exists. No wheel bearing axial play can exist at all.
8. Spread the cotter pin and install the inner hub cap.
9. Installation of the remaining components is in the reverse order of removal.

Rear Wheel Alignment

ADJUSTMENT

200SX and 300ZX

Rear toe-in should be 0.08N–0 in. (2N–0mm)—200SX; 0.059N–0.098P in. (1.5N–2.5Pmm). It can be adjusted by the use of cams on the inside of rear control arm bushing pins.

NOTE: Always set the cams in the same position on the right and left control arm bushing pins.

Maxima (1985 and Later) and 1987 Stanza (All Models)

Rear toe-in should be 0.12P–0.28P in. (3P–7Pmm). It can be adjusted by means of lock nuts and a stud bolt on the rear parallel links. Simply loosen the two lock nuts and then turn the stud bolt to lengthen (decrease toe) or shorten (increase) the link. Be certain that both the left and right links are adjusted to the same length.

NOTE: When tightening the lock nut, or checking its torque, attach a wrench to the "locking" section of the rod in order to prevent the bushing from twisting.

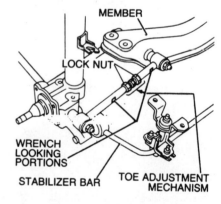

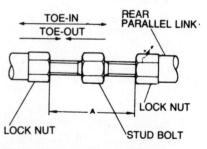

STANDARD LENGTH "A": 49.5mm (1.949 in.)

Adjusting the rear wheel toe—Maxima (1985 and later) and Stanza (1987)

SUSPENSION
Import Models

FRONT SUSPENSION PORSCHE 911, TURBO

Springing is provided by a longitudinal torsion bar at each wheel. A triangular lower arm links the torsion bar to the shock absorber strut and steering knuckle. A permanently lubricated ball joint is located at the bottom of the strut.

Torsion Bars

REMOVAL & INSTALLATION

1. Jack up the front of the car and support it safely with stands.
2. Remove the torsion bar adjusting screw.
3. Take the adjusting lever off the torsion bar and withdraw the seal.
4. Unscrew the retaining bolts from the front mount cover bracket and remove the bracket.
5. Using a drift, carefully drive the torsion bar out of the front of the arm.
6. Check the torsion bar for spline damage and rust. If necessary, replace the bar.
7. Give the torsion bar a light coating of grease before installing it.

NOTE: Torsion bars are marked "L" or "R" to identify them and are not interchangeable.

8. Insert the end cap of the torsion bar, protruding side out, into the control arm. Drive the torsion bar into position with a drift. Carefully.
9. Tighten the retaining bolts on the front mount to 34 ft. lbs.
10. Slide the seal onto the torsion bar from the open side of the crossmember.
11. Using a tire iron, or other suitable lever, pry the control arm down as far as possible. While holding the control arm, slide the adjusting lever onto the splines of the torsion bar. There should only be a slight amount of clearance at the lever adjusting point.

Loosening torsion bar adjusting screw

1. Strut
2. Brake disc
3. Intermediate
4. Universal joint
5. Stabilizer bar
6. Tie-rod
7. Adjusting screw
8. Bellows
9. Control arm
10. Steering column
11. Steering gear
12. Crossmember
13. Bearing support

12. Grease the adjusting screw threads with a moly grease and hand tighten the screw.
13. Check that the end cap is properly seated in the control arm.
14. Install the rubber mount cover bracket. Tighten the retaining bolts to 34 ft. lbs.
15. Lower the car.
16. Check the front wheel alignment.

Shock Absorbers

REMOVAL & INSTALLATION

1. Jack up the front of the car and support it safely on stands. Remove the wheels.
2. Remove the brake line from the clip on the suspension strut. A small amount of brake fluid will run out of the line, plug it so that dirt cannot enter the system.
3. Unscrew the retaining bolts and remove the caliper.
4. Using a soft mallet, tap the hub cap to loosen it.
5. Pry the hub cap off with a small prybar.
6. Loosen the Allen screw in the wheel bearing clamp. Unscrew the clamp nut and remove the nut and washer.
7. Remove the wheel hub along with the brake disc and wheel bearing.
8. Remove the backing plate retaining bolts and remove the plate.
9. Withdraw the cotter pin from the castellated nut on the tie rod end and remove the nut. Using a suitable puller, remove the tie rod joint from the strut.
10. Remove the control arm-to-strut ball joint retaining bolt and pull the ball joint out of the strut by pulling down on the lower control arm.

NOTE: The torsion bar adjusting screw will have to be loosened and the adjusting arm removed.

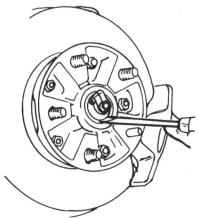

Checking wheel bearing play

Front suspension and steering

2-243

SECTION 2: SUSPENSION
Import Models

11. Remove the keeper for the nut on the top of the strut. Unscrew the nut and remove it, the keeper plate, and washer.

12. Remove the strut from the bottom. It will be necessary to loosen and pull the side of the luggage compartment out for clearance.

13. Check the shock absorber strut for excessive free travel and leaking. Replace the shock absorber if it is at all suspect.

14. Install the strut in a reverse order of the removal.

15. Tighten the top nut to 58 ft. lbs. Use a new keeper plate and ensure that the peg on the plate is pointing up.

16. Tighten the ball joint bolt to 47 ft. lbs.

NOTE: Remember to install the washer between the ball joint seal and strut.

17. Install the torsion bar adjusting lever as described in "Torsion Bar Removal and Installation."

18. Tighten the tie rod nut to 33 ft. lbs. and install a new cotter pin.

19. Torque the backing plate bolts to 18 ft. lbs.

20. Install and adjust the wheel bearings.

21. Tighten the caliper retaining bolts to 50 ft. lbs.

22. Bleed the brakes.

23. Install the wheels and lower the car.

24. Check the wheel alignment.

Stabilizer Bar

REMOVAL & INSTALLATION

1. Jack up the front of the car and safely support it on stands.

2. Loosen the stabilizer clamp bolts and pry the lever ends off their mounts.

3. Remove the stabilizer bar along with the levers.

4. Check the rubber bushings for deterioration and, if necessary, replace them. Lubricate the bushings with glycerine or some other rubber preservative.

5. Install the stabilizer bar in a reverse order of the removal.

6. The square end of the stabilizer should protrude slightly above the clamp. Tighten the clamp nuts to 18 ft. lbs.

Wheel Bearings

REMOVAL & INSTALLATION

NOTE: The inner bearing, seal, and outer bearing may be removed and lubricated once the hub/disc assembly is off the car. If after cleaning, the bearings are noticeably worn or damaged they should be replaced along with their races. If the bearings are satisfactory, skip the race removal steps.

1. Remove the brake disc/hub assembly.

2. Match mark the hub and disc for correct reassembly, remove the five assembly bolts, and separate the hub and disc.

3. Pry the inner seal out of the hub. Remove the inner bearing and outer bearing.

4. Wash the bearings in solvent and blow them dry. Examine the bearings for pitting, scoring, or other damage. Replace the bearing and race as a unit if there is any question as to their condition.

5. Heat the wheel hub to 250°–300°F.

6. Press the inner bearing race out of the hub on a press table, using suitable spacers to prevent damaging the hub.

7. Press out the outer bearing race, using suitable spacers and a support fabricated from the accompanying drawing.

8. Press a new inner bearing race into the hub and then press in a new outer bearing race.

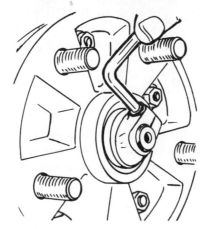

Final tightening of the wheel clamp nut—check play again before installing hub cap

9. Pack the bearings with a lithium multipurpose grease.

10. Align the match marks and install the hub on the disc. Insert the assembly bolts from the inside out and tighten them to 17 ft. lbs.

11. Lightly coat the spindle with grease. Fill the hub with about 2 oz of grease. Lubricate and install the bearings.

12. Grease the sealing edges of a new inner oil seal and carefully tap it into place. The oil seal must be flush with the hub.

13. Install the hub/disc assembly on the car.

14. Adjust the wheel bearings.

ADJUSTMENT

Check and adjust the front wheel bearings after the car has not been run for a few hours. The bearings will be cold then.

1. The front wheel bearings are correctly adjusted when the thrust washer can be moved slightly sideways under light pressure from a small prybar, but no bearing play is evident when the wheel hub is shaken axially.

2. Jack up the front of the car, support it on stands, and remove the wheels. Turn the hub several times to seat the bearings.

3. Pry the hub cap off and perform the check described in Step 1.

NOTE: Don't press the prybar against the hub. Hold it lightly in your hand so you get a better feel.

4. If the bearings require an adjustment, loosen the Allen screw and turn the clamp nut in or out as necessary.

5. Tighten the clamp nut Allen screw to 11 ft. lbs. without altering the adjusted position of the clamp nut.

6. Double check the adjustment and readjust, if necessary.

7. Give the clamp nut and thrust washer a light coating of lithium grease. Tap the hub cap into place with a plastic or rubber mallet.

8. Install the wheels and lower the car.

Adjustments

CAMBER

Camber is adjusted at the top of the strut. Pull back the luggage compartment rug to expose the three mounting bolts. Scrape the undercoating from the bolts and plates. Scribe the positions of the two plates under the bolts. Loosen the bolts and move the strut in or out as necessary to correct the camber angle.

CASTER

Caster is adjusted in the same manner as camber, except that the strut is moved forward or backward to change the caster angle.

Caster and camber adjustment location

TOE-IN

Toe-in is set with the front wheels straight ahead. Tie-rod length is adjusted by loosening the tie-rod clamps and moving them an equal amount in or out to obtain the correct toe-in.

SUSPENSION
Import Models

REAR SUSPENSION PORSCHE 911, TURBO

All models covered in this section have independent rear suspension. The rear suspension is a semi-trailing arm design. Springing is provided by transverse torsion bars located forward of each trailing arm. Telescopic shock absorbers at each wheel provide dampening. A rear stabilizer bar is standard equipment on the 911S Turbo and Carrera and optional on the other 911 models.

Raising the trailing arm

Installing the trailing arm cover

Torsion Bars

REMOVAL & INSTALLATION

1. Jack up the rear of the car and support it safely with stands.
2. Remove the wheel on the side where the torsion bar is being removed.
3. Fabricate a fixture similar to the one shown. The fixture is necessary to hold the trailing arm while it is raised and lowered. The special Porsche tool for this purpose is number P 289.
4. Using a hydraulic jack under the holding fixture, raise the trailing arm.
5. Remove the lower shock absorber bolt.
6. Remove the trailing arm retaining bolts. Remove the toe and chamber adjusting bolts.
7. Remove the four retaining bolts from the trailing arm cover. Withdraw the spacer.
8. Using two small prybars, pry off the trailing arm cover.
9. Remove the holding fixture.
10. Knock out the round body plug and remove the trailing arm.
11. Paint a reference mark on the torsion bar support, matching the location of the "L" or "R" side identification letter, so that the torsion bar may be installed in the same position.

NOTE: The torsion bars are splined to allow adjustment of the rear riding height.

12. Remove the torsion bar. Do not scratch the protective paint on the torsion bar, or it will corrode and possibly develop fatigue cracks.

NOTE: If you are removing a broken torsion bar, the inner end can be knocked from its seat by removing the opposite torsion bar and tapping through with a steel rod. Torsion bars are not interchangeable from side-to-side and are marked "L" and "R" for identification.

13. Check the torsion bar splines for damage and replace it if necessary. If any corrosion is present on the bar, replace it.
14. Coat the torsion bar lightly with a multi-purpose grease. Carefully grease the splines.
15. Apply glycerine or another rubber preservative to the torsion bar support.
16. Install the torsion bar, matching the "L" or "R" with the paint mark you made before removal.
17. Install the trailing arm cover into position and start the three accessible bolts.
18. Raise the trailing arm into place with the holding fixture (or special tool P 289)

Bottom shock absorber mounting

until the spacer and the fourth bolt can be installed.
19. Assemble the remaining components in a reverse order of their removal.
20. Tighten the trailing arm cover bolts to 34 ft. lbs. Tighten the trailing arm retaining bolts to 65 ft. lbs.
21. Tighten the camber adjusting bolt to 43 ft. lbs. and the toe-in adjusting bolt to 36 ft. lbs. Tighten the shock absorber bolt to 45 ft. lbs.
22. Adjust the rear wheel camber and toe-in.

Shock Absorbers

REMOVAL & INSTALLATION

1. Leave the car standing on the ground so that the shock absorber is not tensioned.
2. Open the engine compartment lid and remove the rubber cover from the top of the shock absorber.
3. Hold the shock absorber shaft and remove the nut.
4. On the bottom, remove the retaining nut and bolt.
5. Remove the shock absorber.
6. If the shock exhibits excessive free travel or is leaking, replace it.
7. Install the shock up through the body and screw the nut on hand-tight.
8. Align the shock absorber eye with the hole in the trailing arm and install the nut and bolt.
9. Tighten the top nut and install the rubber cover.
10. Tighten the bottom retaining bolt to 54 ft. lbs.

Top shock absorber mounting

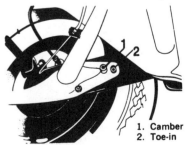

1. Camber
2. Toe-in

Rear wheel alignment adjustment points

SECTION 2
SUSPENSION
Import Models

Stabilizer

REMOVAL & INSTALLATION

1. Raise the rear of the car and safely support it with stands.
2. Using a large prybar, pry the upper eyes of the stabilizer bar off the studs in the trailing arm.
3. Remove the body mounting brackets.
4. Remove the stabilizer.
5. Check the rubber bushings for wear or damage and, if necessary, replace them.
6. Install the stabilizer bar using a reverse of the removal steps.

Adjustments

CAMBER

The rearmost of the two Allen bolts on the trailing arm provides camber adjustment. Tighten the bolt to 43 ft. lbs. after the camber is adjusted to specifications.

TOE-IN

The front Allen bolt on the trailing arm adjusts the toe-in. Tighten the bolt to 36 ft. lbs. after toe-in is adjusted to specification.

FRONT SUSPENSION PORSCHE 924, 928, 944

924, 924 Turbo and 944

The 924 front suspension is a MacPherson strut design. The strut consists of the strut housing, a shock absorber insert in the housing, and a concentric coil spring. The steering knuckle is bolted to the strut assembly. A lower control arm locates the strut at the bottom. A ball joint is riveted to the control arm and bolted to the steering knuckle.

928 and 928S

The 928 front suspension is an independent type with upper and lower control arms, coil springs mounted on the shock absorbers and upper and lower ball joints. Provisions for front end alignment are provided by eccentrics located at the bottom of the lower ball joint mounting plate.

MacPherson Strut

REMOVAL & INSTALLATION

924, 924 Turbo and 944

Front wheel alignment must be reset after a strut is removed.

1. Jack up the front of the car and support it on stands.
2. Remove the brake line from the bracket on the strut.
3. Remove the two through bolts that retain the strut to the steering knuckle.
4. Remove the four retaining nuts from the inner fender in the engine compartment.
5. Pry the lower control arm down and remove the strut from the car.
6. To replace either the spring or shock absorber, place the strut in a spring compressor and remove the large retaining nut at the top.
7. Installation is the reverse of removal.

928 and 928S

1. Remove the self-locking nuts on the upper strut mount, located on the inner fender panel.
2. Remove the front wheel. Remove the flange locknut and press the upper ball joint from the spindle carrier.
3. Remove the inner pivot shaft nuts from the upper control arm.
4. Remove the shock absorber mounting bolts and remove the shock and upper arm as an assembly.

OVERHAUL

For all spring and shock absorber removal and installation procedures, and all strut overhaul procedures, please refer to "Strut Overhaul"

Lower Control Arm

REMOVAL & INSTALLATION

924, 924 Turbo and 944

1. Jack up the front of the car and support it on stands.
2. Remove the thru-bolt at the front that retains the control arm to the suspension crossmember.
3. Detach the stabilizer bar from the control arm.
4. Remove the two bolts that retain the control arm bracket at the rear.
5. Remove the ball joint pinch bolt at the steering knuckle.
6. Pry the control arm down and remove it from the car.

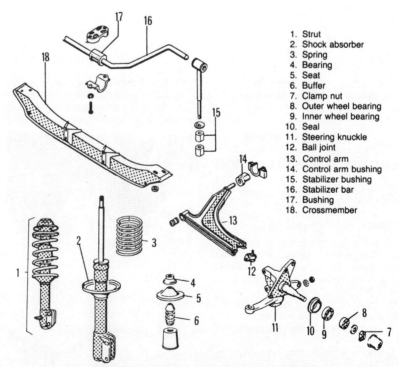

1. Strut
2. Shock absorber
3. Spring
4. Bearing
5. Seat
6. Buffer
7. Clamp nut
8. Outer wheel bearing
9. Inner wheel bearing
10. Seal
11. Steering knuckle
12. Ball joint
13. Control arm
14. Control arm bushing
15. Stabilizer bushing
16. Stabilizer bar
17. Bushing
18. Crossmember

Front suspension of the 924, 924 Turbo, and 944

SUSPENSION
Import Models

7. Installation is the reverse of removal. Caster must be reset after the control arm has been removed.

928 and 928S

NOTE: The front end must be aligned upon completion of the installation.

1. Raise and support the vehicle then remove the wheel.
2. Mark the alignment eccentrics on the lower arm for approximate installation location, if the ball joint is to be removed.
3. Remove the strut bottom link bracket and stabilizer link bolt.
4. Remove the lower ball joint stud nut and press the stud from the spindle. Move the spindle and upper arm upward and block it to gain working clearance.
5. Remove the bolts from the tie-down bracket and control arm bracket. Lower the control arm from the vehicle.
6. The lower ball joint can be replaced, if necessary, while the lower arm is out of the vehicle.
7. Installation is the reverse of removal.

Ball Joint

REMOVAL & INSTALLATION

924, 924 Turbo and 944

1. Remove the lower control arm.
2. Drill out the three rivets retaining the ball joint to the control arm.
3. Install the replacement ball joint using the bolts and nuts supplied in the kit.
4. Reinstall the control arm and align the wheels.

928 and 928S

NOTE: The front suspension must be realigned after the suspension work is done.

The upper ball joint is replaced as a unit with the upper arm assembly. Refer to the "Strut Removal and Installation" section.

The lower ball joint may be replaced by removing the nut from the ball joint stud and pressing the stud from the spindle. The alignment eccentric bolts are removable and the ball joint can be removed from the lower arm assembly.

Front Wheel Bearings

ADJUSTMENT

The front wheel bearings are correctly adjusted when the thrust washer can be moved slightly sideways under light pressure, but no bearing play is evident when the wheel hub is shaken axially.

1. Remove the wheels.
2. Pry the hub cap off and perform the check as described above. Don't press against the hub.

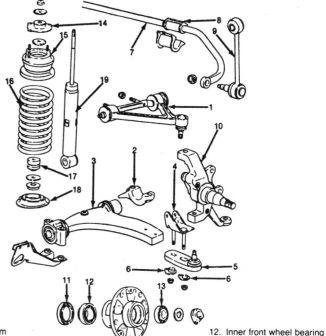

1. Upper control arm
2. Lower control arm support
3. Lower control arm
4. Lower shock mount
5. Lower ball joint
6. Caster eccentric (outer), camber eccentric (inner)
7. Stabilizer bar
8. Bushing
9. Link
10. Steering knuckle
11. Inner front wheel seal
12. Inner front wheel bearing
13. Outer front wheel bearing
14. Upper shock mount
15. Upper spring seat
16. Spring
17. Suspension stop
18. Lower spring seat
19. Shock absorber

Exploded view of the 928 and 928S front suspension

3. If the bearings require an adjustment, loosen the Allen screw and turn the clamp nut. Proper adjustment is achieved when the flat washer can just be moved by finger pressure on a screwdriver.
4. Tighten the clamp nut Allen screw to 11 ft. lbs. without altering the adjusted position of the clamp nut.

Front End Alignment

CAMBER ADJUSTMENT

924, 924 Turbo and 944

Camber is adjusted at the upper strut-to-steering knuckle retaining bolt.

928 and 928S

Camber is adjusted by turning the cam bolts on the inner arm bushings.

CASTER ADJUSTMENT

NOTE: 1983 and later 928 models have the slots for adjusting the caster eccentrics sealed with an elastic sealing compound. This compound must be removed to make camber adjustments, then replaced after adjustment to prevent the entry of dirt which could make adjusting difficult.

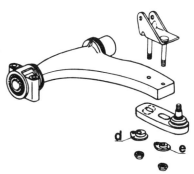

Front suspension caster (e) and camber (d) adjustment eccentrics on the 928 and 928S

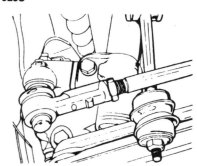

Tie rod ends are adjustable for toe (924 shown, others similar)

2-247

SUSPENSION
Import Models

Caster is adjusted by loosening the two control arms-to-crossmember bolts and moving the control arm laterally.

TOE-IN ADJUSTMENT

924, 924 Turbo and 944

Toe-in is set by loosening the locknuts on the tie rod ends and turning them in or out as necessary.

928 and 928S

Toe-in adjustments are made by turning cam bolts, located at the front of the rear control arms.

Caster is adjusted at the lower control arm mounting on the 924, 924 Turbo, and 944

Camber is adjusted at the upper strut eccentric on the 924, 924 Turbo, and 944

REAR SUSPENSION PORSCHE 924, 928, 944

Torsion Bars

REMOVAL & INSTALLATION

924, 924 Turbo and 944

NOTE: This procedure requires that the rear wheel camber and toe-in be checked and adjusted as the final step.

1. Jack up the rear of the car and support it on stands.
2. Remove the wheel on the side where the torsion bar is being removed.
3. Using a hydraulic jack and a block of wood with a slot cut in it, raise the trailing arm.
4. Remove the lower shock absorber bolt.
5. Remove the trailing arm retaining bolts. Remove the toe and camber adjusting bolts.
6. Remove the four retaining bolts from the trailing arm cover.
7. Pry off the trailing arm cover.
8. Lower the jack.
9. Remove the round body plug and remove the trailing arm.
10. Paint a reference mark on the torsion bar support, matching the location of the L or R side identification letter, so that the torsion bar may be installed in the same position.

NOTE: The torsion bars are splined to allow adjustment of the rear riding height.

11. Remove the torsion bar. Do not scratch the protective paint on the torsion bar, or it will corrode and possibly develop fatigue cracks.

NOTE: If you are removing a broken torsion bar, the inner end can be knocked from its seat by removing the opposite torsion bar and tapping it through with a steel bar. Torsion bars are not interchangeable from side to side and are marked "L" and "R" for identification.

12. Check the torsion bar splines for damage and replace the bar if necessary. If there is any corrosion on the bar, replace it.
13. Coat the torsion bar lightly with grease. Carefully grease the splines.
14. Apply glycerine or another rubber preservative to the torsion bar support.
15. Install the torsion bar, matching the L or R with the paint mark you made before removal.
16. Install the trailing arm cover into position and start the three accessible bolts.
17. Raise the trailing arm into place with a jack and wooden block until the spacer and the fourth bolt can be installed.
18. Assemble the remaining components in the reverse order of their removal.
19. Tighten the trailing arm cover bolts to 25 ft. lbs. Tighten the shock absorber bolt to 50 ft. lbs.
20. Adjust rear wheel camber and toe-in.

Upper Control Arm

REMOVAL & INSTALLATION

928 and 928S

1. Raise the vehicle and support it safely. Remove the rear wheels and support the lower arm assembly with a jack.
2. Loosen and remove the inner and outer bolts from the upper arm ends.
3. Remove the upper arm from the rear crossmember and from the rear flexible mount.

NOTE: The bushings are replaceable.

Installation is the reverse of removal.

Lower Control Arm

REMOVAL & INSTALLATION

928 and 928S

1. Raise and support the vehicle. Remove the rear wheels.
2. Support the hub assembly and the spring strut with a jack.
3. Remove the outer pivot pin nuts and washers. Disconnect the stabilizer bar link.
4. Remove the inner pivot bolts from the arm and pull the front pivot pin from the hub assembly and the spring strut.

NOTE: The bushings are replaceable.

5. Installation is the reverse of removal.

Shock Absorbers

REMOVAL & INSTALLATION

924, 924 Turbo and 944

1. Raise the car on a drive-on hoist or support the wheels on stands for this procedure.
2. Remove the bottom retaining bolt and nut.
3. Remove the top bolt.
4. Remove the shock absorber.
5. Install the replacement shock in the reverse order of removal. Tighten the retaining bolts to 50 ft. lbs.

MacPherson Struts

REMOVAL & INSTALLATION

928 and 928S

1. Remove the locking nuts from the spring strut, located within the trunk area.
2. Raise the vehicle, support safely and remove the wheel.

SUSSPENSION
Import Models

SECTION 2

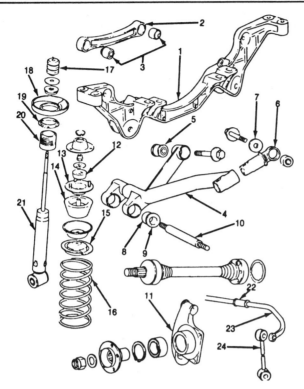

Rear wheel alignment points of the 924 and 924 Turbo. Loosen the fasteners (arrows) to alter the camber and toe-in

3. Remove the front nut on the outer pivot pin rod and remove the pivot rod from the rubber bushings.
4. Disconnect the stablizer bar link from the lower control arm.
5. Remove the spring strut from the vehicle.
6. Installation is the reverse of removal.

NOTE: **The spring can be removed from the shock unit with the use of a spring clamping tool. An adjusting nut and sleeve is used to control the vehicle rear height.**

1. Rear axle crossmember
2. Upper strut
3. Upper strut bushings
4. Lower control arm
5. Lower control arm inner bushing
6. Lower control arm rocker mount
7. Lower control eccentric
8. Lower control arm outer bushing
9. Cone washer
10. Pivot pin
11. Wheel bearing carrier
12. Upper shock mount
13. Shock mount retainer
14. Lower shock mount
15. Upper spring mount
16. Coil spring
17. Shock bumper
18. Lower spring seat
19. Suspension height adjuster
20. Flange
21. Shock absorber
22. Stabilizer bar mount
23. Stabilizer bar
24. Link

Exploded view of the 928 and 928S rear suspension

OVERHAUL

For all spring and shock absorber removal and installation procedures, and all strut overhaul procedures, please refer to "Strut Overhaul" in the Unit Repair section.

Rear Wheel Alignment

CAMBER ADJUSTMENT

Rear wheel camber is adjusted by changing the trailing arm spring plate setting. To increase positive camber, loosen the spring plate-to-trailing arm bolts (with the wheels on the ground). To increase negative camber, do so with car on hoist. Tighten bolts after adjustment.

TOE-IN ADJUSTMENT

Rear wheel toe-in is adjusted by moving the control arm in the slots of the spring plates.

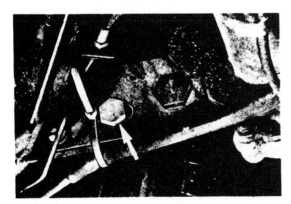

Rear camber adjustment on the 944

Rear toe-in adjustment on the 944

2-249

SECTION 2
SUSPENSION
Import Models

FRONT SUSPENSION RENAULT

Upper Control Arm

NOTE: Alliance and Encore have no upper control arm.

REMOVAL & INSTALLATION

LeCar

1. Remove the cooling system expansion bottle and the ignition coil. Raise the front end.
2. Disconnect the upper ball joint using the removal tool.
3. Screw on and tighten a lock nut on the front end of the arm hinge pin (bolt). Remove the rear lock nut.
4. Drive the pin and lock nut assembly towards the front.
5. Remove the upper arm from the chassis.
6. To install the upper arm, reverse the removal procedure. Tighten the nuts to 70 ft. lb.

18, 18i, Sportwagon and Fuego

1. Raise and support the car on jackstands.
2. Disconnect the castor tie rod.
3. Remove the sway bar.
4. Using a ball joint separator, disconnect the ball joint.
5. Disconnect the tie rod ball joint.
6. Unbolt and remove the upper arm hinge pin.
7. Disconnect the shock absorber lower end.
8. Tilt the arm upward to free the ball joint and free the other end from the shock absorber.
9. Installation is the reverse of removal. Install the arm on the shock absorber first, then the upper ball joint followed by the tie rod ball joint. Hand tighten the nuts. Install the hinge pin, liberally coated with chassis lube and hand tighten the nuts. Install the sway bar; hand tighten the nuts. Attach the castor tie rod. NOW, tighten all nuts.
- Upper ball joint—48 ft. lbs.
- Control arm hinge pin nuts—70 ft. lbs.
- Lower shock absorber nuts—55 ft. lbs.
- Tie rod ball joint—48 ft. lbs.
- Sway bar—20 ft. lbs.
- Caster link nut—60 ft. lbs.

Lower Control Arm

REMOVAL & INSTALLATION

LeCar

1. Remove the torsion bar.
2. Remove the halfshaft.
3. Remove the sway bar.
4. Disconnect the shock absorber lower end and remove the bolt.
5. Disconnect the steering arm ball joint and lower control arm ball joint with a ball joint separator.
6. Unbolt and remove the lower control arm hinge pin.

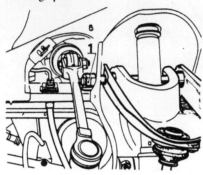

Installing a locknut on the arm hinge pin (bolt)

7. Install a hub puller on the hub end and tighten the puller until the ball joint pulls free of the axle shaft carrier.
8. Installation is the reverse of removal. Observe the following torques:
- Lower control arm hinge pin nuts—75 ft. lbs.
- Lower shock absorber nut—25 ft.lbs.
- Lower ball joint nut—40 ft. lbs.

Alliance and Encore

1. Raise and support the car on jackstands.
2. Remove the sway bar.
3. Remove the wheels.
4. Using a ball joint separator, disconnect the lower control arm ball joint.
5. Remove the control arm hinge pins.
6. Remove the control arm.
7. Installation is the reverse of removal. Don't tighten any nuts completely, until all parts are installed. The nuts are tightened with the car's weight on the suspension. Note the following torques:
- Lower arm hinge pins—55 ft. lbs.
- Lower ball joint nut—40 ft. lbs.
- Sway bar bolts—16 ft. lbs.

18i and Fuego

1. Raise and support the car on jackstands.
2. Using a ball joint separator, disconnect the lower ball joint.
3. Unbolt and remove the control arm hinge pin.
4. Free the lower ball joint from the spindle and remove the arm from the car.
5. Installation is the reverse of removal. Do not tighten any fasteners until all parts are installed. The fasteners must be tightened with the weight of the car on the suspension. Observe the following torques:
- Lower control arm hinge pin—67 ft. lbs.
- Ball joint nut—35 ft. lbs.

Torsion Bars

REMOVAL & INSTALLATION

LeCar

The following procedure requires the use of a special tool.

1. Tilt the front seat forward.
2. There are four bolts on the front face of the seat well. The largest of these is the torsion bar cam bolt. Loosen (but do not remove) the bolt and turn the cam left all the way to zero it.
3. Raise the front end and place jackstands under the frame so that the wheels hang freely.
4. Remove the torsion bar adjusting lever cover.
5. Insert tool Sus.545 in the adjusting lever.
6. In the front seat well, remove the three other bolts mentioned in Step 2. These are the adjusting lever housing attaching bolts.
7. Remove the housing cover cam assembly and gradually ease the pressure on the tool.
8. Mark the position of the adjusting lever on the floor pan.
9. Mark the position of the torsion bar on the lower arm anchor sleeve.
10. Remove the sway bar bushings.
11. Remove the torsion bar from the control arm and check that the marks made previously are lined up with the punch mark on the end of the bar. If not, count and record the number of splines the marks are off.
12. Remove the torsion bar.
13. Apply chassis lube to the ends of the torsion bar.
14. Install the protective cover seal, the housing cover cam assembly and the adjusting lever on the torsion bar.
15. Pass the housing cover cam assembly and the adjusting lever inside the crossmember. Slide the torsion bar into position above the sway bar.
16. Insert the torsion bar into the lower control arm aligning the marks.
17. Position the adjusting lever on the splines, aligning the marks made earlier on the floor crossmember.
18. Position the adjusting lever ⅛–¼ in. to the left of its travel.

SUSPENSION
Import Models

19. With the tool Sus.545 inserted in the adjusting lever, take up the tension on the bar. Center the cover by resetting the cam.
20. Hold the torsion bar assembly against its bushing with a pair of locking pliers and bolt it together.
21. Install the sway bar.
22. Drive the vehicle a short distance, park it on a flat surface and measure the ride height. Ride height is determined by calculating the difference between the distance from the wheel center to the ground and the distance from the underbody frame side member to the ground. The difference should be 1⅞–2⅝ in. To adjust the ride height, loosen the cam adjusting bolt in the seat well and turn the cam to obtain the proper ride height.

Shock Absorber

REMOVAL & INSTALLATION

LeCar

1. Raise and support the car on jackstands.
2. Raise the lower control arm with a floor jack.
3. Disconnect the upper end of the shock.
4. Remove the sway bar.
5. Disconnect the lower end of the shock.
6. Install the shock absorber and sway bar, and tighten the bolts finger tight.
7. Let the lower control arm hang, then raise it 1¾ in., or, about half its travel. Tighten all fasteners.

Ball Joints

REMOVAL & INSTALLATION

All Models

NOTE: The Alliance and Encore have no upper ball joint or control arm.

1. Raise and support the car on jackstands.
2. Remove the ball joint nut. Using a ball joint separator, disconnect the ball joint.
3. Drill out the rivets securing the ball joint to the upper arm and remove the ball joint.

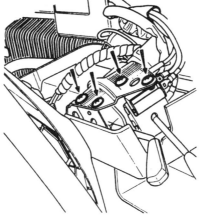

LeCar combination switch removal

4. The new ball joint kit will come with bolts and nuts to replace the rivets and the upper ball joint has a shim which should go on top of the control arm. Install the ball joint and shim. The bolt heads go on the dust cover side. Torque the ball joint large nut to 49 ft. lbs.

MacPherson Strut

REMOVAL & INSTALLATION

All except LeCar

Special tools are needed for this job.

1. Raise and support the car on jackstands placed under the frame. Remove the wheel.
2. Unbolt the strut at the top and bottom. Pull the strut out. It may be necessary to press down on the lower control arm.
3. Installation is the reverse of removal. Observe the following torque specifications for all but the R-18: Shock top mounting nut, 44 ft. lbs.; shock bottom mounting nut, 55 ft. lbs.; Shock top mounting pad nuts, 17 ft. lbs. Observe the following specifications for the R-18: Shock upper mounting nut, 11 ft. lbs.; shock lower mounting nut, 30 ft. lbs.

OVERHAUL

For all spring and shock absorber removal and installation procedures, and any other strut overhaul procedures, please refer to "Strut Overhaul"

REAR SUSPENSION RENAULT

Rear Wheel Bearings

REMOVAL & INSTALLATION

18, 18i, Sportwagon and Fuego

1. Raise the rear of the vehicle and remove the wheels.
2. Remove the rear brake drum;
   a. Release the parking brake.
   b. Slacken the secondary cables of the parking brake so that the lever may be pulled back.
   c. Remove the dust plug in the backing plate so that the automatic adjusting system may be disengaged.
   d. Insert a screwdriver-type tool (tool no. Rou. 370-02) into the hole in the backing plate, through a companion hole in the brake shoe, to the parking brake lever.
   e. Push in with the tool to disengage the catch from the brake shoe. Push the lever to the rear with the tool.

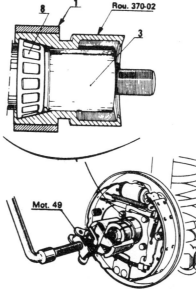

Inner bearing removal—18i and Fuego

   f. Remove the dust cover, the retaining lock nut/cotter pin, the stub axle nut and washer.
   g. Remove the drum, the outer bearing and seal as an assembly.
3. With the use of special pulling tools (No. Rou. 15-01 and B. Vi. 28-01), the inner bearing is removed from the stub axle, along with the deflector.
4. Install the inner deflector, making sure it is mounted correctly.
5. Install the inner bearing on the stub axle by carefully driving it to its seat.
6. Install any of the bearing cups as required. Add lubricant to the bearings and cups.
7. Install the drum assembly in the reverse of its removal procedure.
8. Check the bearing end play, which should be between 0.000 and 0.001 in. (0.00 and 0.03mm).
9. Adjust the foot brake by repeatedly pushing down on the brake pedal.
10. Adjust the parking brake. Install the plug in the rear of the backing plate.

SECTION 2
SUSPENSION
Import Models

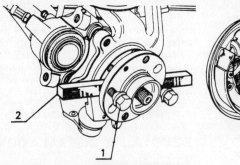

Removing the hub from the 18i or Fuego. (1) is a lug nut; (2) is a metal bar

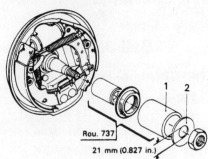

Installing the inner deflector

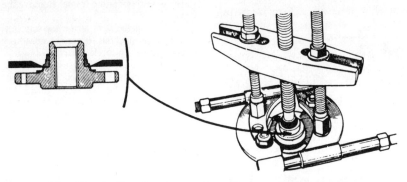

Removing the inner bearing race from the hub on the 18i or Fuego

LeCar

1. Raise the vehicle and support safely.
2. Remove the rear wheel(s) and the brake drum.
3. Remove the outer bearing and the oil seal.

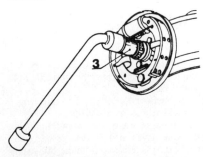

Cold method of installing the LeCar rear axle bearing

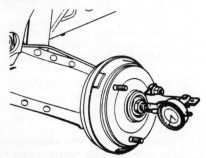

Adjusting the LeCar rear axle bearing; others similar

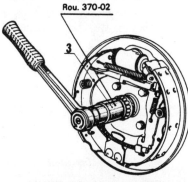

Installing the inner bearing

4. Remove the bearing races as required.
5. With a special puller, remove the inner bearing from the rear axle spindle.

NOTE: Some inner bearings will have a thrust washer that must be removed with the bearing assembly. A different type bearing puller is then used.

There are two methods of bearing replacement, hot or cold. The hot method is preferred and consists of pre-heating the washer so that it can be assembled to the stub axle without the use of tools. The cold method utilizes the special tools to press the components in place. The components should be heated to approximately 200–250 degrees F.

— CAUTION —
Do not use an open flame to pre-heat the components.

Alliance and Encore

NOTE: This procedure requires the use of special tools.

1. Raise and support the car on jackstands.
2. Remove the wheel.
3. Loosen the handbrake cable.
4. Remove the grease cap from the end of the axle shaft. Remove the shaft end nut and washer.
5. Remove the brake drum. It may be necessary to back off the brake shoe adjustment to free the drum. If the drum still won't slide off, it will be necessary to mount a combination slide hammer/hub puller and remove the drum.
6. From the drum, remove the bearing retaining clip and drive the bearing out with a length of 1.93 in. (49mm) diameter pipe.
7. Installation is the reverse of removal. Drive the new bearing in with a length of 2.0 in. (51mm) pipe. Torque the shaft end nut to 118 ft. lb. Adjust the handbrake. Pump the brake pedal several times to adjust the brakes.

ADJUSTMENT

LeCar

An end play of 0.004–0.002 in. (0.01–0.05mm) should exist between the brake drum and the stub axle. Rotate the stub axle nut in or out to obtain this required adjustment. Install the lockplate, nut and cotter pin. Add ⅓ oz. of grease in the dust cover and install.

Alliance and Encore

No adjustment is necessary.

18, 18i, Sportwagon and Fuego

NOTE: A special tool is needed for this procedure.

1. Raise and support the car on jackstands.
2. Remove the wheel.
3. While turning the drum by hand, tighten the axle shaft nut to 22 ft. lb.
4. Tap the drum lightly to make sure the bearing is seated.
5. Back off the shaft nut about ¼ turn.
6. Attach a dial indicator under one of the lug nuts and mount the pointer on the end of the axle shaft.
7. Push and pull on the drum and check the endplay on the indicator. Endplay should be 0–.001 in. Use the shaft end nut to obtain this figure.
8. Assemble all remaining parts.

Torsion Bars

REMOVAL & INSTALLATION

LeCar

A special tool, which can be made at home,

SUSPENSION
Import Models

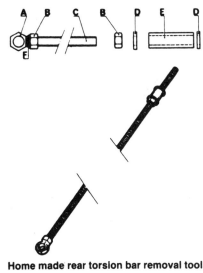

Home made rear torsion bar removal tool

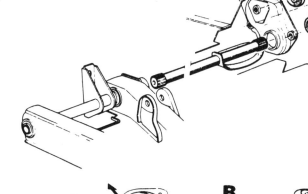

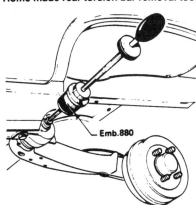

Use a slide hammer to remove the torsion bar on the Alliance and Encore

Installing the LeCar torsion bar

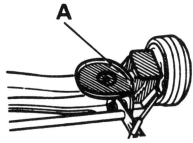

When installing the spring, make sure that the spring end is against the stop (A)

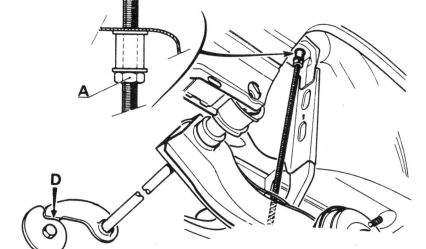

Tool installation adjustment

will aid in the removal of the torsion bar.
1. Raise and support the car on jackstands.
2. Loosen the cam locking nut and zero the cam.
3. Remove the shock absorber.
4. Assemble the torsion bar tool in place of the shock absorber.
5. Tighten nut A until the adjusting lever lifts off of cam D.
6. Remove the torsion bar.
7. Move the adjusting lever until it touches the cam.

8. Apply chassis lube to the torsion bar splines and insert it into the lever and arm.

NOTE: Adjust nut A on the torsion bar remover 23¼ in. for the right arm, or 23⅝ in. for the left arm, prior to torsion bar installation.

9. Tighten the cam locknut. Remove the tool.
10. Install the shock absorber.
11. Lower the vehicle and check the ride height. Ride height is the difference between the distance measured from the wheel center to the ground, and the distance measured from the frame to the ground. The difference should be ¹⁄₁₆–¹³⁄₁₆ in. Turn the cam adjuster to obtain this height.

Alliance and Encore

1. Raise and support the car on jackstands.
2. Remove the sway bar.
3. Remove the shock absorber.
4. Remove the torsion bar with a slide hammer.

2-253

SUSPENSION
Import Models

5. For installation, make a tool identical to that described in the LeCar procedure, above. Install this tool in place of the shock absorber. Turn the nut to obtain a tool length of 24.04 in. for either side.

6. Coat the torsion bar splines with chassis lube and install it. Search around until the easiest point of entry, at each end, is found.

7. Remove the tool. Install the shock and sway bar.

8. Check and adjust the ride height, if necessary. To adjust the ride height, see the LeCar section above. Ride height should be 2.048 in.

Springs

REMOVAL & INSTALLATION

18, 18i, Sportwagon and Fuego

1. Raise and support the car on jackstands placed under the frame.

2. Disconnect the shock absorber lower end.

3. Compress the shock as far as it will go.

4. Remove the brake hose clips from the rear axle.

5. Jack up the rear of the car to allow the axle to drop and the spring to be pulled out and down.

6. Installation is the reverse of removal. Make certain that the lower open coil end is against the stop on the spring pad.

Shock Absorbers

REMOVAL & INSTALLATION

LeCar, Alliance and Encore

1. Working through the trunk, remove the upper nut.

2. Raise and support the car on jackstands.

3. Remove the lower nut and remove the shock absorber.

4. Installation is the reverse of removal. It's best to attach the upper end first. Torque the upper nut to 18 ft. lb.; the bottom nut to 60 ft. lb.

18, 18i, Sportwagon and Fuego

1. Open the trunk and remove the gas tank protection shield.

2. Remove the upper fastener (nut).

3. Raise the vehicle and remove the wheel(s).

4. Remove the lower fastening nut and compress the shock absorber by hand.

5. Remove the retaining clips of the brake hoses from the rear axle.

6. Push the axle assembly downward and remove the shock absorber with the spring. Remove the spacers and the rubber bushings.

Before installing a new shock absorber, pump it up and down manually several times while in an upright position.

1. Place the rubber insulators on the shock absorbers as required on both the top and bottom.

2. Push down on the rear axle assembly and engage the shock absorber with the spring, positioning the spring so that its lower end is placed in the stop of the lower cup.

3. Attach the upper half of the shock absorber with the retaining nut. Be sure to replace the bushings and cups as they were removed.

4. Extend the shock absorber and install its lower retaining nut. Again, be sure to replace the bushings and cups as they were removed.

5. Install the retaining clips of the brake hose to the rear axle.

6. Install the wheels and lower the vehicle to the ground.

FRONT SUSPENSION SUBARU

MacPherson Strut Assembly

REMOVAL & INSTALLATION

For all shock absorber and spring removal and installation procedures, and any other strut overhaul procedures, please refer to "Strut Overhaul" in the Unit Repair section.

1. Raise and support the vehicle. Remove the battery cable from the negative terminal of the battery.

2. Remove the hub caps, loosen the lug nuts, jack up the vehicle until the tire clears the ground and remove the lug nuts and the wheel/tire assembly. Place the jackstands under the vehicle and remove the jack. Perform this operation on the opposite side if the suspension is to be removed from both sides of the vehicle.

3. Remove the handbrake cable bracket and the handbrake cable hanger from the transverse link and the tie-rod end. Remove the handbrake cable end.

4. Remove the axle nut, lockplate, washer, and center piece and remove the front brake drums by using a puller.

5. Disconnect the brake hoses from the brake fluid pipes.

6. With front disc brakes, remove the handbrake cable end from the caliper lever. Remove the outer cable clip from the cable-end support bracket at the caliper. Remove the handbrake cable bracket from the housing mount by loosening the nuts.

7. Drive out the spring pins of the double offset joint by using a drift pin and a hammer. The double offset side of the axle is the side closest to the transaxle.

8. Remove the lower control arm by loosening the self-locking nut which holds it to the inner pivot shaft of the crossmember. Loosen and remove the nuts which clamp the control arm to the stabilizer. Remove the stabilizer rearward from the crossmember by using a lever and pulling the control arm out from the end of the stabilizer.

9. Remove the cotter pin from the castle nut and remove the nuts and ball stud from the knuckle arm of the tie-rod end ball joint housing. Take care not to bend the housing.

10. Disconnect the leading rod from the rear crossmember by removing the self-locking nut, washers, plates, bushings and pipe. Disconnect the stabilizer by removing the bolt at the bracket connecting one end of the stabilizer to the leading rod, and then removing the nuts fixing the bracket to the rear crossmember.

11. Remove the nuts which hold the strut mount to the body (suspension assembly upper mounting nut-top of the shock absorber tower).

12. Pull the double offset joint out of the driveshaft and then remove the suspension assembly from the body.

Install the suspension assembly in the reverse order of removal. Bleed the brakes after installation.

Ball Joints

REMOVAL & INSTALLATION

1980 and Later Models

1. Raise the front of the car and install axle stands. Remove the wheels.

2. Remove the cotter pin and remove the castellated nut. Remove the bolt on the steering knuckle, and remove the ball joint.

3. Inspect the joint, if you're considering reusing it, for damage to the boot that retains grease or stress cracks.

4. To install, insert the ball stud, ungreased, into the steering knuckle, install the bolt, and torque it to 22–29 ft. lb.

5. Connect the joint to the transverse link and install the castellated nut, torquing

SUSPENSION
Import Models — SECTION 2

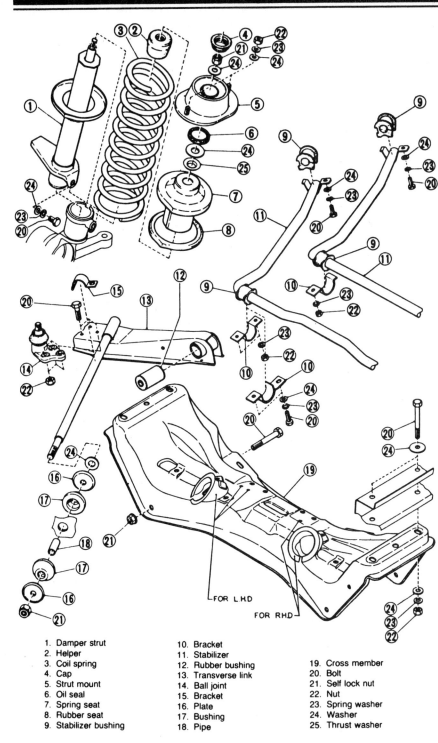

1. Damper strut
2. Helper
3. Coil spring
4. Cap
5. Strut mount
6. Oil seal
7. Spring seat
8. Rubber seat
9. Stabilizer bushing
10. Bracket
11. Stabilizer
12. Rubber bushing
13. Transverse link
14. Ball joint
15. Bracket
16. Plate
17. Bushing
18. Pipe
19. Cross member
20. Bolt
21. Self lock nut
22. Nut
23. Spring washer
24. Washer
25. Thrust washer

Exploded view of the typical front suspension

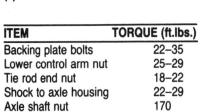

To remove the ball joints, remove the cotter ping (2), castellated nut (1) and bolt (3)

ITEM	TORQUE (ft.lbs.)
Backing plate bolts	22–35
Lower control arm nut	25–29
Tie rod end nut	18–22
Shock to axle housing	22–29
Axle shaft nut	170

2. Jack up the car and support it with jackstands. Block the rear wheels.
3. Remove the lug nuts and the wheel.
4. Remove the parking brake cable clamp from the control arm by unfastening its nut.
5. Unfasten the self-locking nut which attaches the control arm to the crossmember. Be sure to note the installation sequence of the washers.
6. Unfasten the self-locking nut which secures the stabilizer bar to the control arm. Again, note the installation sequence of the washers.
7. Unbolt the leading rod from the rear crossmember.
8. Pry the control arm off the crossmember.
9. Push the control arm forward and detach it from the end of the stabilizer bar.
10. Remove the cotter pin from the castellated nut. Unfasten the nut and remove the ball joint from the axle housing with a puller.
11. Remove the control arm from under the car.
12. Installation is the reverse of removal. Do not grease the upper ball joint stud which fits into the axle housing. Tighten the castellated nut to 30–40 ft. lbs. Use new self-locking nuts on the crossmember and stabilizer bar mounts. Tighten the new self-locking nuts to 73–87 ft. lbs. with the vehicle resting on the wheels.

Front Wheel Bearings

ADJUSTMENT

There is no bearing preload adjustment, other than tightening the axle shaft nut to the proper specifications. See the chart below.

to 29 ft. lb. Then, torque the nut further, just until the castellations are aligned with the hole in the end of the ballstud. Install a new cotter pin and bend it around the nut.

Lower Control Arm

REMOVAL & INSTALLATION

1. Remove the wheel cover and loosen the lug nuts.

2-255

SECTION 2 SUSPENSION
Import Models

Front End Alignment

CASTER AND CAMBER

Caster and camber are not adjustable on these models. If either of these specifications is not within the factory recommended range, this would indicate bent or damaged parts that must be replaced.

TOE-IN

Toe-in is adjusted by loosening the locknuts on the tie-rods, and turning the tie-rods.

NOTE: Before performing the toe-in adjustment, be sure that the steering gear is centered by aligning the marks on it, and that the wheels are straightahead.

Tighten the locknuts after the toe-in adjustment is completed.

REAR SUSPENSION SUBARU

Semi-trailing arms mounted to torque tubes, which act on an internal torsion bar are used. Shock absorbers are mounted to the trailing arm, close to the stub axle.

Torsion Bars

REMOVAL & INSTALLATION

1. Raise the vehicle and support it securely. Remove the rear wheel. Support the brake drum in a position eliminating load from the torsion bar.
2. Remove the lockbolt for the outer bush and the three outer arm to inner arm connecting bolts. Pull the outer arm and torsion bar out of the crossmember. The torsion bar may not be removed from the outer arm.

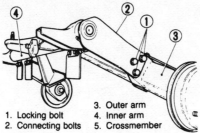

1. Locking bolt
2. Connecting bolts
3. Outer arm
4. Inner arm
5. Crossmember

Torsion bar removal—1980 and later models

3. To install, reverse the removal procedure, keeping the following points in mind:
   a. The torsion bar's splines must be aligned with those in the outer arm and crossmember so that the outer arm lines up with the inner arm as it did during removal, or ride height will be effected.
   b. If you are removing both torsion bars, make sure the markings ("R" or "L") correspond with the side of the car you're installing the bar on.
   c. Use the following torques, in ft. lb.:
   inner arm to outer arm—87–101 ft. lb.
   outer bush lockbolt—23–29 ft. lb.

Shock Absorbers

REMOVAL & INSTALLATION

1. Remove the wheel cover and loosen the lug nuts. Raise the rear of the car and support it with jackstands, after setting the parking brake and blocking the front wheels.

1 Air tank CP
2 Pressure switch
3 Pipe
4 Solenoid valve ASSY
5 Insulator
6 Compressor bracket
7 Compressor
8 Drier
9 Pipe
10 Clip
11 Pipe kit (F. sus. RH)
12 Cap
13 Bush
14 O-ring
15 Strut mount cap
16 Solenoid valve ASSY (RH)
17 Air joint ASSY
18 Pipe
19 Holder
20 Solenoid valve ASSY (LH)
21 Relay ASSY
22 Pipe kit (F. sus. LH)
23 Protector clip
24 Grommet
25 Clip
26 Clip
27 Protector (RH)
28 Protector (LH)
29 Front air suspension ASSY (LH or RH)
30 Strut mount
31 Rear air suspension ASSY (LH or RH)
32 Upper rubber plate
33 Upper rubber
34 Bracket CP (LH or RH)
35 Lower rubber
36 Collar
37 Upper plate
38 Connector
39 O-ring
40 Solenoid valve ASSY

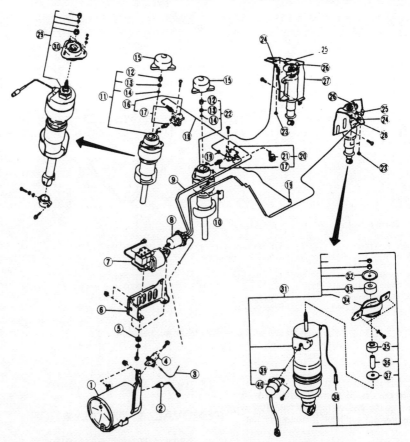

Rear air suspension—exploded view

SUSPENSION
Import Models

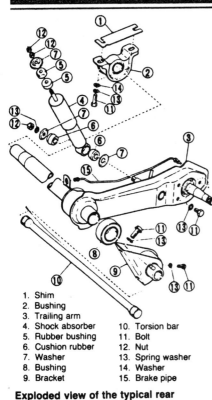

1. Shim
2. Bushing
3. Trailing arm
4. Shock absorber
5. Rubber bushing
6. Cushion rubber
7. Washer
8. Bushing
9. Bracket
10. Torsion bar
11. Bolt
12. Nut
13. Spring washer
14. Washer
15. Brake pipe

Exploded view of the typical rear suspension

2. Remove the lug nuts and the rear wheels.

3. Loosen the two upper shock absorber mounting nuts. Remove the washer and the bushing, being sure to note their correct assembly sequence for installation.

4. Unfasten the nut on the trailing arm pin (nut and bolt on later models) and remove the shock absorber. Note the installing positions of the washers.

Installation is the reverse of removal. Do not fully tighten the upper mounting nuts until the lower shock nut has been installed with the washer and the pin shoulder contacting each other. Tighten the upper nuts to 22–32 ft. lbs. Adjust the ride height bolts, if so equipped.

REAR AIR SUSPENSION ASSEMBLY

Removal & Installation

1. Remove the rear apron protector.
2. Remove the rear solenoid valve from the air suspension assembly.
3. If replacing the solenoid valve, disconnect the air line from the solenoid valve.
4. Remove the vehicle height sensor harness through the access hole located in the body. Disconnect the coupler.
5. Raise the vehicle and support safely. Remove the rear wheel and tire assemblies.
6. Remove the shock absorber mounting nuts. Remove the washer and the bushing.
7. Unfasten the nut on the trailing arm pin. Remove the shock absorber.
8. Installation is the reverse of the removal procedure.

Ride Height

Vehicle height can be adjusted by turning the outer and inner end of the torsion bar by the same number of serration teeth in the opposite direction to the arrow mark on the outer end surface of the torsion bar.

Turning the torsion bar in the direction of the arrow lowers the vehicle height, and changes the height 0.20 in. per tooth shifted.

The torsion bar must be removed from the inner and outer brackets to make the adjustment. 4-wheel drive vehicles have a unique adjusting device that will alter the ride height (in addition to adjusting the torsion bars). See the following procedure.

ADJUSTMENT OF REAR ROAD CLEARANCE

1980 and Later 4 Wheel Drive Vehicles

1. Measure the height of the vehicle from the lowest point of the rear axle crossmember to the ground.
2. To adjust the rear height, remove the access cover from the service hole in the vehicle's floor above the rear axle. Turn the adjusting bolt clockwise to increase the height, counterclockwise to lower it.

FRONT SUSPENSION TOYOTA

MacPherson Struts

REMOVAL & INSTALLATION

All except Van

1. Remove the hubcap and loosen the lug nuts.
2. Raise the front of the car and support it on the chassis jacking plates provided, with jack stands.

— CAUTION —
Do not support the weight of the car on the suspension arm; the arm will deform under its weight.

3. Unfasten the lug nuts and remove the wheel.
4. Detach the front brake line from its clamp.
5. Remove the caliper and wire it out of the way.
6. Unfasten the three nuts which secure the upper shock absorber mounting plate to the top of the wheel arch.

7. Remove the two bolts which attach the shock absorber lower end to the steering knuckle lower arm.

NOTE: Press down on the suspension lower arm, in order to remove the strut assembly. This must be done to clear the collars on the steering knuckle arm bolt holes when removing the shock/spring assembly.

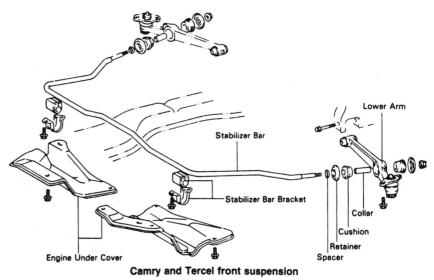

Camry and Tercel front suspension

2-257

SECTION 2: SUSPENSION
Import Models

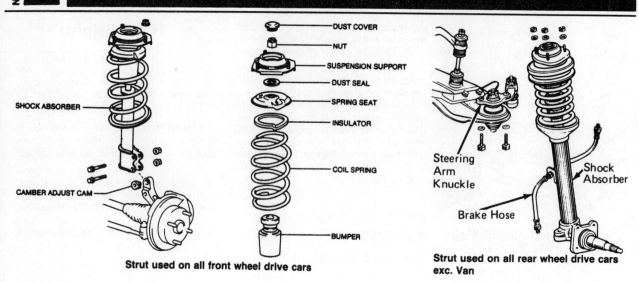

Strut used on all front wheel drive cars

Strut used on all rear wheel drive cars exc. Van

Installation is performed in the reverse order of removal. Be sure to note the following, however:

1. Align the hole in the upper suspension support with the shock absorber piston rod end, so that they fit properly.
2. Always use a *new* nut and nylon washer on the shock absorber piston rod end when securing it to the upper suspension support. Torque the nut to 29–40 ft. lbs.

――― CAUTION ―――
Do not use an impact wrench to tighten the nut.

3. Coat the suspension support bearing with multipurpose grease prior to installation. Pack the space in the upper support with multipurpose grease, also, after installation.
4. Tighten the suspension support-to-wheel arch bolts to the following specification:
Corolla—11–16 ft. lb.
Celica RWD—14–23 ft. lb.
Celica FWD—45–49 ft. lbs.
Camry—45–49 ft. lbs.
Supra—25–29 ft. lbs.
Cressida—25–29 ft. lbs.
Tercel—11–15 ft. lbs.
MR2—21–25 ft. lbs.
5. Tighten the shock absorber-to-steering knuckle arm bolts to the following specifications:
Corolla (RWD): 50–65 ft. lbs.
Corolla (FWD): Gas 105 ft. lb.
　　　　　　　　Diesel 152 ft. lb.
Camry and Celica FWD: 152 ft. lb.
MR2 and Tercel: 105 ft. lb.
Celica RWD and Supra: 72 ft. lb.
Cressida: 80 ft. lbs.
All others: 65 ft. lb.
6. Adjust the front wheel bearing preload as outlined below.
7. Bleed the brake system.

OVERHAUL

For all spring and shock absorber removal and installation procedures, and all strut overhaul procedures, please refer to "Strut Overhaul"

Torsion Bars

REMOVAL & INSTALLATION

Van

1. Raise and support the front end on

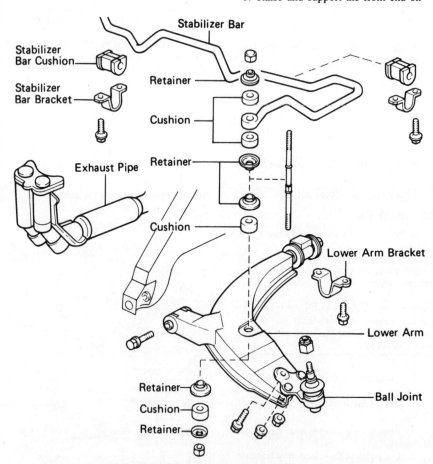

Front suspension components—Corolla FWD: 1986 and later Celica similar

SUSPENSION
Import Models

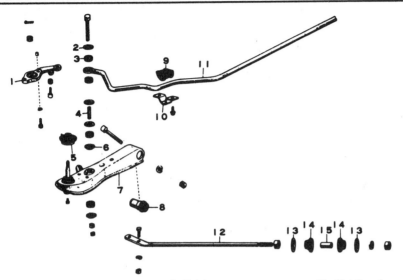

1. Steering knuckle arm
2. Retainer
3. Cushion
4. Collar
5. Dust cover
6. Retainer
7. Lower control arm
8. Bushing
9. Bushing
10. Bracket
11. Stabilizer bar
12. Strut
13. Retainer
14. Cushion
15. Collar

Components of a typical MacPherson strut front suspension

jackstands under the frame.

2. Paint match marks on the torsion bar, anchor arm and torque arm.

3. Remove the locknut and measure the threaded end "A" as shown. Use this figure for an installation reference.

4. Loosen the adjusting nut and remove the anchor arm and torsion bar.

5. Installation is the reverse of removal. Apply a light coating of molybdenum disulphide lithium grease to the splined end of the torsion bar. Align all matchmarks. Tighten the adjusting nut so that the exact length of thread appears as before. The proper length should be 2.76 in.

Lower Control Arm/ Ball Joints

INSPECTION

Corolla (RWD), Cressida, Corona, 1980–81 Celica/Supra and Starlet

Jack up the lower suspension arm (except Corolla, Celica and Cressida). Check the front wheel play. Replace the lower ball joint if the play at the wheel rim exceeds 0.1 in. vertical motion or 0.25 in. horizontal motion. Be sure that the dust covers are not torn and that they are securely glued to the ball joints.

— CAUTION —
Do not jack up the control arm on Corolla, Celica or Cressida models; damage to the arm will result.

Measuring the threaded end of the torsion bar—Van

Tercel, Camry, Corolla (FWD) and 1982 and Later Celica Supra

1. Jack up the vehicle and place wooden blocks under the front wheels. The block height should be 7.09–7.87 inches.
2. Use jack stands for additional safety.
3. Make sure the front wheels are in a straight forward position.
4. Chock the wheels.
5. Lower the jack until there is approximately half a load on the front springs.
6. Move the lower control arm up and down to check that there is no ball joint play.

Van

1. Raise and support the front end with jackstands under the frame.
2. Have someone apply the brakes while you move the lower arm up and down.
3. Vertical play should not exceed 0.09 in.

REMOVAL & INSTALLATION

NOTE: On models equipped with both upper and lower ball joints—if both ball joints are to be removed, always remove the lower and then the upper ball joint.

Corolla (RWD), 1980–81 Celica/ Supra/Cressida and Starlet

The ball joint and control arm cannot be separated from each other. If one fails, then both must be replaced as an assembly, in the following manner:

1. Perform Steps 1–7 of the first "Front Spring Removal and Installation" procedure. Skip Step 6.
2. Remove the stabilizer bar securing bolts.

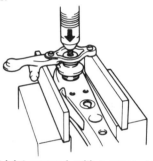

Ball joint removal, with a press, on the Celica, RWD Corolla, Cressida and Starlet

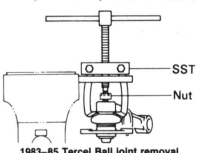

1983–85 Tercel Ball joint removal

3. Unfasten the torque strut mounting bolts.
4. Remove the control arm mounting bolt and detach the arm from the front suspension member.
5. Remove the steering knuckle arm from the control arm with a ball joint puller.

Inspect the suspension components, which were removed for wear or damage. Replace any parts, as required.

Installation is the reverse of removal. Note the following, however:

1. When installing the control arm on the suspension member, tighten the bolts partially at first.
2. Complete the assembly procedure and lower the car to the ground.
3. Bounce the front of the car several times. Allow the suspension to settle, then tighten the lower control arm bolts to 51–65 ft. lbs.

SECTION 2
SUSPENSION
Import Models

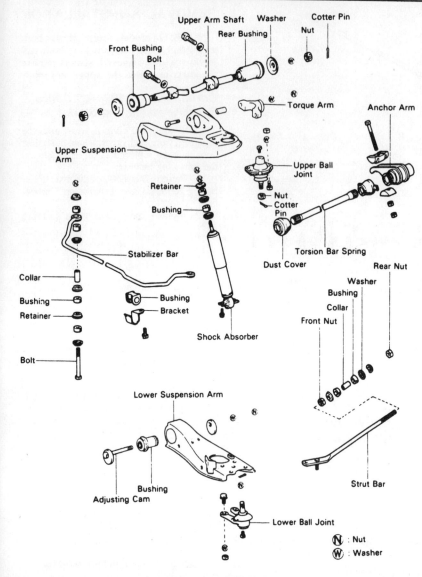

Van front suspension

CAUTION

Use only the bolt which was designed to fit the lower control arm. If a replacement is necessary, see an authorized dealer for the proper part.

4. Remember to lubricate the ball joint. Check front-end alignment.

1980–82 Tercel and 1982–85 Celica

1. Jack up your vehicle and support it with jack stands.

CAUTION

Do not jack up your car on the lower control arms.

2. Remove the front wheels.
3. Remove the tie-rod end.
4. Remove the stabilizer bar end.
5. Remove the strut bar end.
6. Place a jack under the lower control arm for support.
7. Remove the bolt from the bottom of the steering knuckle.
8. Remove the bolt from the lower control arm.
9. Remove the control arm.

NOTE: The lower ball joint cannot be separated from the lower control arm. It must be replaced as a complete unit.

10. The following torques are required: Bottom steering knuckle nut 40–52 ft. lbs.; stabilizer bar 11–15 ft. lbs.; tie rod end 37–50 ft. lbs.; strut bar 29–39 ft. lbs.; lower control arm 51–65 ft. lbs.

1986 and Later Celica (FWD)

1. Raise the front of the vehicle and support it on jackstands. Remove the wheel.
2. Remove the bolt and two nuts and disconnect the lower control arm from the steering knuckle.
3. Remove the nut and disconnect the stabilizer bar from the control arm.
4. On all but the left-side control arm on models with automatic transmissions, remove the control arm front set nut and washer. Remove the rear bracket bolts and then remove the arm.
5. On the left-side arm on models with automatic transmissions, remove the control arm front set nut and washer. Remove the four bolts and two nuts that attach the lower suspension crossmember to the frame and remove the crossmember. Remove the bolt and nut and lift out the lower arm with the lower arm shaft.
6. On all but the left-side control arm on models with automatic transmissions, install the lower control arm shaft washer with the tapered side toward the body. Install the lower arm with the bracket and then temporarily install the washer and nut to the lower arm shaft and bracket bolts.
7. On the left-side arm on models with automatic transmission, position the washer on the lower arm shaft and then install them to the lower arm. Temporarily install the washer and nut to the shaft with the tapered side toward the body. Install the lower arm with the shaft to the body and temporarily install the rear brackets. Install the bolt and nut to the lower arm shaft and tighten them to 154 ft. lbs. (208 Nm). Install the crossmember to the body and tighten the four bolts to 154 ft. lbs. (208 Nm). Tighten the two nuts to 29 ft. lbs. (39 Nm).
8. Connect the lower arm to the steering knuckle and tighten the bolt and two nuts to 94 ft. lbs (127 Nm).
9. Connect the stabilizer bar to the control arm and tighten the nut to 26 ft lbs.
10. Install the wheel, lower the vehicle and bounce it several times to set the suspension.
11. Tighten the front set nut to 156 ft. lbs. (212 Nm). Tighten the rear bracket bolts to 72 ft. lbs. (98 Nm).

MR2

1. Raise the front of the vehicle and support it on jackstands. Remove the wheel.
2. Remove the cotter pin and castle nut and then press the lower arm out of the ball joint.
3. Press the ball joint out of the steering knuckle.
4. Remove the two nuts and disconnect the strut bar from the control arm.
5. Remove the lower control arm-to-body bolt and remove the arm.
6. When installing the lower arm, position it in the strut bar and tighten the nuts finger-tight. Do the same thing with the arm-to-body bolt.

2-260

SUSPENSION
Import Models

7. Connect the control arm to the ball joint and tighten the castle nut to 58 ft. lbs. (78 Nm). Install a new cotter pin.

8. Tighten the strut bar-to-arm bolts to 83 ft. lbs. (113 Nm).

9. Install the tires, lower the car and bounce it several times to set the suspension.

10. Tighten the control arm-to-body bolt to 94 ft. lbs. (127 Nm) and check the wheel alignment.

1982 and Later Supra/Cressida

1. Raise the front of the vehicle and support it on jackstands. Remove the wheel.

2. Remove the two knuckle arm-to-strut bolts, pull down on the control arm and disconnect it and the knuckle arm from the strut.

3. Remove the cotter pin and nut and then press the tie rod off the knuckle arm.

4. Remove the nut attaching the stabilizer bar to the control arm and disconnect the bar.

5. Remove the two nuts and then disconnect the strut bar from the control arm.

6. Disconnect the control arm from the crossmember and remove it and the rack boot protector as an assembly.

7. Remove the cotter pin and nut and then press the knuckle arm off the control arm.

To install:

1. Press the knuckle arm into the control arm and then install the assembly into the crossmember.

2. Connect the stabilizer bar to the control arm and tighten the nut to 13 ft. lbs. (18 Nm).

3. Connect the stut bar to the control arm and tighten the nuts to 48 ft. lbs. (6 Nm).

4. Connect the knuckle arm to the strut housing and tighten the bolts to 72 ft. lbs. (98 Nm).

5. Install the wheel and lower the vehicle. Bounce the car several times to set the suspension and then tighten the control arm-to-body bolt to 80 ft. lbs. (108 Nm).

6. Check the front wheel alignment.

Camry, Corolla (FWD) and 1983 and Later Tercel

1. Raise the front of the vehicle and support it with jackstands. Remove the wheel.

2. Remove the two bolts attaching the ball joint to the steering knuckle.

3. Remove the stabilizer bar nut, retainer and cushion.

4. Jack up the opposite wheel until the body of the car just lifts off the jackstand.

5. Loosen the lower control arm mounting bolt, wiggle the arm back and forth and then remove the bolt. Disconnect the lower control arm from the stabilizer bar.

NOTE: When removing the lower control arm (on the Tercel), be careful not to lose the caster adjustment spacer.

6. On the Tercel and Camry, carefully mount the lower control arm in a vise and then, using a ball joint removal tool, disconnect the ball joint from the arm.

7. Installation is in the reverse order of removal. Please note the following:

a. Tighten the ball joint-to-control

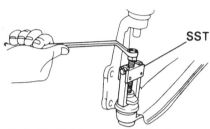

Typical ball joint separation without a press

arm nut to 51–65 ft. lbs. and use a new cotter pin (Tercel) and 67 ft. lbs. (Camry).

b. Tighten the steering knuckle-to-control arm bolts to 59 ft. lbs. on the Tercel; 47 ft. lbs. on the Corolla and 83 ft. lbs. on the Camry.

c. Tighten the stabilizer bar bolt to 13 ft. lbs. on the Corolla.

d. Before tightening the stabilizer bar nuts, on the Tercel and Camry, or the control arm bracket bolts on the Corolla. Mount the wheels and lower the car. Bounce the car several times to settle the suspension and then tighten the stabilizer bolts on the Tercel and Camry to 66–90 ft. lbs.

e. Tighten the arm-to-body bolts on the Tercel and Camry to 83 ft. lbs. On the Corolla, tighten the front arm bolts to 83 ft. lbs. and the rear bolts to 64 ft. lbs.

f. Check the front-end alignment.

Van

1. Raise and support the front end on jackstands under the frame.

2. Remove the hub and caliper.

3. Remove the steering knuckle dust cover.

4. Support the lower arm with a floor jack.

5. Remove the two cotter pins and nuts and disconnect the steering knuckle from the lower ball joint.

6. Disconnect the upper ball joint from the knuckle.

7. Using a ball joint removal tool, remove the ball joint from the arm.

8. Installation is the reverse of removal. Torque the ball joint nut to 50 ft. lb.

Corona

1. Remove the hubcap and loosen the lug nuts.

2. Jack up your vehicle and support it with jack stands.

3. Remove the lug nuts and the wheel.

4. Compress the coil spring by placing a jack underneath the control arm and raising it.

5. Remove the cotter pin and the castelated nut from the ball joint.

6. Use a ball joint puller to detach the lower ball joint from the steering knuckle.

7. Wire the steering knuckle out of the way.

8. Remove the bolt and remove the ball joint.

Installation is the reverse of removal. Tighten the stud nut 51–65 ft. lbs.

Upper Ball Joint

INSPECTION

Disconnect the ball joint from the steering knuckle and check free-play by hand. Replace the ball joint, if it is noticeably loose.

REMOVAL & INSTALLATION

NOTE: On models equipped with both upper and lower ball joints—if both are to be removed, always remove the lower one first.

Corona and Van

1. Remove the steering knuckle as detailed in "Lower Ball Joint Removal & Installation".

2. Suspend the steering knuckle with a wire.

3. Use an open-end wrench to remove the upper ball joint.

Installation is performed in the reverse order from removal. Note the following:

1. Install the upper ball joint dust cover with the escape valve toward the rear.

2. Use sealer on the dust cover before installing it.

3. Tighten the upper ball joint-to-steering knuckle bolt to 40–50 ft. lbs. on the Corona, and 22 ft. lb. on the Van.

Lower Control Arm

REMOVAL & INSTALLATION

Corolla (RWD), 1980–81 Celica/Supra/Cressida and Starlet

1. Raise and support the front end.

2. Remove the wheel.

3. Disconnect the steering knuckle from the control arm.

4. Disconnect the tie-rod, stabilizer bar and strut bar from the control arm.

5. Remove the control arm mounting bolts, and remove the arm.

6. Install in reverse of above. Tighten, but do not torque fasteners until car is on ground.

2-261

SECTION 2: SUSPENSION
Import Models

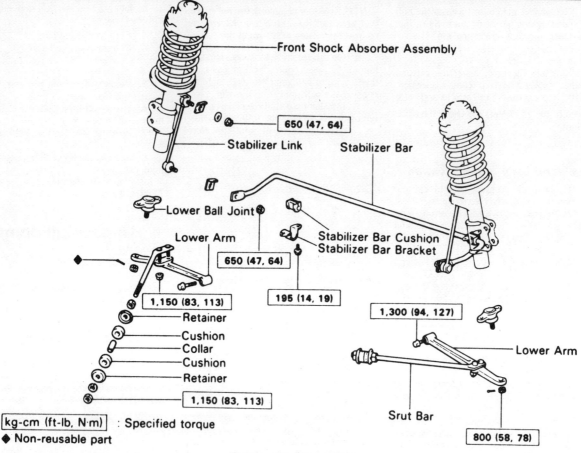

Front suspension—MR2

7. Lower car to ground, rock it from side-to-side several times and torque control arm mounting bolts to 51–65 ft. lbs., stabilizer bar to 16 ft. lbs., strut bar to 40 ft. lbs., and shock absorber to 65 ft. lbs.

Corona

1. Raise and support the vehicle.
2. Remove the front wheel.
3. Remove the shock absorber and disconnect the stabilizer from the lower arm.
4. Install a spring compressor and fully tighten it.
5. Place a jack under the lower arm seat.
6. Disconnect the lower ball joint from the knuckle and lower the jack.
7. Remove the ball joint from the arm, remove the cam plates and bolts and take off the arm.
8. Install in reverse of above. Tighten all fasteners, but do not torque them to specification until vehicle is on ground.
9. Lower vehicle and rock it from side-to-side several times.
10. With no load in vehicle, torque the lower arm mounting bolts to 94–130 ft. lbs.

Van

1. Raise and support the front end on jackstands under the frame.

2. Remove the shock absorber.
3. Disconnect the stabilizer bar from the arm.
4. Disconnect the strut bar from the arm.
5. Remove the ball joint nut and disconnect the ball joint from the knuckle. A ball joint separator is necessary.
6. Place a matchmark on the adjusting cam.
7. Remove the adjusting cam and nut and remove the lower arm.
8. Installation is the reverse of removal. Observe the following torques:

Bolts	ft. lbs.
Ball joint nut	50
Ball joint-to-knuckle	76
Shock absorber upper nut	19
Shock absorber lower bolts	13
Strut bar-to-arm	50
Adjusting cam nut	112

All Others

Please refer to "Lower Control Arm/Ball Joints."

Upper Control Arm

REMOVAL & INSTALLATION

Corona

1. Remove the upper arm mounting nuts from inside the engine compartment, but do not remove the bolts.
2. Raise the vehicle, support the lower arm and remove the wheel.
3. On vehicles equipped with a ball joint wear sensor, remove the wiring from the clamp on the arm.
4. Remove the upper ball joint.
5. Remove the control arm mounting bolts.
6. Pry out the arm with a pry bar.
7. Install in reverse of removal. Do not tighten fasteners until vehicle is on ground.
8. Lower vehicle and torque the control arm mounting bolts to 95–130 ft. lbs.

Van

1. Raise and support the front end with jackstands under the frame.
2. Remove the torsion bar.
3. Remove the air duct.
4. Remove the upper ball joint nut and disconnect the ball joint from the knuckle using a separator.

SUSPENSION
Import Models

5. Unbolt and remove the arm.
6. Installation is the reverse of removal. Observe the following torques:

Bolt	ft. lbs.
Upper arm front bolt	65
Upper arm rear bolt	112
Ball joint arm	22
Ball joint-to-knuckle	58

Front Wheel Bearings

NOTE: For information concerning rear wheel bearings/adjustment on 1982 and later Supra and 1983 and later Celica GTS and Cressida models, refer to the "Axle Shaft Removal and Installation" procedure for that model.

REMOVAL & INSTALLATION

Rear Wheel Drive

1. Remove the disc/hub assembly, as detailed above.
2. If either the disc or the entire hub assembly is to be replaced, unbolt the hub from the disc.

NOTE: If only the bearings are to be replaced, do not separate the disc and hub.

3. Using a brass rod as a drift, tap the inner bearings cone out. Remove the oil seal and the inner bearing.

NOTE: Throw the old oil seal away.

Measuring wheel bearing pre-load with a spring scale

4. Drive out the inner bearing cup.
5. Drive out the outer bearing cup.

Inspect the bearings and the hub for signs of wear or damage. Replace components, as necessary.

Installation is performed in the following order:
1. Install the inner bearing cup and then the outer bearing cup, by driving them into place.

— CAUTION —
Use care not to cock the bearing cups in the hub.

2. Pack the bearings, hub inner well and grease cap with multipurpose grease.
3. Install the inner bearing into the hub.
4. Carefully install a new oil seal with a soft drift.
5. Install the hub on the spindle. Be sure to install all of the washers and nuts which were removed.
6. Adjust the bearing preload.
7. Install the caliper assembly.

PRELOAD ADJUSTMENT

1. With the front hub/disc assembly installed, tighten the castellated nut to the torque figure specified.
2. Rotate the disc back and forth, two or three times, to allow the bearing to seat properly.
3. Loosen the castellated nut until it is only finger-tight.
4. Tighten the nut firmly, using a box wrench.
5. Measure the bearing preload with a spring scale attached to a wheel mounting stud. Check it against the specifications.
6. Install the cotter pin.

NOTE: If the hole does not align with the nut (or cap) holes, tighten the nut slightly until it does.

7. Finish installing the brake components and the wheel.

PRELOAD SPECIFICATIONS

Model/Year	Initial Torque Setting (ft. lbs.)	Preload (oz.)
Tercel '80–'83	22	13–30
Corolla①	19–23	11–25
Celica	19–26	11–25
Corona	19–26	12–31
Supra①	19–23	11–24
Cressida①	22	37–56
Starlet		
'81–'82	22	1–1.5
'83–'84	22	0.8–1.9

① Except models w/IRS

Front Wheel Drive

Please refer to "Front Axle Hub And Bearings."

Front Wheel Alignment

ADJUSTMENT

Front-end alignment measurements require the use of special equipment. Before measuring alignment or attempting to adjust it, always check the following points:
1. Be sure that the tires are properly inflated.
2. See that the wheels are properly balanced.
3. Check the ball joints to determine worn or loose.
4. Check front wheel bearing adjustment.
5. Be sure that the car is on a level surface.
6. Check all suspension parts for tightness.

CASTER

Corolla (RWD), Cressida (1980–84), Celica (RWD), Supra and 1980–82 Tercel

Caster is the tilt of the front steering axis either forward or backward away from the front of the vehicle.

If the caster is found to be out of tolerance with the specifications, it may be adjusted by turning the nuts on the rear end of the strut bar (where it attaches to the body) on all models but the Starlet and the 1983 and later Tercel. The caster is decreased by lengthening the strut bar and increased by shortening it. One turn of the adjusting nut is equal to 8' of tilt on the Corolla (RWD), 9' on the Celica/Supra and Cressida and 7' on the 1980–82 Tercel. 1' is 1/60 of a degree.

Starlet and 1983 and Later Tercel

Caster on the Starlet and the 1983 and later Tercel is adjusted by changing the number of spacers on the stabilizer bar. One space will change the caster 24' on the Starlet and 13' on the 1983 and later Tercel. One minute (1') is equal to 1/60 of a degree.

NOTE: If the caster still cannot be adjusted within the limits, inspect or replace any damaged or worn suspension parts.

Camry and Cressida (1985 and Later)

Increase or decrease the number of spacers on the stabilizer bar. Each spacer changes caster by 30'.

Van

Caster is changed by turning the adjusting cam or strut bar nut. Each graduation of the cam gives 12' of change; each turn of the nut gives 25' change.

MR2

Caster is changed by turning the adjusting nut on the strut bar. Each revolution of the nut changes the caster by 18'.

2-263

SUSPENSION
Import Models

Corolla (FWD) and Celica (FWD)
Caster is not adjustable.

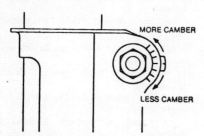

Camber adjusting bolt on the 1983 and later Tercel, front wheel drive Corolla and Camry

CAMBER
Except Corona

Camber is the slope of the front wheels from the vertical when viewed from the front of the vehicle. When the wheels tilt outward at the top, the camber is positive (+). When the wheels tilt inward at the top, the camber is negative (−). The amount of positive and negative camber is measured in degrees from the vertical and the measurement is called camber angle. Camber is preset at the factory, therefore it it is not adjustable on any model but the 1983 and later Tercel, Camry, Van, MR2, FWD Celica and FWD Corolla. If the camber angle is out of tolerance, inspect or replace worn or damaged suspension parts.

Camber on these models is adjustable by means of a camber adjustment bolt on the lower strut mounting bracket. Loosen the shock absorber set nut and then turn the adjusting bolt until the camber is within specifications. Camber will change about 20′ (MR2: 18′) for each gradation on the cam. One minute (1′) is equal to 1/60 of a degree.

CASTER AND CAMBER
Corona

Caster and camber angles are measured in the same way and with the same equipment as all the other models above.

However, the method of adjustment is different:
 1. Measure the camber and adjust it with the *rear* adjusting cam.
 2. Measure the caster and adjust it with the *front* adjusting cam.
 3. Check the caster and camber again.
 4. Tighten the lower control arm mounting bolts to 94–132 ft. lbs.

NOTE: There should be no more than six graduations difference between the front and rear cams; inspect for damaged suspension parts if there is.

Toe

Toe is the amount, measured in a fraction of an inch, that the front wheels are closer together at one end than the other. Toe-in means that the front wheels are closer together at the front of the tire than at the rear; toe-out means that the rear of the tires are closer together than the front.

The wheels must be dead straight ahead. The car must have a full tank of gas, all fluids must be at their proper levels, all other suspension and steering adjustments

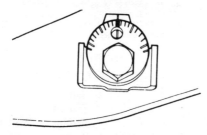

Van caster and camber adjustment cam

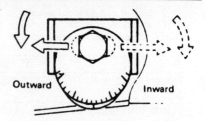

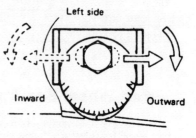

Corona front end alignment adjusting cams

must be correct and the tires must be properly inflated to their cold specification.
 1. Toe can be determined by measuring the distance between the centers of the tire treads, at the front of the tire and the rear. If the tread pattern of your car's tires makes this impossible, you can measure between the edges of the wheel rims, but be sure to move the car and measure in a few places to avoid errors caused by bent rims or wheel run-out.
 2. If the measurement is not within specifications, loosen the four retaining clamp locknuts on the adjustable tie-rods.
 3. Turn the left and right tie rods EQUAL amounts until the measurements are within specifications.
 4. Tighten the lock bolts and then re-check the measurements. Check to see that the steering wheel is still in the proper position. If not, remove it and reposition it.

REAR SUSPENSION TOYOTA

Springs
REMOVAL & INSTALLATION
Leaf Springs

1. Loosen the rear wheel lug nuts.
2. Raise the rear of the vehicle. Support the frame and rear axle housing with stands.
3. Remove the lug nuts and the wheel.
4. Remove the cotter pin, nut, and washer from the lower end of the shock absorber.
5. Detach the shock absorber from the spring seat pivot pin.
6. Remove the parking brake cable clamp.

NOTE: Remove the parking brake equalizer, if necessary.

7. Unfasten the U-bolt nuts and remove the spring seat assemblies.
8. Adjust the height of the rear axle housing so that the weight of the rear axle is removed from the rear springs.
9. Unfasten the spring shackle retaining nuts. Withdraw the spring shackle inner plate. Carefully pry out the spring shackle with a bar.
10. Remove the spring bracket pin from the front end of the spring hanger and remove the rubber bushings.
11. Remove the spring.

CAUTION
Use care not to damage the hydraulic brake line or the parking brake cable.

Installation is performed in the following order:
1. Install the rubber bushings in the eye of the spring.
2. Align the eye of the spring with the spring hanger bracket and drive the pin through the bracket holes and rubber bushings.

SUSPENSION
Import Models

NOTE: Use soapy water as lubricant, if necessary, to aid in pin installation. Never use oil or grease.

3. Finger-tighten the spring hanger nuts and/or bolts.
4. Install the rubber bushings in the spring eye at the opposite end of the spring.
5. Raise the free end of the spring. Install the spring shackle through the bushings and the bracket.
6. Install the shackle inner plate and finger-tighten the retaining nuts.
7. Center the bolt head in the hole which is provided in the spring seat on the axle housing.
8. Fit the U-bolts over the axle housing. Install the lower spring seat.
9. Tighten the U-bolt nuts.

NOTE: Some models have two sets of nuts, while others have a nut and lockwasher.

10. Install the parking brake cable clamp. Install the equalizer, if it was removed.
11. On passenger cars:
    a. Install the shock absorber end at the spring seat. Tighten the nuts.
    b. Install the wheel and lug nuts. Lower the car to the ground.
    c. Bounce the car several times.
    d. Tighten the spring bracket pins and shackles.

Coil Springs

1. Loosen the rear wheel lug nuts.
2. Jack up the rear axle housing and support the frame with jack stands. Leave the jack in place under the rear axle housing.
3. Remove the lug nuts and wheel.
4. If so equipped, disconnect the rear stabilizer bar from the axle housing (or suspension arm, on Supra (1982 and later), Celica GTS (1983–85) and Cressida (1983–85) models.

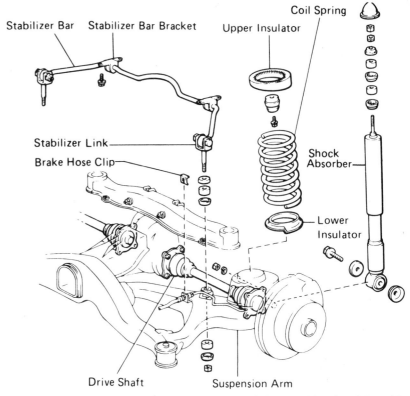

Rear suspension—Supra (1982 and later), Celica GTS (1983 and later) and Cressida (1985 and later)

5. Unfasten the lower shock absorber end. On the Van, Corolla RWD and Tercel 4×4, disconnect the lateral control rod from the axle.

NOTE: On Supra (1982 and later), Cressida (1983 and later) and Celica GTS (1983 and later) models, remove the rear halfshafts.

6. Slowly lower the jack under the rear axle housing until the axle is at the bottom of its travel.
7. Withdraw the coil spring, complete with its insulator.
8. Inspect the coil spring and insulator for wear, cracks, or weakness; replace either or both, as necessary.
9. Installation is performed in the reverse order of removal.

Rear Shock Absorbers

REMOVAL & INSTALLATION

All Exc. Supra (1982 and Later), Cressida (1983 and later) and Celica GTS (1983–85)

1. Raise the rear of the car and support it with jackstands. Position an hydraulic jack under the rear axle.
2. Unfasten the upper shock absorber retaining nuts.

NOTE: Always remove and install the shock absorbers one at a time. Do not allow the rear axle to hang in place as this may cause undue damage.

3. Remove the lower shock retaining nut where it attaches to the rear axle housing.
4. Remove the shock absorber.
5. Inspect the shock for wear, leaks or other signs of damage.

1. Nut
2. Washer
3. Lateral control rod
4. Bushing
5. Bolt
6. Bushing
7. Upper control arm
8. Lower control arm
9. Spring insulator
10. Spring bumper
11. Coil spring
12. Washer
13. Bushing
14. Washer
15. Nut
16. Nut
17. Washer
18. Bushing
19. Shock absorber
20. Bushing

Solid, coil-sprung rear suspension—typical of all models that use rear coil springs (except those with IRS)

SUSPENSION
Import Models

6. Installation is in the reverse order of removal. Please note the following:
—tighten the upper retaining nuts to 16–24 ft. lbs.
—tighten the lower retaining nuts to 22–32 ft.lbs.

Supra (1982 and Later), Cressida (1983 and later) and Celica GTS (1983–85)

1. Jack up the rear end of the car, keeping the pad of the hydraulic floor jack underneath the differential housing. Support the suspension control arms with safety stands.
2. Remove the brake hose clips. Disconnect the stabilizer bar end.
3. Disconnect the drive halfshaft at the CV joint on the wheel side.
4. With a jackstand underneath the suspension control arm, unbolt the shock absorber at its lower end. Using a screwdriver to keep the shaft from turning, remove the nut holding the shock absorber to its upper mounting. Remove the shock.
5. Installation is in the reverse order of removal. Torque the halfshaft nuts to 44–57 ft. lbs.; torque the upper shock mounting nut to 14–22 ft. lbs., and the lower shock mounting nut to 22–32 ft. lbs.

MacPherson Struts

REMOVAL & INSTALLATION

For all spring and shock absorber removal and installation procedures, and all strut overhaul procedures, please refer to "Strut Overhaul" in the Unit Repair section.

Tercel

1. Working inside the car, remove the shock absorber cover and package tray bracket.
2. Raise the rear of the vehicle and support it with jackstands. Remove the wheel.
3. Disconnect the brake line from the wheel cylinder. Disconnect the brake line from the flexible hose at the mounting bracket on the strut tube. Disconnect the flexible hose from the strut.
4. Loosen the nut holding the suspension support to the shock absorber.

CAUTION
Do not remove the nut.

5. Remove the bolts and nuts mounting the strut on the axle carrier and then disconnect the strut.
6. Remove the three upper strut mounting nuts and carefully remove the strut assembly.
7. Installation is in the reverse order of removal. Please note the following:
   a. Tighten the upper strut retaining nuts to 17 ft. lbs.

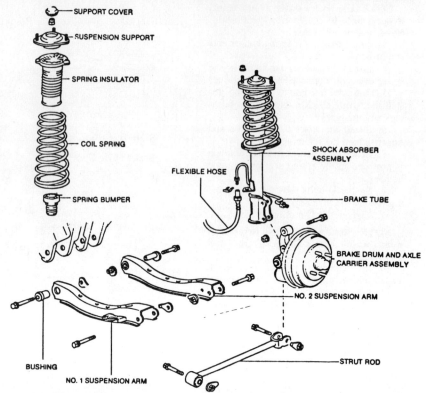

Rear suspension components—1983 and later Tercel (exc 4×4), Camry and front wheel drive Corolla

   b. Tighten the lower strut-to-axle carrier bolts to 105 ft. lbs.
   c. Tighten the nut holding the suspension support to the shock absorber to 36 ft. lbs.
   d. Bleed the brakes.

Corolla FWD and Camry

1. On the 4-door sedan, remove the package tray and vent duct.
2. On the hatchback, remove the speaker grilles.
3. Disconnect the brake line from the wheel cylinder.
4. Remove the brake line from the brake hose.
5. Disconnect the brake hose from its bracket on the strut.
6. Loosen, but do not remove, the nut holding the suspension support to the strut.
7. Unbolt the strut from the rear arm.
8. Unbolt the strut from the body.
9. Installation is the reverse of removal. Torque the strut-to-body bolts to 17 ft. lb.; the strut-to-rear arm bolts to 105 ft. lb. and the suspension support-to-strut nut to 36 ft. lb.
10. Refill and bleed the brake system.

Celica FWD (1986 and Later)

1. Raise the rear of the vehicle and support it with jackstands. Position an hydraulic floor jack underneath the rear hub assembly; raise it just enough to support the assembly.
2. On the lift-back, remove the rear speaker grilles.
3. On the coupe, remove the suspension service hole cover.
4. On the ST and GT models, disconnect and plug the brake line at the backing plate. Remove the clip and E-ring and then disconnect the brake hose and tube from the strut housing.
5. On the GT-S, remove the union bolts and gaskets and disconnect the brake line from the brake cylinder. Remove the clip and E-ring from the strut and then disconnect the brake hose from the strut housing.
6. Loosen, but do not remove, the nut attaching the suspension support to the strut.
7. Disconnect the stabilizer bar at the lower end of the strut housing.
8. Disconnect the strut at the axle carrier.
9. Remove the three strut-to-body bolts and then remove the strut.

To install:

1. Tighten the upper strut-to-body nuts to 23 ft. lbs. (31 Nm).
2. Tighten the lower strut-to-carrier bolts to 119 ft. lbs. (162 Nm).
3. Connect the stabilizer bar to the strut and tighten the bolts to 26 ft. lbs. (35 Nm).
4. Tighten the strut holding nut to 36 ft. lbs. (49 Nm). Install the dust cover onto the suspension support.
5. Reconnect the brake line and hose.

SUSPENSION
Import Models

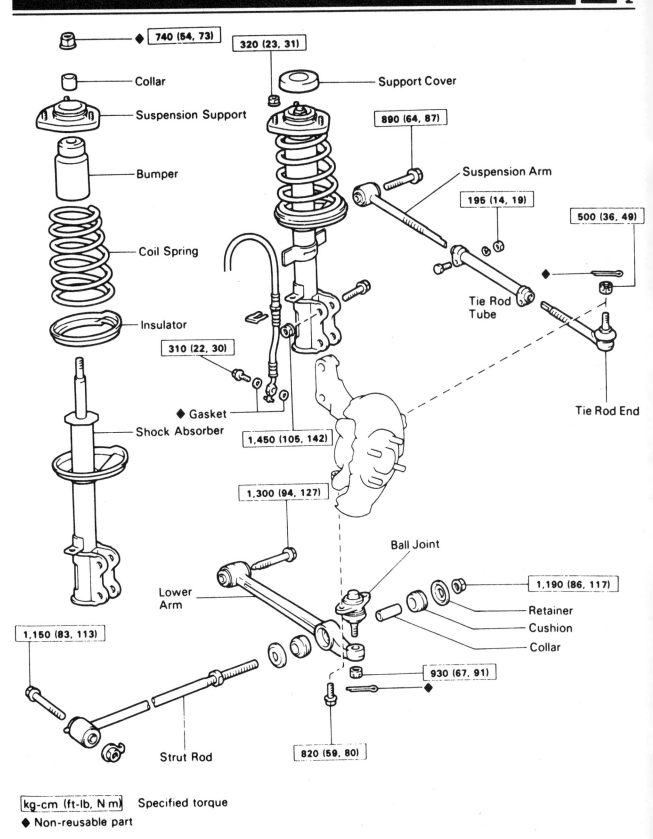

Rear suspension components—MR2

2-267

SECTION 2: SUSPENSION
Import Models

Bleed the system, lower the car and check the rear wheel alignment.

MR2

1. Raise the rear of the vehicle and support it with jackstands. Position an hydraulic floor jack underneath the rear hub assembly; raise it just enough to support the assembly.
2. Remove the union bolts and gaskets and disconnect the brake line from the brake cylinder. Remove the clip and E-ring from the strut and then disconnect the brake hose from the strut housing.
3. Matchmark the lower strut bracket and the camber adjusting cam, remove the two axle carrier bolts and the adjusting cam and disconnect the strut from the carrier.
4. Remove the engine hood side panel.
5. Remove the three upper strut-to-body nuts and then remove the strut.

To install:

1. Position the strut and tighten the upper mounting nuts to 23 ft. lbs. (31 Nm).
2. Install the engine hood side panel.
3. Connect the axle carrier to the lower strut bracket. Insert the mounting bolts from the rear and align the matchmarks made in Step 3. Tighten the nuts to 105 ft. lbs. (142 Nm).
4. Connect the brake line, bleed the system and check the rear wheel alignment.

Rear Axle Hub, Carrier and Bearing

REMOVAL & INSTALLATION

Front Wheel Drive Models
TERCEL (EXCEPT WAGON)

1. Raise the rear of the vehicle and support it with jackstands.
2. Remove the rear wheel and tire assembly.
3. Remove the brake drum.
4. Remove the locknut cap and cotter pin. Pry off the locknut lock and then remove the locknut itself.
5. Pull off the axle hub along with the outer wheel bearing and thrust washer.
6. Disconnect and plug the brake line where it connects to the brake backing plate.
7. Unbolt the rear axle shaft from the carrier and remove it along with the brake backing plate.
8. Remove the bolt and nut attaching the carrier to the strut rod.
9. Remove the bolt and nut attaching the carrier to the No. 1 suspension arm.
10. Remove the bolt and nut attaching the carrier to the No. 2 suspension arm.
11. Unbolt the carrier from the rear strut tube and remove the carrier.
12. Pry the inner bearing oil seal out of the brake drum and then remove the inner bearing.
13. Using a brass drift and a hammer, drive out the bearing races.

To install:

1. Press new outer bearing races into the axle hub and fill it and the bearing cap with grease.
2. Coat your palm with grease and press the bearing into your palm until the grease oozes out the other side.
3. Position the inner bearing into the hub and then drive in a new oil seal. Coat the seal with grease.
4. Position the axle carrier onto the strut tube and tighten the bolts to 105 ft. lb. (142 Nm).
5. Install the bolt and nut attaching the carrier to the No. 2 suspension arm; finger-tighten it only.

NOTE: Make sure that the lip of the nut is on the flange of the arm, not over it.

6. Repeat Step 5 for the No. 1 suspension arm.

NOTE: Make sure that the lip of the nut is in the hole on the arm.

7. Install the strut rod-to-carrier bolt so that the lip of the nut is in the groove on the bracket.
8. Install the axle shaft and brake back-

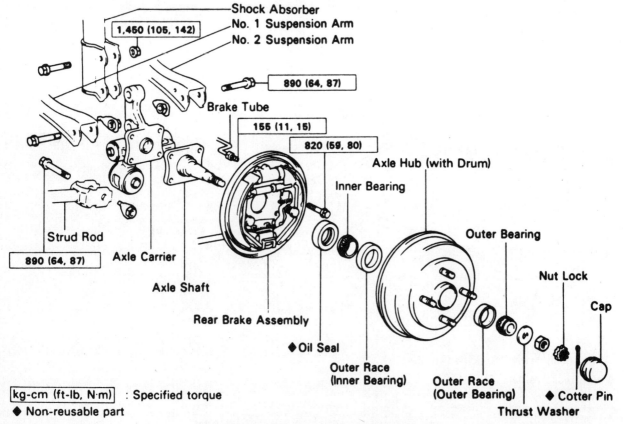

kg-cm (ft-lb, N·m) : Specified torque
♦ Non-reusable part

Rear axle hub and carrier—Tercel (except wagon)

SUSPENSION
Import Models

ing plate. Tighten the four bolts to 59 ft. lbs. (80 Nm).

9. Reconnect the brake line and then slide the axle hub/brake drum onto the axle shaft. Install the outer bearing, fill the hole with grease and position the thrust wsher. Install the bearing locknut and tighten it to 22 ft. lbs. (29 Nm).

10. Spin the axle hub several times to snug down the bearing and then loosen the bearing locknut until it can be turned by hand.

NOTE: There must be absolutely NO brake drag at this time.

11. Retighten the bearing locknut until there is a bearing preload of 0.9–2.2 lbs. (3.9–9.8 N) while turning the wheel.

12. Install the locknut lock, a new cotter pin and the cap. If the cotter pin hole does not line up properly, align the holes by tightening the nut.

13. Bleed the brakes.

14. Lower the vehicle and bounce it a few times to set the rear suspension.

15. Tighten the suspension arm bolts and the strut rod bolt to 64 ft. lbs (87 Nm).

CAMRY, CELICA (1986 AND LATER) AND TERCEL WAGON (EXCEPT 4 X 4)

1. Raise the rear of the vehicle and support it with jackstands.
2. Remove the rear wheel and tire assembly.
3. Remove the brake drum.
4. Disconnect and plug the brake line at the backing plate.
5. Remove the four axle hub-to-carrier bolts and slide off the hub and brake assembly. Remove the O-ring.

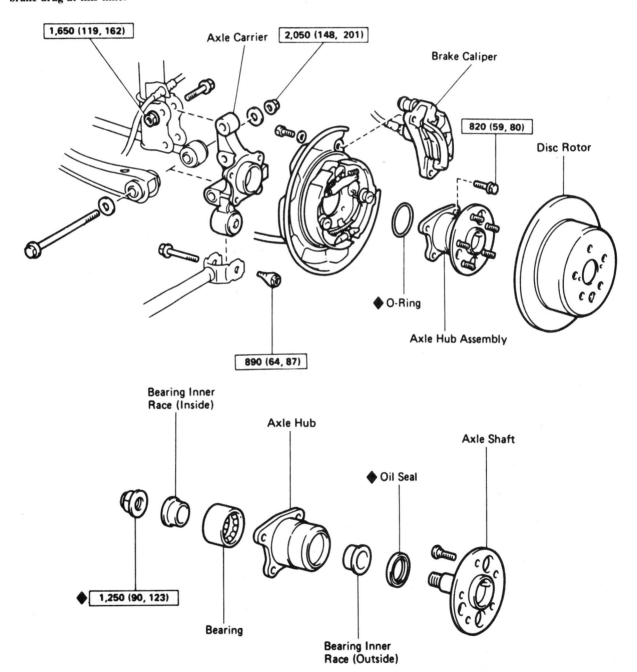

Rear axle hub and carrier—Camry, Celica (1986 and later) and Tercel wagon (except 4 × 4)

SECTION 2: SUSPENSION
Import Models

6. Remove the bolt and nut attaching the carrier to the strut rod.
7. Remove the bolt and nut attaching the carrier to the No. 1 suspension arm.
8. Remove the bolt and nut attaching the carrier to the No. 2 suspension arm.
9. Unbolt the carrier from the rear strut tube and remove the carrier.
10. Using a hammer and cold chisel, loosen the staked part of the hub nut and remove the nut.
11. Using a two-armed puller or the like, press the axle shaft from the hub.
12. Remove the bearing inner race (inside).
13. Using a two-armed puller again, pull off the bearing inner race (outside) from the axle shaft.
14. Remove the oil seal.
15. Position an old bearing inner race (outside) over the bearing and then press it out of the hub.

To install:

1. Position a new bearing inner race (outside) on the bearing and then press a new oil seal into the hub. Coat the lip of the seal with grease.
2. Position a new bearing inner race (inside) on the bearing and then press the inner race with the hub onto the axle shaft.
3. Install the nut and tighten it to 90 ft. lbs. (123 Nm). Stake the nut with a brass drift.
4. Position the axle carrier on the strut tube and tighten the nuts to 119 ft. lbs. (162 Nm).
5. Install the bolt and nut attaching the carrier to the No. 2 suspension arm; finger-tighten it only.

NOTE: Make sure that the lip of the nut is in the hole on the arm.

6. Repeat Step 5 for the No. 1 suspension arm.

NOTE: Make sure that the lip of the nut is in the hole on the arm.

7. Install the strut rod-to-carrier bolt so that the lip of the nut is in the groove on the bracket.
8. Install a new O-ring onto the axle carrier. Install the axle hub and brake backing plate. Tighten the four bolts to 59 ft. lbs. (80 Nm).
9. Reconnect the brake line, install the brake drum and then bleed the brakes.
10. Lower the vehicle and bounce it a few times to set the rear suspension.
11. Tighten the suspension arm bolts and the strut rod bolt to 64 ft. lbs. (87 Nm).

FRONT SUSPENSION
VOLKSWAGEN FRONT WHEEL DRIVE

Ball Joint

REMOVAL & INSTALLATION

1. Jack up the front of the car and support it on stands.
2. Matchmark the ball joint-to-control arm position on the Dasher and Quantum.
3. Remove the retaining bolt and nut from the hub (wheel bearing housing).
4. Pry the lower control arm and ball joint down and out of the strut.
5. Remove the 2 ball joint-to-lower control arm retaining nuts and bolts on the Dasher. Drill out the rivets on 1979–84 Rabbit, Jetta and Scirocco; enlarge the holes to 21/64 in.
6. Remove the ball joint assembly.
7. Install the Dasher or Quantum ball joint in the reverse order of removal. If no parts were installed other than the ball joint, align the matchmarks made in Step 2. No camber adjustment is necessary if this is done. Pull the ball joint into alignment with pliers. Tighten the two control arm-to-ball joint bolts to 47 ft. lbs. (64 Nm) and the strut-to-ball joint bolt to 25 ft. lbs. (34 Nm) (M8 bolt) or 36 ft. lbs. (49 Nm) (M10 bolt).
8. On all other models bolt the new ball joint in place. Torque the bolts to 18 ft. lbs. (25 Nm). Tighten the retaining bolt for the ball joint stud to 37 ft. lbs. (50 Nm).

MacPherson Strut

REMOVAL & INSTALLATION

Dasher and Quantum

1. With the car on the ground, remove the front axle nut. Loosen the wheel bolts.
2. Raise and support the front of the car. Remove the wheels.
3. Remove the brake caliper from the strut and hang it with wire. Detach the brake line clips from the strut.
4. At the tie-rod end, remove the cotter pin, back off the castellated nut, and pull the end off the strut with a puller.
5. Loosen the stabilizer bar bushings and detach the end from the strut being removed.
6. Remove the ball joint from the strut.
7. Pull the axle driveshaft from the strut.
8. Remove the upper strut-to-fender retaining nuts.
9. Pull the strut assembly down and out of the car.
10. Installation is the reverse of removal. The axle nut is tightened to 145 ft. lbs. (196 Nm) (M 18 nut) or 175 ft. lbs. (237 Nm) (M 20 nut). Tighten the ball joint-to-strut nut to 25 ft. lbs. (34 Nm) (M8 nut) or 36 ft. lbs. (49 Nm) (M10 nut), the caliper-to-strut bolts to 44 ft. lbs. (60 Nm) and the stabilizer-to-control arm bolts to 7 ft. lbs. (9 Nm).

1980–84 Rabbit, Jetta and Scirocco

1. Remove the brake hose from the strut clip.
2. Mark the position of the camber adjustment bolts before removing them from the hub (wheel bearing housing). These bolts also serve as the lower strut mounting bolts.
3. Remove the upper mounting nuts and remove the strut from the car.
4. Installation is the reverse of removal. The upper nuts are tightened to 14 ft. lbs. (19 Nm) and the adjusting bolt (upper) to hub to 58 ft. lbs. (80 Nm). Tighten the lower adjusting bolt-to-hub to 43 ft. lbs. (58 Nm). Use new washers on the lower bolts. If the shock absorber was replaced, camber will have to be adjusted.

1985 And Later Golf, Jetta And Scirocco

1. Raise and support the front of the vehicle. Remove the wheel if necessary.
2. Mark the position of the lower strut bolts before removing them from the steering knuckle.
3. Remove the upper mounting nut and the strut from the vehicle.
4. To install, reverse the removal procedures. Torque the strut-to-body nut to 44 ft. lbs. (60 Nm) and the strut-to-steering knuckle bolts to 59 ft. lbs. (80 Nm).

OVERHAUL

For all spring and shock absorber removal and installation procedures, and all overhaul procedures, please refer to "Strut Overhaul"

Lower Control Arm (Wishbone)

REMOVAL & INSTALLATION

All Models Except 1985 And Later Golf, Jetta And Scirocco

Volkswagen refers to the lower control arm as the wishbone.

NOTE: When removing the left side (driver's side) control arm on model years

SUSPENSION
Import Models

SECTION 2

1. Cotter pin
2. Tie rod
3. Axle driveshaft
4. Circlip
5. Retainer nut
6. Brake caliper
7. Wheel bearing
8. Hub
9. Brake disc
10. Axle nut

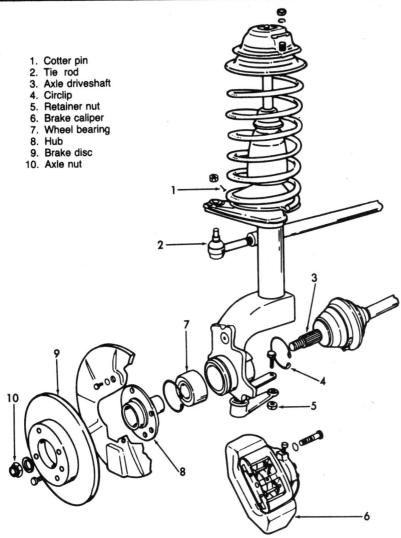

Dasher front suspension components—Quantum similar

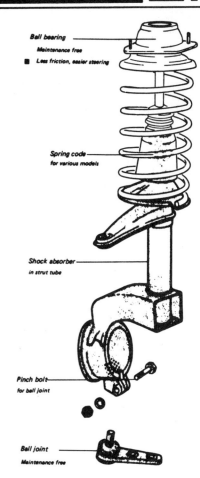

A pinch bolt holds the ball joint to the combination strut and steering knuckle

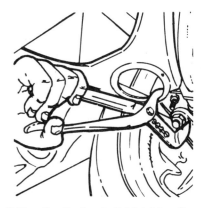

Pulling the Dasher ball joint into alignment on installation

thru 1984 on Rabbit, Jetta and Scirocco equipped with an automatic transmission, remove the front left engine mounting, remove the nut for the rear mounting, remove the engine mounting support and raise the engine to expose the front control arm bolt.

1. Raise the vehicle and support it on jack stands. Remove the wheel.

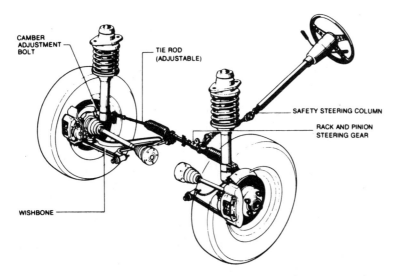

Typical Rabbit, Jetta, Scirocco steering and front suspension components

2-271

SUSPENSION
Import Models

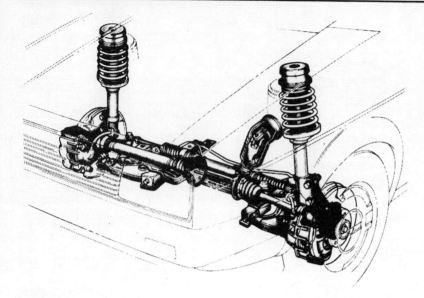

View of the front suspension system for 1985 and later models—except Quantum

2. Remove the nut and bolt attaching the ball joint to the hub (wheel bearing housing) and pry the joint down and out of the hub.

3. Unfasten the stabilizer bar on models so equipped.

4. Unbolt and remove the control arm-to-subframe (crossmember) mounting bolts on the Dasher or Quantum. On the 1980–84 Rabbit, Jetta and Scirocco, remove the control arm mounting bolts from the frame.

5. Remove the control arm. See procedures above for ball joint removal and installation.

6. Installation is the reverse of removal. Tighten the Dasher or Quantum control arm-to-subframe bolts to 44 ft. lbs. (60 Nm), and the 1980–84 Rabbit, Jetta and Scirocco control arm-to-frame front bolt to 50 ft. lbs. (68 Nm), bushing clamp bolts to 32 ft. lbs. (43 Nm). Tighten the ball joint to hub bolt to 21 ft. lbs. (28 Nm) on the Rabbit, Jetta and Scirocco, and to 25 ft. lbs. (34 Nm) (M 8 nut) or 36 ft. lbs. (49 Nm) (M 10 nut).

1985 And Later Golf, Jetta And Scirocco

1. Refer to the "Ball Joint, Removal and Installation" procedures in this section and separate the ball joint from the steering knuckle.

NOTE: If replacing the lower control arm, separate the ball joint from the control arm.

2. If equipped, remove the stabilizer bar from the control arm.

3. Remove the control arm-to-subframe bolts and the control arm from the vehicle.

4. To install, reverse the removal procedures. Torque the control arm-to-subframe

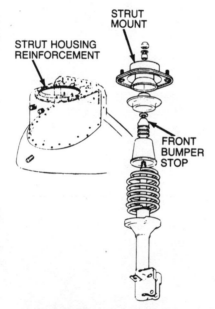

Front strut and mounting—GTI

to 96 ft. lbs. (130 Nm), the stabilizer-to-control arm to 18 ft. lbs. (25 Nm), the ball joint-to-control arm to 18 ft. lbs. (25 Nm) or the ball joint-to-steering knuckle to 37 ft. lbs. (50 Nm).

Front Wheel Bearings

The front wheel bearings are non-adjustable on all models and are sealed, so they should be maintenance-free.

NOTE: Replacement wheel bearings for Quantum models with an enlarged outside diameter require NO moly paste lubrication (Bearing number 321 498 625D).

REMOVAL & INSTALLATION
Dasher And Quantum

1. Remove the grease cup from the wheel spindle.

2. Have an assistant apply pressure to the brake pedal and remove the wheel spindle lock nut.

3. Raise and support the front of the vehicle on jackstands (place under the frame). Remove the wheel assembly.

4. Remove the brake caliper from the steering knuckle and suspend on a wire, DO NOT remove the brake hose.

5. Remove the brake disc from the hub assembly.

6. Loosen the ball joint-to-steering knuckle bolt and separate the ball joint from the steering knuckle by prying down on it with a large pry bar.

7. Remove the tie-rod end-to-steering knuckle nut and use a ball joint puller to separate the tie-rod from the steering knuckle.

8. While supporting the axle shaft, pull the steering knuckle and hub assembly out from the vehicle. After separating the axle shaft from the steering knuckle assembly, support it with a wire.

NOTE: If the Quantum is equipped with a 5 cyl engine, connect a wheel puller to the hub flange, then push the axle shaft from the steering knuckle and strut assembly.

9. Remove the strut assembly from the vehicle.

10. Place a set of parallel rail blocks on an arbor press to support the steering knuckle and strut assembly. Place the steering knuckle and hub assembly on top of tools with the hub facing down.

11. Stack tools VW-295A, VW-420 and VW-412, in order, on top of the hub shaft, them press the hub from the steering knuckle and strut assembly.

12. Secure the hub in a vise. Using tools VW-295A and US-1078, pull the inner race from the hub shaft.

13. Remove the internal snap rings from the steering knuckle.

14. With the steering knuckle assembly in the same pressing position, stack tools VW-519, VW-432 and VW-409, in order, on the bearing, then press the bearing from the steering knuckle and strut assembly.

15. Throughly clean all of the parts. If worn or damaged parts are present, replace them with new ones.

NOTE: Lubricate the new bearing throughly with molybdenum grease. Install the outside internal snap ring in the steering knuckle.

16. To install the new bearing, place a flat plate on the arbor press, then the steering knuckle and strut assembly (in the same position for extracting the bearing). On top of the steering knuckle assembly, stack the new bearing, then tools 40-20 and VW-

SUSPENSION
Import Models

411, in order and press the bearing into the steering knuckle, until it seats against the bottom retaining ring. Install the other retaining ring.

17. Lift the steering knuckle assembly and place the hub on a flat surface followed by the steering knuckle (it must be facing the same direction).

18. Stack tools VW-519, VW-432 and VW-412, in order, on top the bearing, then press the hub into the steering knuckle assembly.

NOTE: When pressing the hub into the steering knuckle, support the inner race of the bearing with tool VW-519.

19. To complete the installation, reverse the removal procedures. Torque the control arm-to-steering knuckle to 37 ft. lbs. (50 Nm), the tie-rod to the steering knuckle to 22 ft. lbs. (30 Nm) and the axle shaft hub nut to 170 ft. lbs. (230 Nm) and the brake caliper-to-steering knuckle to 52 ft. lbs. (50 Nm).

1980–84 Rabbit, Jetta And Scirocco

1. Remove the grease cup from the wheel spindle.

2. Have an assistant apply pressure to the brake pedal and remove the wheel spindle lock nut.

3. Raise and support the front of the vehicle on jackstands (place under the frame). Remove the wheel assembly.

4. Remove the brake caliper from the steering knuckle and suspend on a wire, DO NOT remove the brake hose.

5. Remove the brake disc from the hub assembly.

6. Loosen the ball joint-to-steering knuckle bolt and separate the ball joint from the steering knuckle by prying down on it with a large pry bar.

7. Remove the tie-rod end-to-steering knuckle nut and use a ball joint puller to separate the tie-rod from the steering knuckle.

8. While supporting the axle shaft, pull the steering knuckle and hub assembly out from the vehicle. After separating the axle shaft from the steering knuckle assembly, support it with a wire.

9. Remove the strut-to-steering knuckle bolts and the steering knuckle assembly from the vehicle.

10. Place tools VW-401 and VW-402 on top of parallel rail blocks on an arbor press to support the steering knuckle assembly. Place the steering knuckle and hub assembly on top of tools with the hub facing down.

11. Stack tools VW-418A, VW-421 and VW-409, in order, on top of the hub shaft, then press the hub from the steering knuckle.

NOTE: To press out the hub, if equipped with a drum brake, substitute tools VW-421 and VW-409 with tools VW-431 and VW-411.

12. Secure the hub in a vise. Using tools VW-431 andd US-1078, pull the inner race from the hub shaft.

13. Remove the internal snap rings from the steering knuckle.

14. With the steering knuckle assembly in the same pressing position, stack tools VW-433, VW-420 and VW-412, in order, on the bearing and press the bearing from the steering knuckle.

NOTE: To press out the wheel bearing, if equipped with a drum brake, substitute tools VW-433, VW-420 and VW-412 with tools VW-415A, VW-433 and VW-411.

15. Throughly clean all of the parts. If worn or damaged parts are present, replace them with new ones.

NOTE: Lubricate the new bearing throughly with molybdenum grease.

16. To install the new bearing, use the same set up used to extract the bearing, place the new bearing on the steering knuckle and press it into the steering knuckle, until it seats against the bottom retaining ring. Install the other retaining ring.

17. Lift the steering knuckle assembly and place the hub on a flat surface followed by the steering knuckle (it must be facing the same direction).

18. Stack tools VW-519, VW-432 and VW-412, in order, on top of the bearing, then press the hub into the steering knuckle assembly.

NOTE: When pressing the hub into the steering knuckle, support the bearing with tool VW-519

19. To complete the installation, reverse the removal procedures. Torque the steering knuckle-to-strut to 59 ft. lbs. (80 Nm), the control arm-to-steering knuckle to 37 ft. lbs. (50 Nm), the tie-rod to the steering knuckle to 22 ft. lbs. (30 Nm) and the axle shaft hub nut to 174 ft. lbs. (240 Nm).

1985 And Later Golf, Jetta And Scirocco

1. Remove the grease cup from the wheel spindle.

2. Have an assistant apply pressure to the brake pedal and remove the wheel spindle lock nut.

3. Raise and support the front of the vehicle on jackstands (place under the frame). Remove the wheel assembly.

4. Remove the brake caliper from the steering knuckle and suspend on a wire, DO NOT remove the brake hose.

5. Remove the brake disc and the backing plate from the hub assembly.

6. Loosen the ball joint-to-steering knuckle bolt and separate the ball joint from the steering knuckle by prying down on it with a large pry bar.

7. Remove the tie-rod end-to-steering knuckle nut and use a ball joint puller to separate the tie-rod from the steering knuckle.

8. While supporting the axle shaft, pull the steering knuckle and hub assembly out from the vehicle. After separating the axle shaft from the steering knuckle assembly, support it with a wire.

9. Remove the strut-to-steering knuckle bolts and the steering knuckle assembly from the vehicle.

10. Place tools VW-401 and 3110 on an arbor press. Place the steering knuckle and hub assembly on top of tool 3110 with the hub facing down inside of the tool.

11. Place tools VW-295 and VW-408A on top of the hub shaft, then press the hub from the steering knuckle.

12. Secure the hub in a vise. Using tool 30-11, pull the inner race from the hub shaft.

13. Remove the internal snap ring from the steering knuckle.

14. With the steering knuckle in the same position on tool 3110, place tools VW-433 and VW-407 on the bearing and press the bearing from the steering knuckle.

15. Throughly clean all of the parts. If worn or damaged parts are present, replace them with new ones.

NOTE: Lubricate the new bearing throughly with molybdenum grease.

16. To install the new bearing, place tools VW-401, VW-402 and VW-459/2 on the arbor press. Place the steering knuckle on tool VW-459/2 with large opening facing upward.

17. Place the new bearing on the steering knuckle and tools VW-472/1 and VW-412 on top of the bearing, then press the bearing into the steering knuckle until it seats. Install the internal locking ring and the inner race.

18. Replace tool VW-459/2 with tool VW-519, place the hub on top of bearing and press the hub into the steering knuckle.

NOTE: When pressing the hub into the steering knuckle, support the bearing.

19. To complete the installation, reverse the removal procedures. Torque the steering knuckle-to-strut to 59 ft. lbs. (80 Nm), the control arm-to-steering knuckle to 37 ft. lbs. (50 Nm), the tie-rod to the steering knuckle to 26 ft. lbs. (35 Nm) and the axle shaft hub nut to 170 ft. lbs. (230 Nm).

Front End Alignment

CAMBER ADJUSTMENT

Dasher and Quantum

Camber is adjusted by loosening the two ball joint-to-lower control arm bolts, and moving the ball joint in or out as necessary.

Golf, Rabbit, Scirocco, Jetta

Camber is adjusted by loosening the nuts of the two bolts holding the top of the wheel bearing housing to the bottom of the strut,

2-273

SECTION 2: SUSPENSION
Import Models

and turning the top eccentric bolt. The range of adjustment is 2°.

On models from 1985 the original top bolt can be replaced with a long shank bolt (N903334.01) and adjusted after new bolt is installed—if more adjustment is necessary the lower bolt can also be replaced with a new style.

CASTER

Other than the replacement of damaged suspension components, caster is not adjustable on any model.

Top eccentric bolt provides camber adjustment on Rabbit, Scirocco and Jetta

TOE-IN ADJUSTMENT
Dasher and Quantum

Toe-in is checked with the wheels straight ahead. The left tie-rod is adjustable. Loosen the nuts and clamps and adjust the length of the tie-rod for correct toe-out. If the steering wheel is crooked, remove and align it.

Golf, Rabbit, Scirocco, Jetta

Toe-in is checked with the wheels straight ahead. Only the right tie-rod is adjustable, but replacement left tie-rods are adjustable. Replacement left tie-rods should be set to the same length as the original. Toe-in should be adjusted only with the right tie-rod. If the steering wheel is crooked, remove and align it.

REAR SUSPENSION
VOLKSWAGEN FRONT WHEEL DRIVE

Coil Springs

REMOVAL & INSTALLATION

Dasher Only

1. Raise the car on a lift.
2. Support the axle.
3. Install a spring compressor on the coil spring, and remove.
4. Installation is the reverse of removal.

NOTE: It is not necessary to replace both springs if only one is damaged.

Shock Absorbers

REMOVAL & INSTALLATION

Dasher Only

NOTE: Only remove one shock absorber at a time. Do not allow the rear axle to hang by its body mounts only, as it may damage the brake lines.

This operation requires the use of either special tool VW 655/3 or a suitable spring compressor and floor jack.

1. Raise the car and support it on jack stands. Do not place the jack stands under the axle beam.
2. Remove the wheel.
3. Attach special tool VW 655/3 between the axle beam and a prefabricated hook hung on the body frame above the beam. Jack the tool until you can see the shock absorber compressing. If you are using a spring compressor and a floor jack, compress the spring a little and, placing the floor jack under the beam below the spring, jack it up until you see the shock absorber compress.
4. Unbolt and remove the shock absorber.
5. Installation is the reverse of removal. Tighten the shock absorber bolts to 43 ft. lbs. (58 Nm).

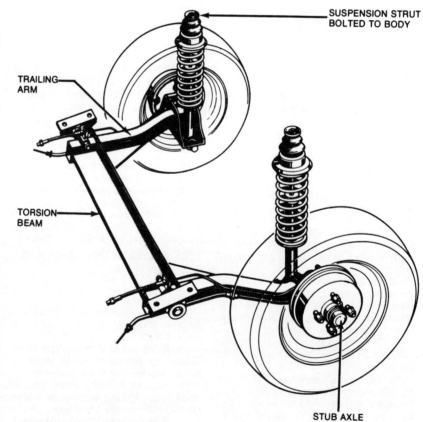

Typical rear suspension (except Dasher). Various models are equipped with a rear stabilizer (not shown)

NOTE: There are two types of shock absorbers for the Dasher and they have different mounts. Make sure you get the correct type for your vehicle.

SUSPENSION
Import Models

SECTION 2

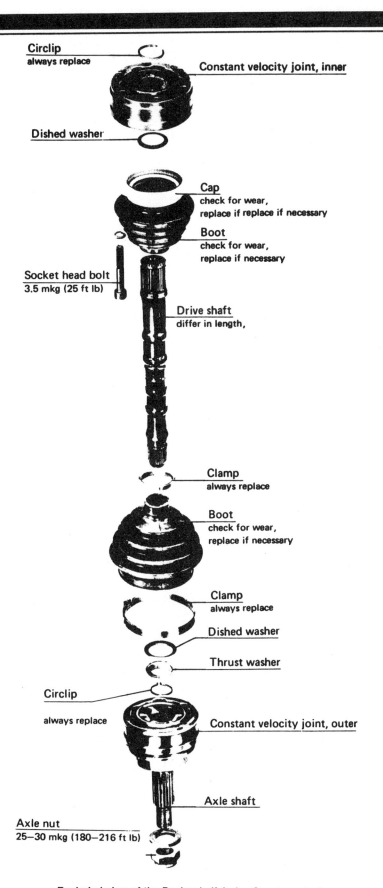

Exploded view of the Dasher halfshaft—Quantum similar

MacPherson Strut

REMOVAL & INSTALLATION

Golf, Rabbit, Jetta, Scirocco

1. Raise and support the rear of the vehicle.
2. Support the axle, but do not put any load on the springs.
3. Remove the rubber guard from inside the car.
4. Remove the nut, washer and mounting disc.
5. Unbolt the strut assembly from the rear axle and remove it.
6. To install reverse the removal procedures. Torque the strut-to-body nut to 44 ft. lbs. (60 Nm) and strut-to-axle body to 59 ft. lbs. (80 Nm).

Quantum

1. Remove the shock strut cover inside car.
2. Unscrew the strut from the body.
3. Slowly lift the vehicle until the wheels are slightly off the ground.
4. Unscrew the strut from the axle.
5. Take the strut out of the lower mounting. Press the wheel down slightly when removing the strut.

--- CAUTION ---
Do not remove both suspension struts at the same time as this will overload the axle beam bushings.

6. Guide the strut out carefully between the wheel and the wheel housing. Do not damage the paint on the spring and wheel housing.

Stub Axle

REMOVAL & INSTALLATION

1. Remove the grease cap, cotter pin, locknut, adjusting nut, spacer, wheel bearing and brake drum.
2. Disconnect and plug the brake line. Remove the brake backing plate with the brakes attached.
3. Unbolt and remove the stub axle.
4. To install, repack the wheel bearings and reverse the removal procedures. Torque the backing plate mounting bolts to 44 ft. lbs. (60 Nm) on the Dasher and Quantum or 52 ft. lbs. (70 Nm) for all other models. Bleed the brake system.
7. To install reverse the removal procedures. Torque the strut-to-body to 26 ft. lbs. (35 Nm) and the strut-to-axle to 52 ft. lbs. (70 Nm).

OVERHAUL

For all spring and shock absorber removal and installation procedures, and all overhaul procedures, please refer to "Strut Overhaul"

SUSPENSION
Import Models

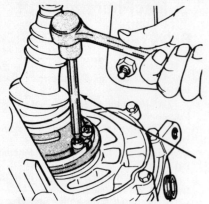
Remove the socket head (Allen) bolts holding the axle shaft to the transaxle

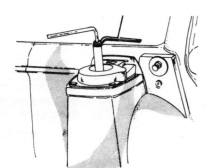
Strut self-locking nut, removal/installation

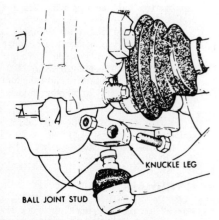
Remove ball joint from knuckle to remove axle shaft—Rabbit, Jetta, Scirocco

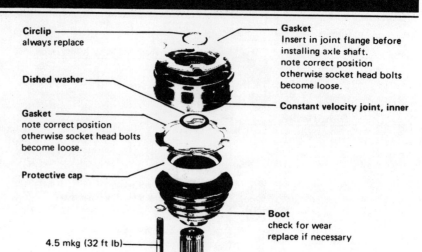

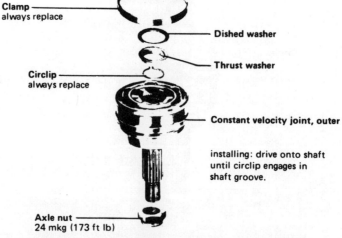
Exploded view of the Rabbit, Jetta and Scirocco halfshaft

SUSPENSION
Import Models

FRONT SUSPENSION
VOLKSWAGEN REAR WHEEL DRIVE

The front suspension consists of upper and lower control arms, a separate upper coil spring/shock absorber mount, steering knuckle and attaching ball joints and a strut arm mounted on the lower control arm for stability.

SHOCK ABSORBER
REMOVAL & INSTALLATION

1. Jack up the front of the vehicle and remove the front wheel.
2. Loosen and remove the single retaining nut at the top of the coil spring/shock absorber upper mount.
3. Remove the through bolt which retains the bottom of the shock absorber to the lower control arm and pull the shock absorber out through the bottom of the lower control arm.
4. Reverse the procedure to install.

Upper Control Arm

REMOVAL & INSTALLATION

1. Jack up the vehicle and remove the front wheel.
2. Place a jack under the lower control arm and raise to put a slight load on the coil spring.
3. Free the upper ball joint from the steering knuckle.
4. Remove the upper control arm to frame mounting bolt and remove the control arm.
5. Reverse procedure to install. Check and adjust wheel alignment.

Lower Control Arm

REMOVAL & INSTALLATION

1. Jack up the vehicle and remove the wheel.
2. Remove the coil spring.
3. Remove the lower control arm to frame mounting bolt and remove the control arm.
4. Reverse procedure to install.

Ball Joint

REMOVAL & INSTALLATION
UPPER BALL JOINT

1. Raise the vehicle and support it on jack stands, then remove the wheel.
2. Place a jack under the lower control arm as close to the steering knuckle as possible and jack up just enough to put a slight load on the coil spring.
3. Loosen the steering knuckle-to-ball joint nut but do not remove completely.
4. Free the ball joint from the steering knuckle using a ball joint removal tool. Then remove the nut.
5. Remove the two upper ball joint to upper control arm bolts and remove the ball joint.
6. Reverse procedure to install. Check wheel alignment.

Lower Ball Joint

1. Jack up the front of the vehicle and support it on stands, then remove the wheel.
2. Place a jack under the lower control arm as close to steering knuckle as possible and put a slight load on the coil spring by jacking up the jack.
3. Disconnect the brake caliper hose form the caliper, and remove the brake caliper and rotor if they are in the way.
4. Loosen the upper ball joint to steering knuckle to ball joint nut, but do not remove it. Free the upper ball joint from the steering knuckle and remove the nut.
5. Remove the lower ball joint to lower control arm nut and free the ball joint from the control arm using a ball joint removal tool. Remove the steering knuckle.
6. Press the ball joint off the steering knuckle.
7. Press a new ball joint in place on knuckle, observing any alignment marks on the ball joint and the knuckle.
8. Reverse the procedure to install. Bleed the brakes.

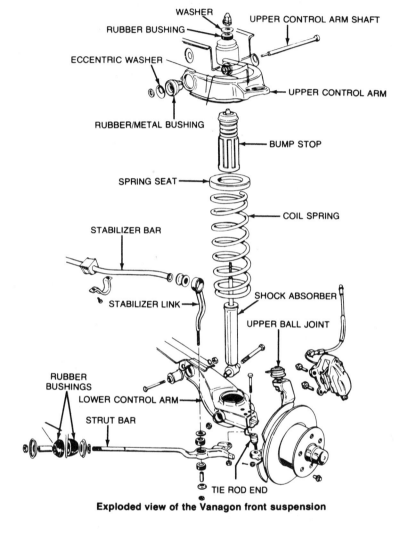

Exploded view of the Vanagon front suspension

SECTION 2
SUSPENSION
Import Models

INSPECTION

1. A quick initial inspection can be made with the vehicle on the ground.
2. Grasp the top of the tire and vigorously pull the top of the tire in and out. Test both sides in this manner.
3. If the ball joints are excessively worn, there will be an audible tap as the ball moves around in its socket. Excess play can sometimes be felt through the tire.
4. A more rigorous test may be performed by jacking the car under the lower torsion arm and inserting a lever under the tire.
5. Lift up gently on the lever so as to pry the tire upward.
6. If the ball joints are worn, the tire will move upward 1/8–1/4 in. or more.
7. If the tire displays excessive movement, have an assistant inspect each joint, as the tire is pried upward, to determine which ball joint is defective.

Front Wheel Bearings
REMOVAL & INSTALLATION

1. Jack up the car and remove the wheel and tire.
2. Remove the caliper and disc (if equipped with disc brakes) or brake drum.
3. To remove the inside wheel bearing, pry the dust seal out of the hub with a screwdriver. Lift out the bearing and its inner race.
4. To remove the outer race for either the inner or outer wheel bearing, insert a long punch into the hub opposite the end from which the race is to be removed. The race rests against a shoulder in the hub. The shoulder has two notches cut into it so that it is possible to place the end of the punch directly against the back side of the race and drive it out of the hub.
5. Carefully clean the hub.
6. Install new races in the hub. Drive them in with a soft faced hammer or a large piece of pipe of the proper diameter. Lubricate the races with a light coating of wheel bearing grease.
7. Force wheel bearing grease into the sides of the tapered roller bearings so that all the spaces are filled.
8. Place a small amount of grease inside the hub.
9. Place the inner wheel bearing into its race in the hub and tap a new seal into the hub. Lubricate the sealing surface of the seal with grease.
10. Install the hub on the spindle and install the outer wheel bearing.
11. Adjust the wheel bearing and install the dust cover.
12. Install the caliper (if equipped with disc brakes).

ADJUSTMENT

The bearing may be adjusted by feel or by a dial indicator.

To adjust the bearing by feel, tighten the adjusting nut so that all the play is taken up in the bearing. There will be a slight amount of drag on the wheel if it is hand spun. Back off fully on the adjusting nut and retighten very lightly. There should be no drag when the wheel is hand spun and there should be no perceptible play in the bearing when the wheel is grasped and wiggled from side to side.

To use a dial indicator, remove the dust cover and mount a dial indicator against the hub. Grasp the wheel at the side and pull the wheel in and out along the axis of the spindle. Read the axial play on the dial indicator. Screw the adjusting nut in or out to obtain 0.001–0.005 in. of axial play. Secure the adjusting nut and recheck the axial play.

Front End Alignment

CASTER ADJUSTMENT

Caster on the coil spring suspension is adjusted by moving the strut bar. Loosen the locknut and then turn the adjusting nut clockwise to increase the caster and counterclockwise to decrease the caster.

CAMBER ADJUSTMENT

Camber is the tilt of the top of the wheel, inward or outward, from true vertical. Outward tilt is positive, inward tilt is negative.

The upper control arm pivots on an eccentric bolt. To adjust the camber, loosen the retaining nut and rotate the bolt.

TOE-IN ADJUSTMENT

Toe-in is the adjustment made to make the front wheels point slightly into the front. Toe-in is adjusted on all types of front suspensions by adjusting the length of the tie-rod sleeves.

REAR SUSPENSION
VOLKSWAGEN REAR WHEEL DRIVE

Trailing Arm
REMOVAL & INSTALLATION

1. Raise the rear of the car and support it with jack stands.
2. Remove the wheel and then disconnect the brake line from the wheel cylinder.
3. Unbolt the driveshaft at the transaxle side.
4. Unscrew the four wheel hub mounting bolts and remove it along with the driveshaft.

NOTE: Removal of the brake drum may provide better access to the wheel hub mounting bolts.

5. Place a floor jack under the trailing arm to hold its position and then remove the upper and lower shock absorber retaining bolts and remove the shock absorber.
6. Lower the trailing arm until you can remove the coil spring. Note the positioning of the upper spring plate and the lower spring seat.
7. Unscrew the two trailing arm mounting bolts and then remove the trailing arm.
8. Installation is in the reverse order of removal. When installing the coil spring, be sure that the contours for the end of the spring in the seat and the trailing arm are aligned. Also, turn the spring plate so that the end of the spring fits into the depression on the plate.
9. Adjust the camber and toe.

Shock Absorber
REMOVAL & INSTALLATION

Procedures for removing the shock absorbers are detailed in the "Trailing Arm Removal and Installation" section.

Coil Spring
REMOVAL & INSTALLATION

Procedures for removing the coil spring are detailed in the "Trailing Arm Removal and Installation" section.

Rear Suspension Adjustments

It is possible to adjust the camber and the toe. To adjust the toe, loosen the INSIDE mounting bolt on the trailing arm and slide it forward or backward in the horizontal slot until the proper toe is achieved. To adjust the camber, loosen the OUTSIDE mounting bolt on the trailing arm and slide it up or down until the proper camber is achieved. Being careful not to move the bolts, tighten them both to 65 ft. lbs. after adjustment.

SUSPENSION
Import Models

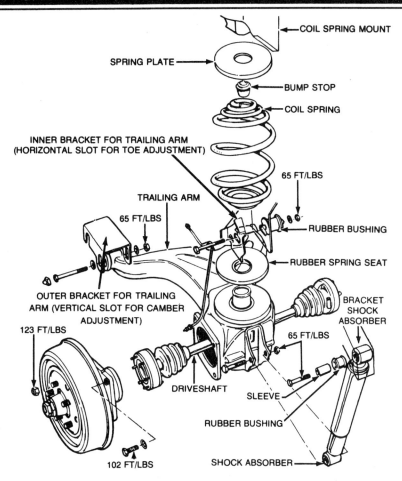

Exploded view of the Vanagon rear suspension

FRONT SUSPENSION VOLVO

MacPherson Strut

REMOVAL & INSTALLATION

1. Remove the hub cap and loosen the lug nuts a few turns.

2. Firmly apply the parking brake and place blocks in back of the rear wheels.

3. Jack up the front of the car with a hoist or using a floor jack at the center of the front crossmember. When the wheels are 2–3 in. off the ground, the car is high enough. Place jack stands beneath the front jacking points. Then, remove the floor jack from the crossmember (if used), and reposition it beneath the applicable lower control arm to provide support at the outer end. Remove the wheel and tire assembly.

4. Using a ball joint puller, disconnect the steering rod from the steering arm.

5. Disconnect the stabilizer bar at the link upper attachment.

6. Remove the bolt retaining the brake

Loosen the upper strut nuts to adjust the camber on 240, 260 (DL, GL) series

line bracket to the fender well.

7. Open the hood and remove the cover for the strut assembly upper attachment.

8. While keeping the strut from turning, loosen and remove the nut for the upper attachment.

9. Before lowering the strut assembly, wire or tie the strut to some stationary component, or use a holding fixture such as SVO 5045, to prevent the strut from traveling down too far and damaging the hydraulic brake lines. Then lower the jack supporting the lower arm and allow the strut to tilt out to about a 60 degree angle. At this angle, the top of the strut assembly should just protrude past the wheel well, allowing removal of the strut from the top.

10. Carefully lift and guide the strut assembly into its upper attachment in the spring tower. Connect the stabilizer bar to the stabilizer link. Guide the shock absorber spindle into the upper attachment and raise the jack beneath the lower control arm. Install the washer and nut on top of the shock absorber spindle. While holding the spindle from turning, tighten the nut to 15–25 ft. lbs. Install the cover.

11. Attach the brake line bracket to its mount. Tighten the nut retaining the stabilizer bar to the link. Connect the steering rod at the steering arm.

12. Install the wheel and tire assembly. Remove the jack stands and lower the car. Jounce the suspension a few times and then road test.

2–279

SUSPENSION
Import Models

OVERHAUL

For all spring and shock absorber removal and installation procedures and any other strut overhaul procedures, please refer to "Strut Overhaul"

Lower Control Arm

REPLACEMENT

1. Jack up car, support on stands and remove wheels.
2. Remove stabilizer bar.
3. Remove ball joint from control arm.
4. Remove control arm front retaining bolt.
5. Remove control arm rear attachment plate.
6. Remove attachment plate from control arm.
7. Remove stabilizer link from control arm.
8. Install in reverse of removal.

NOTE: Right and left bushings are not interchangeable. The right side bushing should be turned so that the small slots point horizontally when installed. Torque the front retaining bolt to 55 ft. lbs., the rear bushing to 4 ft. lbs. and the rear attachment bolts to 30 ft. lbs.

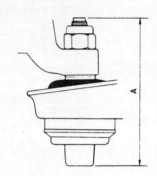

Spring–type lower ball joint maximum allowable length

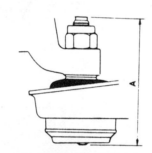

Non-spring type lower ball joint maximum allowable length

Suspending the top of the strut from the body with a wire while removing the lower ball joint

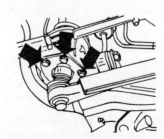

Late type lower ball joint-to-strut retaining bolts

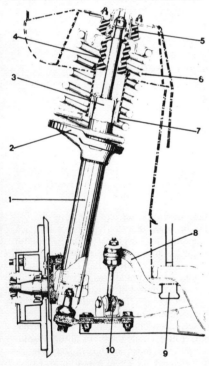

1. Strut assembly
2. Lower spring support
3. Shock absorber
4. Rubber bumper
5. Upper attachment
6. Coil spring
7. Ruber sleeve, protecting the shock absorber
8. Stabilizer bar
9. Stabilizer bar attachment
10. Stabilizer link

Front suspension—240, 260 (DL, GL) series

Lower Ball Joint

REMOVAL & INSTALLATION

All (Except 740 and 760 GLE)

1. Jack up the front of the car and install jack stands beneath the front jacking attachments.
2. Remove the tire and wheel assembly.
3. Reach in between the spring coils and loosen the shock absorber cap nut a few turns.
4. Remove the four bolts (12mm) retaining the ball joint seat to the bottom of the strut.
5. Remove the three nuts (19mm) retaining the ball joint to the lower control arm.
6. Place the ball joint and attachment assembly in a vise and remove the 19mm nut from the ball joint stud. Then, drive out the old ball joint.
7. Install the new ball joint in the attachment and tighten the stud nut to 35–50 ft. lbs.
8. Attach the ball joint assembly to the strut. Tighten to 15–20 ft. lbs.
9. Attach the ball joint assembly to the control arm. Tighten to 70–95 ft. lbs.
10. Tighten the shock absorber cap nut. Install the wheel and tire. Lower the car and road-test.

NOTE: On models with power steering, the ball joints are different for the left and right side.

Compared to previous years, the ball joint is 0.393 in. forward in control rod attachment. It is therefore most important that these ball joints are installed on the correct side.

740 and 760 GLE

1. Jack up the front end of the car. Remove the wheel.
2. Remove the bolt connecting the anti-roll bar link to the control arm.
3. Remove the cotter pin for the ball joint stud and remove the nut.
4. Using a ball joint puller, press out the ball joint from the control arm. Make sure the puller is located directly in line with the stud, and that the rubber grease boot is not damaged by the puller.
5. Remove the bolts holding the ball joint to the spring strut. Press the control arm down and remove the ball joint.
6. Reverse the above procedure for installation. When installing the new ball joint, always use new bolts and coat all threads with a liquid thread sealer. Torque bolts to 22 ft. lbs, checking that the bolt heads sit flat on the ball joint, then angle-tighten (protractor-torque) 90°. Torque the nut holding the control arm ball joint stud to 44 ft. lbs. Use a new cotter pin on the ball joint stud, and install the anti-roll bar link.

SUSPENSION
Import Models

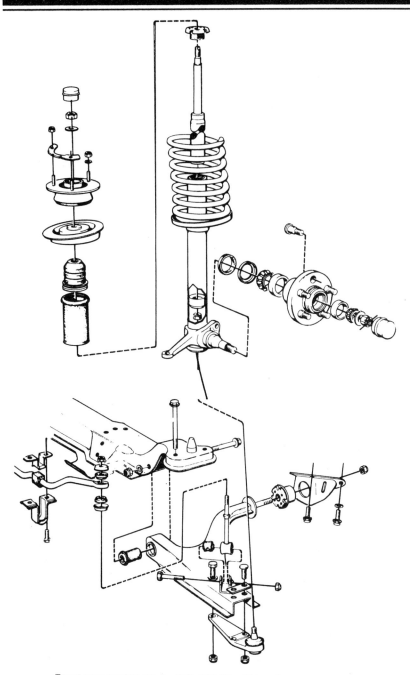

Front suspension strut—240, 260 (DL, GL) series; 760 similar

Front Wheel Bearings

REPLACEMENT AND ADJUSTMENT

1. Remove the hub cap, and loosen the lug nuts a few turns.
2. Firmly apply the parking brake. Jack up the front of the car and place jack stands beneath the lower control arms. Remove the wheel and tire assembly.
3. Remove the front caliper.
4. Pry off the grease cap from the hub. Remove the cotter pin and castle nut. Use a hub puller to pull off the hub. On the 760, remove the brake disc. If the inner bearing remains lodged on the stub axle, remove it with a puller.
5. Using a drift, remove the inner and outer bearing rings.
6. Thoroughly clean the hub, brake disc, and grease cap.
7. Press in the new inner and outer bearing rings with a drift.
8. Press grease into both bearings with a bearing packer. If one is not available, pack the bearings with as much wheel bearing grease as possible by hand. Also coat the outsides of the bearings and the outer rings pressed into the hub. Fill the recess in the hub with grease up to the smallest diameter on the outer ring for the outer bearing. Place the inner bearing in position in the hub and press its seal in with a drift. *The felt ring should be thoroughly coated with light engine oil.*
9. Place the hub onto the stub axle. Install the outer bearing washer, and castle nut.
10. Adjust the front wheel bearings by tightening the castle nut to 45 ft. lbs. to seat the bearings. Then, back off the nut 1/3 of a turn counterclockwise. Torque the nut to 1 ft. lb. If the nut slot does not align with the hole in the stub axle, tighten the nut until the cotter pin may be installed. Make sure that the wheel spins freely without any side play.
11. Fill the grease cap halfway with wheel bearing grease, and install it on the hub.
12. Install the front caliper.
13. Install the wheel and tire assembly. Remove the jack stand and lower the car. Tighten the lug nuts to 70–100 ft. lbs. and install the hub cap.

Front Wheel Alignment

CASTER AND CAMBER ADJUSTMENT

Caster angle is fixed by suspension design and cannot be adjusted. If caster is not within specifications, check front end parts for damage and replace as necessary.

Camber angle, however, may be adjusted. At the strut upper attachment to the body, two of the three bolts holes are eccentric, allowing the upper end of the strut to tilt out or in as necessary. A special pivot lever tool SVO No. 5038, which attaches to the tops of the strut upper attachment retaining bolt threads is recommended for this job. To adjust, loosen the three retaining nuts, install the pivot lever tool, and adjust to specifications. After adjusting, torque the nuts to 15–25 ft. lbs.

TOE-IN ADJUSTMENT

Toe in may be adjusted after performing the caster and camber adjustments. With a wheel spreader, measure the distance (X) between the rear of the right and left front tires, at spindle (hub) height, and then measure the distance (Y) between the front of the right and left front tires, also at spindle (hub) height. Subtract the front distance (Y) from the rear distance (X), and compare that to the specifications table. X−Y = toe-in. If the adjustment is not correct, loosen the locknuts on both sides of the tie rod, and rotate the tie rod itself. Toe-in is increased by turning the tie rod in the normal forward rotation of the wheels, and reduced by turning it in the opposite direction. After the final adjustment is made, torque the locknuts to 55–65 ft. lbs., being careful not to disturb the adjustment.

SECTION 2: SUSPENSION
Import Models

REAR SUSPENSION VOLVO

Springs

REMOVAL & INSTALLATION

1. Remove the hub cap and loosen the lug nuts a few turns. Jack up the car and place jack stands in front of the rear jacking points. Remove the wheel and tire assembly.
2. Place a hydraulic jack beneath the rear axle housing and raise the housing sufficiently to compress the spring. Loosen the nuts for the upper and lower spring attachments.

—— CAUTION ——
Due to the fact that the spring is compressed under several hundred pounds of pressure, when it is freed from its lower attachment, it will attempt to suddenly spring back to its extended position. It is therefore imperative that the axle housing be lowered with extreme care until the spring is fully extended. As an added safety measure, a chain may be attached to the lower spring coil and secured to the axle housing.

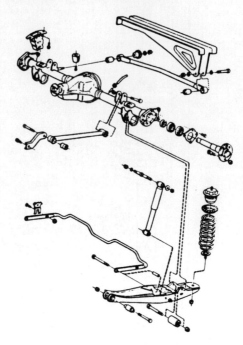

Rear suspension—240, 260 (DL, GL) series

3. Disconnect the shock absorber at its upper attachment. Carefully lower the jack and axle housing until the spring is fully extended. Remove the spring.
4. To install, position the retaining bolt and inner washer, for the upper attachment, inside the spring and then, while holding the outer washer and rubber spacer to the upper body attachment, install the spring and inner washer to the upper attachment (sandwiching the rubber spacer), and tighten the retaining bolt.
5. Raise the jack and secure the bottom of the spring to its lower attachment with the washer and retaining bolt.
6. Connect the shock absorber to its upper attachment. Install the wheel and tire assembly.
7. Remove the jack stands and lower the car. Tighten the lug nuts to 70–100 ft. lbs. and install the hub cap.

Shock Absorbers

REMOVAL & INSTALLATION

1. Remove the hub cap and loosen the lug nuts a few turns. Place blocks in front of the front wheels. Jack up the rear of the car to unload the shock absorbers and place jack stands in front of the rear jacking points. Remove the wheel and tire assembly.
2. Remove the nuts and bolts which retain the shock absorber to its upper and lower attachments and remove the shock absorber. Make sure that the spacing sleeve, inside the axle support arm for the lower attachment, is not misplaced.
3. The damping effect of the shock absorber may be tested by securing the lower attachment in a vise and extending and compressing it. A properly operating shock absorber should offer approximately three times as much resistance to extending the unit as compressing it. Replace the shock absorber if it does not function as above, or if its fixed rubber bushings are damaged. Replace any leaking shock absorber.
4. To install, position the shock absorber to its upper and lower attachments. Make sure that the spacing sleeve is installed inside the axle support (trailing) arm and is aligned with the lower attachment bolt hole. Install the retaining nuts and bolts, and torque to 63 ft. lbs. On 240 and 760 series models, the shock fits *inside* the support arm. On all 260 models, the shock attaches on the *outboard* side of the support arm.
5. Install the wheel and tire assembly. Remove the jack stands and lower the car. Tighten the lug nuts to 70–100 ft. lbs., and install the hub cap.

Light Trucks & Vans

CHEVROLET/GMC TRUCKS
SERIES 10-30 • 1500-9000 • ASTRO • SAFARI
VANS • BLAZER • JIMMY • SUBURBAN • PICK-UPS • S10 • S15

FRONT SUSPENSION

I-Beam Front Axle

This type of front axle is a one-piece steel forging in which dowel pins are installed to locate spring seats. Both ends of the axle are machined to accept the steering knuckle and kingpin assemblies and the kingpin inclination is a built-in angle.

SPRING

Removal and Installation

1. Wire brush all road dirt from threaded areas on U bolts, shock absorbers and stabilizer links, apply a good penetrant on threads.
2. Disconnect shock and stabilizer link at lower bracket.
3. Loosen both spring U bolts.
4. Using jack under I-beam, raise front of vehicle and support at frame side rail with stand. Finish removing U bolts and rebound bumper. Lower jack until spring clears I-beam or tire rests on ground. Remove any caster wedges (shims) and set aside for installation.

NOTE: Thick end of shim goes to rear of vehicle for increased caster.

5. Remove front spring eye bolt.
6. Remove rear shackle (and hanger cam if equipped).
7. Remove spring, inspect hangers and spring seat (center bolt index).
8. Reverse removal procedures to install after placing spring on axle with center bolt head indexed in seat and caster shims in place. Torque all nuts and lubricate.
9. Align front end assembly if the caster wedge had been removed.

STEERING KNUCKLE AND KINGPIN ASSEMBLY F-050, F-055, F-070 AND F-070D AXLES

CAUTION
Steering knuckle bushings are of a split type and are constructed of thermoplastic polyester. Bushings can be cleaned in most conventional solvents, except ketone or chlorinated types.

---

1. Support the frame of the vehicle in a raised position, high enough so that the tires clear the floor.
2. Remove the wheels, hubs and bearings.

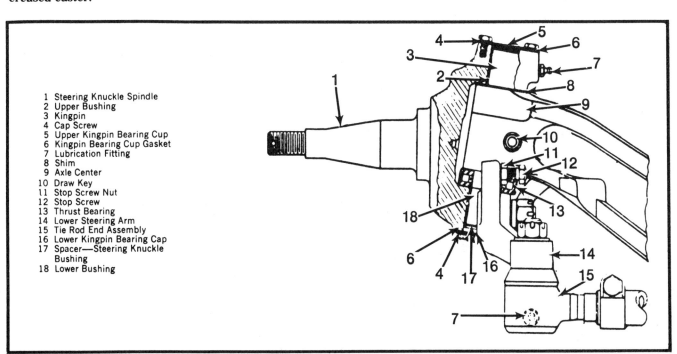

1 Steering Knuckle Spindle
2 Upper Bushing
3 Kingpin
4 Cap Screw
5 Upper Kingpin Bearing Cup
6 Kingpin Bearing Cup Gasket
7 Lubrication Fitting
8 Shim
9 Axle Center
10 Draw Key
11 Stop Screw Nut
12 Stop Screw
13 Thrust Bearing
14 Lower Steering Arm
15 Tie Rod End Assembly
16 Lower Kingpin Bearing Cap
17 Spacer—Steering Knuckle Bushing
18 Lower Bushing

F-050 and F-055 axle (© General Motors Corp.)

SUSPENSION
Light Trucks & Vans

3. Remove brake components as necessary.
4. Disconnect the tie-rod from the steering arm.
5. Remove the two lower steering arm-to-axle flange bolts and swing the steering arm out of the way.
6. Remove the two upper bolts from the axle flange and remove the brake backing plate.
7. Remove the kingpin draw key nut and washer.
8. Thread the draw key nut onto the draw key far enough to protect the threads from damage.
9. Drive the draw key loose by striking the nut with a brass hammer.
10. Finish driving the draw key out with a brass drift.
11. Remove the kingpin bearing cap screws.
12. Remove the kingpin bearing caps and gaskets.
13. Using a brass hammer, drive the kingpin out of the axle.
14. Remove the steering knuckle, thrust bearing, shims and the O-ring.

NOTE: Steering knuckle bushings can be hand pressed into the knuckle bore until flush with the top.

Installation

1. Position the steering knuckle on the axle and insert the thrust bearing.
2. Install a new O-ring seal at the bottom of the upper bushing.
3. Align the steering knuckle yoke, axle end and the thrust bearing to accept the kingpin; start the kingpin through the top of the assembly.
4. With the axle center firmly secured, jack up the steering knuckle until there is zero clearance between the steering knuckle lower yoke, thrust bearing and the axle center.
5. Check the clearance between the top of the axle center and the knuckle upper yoke. Select shims which will provide the correct thrust clearance as indicated in the Front Axle Specification Chart.
6. From the top, insert the kingpin through the steering knuckle yoke, shim, thrust bearing and axle center end. Press the kingpin down until the machined slot in the kingpin aligns with the draw key hole.
7. Insert the draw key into the axle center and install the washer and nut.

--- **CAUTION** ---
If, after tightening the draw key nut, the kingpin is not secured, then replace the draw key.

NOTE: On models using steering knuckle bushing spacer, install spacer at lower end of the kingpin.

8. Using new gaskets, install upper and lower kingpin bearing caps and cap screws.
9. Lubricate the kingpin with chassis lubricant.
10. Secure the brake backing plate to the axle flange with the top two axle flange bolts.
11. Swing the steering arm into position and secure it to the axle flange with two lower flange bolts.
12. Install brake components.
13. Install hubs and bearings.
14. Install wheels.
15. Lower the vehicle and check front end alignment. Make necessary adjustments.

STEERING KNUCKLE AND KINGPIN ASSEMBLY F-090 AXLE (F-120, F-120D AXLE SIMILAR)

NOTE: The steering knuckle is supported on a solid kingpin which is tapered at the center. The kingpin bushing is constructed of steel and the knuckle bushings are constructed of steel backed bronze.

Removal

1. Support the frame of the vehicle in a raised position, high enough so that the tires clear the floor.
2. Remove the wheels, hubs and bearings.
3. Remove brake components as necessary.
4. Disconnect steering linkage components as necessary.
5. Remove the axle flange-to-steering knuckle bolts and remove the brake backing plate.

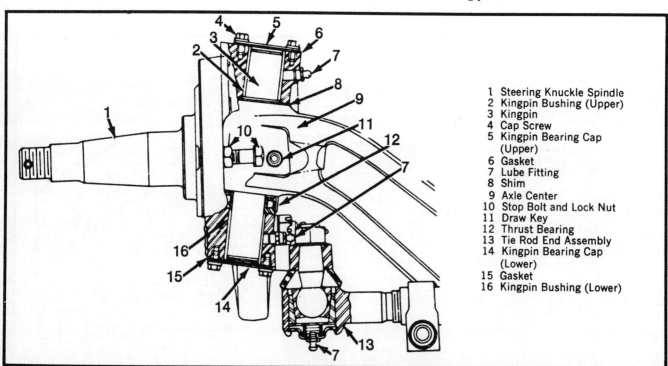

1 Steering Knuckle Spindle
2 Kingpin Bushing (Upper)
3 Kingpin
4 Cap Screw
5 Kingpin Bearing Cap (Upper)
6 Gasket
7 Lube Fitting
8 Shim
9 Axle Center
10 Stop Bolt and Lock Nut
11 Draw Key
12 Thrust Bearing
13 Tie Rod End Assembly
14 Kingpin Bearing Cap (Lower)
15 Gasket
16 Kingpin Bushing (Lower)

F-070 axle (© General Motors Corp.)

SUSPENSION
Light Trucks & Vans

6. Remove the dust cap and gasket.
7. Remove the lower expansion plug retainer and plug.

NOTE: If the plug does not remove freely after the retainer has been removed, it will come out along with the kingpin when the kingpin is driven out.

8. Remove the cotter pin, kingpin nut and the steel washer.
9. Drive the kingpin down and out of the axle with a brass drift.
10. Remove the steering knuckle, thrust bearing and spacers from the axle.

NOTE: Bushing replacement can be accomplished with the use of an arbor press.

Installation

1. With the steering knuckle positioned on the axle center end, insert the thrust bearing assembly between the lower face of the axle center and the steering knuckle lower yoke.

NOTE: Be sure that the retainer is on top of the bearing with the lip of the retainer facing down.

2. Align the knuckle yoke and axle center to accept the kingpin.
3. With the axle center firmly secured, jack up the steering knuckle until there is zero clearance between the steering knuckle lower yoke, thrust bearing and the axle center.
4. Check the clearance between the top face of the axle center end and the face of the upper steering knuckle yoke. Select shims which will provide the correct thrust clearance as indicated in the Front Axle Specification Chart.

NOTE: Kingpin, kingpin bore and component parts must be thoroughly cleaned and dry.

5. Insert the kingpin up through the bottom yoke of the steering knuckle and drive it into place with a soft hammer.
6. Position the kingpin bushing over the kingpin and press into place. Bushing must be flush with the knuckle.
7. Install the kingpin nut and cotter pin.
8. Install a new inverted expansion plug in the lower hole.
9. Install the plug retainer. Retainer must be seated securely in groove.
10. Install the kingpin dust cap and gasket.
11. Connect steering linkage components.
12. Install backing plate to axle flange.
13. Install brake components.
14. Install the hubs, bearings and wheels.

Independent Front Suspension

This suspension consists of upper and lower control arms, pivoting on steel threaded bushings on upper and lower control arm inner shafts which are attached to the crossmember. Control arms are connected to the steering knuckle by ball joints. A coil spring is seated between the upper and lower control arms, thus the lower control arm is the load carrying member.

NOTE: Astro Vans and S-series 2wd front suspension is similar to standard size Vans and Pickups. 4wd models use a torsion bar type front suspension.

COIL SPRING

Removal and Installation

1. Raise vehicle and place stands under frame, allowing control arm to hang free.
2. Disconnect shock absorber (and stabilizer if used) at lower end.
3. Using a floor jack under center of lower control arm inner shaft, raise and remove tension from shaft.

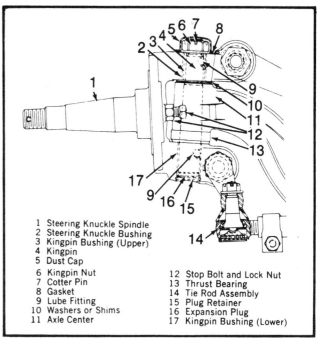

1 Steering Knuckle Spindle
2 Steering Knuckle Bushing
3 Kingpin Bushing (Upper)
4 Kingpin
5 Dust Cap
6 Kingpin Nut
7 Cotter Pin
8 Gasket
9 Lube Fitting
10 Washers or Shims
11 Axle Center
12 Stop Bolt and Lock Nut
13 Thrust Bearing
14 Tie Rod Assembly
15 Plug Retainer
16 Expansion Plug
17 Kingpin Bushing (Lower)

F-090 axle (© General Motors Corp.)

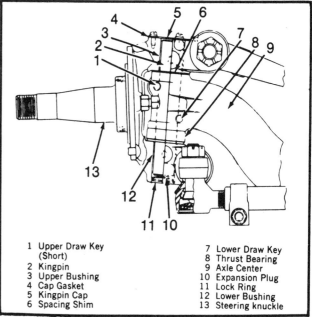

1 Upper Draw Key (Short)
2 Kingpin
3 Upper Bushing
4 Cap Gasket
5 Kingpin Cap
6 Spacing Shim
7 Lower Draw Key
8 Thrust Bearing
9 Axle Center
10 Expansion Plug
11 Lock Ring
12 Lower Bushing
13 Steering knuckle

F-120 axle (© General Motors Corp.)

CAUTION
Install a safety chain through arm and spring.

---

4. Remove both clamps or "U" bolts securing inner shaft to crossmember.
5. Release jack very cautiously, slowly lowering arm with spring until spring is free. Remove safety chain, remove spring.
6. Inspect front end especially at ball joints and both upper and lower control arm inner shaft bushings.
7. Reverse removal steps to install spring. Use a long tapered drift to align holes of inner control arm shaft and crossmember while slowly jacking arm into place.

SUSPENSION
Light Trucks & Vans

FRONT AXLE SPECIFICATIONS
Front Axle Models

Front Axle Component	F-070	F-090	F-120 & F-120D	FF-933	FL-931
KINGPIN					
Length	7.89"	8.750"	9.38"	10.392"	11.212"
Diameter	1.2492"–1.2496"	1.1855"–1.1865" (upper) 1.4330"–1.4340" (lower)	1.7445"–1.7455"	1.7930"–1.7940"	1.9980"–1.9990"
STEERING KNUCKLE					
Bushing Bore Diameter	1.3682"–1.3702"	1.560"–1.562"	1.876"–1.878"	1.919"–1.921"	2.124"–2.126"
STEERING KNUCKLE BUSHING					
Length	1.88"	1.709"–1.729" (upper) 1.990"–2.010" (lower)	2.240"	2.812"	2.625'
I.D. After Reaming or Burnishing	1.2496"–1.2526"	1.4365"–1.4375"	1.7465"–1.7485"	1.7965"–1.7975"	2.001"–2.003"
Installed Depth From Center	Flush	Flush	0.135" Min.	See Text	See Text
KINGPIN SLEEVE					
Length	—	1.9063"	—	—	—
Inside Diameter	—	1.1870"–1.1880"	—	—	—
Outside Diameter	—	1.4330"–1.4340"	—	—	—
STEERING KNUCKLE TO AXLE CENTER CLEARANCE					
In Service Check	0.001"–0.010" (0.03–0.25mm)	0.004"–0.012" (0.10–0.30mm)	0.005"–0.015" (0.13–0.38mm)	0.065" (1.70mm) maximum	0.065" (1.70mm) maximum
At Assembly	0.015"–0.030" (0.38–0.76mm)	0.015"–0.030" (0.38–0.76mm)	0.015"–0.030" (0.38–0.76mm)	0.005"–0.025" (0.12–0.65mm)	0.005"–0.025" (0.12–0.64mm)
SPACING WASHERS AVAILABLE	—	0.114"–0.116" white 0.121"–0.123" yellow 0.128"–0.130" blue	0.0478"	—	—
SHIM WASHERS AVAILABLE	0.005" (0.13mm)	0.005" (0.13mm) 0.010" (0.25mm)	0.010" (0.25mm) 0.015" (0.38mm)	0.005" (0.13mm) 0.010" (0.25mm)	0.005" (0.13mm) 0.010" (0.25mm)
STOP SCREW ADJUSTMENT (ALL MODELS)	Adjustment must provide ⅝" (16mm) minimum clearance from tire to any chassis component.				

UPPER CONTROL ARM

Removal

EXCEPT S–SERIES AND ASTRO VANS

1. Raise the vehicle and support safely. Remove the front wheels.
2. Position an adjustable jackstand under the outboard side of the lower control arm and adjust the height of the jackstand so that the uppermost extremity of the jackstand comes in contact with the metal undersurface of the lower control arm.
3. Remove the cotter pin from the upper control arm ball joint stud and nut.
4. Loosen the stud nut approximately one full turn.
5. If the ball joint stud does not unseat from the steering knuckle, it may be necessary to press the control arm and ball joint stud away from the knuckle, using a tool designed for that purpose.
6. Remove the nut from the ball stud and swing the upper control arm up and away from the steering knuckle.

NOTE: It may be necessary to remove the brake caliper assembly from the steering knuckle in order to facilitate upper control arm removal and installation.

7. Remove the nuts securing the control arm pivot shaft to the frame and remove the control arm.
8. Tape the alignment shims together and tag them in order to properly relocate them during installation.

Installation

NOTE: Special pivot shaft aligning washers must be positioned with the concave and convex sides together.

1. Situate the upper control arm against its normal mounting position and install the pivot shaft nuts. Do not tighten the nuts.
2. Install the alignment shims in their respective positions as noted during removal.

NOTE: Tighten the nut on the thinner shim pack first. This will improve shaft to frame clamping force and torque retention.

4. Insert the ball joint stud into the bore and the steering knuckle and install the nut and cotter pin.
5. Install the brake caliper assembly.
6. Remove the adjustable jackstand from under the lower control arm.
7. Install the wheel and tire assembly.
8. The vehicle may now be positioned to check front end alignment.

NOTE: Ordinarily, a shim pack will leave at least two threads of the bolt exposed beyond the nut. If, in order to properly align the front wheels, it is necessary to build the shim pack beyond the two thread minimum, then check for damaged control arms and related parts. The difference, in thickness, between the front shim pack and the rear shim pack must not exceed 0.03 inches. The front shim pack must be at least 0.24 inches.

LOWER CONTROL ARM

Removal

EXCEPT S–SERIES AND ASTRO VANS

1. Raise the vehicle and support safely. Remove the front coil spring (see Coil Spring/Removal and Installation).
2. Support the disconnected inboard end of the lower control arm after the spring is removed.
3. Remove the cotter pin from the lower ball joint stud and loosen the stud nut approximately one full turn.
4. Press the control arm and ball joint stud away from the steering knuckle, using a tool designed for that purpose.

NOTE: It may be necessary to remove the brake caliper assembly from the steering knuckle in order to facilitate lower control arm removal and installation.

5. Remove the lower control arm.

Installation

1. Insert the lower ball joint stud through the steering knuckle and tighten the nut.
2. Install the coil spring and reattach the inboard end of the lower control arm to the crossmember.
3. Be sure that the ball joint stud nut is properly tightened and install the cotter pin.
4. Install the brake caliper assembly.
5. Remove the vehicle from the hoist.

NOTE: It is always advisable that the front end alignment be checked after any component of the front suspension has been replaced.

UPPER CONTROL ARM

Removal

S–SERIES AND ASTRO VANS

1. Note the location of the shims. Alignment shims are to be installed in the same position from which they were removed. Remove nuts and shims. Raise front of vehicle and support lower control arm with floor stands.

---- **CAUTION** ----

Floor jack must remain under control arm spring seat during removal and installation to retain spring and control arm in position.

Since the weight of the vehicle is used to relieve spring tension on the upper control arm, the floor stands must be positioned between the spring seats and ball joints of the lower control arms for maximum leverage.

2. Remove wheel, then loosen the upper ball joint from the steering knuckle as previously outlined.
3. Support hub assembly to prevent weight from damaging brake hose.
4. It is necessary to remove the upper control arm attaching bolts to allow clearance to remove upper control arm assembly.
5. Remove upper control arm.

Installation

1. Position upper control arm attaching bolts loosely in the frame and install pivot shaft on the attaching bolts.

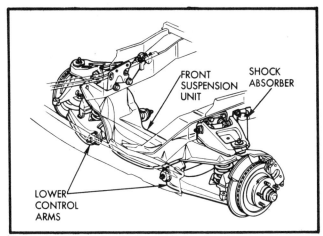

Coil spring suspension (© General Motors Corp.)

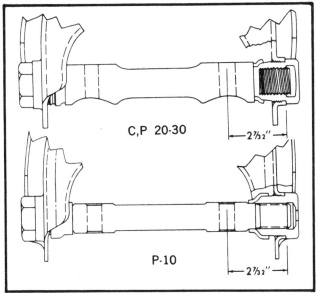

Correctly positioned upper control arm steel bushings
(© General Motors Corp.)

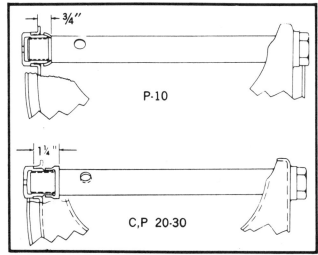

Correctly positioned lower control arm steel bushings
(© General Motors Corp.)

SECTION 2

SUSPENSION
Light Trucks & Vans

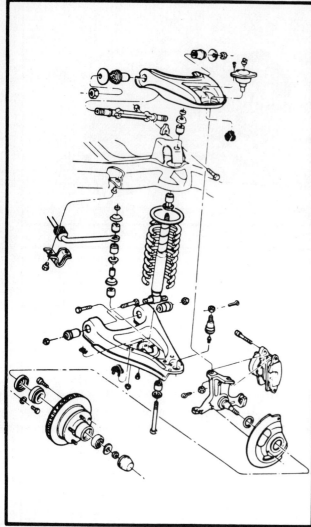

Exploded view of front suspension components, S series (2WD) and Astro van models (© General Motors Corp.)

The inner pivot bolts must be installed with the bolt heads to the front (on the front bushing) and to the rear (on the rear bushing).

2. Install alignment shims in their original position between the pivot shaft and frame on their respective bolts. Torque nuts to 45 ft. lbs

3. Remove the temporary support from the hub assembly, then connect ball joint to steering knuckle as previously outlined.

4. Install wheel, then check wheel alignment and adjust if necessary.

LOWER CONTROL ARM

Removal

S-SERIES AND ASTRO VANS

1. Remove coil spring as described earlier in this section.
2. Remove the lower ball joint stud as previously outlined.
3. After stud breaks loose, hold up on lower control arm. Remove control arm.
4. Guide lower control arm out of opening in splash shield with a putty knife or similar tool.

Installation

1. Install lower ball joint stud into knuckle. Install nut as previously outlined.
2. Install spring.
3. Check front alignment. Reset as required.

STABILIZER BAR

Removal and Installation

S-SERIES AND ASTRO VANS

1. Raise the vehicle and support safely.
2. Disconnect each side of stabilizer linkage by removing nut from link bolt, pull bolt from linkage and remove retainers, grommets and spacer.
3. Remove bracket to frame or body bolts and remove stabilizer shaft, rubber bushings and brackets.
4. To replace, reverse sequence of operations, being sure to install with the identification forming on the right side of the vehicle. The rubber bushings should be positioned squarely in the brackets with the slit in the bushings facing the front of car. Torque stabilizer link nut to 13 ft. lbs and bracket bolts to 24 ft. lbs

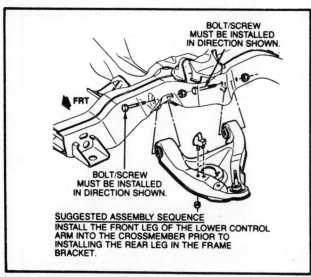

Lower control arm assembly on S-series

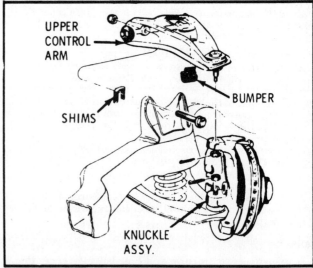

Upper control arm assembly on S-series

SUSPENSION
Light Trucks & Vans

TORSION BAR

Removal and Installation
S-SERIES 4WD

1. Raise the vehicle and support safely.
2. Remove torsion bar adjusting screw.
3. Remove support retainer attaching nuts and bolts.
4. Slide torsion bar forward in lower control arm until torsion bar clears support. Pull down on bar and remove from control arm.
5. Installation is the reverse of removal.

NOTE: Count the number of turns when removing the torsion bar for easy reinstallation. Apply lubricant to top of adjusting arm and adjusting bolt for easy reinstallation. Also apply lubricant to hex ends of torsion bar.

BALL JOINT

Removal and Installation
UPPER

1. Place jack under lower control arm, at coil spring and raise vehicle.
2. Remove tire, wheel and drum or rotor assembly.
3. Remove upper ball stud nut and break the stud taper from the steering knuckle by rapping sides of knuckle flats at stud. Separate stud from knuckle.

NOTE: Support the knuckle assembly to prevent weight of the assembly from damaging the brake hose.

4. Remove rivets and bolt in new ball joint. Rivets can be chiseled off, ground off or drilled out.

─────── CAUTION ───────
Use special hardened bolts only when installing joint (furnished with joint).
─────────────────────

5. Reverse remaining removal steps to complete installation.

LOWER

1. Jack vehicle at lower control arm spring seat and remove tire, wheel and drum or rotor assembly. Support with jack stands under frame, keeping the jack under the spring seat.
2. Remove ball stud nut at knuckle, use stud jack or rap stud loose from knuckle.
3. Press out ball joint. Press new ball joint into arm. Make sure ball joint assembly is fully seated and square with arm. Check inner shaft bushings.
4. Reverse remaining removal steps to complete installation.

WHEEL BEARINGS

Removal, Packing, Installation
WITH LOCKING HUBS

NOTE: This procedure requires snapring pliers and a special hub nut wrench and does not apply to S/T-series trucks. The 4WD S/T series trucks have front sealed wheel bearings, which are pre-adjusted and require no lubrication.

1. Remove the wheel and tire.
2. For ½ and ¾ ton trucks with lock front hubs: lock the hubs. Remove the outer retaining plate Allen head bolts and take off the plate, O-ring and knob. Take out the large snapring inside the hub and remove the outer clutch retaining ring and actuating cam body. This is a lot easier with snapring pliers. Relieve pressure on the axle shaft snapring and remove it. Take out the axle shaft sleeve and clutch ring assembly and the inner clutch ring and bushing assembly. Remove the spring and retainer plate.
3. If the vehicle doesn't have locking front hubs, remove the hub cap and snapring. Next, remove the drive gear and pressure spring. To prevent the spring from popping out, place a hand over the drive gear and use a screwdriver to pry the gear out. Remove the spring.
4. Remove the wheel bearing outer lock nut, lock ring and wheel bearing inner adjusting nut. A special wrench is required.
5. Remove the brake disc assembly and outer wheel bearing. Remove the spring retainer plate if you don't have locking hubs.
6. Remove the oil seal and inner bearing cone from the hub using a brass drift and hammer. Discard the oil seal. Use the drift to remove the inner and outer bearing cups.

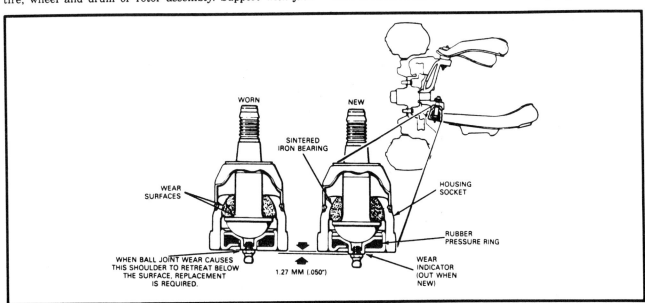

Ball joint wear indicators (© General Motors Corp.)

SECTION 2
SUSPENSION
Light Trucks & Vans

FRONT WHEEL BEARING DIAGNOSIS

CONSIDER THE FOLLOWING FACTORS WHEN DIAGNOSING BEARING CONDITION:

1. GENERAL CONDITION OF ALL PARTS DURING DISASSEMBLY AND INSPECTION.
2. CLASSIFY THE FAILURE WITH THE AID OF THE ILLUSTRATIONS.
3. DETERMINE THE CAUSE.
4. MAKE ALL REPAIRS FOLLOWING RECOMMENDED PROCEDURES.

GOOD BEARING

BENT CAGE

CAGE DAMAGE DUE TO IMPROPER HANDLING OR TOOL USAGE.

REPLACE BEARING.

BENT CAGE

CAGE DAMAGE DUE TO IMPROPER HANDLING OR TOOL USAGE.

REPLACE BEARING.

GALLING

METAL SMEARS ON ROLLER ENDS DUE TO OVERHEAT, LUBRICANT FAILURE OR OVERLOAD.

REPLACE BEARING — CHECK SEALS AND CHECK FOR PROPER LUBRICATION.

ABRASIVE STEP WEAR

PATTERN ON ROLLER ENDS CAUSED BY FINE ABRASIVES.

CLEAN ALL PARTS AND HOUSINGS, CHECK SEALS AND BEARINGS AND REPLACE IF LEAKING, ROUGH OR NOISY.

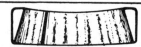

ETCHING

BEARING SURFACES APPEAR GRAY OR GRAYISH BLACK IN COLOR WITH RELATED ETCHING AWAY OF MATERIAL USUALLY AT ROLLER SPACING.

REPLACE BEARINGS — CHECK SEALS AND CHECK FOR PROPER LUBRICATION.

MISALIGNMENT

OUTER RACE MISALIGNMENT DUE TO FOREIGN OBJECT.

CLEAN RELATED PARTS AND REPLACE BEARING. MAKE SURE RACES ARE PROPERLY SEATED.

INDENTATIONS

SURFACE DEPRESSIONS ON RACE AND ROLLERS CAUSED BY HARD PARTICLES OF FOREIGN MATERIAL.

CLEAN ALL PARTS AND HOUSINGS, CHECK SEALS AND REPLACE BEARINGS IF ROUGH OR NOISY.

FATIGUE SPALLING

FLAKING OF SURFACE METAL RESULTING FROM FATIGUE.

REPLACE BEARING — CLEAN ALL RELATED PARTS.

SUSPENSION
Light Trucks & Vans

FRONT WHEEL BEARING DIAGNOSIS

BRINELLING

SURFACE INDENTATIONS IN RACEWAY CAUSED BY ROLLERS EITHER UNDER IMPACT LOADING OR VIBRATION WHILE THE BEARING IS NOT ROTATING.

REPLACE BEARING IF ROUGH OR NOISY.

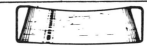

CAGE WEAR

WEAR AROUND OUTSIDE DIAMETER OF CAGE AND ROLLER POCKETS CAUSED BY ABRASIVE MATERIAL AND INEFFICIENT LUBRICATION. CHECK SEALS AND REPLACE BEARINGS.

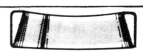

ABRASIVE ROLLER WEAR

PATTERN ON RACES AND ROLLERS CAUSED BY FINE ABRASIVES.

CLEAN ALL PARTS AND HOUSINGS, CHECK SEALS AND BEARINGS AND REPLACE IF LEAKING, ROUGH OR NOISY.

CRACKED INNER RACE

RACE CRACKED DUE TO IMPROPER FIT, COCKING, OR POOR BEARING SEATS.

SMEARS

SMEARING OF METAL DUE TO SLIPPAGE, SLIPPAGE CAN BE CAUSED BY POOR FITS, LUBRICATION, OVERHEATING, OVERLOADS OR HANDLING DAMAGE.

REPLACE BEARINGS, CLEAN RELATED PARTS AND CHECK FOR PROPER FIT AND LUBRICATION.

REPLACE SHAFT IF DAMAGED.

FRETTAGE

CORROSION SET UP BY SMALL RELATIVE MOVEMENT OF PARTS WITH NO LUBRICATION.

REPLACE BEARING. CLEAN RELATED PARTS. CHECK SEALS AND CHECK FOR PROPER LUBRICATION.

HEAT DISCOLORATION

HEAT DISCOLORATION CAN RANGE FROM FAINT YELLOW TO DARK BLUE RESULTING FROM OVERLOAD OR INCORRECT LUBRICANT.

EXCESSIVE HEAT CAN CAUSE SOFTENING OF RACES OR ROLLERS.

TO CHECK FOR LOSS OF TEMPER ON RACES OR ROLLERS A SIMPLE FILE TEST MAY BE MADE. A FILE DRAWN OVER A TEMPERED PART WILL GRAB AND CUT META, WHEREAS, A FILE DRAWN OVER A HARD PART WILL GLIDE READILY WITH NO METAL CUTTING.

REPLACE BEARINGS IF OVER HEATING DAMAGE IS INDICATED. CHECK SEALS AND OTHER PARTS.

STAIN DISCOLORATION

DISCOLORATION CAN RANGE FROM LIGHT BROWN TO BLACK CAUSED BY INCORRECT LUBRICANT OR MOISTURE.

RE-USE BEARINGS IF STAINS CAN BE REMOVED BY LIGHT POLISHING OR IF NO EVIDENCE OF OVERHEATING IS OBSERVED.

CHECK SEALS AND RELATED PARTS FOR DAMAGE.

SUSPENSION
Light Trucks & Vans

7. Check the condition of the spindle bearing. Unbolt the spindle and tap it with a soft hammer to break it loose. Remove the spindle and check the condition of the thrust washer, replacing it if worn. Now you can remove the oil seal and spindle roller bearing.

NOTE: The spindle bearings must be greased each time the wheel bearings are serviced.

8. Clean all parts in solvent, dry and check for wear or damage.
9. Pack both wheel bearings (and the spindle bearing) using wheel bearing grease. Place a healthy glob of grease in the palm of one hand and force the edge of the bearing into it so that grease fills the bearing. Do this until the whole bearing is packed. Grease packing tools are available to make this job easier.
10. To reassemble the spindle: drive the repacked bearing into the spindle and install the grease seal onto the slinger with the lip toward the spindle. It would be best to replace the axle shaft slinger when the spindle seal is replaced.

If you are using the improved seals, fill the seal end of the spindle with grease. If not, apply grease only to the lip of the seal. Install the thrust washer over the axle shaft. On late 1982 models, the chamfered side of the thrust washer should be toward the slinger. Replace the spindle and torque the nuts to 33 ft. lbs for 1980 models and 65 ft. lbs for 1981 and later.

11. To reassemble the wheel bearings: drive the outer bearing cup into the hub, replace the inner bearing cup and insert the repacked bearing.
12. Install the disc or drum and outer wheel bearing to the spindle.
13. Adjust the bearings by rotating the hub and torquing the inner adjusting nut to 50 ft. lbs, then loosening it and retorquing to 35 ft. lbs Next, back the nut off $3/8$ turn or less. Turn the nut to the nearest hole in the lockwasher. Install the outer locknut and torque to a minimum of 80 ft. lbs for 1980, 160–205 ft. lbs for 1981 and later $1/2$ and $3/4$ ton and 65 ft. lbs on 1 ton vehicles. There should be 0.001–0.010 in. bearing end play. This can be measured with a dial indicator.
14. Replace the brake components.
15. Lubricate the locking hub components with high temperature grease. Lubrication must be applied to prevent component failure. For $1/2$ ton and 1980 and later $3/4$ and 1 ton models, install the spring retainer plate with the flange side facing the bearing outer cup. Install the pressure spring with the large end against the spring retaining plate. The spring is an interference fit; when seated, its end extends past the spindle nuts by approximately $7/8$ in. Place the inner clutch ring and bushing assembly into the axle shaft sleeve and clutch ring assembly and install that as an assembly onto the axle shaft. Press in on this assembly and install the axle shaft ring.

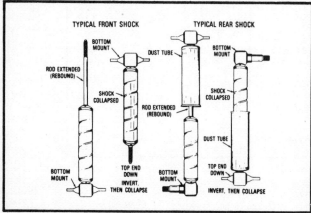

Purging air from shock absorbers (© General Motors Corp.)

Removal & Installation
S/T–SERIES TRUCKS

NOTE: The S/T series trucks have sealed front wheel bearings which are pre-adjusted and require no lubrication. They are replaced as an assembly, should it be required.

1. Raise the front of the vehicle and support safely. Remove the wheel assembly, caliper and rotor.
2. Remove the cotter pin, the locking cap, washer and the axle nut from the axle end.
3. Remove the hub and bearing assembly. It may be necessary to employ a puller to remove the hub from the axle.
4. To install the hub and bearing, reverse the removal procedure. The axle nut should be torqued to 174 ft. lbs

NOTE: Do not back off the axle nut to install the new cotter pin.

DRUM OR ROTOR TYPE ADJUSTABLE WHEEL BEARINGS

Adjustment

1. Check the bearing for a tight or loose fit by gripping the wheel at the top and bottom and moving the wheel in and out on the spindle. The end play should be 0.001–0.005 in..
2. If adjustment is needed, remove the cotter pin and tighten the spindle nut to 12 ft. lbs to fully seat the bearings.
3. Loosen the nut until either hole in the spindle lines up with the slot in the nut.
4. Install the cotter pin and bend the ends against the nut. The end play should be between 0.001 and 0.005 in..
5. Install the dust cover and wheel cover, if equipped.

Shock Absorbers

Shock absorbers are used to dampen the rebound of the two types of springs used: coil and leaf.

LEAF TYPE

The top of the shock absorbers are mounted to the frame and the bottom is usually mounted to the U bolt bracket at the axle area or to a bracket welded to the axle housing.

COIL SPRING

FRONT

The shock absorbers are usually attached to the lower control arm at the bottom and to the frame rail at the top. On some models, the shock absorbers may be mounted through the coil spring.

REAR

The top of the shock absorber is mounted to the body or to a crossmember with the bottom mounted to a stud or bracket welded or mounted on the axle housing.

Removal and Replacement

Removal and replacement is accomplished by the removing of the attaching retainers at the top and bottom of the shock absorber and withdrawing the shock. Replacement is the reverse of removal. Air should be purged from the shock absorber by extending it in the upright position and then inverting and collapsing the shock.

Front Drive Axle

The front axle is a hypoid type gear unit equipped with either

SUSPENSION
Light Trucks & Vans

ball joint or kingpin steering knuckles and is powered through a transfer case which may be one of two types. A full-time four wheel drive unit is used mainly with V8 engines and automatic transmission. The other type is a conventional part-time four wheel drive system.

A yoke and a trunnion universal joint, as part of the drive axle, allows a continuous power flow to each wheel, regardless of the turning angle.

Free-wheeling hubs are available on the front wheels, except those vehicles equipped with the full time four wheel drive transfer case.

For repairs to the hypoid gear unit, refer to the Unit Repair section.

LEAF SPRING

Removal

1. Raise the vehicle and support safely.
2. Position an adjustable jack under the front axle.
3. Situate the axle so that all tension is removed from the spring.
4. Remove shackle retaining bolt (upper).
5. Remove the front eye bolt from the spring.
6. Remove the U-bolt nuts and remove the spring, lower plate and spring pads.
7. Remove the spring-to-shackle bolt and remove the bushings and shackle.

Installation

1. Install the shackle bushings in the spring and attach the shackle. Do not tighten bolt.
2. Place the upper cushion on the spring.
3. Position the front of the spring in its mounted position at the frame and install the bolt. Do not tighten bolt.
4. Position the shackle bushings in the frame and attach the rear shackle. Do not tighten bolt.
5. Install the lower spring pad.
6. Install the spring retainer plate. Tighten bolts.
7. Tighten front and rear spring eye and shackle bolts.
8. Lower the vehicle. Recheck the tightness of the bolts and nuts.

AUTOMATIC LOCKING HUBS

1981 AND LATER

The automatic locking hub engages or disengages to lock the front axle shaft to the hub of the front wheel. Engagement occurs whenever the vehicle is operated in four-wheel drive. Disengagement occurs whenever two-wheel drive has been selected and the vehicle is moved rearward. Disengagement will not occur when the vehicle is moved rearward if four-wheel drive is selected and the hub has already been engaged.

The outer clutch housing is splined to the wheel. The hub sleeve is splined to the front axle shaft. The clutch gear is splined to the hub sleeve. Engagement occurs when the clutch gear is moved on the splines of the hub sleeve to engage the internal teeth of the outer clutch housing.

The cam surface of the steel inner cage forces the cam follower and clutch gear to move outward toward the cover and into engagement with the clutch teeth of the outer clutch housing. A lug on the inside of the drag sleeve retainer washer keys the washer to the axle and two lock nuts retain this washer in position on the axle. Cutouts in the drag sleeve engage the four tabs on the drag sleeve retainer washer to hold the drag sleeve in a fixed position with respect to the axle shaft. The one way clutch spring (called a brake band) is positioned over the serrated portion of the drag sleeve.

Engagement is accomplished (when four-wheel drive is selected) by the movement of the drag sleeve, causing one of the tangs of the brake band to engage the steel outer cage and rotate the cage which will cause the cam ramp to move the clutch gear into mesh with the outer clutch housing. One of the tangs of the brake band is used for engagement when the vehicle is moving forward. The other tang is used to engage reverse when the vehicle is moving rearward.

Disengagement is accomplished (when two-wheel drive has been selected) by the reverse movement of the wheel causing the clutch gear, hub sleeve and cam follower to rotate. The cam follower moves against the lugs of the plastic outer cage, causing the cage to rotate and move to the disengaged condition. The release spring then moves the clutch gear out of mesh with the outer clutch housing to disengage the wheel from the axle shaft.

Preliminary Checking

Before disassembling a unit for complaint of abnormal noise, read the following:

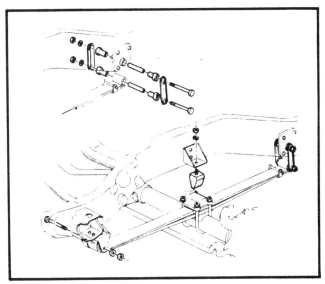

Spring assembly—front drive axle (© General Motors Corp.)

TIGHTENING SEQUENCE

1. INSTALL ALL FOUR NUTS TO UNIFORM ENGAGEMENT ON U-BOLTS TO RETAIN AND POSITION ANCHOR POSITION (PERPENDICULAR TO PLATE IN DESIGN AXIS OF U-BOLTS).
2. TORQUE NUTS IN POSITIONS 1 AND 3 TO 10-25 FT. LBS.
3. TORQUE ALL NUTS TO FULL TORQUE IN FOLLOWING SEQUENCE: 2-4-1-3

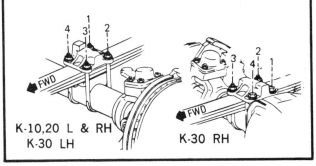

U-bolt tightening sequence—front drive axle (© General Motors Corp.)

SECTION 2
SUSPENSION
Light Trucks & Vans

1. To obtain all-wheel drive, the transfer case lever must be placed in (4L) or (4H), at which time the hub locks will automatically engage.
2. To unlock (free wheel) the hubs, shift the transfer case lever to (2H), then slowly reverse vehicle direction approximately ten feet.
3. Incomplete shift from 2WD to 4WD, or disengagement of only one hub lock may cause an abnormal sound from the front axle. Shift to 4WD to stop the noise, then unlock the hubs as described previously.

Removal Of Hub From Wheel

1. Remove the retaining screws which hold the cover to the outer clutch housing.
2. Remove the cover, seal, seal bridge and the bearing components.
3. Using needle nose pliers, compress the retaining wire ring and pull the remaining components from the wheel.

Disassembly

1. Remove the snap-ring from the groove of the hub sleeve.
2. Turn the clutch gear until it drops into engagement with the outer clutch housing. Lift and cock the drag sleeve to unlock the tangs of the brake band from the window of the inner cage and remove the drag sleeve and brake assembly.

CAUTION

Do not remove the brake band from the drag sleeve. To do so would change the spring tension of the brake band by over expanding the coils, effecting the operation of the hub.

---

3. Remove the snap-ring from the groove in the outer clutch housing.
4. Use a pointed probe to pry the plastic outer cage free of the inner cage while the inner cage is being removed.
5. Pry the plastic outer cage tabs free from the groove in the outer clutch housing and remove the outer cage.
6. Remove the clutch sleeve and attached components from the outer clutch housing.
7. Compress the return spring and hold the spring in the compressed position with fabricated clamps. Position the entire assembly in a bench vise so that the vise holds both ends of the clutch sleeve. Remove the retaining ring.
8. Slowly remove the clamps with the unit still in the vise. Slowly open the vise jaws to permit the release of the return spring tension, in a controlled manner. Remove the retainer seat, spring and spring support washers from the hub sleeve.

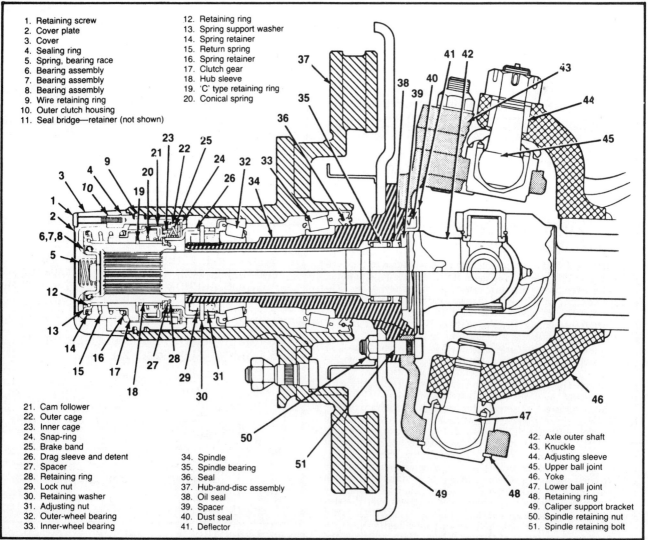

Front axle assembly with automatic locking hubs and ball joints—1981-82 K series (© General Motors Corp.)

SUSPENSION
Light Trucks & Vans

9. Remove the "C" type retaining ring from the clutch sleeve.

NOTE: It is necessary to position the sleeve assembly so that the C-ring ends are aligned with the legs of the cam follower, allowing removal between the two legs.

10. Remove the conical spring from between the cam follower and the clutch gear.
11. Separate the cam follower from the clutch gear.

Assembly

1. Snap the tangs of the cam follower over the flats of the clutch gear.
2. Compress the conical spring and slide it into position with the large diameter of the spring located against the clutch gear.
3. Position the clutch gear assembly over the splines of the hub sleeve. The teeth of the cam follower should be located at the end of the hub sleeve which has no splines. The clutch gear and spring should slide freely over the splines of the hub sleeve.
4. Assemble the "C" shaped retainer ring in the groove of the hub sleeve.
5. Install the spring retainer over each end of the return spring. Place one end of the spring against the shoulder of the clutch gear.
6. Position the spring support washer against the retainer on the end of the spring. Compress the return spring and assemble the retainer ring in the groove of the hub sleeve. Two "C" shaped clamps may be used to retain the return spring while the retainer ring is being assembled.
7. Install the pre-assembled components into the outer housing. The cam follower should be positioned with the two legs directly outboard.
8. Screw three cover screws into three holes of the outer clutch housing.

NOTE: These three screws will support the component to permit the clutch hub to drop down so that the tangs of the brake band can be assembled.

9. Install the plastic outer cage into the outer clutch housing with the ramps facing towards the cam follower. The small external tabs of the plastic cage should be located in the wide groove of the outer clutch housing.
10. Assemble the steel inner cage into the outer cage, aligning the tab of the outer cage with the "window" of the inner cage.
11. Assemble the retaining ring into the groove of the outer clutch housing above the outer cage.
12. Assemble the tangs of the brake band with one tang on each side of the lug on the outer cage, which is located in the "window" of the steel inner cage.

NOTE: The brake band and the drag sleeve are serviced as an assembly.

13. Remove the three screws and rest the end of the hub sleeve on a suitable support. Assemble the washer and snapring above the drag sleeve.
14. As the hub is assembled to the vehicle, assemble the wire retaining ring in the groove in the unsplined end of the outer clutch housing.

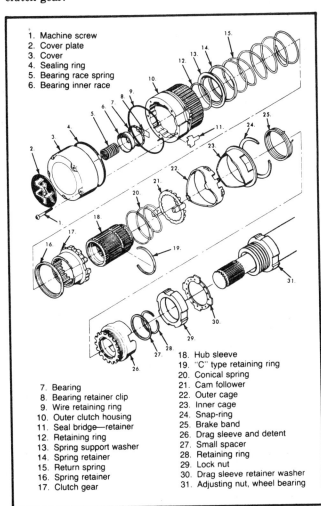

1. Machine screw
2. Cover plate
3. Cover
4. Sealing ring
5. Bearing race spring
6. Bearing inner race
7. Bearing
8. Bearing retainer clip
9. Wire retaining ring
10. Outer clutch housing
11. Seal bridge—retainer
12. Retaining ring
13. Spring support washer
14. Spring retainer
15. Return spring
16. Spring retainer
17. Clutch gear
18. Hub sleeve
19. "C" type retaining ring
20. Conical spring
21. Cam follower
22. Outer cage
23. Inner cage
24. Snap-ring
25. Brake band
26. Drag sleeve and detent
27. Small spacer
28. Retaining ring
29. Lock nut
30. Drag sleeve retainer washer
31. Adjusting nut, wheel bearing

Exploded view of automatic locking hubs, front wheel drive
(© General Motors Corp.)

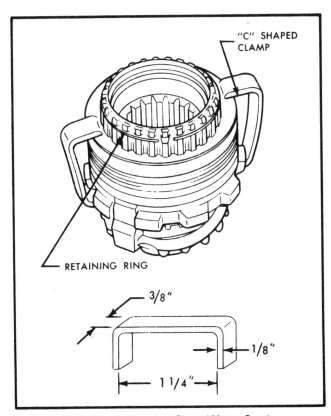

Fabricated "C" shaped clamps (© General Motors Corp.)

SUSPENSION
Light Trucks & Vans

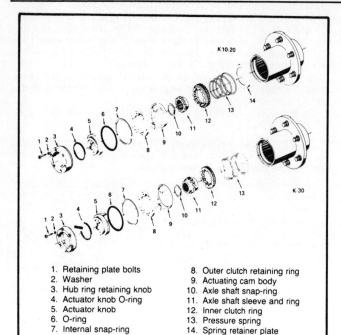

1. Retaining plate bolts
2. Washer
3. Hub ring retaining knob
4. Actuator knob O-ring
5. Actuator knob
6. O-ring
7. Internal snap-ring
8. Outer clutch retaining ring
9. Actuating cam body
10. Axle shaft snap-ring
11. Axle shaft sleeve and ring
12. Inner clutch ring
13. Pressure spring
14. Spring retainer plate

Exploded view of free-wheeling hub (© General Motors Corp.)

NOTE: The tangs of the retainer ring should point away from the splined end of the clutch housing.

15. Hold the tangs of the wire retainer together and assemble the two bent down tabs of the seal bridge over the tangs. The seal bridge will hold the retainer ring in a clamped condition in the groove of the outer clutch housing. Assemble the O-ring in the groove of the outer clutch housing and over the seal bridge.
16. Assemble the bearing over the inner race and lubricate.

NOTE: The steel balls should be visible when the bearing is properly installed.

17. Snap the bearing retainer clip into the hole in the outer race.
18. Assemble the bearing and retainer assembly in the end of the hub sleeve. Assemble the sealing ring over the outer clutch housing.
19. Assemble the bearing race spring into the bore in the cover.
20. Assemble the cover and spring assembly. Align the hole in the cover to the holes in the outer clutch housing and assemble the five screws.
21. Place the O-ring over the seal bridge to prevent it from jumping out of position during the handling prior to the hub bearing being installed in the vehicle.

NOTE: This O-ring may be left on, but is not necessary.

22. The hub sleeve and attached parts should turn freely after the unit has been completely assembled.
23. The five cover screws must be loosened to assemble the hub to the vehicle. After the hub is installed, torque the screws to 40–50 inch lbs.

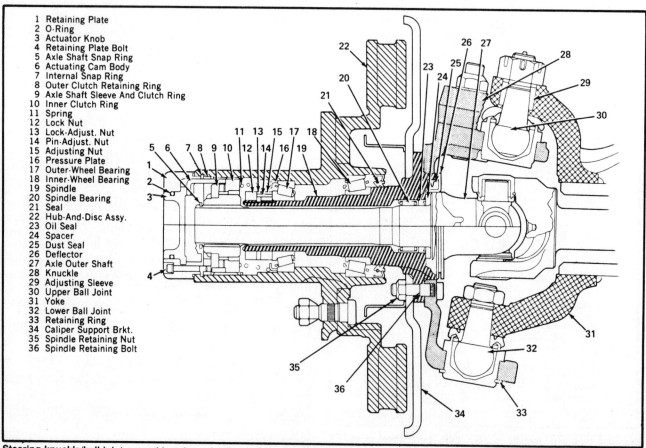

1. Retaining Plate
2. O-Ring
3. Actuator Knob
4. Retaining Plate Bolt
5. Axle Shaft Snap Ring
6. Actuating Cam Body
7. Internal Snap Ring
8. Outer Clutch Retaining Ring
9. Axle Shaft Sleeve And Clutch Ring
10. Inner Clutch Ring
11. Spring
12. Lock Nut
13. Lock-Adjust. Nut
14. Pin-Adjust. Nut
15. Adjusting Nut
16. Pressure Plate
17. Outer-Wheel Bearing
18. Inner-Wheel Bearing
19. Spindle
20. Spindle Bearing
21. Seal
22. Hub-And-Disc Assy.
23. Oil Seal
24. Spacer
25. Dust Seal
26. Deflector
27. Axle Outer Shaft
28. Knuckle
29. Adjusting Sleeve
30. Upper Ball Joint
31. Yoke
32. Lower Ball Joint
33. Retaining Ring
34. Caliper Support Brkt.
35. Spindle Retaining Nut
36. Spindle Retaining Bolt

Steering knuckle/ball joint assembly w/free-wheeling hub (© General Motors Corp.)

SUSPENSION
Light Trucks & Vans

LOCKED HUB (FULL-TIME)

Removal

K10–30, K1500–3500

1. Remove the hub cap and the snap-ring.
2. Remove the drive gear and pressure spring, if equipped.
3. Remove the wheel bearing outer lock nut, lock ring and the wheel bearing inner adjusting nut.
4. Remove the hub and disc assembly.
5. Remove the outer wheel bearing and the spring retainer plate.
6. Drive the inner bearing cone and oil seal from the hub with the use of a brass drift. Discard the oil seal.
7. Using a brass drift, remove the inner and outer bearing cups.

Installation

1. Install the outer wheel bearing cup into the wheel hub.
2. Install the inner wheel bearing cup into the wheel hub.
3. Pack the wheel bearing cone with a suitable wheel bearing grease (high melting point type).
4. Install the cone into the cup.
5. Install a new grease seal into the inboard end of the hub.
6. Lubricate the wheel bearings; install the hub and disc and the bearings on the spindle.
7. While rotating the hub and disc, torque the inner adjusting nut to 50 ft. lbs
8. Back off the inner adjusting nut.
9. While rotating the hub and disc, torque the inner adjusting nut to 35 ft. lbs
10. Again, back off the inner adjusting nut a maximum of $^3/_8$ turn.
    a. K15, K25, K10, K20 models: Assemble the adjusting nut lock by aligning the nearest hole in lock with the adjusting nut pin. Install the outer lock nut torque to 80 ft. lbs.
    b. K30, K35 models: Assemble the lockwasher and outer locknut. Torque the outer locknut to 65 ft. lbs Bend one tab of the lockwasher over the inner nut a minimum of 30 degrees. Bend one tab of the lockwasher over the outer nut a minimum of 60 degrees.

NOTE: End play for all models is 0.001–0.010 in.

11. Install the pressure spring, drive gear, snap-ring and hub cap.

FREE-WHEELING HUB – PART-TIME

CURRENT MODELS

The engagement and disengagement of free-wheeling hubs is a manual operation which must be performed at each front wheel. The transfer case control lever must be in 2-wheel drive position when locking or unlocking hubs. Both hubs must be in the fully locked or fully unlocked position. They must not be in the free-wheeling position when low all-wheeldrive is used as the additional torque output in this position can subject the rear axle to severe strain and rear axle failure may result.

Removal

K10–20, K1500–2500

1. Turn the actuator to the LOCK position.
2. Raise the vehicle on a hoist.
3. Remove the six retaining plate bolts.
4. Remove the retaining plate, actuator knob and O-ring.
5. Remove the internal snap-ring, outer clutch retaining ring and actuating cam body.
6. Remove the axle shaft snap-ring.

NOTE: It may be necessary to first relieve pressure from the axle shaft snap-ring.

7. Remove the wheel bearing outer lock nut and lock ring.
8. Remove the wheel bearing inner adjusting nut.
9. Remove the hub and disc assembly, outer wheel bearing and the spring retainer plate.
10. Drive the inner bearing cone and oil seal from the hub with a brass drift. Discard the oil seal.
11. Using a brass drift, remove the inner and outer bearing cups.

Installation

NOTE: All parts should be lubricated with an ample amount of high speed grease prior to installation.

1. Install the outer wheel bearing cup into the wheel hub.
2. Install the inner wheel bearing cup into the wheel hub.
3. Pack the wheel bearing cone with a suitable wheel bearing grease (high melting point type).
4. Install the cone into the cup.
5. Lubricate the wheel bearings; install the hub and disc and the bearings on the spindle.
6. While rotating the hub and disc, torque the inner adjusting nut to 50 ft. lbs
7. Back off the inner adjustment nut.
8. While rotating the hub and disc, torque the inner adjusting nut to 35 ft. lbs
9. Again, back off the inner adjusting nut a maximum of $^3/_8$ turn.
10. Assemble the adjusting nut lock by aligning the nearest hole in lock with the adjusting nut pin.
11. Install outer lock nut and torque to 50 ft. lbs

NOTE: Hub end play should be 0.001–0.010 in..

12. Install the spring retainer plate over the spindle nuts with the flange side facing the bearing and seat the retainer against the bearing outer cup.
13. Install the pressure spring.

NOTE: The large diameter of the spring seats against the retaining plate. When the spring is seated, it extends past the spindle nuts by approximately $^7/_8$ in..

14. Install the inner clutch ring and bushing into the axle shaft sleeve and clutch ring; install unit as an assembly onto the axle shaft.
15. While pressing in on the assembly, install the axle shaft snap-ring.

NOTE: To facilitate snap-ring installation, thread a $^7/_{16} \times 20$ bolt in the end of the axle shaft and pull outward on the axle shaft.

16. Install the actuating cam body. Cam faces outward.
17. Install the outer clutch retaining ring and the internal snap-ring.
18. Install the O-ring on the retaining plate.
19. Install the actuating knob and retaining plate.

NOTE: Actuating knob should be installed in the LOCK position. The grooves in the knob must fit into the actuator cam body.

20. Install the six cover bolts and seals. Torque bolts to 30 ft. lbs
21. Turn the actuating knob to the FREE position and check free-wheeling operation.
22. Remove vehicle from hoist.

FRONT AXLE ASSEMBLY

NOTE: This procedure does not apply to S-series trucks.

Removal

1. Disconnect the drive shaft from the front axle. Raise the

SUSPENSION
Light Trucks & Vans

vehicle far enough to take the weight off the front springs and place the jack stands under the truck frame.

2. Disconnect the connecting rod at the steering arms.
3. Disconnect the brake hoses at the frame fittings and cover all open ends.
4. Disconnect the shock absorbers at the axle brackets.
5. Disconnect the axle vent tube clip at the differential housing.
6. Unfasten the U-bolts, raise the truck further, as necessary and roll the axle out from underneath.

Installation

1. With truck on axle stands, roll the axle under the truck. Lower the truck until axle and truck are in proper relative positions. Again support the vehicle with axle stands.
2. Attach the shock absorbers to the axle brackets. Connect the brake hoses to the frame fittings and fill and bleed the brake system.
3. Attach the steering connecting rod at the steering arms.
4. Connect the drive axle to the front differential.

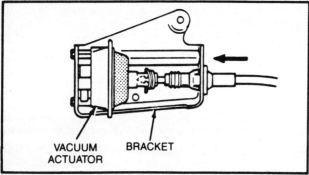

Cable–to–vacuum actuator attachment

AXLE SHAFT ASSEMBLY

Removal

1. Remove the free-wheeling hubs as outlined, if so equipped.
2. Remove the wheel bearing outer lock nut, lock ring and inner adjusting nut.
3. Remove the hub assembly from the spindle.
4. Remove the spindle retaining bolts and tap the end of the spindle with a soft faced hammer, to separate the spindle from the knuckle.
5. Remove the axle shaft and joint assembly by pulling outward on the shaft.
6. Repairs to the wheel hub assembly and to the axle universal joint can be accomplished at this time.

Installation

1. Install a new grease seal onto the slinger of the axle shaft, with the lip of the seal facing toward the spindle.
2. Install the axle shaft into the housing and engage the splines with the pinion side gears of the differential.
3. Place the bronze thrust washer on the axle shaft with the chamfered edge towards the slinger and install the spindle onto the knuckle.
4. Torque the spindle nuts to 45 ft. lbs and assemble the hub to the spindle. Torque the inner adjustment nut to 50 ft. lbs while rotating the hub. Back off the inner nut an additional $\frac{3}{8}$ of a turn maximum.
5. Assemble the lock washer and the outer lock nut to the spindle. Torque the outer lock nut to 50 ft. lbs minimum. The hub should have 0.001–0.010 in. end play.
6. If the vehicle is equipped with free-wheeling hubs, refer to the installation procedure for correct installation and if not equipped, install the hub cap assembly.

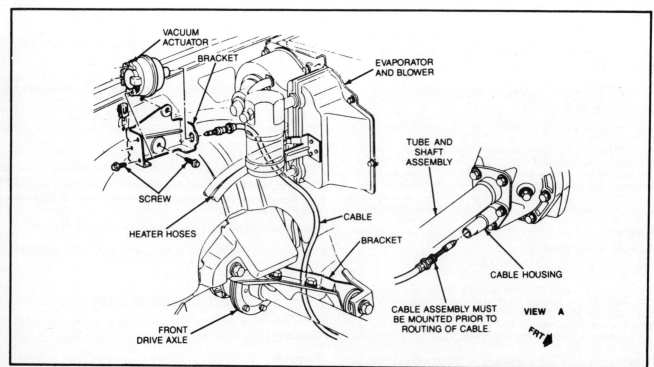

Vacuum actuator assembly

SUSPENSION
Light Trucks & Vans

TUBE/SHAFT ASSEMBLY

Removal

S/T-Series Trucks

1. Disconnect negative battery cable.
2. Disconnect shift cable from vacuum actuator by disengaging locking spring. Then push actuator diaphragm in to release cable.
3. Unlock steering wheel at steering column so linkage is free to move.
4. Raise vehicle. If twin post hoist is used, place jack stands under frame and lower front post hoist.
5. Remove front wheels.
6. Remove engine drive belt shield.
7. Remove front axle skid plate (if equipped).
8. Place support under right hand lower control arm and disconnect right hand upper ball joint, then remove support so control arm will hang free.
9. Disconnect right hand drive axle shaft from tube assembly by removing six bolts.
   Keep axle from turning by inserting a drift through opening in top of brake caliper into corresponding vane of brake rotor.
10. Disconnect four wheel drive indicator light electrical connection from switch.
11. Remove three bolts securing cable and switch housing to carrier and pull housing away to gain access to cable locking spring. Do not unscrew cable coupling nut unless cable is being replaced.
12. Disconnect cable from shift fork shaft by lifting spring over slot in shift fork.
13. Remove two bolts securing tube bracket to frame.
14. Remove remaining two upper bolts securing tube assembly to carrier.
15. Remove tube assembly by working around drive axle. Be careful not to allow sleeve, thrust washers, connector and output shaft to fall out of carrier or be damaged when removing tube.

Installation

1. Install sleeve, thrust washers, connector and output shaft in carrier. Apply sealer #1052357, Locktite 514 or equivalent on tube to carrier surface.
   Be sure to install thrust washer. Apply grease to washer to hold it in place during assembly.
2. Install tube and shaft assembly to carrier and install bolt at one o'clock position but do not torque. Pull assembly down and install cable and switch housing and remaining four bolts. Torque all bolts to 45-60 ft. lb.
3. Install two bolts securing tube to frame and torque.

4. Check operation of four wheel drive mechanism using Tool J-33799. Insert tool into shift fork and check for rotation of axle shaft.
5. Remove tool and install shift cable switch housing by pushing cable through into fork shaft hole. Cable will automatically snap in place.
6. Connect four wheel drive indicator light electrical connection to switch.
7. Install support under right hand lower control arm to raise arm and connect upper ball joint.

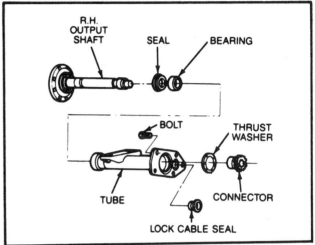

Right side output shaft and tube

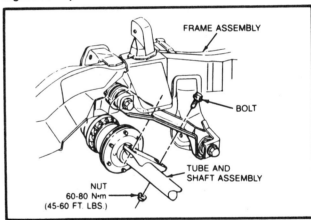

Tube-to-frame attachment

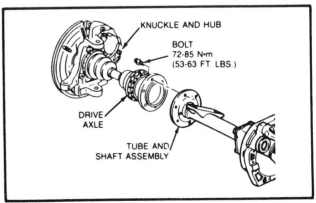

Drive axle bolts

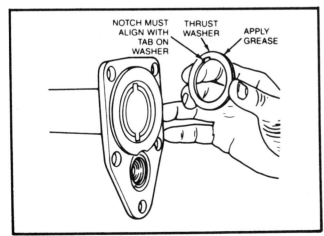

Thrust washer installation

SUSPENSION
Light Trucks & Vans

8. Install right-hand drive axle to axle tube by installing one bolt first, then, rotate axle to install remaining five bolts. Hold axle from turning by inserting a drift through opening in top of brake caliper into corresponding vane of brake rotor. Tighten bolts to 53–63 ft. lb.
9. Install front axle skid plate, if equipped.
10. Install engine drive belt shield.
11. Install front wheels.
12. Lower vehicle.
13. Connect shift cable to vacuum actuator by pushing cable end into vacuum actuator shaft hole. Cable will snap in place automatically.
14. Connect negative battery cable.

SHIFT CABLE REPLACEMENT
S/T–SERIES TRUCKS

1. Disengage shift cable from vacuum actuator by disengaging locking spring, then, push actuator diaphragm in to release cable. Squeeze the two locking fingers of the cable with pliers, then pull cable out of bracket hole.
2. Raise vehicle and remove three bolts securing cable and switch housing to carrier and pull housing away to gain access to cable locking spring. Disconnect cable from shaft fork shaft by lifting spring over slot in shift fork.
3. Unscrew cable from housing.
4. Remove cable from vehicle.
5. Install cable observing proper routing.
6. Install cable and switch housing to carrier using three attaching bolts. Torque mounting bolts to 30–40 ft. lbs
7. Guide cable through switch housing into fork shaft hole and push cable in. Cable will automatically snap in place. Start turning coupling nut by hand, to avoid cross threading, then torque nut to 71–106 in. lbs. Do not overtorque nut as this will cause thread damage to plastic housing.
8. Lower vehicle.
9. Connect shift cable to vacuum actuator by pressing cable into bracket hole. Cable and housing will snap in place automatically.
10. Check cable operation.

DIFFERENTIAL CARRIER RIGHT HALF OUTPUT SHAFT AND TUBE
Disassembly
S/T–SERIES TRUCKS

1. Remove right-hand output shaft from tube by striking inside of flange with a soft face hammer while holding tube.
2. Remove output shaft tube seal by prying out of tube.
3. Remove output shaft tube bearing using J–29369–2.
4. Remove differential shift cable housing seal by driving out with a punch or similar tool.

Assembly

1. Install output shaft tube bearing using tool J–33844. Tool must be flush with tube when bearing is correctly installed.
2. Install output shaft tube seal using tool J–33893. Flange of seal must be flush with tube outer surface when seal is installed.
3. Install output shaft into tube and seat by striking flange with a soft face hammer.
4. Install differential shift cable housing seal using J–33799.

STEERING KNUCKLE
Removal
S–SERIES

1. Raise front of vehicle and support with floor stands under front lift points. Remove wheel.
   Spring tension is needed to assist in breaking ball joint studs loose from steering knuckle. Do not place stands under lower control arm.
2. Remove caliper.
3. Remove hub and rotor assembly.
4. Remove the three bolts attaching shield to knuckle.
5. Remove tie-rod end from knuckle using Tool J–6627 or equivalent.
6. Carefully remove knuckle seal if knuckle is to be replaced.
7. Remove ball studs from steering knuckle using tool J–23742 or equivalent.

— CAUTION —

Floor jack must remain under control arm spring seat during removal and installation to retain spring and control arm in position.

8. Position a floor jack under lower control arm near spring seat and raise jack until it just supports lower control arm.
9. Raise upper control arm to disengage ball joint stud from knuckle.
10. Raise knuckle from lower ball joint stud and remove knuckle.

Inspection

Inspect the tapered hole in the steering knuckle. Remove any

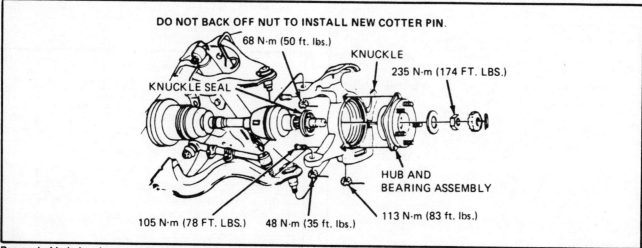

Removal of hub, bearing assembly, knuckle and seal on S-series

SUSPENSION
Light Trucks & Vans

dirt. If out-of-roundness, deformation, or damage is noted, the knuckle MUST be replaced.

Installation

1. Insert upper and lower ball joint studs into knuckle and install nuts.
2. Install shield to knuckle seal and splash shield. Torque attaching bolts to 10 ft. lbs
3. Install tie rod end into knuckle. Install tool J–29193 or equivalent and torque to 15 ft. lbs. Remove tool and install nut to 40 ft. lbs.
4. Replace wheel bearings. Install hub and disc assembly.
5. Adjust wheel bearings. Install caliper.
6. Install remaining parts in reverse order of removal.

STEERING KNUCKLE – WITH BALL JOINTS

Removal

K10, 20, K1500, 2500

1. With the spindle and axle removed, as previously outlined, disconnect the tie rod end from the steering arm.
2. If necessary for working clearance, remove the steering arm from the knuckle.

NOTE: If the steering arm is removed, discard the three self-locking nuts and replace them with new self-locking nuts upon assembly.

3. Remove the upper and lower ball joint retaining nuts.

NOTE: The upper ball joint stud and nut have a cotter pin retainer, while the lower ball point stud and nut have none.

4. With a wedge type tool, separate the lower ball joint stud from the knuckle. Repeat this operation for the upper ball joint stud.
5. Remove the snap-ring retainer from the lower ball joint. With the aid of a C-clamp tool, press the lower ball joint from the knuckle.

NOTE: The lower ball joint must be removed before any service can be performed on the upper ball joint.

6. With the aid of the C-clamp tool, press the upper ball joint from the knuckle. Replacement of the knuckle can be accomplished at this point in the disassembly.

Installation

1. Press the lower ball joint into the knuckle with the aid of the "C" type tool and install the snap-ring retainer on the lower ball joint.
2. Install the upper ball joint into the knuckle with the aid of the C-clamp tool.
3. Position the ball joint studs in their respective openings on the yoke and install the new nuts finger tight.

NOTE: The castellated nut is placed on the upper ball joint stud.

4. Torque the lower ball joint stud nut to 70 ft. lbs while exerting upward pressure on the knuckle.
5. Torque the upper ball joint stud adjusting sleeve to 50 ft. lbs using a spanner type socket.
6. Torque the upper ball joint stud nut to 100 ft. lbs Apply additional torque if necessary to align the cotter pin hole in the nut and stud.
7. Reassemble the steering arm, if removed, tie rod ends, spindle, axle and hub as outlined previously.
8. Torque the steering arm nuts to 90 ft. lbs and the tie rod nut to 45 ft. lbs and install the cotter pin.

STEERING KNUCKLE – WITH KING PINS

Removal

K30, K3500

1. Remove the hub and spindle.

NOTE: It may be necessary to tap lightly on the spindle with a rawhide hammer in order to free it from the knuckle.

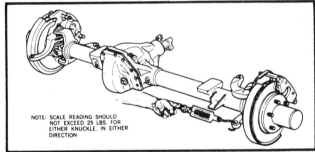
Checking front drive axle ball joint adjustment (© General Motors Corp.)

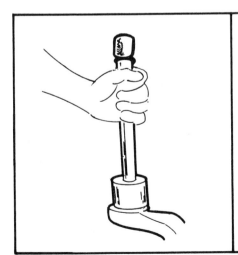
Installing oil seal—front drive yoke w/ king pins

Installing grease retainer—front drive yoke w/king pins

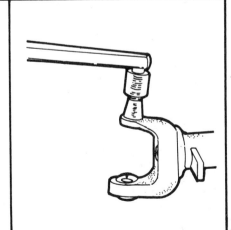
Installing upper king pin—front drive axle

SUSPENSION
Light Trucks & Vans

2. Remove the four cap nuts from the upper king pin.

NOTE: Spring pressure will force the cap up.

3. Remove the cap, spring and gasket. Discard the gasket.
4. Remove the four cap screws from the lower king pin bearing.
5. Remove the cap and the lower king pin.
6. Remove the upper king pin bushing.
7. Remove the knuckle from the yoke.
8. Remove the king pin felt seal.
9. Remove the upper king pin from the yoke with a very large breaker bar and any suitable adapter designed to fit the king pin.

NOTE: Considerable force will be required to remove the king pin from the yoke as the king pin is originally torqued to 500–600 ft. lbs

10. Remove the lower king pin bearing cup, cone, grease retainer and seal. Discard the seal. Discard the grease retainer if damaged.

Installation

1. Install the lower king pin bearing cup and a new grease retainer.
2. Fill the grease retainer with grease.
3. Lubricate the bearing cone with grease. Install the bearing cone.
4. Install a new oil seal for the lower king pin bearing.

NOTE: The oil seal will protrude slightly from the surface of the yoke flange when fully installed.

5. Install the upper king pin and torque to 500–600 ft. lbs
6. Assemble the felt seal to the king pin.
7. Position the steering knuckle over the king pin and install the tapered bushing over the king pin.

8. Install the lower king pin and the lower bearing cap. Torque the cap screws to 70–90 ft. lbs
9. Position the compression spring over the upper king pin bushing.
10. Install the upper bearing cap with a new gasket. Torque the nuts to 70–90 ft. lbs

BALL JOINT

Adjustment

K10, 20, K1500, 2500

1. Raise the vehicle on a hoist.
2. Disconnect the connecting rod and the tie rod to allow the steering knuckle to move freely.
3. Attach a spring scale to the tie rod mounting hole of the steering arm.
4. Move the knuckle to the straight-ahead position.
5. Determine the right angle pull required to keep the knuckle turning after initial movement from standing still. The pull should not exceed 25 lbs. for each knuckle assembly.
6. If the effort required to maintain turning movement is in excess of 25 lbs., then remove the upper ball stud nut and loosen the ball stud adjusting sleeve as required. Tighten the ball stud nut and recheck the turning effort.

WHEEL BEARING

Adjustment

Refer to the Installation procedures under Locked Hub-Full Time or Free-Wheeling Hub-Part-Time for the corresponding wheel bearing adjustment instructions.

Non-driving front wheel axle bearings should have an end play of 0.001–0.005 in..

REAR SUSPENSION

Leaf Spring Type

Removal

LIGHT DUTY TRUCKS AND VANS

1. Jack the vehicle at the frame to relieve tension on the spring.
2. Remove the U-bolt retaining nuts and withdraw the U-bolts.
3. Loosen the shackle bolts and remove the lower bolt.
4. Remove the nut and bolt securing the spring to the front hanger.
5. Remove the spring from the vehicle.

Installation

1. Position the spring assembly and spacers if so equipped, on the axle housing.

NOTE: On springs with metal encased pressed in type bushings the shackle assembly must be attached to the rear spring eye before installing the shackle to the rear hanger.

2. Position the U-bolts and loosely install the U-bolt retaining nuts.
3. Jack as required to align the spring eyes with the front hanger and rear shackle, install the eye bolts.

4. Lower the vehicle.
5. Tighten the U-bolt retaining nuts alternately and evenly to properly seat the spring and tighten the front hanger and rear shackle bolts.

Removal

MEDIUM DUTY TRUCKS

1. Raise vehicle frame to take weight off the spring. Make sure vehicle is supported safely.
2. Remove rear wheels to provide access to spring assembly.
3. Safely support axle on floor jack.
4. Install a C-clamp on radius leaf, to relieve load on radius eye bolt on 45 Series vehicles.
5. On 45 Series at the front and rear hanger, remove rebound pin retainer bolt, then remove retainer. Install suitable puller into tapped hole at end of rebound pin, then remove pin.
6. Remove spring U-bolt nuts, shock absorber bracket (when used), U-bolt anchor plate and U-bolts and U-bolt spacer, then lower axle slightly.
7. Remove spring eye on radius bolt nut and washer, then remove spring eye bolt from spring eye or radius leaf.

NOTE: When tapered shim is used, the position of shim thin and thick edge should be noted so that shim can be installed properly at assembly.

SUSPENSION
Light Trucks & Vans

Installation

1. Set spring assembly and tapered shim or spacer (if used) at axle pad.

NOTE: Tapered shim must be installed on axle in same position that was noted at removal.

2. Install U-bolt spacer over center bolt.
3. Seat U-bolts in spacer grooves, then secure spring to axle by installing anchor plates, shock absorber bracket (when used) and nuts on U-bolts.
4. Lower frame until ends of spring enter the hanger and touch the cam surface of hanger. Compress radius leaf with C-clamp until radius leaf eye and hanger holes are aligned and torque to specifications.
5. Remove C-clamp from radius leaf.
6. Install rebound pin at front and rear hangers. Install rebound pin retainer and secure with retainer bolt.
7. Install wheels.
8. Remove blocking and lower frame to place weight on springs. Check U-bolt nuts for proper torque.

Removal and Installation

SINGLE OR TANDEM AXLES SPRINGS—HEAVY DUTY

1. Raise the rear of the vehicle, place floor jacks under the axle(s) and remove the dual wheels from the hubs to facilitate the removal of the spring eye pin and to expose the other nuts and bolts.
2. Remove the saddle cap stud nuts and/or spring U-bolts.
3. Remove the rebound pin locks or retainers and then remove the rebound pins.
4. Remove the eye bolts or radius lead pin clamp bolts, then remove the lubrication fitting from the inner end of the pin, if equipped.
5. Remove the pins from the springs and lower the axle(s) or raise the frame until the spring will clear the brackets.
6. Remove the spring from the vehicle.
7. The installation is in the reverse of the removal procedure.
8. Torque the U-bolts or saddle cap stud nuts to specifications after the vehicle is lowered to the floor.

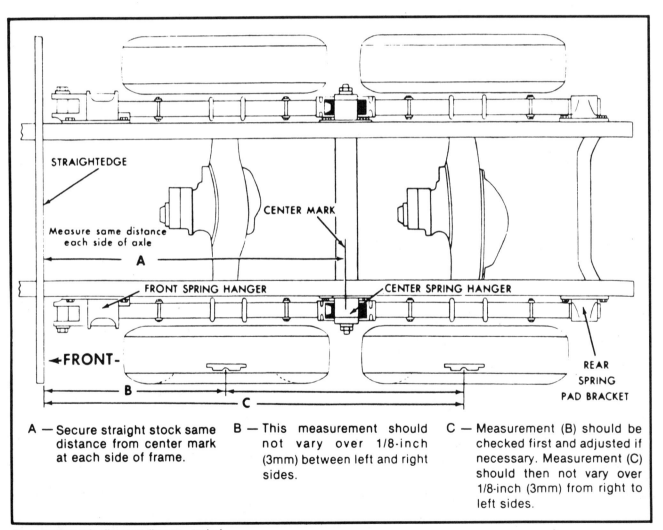

A — Secure straight stock same distance from center mark at each side of frame.

B — This measurement should not vary over 1/8-inch (3mm) between left and right sides.

C — Measurement (B) should be checked first and adjusted if necessary. Measurement (C) should then not vary over 1/8-inch (3mm) from right to left sides.

Reyco tandem axle alignment diagram—typical

SECTION 2

SUSPENSION
Light Trucks & Vans

DODGE/PLYMOUTH TRUCKS
VANS•PICK-UPS•RAMCHARGER
VOYAGER•CARAVAN•TRAILDUSTER•RAMPAGE•SCAMP

FRONT SUSPENSION

WHEEL BEARINGS

Adjustment

D100, D200, AND D300

1. While rotating the wheel, tighten the adjusting nut to 360-480 inch pounds.
2. With the wheel at rest, back off the adjusting nut to completely release the bearing preload.
3. Finger tighten the adjusting nut, and install the lock nut and the cotter key.

NOTE: End-play of 0.0001-0.003 inch is acceptable. If this measurement is obtained, then the wheel bearing adjustment is satisfactory.

Front Drive Axle (4WD)

Removal and Installation

1. Raise the truck and install stands under the frame rails, behind the front springs.
2. Disconnect front driveshaft at drive pinion yoke.
3. Disconnect steering linkage at drag link.
4. Disconnect front shock absorbers and brake line at frame. Disconnect the sway bar link assembly from the spring clip plate.
5. Remove nuts from the spring hold down bolts and remove axle assembly from under vehicle.
6. To install, place axle assembly under vehicle and line up spring center bolts with locating hole in axle housing pad.
7. Install spring clips or spring U-bolts, new lock washer and nuts.
8. Connect the shock absorbers, and the brake line at the frame.
9. Connect the steering linkage to the drag link, and the driveshaft to the pinion yoke. Check lubricants and bleed the brakes.
10. Lower the vehicle and test the operation.

Front Drive Axle Shaft and Joint Assembly

MODEL 44FBJ (W/O LOCKING HUBS)

Removal

1. Remove wheel cover. Remove cotter key and loosen outer axle shaft nut.
2. Raise vehicle and support safely. Remove wheel and tire assembly.
3. Remove caliper retainer and anti-rattle spring assemblies. Remove caliper from disc by sliding out and away from disc. Hang caliper out of the way. Remove inboard shoe.

NOTE: Do not allow caliper to hang or be supported by hydraulic brake hose. Support the brake caliper using wire or other suitable device.

4. Remove outer axle shaft nut and washer.
5. Through hole in rotor assembly remove retainer bolts. Position Puller Tool over wheel studs and install wheel nuts. Tighten main screw of tool to remove hub, rotor, bearings, retainer and outer seal as an assembly.
6. Remove wheel nuts and Puller from hub and rotor assembly.
7. Remove the brake caliper adapter from the knuckle.
8. Position a pry bar behind the inner axle shaft yoke and push the bearings out of the knuckle.

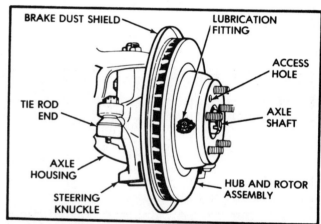

Location of various components on the joint assembly
(© Chrysler Corp.)

9. Remove the "O" ring from the knuckle (if so equipped) and discard.
10. Carefully remove the axle shaft assembly. Remove seal and slinger from shaft.

Installation

1. Inspect the outer axle shaft seal surface for grooving. If the surface is grooved, repair or replace knuckle.
2. Install brake dust shield (if removed), tighten mounting bolts to 160 inch lbs. (18 Nm).
3. Apply RTV sealer to the seal surface of the axle shaft.
4. Using Driver Tool, install the seal slinger onto the outer axle shaft.
5. Install lip seal on the slinger with the lip toward the axle shaft spline.
6. Carefully insert axle shaft into the housing so as not to damage the differential seal at the side gears.
7. Insert a pry bar through the axle shaft "U" joint and wedge it so that the axle shaft is in all the way and cannot be moved out.
8. Using Adaptor Tool and Driver Tool carefully install the seal cup until bottomed in the knuckle.

SUSPENSION
Light Trucks & Vans

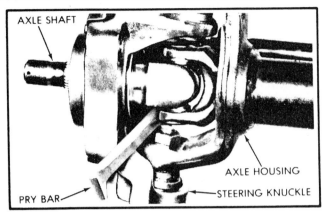

Using pry bar to hold axle (© Chrysler Corp.)

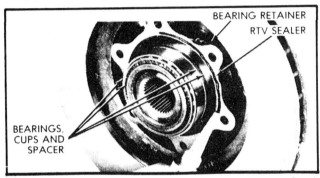

Installing RTV sealer to the bearing retainer face (© Chrysler Corp.)

NOTE: A small amount of wheel bearing grease on the Adaptor face will aid in holding the cup in position. Do not remove the tool at this time.

9. Apply a 1/4 inch (6mm) bead of RTV sealer to the retainer face on the chamfer (this takes the place of the "O" ring discarded at disassembly).
10. Carefully remove seal installing tool from the knuckle bore so that the outer axle shaft remains centered.

NOTE: If the shaft is disturbed, check to make sure that the lip seal is still riding inside the cup. Correct if necessary.

11. Position the bearing retainer on the knuckle so that the lube fitting is facing directly forward. This is extremely important.
12. Install the hub, rotor, retainer and bearing assembly on the knuckle and tighten retainer plate bolts to 30 ft.lb. (41 Nm) using a criss-cross method to position it evenly.
13. Install brake adapter and tighten mounting bolts to 85 ft.lb. (115 Nm).
14. Remove pry bar from "U" joint and install axle shaft washer and nut. Tighten nut to 100 ft.lb. (136 Nm), continue to tighten nut until next slot in nut aligns with cotter key hole in axle shaft. Install cotter key.
15. Insert a grease gun through the access hole in the hub and rotor assembly into the lube fitting. Fill with Multi-Purpose Grease, until grease flows through the new inner seal (observe at the "U" joint area). Remove lube gun and rotate hub and rotor several times. Reinstall grease gun and apply until grease flows from at least 50% of the seal diameter.
16. Locate inboard brake shoe on adapter with shoe flanges in adapter ways. Slowly slide caliper assembly into position in adapter and over disc. Align caliper on machined ways of adapter.

NOTE: Be careful not to pull the dust boot from its grooves as the piston and boot slide over the inboard shoe.

17. Install anti-rattle springs and retaining clips and torque to 180 in.lb. (20 Nm).

NOTE: The inboard shoe anti-rattle spring must always be installed on top of the retainer spring plate.

18. Install wheel and tire assembly. Tighten nuts to 110 ft.lb. (149 Nm). Install wheel covers, lower vehicle and test operation.

MODEL 44FBJ (WITH LOCKING HUBS)

Removal

1. Remove locking hub assembly.
2. Raise vehicle and support safely. Remove wheel and tire assembly.
3. Remove caliper retainer and anti-rattle spring assemblies. Remove caliper from disc by sliding out and away from disc. Hang caliper out of the way. Remove inboard shoe.

NOTE: Do not allow caliper to hang or be supported by hydraulic brake hose. Support the brake caliper using a length of wire or other suitable device.

4. Remove outer axle shaft lock nut washer and nut.
5. Remove rotor and bearing assembly.
6. Remove six nuts which fasten splash shield and spindle to knuckle.
7. Remove splash shield and spindle.
8. Remove the brake caliper adaptor from the knuckle.
9. Carefully remove the axle shaft assembly. Remove seal and stone shield (if equipped) from shaft.

Installation

1. Install lip seal on the axle shaft stone shield with the lip toward the axle shaft spline.
2. Carefully insert axle shaft into the housing so as not to damage the differential seal at the side gears.
3. Install spindle and brake splash shield. Install 6 new nuts and tighten to 25-30 ft.lb. (34-41 Nm).
4. Install the rotor, outer bearing, nut, washer and locknut onto spindle. Tighten as specified.
5. Install brake adapter and tighten mounting bolts to 85 ft.lb. (115 Nm).
6. Locate inboard brake shoe on adapter with shoe flanges in adapter ways. Slowly slide caliper assembly into position in adapter and over disc. Align caliper on machined ways of adapter.

NOTE: Be careful not to pull the dust boot from its grooves as the piston and boot slide over the inboard shoe.

7. Install anti-rattle springs and retaining clips and torque to 180 ft.lb. (20 Nm). The inboard shoe anti-rattle spring must always be installed on top of the retainer spring plate. Install locking hub assembly.
8. Install wheel and tire assembly. Tighten nuts to 110 ft.lb. (149 Nm). Install wheel covers. Remove jack stands, lower vehicle and test operation.

Removal

1. Block brake pedal in the up position.
2. Raise vehicle and support safely.
3. Remove wheel and tire assembly.
4. Remove allen screw holding caliper to adapter.
5. Tap adapter lock and spring from between caliper and adapter.

SUSPENSION
Light Trucks & Vans

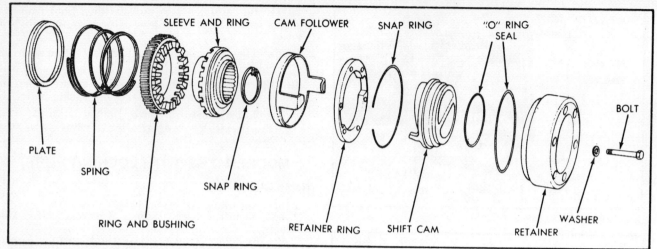

Exploded view of locking hub assembly (© Chrysler Corp.)

6. Carefully separate caliper from adapter. Hang caliper out of the way. Inner brake shoe will remain on adapter.

NOTE: Do not allow caliper to hang or be supported by hydraulic brake hose.

7. Remove hub cap and snap ring.
8. Remove flange nuts and lockwashers. Remove drive flange and discard gasket, or remove locking hub if so equipped.
9. Straighten tang on lock ring and remove outer lock nut, lock ring, inner lock nut and outer bearing. Carefully slide hub and rotor assembly from spindle.
10. Remove inner brake shoe from adapter.
11. Remove nuts and washers holding brake splash shield, brake adapter and spindle to steering knuckle.
12. Remove spindle from knuckle. Slide inner and outer axle shaft complete with bronze spacer, seal and slinger from axle.

Installation

1. Slide axle shaft into position. Position bronze spacer on axle shaft with chamfer facing toward universal joint.
2. Install spindle, brake adapter and brake splash shield. Install washers and nuts and tighten to 50 to 70 ft.lb. (68 to 95 Nm).
3. Position inner brake shoe on adapter.
4. Carefully install hub and rotor assembly onto spindle. Install outer bearing and inner lock nut. Tighten to 50 ft.lb. to seat bearings, back off lock nut and retighten to 30 to 40 ft.lb. (88 Nm) while rotating hub and rotor. Back nut off 135° to 150°. Assemble lock ring and outer lock nut. Tighten lock nut to 65 ft.lb. (88 Nm) minimum. Bend one tang of lock ring over inner lock nut and one tang of lock ring over outer lock nut. Final bearing end play to be 0.001-0.010 inch (0.025-0.254mm).
5. Install new gasket on hub. Install drive flange, lock washers and nuts. Tighten to 30 to 40 ft.lb. (41 to 54 Nm). Install snap ring. Install hub cap, or install locking hub if so equipped.
6. Carefully position caliper onto adapter. Position adapter lock and spring between caliper and adapter and tap into position. Install allen screw and tighten to 12 to 18 ft.lb. (16 to 24 Nm).
7. Install wheel and tire assembly. Tighten nuts to 75 ft.lb. (102 Nm).
8. Remove jack stands, lower vehicle, remove block from brake pedal and test vehicle operation.

Steering Knuckle, Spindle and Ball Joint

MODEL 44FBJ (WITH LOCKING HUBS)

Removal and Disassembly

1. Remove locking hub assembly.

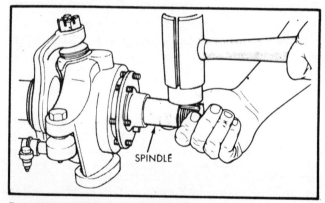

Removing spindle from steering knuckle (© Chrysler Corp.)

Removing or installing spindle and splash shield (© Chrysler Corp.)

SUSPENSION
Light Trucks & Vans

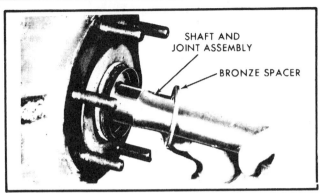

Removing or installing axle shft and U–joint assembly
(© Chrysler Corp.)

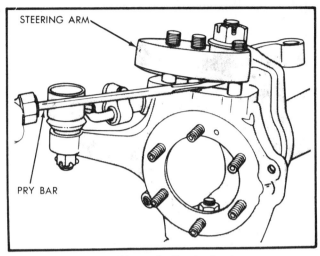

Removing steering knuckle arm (© Chrysler Corp.)

2. Raise vehicle and support safely. Remove wheel and tire assembly.

3. Remove caliper retainer and anti-rattle spring assemblies. Remove caliper from disc by sliding out and away from disc. Hang caliper out of the way. Remove inboard shoe.

NOTE: Do not allow caliper to hang or be supported by hydraulic brake hose. Support the brake caliper with wire.

4. Remove braking disc.
5. Remove brake caliper adapter from the knuckle.
6. Remove the torque prevailing nuts and washers from the spindle to steering knuckle attaching bolts. Remove the brake splash shield.
7. Using a soft faced hammer, hit spindle lightly to break it free from steering knuckle. Upon removal, examine the bronze spacer located between the needle bearing and shaft joint assembly. If wear is evident, replace the spacer.
8. Place spindle in vise having soft jaws. Do not clamp on bearing carrying surfaces. Remove needle bearing grease seal.
9. Remove needle bearings.
10. Carefully remove the axle shaft assembly. Remove seal and stone shield from shaft.
11. Disconnect tie rod from steering knuckle from the left side only. Disconnect drag link from steering knuckle arm.
12. Remove nuts and cone washers from steering knuckle arm. Tap steering knuckle arm to loosen from knuckle. Remove steering knuckle arm.
13. Remove cotter key from upper ball joint nut. Remove upper and lower ball joint nut. Discard lower nut.
14. Using a brass drift and hammer separate steering knuckle from axle housing yoke. Remove and discard sleeve from upper ball joint joke on axle housing.

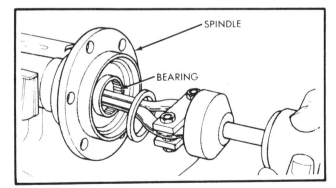

Removing needle bearing from the spindle assembly
(© Chrysler Corp.)

15. Position steering knuckle upside down in a vise and remove snap ring from lower ball joint with snap ring pliers.
16. Press lower ball joint from steering knuckle. Press upper ball joint from steering knuckle. Replace ball joints if any looseness or end play exists.
17. Clean all components with a suitable solvent and blow dry with compressed air. Inspect all parts for burrs, chips, wear or cracks. Replace necessary parts at assembly.

Assembly and Installation

1. Position steering knuckle right side up in a vise. Carefully press lower ball joint into position. Install snap ring.
2. Carefully press upper ball joint into position. Install new boots on both ball joints. Remove steering knuckle from vise.
3. Screw new sleeve into upper ball joint yoke on axle housing leaving about two threads showing at the top.
4. Position steering knuckle on axle housing yoke and install lower ball joint nut. Tighten to 80 ft. lb. (108 Nm).
5. Tighten sleeve in upper ball joint yoke to 40 ft.lb. (54 Nm). Install upper ball joint nut and tighten to 100 ft.lb. (136 Nm). Align cotter key hole in stud with slot in castellated nut, if slot and hole are not aligned continue to tighten nut until aligned. Do not loosen to align. Install cotter key.
6. Position steering knuckle arm over studs on steering knuckle. Install cone washer and nuts. Tighten nuts to 90 ft.lb. (122 Nm). Assemble drag link to steering knuckle arm. Install nut and tighten to 60 ft.lb. (81 Nm). Install cotter key.
7. Assemble tie rod end to steering knuckle. Install nut and tighten to 45 ft. lb. (61 Nm). Install cotter key.
8. Install lip seal on the stone shield with the lip toward the axle shaft spline.
9. Carefully insert axle shaft into the housing so as not to damage the differential seal at the side gears.
10. Assemble new needle bearings into spindle.
11. Install a new grease seal. Fill the seal area of the spindle and thrust face area of shaft and seal with Multi-purpose Grease.
12. Position new bronze spacer on axle shaft, install spindle and brake splash shield.
13. Install new nuts, tighten to 25-35 ft. lbs. (34-41 Nm).
14. Mount braking disc assembly and outer wheel bearing cone onto spindle.
15. Install wheel bearing nuts and locking washer.
16. Install locking hub assembly.
17. Install brake adapter and tighten mounting bolts to 85 ft. lbs. (115 Nm).
18. Locate inboard brake shoe on adapter with shoe flanges in

2–307

SUSPENSION
Light Trucks & Vans

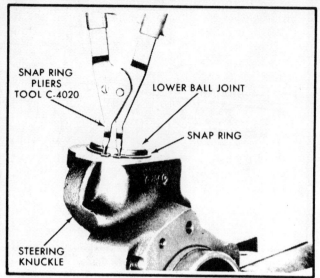

Removing or installing snap ring on the lower ball joint
(© Chrysler Corp.)

adapter ways. Slowly slide caliper assembly into position in adapter and over disc. Align caliper on machined ways of adapter. Be careful not to pull the dust boot from its grooves as the piston and boot slide over the inboard shoe.

19. Install anti-rattle springs and retaining clips and torque to 180 inch lbs. (20 Nm): The inboard shoe anti-rattle spring must always be installed on top of the retainer spring plate.

20. Install wheel and tire assembly. Tighten nuts to 110 ft. lbs. (149 Nm). Install wheel covers. Remove jack stands, lower vehicle and test operation.

Steering Knuckle

MODEL 60

Removal

1. Block brake pedal in the up position.
2. Raise vehicle and support safely. Install jack stands.
3. Remove wheel and tire assembly.
4. Remove allen screw holding caliper to adapter.
5. Tap adapter lock and spring from between caliper and adapter.
6. Carefully separate caliper from adapter. Hang caliper out of the way. Inner brake shoe will remain on adapter.

NOTE: Do not allow caliper to hang or be supported by hydraulic brake hose.

7. Remove hub cap. Remove snap ring.
8. Remove flange nuts and lock washers. Remove drive flange and discard gasket, or remove locking hub if so equipped.
9. Straighten tang on lock ring and remove outer lock nut, lock ring, inner lock nut and outer bearing. Carefully slide hub and rotor assembly from spindle.
10. Remove inner brake shoe from adaptor.
11. Remove nuts and washers holding brake splash shield, brake adapter and spindle to steering knuckle.
12. Remove spindle from knuckle. Slide inner and outer axle shaft complete with bronze spacer, seal and slinger from axle.
13. Remove cotter key and nut from tie rod. Disconnect tie rod from steering knuckle. Remove cotter key and nut from drag link. Disconnect drag link from steering knuckle arm.
14. Remove nuts and upper knuckle cap (left side steering knuckle arm). Discard gasket. Remove spring and upper socket sleeve.
15. Remove capscrews from lower knuckle cap. Work cap free from knuckle and housing.
16. To remove knuckle from housing, swing out at the bottom, then lift up and off upper socket pin.
17. Loosen and remove upper socket pin. Remove seal.
18. Press lower ball socket assembly from axle housing.
19. Clean all components with a suitable solvent and blow dry with compressed air. Inspect all parts for burrs, chips, wear, flat spots or cracks. Replace necessary parts at assembly.

Installation

1. Lubricate lower ball socket assembly with Multi-Mileage Lubricant.
2. Press seal and lower bearing cup into axle housing. Press lower bearing and seal into axle housing.
3. Install upper socket pin and tighten to 500 to 600 ft. lbs. (668 to 813 Nm). Install seal over socket pin.
4. Position steering knuckle over socket pin. Fill lower socket cavity with lubricant. Work lower knuckle cap into place on knuckle and housing. Install capscrews and tighten to 70 to 90 ft. lbs. (95 to 122 Nm).
5. Liberally lubricate upper socket pin with lubricant. Align upper socket sleeve in keyway of steering knuckle and slide into position.
6. Install new gasket over upper steering knuckle studs. Position spring over sleeve. Install cap (left side steering knuckle arm). Install nuts and tighten to 70 to 90 ft. lbs. (95 to 122 Nm).
7. Left side only connect drag link to steering knuckle arm and install nut. Tighten to 60 ft. lbs. (81 Nm). Install cotter key.
8. Connect tie rod to steering knuckle and install nut. Tighten to 45 ft. lbs. (61 Nm). Install cotter key.
9. Slide axle shaft into position. Position bronze spacer on axle shaft with chamfer facing toward universal joint.
10. Install spindle, brake adapter and brake splash shield. Install washers and nuts and tighten to 50 to 70 ft. lbs. (68 to 95 Nm).
11. Position inner brake shoe on adapter.
12. Carefully install hub and rotor assembly onto spindle. Install outer bearing and inner lock nut. Tighten to 50 ft. lbs. (68 Nm) to seat bearings. Back off lock nut and retighten to 30 to 40 ft. lbs. (41 to 54 Nm) while rotating hub and rotor. Back nut off 135° to 150°. Assemble lock ring and outer lock nut. Tighten lock nut to 65 ft. lbs. (88 Nm) minimum. Bend one tang of lock ring over inner lock nut and one tang of lock ring over outer lock nut. Final bearing end play to be 0.001 to 0.010 inch (0.025 to 0.254mm).
13. Install new gasket on hub. Install drive flange, lockwashers and nuts. Tighten to 30 to 40 ft. lbs. (41 to 54 Nm). Install snap ring. Install hub cap, or install locking hub if so equipped.
14. Carefully position caliper onto adapter. Position adapter lock and spring between caliper and adapter and tap into position. Install allen screw and tighten to 12 to 18 ft. lbs. (16 to 24 Nm).
15. Install wheel and tire assembly. Tighten nuts to 75 ft. lbs. (102 Nm).
16. Lubricate at all fittings. Remove jack stands, lower vehicle, remove block from brake pedal and test vehicle operation.

Steering Knuckle Arm

MODEL 60

Removal

1. Raise vehicle and support safely.
2. Turn wheels to the left. Remove cotter key and nut and disconnect drag link.

SUSPENSION
Light Trucks & Vans

3. Remove three steering knuckle arm to steering knuckle mounting nuts and cone washers. Tap steering knuckle arm to loosen. Remove steering knuckle arm.

Installation

1. Position steering knuckle arm on steering knuckle. Install cone washers and nuts. Tighten nuts to 90 ft. lbs. (122 Nm).
2. Install drag link to steering knuckle arm. Install nut and tighten to 60 ft. lbs. (81 Nm). Install cotter key.
3. Remove jack stands. Lower vehicle and test operation.

Servicing Rotor Hub or Bearings

1. Block brake pedal in the up position.
2. Raise vehicle and support safely.
3. Remove wheel and tire assembly.
4. Remove allen screw holding caliper to adapter.
5. Tap adapter lock and spring from between caliper and adapter.
6. Carefully separate caliper from adapter. Hang caliper out of the way. Inner brake shoe will remain on adapter.

NOTE: Do not allow caliper to hang or be supported by hydraulic brake hose.

7. Remove hub cap. Remove snap ring.
8. Remove flange nuts and lockwashers. Remove drive flange and discard gasket, or remove locking hub if so equipped.
9. Straighten tang on lock ring and remove outer lock nut, lock ring, inner lock nut and outer bearing. Carefully slide hub and rotor assembly from spindle.
10. Remove oil seal and inner bearing from hub.
11. Clean bearings and interior of hub, removing all old grease.
12. If bearings and cups are to be replaced. Remove cups from the hub with a brass drift or use suitable remover.
13. Replace bearing cups with a suitable installer.
14. Install inner bearing in grease coated hub and install new seal with a seal installer tool. Exercise extreme care not to damage seals when installing.
15. Carefully install hub and rotor assembly onto grease coated spindle. Install outer bearing and inner lock nut.
16. Tighten to 50 ft. lbs. (68 Nm) to seat bearings, back off lock nut and retighten to 30 to 40 ft. lbs. (41 to 54 Nm) while rotating hub and rotor. Back nut off 135° to 150°. Assemble lock ring and outer lock nut. Tighten lock nut to 65ft. lbs. (88 Nm) minimum. Bend one tang of lock ring over inner lock nut and one tang of lock ring over outer lock nut. Final bearing end play to be 0.001 to 0.010 inch (0.025 to 0.254mm).
17. Install new gasket on hub. Install drive flange, lockwashers and nuts. Tighten to 30 to 40 ft. lbs. Install snap ring. Install hub cap, or install locking hub if equipped.
18. Carefully position caliper onto adapter. Position adapter lock and spring between caliper and adapter and tap into position. Install allen screw and tighten to 12 to 18 ft. lbs. (16-24 Nm).
19. Install wheel and tire assembly. Tighten nuts to 75 ft. lbs. (102 Nm).

Servicing Rotor, Hub or Bearings

MODEL 44 FBJ

Removal and Disassembly

1. Raise vehicle to a comfortable working height and install jack stands. Remove wheel.
2. Remove caliper retainer and anti-rattle spring assemblies. Remove caliper from disc by sliding out and away from disc. Hang caliper out of the way. Do not allow caliper to hang or be supported by hydraulic brake hose. Remove inboard shoe.
3. Remove outer axle shaft nut and washer. Position a wheel puller and remove hub and rotor assembly.
4. Pull the outer bearing cone from the hub and rotor assembly. Discard outer seal. Remove the six retainer bolts and retainer.
5. Remove the brake caliper adapter from the knuckle (if required).
6. Position a pry bar behind the inner axle shaft yoke and push the bearings out of the knuckle.
7. Remove the "O" ring from the knuckle (if so equipped) and discard.
8. Carefully remove the axle shaft assembly.

Inspection

1. Examine the knuckle bore and inner seal surface for evidence of severe wear or damage. Replace the knuckle, if required.
2. Inspect the outer axle shaft seal surface for grooving. If the surface is grooved, repair as follows:
  a. Measure in from the yoke shoulder of the axle approximately $3/8$ inch (9.5mm). Using a center punch, stake at $1/4$ inch (6mm) intervals around the circumference of the shaft. This will size the shaft to insure a tight fit of the inner seal slinger.
3. Check for proper bearing clamp as follows:
  a. Install the bearing cups and spacer into the knuckle bore.
  b. Position the bearing retainer onto the steering knuckle. Install retainer bolts and tighten to 30 ft. lbs.
  c. Insert a 0.004 inch (0.101mm) feeler gauge between the knuckle and the retainer at a point approximately midway between the retainer mounting ears.

NOTE: The brake dust shield may have to be removed to complete this check. If 0.004 (0.101mm) clearance cannot be obtained at each measuring point, the knuckle must be replaced.

  d. Remove retainer plate, bearing cups and spacer. Install brake dust shield (if removed), tighten mounting bolts to 160 inch lbs. (18 Nm).

Assembly and Installation

1. Apply RTV sealer to the seal surface of the axle shaft.
2. Using Driver, Tool C-4398-1, or equivalent install the seal slinger onto the outer axle shaft.
3. Install lip seal on the slinger with the lip toward the axle shaft spline.
4. Carefully insert axle shaft into the housing so as not to damage the differential seal at the side gears.
5. Insert a pry bar through the axle shaft "U" joint and wedge so that the axle shaft is in all the way and cannot be moved out.
6. Using Adapter, Tool C-4398-2 and Driver, Tool C-4398-1 or equivalent carefully install the seal cup until bottomed in the knuckle.

NOTE: A small amount of wheel bearing grease on the Adapter face will aid in holding the cup in position. Do not remove the tool at this time.

7. Install new outer seal in retainer plate. Locate retainer plate over hub of rotor assembly.
8. Pack wheel bearings with Multi-Purpose Grease.
9. Carefully press outer bearing onto hub.
10. Place grease coated outer bearing cup over outer bearing cone followed by spacer, grease coated inner bearing cup and inner bearing cone. Carefully press into position.
11. Apply a $1/4$ inch (6mm) bead of RTV sealer to the retainer face on the chamfer, (this takes the place of the "O" ring dis-

SECTION 2

SUSPENSION
Light Trucks & Vans

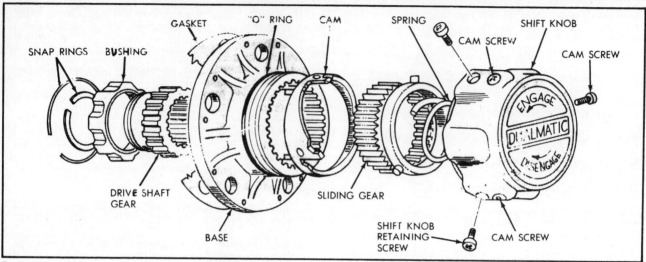
Exploded view of Dualmatic locking hub (© Chrysler Corp.)

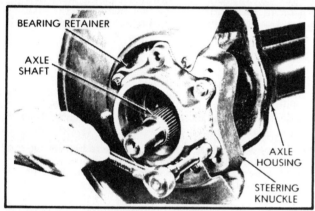
Removing the bearing retainer bolts (© Chrysler Corp.)

Removing the bearing assembly from the steering knuckle (© Chrysler Corp.)

carded at disassembly).
12. Carefully remove seal installing tool from the knuckle bore so that the outer axle shaft remains centered.

NOTE: If the shaft is disturbed, check to make sure that the lip seal is still riding inside the cup. Correct if necessary.

13. Position the bearing retainer in the hub and rotor so that the lube fitting is facing directly forward. This is extremely important.
14. Assemble the hub and rotor to the knuckle and tighten retainer plate bolts to 30 ft. lbs. (41 Nm) using a criss-cross method to position it evenly.
15. Install brake adapter and tighten mounting bolts to 85 ft. lbs. (115 Nm).
16. Remove pry bar from "U" joint and install axle shaft washer and nut. Tighten nut to 100 ft. lbs. (136 Nm), continue to tighten nut until next slot in nut aligns with cotter key hole in axle shaft. Install cotter key.

Front Wheel Locking Hubs

DANA FRONT LOCKING HUBS (MANUAL)
Removal and Disassembly

1. Place hub in lock position. Remove allen head mounting bolts and washers.
2. Carefully remove retainer, O-ring seal and knob. Separate knob from retainer.
3. Remove large internal snap-ring. Slide retainer ring and cam from hub.
4. While pressing against sleeve and ring assembly, remove

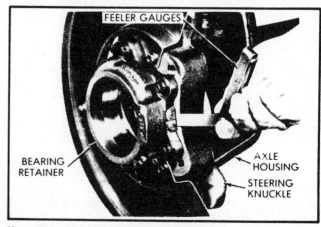
Measuring retainer to steering knuckle clearance (© Chrysler Corp.)

SUSPENSION
Light Trucks & Vans

axle shaft snap-ring. Relieve pressure and remove sleeve and ring, ring and bushing, spring and plate.

5. Inspect all parts for wear, nicks and burrs. Replace all parts which appear questionable.

Assembly and Installation

1. Slide plate and spring (large coils first) into wheel hub housing.
2. Assemble ring and bushing, sleeve and bushing. Slide complete assembly into housing.
3. Compress spring and install axle shaft snap-ring.
4. Position cam and retainer in housing and install large internal snap-ring.
5. Place small O-ring seal on knob, lubricate with waterproof grease and install in retainer at lock position.
6. Place large O-ring seal on retainer. Align retainer and retainer ring and install washers and allen head mounting screws.
7. Check operation.

DUALMATIC LOCKING HUB

Removal and Disassembly

1. Turn shift knob to "engage" position.
2. Apply pressure to the face of shift knob, remove the three screws spaced 120° apart and nearest to the flange.
3. With an outward pull, remove shift knob from mounting base.
4. Remove snap ring from axle shaft.
5. Remove capscrews and lockwashers from mounting base flange.
6. Separate and remove locking hub assembly from rotor hub; remove and discard gasket.

Inspection

Wash parts in mineral spirits and blow dry with compressed air. Examine splines shift knob, cam, sliding gear, drive shaft gear, and mounting base for damage.

Assembly

Lubricate parts lightly with Multi-Purpose lubricant.

1. Position new gasket and locking hub onto rotor hub.
2. Install attaching capscrews and lock washers; tighten to 30ft. lbs. (41-54 Nm).
3. Install axle shaft snap ring.
4. Position shift knob on mounting base. Align splines by pushing inward on shift knob and turning it clockwise to lock it in position.
5. Install and tighten three shift knob retaining screws.

WARN FRONT LOCKING HUBS

Removal and Installation

1. Straighten the lock tabs and remove the six hub mounting bolts.
2. Tap the hub gently with a mallet to remove it.
3. Separate the clutch assembly from the body assembly.
4. Remove the snap ring from the rear of the body assembly. Slip the axle shaft hub out of the body from the front.
5. Remove the allen screw from the inner side of the clutch, and remove the dial assembly from the front side of the clutch housing assembly.
6. Remove the clutch assembly from the rear of the housing complete with the roller pins.
7. Coat the moving parts with a waterproof grease.
8. Slide the axle shaft hub into the body from the front, and replace the snap ring.
9. Replace the dial assembly and the inner clutch. Tighten the allen screw and stake the edge of the screw with a center punch in order to prevent loosening.
10. With the dial in the free position, rotate the outer clutch body into the inner assembly until it bottoms in the housing. Bring it up to the nearest hole and install the roller pins.
11. Position the hub and clutch assembly together with a new gasket in between.
12. Position the hub assembly over the end of the axle and replace the hub mounting bolts and lock tabs.
13. Torque the bolts 30-35 ft. lbs. and bend the lock tabs to anchor the bolts.

WHEEL BEARING ADJUSTING PROCEDURE

1. Raise vehicle and support safely.
2. Remove hub cap. Remove snap ring.
3. Remove flange nuts and lock washers. Remove drive flange and discard gasket, or remove locking hub if so equipped.
4. Straighten tang on lock ring and remove outer lock nut and lock ring.
5. Tighten inner lock nut to 50 ft. lbs. (68 Nm) to seat bearings, back off lock nut and retighten to 30–40 ft. lbs. (41 to 54 Nm) while rotating assembly. Back nut off 135° to 150°. Assemble lock ring and outer lock nut. Tighten lock nut to 65 ft. lbs. (88 Nm) minimum. Bend one tang of lock ring over inner lock nut and one tang of lock ring over outer lock nut. Final bearing adjustment to be 0.001 to 0.010 inch (0.025 to 0.254mm).
6. Install new gasket on hub. Install drive flange, lockwashers and nuts. Tighten to 30 to 40ft. lbs. (41 to 54 Nm). Install snap ring. Install hub cap, or install locking hub if so equipped.
7. Lower vehicle and test vehicle operation.

AUTOMATIC HUBS

Preparing Automatic Hubs for Disassembly

1. Shift transfer case lever into four wheel drive to engage and lock hubs.
2. Shift the transfer case lever into two wheel drive and slowly move the vehicle backwards approximately six feet. Incomplete shift from two wheel drive to four wheel drive or disengagement of only one hub lock may cause an abnormal sound from the axle.
3. Shift to four wheel drive to stop the noise, then shift to two wheel drive to unlock the hub.

Remove Hub from Wheel

1. Using a Torx T-25 driver, remove the five cover screws. Remove cover and bearing race spring assembly, sealing ring, seal bridge retainer, and bearing components.
2. Squeeze the tangs of the wire retaining ring together with a needle nose pliers and pull the remaining components of the automatic locking hub from the wheel.

Assemble Hub to Wheel

1. Make sure that the drag sleeve retainer washer is in position between the wheel bearing adjusting nut and the lock nut. Tighten the lock nut as specified.

NOTE: Before installing the automatic locking hub, make sure that the spacer and retaining ring are in position on the axle shaft.

2. Install the automatic locking hub into the wheel hub aligning the drag sleeve slots with the tabs on the drag sleeve retainer washer.
Align the outer clutch housing splines with the splines of the wheel hub.
3. Loosen the cover screws three or four turns and push in on the cover to allow the retaining ring to expand into the rotor hub groove.

SUSPENSION
Light Trucks & Vans

4. Tighten cover screws to 40-50 inch lbs. (4.52-5.65 Nm) using a Torx T-25 driver.

Disassembling Hub

To disassemble, the hub must be unlocked. If hub is locked when removed from the wheel, hold the hub sleeve and rotate the drag sleeve in either direction to unlock.

1. With the cover removed, turn the clutch gear until it drops to engage with the outer clutch housing.
2. Lift and tilt the drag sleeve and detent to free the brake band tangs from the rectangular opening in the inner cage. Remove the drag sleeve and detent, and brake band assembly.

NOTE: **Never remove the brake band from the drag sleeve and detent as the brake band spring tension may be affected.**

3. Remove retaining ring from outer clutch housing groove.
4. Using a small screw driver, pry plastic outer cage free from inner cage and remove inner cage.
5. Pry free and remove the plastic outer cage from the outer clutch housing.
6. Remove the clutch gear and hub sleeve assembly from the outer clutch housing.
7. Compress the clutch gear return spring and remove the retaining ring from the clutch gear hub sleeve.

---- CAUTION ----
To prevent injury, use a spring compressor to hold spring securely.

8. Carefully release the spring to its normal extended length. Remove the spring support washer, return spring retainers and return spring from the hub sleeve.
9. Remove the "C" type retaining ring from the hub sleeve, remove clutch gear, spring, and cam follower.

Cleaning and Inspection

1. Wash all parts with cleaning solvent; dry with compressed air.
2. Check cover for cracks or porous condition.
3. Check seats of cover screws in cover for pitting, or a tapered, countersunk condition.
4. Examine wire retaining ring for kinks, bends and size (0.088 min.-0.094 max.).
5. Check brake band for distortion or wear.
6. Inspect the teeth on clutch gear and cam follower for wear and broken teeth.
7. Check drag sleeve and drag sleeve retainer washer for cracks or wear.

Lubrication

The automatic locking hub requires lubrication at approximately 24,000 mile intervals.

1. Dip all parts with the exception of the bearing and race assembly, and the brake band and drag sleeve assembly, in automatic transmission fluid.
2. Allow the excess transmission fluid to drip off before proceeding to assemble the unit.
3. Repack the bearing race, and retainer with Lubricant.
4. Lubricate the brake band and drag sleeve using 1.5 grams (0.05 oz.) of lubricant.

Reassembling Hub

1. Snap the cam follower legs over the tooth gaps at the flat surfaces of the clutch gear.

NOTE: **Do not pry the legs of the cam follower apart.**

2. Compress and slide the conical spring into position between the cam follower and the clutch gear having the large diameter of the spring contacting the face of the clutch gear.
3. Install the assembled cam follower, conical spring, and clutch gear assembly onto the hub sleeve having the teeth of the cam follower near the step at the end of the hub sleeve.
4. Position the "C" type retaining ring between the clutch gear and the cam follower, then snap it into place in the groove of the hub sleeve.
5. Place a spring retainer at the end of the return spring and position the spring and retainer in the clutch gear.
6. Install a spring retainer and the spring support washer on the remaining end of the return spring. Compress the return spring, install the retainer ring into the ring groove at the end of the hub sleeve.
7. Install the clutch gear and hub sleeve assembly (spring support washer end first) into the outer clutch housing.
8. For support during the assembling operation, screw three of the cover screws into the outer clutch housing face.
9. Position the plastic outer cage into the outer clutch housing, having the ramps of the plastic outer cage near the cam follower. Carefully work the outer cage into the outer clutch housing until the small external tabs of the plastic cage locates in the wide groove of the outer clutch housing.
10. Install the steel inner cage into the outer cage, aligning the lug of the outer cage with the rectangular opening of the inner cage.
11. Install the retaining ring into the outer clutch housing groove to lock the outer cage into place.
12. Service the brake band and drag sleeve as an assembly. After lubricating, as described previously, position one of the two brake band tangs on each side of the outer cage lug which is located at the rectangular opening of the steel inner cage. If necessary, tilt the drag sleeve while engaging the brake band tangs. When correctly positioned, the brake band tangs will be engaged one on each side of the outer cage lug, and the drag sleeve teeth will be in mesh with the teeth of the cam follower.
13. Remove the three screws from the outer clutch housing which had been previously installed for support during the reassembly procedure. Install the spacer and retaining ring to lock the drag sleeve into position. After making certain that the spacer and retaining ring are in position on axle shaft, install the automatic locking hub onto the vehicle.
14. Position the wire retaining ring in the groove machined in the outer clutch housing surface having the retainer tangs pointing away from the splines.
15. Holding the tangs of the wire retainer together, assemble the seal bridge over the tangs positioned in such a manner that the bent down tabs of the seal bridge clamp the wire retainer tangs together.
16. Install the "O" ring in the outer clutch housing groove and over the seal bridge.
17. Position the bearing over the inner race, lubricate as described previously. Assemble into position, then snap the bearing retainer clip into the hole in the inner race.
18. Install the bearing and retainer assembly into the end of the hub sleeve. Install the sealing ring over the outer clutch housing.
19. Position the bearing race spring into the bore in the cover.
20. Install the spring and cover assembly, install the five screws and tighten to 40-50 inch lbs. (4.5-5.6 Nm) using a Torx T-25 driver.

NOTE: **The "O" ring installed previously to hold the seal bridge in position during assembly, may be left on, but is not required.**

21. After assembling is completed, the hub should turn freely.

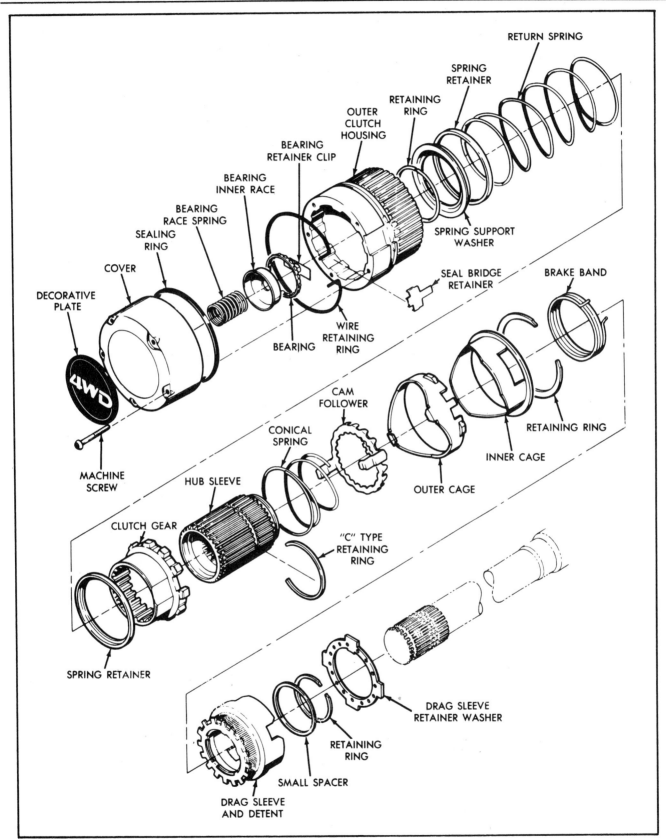

Exploded view of Automatic locking hub (© Chrysler Corp.)

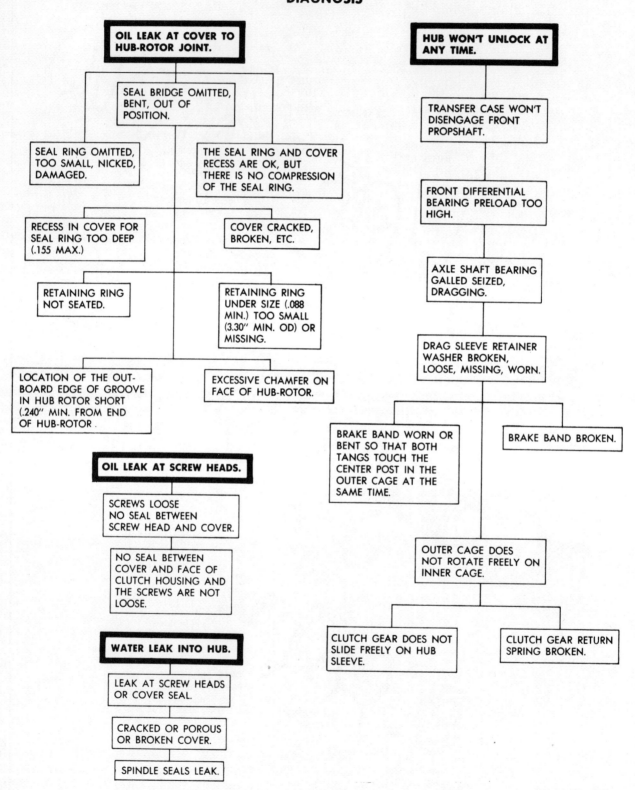

SUSPENSION
Light Trucks & Vans

AUTOMATIC LOCKING HUB DIAGNOSIS

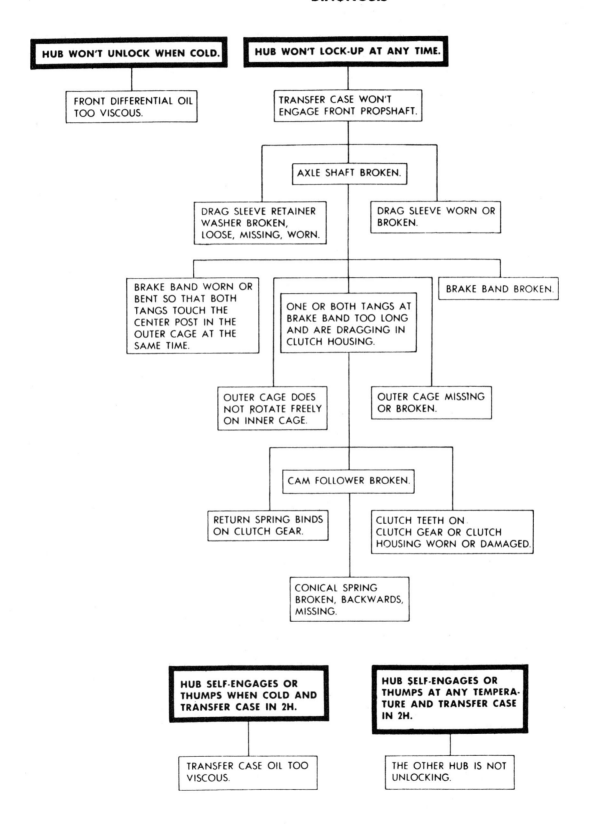

SUSPENSION
Light Trucks & Vans

INDEPENDENT FRONT SUSPENSION

NOTE: These procedures apply to all vans, wagons, and conventional trucks with independent front suspension.

COIL SPRING

Removal and Installation

1. Raise the vehicle and support it safely.
2. Remove the wheel.
3. Remove the shock absorber and upper shock absorber bushing bushing and sleeve.
4. If equipped, remove the sway bar.
5. Remove the strut.
6. Install a spring compressor and tighten finger-tight.
7. Remove the cotter pins and ball joint nuts.
8. Install a ball joint breaker tool and turn the threaded portion of the tool to lock it against the lower stud.
9. Spread the tool to place the lower stud under pressure, then strike the steering knuckle sharply with a hammer to free the stud. Do not attempt to force the stud out of the steering knuckle with the tool.
10. Remove the tool. Slowly release the spring compressor until all tension is relieved from the spring.
11. Remove the spring compressor and spring.
12. Installation is the reverse of removal. Compress the spring until the ball joint can be properly positioned in the steering knuckle.

SHOCK ABSORBER

Removal and Installation

1. Raise and support the vehicle with jackstands positioned at the extreme front ends of the frame rails.
2. Remove the wheel.
3. Remove the upper nut and retainer.
4. Remove the two lower mounting bolts.
5. Remove the shock absorber.
6. Installation is the reverse of removal.

UPPER CONTROL ARM

Removal and Installation

NOTE: Any time the control arm is removed, it is necessary to align the front end.

1. Raise and support the vehicle with jackstands under the frame rails.
2. Remove the wheel.
3. Remove the shock absorber and shock absorber upper bushing and sleeve.
4. Install a spring compressor and tighten it finger-tight.
5. Remove the cotter pins and ball joint nuts.
6. Install a ball joint breaker and turn the threaded portion of the tool, locking it securely against the upper stud. Spread the tool enough to place the upper ball joint under pressure and strike the steering knuckle sharply to loosen the stud. Do not attempt to remove the stud from the steering knuckle with the tool.
7. Remove the tool.
8. Remove the eccentric pivot bolts, after making their relative positions in the control arm.
9. Remove the upper control arm.
10. Installation is the reverse of removal. Tighten the ball joint nuts to 135 ft. lbs. Tighten the eccentric pivot bolts to 70 ft. lbs.
11. Adjust the caster and camber.

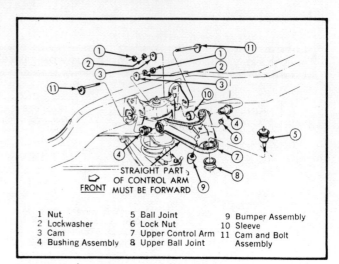

Upper control arm details (© Chrysler Corp.)

1 Nut
2 Lockwasher
3 Cam
4 Bushing Assembly
5 Ball Joint
6 Lock Nut
7 Upper Control Arm
8 Upper Ball Joint
9 Bumper Assembly
10 Sleeve
11 Cam and Bolt Assembly

LOWER CONTROL ARM

Removal and Installation

1. Follow the procedure outlined under "Coil Spring Removal and Installation."
2. Remove the mounting bolt from the crossmember.
3. Remove the lower control arm from the vehicle.
4. Installation is the reverse of removal. After the vehicle has been lowered to the ground, tighten the mounting bolt to 210 ft. lbs.

LOWER BALL JOINT

Removal and Installation

1. Remove the lower control arm.
2. Remove the ball joint seal.
3. Using an arbor press and a sleeve, press the ball joint from the control arm.
4. Installation is the reverse of removal. Be sure that the ball joint is fully seated. Install a new ball joint seal.
5. Install the lower control arm. Be sure to install the ball joint cotter pins.

UPPER BALL JOINT

Removal and Installation

1. Install a jack under the outer end of the lower control arm and raise the vehicle.
2. Remove the wheel.
3. Remove the ball joint nuts. Using a ball joint breaker, loosen the upper ball joint.
4. Unscrew the ball joint from the control arm.
5. Screw a new ball joint into the control arm and tighten 125 ft. lbs.
6. Install the new ball joint seal, using a 2 in. socket. Be sure that the seal is seated on the ball joint housing.
7. Insert the ball joint into the steering knuckle and install the ball joint nuts. Tighten the nuts to 135 ft. lbs. and install the cotter pins.
8. Install the wheel and lower the truck to the ground.

SUSPENSION
Light Trucks & Vans

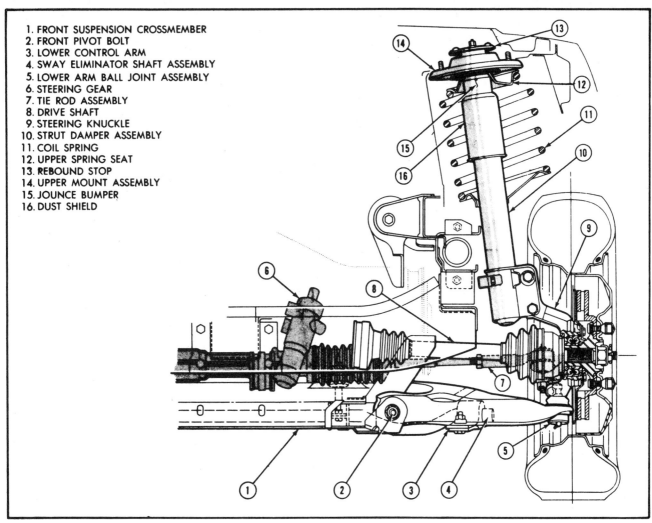

1. FRONT SUSPENSION CROSSMEMBER
2. FRONT PIVOT BOLT
3. LOWER CONTROL ARM
4. SWAY ELIMINATOR SHAFT ASSEMBLY
5. LOWER ARM BALL JOINT ASSEMBLY
6. STEERING GEAR
7. TIE ROD ASSEMBLY
8. DRIVE SHAFT
9. STEERING KNUCKLE
10. STRUT DAMPER ASSEMBLY
11. COIL SPRING
12. UPPER SPRING SEAT
13. REBOUND STOP
14. UPPER MOUNT ASSEMBLY
15. JOUNCE BUMPER
16. DUST SHIELD

Rampage front suspension—front wheel drive with struts (© Chrysler Corp.)

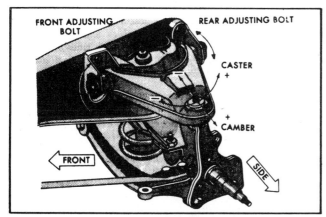

Independent front suspension alignment points (© Chrysler Corp.)

Wheel Bearings

It is recommended that the front wheel bearings be cleaned, inspected and repacked periodically and as soon as possible after the front hubs have been submerged in water.

NOTE: *Sodium based grease is not compatible with lithium based grease. Be careful not to mix the two types. If unsure of the present grease being used in the bearings, clean all of the old grease from the hub assembly before installing any new grease.*

Front Wheel Drive

STRUT ASSEMBLY

Removal and Installation

1. Loosen the wheel lug nuts.
2. Raise the vehicle and remove the wheel assembly.
3. Mark the camber cam to damper bracket position before removing the cam adjusting bolt.
4. Remove the cam adjusting bolt, the through bolt and the brake hose to damper bracket retaining screw.
5. Remove the strut damper to fender shield (strut tower) mounting nut and washer assembly.
6. Install strut assembly into fender (strut tower) and install nut and washer assemblies. Torque the retaining nuts to 20 ft. lbs. (27 Nm).

2-317

SUSPENSION
Light Trucks & Vans

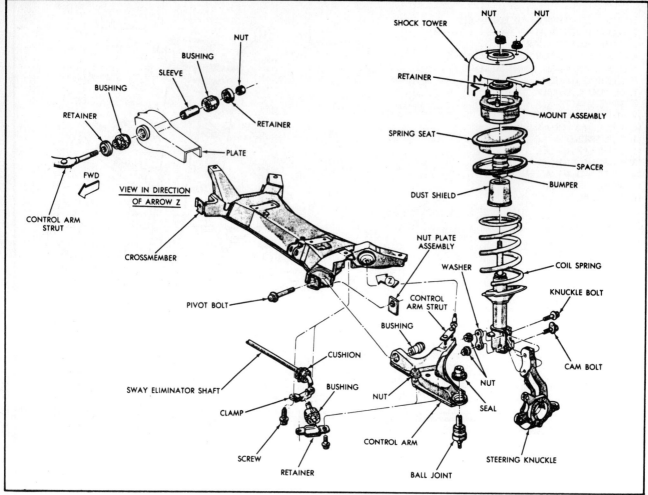

Mini-Van front suspension (© Chrysler Corp.)

7. Position the steering knuckle neck into the strut damper bracket, install the cam adjusting and through bolts.
8. Attach the brake hose retainer to the damper bracket, torque the screws to 10 ft. lbs. (13 Nm).
9. Adjust the camber to the original marked position.
10. Place a 4 inch or larger "C" clamp on the strut and knuckle. Tighten the clamp just enough to eliminate any looseness between the knuckle and the strut. Recheck the alignment marks and tighten the bolts to 45 ft. lbs. (61 Nm). Turn the bolts an additional $\frac{1}{4}$ turn (90°) beyond the specified torque.
11. Install the wheel and tire assembly. Torque the wheel nuts to 80 ft. lbs. (108 Nm).

STRUT TYPE COIL SPRING

Removal and Installation

1. Use a coil spring compressor (strut type) to compress the coil spring.
2. Hold the strut rod while removing the strut rod.
3. Remove the strut damper mount assembly and the coil spring.

NOTE: If both strut springs are to be removed and reinstalled, mark them so they can be returned to their original position.

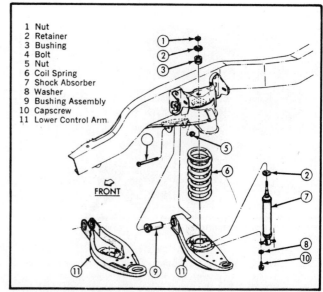

1 Nut
2 Retainer
3 Bushing
4 Bolt
5 Nut
6 Coil Spring
7 Shock Absorber
8 Washer
9 Bushing Assembly
10 Capscrew
11 Lower Control Arm

Lower control details (© Chrysler Corp.)

SUSPENSION
Light Trucks & Vans

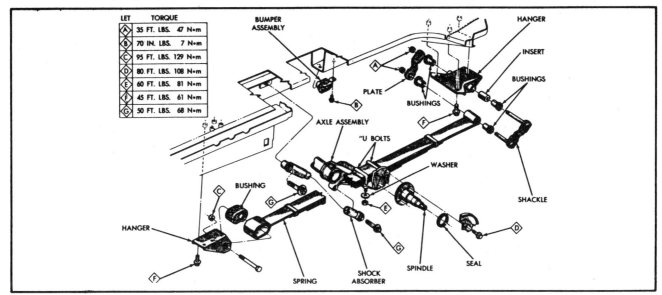

Mini-Van rear suspension (© Chrysler Corp.)

CAUTION
If Chrysler special tool spring compressor tool L-4514 is used, do not open the tool jaws beyond 9 1/4 inches (230mm).

4. Reinstall the strut rod nut and torque it to 60 ft. lbs. (81 Nm) before releasing the spring compressor.

LOWER CONTROL ARM
Removal and Installation

1. Raise the vehicle and support safely.
2. Remove the front inner pivot through bolt, rear stub strut nut, retainer and bushing.
3. Remove the ball joint to steering knuckle clamp bolt.
4. Separate the ball joint stud from steering knuckle.

CAUTION
Do not pull the steering knuckle away from the vehicle, while the ball joint is disconnected, it can separate the inner C/V joint.

5. Remove the sway bar to control arm end bushing retainer nuts and rotate the control arm over the sway bar. Remove rear stub strut bushing, sleeve and retainer.
6. Install retainer, bushing and sleeve on stub strut.
7. Position control arm over sway bar and install rear stub strut and front pivot into crossmember.
8. Install front pivot bolt and loosely assemble the nut.
9. Install the stub strut bushing and retainer and loosely assemble the nut.
10. Install ball joint stud into steering knuckle and install the clamp bolt. Tighten the clamp bolt to 50 ft. lbs. (67 Nm).
11. Position sway bar end bushing retainer to control arm. Install retainer bolts and tighten nuts to 22 ft. lbs. (30 Nm).
12. Lower the vehicle and support the control arm to design height. Tighten the front pivot bolt to 105 ft. lbs. (142 Nm) and the stub strut nut to 70 ft. lbs. (94 Nm).

LOWER BALL JOINT
Inspection

1. With the weight of the vehicle resting on the road wheels, grasp the grease fitting and attempt to move it.
2. No mechanical force is necessary. If the ball joint is worn the grease fitting will move easily. Replace worn ball joints.

LOWER BALL JOINT
Removal and Installation

The lower front ball joints are pressed into the lower control arm and an arbor press will be needed to remove and install them.

1. Pry the seal from the ball joint.
2. Position special tool C-4699-2 to support the lower control arm while receiving ball joint assembly.
3. Use a 1 inch deep wall socket in the arbor press to remove the ball joint.
4. Position the ball joint into the control arm cavity.
5. Position the control arm assembly in the press with special tool C-4699-1 or equivalent supporting the control arm.
6. Align the assembly and press until the ball joint housing ledge stops against control arm cavity down flange.
7. Support the ball joint housing with special tool C-4699-2 or equivalent and position a new seal over the ball joint stud.
8. Using an 1 1/2 inch socket in an arbor press force the seal to seat against the control arm.

WHEEL BEARING
Adjustment

1. Remove the wheel and tire assembly.
2. Remove the hub lock nut cotter pin and lock nut.
3. Loosen the hub nut and apply the brakes.
4. With brakes applied, tighten hub nut to 180 ft. lbs. (245 Nm).
5. Install nut lock and new cotter pin. Wrap cotter pin tightly against nut lock.
6. Install wheel and tire assembly and tighten wheel nuts to 80 ft. lbs. (108 Nm).

DRIVE AXLES A-412 MANUAL TRANSMISSION
Removal and Installation

1. Remove the cotter pin and nut lock.
2. Loosen hub nut and wheel nuts while vehicle is on floor and brakes applied.

SUSPENSION
Light Trucks & Vans

3. Raise vehicle and remove hub nut and wheel and tire assembly.
4. Remove the clamp bolt securing ball joint stud into the steering knuckle.
5. Separate the ball joint stud from the steering knuckle by prying against the knuckle leg and control arm. Do not damage the ball joint or C/V joint boots.
6. Separate out C/V joint splined shaft from hub by holding C/V housing while moving knuckle hub assembly away.
7. Remove the plastic caps from the allen head screws by prying under the cap and against the inner joint flange.

NOTE: The separated outer joint and shaft must be supported during inner joint separation. Tie the assembly to the control arm during the operation.

8. Using special tool L-4550 or equivalent remove the allen head screws attaching the inner C/V joint to the transaxle drive flange. Wipe foreign material from C/V joint and drive flange.
9. Release the outer C/V joint from the control arm.
10. Remove the driveshaft assembly carefully in order to reduce loss of special lubricant from the inner C/V joint. Hold both the inner and outer housings parallel and rotate the outer assembly down while pivoting inner housing up at the drive flange.

NOTE: The drive shaft installed acts as a bolt and secures the hub/bearing assembly. If the vehicle is to be supported or moved on its wheels, install a bolt through the hub to insure that the hub bearing assembly cannot loosen.

11. Replace any lost lubricant with Chrysler part number 4131389 or equivalent.
12. Wipe grease from the joint housing, face, screw holes and transaxle drive flange face before installation.
13. Support assembly vertically inner housing up to retain special lubricant. Do not move inner joint in or out during assembly to drive flange. Pumping action can force lubricant out of the joint.
14. Position inner housing to drive flange and rotate assembly up. Locate the inner housing in the drive flange. Tie the outer end of the drive shaft to the control arm during assembly of the inner end.
15. Install the allen head screws to secure the inner joint to drive axle flange. Torque the screws to 36.6 ft. lbs. (50 Nm).

NOTE: Failure to torque the flange screws to the specified torque may result in loosening during operation.

16. Untie the outer end of the drive shaft and push the hub assembly out far enough to install the outer C/V joint shaft into the hub.
17. Install the knuckle assembly on the ball joint stud.
18. Install and tighten the clamp bolt to 50 ft. lbs. (68 Nm).

NOTE: The steering knuckle clamp bolt is a prevailing torque type and must be replaced with the original and/or equivalent bolt.

19. Install the washer and hub nut. With brakes applied tighten the nut to 180 ft. lbs. (245 Nm).
20. Install lock and cotter pin.

DRIVE AXLES, ALL EXCEPT A-412 TRANSAXLE

Removal and Installation

The inboard C/V joints have stub shafts splined into the differential side gears. The two different types of axles are retained either by spring force or by circlips on the end of the shafts. The clip type axles require the removal of the differential cover to

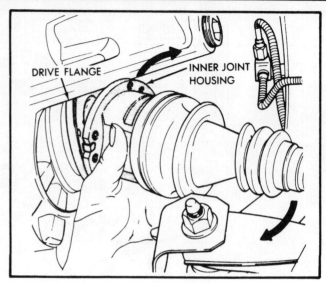
Removing the Rampage drive axle—tilt up to retain lubricant (© Chrysler Corp.)

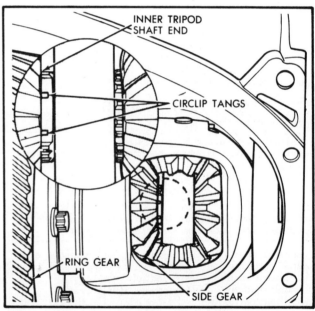
Clip type retained drive axle—Rampage (© Chrysler Corp.)

remove the axle shaft. The spring loaded type shafts do not require removal of the differential cover.

1. On the clip type axles, remove drain plug, drain the lubricant and remove the differential cover.
2. On the right side (passenger side) remove the speedometer pinion.
3. Rotate the driveshaft to expose the circlip tangs. A relief (flat) is machined on the end of the shaft to accommodate the circlip tangs.
4. With needle nose pliers, compress the circlip tangs. Pry the shaft into the side gear splined cavity. The circlip will be compressed into the cavity with the shaft.
5. Remove the clamp bolt securing the ball joint stud into the steering knuckle.
6. Separate the ball joint stud from the steering knuckle by prying against the knuckle leg and control arm. Do not damage

SUSPENSION
Light Trucks & Vans

the ball joint or C/V joint boots.

7. Separate the outer C/V joint splined shaft from hub by holding C/V housing while moving knuckle (hub) assembly away. Do not pry on or otherwise damage the wear sleeve on the outer C/V joint.

8. Support the assembly at the C/V joint housings. Remove by pulling outward on the inner joint jousing.

NOTE: **The drive shaft acts as a bolt and secures the hub/bearing assembly. If the vehicle is to be supported or moved on its wheels, install a bolt through the hub to insure that the hub bearing assembly cannot loosen.**

9. Install new circlips on the inner joint shaft before installation.

NOTE: **If the wear sleeve is bent or damaged, replace it before installing the axle shaft.**

10. Align the tangs on the circlip with the flattened end of the shaft before inserting into transaxle. Failure to align the tangs can cause jamming and component damage.

11. Hold the inner joint assembly at housing, while aligning and guiding the inner joint spline into the transaxle.

12. Check the clip type axle for locking of the circlip in the proper position.

13. Push the knuckle (hub) assembly out and install the splined outer C/V joint shaft in the hub.

14. Install the knuckle assembly on the ball joint stud.

NOTE: **The steering knuckle clamp bolt is the prevailing torque type and the original or exact equivalent must be reinstalled.**

15. Install and tighten clamp bolt to 50 ft. lbs. (68 Nm).
16. Install the speedometer pinion.
17. If the differential cover was removed, apply a $\frac{1}{8}$ inch bead of gasket material, RTV to the gasket surface of the differential cover.
18. Install the differential cover and torque the retaining screws to 13.8 ft. lbs. (19 Nm).
19. Fill the differential to the bottom of the plug hole with Dexron^k type automatic transmission fluid.
20. Install the washer and a new hub nut, apply the brakes and torque the hub nut to 180 ft. lbs. (245 Nm).
21. Install a lock and new cotter pin on the hub nut.
22. Install the wheel and tire assembly and torque the lug nuts to 80 ft. lbs. (108 Nm).

REAR SUSPENSION

Rear Spring

Removal and Installation

CONVENTIONAL TRUCKS, 100–300

1. Raise rear of truck until weight is removed from springs, wheels just touching the floor.

NOTE: **Truck must be lifted by jack or hoist under frame side rail at crossmember behind the axle being careful not to bend flange of side rail.**

2. Place stands under side frame members as a safety precaution.
3. Remove nuts, lockwashers and U-bolts securing spring to axle.
4. Remove spring shackle bolts, shackle and spring front bolt, then remove spring.

5. To install, position spring on axle so spring center bolt enters locating hole in axle housing pad.
6. Line up spring front eye with bolt hole in bracket and install spring bolt and nut.
7. Install the rear shackle, bolts and nuts. Tighten shackle bolt nut until slack is taken up.
8. On headless type spring bolts install the bolts with lock bolt groove lined up with lock bolt hole in bracket. Install lock bolt and tighten lock bolt nut. Install lubrication fittings.
9. Install U-bolts, new lockwashers and nuts, tightening until nuts push lockwashers against axle. Align auxiliary spring parallel with main spring.
10. Remove stands from under vehicle, lower truck so weight is resting on wheels. Tighten U-bolt nuts, spring eye nuts and shackle bolt nuts.
11. Lubricate spring bolts and shackle bolts with chassis lubricant when equipped with lubrication fittings.

SUSPENSION
Light Trucks & Vans

VANS, PICKUPS AND WAGONS

1. Raise the vehicle until springs are accessible. Support safely.
2. Remove U-bolt nuts, U-bolts, and plate.
3. Remove the front pivot bolt nut. Remove the bolt.
4. Remove the rear shackle bolt nuts. Remove the shackle plate.
5. Remove the outer shackle and bolt assembly from the hanger. Remove the spring. On vehicles equipped with one piece shackles, remove the nut, remove the shackle to spring bolt, and remove the spring.
6. To install, position the spring and shackle assembly in the rear hanger.
7. Install the shackle plates and nuts, tightening the nuts to 40 ft. lbs. On vehicles with the one piece shackle, first position the spring to the shackle, and then install the bolt and nut.
8. Position the spring in the front pivot hanger and install the bolt and nut.
9. Position the spring properly on the axle ad install the U-bolt plate.
10. Install the U-bolts and nuts. Make sure that shackled end of the spring is above the shackle bracket pivot.
11. Lower the vehicle to the floor, and tighten all nuts.

Installing New Leaf

1. Clamp spring in a vise, remove center bolt and bend clamp type clips back from spring leafs.
2. Insert long drive in center bolt hole and release vise slowly.
3. Remove assembly from vise and replace broken leaf.
4. Place spring assembly in vise, slowly tightening vise while holding spring leaves in alignment with drift.
5. Remove drift and install new center bolt.
6. Install nut, tightening to 15 ft. lbs. torque.
7. Remove spring from vise.

Removal and Installation
FRONT WHEEL DRIVE – PICKUPS

1. Raise the vehicle and support safely.
2. Using floor stands under the axle assembly, raise axle assembly to relieve the weight of the rear springs.
3. Disconnect the rear brake proportioning valve spring. Disconnect the lower ends of the rear shock absorbers at the axle brackets.
4. Loosen and remove the "U" bolt nuts and remove the "U" bolts and spring plate.
5. Lower the rear axle assembly, allowing the rear springs to hang free.
6. Loosen and remove the front pivot bolt from the front spring hanger.
7. Loosen and remove rear spring shackle nuts and remove shackles from spring.
8. To install, assemble the shackle and bushings in the rear end of the spring. Install the rear spring hanger, start the shackle bolt nuts. Do not tighten.
9. Raise the front of the spring and install the pivot bolt and nut, do not tighten.
10. Raise the axle assembly into correct position with the axle centered under spring center bolts.
11. Install the spring plate and "U" bolts. Tighten the U-bolt nuts to 60 ft. lbs. (81 Nm).
12. Install the shock absorbers and retaining nuts.
13. Lower the vehicle to the floor and with full weight on the wheels, tighten component fasteners as follows:
   a. Front pivot bolt: 115 ft. lbs. (155 Nm).
   b. Shackle nuts: 35 ft. lbs. (47 Nm).
   c. Shock absorber nut: 20 ft. lbs. (27 Nm).
14. Connect the rear brake proportioning valve spring and adjust the valve.

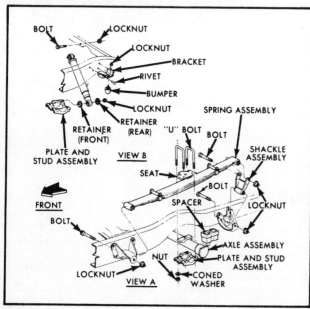

Rear suspension Ramcharger and Trail Duster (© Chrysler Corp.)

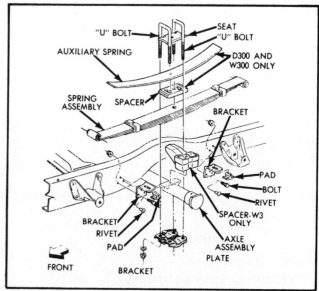

Rear suspension—W200, W300 and D300 shown (© Chrysler Corp.)

SUSPENSION
Light Trucks & Vans

FORD TRUCKS
SERIES F100–300•500–9000
AEROSTAR•VANS•BRONCO•PICK-UPS•RANGER

FRONT SUSPENSION

Solid I Beam

SPRING

Removal and Installation

1980–81 LN 600–700 SERIES

1. Raise the vehicle frame until the weight is off the front springs, with the wheels still touching the floor.
2. Remove the shock absorber.
3. Remove the cotter pin from the support bracket and remove the stud from the front bracket.
4. On vehicles equipped with hydraulic brakes, remove the cotter pin and drive out the spring pin.
5. On vehicles equipped with air brakes, remove the cotter pin and nut from the rear bracket shackle bolt and drive out the bolt.
6. Remove the nuts from the two spring clips (U-bolts) holding the spring on the axle.
7. Position the spring on the spring seat and align the spring eye with the spring bracket.
8. Prior to installation, coat the bushings with lubricant. Drive the stud through the bracket and the eye of the spring.

NOTE: The lubrication opening in the stud should face inward.

9. Install the attaching nut. Tighten the nut to 31–42 ft. lbs. then back it off one castellation. Install a new cotter pin.
10. Install the lubrication fitting.
11. Raise the opposite end of the spring leaf into the spring rear bracket.
12. On vehicles with hydraulic brakes, install the spring pin in the bracket and secure the pin with a new cotter pin.
13. On vehicles with air brakes, install the spring retaining bolt, washer, nut, and new cotter pin.
14. Place the spring clips (U-bolts) in position over the spring clip plate and through the holes in the axle. Make sure that the spring tie bolt is centered in the recess of the axle.
15. Install the nuts on the spring clips. Lower the vehicle to the floor and tighten the spring clip nuts. Lubricate the front pin.

F AND B SERIES, EXCEPT 1983 AND LATER F SERIES WITH 9500 AND 12000 lb. FRONT AXLES

1. Raise the vehicle frame until the weight is off the front spring with the front wheel still touching the floor.
2. Support the front axle with jacks, to remove the weight from the spring U-bolts.
3. Disconnect the shock absorber from the leaf spring.
4. Remove the nut and bolt that retains the spring to the front bracket.
5. Remove the nut and bolt that retains the spring to the rear bracket.
6. Remove the four U-bolt nuts and washers and the two U-bolts. Remove the spring.
7. Position the spring at the spring seat and align the spring eye with the front spring bracket.
8. Prior to installation coat the bushing with Multi-Purpose Lubricant or equivalent. Gently slide the bolt through the bracket and spring eye.
9. Install the washer and nut. Tighten the nut to 220–300 ft. lbs. Install the cotter pin.
10. Raise rear end of spring into the bracket.
11. Install the through bolt and nut and tighten nut to 75–105 ft. lbs. Install cotter pin.
12. Place the U-bolts in position over the spring and through the holes in the axle. Make sure that the spring tie bolt is centered in the recess of the axle. Install the nut and washers on the U-bolts. Lower the vehicle to the floor and tighten the U-bolt nuts to specification. Install the shock absorber. Remove the jack.

1983 AND LATER F SERIES WITH 9500 AND 12000 FRONT AXLES

1. Raise the vehicle until the weight is off the front springs, with the wheels still touching the floor. Remove the shock absorber, if so equipped.
2. Remove the bolts that secure the pin in the front bracket.
3. Remove the retaining pin from the front bracket.
4. Remove the four through bolts and then the two shackle pins securing the shackle assembly. Remove the shackle from the spring and rear bracket.

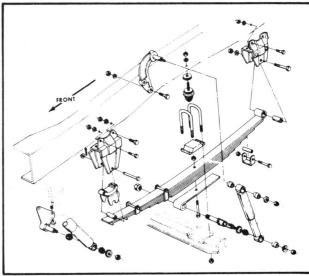

Front spring installation—F-B 600–750 (© Ford Motor Co.)

SUSPENSION
Light Trucks & Vans

5. Remove the two U-bolts that attach the spring to the axle.
6. Lift the spring from the axle noting the position of any caster wedges or spacer.
7. To install, position the spring on the spring seat and align the spring front eye with the spring bracket.
8. Remove the zerk fitting from the retaining pin. Position the pin with the notches aligned with the attaching bolt holes in the front bracket and the lubricator opening facing outward. Drive the pin through the bracket and spring, then install the attaching bolts, washers, and lock nuts. Install the zerk fitting in the pin.
9. Align the shackle assembly upper holes with the rear bracket holes and the shackle lower holes with the spring eye. Drive one shackle pin through the upper hole and bracket and the other pin through the lower hole and spring eye. Be sure that the notches in the shackle pins are aligned with the attaching (pinch) bolt holes in the shackle and that the lube openings face outward.
10. Install the pinch bolts and nuts and tighten to specification.
11. On vehicles equipped with a spacer, place it on the axle.
12. On vehicles equipped with a caster wedge, place it on the axle with the thick edge in the same position as it was originally.
13. Position the U-bolts over the spacer and spring and through the holes in the axle. Make sure the spring tie bolt is centered in the recess of the axle, spacer or caster wedge.
14. Install the shock absorber lower bracket to the underside of the axle with the U-bolt ends entering the holes in the bracket. Install the flat washers and locknuts on the U-bolts.
15. Lower the vehicle to the floor and tighten the spring U-bolt nuts to specifications as listed at toe end of this Section. Lubricate all pins with Multi-Purpose Lubricant, or equivalent.

1980–81 L, LT, LN, LNT, 600, 900, 8000 AND 9000 C AND CT SERIES

1. Raise the vehicle until the weight is off the front springs, with the wheels still touching the floor. Remove the shock absorber, if equipped.
2. On all vehicles except the C-series, remove the bolt securing the spring pin in the front bracket.
3. Remove the retaining pin from the bracket. On C-series vehicles, remove the cotter pin and nut from the spring stud and drive the stud out of the front hanger bracket and spring eye.
4. On all vehicles except C-series, remove the four through bolts and then the two shackle pins securing the shackle assembly. Remove the shackle from the spring and rear bracket. On C-series vehicles, remove the two shackle retaining nuts and slide the shackle assembly out of the shackle bar, hanger and the spring eye.
5. Remove the two U-bolts that attach the spring to the axle.
6. Lift the spring off of the axle, noting the position of any caster wedges or spacers.
7. Position the spring on the spring seat and align the spring front eye with the spring bracket.
8. On all vehicles except C-series, remove the lubricating fitting from the retaining pin. Position the pin with the notches aligned with the attaching bolt holes in the front bracket, and the lubricator opening facing outward. Drive the pin through the bracket and spring, then install the attaching bolt, lockwashers, and nuts. Install the lubricator fittings in the pin. On C-series vehicles, remove the lube fittings from the retaining stud and drive the stud through the front hanger bracket and spring eye. Install the lube fitting on the outer end of the stud and the nut and cotter pin on the inner end.
9. On C-series vehicles, raise or lower the rear end of the spring as required to insert the shackle into the spring eye and rear bracket. Position the shackle bar to the inner side of the rear bracket, and drive the shackle assembly pins through the hanger bracket, spring eye and shackle bar. Install the two shackle retaining nuts and cotter pins.
10. On all vehicles except C-series, align the shackle assembly upper holes with the rear bracket holes and the shackle lower holes with the spring eye. Drive one shackle pin through the upper hole and bracket and the other pin through the lower hole and spring eye.

NOTE: Be sure that the notches in the shackle pins are aligned with the attaching (pinch) bolt holes in the shackle and that the lube openings face outward.

11. Install the pinch bolts and nuts, tighten.
12. On vehicles equipped with a spacer, place it on the axle.
13. On vehicles equipped with caster wedge, place it on the axle in the same position as removed.
14. Position the U-bolts over the spacer and through the holes in the axle.

NOTE: Make sure that the spring tie bolt is centered in the recess of the axle, spacer or caster wedge.

15. Install the shock absorber lower bracket to the underside of the axle, except C-series vehicles, with the spring clip (U-bolt) ends entering the holes in the bracket. Install the flat washers and lock nuts on the spring clip.
16. Lower the vehicle to the floor and tighten the clip nuts. Lubricate the shackle bolts.

1982 AND LATER C AND CT SERIES

1. Raise the vehicle until the weight is off the front springs, with the wheels still touching the floor. Support the front axles with jacks to remove weight from U-bolts. Remove the shock absorber, if so equipped.
2. Remove the cotter pin and nut from the spring stud and drive the stud out of the front hanger bracket and spring eye.
3. Remove the two shackle retaining nuts and slide the shackle assembly out of the shackle bar, the rear hanger bracket and the spring.
4. Remove the two U-bolts that attach the spring to the axle.
5. Lift the spring from the axle noting the position of any caster wedges or spacer.
6. Position the spring on the spring seat and align the spring front eye with the spring bracket.
7. Remove the zerk fitting from the retaining stud and insert the stud through the front hanger bracket and spring eye. Install the zerk fitting on the outer end of the stud and the nut and cotter pin on the inner end.
8. Raise or lower the rear end of spring as required to insert the shackle into the spring eye and rear bracket. Position the shackle bar to the inner side of the rear bracket, and drive the shackle assembly pins through the hanger bracket, spring eye and shackle bar. Install the two shackle retaining nuts and cotter pins.
9. On vehicles equipped with a spacer, place it on the axle.
10. On vehicles equipped with a caster wedge, place it on the axle with the thick edge in the same position as it was originally.
11. Position the U-bolts over the spring and spacer and through the holes in the axle. Make sure the spring tie bolt is centered in the recess of the axle, spacer or caster wedge.
12. Install the shock absorber lower bracket to the underside of the axle, with the U-bolts ends entering the holes in the bracket. Install the flat washers and lock nuts on the U-bolts.
13. Lower the vehicle to the floor and tighten the U-bolt nuts to specifications as listed at the end of this Section. Lubricate the shackle bolts with Multi-Purpose Lubricant, or equivalent.

Spindle

Removal and Installation

1. Loosen the front wheel hub bolt nuts. Raise the front of the vehicle and remove the wheel.

SUSPENSION
Light Trucks & Vans

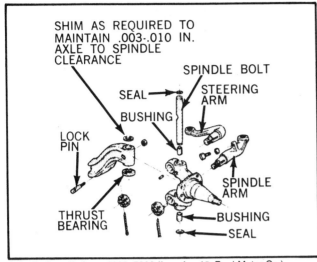

Spindle installation—6000–7000 lb. axles (© Ford Motor Co.)

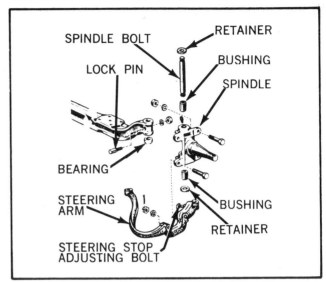

Spindle installation—P-Series (© Ford Motor Co.)

2. Remove the hub and drum and the wheel bearing from the spindle.

3. If the brake backing plate is bolted to the spindle, remove the plate from the spindle. To prevent damage to the brake hose, hang the backing plate from the frame.

4. Disconnect the spindle arms from the spindle.

5. On 5000, 5500, 6000 and 7000 lb axles with hydraulic brakes, remove the seal (O-ring) from the top of the spindle pin. Remove the nut from the spindle bolt lock pin and drive the lock pin out of the axle. Drive the spindle bolt out from the top of the axle with a suitable drift.

6. On 7000 (with air brakes), 8000, 9000, 9500, 12000, 16000 and 18000 lb axles, three separate spindle top mountings are used. Vehicles equipped with either a 7000, 8000, 9000, 9500 or 12000 lb axle utilizes a three-bolt cap/zerk fitting with a gasket. The 16000 and 18000 lb axles have a combination brake hose/zerk cap (Fig. 8). After removing the appropriate cap and gasket, the lower spindle pin seal should be removed on 7000 lb. (with air brake), on 7000, 8000 and 9000 lb. remove the snap ring and expansion plug from the bottom of the spindle bolt. On 9500, 12000, 16,000 and 18,000 lb axles, remove the three bolts, cap, zerk fitting and gasket. Drive out the threaded lock pins with a bronze drift; then, drive out the spindle pin from the top of the axle with a bronze drift.

7. Remove the spindle assembly from the axle.

8. To assemble, place the spindle in position on the axle. Insert the spindle bolt thrust bearing between the bottom of the axle and the spindle. Pack the thrust bearing and coat all mating surfaces of spindle and axle parts with Multi-Purpose Lubricant, or equivalent prior to assembly. On all axles, the thrust bearing must be installed with the retainer lip facing down. The shim (or shims) is placed between the top of the axle and the spindle. A 0.003–0.010 inch axle to spindle clearance must be maintained on all axles.

9. On 6000, 7000 and 8000 lb axles and some 9000 lb. axles, line up the notch in the spindle pin with the spindle pin hole in the axle. The spindle pin must be installed with the end marked T facing upward. Drive the spindle pin with a bronze drift through the axle until the notch and hole are in line. Install a new lock pin from either side of the I-Beam. Seat the lock pin by striking it with a hammer and punch, and then install the nut on the lock pin (certain axles will have two lock pin holes on each side of axle). Do not install both lock pins from the same side of the axle. Install a new seal (O-ring) in the groove at each end (or spindle cap and gasket on top and expansion plug and lock ring on the bottom depending on axle) of the spindle pin.

10. On all other front axles with air or hydraulic brakes, line up the notches in the spindle pin with the lockpin holes in the axle. Drive the spindle pin through the axle, with a drift until the notches and holes are in line. Install the lower (longer) lockpin first and seat by striking prior to installing nut. Install the upper (shorter) lockpin first and seat by striking prior to installing nut. Tighten nuts to 50–70 ft. lbs. Do not install both lockpins from same side of axle.

11. Install the brake backing plate on the spindle. On 6000 lb and larger axles, installing the spindle arms before installing the brake backing plate.

12. Lubricate the spindle bushings until lubricant is visible at the spindle to axle area.

13. Install the hub and drum assembly and the wheel. Adjust the wheel bearings.

14. Check and adjust the toe-in.

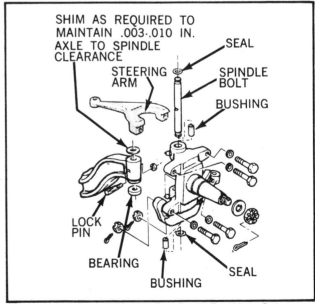

Spindle installation—5000–5500 lb. axles (© Ford Motor Co.)

2-325

SECTION 2

SUSPENSION
Light Trucks & Vans

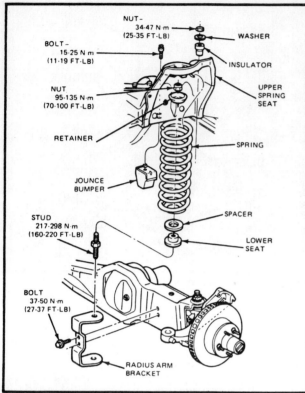

Coil spring and related components—Ranger and Bronco II
(© Ford Motor Co.)

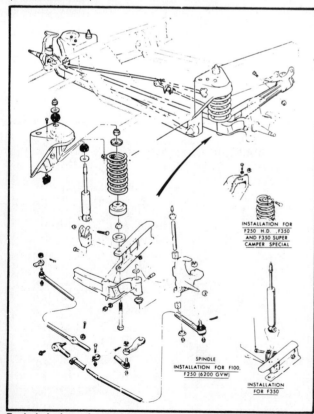

Exploded view of twin I-beam front axle—F-100–350 shown
(© Ford Motor Co.)

SPINDLE BUSHING

Replacement

1. Remove bronze bushings by driving them out with a drift slightly smaller than spindle bore. If a drift is not available, carefully drive a small center punch between the bushing and the spindle bore. Collapse the bushing, then remove.
2. Remove Delrin bushings, light duty vehicles, using a small center punch as described in Step 1.
3. Thoroughly clean spindle bores and make certain that lubrication holes are not obstructed in any way.
4. Place new bushing in spindle bore with lubricating holes properly aligned. Position open end of bushing oil groove toward axle.
5. Drive bronze bushing into spindle bore, using a drift as a pilot.
6. It is not necessary to drive Delrin bushings into spindle bores.
7. Install remaining bushing(s) in the same manner.
8. Ream bronze bushings 0.001 to 0.003 in. larger than spindle bolt diameter.
9. DO NOT ream Delrin bushings.
10. After reaming bronze bushings, clean out spindle bore to remove metal shavings.
11. Apply a light coat of oil to all bushings before spindle assembly and installation.

Twin I Beam

SPRING

Removal and Installation

2WD

1. Raise the front of the vehicle and place jackstands under the frame and a jack under the axle. Remove wheel and tire assemblies. Remove caliper and suspend with wire so that there is no tension on the brake hose.
2. Disconnect the shock absorber from the lower bracket.
3. Remove the rebound bracket. On Ranger and Bronco II, remove the lower spring retainer.
4. On all models except Ranger and Bronco II, remove the two spring upper retainer attaching bolts from the top of the spring upper seat and remove the retainer.
5. Remove the nut attaching the spring lower retainer to the lower seat and axle and remove the retainer.
6. Lower the axle and remove the spring. Some downward pressure using a prybar may be required.
7. Place the spring in position and raise the front axle.
8. Position the spring lower retainer over the stud and lower seat, and install the two attaching bolts.
9. Position the upper retainer over the spring coil and against the spring upper seat, and install the two attaching bolts.
10. Tighten the upper and lower retainer attaching nuts and bolts to 15–25 ft. lbs.
11. Connect the shock absorber to the lower bracket and install the rebound bracket.
12. Remove the jack and safety stands.

4WD BRONCO, F-150

1. Raise the vehicle on a hoist and support it safely. Remove the shock absorber lower attaching bolt and nut.
2. Remove the spring lower retainer nuts from inside of the spring coil.
3. Remove the upper spring retainer by removing the attaching screw.
4. Raise and support the vehicle safely. Position safety stands under the frame side rails and lower the axle enough to relieve tension from the spring.

SUSPENSION
Light Trucks & Vans

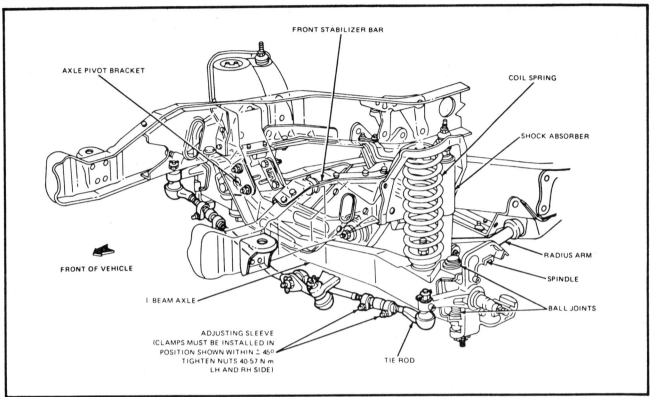

Front suspension—Ranger and Bronco II (2 × 4) (© Ford Motor Co.)

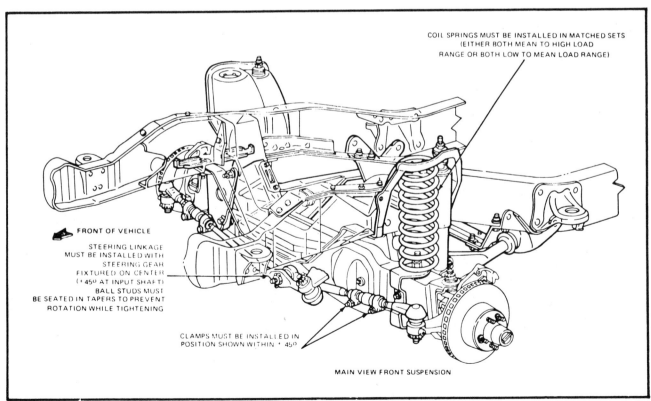

Front suspension—Ranger and Bronco II (4 × 4) (© Ford Motor Co.)

SUSPENSION
Light Trucks & Vans

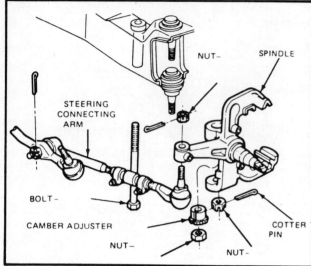

Spindle and related components—Ranger and Bronco II
(© Ford Motor Co.)

NOTE: The axle must be supported on the jack throughout spring removal, and must not be permitted to hang from the brake hose. If the length of the brake hose does not provide sufficient clearance it may be necessary to remove and support the brake caliper.

5. Remove the spring lower retainer and lower the spring from the vehicle.
6. To install place the spring in position and slowly raise the front axle. Make sure the springs are positioned correctly in the upper spring seats.
7. Install the lower spring retainer and torque the nut to 50 ft. lbs.
8. Position the upper retainer over the spring coil and install the attaching screws.
9. Position the shock absorber to the lower bracket and torque the attaching bolt and nut to 53 ft. lbs.
10. lower the vehicle.

4WD F-250–350

1. Raise the vehicle frame until the weight is off the front spring with the wheels still touching the floor. Support the axle to prevent rotation.
2. Disconnect the lower end of the shock absorber from the U-bolt spacer. Remove the U-bolts. U-bolt cap and spacer.
3. Remove the nut from the hanger bolt retaining the spring at the rear and drive out the hanger bolt.
4. Remove the nut connecting the front shackle and spring eye and drive out the shackle bolt and remove the spring.
5. To install position the spring on the spring seat. Install the shackle bolt through the shackle and spring. Torque the nuts to 135 ft. lbs.
6. Position the rear of the spring and install the hanger bolt. Torque the nut to 175 ft. lbs.
7. Position the U-bolt spacer and place the U-bolts to position through the holes in the spring seat cap. Install but do not tighten the U-bolt nuts.
8. Connect the lower end of the shock absorber to the U-bolt spacer.
9. Lower the vehicle and tighten the U-bolt nuts to 100 ft. lbs.

4WD RANGER AND BRONCO II

1. Raise and support the vehicle safely. Position a jack beneath the spring under the axle. Raise the jack and compress the spring.
2. Remove the bolt and nut retaining the shock absorber to the radius arm. Slide the shock out from the bracket.
3. Remove the nut that retains the spring to the axle and radius arm. Remove the retainer.
4. Slowly lower the axle until all spring tension is released and adequate clearance exists to remove the spring. Remove the spring by rotating upper coil out of tabs in upper spring seat. Remove the spacer and seat.

NOTE: The axle must be supported on the jack throughout spring removal and installation, and must not be permitted to hang by the brake hose. If the length of the brake hose is not sufficient to provide adequate clearance for removal and installation of the spring, the disc brake caliper must be removed from the spindle. After removal, the caliper must be placed on the frame or otherwise supported to prevent suspending the caliper from the caliper hose. These precautions are absolutely necessary to prevent serious damage to the tube portion of the caliper hose assembly.

5. If required, remove the stud from the axle assembly.
6. If removed, install the stud in the axle. Tighten to 160–220 ft. lbs.
7. Install the lower seat and spacer over the stud/bolt. Position upper end of spring so end of coil fits into spring stop in upper spring seat and top of coil fits over upper spring retainer.
8. Rotate spring into position. Slowly raise axle until lower end of spring is in position on the lower insulator.
9. Install the retainer and nut on the stud. Tighten nut to 70–100 ft. lbs.
10. Position the shock in the lower bracket. Install nut and bolt and tighten to 42–72 ft. lbs.
11. Remove the jack.

Front Wheel Spindle and King Pin

Removal and Installation

1. Raise and support the vehicle safely.
2. Remove the wheel and tire.
3. Remove the caliper key retaining screw. Drive out the caliper support key and spring with brass drift and hammer. Remove the caliper from the spindle by pushing the caliper downward against the spindle assembly and rotating the upper end of the caliper upward and out of the spindle assembly. It is not necessary to disconnect the brake fluid hose. Wire the caliper to a suspension part to remove the weight of the caliper from the hose. Disconnect the steering linkage from the spindle arm.
4. Disconnect the steering linkage from the integral spindle and spindle arm.
5. Remove the nut and lockwasher from the locking pin, and remove the locking pin.
6. Remove the upper and lower spindle bolt plugs, and drive the spindle bolt out from the top of the axle. Remove the spindle and bearing. Knock out the seal
7. Make sure that the spindle bolt hole in the axle is free of nicks, burrs and dirt. Install a new seal and coat the spindle bolt bushings and bolt hole with oil.
8. Place the spindle in position on the axle.
9. Pack the spindle thrust bearing with chassis lubricant and insert the bearing into the spindle with the open end of the bearing seal facing down into the spindle.
10. Install the spindle pin in the spindle with the locking pin notch in the spindle bolt aligned with the locking pin hole in the axle. Drive the spindle bolt through the axle from the top side until the spindle bolt locking pin notch is aligned with the locking pin hole.
11. Install a new locking pin. Install the locking pin lockwasher and nut. Tighten the nut to 40–55 ft. lbs. Install the spindle bolt plugs at the top and bottom of the spindle bolt.

SUSPENSION
Light Trucks & Vans

12. Position the caliper on the spindle assembly. Be careful to prevent tearing or cutting of the piston boot as the caliper is slipped over the inner brake pad. Use a suitable tool to hold the upper machined surface of the caliper against the surface of the spindle. Install the caliper support spring and key. Drive the key and spring into position with a soft hammer. Install the key retaining screw and tighten the nut to 50–70 ft. lbs. advancing the nut as necessary to install the cotter pin.
13. Install the wheel.
14. Grease the spindle assembly with a grease gun.
15. Check and adjust, if necessary, the toe-in adjustment.

Front Wheel Spindle with Stamped I Beam with Balljoints

Removal and Installation

1. Raise and support the vehicle safely. Remove the tire and wheel assembly.
2. Remove the caliper assembly from the rotor and hold it out of the way with wire.
3. Remove the dust cap, cotter pin, nut retainer, nut washer, and outer bearing, and remove the rotor from the spindle.
4. Remove inner bearing cone and seal. Discard the seal. Remove brake dust shield.
5. Disconnect the steering linkage from the integral spindle and spindle arm by removing the cotter pin and nut and then removing the tie rod end from the spindle arm.
6. Remove the cotter pins from the upper and lower ball joint studs. Remove the nuts from the upper and lower ball joint stud.
7. Strike the inside area of the spindle to pop the ball joints loose from the spindle.

NOTE: Do not use a pickle fork to separate the ball joint from the spindle as this will damage the seal and the ball joint socket.

8. Remove the spindle.
9. Prior to assembly, make sure the upper and lower ball joint seals are in place. Place the spindle over the ball joints.
10. Install the nut on the lower ball joint stud and partially tighten to 30 ft. lbs. Advance the castellated nut as required and install the cotter pin. If the lower ball stud turns while the nut is being tightened, push the spindle up against the ball stud. The lower must not be tightened first.
11. Install the camber adapter in the upper spindle over the upper ball joint stud. Be sure the adapter is aligned properly.
12. Install the nut on the upper ball joint stud. Hold the camber adaptor with a wrench to keep the ball stud from turning. If the ball stud turns, tap the adapter deeper into the spindle. Tighten the nut to 85–100 ft. lbs and contine tightening the castellated nut until it lines up with the hole in the stud. Install the cotter pin.
13. Retighten lower nut. Install the dust shield.
14. Pack the inner and outer bearing cone with C1AZ-19590-B (ESA-M1C75-B) Equivalent bearing grease. Use a bearing packer. If a bearing packer is unavailable, pack the bearing cone by hand working the grease through the cage behind the rollers.
15. Install the inner bearing cone and seal. Install the hub and rotor on the spindle.
16. Install the outer bearing cone, washer, and nut. Adjust bearing end play and install the nut retainer, cotter pin and dust cap.
17. Install the caliper.
18. Connect the steering linkage to the spindle. Tighten the nut to 52–73 ft. lbs and advance the nut as required for installation of the cotter pin.
19. Install the wheel and tire assembly. Lower the vehicle.
20. Check and, if necessary adjust the toe setting.

Camber Adjuster

Removal and Installation

1. Remove the cotter pin and the nut from the upper ball joint stud.
2. Strike the inside of the spindle to pop the upper ball joint from the spindle.
3. If the upper ball joint does not pop loose, remove the cotter pin and back the lower ball joint nut about half way down the lower ball joint stud, and strike the side of the lower spindle.
4. Remove the camber adapter (camber adjustment sleeve) using Ball Joint Removing Tool (D81T-3010-B) or equivalent.
5. Install the correct adaptor in the spindle. On the right spindle the adaptor slot must point forward to make a positive camber change or rearward for a negative camber change. On the left spindle, the adaptor slot must point rearward for a positive camber change and forward for a negative change.
6. If both nuts were loosened, completely remove the spindle, and reinstall. Be sure the lower ball joint stud nut is always tightened before the upper nut. Apply Locktite or equivalent to stud threads before installing nut.
7. If only the upper ball joint stud nut was removed, install the nut and tighten to 85–110 ft. lbs and continue tightening the castellated nut until it lines up with the hole in the upper stud. Install the cotter pin.

Ball Joints

Removal and Installation

1. Raise and support the vehicle safely. Remove the spindle.
2. Remove the snap ring from the ball joints. Assemble the C-frame puller and adapters on the upper ball joint. Turn the forcing screw clockwise until the ball joint is removed from the axle.
3. Assemble the C-frame assembly and receiver cup on the lower ball joint and turn the forcing screw clockwise until the ball joint is removed.
4. Always remove the upper ball joint first. Do not heat the ball joint or the axle to aid in the removal.
5. Installation is the reverse of removal. The lower ball joint must be installed first.

Front Wheel Spindle

Removal and Installtion

4WD

NOTE: This procedure also includes bearing and seal replacement and repacking

1. Raise and support the vehicle safely.
2. If equipped with manual locking hubs as follows:
   a. Remove the six socket head bolts from the cap assembly and separate the cap assembly and separate the cap assembly from the body.
   b. Remove the snap-ring from the end of the axle shaft.
   c. Remove the lock ring seated in the groove of the wheel hub and slide the body assembly out of the wheel hub.
3. If equipped with automatic locking hubs remove the hubs as follows:
   a. Remove the bolts and remove the hub cap assembly from the spindle.
   b. Remove the bolt from the end of the shaft.
   c. Remove the lock ring seated in the groove of the wheel hub.
   d. Remove the body assembly from the spindle. Use a puller if necessary.
   e. Unscrew all three set screws in the spindle locknut until the heads are flush with the edge of the locknut.

SUSPENSION
Light Trucks & Vans

4. Remove the outer spindle locknut with tool T80T-4000V, automatic hub locknut wrench. Remove the front hub grease cap and driving hub snap-ring.
5. Remove the splined driving hub and the pressure spring. Slightly pry off if necessary.
6. Remove the wheel bearing locknut, lock ring and adjusting nut using special tool T59T-1197-B or equivalent.
7. Remove the hub and disc assembly. The outer wheel bearing and spring retainer will slide out as the hub is removed.
8. Remove the spindle nuts and remove the spindle, splash shield and axle shaft assembly.

NOTE: It may be necessary to break the spindle loose with a plastic hammer.

9. Clean all old grease from the needle bearings and wipe clean the spindle face that mates with the spindle bore seal.
10. Remove the spindle bore seal, V-seal, and thrust washer from the outer axle shaft. Clean and replace if necessary.
11. Using Multi-Purpose Lubricant Ford ESA-M1C75-B or equivalent, thoroughly lubricate the needle bearing and pack the spindle face that mates with the spindle bore seal.
12. Position the V-seal in the spindle bore next to the needle bearing. Assemble the spindle bore seal on the axle shaft.
13. Assemble the spindle with the axle shaft on the knuckle studs and tighten the retaining nuts to 75 ft. lbs.
14. Carefully drive the inner bearing cone and grease seal out of the hub using tool T77F-1102-A or equivalent.
15. Inspect the inner bearing cups and if necessary remove with a drift.

NOTE: If new cups are installed, install new bearings.

16. Lubricate the bearings with the lubricant specified earlier and clean all old grease from the hub. Pack the cones and rollers with lubricant. Try to pack as much as possible between the rollers and the cages.
17. Position the inner bearing cone and roller in the inner cup and install the grease retainer.
18. Install the hub and disc assembly on the spindle.
19. Install the outer bearing cone and roller, and the adjusting nut.
20. Using tool T59T-1197-B or equivalent and a torque wrench tighten the bearing adjusting nut to 50 ft. lbs. while rotating the wheel back and forth. Back off the adjusting nut no more than 90 degrees.
21. Assemble the lock ring by turning the nut to the nearest hole and inserting the dowel pin.

NOTE: The dowel pin must seat in the lock ring hole for proper bearing adjustment and wheel retention.

22. Install the outer lock nut and tighten to 65 ft. lbs. Final end play on the wheel and spindle should be 0.001–0.006 inch.
23. Adjust the brake if necessary and lower the vehicle.

Knuckle and Ball Joint

Replacement

4WD

NOTE: A combination ball joint puller/press and a special spanner wrench are needed for this job.

1. Follow the procedures under axle shaft removal.
2. Disconnect the connecting rod end from the knuckle.
3. Remove the cotter pin from the upper ball socket and loosen the upper and lower ball socket nuts. Discard the nut from the lower ball socket after the knuckle breaks loose from the yoke.
4. Remove the knuckle from the yoke. If the upper socket remains in the yoke, remove it by hitting the top of the stud with a soft-faced hammer. Discard the socket and adjusting sleeve.
5. Remove the bottom socket with a ball joint puller, after first removing the snap ring.
6. To install, place the knuckle in a vise and assemble the bottom socket. Place the new socket into the knuckle making sure it isn't cocked, place the driver over the socket, place the forcing screw into the socket and force the socket into the knuckle.
7. Make sure that the socket shoulder is seated against the knuckle. Use a 0.0015 in. feeler gauge between the socket seat and the knuckle.
8. The gauge should not enter the area of minimum contact. Install the snap ring.
9. Assemble the top socket into the knuckle. Assemble the holding plate onto the backing plate screw. Tighten the nuts snugly. Place a new socket into the knuckle. Be sure it is not cocked. Place a driver over the socket and force the socket assembly into the knuckle. Using a 0.0015 in. gauge, check the fit at the shoulder. The gauge should not enter the area of minimum wrench.
10. Install a new adjusting sleeve into the top of the yoke leaving about two threads exposed.
11. Assemble the knuckle and yoke. Install a new nut on the bottom socket and make it finger tight.
12. Place a wrench and step plate over the adjusting sleeve and install the puller so that it grasps the step plate. Tighten the forcing screw to pull the knuckle assembly into the yoke. With torque still applied, tighten the nut to 70–90 ft. lbs. If the bottom stud should turn with the nut, add more torque to the puller forcing screw. Remove the puller, step plate and holding plate.
13. Tighten the adjusting sleeve to 40 ft. lbs. and remove the wrench.
14. Install the top socket nut and torque it to 100 ft. lbs. Line up the cotter pin hole by tightening, not loosening, the nut. Install the cotter pin and test the steering effort with a spring scale attached to the knuckle. Pull should not exceed 26 ft. lbs. If it does, the ball joints will have to be replaced.
15. Connect the steering linkage to the knuckle. Torque it to 40 ft. lbs.
16. Install the axle shaft.

Radius Arm

Removal and Installation

ALL, EXCEPT BRONCO, F-150 WITH 4WD, RANGER AND BRONCO II

1. Raise and support the vehicle safely. Place safety stands under the frame and a jack under the wheel or axle.
2. Disconnect the shock absorber from the radius arm bracket.
3. Remove the two spring upper retainer attaching bolts from the top of the spring upper seat and remove the retainer.
4. Remove the nut which attaches the spring lower retainer to the lower seat and axle and remove the retainer.
5. Lower the axle and remove the spring.
6. Disconnect the steering rod from the spindle arm.
7. Remove the spring lower seat and shim from the radius arm. Then, remove the bolt and nut which attach the radius arm to the axle.
8. Remove the cotter pin, nut and washer from the radius arm rear attachment.
9. Remove the bushing from the radius arm and remove the radius arm from the vehicle.
10. Remove the inner bushing from the radius arm.
11. Position the radius arm to the axle and install the bolt and nut finger-tight.
12. Install the inner bushing on the radius arm and position the arm to the cotter pin.

13. Install the bushing, washer, and attaching nut. Tighten the nut and install the cotter pin.
14. Connect the steering rod to the spindle arm and install the attaching nut. Tighten the radius arm to axle attaching bolt and nut.

F-150 4WD

1. Raise the vehicle and position safety stands under the frame side rails.
2. Remove the shock absorber lower attaching bolt and nut and pull the shock absorber free of the radius arm.
3. Remove the lower spring retaining bolt from the inside of the radius arm.
4. Remove the nut attaching the radius arm to the frame bracket and remove the radius arm rear insulator. Lower the axle and allow the axle to move forward.

NOTE: The axle must be supported on the jack throughout spring removal, and must not be permitted to hang from the brake hose. If the length of the brake hose does not provide sufficient clearance it may be necessary to remove and support the brake caliper.

5. Remove the bolt and stud attaching the radius arm to the axle.
6. Move the axle forward and remove the radius arm from the axle. Then, pull the radius arm from the frame bracket.
7. Installation is the reverse of the removal. Install new bolts and the stud type bolt which attach the radius arm to the axle and tighten to 210 ft. lbs. Tighten the radius arm rear attaching nut to 100 ft. lbs.

BRONCO

1. Raise the vehicle and position safety stands under the frame side rails.
2. Remove the shock absorber to lower bracket attaching bolt and nut and pull the shock absorber free of the radius arm.
3. Remove spring lower retainer attaching bolt from inside of the spring coil.
4. Remove the nut attaching the radius arm to the frame bracket and remove the radius arm rear insulator. Lower the axle and allow axle to move forward

NOTE: The axle must be supported on the jack throughout spring removal and installation, and must not be permitted to hang by the brake hose. If the length of the brake hose is not sufficient to provide adequate clearance for the removal and installation of the spring, the disc brake caliper must be removed from the spindle. After removal, the caliper must be placed on the frame or otherwise supported to prevent suspending the caliper from the caliper hose. Thses precautions are absolutely necessary to prevent serious damage to the tube portion of the caliper hose assembly.

5. Remove the bolt and stud attaching radius arm to axle.
6. Move the axle forward and remove the radius arm from the axle. Then, pull the radius arm from the frame bracket.
7. Position the washer and insulator on the rear of the radius arm and insert the radius arm into the frame bracket.
8. Position the rear insulator and washer on the radius arm and loosely install the attaching nut.
9. Position the radius arm to the axle.
10. Install new bolts and study type bolt attaching radius arm to axle. Tighten to 180–240 ft. lbs.
11. Position the spring lower seat, spring insulator and retainer to the spring and axle. Install the two attaching bolts. Tighten the nuts to 30–70 ft. lbs.
12. Tighten the radius rod rear attaching nut to 80–120 ft. lbs.
13. Position the shock absorber to the lower bracket and install the attaching bolt and nut. Tighten the nut to 40–60 ft. lbs. Remove safety stands and lower the vehicle.

RANGER AND BRONCO II

1. Raise the front of the vehicle and place safety stands under the frame. Place a jack under the axle.

NOTE: The axle must be supported on the jack throughout spring removal and installation, and must not be permitted to hang by the brake hose. If the length of the brake hose is not sufficient to provide adequate clearance for removal and installation of the spring, the disc brake caliper must be removed from the spindle. After removal, the caliper must be placed on the frame or otherwise supported to prevent suspending the caliper from the caliper hose. These precautions are absolutely necessary to prevent serious damage to the tube portion of the caliper hose assembly.

2. Disconnect the lower end of the shock absorber from the shock lower bracket (bolt and nut).
3. Remove the front spring. Loosen the axle pivot bolt.
4. Remove the spring lower seat and stud from the radius arm, and then remove the bolts that attach the radius arm to the axle and front bracket.
5. Remove the nut, rear washer and insulator from the rear side of the radius arm rear bracket.
6. Remove the radius arm from the vehicle, and remove the inner insulator and retainer from the radius arm stud.
7. Installation is the reverse of removal.

Stabilizer Bar

Removal and Installation

BRONCO AND 4WD PICKUPS

1. Remove nuts, bolts and washers connecting the stabilizer bar to connecting links. Remove nuts and bolts of the stabilizer bar retainer.
2. Remove stabilizer bar insulator assembly.
3. To remove the stabilizer bar mounting bracket, the coil spring must be removed as described above under spring removal. Remove the lower spring seat. The bracket attaching stud and bracket can now be removed.
4. To install the stabilizer bar mounting brackets, locate the brackets so that the locating tang is positioned in the radius arm notch (or quad shock bracket notch if vehicle has quad shocks). Install a new stud. Torque to 180–220 ft. lbs. A new stud is required because of the adhesive on the threads. Reposition the spring lower seat and reinstall the spring and retainers.
5. To reinstall the stabilizer bar insulator assembly, assemble all nuts, bolts and washers to the bar, brackets, retainers and links loosely. With the bar positioned correctly, torque retainer nuts to 32–35 ft. lbs. with retainer around the insulator. Then torque all remaining nuts at the link assemblies to 41–50 ft. lbs.

RANGER AND BRONCO II

1. Remove the nuts and U-bolts retaining the lower shock bracket/stabilizer bar bushing to radius arm.
2. Remove retainers and remove the stabilizer bar and bushing.
3. Place stabilizer bar in position on the radius arm and bracket.
4. Install retainers and U-bolts. Tighten retainer bolts to 35–50 ft. lbs. Tighten U-bolts to 48–64 ft. lbs.

FRONT DRIVE AXLE

Axle Shaft

Removal and Installation

F SERIES

1. Remove the axle shaft assembly.
2. Remove the stub assembly by removing the three bolts attaching the retainer plate to the carrier housing.
3. Place the axle shaft in a vise and drill a 1/2 in. hole in the outside of the bearing retaining ring to a depth of 3/4 of the thickness of the ring.
4. Place a chisel across the hole and strike sharply with a hammer to remove the retaining ring. Replace the bearing retaining ring upon assembly.
5. Press the bearing from the axle shaft using an axle bearing remover tool.
6. Remove the steal and retainer plate from the stub shaft.
7. To install place the new seal and retainer plate on the shaft.
8. Place the bearing on the shaft with the large radius on the inner race facing the yoke end of the shaft.
9. Press the bearing onto the shaft using an axle bearing replacer and a pinion bearing cone remover. A 0.0015 in. feeler gauge should not fit between the bearing seat and the bearing.
10. Using the same special tools in Step 9 press the bearing retainer ring onto the stub shaft. A 0.0015 in. feeler gauge should not fit between the ring and the bearing. There must be one point between the bearing and the ring where the feeler

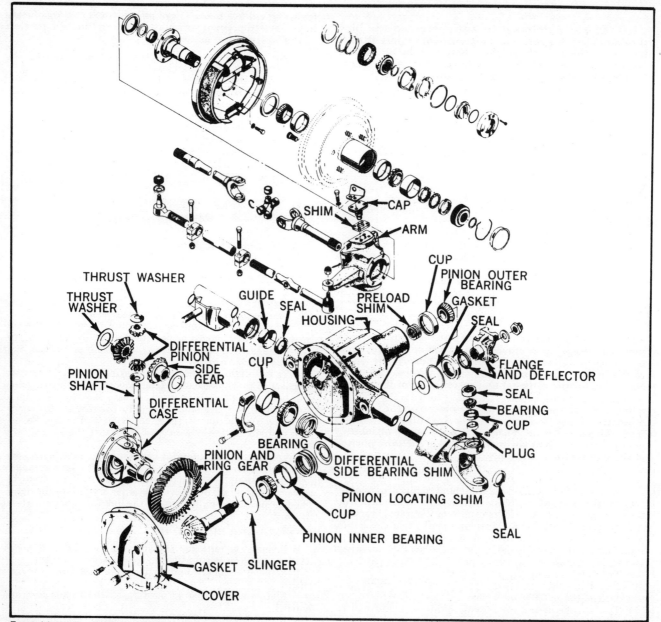

Front drive axle—F-100 and F-150 (© Ford Motor Co.)

SUSPENSION
Light Trucks & Vans

gauge cannot enter. If the feeler gauge enters completely around the circumference press the retainer further onto the shaft.

11. Push the seal and retainer plate away from the bearing to form a space between the seal and the bearing. Fill the space with the proper bearing grease and wrap tape around the space.

12. Pull the seal towards the bearing until it contacts the inner race and forces the grease between the rollers and cup. Remove the tape.

13. Install the stub shaft in the carrier and install the three retainer bolts. Tighten to 35 ft. lbs.

14. Install the right hand axle shaft assembly into the slip yoke.

15. Install the spindle.

BRONCO

1. Remove spindle nuts and remove spindle. It may be necessary to tap the spindle with a rawhide or plastic hammer to break the spindle loose. Remove spindle, splash shield and axle shaft assembly.

2. Place the spindle in a vise with a shop towel around the spindle to protect the spindle from damage. Using a slide hammer and a seal remover tool remove the axle shaft seal and then the needle bearing from the spindle bar.

3. If the tie rod has not been removed, then remove cotter key from the tie rod nut and then remove nut. Tap on the tie rod stud to free it from the steering arm.

4. Remove the cotter pin from the top ball joint stud. Loosen the nut on the top stud and the bottom nut inside the knuckle. Remove the top nut.

5. Sharply hit the top stud with a plastic or rawhide hammer to free the knuckle from the tube yoke. Remove and discard bottom nut. Use new nut upon assembly.

6. Remove camber adjuster with a pitman arm puller.

7. Place knuckle in vise and remove snap ring from bottom ball joint socket if so equipped.

8. Press the bottom ball joint socket from the knuckle with the special removal tools. Remove the top ball joint in the same manner.

NOTE: Always remove bottom ball joint first.

9. Pull out the seal with the appropriate puller tool. Remove and discard seal.

10. Install a new seal on the differential, with the differential seal replacer tool.

11. Slide the seal and tool into the carrier housing bore. Seat the seal with a plastic or rawhide hammer.

12. Place lower ball joint (stud does not have a cotter key hole in stud) in knuckle and press into position using ball joint installation.

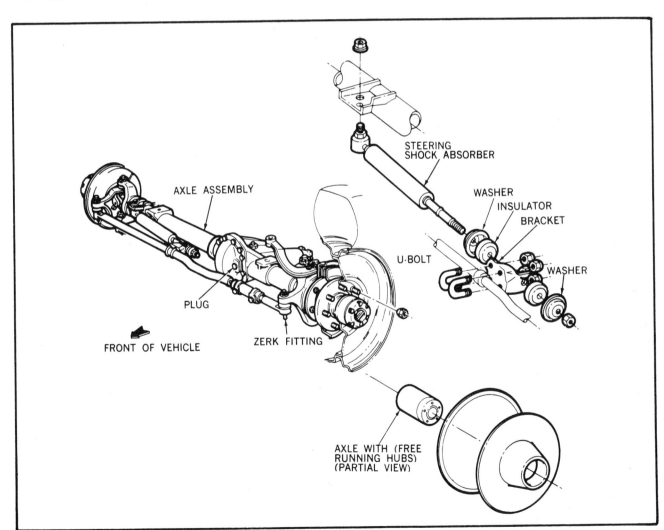

Front axle assembly—F-250 (4 × 4)—typical (© Ford Motor Co.)

SECTION 2

SUSPENSION
Light Trucks & Vans

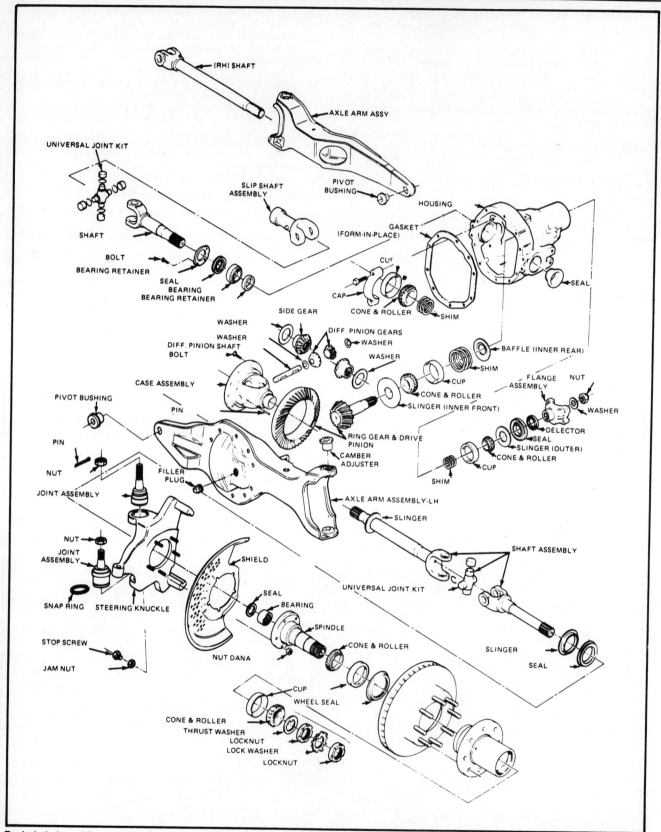

Exploded view of Dana 50 IFS front driving axle—typical of 44 IFS and H/D models (© Ford Motor Co.)

SUSPENSION
Light Trucks & Vans

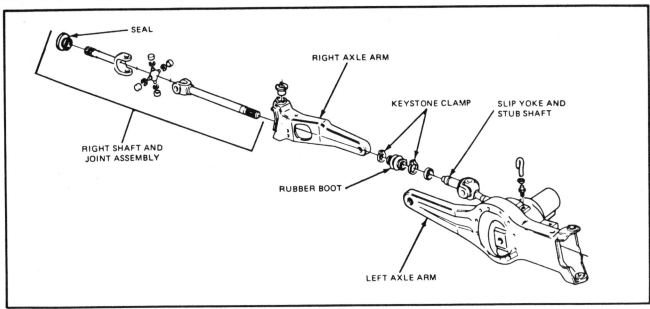

Right axle and joint assembly—Ranger and Bronco II (4 × 4) (© Ford Motor Co.)

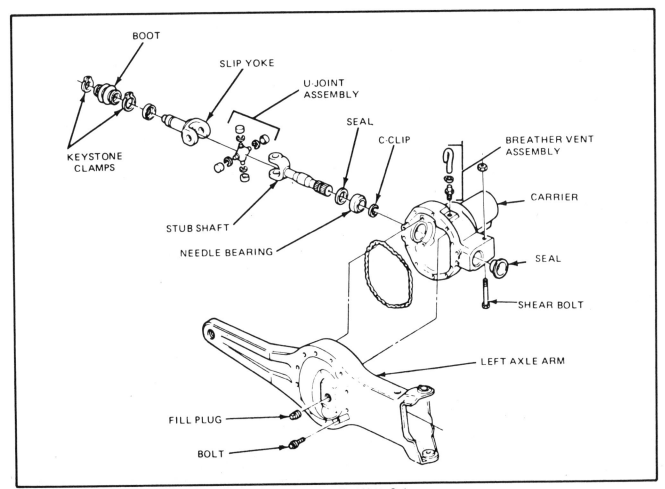

Carrier assembly and related components—Ranger and Bronco II (© Ford Motor Co.)

SUSTENSION
Light Trucks & Vans

13. Install upper ball joint (stud has cotter key hole) in knuckle.

14. Assemble knuckle to tube and yoke assembly. Install camber adjuster on top ball joint stud with the arrow pointing outboard for "positive" camber, pointed inboard for "negative" camber.

15. Install new nut on bottom socket finger tight. Install and tighten nut on top socket finger tight. Tighten bottom nut to 90–110 ft. lbs.

16. Tighten top nut to 100 ft. lbs., then advance nut until castellation aligns with cotter pin hole. Install cotter pin.

NOTE: Do not loosen top nut to install cotter pin.

17. Remove and install a new needle to bearing in the spindle bar using the proper tool.

18. Install the axle shaft assembly into the housing. Install the splash shield and spindle. Install and tighten the spindle attaching nuts.

Axle Shaft and Bearings

Removal and Installation

RANGER AND BRONCO II WITH DANA 28

1. Raise and support the vehicle safely. Remove the wheel and tire assembly. Remove the caliper.

2. Remove hub locks, wheel bearings, and lock nuts.

3. Remove the hub and rotor. Remove the outer wheel bearing cone. Remove the grease seal from the rotor. Remove the inner wheel bearing.

4. Remove the inner and outer bearing cups from the rotor. Remove the nuts retaining the spindle to the steering knuckle. Tap the spindle with a plastic hammer to jar the spindle from the knuckle. Remove the splash shield.

5. Remove the shaft and joint assembly by pulling the assembly out of the carrier.

6. On the right side of the carrier, remove and discard the keystone clamp from the shaft and joint assembly and the stub shaft. Slide the rubber boot onto the stub shaft and pull the shaft and joint assembly from the splines of the stub shaft.

7. Place the spindle in a vise on the second step of the spindle. Wrap a shop towel around the spindle or use a brass jawed vise to protect the spindle. Remove the oil seal and needle bearing from the spindle with slide hammer and seal remover. If required, remove the seal from the shaft, by driving off with a hammer.

8. Clean all dirt and grease from the spindle bearing bore. Bearing bores must be free from nicks and burrs.

9. Place the bearing in the bore with the manufacturer's identification facing outward. Drive the bearing into the bore using spindle bearing replacer, and driver handle tool.

10. Install the grease seal in the bearing bore with the lip side of the seal facing towards the tool. Drive the seal in the bore with spindle bearing replacer, and driver handle. Coat the bearing seal lip with multi-purpose long life lubricant, or equivalent.

11. If removed, install a new shaft seal. Place the shaft in a press, and install the seal with spindle/axle seal installer. On the right side of the carrier, install the rubber boot and new keystone clamps on the stub shaft slip yoke. Since the splines on the shaft are phased, there is only one way to assemble the right shaft and joint assembly into the slip yoke. Align the missing spline in the slip yoke barrel with the gapless male spline on the shaft and joint assembly. Slide the right shaft and joint assembly into the slip yoke making sure the splines are fully engaged. Slide the boot over the assembly and crimp the keystone clamp using clamp pliers.

12. On the left side of the carrier slide the shaft and joint assembly through the knuckle and engage the splines on the shaft in the carrier.

13. Install the splash shield and spindle onto the steering knuckle. Install and tighten the spindle nuts to 35–45 ft. lbs.

14. Drive the bearing cups into the rotor using bearing cup replacer and driver handle.

15. Pack the inner and outer wheel bearings and the lip of the oil seal with Multi-Purpose Long-Life Lubricant, or equivalent.

16. Place the inner wheel bearing in the inner cup. Drive the grease seal into the bore with hub seal replacer, and driver handle. Coat the bearing seal lip with multipurpose long life lubricant, or equivalent.

17. Install the rotor on the spindle. Install the outer wheel bearing into cup.

NOTE: Verify that the grease seal lip totally encircles the spindle.

18. Install the wheel bearing, locknut, thrust bearing, snap ring, and locking hubs.

Right Stub Axle and Carrier

Removal and Installation

RANGER AND BRONCO II WITH DANA 28

1. Remove the nuts and U-bolts connecting the driveshaft to the yoke. Disconnect the driveshaft from the yoke. Wire the driveshaft out of the way, so it will not interfere in the carrier removal process.

2. Remove both spindles and the left and right shaft and U-joint assemblies.

3. Support the carrier with a suitable jack and remove the bolts retaining the carrier to the support arm. Separate the carrier from the support arm and drain the lubricant from the carrier. Remove the carrier from the vehicle.

4. Place the carrier in a holding fixture.

5. Rotate the slip yoke and shaft assembly so the open side of the snap ring is exposed.

6. Remove the snap ring from the shaft. Remove the slip yoke and shaft assembly from the carrier.

7. Remove the oil seal and caged needle bearings at the same time, using a slide hammer, and collet. Discard the seal and needle bearing.

8. Make sure the bearing bore is free from nicks and burrs. Install a new caged needle bearing on the needle bearing replacer, with the manufacturer name and part number facing outward towards the tool. Drive the needle bearing until it is seated in the bore.

9. Coat the seal with Long-Life Multi-Purpose Lubricant, or equivalent. Drive the seal into the carrier using the needle bearing replacer.

10. Install the slip yoke and shaft assembly into the carrier so the groove in the shaft is visible in the differential case.

11. Install the snap ring in the groove in the shaft. Force the snap ring into position with a suitable tool. Remove the carrier from the holding fixture.

NOTE: Do not tap on the center of the snap ring. This may damage the snap ring.

12. Clean all traces of gasket RTV sealant from the surfaces of the carrier and support arm and make sure the surfaces are free from dirt and oil. Apply RTV sealant, in a bead between $1/4$ and $3/8$ in. wide. The bead should be continuous and should not pass through or outside the holes.

NOTE: The carrier must be installed on the support arm within five minutes after applying the RTV sealant.

13. Position the carrier on a suitable jack and install it in position on the support arm using guide pins to align. Install the attaching bolts and hand tighten. Tighten the bolts in a clockwise or counter-clockwise pattern to 40–50 ft. lbs.

14. Install the shear bolt retaining the carrier to the axle arm and tighten to 75–95 ft. lbs.

15. Install both spindles and the left and right shaft and joint

assemblies as described in the removal and installation portion of this section. Connect the driveshaft to the yoke. Install the nuts and U-bolts and tighten to 8–15 ft. lbs.

Axle Shaft Bearing

Removal and Installation
1980–81 MODELS WITH DANA 44IFS, 44IFS-HD or 50IFS AXLES

1. Remove the axle shaft assembly.
2. Remove the stub assembly by removing 3 bolts attaching retainer plate to carrier housing.
3. Place the axle shaft in a vise and drill a 1/4 in. hole in the outside of the bearing retaining ring to a depth 3/4 in. the thickness of the ring.

NOTE: Do not drill through the ring because this will damage the axle shaft.

4. With a chisel placed across the hole, strike sharply with a hammer to remove the retaining ring. Replace bearing retaining ring upon assembly.
5. Press the bearing from the axle shaft with the axle bearing remover and sleeve.

NOTE: Do not use a torch to aid in bearing removal or the stub shaft will be damaged.

6. Remove the seal and retainer plate from the stub shaft. Discard seal and replace with new seal upon assembly.
7. Inspect the retainer plate and stub shaft for distortion, nicks or burns. Replace if necessary.
8. Install retainer plate and new seal on shaft. Coat oil seal with ESA-M175B or equivalent.
9. Place the bearing on the shaft. The large radius on the inner race must face the yoke end of the shaft.
10. Use axle bearing replacer and pinion bearing cone remover to press the bearing onto the shaft until completely seated. A 0.0015 in. feeler gauge should not fit between the bearing seat and bearing.
11. Use the axle bearing replacer and pinion bearing cone remover to press the bearing retainer ring onto the stub shaft. Press the bearing retainer ring until completely seated. A 0.015 in. feeler gauge should not fit between the ring and bearing. There must be one point between the bearing and ring where the feeler gauge cannot enter. If feeler gauge enters completely around the circumference press the retainer further onto the shaft.
12. Push the seal and retainer plate away from the bearing to form a space between the seal and bearing. Fill the space with wheel bearing grease.
13. With the space filled with grease, wrap tape around the space.
14. Pull the seal towards the bearing until it contacts the inner race. This will force grease between the rollers and cup. Remove tape.

NOTE: If grease is not visible on the small end of the rollers, repeat Steps 6 through 8 until grease is visible. Install the slip yoke and U-joint to stub shaft.

15. Install the stub shaft in the carrier and install 3 retainer bolts. Torque to 30–40 ft. lbs. Install right hand axle shaft assembly into slip yoke.
16. Install splash shield and spindle.

Pinion Seal

Removal and Installation
LIGHT DUTY VEHICLES

NOTE: A torque wrench capable of at least 225 ft. lbs. is required for pinion seal installation.

---- CAUTION ----
Some vehicles use a collapsible spacer to set pinion depth and preload. When replacing the pinion seal always install a new spacer. Never tighten the pinion nut more than 225 ft. lbs. or the spacer will be compressed too far.

1. Raise and support the vehicle safely. Position jackstands under the frame rails. Allow the axle to drop to rebound position for working clearance.
2. Mark the companion flanges and U-joints for correct reinstallation position.
3. Remove the drive shaft. Use a suitable tool to hold the companion flange. Remove the pinion nut and companion flange.
4. Use a slide hammer and hook or sheet metal screw to remove the oil seal.
5. If the vehicle uses a collapsible spacer, install new spacer. Install a new pinion seal after lubricating the sealing surfaces. Use a suitable seal driver. Install the companion flange and pinion nut. On models using a spacer, tighten the nut to 225 ft. lbs. On other models, pinion nut torque is 200–220 ft. lbs.

Front Drive Axle Housing

SOLID AXLE HOUSING WITH COIL SPRINGS

Removal and Installation

1. Raise the vehicle and support safely. Remove the wheels from the axle hubs.
2. The hubs, drums or rotors, axle shafts and other components can be removed in the conventional manner.
3. Remove the hydraulic brake line brackets from each end of the axle without breaking the hydraulic connection. Disengage the hydraulic lines from the axle clips. Tie the lines to the frame.
4. Disconnect the steering tie rod at the spindle connecting rod ends. Disconnect the axle stabilizer bar.
5. Disconnect the front drive shaft at the pinion companion flange and universal joint. Secure the drive shaft out of the working area.
6. Lower the vehicle onto the safety stands and place a jack under the axle to support it while disconnecting it from the radius arms.
7. Each radius arm and cap is marked, since they are manufactured as matched pairs. Remove the bolts attaching the radius to the radius arm caps. Remove the rubber insulators and roll the axle from under the vehicle.
8. To install, position the front drive axle under the vehicle, using a floor jack, and install the radius arms, insulators and caps to the axle. Numbers on radius arm and cap should be matched. Torque the attaching bolts to specifications, tightening them diagonally in pairs.
9. Raise the vehicle to working height and install the drive shaft to the pinion companion flange at the universal joint. Torque the universal joint U-bolt nuts to specifications.
10. Connect the axle stabilizer bar. Connect the steering tie rod to the spindle arms by means of the steering connecting rod ends. Torque the attaching nuts to specifications, then install the cotter pins.
11. Install the axle shafts, spindles and brake backing plates.
12. Position the hydraulic brake lines and brackets, then install the retaining clips.
13. Install the hubs, drums or rotors, axle shafts and other components that may have been removed during the removal of the unit from the vehicle.
14. Fill the unit to its proper level with lubricant and lower the vehicle.

SUSPENSION
Light Trucks & Vans

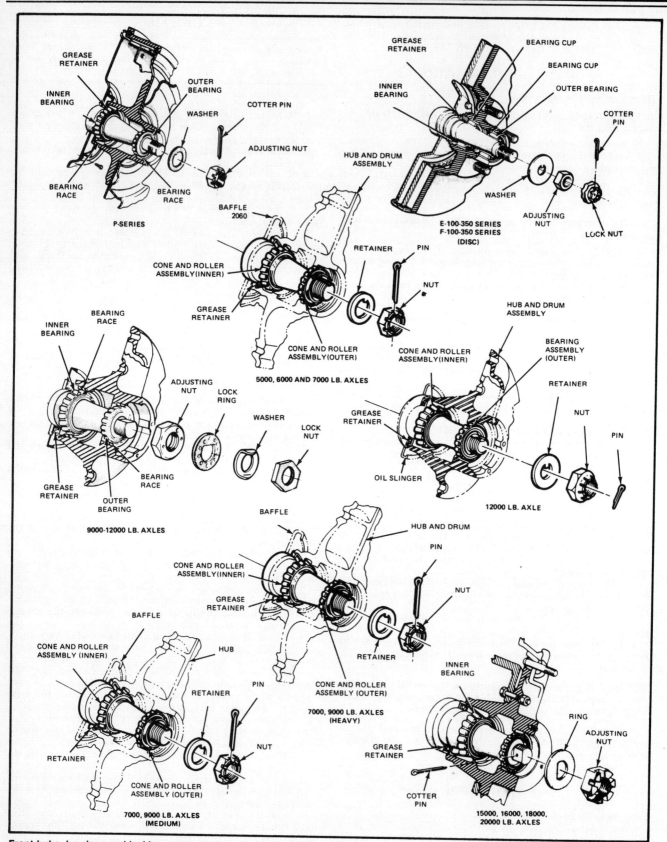

Front hubs, bearings and locking components—varied axles (© Ford Motor Co.)

SUSPENSION
Light Trucks & Vans

SOLID AXLE HOUSING WITH LEAF SPRINGS

Removal and Installation

1. Raise the vehicle and support safely. Remove the wheel assemblies from the hubs.
2. The hubs, rotors, axle shafts and other components can be removed in the conventional manner.
3. Disconnect both front axle shock absorbers at their lower ends.
4. Disconnect the front axle drive shaft at the pinion flange.
5. Support the front axle on a transmission jack, then remove the spring clip (U-bolt) nuts and the spring seats.
6. Lower the axle assembly and roll it from under the vehicle.
7. Installation is the reverse of removal.

Front Wheel Bearings

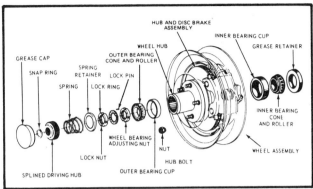

Front hub and wheel bearing components, front wheel drive—typical (© Ford Motor Co.)

Removal and Installation
LIGHT DUTY VEHICLES WITH 2WD

1. Raise and support the vehicle safely.
2. Remove the lug nuts and remove wheel/tire assembly from the hub. Remove the caliper assembly from the rotor and caliper support. Hang the caliper with a length of heavy wire above the hub.
3. Remove the grease cap with a suitable tool.
4. Remove the cotter pin and discard it.
5. Remove the nut lock, adjusting nut, and washer from the spindle.
6. Wiggle the hub so that the outer wheel bearing comes loose and can be removed. Remove the outer bearing.
7. Remove the hub from the spindle.
8. Place a block of wood or drift pin through the spindle hole and tap out the inner grease seal and bearing.
9. Place all of the bearings, nuts, nut locks, washers and grease caps in a container of solvent. Use a light soft brush to throughly clean each part.
10. Clean the inside of the hub, including the bearing races, and the spindle. Remove all traces of old lubricant from these components.
11. Inspect the bearings for pitting, flat spots, rust, and rough areas. Check the races in the hub and the spindle for the same defects and rub them clean with a cloth that has been soaked in solvent. If the races show hair line cracks or worn shiny areas, they must be replaced.
12. Pack the bearings with the proper grade and type wheel bearing grease.

NOTE: Sodium based grease is not compatible with lithium based grease. Be careful not to mix the two types. The best way to prevent this is to completely clean all of the old grease from the hub and spindle before installing any new grease.

13. Turn the hub assembly over so that the inner side faces up, making sure that the race and inner are are clean, and drop the inner wheel bearing into place. Using a hammer and a block of wood, tap the new grease seal in place.
14. Slide the hub assembly onto the spindle. Keep the hub centered on the spindle to prevent damage to the grease seal and the spindle threads.
15. Place the outer wheel bearing in place over the spindle. Place the washer on the spindle after the bearing. Screw on the spindle nut and turn it down until a slight binding is felt.
16. With a torque wrench, tighten the bearings. Install the nut lock over the nut so that the cotter pin hole in the spindle is aligned with a slot in the nut lock. Back off the adjusting nut and the nut lock two slots of the nut lock and install the cotter pin.
17. Bend the longer of the two ends opposite the looped end out and over the end of the spindle. Trim both ends of the cotter pin just enough so that the grease cap will fit, leaving the bent end shaped over the end of the spindle.
18. Install the grease cap, brake caliper if so equipped, and the wheel/tire assembly. The wheel should rotate freely with no noise or noticeable end-play.

4WD FRONT WITHOUT FREE-RUNNING HUBS EXCEPT RANGER AND BRONCO II

NOTE: Sodium based grease is not compatible with lithium based grease. Be careful not to mix the two types. The best way to prevent this is to completely clean all of the old grease from the hub assembly before installing any new grease.

1. Raise and support the vehicle safely.
2. Remove the front hub grease cap and driving hub snap-ring.
3. Remove the splined driving hub and the pressure spring. This may require slight prying with a suitable tool.
4. Remove the wheel bearing locknut, lockring, and adjusting nut.
5. Remove the caliper assembly.
6. Carefully drive out the inner bearing cone and grease seal from the hub.
7. Inspect the bearing cups (races) for cracks and pits. If the cups are excessively worn or there are pits or cracks visible, replace them along with the cones. The cups are removed from the hub by driving them out with a drift pin. They are installed in the same manner.
8. If it is determined that the cups are in satisfactory condition and are to remain in the hub, clean and inspect the cones (bearings). Refer to the bearing diagnosis chart. Replace the bearings if necessary. If it is necessary to replace either the cone or the cup, both parts should be replaced as a unit.
9. Thoroughly clean all components in a suitable solvent and blow them dry with compressed air or allow them to dry while resting on clean paper.

NOTE: Do not spin the bearings with compressed air while drying them.

10. Cover the spindle with a cloth and brush all loose dust and dirt from the brake assembly. Remove the cloth and thoroughly clean the inside of the hub and the spindle.
11. Pack the inside of the hub with wheel bearing grease. Add grease to the hub until the grease is flush with the inside diameter of the bearing cup.
12. Pack the bearing cone and roller assemblies with wheel bearing grease.
13. Position the inner bearing into the inner bearing cup and install the new grease seal.

SUSPENSION
Light Trucks & Vans

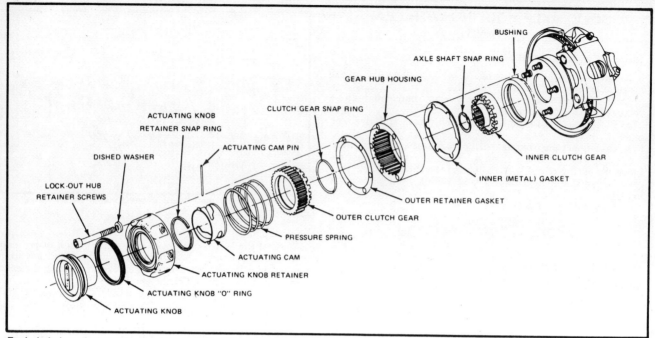

Exploded view of external lock-out hub (© Ford Motor Co.)

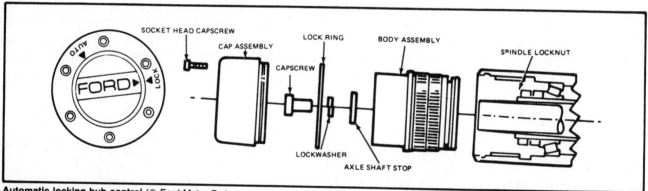

Automatic locking hub control (© Ford Motor Co.)

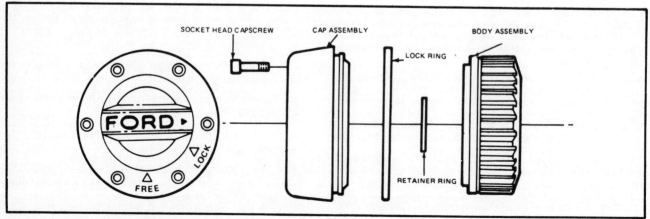

Manual locking hub control (© Ford Motor Co.)

SUSPENSION
Light Trucks & Vans

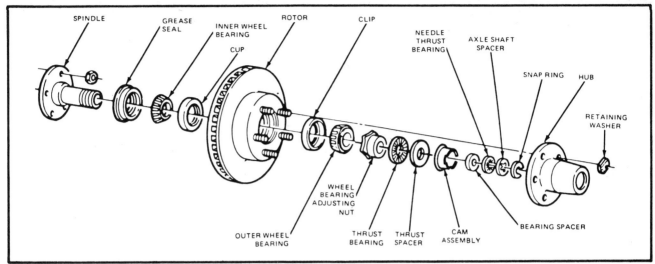

Automatic locking hubs—Ranger and Bronco II (© Ford Motor Co.)

14. Carefully position the hub assembly onto the spindle. Be careful not to damage the new seal. Install the caliper.
15. Place the outer bearing into position on the spindle and into the bearing cup.
16. Install the bearing adjusting nut and tighten it to 50 ft. lbs. while rotating the hub back and forth to seat the bearings.
17. Back off the adjusting nut about 90°.
18. Assemble the lockring by turning the nut to the nearest notch where the dowel pin will enter.
19. Install the outer locknut and torque to 80–100 ft. lbs. The final end-play of the wheel on the spindle should be 0.001–0.010 in.
20. Install the grease cap and adjust the brakes, if they were backed off to remove the hub assembly. Lower the vehicle.

4WD FRONT WITH FREE-RUNNING HUB EXCEPT RANGER AND BRONCO II

1. Raise the vehicle and support it safely. Remove the tire and wheel assembly.
2. To remove hub, first separate cap assembly from body assembly by removing the six (6) socket head capscrews from the cap assembly and slip apart.
3. Remove snap-ring (retainer ring) from the end of the axle shaft.
4. Remove the lock ring seated in the groove of the wheel hub. The body assembly will now slide out of the wheel hub. If necessary, use an appropriate puller to remove the body assembly.
5. Install hub in reverse order of removal. Torque socket head capscrews to 30–35 inch lbs

AUTOMATIC LOCKING HUBS, EXCEPT RANGER AND BRONCO II

1. Raise the vehicle and support it safely. Remove the wheel and tire.
2. Remove capscrews and remove hub cap assembly from spindle. Remove capscrew from end of axle shaft.
3. Remove lock ring seated in the groove of the wheel hub with a knife blade or with a small sharp awl with the tip bent in a hook.
4. Remove body assembly from spindle. If body assembly does not slide out easily, use an appropriate puller.
5. Unscrew all three sets in the spindle locknut until the heads are flush with the edge of the locknut. Remove outer spindle locknut with automatic hub lock nut wrench.
6. Reinstall in reverse order of removal. Tighten the outer spindle locknut to 15–20 ft. lbs. with automatic hub lock nut wrench. Tighten down all three set screws. Firmly push in body assembly until the friction shoes are on top of the spindle outer locknut.
7. Install capscrew into the axle shaft and tighten to 35–50 ft. lbs.
8. Place cap on spindle and install capscrews. Tighten to 35–50 inch lbs Turn dial firmly from stop to stop, causing the dialing mechanism to engage the body spline.

NOTE: Be sure both hub dials are in the same position; "AUTO" or "LOCK."

4WD FRONT HUBS, RANGER AND BRONCO II, EXCEPT WITH AUTOMATIC LOCKING HUBS

1. Raise and support the vehicle safely.
2. Remove the wheel lug nuts and remove the wheel and tire assembly.
3. Remove the retainer washers from the lug nut studs and remove the manual locking hub assembly. To remove the interal hub lock assembly from the outer body assembly, remove the outer lock ring seated in the hub body groove. The internal assembly, spring and clutch gear will now slide out of the hub body.

NOTE: Do not remove the screw from the plastic dial.

4. Rebuild the hub assembly in the reverse order of disassembly.
5. Install the manual locking hub assembly over the spindle and place the retainer washers on the lug nut studs.
6. Install the wheel and tire assembly. Install the lug nuts and tighten to 85–115 ft. lb.

Manual Hub Adjustment

ALL MODELS SO EQUIPPED

1. Raise and support the vehicle safely. Remove the wheel lug nuts and remove the wheel and tire assembly.
2. Remove the retainer washers from the lug nut studs and remove the manual locking hub assembly from the spindle.
3. Remove the snap ring from the end of the spindle shaft.
4. Remove the axle shaft spacer, needle thrust bearing and the bearing spacer.
5. Remove the outer wheel bearing locknut from the spindle using a four-prong spindle nut spanner wrench. Make sure the tabs on the tool engage the slots in the locknut. Remove the locknut washer from the spindle.
6. Loosen the inner wheel bearing locknut using four prong spindle nut spanner wrench. Make sure that the tabs on the tool engage the slots in the locknut and that the slot in the tool is over the pin on the locknut.

SUSPENSION
Light Trucks & Vans

7. Tighten the inner locknut to 35 ft. lbs. to seat the bearings. Spin the rotor and back off the inner locknut $\frac{1}{4}$ turn. Install the lockwasher on the spindle. It may be necessary to turn the inner locknut slightly so that the pin on the locknut aligns with the closest hole in the lockwasher.
8. Install the outer wheel bearing locknut using four prong spindle nut spanner wrench.
9. Tighten locknut to 15 ft. lbs.
10. Install the bearing thrust spacer, needle thrust bearing and axle shaft spacer.
11. Clip the snap ring onto the end of the spindle.
12. Install the manual hub assembly over the spindle. Install the retainer washers.
13. Install the wheel and tire assembly. Install and tighten lugnuts 85–115 ft. lbs.
14. Check the end play of the wheel and tire assembly on the spindle. End play should be 0.001–0.003 inch.

4WD FRONT HUBS, RANGER AND BRONCO II WITH AUTOMATIC LOCKING HUBS.

1. Raise and support the vehicle safely. Remove the wheel lug nuts and remove the wheel and tire assembly.
2. Remove the retainer washers from the lug nut studs and remove the automatic locking hub assembly from the spindle.
3. Remove the snap ring from the end of the spindle shaft.
4. Remove the axle shaft spacer, needle thrust bearing and the bearing spacer. Being careful not to damage the plastic moving cam, pull the cam assembly off the wheel bearing adjusting nut and remove the thrust washer and needle thrust bearing from the adjusting nut.
5. Loosen the wheel bearing adjusting nut from the spindle using a $2\frac{3}{8}$ inch hex socket tool.
6. While rotating the hub and rotor assembly, tighten the wheel bearing adjusting nut to 35 ft. lbs. to seat the bearings, then back off the nut $\frac{1}{4}$ turn (90°).
7. Retighten the adjusting nut to 16 in. lb. using a torque wrench. Align the closest hole in the wheel bearing adjusting nut with the center of the spindle keyway slot. Advance the nut to the next hole if required.
8. Install the locknut needle bearing and thrust washer in the order of removal and push or press the cam assembly onto the locknut by lining up the key in the fixed cam with the spindle keyway.
9. Install the bearing thrust washer. needle thrust bearing and axle shaft spacer. Clip the snap ring onto the end of the spindle.
10. Install the automatic locking hub assembly over the spindle by lining up the three legs in the hub assembly with three pockets in the cam assembly. Install the retainer washers.
11. Install the wheel and tire assembly. Install and tighten lugnuts to 85–115 ft. lbs.
12. Final end play of the wheel on the spindle should be 0.001–0.003 in.

BEARING REPLACEMENT OR REPACKING

1. Raise the vehicle and support on safety stands.
2. If equipped with free-running hubs refer to Free-Running Hub Removal and Installation.
3. Remove the front hub grease cap and driving hub snap-ring.
4. Remove the splined driving hub and the pressure spring. This may require a slight prying assist.
5. Remove the wheel bearing lock nut, lock ring, and adjusting nut using tool T59T-1197-B, or equivalent.
6. Remove the hub and disc assembly. The outer wheel bearing and spring retainer will slide out as the hub is removed.
7. Remove the spindle retaining nuts, then carefully remove the spindle from the knuckle studs and axle shaft.
8. Clean all old grease from the needle bearings and wipe clean the spindle face that mates with the spindle bore seal.
9. Remove the spindle bore seal, V-seal, and thrust washer from the outer axle shaft. Clean any old grease or dirt from these parts and replace those that show signs of excessive wear.
10. Using Multi-Purpose Lubricant, Ford Specification ESA-M1C75-B or equivalent, thoroughly lubricate the needle bearing and pack the spindle face that mates with the spindle bore seal.
11. Assemble the V-seal in the spindle bore next to the needle bearing. Assemble the spindle bore seal on the axle shaft.
12. Assemble the spindle with the axle shaft on the knuckle studs. Adjust the retaining nuts to 50–60 ft. lbs.
13. Carefully drive the inner bearing cone and grease seal out of the hub using Tool T69L-1102-A.
14. Inspect the bearing cups for pits or cracks. If necessary, remove them with a drift. If new cups are installed, install new bearings.
15. Lubricate the bearings with Multi-Purpose Lubricant Ford Specification, ESA-M1C7-B or equivalent. Clean all old grease from the hub. Pack the cones and rollers. If a bearing packer is not available, work as much lubricant as possible between the rollers and the cages.
16. Position the inner bearing cone and roller in the inner cup and install the grease retainer.
17. Carefully position the hub and disc assembly on the spindle.
18. Install the outer bearing cone and roller, and the adjusting nut.
19. Using a torque wrench, tighten the bearing adjusting nut to 50 ft. lbs., while rotating the wheel back and forth to seat the bearings.
20. Back off the adjusting nut approximately 90 degrees.
21. Assemble the lock ring by turning the nut to the nearest hole and inserting the dowel pin. The dowel pin must seat in a lock ring hole for proper bearing adjustment and wheel retention.
22. Install the outer lock nut and tighten to 50–80 ft. lbs. Final end play of the wheel on the spindle should be 0.001 to 0.010 in.
23. Install the pressure spring and driving hub snap-ring.
24. Apply non-hardening sealer to the seating edge of the grease cap, and install the grease cap.
25. Adjust the brake if it was backed off.
26. Lower the vehicle.

REAR SUSPENSION

REAR SPRING (SINGLE AXLE)
Removal and Installation
E SERIES

1. Raise the rear end of the vehicle and support the chassis with safety stands. Support the rear axle with a floor jack or hoist.
2. Disconnect the lower end of the shock absorber from the bracket on the axle housing.
3. Remove the two spring clips (U-bolts) and the spring clip cap.
4. Lower the axle and remove the spring front bolt from the hanger.
5. Remove the two attaching bolts from the rear of the spring. Remove the spring and the shackle.
6. Assemble the upper end of the shackle to the spring with the attaching bolt.
7. Connect the front of the spring to the front bracket with the attaching bolt.
8. Assemble the spring and shackle to the rear bracket with the attaching bolt.
9. Place the spring clip plate over the head of the center bolt.

SUSPENSION
Light Trucks & Vans

10. Raise the axle with a jack and guiding it so that the center bolt enters the pilot hole in the pad on the axle housing.
11. Install the spring clips, cap and attaching nuts. Tighten the nuts snugly.
12. Connect the lower end of the shock absorber to the lower bracket.
13. Tighten the spring front mounting bolt and nut, the rear shackle nuts and spring clip nuts.
14. Remove the safety stands and lower the vehicle.

F SERIES 2WD AND RANGER/BRONCO II

1. Raise the vehicle frame, until the weight is off the rear spring, with the tires still touching the floor.
2. Remove the nuts from the spring U-bolts and drive the U-bolts from the U-bolt plate. If so equipped, remove the auxiliary spring and the spacer.
3. Remove the spring-to-bracket nut and bolt at the front of the spring.
4. Remove the shackle upper and lower nuts and bolts at the rear of the spring. Remove the spring and shackle assembly from the rear shackle bracket.
5. If the bushings in the spring or shackle are worn or damaged, replace them.
6. To install, position the spring in the shackle, and install the upper shackle-to-spring bolt and nut with the bolt head facing outboard.
7. Position the front end of the spring in the bracket and install. Position the shackle in the rear bracket and install.
8. Position the spring on top of the axle with the spring tie bolt centered in the hole provided in the seat. If equipped, install the auxiliary spring and spacer.
9. Install the spring U-bolt plate and nuts. Lower the vehicle. Tighten the spring U-bolt nuts. Tighten the front spring bolt and nut and the rear shackle bolts and nuts.

BRONCO

1. Raise the vehicle by the axles and install safety stands under the frame.
2. Disconnect the shock absorber from the axle.
3. Remove the U-bolt attaching nuts and remove the two U-bolts and the spring clip plate.
4. Lower the axle to relieve spring tension and remove the nut from the spring front attaching bolt.
5. Remove the spring front attaching bolt from the spring and hanger with a drift.
6. Remove the nut from the shackle to hanger attaching bolt and drive the bolt from the shackle and hanger with a drift and remove the spring from the vehicle.
7. Remove the nut from the spring rear attaching bolt. Drive the bolt out of the spring and shackle with a drift.
8. Position the shackle (closed section facing toward front of vehicle) to the spring rear eye and install the bolt and nut.
9. Position the spring front eye and bushing to the spring front hanger, and install the attaching bolt and nut.
10. Position the spring rear eye and bushing to the shackle, and install the attaching bolt and nut.
11. Raise the axle to the spring and install the U-bolts (when an axle cap is not used, the U-bolt shank should contact the leaf edges) and spring clip plate. Align the spring leaves.
12. Tighten the U-bolt nuts and the spring front and rear attaching bolt nuts. The U-bolts should contact the spring assembly edges or axle seat.
13. Connect the shock absorber to the axle and tighten the nut.
14. Remove the safety stands and lower the vehicle.

F SERIES 4WD

1. Raise the vehicle frame until the weight is off the rear springs with the wheels still touching the floor.
2. Remove the nuts from the spring U-bolts.
3. Drive the U-bolts out of the shock absorber lower bracket and the spring cap and remove the U-bolts.
4. Remove the spacer from the top of the spring.

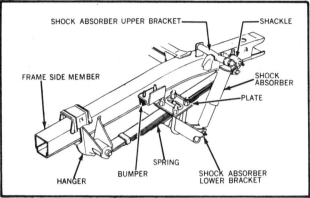

Rear suspension—Bronco (© Ford Motor Co.)

5. If equipped with auxiliary springs, remove the auxiliary spring and spacer.
6. Remove the shackle to bracket bolt and nut from the rear of the spring.
7. Remove the spring-to-hanger bolt and nut from the front of the spring and remove the spring.
8. Remove the shackle-to-spring bolt and nut and remove the shackle from the spring.
9. Position the shackle to the spring and install the attaching bolt and nut. The bolt must be installed so the nut is away from the frame.
10. Position the spring to the spring front hanger and install the attaching bolt and nut.
11. Position the shackle to the bracket and install the attaching bolt and nut.
12. Align the spring toe bolt with the pilot hole in the axle spring seat and, if so equipped, install the auxiliary spring and spacer.
13. Position the spacer on top of the spring and install the U-bolts over the spacer, spring and axle.
14. Position the spring cap and shock lower bracket to the axle and U-bolts. Install the U-bolt attaching nuts.
15. Lower the vehicle and tighten the front spring bracket bolt and nut and the rear shackle bolts and nuts.

C SERIES

1. Raise the vehicle frame until the weight is off the rear springs with the wheels still touching the floor.
2. Remove the nuts from the spring clips (U-bolts) and drive the clips out of the spring seat cap. If so equipped, remove the auxiliary spring and spacer.
3. Remove the shackle pin locking bolts from each spring bracket.
4. A hole is provided in the frame opposite each spring bracket for removing the shackle pin. Insert a drift from the inside of the frame through these holes and drive the shackle pin out of each bracket.
5. Remove the spring and shackle assembly from the truck. Separate the spring from the shackle by removing the locking bolt and driving out the shackle lower pin from the shackle and spring eye.
6. Remove the lubricating fittings from the shackle pins.
7. Align the upper bore of the shackle with the holes in the rear bracket. Drive the shackle upper pin through the shackle and bracket with the pin lubricator hole facing outward.
8. Line up the shackle pin groove with the locking bolt hole in the bracket and install the locking bolt, washer and nut.
9. Install the spring seat and the wedge (if so equipped) between the axle and the spring. Position the spring on the axle, being sure that the spring tie bolt is in the hole provided in the axle or spring seats. If so equipped, install the auxiliary spring and spacer.
10. Drive the shackle lower pin through the shackle and

2-343

SUSPENSION
Light Trucks & Vans

spring rear eye. Install the locking bolt, washer and nut as before. Repeat the operation to install the shackle pin through the spring front bracket and eye.

11. Place the spring clip plate on top of the spring at the tie bolt, and put the spring clips over the spring assembly and the axle.
12. Install the spring seat cap on the spring clips and install the spring clip nuts on the clips.
13. Lower the vehicle to the floor and tighten the spring clip nuts.

L, LN, B, W SERIES AND B-F SERIES WITH CAST SPRING BRACKETS

1. Lift the vehicle until the weight is off of the rear spring but the wheels still touch the ground.
2. Remove the nuts from the U-bolts and drive the bolts out of the spring seat caps.
3. Remove the auxiliary spring and spacer, if equipped.
4. Support the spring and remove the front shackle pins. The lower pin is held in place by a lock bolt and the upper is held by a cotter pin.
5. Remove the cotter pin in the rear shackle pin and remove the pin. Remove the spring from the vehicle.
6. To install place the spring in the rear shackle bracket and install the shackle pin and cotter pin.
7. With a jack or C-clamp press the front eye of the spring up into the bracket until the eye is lined up with the hole in the bracket. Install the shackle pins in the spring and secure them in place with the lock bolt or the cotter pin.
8. Install the spring seat and the wedge (if so equipped) between the axle and spring and install the U-bolts over the axle and install the nuts on the bolts after installing the spring seat cap.
9. Lower the vehicle and tighten the nuts on the U-bolts securely.

F AND B SERIES WITH STAMPED SPRING BRACKETS

1. Raise the vehicle frame until the weight is off the rear springs with the wheels still touching the floor.
2. Remove the nuts and washers from the spring U-bolts (clips) and drive the U-bolts out of the spring cap. Remove the auxiliary spring and spacer, if equipped.
3. Support the spring and remove the front spring bolts and rebound pins. The spring bolts is held in place by a slot nut and cotter pin, and the lower rebound pin is held in place by a self-locking pin and washer.
4. Remove the rear rebound pin which is held in place by a self-locking pin and washer, and remove the spring.
5. Position the spring in the rear bracket and install the rebound pin, self-locking pin and washer.
6. Position the spring in the front bracket. Install the rebound pin, and secure with a self-locking pin and washer.
7. Using a C-clamp or jack, force the front eye of the spring up into the bracket until the eye is aligned with the holes in the bracket. Install the lower spring bolt through the front bracket and spring eye. Install the slot nut and washer. Align the slot opening of the slot nut with the hole on the spring bolt in a tightening direction and secure with a cotter pin.
8. Position the spring seat and wedge (if so equipped) between the axle and spring. If so equipped, install the auxiliary spring and spacer. Place the U-bolt pad on the top of the spring at the tie bolt, and put the spring U-bolt over the spring assembly and axle.
9. Install the spring cap on the spring U-bolt and install the spring U-bolt nuts and washers on the U-bolts.
10. Lower the vehicle to the floor and tighten the spring U-bolt nuts.

REAR SPRING (Hendrickson Tandems)

Removal and Installation

1. Raise the rear of the vehicle and position the blocks under the frame behind the rear axle.
2. Remove the wheels, hub, and drum from the forward rear axle.
3. Remove the support beam bar saddle caps from the lower side of the support beam.
4. Position a jack under the front end of the support beam.
5. Remove the shackle pin lock pin from the spring front bracket and remove the shackle pin.
6. Lower the support beam and spring. Remove the spring from the support beam.
7. RV, RVE, VA and VEA suspension: Remove the U-bolt nuts and remove the saddle and U-bolts from the spring. RT, RTA and RTEA suspension: Remove the spring plate-to-support beam saddle attaching bolts and nuts. Remove the saddle and spring plate from the spring.
8. RV, RVE, VA and VEA suspension: Position the saddle on the spring and install the U-bolts and nuts. Tap the U-bolts with a hammer while tightening the nuts. RT, RTA and RTEA suspension: Position the spring plate and saddle on the spring. Snug up, but do not tighten, the saddle nuts. Tighten the spring plate set screw and lock down, then tighten the saddle nuts.
9. Position the spring and saddle on the support beam and spring rear bracket.
10. Raise the support beam and spring, and position the spring on the front spring bracket.
11. Align the spring with the front bracket and install the shackle pin.
12. Install the shackle pin lock pin and tighten the lock pin nut.
13. Install the hub, drum, and wheel on the forward rear axle.
14. Remove the jack from the support beam and the blocks from the rear of the frame. Lower the saddle on the support beam bar center insulator bushing and install the saddle caps.
15. Install the saddle caps and tighten.

NOTE: The weight of the vehicle must be on the suspension when the saddle cap attaching nuts are tightened.

SHOCK ABSORBER

Removal and Installation
ECONOLINE AND BRONCO

1. Raise the vehicle and support it safely.
2. Remove the shock absorber lower attaching nut and bolt, and swing the lower end free of the mounting bracket on the axle housing.
3. Remove the attaching nut from the upper mounting stud, and remove the shock absorber.
4. To install, position the shock absorber with the rubber bushings and steel washers to the upper mounting bolt.
5. Swing the lower end of the shock absorber into the mounting bracket on the axle housing. Install the attaching washers, mounting bolt, and self-locking nut.
6. Install the self-locking nut on the upper mounting bolt.
7. Lower the vehicle.

ALL OTHERS

1. Remove the self-locking nut, steel washer, and rubber bushings at the upper and lower ends of the shock absorber.
2. Remove the unit from the vehicle.
3. To install, position the shock absorber on the mounting brackets with the large diameter at the top.
4. Install the rubber bushing, steel washer, and self-locking nut.

SUSPENSION
Light Trucks & Vans

NAVISTAR/INTERNATIONAL HARVESTER TRUCKS
SCOUT • TRAVELALL • TRAVELETTE
SERIES 100 • 200 • 500 • 1600-4300

FRONT SUSPENSION

Front I-Beam Suspension

SHOCK ABSORBER

Removal And Installation

1. Raise the vehicle and support safely.
2. Remove the retaining nuts and washers from the upper and lower attaching bolts.
3. Remove the shock absorber from the bolts.
4. Install the new shock absorber on the bolts and position the rubber grommets at the shock absorber eyes.
5. Install the retaining nuts and washers on the upper and lower attaching bolts and tighten securely.
6. Lower the vehicle.

SPRING

Removal and Installation

1. Disconnect shock absorber at lower mount.
2. Raise front of vehicle just enough to take weight off spring.
3. Unbolt U-bolts, tapered caster wedge and U-bolt.
4. Remove lube fittings from spring mountings.
5. Remove spring pins and spring.
6. To install, reverse the above procedure, mounting fixed end of spring first.

KING PIN AND BUSHINGS

Replacement

1. Remove spindle nuts and spindle bearing retaining nuts.
2. Remove wheels, inner bearings and grease retainers from spindles.
3. Remove dirt shields.
4. Remove bolts holding backing plates and place backing plate assemblies over ends of axle I-beam.
5. Remove tapered draw keys holding the knuckle pins.
6. Remove expansion plugs or cap and gasket from the top and bottom of steering knuckles. (Remove expansion plugs by drilling a hole in one of the plugs and driving king pin with a punch to remove the other).
7. Drive out king pin.
8. Remove steering knuckles, thrust bearings and any spacer shims present.
9. Clean all parts thoroughly and inspect for wear and damage.
10. Remove old bushings with an arbor or drift.
11. Install new bushings with an arbor or bushing installing tool, making sure that the grease holes are aligned.
12. Ream or hone bushings to fit king pin with 0.001–0.002 in. clearance.
13. Lubricate and install steering knuckle, thrust bearings, spacer shims and king pins.
14. Install draw key (front side of axle) and tighten securely.
15. Insert expansion plugs or cap and gasket seals in the top and bottom of the steering knuckles.
16. Install brake backing plates, tightening bolts securely.
17. Install dirt shields, their retaining screws, cleaned and repacked wheel bearings and new grease seals.
18. Install wheel and spindle nuts, rotating wheel while tightening nut until slight drag is felt. Back off to the first castellation and install new cotter pin.
19. Lubricate and check and align front wheels if necessary.

Torsion Bar Front Suspension

SHOCK ABSORBER

Removal and Installation

1. Remove the shock absorber upper retaining nut, washer, and rubber grommet from the top of the upper control arm.
2. Raise the vehicle and support it in such a manner so as not to cover the bottom of the lower arm with a floor jack or jack stands.
3. Remove the two shock absorber retaining bolts from the

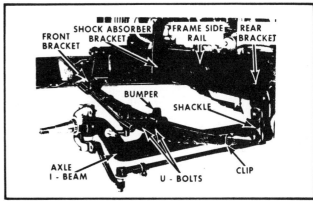

Typical front spring installation (medium only) (© International Harvester Co.)

SECTION 2

SUSPENSION
Light Trucks & Vans

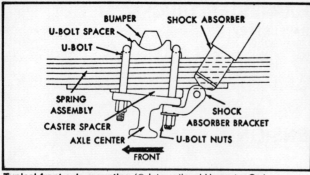

Typical front axle mounting (© International Harvester Co.)

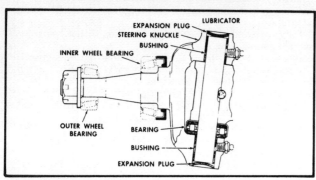

Typical steering knuckle expansion type seal plugs (© International Harvester Co.)

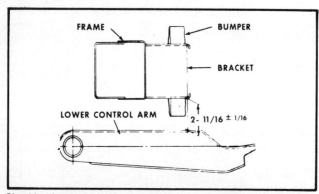

Checking front suspension height (© International Harvester Co.)

bottom of the lower control arm and withdraw the shock absorber.
4. To install new shock absorber, position the washer and rubber grommet on the extended shock absorber rod, and position the shock absorber up through the lower control arm, and engage the hole in the upper control arm with the extended rod.
5. Install the two bottom retaining bolts and tighten the shock absorber securely to the lower control arm.
6. Carefully lower the vehicle, so as not to lose the shock absorber rod from the upper control arm hole.
7. Install the upper rubber grommet, washer, and nut on the rod and tighten until the rubber grommet squeezes out slightly.

NOTE: Follow the manufacturers recommendation concerning the installation of lock nuts or self-locking nuts.

TORSION BAR

Removal and Installation

Right and left torsion bars are not interchangeable. The bars are marked with an "L" or "R" on one end and the bars should always be installed with the marked end towards the rear of the vehicle. There is an arrow indicating the direction of wind-up on the end of the bar.
1. Jack up the vehicle by the frame crossmember and release the load from the torsion bar by loosening the retainer lever adjusting bolt.
2. Remove retainer lever adjusting bolt and slide retainer lever from end of torsion bar.
3. Remove torsion bar by sliding it rearward.

--- CAUTION ---
Do not nick or scratch torsion bars – this may create a fracture.

4. To install torsion bar, position torsion bar in upper control arm, observing right and left side and rearward direction as indicated above.
5. Install retainer lever on end of torsion bar and position bar nut in bracket on frame so that torsion bar adjusting bolt may be installed.
6. Insert bolt in bar washer, then through retainer lever and bracket and thread into bar nut.
7. Adjust height by lowering vehicle to ground (check for correct tire pressure), bouncing front end up and down, then turning bolt on torsion bar adjusting lever until correct height is achieved. Measure height between top and lower control arm and lower edge of rubber bumper frame bracket (vehicle unloaded).

UPPER CONTROL ARM

Removal and Installation

1. Jack up vehicle by front frame crossmember until front wheels are off the ground.
2. Remove wheel and torsion bar (see preceding section).
3. Disconnect top mount of shock absorber.
4. Remove cotter pin, nut and dust seal (cut away seal) from lower ball joint.
5. Drive out lower ball joint stud (do not damage threads) or use special ball stud remover and nut.
6. Remove fender splash panel front shield.
7. Remove nut and washers from front end of upper control arm spindle and carefully drive out spindle with hammer.
8. Remove upper control arm.
9. To install, position upper control arm and install spindle through arm and bracket from rear.
10. Install flat washer, lock washer and nut. Tighten securely.
11. Install fender splash panel front shield.
12. Install new dust seal on ball stud, then line up shock absorber with hole in control arm and position ball stud into steering knuckle. Use jack to raise lower control arm until ball stud is well into steering knuckle.
13. Install nut on ball stud. Tighten securely.
14. Install top mounting of shock absorber, tightening just enough to squash rubber cushion slightly.
15. Install torsion bar on upper control arm as described in preceding section.
16. Mount front wheel.
17. Check alignment (see Unit Repair section) of steering.

LOWER CONTROL ARM

Removal and Installation

1. Raise vehicle by jacking frame crossmember. Remove wheel.

SUSPENSION
Light Trucks & Vans

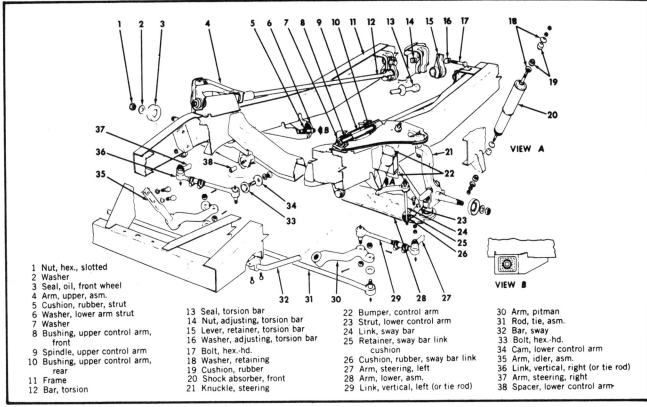

Torsion bar front wheel suspension (© International Harvester Co.)

1 Nut, hex., slotted
2 Washer
3 Seal, oil, front wheel
4 Arm, upper, asm.
5 Cushion, rubber, strut
6 Washer, lower arm strut
7 Washer
8 Bushing, upper control arm, front
9 Spindle, upper control arm
10 Bushing, upper control arm, rear
11 Frame
12 Bar, torsion
13 Seal, torsion bar
14 Nut, adjusting, torsion bar
15 Lever, retainer, torsion bar
16 Washer, adjusting, torsion bar
17 Bolt, hex.-hd.
18 Washer, retaining
19 Cushion, rubber
20 Shock absorber, front
21 Knuckle, steering
22 Bumper, control arm
23 Strut, lower control arm
24 Link, sway bar
25 Retainer, sway bar link cushion
26 Cushion, rubber, sway bar link
27 Arm, steering, left
28 Arm, lower, asm.
29 Link, vertical, left (or tie rod)
30 Arm, pitman
31 Rod, tie, asm.
32 Bar, sway
33 Bolt, hex.-hd.
34 Cam, lower control arm
35 Arm, idler, asm.
36 Link, vertical, right (or tie rod)
37 Arm, steering, right
38 Spacer, lower control arm

2. Disconnect sway bar link from strut and remove two bolts which secure strut to lower control arm.
3. Cut away dust seal from lower ball stud and remove cotter pin and nut from lower ball stud.
4. Either drive out ball stud with a soft hammer while supporting control arm or use special ball stud remover and nut.
5. Disconnect tie rod end from either side of vehicle.
6. Remove nut, lockwasher, cam and bolt from lower control arm and frame bracket.
7. Remove control arm and spacer in bushing.
8. To install, place new dust seal on ball stud and position ball stud into steering knuckle.
9. Tighten nut on ball stud and install cotter pin.
10. Position spacer in bushing and, while holding control arm in position, install bolt from front. Install cam, lockwasher, and nut, tightening to 81–135 ft. lbs. torque.
11. Mount strut to lower control arm tightening bolts securely.
12. Connect tie rod end.
13. Position cushion and retainer on sway bar link into strut with cushion, retainer and nut. Tighten nut until cushion is slightly squashed. Insert cotter pin.
14. Mount front wheel.
15. Check alignment (see Unit Repair section) and tighten nut on strut to 120–150 ft. lbs. torque and camber adjusting bolt nut to 81–135 ft. lbs. torque.

BALL JOINT

Inspection

UPPER BALL JOINT

The upper ball joint is a loose fit when not connected to the steering knuckle.

1. Use a floor jack or position the vehicle on a frame contact lift and raise the vehicle until the wheels fall to the full down position.
2. Grasp the tire at the top and bottom and move the tire in and out. The radial end play should not exceed .180 inch. If so, replace the ball joint.

LOWER BALL JOINT

The lower ball joint is spring loaded in its socket and this minimizes looseness and compensates for normal wear.

1. Locate a floor jack or position the vehicle on a frame contact lift and raise the vehicle until the wheels fall to the full down position.
2. Grasp the tire at the top and bottom and move the tire in and out. Any movement at the ball joint socket and stud indicates wear and the loss of preload, and the ball joint should be replaced.

Removal and Installation

Refer to Independent Front Suspension-Coil Spring, for the procedures necessary to remove and install the ball joints.

Independent Front Suspension — Coil Spring

SHOCK ABSORBER

Removal and Installation

1. Remove the shock absorber upper retaining nut, washer, and rubber grommet from the top of the upper control arm.
2. Raise the vehicle and support it in such a manner so as

SECTION 2

SUSPENSION
Light Trucks & Vans

not to cover the bottom of the lower arm with a floor jack or jack stands.

3. Remove the two shock absorber retaining bolts from the bottom of the lower control arm and withdraw the shock absorber.

4. To install new shock absorber, position the washer and rubber grommet on the extended shock absorber rod, and position the shock absorber up through the lower control arm, and engage the hole in the upper control arm with the extended rod.

5. Install the two bottom retaining bolts and tighten the shock absorber securely to the lower control arm.

6. Carefully lower the vehicle, so as not to lose the shock absorber rod from the upper control arm hole.

7. Install the upper rubber grommet, washer, and nut on the rod and tighten until the rubber grommet squeezes out slightly.

NOTE: Follow the manufacturers recommendation concerning the installation of lock nuts or self-locking nuts.

BALL JOINT

Inspection

UPPER BALL JOINT

The upper ball joint stud is spring loaded in its socket and this minimizes looseness and compensates for normal wear.

1. Locate a floor jack under the lower control arm on the outboard side and raise the vehicle so that the wheels clear the floor.

2. Grasp the tire at the top and bottom and move the tire in and out. If any perceptible lateral or vertical movement is noted, the ball joint should be replaced.

LOWER BALL JOINT

The lower ball joints are a loose fit when not connected to the steering knuckle.

1. Locate a floor jack under the lower control arm on the outboard side and raise the vehicle until the wheels clear the floor.

2. Grasp the tire at the top and bottom and move the tire in and out. The radial play should not exceed .250 inch. If so, the ball joint should be replaced.

COIL SPRING

Removal and Installation

1. Raise the front of the vehicle and support safely on floor stands.
2. Remove the wheels and tires.
3. Remove the caliper from the rotor assembly and support to avoid damage to the brake hose.
4. Remove the hub assembly from the steering knuckle.
5. Remove the brake shield from the knuckle.
6. Disconnect the sway bar link, if equipped.
7. Remove the axle bumper and wheel stop bracket from the lower control arm.

NOTE: This also disconnects the lower control arm rod assembly from the control arm.

8. Remove the shock absorber.
9. Position the spring compressor screw into the shock absorber upper mounting hole in the crossmember. Position the puller hooks under the lower second spring coil and turn the spring compressor screw until coil spring unseats from the lower control arm.
10. Remove the cotter pins from the upper and lower ball joint studs, and loosen the lower nut approximately two turns.

NOTE: The cotter pin is removed from the upper ball joint stud to allow a special tool to be placed over the upper stud and nut.

11. With the aid of special tool SE-2493 (ball joint stud remover) or its equivalent, and by placing it over the upper stud at its hex end, extend the screw to contact the lower ball joint stud.

12. Apply pressure by turning the screw out from the tool. Tap the steering knuckle lightly to loosen the stud from the knuckle.

13. Remove the tool and remove the nut from the lower ball joint stud. Separate the ball joint stud from the knuckle.

14. Loosen the spring compressor and relieve the spring of all tension. Remove the spring from the vehicle.

15. Position new spring into the crossmember and the lower control arm.

NOTE: Turn the coil spring to line up the bottom coil with the seal groove in the lower control arm.

16. Install the hooks of the spring compressor under the second coil of the spring. Tighten the compressor until the lower ball joint stud can be installed into the steering knuckle.

17. Position a hydraulic jack under the lower control arm and raise the arm until the bottom ball joint stud will enter the steering knuckle. Install the nut and torque to specifications.

18. Install the cotter pins in both the upper and lower ball joint studs. Confirm the position of the spring in the lower arm.

19. Remove the spring compressor tool and install the shock absorber.

20. Install the lower control arm rod assembly and axle bumper and wheel stop bracket to the lower control arm.

21. Install the brake shield on the steering knuckle.
22. Install the rotor-hub assembly.
23. Install the caliper on the rotor assembly.
24. Install the wheel and tire assembly.
25. Remove the floor stands and lower the vehicle to the floor.
26. Depress the brake pedal to force the disc pads against the rotor.

UPPER CONTROL ARM AND BALL JOINT

Removal and Installation

Follow the procedure outlined under Coil Spring Removal And Installation, except that the special tool SE-2493 or its equivalent, is used to apply pressure to the upper ball joint stud after loosening the nut approximately two turns. Follow the procedure as outlined.

1. After removing the ball joint stud from the steering knuckle, remove the upper control arm from the frame rail.

2. Place the upper control arm in a vise and with a ball joint remover socket, remove the ball joint from the upper arm.

3. Lubricate the threads in the arm and place the new ball joint into the arm.

4. With the use of the ball joint socket, tighten the ball joint to specifications in the upper arm.

5. Inner upper arm bushings may be replaced by pressing the old bushings out and pressing the new bushings into the upper arm.

6. Install the upper arm onto the frame rail and tighten securely.

7. Follow the procedure outlined under the Coil Spring Removal and Installation to complete the operation.

LOWER CONTROL ARM AND BALL JOINT

Removal and Installation

Follow the procedure outlined under Coil Spring Removal and Installation to remove the coil spring from the suspension. After the coil spring is removed, follow this procedure to remove the lower control arm and ball joint.

SUSPENSION
Light Trucks & Vans

1. After removing the lower ball joint stud from the steering knuckle, remove the lower control arm from the frame bracket.
2. Place the lower control arm in a vise.
3. Remove the ball joint from the lower control arm with the aid of tool SE-2494-2 ball joint remover and installer, or an equivalent tool.
4. Lubricate the threads in the arm and place a new ball joint into the control arm.
5. With the use of tool SE-2494-2 or equivalent, tighten the ball joint into the arm.
6. Inner lower arm bushings may be replaced by pressing the old bushings out and pressing the new bushings into place.
7. Install the lower arm onto the frame rail bracket and tighten securely.
8. Follow the procedure outlined under "Coil Spring Removal and Installation" to complete the operation.

FRONT WHEEL BEARING

Adjustment

ALL NON-DRIVING AXLES

1. While rotating the wheel and hub assembly, adjust the spindle nut to 30 ft.lbs (50 ft. lbs. on Loadstar and Cargostar), then back off nut $\frac{1}{4}$ turn.
2. **Series 100 and 200:** Finger tighten and insert lock so that cotter pin can be inserted with out backing off nut.
Loadstar and Cargostar: If the lock or cotter pin can be installed at this position, do so, if not, tighten to the nearest locking position and insert the cotter pin or lock.

NOTE: Bent type lockwashers must have one tab bent over the adjusting nut. With double locknuts, tighten jam nut to 100–200 ft. lbs. and bend one tab of the lockwasher over the jam nut.

All other series: Finger tighten and if possible, insert cotter pin. If not able to install cotter pin, back off the nut to the nearest hole and insert the cotter pin.

NOTE: When using the cotter pin as a lock, the long tang should be bent over the spindle end. Clip the remaining tang, leaving enough stock to bend down against the side of the nut.

Front Drive Axle

For Service on Transfer Case and Differential, see Unit Repair Section.

The front drive axles incorporate hypoid gears and use both spherical and ball joint wheel-end steering knuckles. The axle shaft assemblies are full floating and may be removed without disassembling the steering knuckles. Two types of axle shafts are used, one, a drive flange arrangement, bolted to the hub, and a second, mated to an internally splined gear, which in turn is splined and mated to the wheel hub, to transmit the driving torque to the front wheels.

LEAF SPRING

Removal and Installation

1. Raise the vehicle and support on the frame rails behind the front springs with floor stands.
2. Remove the shock absorber from the spring.
3. Remove the U-bolts, spring bumpers and retainer, or the U-bolt seat.
4. Remove the lubricators, if used.
5. Remove the nuts from the shackles and bracket pins.
6. Slide the spring off the bracket and shackle pins.
7. Remove the spring from the vehicle.
8. Installation is the reverse of removal. Tighten all nuts and bolts securely.

FRONT DRIVE AXLE

Removal and Installation

1. Jack up truck until load is removed from springs and block up frame to safely hold weight.
2. Drain lubricant from main housing and, if applicable, from wheel end housings.
3. Disconnect brakes.
4. Disconnect drag link from ball stud bracket.
5. Disconnect drive shaft from pinion shaft yoke.
6. Supporting axle with a portable floor jack, remove spring U-bolts.
7. Roll axle assembly out from under truck.
8. To install, reverse the above procedure.

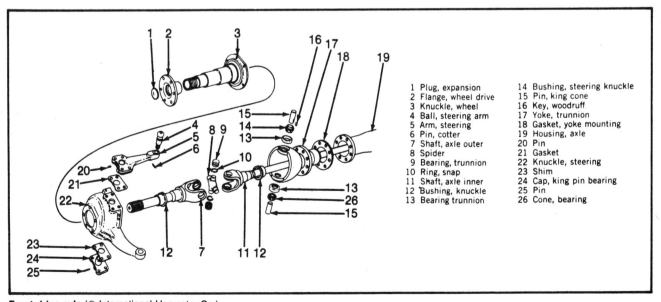

1 Plug, expansion
2 Flange, wheel drive
3 Knuckle, wheel
4 Ball, steering arm
5 Arm, steering
6 Pin, cotter
7 Shaft, axle outer
8 Spider
9 Bearing, trunnion
10 Ring, snap
11 Shaft, axle inner
12 Bushing, knuckle
13 Bearing trunnion
14 Bushing, steering knuckle
15 Pin, king cone
16 Key, woodruff
17 Yoke, trunnion
18 Gasket, yoke mounting
19 Housing, axle
20 Pin
21 Gasket
22 Knuckle, steering
23 Shim
24 Cap, king pin bearing
25 Pin
26 Cone, bearing

Front drive axle (© International Harvester Co.)

SUSPENSION
Light Trucks & Vans

Adjustments

Preload on the knuckle bearings of these front axles must be maintained at all times. Check for looseness each time knuckle is lubricated.

1. Jack up front end of truck until off-center weight of the wheel is relieved (wheel just barely touching ground).
2. Remove wheel and wheel adapter from hub.
3. Disconnect tie rod and drag link.
4. Remove axle shaft.
5. To remove play (check for play by pushing and pulling on top and bottom of knuckle) and increase preload drag, turn adjusting bolt into back of knuckle. Preload should read (spring scale hooked into end of steering arm) 12 lbs.

FRONT WHEEL BEARING

Adjustment

1. Remove wheel and adapter from hub.
2. Remove axle shaft or internal gear, and adjusting nut lock plate.
3. Tighten nut until just against bearing.
4. Rotate the wheel forward and backward until a slight drag can be felt. Turn nut back to the first lock hole to obtain about a ½ hole relief.
5. Bearing adjustment is correct when no play can be felt when pushing and pull at top and bottom of wheel.

Reference for overhaul procedures see International Single Reduction Rear Axle in The Unit Repair section.

AXLE SHAFT AND UNIVERSAL JOINT

Removal

AXLES HAVING DRIVE FLANGE

1. Raise vehicle, support with floor stands and remove wheel from vehicle.
2. Remove grease cap and snap-ring from end of axle shaft.
3. Remove drive flange capscrews, lockwasher, flange and gasket. If equipped with locking hubs, bend up locking tab, take out capscrews and remove clutch body. Remove hub body. Loosen setscrew and unscrew drag shoe from spindle.

NOTE: Lift off clutch body holding it erect so as not to let drive pins fall out of body. If they do fall out, be certain to install them during reassembly.

4. Remove brake drum countersunk setscrews, where applicable and remove drum.
5. Bend the lip on the wheel bearing lockwasher away from the outer wheel bearing nut and remove the nut and lockwasher. Remove wheel bearing adjusting nut (inner) and bearing lockwasher.
6. Remove the wheel hub with wheel bearing.
7. Remove backing plate and wheel spindle retaining bolts and lockwashers. Support backing plate to prevent damage to brake hose if hose is not disconnected.
8. Remove wheel bearing spindle with bushing. If spindle bushing requires replacing, press out bushing using an adapter of correct size. An alternate method of bushing removal is the use of a cape chisel or punch to collapse the bushing.
9. Pull axle shaft and universal joint assembly out of axle housing.

Installation

1. Insert axle shaft and universal joint assembly into axle housing. Position splined end of axle shaft into differential pinion gear and push into place.
2. If wheel bearing spindle bushing was removed, press new bushing into spindle using an installer tool or adapter of proper size. Lubricate ID of bushing with chassis lube when installed to provide initial lubrication. Bushing should be pressed in until bushing flange is seated against shoulder in spindle. Assemble wheel spindle and backing plate to steering knuckle. Secure with six (6) bolts and lockwashers and tighten to specifications. Connect hydraulic brake fluid line if disconnected.
3. Pack wheel bearings using a pressure lubricator or by carefully working lubricant into bearing cones by hand. Slide lubricated inner wheel bearing on spindle until it stops against spindle shoulder.
4. Apply thin coating of lubricant specified for wheel bearings to seal lip and install seal into wheel hub using an adapter of correct diameter. Lip of seal should extend towards wheel (away from backing plate assembly).
5. Assemble wheel hub on spindle. Install lubricated outer wheel bearing cone on spindle. Push cone on spindle until it rests against bearing cup.
6. Install wheel bearing lockwasher and adjusting (inner) nut. Tighten adjusting nut until there is a slight drag on the bearings when the hub is turned; then back-off approximately one-sixth turn.
7. Install tang-type lockwasher and lock nut (outer). Tighten nut and bend lockwasher tang over lock nut. If axle is equipped with locking hubs, install drag shoe on spindle and tighten setscrew.
8. Align splines of drive flange with those of axle shaft and secure drive flange and new gasket to wheel hub with capscrews and lockwashers. Tighten capscrews securely. If equipped with locking hubs, lightly lubricate hub body and clutch using a light grade chassis lubricant and install new gasket, hub body, snap-ring and hub clutch. Be certain that all drive pins are positioned in locking hub clutch when clutch is installed. Secure hub clutch to wheel hub with capscrews and lock. Tighten to specifications and bend tang over head of capscrew.
9. Install snap-ring and grease cup if not equipped with locking hubs.
10. Assemble brake drum and wheels to wheel hub. Bleed and adjust brakes.

CAUTION

Be certain that master cylinder is full of brake fluid after completing bleeding operation.

Removal

AXLES HAVING DRIVE GEAR

1. Raise and support vehicle with floor stands placed under frame rails. Remove wheel from vehicle.
2. Lightly tap alternately around edge of hub cap with hammer and screwdriver or similar tool until hub cap is removed.
3. If axle is equipped with locking hubs, remove the eight (8) socket-head setscrews securing hub clutch assembly to wheel hub assembly.

NOTE: Drive pins may fall out of hub clutch when separated from wheel hub assembly. Be certain to replace them during installation.

4. Remove retaining ring from wheel hub if equipped with locking hubs.
5. Remove snap-ring from axle shaft.
6. Pull drive gear out of wheel hub. If difficulty is encountered in removing drive gear, obtain a screwdriver or similar tool having the end bent approximately 90° with the handle. Insert end of tool into groove in drive gear and withdraw gear. If necessary, move wheel alternately backward and forward to aid removal of gear.
7. Remove retaining ring and locking hub body, if so equipped.
8. Using Wheel Bearing Adjusting Nut Wrench, remove

SUSPENSION
Light Trucks & Vans

wheel bearing outer nut and slide lock ring off of axle shaft. Again using wrench, remove wheel bearing inner nut.

9. Pull drive gear spacer out of wheel hub.
10. Remove brake drum from wheel hub and slide wheel hub assembly off of spindle.

NOTE: Do not allow tapered roller bearings to drop on floor as bearings may be damaged.

11. Remove screws retaining grease guard to backing plate. Take off grease guard and gasket.
12. Remove the six (6) bolts securing wheel spindle and backing plate to steering knuckle. Pull spindle with bushing off of axle shaft. If spindle bushing requires replacing, press or drive out bushing using an adapter of correct size. An alternate method of bushing removal is the use of a cape chisel or punch to collapse the bushing.
13. Pull axle shaft and universal joint assembly out of axle housing.

Installation

1. Proceed with steps 1 through 5 of Axle Shaft and Universal Joint Installation (axles having drive flange).
2. Insert drive gear spacer over spindle and against outer wheel bearing cup.
3. Position wheel bearing inner adjusting nut wheel bearing adjusting nut wrench with pin in nut extending toward handle

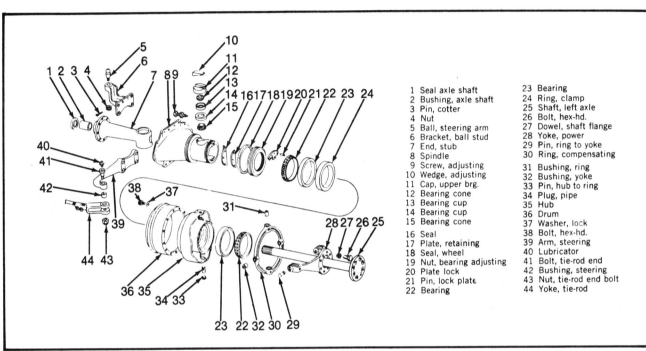

1 Seal axle shaft
2 Bushing, axle shaft
3 Pin, cotter
4 Nut
5 Ball, steering arm
6 Bracket, ball stud
7 End, stub
8 Spindle
9 Screw, adjusting
10 Wedge, adjusting
11 Cap, upper brg.
12 Bearing cone
13 Bearing cup
14 Bearing cup
15 Bearing cone
16 Seal
17 Plate, retaining
18 Seal, wheel
19 Nut, bearing adjusting
20 Plate lock
21 Pin, lock plate
22 Bearing
23 Bearing
24 Ring, clamp
25 Shaft, left axle
26 Bolt, hex-hd.
27 Dowel, shaft flange
28 Yoke, power
29 Pin, ring to yoke
30 Ring, compensating
31 Bushing, ring
32 Bushing, yoke
33 Pin, hub to ring
34 Plug, pipe
35 Hub
36 Drum
37 Washer, lock
38 Bolt, hex-hd.
39 Arm, steering
40 Lubricator
41 Bolt, tie-rod end
42 Bushing, steering
43 Nut, tie-rod end bolt
44 Yoke, tie-rod

Front drive axle (drive gear type) (© International Harvester Co.)

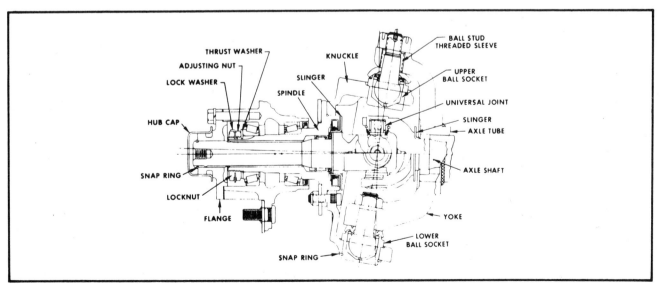

Front drive axle with 40° steer (© International Harvester Co.)

2-351

SUSPENSION
Light Trucks & Vans

end of wrench. Install nut on spindle and tighten until it is snug against outer wheel bearing; then loosen adjusting nut $\frac{1}{4}$ turn. Align tang on adjusting nut lock ring with groove in wheel spindle. Slide ring on spindle and index pin on adjusting nut with hole in lock ring. If pin will not index with hole in lock ring, turn adjusting nut to the left (Loosen) until it will index.

NOTE: When attempting to index pin with hole in lock ring, turn nut very slightly since adjusting nut should be locked with first hole in lock ring past $\frac{1}{4}$ turn lose. Position wheel bearing outer nut in adjusting nut wrench and install on spindle. Tighten nut securely.

4. Align splines on axle shaft and splines in wheel hub with those of drive gear. Insert drive gear on axle shaft. Push gear into hub until it rests against drive gear spacer.

NOTE: Groove on side of gear must be toward hub cap.

5. If axle is equipped with locking hubs, lightly lubricate locking hub body using a light grade chassis lubricant. Align splines and insert hub body into wheel hub.
6. Install snap-ring on end of axle shaft.
7. Place retaining ring in groove in wheel hub, if equipped with locking hub.
8. If applicable, lightly grease hub clutch assembly using a light grade chassis lubricant. Be sure that all eight drive pins are positioned in the locking hub clutch. Assemble hub clutch to hub body and secure with eight socket head setscrews.
9. Position hub cap on wheel hub and lightly tap alternately around cap until flange is against edge of hub.
10. Assemble brake drum and wheel to wheel hub. Bleed and adjust brakes.

―――――― **CAUTION** ――――――
Be certain that the master cylinder is full of brake fluid after bleeding operation.

STEERING KNUCKLE (SPHERICAL)

Removal

1. Remove drag link at steering arm and tie-rod at steering knuckle.
2. Remove oil seal retaining bolts from inner flange of steering knuckle and remove oil seals.
3. Remove bolts and lockwashers securing king pin lower bearing cap. Remove bearing cap and shim pack. Retain shim pack for use during reassembly.
4. Remove capscrews or self-locking nuts, which ever is applicable, securing steering arm or upper bearing cap to steering knuckle.
5. Lift steering arm assembly and knuckle until bronze bearing cone will clear ball yoke. Separate steering knuckle from ball yoke.

NOTE: Do not allow lower tapered roller bearing cone to drop on floor during removal of steering knuckle.

6. Support steering knuckle and with a long brass drift, drive or press king pin out of bronze bearing cone.

NOTE: Be careful not to damage end of king pin during removal of cone.

Installation

1. Assemble steering arm to knuckle using original shim pack. Install self-locking nuts or capscrews and tighten securely.
2. Coat king pin and bronze bearing cone ID and OD with chassis lubricant to prevent galling. Align serrations of new bronze bearing cone with serrations of king pin and press cone on king pin.

NOTE: Make sure the cone is pressed all the way on or against the shoulder.

3. With bronze cone and tapered roller bearing pre-lubricated, place tapered roller bearing cone into cup at lower end of ball yoke. While retaining lower bearing cone in position, assemble steering knuckle to ball yoke. Seat bronze cone into cup at upper end of ball yoke.
4. Lubricate lower king pin with chassis lubricant and using original shim pack, install lower bearing cap to knuckle securing with bolts and lockwasher. Tighten bearing cap bolts securely.
5. Assemble opposite knuckle proceeding with instructions similar to those outlined above.
6. Individually check knuckle bearing preload by placing a torque wrench on any one (1) of the steering arm or bearing cap bolts or nuts. Read the starting torque (not rotating torque). Remove or add shims at the lower bearing cap until the specified preload is obtained. Knuckle bearing preload should be checked without ball joint oil seal, drag link or tie-rod installed.
7. Assemble the knuckle oil seals with the split on top. Knuckle retainer plate must be adjacent to ball yoke; followed by rubber seal, felt seal and metal retainer. Install seal retainer bolts and tighten securely.
8. Connect tie-rod to steering knuckle and tighten nut. Connect drag link to steering arm ball. Install cotter keys.

Cleaning, Inspection

All parts of the wheel end assembly should be thoroughly cleaned and dried with compressed air or a lint-free clean cloth.
Inspect all parts for wear, cracks or other damage. Replace all oil seals, felts and gaskets to prevent lubricant leakage.

STEERING KNUCKLE (BALL JOINT)

Removal

1. With the vehicle safely supported, remove the wheel and brake drum.
2. Remove the backing plate and the spindle from the knuckle.

NOTE: If necessary, tap the spindle lightly with a soft hammer to loosen it from the knuckle bolts. The spindle oil seal, needle bearings and bronze spacer can be removed and replaced at this time.

3. Remove the axle from the housing.

NOTE: The slingers can be removed from the axle by using pullers or tapping the axle through the slingers.

4. Disconnect and remove the tie rod from the steering arm.
5. Remove the cotter pin from the upper ball socket stud and remove the nut.
6. Remove the nut from the lower ball socket stud and discard.

NOTE: This nut is of a special torque design and should only be used one time.

7. Remove the lower ball socket snap-ring (used on 4 × 4 applications only), and unseat the upper and lower ball socket studs with a lead hammer or with a puller tool arrangement, to separate the knuckle from the yoke.

NOTE: If the upper ball socket stud remains in the yoke flange, remove it by striking it on the stud with a soft hammer.

8. With the aid of puller tools or a press and ram, remove the bottom ball socket.
9. Reverse the knuckle and remove the upper ball socket.

SUSPENSION
Light Trucks & Vans

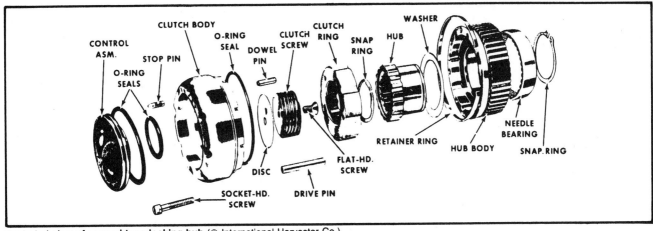

Exploded view of manual type locking hub (© International Harvester Co.)

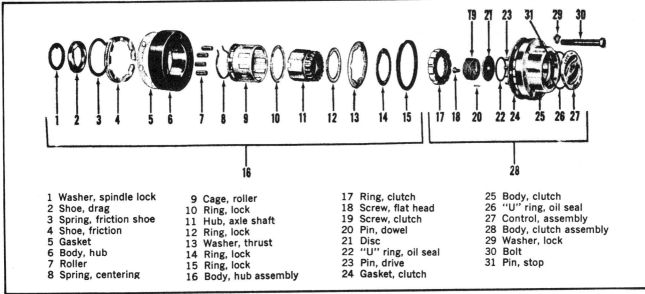

1. Washer, spindle lock
2. Shoe, drag
3. Spring, friction shoe
4. Shoe, friction
5. Gasket
6. Body, hub
7. Roller
8. Spring, centering
9. Cage, roller
10. Ring, lock
11. Hub, axle shaft
12. Ring, lock
13. Washer, thrust
14. Ring, lock
15. Ring, lock
16. Body, hub assembly
17. Ring, clutch
18. Screw, flat head
19. Screw, clutch
20. Pin, dowel
21. Disc
22. "U" ring, oil seal
23. Pin, drive
24. Gasket, clutch
25. Body, clutch
26. "U" ring, oil seal
27. Control, assembly
28. Body, clutch assembly
29. Washer, lock
30. Bolt
31. Pin, stop

Exploded view of lock-o-matic hub (© International Harvester Co.)

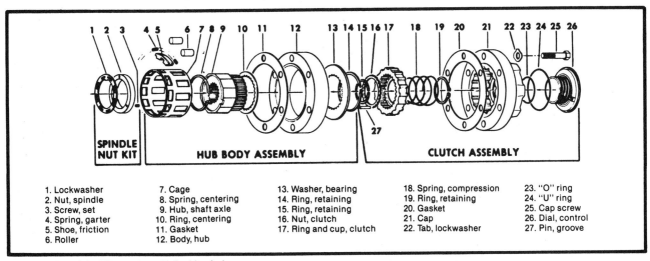

1. Lockwasher
2. Nut, spindle
3. Screw, set
4. Spring, garter
5. Shoe, friction
6. Roller
7. Cage
8. Spring, centering
9. Hub, shaft axle
10. Ring, centering
11. Gasket
12. Body, hub
13. Washer, bearing
14. Ring, retaining
15. Ring, retaining
16. Nut, clutch
17. Ring and cup, clutch
18. Spring, compression
19. Ring, retaining
20. Gasket
21. Cap
22. Tab, lockwasher
23. "O" ring
24. "U" ring
25. Cap screw
26. Dial, control
27. Pin, groove

Exploded view of the Warn automatic locking hub

SUSPENSION
Light Trucks & Vans

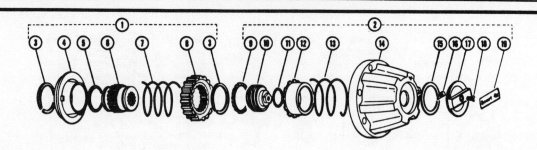

1. Clutch and bearing assy.
2. Cap assy.
3. Ring, retaining
4. Bearing, hub
5. Washer
6. Hub
7. Spring, compression
8. Ring, clutch
9. Nut, clutch
10. Screw, dial
11. Seal "O" ring
12. Cup, clutch
13. Spring, compression
14. Cap, hub
15. Washer
16. Detent, dial
17. Dial, control
18. Screw
19. Label

Exploded view of the Warn manual locking hub

10. With the aid of a special socket, remove the threaded sleeve in the top flange of the yoke.

Installation

1. Assemble the lower ball socket into the knuckle with a press and ram or a puller tool arrangement, making sure that the ball socket is firmly seated against the knuckle. Install the snap-ring on the 4 × 4 application.
2. Assemble the upper ball socket into the knuckle with a press and ram or a puller type tool arrangement, making sure that the ball socket is firmly seated against the knuckle.

NOTE: Use a 0.0015 inch feeler gauge blade between the socket and knuckle. The blade should not enter at the minimum area of contact.

3. Install new threaded sleeve into the top flange of the yoke, leaving approximately two threads exposed.
4. Install the knuckle assembly to the yoke, using a new nut on the lower ball socket stud. Torque the lower nut to 80 ft. lbs.
5. With the use of a special socket, torque the threaded sleeve to 50 ft. lbs. in the upper yoke flange.
6. Install the top ball socket stud nut and torque to 100 ft. lbs. Align the cotterpin holes between the stud and the castellated nut. Do not loosen nut to align the holes. Install the cotter pin.
7. Assemble the tie rod to the steering arm.
8. Assure that slingers are properly installed on the axle shaft and install the shaft into the housing.
9. Position the spindle over the axle end with the bronze bushing in place.
10. Install the backing plate and torque the nuts to 30 ft. lbs.
11. Install the hub and wheel assembly, and lower the vehicle.

CHECKING BALL SOCKETS FOR LOOSENESS

To check the ball sockets for excessive looseness, raise the vehicle and attach a dial indicator to the lower yoke or axle tube and set the indicator against the knuckle or lower ball socket, with a loaded pressure so as to read in both directions. Grasp the wheel at the top and bottom and move the wheel inward and outward. If the total indicator reading exceeds 0.020 inch, both the upper and lower ball sockets should be replaced.

FRONT DRIVE LOCKING HUBS

Two types of locking hubs are used: manual and Lock-O-Matic. Manual locking hubs are either engaged or disengaged, depending on how they are set. Lock-O-Matic hubs, when in "free" position, automatically engaged axle and wheel when forward torque is applied by the axle shaft. Thus, whenever front wheel drive is disengaged at the transmission, the wheels free wheel. "Lock" position is required only when engine braking control on the front wheel is desired.

Removal and Installation

1. Bend up tabs on mounting bolt lock washers.
2. Remove six (eight) mounting bolts using a thin-walled socket or appropriate hex wrench (externally splined type).
3. When clutch body is lifted off, immediately tilt it up so that the drive pins do not fall out.
4. Remove lock ring holding hub body onto axle shaft and pull off hub body.
5. Remove drag shoe (Lock-O-Matic only) from axle spindle by loosening hex-head set screw and unscrew drag shoe.
6. To install, reverse the above procedure.

Steering Linkage

See Unit Repair section for steering alignment procedures. See specifications at the beginning of this section for steering alignment specifications.

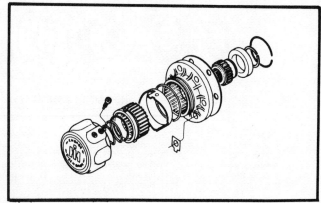

Exploded view of the Dualmatic locking hub

SUSPENSION
Light Trucks & Vans

TIE RODS

Tie rods are of three-piece construction: rod and two end assemblies. The end assemblies are threaded into the end of the tie rod and adjustment is made by turning them either in or out to shorten or lengthen the tie rod. When tightening the clamp it is important to make sure that the end assembly is threaded in far enough so that the clamping action of the clamp is right over the end pieces. Ball studs are integral in the end assemblies.

When disconnecting ball studs, loosen ball stud nut, then strike the nut with one hammer while another larger hammer is backing up the nut.

REAR SUSPENSION

The rear springs used on the IH vehicles are classified as leaf, air, and rubber block types. The heavier the vehicle load requirement, the heavier the spring assemblies would be to carry the load. Care should be exercised in the removal and installation of the spring assemblies so that personal injury can be avoided. The suspensions using shock absorbers will normally be found on the light and medium duty vehicles. The removal and installation of the shock absorbers require only the removal and installation of the retaining nuts or bolts, with the possibility of raising the vehicle for working clearance.

The following procedure can be used as a general outline for the removal and installation of leaf springs. The air and rubber block suspension, along with the tandem leaf springs, are outlined in this section.

REAR SPRING

Removal

1. Place floor jack under truck frame and raise truck sufficiently to relieve weight from spring to be removed.
2. Remove shock absorbers where used.
3. Remove U-bolts, spring bumper and retainer or U-bolt seat.
4. Remove lubricators (not used where springs are equipped with rubber bushings).
5. Remove nuts from spring shackle pins or bracket pins.
6. Slide ring off bracket pin and shackle pin.
7. If spring is rubber bushed, bushing halves may be removed from each side of spring and shackle eye.

SUSPENSION
Light Trucks & Vans

Installation

1. Install pivot end of spring first. Align shackle end to other frame bracket. When installing nuts on spring pins which are welded or pressed in, be sure that washer is tightened against shoulder of pin. Spring pins which are driven in must be installed so that slot for lock bolt is aligned. Spring pins which are threaded in must be installed so that the lubrication hole is facing up. Tighten pin into bracket, then back off one-half turn. Install locknut tightly and install cotter pin. Turn pin out to permit installation of cotter pin.
2. Install lubricators.
3. Install U-bolt seat or retainer and U-bolts. Install U-bolt nuts, but do not tighten.
4. Install shock absorber where used.
5. Lower vehicle.
6. Tighten U-bolt nuts securely.

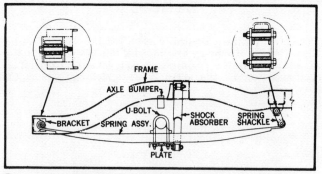

Rear spring installation diagram light and medium duty vehicles
(© International Harvester Co.)

Air Suspension

The air suspension system was developed to improve the ride characteristics of highway transport vehicles. The major components, whether used on single or tandem axles, are as follows: trailing arms, frame hanger brackets, shock absorbers, track bars, air springs, axle connections, air leveling valve.

SPRING HEIGHT

Adjustment

1. Start engine, and wait until the air pressure indicates near maximum pressure of 85–90 lbs.
2. Measure the distance from the top of the trailing arm to the bottom of the frame at the rear of the air spring on one side only.
3. If the distance is greater than $12\frac{1}{4}$ in., ($\pm\frac{3}{8}$ inch), shorten the linkage from the height control valve arm to the axle, wait 15 seconds and remeasure. Repeat the procedure if necessary.
4. If the distance is less than $12\frac{1}{4}$ in., ($\pm\frac{3}{8}$), lengthen the linkage from the height control valve arm to the axle, wait 15 seconds and remeasure. Repeat the procedure if necessary.
5. Repeat the adjustment procedure on the opposite side of vehicle.

AIR SPRING PRESSURE BALANCE

Air spring pressure imbalance on an unloaded IH air suspension system is a normal condition and causes no harm to the suspension components so long as chassis remains unloaded. If, however, the imbalance continues after the chassis is loaded, do not operate the vehicle until the cause is located and corrected. These causes can be incorrect spring height adjustment, air leaks, plugged or pinched air lines, defective leveling control valve, or loose or broken parts.

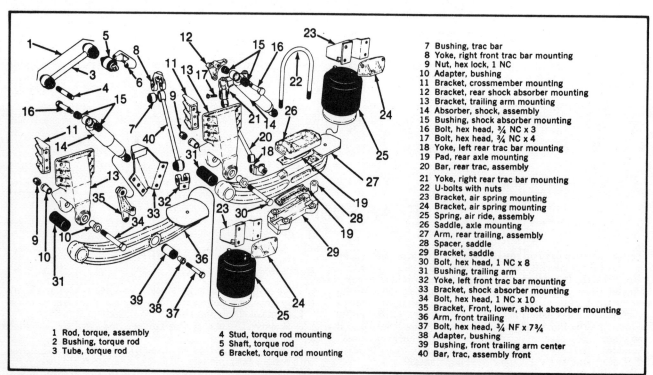

Exploded view of air suspension—typical (© International Harvester Co.)

SUSPENSION
Light Trucks & Vans

AIR SPRING

Removal and Installation

1. Block the vehicle wheels and release the air pressure build-up from the air brake system.
2. Support the vehicle in the raised position to relieve any pressure on the air spring.
3. Remove the front or rear trailing arm, depending upon the air spring affected, by removing the attaching bolts at the front and center of the arm.
4. Lower the arm and the air spring away from the axle.
5. Installation is the reverse of removal.

Equalizing Beam Suspension (Hendrickson)

Tandem drive axles require a special suspension which permits flexibility between the axles, the equalizing beam suspension. Semi-elliptic springs are used mounted on saddle assemblies above the equalizer beams and pivoted at the front end on spring pins and brackets. The rear end of the springs have no rigid attachment to the spring brackets, but are free to move forward and backward to compensate for spring deflection.

NOTE: As options, airspring assemblies and four point rubber mounted suspensions are available in place of the leaf spring type.

There are two approaches to servicing the suspension system. One is the removal and installation of individual parts. Removal and overhaul of the entire unit can also be done.

CAUTION
When complete removal is performed, be careful when disconnecting the torque rods, springs, or rubber cushions from the frame since the axle assemblies will be free to roll or pivot at the equalizer beam ends. Use jacks and other equipment and block the vehicle securely to prevent injury to personnel and damage to the unit.

Four Spring Suspension (Dayton)

The four spring suspension system is used to distribute the load over a greater area of the frame rail. The six torque rods of the Dayton four spring suspension serve a dual purpose. The rods provide a means of suspension alignment as well as permiting the axles to accept complete drive line torque. The torque rods consist of two non-adjustable and four adjustable units.

The removal and installation of parts can be accomplished by the removal of the individual parts or the removal of the complete unit, as with the Equalizing Beam Suspension.

AXLE ALIGNMENT

1. Clamp a straight edge to the top of the frame rail ahead of the forward rear axle. Use a framing square against the straight-edge and the outside surface of the frame siderail to insure the straightedge is perpendicular to the frame.
2. Suspend a plumb bob from the straightedge in front of the tire and on the outboard side of the forward rear axle.
3. Position a bar with pointers that can be engaged in the center holes of the rear axles.
4. Measure the distance between cord of the plumb bob and the pointer on the forward axle and record (dimension A).
5. Position the plumb bob and bar on the opposite side of the vehicle and measure as outlined in paragraph 4. Record the result.
6. Any difference in dimensions from side to side must be equalized if the difference exceeds 0.0625 inch.
7. Equalize the dimensions by loosening the clamp bolts on the lower adjustable torque rod on the forward rear axle and adjusting the length of the torque rod. Tighten the clamp bolts.

NOTE: Remove one end of the left and right upper torque rods on the forward rear axle to relieve any stresses which may be present due to an improperly adjusted torque rod, before adjusting the lower torque rods.

8. Reposition the bar pointers to the axle centers on each side. If any differences exist in the center to center measurement, (dimension B), after the forward rear axle has been squared to the frame, the rear rear axle must also be aligned.
9. To align the rear axle, loosen the clamp bolts on the lower adjustable torque rod and adjust to equalize the center to center distance between the axle ends. Tighten the clamp bolts.
10. Reinstall the upper torque rod ends that were removed in step 7. Tighten the mounting bolts.

AXLE LOAD DISTRIBUTION

The Dayton four spring suspension provides for equal load distribution through the adjustment of the upper torque rod lengths. To adjust, follow this procedure.

1. Disconnect the forward and rear upper torque rods at the frame crossmembers.
2. If the vehicle is equipped with an adjustable fifth wheel, position it in the normal operating location.
3. Apply the maximum rated load on the suspension assembly to obtain the full deflection of the leaf springs when adjusting the torque rods.
4. To settle the suspension to normal operating position, move the vehicle to a level area and bring the vehicle to an easy stop, using the trailer brake, if equipped. Keep the vehicle in a straight ahead position.
5. Loosen the torque clamp bolts and lengthen or shorten the torque rods as required to obtain bolt hole alignment for easy installation of the bolts in the torque rod ends.
6. Tighten the mounting bolts and the torque rod clamp bolts.
7. No further adjustment should be required.

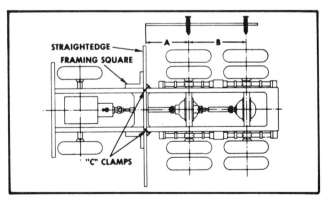

Measurement points of tandem axles for alignment check
(© International Harvester Co.)

JEEP® VEHICLES
CJ SERIES • WAGONEER • CHEROKEE • COMMANCHE

FRONT AXLE AND SUSPENSION

Front Axle Assembly

Removal and Installation

EXCEPT 1984 AND LATER WAGONEER, CHEROKEE AND COMANCHE

1. Raise and support the vehicle safely. Remove the wheel covers and wheels.
2. Index the propeller shaft to the differential yoke for the proper alignment upon installation. Disconnect the propeller shaft at the axle yoke and secure the shaft to the frame rail.
3. Disconnect the steering linkage from the steering knuckles. Disconnect the shock absorbers at the axle housing.
4. If the vehicle is equipped with a stabilizer bar, remove the nuts attaching the stabilizer bar connecting links to the spring tie plates.
5. On vehicles equipped with sway bar, remove nuts attaching sway bar connecting links to spring tie plates.
6. Disconnect the breather tube from the axle housing. Disconnect the stabilizer bar link bolts at the spring clips.
7. Remove the brake calipers, hub and rotor, and the brake shield.
8. Remove the U-bolts and the tie plates.
9. Support the assembly on a jack and loosen the nuts securing the rear shackles, but do not remove the bolts.
10. Remove the front spring shackle bolts. Lower the springs to the floor.
11. Pull the jack and axle housing from underneath the vehicle.
12. Installation is the reverse of the removal procedure. Check the front end alignment, as required.

1984 AND LATER WAGONEER, CHEROKEE AND COMANCHE

1. Raise and support the vehicle safely.
2. Remove the wheels, calipers and rotors.
3. Disconnect all vacuum hoses at the axle.
4. Mark the relation between the front driveshaft and yoke.
5. Disconnect the stabilizer bar, rod and center link, front driveshaft, shock absorbers, steering damper, track bar.
6. Place a floor jack under the axle to take up the weight.
7. Disconnect the upper and lower control arms at the axle and lower the axle from the truck.
8. Installation is the reverse of the removal procedure.

NOTE: Discard the U-joint straps new replacement straps must be used whenever the straps are removed.

Shock Absorbers

The upper ends of the shock absorbers are attached to the frame side rails with mounting brackets and pins. The lower ends are attached to the axle or to the spring by mounting.

Front Leaf Spring

Removal and Installation

1. Raise the vehicle and support it safely.
2. Position a jack under the axle. Raise the axle to relieve the springs of the axle weight.
3. If equipped, disconnect the stabilizer bar. Remove the spring U-bolts and tie plates.
4. Remove the bolt attaching the spring front eye to the shackle.
5. Remove the bolt attaching the spring rear eye to the shackle.
6. Remove the spring from its mounting.
7. Installation is the reverse of the removal procedure.

Front Coil Spring

Removal and Installation

1. Raise the vehicle and support it safely.
2. Remove the wheels.
3. Match-mark and disconnect the front driveshaft.
4. Disconnect the lower control arm at the axle.
5. Disconnect the track bar at the frame. Place a floor jack under the axle.
6. Disconnect the stabilizer bar and shock absorbers.
7. Disconnect the center link at the pitman arm.
8. Lower the axle with a floor jack, loosen the spring retainer and remove the spring.
9. Installation is the reverse of removal.

Selective Drive Hubs

Front drive hubs are serviced as either a complete assembly or a sub assembly, such as the hub body or hub clutch assembly. Do not attempt to disassemble these units. If the entire hub assembly or subassembly has to be replaced it must be replaced as a complete unit.

CJ MODELS

Removal and Installation

1. Remove the bolts and the tabbed lockwashers that attach the hub body to the axle hub. Save the bolts and the washer.
2. Remove the hub body and gasket. Discard the gasket.
3. Do not turn the hub control dial once the hub body has been removed.
4. Remove the retaining ring from the axle shaft. Remove the hub clutch and bearing assembly.
5. Clean and inspect the components for wear and damage. Replace defective components as required.
6. Installation is the reverse of the removal procedure.
7. Turn the control hub dials to the 4X2 position and rotate the wheels. They should rotate freely, if not check the hub installation. Be sure that the controls are in the fully engaged position.

CHEROKEE, GRAND WAGONEER AND TRUCK

Removal And Installation

1. Remove the socket head screws from the hub body assembly.

SUSPENSION
Light Trucks & Vans

2. Remove the large retaining ring from the axle hub. Remove the small retaining ring from the axle shaft.
3. Remove the hub and clutch assembly.
4. Clean and inspect the components for wear and damage. Replace defective components as required.
5. Installation is the reverse of the removal procedure.
6. Turn both controls to the FREE position and rotate the wheels. They must rotate freely, if they drag check the hub installation. Be sure that the control dials are in the fully engaged position.

Axle Shaft

Removal and Installation

CJ AND SCRAMBLER

1. Raise and support the vehicle safely. Remove the tire and wheel. Remove the disc brake caliper.
2. Remove the bolts attaching the front hub to the axle and remove the hub body and gasket.
3. Remove the retaining ring from the axle shaft. Remove the hub clutch and bearing assembly from the axle.
4. Straighten the lip of the lock washer. Remove the outer lock nut, lock washer, inner locknut and tabbed washer. Use tool J-25103, or equivalent, in order to remove the locknut.
5. Remove the outer bearing and remove the disc brake rotor. Remove the disc brake caliper adapter and splash shield. Remove the axle spindle.

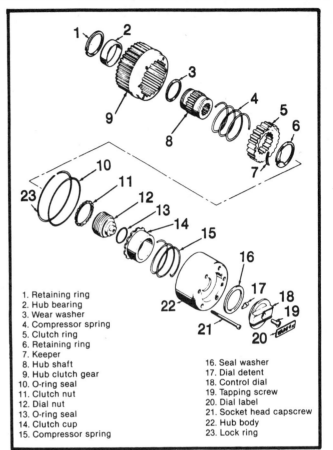

1. Retaining ring
2. Hub bearing
3. Wear washer
4. Compressor spring
5. Clutch ring
6. Retaining ring
7. Keeper
8. Hub shaft
9. Hub clutch gear
10. O-ring seal
11. Clutch nut
12. Dial nut
13. O-ring seal
14. Clutch cup
15. Compressor spring
16. Seal washer
17. Dial detent
18. Control dial
19. Tapping screw
20. Dial label
21. Socket head capscrew
22. Hub body
23. Lock ring

Front drive hubs—Grand Wagoneer and Truck (© Jeep Corp.)

6. Remove the axle shaft and universal joint assembly.
7. Installation is the reverse of the removal procedure.

CHEROKEE, GRAND WAGONEER AND TRUCK

1. Raise and support the vehicle safely. Remove the disc brake caliper.
2. On vehicles without front hubs, remove the rotor hub cap. Remove the axle shaft snap ring, drive gear, pressure spring and spring retainer.
3. On models with front hubs, remove the socket head screws from the hub body. Remove the hub body and the large retaining ring. Remove the small retaining ring from the axle shaft. Remove the hub clutch assembly from the axle.
4. Remove the outer locknut, washer and inner locknut using tool J-6893-03 or equivalent.
5. Remove the rotor. The spring retainer and the outer bearing are removed with the rotor.
6. Remove the nuts and bolts attaching the spindle and support shield. Remove the spindle and the support shield.
7. Remove the axle shaft.
8. Installation is the reverse of the removal procedure.

1984 AND LATER WAGONEER, CHEROKEE AND COMANCHE

1. Raise and support the vehicle safely.
2. Remove the wheels, calipers and rotors.
3. Remove the cotter pin, locknut and axle hub nut.
4. Remove the hub-to-knuckle attaching bolts.
5. Remove the hub and splash shield from the steering knuckle.

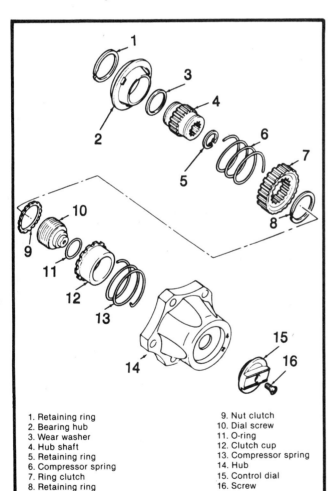

1. Retaining ring
2. Bearing hub
3. Wear washer
4. Hub shaft
5. Retaining ring
6. Compressor spring
7. Ring clutch
8. Retaining ring
9. Nut clutch
10. Dial screw
11. O-ring
12. Clutch cup
13. Compressor spring
14. Hub
15. Control dial
16. Screw

Front drive hubs—CJ and Scrambler (© Jeep Corp.)

2-359

SUSPENSION
Light Trucks & Vans

To remove the left shaft:
6. Remove the axle shaft from the housing.

To remove the right shaft:
7. Disconnect the vacuum harness from the shift motor.
8. Remove the shift motor from the housing.
9. Remove the axle shaft from the housing.
10. To install the right axle shaft first be sure that the shift collar is in position on the intermediate shaft and that the axle shaft is fully engaged in the intermediate shaft end.
11. Install the shift motor, making sure that the fork engages with the collar. Tighten the bolts to 8 ft. lb.
12. On the left side, install the axle shaft in the housing.
13. Partially fill the hub cavity of the knuckle with chassis lube and install the hub and splash shield.
14. Tighten the hub bolts to 75 ft. lb.
15. Install the hub washer and nut. Torque the nut to 175 ft. lb. Install the locknut. Install a new cotter pin.
16. Attach the rotor, caliper and wheel.

Steering Knuckle Service

STEERING KNUCKLE

Removal and Installation

EXCEPT 1984 AND LATER CHEROKEE, WAGONEER AND COMANCHE

1. Remove the axle assembly from the vehicle.
2. Disconnect the tie rod end at the steering knuckle arm.
3. Remove and discard the lower ball stud jamnut.
4. Remove the cotter pin from the upper ball stud. Loosen the stud nut until the top edge of the nut is flush with the top of the stud.
5. Unseat the upper and lower ball studs using a hammer. Remove the upper ball stud nut and the steering knuckle.
6. Remove the upper ball stud split ring seat using tool J-23447 or tool J-25158.
7. Installation is the reverse of the removal procedure.

1984 WAGONEER, CHEROKEE AND COMANCHE

1. Remove the outer axle shaft.
2. Remove the caliper anchor plate from the knuckle.
3. Remove the knuckle-to-ball joint cotter pins and nuts.
4. Drive the knuckle out with a brass hammer.

NOTE: A split ring seat (3) is located in the bottom of the knuckle. During installation, this ring seat must be set to a depth of 5.23mm (.206). Measure the depth to the top of the ring seat (4).

5. Installation is the reverse of removal. Tighten the knckle retaining nuts to 75 ft. lb. and the caliper anchor bolts to 77 ft. lb.

BALL JOINTS

CJ, SCRAMBLER, CHEROKEE, GRAND WAGONEER AND TRUCK

Removal and Installation

LOWER BALL STUD

1. Remove the steering knuckle from the vehicle. Position the assembly in a suitable holding fixture with the upper ball stud pointing downward.
2. Attach tool J-2511-1, or equivalent to the spindle mating surface of the knuckle assembly. Position tool J-25211-3 on the lower ball stud.
3. Assemble and install the puller on the steering knuckle. Hook one arm of the puller in the plate of tool J-25211-1 and the opposite arm of the tool in the steering knuckle.

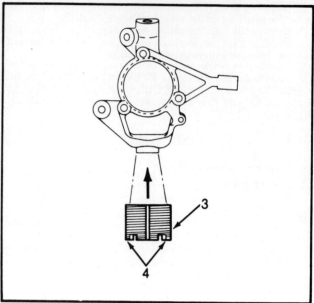

Split retaining ring (© Jeep Corp)

4. Tighten the puller screw to press the lower stud out of the knuckle.
5. Remove the tools from the knuckle.
6. Installation is the reverse of the removal procedure.

UPPER BALL STUD – CJ AND SCRAMBLER

1. Remove the steering knuckle from the vehicle. Position the assembly in a suitable holding fixture with the upper ball stud pointing downward.
2. Remove both arms from tool J-25215. Place button J-25211-3 on the upper ball stud.
3. Install adapter tool J-25211-4 on the nut end of the puller screw so that the adapter shoulder faces the nut end of the screw.
4. Insert the nut end of the puller screw through the upper ball stud hole in the knuckle. Hold the adapter and the frame against the knuckle.
5. Remove the lower ball stud from the knuckle.
6. Installation is the reverse of the removal procedure.

UPPER BALL STUD – CHEROKEE, GRAND WAGONEER AND TRUCK

1. Remove the steering knuckle from the vehicle. Position the assembly in a suitable holding fixture with the upper ball stud pointing downward.
2. Remove both arms from tool J-25215. Place button J-25211-3 on the upper ball stud.
3. Thread the puller frame halfway onto the puller screw. Insert the nut end of the screw through the lower ball stud hole in the steering knuckle. Position the puller frame against the knuckle and the puller screw against tool J-25211-3.
4. Tighten the puller screw and press the upper ball stud out of the steering knuckle.
5. Installation is the reverse of the removal procedure.

1984 AND LATER WAGONEER, CHEROKEE AND COMANCHE

Removal and Installation

UPPER BALL JOINT

1. Remove the steering knuckle from the vehicle.
2. Position the receiver tool (1) over the top of the ball joint.

SUSPENSION
Light Trucks & Vans

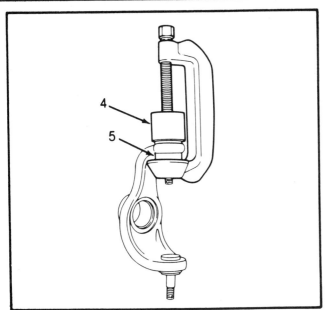

Upper ball joint removal (© Jeep Corp)

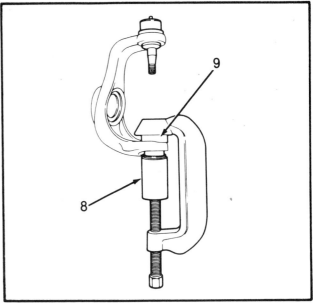

Lower ball joint installation (© Jeep Corp)

Set the adapter tool (2) in a C-clamp (3) and position the clamp so that tightening the clamp screw will remove the ball joint.

3. To install, use the same C-clamp, with adapter tool as illustrated.

LOWER BALL JOINT

1. With the knuckle removed from the vehicle, use a Cclamp, with receiver tool (6) and adapter tool (7) as illustrated, to force out the ball joint.
2. To install, use the same C-clamp, with adapter tools (8 and 9), as illustrated.

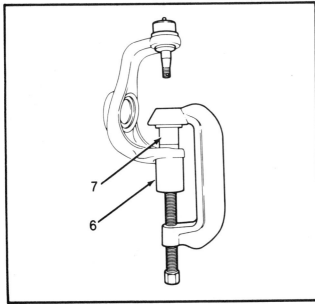

Lower ball joint removal (© Jeep Corp)

Upper Ball Joint Adjustment

Adjustment of the upper ball joint is necessary only when there is excessive play in the steering, persistent loosening of the steering linkage, or abnormal wear of the tires.

Adjustment Procedure

EXCEPT 1984 AND LATER WAGONEER, CHEROKEE AND COMANCHE

1. Raise and support the vehicle safely. Remove the front tires.
2. If the vehicle is equipped with a steering damper, disconnect it at the tie rod and move it aside.
3. Unlock the steering column. Disconnect the steering connecting rod. Disconnect the connecting rod at the right side of the tie rod.
4. Remove the cotter pin and the retaining nut attaching the tie rod to the right side steering knuckle.
5. Rotate both steering knuckles through a complete arc several times. Work from the right side of the vehicle when rotating the knuckles.
6. Install a torque wrench on the tie rod retaining nut and check the torque. The torque wrench must be positioned at a ninety degree angle to the steering knuckle arm in order to obtain a correct reading.
7. Rotate the steering knuckles slowly through a complete arc and measure the torque required to rotate the knuckles.
8. If the reading is less than 25 ft. lbs. turning effort is within specification. If not then procede as follows.
9. Disconnect the tie rod from both steering knuckles. Install a one half by one inch bolt, flat washer and nut in the tie rod stud mounting hole in one of the steering knuckles. Tighten the bolt and nut.
10. Install the torque wrench according to Step six.
11 Rotate the steering knuckles slowly through a complete arc and measure the torque required to rotate the knuckles.
12 If the reading is less than 10 ft. lbs. turning effort is within specification and the defect is not related to the knuckle ball studs.
13. Install components that were removed to accomplish this procedure.
14. Lower the vehicle.

2-361

FRONT WHEEL BEARING

Adjustment

CJ AND SCRAMBLER

1. Raise and support the vehicle safely.
2. Remove the bolts attaching the front hub to the hub rotor. Remove the hub body and gasket.
3. Remove the snap ring from the axle shaft and remove the hub clutch assembly.
4. Straighten the lip of the outer lock nut tabbed washer. Remove the outer locknut and tabbed washer.
5. Loosen and then tighten the inner locknut to 50 ft. lbs. Rotate the wheel while tightening the nut to seat the bearing properly.
6. Back off the inner locknut about one-sixth of a turn as you are turning the wheel. The wheel must rotate freely.
7. Install the tabbed washer and the outer locknut.
8. Torque the outer locknut to 50 ft. lbs.
9. Recheck the bearing adjustment. The wheel must rotate freely. Correct as required.
10. Install components as they were removed.

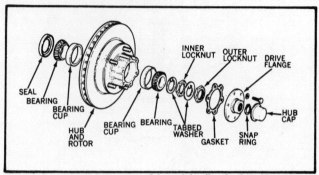

Front wheel attaching parts—CJ model—disc brakes (© Jeep Corp)

CHEROKEE, WAGONEER, AND TRUCK

1. Raise and support the vehicle safely.
2. If the vehicle is not equipped with front hubs, Remove the wheel cover and hubcap. Remove the drive gear snap ring. Remove the drive gear, pressure spring and spring cup.
3. If the vehicle is equipped with front hubs, Remove the socket head screws from the hub body and remove the body from the hub clutch assembly. Remove the large retaining ring from the hub. Remove the small retaining ring from the axle shaft. Remove the hub and clutch assembly. Remove the outer locknut and lock washer.
4. Seat the bearings by loosening and then tightening them to 50 ft. lbs. Back off the inner locknut about one-sixth of a turn while rotating the wheel.
5. Install the lock washer. Align one of the lock washer holes with the peg on the inner locknut and install the washer on the nut.
6. Install and torque the outer locknut to 50 ft. lbs. Recheck the bearing adjustment. Correct as required.
7. On vehicles without front hubs, The spring cup must be installed so the recessed side faces the bearing and the flat side faces the pressure spring. The pressure spring should contact the flat side of the cup only.
8. Install the spring cup and the pressure spring. Install the drive gear and the drive gear snap ring.
9. If the vehicle is equipped with front hubs, Install the clutch assembly. Install the small retaining ring on the axle shaft. Install the large retaining ring on the hub.
10. Install the hub body on the hub clutch. Install the socket head screws in the hub and torque them to 30 inch lbs.
11. Lower the vehicle.

1984 CJ-7, Scrambler, Wagoneer and Cherokee front wheel bearing (© Jeep Corp)

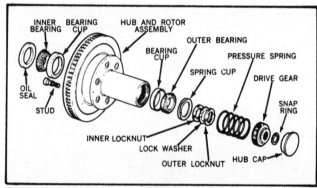

Front wheel attaching parts—Cherokee and Wagoneer (© Jeep Corp)

1984 AND LATER CHEROKEE, WAGONEER AND COMANCHE—TYPE ONE

1. Raise and support the vehicle safely. Remove the wheel. Remove the disc brake caliper as required. Remove the cotter pin, locknut and axle hub nut.
2. Tighten the hub bolts to 75 ft. lbs.
3. Install the hub washer and nut and tighten the hub nut to 175 ft. lbs. Install the locknut and new cotter pins.
4. Install the caliper, if removed. Install the wheel. Lower the vehicle.

1984 and LATER CHEROKEE, WAGONEER AND COMANCHE—TYPE TWO

1. Raise and support the vehicle safely. Remove the wheel. Remove the caliper, as required.
2. Remove the grease cap, cotter pin and nut retainer.
3. Tighten the spindle nut to 25 ft. lbs. while rotating the rotor.
4. Loosen the spindle nut about one half turn while rotating the wheel. Torque the nut to 19 ft. lbs.
5. Install removed components.

SUSPENSION
Light Trucks & Vans

REAR SUSPENSION

Spring

Removal and Installation
MOUNTED BELOW THE AXLE

1. Raise the vehicle and support the axle.
2. Disconnect the shock absorber and stabilizer bar, if so equipped.
3. Remove the U-bolts and tie plates.
4. Disconnect the front and rear ends of the spring and remove the spring.
5. The spring can be disassembled by removing the spring rebound clips and the center bolt.
6. Installation is the reverse of the removal procedure.

MOUNTED ABOVE THE AXLE

1. Raise the vehicle and support the frame ahead of the axle.
2. If the left side spring is being removed, remove the fuel tank skid plate.
3. Remove the wheel.
4. Disconnect the shock absorber.
5. Remove the tie plate U-bolts and the tie plate. Remove the bolt attaching the spring rear eye to the spring shackle.
6. Remove the bolt attaching the spring front eye to the spring hanger on the frame rail.
7. Remove the spring from the vehicle.
8. Installation is the reverse of the removal procedure.

1984 AND LATER WAGONEER, CHEROKEE AND COMANCHE

1. Raise and support the vehicle safely.
2. Raise the axle assembly to relieve the weight.
3. Remove the wheel. Remove the shock absorber at the axle.
4. If equipped, disconnect the stabilizer bar links and the spring tie plate.
5. Remove the spring tie plate U-bolt and the spring tie plate. Remove the rear eye to spring shackle bolt and the front eye to bracket bolt.
6. Lower the spring from the vehicle.
7. Installation is the reverse of the removal procedure.

Shock Absorber

Removal and Installation

1. Raise the vehicle for working clearance and support safely.
2. Place a jack under the axle assembly and raise to relieve the springs of axle weight and to place the shock absorber in its mid stroke.
3. Remove the retaining nuts or bolts and remove the shock absorber from the vehicle.
4. Install the new shock absorber and tighten the attaching nuts or bolts.
5. Lower the vehicle to the ground.

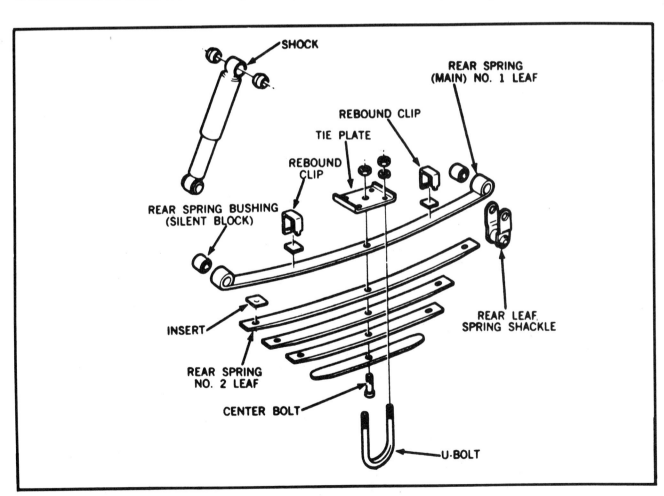

SECTION 2
SUSPENSION
Light Trucks & Vans

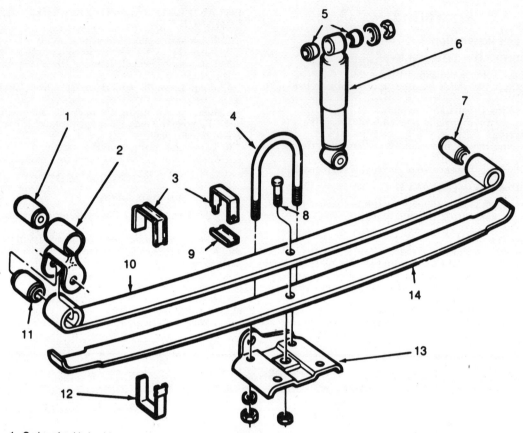

1. Spring shackle bushing
2. Spring shackle
3. Rebound clips
4. U-bolt
5. Bushings
6. Shock absorber
7. Spring eye bushing
8. Center bolt
9. Insulator
10. Main leaf
11. Spring eye bushing
12. Rebound clip
13. Tie plate
14. Tapered leaf

Front suspension used on 1979–83 Wagoneer and Cherokee, 1979–86 J-Series Truck, and 1984–86 Grand Wagoneer. This suspension was also, an option on 1978 models

Suspension Specifications

INDEX

DOMESTIC CARS

American Motors	2-366
Checker Motors	2-366
Chrysler Corporation	
Chrysler	2-366
Dodge	2-367
Plymouth	2-368
Ford Motor Company	
Ford	2-369
Lincoln	2-371
Mercury	2-371
General Motors	
Buick	2-373
Cadillac	2-374
Chevrolet	2-374
Oldsmobile	2-376
Pontiac	2-377

DOMESTIC LIGHT TRUCKS

American Motors/jeep	2-378
Chevrolet	2-378
Dodge	2-381
Ford	2-382
GMC	2-380
International Harvester	2-389
Plymouth	2-381

IMPORT CARS & TRUCKS

Alfa Romeo	2-389
Audi	2-390
BMW	2-390
Buick Opel	2-391
Capri	2-291
Chevrolet	2-391
Chevrolet Luv	2-391
Datsun/Nissan	2-392
Dodge	2-393
Fiat	2-394
Ford Fiesta & Courier	2-394
Honda	2-394
Isuzu	2-396
Jaguar	2-396
Lancia	2-396
Mazda	2-397
Mercedes-Bnz	2-398
MG	2-398
Mitsubishi	2-398
Peugeot	2-398
Plymouth	2-398
Porsche	2-399
Renault	2-400
SAAB	2-400
Subaru	2-400
Toyota	2-402
Triumph	2-404
Volkswagen	2-405
Volvo	2-405
Caster/Camber Refresher Guide	2-409
Adjustment Illustrations Group I	2-406
Adjustment Illustrations Group II	2-407
Adjustment Illustrations Group III	2-408

SPECIAL CANADIAN VEHICLES

Acadian (See Chevrolet Chevette)	2-374
Caravelle	
'84 FWD (See '84 Chrysler LeBaran)	2-367
'84-'86 RWD (See '84-'86 Plymouth Gran Fury)	2-368
'82-'83 (See Plymouth Gran Fury)	2-368
'81 & before (See Chrysler LeBaron)	2-367
Laurentian (See Pontiac Catalina)	2-377
Parisienne '82 & before (See Pontiac Catalina)	2-377

ABBREVIATIONS

ADJ. Ill. No.	Adjustment Illustration Number	NA	Not Available
FWD	Front Wheel Drive	N/A	Not Adjustable
MAX	Maximum	PREF	Preferred
MIN	Minimum	RWD	Rear Wheel Drive

Although information in this manual is based on industry sources and is as complete as possible at the time of publication, the possibility exists that some car manufacturers made later changes which could not be included here. While striving for total accuracy, Chilton Datalog can not assume responsibility for any errors, changes or omissions that may occur in the compilation of this data.

SECTION 2
SUSPENSION
Specifications

MANUFACTURER'S SPECIFIED ALIGNMENT TOLERANCES
U.S. PASSENGER CAR SECTION

All vehicles should be set to the preferred (PREF.) specification when being realigned. The minimum and maximum settings specified are a guide to use when checking alignment. The manufacturers consider alignment within these tolerances acceptable for safe vehicle operation while still limiting abnormal tire wear.

VEHICLE IDENTIFICATION (YEAR MODEL)	ADJ. ILL. NO.	CASTER (Degrees) MIN.	PREF.	MAX.	CAMBER (Degrees) MIN.	PREF.	MAX.	TOE-IN (Inches) MIN.	PREF.	MAX.	TOE-IN (Degrees) MIN.	PREF.	MAX.	TOE-OUT ON TURNS (Degrees) OUTSIDE WHEEL	INSIDE WHEEL	STRG. AXIS INCL. (DEG.)
RESET VEHICLE ALIGNMENT AS CLOSE TO PREFERRED SETTINGS AS POSSIBLE FOR BEST RESULTS.															
AMERICAN MOTORS • Wheels at Full Turning Angle																
83-82 Spirit, Concord	(1)(2)	3½	4½	5				1/16	1/8	3/16	1/8	1/4	3/8	NA	38•	7¾
Left Wheel					1/8	3/8	3/4									
Right Wheel					-1/8	1/8	1/2									
81-80 Spirit, Concord, AMX	(1)(2)	0	1	2½				1/16	1/8	3/16	1/8	1/4	3/8	NA	38•	7¾
Left Wheel					1/8	3/8	3/4									
Right Wheel					-1/8	1/8	1/2									
86-83 Eagle	(1)(2)	2	2½	3½	-1/8	3/8	5/8	1/16	1/8	3/16	1/8	1/4	3/8	NA	38•	11 27/32
82 Eagle	(1)(2)	3	4	5	-1/8	3/8	5/8	1/16	1/8	3/16	1/8	1/4	3/8	NA	38•	11 27/32
81 Eagle	(1)(2)	2	2½	3	-1/8	3/8	5/8									
w/Select Drive								1/16	1/8	3/16	1/8	1/4	3/8	NA	38•	11 27/32
w/o Select Drive								1/16 (out)	1/8 (out)	3/16 (out)	1/8 (out)	1/4 (out)	3/8 (out)	NA	38•	11 27/32
80 Eagle	(1)(2)	2	2½	3	-1/8	3/8	5/8	1/16 (out)	1/8 (out)	3/16 (out)	1/8 (out)	1/4 (out)	3/8 (out)	NA	38•	11½
80-79 Pacer	(3)	1	2	3½	0	1/4	3/4	1/16	1/8	3/16	1/8	1/4	3/8	NA	35•	7¾
79 AMX, Concord, Spirit	(1)(2)	0	1	2½	0	1/4	3/4	1/16	1/8	3/16	1/8	1/4	3/8	NA	35•	7¾
78 Matador, AMX, Concord, Gremlin	(1)(2)	0	1	2				1/16	1/8	3/16	1/8	1/4	3/8	22	25	7¾
Left Wheel					1/8	3/8	5/8									
Right Wheel					0	1/8	1/2									
78 Pacer	(3)	1	2	3				1/16	1/8	3/16	1/8	1/4	3/8	22	25	7¾
Left Wheel					1/8	3/8	5/8									
Right Wheel					0	1/8	1/2									
77 Gremlin, Hornet	(1)(2)	-½	0	½				1/16	1/8	3/16	1/8	1/4	3/8	22	25	7¾
Left Wheel					1/8	3/8	5/8									
Right Wheel					0	1/8	1/2									
77 Pacer	(3)	½	1	1½				1/16	1/8	3/16	1/8	1/4	3/8	22	25	7¾
Left Wheel					1/8	3/8	5/8									
Right Wheel					0	1/8	1/2									
77 Matador, Javelin, Ambassador, AMX	(1)(2)	½	1	1½				1/16	1/8	3/16	1/8	1/4	3/8	22	25	7¾
Left Wheel					1/8	3/8	5/8									
Right Wheel					0	1/8	1/2									
CHECKER MOTORS CORP.																
82-81 All Models	(24)	1½	2	2½	1/4	1/2	3/4	1/16	3/32	1/8	1/8	3/16	1/4	17½	20	7
80-77 All Models	(24)	1½	2	2½	1/2	1	1½	1/16	3/32	1/8	1/8	3/16	1/4	17½	20	7
CHRYSLER CORP.																

(A) Check vehicle suspension height before performing alignment.
(B) Maximum left to right variation in caster not to exceed 1¼ ° when checking alignment. Caster is not adjustable on FWD models.
(C) The engine must be running during toe adjustment of vehicles with power steering.

VEHICLE IDENTIFICATION	ADJ. ILL. NO.	CASTER MIN.	PREF.	MAX.	CAMBER MIN.	PREF.	MAX.	TOE-IN (In.) MIN.	PREF.	MAX.	TOE-IN (Deg.) MIN.	PREF.	MAX.	TOE-OUT OUTSIDE	INSIDE	STRG. AXIS
CHRYSLER (A) (B) (C)																
86-83 Newport, New Yorker 5th Ave. (RWD)	(7)	1¼	2½	3¾	-1/4	1/2	1¼	0	1/8	5/16	0	1/4	5/8	18	20	8
82 New Yorker	(7)	1¼	2½	3¾	-1/4	1/2	1¼	0	1/8	5/16	0	1/4	5/8	18	20	8
81-80 Newport, New Yorker	(7)	-1/4	1	2¼	-1/4	1/2	1¼	0	1/8	5/16	0	1/4	5/8	18	20	8
79 Newport, New Yorker	(7)	-½	3/4	2				1/16	1/8	1/4	1/8	1/4	1/2	18	20	8
Left Wheel					0	1/2	1									
Right Wheel					-1/4	1/4	3/4									
78-77 Newport, New Yorker	(7)	-½	3/4	2				1/16	1/8	1/4	1/8	1/4	1/2	18 5/16	20	9
Left Wheel					0	1/2	1									
Right Wheel					-1/4	1/4	3/4									
83-80 Cordoba	(7)	1¼	2½	3¾	-1/4	1/2	1¼	0	1/8	5/16	0	1/4	5/8	18	20	8
79-77 Cordoba	(7)	-½	3/4	2				1/16	1/8	1/4	1/8	1/4	1/2	18	20	8
Left Wheel					0	1/2	1									
Right Wheel					-1/4	1/4	3/4									

SUSPENSION
Specifications
SECTION 2

U.S. PASSENGER CAR SECTION

VEHICLE IDENTIFICATION YEAR MODEL	ADJ. ILL. NO.	CASTER (Degrees) MIN.	PREF.	MAX.	CAMBER (Degrees) MIN.	PREF.	MAX.	TOE-IN (Inches) MIN.	PREF.	MAX.	TOE-IN (Degrees) MIN.	PREF.	MAX.	TOE-OUT ON TURNS (Degrees) OUTSIDE WHEEL	INSIDE WHEEL	STRG. AXIS INCL. (DEG.)
86 LeBaron, GTS, New Yorker, Laser (FWD)																
Front	(34)		13/16 ■		− 1/4	5/16	3/4	1/8 (out)	1/16 (in)	7/32 (in)	1/4 (out)	1/8 (in)	7/16 (in)	NA	NA	13 5/16
Rear	(48)				− 1 1/4	1/2	1/4	5/16 (out)	0	5/16 (in)	5/8 (out)	0	5/8 (in)			
85-82 LeBaron (FWD), GTS, E-Class, New Yorker, Laser (FWD)																
Front	(4) ★		13/16 ■		− 1/4	5/16	3/4	1/8 (out)	1/16 (in)	7/32 (in)	1/4 (out)	1/8 (in)	7/16 (in)	NA	NA	13 5/16
Rear	(48)				− 1	− 1/2	0	3/16 (out)	0	3/16 (in)	3/8 (out)	0	3/8 (in)			
81-80 LeBaron	(7)	1 1/4	2 1/2	3 3/4	− 1/4	1/2	1 1/4	0	1/8	5/16	0	1/4	5/8	18	20	8
79-77 LeBaron	(7)	1 1/2	2 1/2	3 3/4				1/16	1/8	1/4	1/8	1/4	1/2	18	20	8
Left Wheel					0	1/2	1									
Right Wheel					− 1/4	1/4	3/4									
83-81 Imperial	(7)	1 1/4	2 1/2	3 3/4	− 1/4	1/2	1 1/4	0	1/8	5/16	0	1/4	5/8	18	20	8
DODGE DIV. (A) (B) (C)																
81-80 St. Regis	(7)	− 1/4	1	2 1/4	− 1/4	1/2	1 1/4	0	1/8	5/16	0	1/4	5/8	18	20	8
79 St. Regis	(7)	− 1/2	3/4	2				1/16	1/8	1/4	1/8	1/4	1/2	18	20	8
Left Wheel					0	1/2	1									
Right Wheel					− 1/4	1/4	3/4									
80 Aspen	(7)	1 1/4	2 1/2	3 3/4	− 1/4	1/2	1 1/4	0	1/8	5/16	0	1/4	5/8	18	20	8
79-77 Aspen	(7)	1 1/2	2 1/2	3 3/4				1/16	1/8	1/4	1/8	1/4	1/2	18	20	8
Left Wheel					0	1/2	1									
Right Wheel					− 1/4	1/4	3/4									
83-80 Mirada	(7)	1 1/4	2 1/2	3 3/4	− 1/4	1/2	1 1/4	0	1/8	5/16	0	1/4	5/8	18	20	8
86-80 Diplomat	(7)	1 1/4	2 1/2	3 3/4	− 1/4	1/2	1 1/4	0	1/8	5/16	0	1/4	5/8	18	20	8
79-77 Diplomat	(7)	1 1/2	2 1/2	3 3/4				1/16	1/8	1/4	1/8	1/4	1/2	18	20	8
Left Wheel					0	1/2	1									
Right Wheel					− 1/4	1/4	3/4									
79-77 Charger, Charger SE Magnum XE, Monaco, Royal Monaco																
W/Power Steering	(7)	− 1/2	3/4	2				1/16	1/8	1/4	1/8	1/4	1/2	18	20	8
Left Wheel					0	1/2	1									
Right Wheel					− 1/4	1/4	3/4									
W/Manual Steering	(7)	− 1 3/4	− 1/2	3/4				1/16	1/8	1/4	1/8	1/4	1/2	18	20	8
Left Wheel					0	1/2	1									
Right Wheel					− 1/4	1/4	3/4									
Coronet, Charger																
W/Power Steering	(7)	− 1/2	3/4	1 3/4				1/16	1/8	1/4	1/8	1/4	1/2	18	20	8
Left Wheel					0	1/2	1									
Right Wheel					− 1/4	1/4	3/4									
W/Manual Steering	(7)	− 1 3/4	− 1/2	1/2				1/16	1/8	1/4	1/8	1/4	1/2	18	20	8
Left Wheel					0	1/2	1									
Right Wheel					− 1/4	1/4	3/4									
86 Aries, "400", "600", Daytona, Lancer																
Front	(34)		13/16 ■		− 1/4	5/16	3/4	1/8 (out)	1/16 (in)	7/32 (in)	1/4 (out)	1/8 (in)	7/16 (in)	NA	NA	13 5/16
Rear	(48)				− 1 1/4	1/2	1/4	5/16 (out)	0	5/16 (in)	5/8 (out)	0	5/8 (in)			
85-82 Aries, "400", "600", "600 ES", Daytona, Lancer																
Front	(4) ★		13/16 ■		− 1/4	5/16	3/4	1/8 (out)	1/16 (in)	7/32 (in)	1/4 (out)	1/8 (in)	7/16 (in)	NA	NA	13 5/16
Rear	(48)				− 1	− 1/2	0	3/16 (out)	0	3/16 (in)	3/8 (out)	0	3/8 (in)			
81 Aries																
Front	(4)		13/16 ■		− 1/4	5/16	3/4	5/32 (out)	1/16 (out)	1/8 (in)	5/16 (out)	1/8 (out)	1/4 (in)	NA	NA	13 3/8
Rear	(48)				− 1	− 1/2	0	3/16 (out)	0	3/16 (in)	3/8 (out)	0	3/8 (in)			
86 Omni, Charger																
Front	(4)		1 7/8 ▲		− 1/4	5/16	3/4	1/8 (out)	1/16 (in)	7/32 (in)	1/4 (out)	1/8 (in)	7/16 (in)	NA	NA	13 3/8
Rear	(48)				− 1 1/4	− 3/4	− 1/4	3/16 (out)	3/32 (in)	13/32 (in)	3/8 (out)	3/16 (in)	13/16 (in)			

★ Use ill. no. 34 for 1985-84 models.   ■ Exc. wagon shown; Wagon 7/8°.   ▲ 2 dr. shown; 4 dr. 1 3/8°.

SECTION 2: SUSPENSION
Specifications

U.S. PASSENGER CAR SECTION

VEHICLE IDENTIFICATION YEAR MODEL	ADJ. ILL. NO.	CASTER (Degrees) MIN.	PREF.	MAX.	CAMBER (Degrees) MIN.	PREF.	MAX.	TOE-IN (Inches) MIN.	PREF.	MAX.	TOE-IN (Degrees) MIN.	PREF.	MAX.	TOE-OUT ON TURNS (Degrees) OUTSIDE WHEEL	INSIDE WHEEL	STRG. AXIS INCL. (DEG.)
85-82 Omni, 024, Rampage, Charger																
Front	(4)		1⅞▲		−¼	5/16	¾	⅛ (out)	1/16 (in)	7/32 (in)	¼ (out)	⅛ (in)	7/16 (in)	NA	NA	13⅜
Rear, exc. Rampage	(48)				−1¼	−¾	−¼	5/32 (out)	3/32 (in)	11/32 (in)	5/16 (out)	3/16 (in)	11/16 (in)			
Rear, Rampage	(48)				−1⅛	−⅝	−⅛	5/32 (out)	3/32 (in)	11/32 (in)	5/16 (out)	3/16 (in)	11/16 (in)			
81-78 Omni, 024																
Front	(4)		1⅞▲		−¼	5/16	¾	5/32 (out)	1/16 (out)	⅛ (in)	5/16 (out)	⅛ (out)	¼ (in)	NA	NA	13⅜
Rear	(48)				−½	−1	−1½	5/32 (out)	3/32 (in)	11/32 (in)	5/16 (out)	3/16 (in)	11/16 (in)			
PLYMOUTH (A) (B) (C)																
86-82 Gran Fury	(7)	1¼	2½	3¾	−¼	½	1¼	0	⅛	5/16	0	¼	⅝	18	20	8
81-80 Gran Fury	(7)	−¼	1	2¼	−¼	½	1¼	0	⅛	5/16	0	¼	⅝	18	20	8
77 Gran Fury	(7)	−½	¾	2				1/16	⅛	¼	⅛	¼	½	18	20	9
Left Wheel					0	½	1									
Right Wheel	(7)				−¼	¼	¾									
80 Volare	(7)	1¼	2½	3¾	−¼	½	1¼	0	⅛	5/16	0	¼	⅝	18	20	8
79-77 Volare	(7)	1½	2½	3¾				1/16	⅛	¼	⅛	¼	½	18	20	8
Left Wheel					0	½	1									
Right Wheel					−¼	¼	¾									
86 Reliant, Caravelle																
Front	(34)		1 3/16 ■		−¼	5/16	¾	⅛ (out)	1/16 (in)	7/32 (in)	¼ (out)	⅛ (in)	7/16 (in)	NA	NA	13 5/16
Rear	(48)				−1¼	½	¼	5/16 (out)	0	5/16 (in)	⅝ (out)	0	⅝ (in)			
85-82 Reliant, Caravelle																
Front	(4)★		1 3/16 ■		−¼	5/16	¾	⅛ (out)	1/16 (in)	7/32 (in)	¼ (out)	⅛ (in)	7/16 (in)	NA	NA	13 5/16
Rear	(48)				−1	−½	0	3/16 (out)	0	3/16 (in)	⅜	0	⅜ (in)			
81 Reliant																
Front	(4)		1 3/16 ■		−¼	5/16	¾	5/32 (out)	1/16 (out)	⅛ (in)	5/16 (out)	⅛ (out)	¼ (in)	NA	NA	13⅜
Rear	(48)				−1	−½	0	3/16 (out)	0	3/16 (in)	⅜	0	⅜ (out)			
86 Horizon, Turismo																
Front	(4)		1⅞▲		−¼	5/16	¾	⅛ (out)	1/16 (in)	7/32 (in)	¼ (out)	⅛ (in)	7/16 (in)	NA	NA	13⅜
Rear	(48)				−1¼	−¾	−¼	3/16 (out)	3/32 (in)	13/32 (in)	⅜	3/16 (in)	13/16 (in)			
85-82 Horizon, TC3, Turismo, Scamp																
Front	(4)		1⅞▲		−¼	5/16	¾	⅛ (out)	1/16 (in)	7/32 (in)	¼ (out)	⅛ (in)	7/16 (in)	NA	NA	13⅜
Rear exc. Scamp	(48)				−1¼	−¾	−¼	5/32 (out)	3/32 (in)	11/32 (in)	5/16 (out)	3/16 (in)	11/16 (in)			
Rear, Scamp	(48)				−1⅛	−⅝	−⅛	5/32 (out)	3/32 (in)	11/32 (in)	5/16 (out)	3/16 (in)	11/16 (in)			
81-78 Horizon, TC3																
Front	(4)		1⅞▲		−¼	5/16	¾	5/32 (out)	1/16 (out)	⅛ (in)	5/16 (out)	⅛ (out)	¼ (in)	NA	NA	13⅜
Rear	(48)				−1½	−1	−½	5/32 (out)	3/32 (in)	11/32 (in)	5/16 (out)	3/16 (in)	11/16 (in)			
78-77 Fury																
W/Power Steering	(7)	−½	¾	2				1/16	⅛	¼	⅛	¼	½	18	20	8
Left Wheel					0	½	1									
Right Wheel					−¼	¼	¾									
W/Manual Steering	(7)	−1¾	−½	¾				1/16	⅛	¼	⅛	¼	½	18	20	8
Left Wheel					0	½	1									
Right Wheel					−¼	¼	¾									

★ Use ill. No. 34 for 1985-84 Models. ■ Exc. wagon shown; Wagon ⅞°. ▲ 2 dr. shown; 4 dr. 1⅜°.

SUSPENSION
Specifications

U.S. PASSENGER CAR SECTION

VEHICLE IDENTIFICATION YEAR MODEL	ADJ. ILL. NO.	CASTER (Degrees) MIN.	PREF.	MAX.	CAMBER (Degrees) MIN.	PREF.	MAX.	TOE-IN (Inches) MIN.	PREF.	MAX.	TOE-IN (Degrees) MIN.	PREF.	MAX.	TOE-OUT ON TURNS (Degrees) OUTSIDE WHEEL	INSIDE WHEEL	STRG. AXIS INCL. (DEG.)
FORD MOTOR CO.																
(A) Maximum side to side variation; caster ± ¾°, camber, (Left minus right) within - ½° to 1°.								(E) Caster measurement must be done for each wheel regardless of the equipment being used.								
(B) Maximum side to side variation; caster and camber ± ¾°.								(F) Maximum side to side variation; caster ± ⅞°, camber ± ¾°								
(C) Maximum side to side variation; caster and camber 1°.								(G) Maximum side to side variation; caster and camber ± ⅞°.								
(D) Maximum side to side variation; caster 1° camber ½°.								(H) 86-84 Continental, Mark VII Set Vehicle Ride Height prior to checking alignment.								
FORD DIVISION																
86 Taurus .... (E)(F)																
Front	N/A	3	4	6	−1³⁄₃₂	−½	³⁄₃₂	⁷⁄₃₂ (out)	³⁄₃₂ (out)	¹⁄₆₄ (in)	⁷⁄₁₆ (out)	³⁄₁₆ (out)	¹⁄₃₂ (in)	NA	NA	15³⁄₈
Rear	N/A							¹⁄₁₆ (out)	¹⁄₁₆ (in)	³⁄₁₆ (in)	⅛ (out)	⅛ (in)	⅜ (in)			
Sedan					−1¹¹⁄₁₆	−1	⁵⁄₁₆									
Wagon					−1³⁄₃₂	−1³⁄₃₂	⁵⁄₁₆									
86 Escort .... (E)																
Front	N/A	1¹¹⁄₁₆	2⁷⁄₁₆	3³⁄₁₆				⁷⁄₃₂ (out)	³⁄₃₂ (out)	¹⁄₆₄ (in)	⁷⁄₁₆ (out)	³⁄₁₆ (out)	¹⁄₃₂ (in)			
Left Wheel					⅝	1⅜	2⅛							20	20	14²¹⁄₃₂
Right Wheel					³⁄₁₆	¹⁵⁄₁₆	1¹¹⁄₁₆							18⁷⁄₃₂	20	15³⁄₃₂
Rear					−2¹⁄₃₂	−1³⁄₁₆	−1¹⁄₃₂	0	³⁄₁₆	⅜	0	⅜	¾			
85 Escort, EXP ... (E)																
Front	N/A							0	³⁄₁₆	⅜	0	⅜	¾			
exc. Turbo		⁹⁄₁₆	1⁵⁄₁₆	2¹⁄₁₆												
w/Turbo		¼	1	1¾												
Left Wheel					1¼	1⅞	2⅝							20	20	14²¹⁄₃₂
Right Wheel					¹¹⁄₁₆	1⁷⁄₁₆	2³⁄₁₆							18⁷⁄₃₂	20	15³⁄₃₂
Rear	N/A							¹⁄₁₆ (out)	¹⁄₁₆ (in)	³⁄₁₆ (in)	⅛ (out)	⅛ (in)	⅜ (in)			
Left Wheel					−2⅛	−1¼	−⅜									
Right Wheel					−1½	−⅝	−¼									
84 Escort, EXP ... (E)	N/A															
Front		⅝	1¹³⁄₃₂	2⅛				⁷⁄₃₂ (out)	⅛ (out)	¹⁄₆₄ (in)	⁷⁄₁₆ (out)	¼ (out)	¹⁄₃₂ (in)			
Left Wheel					1⅜	2⅛	2⅞							20	20	14²¹⁄₃₂
Right Wheel					¹⁵⁄₁₆	1¹⁄₁₆	2⁷⁄₁₆							18⁷⁄₃₂	20	15³⁄₃₂
Rear								¹⁄₁₆ (out)	¹⁄₁₆ (in)	³⁄₁₆ (in)	⅛ (out)	⅛ (in)	⅜ (in)			
Left Wheel					−2⅛	−1¼	−⅜									
Right Wheel					−1½	−⅝	−¼									
83-82 Escort, EXP ... (E)	N/A															
Front		⁹⁄₁₆	1⁵⁄₁₆	2¹⁄₁₆				⁷⁄₃₂ (out)	³⁄₃₂ (out)	¹⁄₆₄ (in)	⁷⁄₁₆ (out)	³⁄₁₆ (out)	¹⁄₃₂ (in)			
Left Wheel					1¹³⁄₃₂	2⁵⁄₃₂	2²⁹⁄₃₂							20	20	14²¹⁄₃₂
Right Wheel					³¹⁄₃₂	1²³⁄₃₂	2¹⁵⁄₃₂							17	20	15³⁄₃₂
Rear					−1⁷⁄₁₆	−¹⁹⁄₃₂	¼	0	³⁄₁₆	⅜	0	⅜	¾			
81 Escort, EXP ... (E)	N/A															
Front		⁹⁄₁₆	1⁵⁄₁₆	2¹⁄₁₆				⁷⁄₃₂ (out)	³⁄₃₂ (out)	¹⁄₆₄ (in)	⁷⁄₁₆ (out)	³⁄₁₆ (out)	¹⁄₃₂ (in)			
Left Wheel					1¹³⁄₃₂	2⁵⁄₃₂	2²⁹⁄₃₂							19³¹⁄₃₂	20	14²¹⁄₃₂
Right Wheel					³¹⁄₃₂	1²³⁄₃₂	2¹⁵⁄₃₂							17¹⁄₃₂	20	15³⁄₃₂
Rear					−1⁷⁄₁₆	−¹⁹⁄₃₂	¼	¹⁄₆₄	³⁄₁₆	⅜	¹⁄₃₂	⅜	¾			
86 Tempo .... (E)																
exc. Sport Coupe Front ..	N/A	1¹¹⁄₁₆	2⁷⁄₁₆	3³⁄₁₆				⁷⁄₃₂ (out)	³⁄₃₂ (out)	¹⁄₆₄ (in)	⁷⁄₁₆ (out)	³⁄₁₆ (out)	¹⁄₃₂ (in)	20	20	14²¹⁄₃₂
Left Wheel					¹³⁄₃₂	1⁵⁄₃₂	1²⁹⁄₃₂							18⁷⁄₃₂	20	15³⁄₃₂
Right Wheel					−¹⁄₃₂	²³⁄₃₂	1¹⁵⁄₃₂									
Rear					−1½	−⁹⁄₃₂	¹⁵⁄₃₂	³⁄₁₆ (out)	0	³⁄₁₆ (in)	⅜ (out)	0	⅜ (in)			
Sport Coupe Front	N/A	1⅝	2⅜	3⅛				⁷⁄₃₂ (out)	³⁄₃₂ (out)	¹⁄₆₄ (in)	⁷⁄₁₆ (out)	³⁄₁₆ (out)	¹⁄₃₂ (in)	20	20	14²¹⁄₃₂
Left Wheel					⁷⁄₁₆	1³⁄₁₆	1¹⁵⁄₁₆							18⁷⁄₃₂	20	15³⁄₃₂
Right Wheel					0	¾	1½									
Rear					−1¹⁄₃₂	−⁹⁄₃₂	¹⁵⁄₃₂	³⁄₁₆ (out)	0	³⁄₁₆ (in)	⅜ (out)	0	⅜ (in)			
85 Tempo .... (E)																
Front	N/A	½	1¼	2				⁷⁄₃₂ (out)	⅛ (out)	¹⁄₆₄ (in)	⁷⁄₁₆ (out)	¼ (out)	¹⁄₃₂ (in)			
Left Wheel					1¼	2	2¾							20	20	14⅝
Right Wheel					¾	1½	2¼							18⁷⁄₃₂	20	15⅛
Rear	N/A				−⁹⁄₁₆	−³⁄₁₆	¹⁵⁄₁₆	⁵⁄₃₂ (out)	³⁄₃₂ (in)	1³⁄₃₂ (in)	⁵⁄₁₆ (out)	³⁄₁₆ (in)	1³⁄₁₆ (in)			

2-369

SECTION 2: SUSPENSION Specifications

U.S. PASSENGER CAR SECTION

VEHICLE IDENTIFICATION YEAR MODEL	ADJ. ILL. NO.	CASTER (Degrees) MIN.	PREF.	MAX.	CAMBER (Degrees) MIN.	PREF.	MAX.	TOE-IN (Inches) MIN.	PREF.	MAX.	TOE-IN (Degrees) MIN.	PREF.	MAX.	TOE-OUT ON TURNS (Degrees) OUTSIDE WHEEL	INSIDE WHEEL	STRG. AXIS INCL. (DEG.)
84 Tempo .... (E)																
Front	N/A	9/16	15/16	2 1/16				7/32 (out)	1/8 (out)	1/64 (in)	7/16 (out)	1/4 (out)	1/32 (in)	18 7/32	20	
Left Wheel					1 1/8	1 7/8	2 5/8									14 5/8
Right Wheel					1 1/16	1 1/2	2 3/16									15 1/8
Rear	N/A				−1	−1/4	1/2	5/32 (out)	3/32 (in)	13/32 (in)	5/16 (out)	3/16 (in)	13/16 (in)			
86-85 Mustang .... (B)	(69)	1/4	1	1 3/4	−3/4	0	3/4	1/16	3/16	5/16	1/8	3/8	5/8	19 27/32	20	15 23/32
84-83 Mustang .... (B)	(69)	1/2	1 1/4	2	−3/4	0	3/4	1/16	3/16	5/16	1/8	3/8	5/8	19 27/32	20	15 23/32
82 Mustang .... (B)	N/A	3/8	1 1/8	1 7/8	−1/2	1/4	1	1/16	3/16	5/16	1/8	3/8	5/8	19 27/32	20	15 11/16
81 Mustang .... (B)	N/A	1/4	1	1 3/4	−1/4	1/4	1	1/16	3/16	5/16	1/8	3/8	5/8	19 27/32	20	15 11/16
86-85 Thunderbird ..... (B)	(69)	0	3/4	1 1/2	−1/2	1/4	1	1/16	3/16	5/16	1/8	3/8	5/8	19 23/32	20	15 23/32
84 Thunderbird .... (B)	(69)	1/4	1	1 3/4	−1/2	1/4	1	1/16	3/16	5/16	1/8	3/8	5/8	19 23/32	20	15 23/32
83 Thunderbird .... (B)	(69)	1/2	1 1/4	2	−1/4	1/4	1	1/16	3/16	5/16	1/8	3/8	5/8	19 23/32	20	15 23/32
82-81 Thunderbird .... (B)	(69)	1/8	1	1 7/8	−1/2	3/8	1 1/4	1/16	3/16	5/16	1/8	3/8	5/8	19 3/4	20	15 23/32
83 Fairmont Futura .... (F)	(69)	1/8	1 1/8	2 1/8	−5/16	7/16	13/16	1/16	3/16	5/16	1/8	3/8	5/8	19 27/32	20	15 23/32
82 Fairmont .... (B)	N/A	1/8	1	1 7/8	−5/16	7/16	13/16	1/16	3/16	5/16	1/8	3/8	5/8	19 27/32	20	15 23/32
81 Fairmont																
Sedan .... (B)	N/A	1/8	1	1 7/8	−5/16	7/16	13/16	1/16	3/16	5/16	1/8	3/8	5/8	19 27/32	20	15 23/32
Wagon .... (B)	N/A	−1/8	3/4	1 5/8	−1/4	1/2	1 1/4	1/16	3/16	5/16	1/8	3/8	5/8	19 27/32	20	15 23/32
82-81 Granada Sedan .. (B)	N/A	1/8	1	1 7/8	−5/16	7/16	13/16	1/16	3/16	5/16	1/8	3/8	5/8	19 27/32	20	15 23/32
82 Granada Wagon .... (B)	N/A	−1/8	3/4	1 5/8	−1/4	1/2	1 1/4	1/16	3/16	5/16	1/8	3/8	5/8	19 27/32	20	15 11/16•
86-85 Crown Victoria ... (B)	(47)	2 1/4	3	4	−1/4	1/2	1 1/4	1/16 (out)	1/16 (in)	3/16 (in)	1/8 (out)	1/8 (in)	3/8 (in)	18 1/2	20	NA
84 Crown Victoria ..... (B)	(47)	2 3/8	3 1/8	4 1/8	−1/4	1/2	1 1/4	1/16 (out)	1/16 (in)	3/16 (in)	1/8 (out)	1/8 (in)	3/8 (in)	18 1/2	20	NA
83 Crown Victoria & SW (B)	(47)	2 1/4	3	4	−1/4	1/2	1 1/4	1/16 (out)	1/16 (in)	3/16 (in)	1/8 (out)	1/8 (in)	3/8 (in)	18 1/2	20	11
86-85 LTD Sedan ...... (G)	(69)	1/8	7/8	2 1/8	−3/8	3/8	1 1/8	1/16	3/16	5/16	1/8	3/8	5/8	19 27/32	20	15 23/32
Wagon .... (G)	(69)	0	3/4	2	−5/16	7/16	13/16	1/16	3/16	5/16	1/8	3/8	5/8	19 27/32	20	15 23/32
84 LTD Sedan .... (B)	(69)	1/4	1	2 1/4	−5/16	7/16	13/16	1/16	3/16	5/16	1/8	3/8	5/8	19 23/32	20	15 23/32
Wagon .... (B)	(69)	1/4	1	2 1/4	−1/4	1/2	1 1/4	1/16	3/16	5/16	1/8	3/8	5/8	19 23/32	20	15 23/32
83 LTD																
Sedan .... (G)	(69)	1/8	1 1/8	2 1/8	−5/16	7/16	13/16	1/16	3/16	5/16	1/8	3/8	5/8	19 27/32	20	15 23/32
Wagon .... (G)	(69)	−1/8	7/8	1 7/8	−1/4	1/2	1 1/4	1/16	3/16	5/16	1/8	3/8	5/8	19 27/32	20	15 23/32
82-81 LTD .... (B)	(47)	2 1/4	3	3 3/4	−1/4	1/2	1 1/4	1/16 (out)	1/16 (in)	3/16 (in)	1/8 (out)	1/8 (in)	3/8 (in)	18 1/2	20	10 31/32
80 LTD .... (B)	(47)	2 1/4	3	3 3/4	−1/4	1/2	1 1/4	1/16 (out)	1/16 (in)	3/16 (in)	1/8 (out)	1/8 (in)	3/8 (in)	18 1/2	20	10 7/8
80 Thunderbird .... (A)	N/A	1/8	1	1 7/8	−1/2	3/8	1 1/4	1/16	3/16	5/16	1/8	3/8	5/8	24 29/32	20	9 1/2
80 Fairmont																
Sedan .... (A)	N/A	1/8	1	1 7/8	−5/16	7/16	13/16	1/16	3/16	5/16	1/8	3/8	5/8	19 3/4	20	15 1/4
Wagon .... (A)	N/A	−1/8	3/4	1 5/8	−1/4	1/2	1 1/4	1/16	3/16	5/16	1/8	3/8	5/8	19 3/4	20	15 1/4
80-79 Granada .... (B)	(1)(2)	−1 1/4	−1/2	1/4	−1/2	1/4	1	0	1/8	1/4	0	1/4	1/2			7 1/2
W/Power Steering														18 3/16	20	
W/Manual Steering ...														18 7/16	20	
80-79 Mustang .... (B)	N/A	1/4	1	1 3/4	−1/2	1/4	1	3/16	5/16	7/16	3/8	5/8	7/8	19 3/4	20	15 1/4
80-79 Pinto Sedan ..... (B)	(8)	1/4	1	1 3/4	−1/4	1/4	1 1/4	0	1/8	1/4	0	1/4	1/2	18 7/8	20	10
80 Pinto Wagon .... (B)	(8)	−3/4	1/4	1 1/4	−1/4	1/2	1 1/4	0	1/8	1/4	0	1/4	1/2	18 7/8	20	10
79 Pinto Wagon .... (B)	(8)	−1/2	1/4	1	−1/4	1/4	1 1/4	0	1/8	1/4	0	1/4	1/2	18 7/8	20	10
79 LTD .... (B)	(47)	2 1/4	3	3 3/4	−1/4	1/2	1 1/4	1/16	3/16	5/16	1/8	3/8	5/8	18 1/2	20	11 3/16
79-78 Fairmont .... (A)	N/A	1/8	7/8	1 5/8	−3/8	3/8	1 1/8	3/16	5/16	7/16	3/8	5/8	7/8	19 3/4	20	15 1/4
79-77 LTD II .... (A)	(46)	3 1/4	4	4 3/4				0	1/8	1/4	0	1/4	1/2	18	20	9
Left Wheel					−1/4	1/2	1 1/4									
Right Wheel					−1/2	1/4	1									
79-77 Thunderbird ..... (A)	(46)	3 1/4	4	4 3/4				0	1/8	1/4	0	1/4	1/2	18	20	9 1/2
Left Wheel					−1/4	1/2	1 1/4									
Right Wheel					−1/2	1/4	1									
78-77 Pinto Sedan ..... (B)	(8)	1/4	1	1 3/4	−1/4	1/4	1 1/4	0	1/8	1/4	0	1/4	1/2	18 7/8	20	10
78-77 Pinto Wagon .... (B)	(8)	−1/2	1/4	1	−1/4	1/4	1 1/4	0	1/8	1/4	0	1/4	1/2	18 7/8	20	10
78-77 LTD .... (A)	(8)	1 1/4	2	2 3/4				1/16	3/16	5/16	1/8	3/8	5/8	18 3/4	20	9 3/4
Left Wheel					−1/4	1/2	1 1/4									
Right Wheel					−1/2	1/4	1									
79-77 Granada .... (B)	(1)(2)	−1 1/4	−1/2	1/4	−1/2	1/4	1	0	1/8	1/4	0	1/4	1/2			6 3/4
W/Power Steering														18 3/16	20	
W/Manual Steering ...														18 7/16	20	
78-77 Mustang .... (B)	(8)	1/8	7/8	1 5/8	−1/4	1/4	1 1/4	0	1/8	1/4	0	1/4	1/2	18 7/8	20	9 3/4
77 Maverick .... (B)	(1)(2)	−1 1/4	−1/2	1/4	−1/2	1/4	1	0	1/8	1/4	0	1/4	1/2			6 3/4
W/Power Steering														18 1/8	20	
W/Manual Steering ...														18 3/8	20	
•82 Granada - 15 11/16																

U.S. PASSENGER CAR SECTION

VEHICLE IDENTIFICATION (YEAR MODEL)	ADJ. ILL. NO.	CASTER (Degrees) MIN.	CASTER PREF.	CASTER MAX.	CAMBER (Degrees) MIN.	CAMBER PREF.	CAMBER MAX.	TOE-IN (Inches) MIN.	TOE-IN (Inches) PREF.	TOE-IN (Inches) MAX.	TOE-IN (Degrees) MIN.	TOE-IN (Degrees) PREF.	TOE-IN (Degrees) MAX.	TOE-OUT ON TURNS (Degrees) OUTSIDE WHEEL	TOE-OUT ON TURNS INSIDE WHEEL	STRG. AXIS INCL. (DEG.)
LINCOLN																
(H) 86-84 Continental, Mark VII Set Vehicle Ride Height prior to checking alignment.																
86-85 Continental, Mark VII (G)	(69)	5/8	1 1/2	2 3/4	- 3/4	0	3/4	0	1/8	1/4	0	1/4	1/2	17 1/8	20	11
84 Continental, MarkVII (G)	(69)	7/8	1 3/4	2 15/16	- 7/8	0	7/8	0	1/8	1/4	0	1/4	1/2	17 1/8	20	11
83 Continental (F)	(69)	3/8	1 1/4	2 1/8	- 1/2	3/8	1 1/4	0	1/8	1/4	0	1/4	1/2	19 1/8	20	NA
82 Continental	(47)	1/8	1	1 7/8	- 1/2	3/8	1 1/4	1/16	3/16	5/16	1/8	3/8	5/8	19 1/8	20	NA
86-85 Lincoln Town Car (B)	(47)	2 1/4	3	4	- 1/4	1/2	1 1/4	1/16 (out)	1/16 (in)	3/16 (in)	1/8 (out)	1/8 (in)	3/8 (in)	18 1/2	20	NA
84 Lincoln Town Car (B)	(47)	2 3/8	3 1/8	4 1/8	- 1/4	1/2	1 1/4	1/16 (out)	1/16 (in)	3/16 (in)	1/8 (out)	1/8 (in)	3/8 (in)	18 1/2	20	NA
83 Lincoln Town Car, Mark VI (B)	(47)	2 1/4	3	4	- 1/4	1/2	1 1/4	1/16 (out)	1/16 (in)	3/16 (in)	1/8 (out)	1/8 (in)	3/8 (in)	18 1/2	20	11
82-81 Lincoln Town Car, Mark VI (B)	(47)	2 1/4	3	3 3/4	- 1/4	1/2	1 1/4	1/16 (out)	1/16 (in)	3/16 (in)	1/8 (out)	1/8 (in)	3/8 (in)	18 1/2	20	11
80 Continental, Mark VI (B)	(47)	2 1/4	3	3 3/4	- 1/4	1/2	1 1/4	1/16 (out)	1/16 (out)	3/16 (in)	1/8 (out)	1/8 (in)	3/8 (in)	18 1/2	20	10 7/8
80-77 Versailles (B)	(1)(2)	- 1 1/4	- 1/2	1/4	- 1/2	1/4	1	0	1/8	1/4	0	1/4	1/2			
W/Power Steering														18 3/16	20	6 3/4
W/Manual Steering														18 7/16	20	6 3/4
79-78 Continental (A)	(8)	1 1/4	2	2 3/4				0	1/8	1/4	0	1/4	1/2	18 1/8	20	9 1/2
Left Wheel					- 1/4	1/2	1 1/4									
Right Wheel					- 1/2	1/4	1									
79-78 Mark V (A)	(8)	3 1/4	4	4 3/4				1/16	3/16	5/16	1/8	3/8	5/8	18 1/8	20	9 1/2
Left Wheel					- 1/4	1/2	1 1/4									
Right Wheel					- 1/2	1/4	1									
77 Continental (A)	(8)	1 1/4	2	2 3/4				0	1/8	3/8	0	1/4	3/4	18 5/32	20	9 1/2
Left Wheel					- 1/4	1/2	1 1/4									
Right Wheel					- 1/2	1/4	1									
77 Mark V (A)	(8)	1 1/4	2	2 3/4				1/16	3/16	5/16	1/8	3/8	5/8	18 1/8	20	9 1/2
Left Wheel					- 1/4	1/2	1 1/4									
Right Wheel					- 1/2	1/4	1									
MERCURY																
86 Sable (E)(F)																
Front	N/A	3	4	6	- 1 3/32	- 1/2	3/32	7/32 (out)	3/32 (out)	1/64 (in)	7/16 (out)	3/16 (out)	1/32 (in)	NA	NA	15 3/8
Rear	N/A							1/16 (out)	1/16 (in)	3/16 (in)	1/8 (out)	1/8 (in)	3/8 (in)			
Sedan					- 1 11/16	- 1	5/16									
Wagon					- 1 3/32	- 1 3/32	5/16									
86-85 Merkur Front	N/A	15/16	1 15/16	2 15/16	- 1 7/32	- 17/32	15/32	3/32 (out)	1/32 (in)	5/32 (in)	3/16 (out)	1/16 (in)	5/16 (in)	NA	NA	15 1/8
Rear	N/A				3/16	1 3/16	2 3/16	3/32 (out)	1/16 (in)	3/16 (in)	3/16 (out)	1/8 (in)	3/8 (in)			
86 Lynx (E)																
Front	N/A	1 11/16	2 7/16	3 3/16				7/32 (out)	3/32 (out)	1/64 (in)	7/16 (out)	3/16 (out)	1/32 (in)			
Left Wheel					5/8	1 3/8	2 1/8							20	20	14 21/32
Right Wheel					3/16	15/16	1 11/16							18 7/32	20	15 3/32
Rear					- 2 1/32	- 1 3/16	- 11/32	0	3/16	3/8	0	3/8	3/4			
85 Lynx (E)	N/A															
Front								0	3/16	3/8	0	3/8	3/4			
exc. Turbo		9/16	1 5/16	2 1/16												
w/Turbo		1/4	1	1 3/4												
Left Wheel					1 1/4	1 7/8	2 5/8							20	20	14 21/32
Right Wheel					11/16	1 7/16	2 3/16							18 7/32	20	15 3/32
Rear	N/A							1/16 (out)	1/16 (in)	3/16 (in)	1/8 (out)	1/8 (in)	3/8 (in)			
Left Wheel					- 2 1/8	- 1 1/4	- 3/8									
Right Wheel					- 1 1/2	- 5/8	- 1/4									
84 Lynx (E)																
Front	N/A	5/8	1 13/32	2 1/8				7/32 (out)	1/8 (out)	1/64 (in)	7/16 (out)	1/4 (out)	1/32 (in)			
Left Wheel					1 3/8	2 1/8	2 7/8							20	20	14 21/32
Right Wheel					15/16	1 1/16	2 7/16							18 7/32	20	15 3/32
Rear								1/16 (out)	1/16 (in)	3/16 (in)	1/8 (out)	1/8 (in)	3/8 (in)			
Left Wheel					- 2 1/8	- 1 1/4	- 3/8									
Right Wheel					- 1 1/2	- 5/8	- 1/4									

SECTION 2
SUSPENSION
Specifications

U.S. PASSENGER CAR SECTION

VEHICLE IDENTIFICATION YEAR MODEL	ADJ. ILL. NO.	CASTER (Degrees) MIN.	PREF.	MAX.	CAMBER (Degrees) MIN.	PREF.	MAX.	TOE-IN (Inches) MIN.	PREF.	MAX.	TOE-IN (Degrees) MIN.	PREF.	MAX.	TOE-OUT ON TURNS (Degrees) OUTSIDE WHEEL	INSIDE WHEEL	STRG. AXIS INCL. (DEG.)
83-82 Lynx, LN7 ..... (E)	N/A															
Front		9/16	15/16	21/16				7/32 (out)	3/32 (out)	1/64 (in)	7/16 (out)	3/16 (out)	1/32 (in)			
Left Wheel					1 13/32	25/32	2 29/32							20	20	14 21/32
Right Wheel					31/32	1 23/32	2 15/32							17	20	15 3/32
Rear					−1 7/16	−19/32	1/4	0	3/16	3/8	0	3/8	3/4			
81 Lynx, LN7 .... (E)	N/A															
Front		9/16	15/16	21/16				7/32 (out)	3/32 (out)	1/64 (in)	7/16 (out)	3/16 (out)	1/32 (in)			
Left Wheel					1 13/32	25/32	2 29/32							19 31/32	20	14 21/32
Right Wheel					31/32	1 23/32	2 15/32							17 1/32	20	15 3/32
Rear					−1 7/16	−19/32	1/4	1/64	3/16	3/8	1/32	3/8	3/4			
86-85 Grand Marquis ... (B)	(47)	2 1/4	3	4	−1/4	1/2	1 1/4	1/16 (out)	1/16 (in)	3/16 (in)	1/8 (out)	1/8 (in)	3/8 (in)	18 1/2	20	NA
84 Grand Marquis ..... (B)	(47)	2 3/8	3 1/8	4 1/8	−1/4	1/2	1 1/4	1/16 (out)	1/16 (in)	3/16 (in)	1/8 (out)	1/8 (in)	3/8 (in)	18 1/2	20	NA
83 Grand Marquis ..... (B)	(47)	2 1/4	3	4	−1/4	1/2	1 1/4	1/16 (out)	1/16 (in)	3/16 (in)	1/8 (out)	1/8 (in)	3/8 (in)	18 1/2	20	11
86-85 Marquis Sedan ... (G)	(69)	1/8	7/8	2 1/8	−3/8	3/8	1 1/8	1/16	3/16	5/16	1/8	3/8	5/8	19 27/32	20	15 23/32
Wagon .. (G)	(69)	0	3/4	2	−5/16	7/16	1 3/16	1/16	3/16	5/16	1/8	3/8	5/8	19 27/32	20	15 23/32
84 Marquis Sedan ..... (B)	(69)	1/4	1	2 1/4	−5/16	7/16	1 3/16	1/16	3/16	5/16	1/8	3/8	5/8	19 23/32	20	15 23/32
Wagon ..... (B)	(69)	1/4	1	2 1/4	−1/4	1/2	1 1/4	1/16	3/16	5/16	1/8	3/8	5/8	19 23/32	20	15 23/32
83 Marquis																
Sedan .... (G)	(69)	1/8	1 1/8	2 1/8	−5/16	7/16	1 3/16	1/16	3/16	5/16	1/8	3/8	5/8	19 27/32	20	15 23/32
Wagon .... (G)	(69)	−1/8	7/8	1 7/8	−1/4	1/2	1 1/4	1/16	3/16	5/16	1/8	3/8	5/8	19 27/32	20	15 23/32
82-81 Mercury .... (B)	(47)	2 1/4	3	3 3/4	−1/4	1/2	1 1/4	1/16 (out)	1/16 (in)	3/16 (in)	1/8 (out)	1/8 (in)	3/8 (in)	18 1/2	20	10 31/32
86-85 Capri .... (B)	(69)	1/4	1	1 3/4	−3/4	0	3/4	1/16	3/16	5/16	1/8	3/8	5/8	19 27/32	20	15 23/32
84-83 Capri .... (B)	(69)	1/2	1 1/4	2	−3/4	0	3/4	1/16	3/16	5/16	1/8	3/8	5/8	19 27/32	20	15 23/32
82 Capri .... (B)	N/A	3/8	1 1/8	1 7/8	−1/2	1/4	1	1/16	3/16	5/16	1/8	3/8	5/8	19 27/32	20	15 11/16
81 Capri .... (B)	N/A	1/4	1	1 3/4	−1/2	1/4	1	1/16	3/16	5/16	1/8	3/8	5/8	19 27/32	20	15 11/16
86 Topaz .... (E)																
exc. Sport Coupe Front ..	N/A	1 11/16	2 7/16	3 3/16				7/32 (out)	3/32 (out)	1/64 (in)	7/16 (out)	3/16 (out)	1/32 (in)	20	20	14 21/32
Left Wheel					13/32	15/32	1 29/32							18 7/32	20	15 3/32
Right Wheel					−1/32	23/32	1 15/32									
Rear					−1 1/32	−9/32	15/32	3/16 (out)	0	3/16 (in)	3/8 (out)	0	3/8 (in)			
Sport Coupe Front	N/A	1 5/8	2 3/8	3 1/8				7/32 (out)	3/32 (out)	1/64 (in)	7/16 (out)	3/16 (out)	1/32 (in)	20	20	14 21/32
Left Wheel					7/16	13/16	1 15/16							18 7/32	20	15 3/32
Right Wheel					0	3/4	1 1/2									
Rear					−1 1/32	−9/32	15/32									
85 Topaz .... (E)	N/A															
Front		1/2	1 1/4	2				7/32 (out)	1/8 (out)	1/64 (in)	7/16 (out)	1/4 (out)	1/32 (in)			
Left Wheel					1 1/4	2	2 3/4							20	20	14 5/8
Right Wheel					3/4	1 1/2	2 1/4							18 7/32	20	15 1/8
Rear					−9/16	−3/16	15/32	5/32 (out)	3/32 (in)	13/32 (in)	5/16 (out)	3/16 (in)	13/16 (in)			
84 Topaz																
Front .... (E)	N/A	5/8	1 5/16	2 1/16				7/32 (out)	1/8 (out)	1/64 (in)	7/16 (out)	1/4 (out)	1/32 (in)	18 7/32	20	
Left Wheel					1 1/8	1 7/8	2 5/8									14 5/8
Right Wheel					11/16	1 1/2	2 3/16									15 1/8
Rear		N/A			−1	−1/4	1/2	5/32 (out)	3/32 (in)	13/32 (in)	5/16 (out)	3/16 (in)	13/16 (in)			
83 XR7 .... (B)	(69)	1/2	1 1/4	2	−1/2	1/4	1	1/16	3/16	5/16	1/8	3/8	5/8	19 23/32	20	15 23/32
82-81 XR7 .... (B)	N/A	1/8	1	1 7/8	−1/2	3/8	1 1/4	1/16	3/16	5/16	1/8	3/8	5/8	19 3/4	20	15 23/32
83 Zephyr .... (F)	(69)	1/8	1 1/8	2 1/8	−5/16	7/16	1 13/16	1/16	3/16	5/16	1/8	3/8	5/8	19 27/32	20	15 23/32
82 Zephyr .... (B)	N/A	1/8	1	1 7/8	−5/16	7/16	1 3/16	1/16	3/16	5/16	1/8	3/8	5/8	19 27/32	20	15 23/32
81 Zephyr Sedan ...... (B)	N/A	1/8	1	1 7/8	−5/16	7/16	1 3/16	1/16	3/16	5/16	1/8	3/8	5/8	19 27/32	20	15 23/32
81 Zephyr Wagon ..... (B)	N/A	−1/8	3/4	1 5/8	−1/4	1/2	1 1/4	1/16	3/16	5/16	1/8	3/8	5/8	19 27/32	20	15 23/32
86-85 Cougar .... (B)	(69)	0	3/4	1 1/2	−1/2	1/4	1	1/16	3/16	5/16	1/8	3/8	5/8	19 23/32	20	15 23/32
84 Cougar .... (B)	(69)	1/4	1	1 3/4	−1/2	1/4	1	1/16	3/16	5/16	1/8	3/8	5/8	19 23/32	20	15 23/32
82-81 Cougar Sedan ... (B)	N/A	1/8	1	1 7/8	−5/16	7/16	1 3/16	1/16	3/16	5/16	1/8	3/8	5/8	19 27/32	20	15 23/32•
82 Cougar Wagon ..... (B)	N/A	−1/8	3/4	1 5/8	−1/4	1/2	1 1/4	1/16	3/16	5/16	1/8	3/8	5/8	19 27/32	20	15 11/16
80-79 Capri .... (B)	N/A	1/4	1	1 3/4	−1/2	1/4	1	3/16	5/16	7/16	3/8	5/8	7/8	19 3/4	20	15 1/4
80 Zephyr Sedan ...... (A)	N/A	1/8	1	1 7/8	−5/16	7/16	1 3/16	1/16	3/16	5/16	1/8	3/8	5/8	19 3/4	20	15 1/4
80 Zephyr Wagon ..... (A)	N/A	−1/8	3/4	1 5/8	−1/4	1/2	1 1/4	1/16	3/16	5/16	1/8	3/8	5/8	19 3/4	20	15 1/4

•82 Cougar - 15 11/16

U.S. PASSENGER CAR SECTION

VEHICLE IDENTIFICATION YEAR MODEL	ADJ. ILL. NO.	CASTER (Degrees) MIN.	PREF.	MAX.	CAMBER (Degrees) MIN.	PREF.	MAX.	TOE-IN (Inches) MIN.	PREF.	MAX.	TOE-IN (Degrees) MIN.	PREF.	MAX.	TOE-OUT ON TURNS (Degrees) OUTSIDE WHEEL	INSIDE WHEEL	STRG. AXIS INCL. (DEG.)
80 Bobcat Sedan . . . . . . (B)	(8)	1/4	1	1 3/4	− 1/4	1/2	1 1/4	0	1/8	1/4	0	1/4	1/2	18 7/8	20	10
80 Bobcat Wagon . . . . . (B)	(8)	− 3/4	1/4	1 1/4	− 1/4	1/2	1 1/4	0	1/8	1/4	0	1/4	1/2	18 7/8	20	10
80 Cougar . . . . (A)	N/A	1/8	1	1 7/8	− 1/2	3/8	1 1/4	1/16	3/16	5/16	1/8	3/8	5/8	24 29/32	20	15 3/8
80 Monarch . . . . (B)	(1)(2)	− 1 1/4	− 1/2	1/4	− 1/2	1/4	1	0	1/8	1/4	0	1/4	1/2			
W/Power Steering														18 3/16	20	7 1/2
W/Manual Steering														18 7/16	20	7 1/2
80-79 Mercury . . . . (B)	(47)	2 1/4	3	3 3/4	− 1/4	1/2	1 1/4	1/16	1/16	3/16	1/8	1/8	3/8	18 1/2	20	10 7/8
								(out)	(in)	(in)	(out)	(in)	(in)			
79-78 Zephyr . . . . (A)	N/A	1/8	7/8	1 5/8	− 3/8	3/8	1 1/8	3/16	5/16	7/16	3/8	5/8	7/8	19 3/4	20	15 1/4
79-77 Bobcat Sedan. . . . (B)	(8)	1/4	1	1 3/4	− 1/4	1/2	1 1/4	0	1/8	1/4	0	1/4	1/2	18 7/8	20	10
79-77 Bobcat Wagon . . . (B)	(8)	− 1/2	1/4	1	− 1/4	1/2	1 1/4	0	1/8	1/4	0	1/4	1/2	18 7/8	20	10
79-77 Cougar . . . . (A)	(46)	3 1/4	4	4 3/4				0	1/8	1/4	0	1/4	1/2	18	20	9
Left Wheel					− 1/4	1/2	1 1/4									
Right Wheel					− 1/2	1/4	1									
79-77 Monarch. . . . . (B)	(1)(2)	− 1 1/4	− 1/2	1/4	− 1/2	1/4	1	0	1/8	1/4	0	1/4	1/2			
W/Power Steering														18 3/16	20	6 3/4
W/Manual Steering														18 7/16	20	6 3/4
78-77 Mercury . . . . (A)	(8)	1 1/4	2	2 3/4				1/16	3/16	5/16	1/8	3/8	5/8	18 3/4	20	9 1/2
Left Wheel					− 1/4	1/2	1 1/4									
Right Wheel					− 1/2	1/4	1									
77 Comet . . . . (B)	(1)(2)	− 1 1/4	− 1/2	1/4	− 1/2	1/4	1	0	1/8	1/4	0	1/4	1/2			
W/Power Steering														18 1/8	20	6 3/4
W/Manual Steering														18 3/8	20	6 3/4
•82 Cougar - 15 11/16																

GENERAL MOTORS CORPORATION

BUICK MOTOR DIVISION (A)

(A) Maximum side to side variation; caster and camber 1/2 °.
(B) 1986 Riviera trim heights measured from: front; 22 27/32 " from center of front wheel to rear of car. Rear; 22 5/32 " from center of rear wheel to front of car.
1985-83 Riviera trim heights measured from: front; 24.4" to rear of front wheel center line. Rear; 18.7" to front of rear wheel center line. Trim height is measured at point specified (above) from lower outer edge of rocker panel to ground.
1982-79 Riviera F.W.D. trim height is measured from the edge of the wheel well opening directly over the center of the wheel to the floor. (See block to right for detail).

RIVIERA F.W.D. TRIM HEIGHTS (B) or (C)

Model and Year	Front Suspension Inches	MM	Rear Suspension Inches	MM
86	8.3	210	8.2	208
85-83	9.69	246	9.57	243
82-80	28 1/8	726	27 5/16	694
79	28 1/2	724	28	709

VEHICLE IDENTIFICATION YEAR MODEL	ADJ. ILL. NO.	CASTER MIN.	PREF.	MAX.	CAMBER MIN.	PREF.	MAX.	TOE-IN (Inches) MIN.	PREF.	MAX.	TOE-IN (Degrees) MIN.	PREF.	MAX.	TOE-OUT OUTSIDE WHEEL	INSIDE WHEEL	STRG. AXIS INCL.
86-85 Electra, Park Ave., LeSabre (F.W.D.)																
Front	(5)(67)	2	2 1/2	3				7/32	0	7/32	7/16	0	7/16	NA	NA	NA
								(out)		(in)	(out)		(in)			
Left Wheel					− 1	− 1/2	0									
Right Wheel					0	1/2	1									
Rear (86)	(67)				− 13/16	− 5/16	3/16	0	1/8	7/32	0	1/4	7/16			
(85)	(67)				− 9/16	− 5/16	− 1/16	0	1/8	7/32	0	1/4	7/16			
86-83 Skyhawk (F.W.D.) . . .	(67)	11/16	1 11/16	2 11/16	7/32	23/32	1 7/32	5/32	1/16	1/32	5/16	1/8	1/16	NA	NA	13 1/2
								(out)	(out)	(in)	(out)	(out)	(in)			
82 Skyhawk (F.W.D.).	(67)	11/16	1 11/16	2 11/16	1/16	9/16	1 1/16	5/32	1/16	1/32	5/16	1/8	1/16	NA	NA	13 1/2
								(out)	(out)	(in)	(out)	(out)	(in)			
86-83 Skylark (F.W.D.) and Century (F.W.D.)	(67)	1	2	3	− 1/2	0	1/2	3/32	0	3/32	3/16	0	3/16	NA	NA	NA
								(out)		(in)	(out)		(in)			
82 Skylark, Century (F.W.D.)	(67)	1	2	3	− 1/2	0	1/2	3/32	0	3/32	3/16	0	3/16	NA	NA	14 1/2
								(out)		(in)	(out)		(in)			
86-85 Somerset Regal (F.W.D.)	N/A	11/16	1 11/16	2 11/16	3/16	27/32	1 1/2	1/4	1/8	1/32	1/2	1/4	1/16	NA	NA	NA
								(out)	(out)	(out)	(out)	(out)	(out)			
86-82 Regal (R.W.D.)-All . . .	(10)	2	3	4	− 5/16	1/2	1 5/16	1/16	1/8	1/4	1/8	1/4	1/2	NA	NA	8
86-83 Estate, Electra, LeSabre and LeSabre Wagon (R.W.D.)	(10)	2	3	4	0	13/16	1 5/8	1/16	1/8	1/4	1/8	1/4	1/2	NA	NA	NA
82-77 Electra, Le Sabre. . . .	(10)	2	3	4	0	13/16	1 5/8	1/16	1/8	1/4	1/8	1/4	1/2	NA	NA	9 9/16
82-77 Estate, Electra, & LeSabre Wagon	(10)	2	3	4	0	13/16	1 5/8	1/16	1/8	1/4	1/8	1/4	1/2	NA	NA	10 3/4
86 Riviera Front	(15)(27)	1 15/16	2 5/16	3 5/16	− 13/16	0	13/16	3/32	0	3/32	3/16	0	3/16	NA	NA	NA
								(out)		(in)	(out)		(in)			
Rear								0	3/32	3/16	0	3/16	13/32			
Left Wheel					19/32	23/32	25/32									
Right Wheel					13/32	23/32	1									

U.S. PASSENGER CAR SECTION

VEHICLE IDENTIFICATION YEAR MODEL	ADJ. ILL. NO.	CASTER (Degrees) MIN.	PREF.	MAX.	CAMBER (Degrees) MIN.	PREF.	MAX.	TOE-IN (Inches) MIN.	PREF.	MAX.	TOE-IN (Degrees) MIN.	PREF.	MAX.	TOE-OUT ON TURNS (Degrees) OUTSIDE WHEEL	INSIDE WHEEL	STRG. AXIS INCL. (DEG.)
85-84 Riviera Front	(6)	1½	2½	3½	−13/16	0	13/16	3/32 (out)	0	3/32 (in)	3/16 (out)	0	3/16 (in)	NA	NA	11
Rear	(9)				−13/32	0	13/32	0	3/32	3/16	0	3/16	3/8			
83-79 Riviera																
Front	(6)	1½	2½	3½	−13/16	0	13/16	1/8 (out)	0	1/8 (in)	1/4 (out)	0	1/4 (in)	NA	NA	11
Rear	(9)				−13/32	0	13/32	0	5/32	5/16	0	5/16	5/8			
81-80 Skylark	(34)	−2	0	2	½	1	1½	0	3/32	3/16	0	3/16	3/8	NA	NA	14½
81-78 Century, Regal-All	(10)				−5/16	½	15/16	1/16	1/8	1/4	1/8	1/4	1/2	NA	NA	8
W/Power Steering		2	3	4												
W/Manual Steering		0	1	2												
80-77 Skyhawk	(11)	−1¾	−¾	¼	−½	¼	1	1/16 (out)	1/16 (in)	3/16 (in)	1/8 (out)	1/8 (in)	3/8 (in)	NA	NA	8½
79-77 Skylark	(10)				0	¾	1⅝	1/16	1/8	1/4	1/8	1/4	1/2	NA	NA	10
W/Power Steering		0	1	2												
W/Manual Steering		−2	−1	0				1/16 (out)	1/16 (in)	3/16 (in)	1/8 (out)	1/8 (in)	3/8 (in)	NA	NA	8
77 Century, Regal-All	(10)															
W/Radial Tires		1	2	3												
W/Bias Tires		0	1	2												
Left Wheel					¼	1	1¾									
Right Wheel					−¼	½	1¼									

CADILLAC MOTOR DIVISION (A)(B)

(A) Maximum side to side variation; after reset caster and camber 1½ °.

(B) Check suspension height before performing alignment.

86 Eldorado, Seville																
Front	(15)(27)	1 11/16	2½	3¼	−½	0	½	3/32 (out)	0	3/32 (in)	3/16 (out)	0	3/16 (in)	NA	NA	NA
Rear								0	3/32	3/16	0	3/16	3/8			
86 Fleetwood, Deville (F.W.D.) Front	(5)(67)	1½	2½	3½				3/32 (out)	0	3/32 (in)	3/16 (out)	0	3/16 (in)	NA	NA	NA
Left Wheel					−1	−½	0									
Right Wheel					0	½	1									
Rear	(67)				−13/16	−5/16	3/16	0	3/32	3/16	0	3/16	3/8			
85 Fleetwood, Deville (F.W.D.) Front	(5)(67)	1½	2½	3½	−5/16	½	1¼	3/32 (out)	0	3/32 (in)	3/16 (out)	0	3/16 (in)	NA	NA	NA
Rear	(67)				−¼	3/16	11/16	3/32 (out)	0	3/32 (in)	3/16 (out)	0	3/16 (in)			
86-85 Cimarron	(67)	23/32	1 23/32	2 23/32	3/16	7/8	1½	5/32 (out)	1/16 (out)	1/32 (in)	5/16 (out)	1/8 (out)	1/16 (in)	NA	NA	13½
84-83 Cimarron	(67)	23/32	1 23/32	2 23/32	7/32	23/32	1 7/32	5/32 (out)	1/16 (out)	1/32 (in)	5/16 (out)	1/8 (out)	1/16 (in)	NA	NA	13½
82 Cimarron	(67)	23/32	1 23/32	2 23/32	1/16	9/16	1 1/16	5/32 (out)	1/16 (out)	1/32 (in)	5/16 (out)	1/8 (out)	1/16 (in)	NA	NA	14½
86-80 Cadillac (RWD) exc. Eldorado and Seville	(10)	2	3	4	−5/16	½	15/16	0	1/8	1/4	0	1/4	1/2	NA	NA	10 19/32
85-80 Seville																
Front	(6)	1½	2½	3½	−13/16	0	13/16	1/8 (out)	0	1/8 (in)	1/4 (out)	0	1/4 (in)	NA	NA	11
Rear	(9)				−5/16	1/8	½	0	5/32	5/16	0	5/16	5/8			
85-79 Eldorado																
Front	(6)	1½	2½	3½	−13/16	0	13/16	1/8 (out)	0	1/8 (in)	1/4 (out)	0	1/4 (in)	NA	NA	11
Rear	(9)				−5/16	1/8	½	0	5/32	5/16	0	5/16	5/8			
79-77 Cadillac except Eldorado and Seville	(10)	2	3	4	−¼	½	1¼	1/8 (out)	0	1/8 (in)	1/4 (out)	0	1/4 (in)	NA	NA	10 19/32
79-77 Seville	(10)	1	2	3	−¾	0	¾	1/16 (out)	1/16 (in)	3/16 (in)	1/8 (out)	1/8 (in)	3/8 (in)	NA	NA	10⅝

CHEVROLET MOTOR DIVISION (A)(C)

(A) Maximum side to side variation after reset caster and camber; All except Chevette ½ °, Chevette 2°.

(B) Measure ride height at points 32.5″ to the rear of front wheel center line and 17.2″ forward of rear wheel center line. Height from ground to rocker panel is 8.3″ ± 0.3″ front & rear w/15″ wheels, 7.8″ ± .03″ w/16″ wheels.

(C) Rear alignment measurements are used for complaint vehicles only.

86-82 Chevette	(13)	3	5	7	−½	3/16	29/32	1/64 (out)	1/16 (in)	1/8 (in)	1/32 (out)	1/8 (in)	1/4 (in)	NA	NA	7 9/16

SUSPENSION
Specifications

U.S. PASSENGER CAR SECTION

VEHICLE IDENTIFICATION YEAR MODEL	ADJ. ILL. NO.	CASTER (Degrees) MIN.	PREF.	MAX.	CAMBER (Degrees) MIN.	PREF.	MAX.	TOE-IN (Inches) MIN.	PREF.	MAX.	TOE-IN (Degrees) MIN.	PREF.	MAX.	TOE-OUT ON TURNS (Degrees) OUTSIDE WHEEL	INSIDE WHEEL	STRG. AXIS INCL. (DEG.)
86-85 Cavalier: Front	(67)	23/32	1 23/32	2 23/32	7/32	23/32	1 1/2	5/32 (out)	1/16 (out)	1/32 (in)	5/16 (out)	1/8 (out)	1/16 (in)	NA	NA	13 1/2
Rear (86) .... (C)	(67)				−1 1/32	−1/4	−3/16	1/64	1/16	3/32	1/32	1/8	3/16			
(85) .... (C)	(67)							1/16	5/32	7/32	1/8	5/16	7/16			
84-83 Cavalier	(67)	23/32	1 23/32	2 23/32	7/32	7/8	1 17/32	5/32 (out)	1/16 (out)	1/32 (in)	5/16 (out)	1/8 (out)	1/16 (in)	NA	NA	13 1/2
82 Cavalier	(67)	23/32	1 23/32	2 23/32	1/16	9/16	1 1/16	5/32 (out)	1/16 (out)	1/32 (in)	5/16 (out)	1/8 (out)	1/16 (in)	NA	NA	13 1/2
85 Citation, Celebrity	(67)	15/16	1 15/16	2 15/16	−1/2	0	1/2	3/32 (out)	0	3/32 (in)	3/16 (out)	0	3/16 (in)	NA	NA	14 5/8
86 Celebrity: Front	(67)	15/16	1 15/16	2 15/16	−1/2	0	1/2	3/32 (out)	0	3/32 (in)	3/16 (out)	0	3/16 (in)	NA	NA	14 5/8
Rear ..... (C)					−2 1/2	0	2 1/2	1/32	5/32	1/4	1/16	5/16	1/2			
84 Celebrity	(67)	0	2	4	−1/2	0	1/2	3/32 (out)	0	3/32 (in)	3/16 (out)	0	3/16 (in)	NA	NA	14 5/8
84 Citation	(67)	0	2	4	−1/2	0	1/2	3/32 (out)	0	3/32 (in)	3/16 (out)	0	3/16 (in)	NA	NA	14 1/2
83 Citation, Celebrity	(67)	0	2	4	−1/2	0	1/2	3/32 (out)	0	3/32 (in)	3/16 (out)	0	3/16 (in)	NA	NA	14 1/2
82 Citation, Celebrity	(67)	1	2	3	−1/2	0	1/2	3/32 (out)	0	3/32 (in)	3/16 (out)	0	3/16 (in)	NA	NA	14 1/2
86-85 Camaro exc. Z28 .. (B)	(70)	2 1/2	3	3 1/2	1/2	1	1 1/2	1/16	5/32	1/4	1/8	5/16	1/2	NA	NA	NA
86-85 Camaro Z28 ..... (B)	(70)	3	3 1/2	4	1/2	1	1 1/2	1/16	5/32	1/4	1/8	5/16	1/2	NA	NA	NA
84-83 Camaro exc. Z28.. (B)	(70)	2	3	4	3/16	1	1 13/16	3/32	7/32	5/16	3/16	7/16	5/8	NA	NA	NA
84-83 Z28 .... (B)	(70)	2	3	4	3/16	1	1 13/16	1/16	5/32	1/4	1/8	5/16	1/2	NA	NA	NA
82 Camaro exc. Z28	(70)	2 1/2	3	3 1/2	1/2	1	1 1/2	3/32	7/32	5/16	3/16	7/16	5/8	NA	NA	NA
82 Z28	(70)	2 1/2	3	3 1/2	1/2	1	1 1/2	1/16	5/32	1/4	1/8	5/16	1/2	NA	NA	NA
86-85 Corvette																
Front	(10)	2 1/2	3	3 1/2	5/16	13/16	15/16	0	5/32	1/4	0	5/16	1/2	NA	NA	8 3/4
Rear	(58)(72)				−1/32	13/32	29/32	3/32	5/32	7/32	3/16	5/16	7/16			
84 Corvette																
Front	(10)	2 1/2	3	3 1/2	5/16	13/16	15/16	0	5/32	1/4	0	5/16	1/2	NA	NA	8 3/4
Rear	(58)(72)				−1/2	0	1/2	3/32	5/32	7/32	3/16	5/32	7/16			
86-77 Chevrolet (full size)	(10)	2	3	4	0	13/16	1 5/8	1/16	1/8	1/4	1/8	1/4	1/2	NA	NA	9 25/32
86-78 Malibu, El Camino, Monte Carlo	(10)				−5/16	1/2	1 5/16	1/16	1/8	1/4	1/8	1/4	1/2	NA	NA	7 7/8
W/Power Steering		2	3	4												
W/Manual Steering		0	1	2												
82-80 Corvette																
Front	(10)	1 1/4	2 1/4	3 1/4	0	3/4	1 1/2	1/8	1/4	3/8	1/4	1/2	3/4	NA	NA	7 11/16
Rear	(57)(58)				−1/2	0	1/2	0	1/16	1/8	0	1/8	1/4			
81-80 Camaro	(10)	0	1	2	3/16	1	1 13/16	1/16	1/8	1/4	1/8	1/4	1/2	NA	NA	10 3/8
81-80 Citation	(34)	−2	0	2	1/2	1	1 1/2	0	3/32	3/16	0	3/16	3/8	NA	NA	14 1/2
81-80 Chevette	(13)	2 1/2	4 1/2	6 1/2	−1/2	3/16	29/32	1/64 (out)	1/16 (in)	1/8 (in)	1/32 (out)	1/8 (in)	1/4 (in)	NA	NA	7 9/16
80-77 Monza	(11)	−1 13/16	−13/16	3/16	−5/8	3/16	1	3/16 (out)	1/16 (out)	1/16 (out)	3/8 (out)	1/8 (out)	1/8 (out)	NA	NA	8 9/16
79-78 Camaro	(10)	0	1	2	3/16	1	1 13/16	1/16	1/8	1/4	1/8	1/4	1/2	NA	NA	10 3/64
79-78 Chevette	(13)	2 1/2	4 1/2	6 1/2	−1/2	3/16	7/8	0	3/32	7/32	0	3/16	7/16	NA	NA	7 1/2
79-78 Nova	(10)				0	13/16	1 5/8	1/16	1/8	1/4	1/8	1/4	1/2	NA	NA	10
W/Power Steering		0	1	2												
W/Manual Steering		−2	−1	0												
79 Corvette																
Front	(10)	1 1/4	2 1/4	3 1/4	0	3/4	1 1/2	1/8	1/4	3/8	1/4	1/2	3/4	NA	NA	7 11/16
Rear	(57)(58)				−1	−1/2	0	1/8	3/16	1/4	1/4	3/8	1/2	NA	NA	
78-77 Corvette																
Front	(10)	1 1/4	2 1/4	3 1/4	0	3/4	1 1/2	1/8	1/4	3/8	1/4	1/2	3/4	NA	NA	7 11/16
Rear					−1 1/8	−7/8	−5/8	1/32 (out)	0	1/32 (in)	1/16 (out)	0	1/16 (in)	NA	NA	
77 Chevette	(13)	2 1/2	4 1/2	6 1/2	−1/2	3/16	1	1/32 (out)	1/16 (in)	1/8 (in)	1/16 (out)	1/8 (in)	1/4 (in)	NA	NA	7 1/2
77 Chevelle, El Camino	(10)							1/16 (out)	1/16 (in)	3/16 (in)	1/8 (out)	1/8 (in)	3/8 (in)	NA	NA	9 19/32
Left Wheel					3/16	1	1 13/16									
Right Wheel					−5/16	1/2	15/16									
W/Power Steering																
W/Radial Tires		2	2	3												
W/Bias Tires		0	1	2												
W/Manual Steering		0	1	2												

SECTION 2: SUSPENSION
Specifications

U.S. PASSENGER CAR SECTION

VEHICLE IDENTIFICATION YEAR MODEL	ADJ. ILL. NO.	CASTER (Degrees) MIN.	PREF.	MAX.	CAMBER (Degrees) MIN.	PREF.	MAX.	TOE-IN (Inches) MIN.	PREF.	MAX.	TOE-IN (Degrees) MIN.	PREF.	MAX.	TOE-OUT ON TURNS (Degrees) OUTSIDE WHEEL	INSIDE WHEEL	STRG. AXIS INCL. (DEG.)
77 Monte Carlo	(10)	4	5	6				1/16 (out)	1/16 (in)	3/16 (in)	1/8 (out)	1/8 (in)	3/8 (in)	NA	NA	9 19/32
Left Wheel					3/16	1	1 13/16									
Right Wheel					−5/16	1/2	1 5/16									
77 Camaro	(10)	0	1	2	3/16	1	1 13/16	1/16 (out)	1/16 (in)	3/16 (in)	1/8 (out)	1/8 (in)	3/8 (in)	NA	NA	10 11/32
77 Nova	(10)				0	13/16	1 5/8	1/16 (out)	1/16 (in)	3/16 (in)	1/8 (out)	1/8 (in)	3/8 (in)	NA	NA	10
W/Power Steering		0	1	2												
W/Manual Steering		−2	−1	0												
77 Vega	(11)	−1 13/16	−13/16	3/16	5/8	3/16	1	3/16 (out)	1/16 (out)	1/16 (in)	3/8 (out)	1/8 (out)	1/8 (in)	NA	NA	8 9/16

OLDSMOBILE MOTOR DIVISION (A)

(A) Maximum side-to-side variation after reset; caster and camber 1/2°.
(B) Toronado F.W.D. trim height is measured between the bottom of the rocker moulding to the floor. The measurement positions are:
86: Measure 22 27/32" (580 mm) from center of front wheel toward rear of car and 22 5/32" (563 mm) from center of rear wheel toward front of car.
85-83: Measure trim height from bottom of wheel opening moulding.
82-79: at the front edge of the door, and at 71" (1775 mm) behind the front edge of the door.
78-77: at 6" (152 mm) behind the front edge of the door, and at 66" (1676 mm) behind the front edge of the door. (See chart to right).

TORONADO F.W.D. TRIM HEIGHT (B)

Year	Front Inches	mm	Rear Inches	mm	Year	Max Side-to-Side, Front-to-Rear Deviation Inches	mm
86	8 29/32	226	8 7/8	225	86	3/4	19
85-83	9 5/8	245	9 5/8	245	85-83	7/16	10
82-79	9 1/2	242	9 1/2	242	82-77	3/4	19
78-77	9	229	9 1/4	235			

VEHICLE IDENTIFICATION YEAR MODEL	ADJ. ILL. NO.	CASTER MIN.	PREF.	MAX.	CAMBER MIN.	PREF.	MAX.	TOE-IN (In.) MIN.	PREF.	MAX.	TOE-IN (Deg.) MIN.	PREF.	MAX.	OUT WHEEL	IN WHEEL	STRG. AXIS INCL.
86-85 Firenza	(67)	23/32	1 23/32	2 23/32	3/16	27/32	1 1/2	5/32 (out)	1/16 (out)	1/32 (in)	5/16 (out)	1/8 (out)	1/16 (in)	NA	NA	13 1/2
84 Firenza	(67)	11/16	1 3/16	1 11/16	3/16	11/16	1 3/16	5/32 (out)	1/16 (out)	1/32 (in)	5/16 (out)	1/8 (out)	1/16 (in)	NA	NA	13 1/2
83 Firenza	(67)	23/32	1 23/32	2 23/32	7/32	23/32	1 7/32	5/32 (out)	1/16 (out)	1/32 (in)	5/16 (out)	1/8 (out)	1/16 (in)	NA	NA	13 1/2
82 Firenza	(67)	23/32	1 23/32	2 23/32	1/16	9/16	1 1/16	5/32 (out)	1/16 (out)	1/32 (in)	5/16 (out)	1/8 (out)	1/16 (in)	NA	NA	13 1/2
86-84 Omega, Ciera	(67)	29/32	1 29/32	2 29/32	−1/2	0	1/2	3/32 (out)	0	3/32 (in)	3/16	0	3/16 (in)	NA	NA	14 1/2
83-82 Omega, Ciera	(67)	1	2	3	−1/2	0	1/2	3/32 (out)	0	3/32 (in)	3/16	0	3/16 (in)	NA	NA	14 1/2
81-80 Omega	(34)	−2	0	2	1/2	1	1 1/2	0	3/32	3/16	0	3/16	3/8	NA	NA	14 1/2
86-85 Calais (F.W.D.)	N/A	11/16	1 11/16	2 11/16	3/16	27/32	1 1/2	1/4 (out)	1/8 (out)	1/32 (out)	1/2 (out)	1/4 (out)	1/16 (out)	NA	NA	13 1/2
85-84 Cutlass (R.W.D.)	(10)	2	3	4	−5/16	1/2	15/16	1/16	1/8	1/4	1/8	1/4	1/2	NA	NA	7
83-78 Cutlass (R.W.D.)	(10)				−5/16	1/2	15/16	1/16	1/8	1/4	1/8	1/4	1/2	NA	NA	7
W/Power Steering		2	3	4												
W/Manual Steering		0	1	2												
86-85 88, 98 Regency (F.W.D.) Front	(5)(67)	2	2 1/2	3				3/32 (out)	0	3/32 (in)	3/16 (out)	0	3/16 (in)	NA	NA	12 13/16
Left Wheel					−1	−1/2	0									
Right Wheel					0	1/2	1									
Rear	(67)				−13/16	−5/16	7/32	0	3/32	3/16	0	3/16	3/8			
86 88 & 98 Series, Cutlass (R.W.D.)	(10)	1 13/16	2 13/16	3 13/16	0	13/16	1 5/8	1/16	1/8	1/4	1/8	1/4	1/2	NA	NA	NA
85-84 88 & 98 Series (R.W.D.)	(10)	2	3	4	0	13/16	1 5/8	1/16	1/8	1/4	1/8	1/4	1/2	NA	NA	10 9/16
83 88 & 98 Series	(10)	2	3	4	0	13/16	1 5/8	1/16	1/8	1/4	1/8	1/4	1/2	NA	NA	10 9/16
82-78 88 & 98 Series	(10)	2	3	4	0	3/4	1 5/8	1/16	1/8	1/4	1/8	1/4	1/2	NA	NA	10 1/2
86 Toronado (B) Front	(15)(27)	1 5/16	2 5/16	3 5/16	−13/16	0	13/16	3/32 (out)	0	3/32 (in)	3/16	0	3/16 (in)	NA	NA	NA
Rear					−3/4	−3/8	−1/8	0	3/32	3/16	0	3/16	3/8			
85-79 Toronado (B) Front	(6)	1 1/2	2 1/2	3 1/2	−13/16	0	13/16	1/8 (out)	0	1/8 (in)	1/4	0	1/4 (in)	NA	NA	11
Rear (85-83)	(9)				−13/32	0	13/32	0	3/32	3/16	0	3/16	3/8			11
(82-79)	(9)				−15/16	1/2	5/16	0	3/32	3/16	0	3/16	3/8			
80-77 Starfire	(11)	−1 3/4	−3/4	1/4	−1/2	1/4	1	3/16 (out)	1/16 (out)	1/16 (in)	3/8 (out)	1/8 (out)	1/8 (in)	NA	NA	8 1/2

2–376

SUSPENSION
Specifications

U.S. PASSENGER CAR SECTION

VEHICLE IDENTIFICATION YEAR MODEL	ADJ. ILL. NO.	CASTER (Degrees) MIN.	PREF.	MAX.	CAMBER (Degrees) MIN.	PREF.	MAX.	TOE-IN (Inches) MIN.	PREF.	MAX.	TOE-IN (Degrees) MIN.	PREF.	MAX.	TOE-OUT ON TURNS (Degrees) OUTSIDE WHEEL	INSIDE WHEEL	STRG. AXIS INCL. (DEG.)
79-78 Omega	(10)				0	3/4	1 5/8	1/16	1/8	1/4	1/8	1/4	1/2	NA	NA	10 1/2
W/Power Steering		0	1	2												
W/Manual Steering		-2	-1	0												
78 Toronado	(B)	(6)	-1	0	1				1/8 (out)	0	1/8 (in)	1/4 (out)	0	1/4 (in)	NA	NA	11
Left Wheel					-1/2	5/16	1									
Right Wheel					-1	-5/16	1/2									
77 88 & 98 Series	(10)	2	3	4	0	3/4	1 1/2	0	1/8	1/4	0	1/4	1/2	NA	NA	10 1/2
77 Omega	(10)				0	3/4	1 1/2	1/16 (out)	1/16 (in)	3/16 (in)	1/8 (out)	1/8 (in)	3/8 (in)	NA	NA	10 1/2
W/Power Steering		0	1	2												
W/Manual Steering		-2	-1	0												
77 Cutlass	(10)	1	2	3				1/16 (out)	1/16 (in)	3/16 (in)	1/8 (out)	1/8 (in)	3/8 (in)	NA	NA	10 1/2
Left Wheel					1/4	1	1 3/4									
Right Wheel					-1/4	1/2	1 1/4									
77 Toronado	(B)	(6)	-1	0	1				1/8 (out)	0	1/8 (in)	1/4 (out)	0	1/4 (in)	NA	NA	11
Right Wheel					-1	-1/4	-1/2									
Left Wheel					-1/2	1/4	1									

PONTIAC MOTOR DIVISION (A)

(A) Maximum side to side variation after reset; caster and camber 1/2°.
(B) Measure ride height at points 32.5" to the rear of front wheel centerline and 17.2" forward of rear wheel centerline. Height from ground to rocker panel is 8.3" ± 0.3" front and rear.

VEHICLE IDENTIFICATION YEAR MODEL	ADJ. ILL. NO.	CASTER MIN.	PREF.	MAX.	CAMBER MIN.	PREF.	MAX.	TOE-IN (In) MIN.	PREF.	MAX.	TOE-IN (Deg) MIN.	PREF.	MAX.	OUTSIDE	INSIDE	STRG. AXIS INCL.
86-85 Grand Am	(67)	23/32	1 23/32	2 23/32	7/32	25/32	1 13/32	5/32 (out)	1/16 (out)	1/32 (in)	5/16 (out)	1/8 (out)	1/16 (in)	NA	NA	13 1/2
86-82 T1000	(13)	3	5	7	-1/2	3/16	29/32	1/64 (out)	1/16 (in)	1/8 (in)	1/32 (out)	1/8 (in)	1/4 (in)	NA	NA	7 9/16
86-84 Fiero																
	(13)	3	5	7	-5/16	1/2	15/16	1/16	1/8	3/16	1/8	1/4	3/8	NA	NA	9 3/8
	(67)				-1 1/2	-1	-1/2	1/16	1/8	3/16	1/8	1/4	3/8			
86-84 J2000	(67)	23/32	1 23/32	2 23/32	3/16	11/16	1 3/16	5/32 (out)	1/16 (out)	1/32 (out)	5/16 (out)	1/8 (out)	1/16 (in)	NA	NA	13 1/2
83 J2000	(67)	23/32	1 23/32	2 23/32	7/32	23/32	1 7/32	5/32 (out)	1/16 (out)	1/32 (in)	5/16 (out)	1/8 (out)	1/16 (in)	NA	NA	13 1/2
82 J2000	(67)	23/32	1 23/32	2 23/32	1/16	9/16	1 1/16	5/32 (out)	1/16 (out)	1/32 (out)	5/16 (out)	1/8 (out)	1/16 (in)	NA	NA	13 1/2
86-85 A6000	(67)	29/32	1 29/32	2 29/32	-1/2	0	1/2	3/32 (out)	0	3/32 (in)	3/16 (out)	0	3/16 (in)	NA	NA	14 9/16
84 Phoenix, A6000	(67)	0	2	4	-1/2	0	1/2	3/32 (out)	0	3/32 (in)	3/16 (out)	0	3/16 (in)	NA	NA	14 1/2
83-82 Phoenix, A6000	(67)	1	2	3	-1/2	0	1/2	3/32 (out)	0	3/32 (in)	3/16 (out)	0	3/16 (in)	NA	NA	14 1/2
85-84 Parisenne	(10)	2	3	4	0	13/16	1 5/8	1/16	1/8	1/4	1/8	1/4	1/2	NA	NA	9 3/4
86-85 Grand Prix, Bonneville		2	3	4	-5/16	1/2	15/16	1/16	1/8	1/4	1/8	1/4	1/2	NA	NA	8
84 Grand Prix, Bonneville		3	3 1/2	4	-5/16	1/2	15/16	1/16	1/8	1/4	1/8	1/4	1/2	NA	NA	8
83-82 Grand Prix	(10)	2	3	4	-5/16	1/2	15/16	1/16	1/8	1/4	1/8	1/4	1/2	NA	NA	8
83-82 Bonneville	(10)	2	3	4	-5/16	1/2	15/16	1/16	1/8	1/4	1/8	1/4	1/2	NA	NA	8
86 Firebird (All) (B)	(70)	3	3 1/2	4	1/2	1	1 1/2	1/32	5/32	1/4	1/16	5/16	1/2	NA	NA	NA
85 Firebird (All) (B)	(70)	2 5/16	2 13/16	3 5/16	1/2	1	1 1/2	1/32	5/32	1/4	1/16	5/16	1/2	NA	NA	NA
84 Firebird Trans Am (B)	(70)	2	3	4	3/16	1	1 13/16	1/16	5/32	1/4	1/8	5/16	1/2	NA	NA	NA
84 Firebird exc. Trans Am (B)	(70)	2	3	4	3/16	1	1 13/16	3/32	7/32	5/16	3/16	7/16	5/8	NA	NA	NA
83 Firebird exc. Trans Am (B)	(70)	2	3	4	1/8	1	1 13/16	3/32	7/32	5/16	3/16	7/16	5/8	NA	NA	NA
83 Trans Am (B)	(70)	2	3	4	1/8	1	1 13/16	1/16	5/32	1/4	1/8	5/16	1/2	NA	NA	NA
82 Firebird exc. Trans Am	(70)	2 1/2	3	3 1/2	1/2	1	1 1/2	3/32	7/32	5/16	3/16	7/16	5/8	NA	NA	NA
82 Trans Am	(70)	2 1/2	3	3 1/2	1/2	1	1 1/2	1/16	5/32	1/4	1/8	5/16	1/2	NA	NA	NA
81-80 Phoenix	(34)	-2	0	2	1/2	1	1 1/2	0	3/32	3/16	0	3/16	3/8	NA	NA	14 1/2
81-78 Grand Prix, LeMans, Grand Am					-5/16	1/2	15/16	1/16	1/8	1/4	1/8	1/4	1/2	NA	NA	8
W/Manual Steering	(10)	0	1	2												
W/Power Steering	(10)	2	3	4												
81-78 Firebird	(10)	0	1	2	3/16	1	1 13/16	1/16	1/8	1/4	1/8	1/4	1/2	NA	NA	10 3/8
81-78 Catalina, Bonneville	(10)	2	3	4	0	13/16	1 5/8	1/16	1/8	1/4	1/8	1/4	1/2	NA	NA	10 19/32
80-78 Sunbird	(10)	-1 3/16	-13/16	3/16	-5/8	-3/16	1	3/16 (out)	1/16 (out)	1/16 (in)	3/8 (out)	1/8 (out)	1/8 (in)	NA	NA	8 9/16

SECTION 2: SUSPENSION
Specifications

U.S. PASSENGER CAR SECTION

VEHICLE IDENTIFICATION YEAR MODEL	ADJ. ILL. NO.	CASTER (Degrees) MIN.	PREF.	MAX.	CAMBER (Degrees) MIN.	PREF.	MAX.	TOE-IN (Inches) MIN.	PREF.	MAX.	TOE-IN (Degrees) MIN.	PREF.	MAX.	TOE-OUT ON TURNS (Degrees) OUTSIDE WHEEL	INSIDE WHEEL	STRG. AXIS INCL. (DEG.)
79-78 Phoenix	(10)				0	13/16	1 5/8	1/16	1/8	1/4	1/8	1/4	1/2	NA	NA	10
W/Power Steering		0	1	2												
W/Manual Steering		−2	−1	0												
77 Phoenix, Ventura	(10)				0	3/4	1 1/2	1/16 (out)	1/16 (in)	3/16 (in)	1/8 (out)	1/8 (in)	3/8 (in)	18 1/2	20	10
W/Power Steering		0	1	2												
W/Manual Steering		−2	−1	0												
77 Grand Prix	(10)	4	5	6				1/16 (out)	1/16 (in)	3/16 (in)	1/8 (out)	1/8 (in)	3/8 (in)	NA	NA	10 3/8
Left Wheel					1/4	1	1 3/4									
Right Wheel					−1/4	1/2	1 1/4									
77 LeMans, Grand LeMans																
W/Power Steering	(10)							1/16 (out)	1/16 (in)	3/16 (in)	1/8 (out)	1/8 (in)	3/8 (in)			10 3/8
W/Belted Tires		0	1	2												
W/Radial Tires		1	2	3												
Left Wheel					1/4	1	1 3/4							19 3/16	20	
Right Wheel					−1/4	1/2	1 1/4							18 13/16	20	
W/Manual Steering	(10)	0	1	2				1/16 (out)	1/16 (in)	3/16 (in)	1/8 (out)	1/8 (in)	3/8 (in)			10 3/8
Left Wheel					1/4	1	1 3/4							19 3/16	20	
Right Wheel					−1/4	1/2	1 1/4							18 13/16	20	
77 Firebird	(10)	0	1	2	3/16	1	1 3/4	1/16 (out)	1/16 (in)	3/16 (in)	1/8 (out)	1/8 (in)	3/8 (in)	NA	NA	10 3/8
77 Sunbird, Astre	(11)	−1 3/4	−3/4	1/4	−9/16	3/16	1	3/16 (out)	1/16 (out)	1/16 (in)	3/8 (out)	1/8 (out)	1/8 (in)	NA	NA	8 9/16
77 Catalina, Bonneville, Brougham, Grandville	(10)	2	3	4	0	3/4	1 9/16	1/16	3/16	5/16	1/8	3/8	5/8	NA	NA	10 3/8

U.S. LIGHT TRUCK SECTION

AMERICAN MOTORS:
JEEP: (A) 77 models, 28°-29°; 28°; 75 w/std. tires 31°; w/F78 × 15 tires - 34°; 1981 CJ models 31°-32°; 1980-81 except CJ 37°-38°.

	ADJ. ILL. NO.	CASTER MIN.	PREF.	MAX.	CAMBER MIN.	PREF.	MAX.	TOE-IN (In.) MIN.	PREF.	MAX.	TOE-IN (Deg.) MIN.	PREF.	MAX.	TOE-OUT ON TURNS OUTSIDE	INSIDE	STRG. AXIS INCL.
86-84 Sportwagon, Cherokee, Wagoneer, Comanche	(17)	7	7 1/2	8	−1/2	0	1/2	1/16 (out)	0	1/16 (in)	1/8 (out)	0	1/8 (out)	32-33		NA
83-82 CJ-5	(17)	6	6	7	0	0	1/2	3/64	3/32	3/32	3/16			29		8 1/2
86-82 CJ-7 and Scrambler	(17)	6	6	7	0	0	1/2	3/64	3/32	3/32	3/16			32		8 1/2
86-82 Cherokee, Wagoneer and Pickup Truck	(17)	4	4	5	0	0	1/2	3/64	3/32	3/32	3/16			36-37		8 1/2
81 "CJ" Models	(17)	6	6	7	1 1/2	1 1/2	2	3/64	3/32	3/32	3/16			(A) 31		8 1/2
81-80 Cherokee, Wagoneer and Pickup Trucks	(17)		4			0		3/64	3/32	3/32	3/16			(A) 37		8 1/2
80-77 "CJ" Models	(17)		3			1 1/2		3/64	3/32	3/32	3/16			(A) 31 1/2		8 1/2
79-77 Cherokee, Wagoneer and Pickup Trucks	(17)		4			1 1/2		3/64	3/32	3/32	3/16			37 1/2		8 1/2

GENERAL MOTORS LIGHT TRUCKS
CHEVROLET & GMC VERSIONS

(A) Vehicle ride heights must be checked and corrected before alignment is performed.
(B) With JB8 or JF9 add .3°; with R05 subtract .4° for caster.
(C) Front torsion bar adjusts ride height (see end of GM Truck Section.)

	ADJ. ILL. NO.	CASTER MIN.	PREF.	MAX.	CAMBER MIN.	PREF.	MAX.	TOE-IN (In.) MIN.	PREF.	MAX.	TOE-IN (Deg.) MIN.	PREF.	MAX.	OUTSIDE	INSIDE	STRG. AXIS INCL.
86-85 Astro Van	(10)	1 11/16	2 11/16	3 11/16	1/8	15/16	1 3/4	1/16	5/32	1/4	1/8	5/16	1/2	NA	NA	NA
86-82 S10, S15 4×2	(10)	1	2	3	0	13/16	1 5/8	1/16	1/8	1/4	1/8	1/4	1/2	NA	NA	NA
86-83 S10, S15 4×4	(6)	1	2	3	0	13/16	1 5/8	1/16	1/8	1/4	1/8	1/4	1/2	NA	NA	NA

SUSPENSION
Specifications

GENERAL MOTORS LIGHT TRUCKS
CHEVROLET & GMC VERSIONS

CHART 1

VEHICLE IDENTIFICATION		ADJ. ILL. NO.	CASTER@HEIGHT MEASUREMENT (Degrees) Suspension Height Measurement (M)							CAMBER (Degrees)			TOE-IN (Inches)	TOE-IN (Degrees)
CHEVROLET YEAR MODEL	GMC MODEL		1½	2	2½	3	3½	3¾	4	MIN.	PREF.	MAX.		
1986-82														
C-10	C-1500	(24)	—	—	3⅝	3⅛	2⅝	2⅜	2	0	11/16	1⅜	3/16 ± 1/8	3/8 ± 1/4
C-20, 30	C-2500, 3500	(24)	—	—	1½	15/16	5/16	1/8	0	− 1/2	1/4	1	3/16 ± 1/8	3/8 ± 1/4
1986-85														
K-10, 20, 30	K-1500, 2500, 3500	N/A	—	—	8	8	8	8	8	3/4	1½	2	0 ± 1/8	0 ± 1/4
1984-82														
K-10, 20	K-1500, 2500	N/A	—	—	8	8	8	8	8	5/16	1	1 11/16	3/16 ± 1/8	3/8 ± 1/4
K-30	K-3500	N/A	—	—	8	8	8	8	8	− 3/16	1/2	1 3/16	3/16 ± 1/8	3/8 ± 1/4
1986-82														
G-10, 20	G-1500, 2500	(24)	3½	3⅛	2 11/16	2⅜	2⅛	1 15/16	1⅞	− 3/16	1/2	1 3/16	3/16 ± 1/8	3/8 ± 1/4
G-30	G-3500	(24)	2⅞	2 3/16	1⅝	1	1/2	3/16	0	− 1/2	3/16	7/8	3/16 ± 1/8	3/8 ± 1/4
1986-82														
P-10	P-1500	(24)	—	—	2 5/16	1 11/16	1 3/16	15/16	5/8	− 1/2	3/16	7/8	3/16 ± 1/8	3/8 ± 1/4
P-20, 30 (B)	P-2500, 3500 (B)	(24)	—	—	2 5/16	1 11/16	1 3/16	15/16	5/8	− 1/2	3/16	7/8	3/16 ± 1/8	3/8 ± 1/4
1986-85														
P-30 Motor Home	P-3500 Motor Home	(24)	—	—	5½	5	4⅜	4⅛	3⅞	− 1/2	3/16	7/8	1/4 ± 1/8	1/2 ± 1/4
1984-82														
P-30 Motor Home	P-3500 Motor Home	(24)	—	—	5½	5	4⅜	4⅛	3⅞	− 1/2	3/16	7/8	5/16 ± 1/8	5/8 ± 1/4

CHART 2

VEHICLE IDENTIFICATION		ADJ. ILL. NO.	CASTER@HEIGHT MEASUREMENT (Degrees) Suspension Height Measurement (M)							CAMBER (Degrees)			TOE-IN (Inches)	TOE-IN (Degrees)
CHEVROLET YEAR MODEL (A)	GMC MODEL (A)		1½	2	2½	3	3½	4¾	4½	MIN.	PREF.	MAX.		
1981-80														
G-10, 20	G-1500, 2500	(24)	3½	3⅛	2 11/16	2 13/32	2⅛	1 13/16	—	− 3/16	1/2	1 3/16	3/16 ± 1/8	3/8 ± 1/4
G-30	G-3500	(24)	2 3/16	2 3/16	1⅝	1	1/2	0	—	− 1/2	3/16	7/8	3/16 ± 1/8	3/8 ± 1/4
1981-79														
C-10	C-1500	(24)	—	—	2 13/32	1 13/16	1 3/16	11/16	3/16	− 1/2	3/16	1⅞	3/16 ± 1/8	3/8 ± 1/4
C-20, 30	C-2500, 3500	(24)	—	—	1½	29/32	5/16	0	− 11/16	− 1/2	3/16	7/8	3/16 ± 1/8	3/8 ± 1/4
K-10, 20	K-1500, 2500	N/A	—	—	8	8	8	8	8	5/16	1	1 11/16	0 ± 1/8	0 ± 1/4
K-30	K-3500	N/A	—	—	8	8	8	8	8	− 3/16	1/2	1 3/16	0 ± 1/8	0 ± 1/4
P-10	P-1500	(24)	—	—	2 5/16	1 11/16	1 3/16	5/8	1/8	− 1/2	3/16	7/8	3/16 ± 1/8	3/8 ± 1/4
P-20, 30 (B)	P-2500, 3500 (B)	(24)	—	2 29/32	2 11/16	1 11/16	1 3/16	5/8	3/16	− 1/2	3/16	7/8	3/16 ± 1/8	3/8 ± 1/4
P-30 Motor Home (B)	P-3500 Motor Home (B)		—	—	5½	5	4 13/32	3 13/16	3 5/16	− 1/2	3/16	7/8	5/16 ± 1/8	5/8 ± 1/4
1979														
G-10, 20	G-1500, 2500	(24)	2 29/32	2 5/16	2	1⅝	1 5/16	29/32	—	N/A			3/16 ± 1/8	3/8 ± 1/4
G-30	G-3500	(24)	3 13/32	2 11/16	2⅛	1½	1	13/32	—	N/A			3/16 ± 1/8	3/8 ± 1/4

CHART 3

VEHICLE IDENTIFICATION		ADJ. ILL. NO.	CASTER@HEIGHT MEASUREMENT (Degrees) Suspension Height Measurement (M)							CAMBER (Degrees)			TOE-IN (Inches)	TOE-IN (Degrees)
CHEVROLET YEAR MODEL (A)	GMC MODEL (A)		2½	3	3½	4	4½	4¾	5	MIN.	PREF.	MAX.		
1978-77														
C-10	C-1500	(24)	—	2	1¼	3/4	1/4	0	− 1/2		1/4		3/16 ± 1/8	3/8 ± 1/4
1978-77														
C-20, 30	C-2500, 3500	(24)	1½	1	1/2	0	− 1/2	3/4	− 1		1/4		3/16 ± 1/8	3/8 ± 1/4
1978-77														
K-10, 20, 30	K-1500, 2500, 3500	N/A	8	8	8	8	8	8	8		1/4		0 ± 1/8	0 ± 1/4
1978														
G-10, 20	G-1500, 2500	(24)	3¼	2¾	2½	2	1¾	1½	1½		1/4		3/16 ± 1/8	3/8 ± 1/4
G-30	G-3500	(24)	2¼	1½	1	1/2	0	− 1/4	− 1/2		1/4		3/16 ± 1/8	3/8 ± 1/4
1978-77														
P-10	P-1500	(24)	2½	2	1½	3/4	1/4	0	− 1/4		Chart 4		3/16 ± 1/8	3/8 ± 1/4
P-20, 30	P-2500, 3500	(24)	2½	2	1½	3/4	1/4	0	− 1/4		Chart 4		3/16 ± 1/8	3/8 ± 1/4
1977														
G-10, 20, 30	G-1500, 2500, 3500	(24)	2¼	1½	1	1/2	0	− 1/4	− 1/2		1/4		3/16 ± 1/8	3/8 ± 1/4

SUSPENSION
Specifications

U.S. LIGHT TRUCK SECTION

CHART 4

VEHICLE IDENTIFICATION		ADJ. ILL. NO.	CAMBER @ HEIGHT MEASUREMENT (Degrees) Suspension Height Measurement (MM)									
CHEVROLET YEAR MODEL	GMC MODEL		2½	2½	2¾	3½	3¾	4	4¼	4½	4¾	5	5¼
1978-77 P-10	P-1500	(24)	0	0	¼	¼	¼	¼	0	0	0	−¼	−½
P-20, 30	P-2500, 3500	(24)	0	0	¼	¼	¼	¼	¼	0	0	−¼	−½

NOTE: With vehicle level, measure frame angle with a bubble protractor. Record the suspension height measurement.
 a. Subtract an up-in-rear frame angle from a positive caster specification.
 b. Subtract a down-in-rear frame angle from a negative caster specification.
 c. Add an up-in-rear frame angle to a negative caster specification.
 d. Add a down-in-rear frame angle to a positive caster specification.

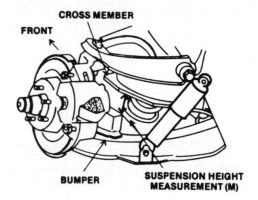

FRONT SUSPENSION HEIGHT
S10/S15 (4x2), Astro Van

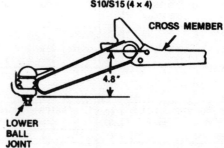

FRONT SUSPENSION HEIGHT
S10/S15 (4 x 4)

SUSPENSION
Specifications

CHRYSLER CORPORATION LIGHT TRUCKS

VEHICLE IDENTIFICATION YEAR MODEL	ADJ. ILL. NO.	CASTER (Degrees) MIN.	PREF.	MAX.	CAMBER (Degrees) MIN.	PREF.	MAX.	TOE-IN (Inches) MIN.	PREF.	MAX.	TOE-IN (Degrees) MIN.	PREF.	MAX.	TOE-OUT ON TURNS (Degrees) OUTSIDE WHEEL	INSIDE WHEEL	STRG. AXIS INCL. (DEG.)
CHRYSLER CORPORATION: DODGE AND PLYMOUTH TRUCKS (A) (B)																
86-84 Caravan, Mini Ram Van, Voyager																
Front ......(C)	(34)		3/8		− 1/4	5/16	3/4	1/8 (out)	1/16 (in)	7/32 (in)	1/4 (out)	1/8 (in)	7/16 (in)	NA	NA	NA
Rear (86)	(48)				− 1 3/8	− 13/16	− 1/4	5/16 (out)	0	5/16 (in)	5/8 (out)	0	5/8 (in)			
(85-84)	(48)				− 1 1/8	− 1/2	− 1/8	1/4 (out)	0	1/4 (in)	1/2 (out)	0	1/2 (in)			
86-81 B100, 200 300; PB100, 200, 300; MB300, 400 CB300, 400	(8)	1 1/4	2 1/2	3 3/4	− 1/4	3/8	1	0	1/8	1/4	0	1/4	1/2	NA	NA	NA
86-85 Ramcharger 4 × 2, D-150, D-250, D350	(6)	− 1/2	1/2	1 1/2	0	1/2	1	0	3/16	1/2	0	3/8	1	33	33	NA
84-83 Ramcharger 4 × 2, D-150, D-250, D350	(6)	− 1/2	1/2	1 1/2	0	1/2	1	1/16	1/8	3/16	1/8	1/4	3/8	33	33	NA
82-79 Ramcharger 4 × 2 Trail Duster 4 × 2, D100, 150, 200, 250, 300, 350	(8)	− 1/2	1/2	1 1/2	0	1/4	1	0	1/8	1/4	0	1/4	1/2	33	33	8 1/2
86-83 Ramcharger 4 × 4, W-150, W-250, W-350	(17)	1/2	2	3 1/2	1/2	1	1 1/2	1/8	1/4	3/8	1/4	1/2	3/4	①	①	8 1/2
82-79 W150, 200, Ramcharger 4 × 4, Trail Duster 4 × 4	(17)		3			1 1/2		1/8 (out)	0	1/8 (in)	1/4 (out)	0	1/4 (in)	NA	NA	8 1/2
82-79 W200, W250 w/extra equip., W300, W350, W400	(17)		3			1/2		1/8 (out)	0	1/8 (in)	1/4 (out)	0	1/4 (in)	NA	NA	8 1/2
80-77 B100, 200, 300, PB100, 200, 300; MB300, 400; CB300, CB400	(8)	1 1/4	2 1/4	3 1/4	0	1/2	1	0	1/8	1/4	0	1/4	1/2	NA	NA	NA
w/Heavy Front Axle Load	(8)	1/4		2 3/4	− 1/4	3/8	1	1/8 (out)	0	1/8 (in)	1/4 (out)	0	1/4 (in)	NA	NA	NA
78 D100, 200, 300	(6)	− 1/2	1/2	1 1/2		1/4		1/8 (out)	0	1/8 (in)	1/4 (out)	0	1/4 (in)	NA	NA	NA
78-77 Ramcharger 4 × 2, Trail Duster 4 × 2	(6)	− 1/2	1/2	1 1/2	0	1/4	1	0	1/8	1/4	0	1/4	1/2	NA	NA	NA
78-77 Ramcharger 4 × 4, Trail Duster 4 × 4 W100, 200	(17)	1 1/2	3	4 1/2	1	1 1/2	2	1/8 (out)	0	1/8 (in)	1/4 (out)	0	1/4 (in)	NA	NA	8 1/2
78-77 W200 w/extra equip., W300, W400	(17)	1 1/2	3	4 1/2	0	1/2	1	1/8 (out)	0	1/8 (in)	1/4 (out)	0	1/4 (in)	NA	NA	8 1/2
77 D100, 200, 300	(6)	− 1/2	1/2	1 1/2	0	1/4	1	0	1/8	1/4	0	1/4	1/2	NA	NA	NA

(A) Engines must be running on vehicles equipped with power steering when centering the steering wheel and setting toe.
(B) For RWD models caster shown is for unladen vehicles. Caster will vary with increased load. If loaded vehicle wanders, increase caster. If loaded vehicle is hard to steer in corners, decrease caster. ①Varies with ride height.
(C) Van caster shown; Wagon use 11/16°. Caster is not adjustable on FWD vehicles. Maximum left to right variation in caster is not to exceed 1 1/2° when checking alignment.

FORD LIGHT TRUCKS

YEAR, MODEL AND AXLE PART NUMBER RIDE HEIGHT MIN.	MAX.	ADJ. ILL. NO.	CASTER (Degrees) MIN.	MAX.	CAMBER (Degrees) MIN.	MAX.	TOE-IN (Inches)	TOE-IN (Degrees)
1986-84 Ranger, 4 × 2 (1)(3)(6)		(40)	(4)	(4)				
W/Forged Axle								
3¼	3½		5¼	8¼	−2	−½	1/32	1/16
3½	3¾		4½	7½	−1⅝	⅛	1/32	1/16
3¾	4		3½	6½	−½	1	1/32	1/16
4	4¼		3	6	½	1¼	1/32	1/16
4½	4¾		1⅞	4⅞	1¼	2¾	1/32	1/16
W/Stamped Axle								
3	3¼		5¼	8¼	−2	−½	1/32	1/16
3¼	3½		4½	7½	−1⅝	⅛	1/32	1/16
3½	3¾		3½	6½	−½	1	1/32	1/16
3¾	4		3	6	½	1¼	1/32	1/16
4¼	4½		1⅞	4⅞	1¼	2¾	1/32	1/16
1983 Ranger, 4 × 2 (1)(3)(6)		(40)	(4)	(4)				
2¾	3¼		4½	7	−1	1	1/32	1/16
3¼	3½		4	6½	−½	1¾	1/32	1/16
3½	4		3⅜	5⅞	0	2⅜	1/32	1/16
4	4¼		2⅝	5⅛	¾	3	1/32	1/16
4¼	4¾		2	4½	1½	3¾	1/32	1/16
1986-84 Ranger & Bronco II 4 × 4 (1)(3)(6)		(40)	(4)	(4)				
2¾	3		5½	8½	−2	−½	1/32	1/16
3¼	3½		4	7	−1	½	1/32	1/16
3½	3¾		3	6	0	1½	1/32	1/16
4	4¼		2	5	1	2½	1/32	1/16
4¼	4½		1	4	2	3½	1/32	1/16
1983 Ranger, 4 × 4 (1)(3)(6)		NA	(4)	(4)	(4)	(4)		
2¾	3¼		5	8	−1	½	1/32	1/16
3¼	3½		4	7	0	1½	1/32	1/16
3½	4		3	6	½	2	1/32	1/16
4	4¼		2½	5½	1¼	2¾	1/32	1/16
4¼	4¾		1¾	5	2	3¾	1/32	1/16
1986 F-250, 350 4 × 2 (1)(2)(A)		(40)	(4)	(4)	(4)	(4)		
3	3¼		5½	7½	−1	0	3/32	3/16
3¼	3½		5	7	−½	½	3/32	3/16
3½	3¾		4¼	6¼	½	1½	3/32	3/16
3¾	4		3½	5½	¾	1¾	3/32	3/16
4	4¼		2¾	4¾	1½	2½	3/32	3/16
1985 F250, 350, 4 × 2 (1)(2)(6)(A)		N/A	(4)	(4)	(4)	(4)		
2½	2¾		5¼	7¼	−1	½	3/32	3/16
2¾	3		5	7	−½	¾	3/32	3/16
3¼	3½		4½	6½	−⅛	1¼	3/32	3/16
3¾	4		4	6	½	1¾	3/32	3/16
4	4¼		3½	5½	¾	2¼	3/32	3/16
1984 F250, 350,4 × 2 (1)(3)(A)		N/A	(4)	(4)				
2½	2¾		5	7	−¾	¾	1/32	1/16
2¾	3		4	6	¼	1¾	1/32	1/16
3¼	3½		3¼	5¼	1	2⅝	1/32	1/16
3¾	4		2½	4½	2	3½	1/32	1/16
4	4¼		1½	3½	3	4½	1/32	1/16
1983 F-250, 350, 4 × 2 (1)(3)(A)		N/A	(4)	(4)				
2½	2¾		5½	7	−½	½	1/32	1/16
2¾	3¼		5	6	−½	1½	1/32	1/16
3¼	3½		4	5	¼	2¼	1/32	1/16
3½	4		3	4	1	3	1/32	1/16
4	4¼		(7)	(7)	2	3½	1/32	1/16
1982-81 F-250, 350, 4 × 2 (1)(3)		N/A	(4)	(4)	(4)	(4)		
2	2¼		5¾	9	−2½	0	1/32	1/16
2¼	2¾		4¾	8	−1½	1	1/32	1/16
2¼	3¼		3¾	7	−¼	1¾	1/32	1/16
3¼	3¾		2¾	6	¼	2¾	1/32	1/16
3½	4		1¾	5	1	3½	1/32	1/16
4	4¼		¾	4	2	4½	1/32	1/16

U.S. LIGHT TRUCK SECTION

YEAR, MODEL AND AXLE PART NUMBER RIDE HEIGHT MIN.	MAX.	ADJ ILL. NO.	CASTER (Degrees) MIN.	CASTER MAX.	CAMBER (Degrees) MIN.	CAMBER MAX.	TOE-IN (Inches)	TOE-IN (Degrees)
1986 F-350 4 × 4 w/Mono Beam Susp. (1)(3)(6)		N/A						
4 3/4	4 3/4		5	5	1 1/2	1 1/2	1/32	1/16
1986-84 F250, 350 4 × 4 (1)(3)(6)(8)		N/A						
5	5 1/4		3	5	− 1 3/4	− 1/4	1/32	1/16
5 1/2	5 3/4		3 1/8	5 1/8	− 3/4	3/4	1/32	1/16
6	6 1/4		3 1/4	5 1/4	1/2	1 1/2	1/32	1/16
6 1/4	6 1/2		3 3/8	5 3/8	1 1/2	3	1/32	1/16
6 3/4	7		3 1/2	5 1/2	2 1/2	4	1/32	1/16
1983 F-250, 350, 4 × 4 (1)(3)(A)		N/A	(4)	(4)	(4)	(4)		
5	5 1/2		3 1/16	5 1/8	− 1 3/4	3/4	1/32	1/16
5 1/2	6		3 1/8	5 1/4	− 3/4	1 3/4	1/32	1/16
6	6 1/4		3 1/4	5 3/8	1/4	3	1/32	1/16
6 1/4	6 3/4		3 3/8	5 1/2	1 1/2	4	1/32	1/16
6 3/4	7		3 1/2	5 1/2	2 1/2	4 1/4	1/32	1/16
1982-81 F-250, 350 4 × 4 (1)(3)		N/A	(4)	(4)				
4 3/4	5		3	5	4 3/4	5	1/32	1/16
5	5 1/2		3 1/8	5 1/8	− 1 3/4	3/4	1/32	1/16
5 1/2	6		3 1/8	5 1/8	− 3/4	1 3/4	1/32	1/16
6	6 1/4		3 1/4	5 1/4	1/4	2 3/4	1/32	1/16
6 1/4	6 3/4		3 3/8	5 3/8	1 1/4	4	1/32	1/16
6 3/4	7		3 1/2	5 1/2	2 1/2	5	1/32	1/16
1986-85 E150 (1)(3)		N/A	(4)	(4)	(4)	(4)		
3 1/2	3 3/4		—	—	—	—	1/32	1/16
4	4 1/2		7 1/2	9 1/2	− 1 1/4	1/4	1/32	1/16
4 1/2	5		6 1/4	8 1/4	− 1/8	1 1/4	1/32	1/16
5	5 1/2		5	7	7/8	2 1/4	1/32	1/16
5 1/2	5 3/4		3 1/4	5 1/4	1 3/4	3 1/4	1/32	1/16
1984 E-150, (1)(3)		N/A	(4)	(4)	(4)	(4)		
3 1/2	3 3/4		—	—	—	—	1/32	1/16
4	4 1/4		4 5/8	6 1/4	− 3/4	1/2	1/32	1/16
4 1/2	4 3/4		3 1/4	5 1/4	1/2	1 3/4	1/32	1/16
5	5 1/4		2	4	1 1/2	2 3/4	1/32	1/16
5 1/2	5 3/4		1/4	2 1/4	2 3/8	3 5/8	1/32	1/16
1983 E-100, 150 (1)(3)		(40)						
3 3/4	4		—	—	− 3/4	3/4	1/32	1/16
4	4 1/2		4 1/2	5 1/4	− 5/8	1 5/8	1/32	1/16
4 1/2	5		3 1/4	4	3/8	2 5/8	1/32	1/16
5	5 1/2		2	2 3/4	1 1/4	3 5/8	1/32	1/16
5 1/2	5 3/4		3/4	2 1/4	2 1/4	4 1/8	1/32	1/16
1982-81 E-100, 150 (1)(3)		N/A						
3 1/4	3 1/2		6 1/4	8	− 1 3/4	− 1/4	1/32	1/16
3 1/2	3 3/4		5 3/4	7 1/4	− 1 1/2	1/4	1/32	1/16
3 3/4	4		5	6 3/4	− 1	3/4	1/32	1/16
4	4 1/4		4 1/2	5 3/4	− 1/2	1 1/4	1/32	1/16
4 1/4	4 1/2		4	5 1/4	0	1 3/4	1/32	1/16
4 1/2	4 3/4		3 1/4	4 1/2	1/2	2 1/4	1/32	1/16
4 3/4	5		2 1/2	4	1	2 3/4	1/32	1/16
5	5 1/4		2	3 1/4	1 1/2	3 1/4	1/32	1/16
5 1/4	5 1/2		1 1/2	2 3/4	2	3 3/4	1/32	1/16
1986-84 F150, 4 × 2 (1)(3)(6)(A)		(40)	(4)	(4)				
3 1/4	3 1/2		5	7	− 3/4	3/4	1/32	1/16
3 1/2	4		4	6	1/4	1 1/4	1/32	1/16
4	4 1/4		3 1/4	5 1/4	1/2	2	1/32	1/16
4 1/4	4 3/4		2 1/2	4 1/2	2	3 1/2	1/32	1/16
4 3/4	5		1 1/2	3 1/2	3	4 1/2	1/32	1/16
1983 F-100, 150, 4 × 2 (1)(3)(A)		(40)	(4)	(4)				
2 3/4	3 1/4		5 1/2	7 1/2	− 1 1/2	3/4	1/32	1/16
3 1/4	3 1/2		5	6	− 3/4	1 1/2	1/32	1/16
3 1/2	4		4 1/4	5 1/4	1/4	2 1/2	1/32	1/16
4	4 1/4		3 1/4	4 1/4	1	3 1/2	1/32	1/16
4 1/4	4 3/4		2 1/2	3 1/2	2	4 1/2	1/32	1/16

(A) Right side height measurement shown; left side height should be "0" to 7/16" higher on any one vehicle.

SECTION 2 SUSPENSION
Specifications

U.S. LIGHT TRUCK SECTION

YEAR, MODEL AND AXLE PART NUMBER RIDE HEIGHT MIN. — MAX.	ADJ. ILL. NO.	CASTER (Degrees) MIN.	MAX.	CAMBER (Degrees) MIN.	MAX.	TOE-IN (Inches)	TOE-IN (Degrees)
1982-81 F-100, F-150 4 × 2 (1)(3)	N/A	(4)	(4)	(4)	(4)		
2¼ — 2¾		6	10	−3	−½	1/32	1/16
2¾ — 3¼		5	9	−2	½	1/32	1/16
3¼ — 3½		4	8	−1¼	1¼	1/32	1/16
3½ — 4		3	7	−¼	2¼	1/32	1/16
4 — 4¼		2	6	½	3	1/32	1/16
4¼ — 4¾		1	5	1½	4	1/32	1/16
1980 F-100, 150 4 × 2 (1)(2)	N/A	(4)	(4)	(4)	(4)		
2¼ — 2¾		6	10	−3	−½	3/32	3/16
2¾ — 3¼		5	9	−2	½	3/32	3/16
3¼ — 3½		4	8	−1¼	1¼	3/32	3/16
3½ — 4		3	7	−¼	2¼	3/32	3/16
4 — 4¼		2	6	½	3	3/32	3/16
4¼ — 4¾		1	5	1½	4	3/32	3/16
1982-81 F-150, Bronco 4 × 4 (1)(3)	(40)	(4)	(4)				
2¾ — 3¼		6	9	−2½	−¼	1/32	1/16
3¼ — 3½		5	8	−1¾	½	1/32	1/16
3½ — 4		4	7	−¾	1½	1/32	1/16
4 — 4¼		3	6	0	2¼	1/32	1/16
4¼ — 4¾		2	5	1	3¼	1/32	1/16
4¾ — 5		1	4	1¾	4	1/32	1/16
1986-84 F150, Bronco 4 × 4 (1)(3)(6)	(40)	(4)	(4)				
3¼ — 3½		6	8	−1¾	−¼	1/32	1/16
3½ — 3¾		5	7	−¾	¾	1/32	1/16
4 — 4¼		4	6	¼	1¾	1/32	1/16
4¼ — 4½		3	5	1¼	2¾	1/32	1/16
1983 F-150, Bronco 4 × 4 (1)(3)	(40)	(4)	(4)				
3¼ — 3½		6	7	−1½	¾	1/32	1/16
3½ — 4		5	6	−¾	1¾	1/32	1/16
4 — 4¼		4	5	¼	2¾	1/32	1/16
4¼ — 4¾		3	4	1¼	3½	1/32	1/16
1980 F-150, Bronco 4 × 4 (1)(2)	(40)	(4)	(4)				
2¾ — 3¼		6	9	−2½	−¼	3/32	3/16
3¼ — 3½		5	8	−1¾	½	3/32	3/16
3½ — 4		4	7	−¾	1½	3/32	3/16
4 — 4¼		3	6	0	2¼	3/32	3/16
4¼ — 4¾		2	5	1	3¼	3/32	3/16
4¾ — 5		1	4	1¾	4	3/32	3/16
1986-84 E250, 350 (1)(3)	N/A	(4)	(4)	(4)	(4)		
3¼ — 3½		—	—	—	—	1/32	1/16
3¾ — 4		7⅝	9⅝	−¾	½	1/32	1/16
4¼ — 4½		6¼	8¼	¼	1½	1/32	1/16
4¾ — 5		5	7	1¼	2½	1/32	1/16
5¼ — 5½		3¾	5¾	2¼	3½	1/32	1/16
1983 E-250, 350 (1)(3)	N/A						
3¾ — 4		—	—	−⅝	⅞	1/32	1/16
4 — 4½		7¼	8	−½	1⅞	1/32	1/16
4½ — 5		6	6¾	½	2⅞	1/32	1/16
5 — 5½		4¾	5½	1½	3⅞	1/32	1/16
5½ — 5¾		3¼	4	2½	4⅜	1/32	1/16
1982-80 E-250, 350 (1)(3)	N/A						
3¼ — 3½		9	10½	−1¾	−¼	1/32	1/16
3½ — 3¾		8½	9¾	−1½	¼	1/32	1/16
3¾ — 4		7⅞	9	−1	¾	1/32	1/16
4 — 4¼		7⅛	8½	−½	1¼	1/32	1/16
4¼ — 4½		6½	7¾	0	1¾	1/32	1/16

2-384

U.S. LIGHT TRUCK SECTION

YEAR, MODEL AND AXLE PART NUMBER RIDE HEIGHT MIN.	MAX.	ADJ. ILL. NO.	CASTER (Degrees) MIN.	MAX.	CAMBER (Degrees) MIN.	MAX.	TOE-IN (Inches)	TOE-IN (Degrees)
1982-80 E-250, 350 (1)(3)		N/A						
4 1/2	4 3/4		5 3/4	7	1/2	2 1/4	1/32	1/16
4 3/4	5		5 1/4	6 1/2	1	2 3/4	1/32	1/16
5	5 1/4		4 5/8	6	1 1/2	3 1/4	1/32	1/16
5 1/4	5 1/2		4	5 1/2	2	3 3/4	1/32	1/16
1980 F-250, 350 4 × 2 (1)(3)		N/A	(4)	(4)	(4)	(4)		
2	2 1/4		5 1/4	9	−2 1/2	0	1/32	1/16
2 1/4	2 3/4		4 3/4	8	−1 1/2	1	1/32	1/16
2 3/4	2 1/4		3 3/4	7	− 3/4	1 3/4	1/32	1/16
3 1/4	3 1/2		2 3/4	6	1/4	2 3/4	1/32	1/16
3 1/2	4		1 3/4	5	1	3 1/2	1/32	1/16
4	4 1/4		3/4	4	2	4 1/2	1/32	1/16
1980 F-250, 350 4 × 4 (1)(2)		(40)	(4)	(4)				
4 1/4	4 3/4		3 1/2	5 3/4	−4	−1 1/4	3/32	3/16
4 3/4	5		3 1/4	5 1/2	−2 3/4	0	3/32	3/16
5	5 1/2		3	5 1/4	−1 1/2	1 1/4	3/32	3/16
5 1/2	6		2 3/4	5	− 1/4	2 1/2	3/32	3/16
6	6 1/4		2 1/2	4 3/4	1	3 3/4	3/32	3/16
6 1/4	6 3/4		2 1/4	4 1/2	2 1/4	5	3/32	3/16
1980 E-100, 150 (1)(5)		N/A						
3 1/4	3 1/2		6 1/4	8	−1 3/4	− 1/4	1/8	1/4
3 1/2	3 3/4		5 3/4	7 1/4	−1 1/2	1/4	1/8	1/4
3 3/4	4		5	6 3/4	−1	3/4	1/8	1/4
4	4 1/4		4 1/2	5 3/4	− 1/2	1 1/4	1/8	1/4
4 1/4	4 1/2		4	5 1/4	0	1 3/4	1/8	1/4
4 1/2	4 3/4		3 1/4	4 1/2	1/2	2 1/4	1/8	1/4
4 3/4	5		2 1/2	4	1	2 3/4	1/8	1/4
5	5 1/4		2	3 1/4	1 1/2	3 1/4	1/8	1/4
5 1/4	5 1/2		1 1/2	2 3/4	2	3 3/4	1/8	1/4

(1) All vehicles with normal operating attitude. **(2)** Nominal toe setting is 2.5 mm (3/32 inch). Range is 0.8 mm (1/32 inch) out to 5.6 mm (7/32 inch) in.
(3) Nominal toe setting is 8 mm (1/32 inch). Range is 2.5 mm (3/32 inch) out to 4 mm (5/32 inch) in. **(4)** Not Adjustable. **(5)** Toe range is 0" to 1/4". 1/8" nominal.
(6) Side-to-side variation: Caster 1 1/2°, Camber 23/32°. **(7)** Vehicle height is too high. **(8)** Except monobeam suspension.

YEAR, MODEL AND AXLE PART NUMBER RIDE HEIGHT MIN.	MAX.	ADJ. ILL. NO.	CASTER (Degrees) MIN.	MAX.	CAMBER (Degrees) MIN.	MAX.	TOE-IN (Inches)	TOE-IN (Degrees)
1979 F-100, F-150, F-250 (D) (6200-6800 GVW) (4 × 2)		N/A						
2 3/4	3		8 3/8	9 5/8	−2	− 1/4	3/32	3/16
3	3 1/4		7 3/4	9	−1 1/2	1/8	3/32	3/16
3 1/4	3 1/2		7	8 3/8	−1 1/8	1/2	3/32	3/16
3 1/2	3 3/4		6 1/4	7 5/8	− 3/4	1	3/32	3/16
3 3/4	4		5 7/8	7 1/8	− 3/8	1 1/4	3/32	3/16
4	4 1/4		5 1/8	6 1/2	0	1 5/8	3/32	3/16
4 1/4	4 1/2		4 1/2	5 7/8	3/8	2	3/32	3/16
4 1/2	4 3/4		3 3/4	5 1/4	3/4	2 3/8	3/32	3/16
4 3/4	5		3 1/4	4 5/8	1 1/4	1 3/4	3/32	3/16
5	5 1/4		2 1/2	4	1 7/8	3 1/8	3/32	3/16
1979 F-250 R/C (7700-7900) (4 × 2) (D) F-250 S/C (6300-7800) (D8TA-BA) (D)		N/A						
2 3/4	3		7 3/4	9	−1 7/8	− 1/4	3/32	3/16
3	3 1/4		7	7 3/8	−1 1/2	1/8	3/32	3/16
3 1/4	3 1/2		6 3/8	7 3/4	−1 1/8	1/2	3/32	3/16
3 1/2	3 3/4		5 7/8	7 1/8	− 3/4	7/8	3/32	3/16
3 3/4	4		5 1/8	6 1/2	− 3/8	1 1/4	3/32	3/16
4	4 1/4		4 1/2	5 7/8	0	1 5/8	3/32	3/16
4 1/4	4 1/2		3 7/8	5 1/4	3/8	2	3/32	3/16
4 1/2	4 3/4		3 1/4	4 5/8	7/8	2 1/2	3/32	3/16
4 3/4	5		2 5/8	4	1 1/4	2 3/4	3/32	3/16
5	5 1/4		2	3 3/8	1 5/8	3 1/8	3/32	3/16

SECTION 2
SUSPENSION
Specifications

U.S. LIGHT TRUCK SECTION

YEAR, MODEL AND AXLE PART NUMBER RIDE HEIGHT MIN.	MAX.	ADJ. ILL. NO.	CASTER (Degrees) MIN.	MAX.	CAMBER (Degrees) MIN.	MAX.	TOE-IN (Inches)						TOE-IN (Degrees)		
1979 F-350, F-250 S/C (3100) (D) (4 × 2) F-250 S/C RPO Suspension (D8TA-DA) (D)		N/A													
2¾	3		9⅝	11	−2⅛	−⅝	3/32						3/16		
3	3¼		8⅞	10⅜	−1¾	−¼	3/32						3/16		
3¼	3½		8⅜	9¾	−1⅜	⅛	3/32						3/16		
3½	3¾		7¾	9	−1⅛	½	3/32						3/16		
3¾	4		7	8⅜	−¾	¾	3/32						3/16		
4	4¼		6⅜	7¾	−1¼	1¼	3/32						3/16		
4¼	4½		5¾	7⅛	0	1⅝	3/32						3/16		
4½	4¾		5⅛	6½	⅜	2	3/32						3/16		
4¾	5		5⅛	6½	¾	2⅜	3/32						3/16		
5	5¼		3⅞	5¼	1¼	2¼	3/32						3/16		
1979 E-100/150 (D7UA-BB) (C) (4 × 2)		N/A													
3¼	3½		6¼	8	−1¾	−¼	1/32						1/16		
3½	3¾		5¾	7¼	−1½	¼	1/32						1/16		
3¾	4		5	6¾	−1	¾	1/32						1/16		
4	4¼		4½	5¾	−½	1¼	1/32						1/16		
4¼	4½		4	5¼	0	1¾	1/32						1/16		
4½	4¾		3¼	4½	½	2¼	1/32						1/16		
4¾	5		2½	4	1	2¾	1/32						1/16		
5	5¼		2	3¼	1½	3¼	1/32						1/16		
5¼	5½		1½	2¾	2	3¾	1/32						1/16		
1979 E-250/350 (D5UA-AB) (C) (4 × 2)		N/A													
3¼	3½		9	10½	−1¾	−¼	1/32						1/16		
3½	3¾		8½	9¾	−1½	¼	1/32						1/16		
3¾	4		7⅞	9	−1	¾	1/32						1/16		
4	4¼		7⅛	8½	−½	1¼	1/32						1/16		
4¼	4½		6½	7¾	0	1¾	1/32						1/16		
4½	4¾		5¾	7	½	2¼	1/32						1/16		
4¾	5		5¼	6½	1	2¾	1/32						1/16		
5	5¼		4⅝	6	1½	3¼	1/32						1/16		
5¼	5½		4	5½	2	3¾	1/32						1/16		
1978 F-100, 150, 250 (4 × 2) 6200-6900 GVW Regular Cab		N/A													
2¾	3		8¼	9⅝	−1	½	1/32 (out)	3/32 (in)	7/32 (in)	1/16 (out)	3/16 (in)	7/16 (in)			
3	3¼		7¾	9	−½	1	1/32 (out)	3/32 (in)	7/32 (in)	1/16 (out)	3/16 (in)	7/16 (in)			
3¼	3½		7	8⅜	−¼	1½	1/32 (out)	3/32 (in)	7/32 (in)	1/16 (out)	3/16 (in)	7/16 (in)			
3½	3¾		6¼	7⅜	0	1¾	1/32 (out)	3/32 (in)	7/32 (in)	1/16 (out)	3/16 (in)	7/16 (in)			
3¾	4		5¾	7	½	2¼	1/32 (out)	3/32 (in)	7/32 (in)	1/16 (out)	3/16 (in)	7/16 (in)			
4	4¼		5	6½	¾	2½	1/32 (out)	3/32 (in)	7/32 (in)	1/16 (out)	3/16 (in)	7/16 (in)			
4¼	4½		4½	5⅞	1¼	3	1/32 (out)	3/32 (in)	7/32 (in)	1/16 (out)	3/16 (in)	7/16 (in)			
4½	4¾		3¾	5	1⅝	3¼	1/32 (out)	3/32 (in)	7/32 (in)	1/16 (out)	3/16 (in)	7/16 (in)			
4¾	5		3⅛	4½	2	3¾	1/32 (out)	3/32 (in)	7/32 (in)	1/16 (out)	3/16 (in)	7/16 (in)			
5	5¼		2½	4	3⅜	4⅛	1/32 (out)	3/32 (in)	7/32 (in)	1/16 (out)	3/16 (in)	7/16 (in)			
1978 F-250 7800-8000 GVW Regular Cab 6350-7800 Super Cab (4 × 2)		N/A													
2¾	3		7⅞	8⅞	−½	1	1/32 (out)	3/32 (in)	7/32 (in)	1/16 (out)	3/16 (in)	7/16 (in)			
3	3¼		6⅞	8⅜	−1¼	1½	1/32 (out)	3/32 (in)	7/32 (in)	1/16 (out)	3/16 (in)	7/16 (in)			
3¼	3½		5⅜	7¾	0	1¾	1/32 (out)	3/32 (in)	7/32 (in)	1/16 (out)	3/16 (in)	7/16 (in)			
3¾	4		5	6½	¾	2½	1/32 (out)	3/32 (in)	7/32 (in)	1/16 (out)	3/16 (in)	7/16 (in)			

SUSPENSION
Specifications
SECTION 2

U.S. LIGHT TRUCK SECTION

YEAR, MODEL AND AXLE PART NUMBER RIDE HEIGHT MIN. / MAX.	ADJ. ILL. NO.	CASTER (Degrees) MIN.	CASTER (Degrees) MAX.	CAMBER (Degrees) MIN.	CAMBER (Degrees) MAX.	TOE-IN (Inches)			TOE-IN (Degrees)		
1978 F-250 7800-8000 GVW Regular Cab 6350-7800 Super Cab (4 × 2)	N/A										
4 / 4¼		4½	5⅞	1¼	3	1/32 (out)	3/32 (in)	7/32 (in)	1/16 (out)	3/16 (in)	7/16 (in)
4¼ / 4½		3⅜	5⅛	1¾	3¼	1/32 (out)	3/32 (in)	7/32 (in)	1/16 (out)	3/16 (in)	7/13 (in)
4½ / 4¾		3⅛	4½	2	3½	1/32 (out)	3/32 (in)	7/32 (in)	1/16 (out)	3/16 (in)	7/16 (in)
4¾ / 5		2½	4	2⅜	4⅛	1/32 (out)	3/32 (in)	7/32 (in)	1/16 (out)	3/16 (in)	7/16 (in)
5 / 5¼		2	3¼	2¾	4½	1/32 (out)	3/32 (in)	7/32 (in)	1/16 (out)	3/16 (in)	7/16 (in)
1978 F-350 All F-250 8100 GVW Super Cab F-250 Super Cab W/RPO Suspension (4 × 2)	N/A										
2¾ / 3		9¾	10⅞	− ⅜	− ¼	1/32 (out)	3/32 (in)	7/32 (in)	1/16 (out)	3/16 (in)	7/16 (in)
3 / 3¼		9	10¼	− 1	− ⅝	1/32 (out)	3/32 (in)	7/32 (in)	1/16 (out)	3/16 (in)	7/16 (in)
3¼ / 3½		8¼	9⅝	− ½	1	1/32 (out)	3/32 (in)	7/32 (in)	1/16 (out)	3/16 (in)	7/16 (in)
3½ / 3¾		7⅝	8⅝	− ¼	1½	1/32 (out)	3/32 (in)	7/32 (in)	1/16 (out)	3/16 (in)	7/16 (in)
3¾ / 4		6⅞	8⅛	0	1¾	1/32 (out)	3/32 (in)	7/32 (in)	1/16 (out)	3/16 (in)	7/16 (in)
4 / 4¼		6⅜	7¼	½	2¼	1/32 (out)	3/32 (in)	7/32 (in)	1/16 (out)	3/16 (in)	7/16 (in)
4¼ / 4½		5¾	7	¾	2½	1/32 (out)	3/32 (in)	7/32 (in)	1/16 (out)	3/16 (in)	7/16 (in)
4½ / 4¾		5⅛	6½	1¼	3	1/32 (out)	3/32 (in)	7/32 (in)	1/16 (out)	3/16 (in)	7/16 (in)
4¾ / 5		4½	5⅞	1¾	3¼	1/32 (out)	3/32 (in)	7/32 (in)	1/16 (out)	3/16 (in)	7/16 (in)
5 / 5¼		3⅞	5⅛	2	3¼	1/32 (out)	3/32 (in)	7/32 (in)	1/16 (out)	3/16 (in)	7/16 (in)
1978 E-100, 150 (4 × 2)	N/A										
4 / 4¼		3¾	6½	− ¾	½	0		¼	0		½
4¼ / 4½		3¼	5¾	− ½	¾	0		¼	0		½
4½ / 4¾		2½	5¼	0	1¼	0		¼	0		½
4¾ / 5		2	4½	½	1¾	0		¼	0		½
5 / 5¼		1¼	4	1¼	2½	0		¼	0		½
5¼ / 5½		¾	3¼	1¾	3¼	0		¼	0		½
5½ / 5¾		0	2¾	1½	3¾	0		¼	0		½
1978 E-250, 350 (4 × 2)	N/A										
4 / 4¼		6¼	9	− 1	¾	3/32 (out)	1/32 (in)	5/32 (in)	3/16 (out)	1/16 (in)	5/16 (in)
4¼ / 4½		5¾	8¼	− ½	1¼	3/32 (out)	1/32 (in)	5/32 (in)	3/16 (out)	1/16 (in)	5/16 (in)
4½ / 4¾		5¼	7¾	0	1¾	3/32 (out)	1/32 (in)	5/32 (in)	3/16 (out)	1/16 (in)	5/16 (in)
4¾ / 5		4½	7¼	½	2¼	3/32 (out)	1/32 (in)	5/32 (in)	3/16 (out)	1/16 (in)	5/16 (in)
5 / 5¼		4	6½	1	2¾	3/32 (out)	1/32 (in)	5/32 (in)	3/16 (out)	1/16 (in)	5/16 (in)
5¼ / 5½		3¼	6	1½	3¼	3/32 (out)	1/32 (in)	5/32 (in)	3/16 (out)	1/16 (in)	5/16 (in)
1977 F-100, 150, 250 6200-6900 GVW Regular Cab (4 × 2)	N/A										
2¾ / 3		7½	10	− 1	½	1/32 (out)	3/32 (in)	7/32 (in)	1/16 (out)	3/16 (in)	7/16 (in)
3 / 3¼		7	9½	− ½	1	1/32 (out)	3/32 (in)	7/32 (in)	1/16 (out)	3/16 (in)	7/16 (in)
3¼ / 3½		6½	9	− ¼	1½	1/32 (out)	3/32 (in)	7/32 (in)	1/16 (out)	3/16 (in)	7/16 (in)

U.S. LIGHT TRUCK SECTION

YEAR, MODEL AND AXLE PART NUMBER RIDE HEIGHT MIN.	MAX.	ADJ. ILL. NO.	CASTER (Degrees) MIN.	MAX.	CAMBER (Degrees) MIN.	MAX.	TOE-IN (Inches)			TOE-IN (Degrees)		
1977 F-100, 150, 250 6200-6900 GVW Regular Cab (4 × 2)		N/A										
3½	3¾		5¾	8½	0	1¾	1/32 (out)	3/32 (in)	7/32 (in)	1/16 (out)	3/16 (in)	7/16 (in)
3¾	4		5¼	7¾	½	2¼	1/32 (out)	3/32 (in)	7/32 (in)	1/16 (out)	3/16 (in)	7/16 (in)
4	4¼		4½	7	¾	2½	1/32 (out)	3/32 (in)	7/32 (in)	1/16 (out)	3/16 (in)	7/16 (in)
4¼	4½		4	6½	1¼	3	1/32 (out)	3/32 (in)	7/32 (in)	1/16 (out)	3/16 (in)	7/16 (in)
1977 F-250 7800-8000 GVW Regular Cab; 6350-7800 Super Cab (4 × 2)		N/A										
2¾	3											
3	3¼		6½	9	− ¼	1½	1/32 (out)	3/32 (in)	7/32 (in)	1/16 (out)	3/16 (in)	7/16 (in)
3¼	3½		5¾	8½	0	1¾	1/32 (out)	3/32 (in)	7/32 (in)	1/16 (out)	3/16 (in)	7/16 (in)
3½	3¾		5¼	7¾	½	2¼	1/32 (out)	3/32 (in)	7/32 (in)	1/16 (out)	3/16 (in)	7/16 (in)
3¾	4		4½	7	¾	2½	1/32 (out)	3/32 (in)	7/32 (in)	1/16 (out)	3/16 (in)	7/16 (in)
4	4¼		4	6½	1¼	3	1/32 (out)	3/32 (in)	7/32 (in)	1/16 (out)	3/16 (in)	7/16 (in)
4¼	4½		3¼	6	1¾	3¼	1/32 (out)	3/32 (in)	7/32 (in)	1/16 (out)	3/16 (in)	7/16 (in)
4½	4¾		2¾	5½	2	3½	1/32 (out)	3/32 (in)	7/32 (in)	1/16 (out)	3/16 (in)	7/16 (in)
1977 E-100, 150 (4 × 2)		N/A										
4¼	4½		3¼	5¾	− ¾	¾	1/32 (out)	3/32 (in)	7/32 (in)	1/16 (out)	3/16 (in)	7/16 (in)
4½	4¾		2½	5¼	− ½	1¼	1/32 (out)	3/32 (in)	7/32 (in)	1/16 (out)	3/16 (in)	7/16 (in)
4¾	5		2	4½	0	1¾	1/32 (out)	3/32 (in)	7/32 (in)	1/16 (out)	3/16 (in)	7/16 (in)
5	5¼		1¼	4	½	2½	1/32 (out)	3/32 (in)	7/32 (in)	1/16 (out)	3/16 (in)	7/16 (in)
5¼	5½		¾	3¼	1½	3¼	1/32 (out)	3/32 (in)	7/32 (in)	1/16 (out)	3/16 (in)	7/16 (in)
5½	5¾		0	2¾	2	3¾	1/32 (out)	3/32 (in)	7/32 (in)	1/16 (out)	3/16 (in)	7/16 (in)
1977 E-250, 350 (4 × 2)		N/A										
4	4¼		6¼	9	−1	¾	3/32 (out)	1/32 (in)	5/32 (in)	3/16 (out)	1/16 (in)	5/16 (in)
4¼	4½		5¾	8¼	− ½	1¼	3/32 (out)	1/32 (in)	5/32 (in)	3/16 (out)	1/16 (in)	5/16 (in)
4½	4¾		5¼	7¾	0	1¾	3/32 (out)	1/32 (in)	5/32 (in)	3/16 (out)	1/16 (in)	5/16 (in)
4¾	5		4½	7¼	½	2¼	3/32 (out)	1/32 (in)	5/32 (in)	3/16 (out)	1/16 (in)	5/16 (in)
5	5¼		4	6½	1	2¾	3/32 (out)	1/32 (in)	5/32 (in)	3/16 (out)	1/16 (in)	5/16 (in)
5¼	5½		3¼	6	1½	3¼	3/32 (out)	1/32 (in)	5/32 (in)	3/16 (out)	1/16 (in)	5/16 (in)
1977 F-350 All F-250 8100 GVW Super Cab (4 × 2) F-250 Super Cab w/RPO Suspension		N/A										
3½	3¾		7	9½	− ¼	1½	1/32 (out)	3/32 (in)	7/32 (in)	1/16 (out)	3/16 (in)	7/16 (in)
3¾	4		6½	9	0	1¾	1/32 (out)	3/32 (in)	7/32 (in)	1/16 (out)	3/16 (in)	7/16 (in)
4	4¼		5¾	8½	½	2¼	1/32 (out)	3/32 (in)	7/32 (in)	1/16 (out)	3/16 (in)	7/16 (in)
4¼	4½		5¼	7¾	¾	2½	1/32 (out)	3/32 (in)	7/32 (in)	1/16 (out)	3/16 (in)	7/16 (in)

SUSPENSION
Specifications

U.S. LIGHT TRUCK SECTION

YEAR, MODEL AND AXLE PART NUMBER RIDE HEIGHT		ADJ. ILL. NO.	CASTER (Degrees)		CAMBER (Degrees)		TOE-IN (Inches)			TOE-IN (Degrees)		
MIN.	MAX.		MIN.	MAX.	MIN.	MAX.						
1977 F-350 All F-250 8100 GVW Super Cab (4 × 2) F-250 Super Cab w/RPO Suspension		N/A										
4½	4¾		4½	7	1¼	3	1/32 (out)	3/32 (in)	7/32 (in)	1/16 (out)	3/16 (in)	7/16 (in)
4¾	5		4	6½	1¾	3¼	1/32 (out)	3/32 (in)	7/32 (in)	1/16 (out)	3/16 (in)	7/16 (in)
5	5¼		3¼	6	2	3¾	1/32 (out)	3/32 (in)	7/32 (in)	1/16 (out)	3/16 (in)	7/16 (in)

VEHICLE IDENTIFICATION		ADJ. ILL. NO.	CASTER (Degrees)			CAMBER (Degrees)			TOE-IN (Inches)			TOE-IN (Degrees)			TOE-OUT ON TURNS (Degrees)		STRG. AXIS INCL. (DEG.)
YEAR	MODEL		MIN.	PREF.	MAX.	MIN.	PREF.	MAX.	MIN.	PREF.	MAX.	MIN.	PREF.	MAX.	OUTSIDE WHEEL	INSIDE WHEEL	
FORD MOTOR CO.																	
79 F-250 (4 × 4), F-150 (4 × 4) SC, F-350 (4 × 4)		N/A	2½	4	5½	0	1½	3		3/32			3/16		NA	NA	NA
79 F-150 (4 × 4), Bronco		N/A	6½	8	9½	1	1½	3		3/32			3/16		NA	NA	NA
78 F-250 (4 × 4)		N/A	2⅛	4	5	½	1	1½	1/16	5/32	¼	⅛	5/16	½	NA	NA	NA
78 F-150, (4 × 4), Bronco		N/A	2¾	3½	4¼	1	1½	2	1/16	5/32	5/16	⅛	3/16	⅝	NA	NA	NA
77 F-250 (4 × 4)		N/A	3	4	5	½	1	1½	1/16	5/32	¼	⅛	5/16	½	NA	NA	NA
77 F-150 (4 × 4) Bronco		N/A	2¾	3½	4¼	1	1½	2	1/16	5/32	5/16	⅛	5/16	⅝	NA	NA	NA

If caster or camber are not with specifications, find the true ride height by installing wooden blocks as pictured below. This will aid in determining which wheel is out of specification.

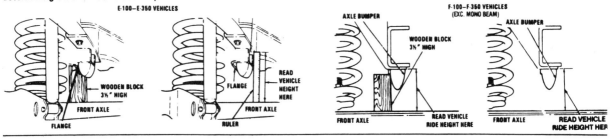

To find ride height on vehicles equipped with front leaf springs: Measure from bottom of frame to top of front axle.

F 350 MONO BEAM VEHICLES

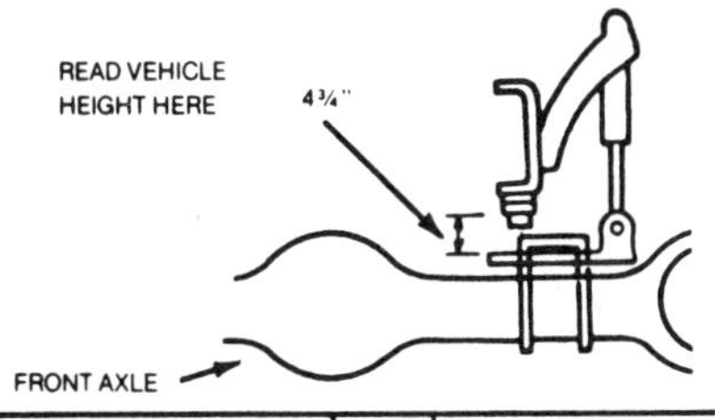

To find ride height on F350 (4 × 4) with Mono Beam front axle: Measure from the bottom of the metal jounce stop to the top of the front spring plate spacer.

INTERNATIONAL HARVESTER		CASTER			CAMBER			TOE-IN			TOE-IN					
80 Scout, All Models	(17)		2½			½		0		3/16	0		⅜	NA	NA	8½
79 Scout, All Models	(17)		0			1		3/32		5/16	3/16		⅝	NA	NA	8½
78-77 Scout, All Models	(17)	−1	0	1	½	1	1½	0	1/16	⅛	0	⅛	¼	NA	NA	8½

Measure suspension height from lower control arm snubber mounting bracket to lower face of front cross member.

IMPORTED CAR AND LIGHT TRUCK SECTION

ALFA ROMEO																
86-78 Sprint Veloce, Sport Sedan, Spider Veloce	N/A	4	4½	5	−5/16	3/16	11/16	0		5/64	0		5/32	NA	NA	NA
77 Alfetta, Alfetta GT	N/A	4	4½	5	−5/16	3/16	11/16	0		5/64	0		5/32	NA	NA	NA
77 Spider Veloce	N/A	1	1½	2	−5/16	3/16	11/16	0		⅛	0		¼	NA	NA	NA

IMPORTED PASSENGER CAR AND LIGHT TRUCK SECTION

VEHICLE IDENTIFICATION	ADJ. ILL. NO.	CASTER (Degrees)			CAMBER (Degrees)			TOE-IN (Inches)			TOE-IN (Degrees)			TOE-OUT ON TURNS (Degrees)		STRG. AXIS INCL.
YEAR MODEL		MIN.	PREF.	MAX.	MIN.	PREF.	MAX.	MIN.	PREF.	MAX.	MIN.	PREF.	MAX.	OUTSIDE WHEEL	INSIDE WHEEL	(DEG.)
AUDI																
86 5000 S Quattro:																
Front	(35)	11/32	1	1 21/32	−1	−1/2	0	3/64 (out)	0	5/64 (in)	3/32 (out)	0	5/32 (in)	NA	NA	NA
Rear	N/A				−1/2	−1/4	0	0	5/64	11/64	0	5/32	11/32			
86-84 4000 S Quattro:																
Front	N/A	15/16	1 7/16	1 15/16	−1 1/4	−3/4	−1/4	0	5/64	11/64	0	5/32	11/32	NA	NA	NA
Rear	N/A				−3/4	−1/2	1/4									
up to Ch. No. FA024249								11/64 (out)	5/64 (out)	0	11/32 (out)	5/32 (out)	0			
from Ch. No. FA024250								5/64	11/64	1/4	5/32	11/32	1/2			
85-83 Quattro 4WD																
Front	(34)	27/32	1 1/2	2 5/32	-1 11/32	−27/32	−11/32	5/64 (out)	0	3/64 (in)	5/32 (out)	0	3/32 (in)	18 3/4	20	NA
Rear	(72)				−1	−1/2	0	0	5/64	11/64	0	5/32	11/32			
86-80 4000, Coupe: Front. (A)	(14)	0	1/2	1	-15/32	−21/32	−5/32	0	5/64	11/64	0	5/32	11/32	NA	NA	NA
Rear	N/A				-1 11/32	−1	−21/32	0	11/64	21/64	0	11/32	21/32			
80 5000 up to VIN No. 43A0016066 and all 79-78 5000																
Front	(35)	−13/16	−5/32	1/2	−1	−1/2	0	5/64 (out)	0	3/64 (in)	5/32 (out)	0	3/32 (in)	NA	NA	NA
Rear	N/A				−1	−1/2	0	5/64	11/64	13/64	5/32	11/32	13/32			
86-84 5000S, Turbo																
Front	(35)	−5/32	27/32	1 1/2	−1	−1/2	0	5/64 (out)	0	3/64 (in)	5/32 (out)	0	3/32 (in)	NA	NA	NA
Rear	N/A				−27/32	−1/2	−5/32									
up to VIN No. EN096669								5/64	1/4	27/64	5/32	1/2	27/32			
from VIN No. EN096670								5/64	11/64	1/4	5/32	11/32	1/2			
83-80 5000 after VIN No. 43A0016066; & 83-80 5000 Turbo, All																
Front	(35)	1/2	1 5/32	1 13/16	−1	−1/2	0	5/64 (out)		3/64 (in)	5/32 (out)		3/32 (in)	NA	NA	NA
Rear	N/A				−1	−1/2	0	5/64	11/64	13/64	5/32	11/32	13/32			
79-78 Fox: Front	(14)	0	1/2	1	0	1/2	1	3/64 (out)	5/64 (in)	13/64 (in)	3/32 (out)	5/32 (in)	13/32 (in)	18 3/4	20	NA
Rear	N/A				−1 5/16	−11/16	0	13/64 (out)	0	13/64 (in)	13/32 (out)	0	13/32 (in)			
									EACH WHEEL			EACH WHEEL				
77 Fox: Front	(14)	0	1/2	1	0	1/2	1	3/64 (out)	5/64 (in)	5/32 (in)	3/32 (out)	5/32 (in)	3/16 (in)	18 3/4	20	NA
Rear	N/A				−1	−1/2	0	13/64 (out)	0	13/64 (in)	13/32 (out)	0	13/32 (in)			
									EACH WHEEL			EACH WHEEL				
77 100 Series	(59)							1/64	3/64	7/64	1/32	3/32	7/32	19 27/32	20	NA
W/Power Steering																
Front		−1/4	0	1/4	−1/2	0	1/2									
W/Manual Steering																
Front		−3/4	−1/2	−1/4	−1/2	0	1/2									
Rear, All	N/A				−1	−1/2	0	1/8 (out)	0	1/8 (in)	1/4 (out)	0	1/4 (in)			
(A) Caster without link rod shown; with link rod 15/16° to 1 15/16°.																
BMW (A)(B)(C)																
86-85 318i, 325e, 325es:																
Front	N/A	8 7/16	8 15/16	9 7/16	−1 3/16	−11/16	−5/32	3/64	5/64	7/64	3/16	11/32	13/32	18 5/16	20	13 15/16
Rear	N/A				-2 11/32	−1 27/32	−1 11/32	3/64	5/64	7/64	3/16	11/32	13/32			
86-85 528e, 535i, 524td																
Front	N/A	7 3/4	8 1/4	8 3/4	−1 3/16	−5/16	3/16	3/64	5/64	7/64	3/16	11/32	13/32	18 3/16	20	12 3/16
Rear	N/A				-2 13/16	−2 5/16	−1 13/16	3/64	5/64	7/64	3/16	11/32	13/32			
86-85 735i: Front	N/A	8 1/2	9	9 1/2	−1/2	0	1/2	0	1/64	1/16	0	3/32	7/32	18 5/16	20	11 9/16
Rear	N/A				-2 11/16	−2 3/16	−1 11/16	3/64	5/64	7/64	3/16	11/32	13/32			
84 318i: Front	N/A	8 1/2	9	9 1/2	−1 3/16	−11/16	−5/16	3/64	5/64	7/64	3/16	11/32	13/32	18 5/16	20	14 1/2
Rear	N/A				−2 5/16	−1 13/16	−1 5/16	3/64	5/64	7/64	3/16	11/32	13/32			
84 528e, 533i, 733i																
Front	N/A	7 3/4	8 1/4	8 3/4	−1 3/16	−5/16	3/16	3/64	5/64	7/64	3/16	11/32	13/32	18 3/16	20	12 3/16
Rear	N/A				−2 1/2	−2	−1 1/2	3/64	5/64	7/64	3/16	11/32	13/32			
86-84 633csi, 635csi: Front	N/A	7 3/4	8 1/4	8 3/4	−1 3/16	−5/16	3/16	3/64	5/64	7/64	3/16	11/32	13/32	18 3/16	20	12 3/16
Rear	N/A				-2 13/16	−2 5/16	-1 13/16	3/64	5/64	7/64	3/16	11/32	13/32			

IMPORTED PASSENGER CAR AND LIGHT TRUCK SECTION

VEHICLE IDENTIFICATION YEAR MODEL	ADJ. ILL. NO.	CASTER (Degrees)			CAMBER (Degrees)			TOE-IN (Inches)			TOE-IN (Degrees)			TOE-OUT ON TURNS (Degrees)		STRG. AXIS INCL.
		MIN.	PREF.	MAX.	MIN.	PREF.	MAX.	MIN.	PREF.	MAX.	MIN.	PREF.	MAX.	OUTSIDE WHEEL	INSIDE WHEEL	(DEG.)
83 528e Front	N/A	7 3/4	8 1/4	8 3/4	– 13/16	– 5/16	3/16	3/64	5/64	7/64	3/16	11/32	13/32	18 3/16	20	12 3/16
Rear	N/A				– 2 5/16	– 1 13/16	– 1 13/16	3/64	5/64	7/64	3/16	11/32	13/32			
83 733i Front	N/A	8 1/2	9	9 1/2	– 1/2	0	1/2	0	1/64	1/16	0	3/32	7/32	18 5/16	20	11 9/16
Rear	N/A				– 2	– 1 1/2	– 1	3/64	5/64	7/64	3/16	11/32	13/32			
83 533i Front	N/A	7 3/4	8 1/4	8 3/4	-13/16	– 5/16	3/16	3/64	5/64	7/64	3/16	11/32	13/32	18 3/16	20	12 3/16
Rear	N/A				-2 13/16	– 2 5/16	– 1 13/16	3/64	5/64	7/64	3/16	11/32	13/32			
82 528e: Front	N/A	7 3/4	8 1/4	8 3/4	– 27/32	– 11/32	5/32	3/64	5/64	7/64	3/16	11/32	13/32	18 3/16	20	12 5/32
Rear	N/A				-2 11/32	– 1 27/32	– 1 11/32	3/64	5/64	7/64	3/16	11/32	13/32			
83-77 320i: Front	N/A	7 13/16	8 5/16	8 13/16	– 1/2	0	1/2	1/32	1/16	3/32	5/32	1/4	3/8	19 1/4	20	10 15/16
Rear	N/A				– 2 1/2	– 2	– 1 1/2	3/64	5/64	7/64	3/16	11/32	13/32			
82-79 528e, 528i: Front	N/A	7 3/16	7 11/16	8 3/16	– 1/2	0	1/2	1/32	1/16	3/32	5/32	1/4	3/8	18 3/16	20	8 1/2
Rear	N/A				– 2 1/2	– 2	– 1 1/2	3/64	5/64	7/64	3/16	11/32	13/32			
82-78 733i: Front	N/A	8 1/2	9	9 1/2	– 1/2	0	1/2	0	1/32	1/16	0	3/32	7/32	18 5/16	20	11 9/16
Rear	N/A				– 2	– 1 1/2	– 1	3/64	5/64	7/64	3/16	11/32	13/32			
82-80 633 CSi: Front	N/A	7 3/16	7 11/16	8 3/16	– 1/2	0	1/2	1/32	1/16	3/32	5/32	1/4	3/8	18 1/2	20	8
Rear	N/A				– 2 1/2	– 2	– 1 1/2	3/64	5/64	7/64	3/16	11/32	13/32			
79-78 530i: Front	N/A	7 3/16	7 11/16	8 3/16	– 1/2	0	1/2	1/32	1/16	3/32	5/32	1/4	3/8	18 3/16	20	8 1/2
Rear	N/A				– 2 1/2	– 2	– 1 1/2	3/64	5/64	7/64	3/16	11/32	13/32			
79-77 630csi: Front	N/A	7 5/16	7 11/16	8 3/16	– 1/2	0	1/2	1/32	1/16	3/32	5/32	1/4	3/8	18 1/2	20	8
Rear	N/A				– 2 1/2	– 2	– 1 1/2	3/64	5/64	7/64	3/16	11/32	13/32			
77 530i: Front	N/A	7 3/16	7 11/16	8 3/16	0	1/2	1	1/32	1/16	3/32	5/32	1/4	3/8	18 3/16	20	8 1/2
Rear	N/A				– 2 1/2	– 2	– 1 1/2	3/64	5/64	7/64	3/16	11/32	13/32			
(A) All models exc. below aligned with 150 lbs. in each front seat, 150 lbs. in rear seat and 46 lbs. in trunk. **(B)** 630, 633 aligned with 150 lbs. in each front seat and 30 lbs. in trunk on left side. **(C)** Toe is measured at wheel rim. N/A - Not Adjustable.																
BUICK OPEL																
79 All	(13)	3	4 1/2	6	– 3/4	1/4	1 1/4	1/32 (out)	1/16 (in)	5/32 (in)	1/16 (out)	1/8 (in)	5/16 (in)	NA	NA	7 7/8
78-77 All	(13)	3 1/2	5	6 1/2	– 1	0	1	1/32	1/8	7/32	1/16	1/4	7/16	NA	NA	7 7/8
CAPRI																
78-77 Capri II	N/A	1	1 1/2	2 1/4	3/4	1 1/2	2 1/4	0		9/32	0		9/16	19•	20	8 +
• Range from 18 1/4 to 19 3/4 acceptable. + Range from 7 1/2° to 8 1/2° acceptable.																
CHEVROLET																
86-85 Nova: Front	N/A	3/32	7/8	1 5/8	– 1	– 1/4	1/2	7/64 (out)	3/64 (in)	13/64 (in)	7/32 (out)	3/32 (in)	13/32 (in)	32 3/4	39 1/2	12 19/32
Rear	N/A				– 1 1/4	– 1/2	1/4	0	5/32	5/16	0	5/16	5/8			
86-85 Spectrum	N/A	1 3/4	2 1/4	2 3/4	– 7/16	11/32	1 1/16	1/16 (out)	0	1/16 (in)	1/8 (out)	0	1/8 (in)	32 1/2	37 1/2	11 27/32
86 Sprint	N/A		3 3/16			1		5/32 (out)	0	5/32 (in)	5/16 (out)	0	5/16 (in)	32	38	12 13/16
85 Sprint	N/A		3 3/16			1		0	1/16	5/32	0	1/8	5/16	32	38	12 13/16
CHEVROLET LUV																
82-81 Series 12, 11 (4 × 2)	(10)	0	1/2	1	0	1/2	1	1/16 (out)	0	1/16 (in)	1/8 (out)	0	1/8 (in)	33	37	7 1/2
82-81 Series 12, 11 (4 × 4)	(18)	– 3/16	5/16	13/16	1/16	9/16	1 1/16	1/16 (out)	0	1/16 (in)	1/8 (out)	0	1/8 (in)	33	35	7 1/2
80-79 Series 10, 9 (4 × 4)	(18)	– 11/16	5/16	15/16	– 3/16	9/16	15/16	1/8 (out)	0	1/8 (in)	1/4 (out)	0	1/4 (in)	30	39	7 15/32
80-77 Series 10,9,8,6 (4 × 2)	(10)	– 1 3/16	– 3/16	13/16	– 1/4	1/2	1 1/4	1/8 (out)	0	1/8 (in)	1/4 (out)	0	1/4 (in)	30	39	7 1/2

LUV SUSPENSION HEIGHT TABLES

Model & Year	Measurement at	Front Suspension Inches	Front Suspension MM	Rear Suspension Inches	Rear Suspension MM
82-81 (4 × 4)		5	127	7 11/16	195
82-81 (4 × 2)	Base Model, Soft Ride	4	102	6 1/8	155
	Base Model, Cab & Chassis Comp.	4	102	7 1/2	190
	Cab & Chassis	4	102	8 5/16	210
	Long Wheel Base Model	4	102	7 1/2	190
80-79 4 × 4		4 13/16	122	7 11/16	195
80-78 4 × 2	Base Model, Soft Ride	4 5/8	116.8	6 1/8	155
	Base Model, Cab & Chassis Comp.	4 5/8	116.8	7 1/2	190
	Cab & Chassis	4 5/8	116.8	8 5/16	210
	Long Wheel Base Model	4 5/8	116.8	7 1/2	190
77 4 × 2	at Curb	4 5/8	116.8	6	152.4
	at G.V.W.	3 5/8	92.0	5 1/32	127.8

IMPORTED PASSENGER CAR AND LIGHT TRUCK SECTION

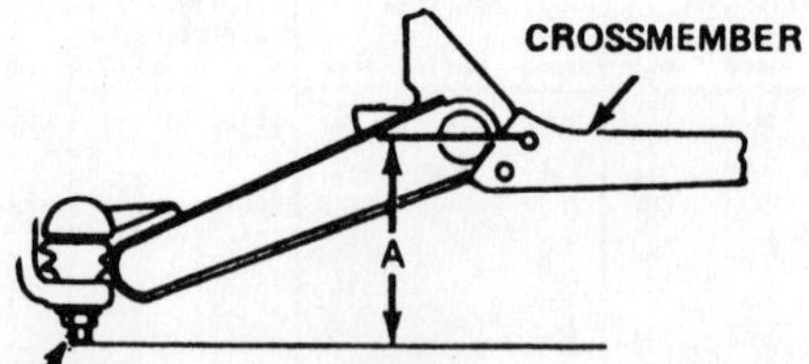

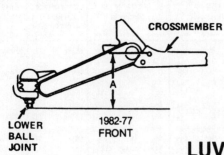

1982-77 FRONT

LUV TRIM HEIGHTS

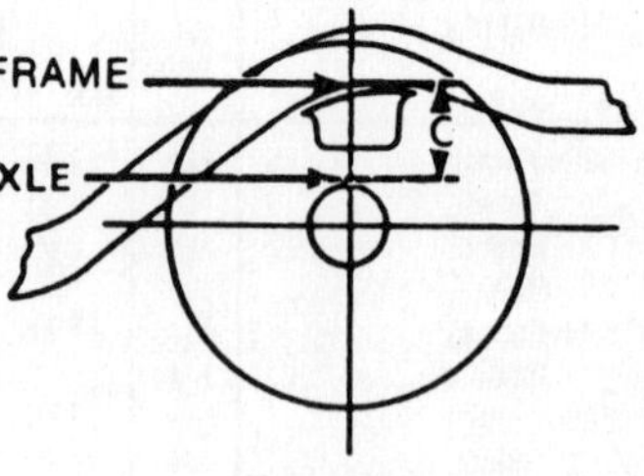

REAR

VEHICLE IDENTIFICATION YEAR MODEL	ADJ. ILL. NO.	CASTER (Degrees) MIN.	PREF.	MAX.	CAMBER (Degrees) MIN.	PREF.	MAX.	TOE-IN (Inches) MIN.	PREF.	MAX.	TOE-IN (Millimeters) MIN.	PREF.	MAX.	TOE-OUT ON TURNS (Degrees) OUTSIDE WHEEL	INSIDE WHEEL	STRG. AXIS INCL. (DEG.)
DATSUN/NISSAN																
86 Pulsar, Sentra: Front . . .	N/A	3/4		2 1/4	– 7/16		1 1/16	1/8		3/16	1/4		3/8	20	22 1/2	12 15/16
Rear					– 1 3/4		– 1/4	1/4		1/4	1/2		1/2			
								(out)		(in)	(out)		(in)			
86 Stanza Wagon (4 × 2)	N/A	3/4		2 1/4	– 1 1/4		– 1/4	1/16		9/64	1/8		9/32	20	21 1/2	12
(4 × 4) Front	N/A	9/16		2 1/16	– 1/2		1 1/16	1/64		1/16	1/32		1/8	20	23	12
Rear					0		1 1/2	5/32		0	5/16		0			
								(out)			(out)					
86-84 300ZX Front	N/A	5 13/16		7 5/16	– 9/16		15/16	1/32		1/8	1/16		1/4	20	22 1/2	13
Rear	N/A				-1 15/16		– 7/16	1/16		3/32	1/8		3/16			
								(out)		(in)	(out)		(in)			
85 Pulsar, Sentra: Front . . .	N/A	3/4		2 1/4	– 7/16		1 1/16	1/8		3/16	1/4		3/8	17 1/2	20	12 15/16
Rear	N/A				– 1 3/4		– 1/4	1/4		1/4	1/2		1/2			
								(out)		(in)	(out)		(in)			
84 Sentra: Front	N/A	3/4		2 1/4	– 7/16		1 1/16	1/8		3/16	1/4		3/8	17 1/2	20	12 15/16
Rear	N/A				– 9/16		1 1/2	3/16		3/16	3/8		3/8			
								(out)		(in)	(out)		(in)			
84 Pulsar: Front	N/A	3/4		2 1/4	– 7/16		1 1/16	1/8		3/16	1/4		3/8	17 1/2	20	12 15/16
Rear	N/A				– 1 3/4		– 1/4	1/32		1/32	1/16		1/16			
								(out)		(in)	(out)		(in)			
83 Pulsar	N/A	3/4		2 1/4	– 9/16		1 1/16	0		5/64	0		5/32	17 1/2	20	12 3/4
86-85 Stanza: Front	N/A	1 1/16		2 3/16	– 7/16		1 1/16	0		1/16	0		1/8	18 1/2	20	14 13/32
Rear	N/A				0		1 1/2	9/32		1/4	9/16		1/2			
								(out)		(in)	(out)		(in)			
84-82 Stanza: Front	N/A	1 1/16		2 3/16	– 3/4		3/4	0		5/64	0		5/32	18 1/2	20	14 13/32
Rear					0		1 1/2	13/64		5/16	13/32		5/8			
								(out)		(in)	(out)		(in)			
86-85 Maxima: Front.	N/A	1 1/4		2 3/4	– 7/16		1 1/16	1/32		1/8	1/16		1/4	20	22 7/8	14 1/2
Rear: Sedan	N/A				– 1/2		1	1/4		1/16	1/2		1/8			
								(out)		(out)	(out)		(out)			
Wagon	N/A				– 5/16		1 3/16	9/32		1/8	9/16		1/4			
								(out)		(out)	(out)		(out)			
84 Maxima Front	N/A	2 15/16		4 7/16	– 5/16		1 3/16	1/32		1/32	1/16		1/16	18 11/16	20	12 1/8
								(out)		(in)	(out)		(in)			
Rear	N/A				1 1/4		2 3/4	7/32		3/16	7/16		3/8			
								(out)		(in)	(out)		(in)			
83-81 Maxima 810: Front . .	N/A	2 15/16		4 7/16	– 5/16		1 3/16	1/32		1/32	1/16		1/16	18 11/16	20	12 1/8
								(out)		(in)	(out)		(in)			
Rear.	N/A				15/16		2 7/16	1/8		9/32	1/4		9/16			
								(out)		(in)	(out)		(in)			
81-79 510 except Wagon . .	N/A	1 1/16		2 9/16	– 1/4		1 1/4	1/32		1/8	1/16		1/4	19 1/2	20	8 27/32
81-78 510 Wagon.	N/A	15/16		2 7/16	1/16		1 9/16	1/32		1/8	1/16		1/4	19 1/2	20	8 5/32
86-84 200 SX Front	N/A	2 3/4		4 1/4	– 3/8		1 1/16	1/64		1/16	1/32		1/8	18 11/16	20	11 11/16
Rear	N/A				– 1 1/4		1/4	5/64		0	5/32		0			
								(out)			(out)					
83-80 200 SX	N/A	1 3/4		3 1/4	– 11/16		13/16	0		5/64	0		5/32	18 45/64	20	8 5/32
80-77 810: Front.	N/A	1 3/16		2 11/16	0		1 1/2	0		3/32	0		3/16	18 29/32	20	7 29/32
Rear	N/A							5/32		9/16	5/16		1 1/8			
82-79 210 except Wagon																
and Canadian	N/A	1 11/16		3 7/16	0		1 1/2	3/64		1/8	3/32		1/4	19 19/64	20	8 19/32
82-79 210 Wagon.	N/A	1 15/16		3 7/16	0		1 1/2	3/64		1/8	3/32		1/4	19 19/64	20	8 19/32
82-80 210 Canadian																
1.2 Litre Mod.	N/A	1 11/16		3 3/16	– 1/4		1 3/32	0		3/32	0		3/16	19 19/64	20	8 13/32

SUSPENSION
Specifications

IMPORTED PASSENGER CAR AND LIGHT TRUCK SECTION

VEHICLE IDENTIFICATION YEAR MODEL	ADJ. ILL. NO.	CASTER (Degrees) MIN.	PREF.	MAX.	CAMBER (Degrees) MIN.	PREF.	MAX.	TOE-IN (Inches) MIN.	PREF.	MAX.	TOE-IN (Degrees) MIN.	PREF.	MAX.	TOE-OUT ON TURNS (Degrees) OUTSIDE WHEEL	INSIDE WHEEL	STRG. AXIS INCL. (DEG.)
83-80 280 ZX	N/A															
W/Power Steering Front		4 3/16		5 11/16	− 9/16		15/16	3/64	1/8	3/32	1/4			18 7/64	20	9 11/32
W/Manual Steering Front		4 3/16		5 11/16	− 9/16		15/16	3/64	1/8	3/32	1/4			18 45/64	20	9 11/32
Rear, All	N/A				− 1/16		17/16	5/64 □		5/32	5/32 □		5/16	18	20	9
86-85 Pickup (4 × 2)	(18)	13/16		1 13/16	0		1									
W/Radial Tires								1/16		5/32	1/8		5/16			
W/Bias Tires								3/16		9/32	3/8		9/16			
84 720 Pickup Truck (4 × 2)																
W/Radial Tires	(18)	13/16		2 3/16	0		1	1/16		5/32	1/8		5/16	18	20	9
W/Bias Tires	(18)	13/16		2 3/16	0		1	3/16		9/32	3/8		9/16	18	20	9
83-82 720 Pickup Truck (4 × 2)	(18)	13/16		1 13/16	0		1	7/32		9/32	7/16		9/16	18	20	9
86-84 720 Pickup Truck (4 × 4)	(18)	15/16		1 15/16	3/16		13/16	1/16		5/32	1/8		5/16	18	20	8
83-81 720 Pickup Truck (4 × 4)	(18)	13/16		2 3/16	0		1	13/64		9/32	13/32		9/16	18	18 1/2	11
81-80 720 Pickup Truck	(18)	13/16		1 13/16	0		1	13/64		9/32	13/32		9/16	18	20	9
82 310: Front	N/A	7/16		1 15/16	1/4		1 3/4	0		3/32	0		3/16	19 ■	20	11 27/32
Rear	N/A				− 1/4		1 3/4	0	0	0	0	0	0			
81-79 310	N/A	7/16		1 15/16	1/4		1 3/4	0		3/32	0		3/16	18 13/32	20	11 27/32
79 280 ZX	N/A															
W/Power Steering Front		4		5 1/2	− 1/2		1	1/16	1/8	1/8	1/4			26 1/2	34 •	9 11/32
W/Manual Steering Front		4		5 1/2	− 1/2		1	1/16		5/32	1/8		5/16	31	35 1/2 •	9 11/32
Rear, All	N/A				− 29/32		1 1/4	1/16		5/32	1/8		5/16			
79 Pick up Truck	(18)	1/2		2	− 1/4		1 1/4	3/32	1/8	3/16	9/32			30 1/2	35 +	NA
79-78 200 SX	N/A	1 1/16		2 9/16	5/16		1 13/16	3/32		5/32	3/16		5/16	30	35	7 13/16
78 620 Pick up Truck	(18)	9/16		2 1/16	− 1/4		1 1/4	3/16		9/32	3/8		9/16	30 1/2	35 +	NA
78 F10 W/Radial Tires	N/A	1/4		1 3/4	13/16		2 1/4	0		3/32	0		3/16	32 1/2	36 1/2	10
W/Bias Tires	N/A							1/8		7/32	1/4		7/16	32 1/2	36 1/2	10
78-77 B210	N/A	1		2 1/2	7/16		1 15/16	3/32		5/32	3/16		5/16	32	38 +	8 1/4
78-77 280Z except (2 + 2)	N/A	2 1/16		3 7/16	5/16		1 13/16	0		1/8	0		1/4	33	34	12
Rear	N/A				− 3/32		1 1/2	7/32 (out)	7/32 (in)	7/16 (out)	7/16 (in)					
78-77 280Z (2 + 2)	N/A	2 1/16		3 7/16	5/16		1 13/16	0		1/8	0		1/4	35 1/2	36 3/4	12
Rear	N/A				− 5/32		1 11/32	7/32 (out)	7/32 (in)	7/16 (out)	7/16 (in)					
77 200 SX	N/A	1 1/32		2 1/4	1/2		1 1/2	3/32		5/32	3/16		5/16	30	35 •	7 13/16
77 620 Pick up Truck	(18)	1 1/16	1 13/16	2 9/16	1/4	1 1/4	2 1/4	3/32		1/8	3/16		1/4	31	36	6 1/4
77 F10 W/Radial Tires	N/A	5/16		1 13/16			2 5/16	0		3/32	0		3/16	32 1/2	36 1/2	10
W/Bias Tires	N/A							7/32		9/32	7/16		9/16	32 1/2	36 1/2	10
77 710, 610																
W/Radial Tires	N/A	1 1/16		2 9/16	1 1/4		2 3/4	5/32		1/4	5/16		1/2	30 1/2	32 1/2 +	7
W/Bias Tires	N/A							1/4		5/16	1/2		5/8	30 1/2 •	32 1/2 +	7

- • Plus or minus 2° is considered acceptable.
- \+ Plus or minus 1° is considered acceptable.
- ▲ Plus or minus 5 1/2° is considered acceptable.
- ■ Plus or minus 1 1/2 is considered acceptable.
- □ 1983-82 280ZX rear toe 0 to 5/64", 0 to 5/32°.

VEHICLE	ADJ. ILL. NO.	CASTER MIN.	PREF.	MAX.	CAMBER MIN.	PREF.	MAX.	TOE-IN (In) MIN.	PREF.	MAX.	TOE-IN (Deg) MIN.	PREF.	MAX.	OUTSIDE	INSIDE	STRG.
DODGE (A) (B)																
86 Conquest: Front	(19)		5 13/16			− 1/2		13/64 (out)	0	13/64 (in)	13/32 (out)	0	13/32 (in)	32	38 1/2	NA
Rear	(38)					0		5/64 (out)	0	5/64 (in)	5/32 (out)	0	5/32 (in)			
86-83 D-50 Pickup Truck (4 × 4)	(10)	1	2	3	1/2	1	1 1/2	5/64		11/32	5/32		11/16	28	30 3/4	8
86-80 D-50 Pickup Truck (4 × 2)	(10)	1 1/2	2 1/2	3 1/2	1/2	1	1 1/2	5/64		11/32	5/32		11/16	30 1/2	37	8
85-84 Conquest: Front	(19)		5 5/16			0		5/64 (out)		13/64 (in)	5/32 (out)		13/32 (in)	31	39	NA
Rear	(38)					5/16		5/64 (out)		5/64 (in)	5/32 (out)		5/32 (in)			
86-84 Colt Vista (4 × 2)																
Front	(19)	5/16	13/16	15/16	− 1/16	7/16	15/16	1/8 (out)		1/8 (in)	1/4 (out)		1/4 (in)	30 11/16	37 3/4	NA
Rear	N/A					− 5/8			0			0				
(4 × 4): Front	(19)	5/16	13/16	15/16	5/16	13/16	15/16	1/8 (out)		1/8 (in)	1/4 (out)		1/4 (in)	30 11/16	37 3/4	NA
Rear	N/A					0			0			0				

SECTION 2 SUSPENSION
Specifications

IMPORTED PASSENGER CAR AND LIGHT TRUCK SECTION

VEHICLE IDENTIFICATION YEAR MODEL	ADJ. ILL. NO.	CASTER (Degrees) MIN.	PREF.	MAX.	CAMBER (Degrees) MIN.	PREF.	MAX.	TOE-IN (Inches) MIN.	PREF.	MAX.	TOE-IN (Degrees) MIN.	PREF.	MAX.	TOE-OUT ON TURNS (Degrees) OUTSIDE WHEEL	INSIDE WHEEL	STRG. AXIS INCL. (DEG.)
86-85 FWD Colt	N/A	3/16	11/16	13/16	−1/2	0	1/2	1/8 (out)	0	1/8 (in)	1/4 (out)	0	1/4 (in)	34 1/8	41 13/16•	NA
84-79 FWD Hatch Back Colt	N/A	1/2	13/16	1 1/8	0	1/2	1	5/64 (out)		5/32 (in)	5/32 (out)		5/16 (in)	29 1/4	35 11/16	12 11/16
83-81 Challenger	(19)	2 3/16	2 11/16	3 3/16	11/16	1 3/16	1 11/16	0		9/32	0		9/16	32	37	9 1/2
80 Challenger	(19)	2 3/16	2 11/16	3 3/16	3/4	1 1/4	1 3/4	5/64		11/32	5/32		11/16	32	37	8 27/32
80 Colt Station Wagon	(19)	2 3/16	2 11/16	3 1/2	3/4	1 1/4	1 3/4	5/64		11/32	5/32		11/16	32	37	8 27/32
79 D-50 Pickup Truck	(10)	2	3	4	1/2	1	1 1/2	5/64		11/32	5/32		11/16	30 1/2	37	8
79-78 Challenger	(19)	2 3/16	2 11/16	3 3/16	1	1 1/2	2	5/64		11/32	5/32		11/16			
W/Power Steering														31	36	8 13/32
W/Manual Steering														32	37	8 13/32
79-78 Colt except Wagon	(19)	1 9/16	2 1/16	2 9/16	1/2	1	1 1/2	5/64		15/64	5/32		15/32	30	35	9
79 Colt Station Wagon	(19)	2 3/16	2 11/16	3 3/16	1	1/2	2	5/64		11/32	5/32		11/16	30 1/2	35	8 27/32
78 Colt Station Wagon	(19)	1 5/16	2	2 11/16	1/2	1	1 1/2	5/64		11/32	5/32		11/16	30 1/2	39	8 13/32
77 Colt Hard Top, GT, Station Wagon	(19)	9/16	1 1/16	1 9/16	3/8	7/8	1 3/8	5/64		15/64	5/32		15/32	30	39	9
77 Colt Coupe & Sedan	(19)	1 9/16	2 1/16	2 9/16	1/4	1	1 3/4	5/64		15/64	5/32		15/32	30	35	9

(A) Variation between wheels 1/2°.
(B) Side to side variation 1/2° or less.
• 86 Premier = 37 1/2° inner, 31 3/4° outer.

VEHICLE IDENTIFICATION	ADJ. ILL. NO.	CASTER MIN.	PREF.	MAX.	CAMBER MIN.	PREF.	MAX.	TOE-IN (In) MIN.	PREF.	MAX.	TOE-IN (Deg) MIN.	PREF.	MAX.	OUTSIDE	INSIDE	S.A.I.
FIAT																
82-80 Strada: Front	(21)	1 1/2		2 1/2	13/16		2 3/16	3/16 (out)		1/8 (out)	3/8 (out)		1/4 (out)	31 3/4	35 1/2	NA
Rear	(22)				1/2		1 1/2	0		5/32	0		5/16			
82-80 X 1/9: Front	(20)	6 5/16		7 5/16	−1		0	1/16		1/4	1/8		1/2	28	32 11/16	NA
Rear					−1 3/4		−3/4	5/32		5/16	5/16		5/8			
82-80 Spider 2000	(23)	2 11/16		3 11/16	−5/16		11/16	5/32		5/16	5/16		5/8	28 1/2	35 13/16	6
81-77 Brava, 131	(21)	3 1/4		4 1/4	7/16		1 7/16	5/32		5/16	5/16		5/8	31	35	NA
79 Strada: Front	(21)	1 1/2	2	2 1/2	13/16	1 11/16	2 3/16	3/16 (out)	1/8 (out)	3/32 (out)	3/8 (out)	1/4 (out)	3/16 (out)	31 3/4	35 1/2	NA
Rear	(22)				1/2	1	1 1/2	0	1/32	1/16	0	1/16	3/16			
79 124 Sport Spider	(23)	2 11/16		3 11/16	5/16		11/16	5/32		5/16	5/16		5/8	NA	NA	NA
79-78 128 Sedan & Wagon Front	(21)	1 1/4		2 1/4	1/2		1 1/2	1/32		1/16	1/16		1/4	31 3/4	35	NA
Rear	(22)				−3 3/4	−3 1/4	−2 3/4	3/32		1/4	3/16		1/2			
79-77 128-3P	(20)	1 3/4	2 1/4	2 3/4	−1/4	−3/4	−1 1/4	3/32 (out)	0	3/32 (in)	3/16 (out)	0	3/16 (in)	31	35 1/4	NA
Rear	(22)				−4	−3 1/2	−3	1/16	5/32	1/4	1/8	5/16	1/2			
79 X 1/9	(20)	6 1/2	7	7 1/2	−1		0	5/64		15/64	5/32		15/32	NA	NA	NA
Rear	N/A				−2 1/4	−1 3/4	−1 1/4	5/32	1/4	5/16	5/16	1/2	5/8			
78-77 X 1/9: Front	(20)	6 1/2	7	7 1/2	−15/16	−1	−5/8	1/32		3/16	1/16		3/8	28	32 11/16	NA
Rear	N/A				−2 5/16	−2	−1 5/8	5/32		1/4	5/16		1/2			
78-77 124 Sport Coupe/Spider	(23)	3	3 1/2	4	0	1/2	1	3/32	1/8	5/32	3/16	1/4	5/16	28 1/2	35 13/16	NA
77 128 Sedan & Wagon Front	(21)	1 3/4	2 1/4	2 3/4	1	1 1/2	2	3/32 (out)	0	3/32 (in)	3/16 (out)	0	3/16 (in)	31 3/4	35	NA
Rear	(22)				−3 3/4	−3 1/4	−2 3/4	3/32		1/4	3/16		1/2			
FORD MOTOR CO.																
80-78 Fiesta	N/A	−7/16	5/16	15/16	1 1/4	2 1/4	3 1/4	15/64 (out)	13/64 (out)	5/32 (out)	15/32 (out)	13/32 (out)	5/16 (out)	NA	NA	NA
82-77 Courier	(24)	3/4		1 1/4	1/2		1 1/2	0		1/4	0		1/2	30 11/16•	32 1/2	NA

•600-14 tire shown; 195 SR-14 tire: Outer wheel 30 11/16°, Inner wheel 33 15/16°.

VEHICLE IDENTIFICATION	ADJ. ILL. NO.	CASTER MIN.	PREF.	MAX.	CAMBER MIN.	PREF.	MAX.	TOE-IN MIN.	PREF.	MAX.	TOE-IN MIN.	PREF.	MAX.	OUTSIDE	INSIDE	S.A.I.
HONDA																
86 Accord: Front	N/A	−1/2	1/2	1 1/2	−1	0	1	1/8 (out)	0	1/8 (in)	1/4 (out)	0	1/4 (in)	30 1/2	39 1/2	6 13/16
Rear	N/A				−1	0	1	0	3/32	5/32	0	3/16	5/16			
86-85 Civic exc. Wagon Front	N/A				−1	0	1	1/8 (out)	0	1/8 (in)	1/4 (out)	0	1/4 (in)	34 1/2	41 1/2	13 1/2
W/OP.S.		1 1/2	2 1/2	3 1/2												
W/P.S.		2	3	4												
Rear	N/A				−1	−3/4	−1/2	0	3/32	5/32	0	3/16	5/16			

SUSPENSION Specifications

IMPORTED PASSENGER CAR AND LIGHT TRUCK SECTION

VEHICLE IDENTIFICATION YEAR MODEL	ADJ. ILL. NO.	CASTER (Degrees) MIN.	PREF.	MAX.	CAMBER (Degrees) MIN.	PREF.	MAX.	TOE-IN (Inches) MIN.	PREF.	MAX.	TOE-IN (Degrees) MIN.	PREF.	MAX.	TOE-OUT ON TURNS (Degrees) OUTSIDE WHEEL	INSIDE WHEEL	STRG. AXIS INCL. (DEG.)
86-85 Civic Wagon (4×2) Front	N/A	1	2	3	−1	0	1	1/8 (out)	0	1/8 (in)	1/4 (out)	0	1/4 (in)	34 1/2	41 1/2	12 1/2
Rear	N/A				−1	−3/4	−1/2	0	3/32	5/32	0	3/16	5/16			
86-85 Civic Wagon (4×4) Front	N/A	15/16	1 15/16	2 15/16	5/16	5/8	15/8	1/8 (out)	0	1/8 (in)	1/4 (out)	0	1/4 (in)	33	40 1/2	11 7/16
Rear					−1	0	1	1/16 (out)	0	1/16 (in)	1/8 (out)	0	1/8 (in)			
84 Civic Coupe: Front	N/A	1 7/16	2 7/16	3 7/16	−1	0	1	3/32 (out)	0	3/32 (in)	3/16 (out)	0	3/16 (in)	34 1/2	41 1/2	12 15/16
Rear	N/A				−1	−3/4	−1/2	0	3/32	5/32	0	3/16	5/16			
84 Civic Hatchback, Sedan: Front	N/A	1 5/16	2 5/16	3 5/16	−1	0	1	3/32 (out)	0	3/32 (in)	3/16 (out)	0	3/16 (in)	34 1/2	41 1/2	12 13/16
Rear	N/A				−1	−3/4	−1/2	0	3/32	5/32	0	3/16	5/16			
84 Civic Wagon: Front	N/A	1 1/16	2 1/16	3 1/16	−1 1/16	5/8	15/8	3/32 (out)	0	3/32 (in)	3/16 (out)	0	3/16 (in)	34 1/2	41 1/2	12
Rear	N/A				−1	−3/4	−1/2	0	3/32	5/32	0	3/16	5/16			
85 Accord: Front	N/A	1/2	1 1/2	2 1/2	−1	0	1	1/8 (out)	0	1/8 (in)	1/4 (out)	0	1/4 (in)	30	38	12 1/2
Rear	N/A					0										
exc. SEi								0	3/32	5/32	0	3/16	5/16			
W/SEi								1/32 (out)	3/32 (in)	3/16 (in)	1/16 (out)	3/16 (in)	3/8 (in)			
84-83 Accord Front	N/A	7/16	1 7/16	2 7/16	−1	0	1	1/8 (out)	0	1/8 (in)	1/4 (out)	0	1/4 (in)	NA	NA	12 1/2
Rear	(45)							0	0	5/32	0	0	5/16			
86-83 Prelude: Front	N/A	−1/2	0	1/2	−1	0	1	1/8 (out)	0	1/8 (in)	1/4 (out)	0	1/4 (in)	NA	NA	6 15/16
Rear	N/A				−1	0	1	0	5/64	5/32	0	5/32	5/16			
83 Civic exc. Wagon: Front	N/A	1 1/16	2 1/16	3 1/16	−1	0	1	5/64 (out)	0	5/64 (in)	5/32 (out)	0	5/32 (in)	NA	NA	12 11/32
Rear	(45)					−1/4		0	5/64	5/32	0	5/32	5/16			
82 Accord: Front	N/A	7/16	1 7/16	2 7/16	−1	0	1	1/8 (out)	0	1/8 (in)	1/4 (out)	0	1/4 (in)	NA	NA	12 1/2
Rear	(45)					0		3/64 (out)	3/64 (in)	1/8 (in)	3/32 (out)	3/32 (in)	1/4 (in)			
81 Accord: Front	N/A	11/16	1 11/16	2 11/16	−1 1/16	5/16	15/16	1/8 (out)	3/64 (out)	1/32 (in)	1/4 (out)	3/32 (out)	1/16 (in)	NA	NA	12 1/2
Rear	(45)					3/16		1/32 (out)	3/64 (in)	1/8 (in)	1/16 (out)	3/32 (in)	1/4 (in)			
82-79 Prelude: Front	N/A	1/2	1 1/2	2 1/2	−1	0	1	1/8 (out)	0	1/8 (in)	1/4 (out)	0	1/4 (in)	NA	NA	12 13/16
Rear	(45)					0		1/16 (out)	1/8 (in)	7/32 (in)	1/8 (out)	1/4 (in)	7/16 (in)			
80-79 Accord: Front	N/A	3/4	1 1/4	1 3/4	0	1/2	1	5/32 (out)	1/32 (out)	3/32 (in)	5/16 (out)	1/16 (out)	3/16 (in)	NA	NA	12 3/16
Rear	(45)				−1/4	1/4	3/4	1/8 (out)	1/32 (in)	1/32 (in)	1/4 (out)	1/16 (out)	1/16 (in)			
82 Civic except Wagon Front	N/A	1 1/2	2 1/2	3 1/2	−1	0	1	1/8 (out)	0	1/8 (in)	1/4 (out)	0	1/4 (in)	NA	NA	12 11/32
Sedan: Rear	N/A					−1/2		0	5/64	5/32	0	5/32	5/16			
Hatchback: Rear	N/A					−1/4		0	5/64	5/32	0	5/32	5/16			
83-82 Civic Wagon	N/A	5/16	1 5/16	2 5/16	−1	0	1	1/8 (out)	0	1/8 (in)	1/4 (out)	0	1/4 (in)	NA	NA	12 11/32
81-80 Civic except Wagon: Front	N/A	3/4	1 3/4	2 3/4	−1	0	1	1/8 (out)	0	1/8 (in)	1/4 (out)	0	1/4 (in)	NA	NA	12 5/16
Rearl (81)	N/A				−1	0	1	0	5/64	5/32	0	5/32	5/16			
Rear (80)	N/A				−1	0	1	5/32 (out)	0	5/16 (out)	0	(in)				
81-80 Civic Wagon	N/A	0	1	2	−1	0	1	1/8 (out)	0	1/8 (in)	1/4 (out)	0	1/4 (in)	NA	NA	12 5/16
79-78 Civic: Front	N/A	1/4	3/4	1 1/4	0	1/2	1	1/8 (out)	3/64 (out)	3/64 (out)	1/4 (out)	3/32 (out)	3/32 (in)	NA	NA	9 5/16
Rear	N/A				0	1/2	1	5/64 (out)	0	5/64 (in)	5/32 (out)	0	5/32 (in)			

SECTION 2: SUSPENSION Specifications

IMPORTED PASSENGER CAR AND LIGHT TRUCK SECTION

VEHICLE IDENTIFICATION YEAR MODEL	ADJ. ILL. NO.	CASTER (Degrees) MIN.	PREF.	MAX.	CAMBER (Degrees) MIN.	PREF.	MAX.	TOE-IN (Inches) MIN.	PREF.	MAX.	TOE-IN (Degrees) MIN.	PREF.	MAX.	TOE-OUT ON TURNS (Degrees) OUTSIDE WHEEL	INSIDE WHEEL	STRG. AXIS INCL. (DEG.)
79 Civic CVCC except Wagon: Front	N/A	1/4	3/4	1 1/4	0	1/2	1	1/8 (out)	3/64 (out)	3/64 (in)	1/4 (out)	3/32 (out)	3/32 (in)	NA	NA	9 5/16
Rear	N/A				0	1/2	1	5/64 (out)	0	5/64 (in)	5/32 (out)	0	5/32 (in)			
79 Civic CVCC Wagon	N/A	0	1/2	1	0	1/2	1	1/8 (out)	3/64 (out)	3/64 (in)	1/4 (out)	3/32 (out)	3/32 (in)	NA	NA	9 5/16
78 Civic CVCC except Wagon: Front	N/A	0	1/2	1	0	1/2	1	1/8 (out)	1/32 (in)	1/16 (in)	1/4 (out)	1/16 (in)	1/8 (in)	NA	NA	9 5/16
Rear	N/A				0	1/2	1	3/32 (out)	0	3/32 (in)	3/16 (out)	0	3/16 (in)			
78 Civic CVCC Wagon	N/A	0	1/2	1	0	1/2	1	1/32 (out)	1/32 (in)	3/32 (in)	1/16 (out)	1/16 (in)	3/16 (in)	NA	NA	9 5/16
78 Accord: Front	N/A	1	2	3	-1/4	3/4	1 3/4	5/32 (out)	3/64 (out)	5/64 (out)	5/16 (out)	3/32 (out)	5/32 (out)	NA	NA	12 3/16
Rear	(45)				-1/4	1/4	3/4	5/64 (out)	0	5/64 (out)	5/32 (out)	0	5/32 (out)			
77 Civic 1200: Front	N/A	3/4	1 3/4	2 3/4	0	1/2	1	5/32 (out)	3/64 (out)	5/64 (out)	5/16 (out)	3/32 (out)	5/32 (out)	NA	NA	8 15/16
Rear	N/A					1/2		3/32 (out)	0	3/32 (in)	3/16 (out)	0	3/16 (in)			
77 Civic CVCC Sedan Front	N/A	1	1 1/2	2	0	1/2	1	1/8 (out)	3/64 (out)	3/64 (in)	1/4 (out)	3/32 (out)	3/32 (in)	NA	NA	9 5/16
Rear					0	1/2	1	5/64 (out)	0	5/64 (in)	5/32 (out)	0	5/32 (in)			
77 Civic CVCC Wagon Front	N/A	1/2	1	1 1/2	0	1/2	1	1/8 (out)	3/64 (out)	3/64 (in)	1/4 (out)	3/32 (out)	3/32 (in)	NA	NA	9 15/16
77 Accord: Front	N/A	1	2	3	-1/2	1/2	1 1/2	5/32 (out)	3/64 (out)	5/64 (out)	5/16 (out)	3/32 (out)	5/32 (out)	NA	NA	12 3/16
Rear	(45)					5/16		1/8 (out)	3/64 (out)	3/64 (out)	1/4 (out)	3/32 (out)	3/32 (in)			

ISUZU

VEHICLE IDENTIFICATION	ADJ. ILL. NO.	CASTER MIN.	PREF.	MAX.	CAMBER MIN.	PREF.	MAX.	TOE-IN (in) MIN.	PREF.	MAX.	TOE-IN (deg) MIN.	PREF.	MAX.	OUTSIDE	INSIDE	STRG. AXIS INCL.
86-85 I-Mark	N/A	1 3/4	2 1/4	2 3/4	-11/16	5/16	15/16	5/64 (out)	0	5/64 (in)	5/32 (out)	0	5/32 (in)	32 7/16	37 11/16	NA
86-85 Trooper II	(18)	0	1/2	1	1/16	9/16	11/16	5/64 (out)	0	5/64 (in)	5/32 (out)	0	5/32 (in)	NA	NA	7 7/16
86-84 Impulse	N/A	3 1/2		6	-1		1/2	0	1/16	1/8	0	1/8	1/4	35	38	8
84-81 I-Mark	(13)	3 11/16	5 3/16	6 3/16	-7/8	1/8	5/8	5/64 (out)	1/8	5/32	5/32	1/4	5/16	35	37 1/2	7 7/8
86-81 Pickup Truck (4×2)	(68)(19)	0	1/2	1	-1	1/2	1	0	5/64	5/32	0	5/32	5/16	33	37	7 1/2
86-81 Pickup Truck (4×4)	(18)	-3/16	5/16	13/16	1/16	9/16	11/16	5/64 (out)	0	5/64 (in)	5/32 (out)	0	5/32 (in)	33	35	7 7/16

JAGUAR

VEHICLE IDENTIFICATION	ADJ. ILL. NO.	CASTER MIN.	PREF.	MAX.	CAMBER MIN.	PREF.	MAX.	TOE-IN (in) MIN.	PREF.	MAX.	TOE-IN (deg) MIN.	PREF.	MAX.	OUTSIDE	INSIDE	STRG. AXIS INCL.
86-84 XJ6, XJS, XJHE Front	(49)	3 1/4	3 1/2	3 3/4	1/4	1/2	3/4	0	0	1/8	0	0	1/4	NA	NA	1 1/2
Rear	(51)				-1	-3/4	-1/2	1/32 (out)	0	1/32 (in)	1/16 (out)	0	1/16 (in)			
83-77 XJ 6: Front	(49)	2	2 1/4	2 1/2	1/4	1/2	3/4	1/16 (out)		1/8	1/8		1/4	NA	NA	1 1/2
Rear	(51)				-1	-3/4	-1/2	1/32 (out)	0	1/32 (in)	1/16 (out)	0	1/16 (in)			
83-79 XJS: Front	(49)	3 1/4	3 1/2	3 3/4	1/4	1/2	3/4	1/16 (out)		1/8	1/8		0	NA	NA	1 1/2
Rear	(51)				-1	-3/4	-1/2	1/32 (out)	0	1/32 (in)	1/16 (out)	0	1/16 (in)			
80-77 XJ 12: Front	(49)	3 1/4	3 1/2	3 3/4	1/4	1/2	3/4	1/16 (out)		1/8	1/8		0	NA	NA	1 1/2
Rear	(51)				-1	-3/4	-1/2	1/32 (out)	0	1/32 (in)	1/16 (out)	0	1/16 (in)			
78-77 XJS: Front	(49)	3 1/4	3 1/2	3 3/4	1/4	1/2	3/4	0		1/8	0		1/4	NA	NA	1 1/2
Rear	(51)				-1	-3/4	-1/2	1/32 (out)	0	1/32 (in)	1/16 (out)	0	1/16 (in)			

LANCIA (A)

VEHICLE IDENTIFICATION	ADJ. ILL. NO.	CASTER MIN.	PREF.	MAX.	CAMBER MIN.	PREF.	MAX.	TOE-IN (in) MIN.	PREF.	MAX.	TOE-IN (deg) MIN.	PREF.	MAX.	OUTSIDE	INSIDE	STRG. AXIS INCL.
82-80 Beta, Zagato: Front	N/A	13/16		1 13/16	3/16		13/16	1/8 (out)		0	1/4 (out)		0	30	36	NA
Rear	N/A				-15/16		-11/16		1/32	5/32		1/16	5/16			

SUSPENSION Specifications

IMPORTED PASSENGER CAR AND LIGHT TRUCK SECTION

VEHICLE IDENTIFICATION YEAR MODEL	ADJ. ILL. NO.	CASTER (Degrees) MIN.	PREF.	MAX.	CAMBER (Degrees) MIN.	PREF.	MAX.	TOE-IN (Inches) MIN.	PREF.	MAX.	TOE-IN (Degrees) MIN.	PREF.	MAX.	TOE-OUT ON TURNS (Degrees) OUTSIDE WHEEL	INSIDE WHEEL	STRG. AXIS INCL. (DEG.)
79-77 Beta, HPE, Zagato																
Front	N/A	1 5/32	1 1/2	1 27/32	5/32	1/2	27/32	5/32 (out)		0	5/16 (out)		0	33 1/2	36	9 3/4
Rear	N/A				− 1 11/32	− 1	− 21/32	1/64		3/32	1/32		3/16			
77 Scorpion: Front	N/A	5 3/4	6	6 1/4	− 5/16	0	5/16	5/64		5/32	5/16		5/16	32	32	13 1/2
Rear	N/A				1 11/16	2	2 5/16	5/32		5/16	5/16		5/8			

(A) Add weight to vehicles before performing alignment: Scorpion requires 375 lbs.; Beta Coupe requires 495 lbs.; Beta HPE & Sedan requires 660 lbs.; Zagato requires 495 lbs.

MAZDA

VEHICLE IDENTIFICATION YEAR MODEL	ADJ. ILL. NO.	CASTER MIN.	PREF.	MAX.	CAMBER MIN.	PREF.	MAX.	TOE-IN (In) MIN.	PREF.	MAX.	TOE-IN (Deg) MIN.	PREF.	MAX.	OUTSIDE WHEEL	INSIDE WHEEL	STRG. AXIS
86 323: Front	(31)	13/16	1 9/16	2 5/16	1/16	13/16	1 9/16	3/64		13/64	3/32		13/32	33	40	12 3/8
Rear	N/A				− 3/4	0	3/4	1/8 (out)	0	1/8 (in)	1/4 (out)	0	1/4 (in)			
86 RX-7: Front	(26)		4 5/8			5/16		0	1/8	1/4	0	1/4	1/2	33	37	13 3/4
Rear	N/A				− 1 1/4	− 3/4	− 1/4	1/8 (out)	0	1/8 (in)	1/4 (out)	0	1/4 (in)			
86 B-2000	(24)				7/16	3/4	1 1/4	0	1/8	1/4	0	1/4	1/2	33	35	8 1/4
W/Manual Steering		1/16	13/16	1 9/16												
W/Power Steering		1 1/16	1 13/16	2 9/16												
86-83 626: Front	(26)	15/16	1 11/16	2 7/16	− 3/16	5/16	1 1/16	0	1/8	1/4	0	1/4	1/2	31	38	12 15/16
Rear	N/A							1/8 (out)	0	1/8 (in)	1/4 (out)	0	1/4 (in)			
83 GLC Wagon	(NA)	13/16	1 9/16	2 5/16	1/4	3/4	1 1/4	0		1/4	0		1/2	31 13/16	42 1/2	8 3/4
85-84 RX7 w/13" Tires	(26)				1/2	1	1 1/2	0	1/8	1/4	0	1/4	1/2	32 1/4	39 11/16	10 3/4
Right Wheel		3 1/16	4 3/16	4 11/16												
Left Wheel		3 3/16	3 11/16	4 3/16												
W/14" Tires	(26)				1/16	9/16	1 1/16	0	1/8	1/4	0	1/4	1/2	32 1/4	39 11/16	11 5/16
Right Wheel		3 11/16	4 3/16	4 11/16												
Left Wheel		3 3/16	3 11/16	4 3/16												
83-81 RX-7					1/2	1	1 1/2	0		1/4	0		1/2	32 1/4	39 11/16	10 3/4
Right Wheel	(26)	3 11/16	4 3/16	4 11/16												
Left Wheel	(26)	3 3/16	3 11/16	4 3/16												
82-81 626	(26)	2 15/16	3 7/16	3 15/16	3/4	1 1/2	1 3/4	0		1/4	0		1/2	33 13/16	40	10 9/16
85-81 GLC except Station Wagon	N/A	13/16	1 15/16	2 11/16	7/16	15/16	1 7/16	1/8 (out)		1/8 (in)	1/4 (out)		1/4 (in)	33	40	12 3/16
82-81 GLC Station Wagon	N/A	3/4	1 1/2	2 1/4	3/4	1 1/4	1 3/4	0		1/4	0		1/2	31 13/16	42 1/2	8 1/4
84-79 B-2000, B2200	(24)	1 1/16	1	1 5/16	7/16	3/4	1 1/4	0	1/8	1/4	0	1/4	1/2	30 11/16	32 1/2	8 1/4
80 RX-7					11/16	13/16	1 11/16	0	1/8	1/4	0	1/4	1/2	32 1/4	39 11/16	10 3/4
Right Wheel	(26)	4	4 1/2	5												
Left Wheel	(26)	3 1/2	4	4 1/2												
80 626	(26)				3/4	1 1/4	1 3/4	0	1/8	1/4	0	1/4	1/2	33 13/16	40	10 11/16
Right Wheel		2 29/32	3 21/32	4 13/32												
Left Wheel		2 7/16	3 3/16	3 15/16												
79 RX-7	(26)	3 1/4	4	4 3/4	11/16	1 3/16	1 11/16	0	1/8	1/4	0	1/4	1/2	32 1/4	39 11/16	10 3/4
79 626	(26)		3 3/4			1 1/4		0	1/8	1/4	0	1/4	1/2	33 13/16	40	10 11/16
80-79 GLC exc. Wagon		15/16	1 11/16	2 7/16	1/4	3/4	1 1/4	0	1/8	1/4	0	1/4	1/2	31	42 1/2	8 3/4
80-79 GLC Wagon	N/A	1	1 3/4	2 1/2	1/2	1	1 1/2	0	1/8	1/4	0	1/4	1/2	31 15/16	42 1/2	8 1/2
78 RX-4 Sedan	(26)	1 1/16	1 13/16	2 9/16	0	1	2	0	1/8	1/4	0	1/4	1/2	30 1/2	41 11/16	9 1/2
Wagon	(26)	1 1/16	1 13/16	2 9/16	1/4	1 1/4	2 1/4	0	1/8	1/4	0	1/4	1/2	30 1/2	41 11/16	9 1/2
78-77 RX-3 SP	N/A	1 7/16	2 3/16	3 15/16	1/16	1 1/16	2 1/16	0	1/8	1/4	0	1/4	1/2	32	43	8 13/16
78-77 GLC	N/A	3/16	1 9/16	2 5/16	− 5/16	11/16	1 11/16	0	1/8	1/4	0	1/4	1/2	31	43	8 13/16
78-77 B-1800 Pick Up	(24)	1 1/16	1	1 5/16	7/16	3/4	1 1/4	0	1/8	1/4	0	1/4	1/2	30 11/16	32 1/2	8 1/4
78-77 Cosmo																
W/Manual Steering	(26)	1 1/16	1 13/16	2 5/16	0	1	2	0	1/8	1/4	0	1/4	1/2	30 1/2	41 11/16	9 3/4
W/Power Steering	(26)	1 1/2	2 1/4	3	0	1	2	0	1/8	1/4	0	1/4	1/2	30 1/2	41 11/16	9 3/4
77 808 (1600) Sedan	N/A	1 1/16	1 13/16	2 9/16	1/16	1 1/16	2 1/16	0	1/8	1/4	0	1/4	1/2	31 1/2	41	8 7/16
Coupe	N/A	15/16	2 1/16	2 13/16	1/16	1 1/16	2 1/16	0	1/8	1/4	0	1/4	1/2	31 1/2	43	8 7/16
Wagon	N/A	1 1/16	1 13/16	2 9/16	1/4	1 1/4	2 1/4	0	1/8	1/4	0	1/4	1/2	31 1/2	43	8 1/4
77 Mizer (808-1300)																
Sedan	N/A	1 1/16	1 7/16	2 3/16	− 3/16	13/16	1 13/16	0	1/8	1/4	0	1/4	1/2	32 11/16	44 1/2	8 1/16
Coupe	N/A	1	1 3/4	2 1/2	1/16	15/16	1 15/16	0	1/8	1/4	0	1/4	1/2	32 11/16	44 1/2	8 1/4
Wagon	N/A	1 1/16	1 7/16	2 3/16	0	1	2	0	1/8	1/4	0	1/4	1/2	32 11/16	44 1/2	8 9/16
77 Rotary Pick up	(24)	1 5/8	1 15/16	2 1/4	− 1/16	1/4	9/16	0	1/8	1/4	0	1/4	1/2	32 9/16	33 5/16	8 3/4
77 RX-4																
Sedan & Hard Top	(26)	1 1/16	1 13/16	2 9/16	0	1	2	0	1/8	1/4	0	1/4	1/2	30 1/2	41 11/16	9 1/2
Wagon (77 only)	(26)	1 1/16	1 13/16	2 9/16	1/4	1 1/4	2 1/4	0	1/8	1/4	0	1/4	1/2	30 1/2	41 11/16	9 1/2

Maximum variation between wheel, caster 1 1/16°, camber 1/2°.

IMPORTED PASSENGER CAR AND LIGHT TRUCK SECTION

VEHICLE IDENTIFICATION YEAR MODEL	ADJ. ILL. NO.	CASTER (Degrees) MIN.	PREF.	MAX.	CAMBER (Degrees) MIN.	PREF.	MAX.	TOE-IN (Inches) MIN.	PREF.	MAX.	TOE-IN (Degrees) MIN.	PREF.	MAX.	TOE-OUT ON TURNS (Degrees) OUTSIDE WHEEL	INSIDE WHEEL	STRG. AXIS INCL. (DEG.)
MERCEDES BENZ (A)																
86 190E 2.3-16	N/A	10	10½	11	– 11/16	– 5/16	– 3/16	1/16	3/32	9/64	5/32	11/32	½	19 11/16	20	NA
86 190 D, E exc. 2.3-16	N/A	10	10½	11	– ½	– 3/16	0	1/16	3/32	9/64	5/32	11/32	½	19 5/16	20	NA
86 300 E, 300 D	N/A	9 11/16	10 3/16	10 11/16	– 5/16	0	3/16	1/16	3/32	9/64	5/32	11/32	½	19 11/16	20	NA
86 300 SDL, 420 SEL, 560 SEC, 560 SEL, 560 SL	N/A	10	10½	11	– ½	– 3/16	0	1/16	3/32	9/64	5/32	11/32	½	19 5/16	20	NA
85-84 190 D, E w/14" wheels	(71)	9 11/16	10 3/16	10 11/16	– 1/16	5/16	9/16	1/16	3/32	9/64	5/32	11/32	½	19 5/16	20	NA
85-81 300SD, 380SEL, 380SEC, 500	(54)	9¼	9¾	10¼	– 3/16	0	3/16	5/64	7/64	5/32	¼	13/32	19/32	18 13/16	20	NA
80-78 300SD	(54)	9½		10½		– 5/16		5/64	7/64	5/32	¼	13/32	19/32	19	20	NA
85-77 300D, 300CD, 300TD	(53)	8¼	8¾	9¼	– 1/16	0	1/16	5/64	7/64	5/32	¼	13/32	19/32	18 13/16	20	NA
81-77 280E, 280CE	(53)	8¼	8¾	9¼	– 3/16	0	3/16	5/64	7/64	5/32	¼	13/32	19/32	18 13/16	20	NA
83-77 280SE	(54)	9½		10½		– 5/16		5/64	7/64	5/32	¼	13/32	19/32	19	20	NA
85-77 240D	(53)	8¼	8¾	9¼	– 1/16	0	1/16	5/64	7/64	5/32	¼	13/32	19/32	18 13/16	20	NA
85-77 380 SL, 380 SLC, 450SL, 450SLC	(52)	3⅜	3 11/16	4	– 5/16	0	5/16	1/32	5/64	7/64	3/32	¼	13/32	19¾	20	NA
80-77 450SE, 450SEL	(54)	9½		10½		– 5/16		5/64	7/64	5/32	¼	13/32	19/32	19	20	NA
79-78 6.9	(54)	9½		10½		– 5/16		1/32	5/64	7/64	3/32	¼	13/32	19	20	NA
79-77 230	(53)	8¼	8¾	9¼	– 1/16	0	1/16	5/64	7/64	5/32	¼	13/32	19/32	18 13/16	20	NA
(A) Toe setting taken at wheel rim.																
MITSUBISHI																
86 Starion: Front	(19)		5 13/16			– ½		13/64 (out)	0	13/64 (in)	13/32 (out)	0	13/32 (in)	32	38½	NA
Rear	(38)					0		5/64 (out)	0	5/64 (in)	5/32 (out)	0	5/32 (in)			
86-85 Mirage: Front	N/A	7/32	23/32	1 7/32	– ½	0	½	⅛ (out)	0	⅛ (in)	¼ (out)	0	¼ (in)			NA
exc. Turbo														34 5/32	41 27/32	
w/Turbo														31¾	37½	
Rear	N/A					– 21/32										
86-85 Galant	N/A	5/32	21/32	1 5/32	0	½	1	⅛ (out)	0	⅛ (in)	¼ (in)	0	¼ (in)	30 19/32	37 5/32	NA
86-83 Cordia, Tredia	(19)	5/16	13/16	1 5/16	– 1/16	7/16	15/16	⅛ (out)	0	⅛ (in)	¼ (out)	0	¼ (in)	30 11/16	37 11/16	NA
86-83 Montero	(24)	2 7/16	2 15/16	3 7/16	½	1	1½	5/64		11/32	5/32		11/16	29	33	8
85 Starion: Front	(19)		5 5/16			0		5/64 (out)	0	13/64 (in)	5/32 (out)	0	13/32 (in)	31	39	NA
Rear	(38)					0		5/64 (out)	0	5/64 (in)	5/32 (out)	0	5/32 (in)			
84-83 Starion: Front	(19)		5 5/16			0		5/64 (out)	0	13/64 (in)	5/32 (out)	0	13/32 (in)	31	39	NA
Rear	(38)					5/16		5/64 (out)	0	5/64 (in)	5/32 (out)	0	5/32 (in)			
86-83 Truck (4WD)	(24)	1	2	3	½	1	1½	5/64		11/32	5/32		11/16	30½	37	8
(2WD)	(24)	1½	2½	3½	½	1	1½	5/64		11/32	5/32		11/16	30½	37	8
MG																
81-77 MGB, GT	N/A	5		7½	– ¼	1	1¼	1/16		3/32	⅛		3/16	19	20	8
79-77 Midget	N/A		3			¾		0		⅛	0		¼	19¼	20	6¾
PEUGEOT																
86-85 505 Turbo Gas	N/A	3	3½	4	– 1¼	– ¾	– ¼	13/64	19/64	⅜	13/32	19/32	¾	NA	NA	NA
86-84 505 exc. Turbo Gas																
Sedan: Front	N/A	3	3½	4	– 1/16	11/16	1 7/16	13/64	19/64	⅜	13/32	19/32	¾	NA	NA	9 5/16
Rear					– 15/16	– 7/16	– 1/16	⅛	¼	21/64	¼	½	21/32			
Wagon: Front	N/A	3	3½	4	5/16	½	11/16	11/64	¼	21/64	11/32	½	21/32	NA	NA	9 5/16
Rear					– 15/16	– 7/16	– 1/16	⅛	¼	21/64	¼	½	21/32			
83-80 505: Front	N/A	3	3½	4	0	¾	1½	13/64	19/64	⅜	13/32	19/32	¾	NA	NA	9 1/16
Rear	N/A				– 15/16	– 7/16	– 1/16	⅛	¼	21/64	¼	½	21/32			
84-77 604: Front	N/A	3	3½	4	– 1¼	– ½	¼	11/64	13/64	19/64	11/32	13/32	19/32	NA	NA	10
Rear	N/A				– 2	– 1½	– 1	5/64	11/64	19/64	5/32	11/32	19/32			
83-77 504	N/A	2 5/64	2 21/32	3 5/64	⅛	⅝	1⅛	11/64	13/64	19/64	11/32	13/32	19/32	NA	NA	8 29/32
PLYMOUTH (A) (B)																
86 Conquest: Front	(19)		5 13/16			– ½		13/64 (out)	0	13/64 (in)	13/32 (out)	0	13/32 (in)	32	38½	NA
Rear	(38)					0		5/64 (out)	0	5/64 (in)	5/32 (out)	0	5/32 (in)			

SUSPENSION
Specifications

IMPORTED PASSENGER CAR AND LIGHT TRUCK SECTION

VEHICLE IDENTIFICATION YEAR MODEL	ADJ. ILL. NO.	CASTER (Degrees) MIN.	PREF.	MAX.	CAMBER (Degrees) MIN.	PREF.	MAX.	TOE-IN (Inches) MIN.	PREF.	MAX.	TOE-IN (Degrees) MIN.	PREF.	MAX.	TOE-OUT ON TURNS (Degrees) OUTSIDE WHEEL	INSIDE WHEEL	STRG. AXIS INCL. (DEG.)
86-85 Colt Vista (4 × 2) Front	(19)	5/16	13/16	15/16	− 1/16	7/16	15/16	1/8 (out)	0	1/8 (in)	1/4 (out)	0	1/4 (in)	30 11/16	37 3/4	NA
Rear	N/A					− 5/8			0			0				
(4 × 4): Front	(19)	5/16	13/16	15/16	5/16	13/16	15/16	1/8 (out)		1/8 (in)	1/4 (out)		1/4 (in)	30 11/16	37 3/4	NA
Rear	N/A					0			0			0				
85-84 Conquest: Front	(19)		5 5/16			0		5/64 (out)		13/64 (in)	5/32 (out)		13/32 (in)	31	39	NA
Rear	(38)					5/16		5/64 (out)		5/64 (in)	5/32 (out)		5/32 (in)			
86-85 FWD Colt	N/A	3/16	11/16	13/16	− 1/2	0	1/2	1/8 (out)		1/8 (in)	1/4 (out)		1/4 (out)	34 1/8	41 13/16•	NA
84-83 Colt FWD Hatchback	N/A	1/2	13/16	1 1/8	0	1/2	1	3/32 (out)		5/32 (in)	3/16		5/16	29 1/4	35 11/16	12 11/16
83-81 Sapporo	(19)	2 3/16	2 11/16	3 3/16	11/16	13/16	1 11/16	0		9/32	0		9/16	32	37	9 1/2
80 Sapporo	(19)	2 3/16	2 11/16	3 3/16	3/4	1 1/4	1 3/4	5/64		11/32	5/32		11/16	32	37	8 27/32
82-80 Arrow Pickup Truck	(10)	1 1/2	2 1/2	3 1/2	1/2	1	1 1/2	5/64		11/32	5/32		11/16	30 1/2	37	8
82-79 Champ F.W.D.	N/A	1/2	13/16	1 1/8	0	1/2	1	5/64 (out)		5/32 (in)	5/32 (out)		5/16 (in)	29 1/4	35 11/16	12 11/16
80-78 Arrow	(19)	1 9/16	2 1/16	2 9/16	1/2	1	1 1/2	5/64		15/64	5/32		15/32	30	35	9
79-78 Sapporo	(19)	2 3/16	2 11/16	3 3/16	1	1 1/2	2	5/64		11/32	5/32		11/16			
W/Power Steering														31	36	8 13/32
W/Manual Steering														32	37	8 13/32
79 Arrow Pickup Truck	(10)	2	3	4	1/2	1	1 1/2	5/64		11/32	5/32		11/16	30 1/2	37	8
77 Arrow	(19)	1 9/16	2 1/16	2 9/16	1/4	1	1 3/4	5/64		15/64	5/32		15/32	30	35	8 27/32

(A) Variation between wheel 1/2°.   **(B)** Side to side variation 1/2° or less.   • 86 Premier = 37 1/2° inner, 31 3/4° outer.

PORSCHE (A)																
86-85 944: Front	N/A	2 1/4	2 1/2	3	− 9/16	− 5/16	− 1/16	3/64	5/64	1/8	3/32	5/32	1/4	19	20	NA
Rear	N/A				− 15/16	− 7/16	1/16	3/64 (out)	0	3/64 (in)	3/32 (out)	0	3/32 (in)			
84-83 944: Front	N/A	2 1/4	2 1/2	3	− 9/16	− 5/16	− 1/16	3/64	5/64	1/8	3/32	5/32	1/4	19	20	NA
Rear	N/A				− 15/16	− 1	− 11/16	3/64 (out)	0	3/64 (in)	3/32 (out)	0	3/32 (in)			
86-85 911 Carrera: Front	(31)	5 13/16	6 1/16	6 5/16	− 3/16	0	3/16	5/64	1/8	11/64	5/32	1/4	11/32	19 1/2	20	NA
Rear	(32)							0	11/64	21/64	0	11/32	21/32			
exc. Turbo Body					13/16	1	1 3/16									
w/Turbo Body					5/16	1/2	11/16									
84-83 911SC: Front	(31)	5 13/16	6 1/16	6 5/16	− 3/16	− 1	13/16	5/64	1/8	11/64	5/32	1/4	11/32	19 1/2	20	NA
Rear	(32)				− 1 3/16	− 1	13/16	0	11/64	21/64	0	11/32	21/32			
84-83 911 Turbo Front	(31)	5 13/16	6 1/16	6 5/16	− 3/16	0	3/16	5/64	1/8	11/64	5/32	1/4	11/32			
Rear	(32)				− 11/16	− 1/2	− 5/16	0	11/64	21/64	0	11/32	21/32			
82 924, 924 Turbo																
Front	(29)	2 1/4	2 1/2	3	− 9/16	− 5/16	− 1/16	3/64	5/64	1/8	3/32	5/32	1/4	19	20	NA
Rear	(30)				− 1 11/16	− 1	− 1/16	5/64 (out)	0	5/64 (in)	5/32 (out)	0	5/32 (in)			
81 924, 924 Turbo	(29)	2 1/4	2 3/4	3 1/4	− 9/16	− 5/16	− 1/16	3/64	5/64	1/8	3/32	5/32	1/4	19	20	NA
Rear Std. Chassis	(30)				− 1 1/2	− 1	− 1/2	5/64 (out)	0	5/64 (in)	5/32 (out)	0	5/32 (in)			
Rear Turbo Chassis	(30)				− 29/32	− 13/32	3/32	5/64 (out)	0	5/64 (in)	5/32 (out)	0	5/32 (in)			
80-79 924, 924 Turbo	(29)	2 1/4	2 3/4	3 1/4	− 9/16	− 5/16	− 1/16	3/64	5/64	1/8	3/32	5/32	1/4	19	20	NA
Rear	(30)				− 1 1/2	− 1	− 1/2	5/64 (out)	0	5/64 (in)	5/32 (out)	0	5/32 (in)			
86-85 928: Front	(43)	3	3 1/2	4	− 11/16	− 1/2	− 5/16	5/64	1/8	11/64	5/32	1/4	11/32	19	20	NA
Rear	(44)				− 13/16	− 11/16	− 1/2	5/64	11/64	1/4	5/32	11/32	1/2			
84-82 928: Front	(43)	3	3 1/2	4	− 11/16	− 1/2	− 5/16	5/64	1/8	11/64	5/32	1/4	11/32	19	20	NA
Rear	(44)				− 7/8	− 11/16	− 1/2	5/64	11/64	1/4	5/32	11/32	1/2			
81-78 928	(43)	3 1/4	3 1/2	3 3/4	− 11/16	− 1/2	− 5/16	5/64	1/8	11/64	5/32	1/4	11/32	19	20	NA
Rear	(44)				− 7/8	− 11/16	− 1/2	5/64	11/64	1/4	5/32	11/32	1/2			
82-78 911 SC	(31)	5 13/16	6 1/16	6 5/16	5/16	1/2	11/16	5/64	1/8	11/64	5/32	1/4	11/32	19 3/4	20	10 15/16
Rear	(32)				− 3/16	0	3/16	0	11/64	21/64	0	11/32	21/32			
79-77 Turbo Carrera	(31)	5 13/16	6 1/16	6 5/16	5/16	1/2	11/16	5/64	1/8	11/64	5/32	1/4	11/32	19 3/4	20	NA
Rear	(32)				− 1/4	0	1/4	0	11/64	21/64	0	11/32	21/32			
78-77 924	(29)	2 1/4	2 3/4	3 1/4	− 1/2	− 5/16	− 1/8	3/64	5/64	1/8	3/32	5/32	1/4	19	20	NA
Rear	(30)				− 1 1/2	− 1	− 1/2	5/64 (out)	0	5/64 (in)	5/32 (out)	0	5/32 (in)			
77 911	(31)	5 13/16	6 1/16	6 5/16	− 3/16	0	3/16	5/64	1/8	11/64	5/32	1/4	11/32	19 3/4	20	10 15/16
Rear	(32)				5/16	1/2	11/16	0	11/64	21/64	0	11/32	21/32			

(A) Do not press wheels to set or check alignment.

IMPORTED PASSENGER CAR AND LIGHT TRUCK SECTION

VEHICLE IDENTIFICATION YEAR MODEL	ADJ. ILL. NO.	CASTER (Degrees) MIN.	PREF.	MAX.	CAMBER (Degrees) MIN.	PREF.	MAX.	TOE-IN (Inches) MIN.	PREF.	MAX.	TOE-IN (Degrees) MIN.	PREF.	MAX.	TOE-OUT ON TURNS (Degrees) OUTSIDE WHEEL	INSIDE WHEEL	STRG. AXIS INCL. (DEG.)
RENAULT																
86-83 Alliance, Encore:																
Front	N/A	1/2		2	– 3/16		5/16	0		1/4	0		1/2	NA	NA	NA
Rear	N/A				– 5/16	13/16	15/16	1/4 (out)	0	1/4 (in)	1/2 (out)	0	1/2 (in)			
84-79 (late) Le Car: Front	(41)	10		13	0	1/2	1	0	3/64	3/32	0	3/32	3/16	NA	NA	NA
Rear	N/A				0		1 1/2	1/16 (out)	0	1/16 (in)	1/8 (out)	0	1/8 (in)	NA	NA	NA
86-81 R-18, Fuego	N/A				– 1/2	0	1/2	0	3/64	3/32	0	3/32	3/16	NA	NA	NA
W/Power Steering		1 1/2		3												
W/Manual Steering		1/2		2												
79 (early) -77 Le Car																
(5 Series): Front	(41)	10		13	0	1/2	1	3/16 (out)		3/64 (out)	3/8 (out)		3/32 (out)	NA	NA	15 1/2
Rear	N/A				0		1/2	0		5/32	0		5/16			
79-77 12, 15, 17 & Gordini																
Front	(23)	4	4	5	1	1 1/2	2	5/32 (out)		3/64 (out)	5/16 (out)		3/32 (out)	NA	NA	18 3/4
Rear					0		1/2	1/16 (out)		0	1/8 (out)		0			
SAAB (A)																
86 9000: Front	N/A	1 1/8	1 5/8	2 1/8	– 1 1/8	– 5/8	– 1/8	1/64	1/16	3/32	3/32	3/16	5/16	20	21	11 5/16
Rear	N/A				– 1/2	– 1/4	0	3/64	3/32	5/32	1/8	5/16	9/16			
86-79 900S, Turbo EMS																
Front					0	1/2	1	3/64	5/64	1/8	1/8	5/16	7/16	20	20 3/4	11 1/2
W/Power Steering	(10)	1 1/2	2	2 1/2												
W/Manual Steering	(10)	1/2	1	1 1/2												
Rear					– 3/4	– 1/2	– 1/4	1/8	5/32	13/64	7/16	5/8	3/4			
78-77 99																
W/Power Steering	(10)	1/2	1	1 1/2	0	1/2	1	3/64 (out)	0	3/64 (in)	1/8 (out)	0	1/8 (in)	20	20 3/4	11 1/2
W/Manual Steering	(10)	1/2	1	1 1/2	0	1/2	1	0	3/64	3/32	0	1/8	5/16	20	20 3/4	11 1/2
Rear					– 1	0	1	5/64 (out)	0	5/64 (in)	5/16 (out)	0	5/16 (in)			
(A) Toe is measured at the wheel rim.																
SUBARU																
86-85 XT 2 WD: Front	N/A	3 5/16	4 1/16	4 13/16	– 3/4	0	3/4	1/8 (out)	0	1/8 (in)	1/4 (out)	0	1/4 (in)	34 1/2	38	NA
Rear	N/A				– 3/4	0	3/4	1/8 (out)	0	1/8 (in)	1/4 (out)	0	1/4 (in)			
86-85 XT 4 WD: Front	N/A	2 5/8	3 3/8	4 1/8	– 1/16	5/8	1 3/8	3/64	5/64	1/8	3/32	5/32	1/4	34 1/2	38	NA
Rear	N/A				– 3/4	0	3/4	1/8 (out)	0	1/8 (in)	1/4 (out)	0	1/4 (in)			
86-85 2 W.D. Sedan exc.																
Air Susp.: Front	N/A	1 3/4	2 1/2	3 1/4	0	3/4	1 1/2	3/64 (out)	5/64 (in)	13/64 (in)	3/32 (out)	5/32 (in)	13/32 (in)	35	39	NA
Rear	N/A				– 3/4		3/4	1/8 (out)		1/8 (in)	1/4 (out)		1/4 (in)			
86-85 Sedan w/Air																
Susp.: Front	N/A	1 7/16	2 3/16	2 15/16	7/16	1 3/16	1 15/16	3/64 (out)	5/64 (in)	13/64 (in)	3/32 (out)	5/32 (in)	13/32 (in)	35	39	NA
Rear	N/A				– 3/4		3/4	1/8 (out)		1/8 (in)	1/4 (out)		1/4 (in)			
86-85 4 W.D. Sedan																
exc. Air Susp.: Front	N/A	1 1/16	1 13/16	2 9/16	15/16	1 11/16	2 7/16	3/64 (out)	5/64 (in)	13/64 (in)	3/32 (out)	5/32 (in)	13/32 (in)	35	39	NA
Rear	N/A				– 3/4		3/4	1/8 (out)		1/8 (in)	1/4 (in)		1/4 (in)			
86-85 2 W.D. Wagon																
Front	N/A	1 5/16	2 1/16	2 13/16	1/4	1	1 3/4	3/64 (out)	5/64 (in)	13/64 (in)	3/32 (out)	5/32 (in)	13/32 (in)	35	39	NA
Rear	N/A				– 3/4		3/4	1/8 (out)		1/8 (in)	1/4 (out)		1/4 (in)			
86-85 4 W.D. Wagon																
Front	N/A	13/16	1 9/16	2 5/16	15/16	1 3/4	2 7/16	3/64 (out)	5/64 (in)	13/64 (in)	3/32 (out)	5/32 (in)	13/32 (in)	35	39	NA
Rear	N/A				– 3/4		3/4	1/8 (out)		1/8 (in)	1/4 (out)		1/4 (in)			

IMPORTED PASSENGER CAR AND LIGHT TRUCK SECTION

VEHICLE IDENTIFICATION YEAR MODEL	ADJ. ILL. NO.	CASTER (Degrees) MIN.	PREF.	MAX.	CAMBER (Degrees) MIN.	PREF.	MAX.	TOE-IN (Inches) MIN.	PREF.	MAX.	TOE-IN (Degrees) MIN.	PREF.	MAX.	TOE-OUT ON TURNS (Degrees) OUTSIDE WHEEL	INSIDE WHEEL	STRG. AXIS INCL. (DEG.)
84-83 4 W.D. exc. Sta. Wagon and Brat: Front	N/A	−1 1/4		1/4	1 11/16		3 3/16	15/64 (out)	13/64 (out)	5/32 (out)	15/32 (out)	13/32 (out)	5/16 (out)	35	36 1/2	NA
Rear	N/A				− 3/4		3/4	1/8 (out)		1/8 (in)	1/4 (out)		1/4 (in)			
84-83 Hardtop, Sta. Wagon and Brat w/Turbo: Front	N/A	−1 1/4		1/4	1 7/16		2 15/16	15/64 (out)	13/64 (out)	5/32 (out)	15/32 (out)	13/32 (out)	5/16 (out)	35	36 1/2	NA
Rear	N/A				−1		1/2	1/8 (out)		1/8 (in)	1/4 (out)		1/4 (in)			
83 4 W.D. Sta. Wagon and Brat exc. Turbo: Front	N/A	−1 7/16		1/16	1 11/16		3 3/16	15/64 (out)	13/64 (out)	5/32 (out)	15/32 (out)	13/32 (out)	5/16 (out)	35	36 1/2	NA
Rear	N/A				− 7/16		1 1/16	1/8 (out)		1/8 (in)	1/4 (out)		1/4 (in)			
84-82 exc. 4 W.D., all 2 W.D. models exc. Sta. Wagons: Front	N/A	−1 3/16		5/16	3/4		2 1/4	0	3/64	5/64	0	3/32	5/32	35	36 1/2	NA
Rear	N/A				− 3/4		3/4	1/8 (out)	0	1/8 (in)	1/4 (out)		1/4 (in)			
81-80 exc. 4 W.D., all 2 W.D. models exc. Sta. Wagons: Front	N/A	−1 3/16	− 7/16	5/16	3/4	1 1/2	2 1/4	5/64		5/16	5/32		5/8	35	36 1/2	NA
Rear	N/A				− 3/4	0	3/4	1/8 (out)		1/8 (in)	1/4 (out)		1/4 (in)			
84-82 Station Wagon except 4 W.D.: Front	N/A	− 13/16		11/16	1		2 1/2	0	3/64	5/64	0	3/32	5/32	35	36 1/2	NA
Rear	N/A				− 3/4		3/4	1/8 (out)	0	1/8 (in)	1/4 (out)		1/4 (in)			
81-80 Sta. Wagon exc. 4 W.D.: Front	N/A	− 13/16	− 1/16	11/16	1	1 3/4	2 1/2	5/64		5/16	5/32		5/8	35	36 1/2	NA
Rear	N/A				− 3/4	0	3/4	1/8 (out)		1/8 (in)	1/4 (out)		1/4 (in)			
82 4 W.D. exc. Sta. Wagon and Brat: Front	N/A	−1 1/4		1/4	1 13/16		3 5/16	3/64	5/64	1/8	3/32	5/32	1/4	35	36 1/2	NA
Rear	N/A				− 3/4		3/4	1/8 (out)	0	1/8 (in)	1/4 (out)	0	1/4 (in)			
81-80 4 W.D. exc. Sta. Wagon & Brat: Front	N/A	−1 1/4	− 1/2	1/4	1 13/16	2 9/16	3 5/16	15/64		15/32	15/32		15/16	35	36 1/2	NA
Rear	N/A				− 3/4	0	3/4	1/8 (out)		1/8 (in)	1/4 (out)		1/4 (in)			
82 4 W.D. Sta. Wagon and Brat: Front	N/A	−1 7/16		1/16	1 13/16		3 5/16	3/64	5/64	1/8	3/32	5/32	1/4	35	36 1/2	NA
Rear	N/A				− 7/16		1 1/16	1/8 (out)	0	1/8 (in)	1/4 (out)	0	1/4 (in)			
81-80 4 W.D. Sta. Wagon: Front	N/A	−1 7/16	− 11/16	1/16	1 13/16	2 9/16	3 5/16	15/64		15/32	15/32		15/16	35	36 1/2	NA
Rear					− 7/16	5/16	1 1/16	1/8 (out)		1/8 (in)	1/4 (out)		1/4 (in)			
81-80 4 W.D. Brat: Front	N/A	−1 9/16	− 13/16	− 1/16	1 7/16	2 3/16	2 15/16	15/64		15/32	15/32		15/16	35	36	NA
Rear	N/A				9/16	1 5/16	2 1/16	5/64		15/64	5/32		15/32			
79-77* all exc. Wagon, 4 W.D. & Brat: Front	N/A	−1 19/32	− 27/32	− 3/32	1	1 3/4	2 1/2	5/64		21/64	5/32		21/32	35	36	NA
Rear	(42)				− 13/32		1 3/32	3/64		13/64	3/32		13/32			
79-77* Wagon exc. 4 W.D.: Front	N/A	− 29/32	− 5/32	19/32	1	1 3/4	2 1/2	5/64		21/64	5/32		21/32	35	36	NA
Rear	(42)				− 13/32		1 29/32	5/64		15/64	5/32		15/32			
79-77* 4 W.D. except Brat: Front	N/A	−1 19/32	− 27/32	− 3/32	1/2	1 1/4	2	15/64		15/32	15/32		15/16	30	31	NA
Rear	(42)				19/32		2 3/32	5/64		15/64	5/32		15/32			
79-77* Brat: Front	N/A	−1 19/32	− 27/32	− 3/32	1/2	1 1/4	2	15/64		15/32	15/32		15/16	35	36	NA
Rear	(42)				19/32		2 3/32	5/64		15/64	5/32		15/32			
77 All exc. Wagon & 4 W.D.: Front	N/A	0	3/4	1 1/2	1	1 1/2	2	5/64		5/16	5/32		5/8	35	36	12 29/32
Rear	(42)				1/4	1	1 1/2	3/64		13/64	3/32		13/32			
77 Wagon: Front	N/A	0	3/4	1 1/2	1	1/2	2	5/64		5/16	5/32		5/8	35	36	12 29/32
Rear	(42)				1	1 1/2	2	5/64		15/64	5/32		15/32			
77 4 W.D.: Front	N/A	0	3/4	1 1/2	2	2 1/2	3	15/64		15/32	15/32		15/16	30	31	12 29/32
Rear	(42)				1 11/32	1 27/32	2 11/32	5/64		15/64	5/32		15/32			

*Oblong fender mounted parking lights on 77 Stage II models.

SECTION 2 SUSPENSION
Specifications

IMPORTED PASSENGER CAR AND LIGHT TRUCK SECTION

VEHICLE IDENTIFICATION YEAR MODEL	ADJ. ILL. NO.	CASTER (Degrees) MIN.	PREF.	MAX.	CAMBER (Degrees) MIN.	PREF.	MAX.	TOE-IN (Inches) MIN.	PREF.	MAX.	TOE-IN (Degrees) MIN.	PREF.	MAX.	TOE-OUT ON TURNS (Degrees) OUTSIDE WHEEL	INSIDE WHEEL	STRG. AXIS INCL. (DEG.)
TOYOTA																
86-85 MR2: Front	(4)(19)	4 13/16	5 5/16	5 13/16	− 1/4	1/4	3/4	0	3/64	5/64	0	3/32	5/32	32	36 1/2	12
Rear	(4)				− 1 1/4	− 3/4	− 1/4	5/32	13/64	15/64	5/16	13/32	15/32			
86-85 Camry: Front	(4)(66)	1/2	1	1 1/2	1/16	9/16	1 1/16	3/64	5/64	1/8	3/32	5/32	1/4	30	38	12 1/2
Rear	(65)				0	1/2	1	5/64	5/32	15/64	5/32	5/16	15/32			
84-83 Camry: Front	(4)(66)				1/16	9/16	1 1/16							30	38	12 1/2
W/Manual Steering		1/2	1	1 1/2				3/64 (out)	0	3/64 (in)	3/32 (out)	0	3/32 (in)			
W/Power Steering		2	2 1/2	3				3/64	5/64	1/8	3/32	5/32	1/4			
Rear	(65)				0	1/2	1	5/64 (out)	0	5/64 (in)	5/32 (out)	0	5/32 (in)			
86-85 Tercel Sedan exc. 4 W.D.: Front	(20)				− 5/16	1/16	11/16	5/64 (out)	3/64 (out)	0	5/32 (out)	3/32 (out)	0			12 1/2
W/Manual Steering		11/16	13/16	1 11/16										32 11/16	36 5/16	
W/Power Steering		2 3/16	2 11/16	3 3/16										32 15/16	35 13/16	
Rear	(45)				− 9/16	− 1/16	7/16	3/64 (out)	0	3/64 (in)	3/32 (out)	0	3/32 (in)			
86-85 Tercel Wagon exc. 4 W.D.: Front	(20)				− 1/2	0	1/2	5/64	3/64	0	5/32	3/32	0			12 9/16
								(out)	(out)		(out)	(out)				
W/Manual Steering		1/4	3/4	1 1/4										32 15/16	36 3/16	
W/Power Steering		1 3/4	2 1/4	2 3/4										32 13/16	35 13/16	
Rear	(45)				− 11/16	− 3/16	5/16	3/64 (out)	0	3/64 (in)	3/32 (out)	0	3/32 (in)			
84 Tercel Sedan exc. 4 W.D.: Front	(20)				− 3/16	5/16	13/16	3/64 (out)	0	3/64 (in)	3/32 (out)	0	3/32 (in)			12 1/2
W/Manual Steering		11/16	13/16	1 11/16										32 11/16	36 5/16	
W/Power Steering		2 3/16	2 11/16	3 3/16										32 15/16	35 13/16	
Rear	(45)				− 9/16	− 1/16	7/16	5/64 (out)	0	5/64 (in)	5/32 (out)	0	5/32 (in)			
84 Tercel Wagon exc. 4 W.D.: Front	(20)				− 1/4	1/4	3/4	3/64 (out)	0	3/64 (in)	3/32 (out)	0	3/32 (in)			12 9/16
W/Manual Steering		1/4	3/4	1 1/4										32 15/16	36 3/16	
W/Power Steering		1 3/4	2 1/4	2 3/4										32 13/16	35 13/16	
Rear	(45)				− 11/16	− 3/16	5/16	5/64 (out)	0	5/64 (in)	5/32 (out)	0	5/32 (in)			
86 Van Wagon	(2)	2	2 1/2	3	− 1/2	0	1/2	3/64 (out)	0	3/64 (in)	3/32 (out)	0	3/32 (in)	33	37	10 1/2
85-84 Van Wagon	(2)	1 7/16	2 1/16	2 9/16	0	1/2	1	3/64 (out)	0	3/64 (in)	3/32 (out)	0	3/32 (in)	34	34 11/16	10
86-85 Cressida: Sedan Front	(19)	4 5/16	4 13/16	5 5/16	− 1/16	7/16	15/16	3/64	5/64	1/8	3/32	5/32	1/4	31	38	10 9/16
Wagon: Front	(19)	3 3/4	4 1/4	4 3/4	− 1/16	7/16	15/16	3/64	5/64	1/8	3/32	5/32	1/4	31	38	10 9/16
Rear (All)	N/A				− 15/16	− 7/16	1/16	3/64	1/8	13/64	3/32	1/4	13/32			
84-83 Cressida Sedan: Front	(19)	2	2 1/2	3	1/4	3/4	1 1/4	5/64	1/8	5/32	5/32	1/4	5/16	31	38	9 1/4
Wagon: Front	(19)	1 11/16	2 3/16	2 11/16	5/16	13/16	15/16	5/64	1/8	5/32	5/32	1/4	5/16	31	38	9 1/16
Rear	N/A				− 3/16	5/16	13/16	5/32 (out)	5/64 (out)	0	5/16 (out)	5/32 (out)	0			
83 Tercel exc. 4 WD: Front	(20)				− 3/8	5/16	13/16	3/64 (out)	0	3/64 (in)	3/32 (out)	0	3/32 (in)			12 1/2
W/Manual Steering		11/16	13/16	1 11/16										32 15/16	36 3/16	
W/Power Steering		2 3/16	2 11/16	3 3/16										32 13/16	35 13/16	
Rear	(45)				− 9/16	− 1/16	7/16	5/64 (out)	0	5/64 (in)	5/32 (out)	0	5/32 (in)			
86-85 Tercel 4 WD	(20)	1 15/16	2 7/16	2 15/16	1/16	9/16	1 1/16	5/64 (out)	3/64 (out)	0	5/32 (out)	3/32 (out)	0			11 13/16
W/Manual Steering														33 1/4	36 1/2	
W/Power Steering														32 13/16	35 3/4	
84-83 Tercel 4 W.D.	(20)	1 15/16	2 7/16	2 15/16	5/16	13/16	15/16	3/64 (out)	0	3/64 (in)	3/32 (out)	0	3/32 (in)			11 13/16
W/Manual Steering														33 1/4	36 1/2	
W/Power Steering														32 13/16	35 3/4	
86-83 Supra: Front	(19)	3 11/16	4 3/16	4 11/16	5/16	13/16	15/16	5/64	1/8	5/32	5/32	1/4	5/16	30 3/4	37 9/16	10 1/2
Rear	(65)				− 11/16	− 3/16	5/16	5/64	0	5/64	5/32	0	5/32			
								(out)		(in)	(out)		(in)			

SUSPENSION Specifications

IMPORTED PASSENGER CAR AND LIGHT TRUCK SECTION

VEHICLE IDENTIFICATION YEAR MODEL	ADJ. ILL. NO.	CASTER (Degrees) MIN.	PREF.	MAX.	CAMBER (Degrees) MIN.	PREF.	MAX.	TOE-IN (Inches) MIN.	PREF.	MAX.	TOE-IN (Degrees) MIN.	PREF.	MAX.	TOE-OUT ON TURNS (Degrees) OUTSIDE WHEEL	INSIDE WHEEL	STRG. AXIS INCL. (DEG.)
84-82 Land Cruiser	N/A				1/4	1	1 3/4							30	31 1/2	9 1/2
W/Bias Tires								5/64	5/32	15/64	5/32	5/16	15/32			
W/Radial Tires								3/64	3/64	1/8	3/32	3/32	1/4			
								(out)	(in)	(in)	(out)	(in)	(in)			
FJ, BJ, HJ-4 Series			1 1/4	1 3/4												
FJ, BJ, HJ-6 Series		9/16	1 1/16	1 9/16												
84-81 Starlet	(66)	1 11/16	2	2 5/16	3/8	1 1/16	1	3/64	5/64	1/8	3/32	5/32	1/4	34	36 27/32	9 3/4
82 Cressida: Sedan	(19)	1 1/16	1 9/16	2 1/16		13/16		5/64	1/8	5/32	5/32	1/4	5/16	31	38	9 9/16
Wagon	(19)	1	1 1/2	2	5/16	13/16	15/16	5/64	1/8	5/32	5/32	1/4	5/16	31	38	9 3/16
81 Cressida: Sedan	(19)	1	1 1/2	2	5/16	13/16	15/16	5/64	1/8	5/32	5/32	1/4	5/16	31	38	9 5/32
Wagon	(19)	1	1 1/2	2	5/16	13/16	15/16	5/64	1/8	5/32	5/32	1/4	5/16	31	38	9 3/32
81 Land Cruiser: FJ40	N/A		1			1		1/8	5/32	13/64	1/4	5/16	13/32	30	32	9 1/2
FJ60	N/A		1 3/32			1		1/8	5/32	13/64	1/4	5/16	13/32	30	32	9 1/2
86 4-Runner	(36)	1 5/8	2 1/8	2 5/8	1/8	5/8	1 1/8	5/64	1/8	5/32	5/32	1/4	5/16	31	32	11 15/16
85 4-Runner	(36)	2	3	4	1/4	1	1 3/4	0	1/32	1/16	0	1/16	1/8	NA	NA	9 1/2
86-85 Pickup (4 × 2)																
1/2 Ton Long Bed	(10)	1 1/16	1 3/16	1 11/16	0	1/2	1							30	34	10
W/Bias Tires								3/16	1/4	9/32	3/8	1/2	9/16			
W/Radial Tires								1/16	1/8	5/32	1/8	1/4	5/16			
1 Ton	(10)	1/16	9/16	1 1/16	0	1/2	1	1/8	5/32	3/16	1/4	5/16	3/8	30	34	10
Cab & Chassis																
W/Single Rear Wheels	(10)	−7/16	1/16	9/16	0	1/2	1	1/8	5/32	3/16	1/4	5/16	3/8	30	34	10
86 Pickup (4 × 2)																
1/2 Ton Extra Long Bed	(10)	1/2	1	1 1/2	0	1/2	1	1/8	5/32	3/16	1/4	5/16	3/8	30	34	10
Cab & Chassis Long																
W/Dual Rear Wheels	(10)	1/16	9/16	1 1/16	0	1/2	1	1/8	5/32	3/16	1/4	5/16	3/8	30	34	10
Cab & Chassis Extra Long																
W/Dual Rear Wheels	(10)	5/16	13/16	15/16	0	1/2	1	1/8	5/32	3/16	1/4	5/16	3/8	30	34	10
86 Pickup (4 × 4)	(36)	13/16	15/16	1 13/16	1/8	5/8	1 1/8	5/64	1/8	5/32	5/32	1/4	5/16	31	32	11 5/16
85-84 Pickup (4 × 4)	(36)	1 1/4	2 1/4	3 1/4	1/4	1	1 3/4							29	30 1/2	9 1/2
W/Radial Tires								0	3/64	5/64	0	3/32	5/32			
W/Bias Tires								1/8	5/32	3/16	1/4	5/16	3/8			
86-84 Pickup (4 × 2)																
1/2 Ton Short Bed		1/16	1 1/16	1 3/16	0	1/2	1							30	34	10
W/Bias Tires	(10)							1/8	5/32	3/16	1/4	5/16	3/8			
W/Radial Tires	(10)							0	3/64	5/64	0	3/32	5/32			
84 Pickup (4 × 2)																
1/2 Ton Long Bed	(10)	1 1/16	1 3/16	1 11/16	0	1/2	1							30	34	10
3/4 Ton	(10)	1/16	1 1/16	1 3/16	0	1/2	1							30	34	10
Cab & Chassis	(10)	1 1/16	1 3/16	1 11/16	0	1/2	1							30	34	10
All W/Bias Tires								13/64	15/64	9/32	13/32	15/32	9/16			
All W/Radial Tires								5/64	1/8	5/32	5/32	1/4	5/16			
83-80 Hilux (4 × 4)	N/A	2 3/4	3 1/2	4 1/4	1/4	1	1 3/4							29	30 1/2	9 1/2
W/Radial Tires								0	3/64	5/64	0	3/32	5/32			
W/Bias Tires								1/8	5/32	13/64	1/4	5/16	13/32			
82-80 Tercel: Front	(19)	1 11/16	2 3/16	2 11/16	0	1/2	1	3/64	5/64	1/8	3/32	5/32	1/4	33	35	11 5/16
Rear	(45)				−1/2	0	1/2	3/64	0	3/64	3/32	0	3/32			
								(out)		(in)	(out)		(in)			
86 Corolla (FWD): Front	(4)	1/8	7/8	1 3/8	−3/4	−1/4	1/4	0	3/64	5/64	0	3/32	5/32			12 9/16
W/Manual Steering														33	40	12 9/16
W/Power Steering														32 1/2	39	
Rear	(65)				−1	−1/2	0	7/64	9/64	3/16	7/32	9/32	3/8			
85-84 Corolla (FWD) Front	(4)	3/8	7/8	1 9/16	−1	−1/2	0	3/64	0	3/64	3/32	0	3/32			12 9/16
								(out)		(in)	(out)		(in)			
Gas W/Manual Steering														33	40	
W/Power Steering														32 1/2	39	
Diesel W/Manual Steering														31 1/2	39 1/2	
W/Power Steering														32	39	
Rear	(65)				−1	−1/2	0	1/16	9/64	7/32	1/8	9/32	7/16			
86-84 Corolla																
2 & 3 Dr. Coupe (RWD)	(19)				−1/4	1/4	3/4	0	3/64	5/64	0	3/32	5/32			8 1/2
W/Manual Steering		2 1/4	2 3/4	3 1/4										33 1/2	38 1/2	
W/Power Steering		3 1/16	3 11/16	4 1/16										33	38 1/2	
83-80 Corolla exc. wagon	(19)	1 1/4	1 3/4	2 1/4	9/16	1 1/16	1 9/16							32	40	8 7/16
W/Radial Tires								0	3/64	5/64	0	3/32	5/32			
W/Bias Tires								5/64	1/8	5/32	5/32	1/4	5/16			
83-80 Corolla Wagon	(19)	1 1/16	1 9/16			1 1/16	1 9/16							32	40	8 5/16
W/Radial Tires								0	3/64	5/64	0	3/32	5/32			
W/Bias Tires								5/64	1/8	5/32	5/32	1/4	5/16			

2-403

IMPORTED PASSENGER CAR AND LIGHT TRUCK SECTION

VEHICLE IDENTIFICATION	ADJ. ILL. NO.	CASTER (Degrees)			CAMBER (Degrees)			TOE-IN (Inches)			TOE-IN (Degrees)			TOE-OUT ON TURNS (Degrees)		STRG. AXIS INCL.
YEAR MODEL		MIN.	PREF.	MAX.	MIN.	PREF.	MAX.	MIN.	PREF.	MAX.	MIN.	PREF.	MAX.	OUTSIDE WHEEL	INSIDE WHEEL	(DEG.)
83-82 Hilux (4 × 2) Pickup																
RN 1/2 Ton	(10)	1/2	1	1 1/2	9/16	1 1/16	1 9/16							29	36	7 5/32
W/Bias Tires								5/32	13/64	15/64	5/16	13/32	15/32			
W/Radial Tires								3/64	5/64	1/8	3/32	5/32	1/4			
RN 3/4 Ton, Cab &																
Chassis	(10)	0	1/2	1	9/16	1 1/16	1 9/16							29	36	7 5/32
W/Bias Tires								5/32	13/64	15/64	5/16	13/32	15/32			
W/Radial Tires								3/64	5/64	1/8	3/32	5/32	1/4			
81-80 Hilux (4 × 2)	(10)	0	1/2	1	9/16	1 1/16	1 9/16							29	36	7 3/16
W/Radial Tires								3/64	5/64	1/8	3/32	5/32	1/4			
W/Bias Tires								5/32	13/64	15/64	5/16	13/32	15/32			
82 Corona Sedan:	(36)	1 1/4	1 3/4	2 1/4	1/4	1	1 3/4	0	3/64	5/64	0	3/32	5/32	31 1/4	38 3/16	7 13/16
Wagon	(36)	1	1 1/2	2	1/4	1	1 3/4	0	3/64	5/64	0	3/32	5/32	31 1/4	38 3/16	7 13/16
81-80 Corona exc. Wagon	(36)	1 1/4	1 3/4	2 1/4	1/2	1	1 1/2	0	3/64	5/64	0	3/32	5/32	31 13/16	38 3/16	7 13/16
81-80 Corona Wagon	(36)	1	1 1/2	2	1/2	1	1 1/2	0	3/64	5/64	0	3/32	5/32	31 13/16	38 3/16	7 13/16
86 Celica: Front	(4)	7/16	1 3/16	1 15/16	−1 1/16	−3/16	5/16	3/64	0	3/64	3/32	0	3/32	30	34	13 1/2
								(out)		(in)	(out)		(in)			
Rear	(65)				−1 1/4	−3/4	−1/4	5/32	13/64	15/64	5/16	13/32	15/32			
85-82 Celica: Front	(19)	2 13/16	3 5/16	3 13/16	7/16	15/16	1 7/16							32	37	9 11/32
W/Manual Steering								7/64	5/32	13/64	7/32	5/16	13/32			
W/Power Steering								5/32	13/64	15/64	5/16	13/32	15/32			
Rear G.T.S.	N/A				−1 1/16	−3/16	5/16	5/64	0	5/64	5/32	0	5/32			
								(out)		(in)	(out)		(in)			
81-80 Celica																
W/Power Steering	N/A	1 3/16	1 11/16			15/16	1 7/16	7/64	5/32	13/64	7/32	5/16	13/32	31 1/4	38 3/16	7 9/16
W/Manual Steering	N/A	1 3/16	1 11/16			15/16	1 7/16	0	3/64	5/64	0	3/32	5/32	31 1/4	38 3/16	7 9/16
82 Supra: Front	(19)	3 11/16	4 3/16	4 11/16	5/16	13/16	1 5/16	5/64	7/64	5/32	5/32	7/32	5/16	30 3/4	37 19/32	7 11/16
Rear	(65)				5 1/2	6	6 1/2	3/64	0	3/64	3/32	0	3/32			
								(out)		(in)	(out)		(in)			
81-79 1/2 Supra	N/A	1 1/4	1 3/4	2 1/4	5/16	13/16	1 5/16	3/64	0	3/64	3/32	0	3/32	31 11/32	37 3/32	7 11/16
80-78 Cressida except								(out)		(in)	(out)		(in)			
78 Wagon	N/A	3/4	1 1/4	1 3/4	5/16	13/16	1 5/16	5/64	7/64	5/32	5/32	7/32	5/16	32	38	7 11/16
80-79 Land Cruiser	N/A	1/4	1	1 3/4	1/4	1	1 3/4							30	32	9 1/2
W/HR78 × 15B								3/64	0	3/64	3/32	0	3/32			
								(out)		(in)	(out)		(in)			
W/H78 × 15B								7/64	5/32	13/64	7/32	5/16	13/32			
79 Hilux (4 × 2)	(10)	0	1/2	1	9/16	1 1/16	1 9/16	5/32	13/64	15/64	5/16	13/32	15/32	29	36	7 5/32
79 Hilux (4 × 4)	N/A		3 1/2			1										9 1/2
W/HR78 × 15B								0	3/64	5/64	0	3/32	5/32	29	36	
W/H78 × 15B								7/64	5/32	15/64	7/32	5/16	15/32	29	30 1/2	
79 Corona	(36)	1 1/4	1 3/4	2 1/4	1/2	1	1 1/2	0	3/64	5/64	0	3/32	5/32	31	38	7 27/32
79 Celica	N/A	1 1/4	1 3/4	2 1/4	9/16	1 1/16	1 9/16	0	3/64	5/64	0	3/32	5/32	31	38	7 7/16
79-77 Corolla	N/A	1 5/16	1 15/16	2 5/16	1/2	1	1 1/2							31 7/16	38	7 27/32
W/Radial Tires								0	3/64	5/64	0	3/32	5/32			
W/Bias Tires								5/64	7/64	5/32	5/32	7/32	5/16			7 27/32
78 Cressida Wagon	N/A	11/16	1 3/16	1 11/16	5/16	13/16	1 5/16	5/64	7/64	5/32	5/32	7/32	5/16	32	37 1/2	7 11/16
78 Celica	N/A	1 1/4	1 3/4	2 1/4	7/16	15/16	1 7/16							31	38	7 9/16
W/Power Steering								0	1/64	1/32	0	1/32	1/16			
W/Manual Steering								3/64	1/16	5/64	3/32	1/8	5/32			
78-77 Corona	(36)	5/16		1 5/16	1/16		1 1/16									
W/Radial Tires								3/64		7/64	3/32		7/32	31	37 1/2	7
W/Bias Tires								7/64		13/64	7/32		13/32	31	37 1/2	7
78-77 Hilux (4 × 2)	(10)	0	1/2	1	1/2	1	1 1/2	13/64	15/64	9/32	13/32	15/32	9/16			
W/Radial Tires														26 1/4	31	7 1/4
W/Bias Tires														27 1/2	33 1/4	7 1/4
78-77 Land Cruiser	N/A		1			1		7/64		13/64	7/32		13/32	30	32	9 1/2
W/900 × 15 Tires								15/64		9/32	15/32		9/16	24	26	9 1/2
77 Celica	N/A	1 1/4		2 1/4	1/2		1 1/2	0	1/64	1/32	0	1/32	1/16	31	38	7 9/16
TRIUMPH																
81-77 TR7, TR8	N/A	2 1/2	3 1/2	4 1/2	−1 1/4	1/4	3/4	0		1/16	0		1/8	NA	NA	11 1/4
80-78 Spitfire: Front	(25)		4			3		1/16		1/8	1/8		1/4	NA	NA	5 3/4
Rear	(60)							1/16		1/8	1/8		1/4			
								(out)		(in)	(out)		(in)			
77 Spitfire: Front	(25)	4	4 1/2	5	1 1/2	2	2 1/2	0		1/16	0		1/8	NA	NA	6 3/4
Rear	(60)				−4 3/4	−3 3/4	−2 3/4	0		1/16	0		1/8			

IMPORTED PASSENGER CAR AND LIGHT TRUCK SECTION

VEHICLE IDENTIFICATION YEAR MODEL	ADJ. ILL. NO.	CASTER (Degrees) MIN.	CASTER PREF.	CASTER MAX.	CAMBER (Degrees) MIN.	CAMBER PREF.	CAMBER MAX.	TOE-IN (Inches) MIN.	TOE-IN (Inches) PREF.	TOE-IN (Inches) MAX.	TOE-IN (Degrees) MIN.	TOE-IN (Degrees) PREF.	TOE-IN (Degrees) MAX.	TOE-OUT ON TURNS (Degrees) OUTSIDE WHEEL	TOE-OUT ON TURNS (Degrees) INSIDE WHEEL	STRG. AXIS INCL. (DEG.)
VOLKSWAGEN																
86-85 Golf, GTI, Jetta																
Front	(4)	1	1 1/2	2	– 3/4	– 7/16	– 1/16	5/64 (out)	0	5/64 (in)	5/32 (out)	0	5/32 (in)	NA	NA	NA
Rear	N/A				– 2	– 1 11/16	– 1 5/16	5/64	13/64	21/64	5/32	13/32	21/32			
86-85 Scirocco, Rabbit Conv. Front	(4)	1 5/16	1 13/16	2 5/16	– 3/16	5/16	13/16	1/4 (out)	1/8 (out)	3/64 (out)	1/2 (out)	1/4 (out)	3/32 (out)	18 1/2	20	NA
Rear	N/A				-1 13/16	– 1 1/4	– 11/16	5/64 (out)	11/64 (in)	27/64 (in)	5/32 (out)	11/32 (in)	27/32 (in)			
86-82 Quantum: Front		0	1/2	1	– 15/32	– 21/32	– 5/32	0	5/64	11/64	0	5/32	11/32	NA	NA	NA
Rear					– 2	– 1 21/32	– 1 11/32	5/64	13/64	21/64	5/32	13/32	21/32			
84-81 Rabbit Pick-up																
Front	(4)	13/16	1 5/16	1 13/16	– 3/16	5/16	13/16	1/4 (out)	1/8 (out)	3/64 (out)	1/2 (out)	1/4 (out)	3/32 (out)	18 1/2	20	NA
Rear	N/A				– 1	0	1	1/2 (out)	0	1/2 (in)	1 (out)	0	1 (in)			
86-80 Vanagon: Front	(62)(63)	7	7 1/4	7 1/2	– 1/2	0	1/2	5/64 (out)	11/64 (in)	27/64 (in)	5/32 (out)	11/32 (in)	27/32 (in)	NA	NA	NA
Rear	(64)				– 1 5/16	– 13/16	– 5/16	11/64 (out)	0	11/64 (in)	11/32 (out)	0	11/32 (in)			
84-79 Rabbit, Scirocco, Jetta: Front	(4)	1 5/16	1 13/16	2 5/16	– 3/16	5/16	13/16	1/4 (out)	1/8 (out)	3/64 (out)	1/2 (out)	1/4 (out)	3/32 (out)	18 1/2	20	NA
Rear	N/A				-1 13/16	– 1 1/4	– 11/16	5/64 (out)	11/64 (in)	27/64 (in)	5/32 (out)	11/32 (in)	27/32 (in)			
80 Rabbit Pickup: Front	(4)	1 5/16	1 13/16	2 5/16	– 3/16	5/16	13/16	1/4 (out)	1/8 (out)	3/64 (out)	1/2 (out)	1/4 (out)	3/32 (out)	18 1/2	20	NA
Rear	N/A				– 9/16	0	9/16	1/2 (out)	0	1/2 (in)	1 (out)	0	1 (in)			
81-78 Dasher: Front	(14)	0	1/2	1	0	1/2	1	3/64 (out)	5/64 (in)	13/64 (in)	3/32 (out)	5/32 (in)	13/32 (in)	NA	NA	NA
Rear	N/A				– 1 3/8	– 11/16	0	13/32 (out)	0	13/32 (in)	13/16 (out)	0	13/16 (in)			
79-77 Type 1, Super Beetle																
Front	(12)	1 7/16	2	2 9/16	5/16	1	1 5/16	1/8	1/4	3/8	1/4	1/2	3/4	NA	NA	NA
Rear	(30)				-1 11/16	– 1	– 5/16	1/8 (out)	0	1/8 (in)	1/4 (out)	0	1/4 (in)			
79-77 Type 2, Bus: Front	(12)	2 5/16	3	3 11/16	3/8	11/16	1	0	1/8	1/4	0	1/4	1/2	17 1/2	20	NA
Rear	(30)				– 1 5/16	– 13/16	– 5/16	5/64 (out)	5/64 (in)	1/4 (in)	5/32 (out)	5/32 (in)	1/2 (in)			
78-77 Rabbit, Scirocco																
Front	(4)	1 5/16	1 13/16	2 5/16	– 3/16	5/16	13/16	1/4 (out)	1/8 (out)	3/64 (out)	1/2 (out)	1/4 (out)	3/32 (out)	18 1/2	20	NA
Rear	N/A				-1 13/16	– 1 1/4	– 11/16	3/32 (out)	5/32 (in)	13/32 (in)	3/16 (out)	5/16 (in)	13/16 (in)			
77 Dasher: Front	(14)	0	1/2	1	0	1/2	0	3/64 (out)	5/64 (in)	13/64 (in)	3/32 (out)	5/32 (in)	13/32 (in)	NA	NA	NA
Rear	N/A				– 1	– 1/2	0	13/64 (out)	0	13/64 (in)	13/32 (out)	0	13/32 (in)			
77 Type 1 Beetle																
Front	(12)	2 11/16	3 5/16	4 5/16	1/8	1/2	7/8	1/8	1/4	3/8	1/4	1/2	3/4	NA	NA	NA
Rear	(30)				– 1 11/16	– 1	– 5/16	1/8 (out)	0	1/8 (in)	1/4 (out)	0	1/4 (in)			
VOLVO																
86-83 740, 760 GLE	N/A		5			13/32		3/32	9/64	11/64	3/16	9/32	11/32	NA	NA	NA
86-85 240 Series	(37)	3		4	1/4	1/2	3/4	1/16	1/8	11/64	1/8	1/4	11/32	20 13/16	20	12
84-80 all models exc. GLT														20 13/16	20	12
W/Power Steering	(37)	3		4	1		1 1/2	1/16	1/8	11/64	1/8	1/4	11/32			
W/Manual Steering	(37)	2		3	1		1 1/2	1/8	13/64	1/4	1/4	13/32	1/2			
84-80 GLT														20 13/16	20	12
W/Power Steering	(37)	3		4	1/4		3/4	1/16	1/8	11/64	1/8	1/4	11/32			
W/Manual Steering	(37)	2		3	1/4		3/4	1/8	13/64	1/4	1/4	13/32	1/2			
79 240, 260 Series					– 1		0							20 13/16	20	12
W/Power Steering	(37)	3		4				1/16	1/8	11/64	1/8	1/4	11/32			
W/Manual Steering	(37)	2		3				1/8	13/64	1/4	1/4	13/32	1/2			
78-77 240, 260 Series	(37)	2		3	1		1 1/2							20 13/16	20	12
W/Power Steering								1/16	1/8	11/64	1/8	1/4	11/32			
W/Manual Steering								1/8	13/64	1/4	1/4	13/32	1/2			

SECTION 2 SUSPENSION
Specifications

ADJUSTMENT ILLUSTRATIONS
Groups I thru III

ADJUSTMENT ILLUSTRATIONS GROUP I

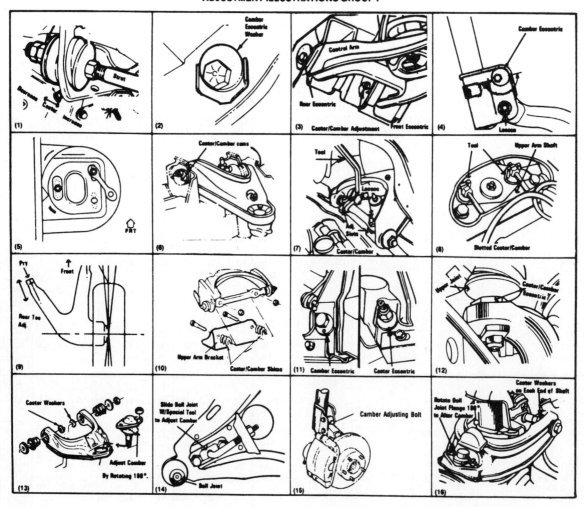

SUSPENSION
Specifications

ADJUSTMENT ILLUSTRATIONS GROUP II

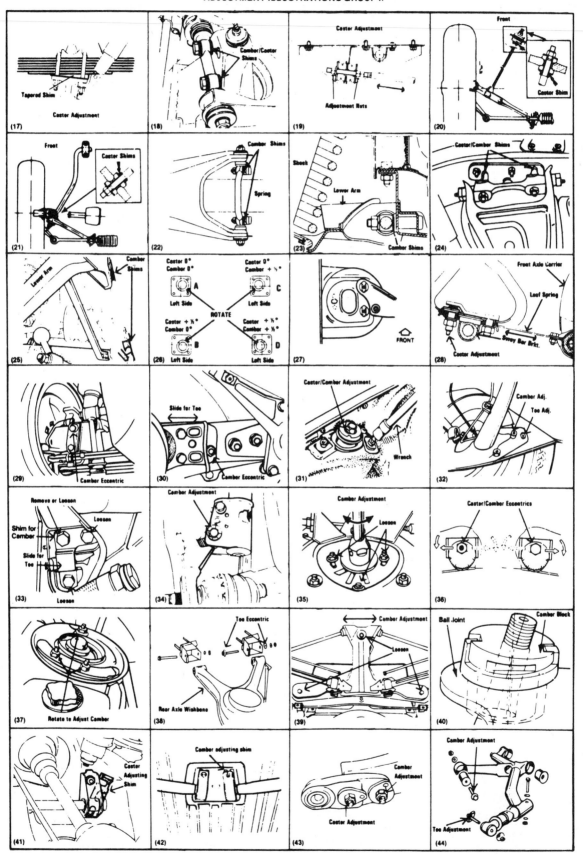

SECTION 2: SUSPENSION
Specifications

ADJUSTMENT ILLUSTRATIONS GROUP III

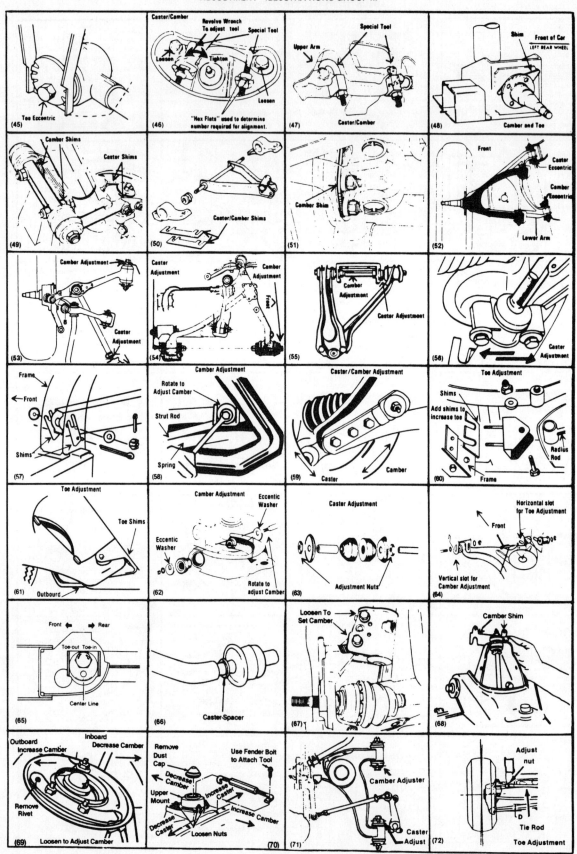

SUSPENSION
Specifications

CASTER/CAMBER REFRESHER GUIDE

MAKE AND MODEL	YEAR	MANNER OF CASTER ADJUSTMENT	MANNER OF CAMBER ADJUSTMENT
American Motors			
All models except Pacer	86-77	Lengthen strut to increase positive caster, shorten strut to decrease positive caster.	Rotate cam to achieve desired measurement. Rotate both eccentric cams to achieve desired measurements.
Pacer	80-77	Rotate rear cam to achieve desired measurement.	
Chrysler Corporation - Slotted mounting type adjustment (see illustration 7)			
Gran Fury, Diplomat	86-82	Increase positive caster by sliding front adjusting bolt away from engine and rear adjusting bolt toward engine. Decrease positive caster by sliding front adjusting bolt toward engine and rear adjusting bolt away from engine.	Increase positive camber by moving both adjusting bolts away from the engine. Decrease positive camber by moving both adjusting bolts toward the engine.
New Yorker (RWD), Caravelle	86-82		
Full sized Chrysler, Dodge and Plymouth vehicles	81-77		
Cordoba, Mirada	83-77		
LeBaron, Diplomat	81-77		
Volare, Aspen, Caravelle	81-77		
Charger, Coronet, Magnum	80-77		
Satellite, Monaco, Fury	78-77		
St. Regis	81-79		
Chrysler Corporation - McPherson strut vehicles.			
All w/FWD	86-83	Caster is not adjustable.	Turn eccentric cam, which is mounted on the strut attaching bracket, as necessary.
Omni, 024, Horizon, TC3	82-78		
Reliant, Aries	82-81		
400, LeBaron (FWD), Rampage	82		
Ford Motor Company - Models that are not adjustable.			
Tempo/Topaz, Escort/Lynx, EXP, Merkur, Taurus, Sable	86-84	Not adjustable.	Not adjustable.
Lincoln Continental	82		
Zephyr/Fairmont	82-78		
Capri/Mustang	82-79		
Cougar XR7/Thunderbird	82-80		
Escort/Lynx, Granada/Cougar	82-81		
Ford Motor Company - Models with front coil spring mounted on lower control arm with McPherson strut and air suspension. (Note: Ride height must be set before checking alignment. For illustrations and procedures for setting ride height refer to page 45).			
Continental, Mark VII	86-84	Not adjustable.	Remove the pop-rivet fastening the fender apron to the upper strut mount. Loosen the three nuts on the camber plate, and move the plate until the desired camber is achieved.
Ford Motor Company - Models with front coil spring mounted on lower control arm with McPherson strut.			
Continental, Thunderbird/Cougar XR7, LTD/Marquis, Fairmont Futura/Zephyr, Mustang/Capri	86-83	Not adjustable.	Remove the pop-rivet fastening the fender apron to the upper strut mount. Loosen the three nuts on the camber plate, and move the plate until the desired camber is achieved.
Mark VII, Continental	86-84		
Ford Motor Company - Models with front coil spring mounted on upper control arm.			
Maverick/Comet	77	Increase positive caster by shortening the lower control arm strut. Decrease positive caster by lengthening the lower control arm strut.	Increase positive camber by rotating lower control arm adjusting cam to move the arm inward. It may be necessary to slightly spread the body bracket at the adjusting cam. Decrease positive camber by reversing the above procedure.
Granada/Monarch	80-77		
Versailles	80-77		
Ford Motor Company - Models with front coil spring mounted on lower control arm and slotted upper control arm adjustment.			
Lincoln Town Car	86-82	Increase positive caster by moving front adjusting bolt away from engine, rear adjusting bolt toward engine. Decrease positive caster by moving rear adjusting bolt away from engine, front adjusting bolt toward engine.	Increase positive camber by moving both adjusting bolts away from the engine. Decrease positive camber by moving both adjusting bolts toward the engine.
Continental Mark VI	83-82		
Full sized Ford/Mercury	86-77		
All Lincoln models	81-77		
Thunderbird, Meteor	79-77		
Pinto/Bobcat	80-77		
Mustang	78-77		
Cougar/LTD II	79-77		
GENERAL MOTORS CORPORATION:			
Buick Motor Division - Models with McPherson strut front suspension without strut mounted eccentric cam.			
Electra, Park Ave., LeSabre (FWD)	86-85	Loosen 2 top strut nuts over slotted holes. Remove nut over oval hole and drill front and rear of hole and file to fully open slot. (see ill. #5). Move strut forward or rearward as necessary.	Loosen strut to knuckle nuts move as necessary.
Buick Motor Division - Models with McPherson strut front suspension without strut mounted eccentric cam.			
Skyhawk (FWD), Skylark (FWD), Century (FWD), Somerset Regal (FWD)	86-82	Caster is not adjustable.	Loosen the lower strut retaining bolts. Grasp the top of the tire firmly and move tire in or out to obtain correct reading. Tighten bolts to final torque specs. (140 ft/lbs.). It may be necessary to modify bottom bolt hole on strut.

2-409

SECTION 2: SUSPENSION
Specifications

CASTER/CAMBER REFRESHER GUIDE

MAKE AND MODEL	YEAR	MANNER OF CASTER ADJUSTMENT	MANNER OF CAMBER ADJUSTMENT
Buick Motors Division - Models with shim pack upper control arm adjustment.			
All (RWD) models except 1980-77 Skyhawk.	86-77	Increase positive caster by adding shims to rear shim pack and removing shims from front shim pack. Decrease positive caster by removing shims from rear shim pack and adding shims to front shim pack.	Increase positive camber by removing shims equally from front and rear shim packs. Decrease positive camber by adding shims equally to front and rear shim pack.
Buick Motor Division - Models with McPherson strut front suspension and strut mounted eccentric cam.			
Skylark (FWD)	81-80	Caster is not adjustable.	Turn eccentric cam on strut mounting bracket as necessary.
Buick Motor Division - Models with cam type adjustment.			
Skyhawk (RWD)	80-77	Turn rear cam as necessary.	Turn front cam as necessary.
Buick Motor Division - Models with McPherson strut front suspension with camber adjustment bolt.			
Riviera (FWD)	86	Remove the 3 strut nuts. Drill front and rear of oval hole to elongate. File to fully open slot. (See ill. #27). Move strut forward or rearward as necessary.	Loosen strut to knuckle bolts. Adjust camber by loosening or tightening camber adjustment bolt. Adjustment bolt must be seated against strut to knuckle mounting bolt after strut to knuckle mounting bolts are tightened.
Buick Motor Division - Models with upper control arm cam adjustment.			
Riviera (FWD)	85-79	Turn front and rear cams as necessary.	Turn front and rear cams as necessary.
Cadillac Motor Division - Models with McPherson strut front suspension with camber adjustment bolt.			
Eldorado, Seville (FWD)	86	Remove the 3 strut nuts. Drill front and rear of oval hole to elongate. File to fully open slot. (See ill. #27). Move strut forward or rearward as necessary.	Loosen strut to knuckle bolts. Adjust camber by loosening or tightening camber adjustment bolt. Adjustment bolt must be seated against strut to knuckle mounting bolt after strut to knuckle mounting bolts are tightened.
Cadillac Motor Division - Models with McPherson strut front suspension without strut mounted eccentric cam.			
Fleetwood, DeVille (FWD)	86-85	Loosen 2 top strut nuts over slotted holes. Remove nut over oval hole and drill front and rear of hole. File to fully open slot. (See ill. #5) Move strut forward or rearward as necessary.	Loosen strut to knuckle nuts move as necessary.
Cadillac Motor Division - Models with McPherson strut front suspension without strut mounted eccentric cam.			
Cimarron (FWD)	86-82	Caster is not adjustable.	Loosen the lower strut retaining bolts. Grasp the top of the tire firmly and move tire in or out to obtain correct reading. Tighten bolts to final torque specs. (140 ft/lbs.) It may be necessary to modify bottom bolt hole on strut.
Cadillac Motor Division - Models with shim pack upper control arm adjustment.			
All (RWD) models except Eldorado, 1986-80 Seville and Cimarron Seville	86-77 79-77	Increase positive caster by adding shims to rear shim pack and removing shims from front shim pack. Decrease positive caster by removing shims from rear shim pack and adding shims to front shim pack.	Increase positive caster by shortening lower control arm strut. Decrease positive caster by lengthening lower control arm strut.
Cadillac Motor Division - Models with upper control arm cam adjustment.			
Eldorado (FWD) Seville (FWD)	85-77 85-80	Turn front and rear cams as necessary.	Turn front and rear cams as necessary.
Chevrolet Motor Division - Models with adjustable upper McPherson Strut Mount			
Camaro (RWD)	86-82	Loosen the uppermost to apron bolts, move the mount rearward to increase caster or foward to decrease caster.	Loosen the upper mount to apron bolts, move the mount outboard to increase camber or inboard to decrease camber.
Chevrolet Motor Division - Models with shim pack upper control arm adjustment.			
All (RWD) models except Vega, Monza, 1986-82 Camaro and Chevette	86-77	Increase positive caster by adding shims to rear shim pack and removing shims from front shim pack. Decrease positive caster by removing shim from rear shim pack and adding shims to front shim pack.	Increase positive camber by removing shims equally from front and rear shim packs. Decrease positive camber by adding shims equally to both front and rear shim packs.
Chevrolet Motor Division - Models with cam type adjustment.			
Vega, Monza (RWD)	80-77	Turn rear cam as necessary.	Turn front cam as necessary.
Chevrolet Motor Division - Models with McPherson Strut front suspension and strut mounted eccentric cam.			
Citation (FWD)	81-80	Caster is not adjustable.	Turn eccentric cam on strut mounting bracket as necessary.

SUSPENSION
Specifications

CASTER/CAMBER REFRESHER GUIDE

MAKE AND MODEL	YEAR	MANNER OF CASTER ADJUSTMENT	MANNER OF CAMBER ADJUSTMENT
Chevrolet Motor Division - Models with McPherson Strut front suspension without strut mounted eccentric cam.			
Celebrity, Citation and Cavalier (FWD)	86-82	Caster is not adjustable.	Loosen the lower strut retaining bolts. Grasp the top of the tire firmly and move tire in or out to obtain correct reading. Tighten bolts to final torque specs. (140 ft/lbs.) It may be necessary to modify bottom bolt hole on strut.
Chevrolet Motor Division - Models with adjustable upper ball joints.			
Chevette (RWD)	86-77	Change the thickness or the configuration of the shims between the control arm locating tube, and the legs of the control arm.	To increase camber, disconnect, and rotate the upper ball joint 180°.
Oldsmobile Motor Division - Models with McPherson Strut front suspension without strut mounted eccentric cam.			
88, 98 Regency (FWD)	86-85	Loosen 2 top strut nuts over slotted holes. Remove nut over oval hole. Drill front and rear of hole. File to fully open slot (see ill. # 5). Move strut forward or rearward as necessary.	Loosen strut to knuckle nuts move as necessary.
Oldsmobile Motor Division - Models with McPherson Strut front suspension without strut mounted eccentric cam.			
Ciera, Omega, Firenza (FWD)	86-82	Caster is not adjustable.	Loosen the lower strut retaining bolts. Grasp the top of the tire firmly and move tire in or out to obtain correct reading. Tighten bolts to final torque specs. (140 ft/lbs.). It may be necessary to modify bottom bolt hole on strut.
Oldsmobile Division - Models with shim pack upper control arm adjustment.			
All (RWD) models except Starfire	86-77	Increase positive caster by adding shims to rear shim pack and removing shims from front shim pack. Decrease positive caster by removing shim from rear shim pack and adding shims to front shim pack.	Increase positive camber by removing shims equally from front and rear shim packs. Decrease positive camber by adding shims equally to front and rear shim packs.
Oldsmobile Division - Models with cam type adjustment.			
Starfire (RWD)	80-77	Turn rear cam as necessary.	Turn front cam as necessary.
Oldsmobile Division - Models with McPherson Strut suspension and strut mounted eccentric cam.			
Omega (FWD)	81-80	Caster is not adjustable.	Turn eccentric cam on strut mounting bracket as necessary.
Oldsmobile Motor Division - Models with McPherson strut front suspension with camber adjustment bolt.			
Toronado (FWD)	86	Remove the 3 strut nuts. Drill front and rear of oval hole to elongate. File to fully open slot. (See ill. #27). Move strut forward or rearward as necessary.	Loosen strut to knuckle bolts. Adjust camber by loosening or tightening camber adjustment bolt. Adjustment bolt must be seated against strut to knuckle mounting bolt after strut to knuckle mounting bolts are tightened.
Oldsmobile Division			
Toronado (FWD)	85-77	Turn front and rear cams as necessary.	Turn front and rear cams as necessary.
Pontiac Motor Division - Models with adjustable upper McPherson Strut Mount.			
Firebird (RWD)	86-82	Loosen the upper mount to apron bolts, move the mount rearward to increase caster or forward to decrease caster.	Loosen the upper mount to apron bolts, move the mount outboard to increase camber or inboard to decrease camber.
Pontiac Motor Division - Models with McPherson Strut front suspension without strut mounted eccentric cam.			
Phoenix, J2000, Sunbird, A6000, Grand Am (FWD)	86-82	Caster is not adjustable.	Loosen the lower strut retaining bolts. Grasp the top of the tire firmly and move tire in or out to obtain correct reading. Tighten bolts to final torque specs. (140 ft/lbs.). It may be necessary to modify bottom bolt hole on strut.
Pontiac Motor Division - Models with shim pack upper control arm adjustment.			
All (RWD) models except Astre, Sunbird, 1986-82 Firebird, Fiero, T1000	86-77	Increase positive caster by adding shims to rear shim pack and removing shims from front shim pack. Decrease positive caster by removing shims from rear shim pack and adding shims to front shim pack.	Increase positive camber by removing shims equally from front and rear shim packs. Decrease positive camber by adding shims equally to front and rear shim pack.
Pontiac Motor Division - Models with McPherson strut front suspension and strut mounted eccentric cam.			
Phoenix (FWD)	81-80	Caster is not adjustable.	Turn eccentric cam on strut mounting bracket as necessary.
Pontiac Motor Division - Models with cam type adjustment.			
Astre, Sunbird	80-77	Turn rear cam as necessary.	Turn front cam as necessary.
Pontiac Motor Division - Models with adjustable upper ball joints.			
Fiero, T1000	86-82	Change the thickness or the configuration of the shims between the control arm locating tube, and the legs of the control arm.	To increase camber, disconnect, and rotate the upper ball joint 180°.